THE

ILLUSTRATED

# FLORA

OF BRITAIN

AND NORTHERN EUROPE

# THE ILLUSTRATED FLORA OF BRITAIN AND NORTHERN EUROPE

Marjorie Blamey
Christopher Grey-Wilson

Hodder & Stoughton
LONDON SYDNEY AUCKLAND TORONTO

First published in Great Britain by Hodder and Stoughton Ltd.

British Museum CIP data
BLAMEY, Marjorie
The illustrated flora of Britain and Northern Europe
1. North-western Europe. Plants.
i. Title ii. Grey-Wilson, Christopher
581.94



ISBN 0 340 40170 2

Devised and produced by Domino Books, Jersey
Design consultant Hermann Heinzel, Gaujac
Systems and programming consultant Michael Mepham, Frome
Typographic consultant Colin Reed, London
Colour separations by J Film Process, Bangkok
Photocomposition by S.C.M., Toulouse
Manufactured by Wm. Collins, Glasgow

# CONTENTS

# PREFACE

Our aim has been to provide an illustrated handbook to our wild flowers with both a fuller – indeed, within our limits a comprehensive – coverage and more numerous and detailed colour illustrations than is possible in a portable field guide. In the interests of keeping the book within reasonable bounds, we have followed other authors in omitting grasses, sedges, rushes and horsetails – all certainly flowering plants, but of rather more specialist interest. Even so, this left over 2500 species and important subspecies to cover, whether native or naturalized or regularly occurring as aliens or casuals. Garden escapes are always a problem, but we have tried to include all those that the reader is likely to encounter, as also a selection on pp. 52-3 of important or conspicuous forestry trees.

The paintings and drawings here are all new, done especially for the book over a period of years, where possible from live plants or failing those, from freshly collected specimens. The text and arrangement of any Flora, however, draw on the past work and writings of other botanists from Linnaeus onwards. In particular, we are indebted to the contributors and editors of that magisterial publication, *Flora Europaea*. Its five unillustrated volumes may be beyond the reach of the public at large in both price and technicality, and with the passage of time sections of it may be overtaken in the scientific literature, but it was a remarkable achievement and will long remain a central work of reference.

Many individuals have helped in the planning and long preparation of this work, but none more than Philip Blamey. His organisational skills and energy, his critical eye, good sense and untiring encouragement, have been a crucial contribution to the whole enterprise.

We are also extremely grateful to botanists in many countries who helped either by supplying live plants, or by checking parts of the text or illustrations – in particular to Dr. K. Callauch, Michael Chinery, Skytte Christiansen, Anders Delin, Richard Fitter, Ted Masson Phillips, Dr. H. van der Meijden, Desmond Meikle, Mats Nettle-bladt, Henry and Anne Noltie, Anne Powell, Michael and Margaret Proctor, R.H. Roberts, John Syms, John White, Goronwy Wynne and Peter Yeo.

We are further most grateful to the Director of the Royal Botanic Gardens, Kew, and to many members of the staff there for particular help: Alan Cook, Philip Cribb, Tony Hall, Brian Halliwell, Nicholas Hind, Charles Jeffrey, Tony Kirkham, Gwilym Lewis, Grenville Lucas, Brian Mathew, Roger Polhill, Alan Radcliffe-Smith and Jeffrey Wood. Finally we thank Douglas Kent for his survey of the nomenclature used in the text; Caron Mitchell for transferring the original manuscript onto disk; Christine Grey-Wilson for proof-reading the text; and Hermann Heinzel for his most valuable guidance on the original design.

Marjorie Blamey

Christopher Grey-Wilson

*June, 1989*

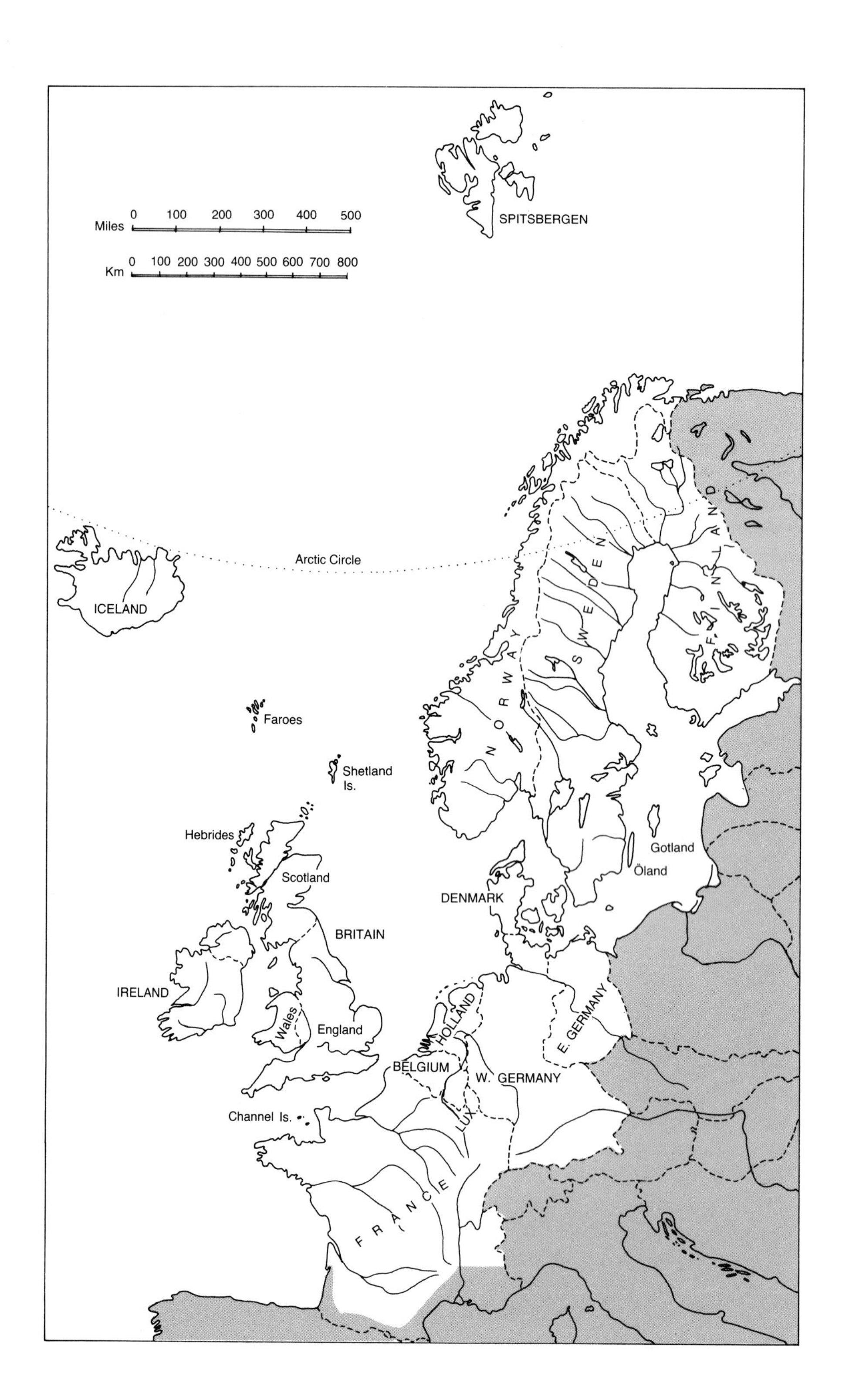

Miles 0 100 200 300 400 500
Km 0 100 200 300 400 500 600 700 800
SPITSBERGEN
Arctic Circle
ICELAND
Faroes
Shetland Is.
Hebrides
Scotland
BRITAIN
IRELAND
Wales
England
Channel Is.
NORWAY
SWEDEN
FINLAND
DENMARK
Gotland
Öland
HOLLAND
BELGIUM
LUX
W. GERMANY
E. GERMANY
FRANCE

# INTRODUCTION

THE AREA covered by this book is north and north-western Europe from the Alps and south-central France northwards. This includes the whole of the Federal Republic of Germany, Belgium, Holland, Denmark, Scandinavia (Norway, Sweden, Finland and Spitsbergen), Great Britain, Ireland, the Faeroes and Iceland. The dividing line in France is roughly where the Mediterranean influence merges into the cooler Atlantic climate covering much of the rest of France – a line more or less from Lyon to Bordeaux northwards, but reaching farher south in the centre-west – almost to the Pyrenees.

Within this region, we cover all the native flowering plants, with the exception of grasses (*Gramineae*), sedges (*Cyperaceae*) and rushes (*Juncaceae*). As far as possible within the limits of space, we have also included both naturalized and casual aliens, whether introduced by man or other agents accidentally or deliberately, or as often, escapes from cultivation in gardens or among crops.

Of the roughly 12,000 native plant species in Europe, only some 2,500 are found in our area. This may seem rather sparse, but the reasons are not hard to see. The warmer climate of the Mediterranean produces the greater wealth of plants; the mountain regions favour greater speciation, with many plants found nowhere else; and in the north the Ice Ages obliterated a large proportion of the native vegetation. The final retreat of the ice was only some 10,000 years ago and most of the plants present in its place today have arrived since then.

However, our flora compensates for its relative paucity in species-numbers with its beauty, variety of forms, and often the profusion of its display. There is more than enough here to sustain a lifetime's interest, and learning more about our flowers affords endless pleasure.

The SCIENTIFIC ORDER – the sequence of families, genera and species – followed here is generally that outlined in the five volumes of *Flora Europaea* (1964-1980), the major reference work for the whole of Europe, including European Russia. The classification of families in it was in turn based upon that established by Engler-Diels, *Syllabus der Pflanzenfamilien*, Ed. 2 of 1936, a system widely followed in Continental Europe, whereas in Britain the system of Bentham and Hooker is more widely used. However, over the years a number of such basic systems of classifying the plant kingdom have been devised, each differing in certain fundamental aspects from the next. It really matters little which is used providing it is used consistently. As *Flora Europaea* has set the precedence of the Englerian system, it would only cause confusion not to follow suit. Except where they have since changed (see below), we also follow it for scientific names.

The PAINTINGS illustrate nearly all species in colour. They show as far as possible the typical form of a species and in some cases a range of variety to be found, particularly as regards leaf-shape and flower-colour. The right-hand pages carry portraits of all major species. The smaller paintings on left-hand pages show a variety of details: aspects of flowers, fruits or leaves; variation within a species; or characters by which species of lesser importance are distinguished from their relatives on the right. The fruit is often the most reliable guide to identification, particularly in the Crucifer and Umbellifer families.

Throughout the book, on both left and right-hand pages, the *scale* of the paintings is life-size except where a reduction is either obvious (as with additional little pictures of whole plants) or specified. A caption '× 1/2' indicates a picture that is half natural size; '× 2', '× 3' etc that the illustration is an enlargement, by two or three times from natural size, to show some detail more clearly.

An asterisk indicates that a plant is illustrated on the facing right-hand page, while the small numbers refer to the illustrations of details on the left. It is important to remember that the texts and illustrations have been designed to complement each other: they should always be studied together.

ENGLISH NAMES by and large follow those in *English Names of Wild Flowers* (Botanical Society of the British Isles, 1980), which set out to standardize usage of common names of British wild flowers. Where there is no generally accepted common name – as of course for many of the non-British plants – none is given. Not surprisingly, established English names often have no connection with the divisions of modern scientific taxonomy. Nonetheless, one should approach the dictation of new ones only with the greatest caution. Names given to birds and flowers are part of a language, incorporated in it through their acceptance and use by past generations. There is something not only arrogant but unrealistic in the recent propensity – in various parts of the world – to decree new 'common' names for one's fellow-countrymen.

SCIENTIFIC NAMES are another matter. These are constantly in a state of flux and under review, since they reflect the latest studies in the genetic relationships between and within the different plant categories – which in turn are central to our understanding of their evolution. As groups of plants and single species are more intensively examined, their current classification is likely to be altered. New names may be given, earlier ones revived, or others found to be invalidated for one reason or another. Infuriating as this may sometimes seem to the casual observer, it is a necessary and continuous process, so that by the time this work is published some of the names included may already have changed. This makes doubly important the inclusion in any book such as this of well-known but now superseded names as SYNONYMS. They are distinguished here by being given always in square brackets, for example:

**Bluebell** *Scilla non-scripta* (L.) Hoffmans. & Link [Endymion non-scriptus] . . .

The current scientific names are immediately followed by their AUTHORITIES. Usually abbreviated, these indicate the author who described the species in question, and they often reveal something of the history of the species name. In the case of the Bluebell the species was first described in 1753 by Linnaeus – shortened to 'L.' He called it *Hyacinthus non-scriptus*. In the nineteenth century it was transferred first to the genus *Endymion* by the Belgian botanist B.C.J. Dumortier and then to *Scilla* by the Germans, J.C. von Hoffmannsegg and J.H.L. Link. In recent times it was moved again, to the genus *Hyacinthoides*, but this has since been considered incorrect, and our Bluebell has been reinstated as a *Scilla*. Admittedly, this is rather an extreme case, but the reader must be prepared for Latin names given here not to match those in older books. We have regrettably not had the space to include authorities for synonyms, but they sometimes make a full name extremely long.

Authority names are particularly important where the same species epithet (the second word in a plant's Latin name) has been used on different occasions for different species. Only by knowing the authors of a plant name can its correct application be checked, against their original description of the plant in the scientific literature. All aspects of the definition and publication of scientific names for plants and animals are controlled by detailed and elaborate Codes of rules, internationally agreed upon and published by their respective committees.

The TEXT DESCRIPTIONS, necessarily concise, stress the characters most important for identification: height; width; habit – whether sprawling, erect, climbing, tufted, cushion or moss-like etc.; months of flowering, habitat, soil-type, altitude and general distribution within our region. Note the following:

**Height**: Even within a family or genus, plants vary a great deal in height from species to species. Many mountain plants above the tree line are low and ground-hugging, adapted to withstand the extremes of temperature, high light intensity and exposure. Small plants also occur in other places with poor thin soils and in exposed coastal habitats such as salt marshes. In contrast, many of the herbs of lowlands grow tall and coarse. Moreover, each species has its own range of heights attained, varying from place to place with different conditions. This information has been simplified in the text as follows:

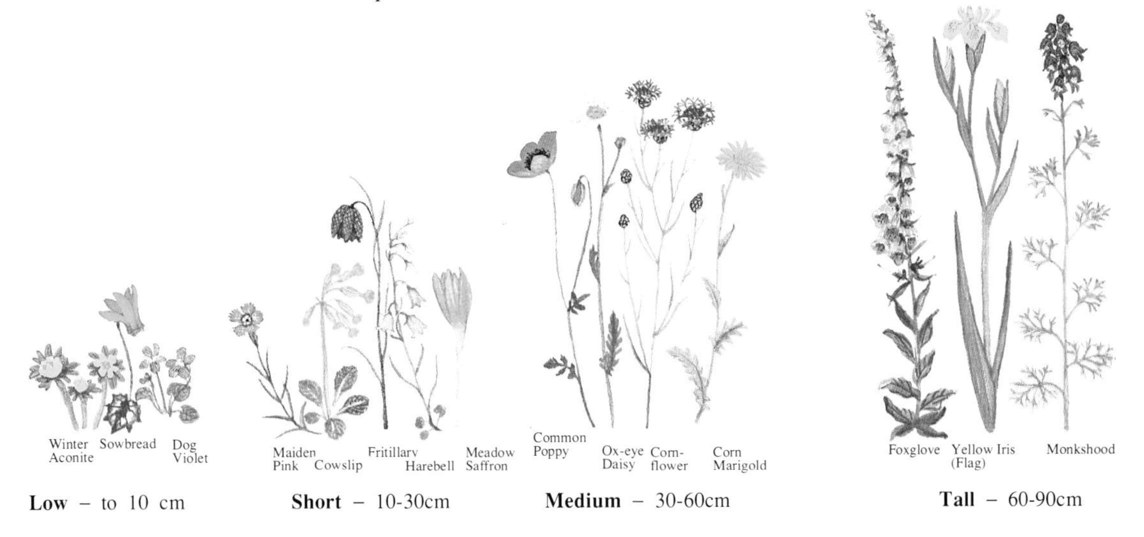


**Low** – to 10 cm **Short** – 10-30cm **Medium** – 30-60cm **Tall** – 60-90cm

For trees and shrubs, as well as some herbs, taller than 90cm, the normal maximum is given: '. . . a tall perennial herb to 1.5m.'

Flower **colour**, unless otherwise stated, refers to the petals (corolla) or, if these are absent or insignificant, to the sepals (calyx). Colour range is described in the text where it is not possible to show the complete range on the colour plates. Flower **size** refers to the diameter of the flower. Where occasionally the measurement is for length, this is specified: '12-18mm long.' **Roots** are not mentioned unless they are of particular interest, such as the intricate root-mass of the Bird's-nest Orchid. But the last thing that anyone should do is to dig up a wild plant merely to examine its roots.

**Habitat**. Some plants have very specific habitats – meadows, marshes, mountain screes – and are not normally found elsewhere. Others tolerate a wide range of habitat types, particularly among the widespread weeds of lowlands. For example, the Sea Sandwort is only found in coastal habitats, especially along shoreline sands or shingle. Away from the coast it is rarely if ever found, presumably unable to exist among the coarser, more competitive plant-communities. In contrast the Common Groundsel is far more widespread and abundant, growing in many types of soil and a wide range of habitats. But no two species are exactly alike in their habitat requirements and tolerances, which constitute valuable clues, both for identification and for knowing what to expect where. Within a given wood, one species will prefer the open glades while others tolerate the deep shade cast by dense trees, some the drier ground and others the damp or waterlogged.

**Flowering time**. The months given indicate the period during which the species can *normally* be expected to be found in flower. This can be greatly affected both by annual fluctuations and by latitude and altitude. After a mild winter a species will often start to flower much earlier in a given area than usual, or later after a cold one. Since the southern part of our region warms up sooner at the beginning of a year, the same species will come into bloom there sooner than it does further north. To some extent the same effect is also found from west to east because of the mild influence of the Atlantic Ocean. A species with a wide altitudinal range will also flower earlier lower down the mountain, often while those higher up are still buried in snow. Even within a small country such as the British Isles a species like the Primrose, *Primula vulgaris*, flowers earlier in Cornwall than in Suffolk, which in turn will be well ahead of the Scottish Highlands. Occasionally some species flower completely 'out of season', but such phenomena are quite unpredictable.

**Distribution** is often another critical guide to a species' identity, for closely related species may occur in different regions. It is indicated here by countries or groups of countries, sometimes by giving the northern limit, as in 'North to Denmark'. For reasons of space, it has not always been possible to specify the Faeroes, Iceland and Spitsbergen as well as parts of Norway, Sweden and Finland north of the arctic circle, and any or all of these are often covered in the general term 'the far north'. Similarly, West Germany, France, Holland, Britain and Ireland are sometimes collectively termed 'W Europe'. 'Continental Europe' in the text indicates that distribution does not extend beyond the English Channel. 'Britain' here covers England, Wales and Scotland. Botanically, Ireland is treated separately and here denotes the geographical island, including both Northern Ireland and the Republic of Ireland.

**British and Irish** distribution and frequency are indicated in more detail, after the symbol '**B**', at the very end of the texts for all species occurring in these islands – except only those which are not found elsewhere, and whose distribution in Britain and/or Ireland will have been already indicated. Otherwise, where there is none, the plant does not occur in Britain or Ireland.

Notes on a species distribution *outside* northern Europe, are only occasionally given. It should be borne in mind that relatively few of our wild plants are indigenous to our region and not found elsewhere. For many of our wild flowers, we are only at the extreme northern or western end of a natural range that may extend south to the Mediterranean and beyond, or far east into Asia.

**Introduced species** include all those species that have been artificially introduced to an area, whether accidentally or deliberately. Introduced species may be *casual*, occurring here and there but not self-maintaining, or *naturalized*, established and self-maintaining in a locality, reproducing themselves like natives. *Alien* species are those believed on good evidence to have been introduced by man.

Furthermore, a species may be native in one area, but naturalized or casual in another. For instance, the Scots Pine is native in parts of Scotland and widely in northern and central Europe down to Greece and southern Spain, but only naturalized in the south of England.

A plant introduced from outside Europe is indicated in this book by '**I**' – followed by the country or region where it is native and the countries in our area in which it has been found.

It is often very difficult to tell, without detailed investigation, where a species is truly native and where naturalized, for it may seem thoroughly 'at home' as both. The movement of plants around the world by man since early times has resulted in many species being established so successfully far from their natural homes, as often to seem a real part of the native landscape. Indeed in the British Isles there are almost as many naturalized species as there are native ones. Naturalized species are constantly turning up in new localities; this needs to be borne in mind when using any wild flower guide or flora.

Casual species often cause confusion and can prove very difficult to name. They are species which occur spasmodically, often in a place for only a year or two. Some may eventually become established (naturalized) aliens. Casuals frequently turn up in coastal regions, especially close to ports, where they are likely to have come into the country with foreign agricultural products or other imports and packaging.

The distribution of some species is very incompletely known, so that the information can only be rather generalized. There is always a chance of any species turning up quite outside its known range. Indeed, new records are constantly being made by both amateur and professional naturalists, and each such discovery, if properly recorded and notified, enhances our knowledge of the species concerned.

**Altitude** is given as completely as possible for species other than wholly lowland plants. 'To 1400m', for instance, indicates that the species may occur from sea level to 1400m above sea level over its range in Europe. Because of local conditions, a given species may show a quite different altitude range in different areas. For example, the Spring Gentian, *Gentiana verna*, rarely occurs below 1000m in the Alps, but in the Burren in Ireland it occurs at sea level.

When the altitude is presented as a range – '1200-2100m' – then this indicates a purely mountain species found only between these altitudes. It should be borne in mind that altitudes for many species are inadequately known, simply because little information on them has been recorded. They need to be read with a degree of caution. Where nothing is said about altitude, the plant may be taken as a lowland species.

**Frequency** can also vary within the total range of a species from very common to 'local' (not widespread or continuous in its distribution but restricted to particular localities) or rare. The populations of many species have been drastically reduced in the second half of this century, the chief causes being destruction of habitat, such as by the drainage of natural wetlands; excessive use of agricultural herbicides; and the more efficient control and greater 'purity' of seed for cereal crops. A few species are increasing, usually exotic introductions that have found a ready-made habitat, such as canal banks and railway embankments. But far more are in decline, some to the point of being threatened with extinction in our area.

Notes of **general information** about a plant are included where space allows. Poisonous plants are always noted as such, but we have not attempted to cover herbal and medicinal uses. For information on this enormous subject – part superstition and fantasy, part undoubted fact – the reader must consult the more substantial of the myriad books upon it, and even there tread with considerable caution. Notes on a plant's methods of **pollination** – by wind, insects, birds, water or other agents, or as often by self-pollination – are sometimes included in the introductions to a family or genus, or within the species texts preceded by '**P**'.

Throughout the text, **technical terms** have been kept to the minimum compatible with precise description. Some – such as bract, calyx, corolla, stipule – are unavoidable. They are all explained, and where appropriate illustrated, in the Glossary on pages 31-38.

Descriptions of closely related species concentrate on features of difference by referring back to the previous species – for instance:

'*Sorbus meinchii* (Lindeb.) Hedl. Like *S. hybrida*, but . . . '

Similarly the descriptions of subspecies and varieties refer back to the species itself (often called the type), again by indicating points of difference without repeating the characters they have in common.

FAMILY AND GENERIC INTRODUCTIONS. Where space allows, brief introductory notes are given for each family – its size, range and common characters

Species within the same genus share a number of characteristics in common. Such details are generally to be found in introductory notes preceding a genus with three or more species.

IDENTIFICATION of wild plants encountered is a facility which grows very rapidly, in both ease and pleasure, with practice. A ×10 magnifying glass (preferably of the small and folding variety, worn round the neck on a string against the otherwise almost inevitable loss) is a great help with small but often crucial features, such as hairs, glands and the nature of the ovary. Many species – perhaps most – can be identified easily enough simply by comparing the plant with the pictures and matching like with like; then checking one's visual judgement against the text. If closely similar species are involved, the salient features of each may need to be considered. With *white* flowers, remember that many species produce an occasional albino form, otherwise identical to the normal plant. Habitat is usually an important factor, and to a lesser degree flowering season and association with other plants. Where fruits are the critical factor, identification will be that much harder early in the year, but in many such cases, particularly with the Umbellifers, the plants obligingly carry both flowers and fruits for a large part of their flowering season.

Where the plant cannot be found, one can resort to **Keys**. These are grouped at the end of the book, with first a general key to families, then individual keys to genera within nine of the largest or most difficult families. Such keys may look a little daunting to the beginner, but their use is quickly learned, and by pointing up the successive categories or 'choices', from the wider to the more particular, using a key to run down a plant or family can itself be most instructive.

## CLASSIFICATION

The modern classification of living forms, as conceived by the great Swedish naturalist Linnaeus in the eighteenth century, aims at reflecting the course of evolution. Of the four major groups of plants only the most recent, and now dominant one, the Spermatophytes or seed-bearers, are our subject here – the others being algae, mosses and liverworts, and ferns. The seed-bearers are divided into two major groups, the Gymnosperms, which contains all the conifers and related plants, and the Angiosperms which are the true flowering plants. In the Gymnosperms the female cells are exposed (Greek *gumnos* = naked). In the Angiosperms they are hidden beneath layers of cells which form a protective wall, the carpel (*angeion* = a vessel).

The Angiosperms are sub-divided into two major divisions. **Monocotyledons** have a single initial seed-leaf (cotelydon). Most have narrow leaves with parallel veins and flower parts generally in multiples of three. Here belong all the members of the lily, iris and daffodil families, many of the water plants, the orchids, grasses and sedges. In contrast, **Dicotyledons** have two seed leaves (below or above the ground), diverging net-veins, and flower parts generally in multiples of four, five or seven, occasionally more. The majority of flowering plants are dicots, from the buttercups and cresses to the vetches, roses, daises and thistles.

Within these two large orders, plants are further divided into smaller groups, notably families, genera (singular, genus) and species. It is worth remembering that only with species are we referring to actual living organisms. The higher groupings have no existence in nature; they are purely human concepts, attempts to understand and arrange in a coherent system diverse forms of life. Subspecies and varieties are more concrete, distinct forms observed within actual species populations.

**Hybrids** between species sometimes occur in the wild, or more often in disturbed or semi-disturbed places. Hybrids are often to be found in the vicinity of the parent species, but not always so, and they generally exhibit intermediate characteristics. Many naturally occuring hybrids are sterile or partially sterile, setting very little seed and usually occuring as isolated individuals. However, hybrids may be fully fertile – hybrids between White and Red Campions, for example. In these instances complex hybrid swarms may result with the hybrids repeately crossing and backcrossing with one or both parents. The possibility of a hybrid should always be considered if a particular plant does not appear to 'fit' within the known range of variability of a species, or if it appears to comes midway between two distinct species growing close together.

**Apomixis** is an extraordinary device of nature – the production of seed without fertilization. The offspring simply inherit the exact genetic material of the parent, whereas in ordinary cross-pollination they inherit half their genetic material from each parent. It occurs particularly in certain genera – notably in lady's mantles (*Alchemilla*), hawkweeds (*Hieracium*), brambles (*Rubus*), whitebeams and rowans (*Sorbus*) and dandelions (*Taraxacum*).

In this process, the ovules develop simply into seeds and contain similar genetic material to all the other cells in the plant. The main difficulty in dealing with the effects of apomixis is that when minor variations occur in the offspring – caused perhaps by simple mutation of a gene, which may result in a slightly different flower shape or colour – this is fixed and thereafter persists from one generation to the next. The result can be a whole series of very similar looking plants, differing only in very minor features from one another. Many botanists used to, and indeed still do, describe these variants as distinct species because the differences are maintained by the process. Thus apomictic groups characteristically have large numbers of 'species' which differ from each other in much finer detail than the average outbreeding species. In *Rubus* alone in the British Isles, more than 300 such apomictic species are recognized. In this book such minute distinctions would be out of place.

Again, certain critical genera have groups or **aggregates** of very closely related species, sometimes referred to as microspecies. Although it is generally easy to identify the group, it is often extremely difficult to determine the exact species without a specialized knowledge of the genus. In these cases, a simplified account is presented here with only the principal species described, particularly in the genera *Alchemilla*, *Hieracium* and *Rubus*.

## THE PARTS OF A FLOWER

The number, shape and arrangement of flower parts is very often a prime factor in identification, so they have to be learned. Whole families can often be recognised by a combination of flower characters – the crucifers, *Cruciferae*, for example, by the combination of 4 separate sepals alternating with 4 separate petals, 6 stamens and 2- celled ovaries. The principal parts are as follows:

Typical Crucifer flower

The **calyx** is the outermost whorl of the flower, composed of **sepals**. As with the petals (see below), there is usually a set number per flower, often 3, 4 or 5. Sepals are usually, but far from invariably, small, green, and less conspicuous than the petals. They may be separate from one another, joined at the base or joined together into a tube. Occasionally, they are large, coloured and petal-like as in *Anemone*, where they replace the petals in their function. Or, as in species of *Potentilla*, there may an extra outer whorl of sepals, often smaller and referred to as an epicalyx.

Sepals are more important to the flower than they may seem. They protect it in bud, and being green they help to disguise it and by photosynthesising supply the developing flower with energy. They may be smooth, inflated, prickly or covered in sticky hairs – all devices to fend off marauders.

Sometimes they are large, coloured and petal-like, as in the Globeflower and Marsh Marigold. Often it is difficult, or even almost impossible, to tell the difference between sepals and petals, as in tulips, where there are two whorls, each with three petal-like organs. In such cases botanists generally give the whole lot a collective name, the *perianth*, or refer to the parts as **tepals**. To add to this confusion other organs may look like sepals or petals. **Bracts**, modified leaves which surround flowers or flowerheads, may become elaborated into large petal-like organs, often far more conspicuous than the actual sepals and petals – and indeed they serve the same function of attracting potential pollinators. This can be seen in dogwoods and some thistles, which have conspicuous bracts and only very reduced sepals and petals.

To the inside of the calyx are the **petals**, which make up the corolla. They are normally the most conspicuous part of the flower, and vary the most in size, shape and colour from species to species. They often equal the sepals in number, generally alternating with them, and they too can be either separate from one another – as in buttercups – or variously fused together. In Campanulas they are united into a bell-shape; in the Pea Family they are differentiated into a standard, two wings

Forms of corolla

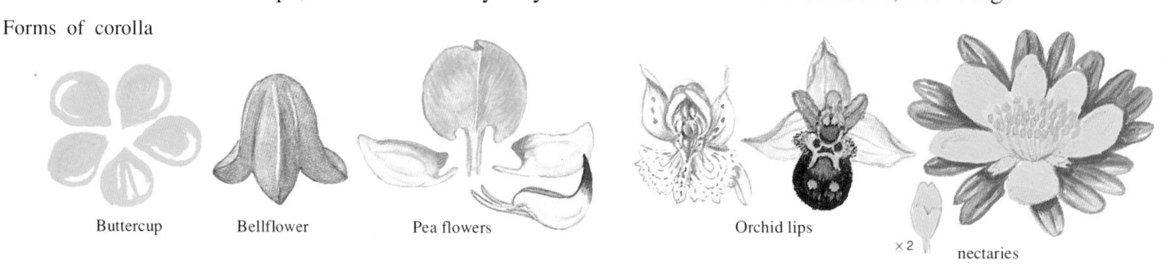

and a keel. In some groups, including most orchids, one petal, generally the lowermost, is elaborated into a conspicuous and often highly specialized lip. Very occasionally the petals may be absent altogether. In the Winter Aconite and Love-in-a-mist the petals are modified into small nectaries on the outside of the stamens, the place of the petals being taken over by coloured, petal-like, sepals.

**Stamens**, the male organs arising within the petals, though often protruding beyond them, carry pollen in specialized processes called anthers, borne on stalks or filaments. Flowers may contain one or many stamens. They may be separate from one another or joined together in various ways, sometimes in bunches, or attached to the corolla or to the rim of a distinctive receptacle – as often in the Rose Family. Flowers of some species may be male, bearing only stamens and no female parts. Sometimes some of the stamens are modified into non-pollen bearing processes called staminodes and in certain female flowers all the stamens may be so modified or absent altogether.

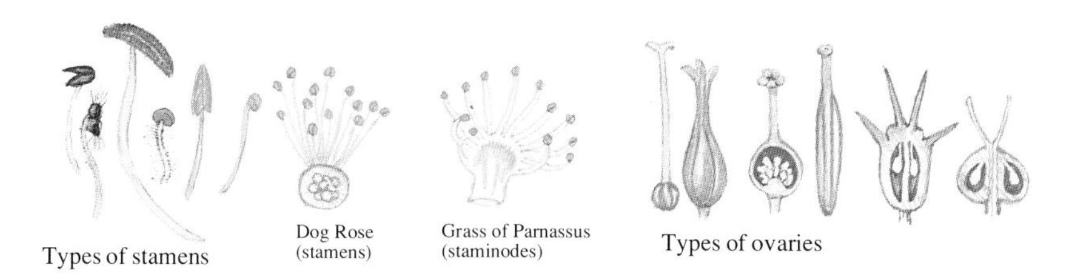

Types of stamens — Dog Rose (stamens) — Grass of Parnassus (staminodes) — Types of ovaries

The female organs of the flower consist of several parts, the **ovary**, **stigma** and **style**, forming the innermost organs of the flower. The ovary may consist of one or several separate or variously fused units or carpels, which contain the ovules; once fertilized the ovules develop into the seeds. The stigma is the receptive surface that receives the pollen and it is generally united to the ovary by a distinctive stalk or style. Flowers may bear a single or several styles.

During the process of evolution, plants have developed more and more specialized adaptations to pollination. The buttercup represents a primitive type, orchids a considerable sophistication. The systematic order aims at following the evolutionary sequence. Hence plants with simple flowers, separate sepals and petals occur in the earlier part of this book, and those with more advanced flowers and joined petals in the later.

The ARRANGEMENT of flowers on plants is a complex subject in itself. In the simplest form flowers are solitary, but more often they are arranged in a specific way. When flowers are aggregated together they rarely open all at the same time but rather in succession to a predetermined pattern, thus ensuring pollination over a prolonged period.

The **spike** is the simplest form of inflorescence with the flowers arranged directly on by a simple unbranched axis. In spikes the flowers usually open from the bottom upwards. A **raceme** is essentially a spike where each flower has a stalk joining it to the main axis. In the **cyme**, each branch terminates in a flower, as do a succession of subsidiary branches below flower, the flower terminating the first branch being the first to open. Cymes may develop bilaterally, producing an even inflorescence, or unilaterally to produce one which is one-sided. In an extreme form of the latter a spiral may result, as in many borages.

Flowers may also be massed in broad or tight heads or **umbels**, as is characteristic of the Cow Parsley family, the umbellifers. All the flowers are borne on stalks arising from one point on the main stem, like the spokes of an umbrella. Each main spoke may be further divided into secondary umbels. The mass effect of numerous small flowers held closely together is irresistible to many insects. Both racemes and cymes can sometimes produce a similar umbel effect with numerous small flowers gathered together in a domed or flat 'head'. In these instances, however, the spokes arise along a contracted axis and unlike a true umbel vary greatly in length so that the lowermost are longer than the upper.

The **composites** or daises display the compactness of an inflorescence better than any other group of flowering plants. In the typical daisy flowerhead numerous small individual flowers, or 'florets', are held closely together on a flattened or domed disk. The outermost florets are often sterile with a long petal-like limb, which attract pollinators and give the daisy flowerhead its characteristic appearance. The central or disk florets have a very reduced corolla. There are several variations on the composite theme. In some, such as the dandelions, all the florets are rayed while in others like the sneezeworts they are all of the disk type. Altogether, composites have solved the problem of successive pollination by cramming in as many flowers into as small an area as possible. The florets mature in succession, generally from the perimeter of the disk inwards.

LEAVES, like flowers, are remarkably variable. Apart from harvesting the sun's energy by photosynthesis – converting carbon dioxide and water trapped in the leaf to sugars and starch – they have other important functions. Their pores or 'stomata' transpire water from the surface, keeping the plant cool, and also effect the exchange of carbon dioxide and oxygen between the leaf cells and the atmosphere. In cool wet regions water loss is unimportant, but in hot dry places it would can result in severe wilt, and leaves have developed a variety of devices to avoid this. The skin may simply be very thick, or the pores few and scattered or indeed all confined to the lower surface away from the direct sunlight. The leaf-margin may be rolled under to reduce the exposed surface. The leaf itself may be small and narrow, or covered in hairs or scales to reduce water loss and reflect excessive sunshine.

The position of leaves on the plant far from haphazard. They may be arranged spirally, so-called 'alternate' leaves, with the spirals in a mathematical progression so that, for instance, every fifth leaf along the stem is in the same position as the one five below; or it may be every third or seventh leaf, and so on. Some have complex systems with two or even three different spirals in combination. Another way of dealing with the 'shadow effect' is for leaves to be borne in pairs, each pair at right angles to the one above and below. Or leaves may be in whorls spaced along the stem, the leaves within each whorl not or only slightly overlapping. Even when a leaf is made up of a number of leaflets, the spatial arrangement of each is important.

The object of all these sytems, together with the inclination of the leaf, is to maximise the total area of leaf exposed to the sun. Of course, in practice not all leaves on a plant can get full sunlight. On large trees those lower down are going to be more shadowed than those high up. The 'shadow effect' is often reflected in larger and thinner leaves, apparently to increase the ratio of leaf-surface to available light.

FRUITS. The fertilized ovary develops into the fruit, which may be simple, being developed solely from the ovary, or of more complex origin, involving various floral organs. In certain groups, particularly among the crucifers and umbellifers, fruit characters may be the only satisfactory way of distinguishing the genera, or indeed individual species. Studying fruits, of which there is an enormous variety and whose characters often show quite well long before they are ripe, can be just as rewarding as examining the flowers.

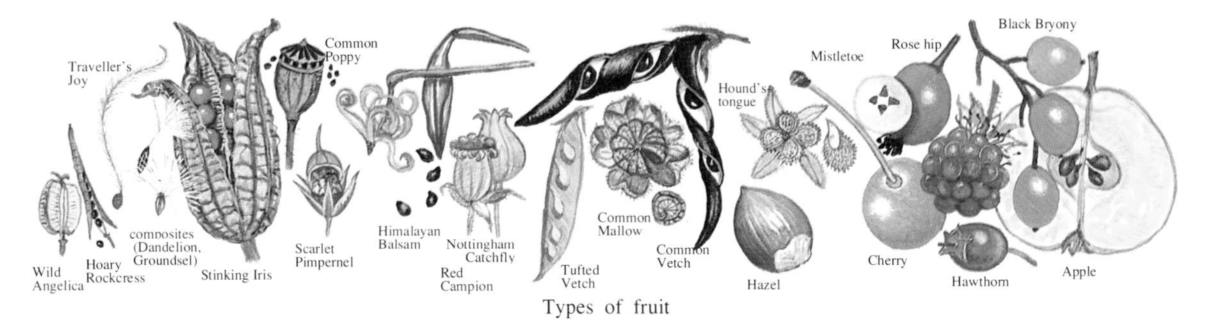

Types of fruit

**Achenes** are single-seeded dry fruits which do not split open when ripe. A flower may develop one or a cluster of achenes. In the Buttercup, a cluster of achenes is normal, whereas in the members of the Daisy Family, each flower (or floret) produces only a solitary achene. In dandelions, *Clematis* and *Pulsatilla* the achenes bear persistent feathery styles, little parachutes with which they can be carried away by wind.

**Capsular fruits**, found in many species, generally consist of a number of fused carpels which may or may not be separated by internal compartments. Capsules vary greatly in size but all split in some manner to release the seeds. The majority split lengthwise as in many monocotyledons – the lily, iris and tulip for instance. Others, like the poppies, have a ring of pores near the top from which the seeds fall when the capsule is buffetted by wind; and in the pimpernels the capsule splits 'equatorially' so that the top half comes off like a cap. Capsules sometimes explode suddenly, to shoot the seeds away, as in balsams, and in the Pink Family they often split at the top into a number of teeth. Capsules vary a good deal in their surface features – they may be smooth or hairy, bristly, prickly or armed with long spines.

**Legumes** are single carpels or pods which split lengthwise into two valves with the seeds attached along one edge only – the typical pod of the Pea Family, Leguminosae. But legumes are highly variable in shape and characteristics: in some the pod splits up into one unit lengths.

**Mericarps**. In fruits in which the carpels are closely fused together the seeds sometimes split off in one-seeded portions or mericarps. This is characteristic of Umbellifers and the Mallow Family.

**Nuts** are of course one-seeded fruits with a hard outer coat or shell. Similar small fruits are often referred to as nutlets: most members of the Labiate Family have fruits with 4 nutlets. **Berries** and **drupes** are both fleshy and often brightly coloured when ripe. Berries are several seeded, the seeds without a stony outer layer – for example in Mistletoe and Black Bryony, while drupes have one or several seeds, each surrounded by a hard stony layer – as in the cherry. In the Raspberry, the fruit consists of a collection of small drupes, or 'drupelets'.

**Hips** (or 'heps'), the typical fruit of the rose, have the seeds attached to the inner wall of a fleshy receptacle, to which the sepals, petals and stamens were also attached when in flower. Hips often form a deep urn-shape with a narrow hole at the top through which the styles protrude. Lastly, a **pome**, the typical fruit of the

Apple and Pear, has a thick fleshy swollen receptacle, with a hardened wall to the ovary, which forms the core surrounding the seeds.

### POLLINATION AND POLLINATORS

The pollination of flowers is a large subject, with its own considerable literature. Involving as it does the observation of pollinators' behaviour patterns it is also far from perfectly understood: there are many species about whose pollination methods little is known.

The transfer of pollen from one flower to another is effected by wind, water – even sea-water, birds, mammals, spiders, but most of all by insects. To them, petals with their bright colours and distinct patterns are often signalling flags as well as landing platforms. Different insects are attracted by different **colours**: flies and butterflies particularly by yellow and blues; bees by blues, purple and reds (in fact any colours that reflect ultra-violet light, and highlighted areas not visible to humans); and moths by white. Lines or patterns on the petals can apparently lead them to the correct part of the flower, the sexual organs. The streaks on the Pansy and the ring in the centre of a Forget-me-not serve the same purpose.

**Scent** cooperates with colour. Bees are attracted by light sweet scents; butterflies and moths by heavy, more sickly ones; wasps, flies and beetles generally by unpleasant, to us often foul smells. Plants even release their scents when the appropriate pollinators are likely to be abroad. Broom and Cowslip are day-scented, but Honeysuckle is most fragrant in the evening, and White Campion at night to attract night-flying moths.

**Pollen** is rich in protein and is a valuable food for small animals. Clearly, if insects were efficient at gathering and keeping it all, no pollination would result. So flowers which produce copious pollen, such as poppies and buttercups, often have open symmetrical flowers that can be approached from a number of directions. Visiting insects will generally rumage around the 'bowl' of the flower, eating or collecting pollen but at the same time being dusted all over with it.

With many 'tubed' flowers, however, especially legumes and two-lipped flowers, insects searching for the sugary nectar at the base of the tube can only enter the bloom from one direction. The stamens are so situated as to brush part of the insect as it enters. In the case of most legumes the stamens are situated beneath so that the underparts gets dusted; in labiates they are above, to brush the creature's head or back.

Not all flowers need to have large petals to attract pollinators. In many groups, composites and umbellifers in particular, small flowers are gathered together in massed heads, often with just one or more larger petals on the outer flowers in the head, are just as effective.

Pollination **strategies** are sometimes highly sophisticated. The wild arum traps insects in an elaborate spathe which enfolds the flowers. During the day the spathe is open and emits a disgusting smell. Flies and beetles attracted by it are funnelled down the spathe, past a ring of downward-pointing hairs which prevent escape, then past a zone of immature male flowers to the female ones at the base. There, if they have picked up pollen from another plant, the insects may transfer some of it to the receptive female flowers. But they are still trapped. The following day the male flowers mature and shed pollen on them, the spathe and ring of hairs quickly start to wither, and the insects can finally escape.

In some orchids not only do the lips closely mimic the appearance of certain bees, spiders or flies, but the flower even emits a smell similar to that of the female. The visiting insect or spider male is fooled into thinking that it is about to mate, and inadvertently picks up pollen (in form of bundles or 'pollinia') before moving on to another supposed female.

In many species self-pollination is still possible if cross-pollination fails for some reason, such as the weather being wet or there being no pollinators about; their fail-safe systems mostly involve the stigma eventually coming into contact with the pollen. In some groups 'cleistogamous' flowers are borne regularly, often either before or after the normally pollinated flowers, in which self-pollination occurs without the flower fully opening. Many species of violet produce them during the summer, after the normal insect-pollinated flowers of spring. Cleistogamy limits genetic diversity but enables a species to build up a large local population.

In general, however, self-pollination is disadvantageous, and plants have developed a number of ways of ensuring cross-pollination. The simplest are when the male and female flowers are on different plants (called **dioecious**) as in hollies and willows; or they may be on the same plant but separate from each other (**monoecious**) as oaks and hazel. Most species, however, have organs of both sexes in the same flower (**hermaphrodite**). In these, often either the male or the female parts ripen first so that in any population there will be some female and some male parts fertile on different plants at the same time. Or the pollen may simply be incompatible with the stigma of a flower on the same plant. Or there may be a mechanical arrangement as in the primrose. Here there are two flower-types, one with a long style and stamens low down in the flower-tube, and the other the reverse. Each plant has only one type of flower and a visiting insect must transfer pollen from one flower type to the other for fertilization to occur. In the Purple Loostrife this system of **heterostyly** is taken to a further extreme, with three basic flower types with stamens and styles set at different levels.

## HABITATS

Within each type of habitat there is a unique association of different plants. This composition of species varies from place to place within habitats according to local factors such as soil acidity, dampness, the amount of shade and so on. Within a particular habitat a particular plant may be either common or rare; in many, one or sometimes several species dominate the vegetation – the beech trees in a beech wood, for instance. Plant ecology is a fascinating though complex subject. Those wishing to learn more can have no better starting point than *A Guide to the Vegetation of Britain and Europe* by Oleg Polunin and Martin Walters, which analyzes the varied habitats and ecology of the whole of Europe, region by region. The following is only an outline of our major habitat types.

### WOODLANDS

It is likely that several thousand years ago one could walk from the south of France to central Scandinavia through an almost uninterrupted forest. Today, most of it is fragmented into scattered patches. The decrease is, of course, the work of man, felling trees for timber and to clear areas for agriculture. What woodland we have today has nearly all been coppiced, where selective felling has removed either the larger trees or chosen species. Except in parts of Scandinavia and in the mountains of central Europe, precious little natural unaffected woodland remains in western Europe.

Plantations, whether of a single or of several interplanted species, generally lack the rich assortment of shrub and herb species found in natural woodland. Hardly any natural forest is left in Britain, except for some isolated patches in Scotland and a few in England and Wales. A little, carefully managed over many centuries, still persists in East Anglia and elsewhere. Over the whole region the different types of natural woodland, reflecting different soil types, rainfall and temperature regimes, are as follows.

**Coniferous Woodland.** Evergreen forests of pine, fir or spruce characterize much of the continental and boreal regions. The harsher climate of the northern and continental regions favours coniferous trees, whereas the milder Atlantic zone favours broad-leaved deciduous species.

The dark and rather sombre forests of the boreal region are generally called TAIGA and are dominated by two species, Norway Spruce and Scots Pine, the spruce often forming extensive almost pure stands, especially in Scandinavia. The Taiga stretches in a broad unbroken band from northern Europe into Russia and Siberia. Because of its dense evergreen habit such forest excludes most other species of trees and shrubs. However, where the canopy is thinner birches and Rowan are often present. On damper ground the conifers may give way to alders. The scrub layer in the Taiga is poorly developed but is composed chiefly of various low shrubs of the heath family, Ericaceae. However, a rich assortment of small herb, ferns, mosses, lichens and fungi are also present, varying with the acidity of the soil.

Further south on the continent, away from the Atlantic, the Norway Spruce is widespread, especially in S Sweden, Finland and Germany, and stretching southwards into the Alps. Much of the lowland SPRUCE woods are uniform stands that have been planted by man. Natural spruce woodland probably only exists in the mountain areas. As in the Taiga the trees are lofty, sometimes reaching more than 30m high and casting a dense shade.

Sub-alpine spruce forest forms the tree-line in the Alps, although often topped by a thin band of deciduous trees, beech and birch in particular. Silver Fir, more sensitive to exposure and frost than the Norway Spruce, occurs mainly in mixed forests in parts of central and east Europe and in parts of France and Germany. It tends to develop a dense canopy so the shrub layer below is only poorly developed, but there is a rich assortment of flowering herbs, ferns, mosses and lichens.

The main species of PINE in our region is the Scots Pine, with a wide altitudinal tolerance, growing typically on acid soils, especially light sandy ones. In the Alps pine woodland usually has a dense scrub layer, often with Alpenrose and other ericaceous shrubs, and a host of small herbs. In the Scottish highlands it is often associated with Juniper, Rowan, birches and various willows, but it is far more restricted than formerly. In parts of southern Britain it has been established for many centuries on heathy soils,though probably not natural in the first instance.

**Deciduous Woodland and Forest.** Broad-leaved deciduous trees, which enjoy long warm summers and mild winters together with a heavy rainfall, are typical of the mild Atlantic region, usually with a well-developed shrub and herb layer.

OAK woodland is typically dominated by two species, Sessile Oak (*Quercus petraea*) and Pedunculate Oak (*Q. robur*). The former predominates in the north and west on drier, often acid, soils while the latter prefers heavier, wetter alkaline soils, especially in the south and east. The composition of oak woods varies from place to place, often mixed with birches, Rowan and Bird Cherry in the north. Pedunculate Oak woodland generally forms a richer community, with birches, Hornbeam, Holly, Apple, Bird Cherry, Wych Elm, and many often colourful herbs.

In central Europe Oak-Hornbeam woodland is frequent, especially in places flanking the Alps and the mountains of central France, to an altitude of about 500m.

BEECH woodland is typical of continental Europe where it forms the natural forest over large areas from the lowlands to the mountains, generally on rather average, often dry, shallow alkaline to slightly acid soils. On the deeper richer soils Beech is unable to compete with Oak. It stretches into the southern part of the Atlantic region, particularly in south-east England and north-west France, but in many areas it has been cleared for agriculture. Beech forms a dense leafy canopy which excludes most direct light from the forest floor. Heavy leaf-fall and slow decomposition result in a deep humus layer. The shrub layer is poorly developed and in the leaf litter saprophytic plants take advantage of the low light intensity, moist humus and lack of competition. On chalky soils in particular, orchids are prominent, though local. Herbs usually form scattered communities in the woodland, though a few like Dog's Mercury and Bluebell may form a continuous carpet, flowering before the Beech comes into full leaf.

In mountain regions Beech is often co-dominant with Silver Fir above 1000m, and at high sub-alpine levels it is dwarfed by altitude and may be only 3-4m tall, though with a richer herb layer than in the lowlands because of the better light penetration.

ASH woodland occurs locally, generally somewhat to the north of the Beech zone, but it may represent a stage in the succession to Beech itself. It is found in river valleys or on limestone rocks and screes, with Whitebeam, Yew, Small-leaved Lime, Wych Elm, and richer shrub and herb layers than in Beech woodland. In some areas of central Europe, but also in south-west England, OAK-HORNBEAM woodland occurs locally; it is far more widespread in southern Europe.

Deciduous woodland also occurs in the far northern regions, often dominated by BIRCHES – Downy Birch, with Silver Birch to the south. On disturbed land it may form a stage in the succession back to conifers. Birch forest is characteristically very open, with a consequently rich assortment of shrubs, herbs, ferns, mosses and lichens.

In parts of southern Sweden, Finland, central and southern Germany, are ALDER – Grey Alder, often with poplars and willows and regularly coppiced. In Britain, Ireland and France there is more Common Alder, forming transitional woodland around the margins of lakes and marshes or along river banks.

GRASSLAND

Grassland is a characteristic feature of large areas of Europe from low altitudes to the higher mountains. It varies greatly with both climate and soil and is also affected by the use of fertilizers, which can promote certain species at the expense of others. Unhappily, ancient pastures have often been ploughed up in recent years and replanted with one or more, often 'improved' species. This destroys the established community and yields a monoculture devoid of wild plants. The composition of meadows, too, may differ markedly if one has been continously grazed and the other used primarily for the production of hay.

In the west and north-west of our region, the **Atlantic** influence encourages rich lush meadows and grassland that can grow almost continuously except in the depths of winter. These are semi-natural, many occupying land areas cleared of forest many centuries ago, and consist primarily of native species formerly present on forest fringes on in clearings. Left to itself, scrub and eventually trees would gradually move in, reconverting the land to woodland.

The richest grassland is found on neutral or slightly alkaline soils. On limestone or chalk hills the soil may be thin and dry, but still harbour a wide range of colourful species. The slopes may be too steep for cultivation, but the flora still affected considerably by the grazing of livestock, particularly sheep, and rabbits. On more acid soils, especially in upland regions or on sandy soils, the composition of the grassland can be very different, generally with fewer species. Common flowers growing on such meadows are Tormentil, Devil's-bit Scabious, Lousewort and Heath Bedstraw.

Wet meadows occur in low lying areas close to rivers and lakes. As they become wetter so they grade into true marshland (see pp. 25-26). A variety of moisture-loving perennials such as Irish Fleabane, Snakeshead Fritillary, Siberian Iris, Marsh Gentian and Ragged Robin are characteristic. Close-grazing or mowing has the effect of eliminating many typical grassland species, but a range of plants has adapted to low sward, often with basal leaf-rosettes and quick maturing fruits which can beat the 'cutters', such as Common Daisy, Broad-leaved Plantain and Dandelion.

Atlantic grassland merges gradually eastwards into **continental** grassland, with a more severe regime of cold winters and hot, relatively dry summers. It too is semi-natural: left ungrazed, especially at lower levels, it will revert gradually to dry woodland. With hot dry summers there is a vigorous flush of growth in the spring and early summer, followed by a long period when the grassland dries out, then as temperatures lower in the late autumn and the ground becomes moister, often another, though lesser, flush of flowers. A prominent feature of the earlier flowering, particularly in the south of the region, is the many species of orchids.

**Mountain** grassland, below the tree-line, is like those of the lowlands only semi-natural, but above the tree-line natural grassland exists, often on poor, thin and rocky, acid or alkaline, soils. Lower meadows are primarily used for hay followed by grazing, but the higher alpine meadows, which often have a low growth, are almost exclusively used for grazing. At these high altitudes small herb communities are typical, often colourful, through the short summer.

### FRESH WATER HABITATS, SWAMPS AND MARSHES

The plants that inhabit watery places, swamps and marshes and river meadows are remarkably similar throughout the region. The most important factors that affect them are the composition of the water, whether acid or alkaline, still or fast-flowing, mineral rich or poor, as well as the actual depth of the water itself. As in most habitats, some plants have a wide range of tolerance, being able to grow and compete in widely different situations, whereas others, the rarer species, are very much restricted by the prevailing conditions.

**River and stream** communities are variable but generally poor in the number of species likely to be found. Fast-flowing water has an obvious limiting affect. Those few species that are able to get a roothold generally have long flexible stems and ribbon leaves to cope with the swirling waters. Slow-moving water is less damaging to plants than swifter waters, plants are able to get a firmer and more permanent roothold, some submerged, others appearing above the water surface.

Along stream banks there is usually a good assort range of species, many flowering during the summer, often with a bordering zone of alders and willows, which are tolerant of occasional inundations.

**Lakes and ponds**, including dykes and ditches, with nutrient-rich waters generally harbour a richer array of species than waters that are acid or poor in nutrients. Still waters are often dominated by water lilies, though these will not tolerate very muddy or polluted water. In recent years, particularly in the north of the region, acid rain has severely affected freshwater communities, especially lakes and ponds, many of which are now biologically dead. In freshwater habitats generally, various submerged plants are important for oxygenating the water. In poor waters bladderwort species capture tiny water organisms in underwater bladders to supplement the general deficiency of nitrogen and other minerals.

**Swamps** are those areas of land that are permanently saturated by shallow water. **Marshes**, on the other hand, are waterlogged below but the soil surface is above the water level, except for occasional flooding. They tend to have similar plant assortments, particularly coarse herbs able to compete with one another.

Many river margin plants are also present on flood meadows, providing they are not heavily grazed by cattle or sheep. Marshy ground in mountains and in the north of our region generally have a layer of lower herbs, often indispersed by willows and birches.

HEATHS, FENS, BOGS AND MIRES

Very distinctive of the Atlantic influence, **heaths** are covered in a dense growth of ericaceous plants – heathers, Ling, Bilberry etc. This habitat is largely man-made, arising where the native woodland has been cleared or burnt, and prevented from returning to woodland by recurrent burning, grazing and tree-clearance. Only in some of the most exposed areas of south-western Britain and south-western Ireland can heathland be considered to be the natural (climax) vegetation.

Heathland requires much moisture and a temperate climate, so it is found from the Atlantic seaboard of Spain and Portugal northwards to SW Scandinavia, and in a small area in NW Germany which comes under the influence of the mild Atlantic weather. Heaths are primarily developed on acid siliceous soils which bear a typical profile, the podsol. The surface layer is a thin dark covering of peat with a pale crust beneath which has been leached of both peat and minerals. Below this the leached particles accumulate in a dark layer.

On the whole, heathland has diminished over much of western Europe in the past hundred years, from changes in land-use and the reafforestation of heathland areas. However, in some areas such as place like Northumberland in N England and the Scottish Highlands large tracts of heath moorland are maintained solely for the purpose of grouse shooting; they are burnt off in a strict rotation over a period of years to encourage the young vigorous growth of Ling upon which the grouse feeds. Gorse is often a conspicuous heathland shrub, forming extensive bright yellow colonies throughout the spring and summer, especially near the coast.

In north-west Germany similar heaths contain a number of species of broom (*Cytisus, Genista* and related genera) and for this reason are sometimes called Broom Heaths. They typically develop on light sandy soils formerly wooded with beech and oak. Ungrazed, these heaths quickly revert back to woodland.

In more northerly latitudes Boreal Heaths occur in the mountains and these may well be natural rather than man-made. They occur in parts of Norway and Iceland and are usually dominated by Ling and Dwarf Birch, Juniper and various ericaceous shrubs. Tiny Arctic alpines may also occur here, adding interest to what is a rather species-poor habitat.

In the mildest heathland areas more specialized heaths are to be found, witness to a period when the climate was much milder over the whole of Europe than it is today – small enclaves persist in SW Britain, SW Ireland, with similar associations in SW France, Spain and Portugal.

In wet areas where peat accumulates on the surface through the incomplete decay of vegetable matter, **mires** may develop. Where the soil is rich in calcium or base-rich salts the resultant plant communities give rise to **fens**. At the other extreme, on acid soils starved of calcium and base-rich soils, **bogs** will result. However, as often where conditions are not one or other extreme a transitional type of habitat will develop.

**Mires** are very extensive in the Boreal and Arctic regions, the peat accumulating on the surface just above the water. The low oxygen level and the continuous deposit of dead organic matter lead to a slow decomposition and the formation of distinct peat layers. Mires may even develop in regions of permafrost, where during the summer months, only the surface layer of soil thaws – peat accumulates on the surface where the conditions are extremely acid and characteristic hummocks, as much as 50m high may develop due to irregular freezing and thawing. Sometimes these raised areas form long ridges with depressions in between. The flora of such places is poor in species, from the mineral deficiencies of the soil as well as the northerly latitude, but Dwarf Birch and various ericaceous shrubs are extensive often with a good deal of lichens and mosses. In the wetter hollows *Sphagnum* mosses are extensive and indeed the prime peat formers. Various sedges and cotton grasses are common as well as the small sundews which entrap small insects on their sticky leaves. In some hollows the mineral content of the surface peats may be higher, which allows a richer association of low herbs to develop.

In the milder Atlantic regions, where the rainfall is considerably higher, both in lowland and upland regions, extensive **bog** formations are to be found – W Scotland, W Ireland and SW Norway in particular. On poorly drained flat or sloping ground, with a very low mineral content, extensive bogs may blanket the ground. These blanket bogs are often dominated by Purple Moor Grass and Deer Grass. In upland areas dwarf ericaceous scrubs, a range of small herbs, as well as cotton grasses, ferns, and an assortment of lichens and mosses are to be found.

In hollows or valleys where the drainage is impeded, shallow water may slowly be covered over by successive intrusions of sphagnum and associated plants, giving rise to a **floating bog**. However, such bogs are generally transitional and normally pass to marsh and finally to scrub and woodland as the water table is successively lowered relative to the accumulating layers of organic matter. But bogs may also develop in areas rich in minerals. As successive layers of peat accumulate so the area is pushed above that of the surrounds. The raised area becomes progressively poorer in mineral contents – such bogs are called raised bogs and generally harbour a similar range of plants to blanket bogs.

**Fens** are formed where the drainage water is rich in calcium and minerals, usually on alkaline or calcareous soils. Again many transitional stages between true fen and woodland can be seen – a natural succession as the region becomes progressively drier. In East Anglia fens may be artificially maintained by regular burning and reed cutting, preventing the establishment of shrubs and trees. Fens are typically richer in species than mires, especially of rushes and sedges, but there is often also a colourful display of summer-flowering perennial herbs as well as a number of orchids.

In central Europe most lowland fens have been destroyed and replaced by good agricultural land. However, in the mountains poor fens are quite frequent, especially in the Alps where they may occur as high as 2400m. Here they are often associated with late-lying snow patch areas and harbour a rich assortment of rushes, sedges and small alpines.

COASTAL AND ESTUARINE COMMUNITIES

The whole of our region has an extremely extensive coastline affording a variety of habitats from cliffs and mud flats to sand-dunes and estuaries. Plants that grow in coastal situations often have to be resistant to extremely exposed and salty conditions. The plant communities are generally very distinctive, though in well sheltered areas inland communities, particularly woodland, may come right down to the coast. At the same time, coastal habitats despite their exposure generally have a reasonably equable climate and species tend to be widespread. Certain groups such as the sea spurreys and glassworts are almost exclusively adapted to salty habitats, while some species like the Sea Pea and the Sea Aster are coastal species of large primarily inland genera.

A number of salt tolerant plants occasionally occur inland where the soils are saline. A number, for instance, occur along motorways inland from the coast where the heavy salting of roads during the winter has produced local saline conditions – this applies particularly to species of glasswort, saltwort and seablite. Many coastal species bear salt tolerant fruits, and seeds or pieces of plant are often distributed by coastal currents. Populations of seabirds also help to distribute seeds along the coast, often over long distances.

**Mudflats and salt-marshes**. These are typical low-lying areas subjected to frequent inundations by the sea. Relatively few flowering plants can tolerate long periods submerged by salt-water, but glassworts and related plants are able to do so and often form extensive communities together with eel-grass. Away from the mudflats the salt-marsh communities are occasionally waterlogged, generally draining through an intricate series of natural channels. They support a range of widespread, often low growing, species of which the sand-spurreys and sea-lavenders are typical. Yet farther from the sea on the landward side there is often an extensive area of rushes.

**Sand-dunes and shingle**. In contrast to mudflats and salt-marshes, dune and shingle provides a generally unstable environment with little soil, excessive drainage and a general lack of mineral nutrients. Such areas are exposed to both vigorous wind and wave action and it is often difficult for plants to get a roothold. Accumulations of shore-line drift litter, which contains various nutrients often helps to anchor seedlings or plant fragments. Thus foreshore communities are able to develop above the high tide mark, but they are extremely vulnerable to storms and entire colonies can be obliterated by a severe gale.

On SAND behind the foreshore, plant communities may help in the creation of sand-dunes by checking wind-blown particles from the foreshore. Here again detritus often accumulates in hollows, together with seeds and plant fragments. Sea-kale, Sea Beet, Radish and species of bistort are common colonizers. Dunes generally form a series moving inland, but eventually the oldest become stationary or 'fixed' as they become increasingly isolated from the sea and colonized by plants, especially woody species. Those dunes closest to the sea are the least stable, but they may reach 7m high. Two grasses, Sand Couch and Lyme Grass, often help to bind the sand and provide a roothold for other plants.

Further inland the dunes are often colonized by Marram Grass, which has extremely vigorous rhizomes that can penetrate several metres of sand and a large colony can bind a big area of dune. Some of the most interesting coastal plants are to be found here, such as Sea Bindweed, Sea Pea and Sea Holly.

Further inland again the dunes become progressively discoloured due to a carpet of mosses and lichens which give them a grey appearance. Such dunes are generally the home for an attractive array of flowering plants – gentians, Bird's-foot Trefoil, several orchids, Grass of Parnassus. As acid peat accumulates on the surface the dunes become brown and represent the oldest in the natural succession from the seashore. On these dunes the most conspicuous feature is the variety of shrubs, which often form dense thickets, much frequented by nesting birds. Ling forms an underlayer and coarser shrubs – Gorse, Juniper, Hawthorn and Sea Buckthorn are common.

SHINGLE presents a harsher environment with little retained moisture and an unstable surface subject to the vagaries of the weather. Pebbles are often graded into ridges following the shoreline, those ridges furthest from the sea offering the best possibility for plants. However, plants are generally few and spaced out. Only those with far-reaching roots, which can sap the moisture well below the surface, can succeed, such as Sea Kale, Horned Poppy, Herb Robert and Sea Pea.

Coastal CLIFFS provide a further habitat for a rich assortment of plants depending on the stability of the cliff and the rock type. The middle and upper parts of the cliff are well out of the reach of the highest tides, but are subjected to salt spray as well as fierce winds. Here many of the plants of salt-marsh and sand-dune communities may find a niche – Thrift, Sea Beet, Sea Lavender and various plantains. Other species are more specific to the cliff environment, their roots able to penetrate deep crevices for support and in their search for moisture and nutrients – Wild Cabbage, Fennel, Hoary Stock and Tree Mallow, are typical.

### ALPINE AND ARCTIC COMMUNITIES

Alpine plant communities occur above the tree-line in mountains. The tree-line itself is greatly restricted by climate, so that as one moves northwards the zones of vegetation occur at increasingly lower altitudes. In the Alps for instance the tree-line is at 2000m in places, while in central Scandinavia it is at about 1000m. Correspondingly, the snow-line also decreases in altitude, the snow lying lower and for longer in Scandinavia than at similar altitudes in the Alps. Alpine and Arctic plant communities have much in common. Plants grow in bleak and generally exposed places subjected to long cold winters and short summers. Yet these communities are relatively rich in species and there are, especially in the Alps, many endemic species occupying specific niches in what is a fairly complex habitat type.

At the end of the Ice Ages, alpine and arctic communities would have widespread. However, as the land warmed up and trees, shrubs and coarse vegetation were able to move in from the south, so these communities were increasingly restricted to higher altitudes or more northerly latitudes which act as refuges. In some cases they were cut off in small refuge areas where they may persist to this day, isolated by large distances from their nearest cousins – the Burren in W Ireland and Upper Teesdale in N England are good examples.

Alpine regions present a variety of habitats from low scrub to mountain heathland, grassland, cliffs, screes and moraines to the bare, very exposed summits of the higher ridges. The plants found in different communities is very much influenced by the rock type and the association of plants can be very different on adjacent acid or alkaline rocks.

**Scrub Communities.** Scrub covered areas may be very extensive above the tree-line. In north-west Europe willow scrub is common and generally associated with a number of coarse herbs such as Angelica and Wood Cranesbill. In the Alps Dwarf Pine generally forms mixed colonies with a scrub of Alpenrose and various other ericaceous shrubs.

**Alpine Heaths**. These are developed above 1500m on acid soils often with a mixture of Juniper, various ericaceous shrubs, Crowberry, smaller alpine herbs, grasses, sedges, ferns, mosses and lichens. Dwarf heaths may develop in exposed and windswept positions where the soils are thin. Here Creeping Azalea may often be seen as well as mountain avens.

**Alpine Grassland**. Though at high altitudes grassland may be natural (see p. 24), lower down in the scrub zone and even below the tree-line grassland it is likely to have been artificially established by tree felling or scrub clearance, and subsequently prevented from reverting to woodland or scrub by summer grazing. Here a rich assortment of colourful herbs brightens the meadows throughout the spring and summer.

**Snow-patch Communities**, where snow lingers in hollows above the tree-line consistently from one year to another, may develop forming a halo around the retreating patch of snow during the spring and summer. In such places plants may be under the snow for lengthy periods, sometimes for as much as eight months in a year. In exceptional seasons such communities may not become clear of snow at all. Small plants such as Alpine Snowbell, Spring Crocus and Dwarf Willow are typical.

**Alpine fens**. Where the water level comes close to the surface, acid alpine fens may develop, dominated by various rushes and sedges. Corresponding calcareous fens, which also occur in the mountains up to 2500m, though considerably lower the further north one goes, have a far richer assortment of plants.

**Rocks and cliffs** provide an extreme environment for plants, depending on the rock types and the stability of the rocks themselves, but for a surprisingly rich assortment - often tufted or cushion-forming perennials or dwarf shrubs, annuals being very rare in such places.

**Screes** are widespread in mountains, but present a very unstable environment, the establishment of plants being continuously hindered by fresh falls of rock detritus from above and by readjustments in the scree itself. Some species are able to cope with such conditions – either by sheltering around the larger rocks or by producing stout or slender stems, beneath the scree and constantly extended to keep the plant close to the surface. Roseroot, Glacier Buttercup and Mountain Sorrel are frequent scree plants.

**Arctic Vegetation**. North of the boreal forests in northern Europe there is an extensive wilderness mainly composed of rocks and dwarf evergreen shrubs of the heath family, Dwarf Birches and various small willows, together with an assortment of rushes, sedges, mosses and lichens, but relatively few grasses. This is the Tundra, forming a narrow zone in northern Norway and Iceland but extending southwards along the mountains of Scandinavia. It is far more extensive in northern Russia, Asia and North America.

Arctic regions are dominated by perma-frost and plant life can only exist in the top few centimetres of crust which may thaw out during the brief summer period. Despite this plant communities can be extensive, though limited in species numbers.

### CULTIVATED AND DISTURBED LAND

Cultivation has had a profound affect on the environment over many centuries, modifying the landscape and often destroying or partly destroying the natural vegetation. In Holland and Britain, for instance, very little remains in the way of natural unmodified vegetation; the woodland has been cleared away and replaced by grassland, arable land or moorland. Yet many of these man- made and semi-natural habitats bear a wealth of wild flowers, either native species that adapted to the changed environment, or introduced species that have become established. Some have become troublesome weeds of cultivation. Weeds like Corn Marigold, Corncockle, Cornflower, Corn Buttercup, Common Poppy and Pheasant's-eye are widespread, but it is difficult to find them anywhere in a truly wild habitat – over the centuries they have become almost exclusively associated with cultivated fields. Today many are in decline from the use of herbicides.

Regular cultivation of land encourages annual and perennial weeds, while discouraging woody plants. Only on waste land or on areas of abandoned cultivation and derelict sites can shrubs and trees become established. On light well-drained soils, brambles, Broom and Silver Birch are frequent invaders, and on the heavier soils Sycamore, Ash, Hawthorn and Sloe. Seeds may lie dormant in the soil for many years awaiting a suitable moment to germinate, or they may simply be blown in and quickly become established. Disturbance or cultivation 'opens up' the land allowing many annual or biennial species to gain a quick roothold that would otherwise fail to compete against more vigorous perennial species. Many of the annual species are quick-growing ephemerals capable of germinating at almost any time of the year; they may produce several generations in a single season and can quickly build up sizeable colonies. Many perennial weeds possess vigorous creeping and invasive rootstocks which are very difficult to eliminate, sprouting anew from the smallest piece left in the soil. Species thriving in such man-made habitats include Bindweed, Creeping Thistle, Enchanters Nightshade and Stinging Nettle.

Roadways and railways can actively assist the spread of some species by wafting seed along the route. The example of the Oxford Ragwort in Britain is well documented. A native of southern Italy and Sicily, this plant was first recorded on walls in Oxford in 1794. Since then it has spread to many parts of Britain, especially along railway lines. Canals and ditches can have a similar effect on alien species which prefer the moister soils, such as Himalayan Balsam and Giant Hogweed from the Caucasus, now widely in Europe along ditches and waterways.

In many man-made and man-maintained environments the semi-native flora is strictly managed by regular cutting, trimming or ploughing. The growth of shrubs and trees in places like railway embankments is often discouraged by regular cutting and firing. The coppicing controls their growth so that those regularly coppiced Ash and Sweet Chestnut, for instance, are prevented from flowering and fruiting. But it often encourages other native flowers, especially herbs – the prolific growth of Wood Anemone, Ramsons, Bluebell and Primrose in recently coppiced woodland follows a regular cycle. As thetrees begin to grow up again they gradually shade out many of the low herbs which decline in consequence. The regular removal of scrub in woodland and on grassland can positively encourage some rare or local native species and is regularly practised by conservation groups, especially to help various species of terrestrial orchids.

In areas where cultivation has been abandoned or where for example a railway has been discontinued it is easy to observe the succession through from coarse herbs and scrub back to woodland. The actual species in the succession is dependent on the surrounding vegetation and environment.

# GLOSSARY

achene

**Achene.** A single-seeded dry fruit, not splitting.

actinomorphic flower

**Actinomorphic.** A regular flower, radially symmetrical and capable of being cut into two equal halves in various directions, e.g. *Ranunculus*.

**Acute.** Forming an angle of less than 90°.

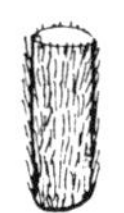

adpressed hairs

**Adpressed.** (or Appressed) Pressed close to another organ – generally refers to hairs when they are pressed close to stems or leaves.

**Aggregated.** In close groups or clusters. An aggregate species is one consisting of a number of very closely related species, often difficult to separate.

**Alien.** A plant which is not native, but which has been introduced, usually by man.

alternate leaves

**Alternate.** Leaves alternating on the stem, arranged in 2 rows or spirally arranged, but not opposite.

**Androecium.** The male organs of the flower – generally consisting of filaments, anthers and the pollen.

**Annual.** A plant that completes its life cycle from germination to fruiting in a single season. Many annuals germinate in the spring, but a few germinate in the autumn and overwinter.

anther

**Anther.** The upper part of the stamen which contains the pollen grains.

**Apomixis.** The production of seed without fertilization (without sexual fusion). Apomictic plants produce genetically similar offspring to the parent plant.

**Aromatic.** Scented with aromatic oils, as for example in many members of the Labiatae.

**Ascending.** Curving or pointing upwards.

auricle

**Auricle.** Small ear-like projections or appendages, usually at the base of leaves.

**Axillary.** Arising in the axil of a leaf or bract.

**Basal.** At the base – in some plants all the leaves occur at the base of the stem.

**Base-rich.** Soils containing large amounts of free basic ions such as calcium and magnesium.

berry

**Berry.** A succulent (fleshy) fruit containing several seeds, but without a stone layer around the seeds.

**Biennial.** A plant that completes its life-cycle in 2 years, germinating in the first and flowering and seeding in the second.

**Biserrate.** Doubly serrate, i.e. with each primary tooth divided into two smaller teeth, as sometimes on a leaf margin.

**Bisexual.** With both stamens and ovary in the same flower – hermaphrodite.

**Bog.** A plant community developed on wet acid peat.

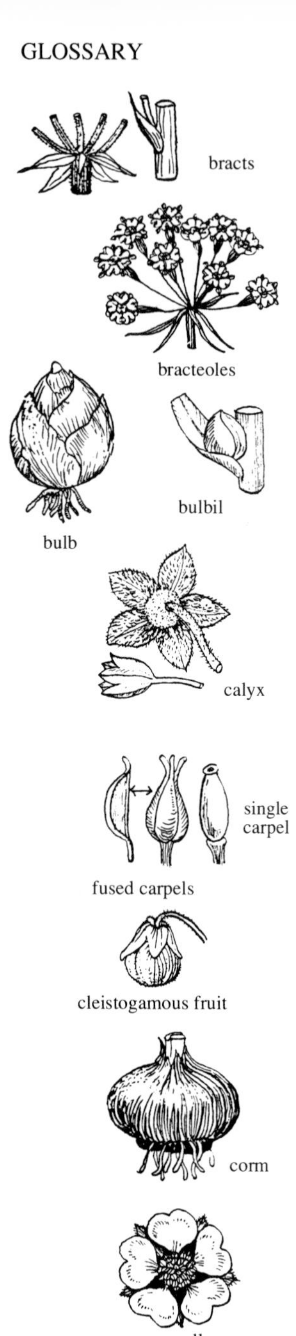


**Bract.** An organ, often small, but sometimes leaf-like, immediately below a flower and located where the flower-stalk joins the stem. In branched inflorescences the bracts on secondary branches are generally referred to as bracteoles.

**Bracteole.** See Bract.

**Bulb.** An underground storage organ developed from the fleshy bases of leaves or scales, with or without an outer skin or tunic.

**Bulbil.** A small bulb-like organ arising in the axils of leaves on aerial stems.

**Calyx.** The sepals – the outer whorl of floral organs in many flowers. Often small and green, sometimes fused together, but extremely variable from species to species.

**Carpel.** One of the units or compartments making up the ovary or gynoecium. Carpels may be separate from one another or variously fused together.

**Casual.** An introduced plant that has not become established, occurring here and there from time to time, often erratic in appearance.

**Catkin.** An erect or pendent tassel of tiny flowers crowded together – often either male or female.

**Chlorophyll.** The green pigment of plants, which harnesses the sun's energy by complex chemical reactions.

**Ciliate.** With hairs projecting from the margin. Often refers to leaves.

**Cleistogamous.** Flowers which fail to open and are self-pollinated. Cleistogamous flowers often occur on the same plant as normal open flowers, but are generally smaller and are often produced early or late in the season, e.g. *Viola.*

**Compound.** Leaves composed of several distinct leaflets or an inflorescence in which the main axis is branched.

**Corm.** A swollen underground storage organ produced from the swollen stem base, erect and generally annual, that of the following season arising on top of the previous, e.g. *Crocus* and *Gladiolus.*

**Corolla.** The petals – the second whorl of floral organs, located inside the sepals. Petals are present in the majority of flowers and are often large and coloured, but they may be as small as the sepals, sometimes absent all together.

**Corymb.** A compound inflorescence in which the lower branches are longer than the upper, bringing all the flowers to more or less the same level, e.g. *Achillea.*

**Cotyledon.** The initial leaf or leaves of a seedling plant, borne above or below the soil surface. Flowering plants are divided into 2 main divisions – those with a single seed leaf (Monocotyledons) and those with two seed leaves (Dicotyledons).

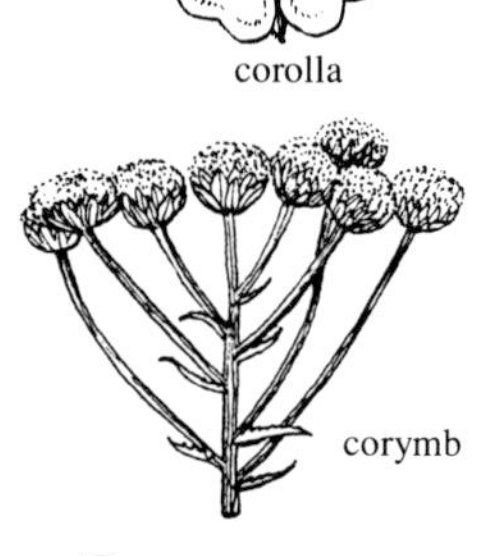


**Cultivated land.** Includes all land that is tilled, including both arable land and gardens.

**Cyme.** An inflorescence in which the main and lateral axis are terminated by a flower, so that continued growth of the inflorescence is from lateral growing points. Cymes may be regularly and symmetrically branched or one-sided and asymmetric.

**Deciduous.** Losing the leaves in the autumn.

**Deflexed.** Bent sharply downwards.

monocotyledon dicotyledon
cotyledons

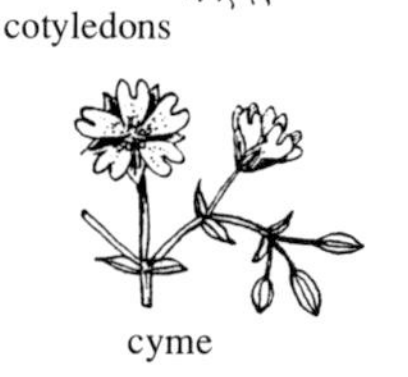

dehiscent fruit

disc florets

**Dehiscent.** Splitting open to release the seeds.

**Disk-floret.** The central floret of a flowerhead which forms a distinctive disk, e.g. in many members of the Daisy Family, Compositae.

**Dioecious.** Having separate male and female flowers borne on different plants, e.g. *Mercurialis.*

**Dominant.** The principal constituent of a particular plant community, e.g. Oak Woodland. Occasionally two or several species may dominate a community and are then co-dominant, e.g. Oak-Ash Woodland.

drupe

**Drupe.** A fleshy fruit with one or sometimes several seeds surrounded by a hard stony layer, e.g. Cherry.

**Elliptic.** Forming an ellipse, widest in the middle and pointed at both ends. Generally used to describe a leaf shape.

**Endemic.** Only native to one country or one area. In the context of this book the term is used to describe those species that only occur in the region, or a part of it, covered by the book.

elliptic leaf

**Epicalyx.** A calyx outside the true calyx – the calyx thus appearing to be composed of two whorls, e.g. *Potentilla.*

**Escape.** Refers to cultivated plants that occur casually outside cultivated areas, often adjacent to them – they have literally escaped, but have not become naturalized.

epicalyx

**Fen.** A community developed on alkaline or neutral, rarely slightly acid, wet peat.

**Filament.** The stalk of the stamen that connects the anther to the receptacle or the corolla.

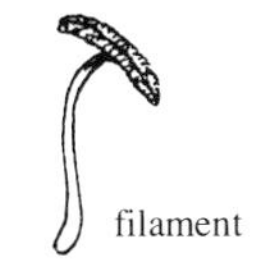
filament

**Filiform.** Slender and parallel-sided, as grass or daffodil leaves.

**Flexuous.** Refers to stems or stalks that are wavy.

**Flowerhead.** Refers to flowers which are aggregated into tight formal heads, as in members of the Daisy and Scabious families (Compositae and Dipsacaceae). The individual flowers of the head, the florets, may be of all the same type or of 2-3 distinct types.

flowerhead

**Floret.** See Flowerhead.

**Flush.** A wet place on or at the base of a hillslope or bank where water seeps or flows but does not form a distinct channel. Such flushes may be permanent or seasonal.

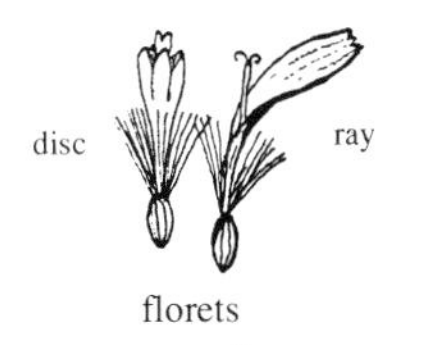

florets

**Follicle.** A dry dehiscent fruit developed from a single carpel. Follicles often occur in small clusters, e.g. *Aquilegia.*

**Free.** Not joined to one another.

**Fruit.** The seeds and the structure that contains them, whether dry or fleshy. Fruits are developed from the ovary and ovules, but may involve other parts of the flower such as the receptacle and the calyx.

follicles

**Gametes.** Simple unisexual organisms, joining to form zygotes.

**Gland.** A small rounded or oblong structure on the plant surface containing oil or some other liquid. Sometimes a gland will occur on the end of a hair – glandular hair.

**Glandular.** Covered with glands – see above.

**Glaucous.** Covered with a waxy bloom, giving a bluish or greyish colouration.

**Globose.** Globe like, rounded.

fruits

gynoecia

indehiscent fruits

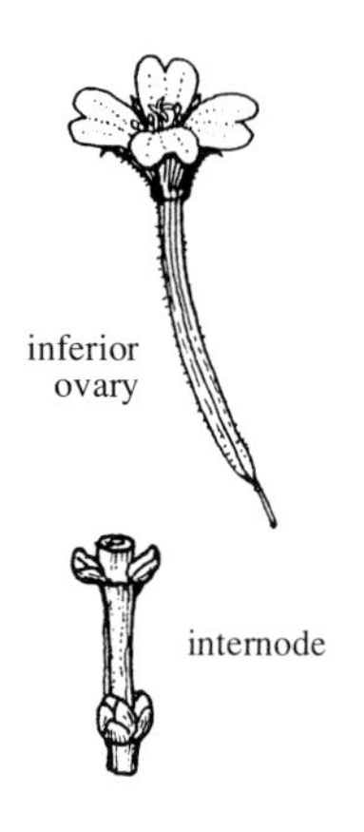
inferior ovary

internode

involucre

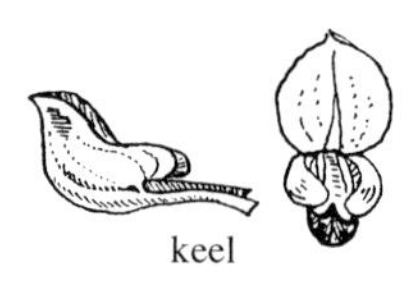
keel

lanceolate leaf

linear leaf

lip

**Gynoecium.** The female organs of the flower – generally consisting of the ovary, ovules, style(s) and stigma(s).

**Heath.** A lowland community dominated by heaths, *Calluna* or *Erica*, developed on thin sandy soils with a shallow peat layer.

**Herb.** Any vascular plant that does not develop woody stems. Those used specifically for flavouring food are referred to as culinary herbs.

**Herbaceous.** Refers to plant organs that are green and with a leaf-like texture. A herbaceous plant is a herb that dies down to ground level each year, overwintering at or below the soil surface.

**Hermaphrodite.** Having both male and female organs in the same flower – the majority of flowers are of this type.

**Heterostyly.** With two or more flower types in the same species – stamens and style set at varying levels so that the stamens in one flower correspond with the stigma in another. Each plant only bears one flower type; thus cross-pollination is assured, e.g. *Primula vulgaris.*

**Hybrid.** A plant originating from a cross between two distinct species. Hybrids are distinguished formally by a multiplication sign in the plant's scientific name, e.g. *Populus* × *canescens.*

**Hybrid swarm.** A group of plants originating as hybrids, but back-crossing through repeated generations with one or both the parent species and with themselves.

**Indehiscent.** Fruits which do not split open to release the seeds.

**Inferior.** Refers to the ovary when it occurs below the other floral organs, being fused with the receptacle, e.g. members of the Iris and Willowherb families (Iridaceae and Onagraceae).

**Inflorescence.** The flowering branch or branches, flowers and bracts above the uppermost stem leaves. Inflorescences are very variable from one species to another – see also Cyme, Panicle, Raceme, Spike and Umbel.

**Internode.** The portion of stem between two adjacent nodes.

**Introduced.** Not native. Thought on good evidence to have been introduced deliberately or accidentally from another country or region.

**Involucre.** Flower-bracts forming a cup-like or collar-like structure surrounding or around the base of a group of flowers or florets, e.g. members of the Daisy Family, Compositae.

**Irregular.** See Zygomorphic.

**Keel.** A petal which has a sharp keel-like edge, as the keel of a boat. Common to members of the Pea Family, Leguminosae.

**Lamina.** The blade of a leaf or petal; generally thin and flat.

**Lanceolate.** Lance-shaped – more or less elliptical, but widest below the middle. Generally refers to leaf-shape.

**Latex.** Milky juice or sap produced by stems, leaves or other organs when cut, e.g. *Taraxacum* species.

**Lax.** Loose.

**Ligule.** A small flap of tissue, often located at the base of a leaf or petal.

**Linear.** Narrow and parallel-sided. Generally refers to leaves, sometimes to fruits, e.g. most grasses have linear leaves.

**Lip.** A petal or petals (sometimes the sepals) which form a distinctive organ sharply divided from the other petals or sepals. Some flowers may

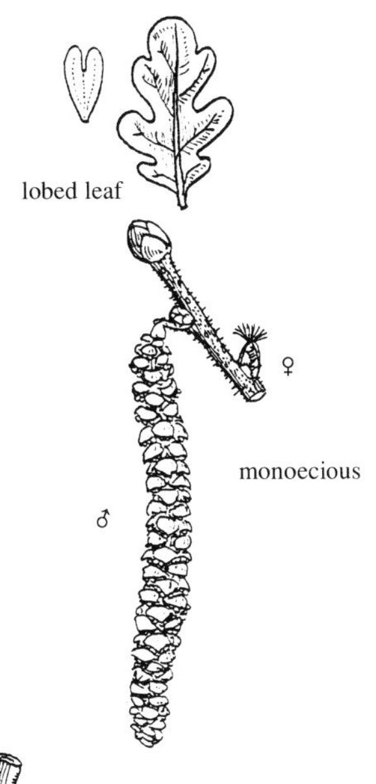

lobed leaf

monoecious

have a single lip (e.g. many orchids), others may have 2, one upper and one lower, and are thus said to be 2-lipped (many members of the Mint Family, Labiatae). The lip is often prominent and acts as a landing platform for visiting pollinators.

**Lobed .** Divided but not separating completely – as in various leaves and petals. Some leaves may be shallowly-lobed, others deeply-lobed.

**Marsh.** A community developed on permanently or seasonally wet (but not peaty) ground.

**Meadow.** A grassy field or area cut for hay or silage. Often grazed after the crop has been cut.

**Membranous.** Thin and dry; often opaque or transparent, like a membrane.

**Mericarp.** A one-seeded portion of a fruit splitting off from the rest, e.g. *Malva* species.

**Monoecious.** Having separate male and female flowers (unisexual flowers), but on the same plant.

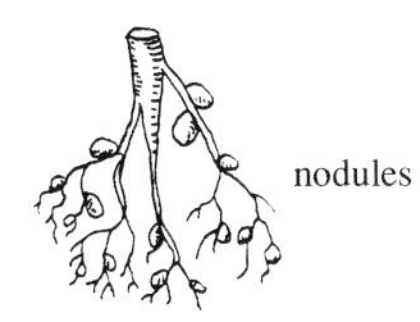
node

nodules

**Moor.** Upland or mountain communities dominated by heathers and developed on dry or damp, but not wet, peaty soils.

**Mycorrhiza.** An association of roots and fungi, dependent or partly dependent on one another. The fungal partner may occur outside the root or it may penetrate the surface tissues. Most orchids have a mycorrhizal association.

**Native .** Occurring naturally in a country or region.

oblong leaf

**Node.** Points on the stem where the leaves arise. Nodes are often regularly spaced, but they may be aggregated along the stem.

**Nodule.** A small regular or irregular swelling, often on roots, but sometimes on the leaves. Root nodules frequently occur on the roots of members of the Pea Family, Leguminosae: these contain bacteria capable of fixing atmospheric nitrogen, which is then usable by the plant.

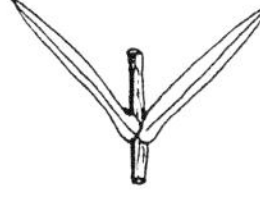
opposite leaves

**Oblong.** Rectangular with rounded ends – used to decribe a leaf or petal shape.

**Ochrea.** A sheath-like extension of the leaf-stalk (petiole) base surrounding the stem above a node, e.g. members of the Polygonaceae.

oval leaf

**Opposite.** Refers to two leaves or other organs arising on opposite sides of the stem at the same level.

**Organ.** Each part of the plant, leaf, petal, anther etc., that serves a function.

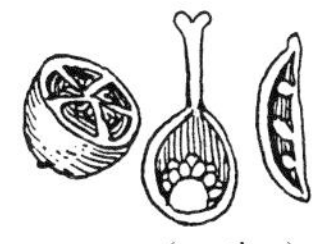
ovary (section)

**Oval.** A broad ellipse with rounded, rather than pointed ends – used to describe a leaf or petal shape.

**Ovary.** The female organ containing the ovules – part of the gynoecium.

**Ovule.** The organ containing the egg, which after fertilization develops into the seed. Seeds may sometimes develop from unfertilized ovules – see Apomixis.

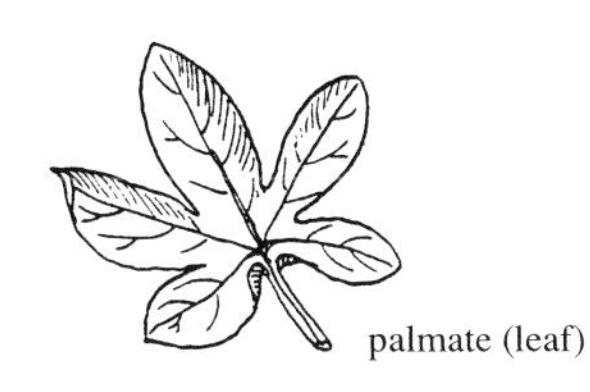
palmate (leaf)

**Palmate.** Hand-like. Generally refers to leaves with more than 3 deep lobes or leaflets arising from the same point.

**Panicle.** Refers to a branched racemose inflorescence, though often loosely applied to various forms of branched inflorescence, e.g.*Aesculus* and various *Verbascum* species.

panicle

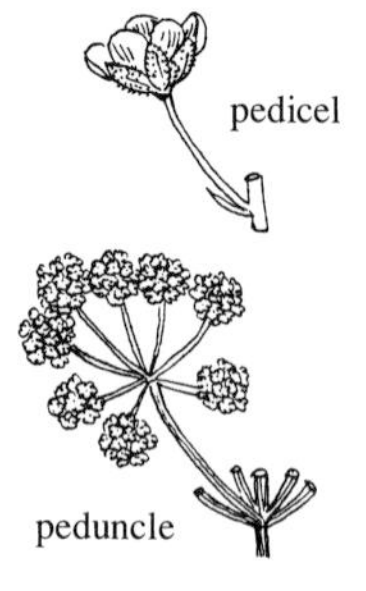


**Pappus** Thistledown – tufts of hairs on achenes and fruits.

**Parasite.** A plant that gains all its sustenance from another living plant to which it is attached. Truly parasitic plants have no chlorophyll, e.g. *Cuscuta* and *Lathraea*. See also Semi-parasite.

**Pasture.** A grassy field used for grazing animals.

**Peat.** Dark brown or black, partly decayed, vegetable matter, often forming thick deposits.

**Pedicel.** The stalk of an individual flower.

**Peduncle.** The stalk of an inflorescence or partial inflorescence.

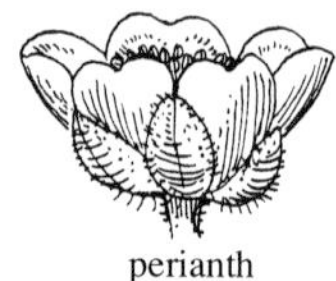


**Peloric.** A strange mutation in which the corolla appears to be turned inside out, e.g. occasionally in *Linaria* species.

**Perennate.** To survive from season to season – used particularly of herbaceous plants.

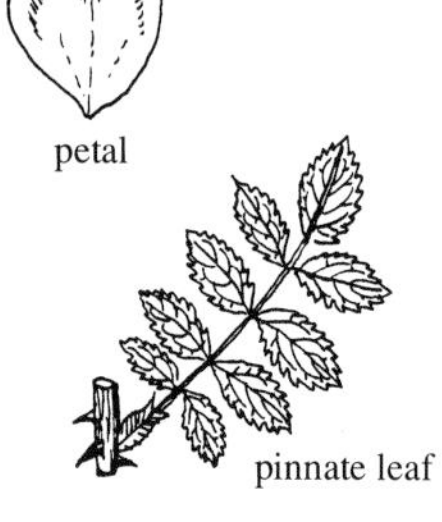


**Perennial.** A plant living for more than 2 years and generally flowering each year. Perennial species often live for many years.

**Perianth.** A collective name for all the floral leaves – petals and sepals when both are present. Often the perianth is not properly distinguished into sepals and petals.

**Petal.** The inner perianth segments when they clearly differ from the outer segments, often by being large and brightly coloured.

**Petaloid.** Brightly coloured and resembling petals.

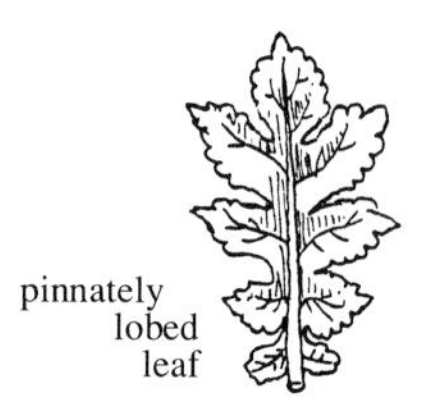


**Petiole.** The leaf stalk.

**Pinnate.** A leaf composed of more than 3 leaflets arranged in 2 rows along a common axis, with or without an end leaflet, e.g. many legumes, *Leguminosae*.

**Pinnately-lobed.** A leaf which is lobed in a pinnate fashion, but not separated into distinct leaflets.

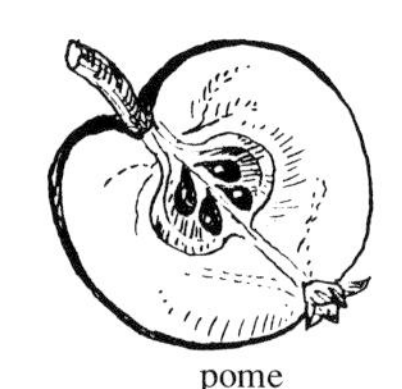


**Pollen.** Tiny particles produced by the anthers and containing the male gametes – often referred to as pollen grains.

**Pollinia.** Pollen grains that are aggregated into regularly shaped masses, e.g. all members of the Orchid Family, Orchidaceae.

**Pome.** A fleshy fruit developed from the ovary wall and receptacle, the flesh surrounding a tough, though not woody, layer that in turn encloses the seeds, e.g. apples and pears.

**Prickle.** A sharp point, often hooked, developed from the outer tissues of stems, or sometimes on the leaf surface. Prickles are generally rather irregularly arranged.

**Process.** A projecting part.

**Prostrate.** Lying close to and along the ground.

**Pungent.** Sharply aromatic.

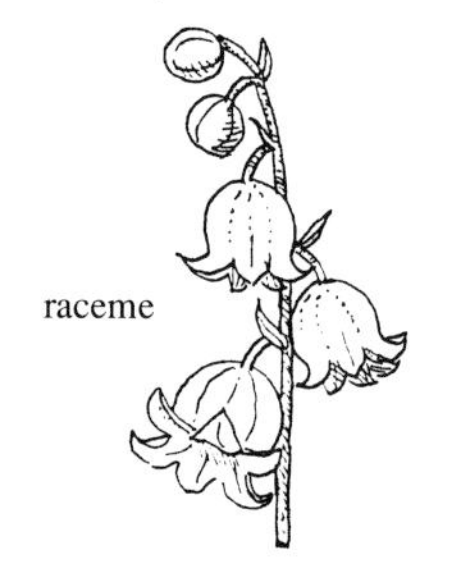


**Raceme.** A spike-like inflorescence in which the individual flowers are clearly stalked. Racemes can be borne in an erect, horizontal or in a pendulous position.

**Ray.** A stalk radiating out from an umbel; like the spokes of an umbrella.

**Ray-floret.** The outer florets of a flowerhead such as a daisy or cornflower, which are often elaborated into a distinctive strap-like or lobed, partly tubular, structure (ligule). Ray-florets generally surround a central disk of shorter florets, the disk-florets.

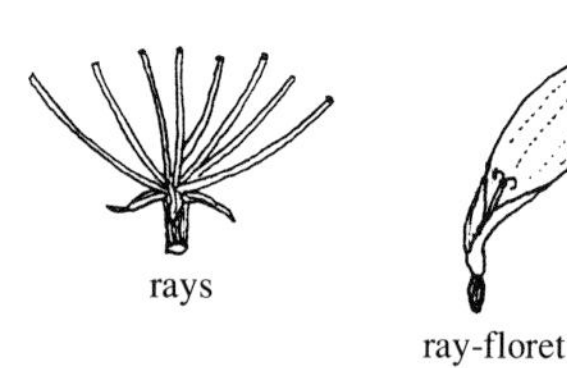

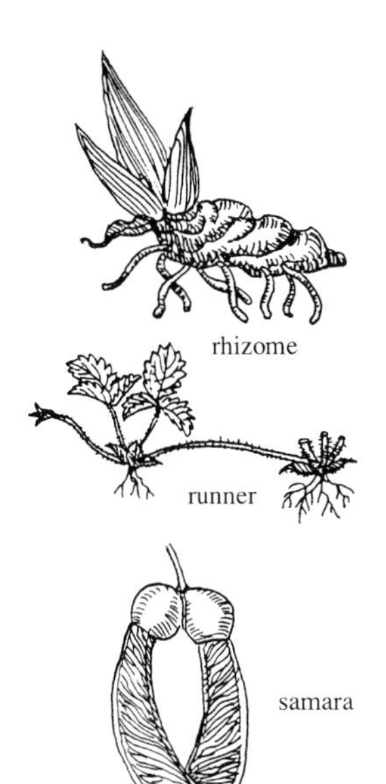


**Receptacle.** The part of the stem from which all the floral organs arise. It may be flat, concave or convex.

**Regular.** See Actinomorphic.

**Rhizome.** An underground or surface stem, often thickly swollen and lasting a number of years, affording the plant a means of perennating from one season to another, e.g. *Iris germanica.* Rhizomes, like stems, bear nodes and buds, though these may be difficult to observe.

**Runner.** A form of aerial stolon which roots down at the node(s)to form a new plant; they eventually become detached from the parent plant.

**Salt-marsh.** Communitities developed on the intertidal zone along coasts and estuaries, often on mud and in sheltered areas.

**Samara.** An indehiscent fruit in which part of the wall forms a distinctive wing, e.g. *Acer* species.

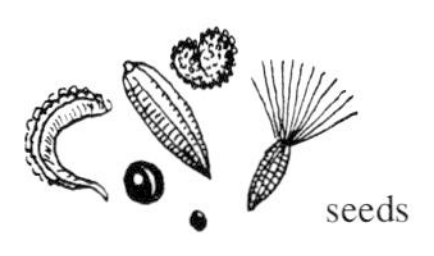


**Saprophyte.** A plant without chlorophyll which lives on humus – decayed vegetation, generally in the darker areas of woodland and forests where there is little competition from other species, e.g. *Neottia nidus-avis.*

**Scape.** A leafless stem bearing flowers, e.g. *Narcissus* species.

**Scrub.** A community dominated by shrubby species.

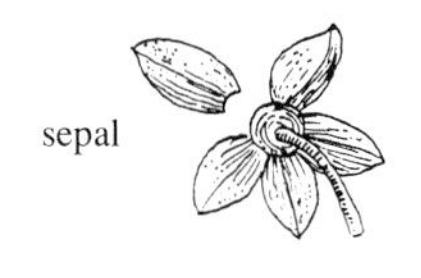


**Seed.** A reproductive unit developed usually from a fertilized ovule, and capable of producing a new individual plant.

**Semi-parasitic.** Only partially dependent on a host plant. Semi-parasitic plants have chlorophyll but cannot survive away from their host species. Many members of the Scrophulariaceae such as *Pedicularis* and *Rhinanthus* are semi-parasitic on grass species.

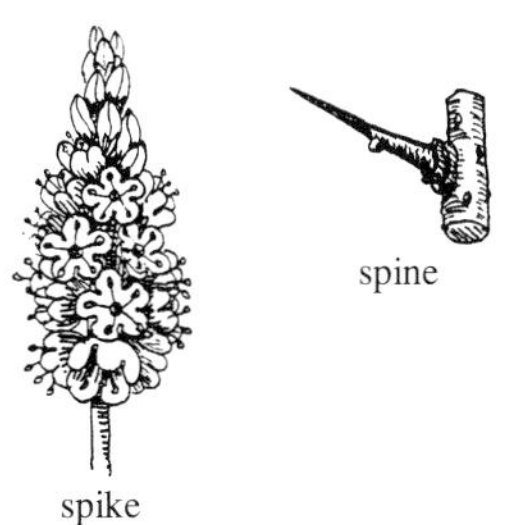


**Sepal.** A member of the outer perianth whorl in most flowers. The sepals collectively make up the calyx.

**Shrub.** A woody plant branching from the base, or close to the base, but not forming a tree.

**Simple.** Not compound (leaves) or unbranched (stems).

**Sinus.** The gap between 2 lobes, especially at the base of a leaf.

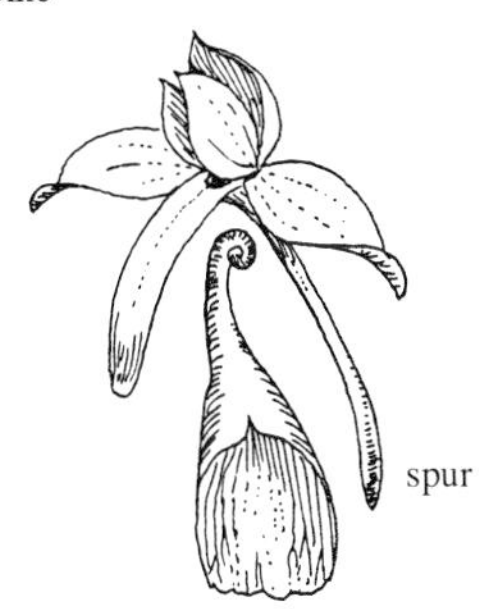


**Spatulate.** Shaped like a paddle.

**Speculum.** A shiny shield-like patch, as on the lip of many *Ophrys* species. The speculum varies greatly from square- to diamond- or horseshoe-shaped.

**Spike.** A simple elongated inflorescence in which the individual flowers are unstalked.

**Spine.** A stiff, sharply-pointed structure, often a modified shoot tip, e.g. *Prunus spinosa.*

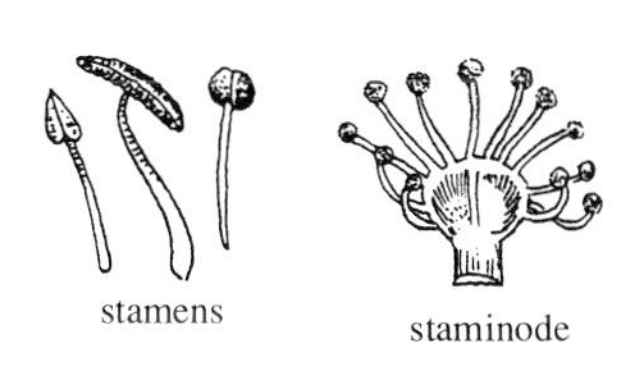


**Spur.** A hollow cylindrical or pouched structure projecting from the calyx or corolla and generally containing nectar at the tip, e.g. *Aquilegia.*

**Stamen.** The male organ of the flower, consisting of a stalk, the filament, and the anther which contains the pollen.

**Staminode.** An infertile, modified stamen. Staminodes can be reduced and inconspicuous, or they may be sometimes elaborate and partially petaloid.

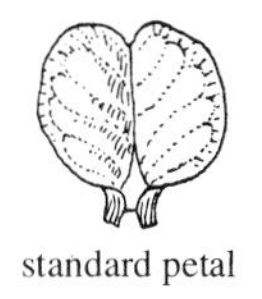


**Standard.** The upper petal of a pea flower, often showy and larger than the other petals.

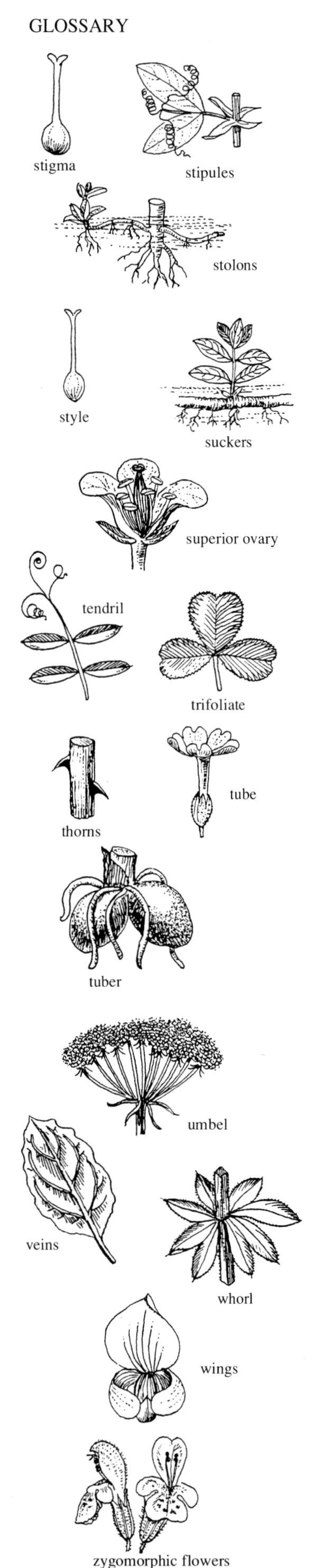


**Stellate.** Star-shaped, usually refers to hairs on various parts of a plant.

**Sterile.** Not producing fertile seeds or pollen.

**Stigma.** The receptive tip of the style to which pollen grains attach themselves and germinate, if both are compatible.

**Stipule.** A leaf-like or scale-like process at the base of the leaf-petiole and sometimes fused to it.

**Stolon.** A creeping stem, below or above ground and produced at the base of erect stems or from basal leaf-rosettes.

**Style.** The part of the gynoecium that links the ovary with the stigma.

**Subspecies.** A subdivision of a species generally separated from the typical plant on several characters such as flower colour and hairiness, but also separated on geographical or ecological criteria.

**Sucker.** A shoot arising from the roots of trees or shrubs and often some distance from the main stem.

**Superior.** Refers to the ovary when it is located above the other floral organs; the ovary being free and located in the centre of the flower, not below it.

**Tendril.** A climbing organ, generally filiform and coiling around supports, sometimes branched. Tendrils are generally modified stems, leaves or leaf-tips.

**Terminal.** Borne at the end of a stem or shoot.

**Trifoliate.** A leaf composed of 3 distinct leaflets, e.g. *Trifolium* species.

**Thorn.** A stiff woody, pointed structure formed from a modified stem.

**Tree.** A woody plant forming a distinct trunk, generally branched well above the ground.

**Trullate.** Angular-ovate; shaped like a brick-layer's trowel.

**Tube.** The fused, often cylindrical part of a corolla or calyx.

**Tuber.** A swollen underground organ, generally only lasting for one or two years and developed from underground stems or from the roots.

**Umbel .** An inflorescence in which all the stalks radiate from the same point like the spokes of an umbrella, e.g. many members of the Cow-Parsley Family, Umbelliferae.

**Unarmed.** Without thorns, spines or prickles.

**Variety.** A subdivision of a species or subspecies differing from the typical plant by one or two characters such as flower colour or leaf-shape, but generally growing in the same general vicinity.

**Vein.** A strand, strengthening or conducting tissue, running through leaves and perianth parts.

**Vittae.** Resin canals; frequent in the fruits of Umbelliferae – they occur (usually 4 in number) between the primary ridges of the fruit.

**Whorl.** More than two organs arising at the same point on the stem. Refers generally to leaves or flowers.

**Wing.** The lateral petals in various flowers, e.g. many orchids, Orchidaceae, and members of the Pea Family, Leguminosae.

**Zygomorphic.** Irregular flowers having only one plane of symmetry.

# THE FLORA

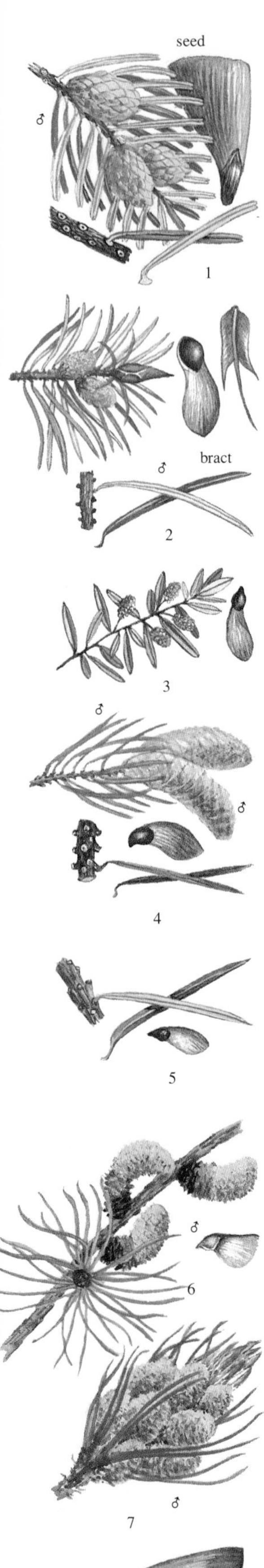


# CLASS – GYMNOSPERMS

Trees and shrubs, usually evergreen. Flowers unisexual. Ovules naked, not enclosed in an ovary. Seeds often borne in a woody cone - in *Taxus*, surrounded by a fleshy aril.

## PINE FAMILY Pinaceae

Trees, occasionally shrubs, often highly resinous; evergreen except *Larix*. Leaves linear, spirally arranged, in bundles of 2, 3 or 5. Male and female flowers borne in separate cones on the same tree; mature female cones woody. Seeds winged. 190 species in 9 genera. North temperate zones to C America, Sumatra and Java.

1* **Silver Fir** *Abies alba* Miller. To 50m. Bark greyish, scaly. Buds not sticky. Leaves 15-30mm, deep green above, whitish beneath, slightly notched at the apex, falling eventually to leave an oval, sucker-like scar. April-May. Cones erect, oblong, 10-20cm, with prominent bracts between the scales; bracts and scales deciduous, leaving erect peg-like projections. Mountain forests to 2100m, occasionally in the lowlands; also in plantations. W France to Corsica, east to the Balkan Peninsula. Occasionally naturalized in Britain, Belgium, Denmark, Norway and Sweden. W France to Corsica, east to the Balkan Peninsula. Trees have a life-span of 200-300 years. Less frequently planted than *A. grandis*, *A. nordmanniana* and *A. procera* – see p. 44. **B**: Fairly widely planted.

2* **Douglas Fir** *Pseudotsuga menziesii* (Mirbel) Franco. To 100m, one of our tallest trees. Bark dark reddish-brown, corky and deeply fissured with age. Leaves 20-35mm, not notched at the apex, rather soft and fragrant when crushed. Late April-May. Cones pendent, oval, 5-10cm, with 3-lobed bracts between the scales; cones fall in one piece, unlike *Abies* where the scales break away individually. **I**: W North America. Planted throughout, except in the far north, for its rapid growth – often over 1m a year; occasionally naturalized. **B**: Frequently planted.

3* **Western Hemlock** *Tsuga heterophylla* (Rafin.) Sarg. To 70m, narrow-pyramidal with irregularly-whorled branches, and drooping shoot tips. Bark brown to purplish-brown, flaking and shredding when old. Leaves flattened, 6-20mm, deep green above, with 2 broad white bands beneath, borne on cushion-like projections, in 2 ranks along the branches, the upper ones shorter than the lateral; when crushed aroma similar to Ground Elder. April-May. Cones small, pendent, egg-shaped, 2-2.5cm, with simple blunt scales. **I**: W North America. Widely planted for timber and ornament. Mainly in Britain, France, Germany and Denmark. Related to *Picea*, but with softer foliage and smaller, more rounded cones. **B**: Frequently planted.

4* **Norway Spruce** *Picea abies* (L.) Karsten. Pyramidal, to 60m, with regularly whorled branches, the shoots hairless or with a few hairs. Bark reddish-brown, scaly. Leaves pointed, 10-25mm, falling to leave a short peg-like projection. May-June. Cones pendent, cylindrical, 10-18cm, without protruding bracts between the scales. Forests and woods, especially in the mountains, forestry plantations, shelter belts, parks and gardens, to 2200m. C & N Europe east to the Balkan mountains and into C Russia. France, Germany and Scandinavia; widely planted and naturalized in Britain, Belgium, Holland and Denmark. The traditional Christmas Tree, sold in vast numbers as small 4-8 year old specimens. Also grown for timber and in parks and gardens.

**Siberian Spruce** *P. a.* subsp. *obovata* (Ledeb.) Hultén [*Picea obovata*]. Like the type, but smaller, rarely over 30m, with densely hairy young shoots. Leaves 10-18mm. Cones 6-8cm. Similar wild habitats. N Europe east to N & C Russia and N Asia.

5* **Sitka Spruce** *Picea sitchensis* (Bong.) Carriere. A bluish-green tree, to 40m. Bark grey to distinctively purplish, scaly and flaking eventually. Leaves like *P. abies*, but flattened and with 2 white bands beneath, rather pungent when crushed. May-June. Cones pendent, narrow-oblong, 6-10cm, with slightly toothed scales. Widely planted for timber and in parks. **I**: W North America. Throughout, except the extreme north, including Iceland. **B**: Much planted.

6* **Larch** *Larix decidua* Miller [*L. europaea*]. Deciduous tree to 35m, with irregularly whorled branches. Bark rough, greyish-brown, becoming fissured with age. Leaves in tufts on short lateral shoots or spirally arranged on leading shoots, rather pale green, 20-30mm. Young shoots pale yellow, often turning bright red; in autumn the needles turn red and yellow before falling. March-April. Cones small, soft and hairy when young, oval or rounded, 2-3.5cm, erect. Mountain forests, but a common forestry tree in the lowlands, also in parks and gardens, along roadsides and shelter belts, to 2500m. Alps to E Europe. E France and S Germany; planted or semi-naturalized further W and N. There are 10 *Larix* species in the north temperate hemisphere. **B**: Sometimes semi-naturalized.

**Pines** *Pinus*. Evergreen trees or shrubs with needle-leaves in clusters of 2-5. Cones woody, ripening in 2-3 years, falling entire and without conspicuous bracts between the thick woody scales. Male cones are borne in masses towards the shoot tips in late spring and early summer; when mature they produce copious quantities of pollen, which is blown in clouds from tree to tree. About 80 species, in the north hemisphere and Java.

7* **Beach Pine** *Pinus contorta* Douglas ex Loudon. A small tree, 6-10m, with sticky buds. Bark dark brown. Needles in pairs, twisted, 30-70mm, yellowish-green. May-June. Cones pendent, egg-shaped, 2-6cm, yellowish-brown and shiny, the scales with a slender recurved spine. Often in extensive plantations, sometimes semi-naturalized. **I**: W North America, coastal. Britain, Iceland and much of Europe, except for the extreme north. **B**: Fairly widely grown in plantations or for shelter.

8* **Maritime Pine** *Pinus pinaster* Aiton. To 40m, with a rounded crown when mature, the buds not sticky. Bark reddish-brown, deeply fissured. Needles very long, in pairs, 10-25cm, green, rigid and curved, sharply pointed. May-June. Cones large and conical, 8-22cm, pale brown and shiny, persistent along the branches. May-June. Coast of C and W Mediterranean. Planted in Britain, Belgium and France for timber and shelter; used especially for stabilizing coastal sand-dunes; naturalized S England and W France.

Silver Fir
♀
♀
Douglas Fir
♀
Western Hemlock
♀
♀
♀
Norway Spruce
Sitka Spruce
Larch
♀
♀
Beach Pine
Maritime Pine

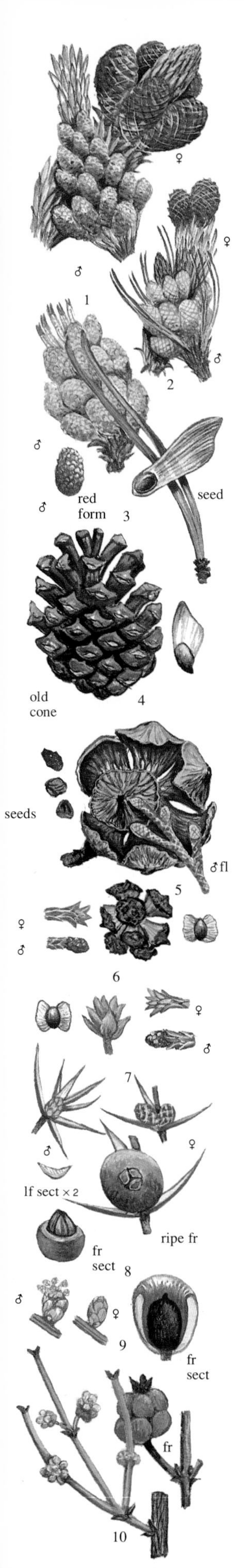


1* **Monterey Pine** *Pinus radiata* D. Don.To 40m. Bark dark greyish brown, deeply cut into thick parallel ridges. Shoots reddish-brown, hairless, with sticky buds. Needles in threes, 10-15cm, slender, pointed, densely crowded. March-May. Cones large, 7-14cm, oval-conical, but markedly asymmetrical, unstalked or short-stalked; persisting on branches and trunks, often for many years. Sometimes planted for timber and shelter in W Europe; occasionally in parks and large gardens. **I**: S California. Britain and France. **B**: Mostly planted near the coast in SW & W; infrequent elsewhere.

2* **Austrian or Black Pine** *Pinus nigra* Arnold. To 50m. Bark coarsely ridged, greyish-black, fissured and flaking. Shoots pale brown, hairless, with sticky buds. Needles in pairs, longer than *P. sylvestris*, 8-16cm, straight or curved. May-June. Cones yellowish-brown, shiny, 3-8cm. **I**: S & SE Europe. Extensively planted to 2300m for timber, shelter or ornament, except in extreme north and Iceland. Favours calcareous soils. The plant usually cultivated is subsp. *larico*. Maire. [*P. larico*]. **B**: Frequent.

3* **Scots Pine** *Pinus sylvestris* L. To 40m, with a dome-shaped crown at maturity. Bark reddish-brown, flaking into small scales. Shoots yellowish-green at first, becoming greyish-brown, hairless, with sticky buds. Needles in pairs, grey and twisted, 3-7cm. May-June. Cones pendent, dull yellowish-brown, egg-shaped, 3-6cm. Forests, woodland, heaths, moors, hedgerows, to 2300m, but widely planted and naturalized; also in parks and gardens. W Europe from Spain to Lapland and Siberia. **B**: C & W Scottish Highlands, but semi-naturalized on heaths in S England.

4* **Mountain Pine** *Pinus mugo* Turra. A shrub to 3.5m with spreading, thickened, rather contorted branches. Shoots green, becoming brown, with sticky buds. Needles in pairs, bright green, 3-8cm. May-June. Cones erect or slightly nodding, pale brown and shiny, 2-5cm. Rocky mountain habitats and open woods, to 2700m. Planted for shelter and for stabilizing dunes, particularly along the coast; frequent in parks and gardens. C & E France & S Germany; naturalized in Denmark.

## CYPRESS FAMILY Cupressaceae

Trees and shrubs with small, paired or whorled, scale-like or bristle-like leaves. Monoecious except for Junipers. Cones small and woody, with separating scales, but fleshy and berry-like in Junipers. Juvenile leaves spreading and linear but become gradually scale-like as the plants begin to mature. 113 species in 17 genera, cosmopolitan.

5* **Macrocarpa, Monterey Cypress** *Cupressus macrocarpa* Hartweg. To 25m with ascending branches making a pyramidal crown when young, later rather flat-topped. Bark brown, shallowly ridged. Shoots 4-angled, hairless. Leaves small, 1-2mm, opposite and at right-angles to the pair above and below. April-May. Cones ripening in second year; male cones yellowish, 3-5mm; female cones rounded, 20-35mm, shiny-brown when mature. **I**: S California. Much planted in Britain and France, often near the coast. **B**: Park, gardens and churchyards, especially in W Britain and SW Ireland.

6* **Lawson Cypress** *Chamaecyparis lawsoniana* (A. Murray) Parl. To 6.5m, forming a columnar shape. Bark reddish-brown. Leading shoots drooping at the tip. Shoots flattened. Leaves opposite and scale-like, all closely pressed together, with white markings beneath, parsley-scented when crushed. Cones generally ripening in first year; male cones pink or reddish; female rounded, small, 7-8mm, yellow-brown when ripe. March-April. Often browned down one side by winter winds. Many cultivars. **I**: W USA. Much planted in W Europe, occasionally naturalized. **B**: Widely planted.

7* **Western Red Cedar** *Thuja plicata* D. Don ex Lambert. To 65m, trunk not branched from the base as in Lawson Cypress. Bark purplish to reddish-brown, widely ridged. Leading shoots erect. Leaves and shoots resin-scented as in Lawson Cypress. April-May. Female cones brown when ripe, conical, 12mm, with thin flexible, overlapping scales (thick and valvate in the previous two species). **I**: W North America. Often planted in NW Europe; sometimes naturalized. Like *Cupressus* but shoots flattened and cones oblong or conical, ripening in the first year.

8* **Juniper** *Juniperus communis* L. Shrub or small tree to 6m, bushy or columnar. Leaves greyish, in whorls of three, spine-tipped, with a single white band beneath. Flowers yellow, male and female on separate plants, May-June. Female cones rounded, berry-like 6-9mm, green at first, later bluish-black. Coniferous woodland, moors, scrub, on calcareous soils, to 1500m. Throughout, except Spitsbergen. The cones ripen in the second or third years. **B**: Widely scattered localities.

* *J. c.* subsp. *alpina* Celak. [subsp. *nana*]. Like the type, but forming a spreading mat-like shrub. Leaves 10-15mm (to 20mm in the type). Mountain habitats and moors, rocky slopes, woodland and scrub, to 3600m, but at low altitudes at the northern end of its range. **B**: Confined to upland localities.

## YEW FAMILY Taxaceae

About 20 species in 5 genera; widely scattered, but primarily in the north hemisphere. Trees and shrubs. Leaves narrow and spirally arranged. Dioecious; male flowers in small cones, female solitary or paired. Fruit enveloped in a fleshy aril.

9* **Yew** *Taxus baccata* L. Evergreen tree or bush to 20m with a wide spreading crown. Bark reddish-brown, flaking. Leaves shiny dark green and pointed, in two rows. Flowers green, males with many yellow stamens, females solitary or 2 together; February-April. Woods, scrub, hedgerows, mostly on calcareous soils, to 1800m. Europe and N Africa. Throughout, except the far north. Widely planted. Huge old trees are often seen in churchyards, particularly in Britain. The leaves and seeds are poisonous, the aril not. **Irish Yew**, *Taxus baccata* 'Fastigiata', is a slender, upright cultivar. **B**: Throughout.

10 **Joint Pine** *Ephedra distachya* L. (Joint Pine Family – Ephedraceae). Mountains of C & S Europe and W Asia. Tough low shrub to 0.5m, much branched with underground rhizomes. Shoots green, appearing leafless and jointed. Leaves reduced to very short sheaths at the stem joints. Flowers small, greenish-yellow, in small clusters, often unstalked, male and female on separated plants; May- June. Fruit globose, 6-7mm, with fleshy red scales. Sandy and dry places, stream banks, rock ledges, to 1100m. NW & C France.

Monterey Pine
Austrian Pine
♀
Scots Pine
Mountain Pine
Macrocarpa
Lawson
Cypress
♂
Western Red Cedar
♀
Juniper
ssp. *alpina*
Yew

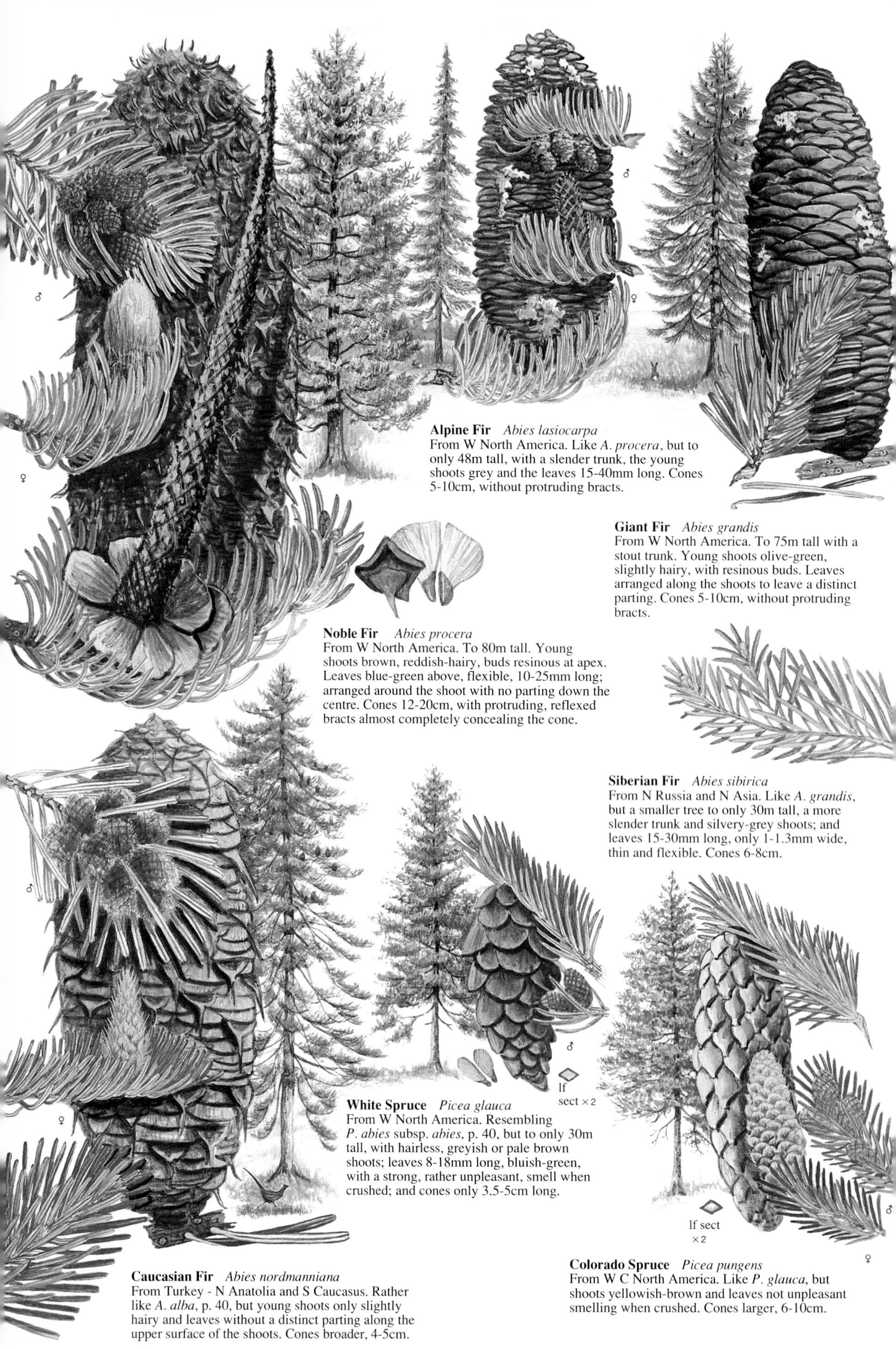


**Alpine Fir** *Abies lasiocarpa*
From W North America. Like *A. procera*, but to only 48m tall, with a slender trunk, the young shoots grey and the leaves 15-40mm long. Cones 5-10cm, without protruding bracts.

**Giant Fir** *Abies grandis*
From W North America. To 75m tall with a stout trunk. Young shoots olive-green, slightly hairy, with resinous buds. Leaves arranged along the shoots to leave a distinct parting. Cones 5-10cm, without protruding bracts.

**Noble Fir** *Abies procera*
From W North America. To 80m tall. Young shoots brown, reddish-hairy, buds resinous at apex. Leaves blue-green above, flexible, 10-25mm long; arranged around the shoot with no parting down the centre. Cones 12-20cm, with protruding, reflexed bracts almost completely concealing the cone.

**Siberian Fir** *Abies sibirica*
From N Russia and N Asia. Like *A. grandis*, but a smaller tree to only 30m tall, a more slender trunk and silvery-grey shoots; and leaves 15-30mm long, only 1-1.3mm wide, thin and flexible. Cones 6-8cm.

**White Spruce** *Picea glauca*
From W North America. Resembling *P. abies* subsp. *abies*, p. 40, but to only 30m tall, with hairless, greyish or pale brown shoots; leaves 8-18mm long, bluish-green, with a strong, rather unpleasant, smell when crushed; and cones only 3.5-5cm long.

**Caucasian Fir** *Abies nordmanniana*
From Turkey - N Anatolia and S Caucasus. Rather like *A. alba*, p. 40, but young shoots only slightly hairy and leaves without a distinct parting along the upper surface of the shoots. Cones broader, 4-5cm.

**Colorado Spruce** *Picea pungens*
From W C North America. Like *P. glauca*, but shoots yellowish-brown and leaves not unpleasant smelling when crushed. Cones larger, 6-10cm.

**Engelmann's Spruce** *Picea engelmannii*
From W North America. Like *P. abies* subsp. *obovata*, p. 40, but shoots pale yellowish-brown, leaves 15-25mm long, and cones 3.5-7.5cm.

**Serbian Spruce** *Picea omorika*
C Yugoslavia. Like *P. sitchensis*, p. 40, but to only 30m tall, with a very slender habit; leaves 8-18mm long, less pungent when crushed; and cones only 3-6cm long.

EXOTIC LARCHES

**Japanese Larch** *Larix kaempferi*. From Japan. Young shoots bluish or greyish-green; leaves 15-25mm long with 2 white bands beneath.

**Siberian Larch** *Larix russica*. From N Russia and Siberia. Cone scales slightly incurved at the apex, about 3 times as long as the cone bracts.

**Dahurian Larch** *Larix gmelinii*. From E Asia. Cone-scales hairless outside, each cone with about 20 scales.

**Jack Pine** *Pinus banksiana*
From NC and NE North America. To 25m tall, with irregularly arranged branches. Bark becoming reddish-brown and forming thick scales; buds very resinous. Leaves in pairs, 20-40mm long, rigid. Cones long-egg-shaped, 2-5cm, curved, erect, shiny and yellowish, persistent.

**Western Yellow Pine** *Pinus ponderosa*
From W North America. Like *P. radiata*, p. 42, but leaves in threes and winter buds resinous. Leaves 100-250mm long. Cones symmetrical, dark grey-green.

**Arolla Pine** *Pinus cembra*
Alps and Carpathians. To 25m tall, with reddish-brown bark becoming scaly on older trees; shoots orange-brown, hairy, the bud-scales with whitish margins. Leaves in groups of five, 50-80mm long. Cone 5-8cm, with large swollen scales, shiny, reddish-brown. Seeds not winged.

**Weymouth Pine** *Pinus strobus*
From E & C North America. To 50m tall. Bark greyish-green, smooth when young, later fissured. Rather like *P. cembra*, but leaves 50-140mm long, slender and flexible. Cones cylindrical, slightly curved near the apex, 8-20cm long. Seeds winged.

# CLASS – ANGIOSPERMS

Trees (usually deciduous in our area), shrubs or herbs, occasionally parasitic. Flowers hermaphrodite, or more arely unisexual. Ovules completely enclosed in an ovary.

**Order - Dicotyledons.** Embryo plants with two seed-leaves (cotyledons), rarely 1 by reduction. Stem vessels usually arranged in a single ring. Leaves with normally diverging, often netted veins. Flowers typically 4- or 5-parted.

## WILLOW FAMILY Salicaceae

Over 300 species of deciduous trees and shrubs in 3 genera, especially in the north hemisphere. Buds with a single scale (willows) or several overlapping (poplars). Leaves usually alternate, often with stipules. Flowers in catkins, generally erect in willows, but drooping in poplars; male and female on separate plants. Catkin scales toothed (poplars) or untoothed (willows). Seeds wind blown, surrounded by a tuft of woolly hairs. Fruits generally ripening quickly after flowering. Hybrids are common. Willows have male flowers with 2, 3 or 5 stamens and are generally insect-pollinated, poplars 4 or more, and are wind-pollinated.

1* **Bay Willow** *Salix pentandra* L. Shrub or small tree, 2-7m, with rugged, fissured, grey-brown bark, hairless. Shoots shiny; stipules soon falling. Leaves elliptical, shiny dark green, glandular near base. Catkins slender, 2-5cm, with the leaves; male catkins bright yellow, female greenish; male flowers with 5-6 stamens; May-June. Damp habitats. From Britain northwards to N Norway. The leaves are sticky and rather fragrant when young. **B**: Throughout, from N Wales and N Midlands north and in Ireland.

2* **Crack Willow** *Salix fragilis* L. A spreading tree to 25m (often pollarded). Bark greyish, fissured. Shoots olive-green, soon hairless, very fragile at the joints. Leaves lanceolate, pointed, toothed, shiny bright green, greyish beneath; stipules soon falling. Catkins 3-7cm, borne with the leaves, drooping, the male catkins yellow, the female green; male flowers with 2 stamens; April-May. Freshwater margins and wet woodland. Throughout, except the Arctic. Hybridizes frequently with *S. pentandra*: *S.* × *meyeriana* Rostkov. ex Willd. **B**: Growing wild up to S Scotland; planted further north and in Ireland.

3* **White Willow** *Salix alba* L. A spreading silvery-grey tree, to 25m (often pollarded), with greyish, non-flaking, bark. Shoots silky when young, later olive-green, not fragile. Leaves lanceolate, pointed, toothed, silky when young, and beneath at maturity. Catkins borne with the leaves, the male catkins yellow, female slenderer, green; male flowers with 2 stamens; April-May. Damp habitats, often planted. Throughout, except the Arctic, to C Asia and N Africa, mostly lowland. Naturalized in parts of Scandinavia. Hybridizes with *S. fragilis*: *S.* × *rubens* Schrank. **B**: Throughout, except NW Scotland.

4 **Cricket-bat Willow** *S. a.* subsp. *coerulea* (Sm.) K.H. Rech. Like the type, but leaves bluish-green and the **Golden Osier** *S. a.* subsp. *vitellina* (L.) Stokes (5) is distinguished by its yellow shoots. Both are commonly planted: *coerulea* is grown beside East Anglian dykes for cricket-bats.

* **Weeping Willow** *Salix* × *sepulcralis* Schrank (*S. alba* var. *vitellina* × *S. babylonica*). A weeping tree to 20m, with long bright green, slender shoots. Leaves narrow lanceolate, pointed, finely toothed, silky at first, bright green above, greyish beneath. Catkins 3-4cm, often curved, borne with the young leaves. April. Widely planted, especially around lakes and large ponds and along river banks, occasionally naturalized. *Salix* × *pendulina* Wenderoth (*S. babylonica* × *S. fragilis*) is like *S.* × *sepulcralis*, but leaves rather broader and practically hairless. Less commonly planted.

6* **Almond Willow** *Salix triandra* L. [*S. amygdalina*]. A shrub or small tree, 4-10m, with smooth, flaking, brown bark. Leaves lanceolate to elliptical, toothed, hairless, dark shiny green, paler beneath; stipules persistent. Catkins 3-5cm, borne with the young leaves; male flowers with 3 stamens; April-May. Riverbanks, ponds, marshes and fens. Throughout, except the Arctic, including the Faeroes and Iceland; widely naturalized. **B**: Locally common in England, Channel Islands and S Ireland, rare in C & N Scotland.

7* **Net-leaved Willow** *Salix reticulata* L. Mat-forming undershrub. Leaves rounded or oval, untoothed, shiny deep green and netted above, whitish beneath. Catkins slender, 1-3cm, erect, borne with the mature leaves, long-stalked. Rocky habitats and ledges in the mountains, generally on basic rocks, to 2500m. June-August. Arctic and sub-Arctic Europe, extending to the Alps – not in Iceland; local and often rare. **B**: Very local in Scotland – Angus and Sutherland to Argyll.

8* **Dwarf Willow** *Salix herbacea* L. Dwarf, patch-forming undershrub, with underground stems and short aerial shoots. Leaves rounded to kidney-shaped, toothed, bright shiny green. Catkins short, 0.5-1.5cm, borne with the leaves. Tundra, rock ledges and mountains tops, to 2800m. June-July. Arctic and sub-Arctic Europe extending south to Britain and the higher mountains of France and Germany. Often locally abundant. **B**: From Wales and N England northwards and in Ireland.

9 **Polar Willow** *Salix polaris* Wahlenb. Very slender shoots and dark green, elliptical, untoothed leaves. Tundra and rocky habitats, to 1700m. July. Arctic Scandinavia, including Spitsbergen.

10* **Whortle-leaved Willow** *Salix myrsinites* L. Low spreading undershrub, to 40cm, developing thick knotty wood; shoots downy at first. Leaves lanceolate to oval, toothed, bright green, shiny on both surfaces, hairy at first, with veins netted; dead leaves persisting into the following season. Catkins fairly large, 2-5cm, purplish, borne with the leaves. Wet rocky habitats to 1750m. May-June. Spitsbergen, Scandinavia and Scotland. **B**: Rare and local in Scotland; Argyll and Perth north to Orkney.

*Salix alpina* Scop. Like *S. myrsinites*, but leaves untoothed, the dead ones not persisting. Rocky mountain habitats, to 2500m. S Germany.

11* **Bluish Willow** *Salix glauca* L. A shrub, 1-2m, with knotty grey-brown branches; shoots shiny, hairy at first. Leaves oval, untoothed, bright green, bluish-green beneath, hairy; no stipules. Catkins rather stout, yellow, appearing with the young leaves; May-June. Tundra, mountains.. N Scandinavia, Iceland, Faeroes.

12 *Salix stipulifera* B. Flod. ex Hayren. Like *S. glauca*, but stipules lanceolate and leaf lateral-veins 5-7 pairs (not 5-6). Similar habitats and flowering time. N Scandinavia.

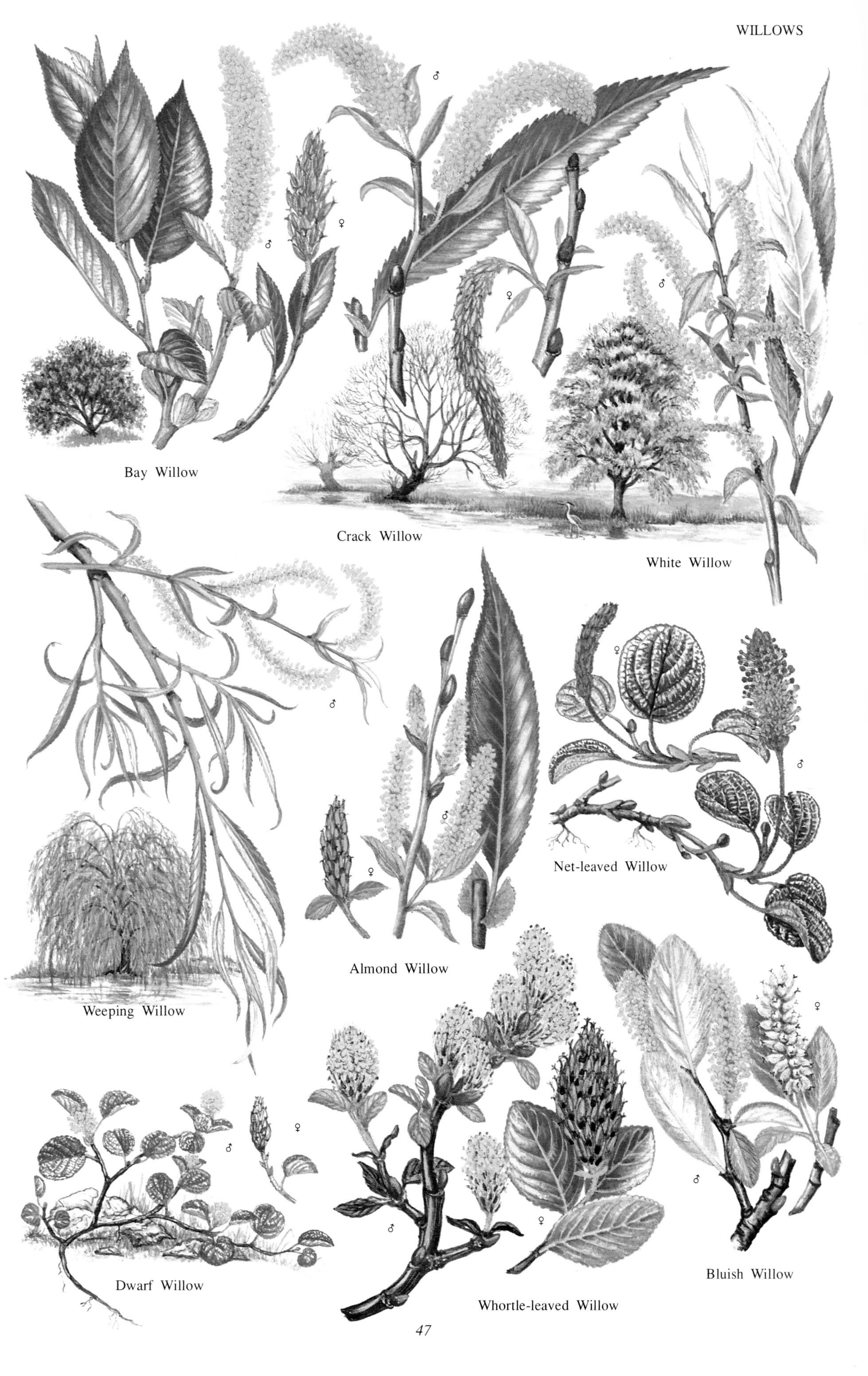

♂
♀
Bay Willow
Crack Willow
White Willow
Weeping Willow
Almond Willow
Net-leaved Willow
Dwarf Willow
Whortle-leaved Willow
Bluish Willow

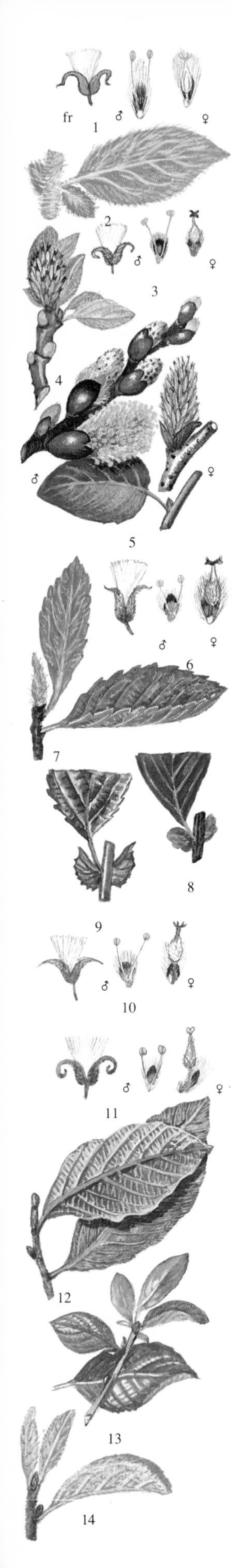


1* **Woolly Willow** *Salix lanata* L. A shrub to 3m with stout branches; shoots thick, with persistent felt and large woolly buds. Leaves broad-oval, pointed, untoothed, downy grey, with distinctive yellowish hairs when young, eventually becoming hairless above; stipules untoothed. Catkins yellow, thick, 2.5-4cm long, appearing before the leaves; May-July. Tundra, damp mountain ledges, generally on basic rocks, to 1750m. From Iceland and arctic Scandinavia, east to the Altai Mountains and extending locally south to Scotland. Hybridizes with *S. hastata* and *S. caprea*. Grown in gardens. **B**: Rare and local in Scotland – Aberdeen, Angus and Perth.

2 *Salix glandulifera* B. Flod. Like *S. lanata* but leaves oblong and glandular and stipules toothed. Similar habitats and flowering time. Arctic Scandinavia.

3* **Tea-leaved Willow** *Salix phylicifolia* L. A shrub to 4m with dark, shiny brown shoots and dark brown buds. Leaves oval to elliptical, pointed, toothed, shiny above, but greyish beneath, rather thick and rigid, not blackening on drying. Catkins 2-4cm, appearing with the young leaves; April-May. Lake and stream margins, wet rocks and stony places, to 800m. Europe east into Arctic Russia. N Britain, Ireland, the Faeroes, Iceland and much of N Europe. Hybridizes with *S. myrsinifolia*, *S.* × *tetrapla* Walker. **B**: Local, Cumbria and Yorkshire north.

4* *Salix bicolor* Willd. Like *S. phylicifolia*, but leaves silky-hairy when young, less shiny above when mature. Buds orange or yellowish. Lake and stream margins. April-May. C France and C Germany southwards.

5 **Irish Willow** *Salix hibernica* Rech. f. Similar to *S. phylicifolia*, but leaves broader, untoothed, rounded (not tapered) at the base. April-May. Moist rocky habitats. NW Ireland – Ben Bulben, not known elsewhere. Rare. Sometimes considered conspecific with *S. phylicifolia*.

6* **Dark-leaved Willow** *Salix myrsinifolia* Salisb. [*S. andersoniana*, *S. nigricans*]. Shrub or small tree to 4m, shoots brownish or greenish, hairy at first. Leaves oval to lanceolate pointed dull green above, greyish beneath except for green tip, toothed, thin and papery, blackening on drying; stipules persistent. Catkins short, 1.5-4cm, appearing with or before the leaves; April-May. Lake and stream margins, damp rock ledges, tundra. Europe east to Siberia. N Britain, Ireland and the mountains of N Europe. Occasionally cultivated in gardens. Hybridizes in the wild with *S. cinerea* and *S. phylicifolia*. Often rather local. **B**: Cumbria and Yorkshire northwards; Ireland – restricted to Antrim and Londonderry.

7 *Salix borealis* Fries. Like *S. myrsinifolia*, but often forming a small tree, with thicker shoots. Mature leaves green beneath, less blackening on drying. Catkins appearing with the young leaves; May-June. Tundra, lake and stream margins. N Scandinavia.

8* **Grey Willow** or **Common Sallow** *Salix cinerea* L. Shrub to 6m with ridged, downy-grey shoots. Leaves oval, wavy, tapered at the base, often with scattered rusty hairs beneath, greyish or dark green above, grey beneath, toothed; stipules toothed, often persistent. Catkins 3.5-5cm, dense, appearing before the leaves, almost unstalked; March-April. Fens, marshes, stream and pond margins, damp woodland, at low altitudes. Throughout, except the Faeroes, Iceland and Spitsbergen. Often dominating damp scrubland. **B**: Mainly E England, local or rare elsewhere.

9* *S. c.* subsp. *oleifolia* Macreight [*S. atrocinerea*, *S. cinerea* subsp. *atrocinerea*]. Like the type, but a taller shrub or small tree with rusty shoots; leaves shiny above, grey or rusty with hairs beneath. Similar habitats and flowering time, to 650m. Britain and W & C France. Also found in the Iberian Peninsular and NW Morocco. Hybridizes with both *S. aurita* and *S. caprea*. **B**: Throughout.

10* **Eared Willow** *Salix aurita* L. A smaller shrub than *S. cinerea*, 1-2m, with numerous spreading branches. Shoots slender, brown and angular, soon hairless. Leaves oval to oblong, dark dull green above, ashy-grey with down beneath, toothed, markedly wrinkled, often with a twisted apex; stipules conspicuous, kidney-shaped, toothed, persistent. Catkins 1.2-5cm, oval, borne before the leaves, with green scales; April-June. Damp habitats, woods, heaths, moors, rocky stream and lake margins, to 1000m. Throughout Europe except the north and the Mediterranean, extending into the C & S USSR and W Asia. Easily identified by its rough-wrinkled leaves and large stipules. **B**: Common throughout.

11* **Goat Willow, Great Sallow, 'Pussy Willow'** *Salix caprea* L. Many parts of Europe, Russia, W & C Asia. Shrub or small tree to 10m. Bark coarsely fissured. Shoots stout, downy at first. Leaves oval to oblong, with a short pointed tip, dark green above, grey and softly downy beneath, toothed; stipules small or large, heart-shaped, sometimes persistent. Catkins 3-7cm, with brownish-black scales, appearing before the leaves, silvery-white in bud; March-April. Damp habitats, woods, scrub and hedgerows, generally at low altitudes but occasionally found as high as 900m. Throughout, except the extreme north and Iceland. The bark is rich in tannin and is used in curing leather – it also contains Salicine, used as a substitute for Quinine. **B**: Common throughout.

12 *S. c.* var. *sphacelata* (Sm.) Wahlenb. [*S. coaetanea*, *S. caprea* subsp. *sericea*]. Like the type, but the leaves and shoots silky-grey, the leaves untoothed; stipules absent. Catkins appearing with the leaves; May-June. Similar habitats. Scottish Highlands and Scandinavia.

13 *Salix starkeana* Willd. [*S. livida*]. Low spreading shrub to 1m, with slender pale, hairless shoots. Leaves broad-oval, slightly hairy and reddish when young, shiny bright green above, paler or greyish beneath, not blackening on drying; stipules well developed. Catkins lax, 1.5-3cm, appearing before the leaves; April-June. Tundra and damp rocky habitats; often local. Scandinavia and Germany.

14 *Salix xerophila* B. Flod. Scandinavia. Like *S. starkeana*, but a larger shrub or small tree to 6m with erect branches; shoots dull and woolly. Leaves oblong, untoothed, grey-woolly; stipules usually absent. Similar habitats and flowering time.

♂
♀
Woolly Willow
fr
Tea-leaved Willow
*Salix bicolor*
♀
♂
Dark-leaved Willow
♂
♀
Grey Willow
fr
♂
*Salix cinerea* ssp. *oleifolia*
Eared Willow
♂
♀
Goat Willow

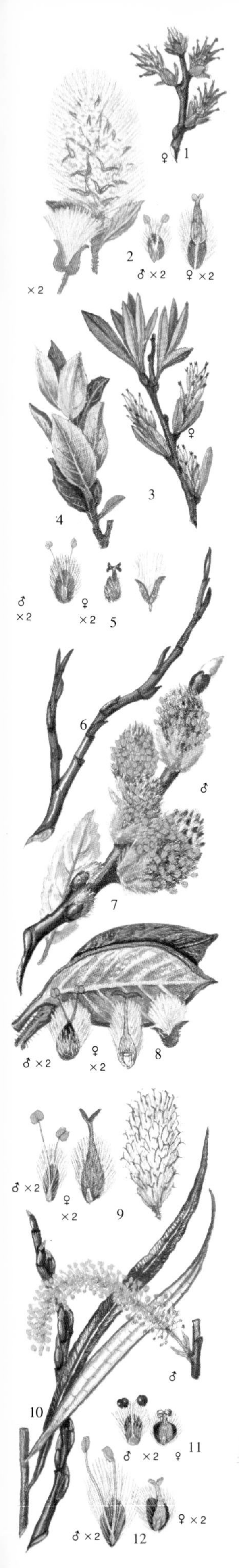

1* *Salix myrtilloides* L. A low shrub, 30-50cm, with a subterranean creeping stem; shoots erect, brown, hairy. Leaves rounded to elliptical, rounded at both ends, dull green, paler beneath, untoothed, the margins often down-turned. Catkins 1.5-2cm: April-May. Swamps and peat bogs. Local, though often forming extensive colonies. Scandinavia and Germany.

2* **Creeping Willow** *Salix repens* L. A spreading shrub with creeping stems to 1.5m tall, with slender shoots, reddish or yellowish-brown, hairy at first. Leaves narrow to broad-oval or elliptical, untoothed, hairless above but silky beneath; no stipules. Catkins short, scales green tipped with purple-brown; April-May. Fruit capsule hairless. Wet heaths, swamps, bogs and fens, to 1000m. Throughout, except Iceland and Spitsbergen. Hybridizes frequently with *S. aurita*: *S.* × *ambigua* Ehrh. **B**: Local throughout.

3 *Salix rosmarinifolia* L. Similar to *S. repens*, but with crowded basal shoots and erect linear leaves. Catkins rounded. Fruit capsules hairy. Similar habitats and flowering time. Belgium, Germany, Denmark and Sweden. Local. Some authorities include this within *S. repens*.

4 *Salix arenaria* L. [*S. repens* subsp. *argentea*]. Similar to *S. repens*, but shoots stouter, dark and hairy. Leaves slightly toothed and spreading, dull green above. Maritime dunes, dune slacks and rocky heaths close to the Atlantic. April-May. Throughout except Iceland. Like the last, this is sometimes included within *S. repens*. **B**: Local throughout.

5* **Mountain Willow** *Salix arbuscula* L. Very local. Much-branched spreading shrub to 1m; shoots ridged, hairless. Leaves elliptical, pointed, slightly toothed, shiny green above, greyish and often hairless beneath. Catkins slender, 1.3-1.8cm, appearing before, or with, the young leaves; scales with rusty hairs; May-June. Mountain slopes and damp ledges, on basic rocks, to 900m. **B**: Perth and Argyll. N Britain – Scotland, and Scandinavia.

6* *Salix hastata* L. Rather erect shrub to 1.5m, with brownish or greyish branches, shiny and hairless. Leaves variable in shape, often rather thin, broad-oval to elliptical, dull green above, paler beneath, hairless at maturity, finely toothed; stipules well developed, persistent. Catkins up to 6cm long. Heaths, moors and mountain slopes. May-June. Germany and Scandinavia.

* subsp. *vegeta* N.J. Anderss. Like the type, but leaves rounded. Lowland habitats, heaths and moors. May. Germany and Finland.

* subsp. *subintegrifolia* (B. Flod.) B. Flod. Like the type, but leaves lanceolate, untoothed. Similar habitats and flowering time. N. Germany and Scandinavia.

7 *Salix pyrolifolia* Ledeb. Very local, also found in the adjacent parts of the USSR. Similar to *S. hastata*, but a taller shrub or a small tree with reddish-brown branches. Leaves whitish and hairless, the midribs and stalks often pinkish beneath; flowers May-June. Moors and bogs at low altitudes. N Finland.

8* **Downy Willow** *Salix lapponum* L. A shrub, much branched, to 1.5m. Shoots rather rigid, grey-downy at first, later hairless and shiny. Leaves oval to lanceolate, dull grey-green and pubescent on both surfaces, untoothed; stipules small, soon falling. Catkins stout, 2-4cm, appearing with or before the leaves; scales brown, with long whitish hairs; May-June. Tundra and wet mountain rocks and cliffs, to 1100m. N Britain and Scandinavia. Cultivated in gardens. Readily recognized by the leaves which are a similar colour on both surfaces. **B**: Restricted to Cumbria and Scotland – Sutherland.

9* **Common Osier** *Salix viminalis* L. A shrub or small tree to 5m. Shoots long and flexible; grey-downy when young, later shiny and yellow-brown. Leaves long and narrow, untoothed, green above, shiny and silky beneath, margin down-turned. Catkins narrow, 1.5-3cm, appearing before the leaves, crowded at shoot tips; March-April. Stream, lake and pond margins, marshes and fens, at low altitudes. Britain, Belgium, Holland, France and Germany. Commonly cultivated – the long flexible shoots are used in basket-making. **B**: Common in lowlands throughout.

10 *Salix elaeagnos* Scop. [*S. incana*]. Shrub or small tree to 6m. Shoots slender, yellowish to reddish-brown, with sparse white hairs when young. Leaves erect, narrow-lanceolate, toothed, tapered, deep green above, grey and downy beneath. No stipules. Catkins up to 6cm long; March-April. Stream, lake and pond margins, occasionally on marshy land, usually at low altitudes. Germany extending west into France; naturalized in Holland. Sometimes grown in gardens.

11* **Purple Willow** *Salix purpurea* L. A variable spreading to erect shrub to 5m, often less. Shoots straight, shiny, hairless, purplish at first. Leaves narrow to broadly oblong, often opposite or sub-opposite, dull bluish-green, paler beneath, scarcely toothed, blackening on drying. Catkins 2-4cm, reddish or purple tinged, appearing before the leaves; March-April. River, lake and pond margins, fens and marshes, occasionally planted as an osier, to 500m. Throughout, except Iceland and Scandinavia; occasionally naturalized in Scandinavia. Occasionally locally dominant, especially on fens. The bark has a very bitter taste. **B**: In most lowlands throughout.

12* **Violet Willow** *Salix daphnoides* Vill. Tall shrub or small tree to 10m. Shoots with a bluish waxy bloom. Leaves oblong, toothed, soon hairless, shiny dark green above, greyish beneath; stipules large and heart-shaped. Catkins 3-4cm, appearing before the leaves; April-May. River and lake margins, moors. Norway and Sweden.

* var. *acutifolia* (Willd.) Doell [*S. acutifolia*]. Like the type, but shoots becoming violet in winter and the leaves narrower and drooping. **I**: USSR. Occasionally cultivated in gardens.

Creeping Willow

*Salix myrtilloides*
Creeping Willow
♀
♂
Mountain Willow
*Salix hastata*
ssp. *vegeta*
ssp. *subintegrifolia*
Downy Willow
♀
♂
Common Osier
Purple Willow
*Salix daphnoides* var. *acutifolia*
Violet Willow

**Poplars** *Populus*. Deciduous trees. Buds with several (not 1 as in *Salix*) outer scales. Catkins pendulous, appearing before the leaves, each flower with 4 or more stamens. Wind-pollinated. 35 species in north temperate regions. A complex genus with many hybrids.

1* **White Poplar** *Populus alba* L. A robust suckering tree to 20m, with a broad crown. Bark white on young stems, later smooth, grey, rough when old. Shoots and buds white-downy. Leaves palmate, 3-5 lobed, toothed, dark green above, white-downy beneath; leaves of short shoots oval, scarcely lobed, greyish beneath. Fruiting catkins 8-10cm; February-March. Frequently planted, but rarely naturalized. S Germany eastwards; widely cultivated elsewhere. **B**: Often planted, but not naturalized.

2 **Grey Poplar** *Populus × canescens* (Aiton) Sm. [*P. alba × P. tremula*]. A much larger tree to 40m or more, freely suckering. Bark smooth, grey or creamy with dark fissures. Shoots white-downy at first. Leaves rounded-triangular, coarsely and irregularly toothed, not lobed, dark green above, grey-downy beneath. Catkin scales deeply cut (shallow in *P. alba*). March-April. Damp woods, freshwater margins. W Europe. **B**: Lowland England, perhaps native in East Anglia and SE England.

3* **Aspen** *Populus tremula* L. A spreading tree to 20m, sometimes only a large shrub, suckering. Bark smooth, grey-brown, rougher when old. Shoots slightly hairy, but buds hairless and slightly sticky. Leaves rounded, with shallow blunt teeth and long flattened stalks, deep green above, pale greyish-green beneath, fluttering in the slightest breeze to show the pale under-surfaces. Fruiting catkins to 12cm; February-March. Damp woods, fens and moors, to 2100m. Throughout, except the far north. Hybridizes with *P. × canescens*, giving *P. × hybrida* (4). **B**: Throughout.

5 *Populus simonii* Carrire. A tree to 12m with a narrow crown. Bark brown, rough and furrowed. Shoots reddish-brown, hairless, with large, sticky, aromatic buds. Leaves elliptic or somewhat diamond-shaped, abruptly pointed, with rounded teeth, hairless, long-stalked. March. **I**: N. China. Planted in France, Holland, Germany, but probably not naturalized.

6* **Balsam Poplar** *Populus candicans* Aiton. [*P. × gileadensis*]. A suckering tree to 20m, with rough fissured bark. Shoots stout, brown, hairy, with sticky buds. Leaves heart-shaped, toothed, densely hairy on veins beneath. Fruiting catkins 7-16cm, but rarely seen; March-early April. Moist habitats beside rivers, streams and roads. W Europe, north to S Sweden. Widely grown for timber and sometimes naturalized. Buds and young leaves strongly aromatic. Origin unknown. **B**: Ocasionally planted.

*Populus trichocarpa* Torrey and Gray ex Hook. Like *P. candicans*, but without suckers and the leaves oval, pointed. March-April. **I**: W North America. Widely planted, rarely naturalized.

7* **Black Poplar** *Populus nigra* L. Robust and spreading tree, to 30m with an uneven crown; rarely suckering; trunk massive, dark brown with large bosses. Shoots yellowish, later greyish, hairless, with sticky buds. Leaves oval-diamond-shaped, to triangular, finely toothed, long-stalked. Fruiting catkins 10-15cm. March-April. River valleys, wet woodland, generally by fresh water. Holland and Germany; widely naturalized elsewhere. **B**: Local in E & C England.

**Lombardy Poplar** *P. n.* var. *italica* Moench. Branches upright, forming a narrow column. Commonly planted. It originated as a fastigiate mutant and is almost invariably male. **B**: Widely planted.

8 *Populus deltoides* Marshall. Tree to 30m, quick-growing and forming a broad crown; branches ascending; shoots angled, greenish then greyish-brown. Leaves oval to almost triangular, pointed, with a hairy, toothed margin; stalk glandular at top. Fruiting catkins 15-20cm; March-April. **I**: SE USA. Occasionally naturalized in W Europe. **B**: Plantations and roadsides, mainly in C & S Britain.

9* **Black Italian Poplar** *Populus × canadensis* Moench (*P. deltoides × P. nigra*). To 30m; trunk often slanting and without bosses. Leaves oval-triangular, red-bronze when young, with 1-2 pimple-like glands at top of the stalk. March-April. Widely planted for timber and shelter, except in the far north. Trunk often slanting. Generally more common than *P. nigra*. **B**: Common.

## WALNUT FAMILY Juglandaceae

About 50 species in seven genera, in temperate and subtropical regions of both hemispheres. Trees with alternate and pinnate leaves. Monoecious; male flowers with 3 or more stamens, female enclosed in a fleshy cupule. Fruit a nut or drupe.

10* **Walnut** *Juglans regia* L. Spreading tree to 30m with smooth greyish bark, eventually fissured. Leaves pinnate, alternate, with 7-9 elliptical, untoothed, leaflets, aromatic when crushed. Male catkins pendent, yellowish-green, 5-15cm, female flowers in short erect spikes. April-May. Fruit oval, fleshy, deep green, aromatic and staining, 4-5cm, containing the walnut. **I**: SE Europe. Long cultivated for its nuts and fine-grained timber; occasionally naturalized in hedgerows or along woodland margins in S Britain and France. **B**: Naturalized locally in S England.

11 **Black Walnut** *Juglans nigra* L. Leaves with 11 or more toothed leaflets; stone of fruit not splitting. E North America. **I**: Planted for timber in Denmark, Germany, occasionally Britain; not naturalized.

## BOG MYRTLE FAMILY Myricaceae

About 40 species in 4 genera, widely scattered throughout the world. Shrubs or small trees with alternate leaves, often gland-dotted. Monoecious, flowers in catkins. Fruit a small nut or berry. Wind-pollinated. Most have aromatic foliage.

12* **Bog Myrtle, Sweet Gale** *Myrica gale* L. Shrub or small tree to 2.5m; shoots reddish-brown, with scattered yellowish glands. Leaves elliptical, strongly aromatic, toothed at the top, hairy beneath, glandular. Catkins on leafless branches, orange or reddish; April-May. Fruit a dry flattened nut. Bogs and wet heaths. Throughout, except the far north. Plants may change sex from one year to another. **B**: Mainly in N & W, but scattered localities in Cornwall, S England and East Anglia.

13 **Bayberry** *Myrica caroliniensis* Miller. Like *M. gale*, but more shrub-like and with downy shoots. Catkins borne among the current season's leaves. Fruit a small waxy berry. **I**: E North America. Naturalized in S England and Holland.

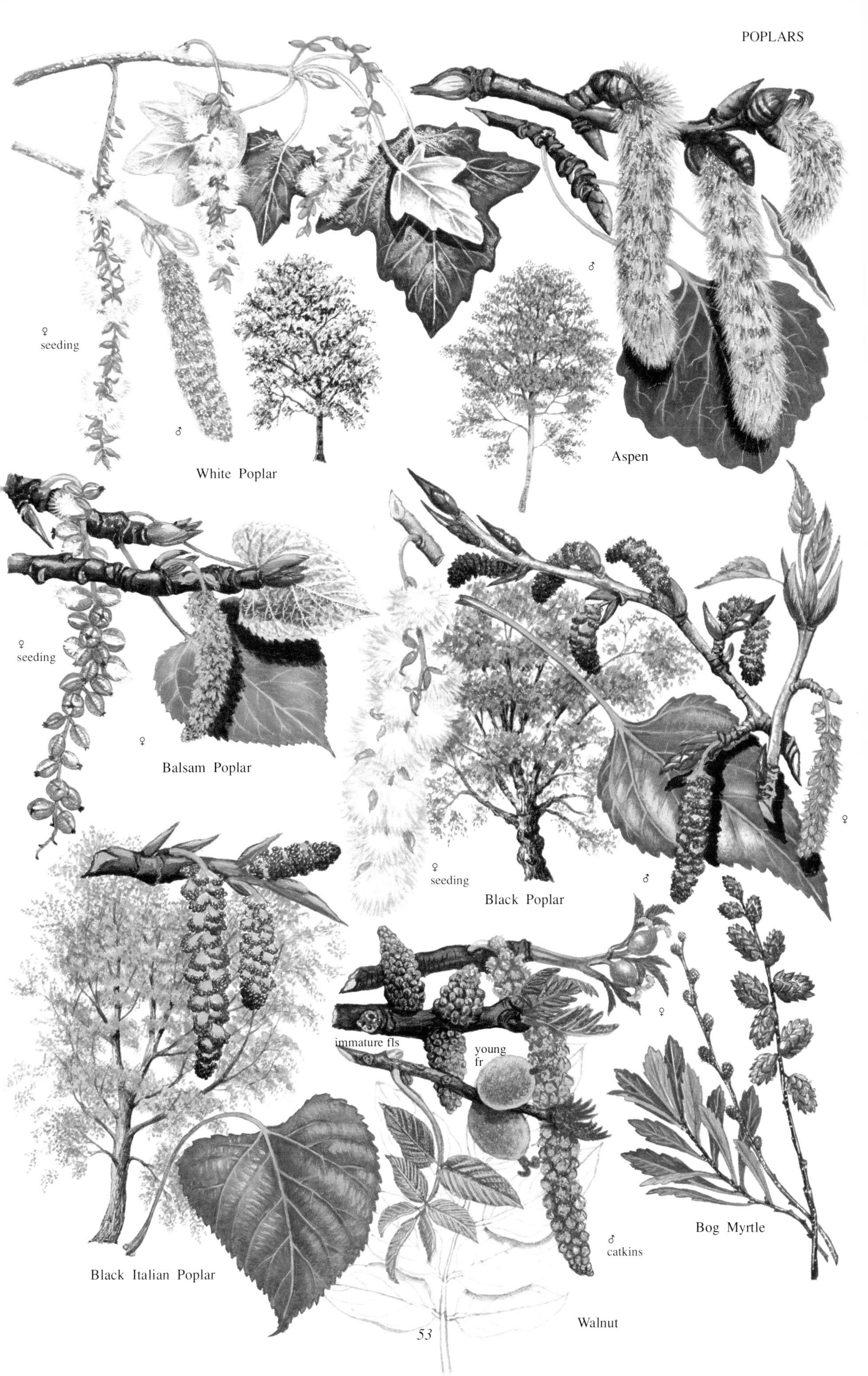

♀
seeding
♂
White Poplar
♂
Aspen
♀
seeding
♀
Balsam Poplar
♀
♀
seeding
♂
Black Poplar
immature fls
young
fr
♀
Black Italian Poplar
♂
catkins
Bog Myrtle
Walnut

## BIRCH FAMILY Betulaceae

Deciduous trees and shrubs; stipules quickly falling. Flowers in tight catkins, male and female separate but on the same plant, male pendent, female erect; cone scales 3-lobed (*Betula*) or 5-lobed and woody (*Alnus*). Fruit a small winged nutlet. *Betula* has about 40 species – in Europe, Asia south to the Himalaya, and N America. Wind-pollinated.

1* **Silver Birch** *Betula pendula* Roth. [*B. verrucosa*]. An erect tree to 30m. Bark reddish-brown at first, later silvery-white, peeling, with dark bosses when older. Shoots slender, hairless, pendent, with resin glands. Leaves oval-triangular, irregularly toothed. Catkins yellowish, the male present and unprotected from the winter, maturing with the leaves; April-May. Woods and heaths, often on poor soils, to 2000m. Throughout, except the far north. Shallow-rooted and rather short-lived. The fruiting catkins break up during the winter. Hybridizes with *B. pubescens*, *B* × *aurata* Borkh. **B**: Throughout.

2* **Downy Birch** *Betula pubescens* Ehrh. A shrub or small tree to 20m; bark brownish or greyish, without bosses. Young shoots finely hairy, without resin glands, not or scarcely pendent. Leaves evenly toothed, hairy on angles beneath. Catkins as *B. pendula*; April-May. Generally on rather wet soils and often forming extensive woods, to 2000m. Throughout, except Holland. Generally replaces *B. pendula* on badly drained soils. Rarely cultivated, unlike Silver Birch. Europe to C Asia. **B**: Common throughout, especially in the Scottish Highlands. *B.* × *intermedia* is a hybrid *B. nana*.

subsp. *carpatica* (Willd.) Aschers. & Graebn. Like the type, but shoots hairless; a shrub or small tree to 5m. Primarily in mountain habitats and on Arctic moors. N & C Europe.

subsp. *tortuosa* (Ledeb.) Nyman. Like the type, but with distorted, interlaced branches. Arctic Europe, including Iceland and the mountains of Scandinavia and Scotland.

3 *Betula humilis* Schrank. Small, much-branched shrub to 2m with finely hairy shoots. Leaves oval, pointed, thick, coarsely toothed. Young male catkins protected by bud scales during the winter; fruiting catkins 8-15mm; June-July. Bogs and fens and other wet habitats, mainly in the mountains. N Germany to NE Switzerland. Occasionally cultivated in gardens.

4* **Dwarf Birch** *Betula nana* L. Dwarf shrub, rarely exceeding 1m, with spreading branches and hairless shoots. Leaves rounded, coarsely toothed, downy when young. Fruiting catkins 5-10mm; June-July. Moors, bogs and tundra, to 2200m. N Britain, Iceland, Scandinavia and the mountains of France and Germany. **B**: Scattered localities in N England and Scotland.

**Alders** *Alnus*. Deciduous trees. Female flowers 2 to each bract, male with 4 stamens. Fruiting catkins cone-like, woody, persisting long after the seeds have been released. About 30 species, mainly in the north hemisphere.

5* **Green Alder** *Alnus viridis* (Chaix) DC. A shrub to 2-5m, often less; shoots hairless or finely hairy. Leaves sticky when young, elliptical or oval, 2-serrate, hairy on veins beneath. Catkins on short shoots with 2-3 leaves beneath. Fruiting catkins 8-15mm; April-May. Moist mountain habitats, freshwater margins, open woodland, gullys and rocky places, 1500-2300m. E France and S Germany.

6* **Common Alder** *Alnus glutinosa* (L.) Gaertner. A tree or shrub to 20mm; bark dark grey-brown, fissured. Young shoots sticky, hairless. Leaves roundish, blunt, obovate, green beneath, 2-serrate. Male catkins yellow, female purplish, appearing before the leaves. Fruiting catkins cone-like, 10-30mm, stalked; February-April. By fresh water, to 1800m. Throughout, except the far north. Europe east to Siberia. Hybridizes with *A. incana*: *A.* × *pubescens* Tausch. **B**: Throughout.

7 *Alnus rugosa* (Duroi) Spreng. Like *A. glutinosa*, but a shorter shrub, the reddish-brown shoots hairy at first. Leaves narrower and more pointed, with reddish-brown hairs beneath at the vein angles. Fruiting cones 10-15mm. February-April. **I**: North America. Frequently planted and naturalized in moist habitats in Denmark, Germany and S Sweden.

8* **Grey Alder** *Alnus incana* (L.) Moench. Tree or shrub to 10m, sometimes taller, with smooth opaque grey bark. Young shoots hairy. Leaves oval-lanceolate, pointed, downy and grey-green beneath. Fruiting catkins cone-like, 11-17mm. Growing in drier habitats than *A. glutinosa*, often in mixed deciduous woodland, to 1600m. February-March. France, Germany and Scandinavia, east to the Caucasus. **B**: Plantations mainly in Scotland and Ireland, occasionally naturalized.

subsp. *kolaensis* (Orlova) Á. & D. Löve. Like the type, but with yellowish translucent bark; leaves sometimes hairless. Arctic habitats in Europe.

## HAZEL FAMILY Corylaceae

Deciduous trees and shrubs with alternate leaves. Monoecious; male flowers in slender pendent catkins, female in clusters (drooping catkins in *Carpinus*, but in small erect clusters in *Corylus*). Fruit a nut, surrounded by a ruff of bracts. A small family: *Carpinus* has about 26 species and *Corylus* 15, in northern temperate regions. Wind-pollinated.

9* **Hornbeam** *Carpinus betulus* L. Deciduous tree to 25m with smooth silvery-grey bark striped with brown; trunk fluted. Leaves oval, pointed, somewhat pleated, sharply toothed, hairy on veins beneath. Catkins greenish to 5cm, with the leaves; April-May. Fruit in pendent clusters, surrounded by 3-lobed bracts; nuts small. Woods, hedgerows and scrub, occasionally pollarded, mostly at low altitudes. S Britain and Europe north to S Sweden. Occasionally occurs as pure woods. **B**: Mainly in SE England and East Anglia; scattered localities elsewhere.

10* **Hazel** *Corylus avellana* L. Deciduous shrub to 6m, rather upright with smooth brown bark. Leaves rounded, short-pointed, toothed, downy. Flowers appearing before the leaves, male catkins yellow, pendent; female flowers tiny, bud-like, with reddish styles; January-March. Fruit a nut, solitary or several together, each enclosed in a leafy husk, edible. Woods and hedgerows, to 1800m. Throughout, except the far north. The wood is used for fencing and basket making. The species cultivated for its nuts is not this but *C. maxima* Miller. **B**: Throughout, except for the Shetland Islands.

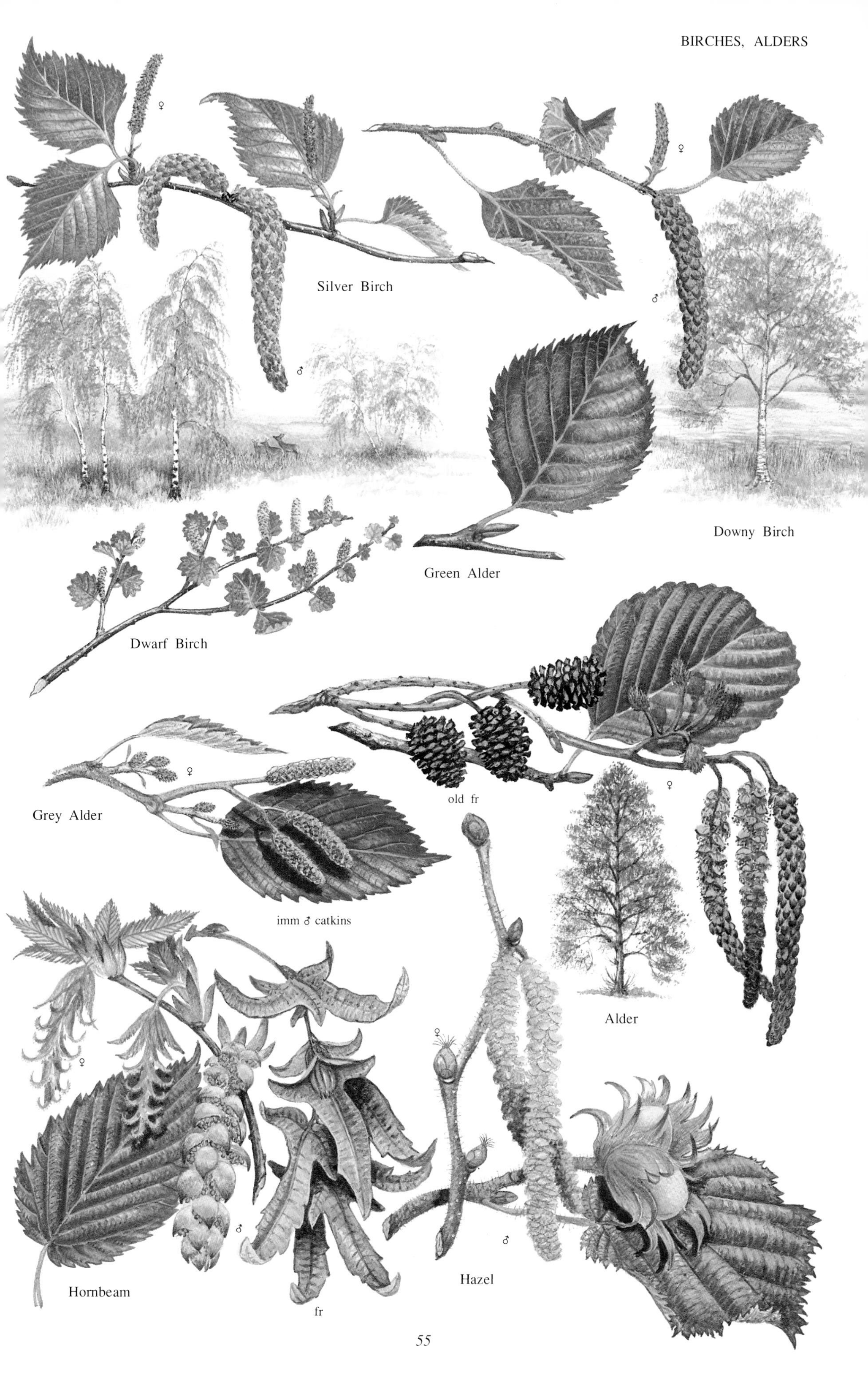

♀
Silver Birch
♂
♀
♂
Downy Birch
Green Alder
Dwarf Birch
Grey Alder
♀
old fr
♀
imm ♂ catkins
Alder
♀
♀
Hornbeam
♂
fr
Hazel
♂

# OAK FAMILY Fagaceae

Trees or shrubs with simple alternate leaves. Monoecious, male flowers in long slender catkins or heads, each flower with 8-20 stamens; female 1-3, surrounded by a close cluster of scales, each with 3 or 6 styles. Some 900 species in 6 genera, in both hemispheres.

1* **Beech** *Fagus sylvatica* L. Large spreading tree to 30m. Bark smooth and greyish. Shoots with long elliptical reddish-brown buds. Leaves elliptical, pointed, shallowly toothed, with parallel veins running to the leaf margin, silky at least when young. Flowers appearing with the young leaves, male in drooping tassels, female separate and erect; April-May. Fruit 1-3 triangular nuts in a bristly, woody splitting husk. Woods, often on calcareous but also on sandy soils, to 1900m. Europe north to SE Norway. Trees are frequently pollarded in ancient woodlands. **P**: Wind. Heavy 'mast' years, when abundant fruits are produced, often follow a hot summer of the previous year. **B**: SE England, Midlands and S Wales, but extensively planted and naturalized elsewhere.

2* **Sweet Chestnut** *Castanea sativa* Miller. Large spreading tree to 30m. Bark brownish-grey, often with spiralled fissures. Leaves oblong, pointed, toothed, scaly beneath. Flowers yellowish-green, in erect or spreading catkins, female at the base, male above; July. Fruit 1-3 smooth brown shiny nuts in a very spiny splitting husk. Woodland, particularly on acid soils but widely planted. **P**: Insects. **I**: S Europe. 10 species in north temperate regions. **B**: Widely naturalized, especially in SE England.

**Oaks** *Quercus*. Leaves often toothed or with deep rounded lobes, deciduous or evergreen. Wind-pollinated, male in clusters in slender drooping catkins, female in smaller, shorter catkins. Fruit an oblong nut (acorn) held in a cluster of fused bracts, the cup or cupule. 800 species, both deciduous and evergreen, mainly in northern temperate regions, but also at higher altitudes in the tropics.

3* **Red Oak** *Quercus rubra* L. [*Q. borealis*]. Deciduous tree to 25m; bark smooth, silvery-grey, sometimes warted and fissured; shoots hairless, dark red. Leaves oval, deeply-lobed and sharply toothed, stalked. April-May. Cup shallow, 18-25mm wide, with thin scales, the acorns ripening in second year. **I**: E North America. Sometimes naturalized. The leaves turn brilliant red in the autumn. **B**: Planted throughout, except the north and Ireland. Sometimes regenerates.

4 **Pin Oak** *Quercus palustris* (L.) Moench. Like *Q. rubra*, but leaves narrower and more deeply lobed, with jagged teeth. Cup 10-15mm wide. Planted for timber in Denmark and Germany. **I**: E North America. Planted occasionally in parks and large gardens in S England.

5* **Holm Oak, Evergreen Oak** *Quercus ilex* L. An evergreen tree to 25m forming a dense rounded head, with greyish or blackish-brown scaly bark; shoots and buds downy-grey. Leaves thick and leathery, deep green, oval, finely toothed or with shallow spines, grey-downy beneath. Catkins greenish-yellow; May-June. Acorns ripening in the first year, small. Warm, often rocky, habitats on well-drained soils. Primarily Mediterranean. W France, especially near the coast. **B**: Commonly planted, sometimes naturalized.

6* **Turkey Oak** *Quercus cerris* L. Large deciduous tree to 35m with rough downy young shoots; bark dull grey, fissured. Leaves oblong, rather unevenly lobed and toothed, dull green and slightly rough above, downy beneath, stalked. May-June. Acorns ripening in the second year, with a bristly cup, short-stalked. Woodland and hedgerows, widely planted. France and Germany; naturalized in Britain and Belgium. **B**: Widespread north to Ross-shire; spreading in some areas.

7* **Sessile Oak, Durmast Oak** *Quercus petraea* (Mattuschka) Liebl. [*Q. sessiliflora*, *Q. sessilis*]. Large deciduous tree to 40m with grey fissured bark; shoots dark grey, hairless. Leaves oval with rounded lobes, narrowed at base, hairy beneath, stalked. Flowers appearing with young leaves; April-May. Acorns with a hairy cup, scarcely stalked. Woodland and hedgerows often on rather poor soils, sometimes planted, to 1800m. Throughout, except the far north. Unlike *Q. robur* and other oaks, the leaves are rarely eaten by insects or galled. May hybridize with *Q. robur*: *Q.* × *rosacea* Bechst. **B**: Throughout.

8* **Pedunculate Oak, English Oak** *Quercus robur* L. [*Q. pedunculata*]. Large spreading tree to 45m, with pale grey, densely fissured bark; shoots usually hairless. Leaves oval, with rounded lobes, rounded at base and with a very short stalk; April-May. Acorns with a finely hairy, grey-green cup, long-stalked. Woods, hedgerows on a variety of soils, especially basic loams and clays, to 1400m. Widely planted, especially in parks. Throughout, except the north, the Faeroes and Iceland. Europe to NE Russia, SW Asia and N Africa. Older trees are frequently pollarded, especially in Britain where the boles may reach 2-3m diameter. The leaves are often holed, and galled beneath (spangle gall). **B**: Throughout, especially in the lowlands.

9 **White Oak, Downy Oak** *Quercus pubescens* Willd. [*Q. lanuginosa*]. A shrub or tree to 25m, deciduous, with dark grey, finely cracked bark; shoots densely hairy. Leaves, elliptical with rounded lobes, grey-green, hairy beneath, stalked. April-May. Acorns short-stalked; cups downy, with narrow scales. Woods, sometimes hedgerows, generally on light well-drained soils, to 1500m. France, Belgium and S Germany. Mainly from S Europe to W Asia and the Caucasus. **B**: Occasionally grown in parks, but not a fine tree.

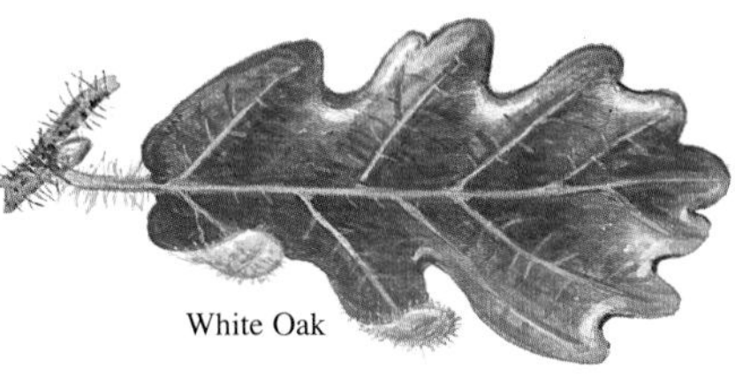

White Oak

Beech in autumn

Oak in autumn

♀
♂
developing
fr
spring
foliage
Beech
♂
♀
Sweet Chestnut
♀
♂
Holm Oak
Red Oak
♂
♂
♂
first year
acorn
Turkey Oak
winter
Sessile Oak
winter
Pedunculate Oak

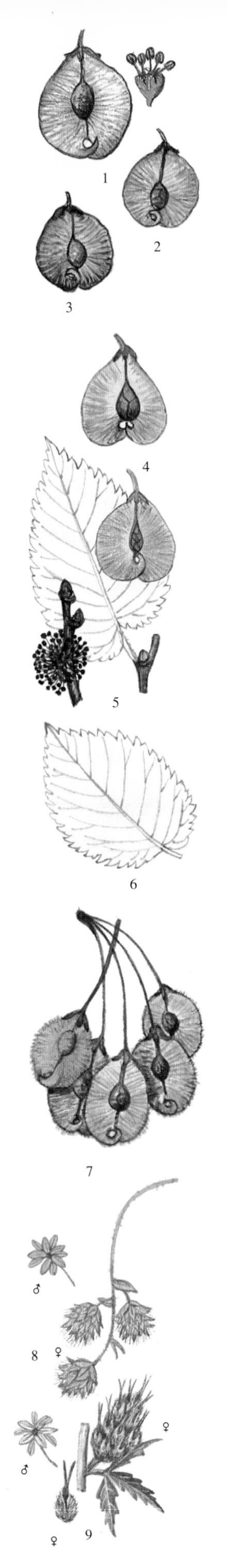


## ELM FAMILY Ulmaceae

Deciduous trees or shrubs with simple, alternate leaves, often asymmetrical at the base; margin twice serrate. Size and shape of leaves can vary on different parts of the same tree. Flowers in small lateral clusters, hermaphrodite or male, appearing before the leaves. Fruit a winged nut, notched at the apex. 15 genera and some 200 species. The genus *Ulmus* is a complex one in Europe with numerous intermediates and local populations. In W Europe all species have been severely affected in recent years by Dutch Elm Disease, *Ceratocystis ulmi*, carried from tree to tree by small beetles. Wind-pollinated.

1* **Wych Elm** *Ulmus glabra* Huds. Large spreading tree to 40m, without suckers; bark with smooth silvery-grey, but eventually fissured and brownish; twigs stout, rough when young. Leaves rounded to elliptical with 12-18 pairs of lateral veins, rough hairy above, scarcely stalked. Fruit oval, 15-20mm, with a central seed. Woods and hedgerows. March- April. Throughout, except the Faeroes, Iceland and Spitsbergen. N & C Europe to W Asia. Occasionally planted in parks and churchyards. **B**: Local in England and Wales, north to Yorkshire; most abundant in the west and in Ireland.

2* **English Elm** *Ulmus procera* Salisb. [*U. campestris*] Large suckering domed tree, to 30m, with dark brown bark 'cracked' into squares; twigs fairly slender, hairy. Leaves like *U. glabra* but smaller, with 10-12 pairs of lateral veins. Fruit rounded, 10-17mm, winged, with the seed above the middle. Hedgerows and fields on well-drained soils, rarely woodland. March-April. C & S Britain. Also found in SE France. The fruit is borne irregularly from one season to another and often proves to be sterile. **B**: Midlands, Bristol Channel and Devon mainly; rare and planted in Ireland.

3* **Smooth-leaved Elm** *Ulmus minor* Miller [*U. carpinifolia*]. Large suckering tree to 30m, with grey-brown bark with deep vertical fissures; twigs slender, quickly hairless. Leaves elliptical to oval, smooth and bright shiny green above, with 7-12 pairs of lateral veins, stalked. Fruits elliptical with the seed above the centre. Hedges, fields and woodland margins. March-April. Throughout, except the north and most of the islands. The common elm in Europe, less common in the extreme west. Occasionally planted in parks. **B**: Mainly confined to the Midlands and SE England.

4* **Lock Elm, Small-leaved Elm** *U. m.* var. *lockii* (Druce) Richens. [*U. coritana*, *U. plotii*]. Like the type, but with a rather thin, arching, one-sided crown; leaves less shiny. Similar habitats and flowering time. **B**: Confined to the English N Midlands.

5 **Cornish Elm** *U. m.* var. *cornubiensis* (Weston) Richens. [*U. stricta*]. Like the type, but branches steeply ascending, producing a dense, leafy, fan-shaped crown; leaves bright green. Similar habitats and flowering time. **B**: Cornwall and W Devon where it replaces *U. procera*; also in SW Ireland.

6 **Jersey Elm** *U. m.* var. *sarniensis* (Loud.) Rehder [*U. angustifolia*, *U. wheatleyi*]. Like var. *lockii*, but the central axis persisting to the top of the tree to form a dense symmetrical crown, not fanning out; leaves dark green, rather rounded. Parks, roadsides, streets, occasionally in hedgerows. Usually planted. Britain only, mainly in C & N England and S Scotland; rare in Ireland.

**Dutch Elm** Hybrids between *U. glabra* and *U. minor*, *U.* × *hollandica* Miller. Common in England, particularly in the south and East Anglia. This hybrid exists in several forms, generally vigorously suckering and with radiating branches with their long straight, rather than twisted, divisions.

7* **Fluttery Elm, European White Elm** *Ulmus laevis* Pallas [*U. effusa*]. Large tree to 35m with dull grey or pale brown, smoothly-ridged bark; twigs reddish-brown, hairy. Leaves broad-oval, with 12-19 pairs of lateral veins, finely hairy on both surfaces. Fruit oval, 10-12mm, with a central seed, hairy on the margin and long-stalked. Woodland, hedgerows and fields. March-April. C Europe extending locally west to Belgium and France and north to S Sweden; occasionally planted elsewhere, but generally rare in cultivation. C Europe to W Asia.

## HEMP FAMILY Cannabaceae

Herbs with opposite or alternate leaves. Dioecious; male fowers stalked with a 5-parted perianth, female unstalked with an undivided perianth. Fruit an achene within the persistent perianth. Only 3 species in 2 genera, from north temperate regions.

8* **Hop** *Humulus lupulus* L. Climbing and twining herb to 6m; stem square, rough-hairy. Leaves opposite palmately 3-5-lobed, coarsely toothed. Male flowers green, 4-5mm, in lax lateral panicles; female flowers pale green, in rounded clusters. Fruit cone-like, 25-30mm, with closely overlapping bracts. Hedgerows, scrub, sometimes clambering on walls, on loamy and peaty soils. July-September. Throughout, except the far north. The ripe female cones have been used for flavouring beer since as early as the ninth century. **B**: Throughout, though mostly naturalized in much of Scotland and in Ireland.

9* **Cannabis, Hemp** *Cannabis sativa* L. Tall erect, strong-smelling, herb to 2.5m, though often less. Leaves alternate, palmate, lobed to the base into 3-9 narrow, toothed lobes. Flowers small, greenish, male in much- branched clusters, female in slender stalked spikes, with resinous glands. Nut smooth. A weed of disturbed land, waysides, rubbish tips. July-September. **I**: Asia. Naturalized in Belgium, France and Germany; casual elsewhere. Formerly cultivated for its fibre, oil, and medicinally; now illegally for marijuana. **B**: Casual.

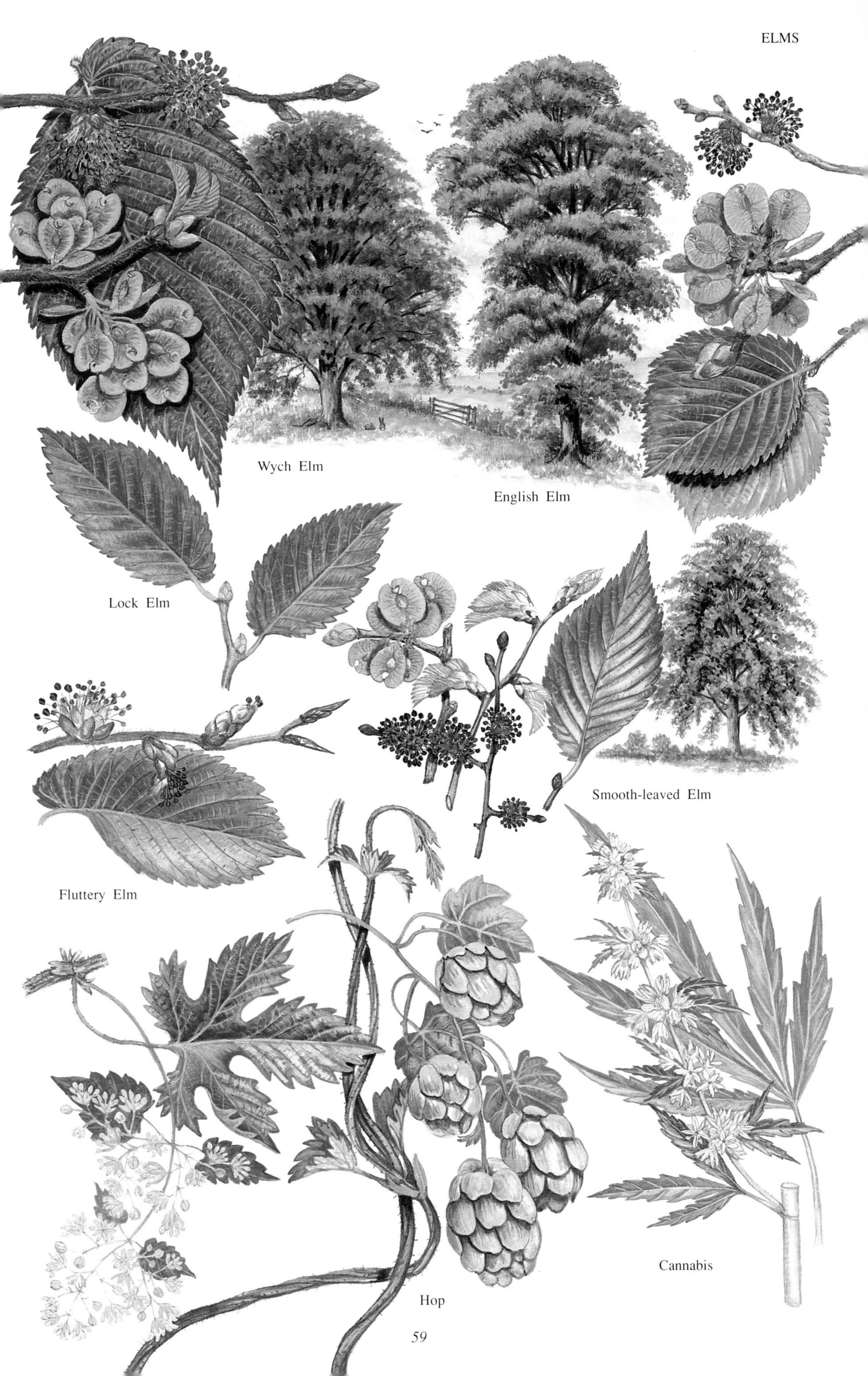
Wych Elm
English Elm
Lock Elm
Smooth-leaved Elm
Fluttery Elm
Cannabis
Hop

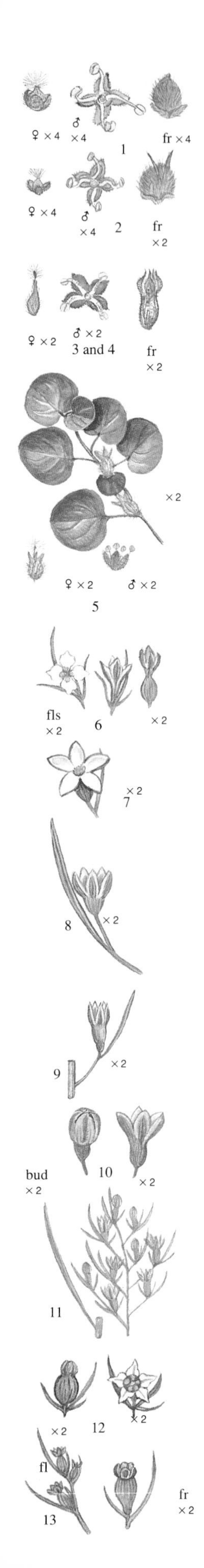


# NETTLE FAMILY Urticaceae

Herbs without latex. Leaves opposite or alternate, simple, with stipules. Flowers usually unisexual, sometimes on separate plants; perianth 4-5 parted, often persistent. Fruit a small achene. 550 species in 45 genera, in temperate and tropical regions of the world.

1* **Stinging Nettle** *Urtica dioica* L. Very variable, vigorous, medium-tall perennial with stout stolons; stems square. Leaves opposite, heart-shaped to lanceolate, armed with stinging hairs, toothed. Male and female flowers greenish, on separate plants; male in long drooping catkin-like spikes, female in small clusters. Widely varied habitats, often near habitation and on rich soils, often forming extensive colonies, to 3150m. June-September. **P**: Wind. Throughout, except Spitsbergen. The stem-fibres were used for making cloth from the Bronze Age onwards, and the young shoots are edible. **B**: Ubiquitous.

2* **Annual Nettle** *Urtica urens* L. Short to medium annual, with weakly stinging hairs. Leaves as in *U. dioica*, but more coarsely toothed, the lower short- not long-stalked. Male and female flowers on the same plant, many female and a few male in ascending spikes. Cultivated, waste and disturbed ground, gardens, often on slightly acid and well-drained soils, generally at low altitudes, but to 2700m in C Europe. June-October. Throughout, except Spitsbergen. **B**: Throughout – commonest in the east.

* **Roman Nettle** *Urtica pilulifera*. Like *U. urens*, but female flowers in rounded heads, male in spikes. **I**: S Europe. A rare casual in Britain, Belgium and France.

3* **Parietaria** *Parietaria officinalis* L. [*P. erecta*]. Medium to tall, erect, hairy perennial; stems scarcely branched. Leaves alternate, elliptical, untoothed, short-stalked. Flowers hermaphrodite, small, greenish with yellow stamens, in clusters at base of leaves, each 4-parted. Achenes black. Open habitats, rocky places, banks. June-October. C & E France and Germany; naturalized in Belgium, Denmark and Holland.

4* **Pellitory of the Wall** *Parietaria judaica* L. [*P. diffusa*]. Similar to *P. officinalis*, but a short to medium much branched herb with spreading stems. Leaves generally smaller, up to 5cm (instead of 3-12cm). Walls, hedge banks and stony places on dry, well-drained soils. June-October. Confined to W & S Europe. **B**: Throughout much of England and Ireland, mostly coastal in Wales; rare in Scotland.

5* **Mind Your Own Business** *Soleirolia soleirolii* (Req.) Dandy [*Helxine soleirolii*]. Low creeping, hairy, evergreen, mat-forming perennial with slender stems rooting at the nodes. Leaves small, rounded, 2-6mm, 3-veined. Flowers pinkish, unisexual, solitary at base of the leaves. May-August. **I**: S Europe. Widely naturalized in damp places, on shady walls and in greenhouses: Britain, France, Belgium and Holland, possibly elsewhere. Also cultivated as a small pot plant, including forms with variegated leaves. **B**: Especially SW Britain and Ireland.

# SANDALWOOD FAMILY Santalaceae

Herbs or shrubs with narrow, alternate, untoothed leaves; usually semi-parasitic on the roots of small herbs or shrubs. Flowers hermaphrodite (in *Thesium*) bell-shaped or tubular, 4-5 parted. Fruit a small nut. 30 genera and 400 species in temperate and tropical regions. They are sometimes confused with members of the Caryophyllaceae, but the leaves are alternate and the perianth consists of only one whorl – not differentiated into sepals and petals. Species of *Thesium* are often difficult to separate and details of the leaves and inflorescence should be examined with care.

6* **Alpine Bastard Toadflax** *Thesium alpinum* L. [*T. tenuifolium*]. Short ascending to erect perennial; stems often simple. Leaves linear, 1-veined. Flowers usually in one-sided, leafy, racemes, white, 4-lobed each flower with 2 bracteoles. Nut shorter than persistent perianth. Dry meadows and stony habitats, mainly in the mountains, to 2600m. May-August. France and Germany.

7* *Thesium pyrenaicum* Pourret [*T. pratense*]. Like *T. alpinum*, but flower 5-lobed, in spikes or panicles, not one-sided, rather zig-zagged. Dry habitats, both grassy and rocky, to 2600m. May-August. Belgium, Holland, France and Germany.

8* *Thesium ebracteatum* Hayne. Short perennial with a slender stoloniferous stock; stems erect or ascending. Leaves linear-oblong, 1-veined. Flowers 5-lobed, bell-shaped, in slender racemes. Nut longer than the persistent perianth; no bracteoles. Dry open habitats, at low altitudes. June-August. E Denmark and N Germany. Very local.

9 *Thesium rostratum* Mert. & Koch. Like *T. ebracteatum*, but plants with a woody, non-stoloniferous stock. Flowers tubular, the perianth longer than the nut. C & S Germany.

10* **Bavarian Bastard Toadflax** *Thesium bavarum* Schrank [*T. montanum*]. Short to medium erect perennial herb, without stolons; stems usually simple. Leaves lanceolate, 3-5-veined, dark green. Flowers whitish, in a lax panicle, each with 2 bracteoles. Dry meadows and scrub, to 1250m. June-August. E France and S Germany.

11 *Thesium linophyllon* L. [*T. linifolium*]. Like *T. bavarum* but shorter and usually stoloniferous. Leaves narrower, stiff, yellowish-green, usually 3-veined. Sandy habitats on acid soils, to 1200m. June-August. E France, C & S Germany.

12* **Bastard Toadflax** *Thesium humifusum* DC. Short, often prostrate, perennial, not stoloniferous, weakly branched. Leaves linear, 1-veined. Flowers 5-lobed, yellowish-green outside, dull white inside, in branched racemes; each flower with 2 bracteoles, the persistent perianth shorter than nut. Dry grassy places, on calcareous soils, at low altitudes. June-August. Britain, France and Holland, confined mainly to W Europe. Rare and decreasing in most areas. **B**: C & S England and East Anglia, from Shropshire southwards; rare.

13 *Thesium divaricatum* Jan ex Mert. & Koch. Like *T. humifusum*, but an erect plant with more robust and rigid stems, much-branched. Dry meadows and scrub, to 2150m. June-August. C & S France.

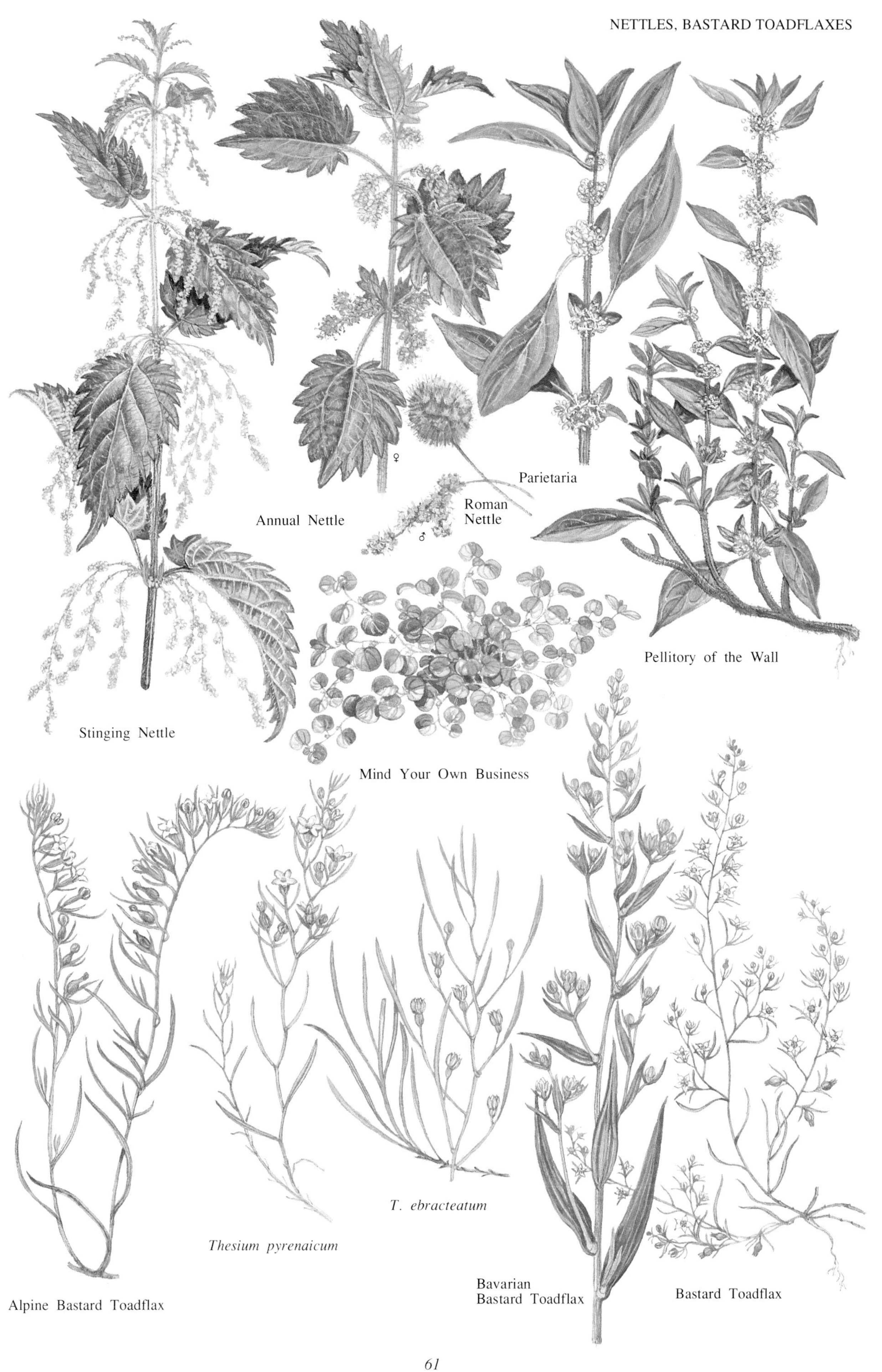
Stinging Nettle
Annual Nettle
♀
Roman
Nettle
♂
Parietaria
Pellitory of the Wall
Mind Your Own Business
Alpine Bastard Toadflax
*Thesium pyrenaicum*
*T. ebracteatum*
Bavarian
Bastard Toadflax
Bastard Toadflax

## MISTLETOE FAMILY Loranthaceae

Parasitic shrubs with opposite untoothed leaves. Flowers mostly 4-parted, unisexual. Fruit a fleshy berry. The sticky berries are eaten by a variety of birds, which wipe the seeds off onto other trees. Over 900 species in 70 genera, mostly tropical.

1* **Mistletoe** *Viscum album* L. Regularly branched, yellowish-green, parasitic shrub, hairless, forming rounded tufts up to 1m across on tree branches. Leaves oblong, leathery, opposite, untoothed. Flowers inconspicuous, unisexual, male and female on separate plants, 4-parted, in small stalkless clusters. Fruit a white berry, 6-10mm, during winter. Parasitic on deciduous trees, especially *Malus* and *Populus*. February-April. Throughout, except Ireland and the far north. **B**: England and Wales north as far as the Humber. subsp. *abietis* (Wiesb.) Abromeit has longer leaves, up to 8cm. Parasitic on conifers, especially *Abies*. France and Germany. subsp. *austriacum* (Wiesb.) Vollmann has usually smaller leaves, 2-4cm and yellow berries. Parasitic on *Pinus* and *Larix*. France and Germany.

## BIRTHWORT FAMILY Aristolochiaceae

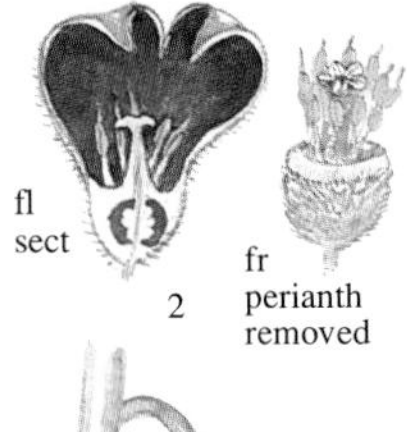


Herbs with alternate untoothed leaves. Flowers solitary or clustered at the leaf axils, 3-parted, the perianth forming a tube; stamens 6 or 12; styles usually 6. Fruit a capsule. 7 genera and 400 species, mainly tropical and subtropical.

2* **Asarabacca** *Asarum europaeum* L. Low, creeping, hairy, evergreen perennial. Leaves kidney-shaped, shiny deep green, long-stalked. Flowers purplish-brown, solitary, hidden beneath the leaves, bell-shaped, 12-15mm, with a 3-lobed perianth. Fruit capsule small, rounded, splitting irregularly. Woodland, scrub and copses, on calcareous soils, to 1300m. March-August. Belgium, France, Germany, S Norway and S Sweden; naturalized, possibly native, in Britain, Denmark and Holland. **B**: Rare and declining; scattered localities in England, S Wales and S Scotland.

3* **Birthwort** *Aristolochia clematitis* L. Medium to tall tufted perennial, hairless, evil-smelling and stinking; stems erect, unbranched. Leaves oval, blunt, stalked. Flowers irregular, partly tubular, 20-30mm, dull yellow with a yellowish-brown flattened lip. Fruit a large oval capsule, splitting symmetrically. Damp habitats. June-September. Probably native to C & S Europe. Naturalized in Britain, Belgium, Holland, France and Germany. **B**: Rare and declining; C & E England and S Scotland.

## DOCK FAMILY Polygonaceae

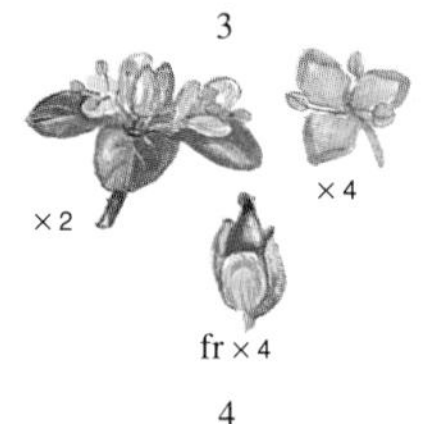


Herbs, shrubs or climbers, without latex. Leaves alternate; stipules often a membranous sheath (ochrea) around the stem. Flowers hermaphrodite or unisexual, 3-6 parted, the perianth often persisting and enlarging in fruit; stamens (3) 6-9; styles 2-4. Fruit an elliptical or triangular nut. 800 species in 40 genera, worldwide.

4* **Iceland Purslane** *Koenigia islandica* L. Tiny insignificant annual with erect reddish stems. Leaves 3-5mm only, broad-oval, unstalked. Flowers pale green with reddish anthers, 3-parted, in small clusters; stamens 3. Open bare moist ground or among sparse low vegetation, to 1500m. June-September. Britain, N Europe, the Faeroes and Iceland. **B**: Very rare, confined to mountains on Skye and Mull.

**Knotgrasses** *Polygonum*. Herbs, often with rather narrow leaves. Flowers with a 5-parted perianth; stamens often 8; stigmas 2 or 3. Nut not winged. Over 300 species.

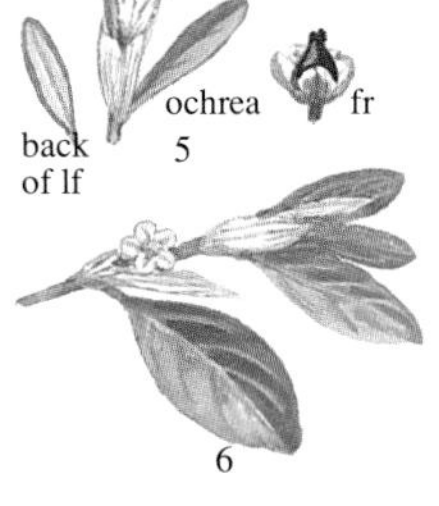


5* **Sea Knotgrass** *Polygonum maritimum* L. Short to medium, branched, generally prostrate perennial, woody at the base. Leaves narrow-elliptical, bluish-green, with downrolled margins; ochrea silvery, many-veined. Leaves blackening on drying. Flowers pink or whitish, solitary or in small clusters. Nut glossy. Maritime sands and shingle, occasionally among rocks. July-September. Primarily Mediterranean. France, Britain. **B**: Very rare in S Ireland and the Channel Islands.

6 *Polygonum cognatum* Meissner. Like *P. maritimum*, but with oblong leaves and weakly veined ochrea. Naturalized on waste ground. June-September. **I**: SW Asia. France; formerly in Britain.

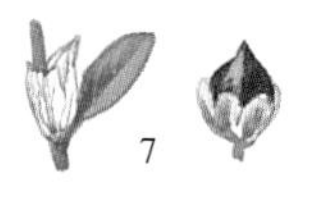


7* **Ray's Knotgrass** *Polygonum oxyspermum* Meyer & Bunge ex Ledeb. Low annual or biennial with sprawling stems, often woody at base. Leaves narrow-lanceolate, the margin scarcely down-rolled; ochrea short, few-veined. Flowers in small lateral clusters, greenish edged with red. Nut glossy, longer than the dead perianth. Coastal habitats, sand and shingle coasts, often associated with other drift-line species and rather erratic in appearance. June-November. Baltic Norway. Very local.

subsp. *raii* (Bab.) D.A. Webb & Chater [*P. raii*] has broader blue-green leaves; flowers with white or pink margins. Shores of Europe north to parts of Scandinavia and N Germany. **B**: Rare north to Harris.

*Polygonum patulum* Bieb. Like *P. oxyspermum*, but with upper bracts of the spike shorter than the flowers. France and Germany.

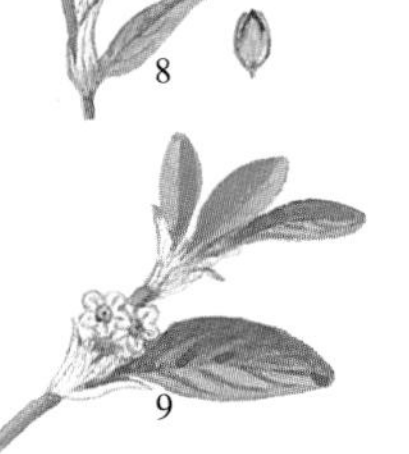


8* **Knotgrass** *Polygonum aviculare* L. [*P. heterophyllum*]. Variable short to medium, erect to sprawling, hairless annual. Leaves lanceolate to oval, larger on main stem; ochrea silvery, ragged, few-veined, concealing the leaf-stalk. Flowers solitary or in small clusters, greenish with pink or whitish margins; perianth-segments fused only at base. Nut dull, enclosed within the persistent perianth. A weed of waste places, bare ground, roadsides and seashores, to 2300m. June-November. Throughout, though alien in the extreme north. **B**: Common except in the far north, where it is replaced by *P. boreale*.

9 *Polygonum boreale* (Lange) Small. Like *P. aviculare*, but leaves smaller and broadest above the middle; leaf-stalks extending beyond ochrea. Flowers with bright pink margins. Similar habitats and flowering time, particularly seashores, local. N Britain, the Faeroes, Iceland, Scandinavia. **B**: Local in N Scotland, Orkney, Shetland, Outer Hebrides.

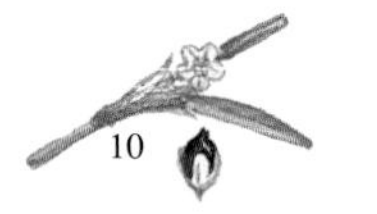


10* *Polygonum rurivagum* Jordan ex Boreau. Like *P. aviculare*, but leaves linear to linear-lanceolate with reddish-brown ochrea. Flowers usually reddish. Nut protruding beyond the persistent perianth. Dry habitats on light calcareous soils; a common weed on arable land. June-October. Europe north to S Sweden. Declining. **B**: Mainly confined to S & E England.

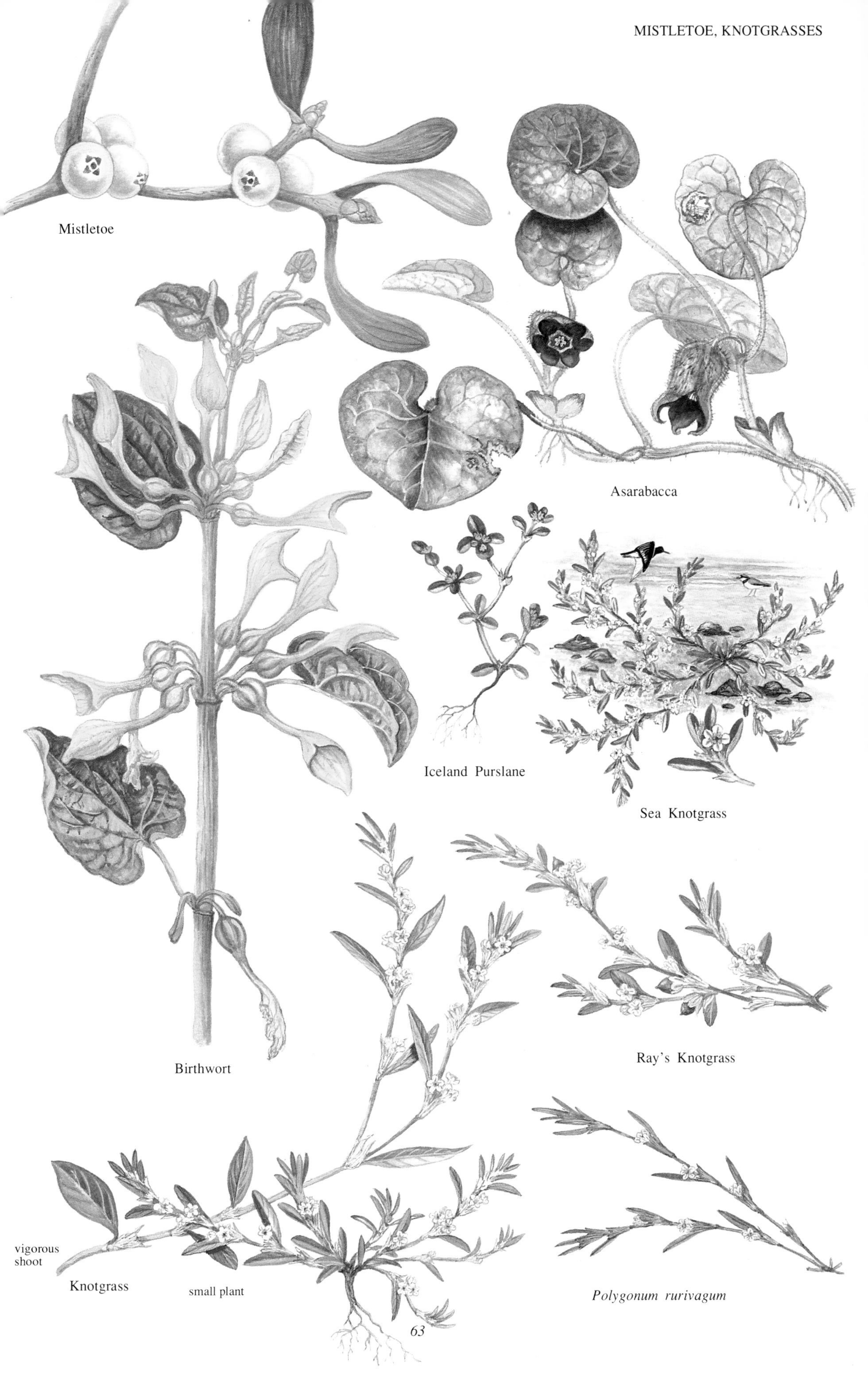
Mistletoe
Asarabacca
Birthwort
Iceland Purslane
Sea Knotgrass
Ray's Knotgrass
vigorous shoot
Knotgrass
small plant
*Polygonum rurivagum*

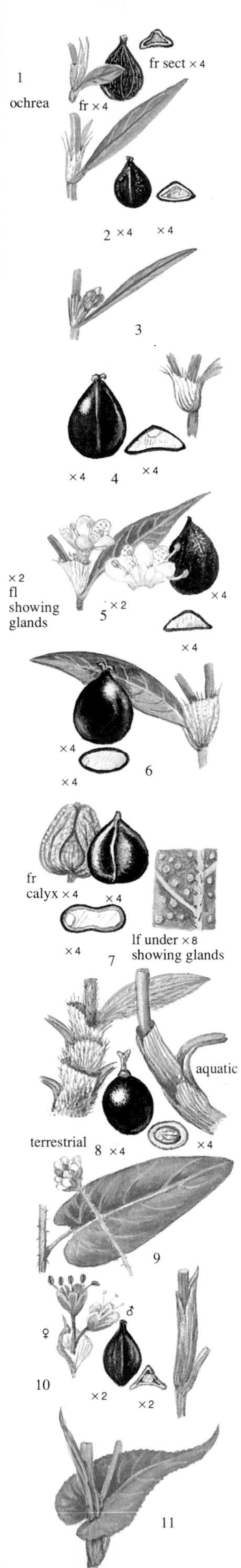


1* **Small-leaved Knotgrass** *Polygonum arenastrum* Boreau [*P. aequale*]. Low mat-forming annual. Leaves elliptical to lanceolate, uniform; ochrea silver. Flowers lateral, solitary or in small clusters, greenish-white or pinkish. Nut dull. Open ground, cultivated land, waste places, pathsides and roadsides. July-September. Throughout, except the extreme north. A widespread weed in Europe and North America. **B**: Common throughout, though less so than *P. aviculare*.

2* **Small Water-pepper** *Polygonum minus* Hudson. Low to short hairless annual with sprawling or ascending stems; ochrea brownish, fringed. Leaves narrow-lanceolate, pointed, scarcely stalked; flowers in lax slender erect spikes, deep pink, rarely white. Nut black, glossy. Damp habitats, marshy ground, lake, pond and river margins, at low altitudes. July-September. Throughout, except Iceland and the extreme north. Generally declining, especially in W Europe, due primarily to land drainage. Often grows in association with *P. mite*, though a smaller plant. **B**: Uncommon, except C & N Scotland.

3 *Polygonum foliosum* H. Lindb. Like *P. minus*, but with linear leaves; ochrea scarcely fringed. Similar habitats and flowering time. Scandinavia.

4* **Tasteless Water-pepper** *Polygonum mite* Schrank. Short to medium, slender, hairless annual with erect stems. Leaves lanceolate, scarcely stalked, tasteless; ochrea brown, fringed. Flowers in a lax slender, undotted spike, often branched, pink, rarely white. Nut black, glossy. Damp habitats, river, stream and pond margins, ditches and dykes, at low altitudes. July-September. Throughout, except Iceland and Scandinavia. **B**: Lowland England and Wales, W Ireland; rare and declining.

5* **Water-pepper** *Polygonum hydropiper* L. Medium to tall, erect hairless annual, with a burning peppery taste. Leaves narrow-lanceolate, pointed; scarcely stalked; ochrea brown, shortly fringed. Flowers in a lax slender yellow dotted spike, nodding at tip, pink or greenish- white. Nut brown or black, dull. Damp habitats, meadows and shallow water, on acid soils. June-September. Lowlands throughout, except the Faeroes, Iceland and Spitsbergen. Cleistogamous flowers are usually present at the leaf axils. **B**: Common throughout.

6* **Redshank** *Polygonum persicaria* L. Short to medium, erect to rather sprawling, hairless, branched annual. Leaves lanceolate, tapered at the base, often with a large blackish blotch; ochrea brown, shortly fringed. Flowers crowded, in stout short spikes, pale to bright pink. Nut black, glossy. Waste, bare and cultivated land, railway embankments, often close to water especially along rivers, streams and canals, sometimes abundant on heavy non-calcareous soils. June-October. Throughout, except Spitsbergen. **P** A range of small insects, but also self-pollinated. Frequent and often troublesome weed. **B**: Widespread.

7* **Pale Persicaria** *Polygonum lapathifolium* L. [*P. nodosum*, *P. scabrum*]. Stouter plant than *P. persicaria*, a variable medium-tall, slightly hairy annual, sometimes red-spotted. Leaves oval to linear-lanceolate, occasionally with a pale blackish blotch, with yellowish dots (glands) beneath; ochrea brown, fringed or not. Flowers in dense spikes, dull pink or greenish-white; stalks with yellowish glands. Nut black, glossy. Waste places, pond and river margins, cultivated land. June-October. Throughout, except Spitsbergen. **P** Mostly self-pollinated. A common weed of cultivated land. Often confused with *P. persicaria*, the two frequently growing together at low altitudes; *P. lapathifolium* can be distinguished by the yellowish glands on the flowers and flower stalks. A small more procumbent form was distinguished in the past as *P. nodosum*. **B**: Widespread.

8* **Amphibious Bistort** *Polygonum amphibium* L. Short to medium creeping perennial, terrestrial or aquatic, rooting at the nodes. Floating leaves oblong, hairless, rounded at the base or truncate, long-stalked, those of terrestrial plants smaller and narrower, slightly hairy; ochrea not fringed. Flowers in short dense, broad, often solitary spikes, deep pink. In or close to still or slow-moving water, particularly pools, canals, ditches and dykes, at low altitudes. June-September. Throughout, except Spitsbergen. Exists in 2 forms, a terrestrial and an aquatic, the former often shorter and with smaller thicker leaves. Distinguished in leaf by its characteristic rounded leaf-bases. **B**: Throughout, but rather rare in Scotland.

9 *Polygonum sagittatum* L. Weak medium to tall, scrambling annual; stems armed with small hooked bristles. Leaves oblong-heart-shaped, hairless. Flowers in long-stalked rounded 'heads', whitish. Nut glossy. June-September. **I**: East Asia and N America. **B**: Ireland – Kerry, where it has naturalized in ditches but is almost extinct.

10* **Bistort, Snake-root** *Polygonum bistorta* L. Short to medium patch-forming perennial with creeping rhizomes and simple erect stems. Leaves oblong, blunt, with long, winged stalks, the upper narrower, triangular-lanceolate and pointed. Flowers in dense broad spikes, bright pink. Meadows and woods, alluvial pastures, damp places often close to water, roadsides, alder carr, on humus-rich, usually non-calcareous soils, to 2500m. June-October. Throughout, except the extreme north and Iceland; naturalized in Ireland and Scandinavia. The young leaves can be eaten and in some regions, particularly N England, much folklore surrounds the plant. Frequently cultivated in gardens. **B**: Throughout, but commonest in NW England.

11 *Polygonum amplexicaule* D. Don. Like *P. bistorta*, but with larger heart-shaped leaves, the upper clasping the stem. Flowers in dense spikes, deep red. **I**: Himalaya. **B**: Naturalized in damp habitats in England and W Ireland. Widely cultivated in gardens.

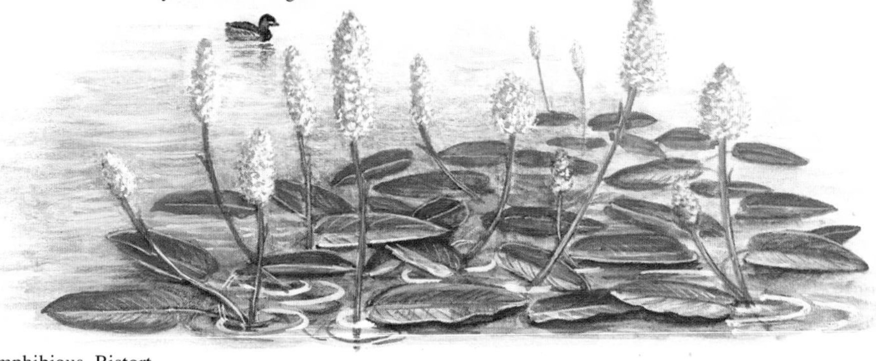

Amphibious Bistort

Small-leaved Knotgrass
Small Water-pepper
Tasteless
Water-pepper
Pale Persicaria
Bistort
Water-pepper
Redshank
Amphibious Bistort

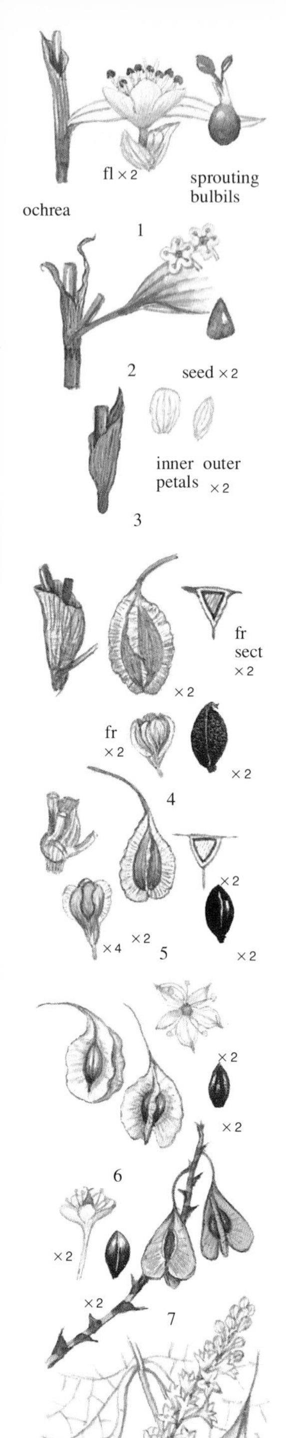


1* **Alpine Bistort** *Polygonum viviparum* L. Low to short, tufted, hairless perennial with erect, unbranched stems. Leaves linear-lanceolate, tapered at base, the margins rolled under. Flowers pale pink or white, in long slender spikes, the lower flowers replaced by small brown or purplish-red bulbils. Nut rarely produced. Moist grassland, pastures, screes and wet rocks, to 2300m, only on the mountains in the south of its range. June-August. Throughout, except Iceland, Belgium and Holland. The bulbils fall off eventually, rooting down to form new plants – they often start to grow while still attached to the parent plant. **B**: Mountains of N England and Scotland but rare in Wales and Ireland.

2* **Alpine Knotgrass** *Polygonum alpinum* All. Medium to tall, erect perennial. Leaves oblong to narrow-lanceolate, tapered; ochrea membranous, quickly disappearing. Flowers white or pale pink, in a lax-branched panicle. Nut triangular pale brown, glossy. Damp mountain habitats, particularly meadows, stream banks and among rocks, to 2200 m. June-August. C & E France; naturalized in Britain and Denmark. Occasionally cultivated.

3* **Himalayan Knotgrass** *Polygonum polystachyum* Wall. ex Meissner. Stout tall, patch-forming perennial with erect shoots to 1.2m. Leaves oblong to lanceolate, slightly cordate at base, pointed, often hairy beneath; ochrea brown, thick and persistent. Flowers white in lax leafy panicles, with red stalks. Nut rarely produced in Europe. Moist waste ground, river and stream margins. July-October. **I**: Himalaya. Widely naturalized in Britain and parts of W & C Europe. Long cultivated in gardens. Plants produce large leafy colonies but are often rather shy flowering. **B**: Naturalized especially in SW England and W Ireland.

**Fallopia**. Like *Polygonum*, but plants with twining stems. Perianth-segments winged in fruit. 9 species from north temperate regions; formerly included in *Polygonum*.

4* **Black Bindweed** *Fallopia convolvulus* (L.) Á. Löve [*Bilderdykia convolvulus*, *Polygonum convolvulus*]. Medium to tall, finely downy, annual twiner, occasionally spreading on the ground. Leaves heart-shaped, mealy beneath, untoothed. Flowers greenish-pink or greenish-white, in lax leafy spikes from the leaf axils. Fruit-stalks to 3mm. Nut, dull black, finely granular. Waste places, cultivated land, especially arable land, roadsides, rubbish tips. July-October. Throughout, except Spitsbergen. The seeds were eaten by early man and have been found in Bronze Age deposits. A spring or autumn germinating annual. **B**: Throughout, except parts of N & W Scotland.

5 **Copse Bindweed** *Fallopia dumetorum* (L.) Holub [*Bilderdykia dumetorum*, *Polygonum dumetorum*]. Like *F. convolvulus* but taller. Fruit stalks 5-8mm. Nut black, smooth and glossy. Warm sunny places, often in hedgerows and scrub, and recently coppiced land, on well-drained calcareous soils, at low altitudes. July-September. Throughout, except Iceland and the extreme north. Local, often rather erratic in appearance. **B**: Rare and decreasing; S England north as far as Buckinghamshire, very rare in the Midlands and S.Wales.

6* **Russian Vine** *Fallopia aubertii* (Louis Henry) Holub [*Bilderdykia aubertii*]. Vigorous woody climber, to 5m or more. Leaves heart-shaped, blunt. Flowers white or pale green, becoming pinkish in fruit, in spreading lax terminal or lateral panicles. Nut black, glossy. Occasionally naturalized in hedgerows and waste places, sometimes on fences and old or abandoned buildings. July-October. **I**: W China. Naturalized in Britain. Widely cultivated in gardens. An invasive plant forming dense entanglements.

*Fallopia baldschuanica* (Regel) Holub [*Bilderdykia baldschuanica*, *Polygonum baldschuanicum*]. Like *F. aubertii*, but flowers pale pink. **I**: S USSR. Perhaps naturalized in W Europe. Occasionally cultivated, but not nearly as commonly grown as *F. aubertii*. The plant grown in gardens under this name often proves to be *F. aubertii*.

7* **Japanese Knotweed** *Reynoutria japonica* Houtt. [*Polygonum cuspidatum*]. Vigorous, spreading, tall perennial with stout erect stems to 1-2m; stems bluish-green, often reddish, branched above. Leaves broadly oval-triangular, stalked. Flowers white, in narrow leafy panicles from upper leaf-joints. Nut black, glossy. Moist habitats, grassy banks, waste ground, railway embankments, margins of thickets and woodland, ditches and riverbanks. August-October. **I**: N China, Japan and Taiwan. Naturalized throughout, generally as a garden escape. A popular garden plant in Victorian times, but a serious weed pest today in some places – difficult to eliminate. **B**: Throughout, except the Orkneys.

8 **Giant Knotweed** *Reynoutria sachalinensis* (Friedrich Schmidt Petrop.) Nakai [*Polygonum sachalinensis*]. Like *R. japonica*, but even stouter, often 3m or more. Leaves more heart-shaped, pointed. Flowers greenish, in denser panicles; flower cluster 4-7 (not 2-4). Similar habitats and flowering time to *R. japonica*. Also a garden escape, but less frequently naturalized. **I**: E Asia. Throughout, except for Iceland and most of Scandinavia. Generally a shy flowerer. **B**: Widely naturalized in England, local in Wales and less frequent in Scotland; rare in Ireland.

9* **Buckwheat** *Fagopyrum esculentum* Moench [*F. vulgare*, *Polygonum fagopyrum*]. Medium hairless annual with hollow, erect stems. Leaves arrow-shaped, the lower stalked, untoothed; ochrea short, not fringed, often reddish. Flowers greenish-white or pinkish, in lax terminal or lateral panicles, perianth segments not winged or keeled in fruit. Nut dark dull brown, 5-6mm. Cultivated occasionally but frequently naturalized in open or waste places, field borders, and rubbish tips, often as a relict of former cultivation. June-September. Throughout, except Ireland, the Faeroes, Iceland and Spitsbergen (probably a native of C Asia). Long cultivated as a 'grain crop' for milling into flour or as a cattle feed. Less frequently cultivated today except in remote mountain districts. However, it is still extensively grown in parts of Asia, especially in the Himalaya. **B**: Casual throughout, occasionally well established, especially in East Anglia.

*Fagopyrum tataricum* (L.) Gaertner [*Polygonum tataricum*]. Like *F. esculentum*, but taller, seldom red-tinged and leaves paler green. Flowers smaller, 2mm, entirely green. Nut rough. Waste places; a weed, often associated with *F. esculentum*. Widespread throughout. **B**: Widespread.

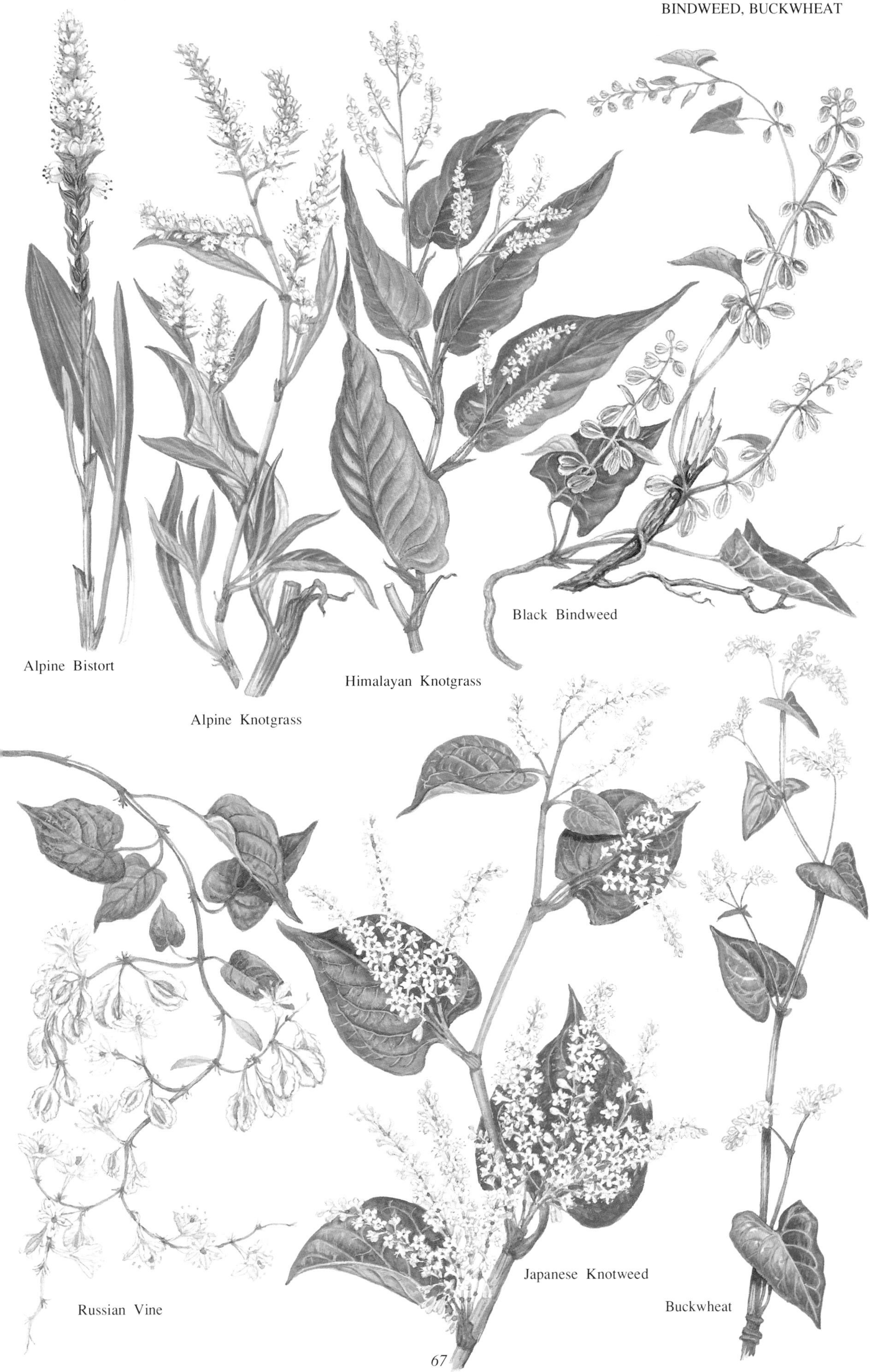

Alpine Bistort

Alpine Knotgrass

Himalayan Knotgrass

Black Bindweed

Russian Vine

Japanese Knotweed

Buckwheat

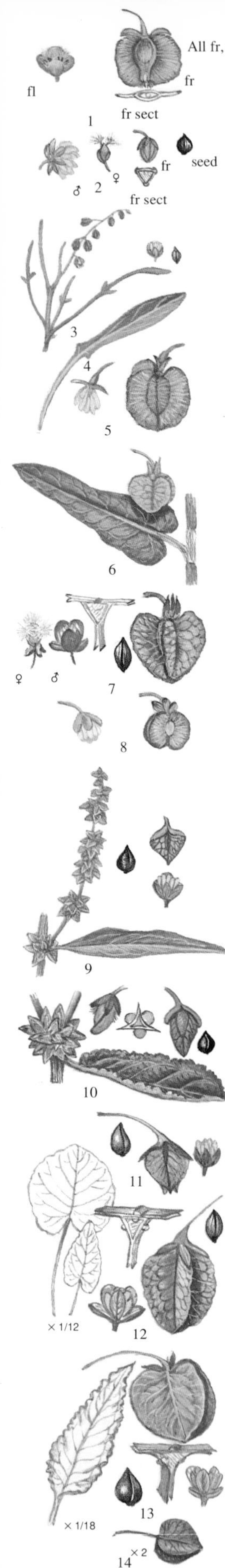


1* **Mountain Sorrel** *Oxyria digyna* (L.) Hill. Short, tufted, hairless perennial, with erect stems. Leaves kidney-shaped, untoothed, long- stalked, mostly basal. Flowers tiny, greenish, 4-petalled, in branched spikes. Fruit a drooping, winged, nut. Damp rocky places, wet ledges, streamsides, often in granite areas, only on the mountains in the south of its range, to 3500m. June-August. Throughout, except Holland and Denmark. Locally common. In mountain regions seeds are often washed down streams and plants may occur at lower altitudes than is normal. **B**: Locally common in Wales, Lake District, Scottish Highlands and in Ireland.

**Docks** *Rumex*. Mostly perennials, generally hairy, with tubular ochrea. Flowers hermaphrodite or unisexual, with 6 perianth segments (5 in *Polygonum*), in a simple or branched inflorescence. Fruit a triangular nut, often winged. 200 species, widespread, mostly in north temperate regions. A complicated genus; fruit characters are especially important. Pollinated by wind.

2* **Sheep's Sorrel** *Rumex acetosella* L. Variable low to short, slender, tufted perennial with erect stems. Leaves arrow-shaped with basal lobes spreading or pointing forward. Flowers greenish, unisexual, male and female on separate plants. Nut 1.3-1.5mm, longer than wide, scarcely winged. Dry meadows, grassy heaths and commons, bare places, coastal habitats, on sandy, usually acid, well-drained soils, to 2400m. May-August. Throughout. **B**: Widespread, particularly in the west.

3 *Rumex tenuifolius* (Wallr.) Á. Löve [*R. acetosella* var. *tenuifolius*]. Like *R. acetosella*, but with narrow leaves, those on the stem not more than 2mm wide; stems branched from below the middle (not above). May-September. Britain and France northward. **B**: Widespread but uncommon.

4* *Rumex graminifolius* Rudolph ex Lamb [*R. angustissimus*]. Like *R. acetosella*, but generally with sprawling stems and narrow leaves with indistinct basal lobes. June-September. Arctic Finland.

5* **French Sorrel** *Rumex scutatus* L. Short to medium, often greyish, tufted, hairless perennial, branched from the base. Leaves arrow-shaped with spreading basal lobes. Flowers greenish in branched spikes, branches erect. Nut yellowish-grey, 3-3.5mm; frujt wings without pimples. Mountain screes, cliffs, old walls, rocky roadside and waste places, to 2500m. June-July. Belgium, France and Germany; naturalized in Britain, Holland and Sweden. Introduced into some areas as a culinary herb – the leaves have a cool acid taste. **B**: Established but rare – mostly in N England and Scotland.

6 *Rumex arifolius* All. [*R. montanus*]. Tall stout perennial to 1.2m, with very leafy erect stems. Leaves heart-shaped; ochrea not fringed. Flowers in a lax panicle with flexuous branches. Nut 2.5-3mm, yellowish-grey or brownish. Meadows, especially in mountain areas, to 2500m. July-August. Iceland, the Faeroes, France and Germany.

7* **Common Sorrel** *Rumex acetosa* L. Very variable short to tall, acid tasting perennial. Leaves arrow-shaped, the basal lobes pointing backwards, upper clasping the stem; ochrea fringed. Flowers in few- branched or unbranched spikes, male and female flowers on separate plants. Nut 1.8-2.2mm, black, glossy. Mainly a meadow plant, sometimes in open woodland, road verges, riverbanks, coastal shingle or mountain ledges, on neutral or mildly acid soils, to 2100m. May-June. Throughout. Long used in salads and sauces, often to accompany fish dishes. **B**: Common throughout.

*Rumex hibernicus* Rech.f. Like *R. acetosa*, but only 10-15cm tall, with few stem leaves. Sand-dunes. **B**: W Ireland and W Scotland.

8* *Rumex thyrsiflorus* Fingerh. Like *R. acetosa*, but leaves narrower, the basal lobes spreading. Panicle dense, much-branched. Nut dark brown. Dry open habitats. June-September. C & N Europe – Holland, Germany, Denmark and Scandinavia; naturalized in NE France.

9 **Willow-leaved Dock** *Rumex triangulivalvis* (Danser) Rech. f. Medium erect perennial. Leaves linear-lanceolate, pointed, pale green, short-stalked. Flowers in a wide-spreading panicle. Fruit valves triangular, untoothed, swollen, without pimples. Waste ground, rubbish tips, often near sea ports. July-August. **I**: North America. Naturalized or casual in several countries. **B**: Mostly casual.

10 **Argentine Dock** *Rumex frutescens* Thouars [*R. cuneifolius*]. Short rhizomatous, creeping perennial; stems erect, branched. Leaves thick, oblong, wavy-edged. Flowers in crowded leafless, panicles. Fruit valves triangular, untoothed, pimpled. Naturalized on dune slacks, often close to sea-ports, occasionally alien or casual elsewhere. July-August. **I**: South America. Britain and Denmark. **B**: Local in SW England and S Wales.

11* **Monk's Rhubarb** *Rumex alpinus* L. Stout, medium to tall, rhizomatous perennial. Leaves large round-heart-shaped, as long as broad, long- stalked, margin wavy; ochrea papery. Flowers in crowded panicles, yellowish-green, hermaphrodite. Fruit valves untoothed, not swollen, without pimples. Mountain meadows, often by cattle sheds and other farm buildings, to 2500m. June-August. France and Germany; naturalized in Britain. Formerly grown for eating and for medicinal or veterinary uses. In the Alps the large leaves are used to wrap butter in. **B**: Locally naturalized from Staffordshire northwards.

12 **Scottish Dock** *Rumex aquaticus* L. Stout, tall, erect perennial, 1-2m. Basal leaves triangular-heart-shaped, long-stalked. Flowers in large dense panicles. Fruit valves oval, untoothed, not swollen. Wet habitats, marshes and swamps. July-August. Throughout, except Ireland, Iceland and the Faeroes. Hybridizes with *R. hydrolopathum* and *R. longifolius*. **B**: Rare in N Britain – Loch Lomond region.

13 **Northern Dock** *Rumex longifolius* DC. [*R. domesticus*]. Medium to tall stout perennial. Leaves broad-lanceolate with margins wavy. Flowers greenish, in dense branched spikes. Fruit valves kidney-shaped, with wavy untoothed margins. Damp grassy habitats, meadows, river and lake margins, ditches, to 2500m. June-July. Throughout, except Holland and Spitsbergen. Like *R. aquaticus* in fruit, but the leaves long and narrow with crisped margins. **B**: Lancashire and Yorkshire northwards.

14 *Rumex pseudonatronalis* Borbás [*R. fennicus*]. Like *R. longifolius*, but leaves much narrower, linear-lanceolate. Seashores and river margins. June-August. Finland and Sweden. Rare.

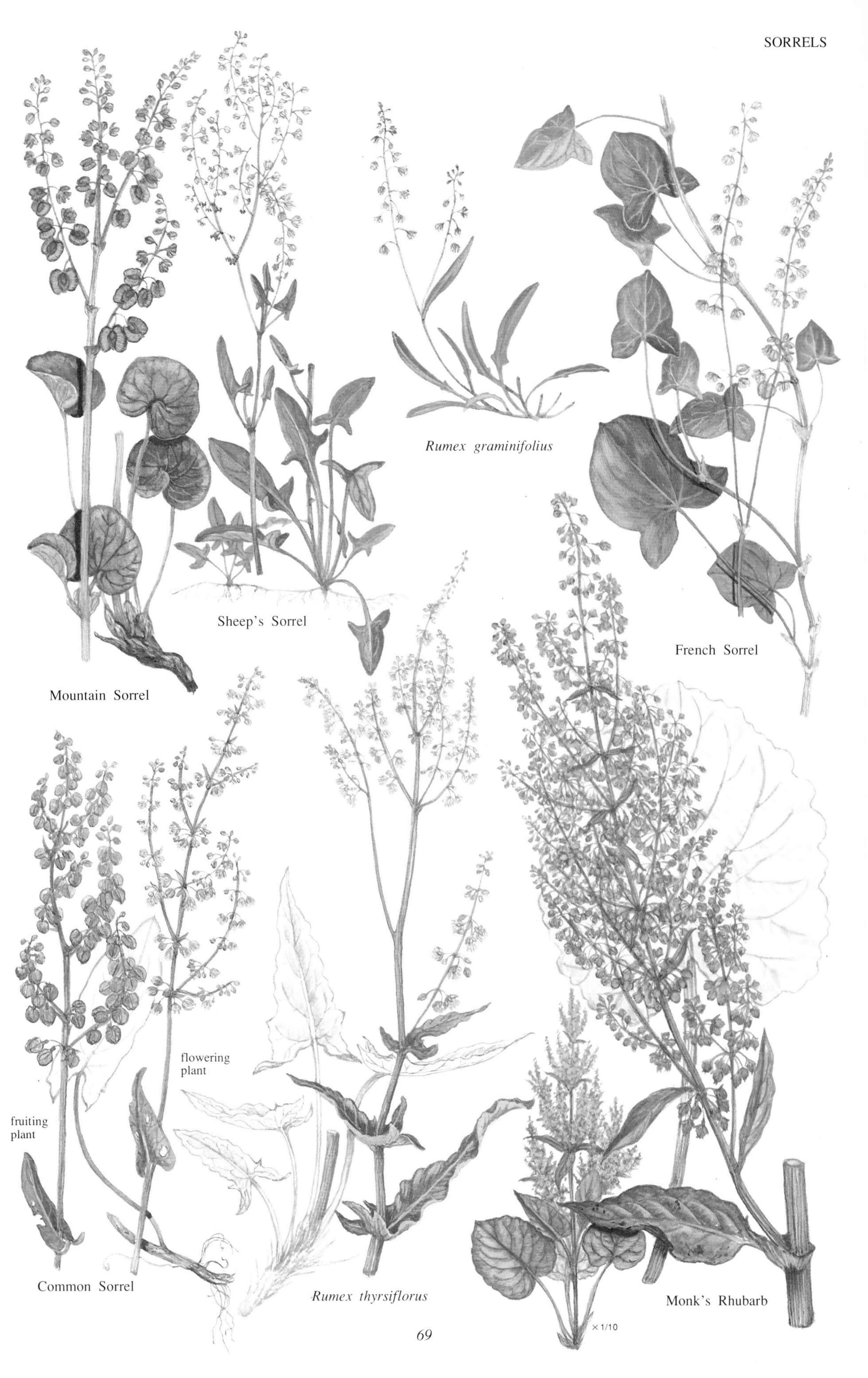
Mountain Sorrel
Sheep's Sorrel
*Rumex graminifolius*
French Sorrel
Common Sorrel
flowering plant
fruiting plant
*Rumex thyrsiflorus*
Monk's Rhubarb
× 1/10

All fruits and sections ×2

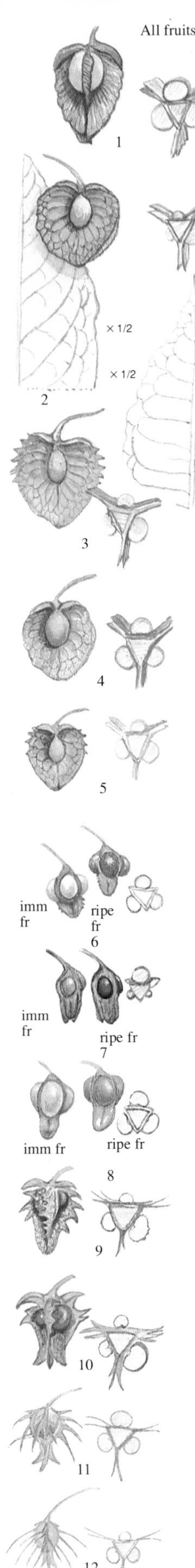


1* **Water Dock** *Rumex hydrolapathum* Hudson. Stout, tall perennial to 2m, much branched. Leaves large, rigid, broad-lanceolate, tapered at both ends, lateral veins at right angles to the midrib. Inflorescence dense; flowers in much-branched panicles, leafy below; perianth segments without teeth. Fruit valves triangular, untoothed, swollen, each with an elongated pimple. Marshes, river, stream and canal banks, lake and pond margins, ditches, dykes, fen peats, often growing in shallow water. July-September. Throughout, except for Iceland and N Scandinavia. Hybridizes with *R. aquaticus*, *R.* × *heterophyllus* C.F. Schultz, where the two parent species grow in close proximity. **B**: Throughout England, scarcer elsewhere; rare in Scotland.

2 **Patience Dock** *Rumex patientia* L. subsp. *orientalis* (Bernh.) Danser. Tall perennial to 2m. Leaves pale green, oval to oblong, pointed, the lateral veins at an angle of 50-60° to the midrib. Flowers crowded, in branched spikes. Fruit valves untoothed, rounded, often only one swollen. Waste and cultivated land, especially close to docks, breweries and along canal and river banks. May-July. **I**: C & SE Europe and W Asia. Locally naturalized in Britain and parts of NW Europe . Formerly cultivated as a green vegetable. **B**: Scattered localities in S Britain, particularly around Bristol and London.

3 **Greek Dock** *Rumex cristatus* DC. Like *R. patientia*, but leaves with vein angles of 60-90°; fruit valves toothed, generally all the valves swollen. Similar habitats and flowering time. **I**: SE Europe & SW Asia. Naturalized locally in S Britain. **B**: Scattered localities in S England and S Wales.

4* **Curled Dock** *Rumex crispus* L. Medium to tall perennial. Leaves narrow- lanceolate with wavy margins, short-stalked. Flowerheads dense, in branched spikes, branches rather upright. Fruit valves heart-shaped, all 3 swollen and untoothed. Disturbed and cultivated land, waste places, hedgebanks, river, lake and pond margins, river gravels and seashores, on a variety of soils. June-October. Throughout, except Spitsbergen; naturalized in Iceland. A cosmopolitan weed; a serious agricultural pest in some places. **P**: Wind. Hybridizes with *R. obtusifolius*. **B**: Throughout, though less common in N Britain – scheduled as an injurious weed.

5 *Rumex stenophyllus* Ledeb. Like *R. crispus*, but a shorter plant, rarely exceeding 50cm tall. Leaves less wavy. Fruit valves distinctly toothed. Saltmarshes, river estuaries. June-September. Germany; naturalized in Denmark and Sweden.

6* **Clustered Dock** *Rumex conglomeratus* Murray. Medium to tall, erect perennial; stems zig-zagged. Leaves oblong, rounded at base, long- stalked; upper leaves narrower, pointed. Flowers in rather distant clusters, forming a much-branched panicle; flowering branches leafy almost to the top. Fruit valves oval, untoothed, all swollen. Damp grassy habitats, marshy meadows, river, stream and pond margins, ditches and streambanks, waste places; often areas subjected to winter flooding. June-October. Throughout, except the Faeroes, Iceland and Spitsbergen. **B**: Throughout, but local in much of the north and in Wales.

7* **Wood Dock** *Rumex sanguineus* L. Like *R. conglomeratus*, but slenderer and with straighter stems. Panicles less leafy. Fruit with only one valve swollen. Damp, generally shaded habitats, woodland ridges, pathsides, road verges and grassy waste ground, generally on heavy soils. June-September. Like *R. conglomeratus* in general appearance, but with more acute branching and with few leaves in the inflorescence. The fruit have only one (not all three) valves swollen. **B**: Common in S England, Wales and Ireland, rare elsewhere.

8* **Shore Dock** *Rumex rupestris* Le Gall. Similar to *R. conglomeratus*, but shorter and stouter, with blue-green leaves and straighter stems. Inflorescence interrupted, only the lower flower clusters with a leaf. Seashores, dune slacks, coastal rocks. June-August. N Spain and France north to S Wales; absent from Ireland. **B**: Very rare, confined to a few localities in S Wales, SW England, Isles of Scilly and the Channel Islands.

9* **Fiddle Dock** *Rumex pulcher* L. Short to medium spreading perennial. Leaves small and fleshy, oblong but often waisted and fiddle-shaped. Flowers in branched leafy spikes; branches flexuous, often entangled. Fruit valves oval-triangular, unevenly swollen. Dry, often waste places, short grassy commons and greens, churchyards and roadsides, generally on sandy calcareous soils, at low altitudes. June-August. Britain and France; naturalized in Ireland, Denmark and Sweden. **B**: Lowland Britain south of the Wash, often coastal.

10* **Broad-leaved Dock** *Rumex obtusifolius* L. Medium to tall erect perennial, to 1m. Leaves oblong, the lower heart-shaped at base, usually downy beneath; margins wavy. Flowers in spreading branches, leafy below. Fruit valves triangular, toothed, often only one valve swollen. Waste and disturbed ground, poor pastures and meadows, lake and pond margins, hedgerows and field boundaries in sunny or partially shaded sites. June-October. Throughout, except Spitsbergen. **P**: Wind. Frequently hybridizes with *R. crispus*. An often serious weed pest – difficult to eradicate once established. **B**: Throughout; a scheduled injurious weed.

11* **Marsh Dock** *Rumex palustris* Sm. Medium erect, much-branched, annual or biennial, yellowish-brown in fruit; stems wavy. Leaves lanceolate, tapered at the base; upper leaves linear. Flowers greenish-yellow, in branched spikes, dense and leafy towards tips. Fruit valves toothed and swollen, long, blunt-tipped. Bare muddy habitats, marshes, pond and river margins, reservoirs and gravel pits. June-September. Europe north to S Sweden. Often growing in places subjected to flooding in winter. **B**: Local in E & S England north to the Humber.

12* **Golden Dock** *Rumex maritimus* L. Short to medium erect annual rather like *R. palustris*, but plant golden-yellow in fruit. Leaves narrowly elliptical. Fruits with slender stalks longer than valves (shorter and thick in *R. palustris*). Bare muddy ground, seashores, river, pond and reservoir margins, clay pits, ditches and dykes, particularly along, or close to, the coast. June-September. Throughout, except Iceland and the extreme north, often local. Like *R. palustris* in general appearance, but they can be distinguished in fruit – those of *R. maritimus* have sharply pointed, not blunt, valves. **B**: Scattered localities throughout, especially in England, but rare and declining.

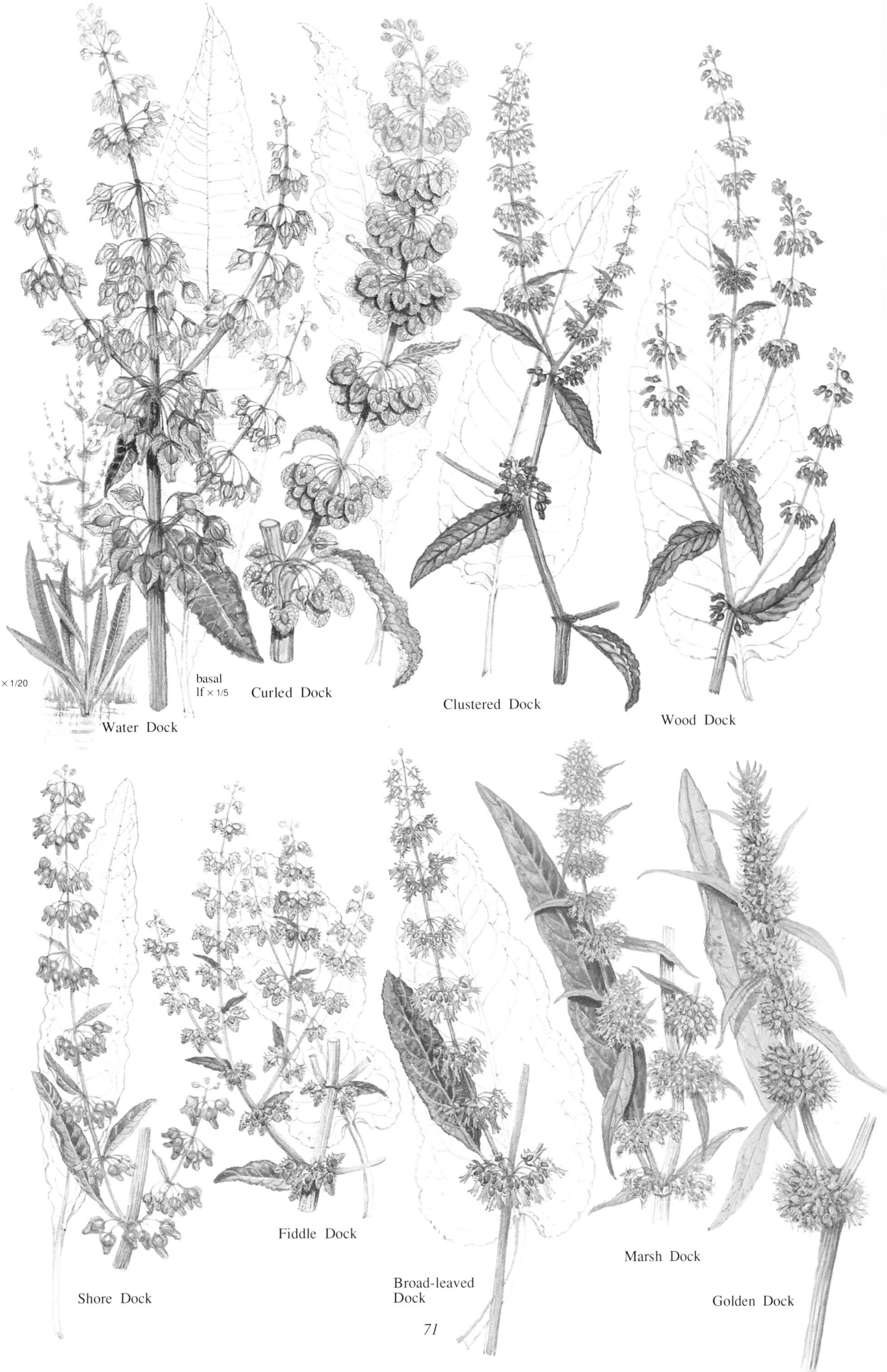

× 1/20
Water Dock
basal
lf × 1/5
Curled Dock
Clustered Dock
Wood Dock

Shore Dock
Fiddle Dock
Broad-leaved
Dock
Marsh Dock
Golden Dock

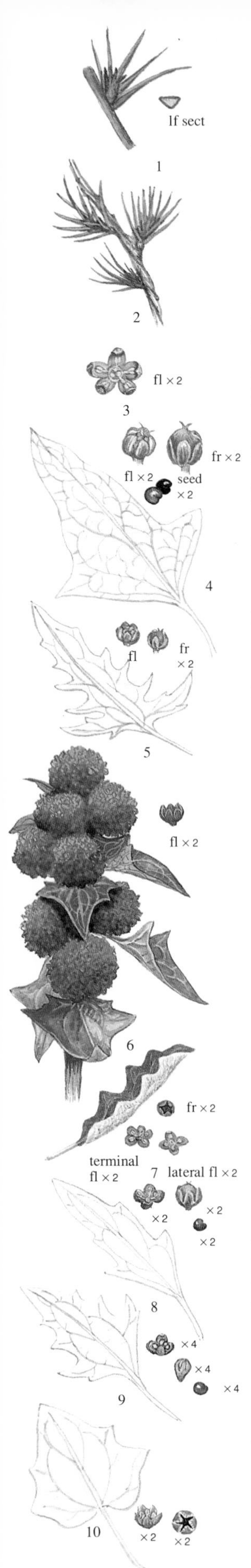


# FAT HEN FAMILY Chenopodiaceae

Herbs or shrubs, often with succulent leaves and stems and alternate leaves, occasionally opposite. Flowers small and greenish, hermaphrodite or unisexual, regular. Perianth 3-5 lobed usually, persistent and often enlarging in fruit. Stamens generally the same number as the perianth segments. Fruit indehiscent, 1-seeded. 1400 species in 102 genera, in many parts of the world. Many are characteristic of dry sandy and saline habitats but a number have become widespread weeds of cultivated land – especially species of *Chenopodium.*

1* **Polycnemum** *Polycnemum majus* A. Braun. Short hairless, often sprawling, sometimes erect, annual. Leaves linear, spine-tipped, triangular in section. Flowers tiny, greenish, solitary in axils of upper leaves, petalless, but with 5 segments. Dry, often bare habitats, frequently a weed. July-October. Belgium and France east into Germany.

2 *Polycnemum arvense* L. Like *P. majus*, but a taller plant with warted, spirally-twisted branches. Leaves with a soft-tip. Sandy soils, often as a weed of cultivated or disturbed land. July-September. C & E France and Germany.

3* **Sea Beet** *Beta vulgaris* L. subsp. *maritima* (L.) Arcangeli [*B. maritima, B. perennis*]. Very variable medium to tall hairless, often red-tinged biennial, occasionally perennial. Basal leaves dark green and leathery, oval to heart-shaped, or lanceolate, untoothed. Flowers green, in dense leafy clusters forming a long narrow spike. Fruits with swollen flower segments, becoming corky and sticking together in small clusters. Coastal habitats, drift-line of shingle beaches, the drier areas of salt marshes, grassy banks and sea walls; inland on drainage dykes. June-September. Europe north to S Sweden. **P**: Wind. The ancestor of the cultivated beet, mangold etc. **B**: Coasts, except for the extreme north.

*Chenopodium.* Mainly annual plants, with alternate, flat leaves. Flowers usually hermaphrodite, borne in clusters, often forming large spikes or panicles. The perianth-segments persist, but are scarcely altered in the fruiting stage. 100 species, mainly in temperate regions. Many have become widespread weeds of cultivated land.

4* **Good King Henry, Mercury** *Chenopodium bonus- henricus* L. Short to tall, often reddish, erect annual, mealy when young. Leaves triangular to diamond-shaped, generally large, poorly toothed. Flowers green, often red-tinted, in an almost leafless, tapering spike; stigmas long-protruding. Waste places, cultivated land, farmyards, manure heaps, hedgebanks and old walls, often on organically enriched soils. May-August. Throughout, except the Faeroes, Iceland and Spitsbergen, probably naturalized in much of the north and north-west. Formerly widely grown as a green vegetable. The fruits are red but eventually ripen black. **B**: Throughout, but most frequent in England.

5* *Chenopodium foliosum* Ascherson. Medium to tall erect, almost hairless annual. Leaves diamond-shaped, coarsely toothed, with a projecting lobe on either side near the base; upper leaves narrow-lanceolate. Flowers reddish, in many clusters along the main stem; perianth becoming fleshy in fruit. Cultivated and disturbed ground, waysides. July-September. Scattered localities in France and Germany; casual or locally naturalized elsewhere, especially in Holland, Denmark and S Sweden.

6 **Strawberry-blite** *Chenopodium capitatum* (L.) Ascherson. Similar to *C. foliosum*, but lower leaves less toothed and upper leaves broader, diamond-shaped to narrow lanceolate. Flower clusters fewer and more scattered, scarlet when mature. Cultivated and waste land, rubbish tips. July-August. Naturalized in much of Europe, except for Scandinavia. **B**: Casual.

7* **Oak-leaved Goosefoot** *Chenopodium glaucum* L. Low to short sprawling to erect, much-branched annual, hairless except the leaves beneath. Leaves lanceolate to elliptical, sinuate to toothed, like an oak-leaf, green above but grey-green and mealy beneath. Flowers greenish, in shorter slender spikes or clustered below. Waste and disturbed ground, refuse tips, arable land, occasionally in coastal habitats, on rich fertile soils. June-September. Throughout, except Ireland, the Faeroes, Iceland and the extreme north. The seeds are thought to be distributed in horse manure. **B**: Local, north to Northumberland, generally decreasing.

8* **Red Goosefoot** Chenopodium rubrum L. Variable short to tall hairless, spreading to erect annual, usually much-branched, often red-tinged. Leaves diamond-shaped to lanceolate, irregularly coarsely toothed to almost toothed, green, often reddish beneath. Flowers in simple or branched leafy spikes, clusters crowded; flower segments free for at least half-way. Waste places, disturbed ground, dried up pond and reservoir margins, manure heaps, arable land, marshes, often in coastal habitats, on rich fertile soils. July-October. Throughout, except the Faeroes, Iceland and Spitsbergen. **P**: Insects, wind, or self-pollinated. **B**: Throughout lowlands.

9* **Small Red Goosefoot** *Chenopodium chenopodioides* (L.) Aellen [*Blitum chenopodioides, Chenopodium botryodes*]. Similar to *C. rubrum*, but generally spreading from the base. Leaves triangular to diamond- shaped, scarcely toothed. Flower segments fused almost to the apex and forming a close sac around the fruit. Waste land, salt marshes and other coastal habitats, ditches. July-September. Britain, Belgium, Holland, France, Germany and Denmark. **B**: Very local; mostly restricted to the E & SE coasts of England.

10* **Sowbane, Maple-leaved Goosefoot** *Chenopodium hybridum* L. Short to medium, erect, slightly mealy, hairless annual. Leaves oval-heart- shaped to triangular, with a few large angular teeth. Flower clusters lax, leafless and scattered, forming a branched panicle. Disturbed ground and waste places, arable land, farmyards and manure heaps. July- October. Throughout, except Ireland, the Faeroes, Iceland and Spitsbergen. **B**: Rare; scattered localities in England south of the Humber, but not in SW England.

× 1/3
Polycnemum
Sea Beet
× 1/3
Good King Henry
*Chenopodium foliosum*
Oak-leaved Goosefoot
Red Goosefoot
Small Red Goosefoot
Sowbane

Flowers, seed capsules and seeds ×6 except where shown

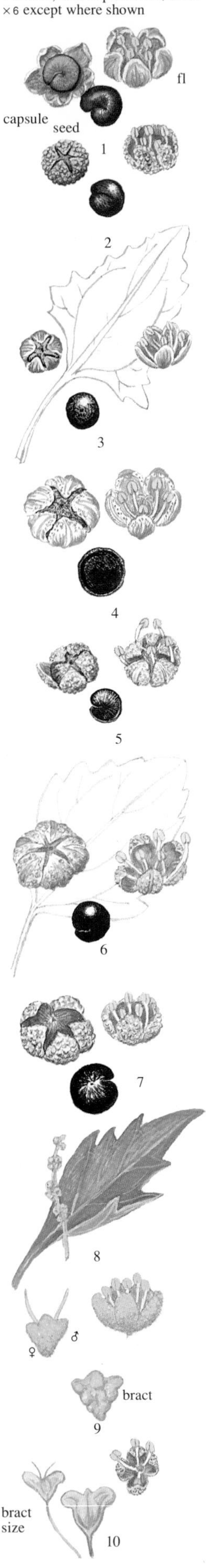


1* **Many-seeded Goosefoot, All-seed** *Chenopodium polyspermum* L. Variable short to tall, erect to spreading hairless annual. Stems square, often reddish. Leaves oval to elliptic, untoothed but occasionally with an angle on one or both sides, narrowed at the base. Flowers in long lax spikes, leafy, branched below. Cultivated land, particularly among arable crops, waste places, on a variety of soils. July-October. Throughout, except the Faeroes, Iceland and Spitsbergen. The almost square stems and usually untoothed leaves serve to distinguish this species. **B**: Throughout lowland England north to the Humber, rare elsewhere.

2* **Stinking Goosefoot** *Chenopodium vulvaria* L. Spreading and ascending, short to medium, much-branched, grey-mealy annual. Leaves diamond-shaped to oval, untoothed. Flowers in lateral and terminal leafy clusters. Coastal salt-marshes and shingle, waste places and rubbish tips inland. July-September. Throughout, except Ireland, the Faeroes, Iceland, Norway and Spitsbergen. The strong smell of decaying fish is due to the presence in the plant of Trimethylamine. **B**: Rare and declining; restricted to S & SW England and the Channel Islands.

3 **Upright Goosefoot** *Chenopodium urbicum* L. Short to tall erect, practically hairless, annual. Leaves triangular to oval, tapered at base, coarsely toothed, the teeth often hooked. Flowers in a crowded panicle, with rather upright branches; perianth segments and stamens 5. Arable land, waste ground, farmyards and manure heaps, on rich fertile soils. July-September. Throughout, except Ireland, the Faeroes, Iceland, Holland and Spitsbergen. **B**: Very rare: a few localities near the S coast of England and the Channel Islands.

4* **Nettle-leaved Goosefoot** *Chenopodium murale* L. Fairly stout medium to tall slightly mealy annual. Leaves diamond-shaped, usually coarsely toothed, teeth incurved. Inflorescence slightly mealy. Flowers in a crowded leafy panicle. Light disturbed ground, waste places, sand-dunes and rubbish tips, generally on light, well-drained soils. July-October. Throughout, except Ireland, the Faeroes, Iceland, Norway and Spitsbergen. The inflorescence is mealy and leafy almost to the top. **B**: Scattered localities in England and Wales, casual elsewhere.

5* **Fig-leaved Goosefoot** *Chenopodium ficifolium* Sm. An erect or spreading, medium to tall, green, mealy annual. Leaves lanceolate, 3-lobed, the central lobe much the largest, parallel-sided; uppermost leaves generally narrower and unlobed. Flowers in a slender-branched panicle. The seed has a distinctly sculptured seed-coat. Waste ground, arable land, farmyards and manure heaps, particularly on rich fertile soils. July-September. Throughout except Ireland, the Faeroes, Iceland and most of Scandinavia. Sometimes confused with *C. album*, but distinguished by its leaf-shape – the terminal lobe of the leaf has long parallel margins. **B**: Local in England, especially the SE, rare elsewhere.

6 **Grey Goosefoot** *Chenopodium opulifolium* Schrader ex Koch & Ziz. Medium to tall, much-branched, erect mealy annual. Leaves broad diamond-shaped, often broader than long and with a lobe on each side, sometimes toothed, bluish-green, mealy beneath, especially when young. Flowers in a grey-blue, mealy panicle. Disturbed and waste places, rubbish tips. July-October. Belgium, France and Germany; casual in S Britain. Widespread in C & E Europe and parts of temperate Asia. **B**: Casual in S England.

7 **Fat Hen** *Chenopodium album* L. Very variable medium to tall, erect, deep green mealy annual; stems often red-tinged. Leaves very variable, diamond-shaped to lanceolate, pointed, toothed, white-mealy especially beneath. Flowers in a leafy spike or panicle. Disturbed and waste ground, arable land, farmyards, manure heaps and roadsides, often abundant. June-October. Throughout, except Spitsbergen; naturalized in the Faeroes and Iceland. An often troublesome weed of arable land, especially among rootcrops and barley, generally on rich fertile, non-calcareous soils. Formerly cultivated as a green vegetable. The seeds were at one time ground into flour. **P**: Wind and insects. **B**: Throughout, less frequent in the north.

* *C. a.* subsp. *striatum* (Krašan) J. Murr. Like the type but with blunt oblong leaves and red-striped stems. Similar habitats and flowering time. Belgium and France, east into Germany.

8 **Green Goosefoot** *Chenopodium suecicum* J. Murr. Like *C. album*, but with bright blue-green leaves, mealy only when young; stem not red-tinged. Leaves with sharp, forward-pointing teeth. Waste places, rubbish tips and disturbed ground. July-September. N and E Europe to temperate Asia.**B**: Casual.

**Halimione** *Halimione*. Like *Atriplex*, p. 76, but fruit 'bracts' united almost to the apex (not separated).

9* **Sea Purslane** *Halimione portulacoides* (L.) Aellen [*Atriplex portulacoides, Obione portulacoides*]. Short to medium, spreading, silvery or greyish subshrub; stems often rooting down, brownish. Leaves mostly opposite, thick and fleshy, oblong, untoothed. Flowers greenish- yellow in short, practically leafless panicles. Fruit held within 3-lobed bracts, stalkless. Maritime habitats, particularly salt marshes, pools and channel margins, often forming a conspicuous zone on the upper levels of estuarine marshes. July-October. Coasts of W Europe, Britain, Belgium, Holland and France. **B**: South of the Solway Firth and Tweed and in E Ireland.

10* **Stalked Orache** *Halimione pedunculata* (L.) Aellen [*Atriplex pedunculata, Obione pedunculata*]. An erect, short to medium hairless annual, silvery-grey mealy, especially when young. Leaves alternate, oblong to elliptic, untoothed. Flowers greyish-green in clusters or short spikes. Fruit enclosed within 2-lobed bracts, long-stalked. Coastal habitats, particularly muddy salt marshes. June-September. Europe north to S Sweden. **B**: Extinct, last seen here in 1935.

Sea Purslane

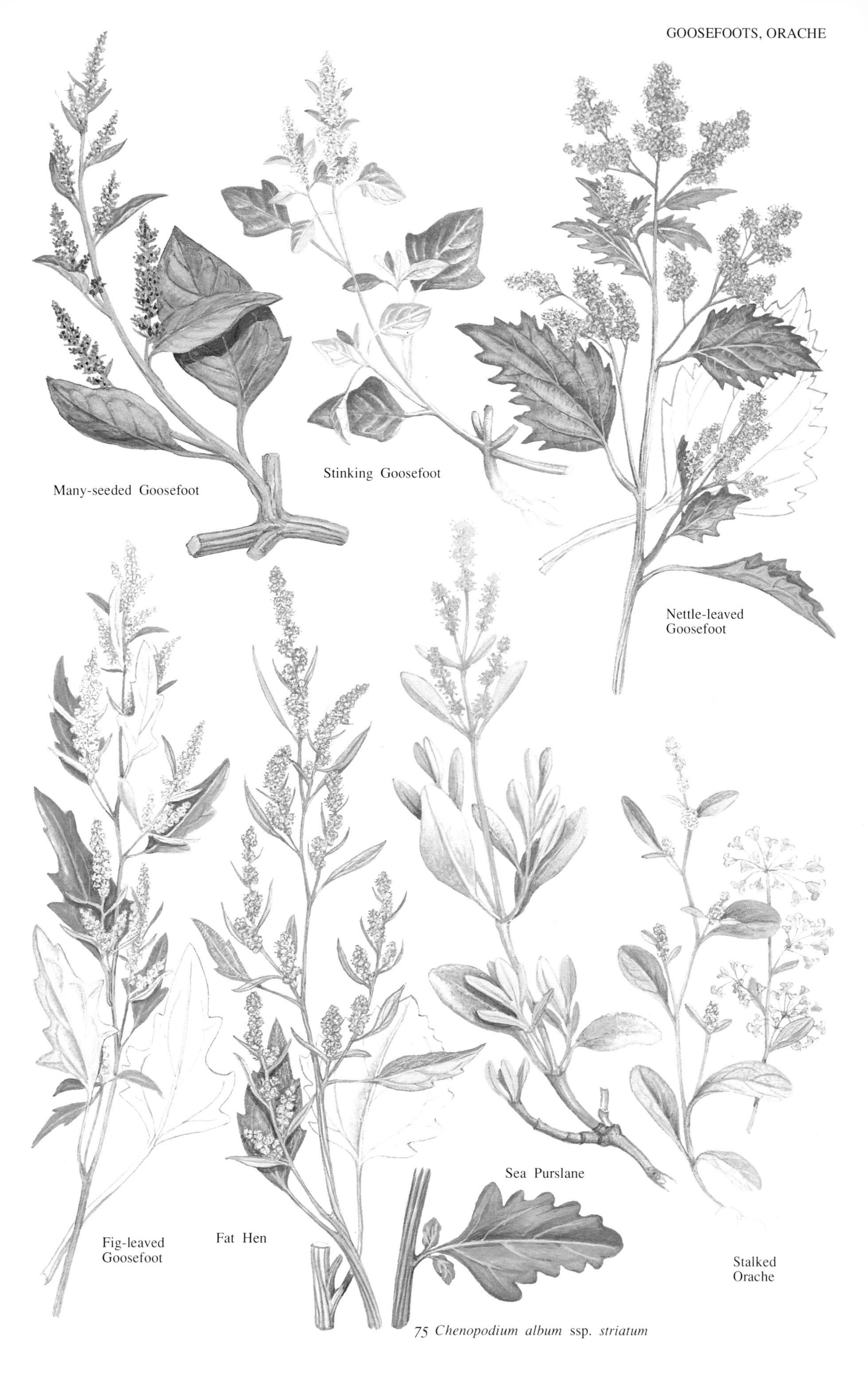


75 *Chenopodium album* ssp. *striatum*

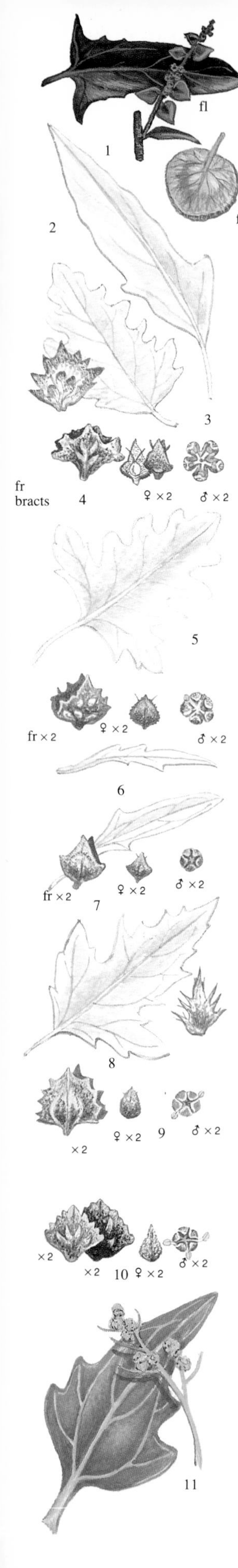


**Oraches** *Atriplex*. Annual herbs or small shrubs, frequently mealy. Rather like *Chenopodium*, but flowers unisexual and fruits enclosed by two small erect bracts; monoecious. 200 species scattered in temperate and subtropical regions of the world.

1* **Orache** *Atriplex hortensis* L. Tall erect annual, green or heavily flushed with purple-brown. Leaves heart-shaped to triangular, slightly toothed, hairless eventually. Flowers in a terminal spike. Disturbed ground and waste places, cultivated land. July-October. Naturalized in Holland and Germany, often casual elsewhere. Frequently cultivated in gardens, including a form with reddish-purple foliage and fruits. A frequent garden escape. **B**: Casual.

*Atriplex nitens* Schkuhr. Like *A. hortensis*, but leaves more deeply toothed, mealy-white beneath. Similar habitats and flowering time. S Germany.

2 *Atriplex oblongifolia* Waldst. & Kit. Tall erect annual. Leaves narrow-oval to triangular, toothed; upper leaves narrower and untoothed. Flowers in a long 'broken' spikes, or in lateral clusters. Waste places and disturbed ground. July-October. E France and Germany. To W and C USSR.

*Atriplex heterosperma* Bunge. Like *A. oblongifolia*, but leaves arrow-shaped, untoothed or slightly toothed. France and Germany.

3 *Atriplex rosea* L. Much-branched medium to tall annual. Leaves diamond-shaped to triangular, whitish, coarsely toothed. Flowers mostly in lateral leafy clusters. Disturbed ground and waste places, coastal and inland. July-September. France and Germany; casual in Britain, Belgium and Holland. **B**: Casual.

4* **Frosted Orache** *Atriplex laciniata* L. [*A. sabulosa*]. Low to short spreading, much-branched, silvery annual; stems often reddish, and rather weak. Leaves 1.5-2cm, oval to diamond-shaped, with deep blunt teeth. Flowers in lateral leafy clusters. Fruit enclosed within heart-shaped, lobed bracts. Coastal dunes and sandy beaches, sometimes on fine shingle, often close to the high tide mark. July-September. Throughout, except the Faeroes, Iceland, Finland and Spitsbergen. Forms a distinctive zone at high level water mark, often in association with *Salsola kali* (p. 78). Easily recognized by its silvery-frosted appearance. **B**: Local, coastal.

5* *Atriplex tatarica* L. [*A. laciniata*]. Very variable, medium to tall, spreading to erect whitish annual. Leaves up to 10cm, silvery, triangular to almost arrow-shaped, with irregular blunt teeth. Flowers in long, leafless, terminal panicles or spikes; bracts rounded to oblong. Disturbed and waste places. July-September. C & S Germany; naturalized in Belgium.

6* **Grass-leaved Orache** *Atriplex littoralis* L. Medium to tall, much-branched, erect annual, with ridged stems. Leaves linear to linear-oblong, toothed or not, the uppermost unstalked. Flowers in a long spike, leafy only at the base. Fruit enclosed within triangular, toothed bracts. Maritime habitats, especially the upper parts of salt marshes, rough grassland by the sea, but also on saline habitats inland, often on muddy soils. July-October. Throughout, except the extreme north, the Faeroes and Iceland. The narrow untoothed or slightly toothed leaves are distinctive. **B**: Local, coastal, N to Moray Firth; very rare in Ireland.

7* **Common Orache** *Atriplex patula* L. Very variable, medium to tall, much-branched annual, somewhat mealy, with ridged stems. Leaves rather mealy, 3-14cm, arrow-shaped or almost diamond-shaped, narrowed at the base; upper leaves much narrower. Flowers in lateral clusters or long spikes. Fruit enclosed within diamond-shaped bracts. Disturbed ground and waste places, cultivated land, sometimes along sea-shores. July-October. Throughout, except Spitsbergen. The seeds have been shown to remain viable for 30 years. Sometimes confused with *A. glabriuscula*, but leaves are a rather different shape and somewhat mealy on both surfaces. **B**: Common throughout, except for parts of the north.

8 *Atriplex calotheca* (Rafn) Fries. Medium to tall hairless annual. Leaves arrow-shaped, irregularly toothed. Flowers in pointed spikes; bracts deeply toothed. Coastal habitats. July-October. Denmark, N Germany and Scandinavia.

9* **Spear-leaved Orache** *Atriplex prostrata* Boucher ex DC [*A. hastata*]. Very variable spreading, prostrate to erect, medium to tall annual, green, mealy when young, often reddening with age. Leaves arrow-shaped, generally not mealy, with the basal lobes at right angles to the stalk. Flowers in fairly dense panicles, leafy below. Fruit enclosed within triangular, toothed, bracts. Coastal habitats, brackish marshes, sand and shingle above high tide mark, arable land, roadsides, gardens and waste places. July-October. Throughout, except the extreme north, the Faeroes and Iceland. A coastal variant found on the coasts of England has strong red and purple tinges to the stems and leaves. **B**: Common south of the Humber; coastal elsewhere, including Ireland.

10* *Atriplex glabriuscula* Edmondston. Similar to *A. prostrata*, but a short spreading, prostrate, often reddish plant. Leaves arrow-shaped, but with a truncated base, less toothed. Maritime habitats, shingle and sandy shores. July-October. Throughout, except Iceland and the extreme north. An early colonizer of shingle together with *Lathyrus japonicus*. The spongy inflated bracteoles distinguish it from its closest allies. **B**: Throughout, coastal.

*Atriplex praecox* Hülphers. Like *A. glabriuscula*, but a low plant not exceeding 10cm tall and bracteoles not spongy, fused only at the base (not to the middle). Coastal habitats, meadows and short herb vegetation, along shingle beaches. July-October. **B**: Rare and local in NW Scotland.

11 *Atriplex longipes* Drejer [*A. patula* var. *bracteosa*]. Similar to *A. prostrata*, but always erect, to 80cm, with variable colour, some bracts distinctly stalked, generally untoothed (unstalked and toothed in *A. prostrata*). Coastal habitats, brackish marshes in particular. July-October. Britain and Scandinavia. Frequently confused with *A. prostrata* and in Britain only recently recognized as a distinct species – differs in its more upright and taller habit and in some of the bracteoles being distinctly stalked. Generally grows among tall marsh communities with e.g. *Phragmites australis* and *Scirpus maritimus*. **B**: Very local in saltmarshes in S & NW England and S Scotland.

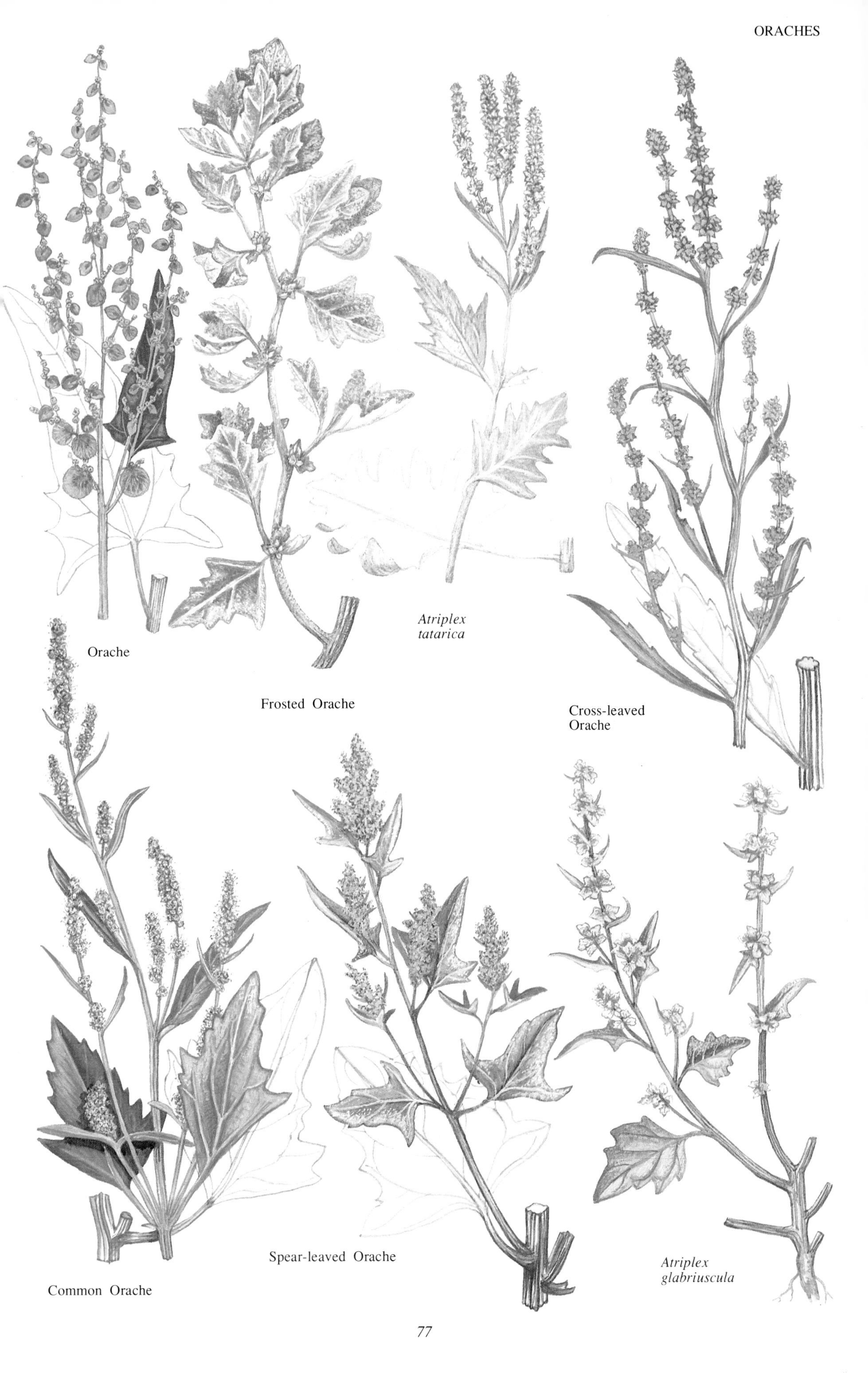
Orache
Frosted Orache
*Atriplex tatarica*
Cross-leaved Orache
Common Orache
Spear-leaved Orache
*Atriplex glabriuscula*

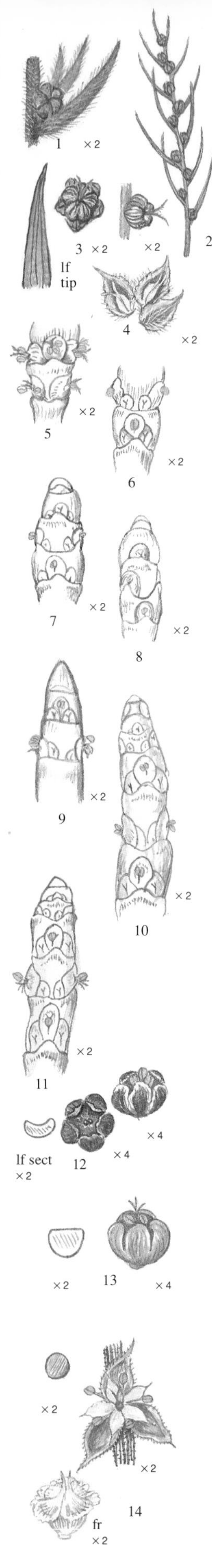


1* **Hairy Seablite** *Bassia hirsuta* (L.) Ascherson [*Echinopsilon hirsutum, Kochia hirsuta*]. Short to medium prostrate to spreading, often slightly hairy annual. Leaves linear and fleshy to 15mm, borne on flexuous branches. Flowers small, green, in the leaf axils; flowers with both sexes or just female. Ripe fruit spiny. Coastal habitats, saline soils. July-October. Europe north to S Sweden.

2 **Kochia** *Kochia laniflora* (S.G. Gmelin) Borbás. Medium to tall erect annual. Leaves thread-like, to 25mm, soft, 1-veined. Flowers small, greenish, borne in the leaf axils. Fruit winged, spineless, 4-6mm. Sandy dry habitats. August-October. France and Germany. Local.

3* *Kochia scoparia* (L.) Schrader. Like *K. laniflora*, but often taller, the leaves flat, lanceolate, 3-veined, to 50mm long. Fruit scarcely winged, 3-4mm. Sandy and waste places. July-October. **I**: Temperate Asia. Naturalized in Belgium, Holland, France, Denmark and Germany; casual in Britain.

4* **Corispermum** *Corispermum leptopterum* (Ascherson) Iljin. Short to medium branched, usually hairless, annual. Leaves narrow-lanceolate, flat, untoothed. Flowers small greenish, borne in the leaf axils and forming dense spikes, rather like *Kochia* but flower segments 1, not 5. Sandy habitats. July-October. **I**: S Europe. Naturalized in Belgium, Holland and Germany.

5* **Perennial Glasswort** *Arthrocnemum perenne* (Miller) Moss [*Salicornia perennis, S. radicans*]. Dwarf subshrub, creeping, forming mats up to 1m across; stems jointed, short and erect, becoming orange, woody with age. Leaves opposite, scale-like. Flowers tiny, borne in threes, each with 2 yellow stamens, forming branched spikes. Coastal habitats, particularly salt marshes, also on gravelly soils. August-October. France, Britain. **B**: S & SE England; Ireland; isolated localities in N Wales and NE England.

*Arthrocnemum fruticosum* (L.) Moq. Like *A. perenne*, but a stouter plant to 1m; stems bluish-green, not rooting down. Coastal habitats. July-October. W France.

**Glassworts** *Salicornia*. Annual herbs with succulent, jointed stems and opposite pairs of scale-like leaves fused across the stem. Flowers tiny, usually in threes at each leaf-scale, forming slender, often branched spikes, each flower with 3-4 segments and 1-2 stamens. A complex and difficult group of saline habitats. Distribution is imperfectly known. Identification is easiest in the autumn, when the plants become pigmented.

6* **Glasswort, Marsh Samphire** *Salicornia europaea* L. Very variable low to medium hairless annual. Stems translucent, erect, with long ascending branches, blue- or grass-green but flushed with red or pink at flowering time. Terminal flower spikes 10-50mm long; flowers in threes. Coastal muddy salt marshes, low down on the shore line, occasionally sandy ground inland. August-September. Throughout, except the far north. The young stems are edible. **B**: Throughout.

7* *Salicornia ramosissima* J. Woods. Like *S. europaea*, the stems erect or spreading, much branched or almost unbranched, dark green but deep purplish-red around the flowers; the leaves have a distinctive colourless border to the upper edge. Terminal flower-spikes 5-30mm long. Upper levels of coastal salt marshes, along drift line, estuaries and creeks. August-September. Britain, Ireland, Belgium, Holland, Denmark and Germany. **B**: E & S England, S Wales, S & E Ireland.

8 *Salicornia pusilla* J. Woods. A smaller plant, generally erect, with rather short, stubby branches, yellowish-green, becoming brownish or pinkish. Terminal flower spikes up to 6mm long; flowers solitary, not in threes. Upper level of salt marshes, estuaries, often along the drift line and below sea walls. August-September. Local in Britain, Ireland and NW France. **B**: SE England, East Anglia, S Wales, S Ireland.

9 *Salicornia nitens* P.W. Ball & Tutin. Low to short plant with erect stems with short upcurved branches, green to yellowish-green, soon tinged with brown or orange-purple. Terminal flower-spike 10-40mm long; flowers with protruding stamens (not or scarcely so in *S. europaea*). Coastal muddy and salt marsh habitats, salt pans. August-September. England and S Ireland. The lateral flowers are equal in size to the central one in each cluster of 3; in *S. europaea* they are smaller. **B**: Common in parts of S England and East Anglia.

10 *Salicornia fragilis* P.W. Ball & Tutin. Like *S. nitens*, but stouter, much branched and becoming bright yellow. Terminal flower-spikes 25-80mm long. Many habitats based on sandy soils, lower levels of salt marshes and bare muddy ground. August-September. C & S England and S Ireland; possibly also in W France.

11* *Salicornia dolichostachya* Moss. Like *S. fragilis*, but a dull green plant becoming dull yellow or brownish. Terminal flower-spikes 50-120mm, tapered. Muddy and sandy habitats at lower tidal levels, coasts and estuaries. July-August. W & N Europe. **B**: Throughout, except for W & N Scotland; coastal.

12* **Annual Seablite** *Suaeda maritima* (L.) Dumort. Very variable, short to medium, usually erect, branched annual, blue-green, becoming flushed with purple, then bright red. Leaves fleshy, linear, pointed, 10-50mm. Flowers small, green, 1-3 at base of upper leaves; stigmas 2. Coastal habitats, particularly muddy saltmarshes. July-September. Throughout, except the Faeroes, Iceland and Spitsbergen. Very variable. **B**: Common throughout, coastal.

13* **Shrubby Seablite** *Suaeda vera* Forskål ex J.F. Gmelin. Greyish hairless subshrub to 1.2m. Leaves linear, blunt, 5-18mm. Flowers small, yellowish-green, 1-3 at base of upper leaves; stigmas 3. Coastal habitats, shingle, rocks and sandy places, edge of salt marshes, often forming a distinct zone along the drift line. Late June-October. S Britain and W France. Extremely resistant, adapting to changing levels of sand and shingle. **B**: Scattered in East Anglia, SE England W as far as Chesil Beach.

14* **Prickly Saltwort** *Salsola kali* L. Variable short to medium stiff, prickly, hairy, often semi-prostrate annual. Leaves linear, spine-tipped, bluish-green. Flower solitary, green, sometimes tinged pink, at base of leaf-like bracts; stamens 5, stigmas 2. Fruit covered by persistent flower parts. Sandy coastal habitats, often non-saline, sometimes growing on waste places inland. July-October. Throughout, except the Faeroes, Iceland and Spitsbergen. Often associated with *Atriplex laciniata* and *Cakile maritima* along the shore drift line and along the edge of developing dunes. **B**: Common throughout, coastal.

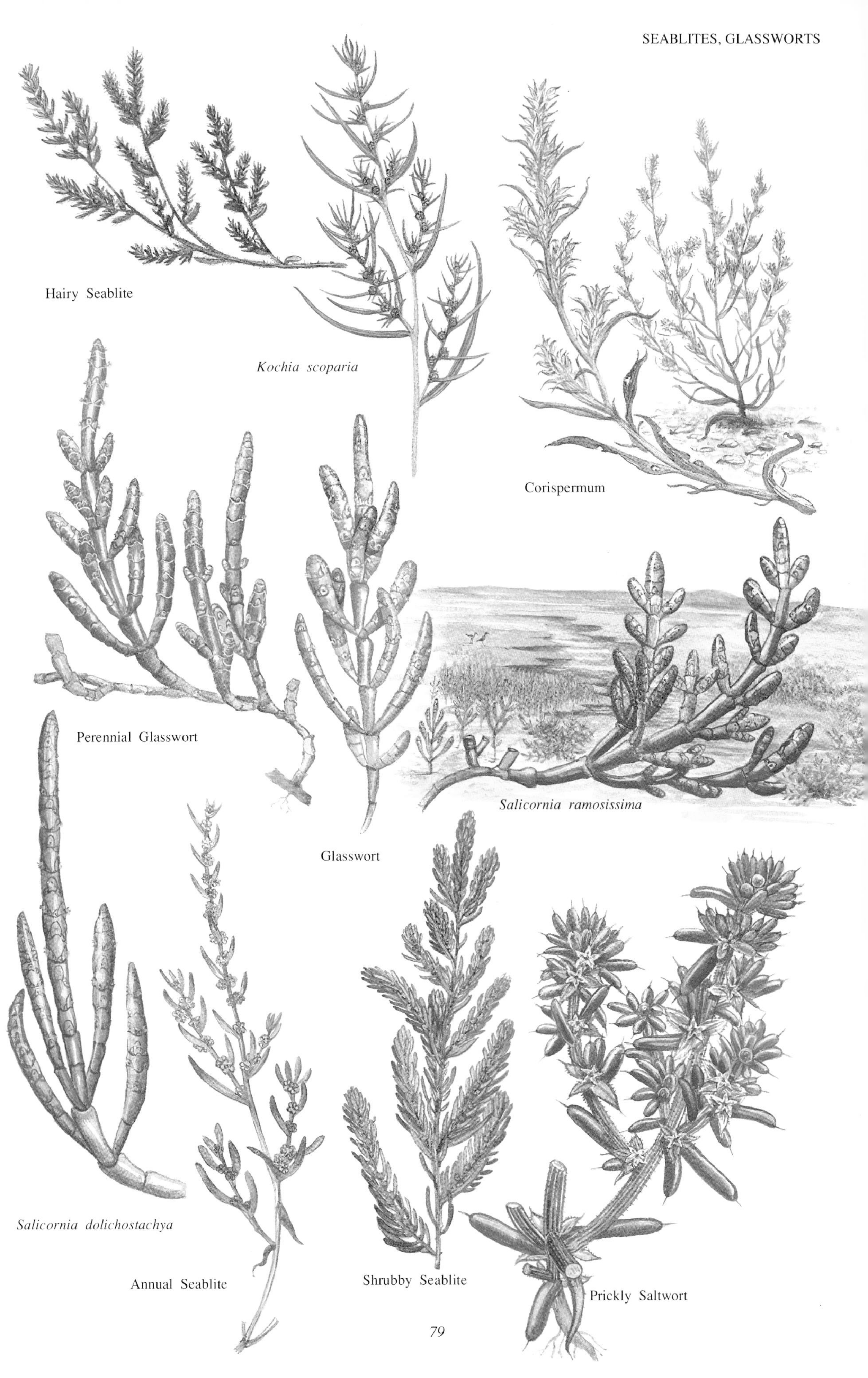
Hairy Seablite
*Kochia scoparia*
Corispermum
Perennial Glasswort
Glasswort
*Salicornia ramosissima*
*Salicornia dolichostachya*
Annual Seablite
Shrubby Seablite
Prickly Saltwort

# AMARANTH FAMILY Amaranthaceae

Herbs, generally with alternate, untoothed leaves. Flowers usually hermaphrodite, in dense heads or spikes. Perianth dry and membranous, 4-5-parted. Stamens 5. Styles 2-3. Fruits dry with a membranous wall. 900 species in 65 genera, cosmopolitan, but particularly in African and American tropics. Some are cultivated, for food or ornament.

1* **Green Amaranth** *Amaranthus hybridus* L. [*A. chlorostachys*]. Medium to tall erect, yellowish-green, hairless, or shortly hairy annual. Leaves oval, stalked. Flowers tiny, greenish or reddish, in dense, often branched, spikes, mixed with pointed bracts. Waste cultivated land. July-October. **I**: Tropical and sub-tropical America. Naturalized or casual in W Europe. **B**: Casual.

*Amaranthus cruentus* L. Like *A. hybridus*, but flowers in denser spikes with short basal branches. Waste and cultivated land. July-October. **I**: Tropical and subtropical America. Naturalized locally in France and Germany.

2* **Common Amaranth** *Amaranthus retroflexus* L. Short to tall erect, rather greyish, annual, hairy above. Leaves oval, stalked. Flowers always green, in a short dense spike, almost leafless; bracteoles stiffly pointed. Waste and cultivated land. July-October. **I**: North America. Naturalized or casual in much of Europe except the far north. **B**: Casual, occasionally established for a year or so.

3 *Amaranthus blitoides* S. Watson. Short hairy or hairless annual with spreading whitish stems. Leaves oblong to lanceolate, blunt, with membranous margins. Flowers greenish, in short lateral clusters; flower segments 4-5. Waste and cultivated land. July-September. **I**: C & W North America. Naturalized or casual in W Europe. **B**: A rare casual.

4 **White Amaranth** *Amaranthus albus* L. Short to medium, hairless annual with erect or spreading stems. Leaves oblong or spoon-shaped, blunt, with a wavy margin. Flowers greenish, in lateral clusters mixed with bristly bracts; flower segments 3. Waste places, cultivated land. July-September. **I**: S North America. Naturalized or casual in W Europe. **B**: A rare casual.

5* **American Pokeweed** *Phytolacca acinos* Roxb. [*P. americana*, *P. decandra*]. (Pokeweed Family – Phytolaccaceae). Stout tall, hairless, perennial herb to 3m, frequently less, often with reddish stems. Leaves large, oval, untoothed. Flowers greenish-white, 5-6mm, in long racemes. Fruit ribbed and fleshy, purplish-black when ripe. Cultivated land, waste places, waysides and rubbish tips, local. July-September. **I**: USA. Naturalized or casual in Britain, Belgium, Holland, France and Germany. A dye is made from the purple juice of the poisonous fruits. Sometimes grown (or a weed) in gardens. **B**: Occasionally naturalized.

6* **Hottentot Fig** *Carpobrotus edulis* (L.) N.E. Br. (Mesembryanthemum Family – Aizoaceae). Low spreading perennial with woody stems and oblong fleshy leaves, triangular in section, borne in pairs. Flowers large, bright magenta-pink or pale yellow, 80-90mm, with numerous linear petals. Fruit fleshy, edible. Coastal cliffs and sand-dunes. May-August. **I**: South Africa. Naturalized in SW Britain, S Ireland, coastal Belgium and France. The flowers open in bright sunshine. Plants often form extensive colonies and both colour forms may occur in the same population. **B**: Local, coastal SW Britain and S Ireland.

# PURSLANE FAMILY Portulacaceae

Annual or perennial herbs, usually hairy and fleshy. Leaves simple, untoothed. Flowers solitary or in cymose clusters, hermaphrodite. Sepals 2. Petals 4-6, free or joined below. Stamens numerous (*Portulaca*) or 3-5 (*Montia*). Fruit a capsule. 500 species in 19 genera, cosmopolitan but mostly American or South African.

7* **Spring Beauty** *Montia perfoliata* (Donn ex Willd.) Howell [*Claytonia perfoliata*]. Short, hairless annual. Basal leaves in a rosette, oval, stalked; stems with one pair of leaves united in a whorl below the flowers. Flowers white, 4-5mm, the petals slightly notched or not. Dry light sandy soil, waste and disturbed ground, stabilized sand-dunes, on acid soils. Late April-July. **I**: W North America, from Alaska to Mexico. Naturalized or casual in Britain, Belgium, Holland, France and Germany. Often forming extensive colonies. **B**: Introduced in 1749, now widespread, but scarcer in the north and west.

8* **Pink Purslane** *Montia sibirica* (L.) Howell [*Claytonia sibirica*]. Low hairless annual. Basal leaves in a lax rosette, oval, stalked; stem leaves not closely united. Flowers pink, rarely white, 8-10mm, the petals deeply notched. Damp habitats, woods, shady stream banks, on acid sandy soils. April-July. **I**: W North America and NE Asia. **B**: Naturalized in N & W Britain, casual elsewhere.

9* **Blinks** *Montia fontana* L. [*M. minor*, *M. rivularis*, *M. verna*]. Very variable, low to short, straggling annual or perennial; stems often reddish. Leaves narrow, oval. Flowers small and rather inconspicuous, white, 2mm, in small lax clusters. Wet habitats, sometimes submerged, generally on acid soils. April-October. Throughout. Variable, with several subspecies. **B**: Throughout.

10* **Purslane** *Portulaca oleracea* L. Low spreading, hairless annual with branched stems. Leaves spoon-shaped, crowded and sub-opposite below flowers. Flowers yellow, 8-12mm, solitary, lateral or at stem forks. Cultivated land and waste places. June-September. **I**: Tropical regions. Naturalized or casual in Britain, Belgium, Holland, France, Germany and S Denmark. The form cultivated as a salad vegetable is subsp. *sativa* (Haw.) Celak.

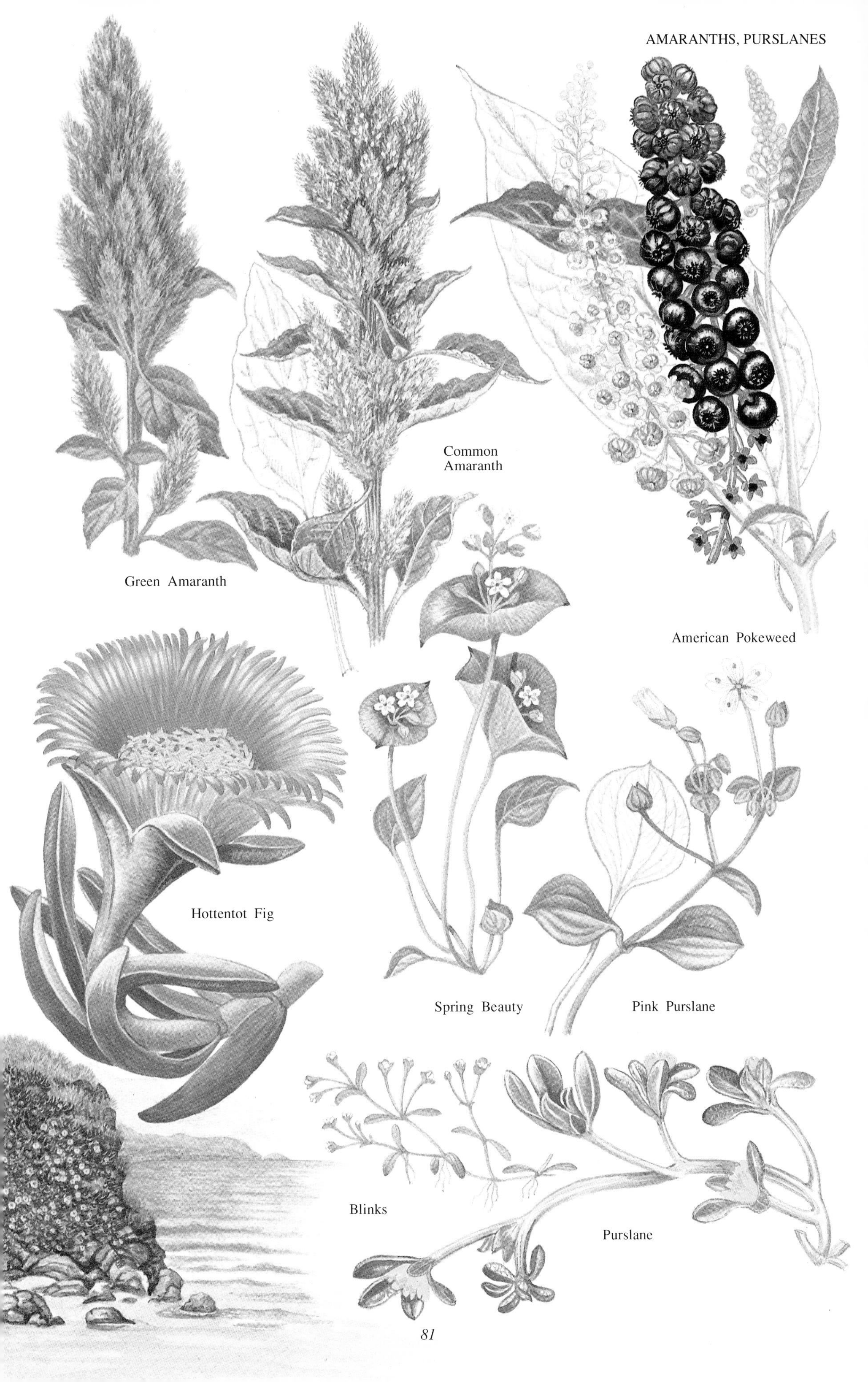
Green Amaranth
Common Amaranth
American Pokeweed
Hottentot Fig
Spring Beauty
Pink Purslane
Blinks
Purslane

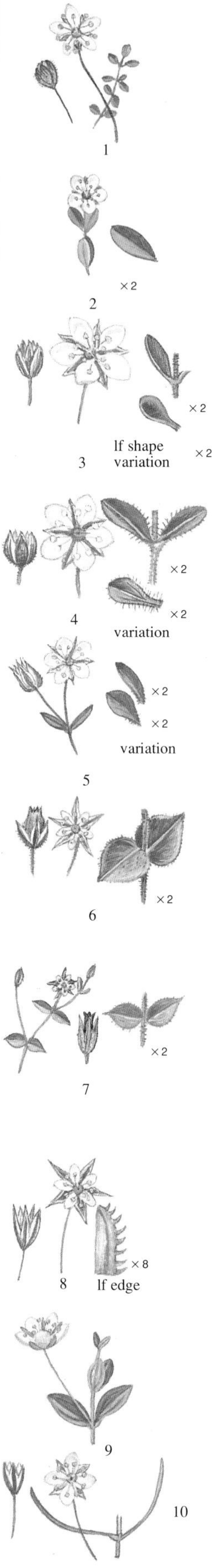

# PINK FAMILY Caryophyllaceae

Herbs with opposite, rarely alternate leaves, each pair at right angles to the one below; occasionally with membranous stipules. Flowers regular, generally hermaphrodite. Sepals 4-5, free or fused. Petals 4-5, sometimes absent, clawed or not. Stamens often 8-10, sometimes fewer. Stigmas 2-5. Fruit a capsule splitting with as many or twice as many teeth as styles, (a 'berry' in *Cucubalus*). 2000 species in about 80 genera, in temperate regions of the world, particularly in Mediterranean areas.

**Sandworts** *Arenaria*. Annual or perennial herbs, mostly with oblong to lanceolate leaves. Flowers in small cymes or solitary. Sepals separate from one another. Petals 5, not notched. Stamens 10. Styles 3. Capsule splitting by 6 teeth. 150 species widespread, but more especially in Arctic and temperate regions of the north hemisphere.

1* **Balearic Sandwort, Mossy Sandwort** *Arenaria balearica* L. Low mat-forming, bright green perennial. Leaves small oval to rounded, slightly hairy. Flowers solitary, white, 7-9mm, on slender erect stalks; sepals hairy. Damp shady rocky place and walls. March-August. **I**: W Mediterranean. Naturalized in Britain. **B**: Widely naturalized, as a garden escape.

2* *Arenaria humifusa* Wahlenb. Low mat-forming perennial with runners, hairless. Leaves small, elliptical. Flowers solitary or two together, white, 5-6mm, with pale purple anthers; sepals hairless. Open mountain habitats on calcareous soils, to 1500m. June-July. Norway, Sweden.

3* **Arctic Sandwort** *Arenaria norvegica* Gunnerus. Low, tufted, scarcely hairy perennial. Leaves dark green, oblong, rather fleshy with a single, rather obscure, vein. Flowers solitary or two, 8-12mm, white with pale yellowish-white anthers, petals longer than the sepals. Stony and bare habitats and screes, river shingles, on calcareous soils, to 1450m. May-July (October). Britain, Iceland, Scandinavia. Local. **B**: Rare, W Highland Scotland, Rhum, Shetland; a single Irish locality in Co Clare.

**English Sandwort** *A. n.* subsp. *anglica* Halliday. Like the type, but an annual or biennial with a laxer habit, pale green leaves and slightly larger flowers. Limestone pavement. June-July. N England – W Yorkshire, very rare. A winter germinating plant.

**Alpine Sandwort** *Arenaria montana* L. Low to short, grey-green, hairy perennial. Leaves linear-lanceolate, finely pointed, 1-veined, hairy on edges. Flowers large,white, 16-24mm, in lax clusters, the petals at least twice as long as the sepals. Rocky and waste places. May-July. **I**: SW Europe. Casual or locally naturalized in Britain.

4* **Fringed Sandwort** *Arenaria ciliata* L. Low rough-stemmed perennial, not tufted. Leaves elliptical to spoon-shaped, hairy on edges towards the base, clearly 1-veined. Flowers in a small cluster or solitary, white, 12-16mm, anthers white, petals just exceeding sepals. Meadows and rocky habitats, on calcareous soils, to 3200m. July-August. Ireland, Scandinavia, Spitsbergen, mountains of C & E France and S Germany. The Irish plant is sometimes distinguished as (subsp. *hibernica* Ostenf. & O.C. Dahl.). **B**: Very rare: cliffs on Ben Bulben, Co Sligo

5 *Arenaria gothica* Fries. Taller, to 12cm, annual or biennial. Flowers slightly smaller. Limestone pavements and lakeshores. July-August. S Sweden – Gotland, France – Jura. Very local.

6* **Thyme-leaved Sandwort** *Arenaria serpyllifolia* L. Low to short, prostrate or bushy, rough-hairy, greyish annual, rarely biennial, much-branched. Leaves oval, pointed. Flowers in a spreading cluster, white with yellowish anthers, 5-8mm, petals about half the length of the sepals. Dry sandy, often bare places, fields, heaths, walls. April-September. Throughout, except the far north.

7 **Lesser Thyme-leaved Sandwort** *Arenaria leptoclados* (Reichenb.) Guss. Like *A. serpyllifolia*, but more delicate, the sepals less than 3mm long (not 3-4.5mm) and capsule equalling sepals (not longer). June-August. Throughout, except the far north. Sometimes growing together with *A. serpyllifolia*, though generally less common. **B**: Scattered localities throughout.

**Moehringia** *Moehringia*. Like *Arenaria*, but seeds with an appendage (strophiole). Styles 2-3. Fruit capsule with 4 or 6 teeth. 28 species in the north hemisphere.

8* **Three-veined Sandwort** *Moehringia trinervia* (L.) Clairv. [*Arenaria trinervia*]. Low to short, rather straggling annual, sometimes overwintering, shoots hairy. Leaves oval, 3-5-veined, the lower stalked. Flowers white, 4-7mm, 5-parted, solitary or several on slender hairy stalks, petals about half the length of the hairy-margined sepals. Woods and other shady habitats, occasionally in more open situations, on rich acid or neutral soils. May-June. Throughout, except the far north. Often indicative of ancient woodland. **B**: Throughout, though scarce in the extreme north.

9 *Moehringia lateriflora* (L.) Fenzl. Like *M. trinervia*, but always perennial, the leaves 1-3-veined. Flowers 6-9mm, the sepals hairless and petals longer than sepals. Wooded habitats, coastal shingle and river and stream margins. June-August. Scandinavia. Circumpolar in the northern hemisphere.

10* **Mossy Sandwort** *Moehringia muscosa* L. Low to short, very variable, weak-stemmed, hairless moss-like perennial, occasionally annual. Leaves linear. Flowers 4-parted, white, 5-7mm, the petals longer than sepals; stamens 8. Shady damp rocks and mossy places, to 2350m. May-September. Mountains of C & E France and S Germany.

**Ciliate Sandwort** *Moehringia ciliata* (Scop.) Dalla Torre. Like *M. muscosa*, but forming creeping mats with fleshy linear leaves. Flowers small, 4-5mm, 5-parted usually, solitary or 2-3 together, sickly scented. Limestones screes and rocks, to 3000m. June-August. Mountains of E France and S Germany – Alps. Occasionally hybridizes with *M. muscosa*.

Balearic Sandwort
*Arenaria humifusa*
Arctic Sandwort
Fringed Sandwort
Thyme-leaved Sandwort
Three-veined Sandwort
Mossy Sandwort

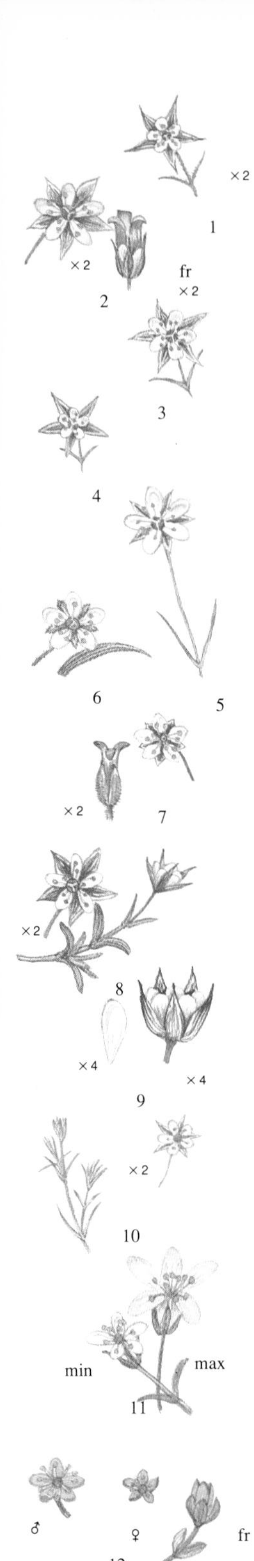


**Sandworts** contd. *Minuartia*. Like *Arenaria* but leaves narrow, often linear or bristle-like. Fruit capsule with as many teeth as styles.

1* **Sticky Sandwort** *Minuartia viscosa* (Schreber) Schinz & Thell. [*Alsine viscosa*]. Low, slender, stickily-hairy, annual; stems branched above the middle. Leaves linear, pointed. Flowers white, small, 2-3mm, in lax clusters, petals shorter than sepals. Dry sandy habitats. May-September. E France, Germany, Denmark and S Sweden.

2* **Fine-leaved Sandwort** *Minuartia hybrida* (Villars) Schischkin [*M. tenuifolia, Alsine tenuifolia*]. Low slender, much-branched, erect, annual, stickily-hairy above, without sterile shoots. Leaves linear. Flowers white with yellow anthers, 6mm, in lax clusters; petals slightly shorter than sepals, white-edged. Dry, sandy and stony habitats, arable fields, waste ground, old walls. May-September. Britain, Ireland, France to C Germany. Primarily in S and W Europe. **B**: Local, scattered through England, Wales and Ireland.

3 *Minuartia mediterranea* (Link) K. Maly. Similar, but a taller plant and hairless. Flowers in denser clusters; petals sometimes absent. Coastal habitats. May-August. France north to Normandy, rare.

4* *Minuartia rubra* (Scop.) McNeill [*M. fastigiata, Alsine fasciculata*]. Short erect biennial, hairless or slightly hairy above. Leaves bristle-like. Flowers in dense lateral or terminal clusters, 6-8mm, petals only one third the length of the sepals. Dry sandy, bare and waste habitats. June-September. C France to C & S Germany.

5* *Minuartia setacea* (Thuill.) Hayek [*Alsine setacea*]. Low to short, tufted or cushion-forming perennial with numerous erect flowering shoots, hairy only below. Leaves bristle-like, usually erect, rough. Flowers small, white, 4-5mm, 3 or more in a lax cluster, petals slightly longer than the sepals. June-September. C & E France and C & S Germany.

6* **Curved Sandwort** *Minuartia recurva* (All.) Schinz & Thell. [*Alsine recurva*]. Low densely tufted perennial, slightly stickily-hairy. Leaves narrow, sickle-shaped. Flowers white, 7-9mm, solitary or in clusters of up to 8, petals slightly longer than the 5-7-veined sepals. Grassy and stony habitats, usually on acid soils, to 1900m. June-October. Mountains of France.

7 **Spring Sandwort** *Minuartia verna* (L.) Hiern [*Alsine verna*]. Variable low to short, loose cushion-forming perennial, often somewhat stickily-hairy. Leaves linear, pointed, keeled beneath, 3-veined. Flowers white, 6-8mm, with purplish anthers, in clusters of 2-7, petals longer than sepals. Dry rocky places, screes, mine spoil heaps, often on calcareous soils, to 3000m. May-September. Britain, France and Germany. Differs from *M. stricta* in its 3-veined leaves. **B**: Local, N England, Wales and W Ireland, few localities in S England and Scotland.

8 **Mountain Sandwort** *Minuartia rubella* (Wahlenb.) Hiern. Like *M. verna*, but smaller and more densely tufted, flowers generally only 1-2, the petals shorter than the sepals. Mountain rocks and ledges, on mineral-rich soils. Mainly arctic and sub-arctic. July-August. Britain, Iceland and Scandinavia. **B**: Rare; only in Perthshire.

9* **Teesdale Sandwort** *Minuartia stricta* (Swartz) Hiern. Low tufted, hairless, perennial. Leaves mostly basal, linear, 1-veined. Flowers white, 7-8mm, often in threes, petals slightly shorter than the sepals. June-July. Wet mountain habitats, wet flushes, rocks, low sward and screes, to 530m. June-July. Arctic and sub-Arctic Europe, N Britain, Scandinavia, E France and S Germany, local and often rare. **B**: Very rare; only in upper Teesdale.

10 *Minuartia rossii* (R. Br.) Graebner. Like *M. stricta*, but stems more leafy with detachable vegetative shoots in the upper leaf-angles; flower stems spreading. Rocky habitats. July-August. Spitsbergen. Also found in Arctic North America and Greenland.

11* *Minuartia biflora* (L.) Schinz & Thell. [*Alsine biflora*]. Low slender, tufted, slightly hairy perennial. Leaves linear, blunt, 1-veined. Flowers white, rarely pale lilac, 7-12mm, solitary or 2-3 together; petals slightly longer than the sepals. Damp open mountain habitats, often by snow patches, to 2800m. July-August. Iceland and Scandinavia. Also found in the Alps.

12* **Cyphel** *Minuartia sedoides* (L.) Hiern [*Alsine sedoides, A. cherleri*]. Low dense cushion perennial, hairless. Leaves somewhat fleshy, linear-lanceolate, closely overlapping. Flowers yellowish-green, 4-7mm, solitary, borne almost on the cushion, often without petals, sometimes only male or female. Rocky mountain ledges, exposed ridges and moraines, shingle, to 3000m. June-August. N Britain, E France and S Germany. **B**: Scotland, local; C & W Highlands and the Inner Hebrides.

13* **Sea Sandwort** *Honkenya peploides* (L.) Ehrh. [*Ammadenia peploides*]. Low creeping, rather yellowish perennial, rooting at the nodes. Leaves succulent, oval, pointed, yellowish-green. Flowers greenish-white, 6-10mm, solitary or in small leafy clusters, sometimes only male or female; petals slightly shorter than sepals. Maritime habitats, especially on sand and shingle. May-August. Throughout. Plants often colonize the youngest sand-dunes and are very tolerant of salt spray. The flowers are hermaphrodite or unisexual. **B**: Throughout.

14* *Bufonia paniculata* F. Dubois [*B. macrosperma*]. Short, much-branched, annual with erect flowering stems. Leaves linear, rather bristly, closely pressed to the stem. Flowers white, 4-parted, 4-5mm long in lax panicles; styles 2. Fruit capsule splitting by 2 teeth only. July-October. Dry sandy or rocky places. C & S France.

Fine-leaved Sandwort

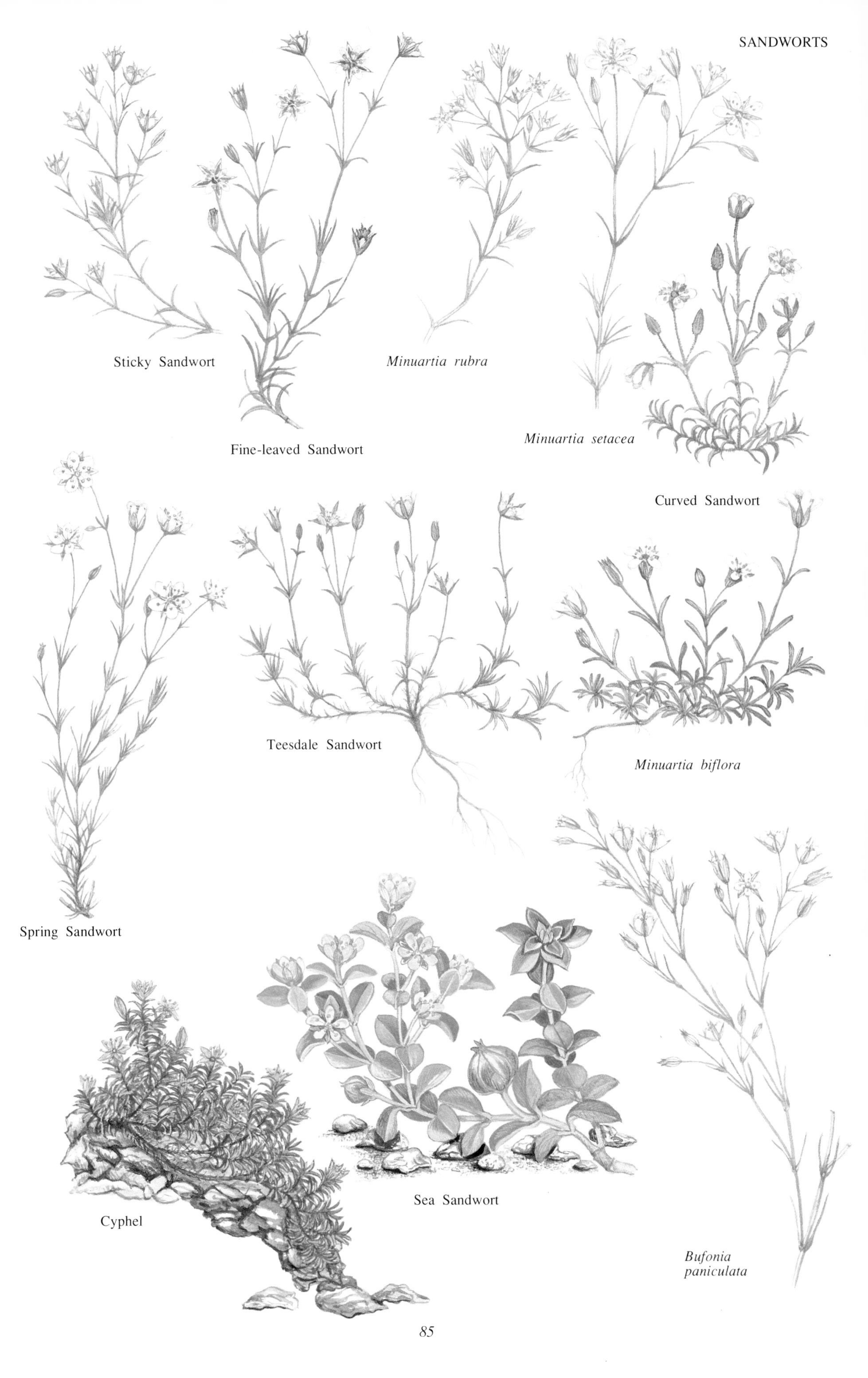

Sticky Sandwort

Fine-leaved Sandwort

*Minuartia rubra*

*Minuartia setacea*

Curved Sandwort

Teesdale Sandwort

*Minuartia biflora*

Spring Sandwort

Sea Sandwort

Cyphel

*Bufonia paniculata*

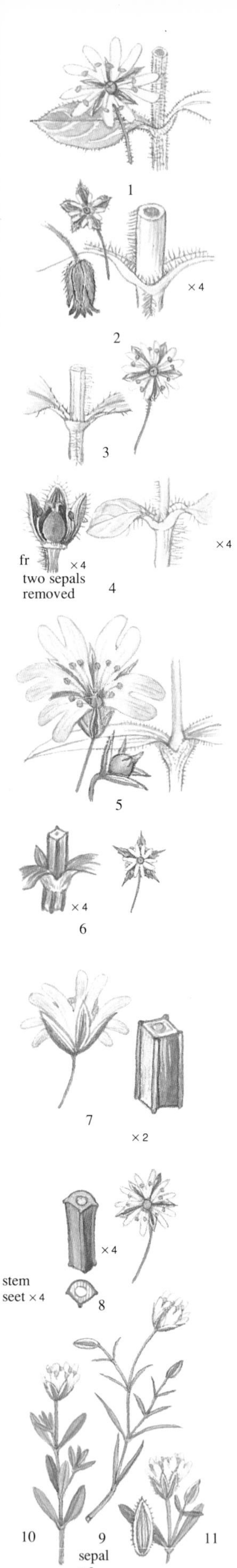


**Stitchworts & Chickweeds** *Stellaria*. Annual or perennial herbs. Inflorescence usually a branched cyme (dichasium), occasionally the flowers very few or solitary. Sepals 5, separate. Petals often 5, sometimes fewer or absent, white, deeply cleft. Stamens 10, occasionally fewer. Styles 3. Fruit capsule splitting with 6 teeth. 90 species widespread in many regions of the world. A number have become widespread weeds of cultivated land.

1* **Wood Stitchwort** *Stellaria nemorum* L. Short to medium stoloniferous perennial; stems hairy all round. Leaves oval, pointed, long-stalked, the upper narrower. Flowers 20-24mm, the petals cleft almost to the base, twice as long as the hairless sepals. Damp wooded habitats, generally deciduous, streamsides, usually on base-rich soils, to 2400m. May-July. Throughout, except the extreme north, Ireland, the Faeroes and Iceland. **P**: Flies and small beetles. **B**: Upland Britain from N England to C Scotland and Wales.

subsp. *glochidisperma* Murb. Like the type, but with bracts decreasing rapidly in size after the first branch of the inflorescence (gradually so in subsp *nemorum*). Similar habitats and flowering time. Scattered in W & C Europe from Britain to S Sweden. **B**: Wales.

2* **Common Chickweed** *Stellaria media* (L.) Vill. Low to short variable, often semi-prostrate annual; stems with one line of hairs down each internode. Leaves oval, pointed, long-stalked, the uppermost unstalked. Flowers few to many, 8-10mm; petals equalling sepals; stamens 3-8 usually. Flower stalks glandular-sticky, spreading to down-curved in fruit. Cultivated land, farmyards, road verges, bare and waste ground, rubbish tips, beaches – especially on shingle, to 2500m. Flowering throughout the year. Throughout Europe. Widespread and often persistent weed, resistant to many common herbicides. Like many successful weeds, fast-growing, self-pollinated and flowering over a long period. Larger plants from woodland habitats can be easily mistaken for *S. neglecta*. **B**: Throughout.

3* **Great Chickweed** *Stellaria neglecta* Weihe. Like *S. media*, but a larger plant; flowers 10-13mm usually with 10 stamens. Flower stalks not glandular, straight but deflexed in fruit. Damp shaded habitats, woodland, streamsides, hedgerows. April-July. Europe north to S Sweden. **B**: Local in England and Wales, rare elsewhere.

4* **Lesser Chickweed** *Stellaria pallida* (Dumort.) Piré. Like *S. media* but a more slender plant, all the leaves short-stalked. Flowers smaller, 4-5mm, petals generally absent or very small; stamens usually 1-3, violet. Arable land, heathy grassland, road verges and waste or bare habitats, often on light sandy soils. March-May. Britain, Belgium, Holland, France, Germany and Denmark. **P**: Self-pollinated. **B**: Scattered throughout, though rare in Scotland and Ireland.

5* **Greater Stitchwort** *Stellaria holostea* L. Short to medium rather straggly perennial; stems weak, square and rough. Leaves narrow lanceolate, pointed, rough on margins. Flowers 18-30mm, in lax clusters; petals cleft to half-way, twice as long as the sepals. Woods and hedgerows, road verges and banks, on base-rich or slightly acid, often heavy soils, to 2000m. April-June. Throughout, except the extreme north, the Faeroes and Iceland. **P**: Bees, beetles, butterflies, moths and other insects; perhaps also self-pollinated. **B**: Throughout.

6* **Bog Stitchwort** *Stellaria uliginosa* Murray [*S. alsine*]. Variable, short, creeping, bluish-green perennial; stems spreading to ascending, square, hairless and smooth. Leaves elliptical to lanceolate, pointed, unstalked (stalked only on overwintering shoots). Flowers 5-7mm, sepals longer than the deeply cleft petals. Damp or wet habitats, streamsides, marshes, woodland rides, woodland flushes and spring lines, on acid soils, to 2300m. **P**: small insects, perhaps also self-pollinated. May-June. Throughout. **B**: Widespread, except in extreme north.

7* **Marsh Stitchwort** *Stellaria palustris* Retz. Short to medium greyish, creeping perennial with erect, square, smooth stems. Leaves linear-lanceolate. Bracts without a hairy margin. Flowers few in a lax cluster, 12-18mm; petals cleft almost to the base, up to twice as long as the sepals. Wet grassy habitats, marshes and fens, usually lowland on base-rich or calcareous soils. May-July. Throughout, except Ireland, the Faeroes and Iceland. **P**: Small flies. Often local and declining in some areas through habitat destruction. **B**: Local, but decreasing, throughout, except the extreme north.

*Stellaria fennica* (Murb.) Perf. Like *S. palustris*, but stem angles and leaf-margins rough; flowers smaller. Similar habitats. June-August. Scandinavia.

8* **Lesser Stitchwort** *Stellaria graminea* L. Short to tall creeping perennial with weak straggly, much-branched stems, square and smooth. Leaves linear to lanceolate, pointed, smooth. Bracts membranous with a hairy margin. Flowers in lax clusters of 10 or more, 5-12mm, petals equalling or longer than the sepals. Woods, rough grassland, hedgebanks, heaths, on light, generally calcareous soils, to 2000m. May-August. Throughout, except the Faeroes and Spitsbergen. The flowers are either hermaphrodite or functionally female. **P**: Primarily flies. **B**: Throughout.

9 *Stellaria longipes* Goldie. Low to short creeping or tufted perennial; stems square ascending, occasionally with lax non-flowering shoots in axils of upper leaves. Leaves linear-lanceolate, unstalked, finely pointed. Flowers few, 7-15mm, the petals longer than the hairless sepals. June-August. Spitsbergen, extending to N Russia and Arctic Asia.

10 *Stellaria crassipes* Hultén. Like *S. longipes*, but leaves oval to lanceolate, thick and rigid, generally with short condensed shoots in axils of upper leaves. Flowers usually solitary. June-July. NW Finland and C Norway, local.

11 *Stellaria ciliatisepala* Trautv. Like *S. crassipes*, but without condensed shoots. Flowers generally several together; bracts and sepals with hairy margins. June-August. Spitsbergen, extending to Arctic Asia and North America.

Common Chickweed
Wood Stitchwort
Great Chickweed
Lesser Chickweed
Bog Stitchwort
Greater Stitchwort
Marsh Stitchwort
Lesser Stitchwort

1* *Stellaria longifolia* Muhl. ex Willd. Low to short creeping perennial; stems square, ascending, branched and rough. Leaves linear-lanceolate to linear, rough on margins. Flowers small, 4-8mm, many in lax clusters, petals equalling hairless sepals. Damp woodland. June-July. Scandinavia and Germany, often rather local.

2 *Stellaria calycantha* (Ledeb.) Bong. Short creeping or tufted perennial with square, smooth stems. Leaves oval to linear-lanceolate, pale yellowish-green with a hairy margin. Flowers few or solitary, 8-12mm, petals cleft almost to the base, shorter than the sepals, or absent. Rocky habitats and moraines. June-July. Arctic and sub-Arctic Europe, including Iceland.

3* *Stellaria crassifolia* Ehrh. Low to short creeping perennial; stems ascending, square, smooth, loosely branched. Leaves ovate to linear-lanceolate, pointed, unstalked, somewhat fleshy, not hairy. Flowers few or solitary, 5-8mm, petals cleft almost to the base, longer than the sepals. Rocky habitats. June-August. Iceland and Scandinavia extending into N Germany.

4 *Stellaria humifusa* Rottb. Like *S. crassifolia*, but generally prostrate with matted stems. Leaves crowded, often with vegetative buds in leaf angles. Flowers 8-16mm. June-August. Iceland, Finland, Norway and Spitsbergen.

5* **Umbellate Chickweed** *Holosteum umbellatum* L. Low to short rather sticky annual, greyish-green towards the base; stems erect. Leaves oblong, pointed, the upper narrower and unstalked. Flowers in simple umbels, with slender stalks of varying lengths, white to pale pink, petals irregularly toothed, twice the length of the sepals. Dry, often sandy, habitats, disturbed ground or walls, usually at low altitudes. March-May. Europe north to S Sweden; extinct in Britain.

**Mouse-ears** *Cerastium*. Herbs, occasionally woody at the base, usually hairy. Flowers in a cymose inflorescence, occasionally solitary. Sepals separate. Petals white, notched. Stamens 5-10. Styles usually 5, occasionally 3. Fruit an oblong capsule splitting with twice as many teeth as styles. Cosmopolitan; 100 species, although most occur in northern temperate regions of the Old World. A number are widespread weeds of cultivated land.

6* **Starwort Mouse-ear** *Cerastium cerastoides* (L.) Britton. Low loosely matted perennial, hairless except for a line of hairs down each internode; flowering stems ascending, but vegetative stems creeping and rooting down. Leaves pale green, oblong to linear, generally curving to one side. Bracts leaf-like. Flowers 1-3, 9-12mm, on sticky stalks; petals deeply notched. Styles usually 3. Damp rocky and grassy mountain habitats, screes, to 3000m. June-August. Arctic and sub-Arctic Europe; N Britain, Scandinavia, the Faeroes and Iceland. **P**: Bees, but often self-pollinated. **B**: Local on siliceous rocks; Scottish Highlands.

7* **Dusty Miller, Snow-in-summer** *Cerastium tomentosum* L. Low to short, densely matted greyish-white, hairy perennial. Leaves lanceolate to linear-lanceolate, the margins rolled under. Flowers in lax clusters, 15-25mm, petals deeply notched. Capsule teeth spreading. Banks, hedgerows, rocky habitats and walls. May-July. **I**: C & S Apennines. Naturalized in Britain, France, Belgium and Holland. Widely cultivated in gardens and a frequent escape, favouring dry sunny places. **B**: Scattered localities, mainly in C & S Britain.

*Cerastium biebersteinii* DC. Like *C. tomentosum*, but leaves with flat margins; capsule teeth erect. Similar habitats. May-July. **I**: Crimea. Britain and parts of W Europe. Less common than *C. tomentosum*.

8* *Cerastium dubium* (Bast.) O. Schwarz [*C. anomalum*]. Resembles *C. cerastoides*, but a rather taller annual with somewhat stickily-hairy stems. Leaves linear to linear-oblong, hairless or glandular-sticky. Bracts leaf-like. Flowers in lax clusters, 11-15mm, with 3 styles. Damp habitats, grassy and rocky places. May-July. France and Germany.

9* **Field Mouse-ear** *Cerastium arvense* L. Very variable, low, slightly hairy perennial, loosely matted with ascending flowering stems; vegetative shoots spreading and rooting at the nodes. Leaves linear-lanceolate. Bracts with membranous margins, hairy only on the edge. Flowers in lax clusters, 12-20mm, petals deeply notched. Dry open habitats, grassland, hedgebanks and road verges, on calcareous or slightly acid soils, to, 3100m. April-August. Throughout, except the extreme north, the Faeroes and Iceland; naturalized in Finland and Norway. **P**: Bees and flies. **B**: Throughout, though rare in S Wales and Ireland.

Dusty Miller

*Stellaria longifolia*
Umbellate Chickweed
Starwort Mouse-ear
*Stellaria crassifolia*
Dusty Miller
Field Mouse-ear
*Cerastium dubium*

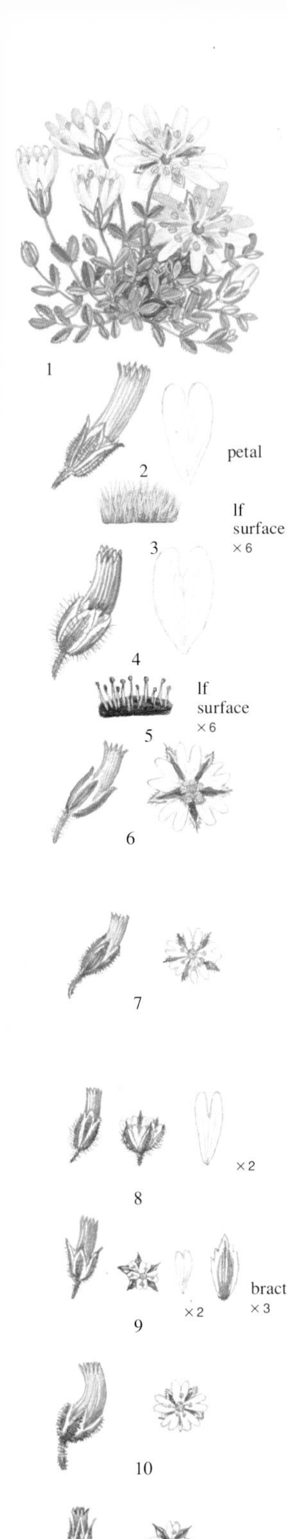


1 *Cerastium regelii* Ostenf. Low dense cushion-forming perennial, practically hairless, often non-flowering and increasing by thin underground runners with deciduous bulbil-like buds. Leaves roundish to elliptical, yellowish-green, shiny, hairy on edges. Flowers, when produced, solitary, 18-24mm, petals deeply cleft, twice as long as sepals. Arctic tundra. July-August. Spitsbergen and Arctic Russia.

2* **Alpine Mouse-ear** *Cerastium alpinum* L. Low, mat-forming, greyish-green, hairy, perennial. Leaves oblong to elliptical, narrowed at base, with soft, whitish hairs; bracts with white margins. Flowers 1-5, 18-25mm, petals deeply cleft, twice the length of sepals. Mountain rocks and ledges, 1800-2850m. June-August. E France, S Germany and Scandinavia. A typical Arctic-alpine, circumpolar.

3* subsp. *lanatum* (Lam.) Ascherson & Graebner. Like the type, but the whole plant white-woolly, with glandular hairs. Habitat and flowering time similar. Distribution similar but including N Britain and Iceland. **B**: Local in N Wales, Cumbria and the Scottish Highlands.

subsp. *glabratum* (Hartman) Á. & D. Löve. Like the type but stems and leaves hairless; habit more erect. Similar habitats and flowering time. Iceland and Finland.

4* **Arctic Mouse-ear** *Cerastium arcticum* Lange. Variable, low, mat-forming perennial, with underground stolons. Leaves elliptical, yellowish-green, covered as stems in short white hairs; bracts leaf-like. Flowers 1-3, 18-30mm, petals shallowly notched, twice the length of the sepals. Rocky mountain rocks, ledges and screes, to 1700m, often local. June-August. N Britain, Scandinavia and Iceland. **B**: Scattered localities in N Wales and the Scottish Highlands.

5* *C. a.* subsp. *edmondstonii* (H.C. Watson) Á. & D. Löve. More compact and with rounded, dark purplish-green leaves, densely glandular. Confined to Shetland – Unst.

6* **Common Mouse-ear** *Cerastium fontanum* Baumg. subsp. *holosteoides* (Fries) Salman, van Ommerang & de Voogd [*C. holosteoides*]. Variable low-short, slightly hairy, laxly tufted perennial, with short basal, non-flowering shoots. Leaves lanceolate, unstalked. Bracts mostly white-edged. Flowers in lax clusters, 6-10mm, petals deeply notched, equalling sepals. Capsule distinctly curved. Grassy habitats, shingle, sand-dunes, on calcareous or neutral soils. Low altitudes. April-November. **B**: Throughout.

subsp. *scandicum* H.Gartner. Rather larger fruit capsules and flowers with sepals 6-9mm long (rather than 3-5mm). Rocky habitats. Scandinavia, the Faeroes and Iceland.

7* **Grey Mouse-ear** *Cerastium brachypetalum* Pers. Low to short, hairy annual. Leaves oblong to elliptical. Bracts all leaf-like. Flowers in lax clusters, 6-9mm, the petals deeply notched, shorter or longer than the sepals. Dry open habitats, on calcareous soils, at low altitudes. April-November. W Europe north to Denmark. **B**: Only on railway cuttings in Bedfordshire and Northamptonshire. Doubtfully native.

subsp. *tauricum* (Sprengel) Murb. Like the type, but the flower stalks and sepals with both simple, as well as glandular, hairs. Similar habitats and flowering time. Continental Europe north to S Sweden.

8* **Sticky Mouse-ear** *Cerastium glomeratum* Thuill. Low to short, yellowish-green, stickily-hairy annual. Leaves oval to elliptic. Bracts leaf-like. Flowers in compact clusters, 5-8mm, not fully opening, petals shallowly notched, equalling sepals. Arable land, pathways, walls and sand-dunes, at low altitudes. April-October. Throughout, except Finland and Spitsbergen. A common weed. **B**: Throughout.

9* **Little Mouse-ear** *Cerastium semidecandrum* L. Low to short stickily-hairy annual, with semi-prostrate to ascending stems. Leaves oblong, the lower stalked. Bracts with white edges near the top. Flowers 5-9mm, in small clusters, petals slightly notched, shorter than sepals. Dry open habitats on sandy and calcareous soils, occasionally on walls, at low altitudes. March-May. Throughout, except the Faeroes, Iceland and Spitsbergen. **B**: Scattered localities throughout, particularly in East Anglia; rare in Scotland.

subsp. *macilentum* (Aspegren) Moschl. Similar to the type, but the whole plant hairless. S Sweden.

10* **Dwarf Mouse-ear** *Cerastium pumilum* Curtis. Low stickily-hairy, more or less erect annual, often red-tinged. Leaves oblong to elliptic, the lower stalked. Bracts with white margins. Flowers 6-9mm, in lax clusters, the petals often purple tinged, equalling the sepals. Dry grassy habitats and bare places, on calcareous soils. April-June. Throughout, except the extreme north, the Faeroes and Norway. **B**: Rare and local; C & S England.

subsp. *pallens* (F.W.Schultz) Schinz & Thell. Like the type but a pale green plant with broader petals and 6-10 (rather than 5) stamens. Habitat and distribution as above.

11* **Sea Mouse-ear** *Cerastium diffusum* Pers. [*C.atrovirens*]. Low to short stickily-hairy annual. Leaves oval to elliptical, the lowermost spoon-shaped. Bracts leaf-like. Flowers in lax, few-flowered, clusters, 4-parted, petals shallowly notched, shorter than the sepals. Dry grassy or stony places, mainly coastal in W Europe, sometimes growing on railway ballast inland. March-July. Throughout, except the extreme north.

subsp. *subtetrandrum* (Lange) P.D. Sell & Whitehead. Flowers usually 5-parted and sepals 7-9mm (not 4-7). Similar habitats and flowering time. Germany and Sweden.

12* **Upright Chickweed** *Moenchia erecta* (L.) P. Gaertner, B. Meyer & Scherb. [*Cerastium erectum*]. Low greyish, hairless annual, erect. Leaves linear to linear-lanceolate, rigid and pointed. Flowers white, 1-3 usually, 7-8mm, usually 4-parted; petals narrow, not notched, shorter than the sepals. Capsule splitting with twice as many teeth as styles. Bare sandy habitats at low altitudes. April-June. S Britain, Belgium, Holland, France and Germany. Generally declining. **B**: Restricted to S England and Wales.

13* **Water Chickweed** *Myosoton aquaticum* (L.) Moench [*Cerastium aquaticum, Stellaria aquatica, Malachium aquaticum*]. Very variable medium, straggly perennial; stems stickily hairy above. Leaves ovate with a truncated or heart-shaped base, hairy or not. Flowers in lax clusters, white, 14-15mm, the 5 petals cleft to the base, much longer than the sepals. Wet habitats, marshes, fens, pond and river margins, on fertile, often base-rich soils. June-October. Throughout, except Ireland, the Faeroes, Iceland and Spitsbergen. **B**: C & S England, rare in Wales, absent elsewhere.

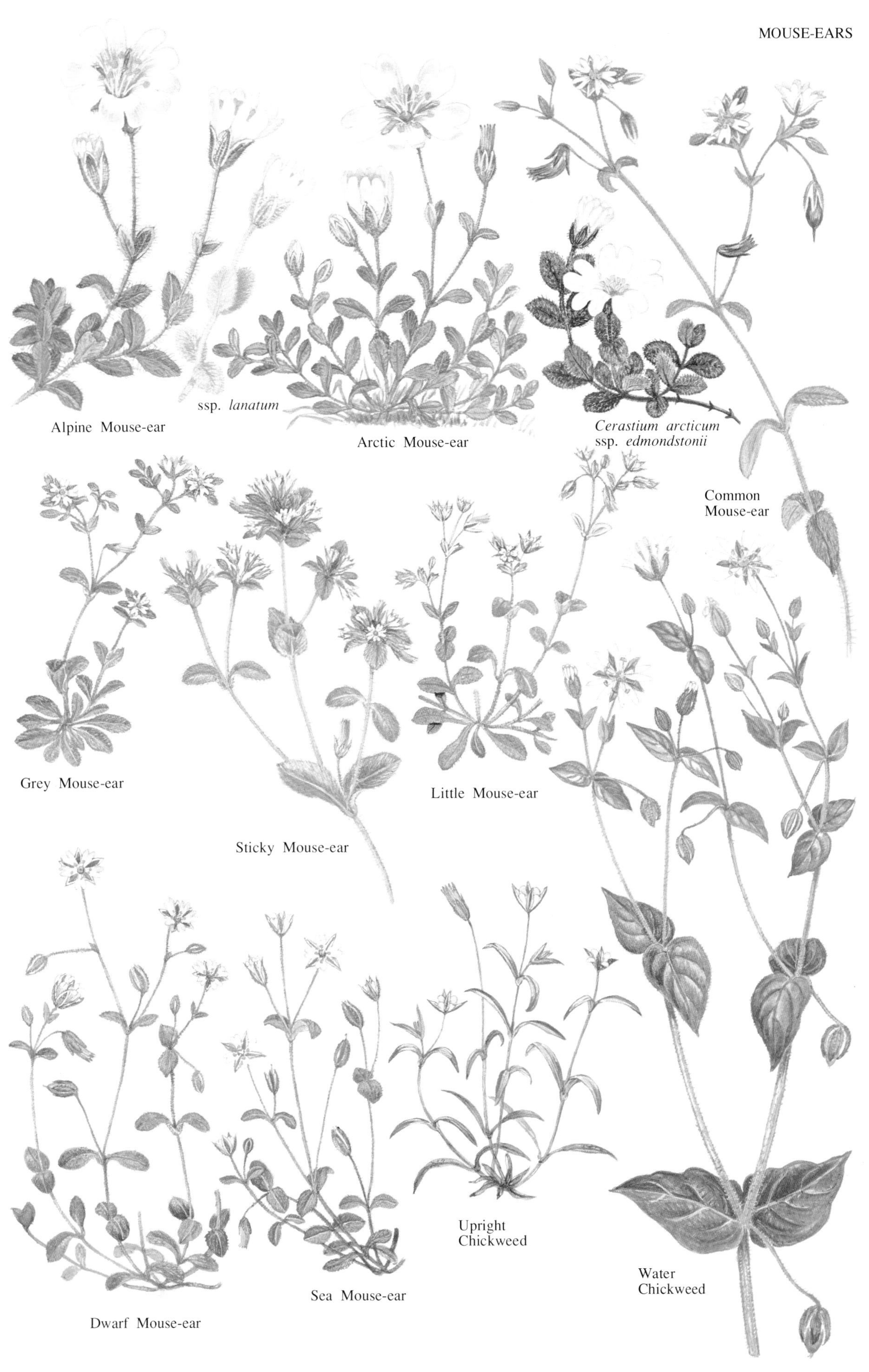
Alpine Mouse-ear
ssp. *lanatum*
Arctic Mouse-ear
*Cerastium arcticum*
ssp. *edmondstonii*
Common
Mouse-ear
Grey Mouse-ear
Sticky Mouse-ear
Little Mouse-ear
Upright
Chickweed
Water
Chickweed
Sea Mouse-ear
Dwarf Mouse-ear

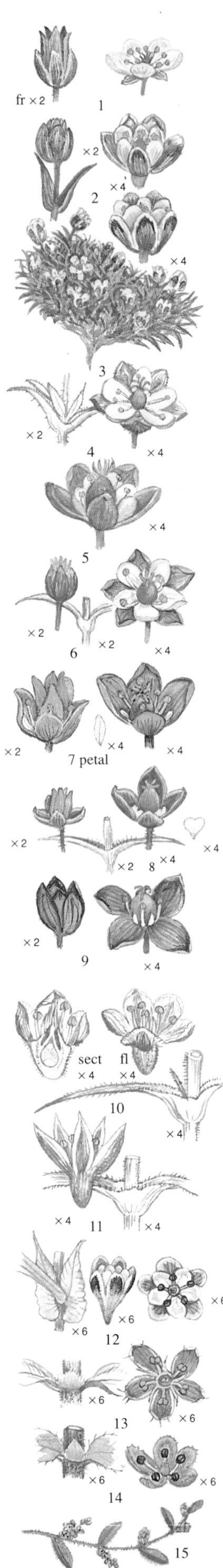


**Pearlworts** *Sagina*. Small annual or perennial herbs, with slender flowering stems. Leaves opposite, linear, fused together at the base. Flowers 4-5-parted, in lax, few-flowered cymes, or solitary; stamens 4,5,8, or 10; styles 4-5. Fruit capsule splitting to base into 4-5 valves.

1* **Knotted Pearlwort** *Sagina nodosa* (L.) Fenzl. Short, tufted perennial, with a central leaf-rosette giving rise to lateral flowering stems. Leaves pointed, with small leafy tufts in the angles. Flowers 1-3, 5-10mm, petals two or three times the sepals, 5-parted. Damp habitats, short wet turf, to 2000m. July-September. Throughout. **B**: Local throughout. var. *moniliformis* (G.F.W. Meyer) Lange is not prostrate and has bulbil-like growths at the leaf-angles. Similar habitats, rarely flowering. N Europe.

2* **Snow Pearlwort** *Sagina nivalis* (Lindblad) Fries [*S. intermedia*]. Low compact, tufted perennial. Basal leaf-rosette only present during first year, otherwise leaves linear with a short point, hairless. Flowers solitary, 3-6mm, petals narrower, slightly shorter than the violet-edged sepals, 4-5-parted; stamens 8-10. Fruit with erect sepals. Broken rocky ground and fine screes, to 1700m. June-September. Arctic and sub-Arctic Europe, southwards to N Britain. Local. **B**: A few localities in the Scottish Highlands, very rare.

3 *Sagina caespitosa* (J. Vahl) Lange. Like *S. nivalis*, but tufts with persistent dead leaves. Flowers 1-2 together, held almost on the leaf cushion; petals longer than the sepals. Mountains, damp rocky places, gravelly ground, often by snow patches, to 1670m. Norway & NW Sweden, Spitsbergen and Iceland.

4* **Heath Pearlwort** *Sagina subulata* (Swartz) C. Presl. Low, slightly hairy, mat-forming perennial with a central leaf rosette and lateral ascending flowering stems. Leaves long, pointed. Flowers usually solitary, 4-5mm, 5-parted, usually with stickily-hairy stalks, petals equalling sepals; stamens 10. Dry sandy or gravelly habitats, heathland. May-August. Throughout, except Finland and Spitsbergen. **B**: Local in Scotland, W & S Britain; N & W coastal Ireland.

5* **Alpine Pearlwort** *Sagina saginoides* (L.) Karsten (*S. linnaei* C. Presl nom illegit). Low laxly tufted, hairless perennial. Leaves bristle-tipped, rosetted. Flowers solitary or 2, borne directly from leaf-rosettes, petals equalling sepals. Fruit on recurved stalks. Damp rocky places, to 2750m. June-September. Arctic and sub-Arctic Europe, south to the Alps. **B**: Rare, confined to the Scottish Highlands and Skye.

6 *Sagina* × *normaniana* Lagerh. Natural hybrid and intermediate between *S. procumbens* and *S. saginoides*, rarely producing seeds. Seeds are rarely produced. Very local and often growing close to the parent species. N Britain and Scandinavia. **B**: Rare, Scottish Highlands.

7* **Procumbent Pearlwort** *Sagina procumbens* L. Low hairless, mat-forming perennial, spreading outwards from central leaf-rosettes, rooting at the nodes. Leaves bristle-tipped. Flowers solitary, 4-parted, 2-4mm, petals minute or absent, fruit with spreading sepals. Damp, generally bare, often shaded places, walls, to 2600m. May-September. Throughout, except Spitsbergen. **B**: Throughout.

8* **Annual Pearlwort** *Sagina apetala* Ard. Variable low, lax, annual, with ascending stems. Leaves long-pointed, not forming a rosette. Flowers solitary, 4-parted, 2-4mm, on stickily-hairy or hairless stalks, petals minute, soon falling. Dry sandy and grassy habitats and walls, to 2200m. April-August. Throughout, except the far north. **B**: Throughout, though rare in N Scotland.

9 **Sea Pearlwort** *Sagina maritima* G. Don. Variable low, dark green, tufted annual, usually hairless. Leaves linear-lanceolate, rather fleshy and blunt. Flowers 4-6mm, 4-parted, petals minute or absent. Fruit with erect purple-edged hooded sepals. Bare open habitats, often coastal, occasionally on mountains inland, to 1500m. May-September. Throughout, except Germany and the far north. **B**: Local, coastal, rare; E Highlands of Scotland.

10* **Perennial Knawel** *Scleranthus perennis*. L. Low to short rather stout, erect perennial, woody at the base. Flower clusters with short bracts, sepals blunt with wide membranous margins. Fruit 2-6.5mm. Dry and waste places, on thin, usually sandy soils, also on cliffs, to 2250m. May-October. Throughout, except the far north. **B**: Very rare – now a solitary locality in Wales. subsp. *prostratus* P.D. Sell. A prostrate plant with smaller leaves, 3-5mm (not 3-9mm). Endemic to E England – Suffolk, on Breckland soils.

11* **Annual Knawel** *Scleranthus annuus* L. Variable low, spiky-looking annual or biennial with ascending stems. Flower clusters exceeded by bracts, sepals pointed, with narrow membranous margins. Fruit 3.2.-5.5mm. Dry sandy, grassy and gravelly habitats, on open acid soils. May-October. Throughout, except the Faeroes, Iceland and Spitsbergen. **B**: Throughout, though scarce in NW. subsp. *polycarpos* (L.) Thell. has smaller leaves and fruits only 2.2-3mm. Habitat and distribution similar.

12* **Strapwort** *Corrigiola litoralis* L. Low, spreading, greyish, annual, often with reddish stems. Leaves alternate (unique in the Family), narrow-oblong, blunt; stipules small, membranous. Flowers minute, 5-parted, white, in leaf-opposed clusters, sepals often reddish, slightly exceeding the petals. Seasonally wet habitats, on sandy or gravelly soils. June-October. W Europe. Local, populations fluctuating from year to year. **B**: Rare; S Devon shoreline and a few places inland.

**Ruptureworts** *Herniaria*. Herbs, often woody at base. Leaves opposite, or apparently alternate; stipules membranous, conspicuous. Flowers small 4-5-parted; petals minute or absent. Stamens 5. Styles 2. Fruit an achene in a membranous pericarp.

13* **Glabrous Rupturewort** *Herniaria glabra* L. Variable low, bright green, prostrate annual to perennial, slightly hairy. Leaves elliptical. Flowers greenish, 1mm, 5-parted, in crowded clusters on leafy lateral branches. Dry bare habitats, excavations, on coarse gravelly or sandy soils. May-October. Throughout, except Ireland and the far north. **B**: Lincolnshire and East Anglia; rare.

14 **Ciliate Rupturewort** *Herniaria ciliolata* Melderis [*H. ciliata*]. Low dwarf, woody perennial subshrub; stems hairy along one side. Leaves rounded to elliptical, hairy on margin. Flowers 1.5-2mm, the clusters not crowded. Rocky and grassy coastal habitats. June-August. France and SW Britain. **B**: Very rare; Lizard Peninsular and the Channel Islands (latter called subsp. *subciliata* (Bab.) Chaudhri [*H. glabra*]).

15 **Hairy Rupturewort** *Herniaria hirsuta* L. Low prostrate, hairy annual, with regularly alternating branches. Leaves elliptical, stiffly-hairy; stipules fringed. Flowers tiny, 1-1.5mm, in dense leaf-opposed clusters, sepals very hairy. Sandy, coastal rocks. June-August. W Europe. **B**: Casual in S England.

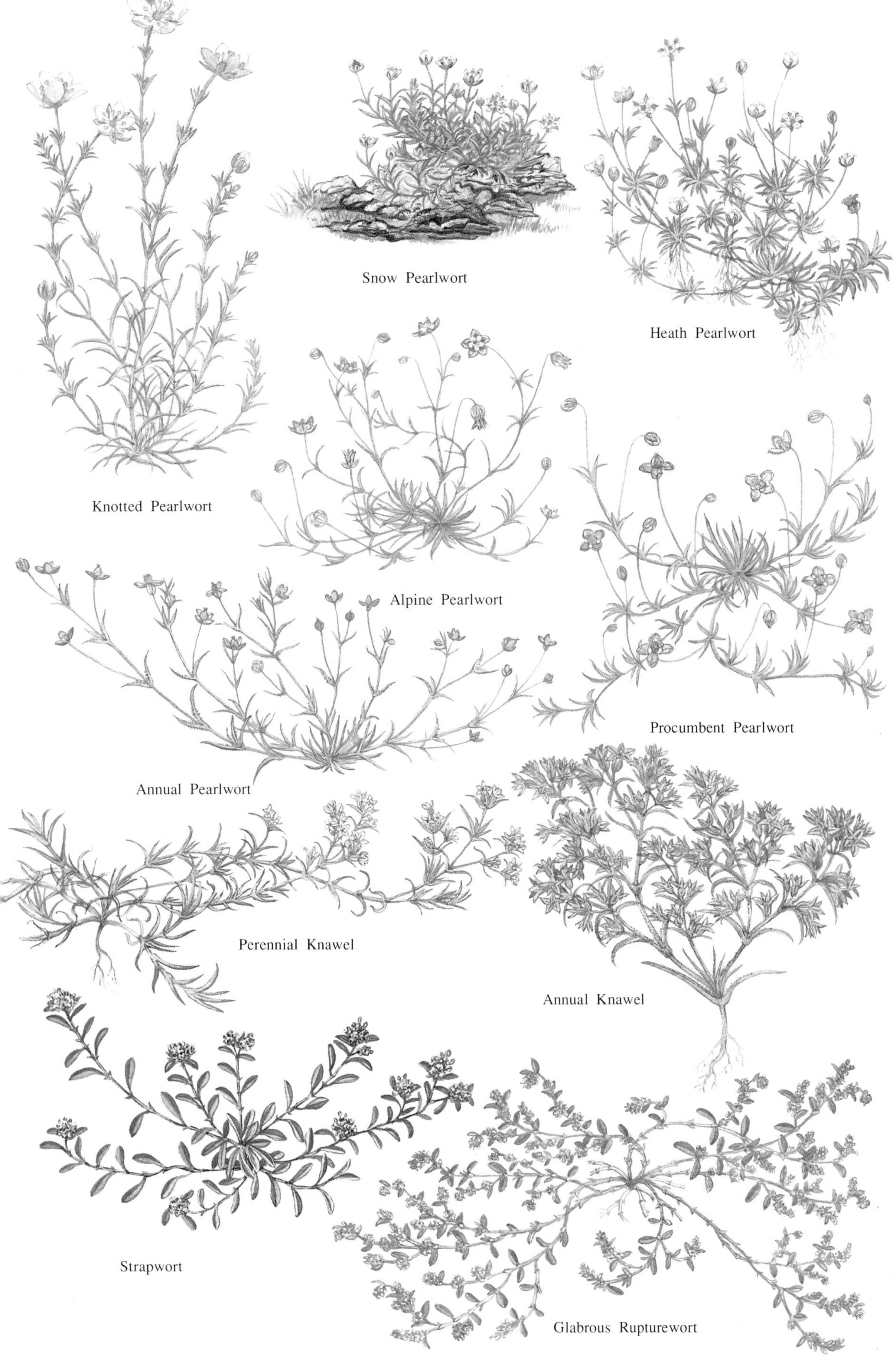

Snow Pearlwort

Heath Pearlwort

Knotted Pearlwort

Alpine Pearlwort

Procumbent Pearlwort

Annual Pearlwort

Perennial Knawel

Annual Knawel

Strapwort

Glabrous Rupturewort

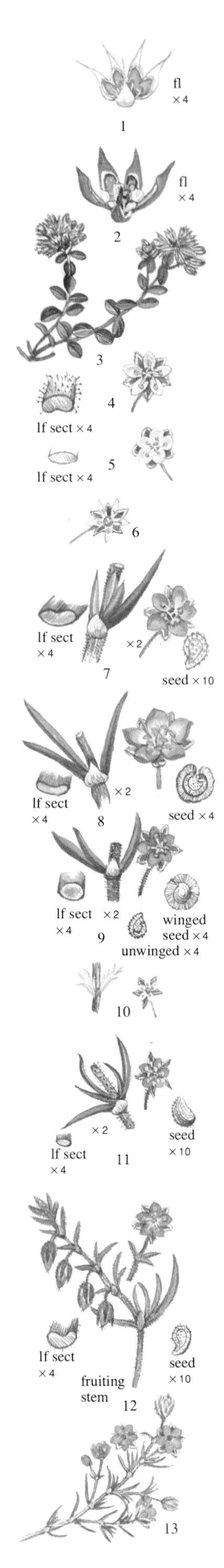


1* **Coral Necklace** *Illecebrum verticillatum* L. Variable, low, prostrate hairless annual, with slender square stems rooting near the base. Leaves opposite, narrow-spatular-shaped to oblong, blunt; stipules small. Flowers in small dense clusters at the leaf-bases, very white, with spongy persistent sepals and silvery bracts. Seasonally damp, sandy or gravelly habitats, on non-calcareous soils. June-October. W Europe. Local. **B**: Rare, restricted to Cornwall and Hampshire – New Forest.

2* **Four-leaved Allseed** *Polycarpon tetraphyllum* (L.) L. [*Mollugo tetraphylla*]. Low hairless annual, rarely perennial, with much-branched stems and conspicuous silvery stipules and bracts. Leaves green, mostly in fours, oval. Flowers tiny, whitish, in lax, much-branched clusters. Sandy or rocky habitats, sand-dunes, bulb fields, bare ground, occasionally on walls, usually inland. June-August. SW Britain, France and Germany. **B**: Restricted to Devon, Cornwall and the Channel Islands.

3 **Two-leaved Allseed** *Polycarpon diphyllum* Cav. Low to short unbranched annual. Leaves mostly opposite, always purple-tinged. Flowers whitish, in dense, few-flowered, clusters; petals shorter than the sepals, not notched. Sandy habitats, mostly coastal. June-August. France; naturalized in Britain, although the status of the British plants is uncertain.

**Spurreys** *Spergula*. Annual herbs, much-branched at the base. Leaves linear, blunt, in pairs, but with leafy tufts in the axils giving the leaves a whorled appearance; stipules membranous, not united around each node. Flowers 5-parted, in lax cymose clusters; petals white, not notched; stamens 5-10; styles 5. Fruit capsule splitting by 5 valves. 5 species in northern temperate regions.

4* **Corn Spurrey** *Spergula arvensis* L. [*S. vulgaris, S. sativa, S. maxima, S. linicola*]. Variable short to medium straggly annual; stems stickily-hairy above. Leaves linear, fleshy, channelled beneath. Flowers white, 4-8mm, the petals slightly longer than the sepals. Weed of arable and disturbed land, road verges, also along seashores, on sandy acid soils, at low altitudes. May-September. Throughout, except Spitsbergen. **P**: The rather unpleasant smell of the flowers, which open in the afternoon, attracts various insects, but despite this they are generally self-pollinated. An often persistent weed. **B**: Throughout.

5* **Pearlwort Spurrey** *Spergula morisonii* Boreau. Short stiffly erect annual, sometimes slightly hairy. Leaves linear, not channelled beneath. Flowers white, 6-8mm, petals overlapping, equalling sepals. Bare sandy habitats. April-June. Europe north to Central Sweden. **B**: Naturalized and rare.

6 *Spergula pentandra* L. Like *S. morisonii* has narrower non-overlapping petals; stamens usually 5 (instead of 5-10). May-August. Belgium, France and Germany.

**Sand-spurreys** *Spergularia*. Small herbs, sometimes woody at the base. Leaves opposite, linear, with leaf fascicles on one side only at each node; stipules partly united around each node. Flowers 5-parted, petals not notched; stamens 1-10; styles 3. Fruit capsule splitting by 3 valves. 20 species in many parts of the world, mostly favouring salty habitats.

7* **Cliff Sand-spurrey** *Spergularia rupicola* Lebel ex Le Jolis [*S. lebeliana*]. Low to short, rather robust, stickily-hairy perennial with a woody stock. Leaves narrow-linear, pointed; stipules somewhat silvery. Flowers pink, 8-10mm, petals equalling sepals; stamens 10. Coastal rocks and cliffs, short turf and screes, at low altitudes. May-September. Atlantic coasts of Britain and France. Local. **B**: W & S coasts of Britain, east to the Isle of Wight.

8* **Greater Sand-spurrey** *Spergularia media* (L.) C. Presl [*S. marginata*]. Low to short slender, almost hairless perennial. Leaves linear, awned. Flowers white or pale bluish-pink, 7-12mm, the petals slightly longer than sepals; stamens 10. Saline habitats, salt marshes, sandy or gravelly shores, sometimes inland, at low altitudes. May-September. Coastal Europe, except the far north. Occasionally growing along road verges inland from heavy use of salt during the winter months causing local saline conditions. **B**: Throughout, coastal.

9* **Lesser Sand-spurrey** *Spergularia marina* (L.) Griseb. Less robust than *S. media*, generally annual or biennial. Stipules short, sheath-like. Flowers pink, white in the centre, 5-8mm, the petals shorter than sepals; stamens 1-5 usually. Coastal habitats, drier parts of salt marshes, seashores, sometimes on saline soils inland. May-September. Coastal Europe, except the Faeroes and Spitsbergen. Less common than *S. media*. **B**: Local, coastal; inland in Cheshire and Worcestershire.

10* *Spergularia segetalis* (L.) G. Don fil. [*Alsine segetalis*]. Low slender annual with ascending stems. Leaves linear, awned; stipules silvery, fringed at apex. Flowers white, tiny, 2-3mm; petals shorter than the sepals. Arable fields and disturbed ground. May-July. Belgium, Holland, France and Germany.

11* **Sand-spurrey** *Spergularia rubra* (L.) J. & C. Presl [S. campestris]. Low to short stickily-hairy annual or perennial; stems spreading. Leaves linear, bristle-tipped, the upper markedly tufted; stipules lanceolate, silvery. Flowers uniformly pale pink 3-6mm; stamens usually 10. Heathland, commons, disturbed ground, cliff tops and sand-dunes, usually on dry sandy, non-saline soils. June-September. Throughout, except the far north. **B**: Throughout, though rare in both Scotland and Ireland.

12 **Boccone's Sand-spurrey** *Spergularia bocconii* (Scheele) Ascherson & Graebner. Low to short, densely stickily-hairy annual or biennial. Leaves linear, the upper not tufted; stipules not silvery. Flowers white or pink with a white centre, 5-6mm, the petals shorter than, or equalling, the sepals. Coastal sands and rocks, waste places, dry sandy and gravelly paths. May-September. S Britain, Holland and Germany. Primarily SW and Mediterranean Europe. **B**: Very rare, restricted to Cornwall, Guernsey and Jersey.

13 *Spergularia echinosperma* Celak. Low annual. Leaves linear, slightly tufted above. Flowers few, pale pink, 4-6mm, petals shorter than or equalling sepals; stamens 2-5. Sandy and rocky habitats, disturbed ground, primarily along the margins of streams and lakes. May-August. France and Germany.

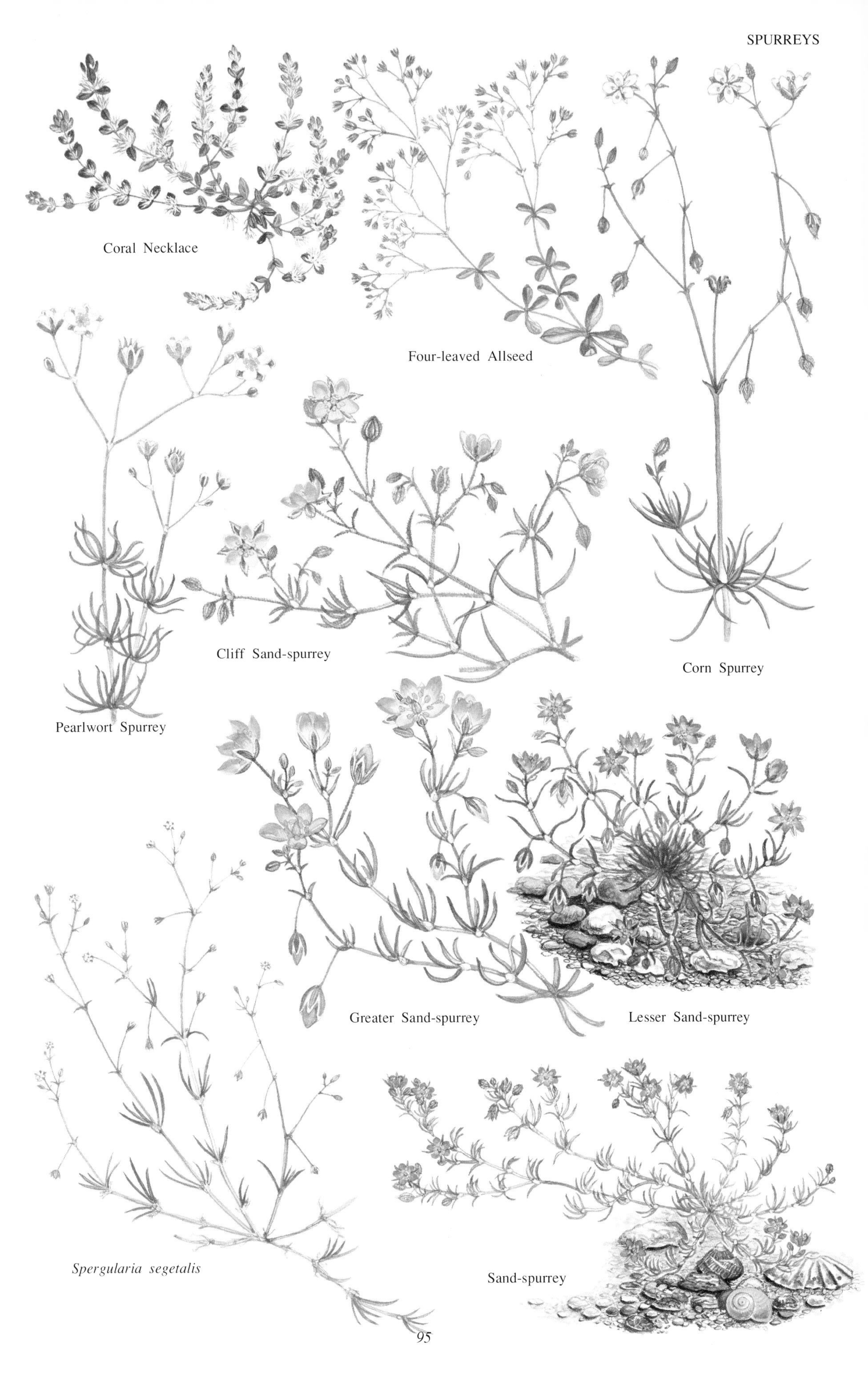
Coral Necklace
Four-leaved Allseed
Corn Spurrey
Pearlwort Spurrey
Cliff Sand-spurrey
Greater Sand-spurrey
Lesser Sand-spurrey
*Spergularia segetalis*
Sand-spurrey

**Lychnis** *Lychnis*. Erect, often tufted perennials. Flowers 5-parted, the calyx tube 10-veined, with 5 short teeth. Petals with a distinct claw, as in *Silene*, usually with scales at the base of the limb. Stamens 10. Styles 5. Fruit capsule splitting with 5 teeth. 15 species in the northern hemisphere. Several are cultivated in gardens.

1* **Ragged Robin** *Lychnis flos-cuculi* L. [*Coronaria flos-cuculi*] Medium to tall perennial, rough-hairy, often branched. Basal leaves oblong to spoon-shaped, stalked; stem leaves linear-lanceolate. Flowers pale to bright purplish-pink, occasionally white, 30-40mm, the petals cut into four narrow pointed segments. Wet meadows, marshes, fens, wet woodland and other damp habitats, on peaty or mineral-rich soils, to 2500m. May-August. Throughout, except Spitsbergen. **P**: Various butterflies, long-tongued bees and flies. **B**: Widespread.

2* **Sticky Catchfly** *Lychnis viscaria* L. [*Viscaria viscosa, V. vulgaris*]. Variable medium, tufted perennial, hairless or slightly hairy, sticky below each upper node. Leaves mostly basal, lanceolate, stalked. Flowers bright rosy-purple, rarely white, 18- 22mm, in apparent whorls forming a narrow panicle, petals notched or not. Dry rocky habitats, cliffs and rock debris, on acid soils usually, to 1800m. May-August. Throughout, except Ireland and the extreme north. **P**: Butterflies and long-tongued bees. **B**: Rare, confined to N Wales and Scotland.

3* **Alpine Catchfly** *Lychnis alpina* L. [*Viscaria alpina*]. Short, tufted, hairless perennial; stem simple, not sticky. Leaves mostly basal, linear to spoon-shaped. Flowers pale purple, rarely white, 8-12mm, in a small dense head, petals deeply notched; smaller flowers often only female. Mountain meadows, rocky and stony places, often on mineral rich soils, to 3100m. June-August. N Britain, Scandinavia, C & E France, Iceland. Often local. Frequently depleted by over-collecting. **P**: Butterflies and long-tongued bees, or self-pollinated. **B**: Lake Distrtict and Angus, very rare. Widely cultivated in gardens.

4* **Corncockle** *Agrostemma githago* L. [*Lychnis githago*]. Medium to tall greyish, hairy, erect annual. Leaves narrow-lanceolate, pointed. Flowers dull purple, occasionally white, 30-50mm, with long leaf-like sepals much exceeding the petals, borne on long slender stalks. A weed of cornfields, occasionally in waste places, to 2000m. May-August. **I**: Mediterranean. Naturalized throughout except Spitsbergen. **P**: Butterflies or self-pollinated. Greatly declined in recent years due to cleaner agricultural seed and the use of herbicides. The seeds are poisonous. **B**: Local in East Anglia and Morayshire.

**Catchflies & Campions** *Silene*. Annual or perennial herbs. Flowers 5-parted. Calyx tube 10-30-veined, with 5 short teeth. Petals with a distinct claw, often with scales at the base of the limb. Stamens 10. Styles 3, rarely 5. Fruit capsule splitting by 6 teeth. 300 species widespread in temperate regions of Europe and Asia, Africa and North America. Several have become widespread weeds of cultivated land.

5* **Italian Catchfly** *Silene italica* (L.) Pers. Medium to tall, stickily-hairy perennial with branched stems. Leaves lanceolate to oblong with wavy-margins. Flowers creamy-white, greenish or reddish underneath, erect, petals deeply cleft; calyx 14-21mm, stickily-hairy. Dry grassy habitats, banks, road verges, quarries and cliffs. June-July. France and Germany; naturalized in Britain. The flowers open in the evening. **P**: Moths. **B**: Rare, a few localities in England and Scotland, especially in Kent.

6* **Nottingham Catchfly** *Silene nutans* L. Very variable, medium, hairy perennial; stems sticky above, usually unbranched. Leaves oblong, broadest above the middle, the upper linear-lanceolate. Flowers whitish, greenish or reddish beneath, drooping in a lax one-side panicle; petals deeply cleft, the lobes rolled inwards; calyx 9-12mm, stickily-hairy. Dry disturbed habitats, rocky places, shingle, grassland and field boundaries, often on calcareous soils. May-August. Throughout Europe, except the extreme north. The flower open fully and are fragrant during the evening, the petals being curled up during the day. **P**: Moths and bees. **B**: Rare and local, throughout.

7* **White Sticky Catchfly** *Silene viscosa* (L.) Pers. [*Melandrium viscosum*]. Medium, robust, stickily-hairy biennial, occasionally perennial. Lower leaves oval to lanceolate, pointed, undulate. Flowers large, white, 18-22mm, in whorls forming an erect, short-branched panicle; petals without scales. Dry grassy habitats, banks, road verges and waysides. June-July. Denmark, Germany, S Sweden and Finland.

8 *Silene tatarica* (L.) Pers. Medium, almost hairless, perennial, often with short non-flowering shoots. Leaves lanceolate to oblong. Flowers white, cream or greenish-white 14-18mm, inclined; petals deeply cleft with narrow lobes but no scales. Grassy and rocky habitats. July-August. N Germany, Finland and S Norway. **P**: Moths and bees.

9 **Northern Catchfly** *Silene wahlbergella* Chowdhuri [*Lychnis apetala, Melandrium apetalum*]. Short, unbranched, slightly hairy perennial. Leaves narrow oval. Flowers solitary, nodding at first, with a greenish-white inflated calyx, 14-18mm long; petals purplish, inconspicuous. Mountain marshes and bare places, on calcareous soils, to 1900m. June-August. Arctic Scandinavia. Small and rather inconspicuous. **P**: Mainly bees.

10 *Silene furcata* Rafin [*Lychnis affinis*]. Like *S. wahlbergella*, but stems often branched, stickily-hairy and flowers smaller and erect; calyx scarcely inflated, 10-12mm; petals whitish. Arctic habitats. June-August. Scandinavia.

11* **Spanish Catchfly** *Silene otites* (L.) Wibel. Variable, medium, erect biennial or short-lived perennial, stickily-hairy towards the base. Leaves elliptical, the lowermost stalked, in lax rosettes. Flower small, greenish-yellow, 3-4mm, many in lax-branched clusters; petals linear. Grassland, banks, heaths, road verges, on dry calcareous or sandy soils, to 2000m. June-September. E Britain, Holland, France, Denmark and Germany; naturalized in Belgium and Finland. Widespread in C, S and E Europe. **P**: Butterflies and wind. Often dependent on disturbed ground for seed germination. **B**: Local, confined to East Anglian – Breckland.

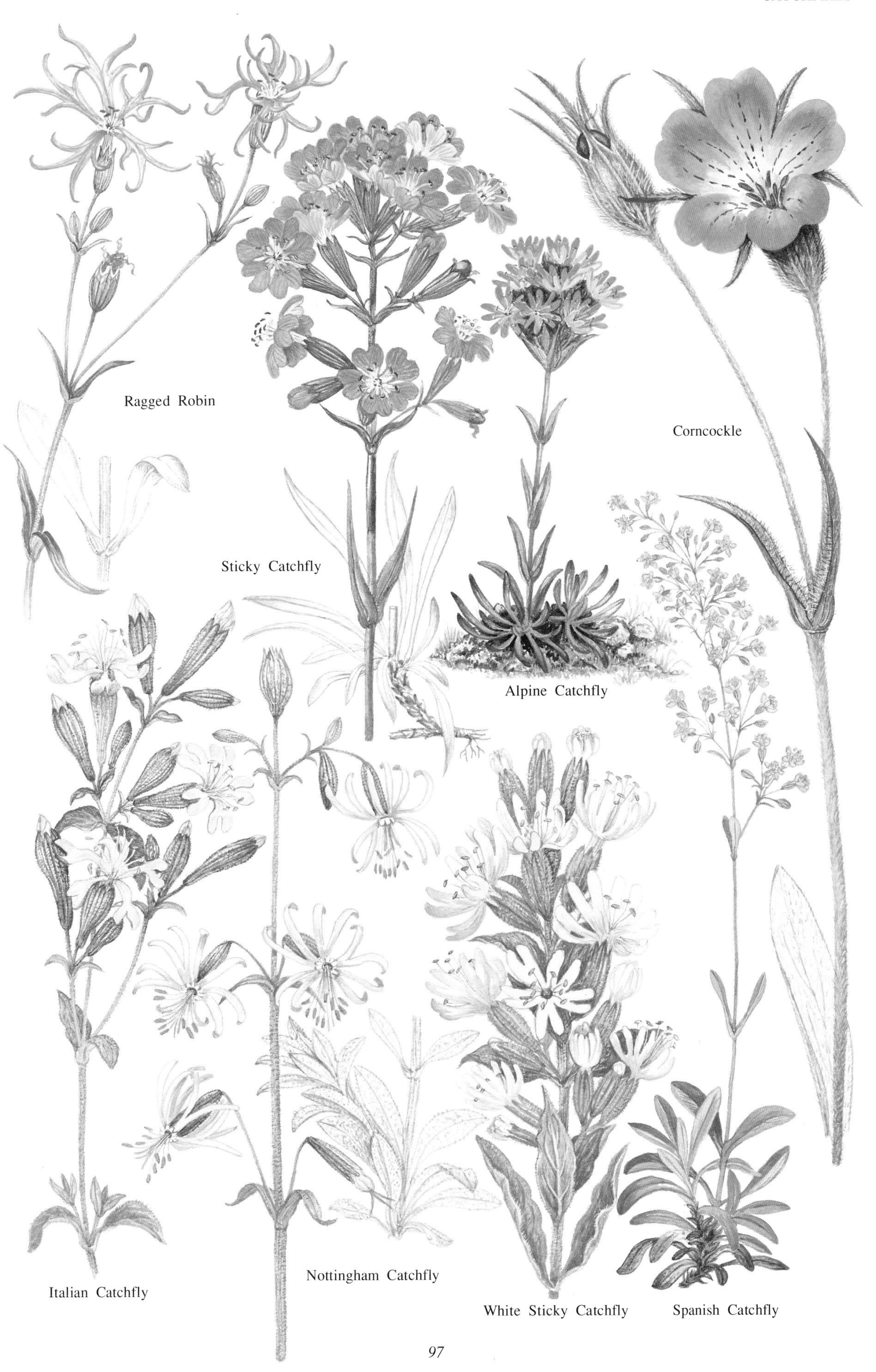
Ragged Robin
Corncockle
Sticky Catchfly
Alpine Catchfly
Italian Catchfly
Nottingham Catchfly
White Sticky Catchfly
Spanish Catchfly

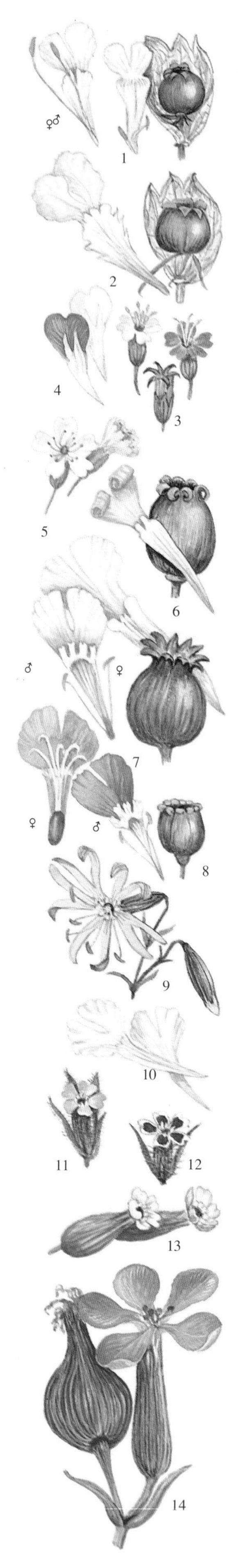


1* **Bladder Campion** *Silene vulgaris* (Moench) Garcke. Medium to tall, generally hairless, greyish perennial with a woody base and stout, erect stems; all shoots flowering. Leaves oval to elliptical, the lowermost stalked, the upper unstalked, sometimes with a hairy margin. Flowers white, 16-18mm, fragrant, petals deeply notched, the petals not overlapping, styles 3; calyx inflated and bladder-like; constricted at the mouth. Seed capsule with 6 erect teeth. Grassy and waste places, preferring rather dry calcareous soils. May-September. Throughout except for some northerly islands. **B**: Often absent in N and W Scotland.

2* **Sea Campion** *Silene uniflora* Roth [*S. maritima*, *S. vulgaris* subsp. *maritima*]. Low to medium, generally hairless, greyish perennial, forming tufts or mats with a mixture of flowering and non-flowering shoots. Leaves elliptical to lanceolate, mostly unstalked, rather fleshy. Flowers white, 20-25mm, few to an inflorescence, petals deeply notched and overlapping one another, styles 3; calyx inflated and bladder-like. Fruit capsule with 6 erect teeth. Sea cliffs, rocks and shingle, sometimes inland, to 1000m. June-August. W & NW Europfe. Often locally abundant. **B**: Throughout; coastal, occasionally inland.

3* **Moss Campion** *Silene acaulis* (L.) Jacq. Low, hairless cushion or mat-forming perennial, bright green and moss-like. Leaves small, linear, pointed. Flowers pale to deep pink, 6-10mm, solitary on short stalks, the petals notched, stamens protruding ; calyx almost bell-shaped, generally flushed with purple or red; usually hermaphrodite.. Damp rocks and short turf, to 3700m. June-August. Throughout, except Belgium, Holland and Denmark, but often local and only on the mountains in the south. **B**: Scattered localities in the Scottish Highlands, Lake District, Snowdonia and in W Ireland (Sligo and Donegal).

4* **Sweet William Catchfly** *Silene armeria* L. A short, erect, hairless annual or biennial, the stems sticky above, especially at the nodes. Leaves lanceolate, often greyish, the lower stalked, the upper unstalked and clasping the stem. Flowers bright pink, 14-16mm, in rather flat-topped clusters, the petals shallowly notched; calyx reddish, 10-veined. Dry, generally shaded places, often a garden escape. June-September. France, Germany, S Europe, naturalized elsewhere. **B**: Scattered localities in S England.

5. **Rock Campion** *Silene rupestris* L. A short erect, hairless, grey-tufted perennial. Basal leaves spoon-shaped, stalked; stem leaves lanceolate, mostly unstalked. Flowers white or pink, 7-9mm, in lax, branched clusters, the petals deeply notched; calyx hairless. Seed capsule with 6 teeth. Poor dry pastures, screes, generally on acid rocks, to 2900m. June-September. C & E France, S Germany, Norway and Sweden.

6* **Night-flowering Catchfly** *Silene noctiflora* L. [*Melandrium noctiflorum*]. Short to medium hairy annual, sticky with glands in the upper half. Leaves oval to lanceolate, the lower stalked, the upper generally narrower and unstalked. Flowers pink, yellowish beneath, 17-19mm, few to a cluster, the petals deeply notched, inrolled during the day, but expanded and strongly fragrant by evening; styles 3. Seed capsule with 6 recurved teeth. Arable fields and dry waste places, often on sandy soils. June-August.C & S Britain, Belgium, France and Germany; naturalized in Ireland, Iceland and parts of Scandinavia. **B**: Local and decreasing in C & S England from the Tees southwards, rare elsewhere including SW England.

7 **White Campion** *Silene latifolia* Poiret subsp. *alba* (Miller) Greuter & Burdet [*S. alba*, *Melandrium album*]. Variable medium to tall, stickily-hairy, short-lived perennial, sometimes annual. Leaves oval to lanceolate, the lowermost stalked, the upper usually unstalked. Flowers white, 25-30mm, in lax, branched clusters, the petals deeply notched; male and female flowers on separate plants; calyces slightly inflated, those of male flowers smaller and 10-veined, those of the female larger and 20-veined; styles 5; expanded and fragrant by evening. Seed capsule with 10 erect teeth. Arable and waste land. and waysides, especially on dry calcareous soils. May-October. Throughout, except for the far north. Common. **B**: Throughout, but scarce in the west and local in Ireland.

8* **Red Campion** *Silene dioica* (L.) Clairv. [*Melandrium dioicum*]. Medium to tall, hairy but not sticky, biennial or perennial, with a creeping stock and erect flowering stems. Leaves oblong, the lowermost stalked, the upper short-stalked or unstalked. Flowers bright rose-pink, 18-25mm, not scented, numerous, in lax, branched clusters, styles 5; monoecious, male calyces 10-veined, female 20-veined. Seed capsule with 10 recurved teeth. Deciduous woodland and hedgerows, sometimes on rocky slopes, generally on calcareous or base-rich soils, to 2400m. May-November. Throughout. Hybridizes with *S. latifolia* subsp. *alba*; hybrids are fully fertile and vigorous with pale or mid-pink flowers. In some places they may completely replace the parent species. **B**: Throughout, but scarcer in E Anglia, N Scotland and Ireland.

9 **Flaxfield Catchfly** *Silene linicola* C.C. Gmelin. Medium, rough-stemmed, branched annual. Leaves narrowly lanceolate to linear, the lowermost stalked. Flowers pink, with reddish-purple veins, 14-16mm, in lax, flat-topped clusters, the petals deeply notched; styles 3. Seed capsule with 6 teeth. Flaxfields, local. June-July. France and Germany. Declined from drop in both flax-growing and seed impurities.

10* **Forked Catchfly** *Silene dichotoma* Ehrh. Medium to tall erect, hairy annual, the stems branched above, not sticky. Leaves lanceolate, the lower stalked, the upper generally narrower and unstalked. Flowers white, rarely pale pink, 15-18mm, half-nodding, in characteristically forked, one-sided, spike-like racemes, petals deeply notched; styles 3; often only fully open in the evenings or dull weather. Seed capsule with 6 teeth. Arable, waste and bare places. May-August. **I**: E Europe and W Asia. Naturalized in France and Germany, casual elsewhere. Imported with grain. **B**: Casual, particularly in the south.

11* **Small-flowered Catchfly** *Silene gallica* L. [*S. anglica*]. Variable short to medium, stickily-hairy annual, often overwintering, with erect branched stems. Leaves lanceolate to almost linear, the lower stalked, the upper unstalked. Flowers white or pink, 6-10mm, short-stalked, in one-sided spike-like racemes, the petals slightly notched or unnotched, styles 3; calyx stickily-hairy, 10-veined. Arable land, waste places, on sandy or gravelly soils. June-October. W Europe; sometimes casual elsewhere. var. *quinquevulnera* (L.) Koch (12), with a crimson spot on each petal, W France and Channel Islands, is casual in S England and near Edinburgh. **B**: Local in S England, E Anglia and S Wales, rare in Ireland.

13* **Sand or Striated Catchfly** *Silene conica* L. Short, stickily-hairy, rather greyish annual. Leaves lanceolate, the lowermost stalked, the upper unstalked. Flowers rose-pink or whitish, 4-5mm, solitary or few to a cluster, petals slightly notched, styles 3; calyx flask- shaped, strongly ribbed, greatly inflated in fruit. Sandy and waste places, often on calcareous soils, particularly near the coast. May-August. S Britain, France, Holland and Germany; naturalized in Denmark. Often introduced, but not always persisting for long. **B**: Local along E & S coasts of England, E Anglia and Channel Is; casual further north.

14 *Silene conoidea* L. Like *S. conica*, butrather taller plant, less hairy. Flowers larger, pink, 12-18mm; calyx 15-25mm long in fruit (not 10-15mm). Casual on waste or sandy ground near the coast. June-August. **I**: Southern. Local, coasts of W France.

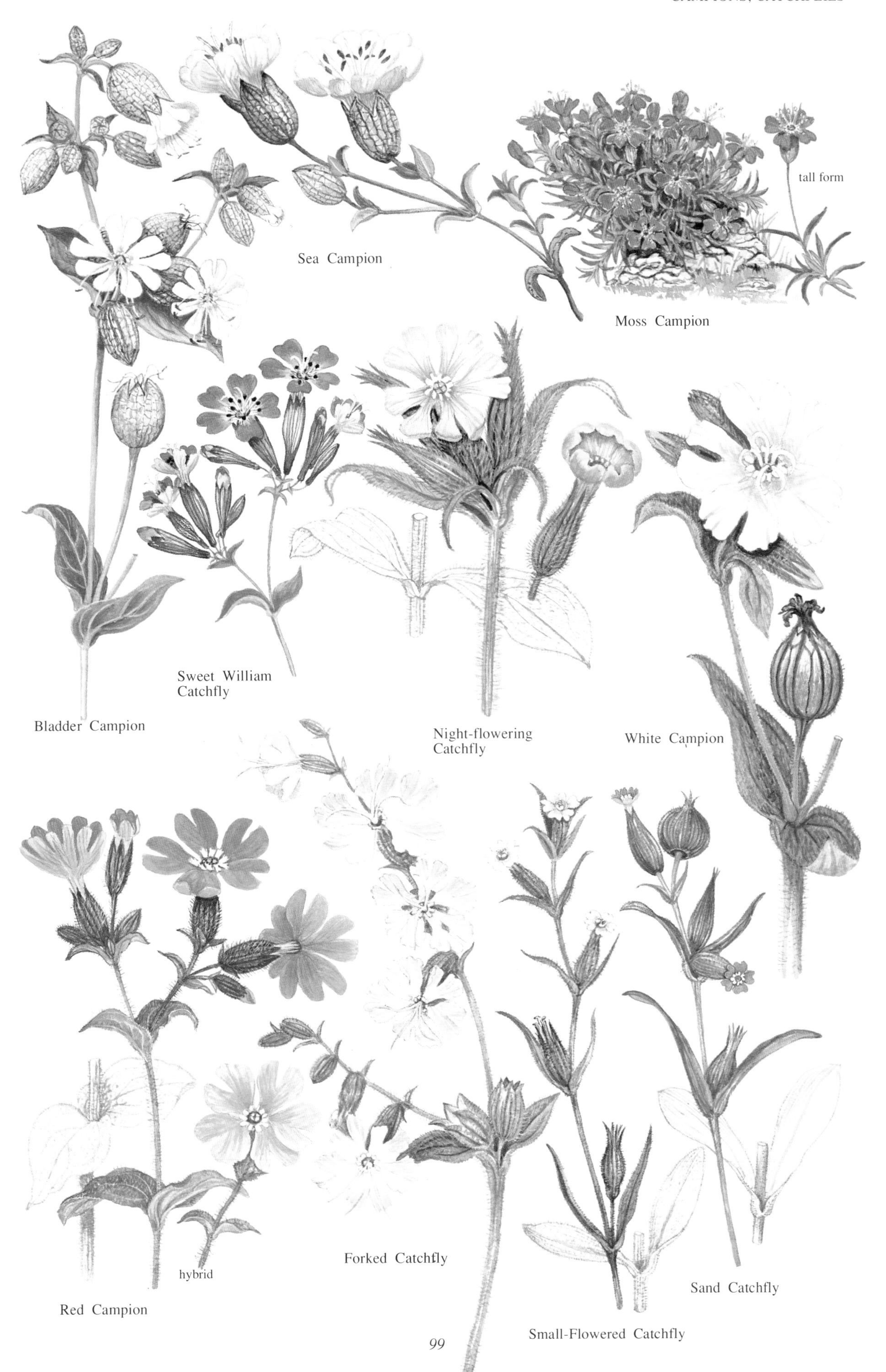
Sea Campion
Moss Campion
tall form
Bladder Campion
Sweet William Catchfly
Night-flowering Catchfly
White Campion
Red Campion
hybrid
Forked Catchfly
Small-Flowered Catchfly
Sand Catchfly

1* **Berry Catchfly** *Cucubalus baccifer* L. Medium to tall, weak-stemmed, hairy perennial herb, with brittle branches. Leaves oval, pointed, untoothed, with short stalks. Flowers greenish-white, 16-18mm, drooping in lax leafy clusters; petals deeply notched. Fruit black, berry-like. Shaded habitats, woodland and scrub. July-September. Belgium, Holland, France and Germany; naturalized in S Britain. The flowers look very similar to those of various species of *Silene* but the berry-like fruit quickly distinguishes the genus *Cucubalus*. Widely distributed from Britain to Asia and Japan. **B**: Local, S England.

**Gypsophilas** *Gypsophila*. Annual or perennial herbs with lax or crowded inflorescences. Calyx tubular with 5 teeth, 5-veined. Petals clawed, without scales. Stamens 10. Styles 2. Fruit splitting by 4 teeth. 10 species widespread mainly in SE Europe and W Asia; several are cultivated in gardens.

2* **Alpine Gypsophila** *Gypsophila repens* L. Low to short hairless, sprawling perennial. Leaves linear-lanceolate, greyish-green, often curved. Flowers in lax clusters, white, pale pink or lilac-purple, 8-10mm, petals notched. Mountain habitats, rocky places, ledges, cliffs and stabilized screes, sometimes in short rocky turf, generally on limestone, to 2900m. May-September. C & E France and S Germany. Confined to the mountains of C and S Europe. Widely cultivated.

3* **Fastigiate Gypsophila** *Gypsophila fastigiata* L. Variable low to tall erect perennial, stickily-hairy above. Leaves linear, 1-veined. Flowers in rather dense, flat-topped clusters, white or pale purple, 5-8mm, petals slightly notched. Dry rocky habitats. June-September. Germany & S Sweden.

4* *Gypsophila paniculata* L. Tall greyish, hairless perennial with a stout rhizome; stems much-branched. Leaves narrow-lanceolate, pointed. Flowers small, 3-4mm, white or pink, in lax clusters, petals narrow. Dry sandy and stony habitats. July-September. **I**: E Europe and W Asia. Naturalized in Belgium and Germany. Widely cultivated as an ornamental species and as a cut flower. It has a thick, somewhat fleshy rootstock.

5* **Annual Gypsophila** *Gypsophila muralis* L. Low to short, slightly hairy annual. Leaves linear, greyish-green. Flowers white or pink with darker veins, 4mm, in lax flat-topped clusters; petals notched or not. Damp woods and meadows. June-October. Belgium, Holland, France, Denmark, Germany and S Sweden.

**Soapworts** *Saponaria*. Perennial herbs. Calyx tubular, with 5 teeth, 15-25-veined. Petals clawed, with scales at the base of the limb. Stamens 10. Styles usually 2. Fruit capsule splitting by 4 teeth. 20 species, mainly in the Mediterranean region.

6* **Rock Soapwort** *Saponaria ocymoides* L. Low sprawling, much branched, hairy perennial; stems often reddish. Leaves lanceolate, blunt. Flowers pink or purplish, 6-10mm, in lax spreading clusters, petals blunt. Grassy, rocky and stony places, to 2000m. March-October. **I**: S Europe. Naturalized in Britain and Denmark. Widely cultivated in gardens. **B**: Naturalized in a few localities – Cornwall and Northumberland.

7* **Soapwort** *Saponaria officinalis* L. Medium to tall, rather straggling, generally hairless perennial, with stout runners. Leaves oval, 3-veined, the lower stalked. Flowers flesh-pink, 25-38mm, in dense branched clusters; petals not notched, the claws standing clear of the greenish or reddish calyx. Hedgebanks, open woodland, roadsides, waysides, streambanks and waste places; often close to habitation. June-September. Widely naturalized in much of W & N Europe (probably native in C & S Europe). Widely cultivated in gardens, including a double-flowered form 'Flora Plena'. Long used as a soap, the leaves producing a lather when rubbed in water. **P**: Moths. **B**: Throughout, rather rare in Scotland.

8* **Cow Basil** *Vaccaria hispanica* (Miller) Rauschert [*V. pyramidata*, *V. vulgaris*, *V. segetalis*, *Saponaria vaccaria*]. Medium, hairless, erect annual, regularly branched. Leaves oval to lanceolate, grey-green. Flowers pink, 10-15mm, petals notched or not; calyx inflated, winged on the angles; styles 2. A weed of cultivated land, especially arable fields on calcareous soils, at low altitudes. June-July. Belgium, France and Germany; casual in Britain and parts of NW Europe. W Europe to W Asia.

**Petrorhagias** *Petrorhagia*. Annual or perennial herbs. Calyx 5-15-veined, membranous between teeth. Petals 5. Stamens 10. Styles 2. Fruit capsule splitting by 4 teeth. 5 species in C and S Europe closely related to *Dianthus*, but differing primarily in the prominent whitish membranous areas between the calyx teeth.

9* **Tunic Flower** *Petrorhagia saxifraga* (L.) Link [*Kohlrauschia saxifraga*, *Tunica saxifraga*]. Short to medium, hairless or rough perennial. Leaves linear. Flowers white or pale pink, 5-6mm, solitary on long stalks, forming a lax cluster; petals notched. Dry sandy habitats, banks, fields, occasionally on walls, to 1300m. June-August. Germany, NE & C France, naturalized, but rare, in Britain and Holland. **B**: Only in Pembrokeshire.

10 **Proliferous Pink** *Petrorhagia prolifera* (L.) P.W. Ball & Heywood [*Dianthus prolifer*, *Tunica prolifera*, *Kohlrauschia prolifera*] Short to medium, erect, hairless or rough annual. Leaves linear-lanceolate, 3-veined. Flowers pink or white, 6-7mm, in a small dense head surrounded by brownish, papery bracts; petals notched. Dry open habitats, often on calcareous soils, to 1200m. May-September. E Belgium, Holland, France, Germany and S Sweden. **P**: Butterflies.

11 **Childing Pink** *Petrorhagia nanteulii* (Burnat) P.W. Ball & Heywood. Like *P. prolifera*, but centre of stem hairy and leaf-sheaths twice as long as wide (not as long as wide). Sandy and gravelly habitats, often close to the sea. July-August. Britain and France. Local and often rare. **P**: Butterflies. From W and SW Europe to N Africa and the Canary Isles.

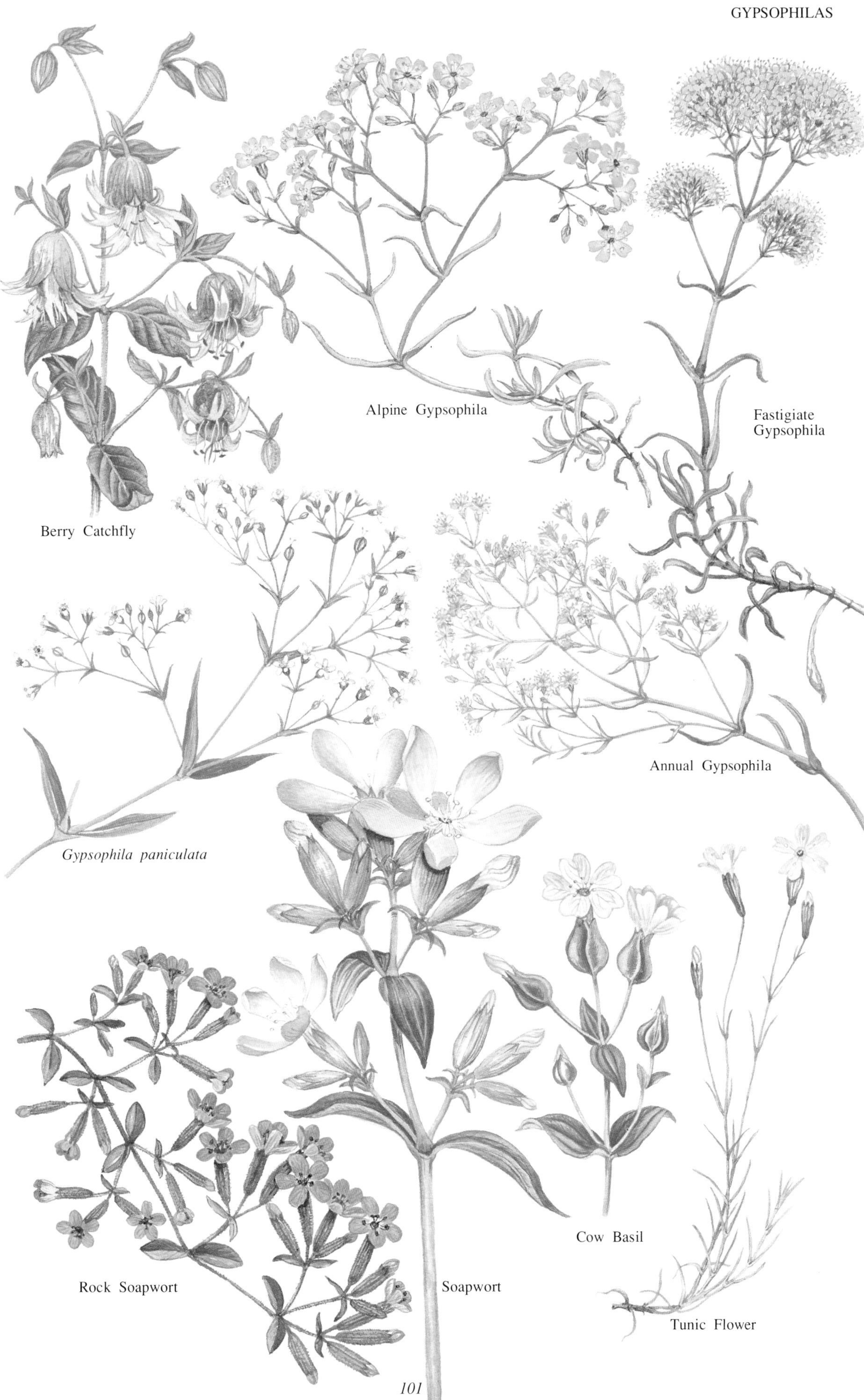
Berry Catchfly
Alpine Gypsophila
Fastigiate Gypsophila
Annual Gypsophila
*Gypsophila paniculata*
Cow Basil
Rock Soapwort
Soapwort
Tunic Flower

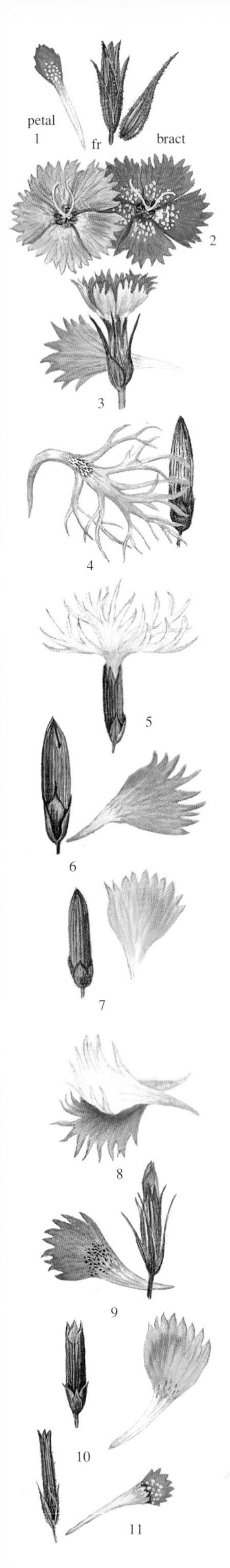


**Pinks** *Dianthus*. Tufted perennial or annual herbs, generally with slender, parallel-veined, leaves. Flowers solitary or clustered, often fragrant. Calyx with opposing pairs of scales at the base. Petals toothed or deeply fringed, long-clawed. Stamens 10; styles 2. Capsule splitting with 4 teeth. **P**: Mostly butterflies. 300 species widespread in Europe, Asia, Africa and North America. Several are grown in gardens, including the hybrid pinks and carnations.

1* **Deptford Pink** *Dianthus armeria* L. Short, stiffly-branched, hairy annual or biennial. Leaves dark green. Flowers bright reddish-pink, 8-15mm, in small rather flat clusters; petals toothed, bearded; flower clusters surrounded by long leafy bracts; scales pointed, equalling the calyx. Dry well-drained grassland and banks, road verges, hedgerows, on sandy, rocky or calcareous soils, to 1250m. June-August. Throughout, except the extreme north. **P**: Self-pollinated. **B**: Local and rare, scattered localities in England, Wales and S Scotland.

2* **Sweet William** *Dianthus barbatus* L. Medium, almost hairless, biennial or short-lived perennial. Leaves green, lanceolate. Flowers purple, pink or reddish, often spotted, 20-30mm, in dense, flat-topped, clusters; petals toothed; scales green, longer than the calyx. Woodland margins, grassy places, banks and hedgerows, to 2500m. June-August. Frequently cultivated in gardens and widely naturalized in N & NW Europe, except the extreme north. Often sold as a cut flower.

3* **Carthusian Pink** *Dianthus carthusianorum* L. Variable medium, hairless, perennial, with stiff stems. Leaves green, up to 5mm wide, flat. Flowers deep pink to purple, rarely white, 18-20mm, in small dense clusters surrounded by brown bracts; petals toothed, bearded; scales long-pointed, half the length of the calyx. Dry grassy and rocky habitats and open woodland, to 2500m. May-August. France and Germany.

4* **Large Pink** *Dianthus superbus* L. Medium to tall, hairless, branching perennial. Leaves green, narrow-lanceolate. Flowers pink or purplish, 30-50mm, in branched clusters, fragrant; petals deeply cut into narrow, pointed segments; scales pointed, one third the length of the calyx. Dry, generally semi-shaded, habitats, to 2400m. June-September. Europe north to Sweden and S Norway.

subsp. *speciosus* (Reichenb.) Pawl. Like the type, but the leaves greyish and flowers rather larger. Generally only on the higher mountains, the Alps in particular.

5 *Dianthus arenarius* L. Like *D. superbus*, but leaves narrower and flowers white, 20-30mm; petals often green-marked at the base and with a purplish margin. Grassy and rocky habitats. July-August. Germany and E Scandinavia.

6* **Jersey Pink** *Dianthus gallicus* Pers. Short to medium, laxly-tufted, downy perennial. Leaves greyish. 1.5-3mm wide, short and rather stiff. Flowers pink or mauvish, 20-30mm, solitary or 2-3 together; petals toothed, bearded at the base. Coastal grassy sand-dunes. June-August. W France. Also known from parts of Spain and Portugal. **B**: Only on Jersey.

7* **Clove Pink** *Dianthus caryophyllus* L. Like *D. plumarius*, but leaves 2-4mm wide, smooth-edged. Flowers pink or purplish, 30-40mm, usually solitary, very fragrant; petals toothed; scales short, one quarter the length of the calyx. Rocky habitats and walls. July-August. **I** S Europe. Naturalized in many places in W & NW Europe but absent from most of the north. The ancestor of the garden carnation. **B**: Occasionally found on old walls.

8* **Common Pink** *Dianthus plumarius* L. Short to medium, laxly tufted, hairless perennial. Leaves rough-edged, greyish, only 1mm wide. Flowers pink or white, 25-35mm, fragrant; petals deeply toothed; scales short, pointed, one quarter the length of the calyx. Old walls and dry calcareous banks. June-August. **I**: SE Europe. Widely naturalized in W & NW Europe, but absent from most of the north. Many forms and hybrids are cultivated. **B**: Naturalized in places.

9* **Sequier's Pink** *Dianthus seguieri* Vill. Medium hairless perennial, laxly tufted. Leaves green, only 1-2mm wide. Flowers pink or purplish, white-spotted in centre, bearded, 14-22mm, solitary or paired, petals toothed; scales often as long as the calyx. Dry meadows, stony and rocky places, to 1600m. June-September. E France. Local.

subsp. *glaber* Celak. Like the type, but leaves 2-4mm wide; scales one third the length of the calyx. Similar habitats and flowering time. S Germany.

10* **Cheddar Pink** *Dianthus gratianopolitanus* Vill. Low to short, hairless, densely tufted perennial, with long, creeping, sterile shoots. Leaves greyish, 1-2mm wide, rough-edged. Flowers plain pink, 20-30mm, solitary; petals toothed, bearded at the base; scales short, one quarter the length of the calyx. Sunny rocky habitats, cliffs and screes, generally on limestone, to 2200m. May-July. S Britain, Belgium, France and C & S Germany. Local. Much reduced by overcollecting, especially in Britain. **P**: Butterflies and moths. **B**: Rare, only found in Cheddar Gorge – protected.

11* **Maiden Pink** *Dianthus deltoides* L. Short to medium, loosely-tufted, perennial, downy, with numerous sterile shoots and short blunt bluish- green leaves. Flowers small, deep pink, spotted in the centre, rarely white, 15-20mm; petals toothed, bearded; scales long-pointed, half the length of the calyx. Meadows and other grassy habitats, banks and rocky slopes, often on light sandy soils, calcareous or slightly acid, to 2000m. June-September. Throughout, except the extreme north, Ireland, Iceland and most of the smaller islands. **B**: Scattered localities through England, Wales and S Scotland.

Cheddar Pink

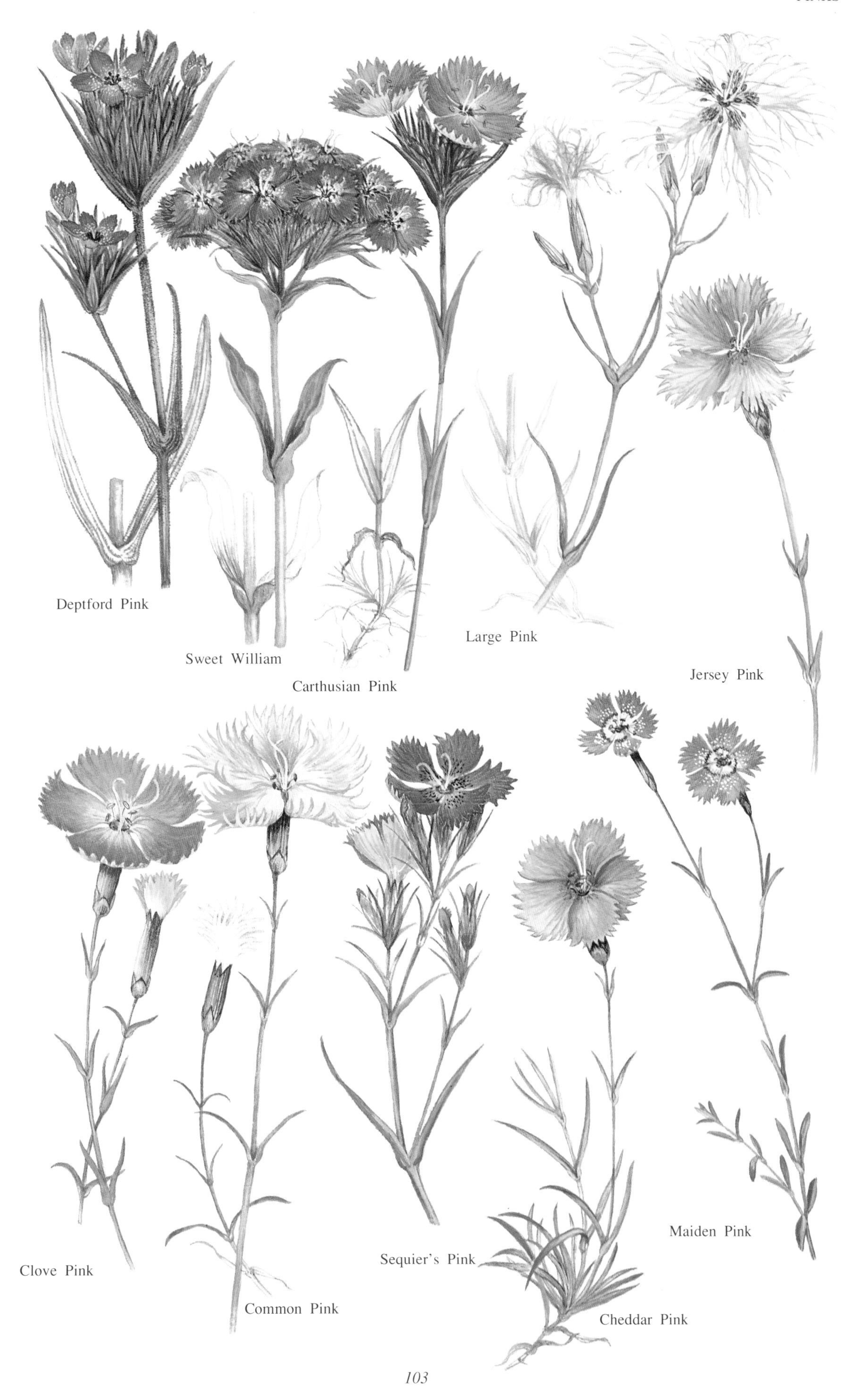
Deptford Pink
Sweet William
Carthusian Pink
Large Pink
Jersey Pink
Clove Pink
Common Pink
Sequier's Pink
Cheddar Pink
Maiden Pink

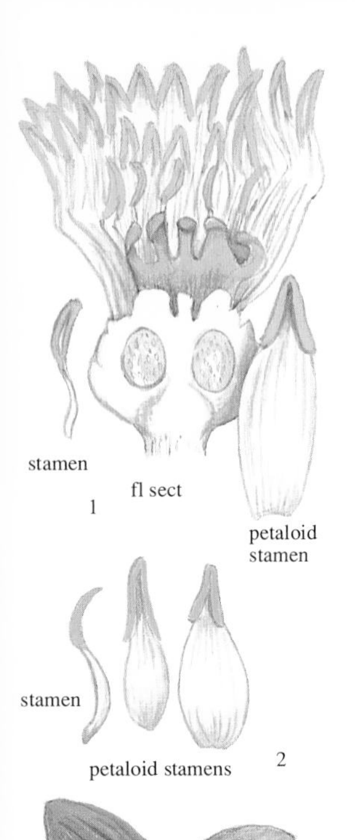


## WATER-LILY FAMILY Nymphaeaceae

Aquatic, hairless perennials of slow-moving or stagnant fresh water. Stems rooting in the mud, the leaves long-stalked, floating or submerged, oval or circular in outline. Flowers at or above the water level, with 4-6 sepals and 8 or more petals and numerous stamens. Fruit capsule ripening above the water (*Nuphar*) or below (*Nymphaea*). 60 species in 6 genera, in both temperate and tropical regions. The largest genus is *Nymphaea*, many species of which are cultivated in gardens and hot-houses.

1* **White Water-lily** *Nymphaea alba* L. Rhizomatous perennial. Leaves circular in outline, the basal lobes parallel or diverging, all floating. Flowers white with a mass of pale yellow anthers, 100-200mm, held at water level, with 20 or more petals, fragrant; buds rounded. Stigmas greenish-yellow. Fruit a spongy capsule. Lakes, ponds, dykes and deep ditches, slow-moving streams and rivers, in fresh, often nutrient-rich waters, rooting on muddy bottoms to a depth of up to 3m. June-September. Throughout, except the Faeroes, Iceland and Spitsbergen. Widely cultivated. The flowers open in bright daylight. **P**: Various small beetles. **B**: Throughout.

2* *Nymphaea candida* C. Presl. Like *N. alba*, but the leaves with touching or overlapping lobes. Flowers somewhat smaller, 75-130mm, with pointed buds, 15-18 petals and deep yellow anthers. Stigmas reddish. Lakes, ponds and slow-moving rivers. July-September. Germany and Scandinavia. Local.

3* *Nymphaea tetragona* Georgi. Altogether smaller than either *N. alba* or *N. candida*, with oval leaves, with diverging lobes; all leaves floating. Flowers white with pale yellow anthers, 35-60mm, with only 8-17 petals. Sepals thick and fleshy, persisting in fruit. Shallow lakes and ponds. July-September. Finland, extending east into the USSR. Far smaller than the previous two. Occasionally cultivated in gardens.

4* **Yellow Water-lily** *Nuphar lutea* (L.) Sm. Stout rhizomatous perennial. Floating leaves oval, deeply cleft to the stalk; submerged leaves shorter-stalked, rounded, thin and translucent. Flowers bright yellow, 12-40mm, rising out of the water, with 5-6 large overlapping sepals and numerous small petals. Fruit flask-shaped. Slow moving rivers and streams, canals, deep ditches, lakes and ponds, rooting on muddy bottoms in depths of up to 5m, in fresh, generally nutrient-rich waters. June-September. Throughout, except the Faeroes, Iceland and Spitsbergen. The flowers, which are held high above the water, smell of alcohol and attract numerous small flies. More shade tolerant than *Nymphaea alba*, but more susceptible to polluted water. **B**: Throughout, except the extreme north.

5* **Least Water-lily** *Nuphar pumila* (Timm) DC. Like *N. lutea*, but smaller, flowers 15-35mm, with narrower, scarcely touching, petals. Shallow lakes, mires and pools, often in upland areas, generally in acid, nutrient-poor, water. June-July. Throughout, though absent from most of the smaller islands. Hybridizes with *N. lutea*, *N.* × *spenneriana* Gaudin; hybrids occur where the parent species grow in close proximity to one another. **B**: Rare, Shropshire and Scottish Highlands.

## HORNWORT FAMILY Ceratophyllaceae

Aquatic, submerged, perennial herbs. Leaves in whorls. Flowers unisexual, tiny, male and female at different nodes on the same plant. Plants of stagnant or slow-moving fresh water. Three species in a single genus, in many parts of the world.

6* **Rigid Hornwort** *Ceratophyllum demersum* L. Stems flexuous with rigid leaves in whorls of 3-8. Leaves dark green, 1-2 forked, divided into slender toothed segments. Flowers solitary, with 8 or more linear petals; female flowers green, male whitish. Fruit a small nut, 4-5mm, with basal spines, otherwise smooth. Fresh water habitats, ponds, ditches, slow-moving streams and rivers in nutrient-rich water, occasionally in brackish dykes, rooting in the muddy bottoms. July- September. Throughout, except the Faeroes, Iceland and Spitsbergen. **P**: Under water. A rapid colonizer, frequently choking areas of water. **B**: Lowland England, rare elsewhere.

7 **Soft Hornwort** *Ceratophyllum submersum* L. Like *C. demersum*, but a smaller plant; leaves soft, bright green, 3-4-forked. Fruit warty, spineless. Ponds and ditches, very slow moving streams and canals, often in rather brackish or nutrient-rich water. July-September. Throughout, except the extreme north. **B**: Coastal SE England, also in Somerset and S Wales.

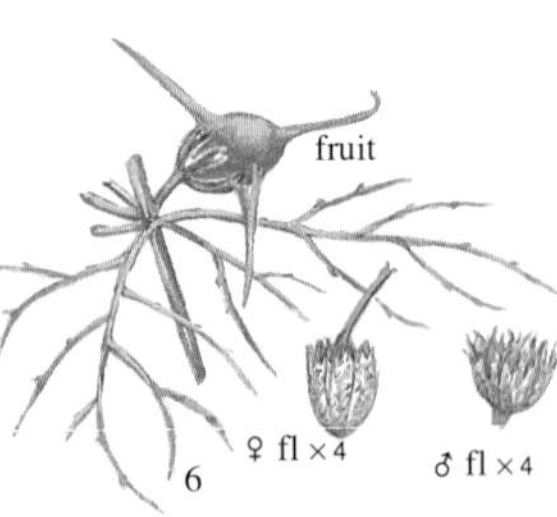


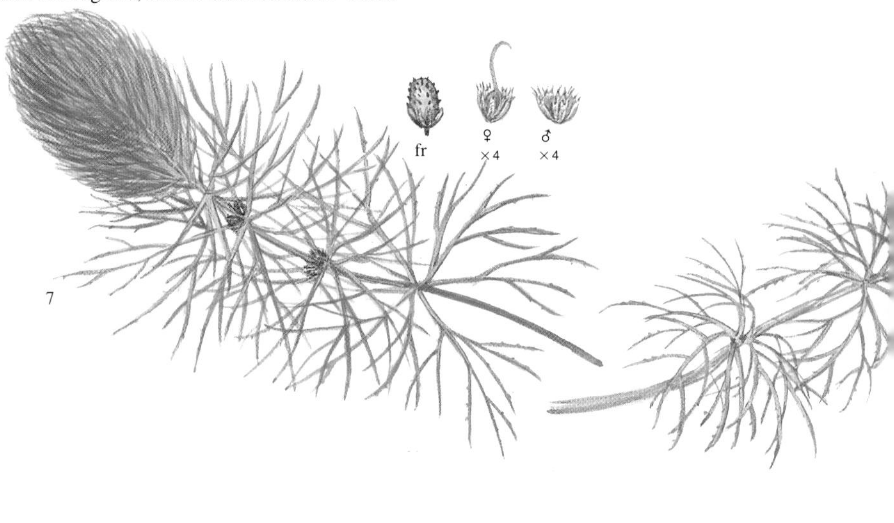

White Water-lily
Nymphaea candida
Nymphaea tetragona
Rigid Hornwort
Least Water-lily
Yellow Water-lily

# BUTTERCUP FAMILY Ranunculaceae

A large family of perennial herbs (woody climbers in *Clematis*). Leaves alternate, but opposite in *Clematis*, sometimes mostly basal; simple to pinnately or palmately lobed. Flowers hermaphrodite, often large and showy. Sepals usually 5 but often petal-like. Petals 5 or more, sometimes absent or reduced to small nectaries or spurred. Stamens numerous. Fruit a collection of achenes or follicles, rarely a berry (*Actaea*). Some 1750 species in 58 genera, primarily in temperate and boreal regions of the world.

1* **Stinking Hellebore** *Helleborus foetidus* L. Medium to tall, rather robust, foetid perennial with leafy overwintering stems. Leaves palmate, deep green, lobes simple, toothed. Uppermost leaves and bracts unlobed or with a very reduced blade. Flowers nodding, yellowish-green with a purple rim, bell-shaped, 10-30mm, in large clusters. Fruit a cluster of 3 follicles. Woodland, often of beech and yew, scrub and rocky places, generally on rather dry, shallow calcareous soils, to 1600m. January-May. Britain, Belgium, Holland, France and Germany. **P**: Honey- and bumble-bees. Widely cultivated in gardens, All parts of the plant are poisonous. **B**: From Yorkshire southwards, rather rare.

2* **Green Hellebore, Bear's-foot** *Helleborus viridis* L. Short, tufted perennial, generally with two basal leaves. Leaves palmately-lobed, dull green, the lobes further lobed and toothed, hairless beneath, not overwintering. Flowers half-nodding, yellowish-green, 40-50mm, with spreading sepals. Damp calcareous woodland and scrub, growing in deep leaf-mould, to 1600m. February-April. E France and Germany. **P**: Early flying bees. Sometimes grown in gardens. All parts of the plant are extremely poisonous, with the glycosides Helleborin and Helleborein.

subsp. *occidentalis* (Reuter) Schiffner. Like the type, but the leaves more coarsely toothed, hairy beneath. Similar habitats and flowering time. Britain, Belgium and France. **B**: SE England and scattered localities north to Lancashire.

3* **Winter Aconite** *Eranthis hyemalis* (L.) Salisb. Low hairless, tuberous-rooted, perennial. Leaves basal, circular in outline, palmately-lobed, shiny green. Flowers solitary, nodding in bud, buttercup-like, yellow, 20-30mm, usually with 6 oval sepals, with a ruff of three stalkless, bract-like leaves below. Fruit a cluster of about 6 follicles. Woods and scrub, in moist humus-rich soils, to 1500m. January-March. **I**: S Europe and W Asia. Naturalized in Britain, Belgium, Holland, France and Germany. **P**: Bees and flies. Widely cultivated in gardens and frequently becoming naturalized or semi-naturalized. **B**: Widely naturalized north to S Scotland.

4* **Nigella** *Nigella arvensis* L. Short erect, branched, hairless annual, with soft, finely divided, feathery leaves. Flowers solitary, bluish, 20-30mm, the sepals often green-veined, without a ruff of leaves immediately below. Fruit a collection of partly fused follicles, each 3-veined. Arable fields on calcareous soils, generally less common than formerly, to 1200m. June-September. Throughout Europe, except the extreme north; casual in Britain. Unlike the next species the fruits are not or scarcely inflated. Occasionally cultivated in gardens.

5 **Love-in-a-mist** *Nigella damascena* L. Similar to *N. arvensis*, but the upper leaves in a close whorl just below the flower. Flowers pale to mid-blue, 25-40mm. Follicles closely united. Fruit capsule inflated, papery when ripe; seeds black. Dry open habitats, fields and waste places, disturbed ground. June-July. **I**: S Europe and West Asia. Naturalized in Belgium, Holland, France and Germany; occasionally casual in Britain. Frequently cultivated in gardens; forms include blue, pink and white flowers.

6* **Globeflower** *Trollius europaeus* L. Short to medium tufted perennial, hairless. Leaves palmately-lobed, the basal long-stalked. Flowers globular, lemon-yellow, 30-50mm, with about 10 incurved sepals. Fruit a cluster of follicles. Rich moist meadows, alpine pastures, fens, fen carrs, moist open woodland, occasionally on cliff ledges to 2800m. May-August. Throughout, mainly upland. **P**: Various small insects. Widely cultivated in gardens including forms with orange flowers. Very poisonous. **B**: Local in Wales, Scotland and N England, as well as Ireland.

7* **Isopyrum** *Isopyrum thalictroides* L. Short, slender, tufted, hairless perennial. Leaves bluish-green, trifoliate, with 3-lobed leaflets. Flowers solitary, white, 10-20mm, buttercup-like, with 5 sepals. Fruit generally of 2 follicles. Damp shady upland deciduous woodland, rock crevices, to 1200m. March-May. S Germany. Occasionally cultivated in gardens.

8* **Baneberry, Herb Christopher** *Actaea spicata* L. Medium, hairless, strong smelling perennial. Basal leaves large, 2-trifoliate or pinnate, with toothed leaflets; stem leaves smaller. Flowers small, white, in short oblong racemes; sepals and petals 4-6. Fruit a shiny black berry, 10-11mm. Woods over limestone and limestone pavement, growing in the grikes, 1900m. May-July. Throughout, but often local. A very poisonous plant. often grown in gardens. **P**: Various small insects. **B**: Rare, confined to Cumbria, Yorkshire and Lancashire.

9 **Red Baneberry** *Actaea erythrocarpa* Fischer. Like *A. spicata*, but leaflets more deeply divided and berries red, smaller, 8-9mm. Wooded and rocky habitats. May-July. Finland and Sweden. Poisonous.

10* **Marsh Marigold, Kingcup** *Caltha palustris* L. Very variable short, tufted, hairless, perennial; stems upright to creeping and rooting at the nodes. Leaves heart-shaped, toothed, the lower mostly long-stalked. Flowers bright shiny golden-yellow, 15-50mm, often greenish beneath. Fruit a cluster of pod-like follicles, with nectaries at the base. Wet meadows, marshes, fens, ditches and wet woodland, to 2500m. March-August. Throughout, except Spitsbergen. **P** Visited by various insects for nectar as well as pollen. Poisonous, but widely cultivated in gardens. Dwarf plants from N Europe with creeping stems, rooting at the nodes, are often referred to subsp. *minor* (Miller) Asch. & Graebner. **B** Throughout.

Stinking Hellebore
× 1/10
Nigella
Green Hellebore
Winter Aconite
× 1/4
× 1/6
Baneberry
Globeflower
Isopyrum
Marsh Marigold

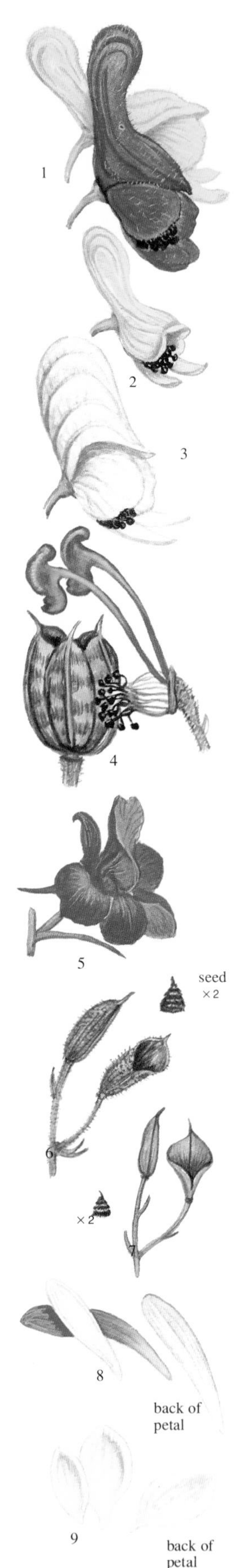


**Monkshoods** *Aconitum*. Herbaceous perennials with stout erect stems and tuberous rootstocks. Leaves palmately-divided, with lobed divisions. Flowers irregular, in spike-like racemes, occasionally branched. Perianth segments 5, petal-like, the uppermost forming a prominent hood or helmet; staminodes usually 10, the upper two large and spurred and with nectaries contained within the hood, others small. Stamens many. Fruit a cluster of 2-5 follicles. 100 species widespread in temperate regions of the northern hemisphere. A number are cultivated in gardens. All are poisonous to some extent.

1 **Northern Wolfsbane** *Aconitum septentrionalis* Koelle [*A. lycoctonum*]. Medium to tall, partly hairy perennial. Leaves dark green, palmately 4-6-lobed to the midrib. Flowers dark violet, very rarely yellow; helmet long-tapered, 18-25mm long, widening at the base; inflorescence hairy. Meadows, woods and scrub, generally in moist semi-shaded habitats, to 1300m. July-August. Scandinavia, extending east into the USSR. **P**: Bees.

2* **Wolfsbane** *Aconitum vulparia* Reichenb. Medium to tall, partly hairy perennial. Leaves dark green, palmately 7-8-lobed, divided to the midrib. Flowers generally rather few, pale yellow; helmet narrow-oblong, 15-22mm long, not widened much at the base; inflorescence branched, hairy. Damp, generally shaded, habitats, woodland, among rocks and along stream margins, to 2400m. Belgium, Holland, France and Germany. **P**: Bees. Very similar in general appearance to the preceding species, but the two do not overlap in distribution or area and, besides, usually have different flower colours. Cultivated in gardens.

3* **Variegated Monkshood** *Aconitum variegatum* L. Medium to tall perennial. Leaves palmately 5-7-lobed, divided almost to the midrib, the divisions broad and deeply lobed. Flowers blue streaked with white, rarely all white; helmet broadly-oblong, twice as high as broad, 8-15mm; inflorescence hairless or slightly hairy. Meadows, woodland and woodland clearings, in sunny or semi-shaded habitats, to 2000m. Mountains of France and S Germany; naturalized in Denmark. **P**: Bees.

4* **Monkshood** *Aconitum napellus* L. [*A. anglicum*]. Tall, hairless perennial. Leaves palmately 5-7-lobed, divided almost to the midrib, the divisions narrow and deeply lobed. Flowers deep reddish-violet or blue; helmet rounded, as broad as high, 10-18mm; inflorescence dense, generally branched. Damp habitats, woodland, meadows, scrub and stream margins, to 2500m. June-September. S Britain, Belgium, France and Germany. A very poisonous plant containing the alkaloid Aconitine. **P**: Bumble-bees, seeking nectar from the two nectaries hidden below the helmet. Cultivated in gardens. **B**: Local in SW England and S Wales, elsewhere as a garden outcast.

**Larkspurs** *Consolida*. Erect, generally branched annuals with finely cut leaves, rounded in outline. Flowers in racemes with 5-petaloid segments and a backward pointing spur. Fruit a solitary follicle. 50 species in C and S Europe and temperate Asia. Closely related to *Delphinium*, but distinguished by being annual in habit, rather than perennial, and by the fruit having a solitary follicle, rather than a cluster of follicles.

5 **Eastern Larkspur** *Consolida orientalis* (Gay) Schrödinger. Medium to tall, stickily-hairy annual. Leaves with narrow-oblong or linear lobes. Flowers purplish-violet, in a fairly dense raceme, spur 10-12mm; flower-stalks shorter than the lower dissected bracts. Fruit hairy. Arable fields, waste places and cultivated land, generally on rather dry well-drained soils. June-August. **I**: SE Europe and W Asia. Naturalized in S Britain, France and Germany. Cultivated in gardens. Naturalized or casual C & S Britain.

6* **Larkspur** *Consolida ambigua* (L.) P.W. Ball & Heywood [*Delphinium ambiguum*, *D. ajacis*]. Medium to tall annual like *C. orientalis*, but flower-stalks longer than the lower dissected bracts; flowers deep blue, spur 13-18mm; racemes occasionally branched. Arable and waste land, on light sandy soils. June-August. **I**: Mediterranean region. Naturalized or casual in Britain, France and Belgium, possibly elsewhere. The basal petal is marked with lines resembling the letters AIA. Widely cultivated in gardens. **B**: Casual, scattered localities in C & S Britain.

7* **Forking Larkspur** *Consolida regalis* S.F. Gray [*Delphinium consolida*]. Medium, downy annual, widely branched. Leaves with linear lobes. Flowers violet-blue, in lax-branched panicles, spur 12-25mm; all bracts unlobed. Fruit hairless. Arable fields, waste or disturbed land, on calcareous soils. June-August. Throughout except the extreme north. W Europe to C Asia. Widely cultivated in gardens. **P**: Bees.

**Anemones** *Anemone*. Perennial herbs, often tufted. Leaves palmately-lobed, the basal usually long-stalked; stem leaves 3, in a whorl beneath the flowers. Perianth segments 5 or more, petal-like. Stamens numerous. Fruit a tight cluster of single-seeded achenes. 120 species widespread in many parts of the world, but with the majority in northern temperate, alpine and Arctic regions. Many are cultivated in gardens. Plants are poisonous, especially to livestock, due to the presence of the narcotic Anemonin.

8* **Blue Anemone** *Anemone apennina* L. Low to short, somewhat hairy, rhizomatous perennial. Leaves 2-trifoliate, stalked, the segments elliptical, toothed, often tinged with purple, the basal leaves developing after the flowers fade. Flowers deep blue, 20-35mm, nodding in bud, 'petals' 10-15, narrow. Woodland, woodland margins and hedgerows and banks. April-early May. **I**: S Europe. Locally naturalized in Britain and parts of W Europe. Widely cultivated in gardens.

9* **Narcissus-flowered Anemone** *Anemone narcissiflora* L. Low to medium, hairy, tufted perennial. Leaves palmately-lobed to the middle, with lobed divisions. Flowers white, often flushed with pink beneath, 20-30mm, in umbels of 3-8; 'petals' 5-6. Meadows, scrub, rocky slopes and stabilized screes, generally in open sunny habitats,1600- 2600m. June-July. Mountains of S Germany and E France. Rather local. Sometimes cultivated in gardens.

Variegated Monkshood
× 1/6
Wolfsbane
Monkshood
Larkspur
Forking Larkspur
Blue Anemone
Narcissus-flowered Anemone

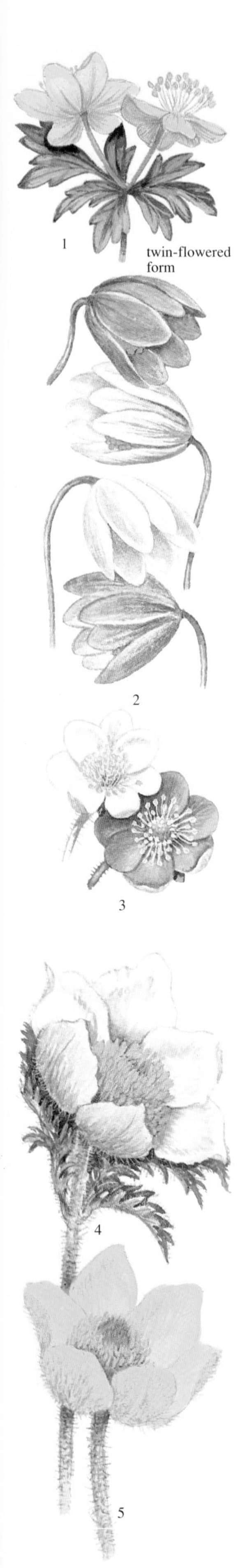


1* **Yellow Wood Anemone** *Anemone ranunculoides* L. Short, hairy, rhizomatous perennial. Leaves deeply divided into oblong, toothed, lobes. Flowers yellow, buttercup-like, 15-20mm, with 5-8 'petals', solitary or 2 together. Deciduous woodland, copses and scrub, occasionally in hedgerows, to 1500m. Continental Europe, except the extreme N; naturalized locally in Britain. W Europe south to Spain and east to Russia. Cultivated in gardens. **B**: Scattered localities in England.

subsp. *wockeana* (Ascherson & Graebner) Hegi. Like the type, but plants with short rhizomes, more patch-forming rather than widely spreading. Similar habitats and flowering time. S Germany.

2* **Wood Anemone, Wind Flower** *Anemone nemorosa* L. Low to short, hairless, rhizomatous perennial. Leaves with deeply-lobed, toothed divisions, the basal generally appearing after the flowers. Flowers white, often flushed with pink or blue outside, solitary, 20-40mm, half-nodding to erect; 'petals' 6-12. Woodland, woodland rides and clearing, copses, scrub, hedgerows, sometimes in more open habitats such as grassland, usually on calcareous or neutral soils occasionally on acid soils, to 1800m. March-May. Throughout, except the Faeroes, Iceland and Spitsbergen. Tends to grow where there is not too much competition from coarser vegetation. **P**: Bees and flies. Widely cultivated in gardens, including white, pink and blue-flowered forms. **B**: Throughout, but absent from Orkney and Shetland.

* **Snowdrop Windflower** *Anemone sylvestris* L. Short to medium, hairy perennial. Leaves palmately-lobed almost to the middle, with oval, toothed divisions; stem leaves similar but shorter-stalked. Flowers white, large, 40-70mm, solitary or 2 together, erect, with 5 broad oval 'petals'. Dry woods and scrub on calcareous soils, to 1200m. April-June. France (except the W), C & S Germany and SE Sweden. Cultivated occasionally in gardens. Distinguished from the previous species by its large flowers which have only 5 'petals'.

3* **Hepatica** *Hepatica nobilis* (L.) Miller [*Anemone hepatica*, *Hepatica triloba*]. Low, generally evergreen, hairy, tufted perennial. Leaves basal, 3-lobed, with a heart-shaped base, deep green or mottled, purplish beneath. Flowers solitary, purple, bluish-violet, occasionally pinkish or white, 15-25mm, with 6-10 narrowly oval 'petals' and 3 sepal-like bracts immediately below; anthers whitish. Woods, scrub, rocky and grassy habitats, mainly in mountain regions, to 2200m. March-April. Continental Europe except the extreme north, Belgium and Holland. **P**: Bees and flies. Widely cultivated in gardens, including both single- and double-flowered forms.

**Pasque Flowers** *Pulsatilla*. Like *Anemone*, but the leaves pinnately-divided, often ferny and the fruits with long feathery, persistent, styles. Flowers with 6-7 petal-like segments, hairy outside; flower stalk much elongating in fruit. 15 species in temperate regions of the northern hemisphere, especially in mountain regions. Most of the species are cultivated in gardens and are poisonous like *Anemone*.

* **White Pasque Flower** *Pulsatilla alba* Reichenb. Short, hairy perennial. Basal leaves long-stalked, 2-pinnate, with narrow pointed divisions, practically hairless at maturity; stem leaves smaller but short-stalked. Flowers white, sometimes with a bluish flush, 25-45mm erect, bowl-shaped. Mountain meadows, generally over acid rocks, to 2200m. May-July. C & S France and S Germany. The basal leaves are only partly developed at flowering time. Rarely cultivated in gardens.

4 **Alpine Pasque Flower** *Pulsatilla alpina* (L.) Delarb. Like *P. alba*, but with hairy leaves; flowers white flushed with bluish-purple outside, longer, 40-60mm. Mountain meadows over limestone, to 2700m. May-July. C & E France and S Germany. Widely distributed in the mountains of C & S Europe. Very rarely growing in association with subsp. *apiifolia*. Cultivated in gardens.

5 subsp. *apiifolia* (Scop.) Nyman [*P. sulphurea*]. Like the type, but flowers yellow. Meadows and scrub, generally over acid rocks, to 2700 m. Distribution and flowering time as for the species. Sometimes cultivated in gardens.

* **Eastern Pasque Flower** *Pulsatilla patens* (L.) Miller [*Anemone patens*]. Short hairy perennial. Basal leaves palmately-divided, with broad, toothed leaflets; stem leaves with narrow divisions. Flowers bluish-violet, 50-80mm, erect, with widely spreading 'petals'. Lowland meadows. April-May. Germany, Finland and Sweden. N Europe to Russia. Sometimes confused with *P. pratensis*, but flowers larger and with widely spreading 'petals', not forming a close bell-shape.

* **Spring Pasque Flower** *Pulsatilla vernalis* (L.) Miller [*Anemone vernalis*]. Short, hairy perennial. Basal leaves evergreen, pinnately-lobed, the oblong leaflets toothed; stem leaves unstalked, with linear divisions. Flowers white inside, but flushed with pink, violet or blue outside and covered in golden hairs, 40-60mm, nodding at first, later erect. Mountain meadows and rocky habitats, often close to melting snow, 1300-3600m. April-June. C & E France, C & S Germany, C & S Norway and Sweden. **P**: Bees. Widely cultivated in gardens – the smallest European species of *Pulsatilla*.

* **Common Pasque Flower** *Pulsatilla vulgaris* Miller [*Anemone pulsatilla*]. Low, hairy perennial. Leaves 2-pinnate, feathery, covered with long hairs at first; stem leaves unstalked, with linear divisions. Flowers large, dark to pale purple, bell-shaped, 55-85mm, erect at first, then half-drooping, the 'petals' twice as long as the stamens. Meadows, generally over limestone, particularly on short turfy slopes, often grazed, to 1200m. April-June. C & E Britain, Belgium, Holland, W France, Germany, Denmark and S Sweden. Often growing in short turf on thin poor soils. Declining. Poisonous. **B**: Local, generally declining. Cotswolds, Chilterns and East Anglia.

subsp. *gotlandica* (K. Joh.) Zamels & Paegle. Like the type, but with broader 'petals', the flowers appearing before the basal leaves. Meadows. S Sweden – Germany.

* **Small Pasque Flower** *Pulsatilla pratensis* (L.) Miller [*P. nigricans*, *Anemone pratensis*]. Similar in height and leaf to *P. vulgaris*. Flowers dark purple in the north of its range but pale greenish-violet in the south, 30-40mm, narrow nodding bells, the 'petals' scarcely longer than the stamens, the tips recurved. Meadows and scrub, to 2100m. April-May. Germany, extending to S Sweden, SE Norway and W Denmark.

**Red Pasque Flower** *Pulsatilla rubrum* (Lam.) Delarbre. Like *P. pratensis*, but flowers reddish-brown, purplish-brown or reddish-black, the 'petals' twice the length of the stamens. Rocky and grassy meadows, to 1500m. April-June. C & S France.

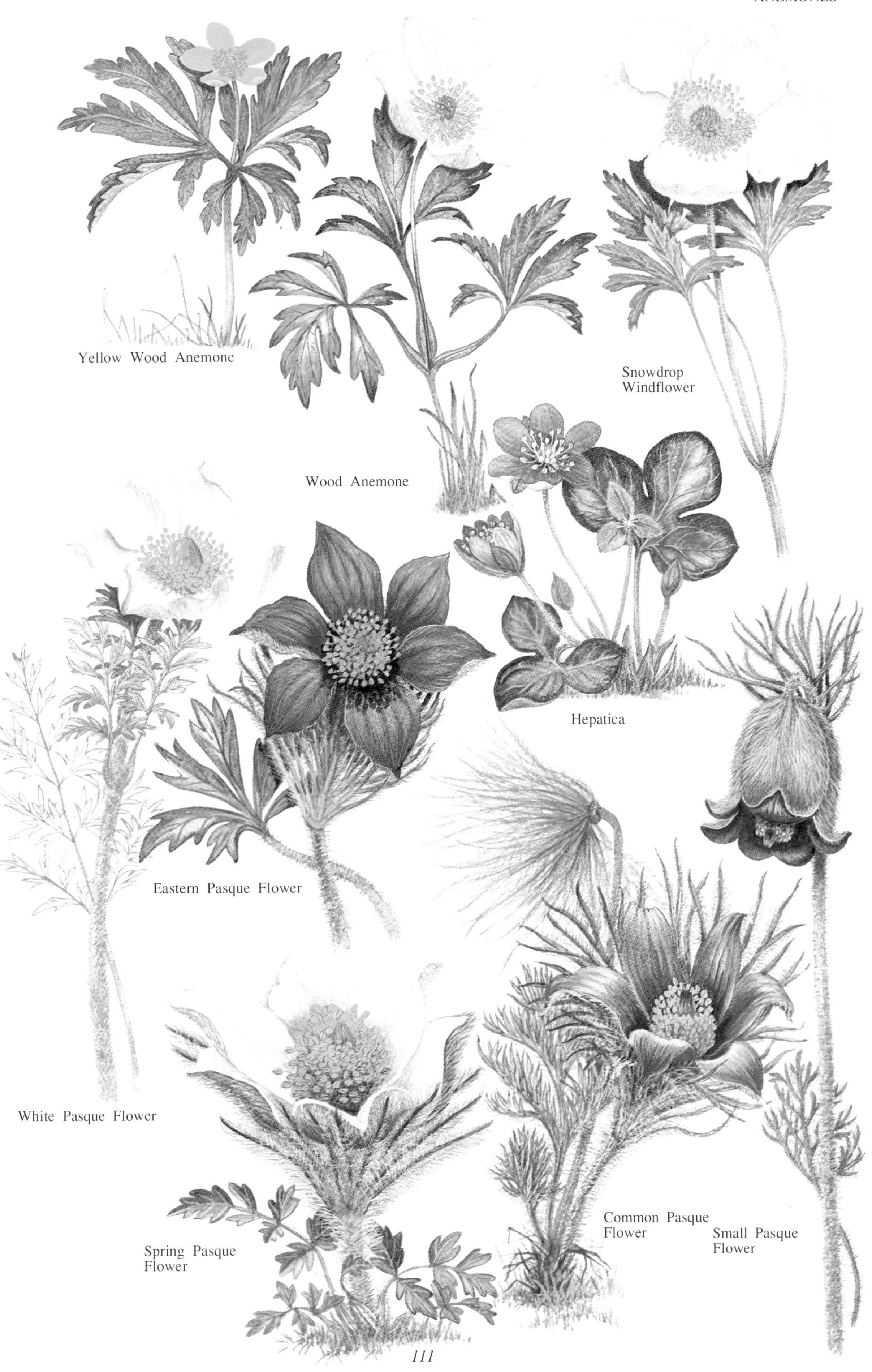
Yellow Wood Anemone
Wood Anemone
Snowdrop Windflower
Hepatica
Eastern Pasque Flower
White Pasque Flower
Spring Pasque Flower
Common Pasque Flower
Small Pasque Flower

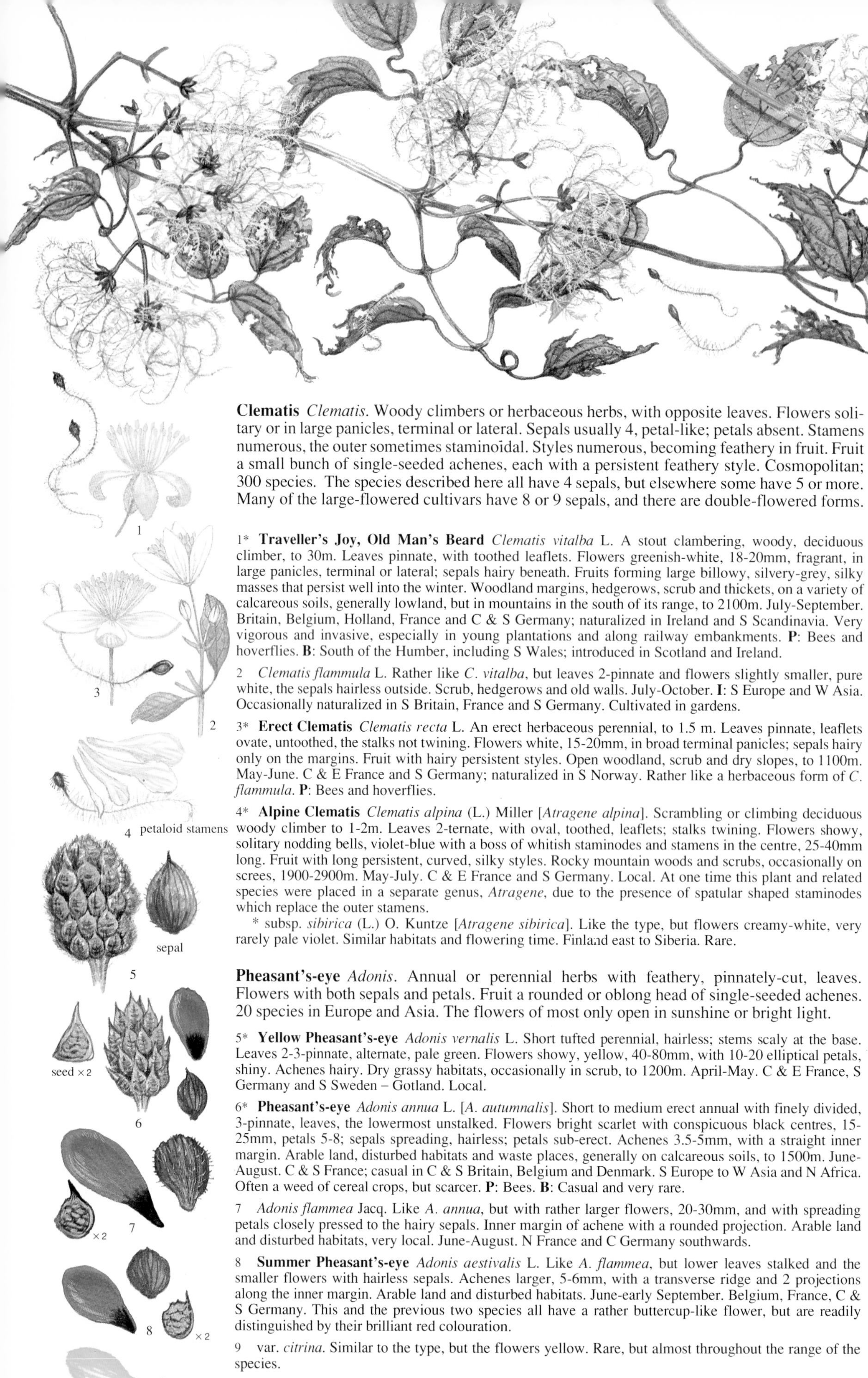


**Clematis** *Clematis.* Woody climbers or herbaceous herbs, with opposite leaves. Flowers solitary or in large panicles, terminal or lateral. Sepals usually 4, petal-like; petals absent. Stamens numerous, the outer sometimes staminoïdal. Styles numerous, becoming feathery in fruit. Fruit a small bunch of single-seeded achenes, each with a persistent feathery style. Cosmopolitan; 300 species. The species described here all have 4 sepals, but elsewhere some have 5 or more. Many of the large-flowered cultivars have 8 or 9 sepals, and there are double-flowered forms.

1* **Traveller's Joy, Old Man's Beard** *Clematis vitalba* L. A stout clambering, woody, deciduous climber, to 30m. Leaves pinnate, with toothed leaflets. Flowers greenish-white, 18-20mm, fragrant, in large panicles, terminal or lateral; sepals hairy beneath. Fruits forming large billowy, silvery-grey, silky masses that persist well into the winter. Woodland margins, hedgerows, scrub and thickets, on a variety of calcareous soils, generally lowland, but in mountains in the south of its range, to 2100m. July-September. Britain, Belgium, Holland, France and C & S Germany; naturalized in Ireland and S Scandinavia. Very vigorous and invasive, especially in young plantations and along railway embankments. **P**: Bees and hoverflies. **B**: South of the Humber, including S Wales; introduced in Scotland and Ireland.

2 *Clematis flammula* L. Rather like *C. vitalba*, but leaves 2-pinnate and flowers slightly smaller, pure white, the sepals hairless outside. Scrub, hedgerows and old walls. July-October. **I**: S Europe and W Asia. Occasionally naturalized in S Britain, France and S Germany. Cultivated in gardens.

3* **Erect Clematis** *Clematis recta* L. An erect herbaceous perennial, to 1.5 m. Leaves pinnate, leaflets ovate, untoothed, the stalks not twining. Flowers white, 15-20mm, in broad terminal panicles; sepals hairy only on the margins. Fruit with hairy persistent styles. Open woodland, scrub and dry slopes, to 1100m. May-June. C & E France and S Germany; naturalized in S Norway. Rather like a herbaceous form of *C. flammula*. **P**: Bees and hoverflies.

4* **Alpine Clematis** *Clematis alpina* (L.) Miller [*Atragene alpina*]. Scrambling or climbing deciduous woody climber to 1-2m. Leaves 2-ternate, with oval, toothed, leaflets; stalks twining. Flowers showy, solitary nodding bells, violet-blue with a boss of whitish staminodes and stamens in the centre, 25-40mm long. Fruit with long persistent, curved, silky styles. Rocky mountain woods and scrubs, occasionally on screes, 1900-2900m. May-July. C & E France and S Germany. Local. At one time this plant and related species were placed in a separate genus, *Atragene*, due to the presence of spatular shaped staminodes which replace the outer stamens.

* subsp. *sibirica* (L.) O. Kuntze [*Atragene sibirica*]. Like the type, but flowers creamy-white, very rarely pale violet. Similar habitats and flowering time. Finland east to Siberia. Rare.

**Pheasant's-eye** *Adonis.* Annual or perennial herbs with feathery, pinnately-cut, leaves. Flowers with both sepals and petals. Fruit a rounded or oblong head of single-seeded achenes. 20 species in Europe and Asia. The flowers of most only open in sunshine or bright light.

5* **Yellow Pheasant's-eye** *Adonis vernalis* L. Short tufted perennial, hairless; stems scaly at the base. Leaves 2-3-pinnate, alternate, pale green. Flowers showy, yellow, 40-80mm, with 10-20 elliptical petals, shiny. Achenes hairy. Dry grassy habitats, occasionally in scrub, to 1200m. April-May. C & E France, S Germany and S Sweden – Gotland. Local.

6* **Pheasant's-eye** *Adonis annua* L. [*A. autumnalis*]. Short to medium erect annual with finely divided, 3-pinnate, leaves, the lowermost unstalked. Flowers bright scarlet with conspicuous black centres, 15-25mm, petals 5-8; sepals spreading, hairless; petals sub-erect. Achenes 3.5-5mm, with a straight inner margin. Arable land, disturbed habitats and waste places, generally on calcareous soils, to 1500m. June-August. C & S France; casual in C & S Britain, Belgium and Denmark. S Europe to W Asia and N Africa. Often a weed of cereal crops, but scarcer. **P**: Bees. **B**: Casual and very rare.

7 *Adonis flammea* Jacq. Like *A. annua*, but with rather larger flowers, 20-30mm, and with spreading petals closely pressed to the hairy sepals. Inner margin of achene with a rounded projection. Arable land and disturbed habitats, very local. June-August. N France and C Germany southwards.

8 **Summer Pheasant's-eye** *Adonis aestivalis* L. Like *A. flammea*, but lower leaves stalked and the smaller flowers with hairless sepals. Achenes larger, 5-6mm, with a transverse ridge and 2 projections along the inner margin. Arable land and disturbed habitats. June-early September. Belgium, France, C & S Germany. This and the previous two species all have a rather buttercup-like flower, but are readily distinguished by their brilliant red colouration.

9 var. *citrina*. Similar to the type, but the flowers yellow. Rare, but almost throughout the range of the species.

Traveller's Joy
Erect Clematis
Alpine
Clematis
Pheasant's Eye
Yellow
Pheasant's Eye
Clematis alpina
ssp. sibirica

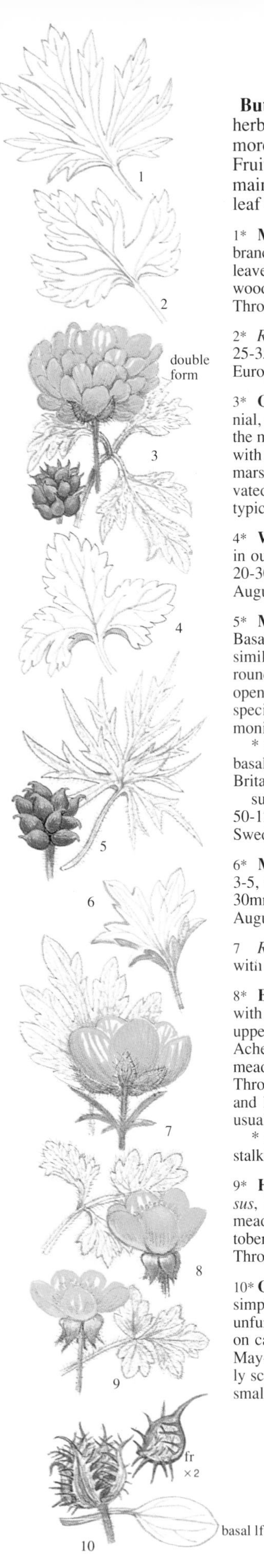


**Buttercups** and **Crowfoots** *Ranunculus*. Terrestrial or semi-aquatic annual or perennial herbs. Flowers solitary or in spreading cymose clusters. Sepals and petals often 5, sometimes more or fewer; petals often shiny, yellow or white, with basal nectaries. Stamens numerous. Fruit a head of single-seeded achenes. Cosmopolitan; 300 species. In the warmer climates mainly restricted to mountain habitats. Many are difficult to distinguish, though details of the leaf shape, sepals and fruit characteristics are often important for accurate identification.

1* **Multiflowered Buttercup** *Ranunculus polyanthemos* L. [*R. meyeranus*]. Short to medium, much-branched, hairy perennial. Basal leaves rounded in outline, the 5 segments cut into narrow lobes; stem leaves similar but smaller. Flowers golden-yellow, 18-25mm, numerous usually; sepals erect. Damp woods and meadows and other grassy habitats, on calcareous soils, to 1250m. May-July (-October). Throughout Continental Europe, except the extreme north.

2* *Ranunculus nemorosus* DC. Like *R. polyanthemos*, but leaves with 3 broadly-oval segments. Flowers 25-35mm; sepals erect. Meadows and other grassy habitats, to 2000m. May-July (-October). Continental Europe north to Denmark and S Sweden.

3* **Creeping Buttercup** *Ranunculus repens* L. Very variable short to medium, generally hairy perennial, with long creeping and rooting runners. Leaves triangular in outline with 3 deeply toothed segments, the middle one stalked. Flowers golden-yellow, 20-30mm, in lax irregular clusters; sepals erect. Achenes with a curved beak. Grassy habitats, more especially on wet, heavy calcareous or slightly acid clay soils, marshes, fens, road verges and pathways, occasionally along woodland rides or dune slacks, or on cultivated land, to 2500m. May-September. Throughout. More tolerant of trampling and grazing than other typical meadow *Ranunculus*. **B**: Throughout.

4* **Woolly Buttercup** *Ranunculus lanuginosus* L. Medium, very hairy, perennial. Basal leaves rounded in outline, the 3 segments, oval, toothed; stem leaves similar but smaller. Flowers deep orange-yellow, 20-30mm. Achenes with a strongly recurved beak. Damp woodland, meadows and scrub, to 1600m. May-August. Belgium, Holland, France, Denmark and Germany.

5* **Meadow Buttercup** *Ranunculus acris* L. Very variable medium to tall, usually hairy, perennial. Basal leaves deeply divided into 3-7 narrow wedge-shaped segments, each toothed or lobed; stem leaves similar but smaller. Flowers golden-yellow, 15-25mm, with erect sepals. Achenes with a hooked beak, in rounded heads. Damp meadows and pastures, especially hay meadows or on grazed land, road verges, open grassy woodland and ditches, to 2500m. April-September. Throughout. Often abundant. Like most species of *Ranunculus* the leaves are poisonous and are avoided by livestock; the presence of Protoanemonin in the sap gives the leaves a bitter taste. **P**: Various short-tongued insects. **B**: Throughout.

* subsp. *borealis* (Trautv.) Nyman [*R. borealis*]. Like the type, but seldom more than 20cm tall, the basal leaves with only 3 main segments. Flowers often brown-veined. Grassy habitats. June-September. N Britain, the Faeroes, Iceland and Scandinavia.

subsp. *friesianus* (Jordan) Rouy & Fouc. Like the type, but leaves softly-hairy beneath; a taller plant, 50-120cm usually. Grassy habitats. April-September. France; naturalized in Belgium, Germany and Sweden.

6* **Mountain Buttercup** *Ranunculus montanus* Willd. Short, slightly hairy perennial. Basal leaves with 3-5, oval, toothed segments; stem leaves narrower, half-clasping the stem. Flowers golden-yellow, 10-30mm, with erect hairy sepals. Meadows, open woodlands, screes and by snow patches, to 2800m. May-August. Mountains of E France and S Germany. Local. Confined to the mountains of C and S Europe.

7 *Ranunculus oreophilus* Bieb. Similar to *R. montanus*, but stem leaves linear and flower rather smaller, with a hairy (not hairless) receptacle. Similar habitat and distribution. Local .

8* **Bulbous Buttercup** *Ranunculus bulbosus* L. Short to medium, very variable, hairy perennial; stem with a swollen, corm-like base. Basal leaves with 3 lobed and toothed, segments, the central one stalked; upper leaves with narrow lobes. Flowers bright yellow, 20-30mm, on furrowed stems; sepals downturned. Achenes in a rounded head. Grassy habitats on well-drained, calcareous or somewhat acid soils, meadows, pastures, road verges, banks, grassy mountain slopes and fixed dunes, to 2500m. March-July. Throughout, except the Faeroes, Iceland and Spitsbergen. Often abundant, especially in grazed meadows and hayfields. Often growing in association with *R. acris*, but generally prefering the drier soils and usually coming into flower earlier. **P**: Several insects, though more especially honey-bees. **B**: Throughout.

* subsp. *bulbifer* (Jordan) Neves. Like the type, but the central leaf-segment unstalked and flower-stalks scarcely furrowed. Similar habitats and flowering time. England and France.

9* **Hairy Buttercup** *Ranunculus sardous* Crantz. Short to medium, slightly hairy annual, like *R. bulbosus*, but the stem base scarcely swollen. Leaves often shiny. Flowers pale yellow, 12-25mm. Grassy meadows, usually grazed, arable land and waste places, on damp, often alluvial soils, to 2500m. May-October. Throughout, except Ireland, the Faeroes, Iceland and much of Scandinavia. **P**: Bees and flies. **B**: Throughout lowlands, especially East Anglia – most common near the coast.

10* **Corn Crowfoot** *Ranunculus arvensis* L. Short to medium, often hairless annual. Lowermost leaves simple, others with 3 segments, each further lobed or toothed. Flowers pale greenish-yellow, 4-12mm, on unfurrowed stalks; sepals downturned. Achenes large, 6-8mm, spiny. Arable land, especially cornfields, on calcareous or clay soils, occasionally on disturbed land or in waste places, usually at low altitudes. May-July. Throughout, except Ireland, the Faeroes, Iceland and Spitsbergen. Local. Becoming increasingly scarce due to the use of herbicides on arable fields. Easily recognized by its spiny fruits. **P**: Probably small flies. **B**: South of the Humber, scattered localities elsewhere.

Ranunculus nemorosus
Multi-flowered Buttercup
Creeping Buttercup
Woolly Buttercup
subsp. borealis
Meadow Buttercup
Mountain Buttercup
subsp. bulbosus
basal leaves
Bulbous Buttercup
Hairy Buttercup
Corn Crowfoot

1* **Small-flowered Buttercup** *Ranunculus parviflorus* L. Short, spreading, hairy annual. Lower leaves with 3-5 toothed lobes; upper leaves simple or with oblong lobes. Flowers small, pale yellow, 3-6mm, on furrowed stalks; sepals downturned. Achenes small, 3mm, spiny. Dry grassy banks, pathsides, roadsides, banks, occasionally a weed of cultivated land such as bulb fields and gardens, generally on exposed light sandy soils. May-July. Britain and W France. Rather inconspicuous and generally decreasing, most often occurring close to the sea. **B**: Mostly in SW England, S Wales, Isles of Scilly and SW Ireland.

2* *Ranunculus illyricus* L. Short, silkily-hairy, tuberous-rooted, perennial. Basal leaves with 3 narrow segments, these sometimes further lobed. Flowers yellow, 20-35mm; sepals downturned. Achenes triangular, in a cylindrical head, beak straight. Dry grassy habitats, banks and scrub. May-July. S Germany. Primarily in C and SE Europe. Cultivated in gardens. **P**: Various insects, particularly bees.

3* **Fan-leaved or Jersey Buttercup** *Ranunculus paludosus* Poiret [*R. flabellatus*]. Very variable short to medium, hairy, tuberous-rooted perennial. Basal leaves shallowly 3-lobed, the others with 3 narrow, toothed segments, the central one stalked. Flowers yellow, 20-32mm, sepals erect. Achenes in a cylindrical head, beak hooked. Seasonally waterlogged ground which dries out during the summer, when the plants quickly die down. May-June. W and NW France. Local. **B**: Only on Jersey.

4* **Goldilocks** *Ranunculus auricomus* L. Short, variable, slightly hairy perennial. Basal leaves and lower stem leaves are only slightly lobed, others with 3-5 narrow segments. Flowers yellow, 15-25mm, petals frequently imperfect, one or more often absent. Achenes downy, in a rounded head. Meadows, woodland, hedgerows and rocky habitats, mainly in the lowlands, but to 2100m in places. April-May. Throughout. Often forming large colonies, particularly on heavy, fertile, calcareous or neutral soils. Some flowers are petaless, while others may have 1-5 petals of varying sizes. Widespread in Europe and W Asia. **B**: Most frequent in the east and south.

5 *Ranunculus affinis* R. Br. [*R. auricomus* var. *glabratus*]. Like *R. auricomus*, but the basal leaves deeply 3 to 5-lobed; stem leaves with linear segments. Achenes hairless. Arctic habitats, grassy places, open woodland or scrub. June-August. Arctic Europe, including Iceland.

6 *Ranunculus pedatifidus* Sm. Short, practically hairless perennial with unbranched stems. Basal leaves heart-shaped with 5-9 lobes; stem leaves deeply cut into linear lobes, unstalked. Flowers yellow, 16-18mm, sepals downturned. Achenes finely downy. Confined to rocky habitats, tundra and moraines. June-July. Spitsbergen.

7* *Ranunculus pygmaeus* Wahlenb. Low, hairless perennial, often tiny; stems unbranched. Leaves kidney-shaped with 3 blunt lobes, the outer two often further lobed; upper leaves unstalked. Flowers yellow, 5-10mm, sepals hairy. Achenes smooth with a short beak. Mountain habitats, short turf and snow patches, 1800-2800m. June-August. Arctic and sub- Arctic Europe, including the Faeroes and Iceland.

8* *Ranunculus nivalis* L. Like *R. pygmaeus*, but taller, to 15cm, the basal leaves more deeply lobed; flowers larger, 15-20mm. Rocky habitats, tundra and moraines, snow patches, to 1550m. June-August. Arctic Scandinavia.

9 *Ranunculus sulphureus* C.J. Phipps. Like *R. nivalis*, but basal leaves wedge-shaped, shallowly-lobed; sepals with dense brown hairs. Arctic habitats, tundra and moraines. June-August. Arctic Scandinavia.

10 *Ranunculus hyperboreus* Rottb. Low slender creeping or floating perennial, practically hairless. Leaves oval, 5-lobed, short-stalked. Flowers small, yellow, 5mm, solitary; sepals and petals 3. Achenes with a short blunt beak. Mountain habitats, rocky place, moraines and tundra, to 2100m. July-August. Arctic and sub-Arctic Europe, including the Faeroes, Iceland and Spitsbergen.

11* **Celery-leaved Crowfoot** *Ranunculus sceleratus* L. Medium, rather stout, mostly hairless annual. Basal leaves with 3 segments, these further lobed or toothed, upper leaves 3-lobed, unstalked; all leaves rather shiny green. Flowers yellow, 5-10mm, numerous in branched clusters; petals equalling the downturned sepals. Achenes hairless with a short blunt beak, in elongated heads. Wet meadows, marshes, shallow pools, the margins of slow-moving streams, rivers and lakes in muddy places, often near the coast, on fertile mineral-rich or alluvial soils, at low altitudes. May-September. Throughout, except the Faeroes, Iceland and Spitsbergen. Local, sometimes abundant. Plants generally germinate in the autumn and overwinter. **P**: Mainly small flies. **B**: Mainly in the lowlands and most common in the south and east.

*Ranunculus nivalis*

Small-flowered Buttercup

*Ranunculus illyricus*

Fan-leaved Buttercup

Goldilocks

*Ranunculus pygmaeus*

*Ranunculus nivalis*

Celery-leaved Crowfoot

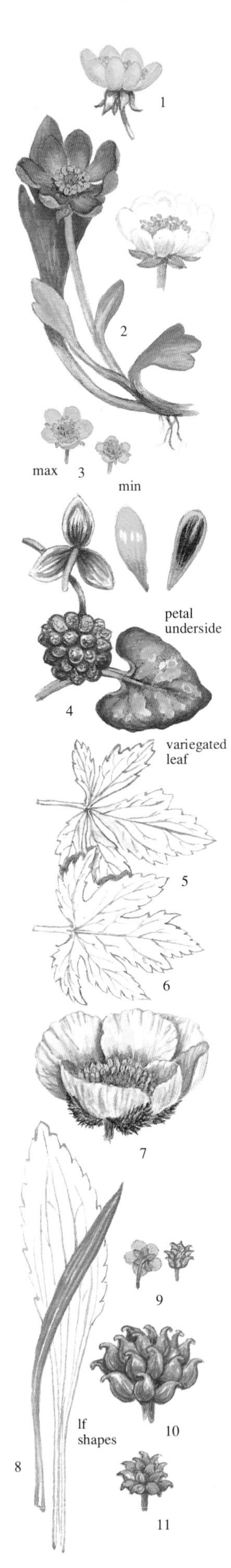


1* **Lapland Buttercup** *Ranunculus lapponicus* L. Low, slender, creeping perennial, rooting at the nodes. Leaves kidney-shaped, deeply 3-lobed. Flowers yellow, 8-13mm, solitary, long-stalked, with 6-8 petals and downturned sepals. Achene hairless. Mountain habitats, rocky places, moraines and tundra. June-August. Arctic Europe with the exception of Iceland.

2 *Ranunculus pallasii* Schlecht. Like *R. lapponicus*, but leaves more shallowly-lobed and flowers larger, with 6-12 reddish-violet or white petals. Rocky habitats and moraines. June-August. Spitsbergen. Hybridizes with *R. lapponicus*, *R.* × *spitzbergensis* Hadac.

3* *Ranunculus cymbalaria* Pursh. Low to short, slender perennial with ascending stems and creeping stolons. Leaves kidney-shaped, blunt, toothed. Flowers bright yellow, 6-9mm, 5-petalled, solitary or several on branched stalks. Achenes in an elongated head. Damp habitats, pool and stream margins, marshy ground. **I**: North America. Naturalized in Scandanavia.

4* **Lesser Celandine** *Ranunculus ficaria* L. [*Ficaria verna*, *F. ranunculoides*]. Variable low to short, tuberous-rooted, hairless perennial. Leaves heart-shaped, rather fleshy, dark green, sometimes with dark markings. Flowers bright glistening yellow, turning white with age, 20-30mm, with 8-12 petals, narrow elliptical; sepals 3. Deciduous woodland, hedgerow, road verges, grassy banks, ditches, damp grassy habitats, stream and river margins, often in seasonally flooded habitats, in both open and shaded habitats, to 900m. March-May. Throughout, except the Faeroes, Iceland and Spitsbergen. An often troublesome weed, especially in gardens. The typical species is a diploid, but subsp. *bulbilifer* is tetraploid and practically sterile. **P**: Various insects, including bees and flies. **B**: Throughout, often abundant.

* subsp. *bulbilifer* Lambinon [*R. ficaria* var. *bulbifer*] Like the type, but more slender and with small bulbils in the leaf angles; flowers smaller. Achenes mostly sterile. Often in more shaded habitats or on disturbed ground. Throughout the range of the species.

5* **Aconite-leaved Buttercup** *Ranunculus aconitifolius* L. Medium, slightly hairy, tufted perennial with erect stems. Leaves palmately 3-5-lobed, the lobes lanceolate, toothed, the middle one free to the base. Flowers white, with reddish or purplish sepals, 10-20mm, in laxly branched clusters; petals 5. Damp mountain habitats, meadows, open woodland, stream margins, marshy ground and ditches, often forming extensive colonies, to 2600m. May-August. C & E France and S Germany. Confined to the mountains of C & S Europe. **P**: Various small insects. Cultivated in gardens.

6 *Ranunculus platanifolius* L. Like *R. aconitifolius*, but taller, to 1.3 m. Leaves 5-7-lobed, the middle lobe not free to the base. Damp woodland, ditches and marshy areas, mainly lowland, but up to 1600m in the south of its range. May-August. North to W Norway and S Sweden.

7* **Glacier Crowfoot** *Ranunculus glacialis* L. [*Oxygraphis vulgaris*]. Low to short mostly hairless perennial. Basal leaves thick, rounded in outline but deeply-lobed, stalked; stem leaves few-lobed, unstalked. Flowers large, white becoming pink or purplish tinted, 25-40mm, solitary or 2-3; sepals with purple-brown hairs. Mountain habitats, bare rocky places, moraines, screes, often near late snow patches, 2300-4250m. June-October. Mountains of France and Germany, Scandinavia, the Faeroes and Iceland. The pinking of the ageing flowers is very characteristic.

8* **Lesser Spearwort** *Ranunculus flammula* L. Very variable short to medium, hairless tufted perennial, erect or prostrate and then rooting at some nodes, but without runners; stems often reddish. Basal leaves lanceolate to oblong, thin sometimes toothed. Flowers yellow, 7-20mm, several together, rarely solitary, on furrowed stalks. Wet habitats, meadows, marshes, fens, woodland flushes, pool and stream margins, on calcareous or neutral soils usually, to 2000m. May-September. Throughout, except Spitsbergen, mainly in the mountains in the south of its range. **P**: Small flies and bees. **B**: Throughout.

subsp. *minimus* (Ar. Benn.) Padmore. Like the type, but stems more prostrate, not rooting at the nodes; basal leaves heart-shaped at the base, thick and short-stalked. Exposed coastal habitats. June- September. **B**: W Ireland, N & W Scotland. Very local.

subsp. *scoticus* (E. S. Marshall) Clapham. Like the type, but always erect, the basal leaves much reduced and soon withering, other leaves linear or with a reduced lanceolate blade; flowers often solitary. Lakeshore habitats, similar flowering time. **B**: N Scotland. Very local.

9* **Creeping Spearwort** *Ranunculus reptans* L. Like small slender forms of *R. flammula*, but with creeping runners rooting at each node. Leaves elliptical to paddle-shaped, all stalked. Flowers yellow, 5-10mm, solitary. Gravelly lake margins, very local and often rare. June-August. The Faeroes, Iceland and parts of Scandinavia. Hybrids with *R. flammula* frequently occur and many so-called *R. reptans* prove to be of hybrid origin, particularly in Britain. Hybrid plants produce very little fertile pollen. **B**: Extinct in Britain, where replaced by hybrids, *R. reptans* × *R. flammula*.

10* **Great Spearwort** *Ranunculus lingua* L. Medium to tall, robust, stoloniferous, hairless perennial, with erect hollow stems, often reddish. Basal leaves oval-heart-shaped, long-stalked or unstalked, toothed. Flowers large, bright glossy yellow, 30-50mm, on long, branched stalks. Marshy ground, fens, stream margins, ponds and ditches, often growing in shallow water, on mineral-rich or organic, often calcareous, soils, to 1200m. June-September. Throughout, except the Faeroes, Iceland and Spitsbergen. **P**: Various small flies. A vigorous but rather local species often planted in water gardens. **B**: Throughout, though rare in Scotland.

11* **Adder's-tongue Spearwort** *Ranunculus ophioglossifolius* Vill. Short, sometimes slightly hairy annual. Basal leaves oval to rounded-heart-shaped, long-stalked, the upper small leaf often unstalked. Flowers yellow, 5-9mm, numerous in branched clusters; petals 5. Marshy habitats. May-August. S Britain, France and S Sweden – Gotland. Local and often rare. Seeds germinate in the autumn and young plants overwinter to flower the following year. Very vulnerable; susceptible to land-drainage and to competition from coarse vegetation. **B**: Very rare, confined to 2 sites in Gloucestershire; a protected species.

Ranunculus cymbalaria
Lesser Celandine
ssp. bulbifer
Lapland Buttercup
Lesser Spearwort
Glacier Crowfoot
Adder's-tongue Spearwort
Aconite-leaved Buttercup
Creeping Spearwort
Great Spearwort

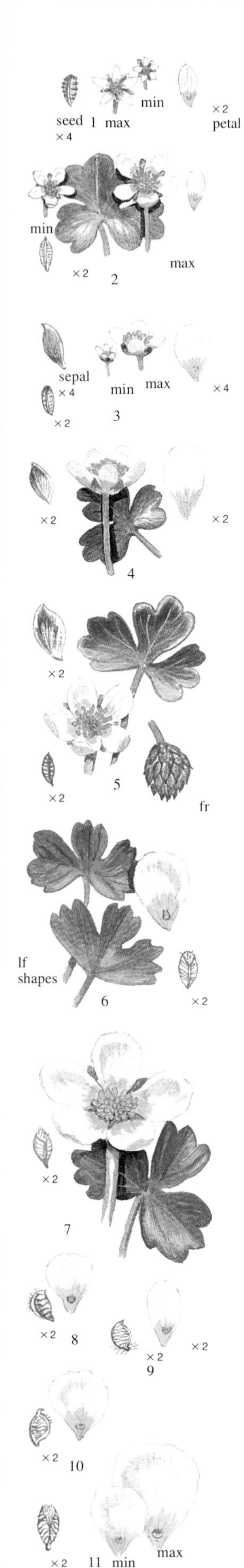


**Water Crowfoots** (*Ranunculus* subgenus *Batrachium*). Aquatic plants, occasionally growing on mud. Leaves of 2 kinds, floating and leaf-like or submerged and composed of numerous linear segments, either or both on the same plant. Flowers solitary, white. Widespread in the temperate northern hemisphere. Those adapted to faster-flowing water tend to have no floating leaves and long parallel divisions on the submerged leaves. In all species the flowers are held just above the water surface.

1* **Ivy-leaved Crowfoot** *Ranunculus hederaceus* L. Trailing annual or biennial of mud or shallow water. Leaves all laminate, kidney or heart-shaped, with 3-5 shallow lobes, the lobes widest at the base. Flowers small, 3-6mm, petals scarcely longer than the sepals. Muddy habitats and shallow fresh water, often in temporary pools, rare in limestone regions. May-September. Britain, Ireland, Belgium, Holland, France and Germany. This and the following species never have submerged leaves with slender segments. **B**: Throughout, less frequent in the north.

2 **Round-leaved Water Crowfoot** *Ranunculus omiophyllus* Ten. [*R. lenormandii*]. Similar to *R. hederaceus*, but leaf-lobes narrowed at the base and flowers larger, 8-12mm, the petals at least twice as long as the sepals. Muddy habitats, ditches, dykes and slow moving streams, on acid, non-calcareous soils, always in open habitats. June-August. Britain, Ireland, France and Holland. **B**: S & W Britain and S Ireland, scarce elsewhere.

3* **Three-lobed Water Crowfoot** *Ranunculus tripartitus* DC. [*R. lutarius*]. An annual or perennial. Floating leaves round with 3 shallow lobes; submerged leaves (absent in terrestrial plants) with compressed segments. Flowers 3-10mm, the petals to 4mm long, up to twice the length of the sepals which have a blue band near the tip. Muddy habitats and shallow, often temporary pools, ditches and dykes in open habitats, on acid or non-calcareous soils. March-July. Britain, Ireland, Holland, France and Germany. Often behaves as an annual in temporary pools and muddy areas. Declining generally, particularly in Britain. **B**: SW England, Wales, Cheshire, Ireland – Cork.

4 *Ranunculus ololeucos* Lloyd. Like *R. tripartitus*, but flowers larger, the petals 6mm long or more, more than twice the length of the sepals. Shallow water habitats. May-July. Belgium, Holland, France and Germany.

5 **Brackish Water Crowfoot** *Ranunculus baudotii* Godron [*Batrachium marinum*]. An annual or perennial, with or without floating leaves. Floating leaves deeply 3-lobed; submerged leaves yellowish-green with spreading rigid segment not collapsing out of water. Flowers 12-18mm, the sepals blue-tipped. Fruit head elongating, not rounded. Brackish water ditches and pools, ditches and dykes, often close to the coast, on nutrient rich and alluvial soils. May-September. Throughout Europe, except the extreme north. Usually behaving as an annual in temporarily wet habitats. **B**: S & E England, local or rare elsewhere.

6* **Pond Water Crowfoot** *Ranunculus peltatus* Schrank [*Batrachium dichotomum*, *B. langei*, *B. triphyllos*]. An annual or perennial with floating or submerged leaves, or both. Floating leaves rounded, with 3-7 shallow lobes; submerged leaves shorter than internodes, with rigid segments. Flowers 15-20mm, petals with pear-shaped nectaries; fruiting stalks 5-15cm long. Shallow water habitats, lakes, ponds, slow-moving streams, ditches and dykes, generally in nutrient-rich waters. May-August. Throughout, except the Faeroes, Iceland and Spitsbergen. Can be an annual in temporarily wet habitats. **B**: Generally frequent.

7 *Ranunculus penicellatus* (Dumort.) Bab. [*R. pseudofluitans*, *R. peltatus* subsp. *pseudofluitans*]. More robust than *R. peltatus*, the submerged leaves less rigid, longer than the internodes and the flowers larger; fruit-stalks 20-30mm. Fast-flowing rivers and streams, particularly in limestone regions, generally in nutrient-rich waters, especially on harder rocks. June-July. Throughout Europe, except the extreme north. **B**: Throughout, except parts of the north.

8* **Common Water Crowfoot** *Ranunculus aquatilis* L. [*Batrachium gilibertii*]. Like *R. peltatus*, but petals with circular nectaries and fruit-stalks up to 5cm long. Shallow, fresh-water habitats, still or slow-moving, ponds, ditches, dykes and streams, often over rather hard rocks. April-September. Throughout, except the Faeroes, Iceland and Spitsbergen. The commonest species in many regions. **B**: Throughout, scarcer in the north.

9* **Thread-leaved Water Crowfoot** *Ranunculus trichophyllus* Chaix [*R. paucistamineus*, *Batrachium divaricatum*]. Dark green annual or perennial with submerged leaves only. Submerged leaves short, up to 4cm, with spreading, sometimes rigid, segments. Flowers 5-10mm, petals with a moon-shaped nectary at the base; fruit-stalks usually less than 4cm. Shallow fresh-water habitats, moderately fast-moving streams, canals, ditches, dykes and pools, occasionally in seasonally flooded habitats (then acting as an annual). May-July. Throughout, except Spitsbergen and Arctic-alpine regions. **B**: Throughout, most frequent in the south and east.

subsp. *lutulentus* (Perr. & Song.) Vierh. More delicate than the type, the stems rooting at most internodes. Arctic alpine habitats. Scandinavia.

10* **Fan-leaved Water Crowfoot** *Ranunculus circinatus* Sibth. [*R. divaricatus*]. Perennial with rigid submerged leaves, circular in outline, segments all in one plane; no floating leaves. Flowers 8-18mm, petals scarcely touching. Lakes, ponds, canals, ditches and slow-moving streams, generally in mineral-rich waters, often silty, locally common. June-August. Throughout, except the Faeroes, Iceland and Spitsbergen. Distinguished by the absence of floating leaves and by the submerged leaves all being in one plane. **B**: Britain from the Midlands southwards, local or rare elsewhere, including Ireland.

11* **River Water Crowfoot** *Ranunculus fluitans* Lam. Perennial, rather greenish-black overall. Submerged leaves with a few very long parallel segments easily collapsing out of water; without floating leaves. Flowers large, 20-30mm, petals overlapping. Fast-flowing streams and rivers, generally over hard rocks, often growing in depths of 1m, sometimes more, in nutrient-poor or calcium-rich waters. June-August. Europe north to S Scandinavia. **B**: S Scotland southwards, but rare in Wales and SW England; Ireland – confined to Antrim.

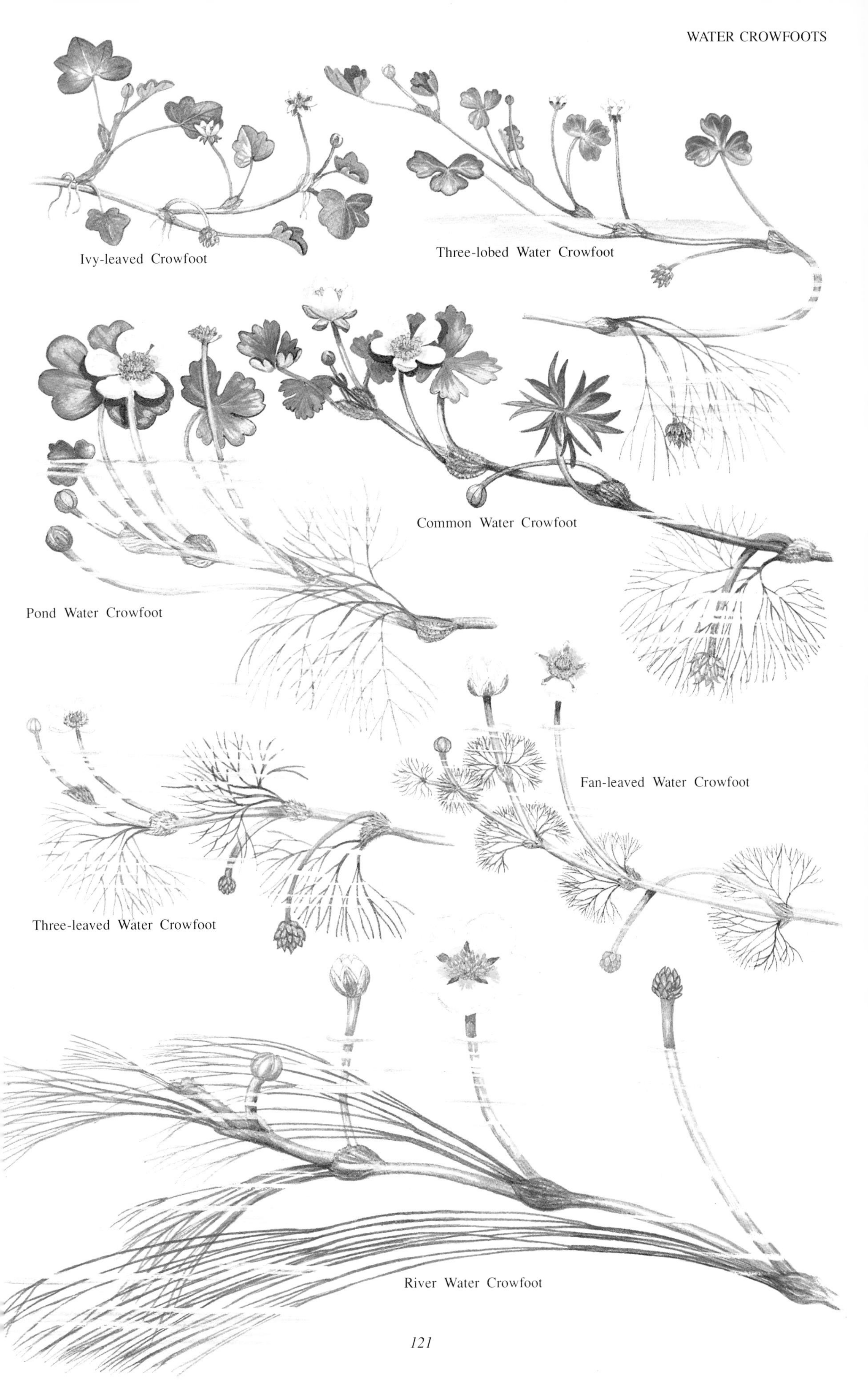
Ivy-leaved Crowfoot
Three-lobed Water Crowfoot
Common Water Crowfoot
Pond Water Crowfoot
Fan-leaved Water Crowfoot
Three-leaved Water Crowfoot
River Water Crowfoot

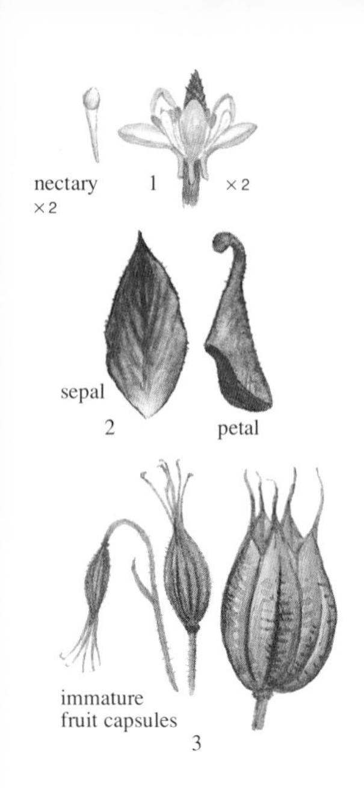


1* **Mousetail** *Myosurus minimus* L. Low to short hairless, rather fleshy, annual. Leaves in a basal rosette, linear, untoothed. Flowers pale greenish-yellowish, 4-5mm, solitary on long stalks; sepals 5-7, petal-like. Fruit a long slender, 'mousetail-like' spike of achenes. Bare and waste habitats, arable fields, margins of paths, always local, to 1400m. March-June. Throughout Europe, except the extreme north. Increasingly scarce due to modern farming practices. **P**: Small flies, often self-pollinated. An inconspicuous little plant. **B**: Rare and decreasing, mostly confined to S England.

## **Columbines** *Aquilegia*. Tufted perennial herbs with ternately divided leaves. Flowers usually nodding, 5-parted, sepals petal-like, the petals each with a long backward pointing spur. Fruit a small cluster of follicles. 60 species scattered right across the northern hemisphere. Many are cultivated in gardens, where they hybridize readily.

2* **Common Columbine** *Aquilegia vulgaris* L. Medium to tall, branched, hairy perennial. Leaves 2-ternate, dull green, the leaflets usually 2 to 3-lobed. Flowers purple-blue or violet, rarely white, 30-50mm, with hooked spurs; stamens scarcely protruding beyond the petals. Fruit stickily-hairy. Open woods, scrub, grassy habitats and banks, sometimes on marshes or fens, to 2000m. May-July. **B**: Throughout Europe, except the extreme north. Frequently cultivated in gardens and there usually hybridizing with other species. These and double-flowered forms are cultivated and frequently escape to become naturalized. **P**: Long-tongued bumble-bees. **B**: S Scotland southwards, local.

3* **Dark Columbine** *Aquilegia atrata* Koch. Like *A. vulgaris*, but more densely hairy above. Leaflets hairless (not hairy) beneath. Flowers dark purplish-violet, 30-40mm; stamens protruding well beyond the petals. Woodland clearing, scrub and rocky habitats, on calcareous soils, to 2000m. May-July. E France and S Germany. Confined to the mountains of C and S Europe. Hybridizes with *A. vulgaris* in gardens.

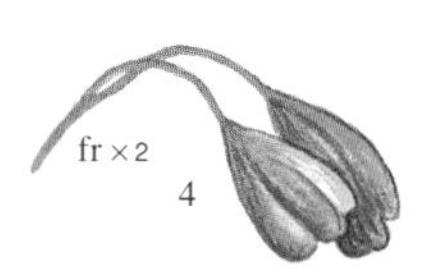


## **Meadow-rues** *Thalictrum*. Perennial herbs with 2-3-pinnate or 2-3-ternate leaves. Flowers small, in panicles or racemes. Perianth segments 4-5, sepal-like, usually falling quickly. Stamens many, conspicuous. Fruit a small cluster of single-seeded achenes.

4* **Greater Meadow-rue** *Thalictrum aquilegifolium* L. Tall, hairless perennial, to 1.5m. Leaves 2-3-ternate, with broad, toothed, leaflets. Flowers in a much-branched, 'mop-headed' panicle, with whitish or lilac stamens. Achenes drooping, winged, long-stalked. Damp woodland and meadows, to 2500m. June-July. France and Germany, Finland and S Sweden. Commonly cultivated in gardens. The fruit is very distinctive.

5* **Alpine Meadow-rue** *Thalictrum alpinum* L. Low to short, rather delicate perennial. Leaves 2-ternate, with small rounded leaflets. Flowers purplish-green or violet, with yellow anthers, pendent at first, in a simple raceme. Fruit pendent, achenes 2-3, ribbed. Damp mountain turf, rocky habitats, ledges, on calcareous or slightly acid soils, to 2900m. May-July. Britain, Iceland, France and Scandinavia. Sometimes grows close to sea-level in the north of its range. A rather inconspicuous little plant.**P**: Wind. **B**: N England and N Wales northwards, scarce in Ireland.

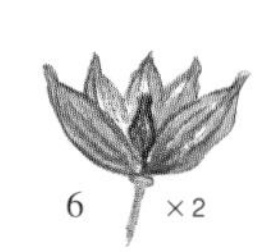


6* **Lesser Meadow-rue** *Thalictrum minus* L. Very variable, sometimes glandular, medium, perennial. Leaves 3-4-ternate, leaflets rounded to oval, irregularly lobed or toothed. Flowers yellowish, in lax long-branched panicles. Fruit erect, achenes 3-15, unstalked, weakly ribbed. Damp to rather dry grassland, rocky places and sand-dunes, sometimes along stream margins, generally on calcareous soils, to 2850m. June-August. Throughout, except the extreme north, the Faeroes, Iceland and Spitsbergen. **P**: Usually wind. **B**: Throughout England, Scotland, Wales and N & E Ireland, local.

subsp. *majus* (Crantz) Hook. f. [*T. elatum*]. Like the type, but taller, to 1.2m; leaflets larger, mostly 10-30mm (not 5-15mm). Similar habitats and distribution.

subsp. *kemense* (Fries) Tutin. Like the type, but the inflorescence narrow and few-flowered. Rocky places, moraines, tundra. June-August. Arctic Europe.

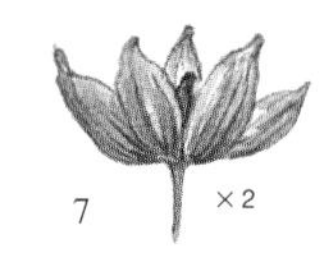


7* **Small Meadow-rue** *Thalictrum simplex* L. Medium to tall hairless perennial. Basal leaves 2-3-pinnate, the leaflets oblong-wedge-shaped, toothed. Flowers yellowish, in a leafy, branched, rather narrow panicle, short-stalked, nodding at first. Achenes elliptical, ribbed. Damp meadows, lake and stream margins, occasionally in rocky habitats, to 2000m. June-August. Continental Europe north to S Scandinavia. Local.

subsp. *boreale* (F. Nyl.) Tutin. Like the type, but leaflets more oval; flowers fewer, on long stalks. Similar habitats. July-August. Finland.

subsp. *gallicum* (Rouy & Fouc.) Tutin. Like the type, but leaflets mostly 3-lobed, with recurved margins; flowers numerous, long-stalked. Similar habitats and flowering time. E France.

subsp. *bauhinii* (Crantz) Tutin. Like subsp. *gallicum*, but leaflets linear-lanceolate, untoothed to 2-3-lobed; flowers short-stalked. Similar habitats and flowering time to *T. simplex*. Germany and Sweden.

8 *Thalictrum lucidum* L. Like *T. simplex*, but tufted, the leaves mostly unstalked and with narrower leaflets, untoothed to 2-3-lobed. Flowers in dense rounded clusters; stamens erect. Grassy habitats, particularly meadows. June-July. C & S Germany.

9* **Common Meadow-rue** *Thalictrum flavum* L. Tall, almost hairless perennial, with a far-creeping rhizomatous stock. Leaves 2-3-pinnate, leaflets oblong-wedge-shaped, 3-4-lobed. Flowers yellow, in dense oblong panicles, stamens erect. Achenes round, 6-ribbed. Wet grassy habitats, meadows, fens and marshes, river and stream margins, on base-rich or calcareous soils, generally at low altitudes. June-August. Throughout, except the Faeroes, Iceland and Spitsbergen. **P**: Bees and flies, also wind. **B**: North to Inverness, including Ireland.

10 *Thalictrum morisonii* C.C. Gmelin [*T. exaltatum*]. Like *T. flavum*, but often taller and with shiny stems. Flower panicles rounded and less crowded. Achenes with 8-10 ribs. Meadows and open scrub. June-July. C & E France and SW Germany. Local and rather rare.

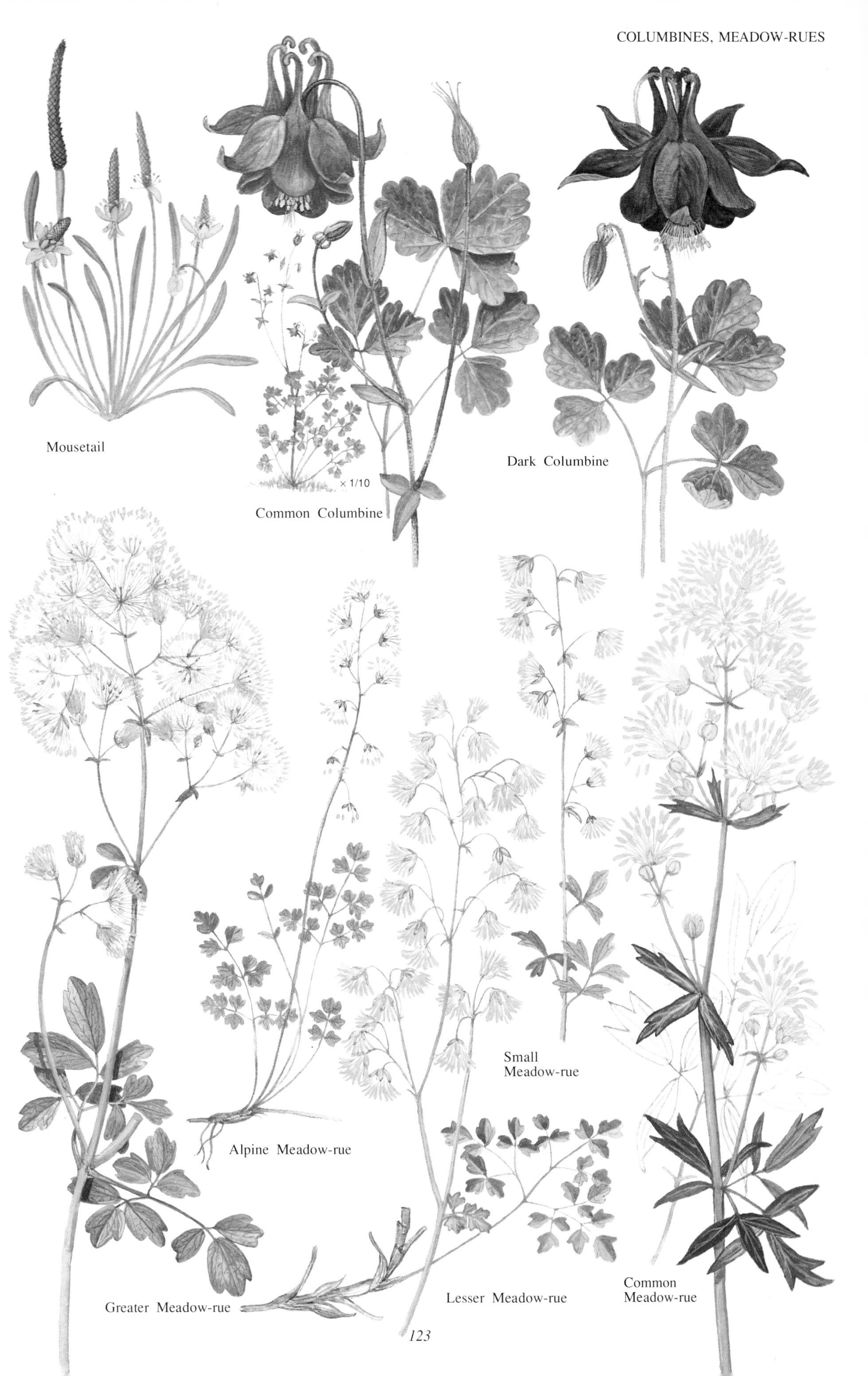
Mousetail
× 1/10
Common Columbine
Dark Columbine
Small Meadow-rue
Alpine Meadow-rue
Greater Meadow-rue
Lesser Meadow-rue
Common Meadow-rue

1* **Peony** *Paeonia mascula* (L.) Miller [*P. corallina*, *P. caucasica*]. (Peony Family – Paeoniaceae – 33 species). Medium to tall, stout, herbaceous perennial. Leaves alternate, large, 2-ternate, the elliptical leaflets occasionally with 1-2 lobes, untoothed, hairless. Flowers large, red, 8-14cm, solitary, with 5-9 glossy petals. Fruit a group of 3-5 furry follicles, containing large red or black seeds. Scrub covered slopes, rocky habitats and cliffs, generally on calcareous soils. April-May. NC & E France south; naturalized in W Britain. Poisonous. **B**: Very locally naturalized – Steep Holme in the Bristol Channel.

## BARBERRY FAMILY Berberidaceae

Shrubs or herbs with leathery, often spiny-edged leaves. Flowers hermaphrodite, generally 3-4-parted, the inner segments petaloid, with nectaries. Stamens 4-6, opposite the inner petals; anthers opening by apical valves. Fruit a capsule or berry. 575 species in 14 genera.

2* **Barren-wort** *Epimedium alpinum* L. Short tufted perennial. Leaves 2-ternate, leaflets heart-shaped, finely spine-edged. Flowers dull red or pinkish-grey, with yellow petals inside, 9-13mm, in lax racemes. Fruit a capsule. Damp shaded places, woods, mountain slopes, to 1300m. April-July. **I**: SE Europe. Naturalized W Europe.. Young leaves often flushed red. **B**: Locally naturalized, often near buildings.

3* **Barberry** *Berberis vulgaris* L. Densely branched spiny deciduous shrub, 1.5-3m, with yellowish, ridged twigs; spines 3-forked. Leaves oval, finely spine-edged. Flowers bright yellow, 6-8mm, in pendent racemes. Berry oblong, bright orange-red when ripe, edible. Hedgerows, dry woodland and scrub, rocky places, often on calcareous soils, to 2300m. May-June. Throughout, except the far north; naturalized in Britain and Scandinavia. A host for Wheat Rust, and eliminated in some areas. **B**: Local and rare throughout.

4* **Oregon Grape** *Mahonia aquifolium* (Pursh) Nutt. [*Berberis aquifolium*]. Stoloniferous evergreen shrub 0.5-1m, with stout, scarcely branched stems. Leaves pinnate with 5-9 oval, deep shiny-green, spine-edged. Flowers yellow, in clustered terminal racemes, fragrant. Berry rounded, black with a bluish bloom. Woods, hedgebanks, usually in shaded or semi-shaded places. March-May. **I**: W North America. Locally naturalized in Britain, Holland and Germany. **B**: Naturalized, particularly in C & E England.

## POPPY FAMILY Papaveraceae

Herbs with latex. Leaves usually alternate. Flower hermaphrodite with 2 deciduous sepals. Petals 4, occasionally more. Stamens numerous. Fruit a capsule with pores or splitting lengthwise. 200 species in 26 genera, chiefly confined to the north temperate hemisphere.

5* **Opium Poppy** *Papaver somniferum* L. Medium to tall, erect greyish-green annual. Leaves oblong, but lobed and undulate, the upper clasping the stem. Flowers large 8-18cm, purple, often with a deep blotch at the base of each petal, pale lilac or white, sometimes red; anthers pale yellow. Capsule rounded, hairless, variable. Cultivated and waste land, waysides. June-August. **I**: E Europe and W Asia. Naturalized or casual throughout, except the far north. The edible seeds are used in bread and cakes. Opium from the capsule is refined to heroin and some 30 different alkaloids, including Codeine and Morphine. **B**: Throughout.

6* **Common Poppy** *Papaver rhoeas* L. Medium, rough-hairy annual, usually branched. Leaves pinnately-lobed, with narrow, toothed divisions. Flowers scarlet, 7-10cm, the petals overlapping, often with a black blotch at the base, anthers bluish; flower stalks with spreading hairs. Capsule rounded, hairless. Cultivated fields, waste places, field boundaries, roadsides, to 1800m. June-October. Throughout, except the far north. Declining in many placese **B**: Throughout, local in Wales and Scotland.

7* **Long-headed Poppy** *Papaver dubium* L. [*P. obtusifolium*, *P. modestum*]. Short to medium hairy annual with white latex. Leaves pinnately-lobed, with broad, blunt, divisions, somewhat greyish-green. Flowers pale scarlet, 3-7cm, the petals occasionally blotched at the base, anther violet; flower stalks with hairs pressed close to the stem. Capsule oblong, hairless. A widespread weed, to 1800m. June-August. Throughout, except Iceland and Spitsbergen. **B**: Throughout, but scarce in the far north of Scotland.

8 **Babington's Poppy** *Papaver lecoqii* Lamotte. Like *P. dubium*, but latex turning yellow on exposure to air and flowers more orange-yellow, with yellow anthers. Cultivated and waste land, waysides. June-August. Britain, Belgium and France. **B**: Scattered localities, particularly in C & S England.

9* **Prickly Poppy** *Papaver argemone* L. Short to medium, stiffly-hairy annual. Leaves pinnately-lobed with linear or oblong, bristle-tipped, divisions. Flowers pale scarlet, 2-6cm, the petals not overlapping, sometimes with a dark blotch at the base, anthers bluish. Capsules oblong, with a few erect bristles. Arable fields, disturbed ground, roadsides, generally on light dry sandy soils, to 1700m. May-July. Throughout, except Iceland and the extreme north. **B**: Scattered localities throughout.

10 **Rough Poppy** *Papaver hybridum* L. [*P. hispidum*]. Rather like *P. argemone*, but flowers crimson, the petals with a dark blotch at the base and the capsule rounded, with dense yellowish bristles. Arable fields, disturbed land, roadsides, generally on well-drained calcareous soils, to 1800m. June-August. Throughout, except the north, Ireland and Iceland. **B**: Rare, S and E England, sometimes casual elsewhere.

Peony
Barren-wort
Barberry
Oregon Grape
Opium Poppy
Common Poppy
Prickly Poppy
Long-headed Poppy

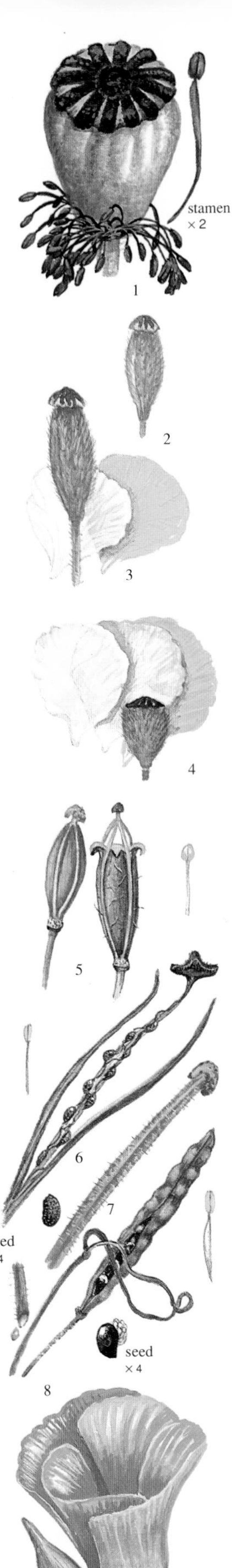


1* **Oriental Poppy** *Papaver orientale* L. Tall, robust, bristly perennial forming large spreading tufts. Leaves oblong, pinnately-lobed, rough, the basal long-stalked, the upper unstalked. Flowers very large, 10-18cm, orange-red or crimson, the petals often with a large dark blotch at the base; anthers deep violet. Capsule rounded, hairless. Scrub and cultivated land. May-July. **I**: SW Asia. Locally naturalized in Britain, France, Holland and Finland. Widely cultivated in gardens, including forms with white, pink and orange flowers – some cultivars represent hybrids between this and the following species. **P**: Bees. **B**: Locally naturalized from gardens in a few places.

*Papaver bracteatum* Lindley. Like *P. orientale* and much confused with it. Distinguished by having one or several short leaves immediately below the flower. **I**: W Asia. Possibly locally naturalized from gardens in a few places in Britain and W Europe. May-June.

2* **Lapland Poppy** *Papaver lapponicum* (Tolm.) Nordh. Low-short, tufted, bristly perennial with yellow latex. Leaves pinnately-lobed, with narrow divisions. Flowers yellow, 2.6-4.6mm; flower stalks with hairs pressed close to the stem. Capsule narrow-oblong, broadest towards the top, with a few closely adpressed bristles. Rocky habitats, moraines, stabilized screes and tundra. July-August. Arctic Norway. Occasionally cultivated in gardens.

3 *Papaver dahlianum* Nordh. Like *P. lapponicum*, but the leaves less finely dissected, the latex white; flowers yellow or white, slightly larger. Similar habitats and flowering time. Arctic Norway.

4* **Arctic Poppy** *Papaver radicatum* Rottb. Low-short, tufted, hairy perennial with persistent leaf-bases; latex white or yellow. Leaves pinnately-lobed, with oval divisions. Flowers usually yellow, sometimes white or pale pink, 30-40mm; flower-stalks with spreading hairs. Capsule elliptical to almost rounded, covered in dense soft hairs. Bare rocky habitats and tundra, moraines, screes, gravelly places, to 1850m. June-August. NW Europe – the Faeroes, Iceland and Scandinavia. Sometimes cultivated in gardens.

5* **Welsh Poppy** *Meconopsis cambrica* (L.) Vig. Medium, slightly hairy, loosely tufted perennial with yellow latex. Leaves pinnately-lobed with rounded or oblong divisions; stem leaves similar but short-stalked. Flowers yellow, rarely orange, 4-8cm, solitary on long stalks. Capsule elliptical, hairless, with a short style at the apex. Moist shaded habitats, woodland, hedgebanks, old walls, streamsides and rocky places, to 2000m. June-August. Britain, Ireland, W France; naturalized in Holland and Germany. Orange and double-flowered forms sometimes escape from, gardens. **B**: Restricted to SW England, Wales and parts of S & W Ireland; naturalized elsewhere, especially in the Cotswolds and the Lake District.

6* **Yellow Horned-poppy** *Glaucium flavum* Crantz [*G. luteum*, *Chelidonium glaucium*]. Medium to tall, greyish, somewhat fleshy biennial or short-lived perennial, with yellow latex. Leaves pinnately-lobed, undulate, the basal long-stalked, the upper clasping the stem, roughly hairy. Flowers yellow, 6-9cm, solitary. Capsule very long, to 30cm, linear, sickle-shaped, splitting lengthwise when ripe, hairless. Sandy habitats, especially coastal dunes, shingle and cliffs, waste ground inland, always on well drained-soils, especially calcareous ones. June-September. Europe north to S Norway. **P**: Bees and flies. Poisonous. **B**: Coasts north to Argyll in Scotland.

7* **Red Horned-poppy** *Glaucium corniculatum* (L.) J.H. Rudolph [*G. phoeniceum*, *Chelidonium corniculatum*]. Like *G. flavum*, but flowers orange or reddish, smaller, 3-5cm. Capsule straight and hairy. Bare ground and waste places. June-August. **I**: S Europe and SW Asia. Naturalized in Europe north to S Sweden. **B**: A very rare casual.

8* **Greater Celandine** *Chelidonium majus* L. Medium to tall, rather brittle, tufted perennial with orange latex, practically hairless. Leaves pale greyish-green, pinnately-lobed with rounded divisions. Flowers small, bright yellow, 1.5-2.5cm, in small umbels of 2-6. Capsule linear, hairless. Bare and waste ground, open woodland, hedgebanks, beneath old walls, often close to habitation, in moist semi-shaded habitats, to 1700m. April-October. Throughout, except Iceland and most of the smaller islands. Cultivated since medieval times, the latex being used for curing warts and as an eye ointment. The plant is, however, extremely poisonous. **P**: Bees and flies. **B**: Throughout, though scarcer in N Scotland and parts of Ireland.

9 **Californian Poppy** *Eschscholzia californica* Cham. Short-medium, hairless, greyish, short-lived perennial or annual; stems erect to sprawling. Leaves finely divided into linear segments. Flowers yellow or orange, 3-7cm, with satiny petals, solitary on long slender stems; sepals forming a pointed 'cap' which falls off in one piece as the petals expand. Capsule linear, slightly curved, splitting lengthwise. Open sunny habitats on light well-drained soils. July-September. **I**: SW USA. Naturalized from gardens in parts of Britain, France, Holland and Germany. Widely cultivated in gardens – forms range in colour from pale cream to yellow, orange, red and pink; double-flowered forms are also grown. The yellow and orange single-flowered forms are normally the ones which become naturalized. **B**: Scattered localities, especially in C & S Britain.

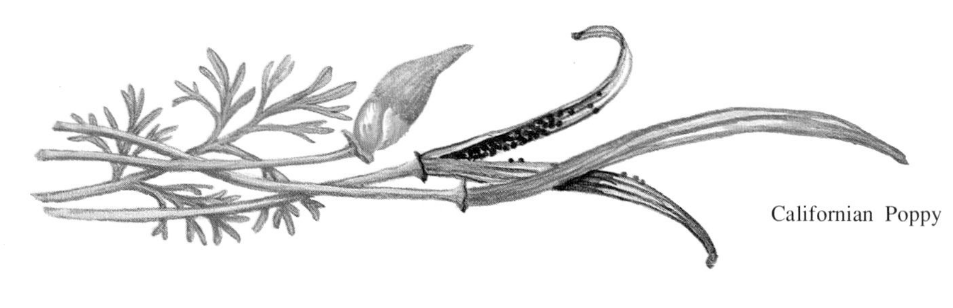

Californian Poppy

Oriental Poppy
Lapland Poppy
Arctic Poppy
Greater Celandine
Yellow Horned-poppy
Red
Horned-poppy
Welsh Poppy

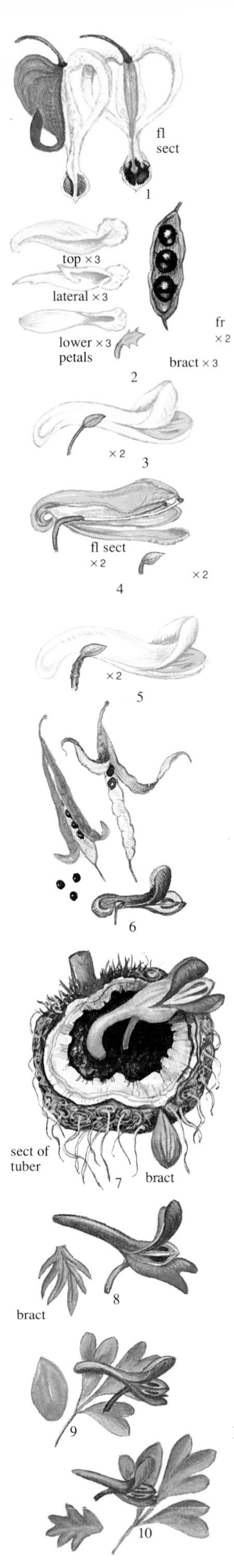


# FUMITORY FAMILY Fumariaceae

Included within the Papaveraceae by some authorities but distinguished in our area at least by having small, irregular, 2-lipped flowers. Most are hairless herbs with watery sap. Sepals 2, small and often inconspicuous. Petals 4, the outer 2 larger, one or both pouched or spurred at the base. Stamens generally 2. Fruit a capsule splitting lengthwise or single-seeded and indehiscent. 450 species in 16 genera, mainly in the north temperate hemisphere, but also in the mountains of E and S Africa. A number are cultivated, particularly *Dicentra* and *Corydalis*.

1* **Bleeding Heart** *Dicentra spectabilis* (L.) Lemaire. Medium hairless, leafy perennial with reddish, succulent stems. Leaves 2-ternate, the leaflets often 3-parted, irregularly toothed. Flowers heart-shaped, pendent, 14-18mm, the outer petals red, pouched at the base, the inner white, in lax almost horizontal racemes. Naturalized in moist semi-shaded habitats. June-July. **I**: E Asia. Naturalized in Finland. A garden form has pure white flowers.

**Corydalis** *Corydalis*. Annual or perennial herbs, hairless, with compound, often ferny-looking leaves. Flowers 2-lipped, the upper petal with a pronounced spur at the base. Fruit an oblong or linear 2-valved capsule splitting lengthwise and containing several to many seeds. About 320 species, mainly in north temperate regions, especially in the Himalaya and W China. A number are commonly cultivated in gardens. Similar to *Fumaria* in general appearance, but plants mostly perennial and with long, dehiscent fruit pods.

2* **Climbing Corydalis** *Ceratocapnos claviculata* (L.) Lidén [Corydalis claviculata]. Medium to tall, rather delicate, pale green, much-branched climbing annual. Leaves 2-pinnate, terminating in a branched tendril. Flowers cream, in lax racemes of 6-8, short-spurred. Fruit 9-10mm long. Rocky and wooded habitats, scrub and heath, generally on acid soils, occasionally on calcareous peat. June-September. Europe north to S Norway. **P**: Bees, often self-pollinated. Often local and frequently associated with ancient woodland. **B**: Throughout, but rare in Ireland.

3* *Corydalis capnoides* (L.) Pers. Short to medium, rather robust annual or biennial with ascending or erect stems. Leaves 2-ternate, stalked, with dissected divisions. Flowers cream or white with a yellow apex, 11-16mm, in lax racemes of 5-8; lower bracts leaf-like, upper undivided. Fruit 20-30mm with a persistent style. Shaded habitats, particularly deciduous woodland. June-August. **I**: C & E Europe. Finland.

4* **Yellow Corydalis** *Corydalis lutea* (L.) DC. [*Pseudofumaria lutea*]. Short to medium densely tufted, rather greyish-green perennial. Leaves 2-3-pinnate with narrow-oblong divisions. Flowers golden-yellow, 12-20mm, 6-16 in racemes opposite the upper leaves. Fruit pendent, with a deciduous style. Walls and rocky places and other shaded habitats, to 1700m. May-October. **I**: C & E Alps. Widely naturalized in Europe, except for Scandinavia. Frequently cultivated in gardens. In the wild this is a species of limestone rocks and screes, up to 1700m. **P**: Bees. **B**: Throughout most of country.

5* *Corydalis ochroleuca* Koch. Similar to *C. lutea* but leaves more greyish and flowers cream with a yellow apex, 14-15mm, in dense racemes. Fruit erect. Walls and rocky places. May-September. **I**: SE Europe. Naturalized in Britain, Belgium, Holland, France and Germany.

6* *Corydalis sempervirens* (L.) Pers. Short to tall erect annual or biennial, branched above. Leaves 1-3-pinnate, greyish, mostly stalked. Flowers pale purple with a yellow apex, in lax racemes or few-branched panicles. Fruit 30-50mm with a persistent style. Rocky and waste places. June-August. **I**: North America. Locally naturalized in Norway.

7* **Bulbous Corydalis** *Corydalis cava* (L.) Schweigg. & Koerte [C. *bulbosa*]. Short erect perennial with a hollow tuberous rootstock; stem without a scale near the base. Leaves 2-ternate, often greyish. Flowers purplish or white, 18-30mm, spur long, down-curved; bracts scale like, undivided. Fruit 20-25mm, pendent when ripe. Woods, hedgerows and cultivated land, to 2000m. France and Germany north to S Sweden; naturalized in Britain and Belgium. Cultivated in gardens, including forms with white flowers. Often confused with *C. solida* but distinguished by the absence of a bract-like scale at the base of the stem and by the undivided bracts. **B**: Very local, not nearly so frequent in Britain as *C. solida*.

8* **Solid-tubered Corydalis** *Corydalis solida* (L.) Clairv. [*C. halleri*, *C. tenella*]. Variable short perennial with a solid tuberous rootstock; stem with an oval scale near the base. Leaves 1-3, mostly 2-ternate, with narrow segments. Flowers purplish, 15-30mm, in rather dense racemes; spur long, slightly curved; lower bracts lobed. Fruit 10-25mm, pendent when ripe. Woodland habitats and hedgerows, to 2000m. April-May. Throughout except the extreme north and Iceland; naturalized in Britain, Denmark and Norway. Widely cultivated in gardens. **P**: Bumble-bees. **B**: An occasional garden escape, particularly in S England.

subsp. *laxa* (Fries) Nordst. Like the type, but racemes laxer and leaf segments broader. Similar habitats and flowering time. Sweden and SW Finland.

9 *Corydalis intermedia* (L.) Mérat [*C. fabacea*]. Like *C. solida*, but all bracts unlobed; flowers smaller, rarely white, 10-15mm, only 2-8 (not 10 or more) in a raceme. Woodland and pastures, generally on limestone, to 2000m. March-April. Throughout Europe except for much of the north. Occasionally cultivated in gardens.

10 *Corydalis pumila* (Host) Reichenb. Like *C. intermedia*, but bracts deeply lobed and flower stalks less than 5mm long (not 5mm long or more). Woodland and pastures. March-April. E France, S Germany and S Scandinavia.

Bleeding Heart

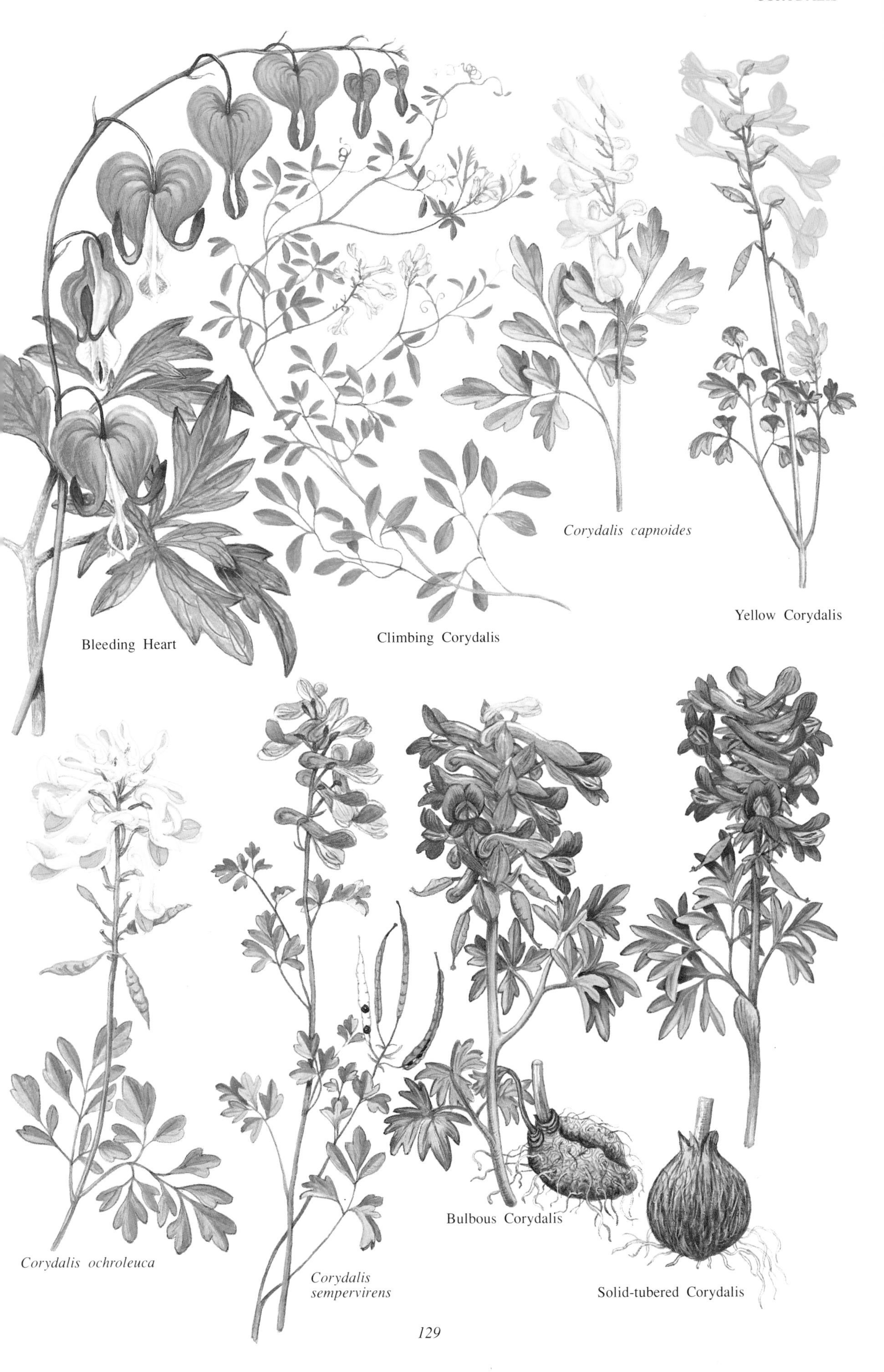
Bleeding Heart
Climbing Corydalis
*Corydalis capnoides*
Yellow Corydalis
*Corydalis ochroleuca*
*Corydalis sempervirens*
Bulbous Corydalis
Solid-tubered Corydalis

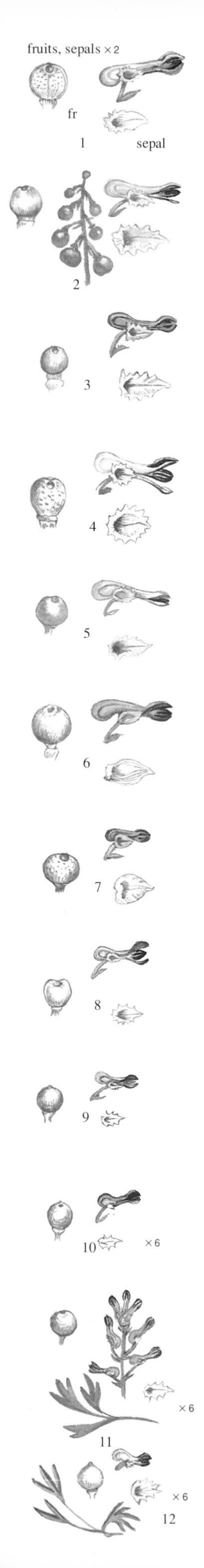


**Fumitories** *Fumaria*. Like *Corydalis* but always annual, often delicate with spreading or climbing stems. Leaves 2-4-pinnately divided, usually with small, toothed or lobed divisions. Flowers in lateral or terminal racemes. Fruit rounded, 1-seeded, not splitting. A difficult group separated on fine details of the flowers and fruit; many are weeds of arable land. About 55 species in Europe, C Asia and the Himalaya. Mostly self-pollinated. Several have become widespread weeds of cultivated land.

1* **Cornish Fumitory** *Fumaria occidentalis* Pugsley. Medium, rather robust clambering annual. Leaf-segments flat, oblong. Racemes long-stalked. Flowers white at first, turning bright pink, with a dark purple apex, 12-14mm long, lower petal with a broad flat margin; sepals toothed. Fruit 3mm, notched at apex. Arable fields (particularly bulb fields), walls and waste places. May-October. SW Britain – not known elsewhere. **B**: Confined to SW Cornwall and the Isles of Scilly.

2* **Ramping-fumitory** *Fumaria capreolata* L. Medium-tall robust, clambering annual. Leaf-segments oblong or wedge-shaped. Racemes long-stalked. Flowers creamy-white with a reddish-black apex, 10-14mm long, upper petal compressed with upturned margins; sepals toothed. Fruit on recurved stalks, 2mm, smooth, with a faint keel. Hedges, scrub, arable land and waste places. May-September. Belgium, France and Germany; naturalized in parts of Holland, Denmark and S Norway. subsp. *babingtonii* (Pugsley) P. D. Sell. Like the type, but bracts equalling fruit-stalks (not shorter) and fruit 2-5mm. Similar habitats and flowering time. **B**: Scattered localities throughout, but particularly coastal Wales and SW England.

3* *Fumaria purpurea* Pugsley. Similar to *F. capreolata*, but flowers purple, 10-13mm long, the upper petal not compressed. Fruit squarish, the bracts equalling or longer than the fruit-stalks. Hedgerows, cultivated and waste ground. July-October. Britain and Ireland – not known elsewhere. Always local. **B**: Throughout, but rare and local in Scotland and Ireland.

4* **Tall Ramping-fumitory** *Fumaria bastardii* Boreau [*F. confusa*]. Short to medium, rather robust annual. Leaf-segments oblong. Racemes longer than their stalks. Flowers numerous, pale pink with a dark purple apex, 9-12mm, upper petal compressed; sepals toothed. Fruit 2-2.5mm, rough when dry. A weed of cultivated and waste land. April-October. Britain, Ireland and France. **B**: W Britain and Ireland, scarce elsewhere.

5* **Common Ramping-fumitory** *Fumaria muralis* Sonder ex Koch. Variable, rather slender spreading or clambering annual. Racemes lax, equalling their stalks. Flowers few, up to 15, pink with a reddish-black apex, 9-10mm, the lower petal with erect margins; sepals toothed. Fruit 2-2.5mm, smooth when dry. Cultivated ground, particularly arable land and waste places, banks and walls. April-October. W Europe north to S Norway, but absent from most of Scandinavia. Often confused with *F. martinii* and *F. bastardii* but generally with fewer flowers to each raceme and the lower petal with erect, not spreading, margins. **B**: Rare, possibly naturalized, S & W England and Wales. subsp. *boraei* (Jordan) Pugsley is more robust, with flowers 10-12mm long. Fruit slightly rough. Habitat and distribution the same. **B**: Throughout except W coast Ireland. subsp. *neglecta* Pugsley, also more robust, has racemes up to 20-flowered; sepals scarcely toothed. Arable fields. **B**: England – W Cornwall, not known elsewhere.

6* *Fumaria reuteri* Boiss. [*Fumaria martinii*, *F. paradoxa*]. Rather robust, medium, often clambering annual. Racemes lax, much longer than their stalks. Flowers pink with a reddish-black apex, 11-13mm long, with a broad upper petal, not compressed; sepals practically untoothed. Fruit 2-5mm, almost smooth when dry. Arable land. May-October. Britain and W France. Rare, extinct in many of its former localities. **B**: Probably extinct.

7* *Fumaria densiflora* DC. [*F. micrantha*] Short to medium rather robust annual. Leaf-segments linear, channelled. Racemes dense, much exceeding the very short stalks; bracts longer than the flower-stalks. Flowers small, pink with a reddish-black apex, 6-7mm long; sepals untoothed or toothed at the base. Fruit 2-2.5mm, rough when dry. Arable land and waste places, generally on light sandy or chalky soils. June-October. Britain, France and Belgium; occasionally naturalized elsewhere. Local but widely scattered. Europe east to the Caucasus Mts and the Canary Islands and Azores. **B**: Restricted mainly to S and E England and E Scotland, scarce elsewhere.

8* **Common Fumitory** *Fumaria officinalis* L. Weak, low to short scrambling annual. Leaf segments flat. Racemes longer than their stalks. Flowers numerous, 120 or more, purplish-pink with a reddish-black apex, 7-9mm long; sepal with irregular teeth. Fruit 2-2.5mm, straight or notched at the apex, rough when dry. Cultivated land, waste places, on light, well-drained, acid or calcareous soils, to 1500m. May-October. Throughout, except Spitsbergen; casual in the Faeroes and Iceland. An often common and gregarious annual weed. **B**: Throughout.

9 *Fumaria caroliana* Pugsley. Like *F. officinalis* but the racemes with only 10-15 flowers, short-stalked; flowers 6-7mm long. Fruit rounded at the apex. N France – Arras. Rare and very local.

10* *Fumaria vaillantii* Loisel. Low to short, much-branched, greyish annual. Leaf-segments flat. Racemes only 6-12 flowered, longer than the short stalks. Flowers small, pale pink, often with a reddish-black apex, 5-6mm; sepals toothed. Fruit 2mm, rounded, granular when dry. Arable land on calcareous soils, to 2100m. June-September. Throughout, except for much of northern Europe and Iceland. Sometimes confused with *F. officinalis*, but the sepals are tiny, never more than 1mm long (2mm long or more in *F. officinalis*). Rare. **B**: Confined to SE England and East Anglia.

11 *Fumaria schleicheri* Soyer-Willemet. Like *F. vaillantii*, but racemes 12-20 flowered, equalling or longer than their stalks; flowers deep pink with a dark purple apex. Arable and waste land, to 1700m. May-September. France and Germany; naturalized in Denmark and Holland.

12 *Fumaria parviflora* Lam. Like *F. vaillantii*, but leaf-segments linear, channelled. Racemes scarcely stalked; flowers white, occasionally pink-tinged. Fruit often pointed. Arable land, usually on calcareous soils. June-September. Britain, Belgium, Holland, France and Germany. **B**: Confined to S England and East Anglia and a few scattered localities in C & N England.

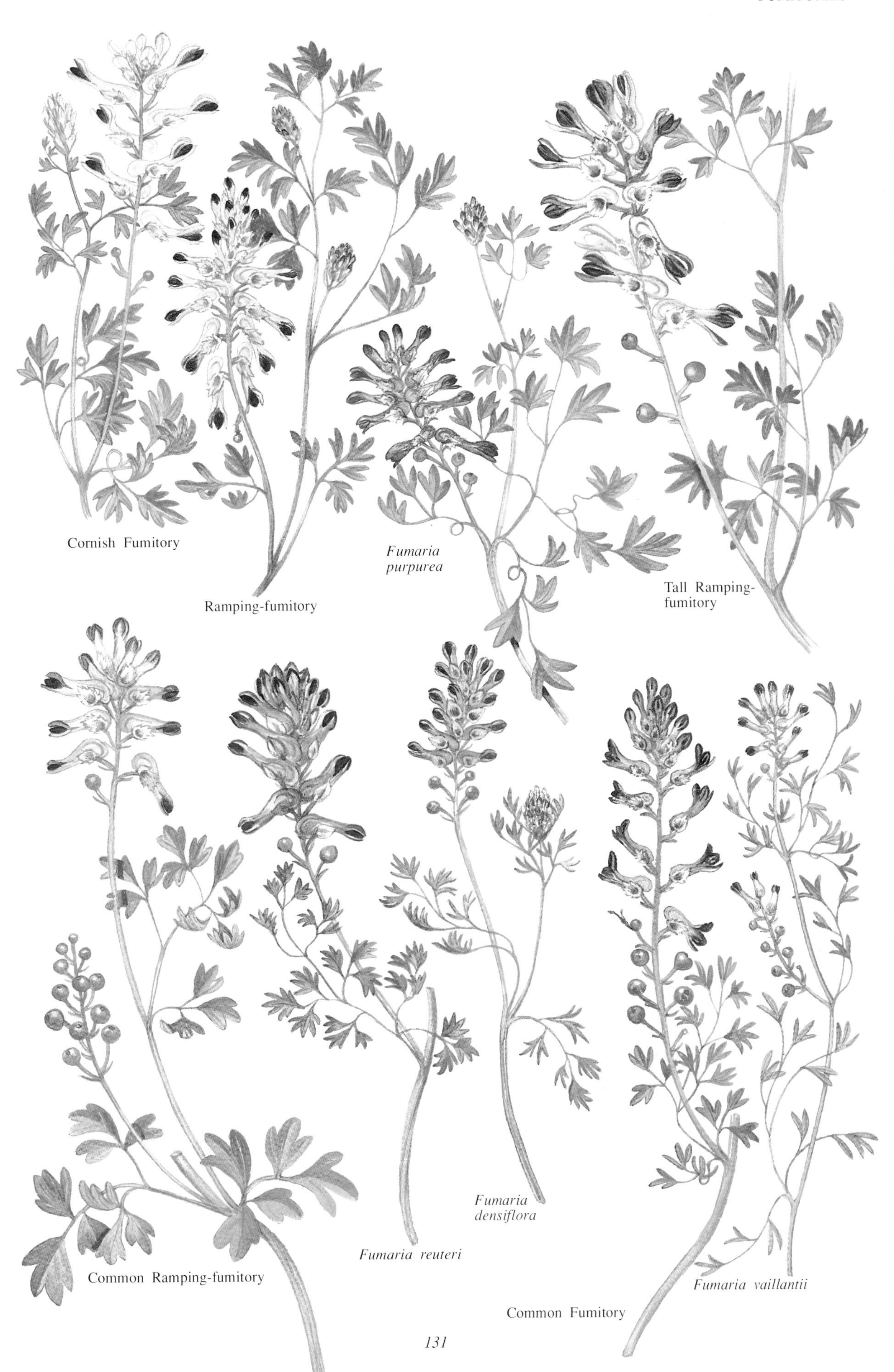
Cornish Fumitory
Ramping-fumitory
*Fumaria purpurea*
Tall Ramping-fumitory
Common Ramping-fumitory
*Fumaria reuteri*
*Fumaria densiflora*
Common Fumitory
*Fumaria vaillantii*

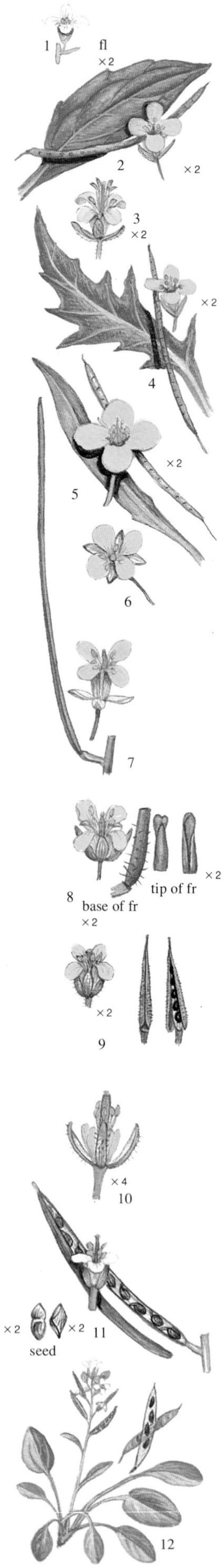

# CRESS FAMILY Cruciferae

Annual or perennial herbs with alternate leaves. Flowers generally borne in heads or racemes, with 4 sepals and 4 petals, all separate; stamens 6; ovary usually with 2 compartments separated by a membranous septum – with a solitary style. Fruit very variable in shape, but either long and linear (a siliqua) or short, less than three-times as long as broad (a silicula), usually opening by 2 valves from the base, but occasionally indehiscent or breaking off into 1-seeded segments. Over 2000 species in 220 genera, throughout the world but especially in north temperate regions. The fruit is particularly important. Flowers and fruits can often be found on the plant at the same time.

**Sisymbrium** *Sisymbrium.* Annual or perennial herbs with simple, unbranched hairs. Flowers in simple or branched racemes. Flowers yellow, occasionally white with unnotched petals. Fruit a siliqua, linear, the valves usually 3-veined. Mostly weedy lowland species. 90 species widespread from Europe to Asia, S Africa and the Americas. Differs from similar looking crucifers by having unbeaked fruits; the lateral 2 veins of the fruit valves are fainter than the central vein.

1* *Sisymbrium supinum* L. Short, hairy, annual. Leaves pinnately-lobed with an oblong terminal lobe. Flowers white 3mm, short-stalked; bracts present below each flower. Fruit 10-30mm, hairy, valves 1-veined. Cultivated ground, waste places and rocky habitats. June-August. Continental Europe, except the extreme north and Denmark; naturalized in Finland and Norway.

2 *Sisymbrium strictissimum* L. Tall, hairless or slightly hairy perennial. Leaves not lobed, the lower oval, pointed, toothed or untoothed. Flowers bright yellow, 4-6mm. Fruits 30-80mm, curving upwards on long slender stalks, hairless. Waste places and cultivated land. June-August. France and Germany; naturalized locally in Britain and Belgium. **B**: Established in a few scattered localities.

3* **London Rocket** *Sisymbrium irio* L. Short to medium hairy or hairless annual, often over-wintering. Lower leaves pinnately-lobed with a large terminal lobe; stem leaves short stalked. Flowers pale yellow, 3-4mm, in condensed racemes over-topped by young fruits. Mature fruits 30-50mm, slightly constricted between the seeds, hairless. Waste places, road verges and old walls, to 1400m. June-August. **I**: S Europe. Naturalized throughout, except much of the north. **B**: Rare, established in the London area; occasionally casual elsewhere. Flourished on ruins following the Great Fire of London in 1666.

4 **False London Rocket** *Sisymbrium loeselii* L. Like *S. irio*, but hairier, especially below, and with brighter yellow flowers, 4-6mm, not overtopped by the young fruits. Cultivated lands, waste and bare places. June-August. Germany; casual or naturalized throughout, except Ireland, Iceland and much of the north. Generally an overwintering annual. **B**: Naturalized in the London area and elsewhere.

5 *Sisymbrium volgense* Beib. ex E. Fourn. Like *S. loeselii*, but perennial and flowers larger, 7-9mm; upper leaves linear-lanceolate (not lobed or arrow-shaped as in the previous 2 species). Disturbed and waste ground. **I**: SE USSR. Naturalized throughout, except Ireland, Iceland and the extreme north. Often introduced with foreign grain. **B**: Rare casual, mainly around docks and flour mills.

6* *Sisymbrium austriacum* Jacq. [*S. pyrenaicum*]. Short to tall, very variable, hairy or hairless biennial or perennial. Lower leaves untoothed to lobed, upper arrow-shaped or pinnately-lobed. Flowers golden-yellow, 7-10mm. Fruits 20-60mm with curved and twisted stalks. Disturbed and rocky ground, to 1600m. June-September. France and Germany; naturalized or casual in Holland and Scandinavia.

7* **Tall Rocket** *Sisymbrium altissimum* L. [*S. sinapistrum*]. Tall, branched, often overwintering annual, rough-hairy below. Lower leaves pinnately-lobed, soon withering, upper unstalked, with linear lobes. Flowers pale yellow 10-11mm; outer, sepals with a short horn at the apex. Fruit long, 50-100mm. Waste places and disturbed ground. June-August. S Germany; naturalized or casual throughout, except for Ireland and Iceland. C and S Europe to the Near East. **B**: Commonest in the south-east.

8* **Eastern Rocket** *Sisymbrium orientale* L. [*S. columnae*]. Like *S. altissimum*, but stems softly hairy and upper leaves unlobed or only 3-lobed; fruit hairy at first, 5cm long or more. Waste ground. June-August. **I**: S Europe, N Africa and W Asia. Naturalized or casual throughout much of Europe. Generally an overwintering annual. **B**: SW England, rare and casual elsewhere.

9* **Hedge Mustard** *Sisymbrium officinale* (L.) Scop. [*Chamaeplium officinale*]. Medium to tall, rough-hairy or hairless, annual or biennial with spreading branches. Lower leaves deeply pinnately-lobed with a large terminal lobe. Flowers pale yellow, small, 3mm. Fruits short 10-20mm, closely pressed to the stem. Waste places, hedgerows, occasionally arable land. May-September. Throughout, except the far north. Often an overwintering annual. Europe to W Asia and N Africa. A widespread weed in many parts of the world. **B**: Throughout, scarce in C & N Scotland.

10* **Flixweed** *Descurainia sophia* (L.) Webb ex Prantl [*Sisymbrium sophia*]. Medium To tall hairy annual or biennial. Leaves grey-green, 2-3-pinnately-lobed, ferny looking. Flowers pale yellow, small, 3mm; petals shorter than sepals. Fruit very slender, 8-45mm, curving, almost erect. Waste places, roadsides, arable land, often on light sandy soils. June-August. Throughout, except the far north. **B**: Casual, scattered localities, mainly in E England and E Scotland.

11* **Garlic Mustard, Jack-by-the-hedge** *Alliaria petiolata* (Bieb.) Cavara & Grande [*A. officinalis*, *Sisymbrium alliaria*]. Variable, short to tall, usually unbranched biennial, hairy. Leaves pale green, kidney-shaped to heart-shaped, toothed, smelling of garlic when crushed. Flowers white, 3-5mm. Fruit 20-70mm, erect. Hedgerows, woodland margins, scrub, roadsides, and waste ground, generally on calcareous or base-rich soils. April-June. Throughout, except the Faeroes, Iceland and Spitsbergen. Often gregarious. **B**: Throughout, though scarce in N Scotland and much of Ireland.

12 *Eutrema edwardsii* R. Br. Short, generally hairless, rhizomatous perennial. Basal leaves oval, often heart-shaped at base, fleshy, the upper narrow unstalked. Flowers white, 3mm. Fruit linear-oblong, 6-20mm, erect. Rocky habitats. July-August. Spitsbergen extending into parts of Arctic Russia.

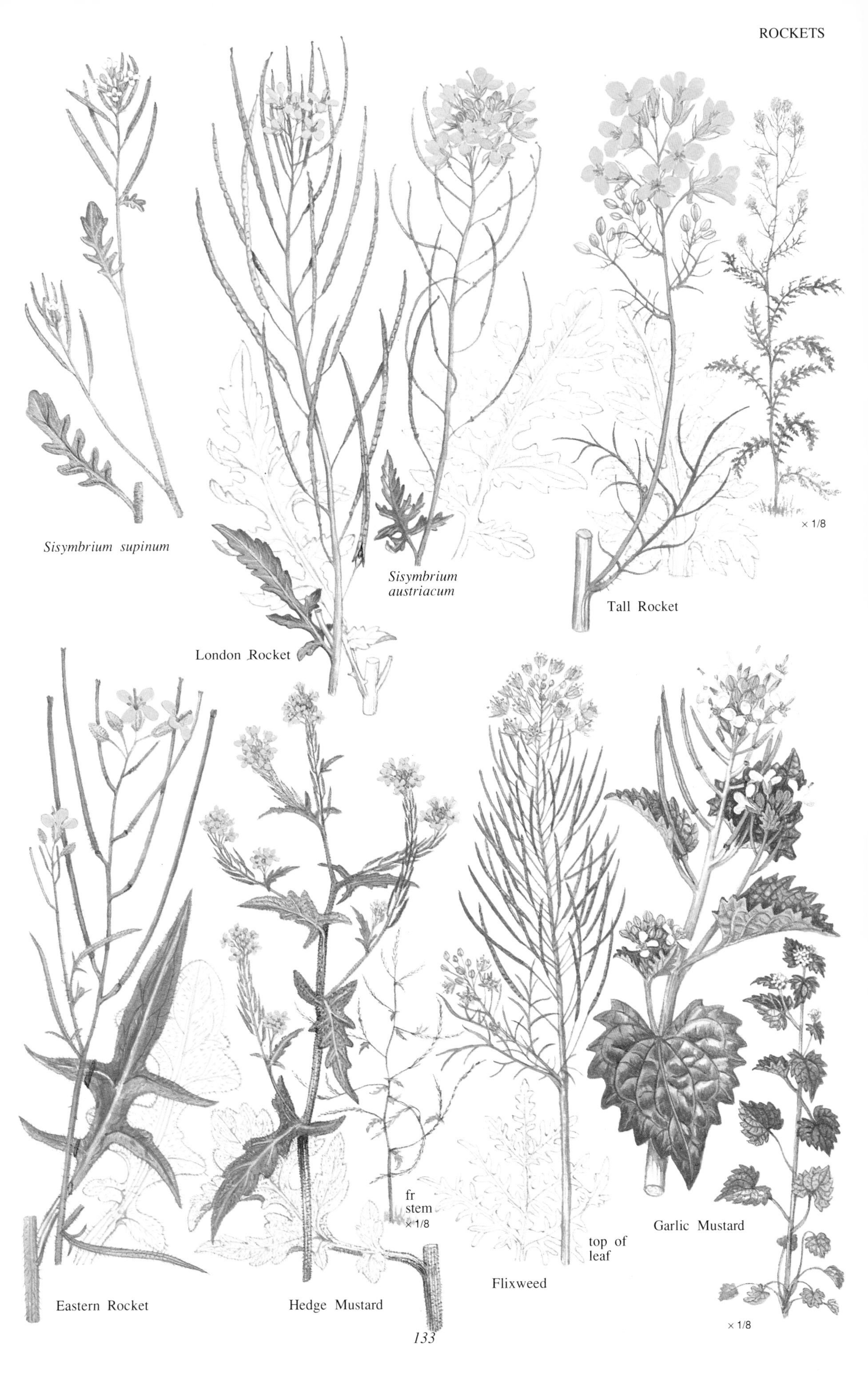
*Sisymbrium supinum*
London Rocket
*Sisymbrium austriacum*
Tall Rocket
× 1/8
Eastern Rocket
Hedge Mustard
fr stem × 1/8
Flixweed
top of leaf
Garlic Mustard
× 1/8

1* **Thale Cress** *Arabidopsis thaliana* (L.) Heynh. [*Sisyrinchium thalianum*]. Short to medium, slightly hairy annual or biennial. Basal leaves forming a rosette, elliptical toothed or untoothed; stem leaves usually untoothed, unstalked. Flowers white, 3mm. Fruits linear, 5-20mm, hairless, not flattened. Arable and cultivated land, waste places, banks, walls and hedgerows, often on rather dry sandy soils. March-October. Throughout, except for much of the north, the Faeroes and Iceland. The plant exists in a number of physiological races requiring different germination regimes. Some can produce seed within 4-5 weeks of germination. Often abundant. **B**: Throughout, particularly in E Britain.

2 *Arabidopsis suecica* (Fries) Norrlin. Like *A. thaliana*, but the basal leaves are strongly toothed or pinnately-lobed; flowers larger, 4-5mm; fruit 20-30mm. Stony and gravelly places. May-October. Scandinavia. Local.

3* **Braya** *Braya linearis* Rouy. Low to short, loosely tufted, slightly hairy perennial. Leaves linear, sometimes slightly toothed. Flowers white or purplish, 2-3mm, petals truncated. Fruit linear, 8-15mm, with 1-veined valves. Calcareous screes and gravels. July-August. Arctic Norway and Sweden.

4 *Braya purpurascens* (R.Br.) Bunge. Like *B. linearis*, but the stems with no, or only one, stem leaf (not 1 or more); leaves linear-oblong; fruit 4-10mm linear-oblong. Calcareous rocks. June-August. Arctic Norway and Spitsbergen.

5 **Mitre Cress** *Myagrum perfoliatum* L. Short to tall, variable, greyish, hairless, annual. Basal leaves deeply toothed to pinnately-lobed, stalked; stem leaves unstalked, clasping the stem. Flowers yellow, 3-4mm. Fruit a silicula, mitre-shaped, 5-8mm, with a thick stalk. Waste places and disturbed land, waysides. June-August. C & S France; a rare casual in Britain, naturalized in Belgium and S Scandinavia.

6* **Woad** *Isatis tinctoria* L. [*I. canescens*, *I. littoralis*, *I. taurica*]. Medium to tall, mostly hairless, greyish biennial. Stem leaves arrow-shaped, clasping the stem. Flowers yellow, 3-4mm in much branched racemes. Fruit an oblong, flattened, pendent silicula, 11-27mm, dark brown when ripe. Dry habitats, cultivated land, waste places, cliffs and rocks, to 2000m. July-August. **I**: E Europe, W Asia. Throughout, except Ireland, the Faeroes, Iceland and Spitsbergen, naturalized probably in most places and persisting as a relict of former cultivation. Used by the Greeks and Romans as a medicinal plant. During the Middle Ages it was widely cultivated for the production of a blue dye, Indigotine, produced by crushing and fermenting the leaves. Still cultivated in gardens today. **B**: Rare, persisting in isolated localities in Surrey and Gloucestershire.

7* **Bunias** *Bunias erucago* L. Medium, rough-hairy, glandular annual or biennial. Lower leaves pinnately-lobed, upper toothed or untoothed. Flowers yellow, 8-10mm, petals notched. Fruit a silicula, 10-12mm, square in section with irregular wings and ridges on the corners, not splitting. Waste ground. May-August. C & S France; naturalized or casual in Germany, particularly in the vicinity of ports. **P**: Bees or self-pollinated. The roots and young shoots can be eaten as a salad vegetable.

8* **Warty Cabbage** *Bunias orientalis* L. Similar to *B. erucago*, but a hairless biennial or perennial with slightly smaller flowers. Fruits shiny, covered in irregular warty humps. Waste ground, roadsides and rubbish tips. May-August. **I**: E Europe and W Asia. Naturalized or casual throughout much of Europe and Britain. **P**: Various small insects, often self-pollinated. The leaves can be used as a salad vegetable or for fodder. **B**: Local, particularly occurring in SE England.

**Treacle-mustards** *Erysimum*. Annual or perennial herbs with branched hairs. Flowers yellow with long-clawed petals; sepals erect, the inner 2 pouched at the base. Fruit a siliqua, linear, valves 1-veined. A difficult group, with many elements of uncertain status, particularly in S Europe. 100 species, primarily in the Mediterranean region, Europe, Asia and the Azores. Several are garden plants.

9 *Erysimum crepidifolium* Reichenb. Short to medium biennial or perennial. Basal leaves oblong, toothed to pinnately-lobed; stem leaves deeply toothed or untoothed. Flowers yellow, 16-18mm. Fruit 25-80mm, square in section, erect. Dry grassland, generally on calcareous soils. June-September. Germany.

10 *Erysimum odoratum* Ehrh. [*E. pannonicum*, *E. erysimoides*]. Like *E. crepidifolium*, but always biennial, the basal leaves withered by flowering time; stem leaves comb-like with narrow pinnate divisions. Flowers bright yellow. Grassy and rocky habitats. June- September. E France and Germany.

11* *Erysimum hieracifolium* L. [*E. strictum*, *E. marschallianum*]. Medium to tall biennial or perennial, with branched, star-shaped, hairs. Leaves linear to oblong, deeply toothed, green or grey. Flower yellow, 8-10mm. Fruit erect, parallel to the stem, 30-55mm. Waste places and rocky habitats, woods, on limestone, to 1700m. June-September. Belgium, E France, Germany, Denmark and S Finland. Locally common.

12 *Erysimum repandum* L. Rather like *E. hieracifolium*, but annual, with 2-branched hairs and leaves always green. Fruit long, 45-100mm, circular (not square) in section. Waste places. June-August. Germany, but casual or naturalized throughout much of Europe, including Britain. **B**: Casual.

13* **Treacle-mustard** *Erysimum cheiranthoides* L. Variable, short to medium, hairy annual with square stems. Lower leaves oblong or lanceolate, untoothed to deeply toothed but soon withering; stem leaves pointed, ascending. Flowers small, yellow, 5-6mm. Fruit 10-50mm, slightly curved, square in section. Waste places and cultivated land, road verges, especially on sandy soils. June-August. Throughout, except the extreme north, the Faeroes and Spitsbergen. A common weed of arable fields and gardens. **P**: Self-pollinated. **B**: Naturalized and common in lowland England, scarce elsewhere including Ireland.

subsp. *altum* Ahti. Like the type, but biennial and larger, more densely hairy; stem leaves closely pressed to the stem, not spreading. Similar habitats and flowering time. Finland.

Thale Cress

Braya

Woad

Bunias

Warty Cabbage

*Erysimum hieracifolium*

Treacle-mustard

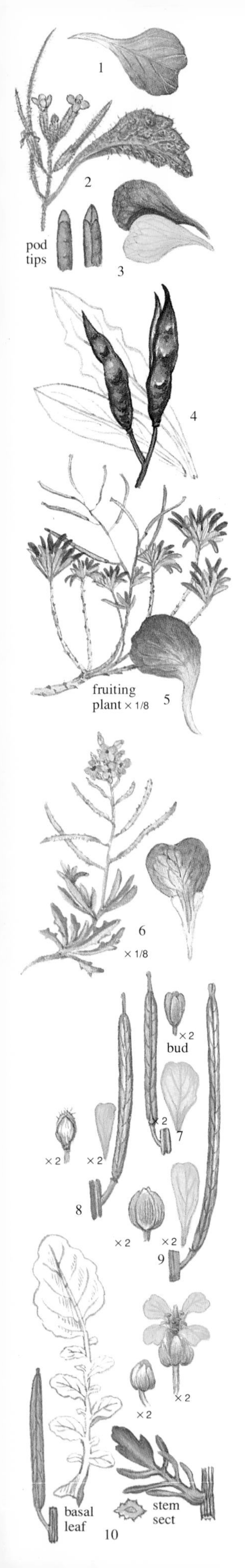


1* **Dame's Violet** *Hesperis matronalis* L. [*H. sibirica*]. Medium to tall erect, hairy, biennial or perennial; stems unbranched. Leaves lanceolate, short-stalked. Flowers large, white or purple, 15-20mm, in broad panicles, fragrant. Fruit a slender siliqua, 25100mm, curving upwards. Damp or semi-shaded habitats, field boundaries, hedgerows, banks and road verges, often close to habitations, to 1400m. May-August. C & S France; naturalized widely throughout, including Iceland. Cultivated for many centuries as a garden plant. **P**: Various insects, especially butterflies. Native over much of S Europe. **B**: Local throughout.

2 **Malcolmia** *Malcolmia africana* (L.) R. Br. Short to medium hairy annual. Leaves lanceolate, untoothed to deeply toothed. Flowers violet, 9-10mm. Fruit a rigid siliqua, 25-65mm, spreading, 4-angled and bristly. Waste and disturbed ground. June-August. **I**: S Europe & N Africa, W Asia. Casual in parts of W & N Europe. Probably not naturalized in the area covered by this book.

**Virginia Stock** *Malcolmia maritima* (L.) R. Br. Like *M. africana*, but the flowers pink or violet, larger, the petals 12-25mm long (not 8-10mm) and the outer sepals pouched at the base. Sandy and waste places at low altitudes. May-August. **I**: S & W Greece and S Albania. Casual in Britain and France and possibly elsewhere. Widely cultivated in gardens.

3* **Wallflower** *Cheiranthus cheiri* L. Variable medium to tall perennial, covered with flattened hairs attached from the middle. Leaves oblong-lanceolate, untoothed, the uppermost narrower. Flower yellow or orange-brown, large, 20-25mm, in elongating racemes, very fragrant. Fruit a slender siliqua, 25-75mm, erect, flattened. Rocky habitats, cliffs and walls, generally at low altitudes. March-June. **I**: SE Europe. Widely naturalized in Britain, Belgium, Holland, France and Germany. Widely cultivated in gardens as an ornamental and as a cut flower. **P**: Butterflies. **B**: Throughout most of Britain.

4* **Parrya** *Parrya nudicaulis* (L.) Boiss. Low to short hairless or stickily-hairy, rhizomatous perennial. Leaves spoon-shaped to linear-oblong, toothed or not. Flowers large, white or purplish, 16-22mm, the petals slightly notched. Fruit a slender, flattened, siliqua, 20-65mm. Rocky habitats, moraines and screes. June-August. Spitsbergen, extending into Arctic N Asia and North America. Occasionally cultivated in alpine houses.

5* **Hoary Stock** *Matthiola incana* (L.) R. Br. Stout, short to tall, densely white-downy perennial, woody at the base. Leaves lanceolate, mostly unlobed. Flowers white, purple or pink, 25-30mm, in lax racemes, fragrant. Fruit a slender downy siliqua, 45-160mm, ascending, with a pair of 'horns' at the apex. Coastal habitats, particularly cliffs, generally on calcareous rocks. Late May-August. S Britain and W France, occasionally locally naturalized from garden escapes. Local. **P**: Butterflies. **B**: Rare, S England, especially Sussex, the Isle of Wight and the Channel Islands.

6* **Sea Stock** *Matthiola sinuata* (L.) R. Br. Like *M. incana*, but less bushy, the leaves with a wavy, toothed or lobed, margin, the uppermost untoothed. Fruit downy, 50-150mm, *sticky* with conspicuous yellow or blackish glands. Coastal habitats, sea cliffs and sand-dunes. June-August. Local in S Britain and W France; formerly in Ireland. **P**: Butterflies. **B**: Restricted to W Europe and the W Mediterranean region. **B**: Rare, restricted to a few localities in N Devon, Glamorgan and the Channel Islands.

**Wintercresses** *Barbarea*. Biennial or perennial herbs, hairless or with unbranched hairs. Leaves pinnately-lobed or divided; stem leaves usually clasping the stem. Flowers yellow, in racemes, often branched. Fruit a siliqua, 4-angled, the valves with a strong mid-vein. 20 species from north temperate regions. Mostly wet or damp habitats. Several have become widespread weeds.

7* **Common Wintercress** *Barbarea vulgaris* R. Br. Medium to tall hairless biennial or perennial. Basal leaves with 2-5 pairs of lateral lobes; uppermost leaves unlobed, toothed. Flowers bright yellow, 7-9mm; buds hairless. Fruit 15-30mm, stiffly erect. Damp roadside habitats, hedgerows, ditches and riversides. May-August. Throughout, except the Faeroes and Spitsbergen; naturalized in Iceland. **P**: Various short-tongued bees, bumble bees, moths, flies and beetles, sometimes self-pollinated, especially during dull or wet weather. **B**: Throughout, most frequent in the south and east.

8* **Small-flowered Wintercress** *Barbarea stricta* Andrz. Medium to tall, practically hairless, biennial. Basal leaves with only 1-2 pairs of lateral lobes; uppermost leaves unlobed, toothed. Flowers yellow, 5-6mm; buds hairy. Fruit 18-30mm. Damp habitats, roadsides, waste places and river-banks. May-August. Britain, Holland, Germany and Scandinavia; naturalized in France and Ireland. N & E Europe and N Asia. **B**: Rare, restricted mainly to C & N England and E Wales.

9* **American Wintercress** *Barbarea verna* (Miller) Ascherson [*B. praecox*]. Short to tall hairy biennial. Basal leaves with 6-10 pairs of lateral lobes; uppermost leaves pinnately-divided. Flowers mid-yellow, 7-10mm. Fruit long, 30-70mm. Damp habitats, waste places and cultivated ground. April-July. C & S France; widely naturalized throughout, except for most of Scandinavia, the Faeroes and Iceland. Widely cultivated as a salad vegetable, often as an alternative to watercress. A native of S & SW Europe. Widespread weed in North America, South Africa and Australasia. **B**: Throughout, except for much of N Britain.

10 **Medium-flowered Wintercress** *Barbarea intermedia* Boreau. Short-medium, hairy or hairless, biennial. Basal leaves with 3-5 pairs of lateral lobes; uppermost leaves pinnately-lobed. Flowers yellow, 5-6mm. Fruit short, 10-30mm. Damp habitats, arable and waste land, waysides and riversides. May-August. C & S France & S Germany; naturalized throughout, except for most of Scandinavia, the Faeroes and Iceland. A native of C & S Europe. **B**: Locally established throughout.

Dames' Violet

Dame's Violet
Wallflower
Parrya
Hoary Stock
Sea Stock
×1/3
Common Wintercress
Small-flowered Wintercress
American Wintercress

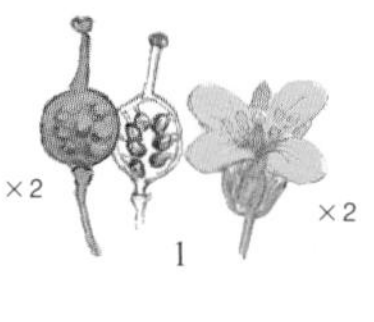


**Yellowcresses** *Rorippa.* Annual to perennial herbs, hairless or with unbranched hairs. Leaves simple to pinnately-lobed. Flowers yellow, the inner sepals pouched at the base. Fruit a short siliqua or a silicula, the valves veinless or with a weak mid-vein. Cosmopolitan; 70 species in temperate and tropical regions. Several have become widespread weeds of cultivated land.

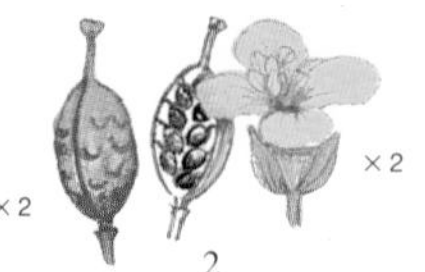


1* **Austrian Yellowcress** *Rorippa austriaca* (Crantz) Besser. Medium to tall, more-or-less hairless perennial with runners. Leaves elliptical, toothed but not lobed, the lower stalked but the upper clasping the stem. Flowers yellow, 3-4mm, petals slightly longer than the sepals. Fruit rounded, on long slender stalks. Wet habitats, riverbanks, ditches, dykes and railway embankments. June-August. Germany; widely naturalized throughout, except for Ireland, the Faeroes and Iceland. Europe to W Asia. **B**: Rare, scattered localities in England and Wales, north to Lancashire.

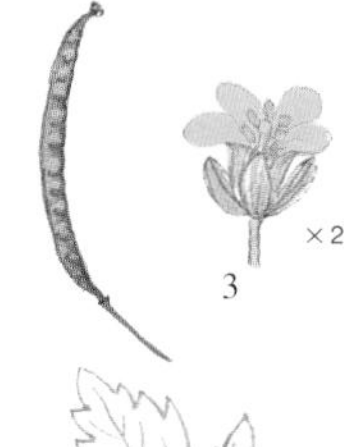


2* **Great Yellowcress** *Rorippa amphibia* (L.) Besser [*Nasturtium amphibium*]. Medium to tall, usually hairless perennial, with runners. Lower leaves variable, toothed or pinnately-lobed, short-stalked; upper leaves lanceolate, unstalked. Flowers bright yellow, 5-6mm, petals twice as long as sepals. Fruit oval, 3-6mm, on slender spreading or deflexed stalks. Wet habitats on fertile soils, river and stream margins, ditches, dykes and shallow ponds. June-August. Throughout, except the north, the Faeroes, Iceland and Spitsbergen. **P**: Bees or may be self-pollinated. The stem leaves do not clasp the stem as in the previous species. Europe to Siberia and N Africa. **B**: Scattered localities in England north to the Humber and in Ireland.

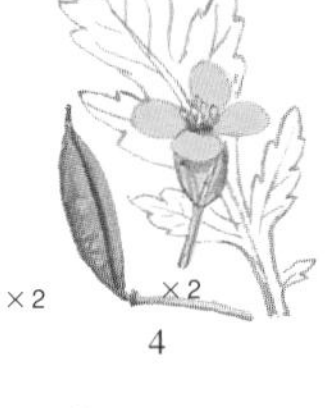


3* **Creeping Yellowcress** *Rorippa sylvestris* (L.) Besser [*Nasturtium sylvestris*]. Short to medium, rather straggling, usually hairless perennial, with runners. Leaves all stalked, mostly pinnately-lobed. Flowers yellow, 5mm, petals twice as long as sepals. Fruit linear, 8-18mm, ascending. Damp and bare habitats, stream and pool margins, dykes, especially where water stands only during the winter. June- September. Throughout, except the Faeroes and N Scandinavia; naturalized in Iceland. Often growing in association with *Rorippa palustris*, *Ranunculus sceleratus* and *Chenopodium rubrum*. **P**: Small bees and flies. Confined to Europe and parts of N Africa. **B**: Throughout, but rare in Scotland.

4 *Rorippa prostrata* (J.P. Bergeret) Schinz & Thell. [*Nasturtium anceps*]. Medium to tall, hairless perennial. Leaves very variable, mostly deeply pinnately-lobed. Flowers yellow, 4mm. Fruit elliptical to almost linear, 5-6mm, on horizontal or ascending stalks. Damp and bare habitats. June-September. Holland and Germany, S Scandinavia; naturalized or casual in Britain and Belgium. Intermediate between *R. amphibia* and *R. sylvestris* and probably of hybrid origin between these two.

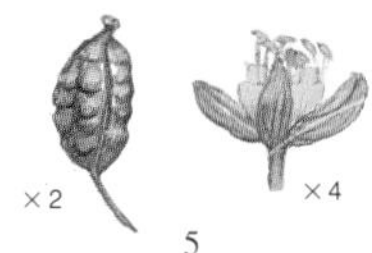


5* **Iceland Yellowcress** *Rorippa islandica* (Oeder ex Murray) Borbás [*Nasturtium palustre*, *Rorippa palustris*]. Short to medium, practically hairless annual or biennial. Lower leaves pinnately-lobed, often rather lyre-shaped; upper leaves unstalked, pinnately-lobed to unlobed, with slight 'ears' at the base. Flowers pale yellow, small, 3mm, the petals shorter than or only just equalling sepals. Fruit elliptical-oblong, 4-7mm, slightly curved. Bare ground subjected to winter flooding, temporary pools, margins of streams and lakes, damp pathways. June-August. Throughout. Also native to Greenland. It has been suggested that the seeds are distributed along coasts by grazing geese. **B**: Coastal localities in Scotland, Isle of Man and W Ireland.

**Marsh Yellowcress** *Rorippa palustris* (L.) Besser [*Nasturtium palustre*]. Very like *R. islandica*, but the sepals more than 1.6mm long and the fruit shorter, rarely much longer than the stalk (2-3 times as long as the stalk in *R. islandica*). Similar habitats and distribution. Generally more common than *R. islandica*. **B**: Throughout.

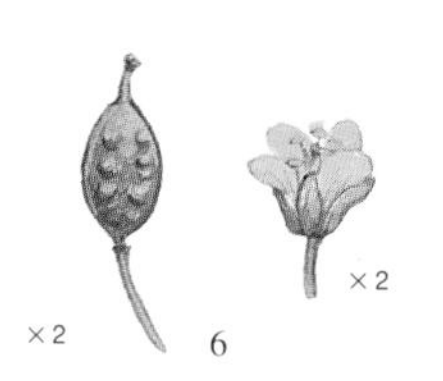


6* *Rorippa pyrenaica* (Lam.) Reichenb. [*Nasturtium pyrenaicum*]. Low to short perennial, hairy below. Lowest leaves in a rosette, long-stalked, oval, lobed or unlobed; stem leaves unstalked, clasping the stem, narrowly-lobed. Flowers yellow, 4mm. Fruits elliptical, 2.5-6mm. Damp and waste habitats. June-August. S Germany, C & S France; naturalized in Belgium. A native of C & S Europe.

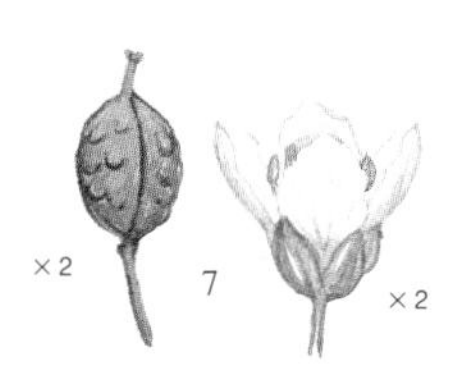


7* **Horse-radish** *Armoracia rusticana* Gaertn., Mey. & Scherb. Tall, stout, hairless stoloniferous perennial. Basal leaves large, oblong, long-stalked, toothed; stem leaves small and short-stalked, the lower often lobed. Flowers white, 8-9mm, many in broad panicles. Fruit rounded, inflated, 4-6mm, on slender stalks. Fields, roadsides, railway embankments, streamsides, waste land, at low altitudes. May-June. **I**: S Russia. Widely naturalized throughout except much of the north and Iceland. Widely cultivated since the Middle Ages. Horse-radish is prepared by grating the thick parsnip-like rootstock into soured cream. **P**: Various small insects. Fruits are rarely ripened in most of N and NW Europe, indeed some colonies flower very sparsely. **B**: Widely naturalized; rarer in Ireland.

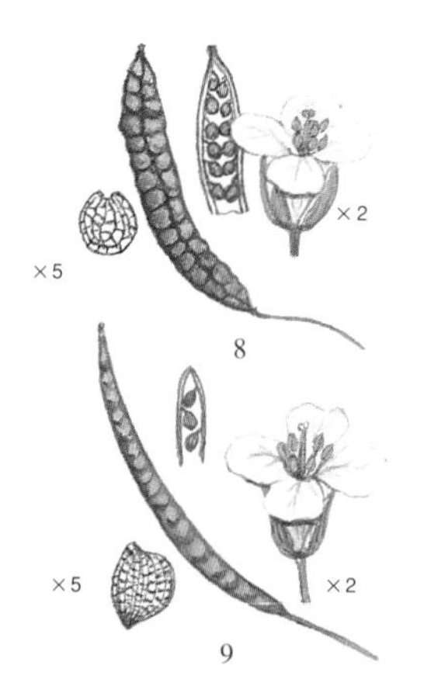


8* **Watercress** *Nasturtium officinale* R. Br. [*Rorippa nasturtium-aquaticum*]. Short to medium hairless perennial, the stems thick and succulent, creeping and rooting, sometimes floating. Leaves pinnate, deep glossy green, leaflets rounded or elliptical, usually unlobed. Flowers white, 4-6mm, in congested racemes. Fruit a narrow oblong siliqua, 13-18mm, straight or curved, with 2 rows of seeds within each valve. Shallow, clear, freshwater habitats, rivers and streams, ditches, dykes and wet flushes, May-October. Throughout, except the extreme north, the Faeroes and Iceland. Widely cultivated as a salad vegetable. Plants are very susceptible to polluted water. **P**: Various small insects, bees and flies. Europe to W Asia and N Africa. **B**: Throughout, except for much of C & N Scotland.

9 *Nasturtium microphyllum* (Boenn.) Reichenb. [*Rorippa microphylla*]. Like *N. officinale*, but with rather larger flowers and more slender, 16-24mm fruits, the valves with only a single row of seeds. Leaves and stems turning purplish-brown in autumn (not remaining green). Similar habitats and flowering time to *N. officinale*. Britain, Belgium, Holland, France and Germany. Hybrids with *N. officinale*, *N. × sterile* (Airy-Shaw) Oefel., frequently occur where the parent species grown in close proximity. **B**: Throughout, but commoner in the north than *N. officinale*.

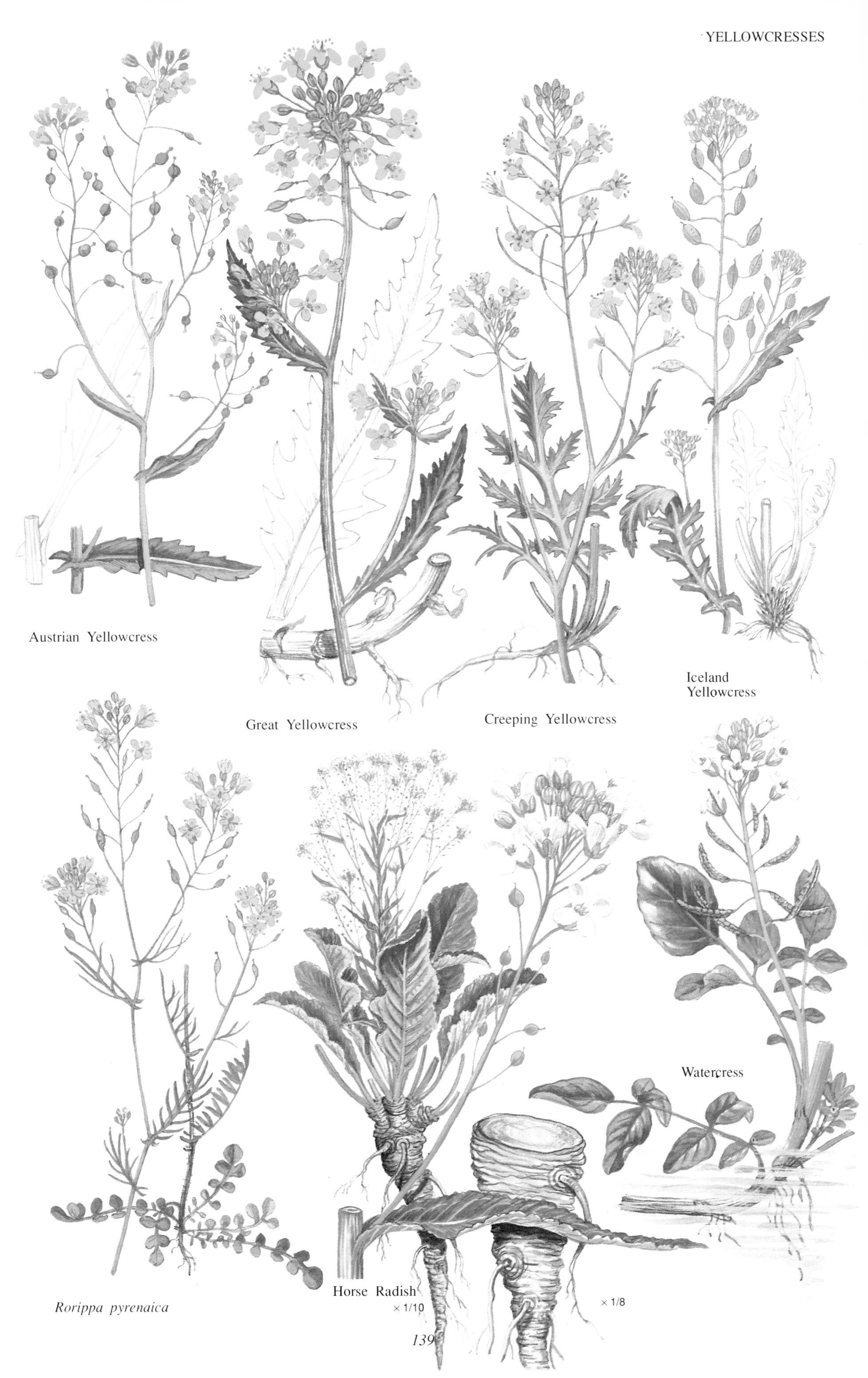
Austrian Yellowcress
Great Yellowcress
Creeping Yellowcress
Iceland Yellowcress
Rorippa pyrenaica
Horse Radish
× 1/10
× 1/8
Watercress

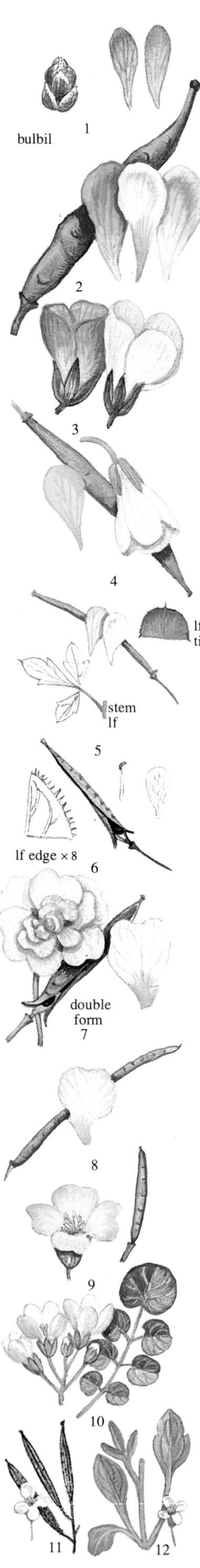


**Bittercresses** *Cardamine*. Annual or perennial herbs, often hairless, or with simple unbranched hairs. Leaves simple to pinnate. Flowers in racemes, sometimes branched. Fruit a flattened siliqua, the valves coiling spirally from the base as the fruit splits. Cosmopolitan; 100 species, those in the tropics confined to the mountains. Several are cultivated in gardens.

1* **Coralroot Bittercress** *Cardamine bulbifer* (L.) Crantz [*Dentaria bulbifer*]. Medium, hairless, rhizomatous perennial, without root-leaves. Stem leaves pinnate with 1-3 pairs of lateral leaflets, toothed; uppermost leaves simple. Most upper leaves with small purplish-brown bulbils at the base. Flowers pale purple or lilac, 12-18mm. Fruit 20-35mm, not always produced. Woodland, often of beech, woodland banks and stream margins, on calcareous soils, to 1500m. May-June. Throughout, except the extreme north, Ireland and Iceland, often grows in association with *Allium ursinum* and *Scilla non-scripta*. The bulbils are readily detached and grow into new plants in suitable habitats. Europe to SE Russia and the Caucasus Mountains. **B**: Confined to S England – Chilterns and the Weald of Kent and Sussex.

2* *Cardamine heptaphylla* (Vill.) O.E. Schulz [*C. pinnata*]. Medium, almost hairless perennial. Leaves all pinnate, the lower with 3-5 pairs of lanceolate, toothed, leaflets. Flowers 19-21mm, white, pink or purplish. Fruit 40-80mm. Mountain woods and road verges, particularly beneath beech trees, to 1800m. May-July. France & S W Germany; naturalized in Belgium.

3 *Cardamine pentaphyllos* (L.) Crantz [*Dentaria pentaphyllos, D. digitata*]. Like *C. heptaphylla* but all leaves palm-like or trifoliate, not pinnate; flowers slightly larger, white or pale purple. Mountain woodlands and rocky habitats, to 2200m. May-July. France & S Germany.

4* **Drooping Bittercress** *Cardamine enneaphyllos* (L.) Crantz. Short to medium, hairless perennial. Leaves in a whorl of 2-4, 2-ternate, with lanceolate, toothed leaflets. Flowers pale yellow, 12-18mm, drooping in clusters. Mountain woods, growing in deep loamy soils in shaded and semi-shaded places, to 2150m. April-July. S Germany. **P**: Bees. Occasionally cultivated in gardens.

5* **Trifoliate Bittercress** *Cardamine trifolia* L. Short, slightly hairy, rhizomatous perennial. Leaves mostly basal, trifoliate, leaflets oblong or rounded, purplish beneath. Flowers white or pink, 10-14mm; anthers yellow. Fruit 20-25mm. Moist woodland and other shady habitats, to 1400m. April-June. E France & S Germany; naturalized in Britain. Cultivated in gardens and sometimes escaping. **B**: Occasionally naturalized.

6* **Large Bittercress** *Cardamine amara* L. Medium hairless perennial, with angular stems and slender stolons. Leaves pinnate, not in a basal rosette, the lower with 2-4 pairs of oval leaflets, the upper with narrower leaflets. Flowers white, occasionally purplish, 11-12mm, 1; anthers blackish-violet. Damp habitats, pastures, woods, woodland flushes, marshes and fens, streamsides, to 2000m. April-June. Throughout, except the extreme north, the Faeroes and Iceland. Generally growing in places where there is moving ground water – tolerant of shade, often abundant. Throughout much of Europe east to W & C Asia. **B**: Throughout, except for much of S W Britain.

7* **Cuckoo Flower, Lady's Smock** *Cardamine pratensis* L. Medium, hairless perennial with erect stems. Basal leaves in a rosette, pinnate with a large end leaflet; upper leaves with more numerous, narrower, leaflets. Flowers pale lilac-pink, rarely white, 12-18mm, petals notched, anthers yellow. Fruit 25-40mm. Damp habitats, meadows and pastures, open woodland, marshes, hedgerows, roadsides, stream and pond margins. April-June. Throughout Europe except for the extreme north. Often forming extensive colonies. **P**: Bee-flies and long-tongued hoverflies. **B**: Common throughout.

8* *Cardamine palustris* (Wimmer & Grab.) Peterm. Like *C. pratensis*, but often smaller, with 5 or more stem leaves (not 2-3), with broader, stalked leaflets. Flowers larger, 18-24mm. Fruit 30-55mm. Similar habitats and flowering time to *C. pratensis*. Throughout Europe except the extreme north; possibly also in Britain, though there is little evidence to substantiate this. Generally readily separated from *C. pratensis*, but in the west of its range, particularly in Britain, the two species overlap to some extent. For this reason it is sometimes treated as a morphological variant of *C. pratensis*.

9 *Cardamine nymanii* Gand. Like *C. pratensis*, but a low, short, tufted plant with rather thick leaves, mainly in a basal rosette; leaflets with impressed veins. Flowers generally slightly smaller, lilac. Fruit 10-18mm. Damp meadows and stream margins. May-July. N Europe, the Faeroes and Iceland.

10 *Cardamine matthiola* Moretti [*C. hayneana*]. Like *C. pratensis*, but stems generally much-branched from the base with many lateral inflorescences. Flowers smaller, 9-13mm, white, the petals not notched. Fruit 18-30mm. Moist habitats, particularly meadows. April-June. S Germany.

11* **Alpine Bittercress** *Cardamine bellidifolia* L. Low hairless perennial. Leaves thick, spoon-shaped, mostly in a basal rosette, untoothed. Flowers white, 7-10mm in clusters of 2-5, fruit purplish-brown, 12-25mm. Damp habitats, meadows, gravelly places, stream margins, usually on acid soils, to 2100m. Arctic and sub-Arctic Europe.

subsp. *alpina* (Willd.) B.M.G. Jones [*C. alpina*]. Like the type, but usually with 2-3 stem leaves and flowers in clusters of 3-8; fruit 10-15mm. Mountain habitats, grassy and rocky places, to 3000m. C & E France and S Germany.

12 *Cardamine resedifolia* L. Like *C. bellidifolia*, but generally taller, the leaves mostly lobed or toothed, eared at the base of the stalk close to the stem. Flowers slightly larger, 9-12mm. Damp rocky mountain habitats, to 3500m. June-August. C & E France and S Germany.

Cuckoo
Flower

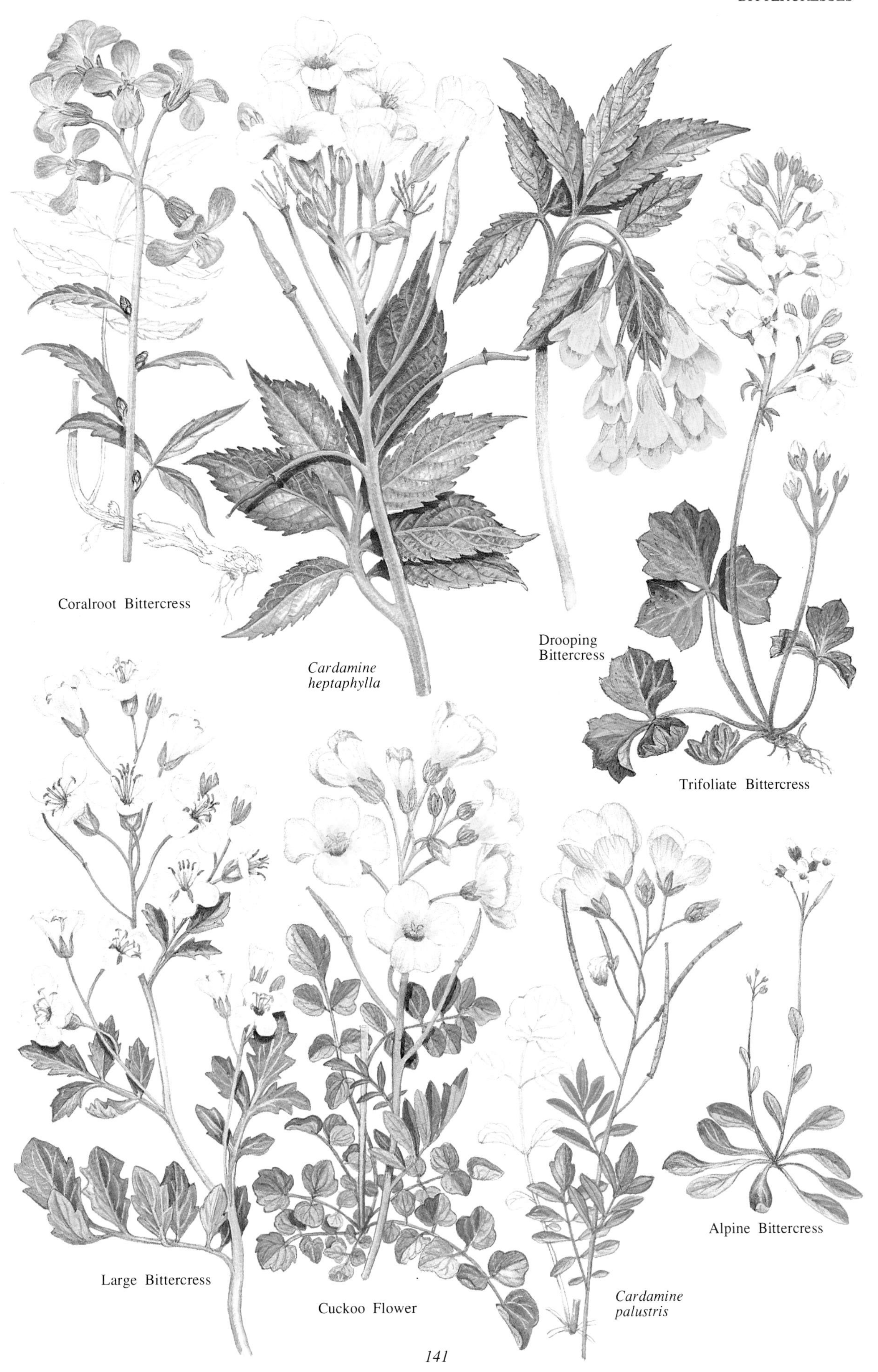
Coralroot Bittercress
*Cardamine heptaphylla*
Drooping Bittercress
Trifoliate Bittercress
Large Bittercress
Cuckoo Flower
*Cardamine palustris*
Alpine Bittercress

1* *Cardamine parviflora* L. Short hairless annual. Leaves pinnate, the lower with 3-5 pairs of leaflets; upper leaves with up to 8 pairs of linear-oblong, untoothed, leaflets. Flowers white, small, 2-3mm. Fruit 12-20mm, erect. May-August. Continental Europe, except for most of the extreme north, Holland and Denmark.

2* **Narrow-leaved Bittercress** *Cardamine impatiens* L. Medium hairless annual, sometimes overwintering. Leaves mostly stalked, pinnate, with 5-9 pairs of usually 3-lobed leaflets, eared at the base of the leaf stalk, margin finely bristled. Flowers inconspicuous, white or greenish, 3-4mm, petals sometimes absent; anthers greenish. Fruit 18-30mm, spreading. Damp shaded habitats, woodland, often beech or ash, rocky places and screes, generally on calcareous soils. May-August. Throughout, except the extreme north. **B**: Local, scattered localities N & W England, rare elsewhere including Wales and SE England.

3* **Hairy Bittercress** *Cardamine hirsuta* L. Low to short annual, branched only at the base; stems straight, hairless. Leaves pinnate, mostly in a basal rosette; leaflets angular, hairy above. Flowers small, white, 3-4mm, petals equalling sepals or absent; stamens 4 usually. Fruit 18-25mm, greatly overtopping the flower buds. Bare open ground, cultivated land, rocky places and old walls. February-November. Throughout, except Spitsbergen. A common weed of cultivation, particularly gardens. The seeds are flung from the plant by the explosive mechanism of the fruit capsule. Throughout most of the northern hemisphere. **B**: Widespread.

4 **Wavy Bittercress** *Cardamine flexuosa* With. [*C. sylvatica*]. Like *C.hirsuta*, but often taller biennial or perennial, the stem flexuous and branched above, hairy at the base. Flowers similar but usually with 6 stamens. Fruit 12-25mm, scarcely overtopping flowerbuds. Damp shaded habitats, ditches and areas subject to winter flooding. April-September. Throughout, except Iceland and Spitsbergen. Normally perennial but can be annual. **B**: Throughout.

**Cardaminopsis** *Cardaminopsis*. Like *Cardamine* but fruit valves not coiling elastically on dehiscence, and with a strong midvein (not weak or indistinct).

5* **Tall Rockcress** *Cardaminopsis arenosa* (L.) Hayek. [*Arabis arenosa*]. Variable short to tall, much-branched, rather robust, hairy annual or perennial. Basal leaves pinnately-lobed; stem leaves pinnate to simple and toothed. Flowers many, white or lilac (especially in the south) 7-9mm. Fruit 10-45mm, spreading. Sandy, generally calcareous soils, to 1400m. April-June. Throughout, except Ireland, Iceland, the Faeroes and Spitsbergen.

6* **Northern Rockcress** *Cardaminopsis petraea* (L.) Hiitonen [*C. hispida*]. Short, hairy or hairless, stoloniferous perennial. Basal leaves lanceolate long-stalked, pinnately-lobed to toothed; stem leaves few, lanceolate, untoothed. Flowers white or purplish, 5-8mm, few. Fruit 14-45mm, spreading. Rocky mountain habitats, screes and river gravels, on basic soils usually, to 1400m, occasionally at sea level. June-August. Britain, the Faeroes, Germany, Scandinavia and Iceland. Variable – some forms are more compact and may have pure white or pink flowers. N & NW Europe to Siberia and North America. **B**: Local – N Wales, Scotland, the Scottish Islands and Ireland (Galtee & Glenade Mts).

7 *Cardaminopsis halleri* (L.) Hayek. Like *C. petraea*, but basal leaves rounded, or if pinnately-lobed then with a rounded end leaflet. Flowers numerous, white or lilac. Fruit 10-25mm. Mountain habitats. April-May. S Germany and the neighbouring parts of France.

**Rockcresses** *Arabis*. Annual or perennial herbs, hairs unbranched to starry. Leaves simple. Flowers often white; inner sepals slightly pouched at the base. Fruit a slender, flattened siliqua, generally without a distinct mid-vein. About 120 species in temperate regions of Europe and Asia, North America and the mountains of E Africa.

8* **Tower Mustard** *Arabis glabra* (L.) Bernh. [*A. perfoliata*, *Turritis glabra*]. Tall erect, unbranched biennial, hairy only at the base. Leaves grey-green, the lower deeply toothed, starry-haired; stem leaves arrow-shaped, hairless. Flowers pale yellow or greenish, 5-6mm. Fruit erect, 40-70mm. Dry habitats, open woods, heaths. May-July. Throughout, except the extreme north, Ireland, the Faeroes and Iceland. **B**: Scattered localities in C, S & E England – generally declining.

9 *Arabis pauciflora* (Grimm) Garcke [*A. brassicaeformis*]. Medium to tall, hairless, grey-green perennial. Basal leaves oval, untoothed, long-stalked; stem leaves mostly guitar-shaped, unstalked. Flowers white or pink, 5-7mm. Fruit erect, 30-80mm. Dry rocky meadows and woods, on calcareous soils. May-July. France and S Germany. Local.

10* *Arabis planisiliqua* (Pers.) Reichenb. [*A. gerardii*]. Medium to tall biennial, stems branched above, frequently reddish, hairy below. Stem leaves arrow-shaped, closely clasping the stem at the base. Flowers white 4-5mm. Fruits 30-50mm, on erect 3-7mm stalks, constricted between each seed. Dry rocky habitats and waste ground. May-July. Holland, France and Germany; naturalized in Sweden.

*Arabis sagittata* (Bertol.) DC. Like *A. planisiliqua*, but stem hairy at the base, leaves with spreading basal lobes and flowers slightly larger. Dry calcareous soils. May-July. Holland and Germany southwards.

*Arabis borealis* Andrz. ex Ledeb. Like *A. planisiliqua*, but plant hairy all over; fruit stalks 5-20mm. Dry calcareous slopes. **I**: N USSR. Naturalized in Finland.

11* **Hairy Rockcress** *Arabis hirsuta* (L.) Scop. Very variable short-medium biennial or short-lived perennial; stems branched above, hairy. Basal leaves in a rosette; stem leaves lanceolate, held close to the stem. Flowers white, 3-5mm. Fruit 15-35mm, erect and held close to the stem. Dry habitats, grassland, dunes, dry banks, rocks and walls on calcareous soils, occasionally on wet mountain rocks, to 2050m. May-August. Throughout, except Iceland and Spitsbergen. Widespread from Europe and N Africa, to N Asia, Japan and North America. **B**: Throughout, local.

12 **Fringed Rockcress** *A. h.* var. *brownii* (Jordan) Titz [*A. brownii*]. Like *A. hirsuta*, but shorter, to 20cm only, and hairless except for the leaf-margins; stem leaves few. Fruits in a compact cluster. Sandy coastal dunes. July-August. W Ireland, not known elsewhere. Plants often regarded as hybrids between this variety and *A. hirsuta* are sometimes found in W Ireland.

*Cardamine parviflora*
Narrow-leaved Bittercress
Hairy Bittercress
Tall Rockcress
Northern Rockcress
Tower Mustard
*Arabis planisiliqua*
Hairy Rockcress

1* **Tower Cress** *Arabis turrita* L. Medium, softly hairy biennial or perennial; stems often reddish. Basal leaves in a lax rosette, oval, long-stalked, grey-hairy; stem leaves narrow heart-shaped, clasping the stem. Flowers pale yellow, 7-9mm. Fruits very long, 100-140mm, arching over in a one-sided spray. Rocky habitats, old walls, to 1500m. April-July. Belgium, France and Germany; naturalized in Britain. Readily recognized by its extremely long arching fruits. **P**: Probably self-pollinated. A native of C & S Europe, NW Africa and W Asia. **B**: Established only on old walls in Cambridge.

2* **Annual Rockcress** *Arabis recta* Vill. [*Arabis auriculata*]. Short, softly hairy annual; stem unbranched or branched above. Basal leaves oval, untoothed, soon withering; stem leaves narrow arrow-shaped, untoothed, blunt, clasping the stem. Flowers white, 5mm, with erect petals. Fruit 10-35mm, spreading, hairy or not. Mountain rocks and screes, to 1500m. April-June. France and Germany; naturalized in Belgium.

3* **Bristol Rockcress** *Arabis scabra* All. [*A. stricta*]. Low to short, rough-hairy perennial; stems unbranched. Leaves dark glossy green, mostly in a basal rosette, oblong, toothed and stalked; stem leaves few, clasping the stem. Flowers few, cream, 6-8mm. Fruit 35-50mm, erect. Rocky habitats, cliffs and screes, generally on calcareous soils, to 1400m. March-May. SW England, French Alps and Jura Mts. Local. **P**: Probably self-pollinated. Restricted to parts of W and SW Europe. **B**: Very rare, confined to the Avon Gorge.

4* **Alpine Rockcress** *Arabis alpina* L. Low to short grey-green, hairy, perennial, creeping and forming low mats. Basal leaves in rosettes, elliptical, coarsely toothed; stem leaves heart-shaped pointed, with rounded basal lobes. Flowers white, 6-10mm, petals spreading. Fruit 20-35mm, spreading. Mountain habitats, wet rocks and rock ledges, gravels and screes, often on calcareous soils, to 3000m. May-July. Britain and most of Europe, except Belgium, Holland and Denmark. **P**: Self-pollinated. Europe east to Siberia, the Himalaya and Alaska. **B**: Very rare, confined to the island of Skye.

5 **Garden Arabis** *Arabis caucasica* Schlecht. [*A. albida*]. Like *A. alpina*, but with greyish-white denser rosettes and stem leaves arrow-shaped with pointed basal lobes. Flowers larger, white, 14-16mm. Fruit 40-70mm. Rocky habitats and rock ledges, gravels and screes, sometimes naturalized on banks and walls, to 3000m. May-July. **I**: W Asia. Naturalized in Britain, Belgium, Holland and Germany. Widely cultivated in gardens, including forms with pale pink and double flowers. **B**: Locally naturalized.

6* **Aubrieta** *Aubrieta deltoides* (L.) DC. Low to short hairy, mat-forming or straggling perennial. Leaves small, spoon-shaped to rhombic, few-toothed or untoothed. Flowers reddish-purple, violet or occasionally white, 13-18mm, petals spreading. Fruit a narrow oblong siliqua, 6-16mm, downy. Rocky places, banks and walls. April-June, occasionally later. **I**: S Greece, Crete and Sicily. Naturalized in Britain, France and Holland. Widely cultivated in gardens, especially on rock gardens and along path edges. **P**: Bees and butterflies. **B**: Locally naturalized on old walls.

**Honesty** *Lunaria*. Biennial or perennial herbs with unbranched hairs. Fruit large and flat, splitting to leave a silvery, conspicuous, persistent membrane. 3 species in C and SE Europe. The fruits are often used for dried flower arrangements.

7* **Perennial Honesty** *Lunaria rediviva* L. Tall rather stout perennial herb spreading by underground runners; stems erect, hairy. Leaves oval, pointed, toothed, all stalked. Flowers pale purple to violet, 20-25mm, in branched racemes, fragrant. Fruit elliptical, pointed, 35-90mm. Damp wooded habitats and other shaded places, primarily on calcareous soils, to 1400m. May-July. Belgium, Holland, Denmark, France and Germany. W Europe to Russia. Cultivated in gardens. **P**: Bees and butterflies.

8* **Honesty** *Lunaria annua* L. [*L. biennis*]. Medium to tall biennial forming a single lax rosette of leaves in the first season. Leaves oval to lanceolate, coarsely toothed, only the lower stalked. Flowers reddish-purple or mauve, rarely white, 25-30mm, in branched racemes usually, not noticeably scented. Fruit subrounded, 30-55mm. Cultivated land, waysides and waste places, to 1200m. April-June. **I**: SE Europe. Naturalized or casual in Britain and much of Europe. Widely cultivated in gardens, primarily for its attractive fruits – cultivated forms include purple and white-flowered forms as well as those with leaf variegations. The roots are partly tuberous. **P**: Butterflies and long-tongued bees, but may also be self-pollinated. **B**: A frequent escape from gardens in Britain, though rarely persisting for very long.

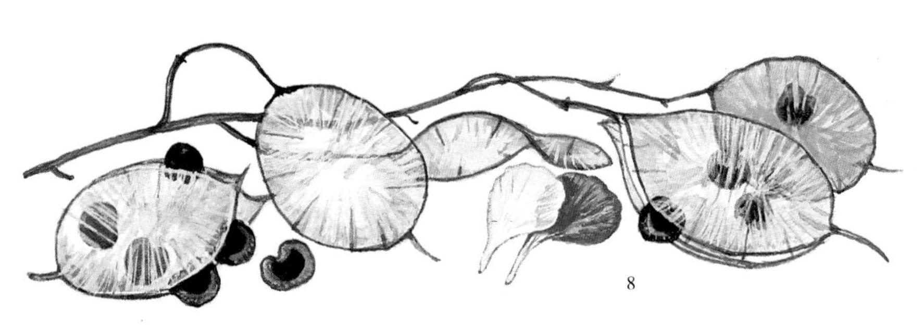

Tower Cress
Annual Rockcress
Bristol Rockcress
Alpine Rockcress
Aubrieta
Perennial Honesty
Honesty

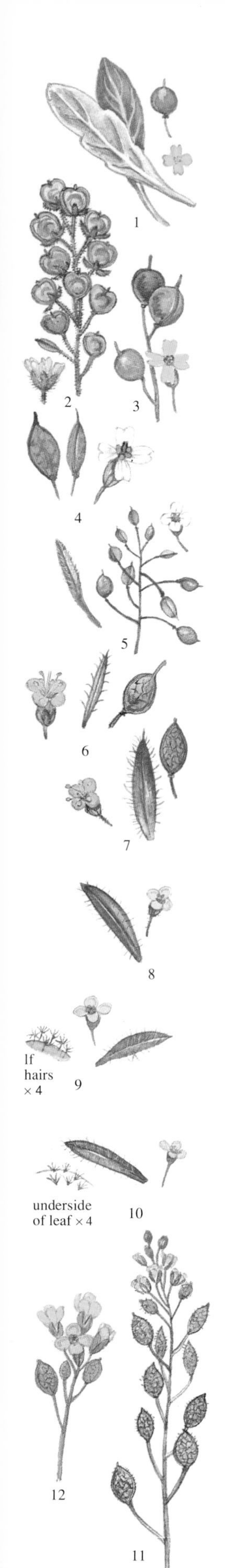


**Alisons** *Alyssum*. Annual or perennial herbs with branched or starry hairs. Flowers yellow (in the species included); sepals not pouched at the base. Fruit a small silicula, often oval or rounded, without conspicuous veins. 150 species in Europe and the Mediterranean region to W Asia and Siberia. Several are grown in gardens.

1* **Golden Alyssum** *Alyssum saxatile* L. Short, laxly tufted, grey-green perennial, becoming woody at the base with rather straggling stems. Leaves mostly basal, oblong, untoothed to lobed. Flowers bright yellow, 4-6mm, many in lax, branched clusters, petals slightly notched. Fruit oval, 3-5mm, hairless. Rocky habitats, walls and embankments. April-July. **I**: C & SE Europe. Naturalized in Britain and Germany. Cultivated in gardens – the normal form has bright yellow flowers but there is also a pale creamy yellow form.

2* **Small Alison** *Alyssum alyssoides* (L.) L. [*A. calycinum*]. Low to short downy, generally erect, greyish annual or biennial. Leaves small, lanceolate, the upper narrower. Flowers pale yellow, 3-4mm, in dense racemes; petals notched. Fruit rounded, 3-4mm, downy, with persisting sepals. Arable land, grassy habitats, waste places, generally on sandy and calcareous soils at low altitudes. April-June. Throughout, except Ireland, the Faeroes, Iceland and Spitsbergen. Populations vary from one year to another, especially on agricultural land. **B**: Very local in S & E England and E Scotland – declining.

3* **Mountain Alison** *Alyssum montanum* L. Low to short spreading to erect, green or whitish, perennial, with numerous non-flowering leaf rosettes. Basal leaves oblong, untoothed; upper leaves narrower. Flowers bright yellow, 5-6mm, in long racemes; petals notched. Fruit rounded, 3.5-5.5mm, slightly inflated, without persistent sepals. Rocky and stony habitats, mostly in mountains, to 2500m. April-June. France and Germany. Cultivated in gardens, especially on walls and rock gardens. **P**: Mainly bees and buterflies.

subsp. *gmelinii* (Jordan) Hegi & E. Schmid. Like the type, but with very few non-flowering leaf-rosettes and smaller flowers. Fruit oval, 3-5mm. Sandy lowland habitats. April-July. N & C France & S Germany.

4* **Hoary Alison** *Berteroa incana* (L.) DC. [*Alyssum incanum, Farsetia incana*]. Medium grey-hairy annual or perennial. Leaves oblong to lanceolate, usually untoothed. Flowers white, 5-8mm; petals deeply notched. Fruit elliptical 4.5-8mm, inflated, downy. Arable and waste ground, on light sandy soils. June-October. Europe north to S Norway and Finland; naturalized in Britain. Sometimes imported as an impurity in clover seed. The leafy stems and deeply cleft white petals make this easy to recognise. N & C Europe to W Asia. **B**: Scattered localities, especially in E England.

5* **Sweet Alison** *Lobularia maritima* (L.) Desv. [*Koniga maritima*]. Short hairy, greyish-white, spreading annual, branched at the base. Leaves linear-lanceolate, pointed, untoothed. Flowers white, 5-6mm, sweetly scented, petals not notched, in dense clusters at first, but elongating in fruit. Fruit an oval silicula, 2-3.5mm, downy often. Rocky and sandy habitats, waste land, especially near the coast. June-September. **I**: S Europe – Mediterranean region. Naturalized in Britain and much of Europe except for the north. Widely cultivated in gardens, including forms with pink or reddish flowers, often used for edging or bedding in gardens. **P**: Various small insects. **B**: Scattered localities in Britain – more especially in the SW.

**Whitlow Grasses** *Draba*. Annual or perennial herbs with simple, toothed or untoothed leaves. Flowers white or yellow. Fruit a silicula, valves flat with a mid-vein in the lower half. Mostly Arctic or mountain plants of rocky and gravelly places. 300 species scattered in northern temperate and Arctic regions, Central and South America.

6* **Yellow Whitlow Grass** *Draba aizoides* L. Variable low tufted, hairless perennial. Leaves all in basal rosettes, linear-lanceolate, edged with stiff bristles. Flowers yellow, 5-7mm, in dense clusters, elongating in fruit; stamens equalling petals. Fruit elliptical, 6-12mm, usually hairless. Rocky mountain habitats, cliffs, crevices, screes, on calcareous soils, and occasionally on old walls, to 3600m. March-May. W Britain, Belgium, France and Germany. Cultivated occasionally in gardens. **P**: Bees and flies, also self-pollinated. **B**: Rare – confined to Wales, SW Glamorgan.

7* **Alpine Whitlow Grass** *Draba alpina* L. Low to short, densely tufted, hairy perennial. Leaves all in basal rosettes elliptical. Flowers bright yellow, 4-6mm, in a dense cluster, scarcely elongating in fruit. Fruit 4-8mm, hairless. Rocky and gravelly habitats, to 1650m. June-July. Arctic and sub-Arctic Europe; including Iceland. Occasionally cultivated in gardens. *D. alpina* represents a complex of species of which the following (in our area) are fairly distinct.

8 *Draba gredinii* Elis. Ekman. Like *D. alpina*, but flowers pale yellow and leaves sparsely hairy. Rocky habitats. Spitsbergen.

9 *Draba bellii* Holm. Like *D. alpina*, but leaves with long rigid hairs on the upper surface; flowers dull yellow. Fruit hairless. Rocky habitats. Spitsbergen.

10 *Draba kjellmanii* Lid. ex Elis. Ekman. Like *D. bellii*, but the star-shaped hairs on the leaf undersurface unstalked (not stalked). Rocky habitats. Spitsbergen.

11 *Draba oblongata* R. Br. ex DC. Like *D. alpina* but flower-clusters greatly elongating as the fruits develop – to 20cm long. Flowers pale yellow or cream, 3-4mm; fruit hairy. Rocky habitats. Spitsbergen.

12 *Draba glacialis* Adams. Like *D. oblongata*, but more densely hairy and flowers bright yellow, larger, 6mm; fruit scarcely hairy. Rocky habitats, moraines and screes. June-August. Arctic Norway and Spitsbergen, extending into the USSR.

Yellow Whitlow Grass

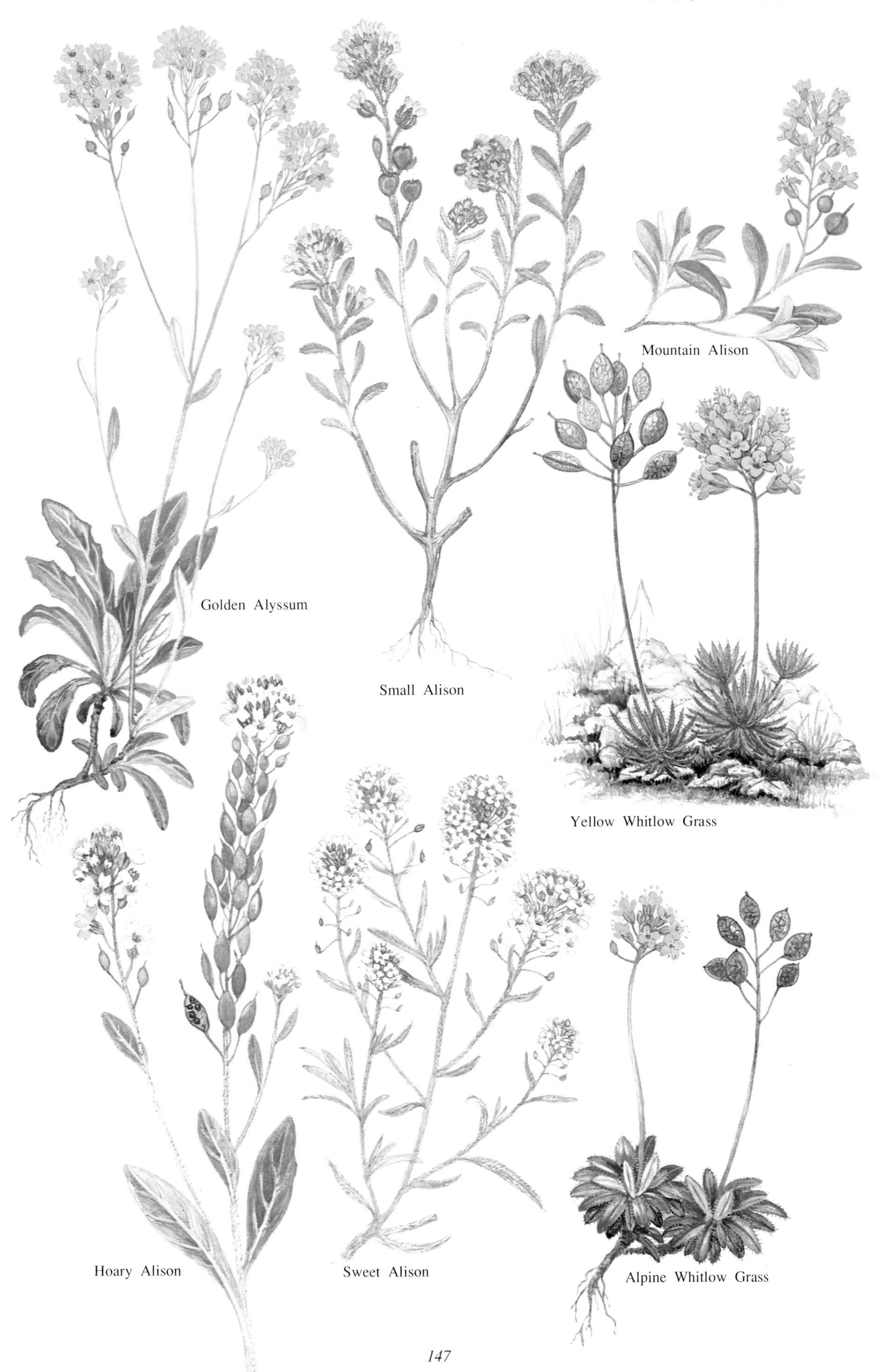
Mountain Alison
Golden Alyssum
Small Alison
Yellow Whitlow Grass
Hoary Alison
Sweet Alison
Alpine Whitlow Grass

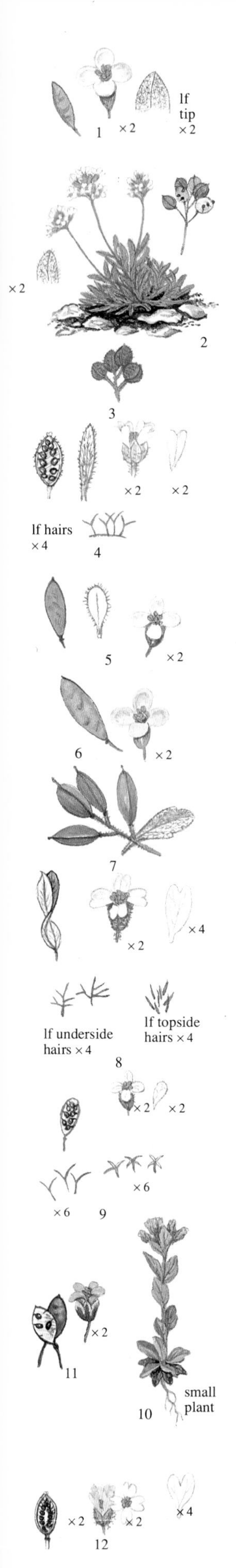


1* *Draba nivalis* Liljeblad. Low densely tufted, hairy, bluish-green perennial. Leaves in dense basal rosettes, oblong, narrowed to the base, untoothed, densely starry-hairy; stem leaves usually absent. Flowers white or cream, 5-6mm, in dense, few-flowered racemes. Fruit 4-9mm, hairless. Mountain habitats, rocks, screes and moraines, to 1920m. June-July. Scandinavia, including Spitsbergen, and Iceland.

2 *Draba subcapitata* Simmons. Like *D. nivalis*, but more dwarf, the leaves with star-shaped hairs as well as long silky hairs. Fruit in a congested head (not elongated). Rocky habitats, screes and moraines. Iceland and Spitsbergen, extending E into the USSR.

3 *Draba cacuminum* Elis. Ekman. Like *D. subcapitata*, but fruit hairy. Rocky mountain habitats. Only in Norway.

4* **Rock Whitlow Grass** *Draba norvegica* Gunn. (incl. *D. rupestris* R. Br.). Very variable, low-short, somewhat hairy, tufted perennial with slender, often flexuous, hairy stems. Leaves mostly in basal rosettes, oblong-lanceolate, with a hairy margin and occasionally a few teeth; stem leaves up to 3. Flowers white or cream, 4-5mm, often with an odd flower below the main cluster; petals notched. Fruit elliptical, 5-7mm, erect, hairy or not. Mountain rocks, screes and stony grassland, to 1250m. July-August. Arctic and sub-Arctic Europe south to Scotland and including Iceland. Local, generally growing on base-rich rocks. **B**: Very rare, confined to the C Highlands of Scotland.

5* **Bald White Whitlow Grass** *Draba fladnizensis* Wulfen [*D. wahlbergii*]. Low tufted perennial, hairless except for the leaf margins. Leaves oblong, in basal rosettes; stem leaves often absent, sometimes up to 2. Flowers white, 3-4mm, in small clusters. Fruit elliptical 5-8mm. Rocky and grassy mountain slopes, preferring acid soils, 1600-3400m. June-August. France, Germany, Scandinavia and Iceland. Arctic plants without stem leaves and with hairs on the rosette leaves are sometimes separated as *D. lactea* Adams.

6* *Draba daurica* DC. Like *D. fladizensis*, but more robust, to 25cm, with hairy, often flexuous, stems. Leaves narrow-lanceolate, toothed or not, hairy; stem leaves 1-4. Flowers cream, 5-6mm. Fruit larger, 6-12mm, hairless. Mountain habitats, rocks, screes and moraines. June-July. Scandinavia and the Arctic mountains, including Iceland.

7 *Draba cinerea* Adams [*D. arctica*]. Like *D. daurica*, but more densely hairy; fruit inflated (not flat). Rocky habitats, screes and moraines. Arctic Europe – Finland, Norway and Spitsbergen.

8* **Hoary Whitlow Grass** *Draba incana* L. Very variable short, rather robust, hairy biennial, occasionally perennial, with erect leafy stem, sometimes branched. Basal leaves lanceolate, in a lax rosette, hoary, toothed; stem leaves numerous. Flowers white, 4-5mm, in a lax raceme, petals notched. Fruit oblong, 7-9mm, twisted when ripe, hairless. Mountain habitats, rocks, screes, moraines and cliffs, generally on limestone, to 2600m, occasionally at low altitudes on calcareous dunes (especially N Scotland, Outer Hebrides and the Shetland Is.). May-July. Throughout, except parts of the north, Belgium, Holland and Germany. **P**: Self-pollinated. N Europe to the mountains of C Europe, east to C Asia – also native in Greenland. **B**: N England, N Wales and Scotland, north to the Shetland Is.

9* **Wall Whitlow Grass** *Draba muralis* L. Short, slightly hairy, erect annual; stem branched or not, leafy. Leaves broad-oval, the upper partly clasping the stem. Flowers tiny, white, 2-3mm, petals not notched, in a lax raceme, elongating in fruit. Fruit oblong-elliptical, 3-6mm, hairless. Rocky habitats, walls, on well-drained soils, mainly over limestone, to 1300m. April-June. Throughout, except for much of the north, the Faeroes and Iceland; naturalized or casual in many areas, especially in Britain, as a garden weed. **P**: Automatically self-pollinated. From Europe to N Africa and W Asia. **B**: Rare, mainly confined to the Pennines and the Lake District, and a few isolated localities in SW England.

10 *Draba nemorosa* L. Like *D. muralis*, but stems hairless above and flowers pale yellow, fading to white; fruit hairy. Rocky habitats, waste and disturbed ground. April-July. Belgium, Holland, France and Germany, C & S Scandinavia.

11* *Draba crassifolia* R.C. Graham. Short tufted perennial; stems hairless and leafless. Leaves thick, spoon-shaped. Flowers pale yellow, 3-4mm, few in a condensed raceme; petals scarcely longer than the sepals. Fruit elliptical, 4-7mm, hairless. Arctic habitats. June-July. Norway and Sweden, extending to Greenland and North America.

12* **Whitlow Grass** *Erophila verna* (L.) Chevall. [*Draba verna*]. Very variable, low, slightly hairy annual, sometimes overwintering; stems leafless. Leaves lanceolate to elliptical, toothed, in a basal rosette. Flowers white or pinkish, 3-5mm, the petals deeply cleft. Fruit narrow-elliptical, 6-10mm, hairless, on long stalks. Dry rocks, walls, sandy and stony ground, dry sandy heaths, both coastal and inland, to 1700m. March-May. Throughout, except the Arctic and the Faeroes. **P**: Occasionally small insects, but normally the flowers are self-pollinated. Europe to Asia and N Africa – widely naturalized in parts of North America. **B**: Throughout, scarcer in the W and in Ireland. A form with shorter fruits containing up to 40 ovules (not 40-60) is sometimes distinguished as subsp. *spathulata* (Láng) Walters, confined mostly to sand-dunes in E & S Britain, rarer elsewhere, however, this species represents a complex which requires further investigation.

Hoary Whitlow Grass

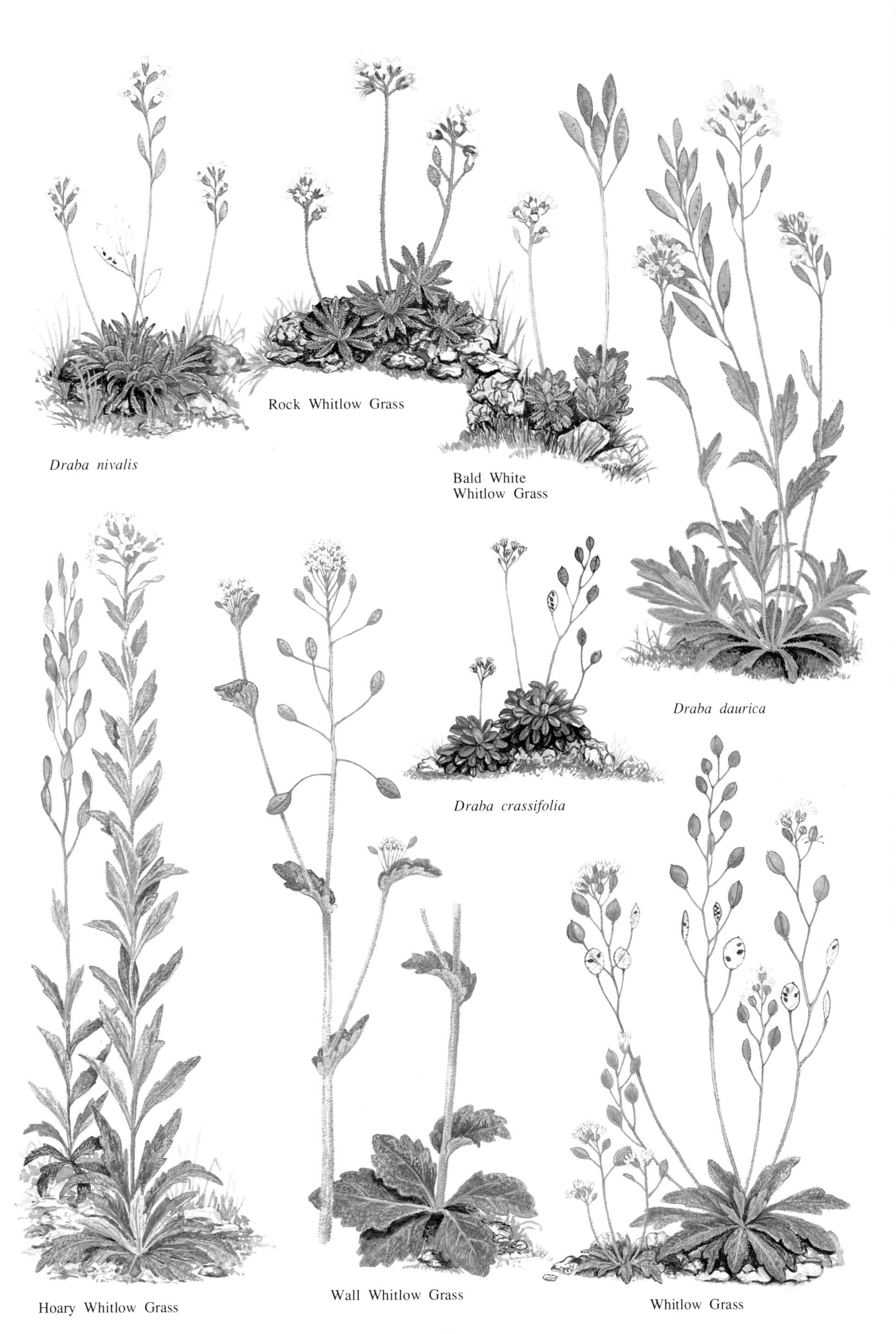
Rock Whitlow Grass
*Draba nivalis*
Bald White
Whitlow Grass
*Draba crassifolia*
*Draba daurica*
Hoary Whitlow Grass
Wall Whitlow Grass
Whitlow Grass

**Scurvy-grasses** *Cochlearia.* Annual or perennial herbs with unbranched hairs, or hairless. Leaves simple. Flowers in racemes, with short-clawed petals. Fruit a swollen rounded or oval silicula, with convex valves. 25 species in N temperate regions, the Himalaya and the mountains of Java (where probably introduced). Associated with coastal and mountain habitats.

1* **Danish Scurvy-grass** *Cochlearia danica* L. Variable low-short annual, sometimes overwintering, not fleshy. Basal leaves long-stalked, rounded to triangular-heart-shaped; upper leaves stalked, the lower stem leaves with 3-7 lobes. Flowers white or lilac-purple, 4-5mm. Fruit oval, 3-6mm. Coastal habitats, cliffs, sandy and rocky shores, walls, banks, occasionally on disturbed ground inland. January-September. Throughout, except Denmark, the Faeroes and Iceland. **P**: Various small insects, but generally automatically self-pollinated. **B**: Frequent, more common in the W and SW.

2* **Common Scurvy-grass** *Cochlearia officinalis* L. Very variable, low-medium, fleshy biennial or perennial. Basal leaves generally kidney-shaped, long-stalked; stem leaves mostly unstalked, loosely clasping. Flowers white, 8-10mm. Fruit rounded, 4-7mm; stalks longer than the capsule. Salty and brackish marshes, grassy cliffs, banks and sea walls, occasionally on mountains inland, to 2200m. April-August. Throughout, except the extreme north, Iceland and Finland. Scurvy-grass is rich in vitamin C (Ascorbic Acid) and was widely used in the past as a prevention for scurvy, especially by sailors. **B**: Throughout, except much of S and SE England.

3* *Cochlearia pyrenaica* DC. [*C. officinalis* subsp. *pyrenaica*]. Like *C. officinalis*, but leaves oval-elliptical. Fruits narrowed at both ends, their stalks shorter than the capsule. Rocky ledges and stream margins, to 2200m. June-September. Mountains of C Europe, E France and S Germany, also in Britain, and Belgium. The British plants are sometimes referred to *C. alpina* (Bab.) H.C. Watson. **P**: Flies and beetles. **B**: Very local in N England and Ireland, on carboniferous limestone.

4 *Cochlearia aestuaria* (Lloyd) Heywood. Like *C. officinalis*, but capsule truncated or notched at the apex. Muddy shores and estuaries. Coastal W France. June-August. Hybridises with *C. officinalis*.

5* **English Scurvy-grass** *Cochlearia anglica* L. Robust short biennial or perennial, with erect stems. Basal leaves oblong to oval, narrowed into a long stalk, slightly toothed or untoothed, the upper leaves clasping the stem. Flowers usually white, 9-14mm. Fruit elliptical, 8-15mm. Muddy shores of salt marshes and tidal estuaries, sandy and shingle shores. April-July. NW Europe. **P**: Various small insects, especially flies and wasps. The tapering leaf-bases distinguish this from *C. danica* and *C. officinalis*. Hybridises with *C. aestuaria*. **B**: Coasts, but rare in Scotland.

6 *Cochlearia fenestrata* R.Br. (incl. *C. arctica* Schlecht.). Like *C. anglica*, but less robust and leaves truncated or heart-shaped at the base. Flowers smaller, 6-9mm, white or occasionally purplish. Fruit narrow-elliptical, 5-8mm. Rocky habitats. June-August. Arctic and sub-Arctic Europe; including Iceland.

7 *Cochlearia groenlandica* L. Like *C. fenestrata*, but stems leafless. Fruit rounded, 3.5-5mm. Similar habitats and flowering time. Iceland and Spitsbergen.

8* **Scottish Scurvy-grass** *Cochlearia scotica* Druce. Low compact perennial, rather like dwarf forms of *C. officinalis*. Flowers smaller, 5-6mm, purplish, sometimes white, petals rather square. Fruit oval, 3mm. Coastal habitats, especially sandy and rocky shores. June-August. Only in Britain. **B**: NW Scotland, Isle of Man, N & W Ireland.

9* **Kernera** *Kernera saxatilis* (L.) Reichenb. Variable, low-short hairless perennial, usually branched. Basal leaves lanceolate to spoon-shaped, toothed or not, stalked, in a rosette; upper leaves clasping the stem. Flowers white, 3-5mm, in lax racemes; petals rounded. Fruit elliptical, 2-4.5mm, long-stalked. Rocky and grassy habitats, over limestone, to 2000m. June-August. Mountains of France and Germany. Generally very local.

**Camelina** *Camelina.* Annual or biennial herbs with simple or branched. Stem leaves often arrow-shaped, unstalked. Inflorescence without bracts. Flowers with erect sepals and yellow or white petals. Fruit a somewhat inflated silicula. 10 species in Europe, the Mediterranean region and W & C Asia.

10* **Gold-of-pleasure** *Camelina sativa* (L.) Crantz [*C. glabrata*, *C. pilosa*]. Short to medium, erect, usually hairless, annual. Leaves narrow arrow-shaped, toothed or not. Flowers yellow, 3-4mm. Fruit elliptical 7-9mm, becoming yellowish; stalks 10-20mm. Arable land and waste ground. May-July. Throughout, except the Faeroes, Iceland and Spitsbergen, often casual. Cultivated since Neolithic time in parts of Europe for its oil bearing seed and as a source of fibre; today it is becoming increasingly scarce. **P**: Bees or self-pollinated. An overwintering annual usually. *C. sativa* is probably only truly native in E Europe and W Asia. **B**: Local and casual throughout.

11 *Camelina microcarpa* Andrz. ex DC. Like *C. sativa*, but less robust, the stems hairy and the flowers pale yellow. Fruit grey-green, 5-7mm. Cultivated land, waste and bare places. **I**: C & S Europe, W Asia. Casual and naturalized throughout, except Ireland, the Faeroes and Spitsbergen. Europe to N Africa and W Asia. A rare weed of Lucerne fields, probably introduced with imported seed.

12 *Camelina alyssum* (Miller) Thell. [*C. foetida*, *C. linicola*]. Like *C. sativa*, but more slender and leaves more strongly toothed or lobed. Fruit truncated or notched at the top. A weed of flaxfields (*Linum usitatissimum*), and waste places. June-August. **I**: C & S Europe and parts of N Europe. Naturalized throughout, except the Faeroes, Iceland and Spitsbergen. **B**: a rare casual, occasionally naturalized.

13 *Camelina macrocarpa* Wierzb. ex Reichenb. Like *C. sativa*, but fruits larger, 10-12mm, oval to pear-shaped, on stalks 15-30mm long. A weed of flaxfields. June-August. Local in N & C Europe – Germany & S Scandinavia.

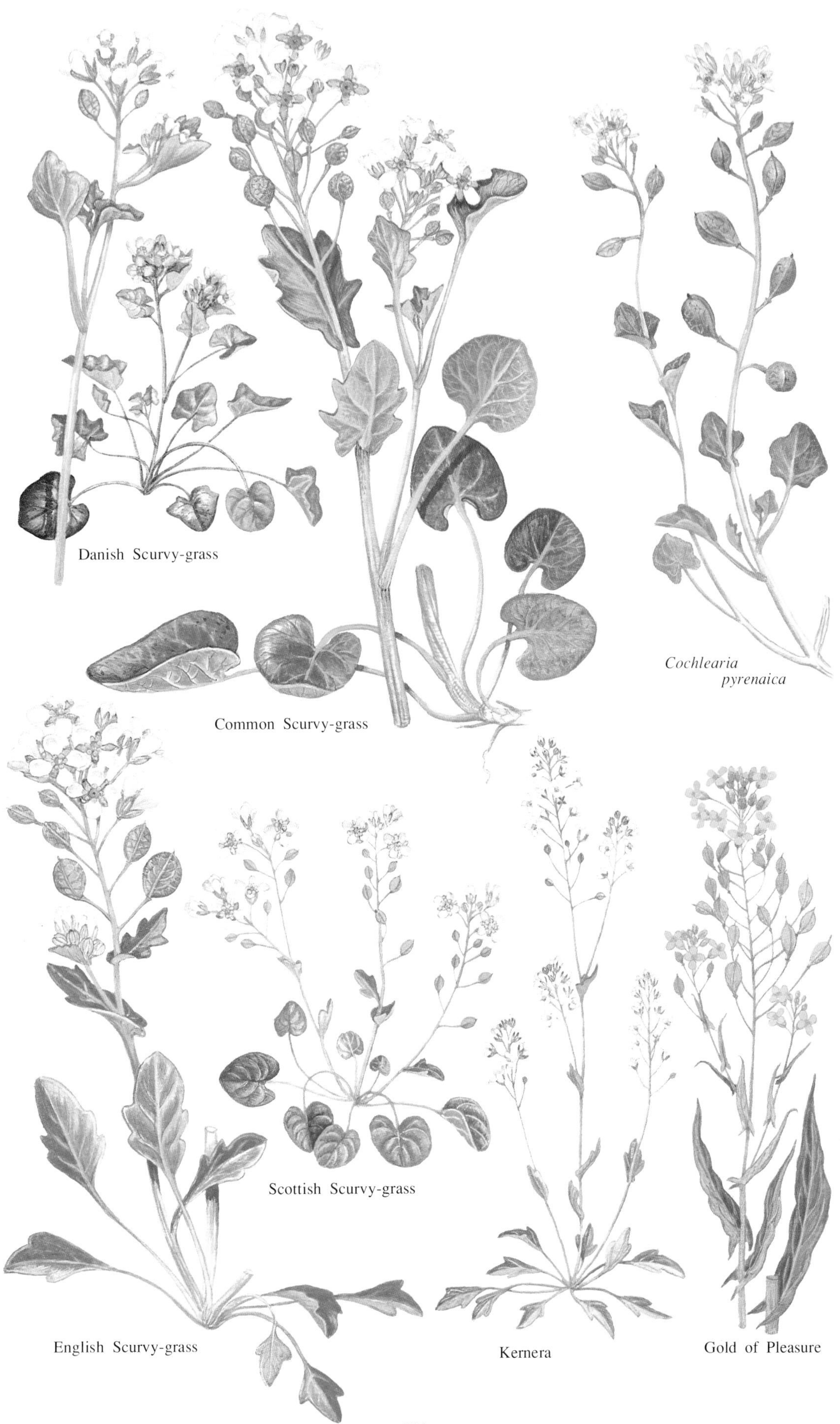
Danish Scurvy-grass
Common Scurvy-grass
*Cochlearia pyrenaica*
English Scurvy-grass
Scottish Scurvy-grass
Kernera
Gold of Pleasure

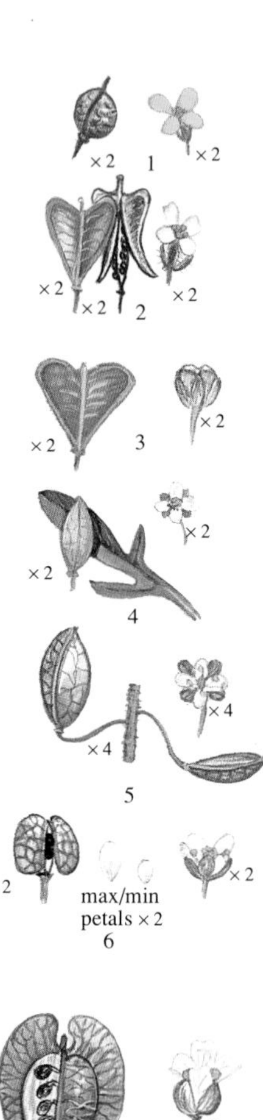


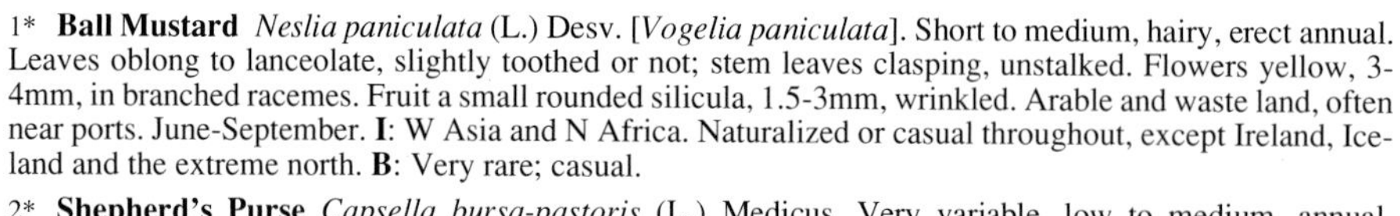

1* **Ball Mustard** *Neslia paniculata* (L.) Desv. [*Vogelia paniculata*]. Short to medium, hairy, erect annual. Leaves oblong to lanceolate, slightly toothed or not; stem leaves clasping, unstalked. Flowers yellow, 3-4mm, in branched racemes. Fruit a small rounded silicula, 1.5-3mm, wrinkled. Arable and waste land, often near ports. June-September. **I**: W Asia and N Africa. Naturalized or casual throughout, except Ireland, Iceland and the extreme north. **B**: Very rare; casual.

2* **Shepherd's Purse** *Capsella bursa-pastoris* (L.) Medicus. Very variable, low to medium, annual, sometimes overwintering, hairy or not. Basal leaves in a lax rosette, lanceolate, pinnately-lobed. Flowers small, white, 2-3mm, in lax, often branched, racemes; sepals often hairy. Fruit triangular-heart-shaped, 6-9mm. A common weed of cultivated land, bare places, waysides and embankments, to 2000m. Throughout. Highly successful, germinating almost throughout the year. **B**: Abundant.

3 *Capsella rubella* Reuter. Like *C. bursa-pastoris*, but flowers smaller, the reddish petals scarcely exceeding the hairless sepals. Fruit more deeply notched. Britain and France; naturalized in Germany. Mainly in S Europe. **B**: Local, England north to Buckinghamshire.

4 **Hymenolobus** *Hymenolobus procumbens* (L.) Nutt. ex Torrey & A. Gray [*Hornungia procumbens, Hutchinsia procumbens*]. Low to short, slightly hairy, annual or biennial; stems spreading to erect. Leaves deeply pinnately-lobed to unlobed. Flowers small, white, 2-3mm, many in lax racemes, petals equalling or slightly longer than sepals. Fruits elliptical, 2-5mm. Bare and sandy habitats, often coastal. April-June. France and Germany. A rather insignificant plant.

*Hymenolobus pauciflorus* (Koch) Schinz & Thell. Like *H. procumbens*, but a lower plant, not more than 5cm. Leaves 3-lobed to unlobed. Flowers few to a raceme. Rocky and stony habitats, to 1600m. May-August. Mountains of France.

5* **Hutchinsia** *Hornungia petraea* (L.) Reichenb. [*Hutchinsia petraea*]. Low, hairless, slender annual. Leaves all pinnate. Flowers tiny, greenish-white, 1-1.3mm, in elongating racemes, petals not notched, equalling or slightly longer than sepals. Fruit elliptical, 2-2.5mm. Rocky and sandy places, mainly on calcareous soils. March-May. Throughout, except the north. **B**: Rare, Wales, Mendips, N England.

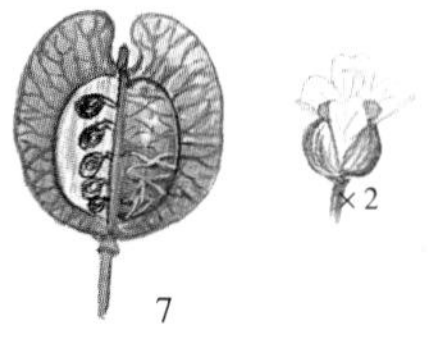


6* **Shepherd's Cress** *Teesdalia nudicaulis* (L.) R. Br. Low to short, hairless annual. Leaves pinnately-lobed, mostly basal, the end segment often 3-lobed. Flowers white, 2mm, petals unequal, not notched, in lax racemes. Fruit oval, 3-4mm, blunt or notched. Sandy and gravelly habitats, heaths and disturbed places, generally on acid soils, often coastal. May-October. Throughout. Often associated with *Ornithopus perpusillus, Rumex acetosella* and *Trifolium arvense*. Europe and N Africa. **B**: Throughout, but scarce in Scotland; in Ireland confined to the north-east.

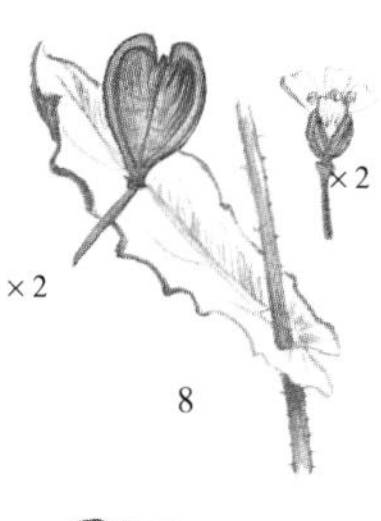


**Pennycresses** *Thlaspi*. Annual or perennial herbs with unstalked stem leaves, often clasping the stem; hairs when present unbranched. Flowers white or sometimes purplish, in racemes; petals short-clawed. Fruit a silicula, often notched at the top; valves usually winged. 60 species in Europe, Asia and North America, extending to South America.

7* **Field Pennycress** *Thlaspi arvense* L. Short to medium erect, hairless, foetid annual. Stem leaves oblong, clasping the stem closely; lower leaves stalked, not in a rosette. Flowers white, 4-6mm; anthers yellow. Fruit quite large, 10-15mm, rounded with a deep notch, broadly winged. A weed of arable land and waste places, disturbed ground, generally on rather fertile soils. May-August. Throughout, except the extreme north and the Faeroes. **B**: Throughout, often abundant.

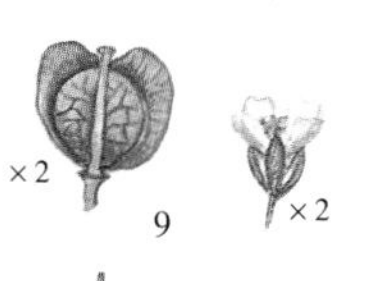


8 **Garlic Pennycress** *Thlaspi alliaceum* L. Like *T. arvense*, but stem hairy below, the whole plant smelling of garlic; flowers slightly smaller. Fruit heart-shaped, with a shallow notch. Similar habitats. April-June. France and Germany; naturalized in Britain. Primarily a native of C and S Europe. **B**: Rare in Kent; an occasional casual elsewhere.

9* **Perfoliate Pennycress** *Thlaspi perfoliatum* L. Low to short, greyish, hairless, overwintering annual. Basal leaves in a lax rosette, oblong, toothed or not; stem leaves heart-shaped, clasping the stem. Flower white, 3-4mm; anthers yellow. Fruit broadly heart-shaped, with wide wings. Arable land, bare and waste places, walls and railway embankments. May-August. Throughout, except much of the north. **B**: Rare, confined to oolitic limestone in Gloucestershire, Oxfordshire and Wiltshire; casual further south, also rare in Ireland.

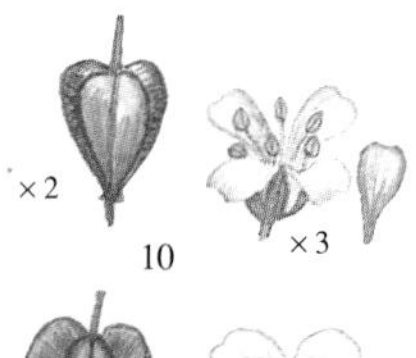


10* **Alpine Pennycress** *Thlaspi caerulescens* J. & C. Presl [*T. alpestre*]. Variable short to medium biennial; greyish, with crowded non-flowering leaf-rosettes. Basal leaves elliptical, stalked, untoothed; stem leaves heart-shaped, half-clasping. Flower white or purplish, 3-4mm, in elongating racemes; anthers violet. Fruit heart-shaped, 5-9mm, winged in the upper half. Limestone rocks, screes and lead-mine debris, mountain woods, to 2000m. April-July. Throughout, except Ireland; naturalized in Scandinavia and Iceland. Tolerant of soils rich in minerals such as lead and zinc. **B**: N England, Mendips, Snowdonia, rare in Scotland.

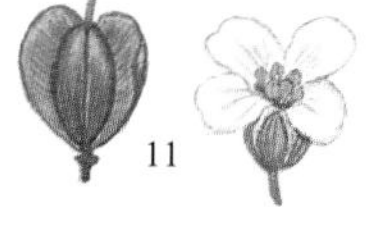


11* **Mountain Pennycress** *Thlaspi montanum* L. Short, hairless mat-forming perennial, with erect stems. Basal leaves oval or rounded, long-stalked, untoothed; stem leaves heart-shaped, occasionally finely toothed. Flowers white, 9-12mm; anthers pale yellow. Fruit heart-shaped, 7-8mm, broadly winged in the upper half. Grassy slopes, cliff ledges, screes and rocks, generally on limestone, to 2500m. Mountains of Belgium, France and Germany. Confined to the mountains of C & E Europe.

12* **Wild Candytuft** *Iberis amara* L. Short, slightly hairy, erect, leafy annual, branched above. Leaves elliptical lobed or toothed. Flowers white or purplish, 6-8mm, petals unequal. Fruit a rounded silicula, 3-5mm, with triangular lobes. Cornfields, dry hillslopes, open woods, disturbed ground, often around rabbit-warrens, on calcareous or dolomitic soils. May-September. Britain, Belgium, Holland, France and Germany. **B**: Rare, local in the Chilterns, Cambridgeshire and Surrey.

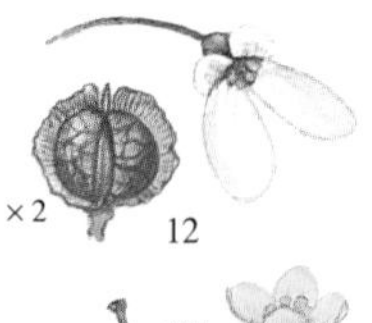


13* **Buckler Mustard** *Biscutella laevigata* L. Variable short to medium, tufted hairy perennial. Basal leaves linear to lanceolate, lobed or toothed often; stem leaves up to 2. Flowers yellow, 6-10mm, in branched racemes. Fruit a flattened silicula with 2 rounded lobes, 4-8mm. Dry limestone rocks, occasionally in waste places, to 2600m. May-July. Belgium, France and Germany. A very difficult complex which includes *B. apricorum* Jordan, *B. divionensis* Jordan, *B. guillonii* Jordan and *B. neustriaca* Bonnet, all from France and maintained as distinct species by some authorities.

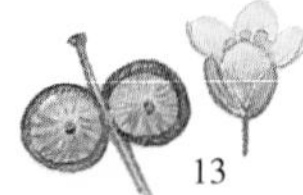

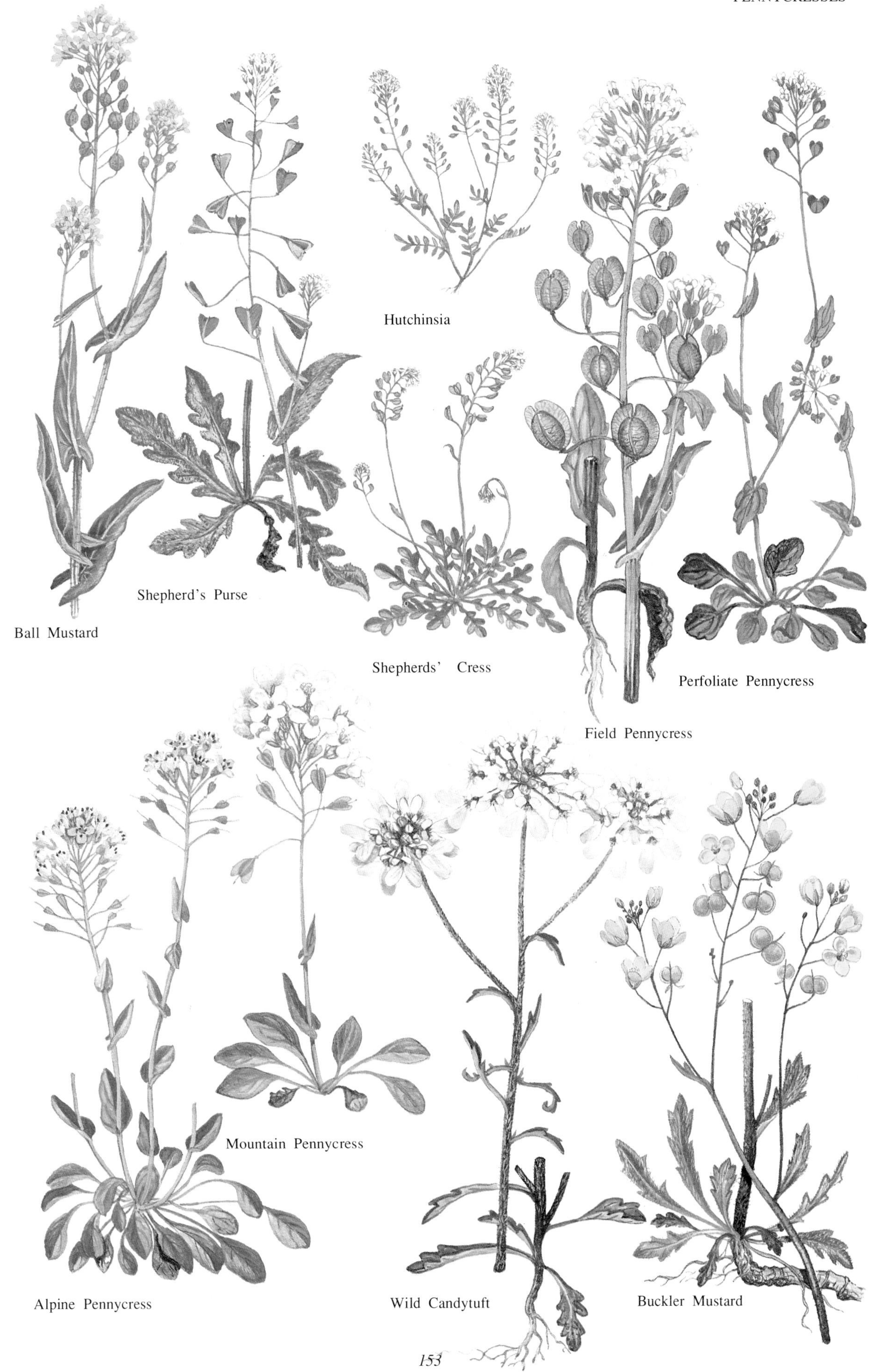
Ball Mustard
Shepherd's Purse
Hutchinsia
Shepherds' Cress
Field Pennycress
Perfoliate Pennycress
Alpine Pennycress
Mountain Pennycress
Wild Candytuft
Buckler Mustard

**Pepperworts** *Lepidium*. Annual to perennial herbs, usually with unbranched hairs. Flowers small, in terminal, often lax, racemes; petals white, sometimes yellow or absent. Stamens 2, 4 or 6. Fruit a flattened silicula, often blunt-ended or notched, valves keeled or winged. 150 species in the northern hemisphere. Sometimes confused with *Thlaspi* but distinguished by has only a single ovule in each compartment of the fruit, rather than 2 or more.

1* **Field Pepperwort** *Lepidium campestre* (L.) R.Br. Medium, greyish, hairy annual or biennial. Basal leaves oval, untoothed, sometimes slightly lobed; upper stem leaves numerous, narrow arrow-shaped, often toothed. Flowers white, 2mm, petals scarcely longer than sepals; anthers yellow. Fruit 5-6mm, notched, covered in scales. Arable land, dry fields, waysides, banks, road and waste places, occasionally on walls. May-August. Throughout; introduced and naturalized in the Faeroes and Iceland. A typical and rather common weed of dry open places. From Europe to W Asia as far as the Caucasus Mountains. An introduced weed in many parts of North America. **B**: Throughout except NW Scotland, rather rare in Ireland.

2* **Smith's Cress** *Lepidium heterophyllum* Bentham. Short to medium, hairy, branched perennial. Basal leaves elliptical, untoothed, withering before flowers appear; stem leaves narrowly triangular, clasping the stem with narrow lobes. Flowers white, 3-4mm, petals longer than sepals; anthers violet. Fruit 4-7mm, with a small notch, smooth, winged in the upper third. Arable and waste land, waysides, in dry open habitats. May-August. Britain, Ireland and France; naturalized in Belgium, Holland, Germany and parts of Scandinavia. Close to *L. campestre*, but a perennial with smooth fruits with the style projecting beyond the notch. Confined to W Europe, but is a widespread weed in parts of North America and New Zealand. **B**: Throughout but most common in the S & W; Ireland except for the north-west.

*Lepidium villarsii* Gren. & Godron. Like *L. heterophyllum*, but with hairless fruit stalks. Similar habitats and flowering time. E France. W Alps and E Pyrenees.

3 *Lepdium hirtum* (L.) Sm. Like *L. heterophyllum*, but a grey-hairy plant with yellow anthers. Fruit hairy, at least when young. Similar habitats and flowering times. C & S France; naturalized or casual in Britain. Primarily Mediterranean. Sometimes considered a form of *L. heterophyllum*.

4* **Garden Cress** *Lepidium sativum* L. Short to medium, strong-smelling, greenish, hairless annual with a single erect stem. Basal leaves pinnately-lobed, soon falling; stem leaves lobed, not clasping, the uppermost linear, unlobed. Flowers white, occasionally reddish, 3-5mm, fragrant, petals twice as long as sepals. Fruit 5-6mm oval, notched, not hairy. Cultivated land, waste and disturbed ground. June-July. **I**: W Asia. Naturalized or casual throughout, except Ireland, Iceland and Spitsbergen. **B**: Occasional garden escape; also in birdseed.

5 *Lepidium virginicum* L. Like *L. sativum*, but finely hairy, the upper leaves toothed. Fruit 3-4mm. Disturbed and waste places. June-August. **I**: N America. Naturalized throughout, except Ireland, Iceland and much of the north.

6 *Lepidium bonariense* L. Like *L. virginicum*, but all leaves pinnate and petals linear, shorter than the sepals. Disturbed and waste places. June-August. **I**: N America. Naturalized or casual in Belgium, Holland, France, Germany and Norway.

7* *Lepidium densiflorum* Schrader. Medium, finely hairy, grey annual or biennial, with a single erect stem. Leaves elliptical, deeply toothed, long-stalked, uppermost linear-lanceolate, all with a hairy margin. Flowers greenish, 2mm, the petals linear, shorter than the sepals or absent. Fruit 3-4mm, oval with a shallow notch. Cultivated, disturbed and waste land. May-July. **I**: N America. Naturalized or casual throughout, except the far north. An introduced weed in many parts of the world. **B**: Casual.

8 **Least Pepperwort** *Lepidium neglectum* Thell. Like *L. densiflorum*, but upper stem leaves linear, 1-veined. **I**: N America. Similar habitats and distribution; sometimes casual in Britain. Often regarded as a form of *L. densiflorum*.

9* **Narrow-leaved Pepperwort** *Lepidium ruderale* L. Short, scarcely hairy, strong-smelling annual or biennial with a single erect stem. Basal leaves pinnately-lobed, upper linear, untoothed. Flowers tiny green, 1mm, usually petalless. Fruit 2-2.5mm elliptical, deeply notched, with narrow wings on hairy stalks. Waste places, dry open coastal habitats, sea walls, rubbish tips, salt marshes, waysides. May-July. Throughout, except Ireland and Spitsbergen. Europe to SW Asia, but a widespread weed in North America and Australia. **B**: Throughout England, especially in the east, occasionally elsewhere; often coastal, where it may be native.

10 *Lepidium divaricatum* Solander. Like *L. ruderale*, but stems finely hairy and upper leaves oblong, toothed. Petals present, shorter than the sepals. Disturbed and waste places, waysides. May-July. **I**: Africa. Naturalized or casual in Belgium, Holland and France, occasionally elsewhere in NW Europe. **B**: Casual.

11* **Perfoliate Pepperwort** *Lepidium perfoliatum* L. Short annual or biennial with a single erect stem, somewhat hairy usually. Basal leaves pinnately-lobed; upper leaves oval, completely encircling the stem. Flowers pale yellow, tiny, 1mm, petals slightly longer than sepals. Fruit 4mm, rounded, scarcely notched. Cultivated land, waste and disturbed places. May-June. **I**: E Europe & W Asia. Widely naturalized or casual throughout, except Ireland and Finland. Often introduced in grass seed mixtures. **B**: Casual, rare.

12* **Dittander** *Lepidium latifolium* L. Rather stout, tall, hairless, stoloniferous perennial; stems erect, much-branched. Leaves oval, toothed, the lower stalked, the upper narrower and unstalked. Flowers white, 2-3mm, in broad panicles; petals longer than the white-edged sepals. Fruit 2mm, rounded, notched or not. Damp sandy habitats, salt marshes among coarse vegetation. June-August. Throughout, except the Faeroes, Iceland, Finland and Norway. Used for flavouring food before the discovery of pepper. Europe to N Africa and SW Asia. **B**: Scattered coastal localities in E & SE England, declining; also along the Severn Estuary and along canals and in gravel pits in the London area.

13* **Tall Pepperwort** *Lepidium graminifolium* L. Like *L. latifolium*, but not stoloniferous and with narrower leaves, the uppermost linear. Fruit oval, 2.5-4mm, pointed, not notched. Disturbed and waste habitats. June-July. C & S France, S Germany; casual in Britain, Belgium and Holland. C & S Europe to W Asia.

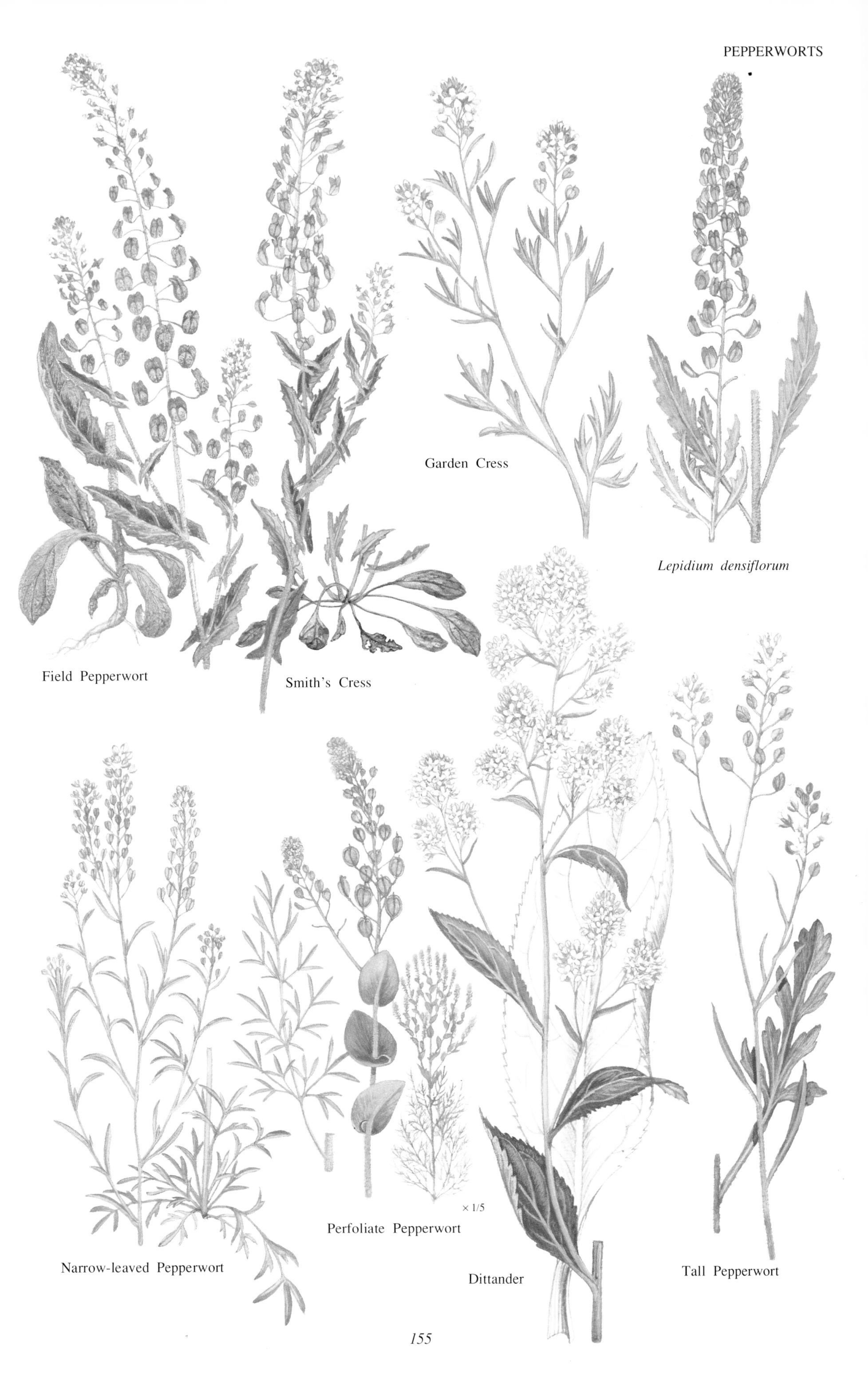
Field Pepperwort
Smith's Cress
Garden Cress
*Lepidium densiflorum*
Narrow-leaved Pepperwort
× 1/5
Perfoliate Pepperwort
Dittander
Tall Pepperwort

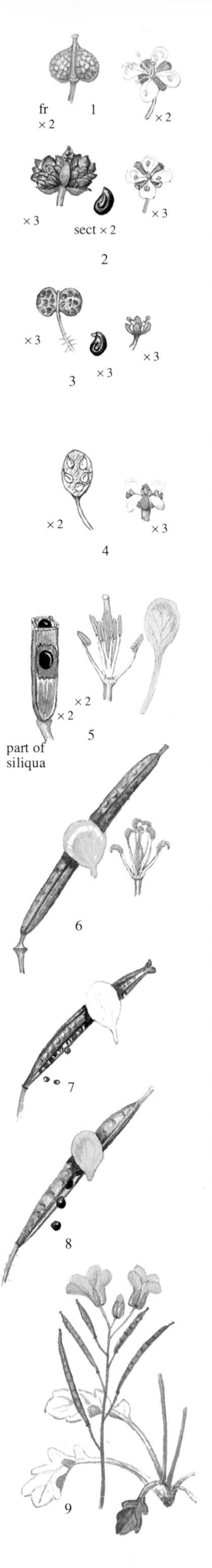


1* **Hoary Cress** *Cardaria draba* (L.) Desv. [*Lepidium draba*]. Medium to tall, greyish, hairless or slightly hairy perennial. Basal leaves oblong, pointed, deeply toothed, long-stalked; stem leaves clasping. Flowers white, 5-6mm, in dense umbel-like clusters. Fruit kidney-shaped 3-4.5mm, inflated, not splitting. Waste and disturbed ground, field boundaries, roadsides and hedgerows on calcareous or neutral soils. May-June. **I**: S Europe & W Asia. A common weed throughout, except Iceland and the Faeroes. **P**: Mostly self-pollinated. **B**: Throughout, but scarce in Scotland and Ireland.

2* **Swinecress** *Coronopus squamatus* (Forsk.) Aschers. [*C. procumbens*, C. *ruellii*, *Senebiera coronopus*]. Low to short prostrate annual or biennial. Leaves mostly 1-2-pinnately-lobed. Flowers small, white, 2-3mm, in short crowded racemes; petals longer than sepals; fertile stamens 6. Fruit kidney-shaped, 2-3mm, net-veined. Waste ground, pathways and other trampled ground, avoiding well-drained acid soils. June-September. Throughout, except the Faeroes, Iceland and Spitsbergen. mostly self-pollinated. Throughout most of Europe and N Africa. A widely introduced weed in North America, South Africa and Australia. **B**: Almost throughout, mostly coastal in the north.

3* **Lesser Swinecress** *Coronopus didymus* (L.) Sm. [*Senebiera didyma*]. Like *C. squamatus*, but more delicate, with more feathery, pungent leaves. Flowers smaller, 1mm, sepals longer than petals; fertile stamens only 2. Fruit 1.5-2mm. Cultivated land and waste places. June-September. **I**: Possibly native to S America. Naturalized throughout, except the extreme north, the Faeroes and Iceland. A cosmopolitan weed. **P**: Self-pollinated. **B**: Mainly confined to S England, S Wales and S Ireland.

4* **Awlwort** *Subularia aquatica* L. Low hairless, aquatic annual. Leaves all basal, grasslike, round in section. Flowers white, 2-2.5mm, in lax racemes, petals sometimes absent. Fruit an oval, swollen silicula, 2-5mm. Shallow base-poor pools and lakes, in hilly and mountain habitats, often in fine gravelly places. June-September. Throughout, except Holland. **P**: Underwater flowers remain closed and are self-pollinated, those above the water surface open fully and are pollinated by various small insects. **B**: Local in N Wales, NW England, N & W Scotland and parts of W Ireland.

5* **Hare's-ear Cabbage** *Conringia orientalis* (L.) Dumort. Short to medium hairless annual. Leaves greyish, the lower oblong, stalked; stem leaves narrow, with a broad clasping base. Flowers yellowish-white or greenish-white, 9-12mm, in a terminal raceme. Fruit a long 4-angled siliqua, 6-14cm. Arable land, waste land, and sea cliffs, on clay and calcareous soils. May-July. Germany; widely naturalized or casual elsewhere in W & NW Europe. **B**: Casual in parts of Britain, mostly coastal.

**Wall Rockets** *Diplotaxis*. Annual or perennial herbs with pinnately-lobed leaves usually. Sepal almost erect. Petals yellow or violet, clawed. Fruit a slender siliqua with a short beak; valves with a strong midvein. Distingushed from similar looking Crucifers by a combination of the beaked, slightly flattened, fruits with a single midvein along each valve and 2 rows of ovules within each compartment. About 27 species in Europe and the Mediterranean region.

6* **Perennial Wall Rocket** *Diplotaxis tenuifolia* (L.) DC. Medium to tall greyish perennial, woody at the base, stems leafy. Leaves not in a basal rosette, mostly with 4-8 segments, hairless except at margins, strong-smelling when crushed. Flowers sulphur-yellow, 15-30mm. Fruit 20-60mm, stiffly erect. Waste places, rocky habitats and old walls. May-September. Belgium, Holland, France, Germany and S Sweden; naturalized elsewhere in Scandinavia and in Britain. Native in many parts of C & S Europe. **B**: Naturalized in SE England, especially about railways; casual further north.

7* **White Wall Rocket** *Diplotaxis erucoides* (L.) DC. [*Sinapis erucoides*, incl. *D. valentina*]. Medium, leafy erect annual, sometimes overwintering. Lower leaves in a lax rosette, with 6-10 segments, and a large terminal lobe, sparsely hairy; upper leaves clasping the stem. Flowers white with violet veins, 14-15mm. Fruit 18-45mm, ascending. Arable land, waste places and disturbed ground. May-September. C & S France; casual in parts of NW Europe, formerly in Britain. A mainly Mediterranean species that has become a troublesome weed in parts of Europe, especially in the south.

8* **Annual Wall Rocket** *Diplotaxis muralis* (L.) DC. [*Sisymbrium murale*, incl *Diplotaxis scaposa*]. Short to medium, much-branched yellow-green foetid annual to perennial; stem hairless or bristly below. Leaves mostly in a basal rosette, elliptical, toothed or lobed, long-stalked. Flowers sulphur-yellow, occasionally becoming violet, 10-15mm. Fruit 18-45mm, ascending. Rocky habitats, cliffs, quarries and old walls, arable land and waste places, on dry calcareous soils usually. June-September. France and S Germany; naturalized in Britain, Holland and Sweden. **P**: Bees and flies. A native mainly of C & S Europe.The plant stinks; the smell has been likened to foxes. **B**: Naturalized or casual, but scarce in much of the north and west.

9 *Diplotaxis viminea* (L.) DC. Like *D. muralis*, but a shorter, more slender, plant. Flowers smaller, only 7-9mm. Fruit 10-35mm. Rocky habitats, waste and cultivated land. June-September. C & S France; naturalized in Holland and Germany, occasionally casual elsewhere. Mainly Mediterranean.

Annual Wall Rocket

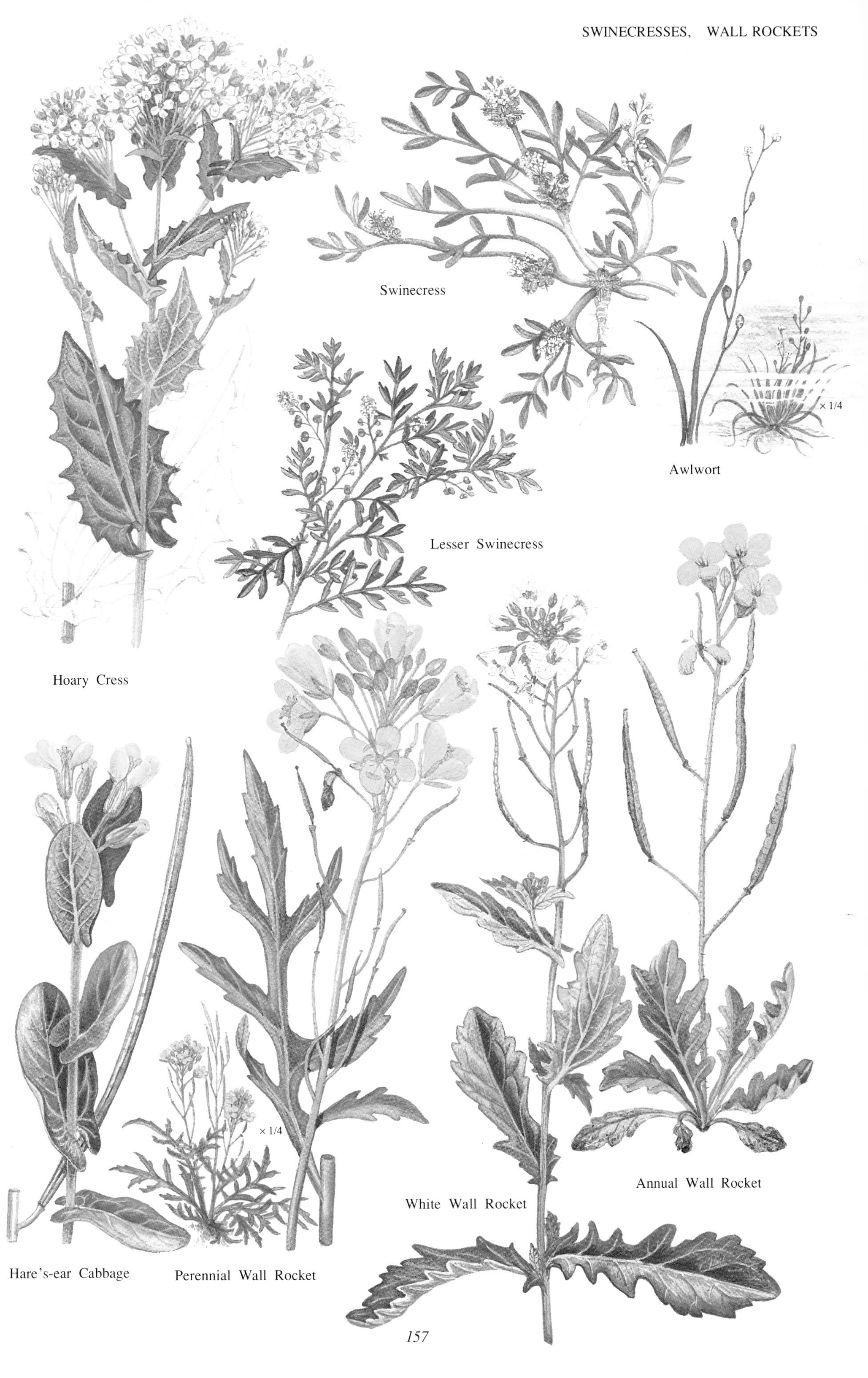
Swinecress
Awlwort
× 1/4
Lesser Swinecress
Hoary Cress
× 1/4
Annual Wall Rocket
White Wall Rocket
Hare's-ear Cabbage
Perennial Wall Rocket

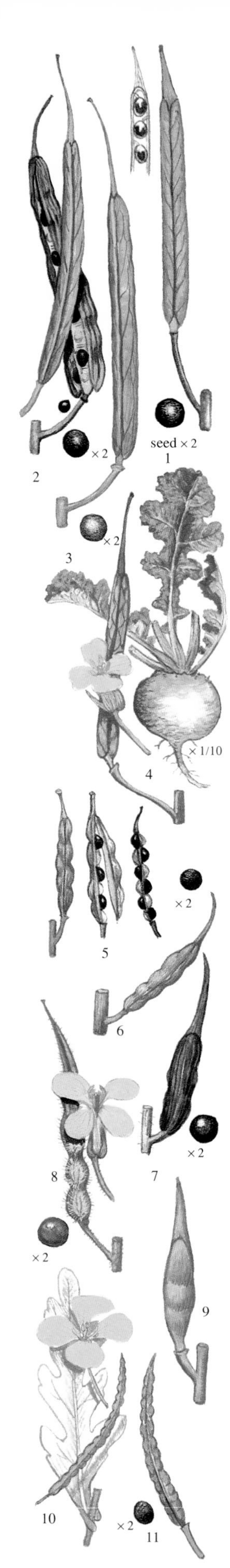


**Brassicas** *Brassica*. Annual or perennial herbs with entire or pinnately lobed leaves. Flowers in racemes, often branched; sepals erect to spreading; petals yellow or white, clawed. Fruit a narrow siliqua with a distinctive beak; valves with a prominent midvein.

1* **Wild Cabbage** *Brassica oleracea* L. [*B. sylvestris*]. Medium to tall, grey, hairless biennial or perennial with a thick stem becoming woody and leaf-scarred below. Basal leaves pinnately lobed, undulate, stalked; upper stem leaves clasping, unlobed. Flowers yellow, 30-40mm, overtopped by buds, in branched racemes. Fruit 50-70mm, linear, with a conical beak. Maritime habitats, particularly calcareous sea cliffs. May-September. Britain and France; naturalized in Holland and Germany. Cultivated since Greek and Roman times, giving rise to our cabbage, cauliflower, brussel sprouts etc. **B**: long established in scattered coastal localities, most in S & W Britain.

2* **Rape** *Brassica napus* L. Tall, usually slightly bristly, annual or biennial, with a carrot-like root. Basal leaves stalked, upper clasping the stem; all leaves grey. Flowers pale yellow, 14-25mm; buds slightly overtopping the open flowers. Fruit 50-100mm, suberect, with a long slender beak to 25mm. Cultivated extensively and naturalized or casual on waste and disturbed ground, along the banks of streams and ditches. May-August. Throughout, except the extreme north. Probably originated as a cross between *B. oleracea* and *B. rapa*. **P**: Bees. **B**: Naturalized or casual.

**Oil Seed Rape, Cole** *B. n.* subsp. *oleifera* (DC.) Metzger. Like the type, but a biennial without a carrot-like root. Naturalized or casual throughout. Widely cultivated for its rich oil-bearing seeds.

**Swede** *B. n.* subsp. *rapifera* Metzger var. *napobrassica* Reichenb. Like the type, but with a swollen stem base and buff-coloured flowers. Widely cultivated in lowland Europe, sometimes naturalized or casual. June-August. The root is eaten as a vegetable or used for animal feed.

3* **Wild Turnip** *Brassica rapa* L. subsp. *campestris* (L.) Clapham [*B. campestris*]. Similar to *B. napus*, but basal leaves bright green, bristly; upper leaves grey. Flowers rather smaller; open flowers overtopping the buds. Arable land and waste places, banks of streams and ditches. May-August. Throughout, except the extreme north. **P**: Bees and hoverflies.

subsp. *sylvestris* (L.) Janchen. Like the type, but biennial (not annual) with a well defined basal leaf rosette. Often growing along stream banks and also in other damp places. May-August. The cultivated turnip with a white fleshy root is generally assigned to subsp. *rapa* (4).

5* **Black Mustard** *Brassica nigra* (L.) Koch. Tall, rather greyish annual, stems bristly below. Lower leaves pinnately lobed, upper narrow-oblong, narrowed at the base, not clasping the stem. Flowers yellow, 12-15mm. Fruit short, 10-20mm, erect and pressed close to the stem. Arable and waste land, waysides, banks of streams and ditches, sea cliffs. June-August. Throughout, except Iceland and Spitsbergen. Long cultivated for its edible seed – the source of black mustard. **B**: Throughout except N Scotland, rare in Ireland.

6 *Brassica juncea* (L.) Czern. Like *B. nigra*, but with often very large, greyish, lower leaves and fruits ascending, not pressed close to the stem, constricted at intervals. Waste and disturbed ground, ditches and stream banks. June-August. **I**: S & E Asia. Naturalized in Holland and Germany; casual elsewhere especially in the vicinity of sea ports.

**Sinapis** *Sinapis*. Like *Brassica*, but fruit valves 3-7-veined. 10 species in Europe and the Mediterranean region.

7* **Charlock, Wild Mustard** *Sinapis arvensis* L. [*S. orientalis*, *S. schkuhriana*]. Medium to tall, bristly annual. Lower leaves large, lyre-shaped; upper lanceolate, not clasping the stem. Flowers yellow, 15-20mm. Fruit 25-45mm, beaded, sometimes bristly; seeds dark reddish-brown. Arable fields, waste and disturbed ground, roadsides and banks, often on calcareous soils. April-October. Throughout, except Spitsbergen. **P**: Various bees and flies. The seeds can be ground for a type of mustard. A pernicious weed in some areas. From Europe to SW Asia, Siberia and N Africa. Widely introduced weed in parts of North and South America, Southern Africa, Australia and New Zealand. **B**: Almost throughout.

8 **White Mustard** *Sinapis alba* L. [*Brassica hirta*, *B. alba*]. Medium to tall bristly annual. Leaves all pinnately lobed and stalked. Flowers yellow, 18-25mm. Fruit 20-40mm; seeds greyish-brown. Arable land and waste places, especially on calcareous soils. June-August. **I**: Mediterranean region. Naturalized throughout, except Spitsbergen. Widely cultivated for fodder, green manure and the seeds which are ground to produce white mustard. **P**: Bees and flies. A widely introduced weed in North and South America and New Zealand. **B**: Throughout, especially in the S & E.

subsp. *dissecta* (Lag.) Bonnier [*S. dissecta*]. Like the type, but leaves 2-pinnately lobed and fruit 25-30mm, slightly hairy. Flaxfields and arable land. June-August. Naturalized in Germany, France, occasionally in Belgium and Holland. **B**: A rare casual.

9* **Rocket, Eruca** *Eruca sativa* (L.) Cav. Medium to tall, bristly annual. Leaves pinnately lobed, with a large terminal lobe. Flowers whitish or yellowish with deep violet veins, 20-32mm. Fruit a thick siliqua, 12-25mm, with a saber-shaped beak; valves 1-veined. Waste and disturbed ground, waysides. May-August. **I**: Mediterranean region. A frequent casual, occasionally naturalized in N & NW Europe, including Britain. Long since cultivated as a salad vegetable, so that its origins are rather obscure. Sometimes included under *E. vesicaria* (L.) Cav. as a subspecies. **B**: Casual.

10 *Erucastrum nasturtiifolium* (Poiret) O.E. Schulz [*E. baeticum*, *E. pseudosinapis*, *Sinapis laevigata*]. Medium to tall bristly biennial or perennial. Basal leaves pinnately lobed; stem leaves smaller, with the basal lobes clasping the stem. Flowers bright yellow, 12-20mm. Fruit a beaded siliqua, 23-45mm, with a short beak. Waste and disturbed land, often near ports. May-September. Holland, France and Germany; naturalized in Britain. Native of C and SW Europe. **B**: A rare casual near ports.

11* **Hairy Rocket** *Erucastrum gallicum* (Willd.) O.E. Schulz. Like *E. nasturtiifolium*, but basal lobes of stem leaves not clasping. Flowers pale yellow or whitish, 10-15mm; sepals erect (not spreading). Dry rocky habitats and waste places, often close to ports. May-September. Holland, France and Germany; naturalized or casual in Britain and S Scandinavia. Native of W & C Europe. **B**: Locally naturalized, or casual, especially near ports.

Wild Cabbage
Rape
Wild Turnip
Black Mustard
Charlock
Rocket
Hairy Rocket

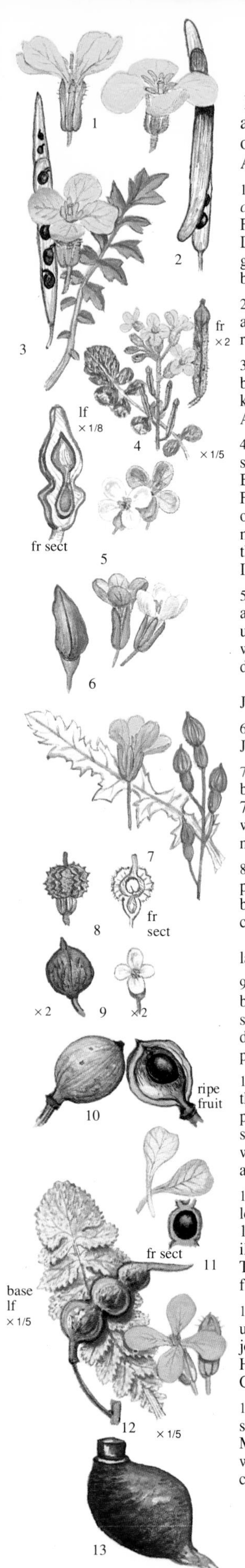


**Rhynchosinapis** *Rhynchosinapis.* Annual or perennial herbs with a carrot-like rootstock; hairs absent or simple. Inflorescence without bracts. Sepals erect. Petals yellow, sometimes with violet or reddish veins. Fruit a linear siliqua with a pronounced beak. 8 species in W Europe, NW Africa and the Canary Islands.

1* **Wallflower Cabbage** *Rhynchosinapis cheiranthos* (Vill.) Dandy [*Brassicella erucastrum, Brassica cheiranthos*]. Medium annual, or short-lived perennial; stem bristly below. Leaves pinnately lobed, bristly. Flowers yellow, sometimes with reddish veins, 20-25mm. Fruit 30-80mm long, ascending, only 2mm wide. Dry habitats, waste places, generally on calcareous soils, mainly in mountain regions, to 1500m. June-August. A native of parts of C & SW Europe. France and Germany; naturalized in Britain and Holland. **P**: butterflies. **B**: more or less confined to England and S Wales.

2 **Lundy Cabbage** *Rhynchosinapis wrightii* (O.E. Schulz) Dandy ex Clapham. Like *R. cheiranthos*, but a stout, densely hairy, perennial; fruits 3-4mm wide. Cliff habitats. June-August. Britain – Lundy Island, rare. Not known elsewhere. Declining.

3 **Isle of Man Cabbage** *Rhynchosinapis monensis* (L.) Dandy ex Clapham. Like *R. cheiranthos*, but biennial, usually with a hairless stem. Leaves mostly basal, greyish and hairless. Flowers yellow, with darker veins. Sandy coastal habitats. June-August. **B**: Only in W Britain and adjacent islands – Lancashire to Ayrshire, Arran Is and Isle of Man.

4 **Hoary Mustard** *Hirschfeldia incana* (L.) Lagrèze-Fossat. [*Brassica incana*]. Medium to tall annual, sometimes overwintering; stem simple or branched, with downward pointing stiff hairs in the lower half. Basal leaves grey-hairy, pinnately lobed, coarsely toothed; upper leaves narrowly lanceolate, usually hairy. Flowers pale yellow, 6-8mm, the petals often with darker veins, borne in racemes with the open flowers overtopping the buds. Fruit a siliqua, 8-12mm, with a short swollen beak, erect and closely pressed to the main stem. Fields, sandy and waste places, at low altitudes. June-September. **I**: Mediterranean region and the Middle East. Naturalized north to Denmark. A weed of cultivated land. **B**: naturalized in the Channel Islands; casual in S Britain.

5* **Sea Rocket** *Cakile maritima* Scop. Short to medium greyish, hairless annual. Leaves succulent, variable from linear to pinnately lobed. Flowers lilac, pink or white, 6-14mm. Fruit 10-25cm, 2-parted, the upper egg-shaped, the lower with 2 projections. Sandy coastal habitats. June-September. North to S Norway. Tolerant of salt spray and often growing just above the tide line. An early colonizer of young sand-dunes. **B**: Throughout, coastal.

subsp. *baltica* Jordan ex Rouy & Fouc. Like the type, but leaves 1-2-pinnately lobed. Coastal habitats. June-August. Baltic coasts.

6 *Cakile edentula* Hyl. ex P.W. Ball. Like *C. maritima*, but the fruit without projections. Coastal habitats. June-August. Faeroes, Iceland and N Norway, also N Russia.

7 **Perennial Bastard Cabbage** *Rapistrum perenne* (L.) All. Medium to tall biennial or perennial, bristly below. Leaves pinnately lobed, toothed, the upper often unstalked. Flowers bright yellow, 6-8mm. Fruit 7-10mm, a 2-parted silicula, the upper part rounded and ribbed with a short beak. Arable and waste land, waysides. June-September. Germany; casual or sometimes naturalized in W Europe, including Britain, but not Ireland. **B**: casual.

8* **Bastard Cabbage** *Rapistrum rugosum* (L.) All. Medium annual, bristly at least below. Lower leaves pinnately lobed, upper toothed, all stalked. Flower pale yellow, 8-10mm. Fruit 3-10mm, like *R. perenne*, but the upper segment oval. Arable land and waste places. May-September. C & S France; naturalized or casual in Britain, Belgium, Holland and Germany. Native of S Europe and Turkey.

subsp. *orientale* (L.) Arcangeli. has the upper fruit segment globose. Disturbed, waste and cultivated land. May-September. A frequent casual in France, Belgium, Holland and Germany.

9* **White Ball Mustard** *Calepina irregularis* (Asso) Thell. [*C. corvini*]. Medium to tall hairless annual or biennial. Basal leaves pinnately lobed or not; stem leaves clasping with spreading lobes. Flowers white, small, 2-4mm, with unequal petals. Fruit 2.5-4mm, with a short thick beak, no lower segment, rough when dry. Waste ground, mainly on calcareous soils. May-June. Belgium and France, often local; naturalized in parts of Britain, Holland and Germany.

10* **Sea Kale** *Crambe maritima* L. Medium to tall stout hairless perennial forming large clumps; stem thick. Leaves fleshy, undulate and pinnately lobed, waxy. Flowers white, 10-15mm, in broad billowy panicles. Fruit 8-14mm, 2-parted, the upper segment globose. Coastal habitats, shingle and sandy beaches, sea cliffs. June-August. Atlantic and Baltic coast of Europe, including Britain and Ireland. Often growing where there is a rich accumulation of seaweed humus. The young shoots used to be blanched and eaten as a vegetable. **B**: Coasts, especially in the S & SE; declining.

11* **Wild Radish** *Raphanus raphanistrum* L. Medium to tall, rough bristly annual or biennial. Lower leaves pinnately lobed, upper unlobed; blunt-toothed. Flowers white or yellow, usually with violet veins, 15-30mm. Fruit 30-90mm, a jointed, beaded, siliqua, strongly veined with a prominent beak, breaking easily at the joints. Cultivated and bare ground, waste places, mainly on clay soils. May-September. Throughout, but doubtfully native in the north. An often persistent weed of agricultural land. The yellow-flowered form is more common in the south of the range. **B**: Throughout; probably an alien.

12 **Sea Radish** *R. r.* subsp. *maritimus* (Sm.) Thell. [*R. maritimus*]. Taller and bushier than the type, usually with a carrot-like root. Fruit pods 5-8mm wide (not less than 5mm), not breaking easily at the joints. Coastal habitats, sandy and rocky places, rough grassland and cliffs. May-August. Britain, Belgium, Holland and W & N France. Native along the coast of W Europe, the Mediterranean and the Black Sea. **B**: Coasts, mainly in the S & W.

13* **Garden Radish** *Raphanus sativus* L. Like *R. raphanistrum*, but with a swollen, edible, fleshy rootstock. Flowers white or lilac. Fruit 20-90mm, scarcely beaded. Cultivated, waste and disturbed ground. May-September. Origin unknown. Widely cultivated for its edible roots – many cultivars are known, with white, red or blackish roots. The Wild Radish does not have a swollen root. **B**: Occasionally a garden escape.

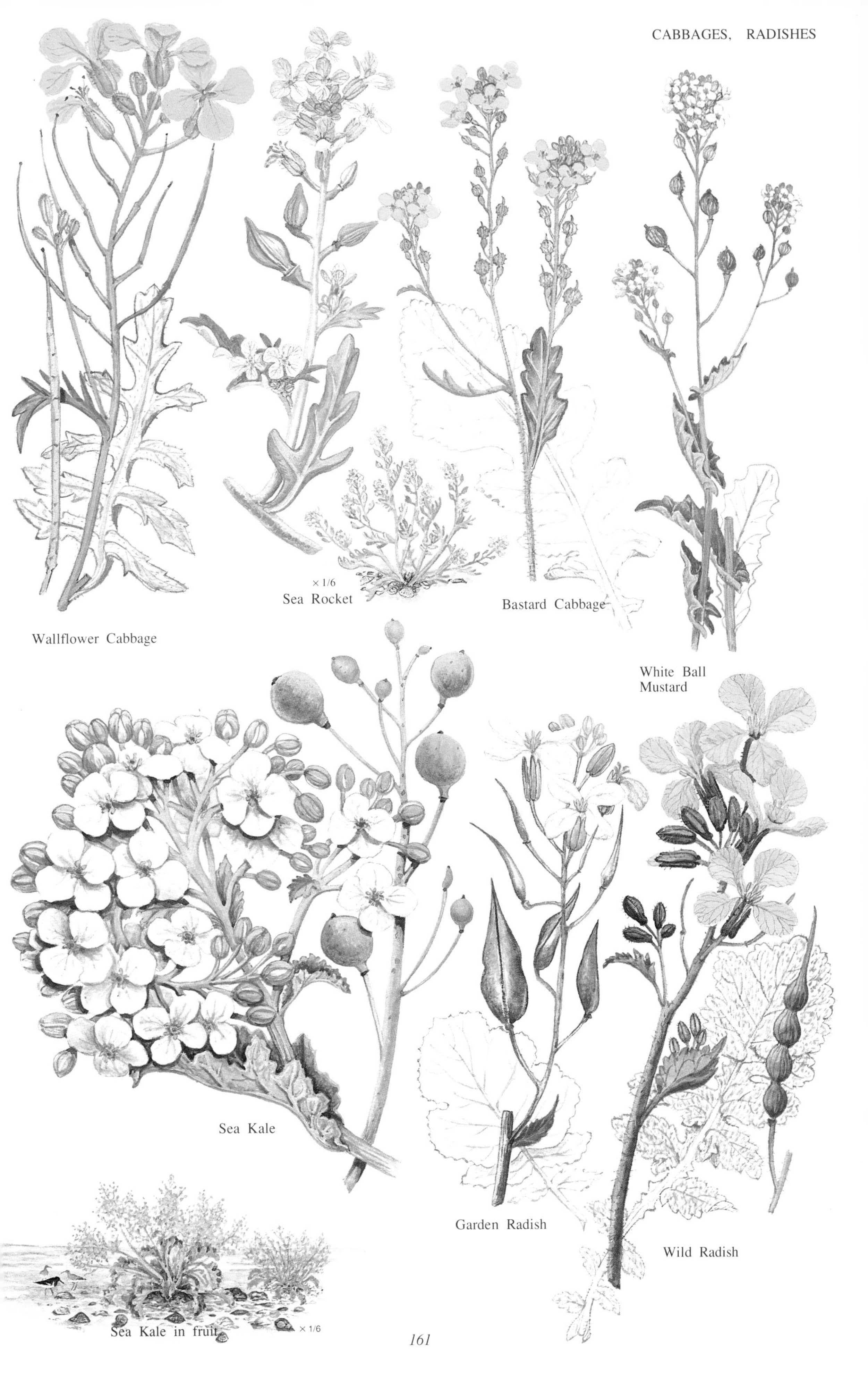
× 1/6
Sea Rocket
Bastard Cabbage
Wallflower Cabbage
White Ball
Mustard
Sea Kale
Garden Radish
Wild Radish
Sea Kale in fruit
× 1/6

## MIGNONETTE FAMILY Resedaceae

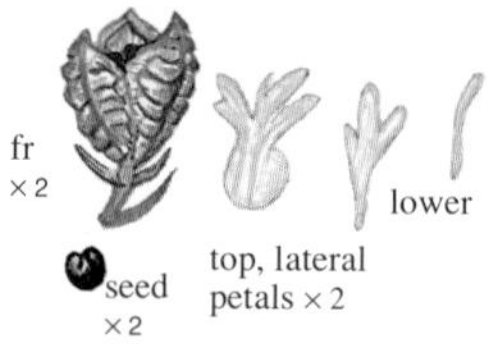

Annual or perennial herbs with simple or pinnately lobed leaves. Flowers in terminal racemes or spikes; bracts present. Sepals and petals 4-8, the petals separate, often deeply cut. Stamens 7-25. Fruit of 3-7 separate carpels (*Sesamoides*) or united into an open-ended capsule (*Reseda*). 70 species in Europe and W & C Asia; especially Mediterranean.

1* **Weld, Dyer's Rocket** *Reseda luteola* L. Tall hairless biennial, to 130cm. Leaves lanceolate, unlobed, wavy-edged. Flowers yellowish-green, 4-5mm; sepals and petals 4. Fruit capsule 3-4mm, 3-parted, globular, erect. Stony and sandy habitats, field margins, pathsides, waste places and old quarries, on calcareous soils, to 2000m. June-September. Throughout, except the far north; naturalized in Germany. Grown in the past for a yellow dye. **B**: throughout, scarce in the north.

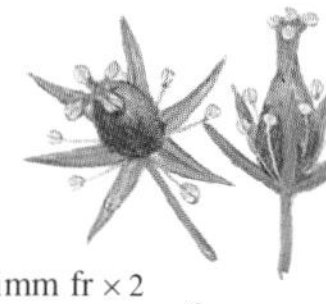

2* **White Mignonette** *Reseda alba* L. Medium to tall erect perennial; stems branched above. Leaves pinnately lobed, with 10 or more lobes. Flowers white, 8-9mm; sepals and petals 5-6. Fruit capsule 8-15mm, 4-parted, narrow-elliptical with a narrowed apex, with persistent filaments around the fruit. Dry rocky and waste places, old walls, often near ports. June-August. France. Naturalized in Britain, Holland and Germany, casual elsewhere. **B**: Naturalized or casual, especially in the SW.

3* **Corn Mignonette** *Reseda phyteuma* L. Short to medium hairy annual or biennial; stems branched near the base. Leaves oblong, unlobed or with 1-2 lobes on each side. Flowers white, 6-10mm; sepals and petals 6. Fruit capsule nodding, 13-14mm, oblong. Disturbed, waste and cultivated land, to 1600m. June-August. C & S France; naturalized or casual in Britain, Holland and Germany. **B**: Rare, casual in several places in S England.

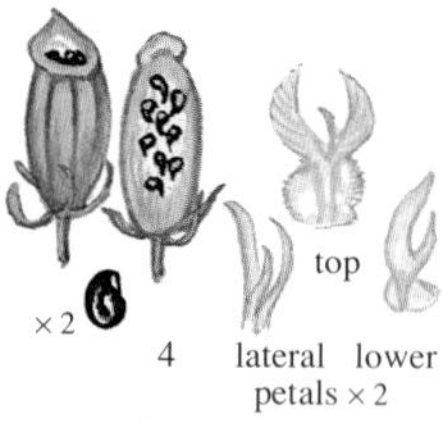

4* **Wild Mignonette** *Reseda lutea* L. [*R. ramosissima*]. Medium to tall, branched, bushy, hairless biennial or perennial. Leaves mostly small, pinnately lobed with 1-2 pairs of lobes on each side. Flowers yellow; sepals and petals 6. Fruit capsule 7-12mm, 3-parted, oblong, erect. Disturbed, waste and cultivated land, road verges, pathsides and field edges, on calcareous soils, to 2000m. June-September. Britain, Ireland, Belgium, Holland and France; naturalized further N and in Germany. **P**: Bees or self. **B**: Widespread in S & E England, scarcer elsewhere.

5* **Sesamoides** *Sesamoides canescens* (L.) O. Kuntze [*Astrocarpus purpurascens*, *A. clusii*]. Low to short, hairless biennial or perennial. Leaves lanceolate, unlobed, the basal ones in dense rosettes. Flowers whitish, 4-5mm, in long spikes; sepals and petals 6. Fruit star-shaped, 4-5 parted. Grassy and rocky habitats, roadsides. May-September. Spain and Portugal north to NW & C France, rather rare.

6 *Sesamoides pygmaea* (Scheele) O. Kuntze [*Reseda sesamoides*, *Astrocarpus sesamoides*]. Like *S. canescens*, but flowers with 7-9 (not 12-14) stamens. Fruit 4-6-parted. Mountain habitats, meadows, rocky habitats and damp screes, roadsides, to 2000m. June-September. C & S France.

## PITCHER PLANT FAMILY Sarraceniaceae

Insectivorous plants with basal rosettes of leaves modified into pitchers to entrap insects and other small animals. Flowers usually 5-parted , nodding on long stalks, hermaphrodite. An American family with 3 genera and 17 species.

7* **Pitcher Plant** *Sarracenia purpurea* L. Short stout, tufted perennial. Leaves all basal, narrow inverted cones (pitchers) topped by a broad hood, with a fluid reservoir inside which drowns trapped insects. Flowers large, solitary, 40-50mm, with 3 purplish-red sepals and 5 purplish-red petals, pale green on the inside. Styles expanded to form a 5-lobed 'umbrella'. Peat bogs. June-July. **I**: E North America. Naturalized in C Ireland. Frequently cultivated. Fluid within the pitchers contains various enzymes which absorb trapped organic matter, the soft tissue of insect bodies in particular.

## SUNDEW FAMILY Droseraceae

Small insectiferous herbs. Leaves in basal rosettes, the blade covered in sticky dew-tipped, reddish, translucent hairs. Flowers in lax spikes, 5-8-parted. Fruit a capsule. 105 species in 4 genera, *Drosera* being the largest with 100 species in tropical and temperate regions. In *Drosera* insects are trapped by the sticky, long-stalked glands on the leaf surface. Once trapped the surrounding leaf-glands curve in around the insects body which is then digested by the proteolytic enzymes secreted by glands in the centre of the leaf.

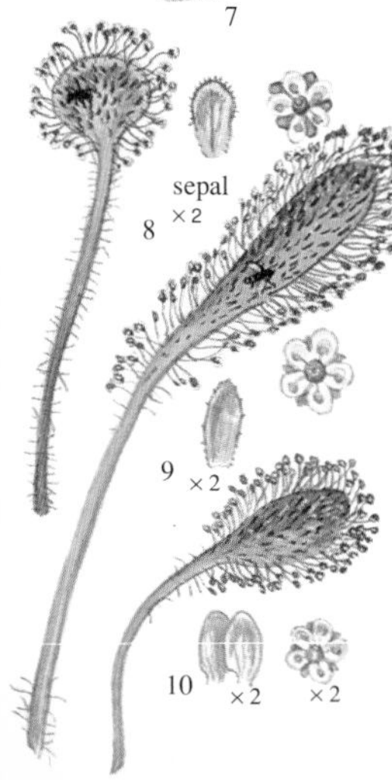

8* **Common Sundew** *Drosera rotundifolia* L. Low perennial; plants usually, solitary. Leaves rounded, long-stalked, horizontal or ascending; stalks hairy. Flowers white, 5mm, from the centre of the leaf-rosette, much overtopping the leaves. Wet heaths, moors and *Sphagnum* bogs, especially around margins of bog pools, on acid soils, to 2000m. June-August. Throughout, except Spitsbergen. **P**: Often self-pollinated, only opening properly in bright sunshine. Europe to N Asia and North America. **B**: Throughout N Britain and Ireland; scattered localites in S Britain.

9 **Great Sundew** *D. longifolia* L. [*Drosera anglica*]. Like *D. rotundifolia*, but leaves narrow oblong, erect to ascending, stalks hairless. Bogs, especially of *Sphagnum*; more tolerant of basic habitats than the previous species, to 1900m. July-August. Throughout, except the far north. **B**: Common in W & NW Scotland & W Ireland; scattered localities elsewhere, decreasing. Hybrids between *D. anglica* and *D. rotundifolia*, *D. × obovata* Mert. & Koch, occur frequently and are sterile.

10 **Long-leaved Sundew** *Drosera intermedia* Hayne [*D. longifolia*]. Low perennial; plants usually gregarious, often forming floating mats. Leaves erect, oblong, narrowed into a long hairless stalk. Flowers white, 5mm; spikes scarcely exceeding the leaves, borne to one side of the rosette. Damp acid peaty habitats, heaths and bogs, sometimes in drier peaty places, to 1900m. June-August. Throughout, except the far north. **B**: Local, scattered throughout, especially NW Scotland, the Lake District and W Ireland.

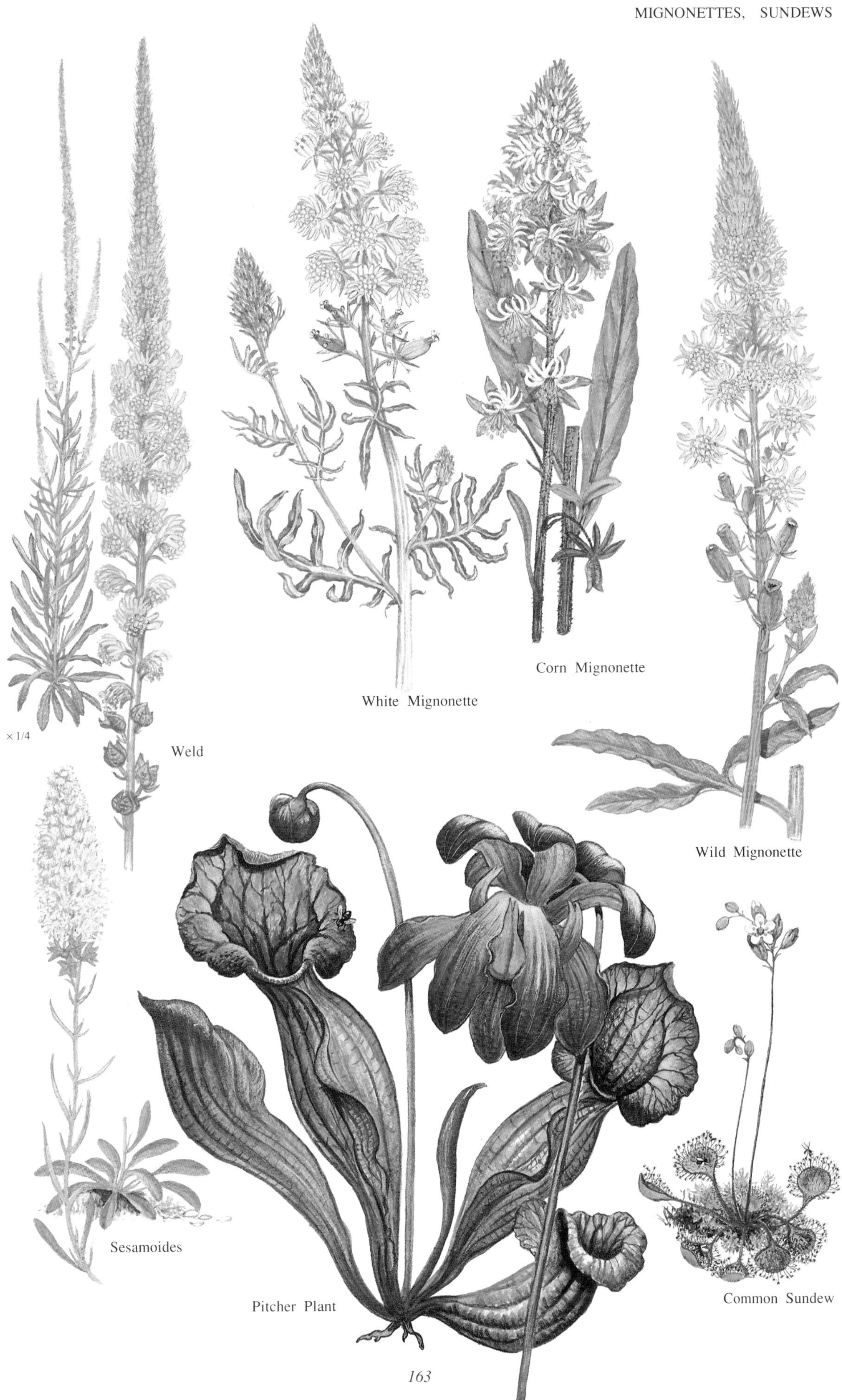
× 1/4
Weld
White Mignonette
Corn Mignonette
Wild Mignonette
Sesamoides
Pitcher Plant
Common Sundew

# STONECROP FAMILY Crassulaceae

Succulent annual or perennial herbs. Flowers regular, starry, or bell-shaped, often in cymes; sepals and petals 3-20, united or separate. Stamens equal in number, or twice as many as the petals. Fruit a cluster of carpels, usually equal in number to the petals. 1500 species in 35 genera, cosmopolitan but especially in South Africa.

**Crassula** *Crassula*. Small hairless annuals, rarely perennials, with opposite fused leaves, often reddish. Flowers 3-4-parted; stamens equal in number to the petals. Over 300 species, primarily in southern Africa.

1* **Mossy Stonecrop** *Crassula tillaea* Lester-Garland [*C. muscosa*, *Tillaea muscosa*]. Low minute moss-like plant, often reddish. Leaves tiny, oval, crowded. Flowers whitish or pale pink, 1-2mm, in small clusters, 3-parted. Damp sandy or gravelly habitats, often where the surface is compacted, such as pathways and tracks, which are frequently wet during the winter. June-September. Britain, Belgium, Holland, France and Germany. **P**: Self-pollinated. A tiny annual, often overlooked. **B**: Very local in East Anglia, Hampshire, Dorset and the Channel Islands.

2 **Water Tillaea** *Crassula aquatica* (L.) Schönl. [*Tillaea aquatica*]. Like *C. tillaea*, but leaves 3-5mm and flowers 4-parted; petals longer than sepals (not shorter). Damp habitats, margins of muddy pools. June-September. Scandinavia, N Germany and Iceland, rare in Britain – possibly introduced. N & C Europe to N Asia and North America. **B**: Very rare – probably confined to Inverness.

3 *Crassula vaillantii* (Willd.) Roth. Like *C. aquatica*, but flowers on long slender stalks. Similar habitats and flowering time. France – often very local. Mainly in S Europe.

*Crassula helmsii* (T. Kirk) Cockayne. Like *C. aquatica*, but perennial and with stalked flowers. Shallow pools and dykes. June-August. **I**: Australia and New Zealand. Naturalized in Britain. **B**: Mostly in S, SE & E England.

4* **Navelwort** *Umbilicus rupestris* (Salisb.) Dandy [*U. pendulinus*, *Cotyledon pendulina*]. Variable low to medium, hairless perennial. Leaves rounded with a central 'navel' and a long stalk; stem leaves progressively smaller. Flowers greenish-white or straw-coloured, bell-shaped, pendent, 8-10mm, borne in tapered spikes. Walls, cliffs, rocks, roof tops, grassy habitats and banks, often in mountain regions, generally on acid soils. June-August. Britain, Ireland and France. **P**: Probably self-pollinated. **B**: Widespread in W Britain and Ireland, scarce elsewhere.

5* **Houseleek** *Sempervivum tectorum* L. Variable, low to short perennial with dense rounded, open, rosettes, 3-8cm across. Leaves lanceolate, pointed, bluish-green tipped with dull red; margin bristly. Flowers dull reddish-purple, 20-30mm, in broad, stout-stemmed, clusters; petals 8-18, often 12. Walls, roof tops, grassy and rocky habitats, primarily in the mountains, to 2800m. June-July. Mountains of France and Germany; cultivated on old walls, roof tops and rock gardens for centuries, but scarcely ever truly naturalized. **P**: Various insects. **B**: Occasionally on roof tops.

6* **Hen and Chickens Houseleek** *Jovibarba sobolifera* (J. Sims) Opiz [ *Sempervivum soboliferum*]. Rather like the previous species, but leaf rosettes closed, only 2.5-3cm; leaves, often red-tipped. Flowers pale yellow, 15-17mm, erect bells, petals toothed. Sandy ground, dry grassland, usually on acid soils and at fairly low altitudes, to 1500m. C & S Germany. Sometimes grown in gardens.

**Stonecrops** *Sedum*. Leaves seldom in rosettes, usually alternate. Flowers hermaphrodite, usually 5-parted; stamens twice as many as petals. 600 species, mainly in N temperate regions of the world, but also on some mountains in tropical Africa and South America.

7* *Sedum aizoon* L. Medium erect perennial; stems erect. Leaves alternate, flat, lanceolate, pointed and toothed. Flowers golden yellow, 15-20mm, in dense clusters with leaves immediately below. Rocky places. July-September. **I**: N Asia. Locally naturalized from gardens in France, Germany and Scandinavia.

8 *Sedum hybridum* L. Like *S. aizoon*, but a shorter plant with low non-flowering shoots. Leaves only 2-3cm (not 5-8cm), with reddish teeth. Rocky and waste places. June-August. **I**: N Asia. Naturalized from gardens in Britain, Scandinavia, France and Germany.

9* **Orpine** *Sedum telephium* L. Variable short to tall, rather greyish perennial, often red-tinged; all shoots flowering. Leaves rounded to oblong, 5-8cm, toothed. Flowers purplish-red, greenish-yellow or whitish, 8-10mm, in dense clusters. Shaded habitats, woods, hedgebanks, scrub, generally on light sandy soils, to 2500m. July-September. Throughout, except the Faeroes, Iceland and Spitsbergen. **P**: various bees and flies. Widespread from Europe to W Turkey, temperate Asia and North America. **B**: Throughout; probably introduced in Ireland.

10 *Sedum ewersii* Ledeb. Like *S. telephium*, but a lower plant, not exceeding 20cm and with non-flowering shoots. Leaves oval or rounded, 1.5-2cm, opposite, grey. Flowers pink or mauve, 9-12cm. Rocky places. June-August. **I**: C Asia and the N Himalaya. Naturalized in Scandinavia.

11 **Caucasian Stonecrop** *Sedum spurium* Bieb. Low to short creeping, mat-forming, perennial with short non-flowering shoots. Leaves to 2.5cm, oval to rhombic, alternate, toothed towards the apex. Flowers reddish-purple, rarely pink or white, 9-12mm long, in dense clusters, petals erect. **I**: Caucasus. Naturalized from gardens, except in the far north. July-August.

12* **Reflexed Stonecrop** *Sedum rupestre* L. [*S.reflexum*]. Short, creeping greyish, evergreen perennial forming mats, with erect flowers stems. Leaves linear-cylindrical, 8-20mm, pointed, falling when dead. Flowers bright or pale yellow, 14-15mm, drooping in bud; petals often 7. Rocky places, walls, stony pathways, to 2000m. June-August. North to S Scandinavia; naturalized in Britain and Ireland. W Europe to the Ukraine. **B**: Mainly in S, rare in most of Scotland and in Ireland.

13* **Rock Stonecrop** *Sedum forsteranum* Sm. [*S. elegans*, *S. pruinatum*]. Like *S. reflexum*, but leaves of non-flowering shoots aggregated into cone-like clusters, with persistent dead leaves below. Flowers 11-12mm. Rocks and screes, old walls, rocky woodland, at rather low altitudes. June-July. Europe north to S Scandinavia. **B**: Locally naturalized in England, Wales and (very rare) Ireland.

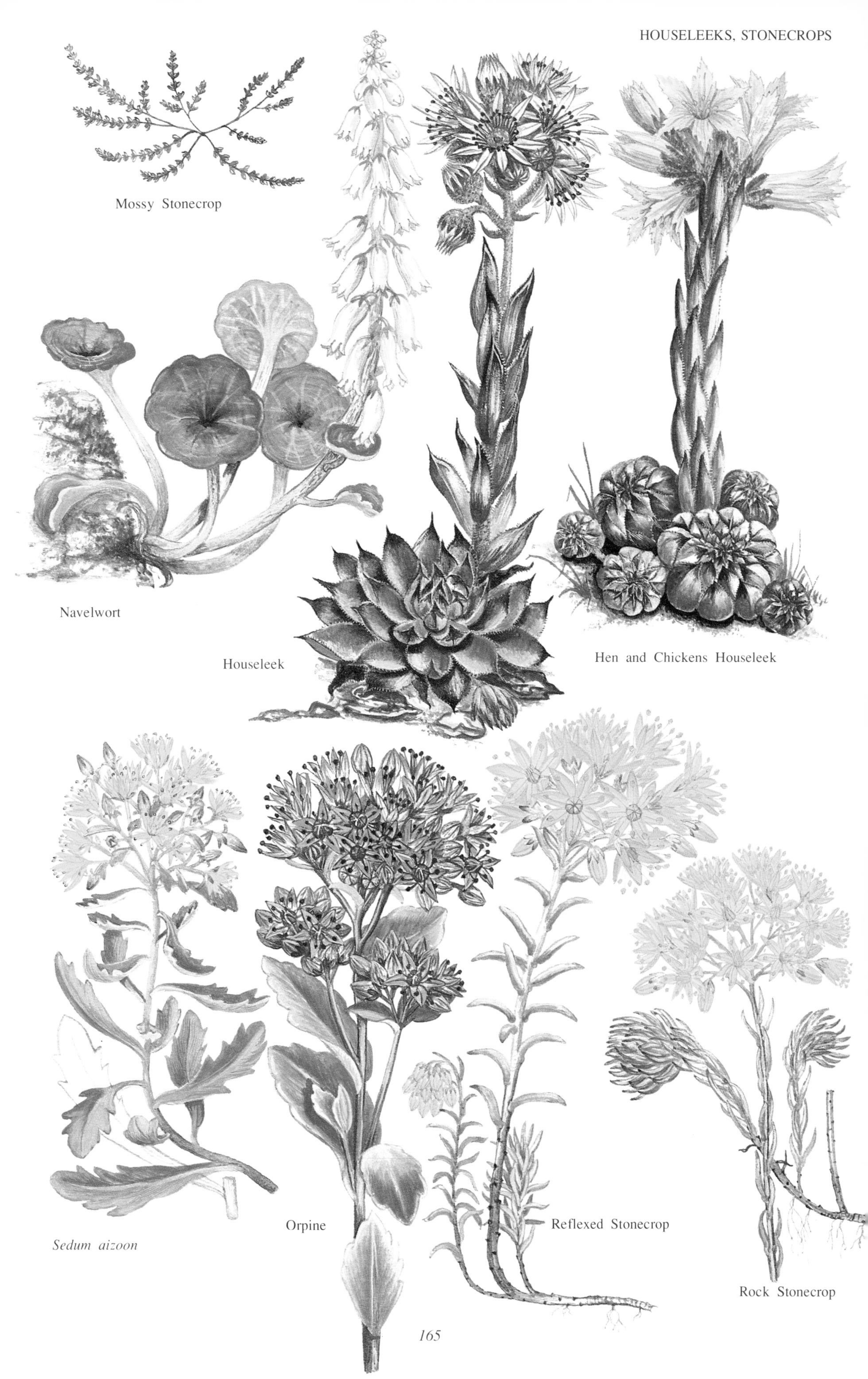
Mossy Stonecrop
Navelwort
Houseleek
Hen and Chickens Houseleek
*Sedum aizoon*
Orpine
Reflexed Stonecrop
Rock Stonecrop

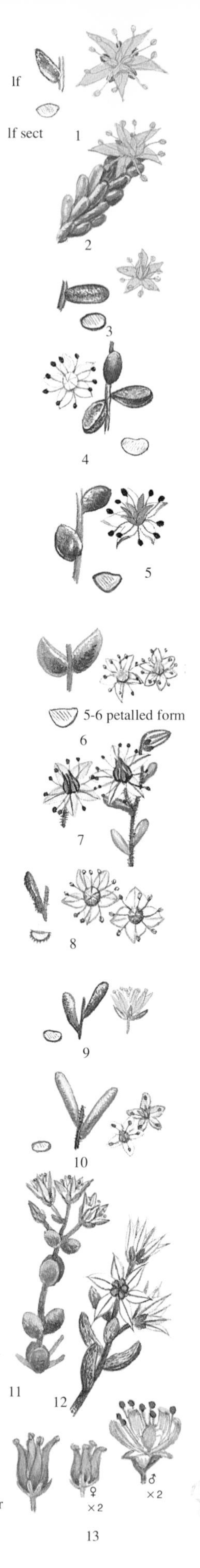


1* **Biting Stonecrop, Wallpepper** *Sedum acre* L. Low hairless evergreen, perennial forming deep yellowish-green mats. Leaves 3-6mm, oval-cylindrical, broadest near base, densely overlapping on vegetative shoots. Flowers bright yellow, 10-12mm, 5-parted, in small clusters. Rocky and sandy places, especially sand-dunes, shingle beaches, heaths, banks and walls, to 2300m. May-July. Throughout, except Spitsbergen. Widely cultivated in gardens – often seen on old walls and root tops. The pale green shoots have a peppery taste. **B**: Common throughout, except parts of NW Scotland.

2 **Insipid Stonecrop** *Sedum sexangulare* L. [*S. boloniense, S. mite*]. Like *S. acre*, but leaves cylindrical, parallel-sided or broadest above the middle, spreading, bright green, in 5-6 regular rows, not peppery-tasting. Rocky habitats and old walls. July-August. C & E France, Germany and S Finland; naturalized in Britain – England and Wales. Cultivated in gardens. Local on old walls in England and Wales.

3* **Alpine Stonecrop** *Sedum alpestre* Vill. Somewhat like *S. acre*, but forming low evergreen mats with very short vegetative shoots. Leaves rather flattened, parallel-sided, often streaked with red. Flowers dull yellow, 6-8mm; follicles dark red. Mountain rocks and moraines, on non-calcareous soils, to 3500m. June-July. C & S Germany, C & S France.

4* **White Stonecrop** *Sedum album* L. Low to short mat-forming, often rather straggling, perennial. Leaves alternate, bright green, often tinged with red, 4-12mm, oval-cylindrical. Flowers white, 6-9mm, in much-branched flat-topped clusters on erect stems; follicles pink, erect. Rocky habitats, screes, moraines and ledges, roadsides and old walls, to 2500m. June-August. Throughout, except the Faeroes, Iceland and Spitsbergen; naturalized in Ireland. Sometimes cultivated in gardens. **B**: Throughout, perhaps only native in the south; naturalized in Ireland.

5* **English Stonecrop** *Sedum anglicum* Hudson. Low, lax, mat-forming, hairless perennial, with short vegetative shoots. Leaves alternate, 3-5mm, cylindrical-oval, blue-green, often tinged with pink. Flowers white, pink below, with blackish anthers, 9-12mm, few to a cluster; follicles red. Rocky habitats, banks and dry grassland, dunes and shingle beaches, usually on acid soils, to 1800m. June-September. Britain, Ireland, Belgium, Holland, France, Denmark, Norway and Sweden. The inflorescence normally has 2 main branches (in *S. album* there are several branches). **B**: W Britain, S & W Ireland; coastal elsewhere.

6* **Thick-leaved Stonecrop** *Sedum dasyphyllum* L. Like *S. anglicum*, but leaves mostly opposite, stickily hairy and greyish-pink. Flowers 5-6mm, in more-branched clusters. Acid rocks, walls, banks and roadsides, to 2500m. June-September. France and Germany; naturalized in Britain, Ireland and various parts of NW Europe. Widespread in C & S Europe. **P**: Flies and bees. **B**: Scattered localities in England, Wales and Ireland.

7 *Sedum hirsutum* All. Low mat-forming, densely hairy, sticky perennial. Leaves alternate, 5-15mm, cylindrical-oval. Flowers white or pinkish, 11-13mm, the petals with a red mid-vein, nodding in bud. Dry, generally acid, rocky habitats. June-August. France. Locally in parts of W and SW Europe.

8* **Hairy Stonecrop** *Sedum villosum* L. Low reddish, hairy, perennial or biennial with short vegetative shoots and erect flowering stems. Leaves alternate, 4-7mm, linear-oblong, glandular. Flowers lilac or pale pink, 6-8mm, long-stalked, in lax clusters; follicles green or purplish. Wet stony ground, wet pastures, river and stream banks, generally on calcareous or base-rich soils. June-August. Britain, the Faeroes, Iceland and most of Europe. **B**: Local in England and Scotland, except the north.

var. *glabratum* Rostrup. Like the type, but almost hairless. Confined to sub-Arctic regions.

9* **Annual Stonecrop** *Sedum annuum* L. Low hairless annual or biennial, rather straggly and branched from the base, often red-spotted or streaked. Leaves alternate, linear-oblong, 6mm. Flowers yellow, 3-5mm, short-stalked, in lax, few-flowered, clusters. Rocky habitats, screes, moraines and ledges, only on the mountains in the south of its range, to 2900m. June-August. Scandinavia, Iceland, France and Germany.

10* **Reddish Stonecrop** *Sedum rubens* L. [*Crassula rubens*] Low erect annual, greyish and flushed with red; stems hairy above, often sticky. Leaves linear, alternate, 10-20mm. Flowers white or pink, 9-11mm, unstalked; stamens usually only 5. Rocky and stony habitats. May-July. Belgium, France and SW Germany.

11 *Sedum andegavense* (DC.) Desv. Like *S. rubens*, but hairless, the leaves oval and the flowers 4-5-parted, with 4-5 stamens. Similar habitats and flowering time. France. Only in France and SW Europe.

12 *Sedum hispanicum* L. [*S. glaucum, S. sexfidum*]. Like *S. rubens*, hairy or hairless, occasionally biennial. Flowers white, the petals with a pink mid-vein, usually 6-parted. **I**: C & SE Europe. Naturalized in Britain, Germany and S Sweden.

13* **Roseroot** *Rhodiola rosea* L. [*Sedum roseum, S. rhodiola*]. Variable low to short, tufted, hairless, erect perennial, often tinged with purple. Leaves flattish, oval to linear-oblong, 10-40mm, alternate, greyish and toothed. Flowers dull yellow, sometimes petaless, 5-8mm, 4-parted, in dense clusters, male and female flowers on separate plants; anthers purple; follicles reddish-orange. Rocks, sea cliffs, moraines and stabilized screes, on acid or calcareous soils, to 3000m, at sea level in the north of its range. May-June. Throughout, except Belgium, Holland, Denmark and Germany. Cultivated in gardens. **B**: N Britain, Wales and W Ireland. The English name derives from the rose-like fragrance of the plant.

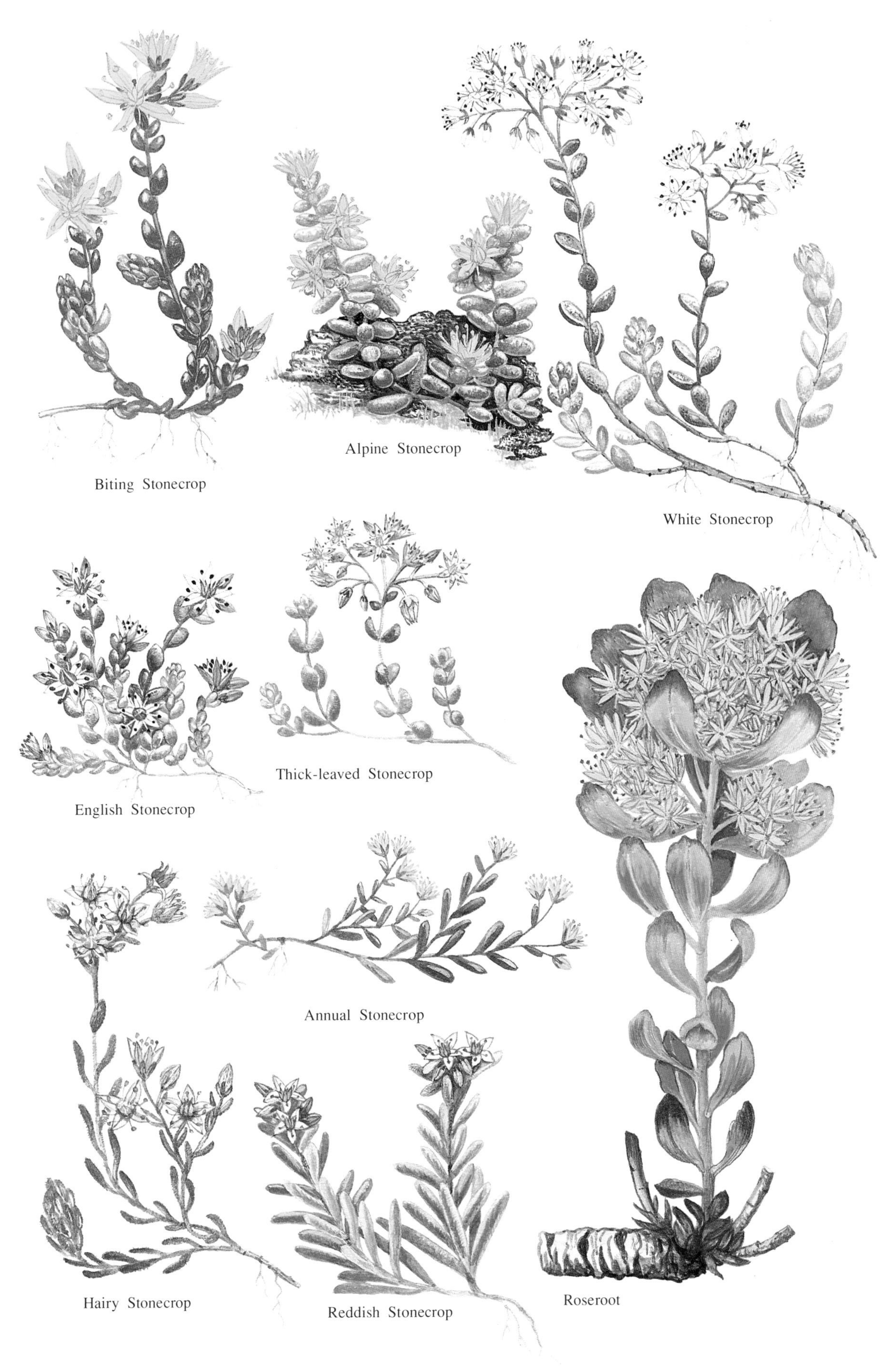
Biting Stonecrop
Alpine Stonecrop
White Stonecrop
English Stonecrop
Thick-leaved Stonecrop
Annual Stonecrop
Hairy Stonecrop
Reddish Stonecrop
Roseroot

# SAXIFRAGE FAMILY Saxifragaceae

Herbs, usually perennial. Flowers in branched cymes, sometimes solitary or in racemes; petals 4-5. Stamens twice as many as sepals, sometimes fewer. Style 2. Fruit a 2-parted capsule containing many seeds. 580 species in 30 genera, mainly in temperate regions. Many are cultivated in gardens, particularly in the genus *Saxifraga* and *Bergenia*; *Saxifraga* alone contain about 370 species, mainly from temperate, alpine and Arctic regions.

1* **Hawkweed Saxifrage** *Saxifraga hieracifolia* Waldst. & Kit. Low to short hairy perennial with an underground stem. Leaves all in a basal rosette, oval, usually toothed, hairy beneath and on margin, stalk broad; bracts large and leaf-like. Flowers green tinged with reddish-purple, 3-5mm, in slender interrupted spikes; petals as long as sepals. Damp mountain rocks, moraines and streamsides, to 2400m. Arctic Europe and C France – Auvergne. Local. Easily identified by its slender spikes of rather dull little flowers.

2* **Arctic Saxifrage** *Saxifraga nivalis* L. Low rhizomatous, stickily-hairy, perennial. Leaves all in basal rosettes, thick, oblong to rhombic, toothed, reddish-purple beneath. Flowers white or pink, unspotted, 5-6mm, scarcely stalked, in a crowded head; sepals erect. Mountain habitats, especially shaded rock ledges, generally on basaltic rocks, screes and moraines, to 2100m. July-August. N Britain, Ireland. Arctic and sub-Arctic Europe. Rather rare in cultivation. Local in the wild. **B**: Rare and local, N Wales, Lake District, W & C Highlands of Scotland; Ireland – confined to Co Sligo.

3 *Saxifraga tenuis* (Wahlenb.) H. Smith ex Lindman. A more slender plant, the flowers distinctly stalked, borne in lax clusters. Similar habitats and flowering time. Arctic and sub-Arctic Europe.

4* **Starry Saxifrage** *Saxifraga stellaris* L. Low to short, slightly hairy, densely tufted perennial. Leaves in basal rosettes, oblong, toothed, scarcely stalked. Flowers white, each petal with 2 yellow spots near the base, 10-15mm, in lax, branched clusters; sepals downturned; anthers pink. Damp mountain habitats, rock ledges, stony ground, wet flushes, stream bank and marshes, to 1900m. June-August. Throughout except Belgium, Holland and Denmark. **P**: Flies. **B**: Mountains of Wales, N England, Scotland and Ireland.

subsp. *alpigena* Temesy. Like the type, but a laxer plant, the leaves almost hairless; flowers smaller, 7-10mm. Similar habitats and flowering time. S Germany, from the Vosges southwards. C & S Europe.

5 *Saxifraga foliolosa* R.Br. Like *S. stellaris*, but leaf-rosettes usually solitary (not clustered). Inflorescence with a terminal flower, those lower down generally replaced by small reddish bulbils. Damp rocky habitats. July-August. Arctic and sub-Arctic Europe.

6* **St Patrick's Cabbage** *Saxifraga spathularis* Brot. Low to short laxly tufted perennial. Leaves thick, ascending, mostly in basal rosettes, spoon-shaped, toothed, with a narrow translucent margin; stalk longer than blade. Flowers white, the petals with 1-3 basal yellow spots, with smaller red spots above, 7-9mm; sepals downturned; anthers and ovary pink. Damp Rocky habitats, woodland and stream margins, on acid soils, to 1035m. June-August. Ireland. Also found in Spain and Portugal. Sometimes cultivated in gardens, but largely replaced by *S.* × *urbium*, of which it is one of the parents. **B**: Widespread in mountains of W Ireland.

7* **Pyrenean Saxifrage** *Saxifraga umbrosa* L. Low to short, tufted perennial, rather like *S. spathularis*, but leaves spreading, with a broad stalk, bristle-edge, shorter than the blade, blunt-toothed. Flowers white, the petals with a yellow spot at base and tiny red spots above, 5-8mm. Fruit reddish. Naturalized in damp semi-shaded places. June-July. **I**: W & C Pyrenees. N Britain. Cultivated in gardens. **P**: Various small insects. **B**: Naturalized in Yorkshire – Haseldon Gill.

8* **Kidney Saxifrage** *Saxifraga hirsuta* L. Like *S. spathularis*, but leaves less leathery, kidney to heart-shaped, hairy on both surfaces, bright green. Flowers like *S. umbrosa*, but sometimes without red spots. Fruit green. Shady mountain habitats, damp rocks, stream banks, to 915m. May-July. SW Ireland; occasionally naturalized in Britain. The following hybrids, between the previous 3 species, are widely cultivated and naturalized in W Europe. **B**: Very local, confined to uplands of SW Ireland.

9* **London Pride** *Saxifraga* × *urbium* D.A.Webb (*S. spathularis* × *S. umbrosa*). Intermediate in character between the parent species. Plants generally sterile. Naturalized in Britain and France. Not known naturally in the wild. Widely grown in gardens.

10 *Saxifraga* × *polita* (Haw.) Link (*S. hirsuta* × *S. spathularis*). Like *S. spathularis*, but leaves more deeply toothed and without a translucent margin. Occurs naturally where the parent species grow in close proximity to one another. May-July. W Ireland.

11 *Saxifraga* × *geum* L. (*S. hirsuta* × *S. umbrosa*). Intermediate in character between the parent species. Leaves with a translucent margin, stalks narrow, considerably longer than the blade. Widely cultivated in gardens. June-July. Naturalized in Belgium, France and Germany; occasionally casual elsewhere.

12* **Round-leaved Saxifrage** *Saxifraga rotundifolia* L. Variable short to medium, loosely tufted, slightly hairy perennial with leafy stems. Leaves round to kidney-shaped, sharply toothed the lower long-stalked. Flowers white, each petal yellow-spotted near the base, red-spotted near the tip, in laxly-branched clusters; sepals erect. Damp or shaded habitats, woodland, stream margins and rocks, to 2500m. June-October. France and Germany; naturalized in Britain and Belgium. Cultivated in gardens.

London Pride

Arctic Saxifrage
Hawkweed Saxifrage
St Patrick's Cabbage
Starry Saxifrage
Kidney Saxifrage
London Pride
Round-leaved Saxifrage
Pyrenean Saxifrage

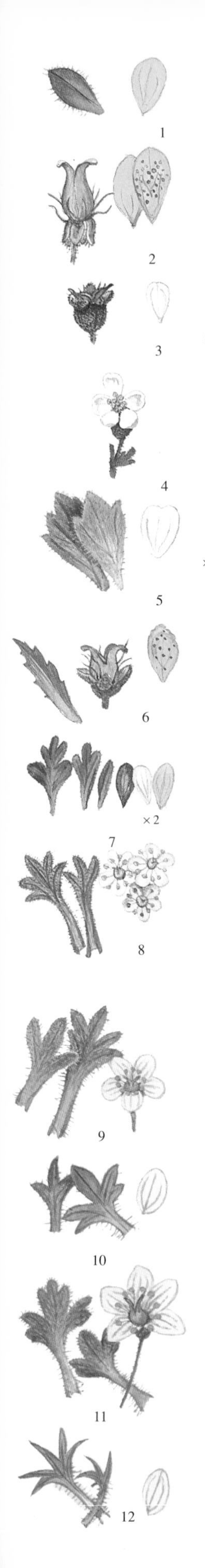


1* Saxifraga platysepala (Trautv.) Tolm. Low, slightly hairy, creeping perennial. Leaves in dense basal rosettes, oblong, pointed; rosettes producing long slender, reddish runners, to 15cm, each forming a new leaf-rosette on the end. Flowers yellow, 10-14mm, few in a lax cluster. Rocky habitats, screes and moraines. July-August. Spitsbergen. Very local. Also found in the N USSR.

2* **Yellow Marsh Saxifrage** *Saxifraga hirculus* L. Short loosely-tufted perennial, stems erect, leafy below. Leaves lanceolate, untoothed, hairy towards the base. Flowers bright yellow, sometimes with reddish spots, 20-30mm, solitary or 2-4 clustered. Wet grassy moors and bogs, only on the mountains in the south of its range, to 1500m. Throughout, except the Faeroes, Belgium, Holland. Occasionally cultivated in gardens. **P**: Bees and butterflies. **B**: Confined to N England, Scotland and Ireland; rare, local and decreasing rapidly.

3* **Rue-leaved Saxifrage** *Saxifraga tridactylites* L. Low to short, stickily-hairy annual, often reddish. Lowest leaves spoon-shaped, unlobed, withered at flowering time, others 3-5-lobed. Flowers white, 4-6mm, in lax, leafy clusters; petals slightly notched; flowerstalks much longer than the flowers. Walls, rocky and bare places, field margins, sand- dunes, grassy heaths, usually on limestone and often lowland, but to 1550m in the south of its range. June-September. Throughout, except the extreme north, the Faeroes and Iceland. **P**: Self-pollinated. **B**: Widespread throughout, but very rare in Scotland.

4 *Saxifraga osloensis* Knaben. Like *S. tridactylites*, but flowers larger, 6-10mm, and with shorter stalks. Similar habitats and flowering time. Sweden and E Norway, rare. Thought to be of hybrid origin between *S. adscendens* and *S. tridactylites*.

5* **Biennial Saxifrage** *Saxifraga adscendens* L. [*S. controversa*]. Rather robust, short, hairy biennial; stem simple or branched. Leaves mostly in a basal rosette, wedge-shaped, 2-5-lobed, unstalked. Flowers white, occasionally yellowish, 6-10mm, in lax clusters; petals slightly emarginate. Mountain pastures, rocks and screes, 1800-3500m. June-August. Scandinavia and France. Local.

6* **Yellow Mountain Saxifrage** *Saxifraga aizoides* L. Low to short, leafy, slightly hairy perennial. Leaves fleshy, linear-oblong, often slightly toothed. Flowers bright yellow or orange, sometimes red-spotted, 5- 10mm, borne in lax clusters. Damp and stony or rocky habitats, along streamsides and wet flushes, to 3150m. June-September. Throughout, except the Faeroes and most of lowland Europe. **P**: Flies. **B**: Widespread in N England and the Scottish Highlands, as well as the mountains of Ireland.

7* **Musky Saxifrage** *Saxifraga moschata* Wulfen [*S. varians*]. Variable low, slightly hairy perennial, forming flattish cushions. Leaves often 3-lobed, but sometimes 5-lobed or unlobed, lobes not grooved. Flowers dull yellow or cream, 5-8mm, solitary or 2-7 on almost leafless stems; petals narrow, not touching. Rocky and stony mountain habitats, cliffs and moraine, 1200-4000m. July-August. C & E France and S Germany. Widespread in the mountains of C Europe. Sometimes cultivated.

8 **White Musky Saxifrage** *Saxifraga exarata* Vill. Like *S. moschata*, but leaves 3-5-lobed, the lobes grooved above; flowers white or pale cream, rarely pink, with touching petals. Similar habitats. June-August. C & E France and S Germany.

9* **Tufted Saxifrage** *Saxifraga cespitosa* L. Low, loose, cushion-forming perennial, stickily-hairy. Leaves mostly 3-lobed, in basal rosettes, rather incurved when young. Flowers dull white or creamish, 7-9mm, solitary or 2-3 on a slightly leafy stem. Mountain rocks, cliffs and moraines, to 2800m. May-July. N Britain, the Faeroes, Iceland and Scandinavia. Usually local. **B**: Rare; confined to Scotland and N Wales.

10* **Irish Saxifrage** *Saxifraga rosacea* Moench [*S. decipiens*]. Variable low to short perennial forming compact cushions or a loose mat, usually hairy. Leaves wedge-shaped or rhombic with 3-5-lobes, usually without glandular hairs. Flowers pure white, 12-20mm, erect in bud, on almost leafless stems. Mountains and sea-cliffs, among boulders and along stream margins, gullies and stabilized screes, to 2400m. June-August. Ireland, the Faeroes, Iceland, France, C & S Germany. **B**: Rare; mountains and sea cliffs in S & W Ireland; formerly in N Wales, but now extinct.

11 *Saxifraga hartii* D.A. Webb. Intermediate between *S. cespitosa* and *S. rosacea*, but with rather fleshy, flatter leaf-rosettes and 5-7-lobed leaves; flowers pure white. Grassy sea cliffs. May-June. **B**: NW Ireland; very rare, confined to Co Donegal.

subsp. *sponhemica* (C.C. Gmelin) D.A. Webb. Like the type, but leaf-lobes with a finely pointed tip. Similar habitats and flowering time. E France and W Germany.

12* **Mossy Saxifrage, Dovedale Moss** *Saxifraga hypnoides* L. [*S. hypnoides* subsp. *boreali-atlantica*]. Low to short, mat-forming, hairy perennial with long slender prostrate leafy shoots; bulbs often present at the axils of lower leaves. Leaves linear-lanceolate to 3-lobed, often widely spaced; rosette leaves at shoot tips generally 3-, sometimes 5- or 7-lobed. Flowers pure white, 14-20mm, nodding in bud. Damp, often grassy habitats, in hilly and mountain regions, stream margins, cliffs, screes and morains, on calcareous or base-rich soils, to 1215m. May-July. Britain, Ireland, the Faeroes, Iceland, Belgium, France - Vosges, and Norway. **B**: Scattered localities (except in the extreme south and in Ireland).

*Saxifraga hartii*

*Saxifraga platysepala*
Yellow Marsh Saxifrage
Rue-leaved Saxifrage
Tufted Saxifrage
Biennial Saxifrage
Yellow Mountain Saxifrage
Irish Saxifrage
Mossy Saxifrage
Musky Saxifrage

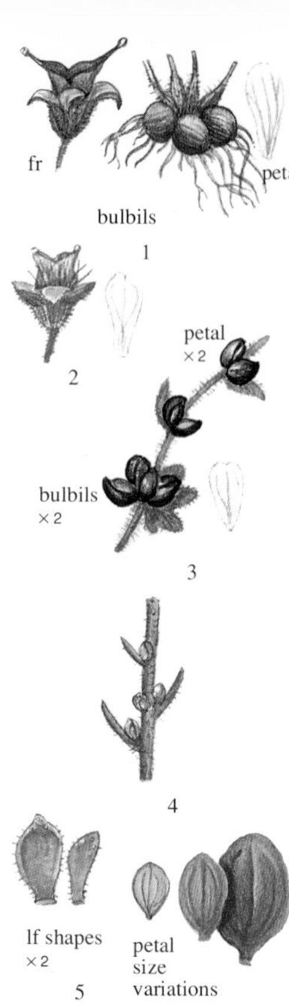


1* **Meadow Saxifrage** *Saxifraga granulata* L. Variable, short to medium, hairy perennial; stem erect, branched above. Leaves mostly basal with small bulbils at base of stalks (below ground), kidney-shaped, lobed or deeply blunt-toothed. Flowers white, 18-30mm, in loose, branched clusters. Meadows and other grassy habitats, pastures, bank and road verges, rocky places, generally on well-drained neutral or calcareous soils, to 2200m. April-June. Throughout, except the Faeroes, Iceland and Spitsbergen. **P**: Various insects. Occasionally grown in gardens, including a double-flowered form. **B**: Local in Britain, not in N Scotland or SW England, rare in Ireland.

2* **Highland Saxifrage** *Saxifraga rivularis* L. Low, tufted, hairless perennial with creeping stolons. Basal leaves 3-5-lobed, with bulbils at the base of the stalks (below ground); stem leaves present, without bulbils. Flowers white, veined or flushed with pink, 8-10mm, few to a cluster. Wet rocky mountain habitats, to 1220m. July-August. Arctic and sub-Arctic Europe to S Norway, Iceland and N Britain. **P**: Probably flies. **B**: Very rare, known from a few scattered localities in the C & W Highlands of Scotland.

3* **Drooping Saxifrage** *Saxifraga cernua* L. Low to short, erect, thin-stemmed perennial. Basal leaves kidney-shaped, 5-7-lobed, long-stalked; stem leaves numerous, few lobed or unlobed, mostly unstalked, with small reddish bulbils in the axils. Flowers white, 12-18mm, often solitary, or one flower per branch, petals not always fully developed. Rocky, often shaded, habitats, on basic rocks, 900-2500m. June-July. Arctic and sub-Arctic Europe, Iceland and Scotland, mountains of France and Germany. A local species which apparently reproduces mainly by axillary bulbils, rarely setting viable seed. **B**: Very rare, C Highlands of Scotland; protected.

4 *Saxifraga* × *opdalensis* Blytt (*S. cernua* × *S. rivularis*) is stouter than the parent species, with greenish bulbils in the axils of stem leaves; flowers intermediate in character. Confined to the Opdal region of Norway.

5* **Purple Saxifrage** *Saxifraga oppositifolia* L. Low mat-forming perennial, with long trailing stems. Leaves small, oval, opposite, green or greyish, often lime-encrusted, with 1-5 lime pores near the apex. Flowers pale pink to deep purple, 10-20mm, solitary, almost stalkless; anthers bluish. Damp rocky and stony mountain habitats, screes, moraines and cliff-ledges, often on limestone or base-rich soils, sea level to 3800m, but only on the mountains in the south of its range. March-August. Throughout. Widely cultivated in gardens. Widespread in the mountains of Europe, where various local forms exist. **P**: Bees and butterflies, but may also be self-pollinated. **B**: N England, N Wales, Scotland and N & NW Ireland.

6 *Saxifraga biflora* All. Like *S. oppositifolia*, but the leaves larger, 5-9mm long (not 2-6mm), with a single lime pore; flowers reddish-purple, in clusters of 2-5, the anthers pink to orange. Damp mountain screes, moraines and river gravels, 2000-3200m. E France & S Germany. Cultivated in gardens.

7* **Pyramidal Saxifrage** *Saxifraga cotyledon* L. A short to medium perennial with large flattish leaf-rosettes. Leaves grey-green, broad-oblong with a point at the blunt tip, 20-60mm, finely toothed, not lime encrusted. Flowers white, 11-18mm, sometimes purple-spotted, in broad pyramidal panicles, branched from the base. Acid mountain rocks, especially cliff crevices, but also on stabilized moraines, 1500-2600m. July-August. S Germany, Norway, Sweden and Iceland. Widely cultivated in gardens, particularly in frames and alpine houses. The large basal leaf-rosettes are very distinctive.

8* **Live-long Saxifrage** *Saxifraga paniculata* Miller [*S. aizoon*]. Variable, short to medium tufted perennial with numerous rather small stiff leaf-rosettes. Leaves narrow-oblong to narrow-oval, broadest above the middle, 12-40mm, greyish-green, lime encrusted, finely toothed. Flowers white or cream, rarely spotted with pink or red, 8-11mm, in loose narrow panicles, branched only in the upper third. Rocky and stony habitats, cliffs moraines and stabilized screes, to 2700m. Iceland, local in S Norway, mountains of France and Germany from the Vosges south. Cultivated in gardens. Variable in the wild.

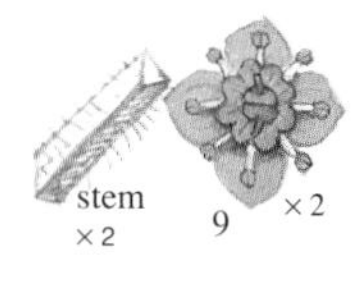


**Golden Saxifrages** *Chrysosplenium*. Perennial herbs with stalked leaves. Flowers small, 4-parted, in flat leafy clusters; petals absent; sepals 4; stamens 4 or 8. 55 species from N temperate and Arctic regions, but also represented in N Africa and South America. Several are cultivated in gardens.

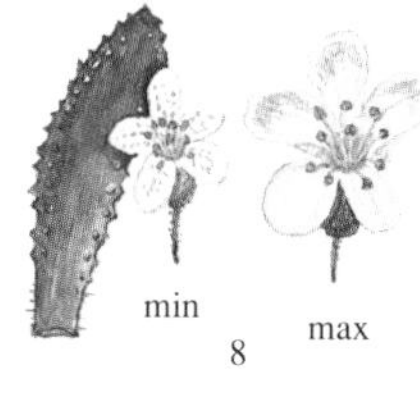


9* **Alternate-leaved Golden Saxifrage** *Chrysosplenium alternifolium* L. Low to short, slightly hairy, creeping perennial with triangular stems. Leaves mostly basal, heart or kidney-shaped, toothed, alternate. Flowers small, yellowish, 2-3mm, surrounded by yellowish leafy-bracts; stamens 8. Damp and shaded habitats, marshy ground, streamsides, damp open woodland and seepage areas, generally on calcareous soils where there is a continuous movement of water. April-July. Throughout, except the extreme north, Ireland, the Faeroes, and Iceland. **B**: Local, except for the northern Isles.

10 *Chrysosplenium tetrandrum* (N. Lund) Th. Fries. Like *C. alternifolium*, but hairless and with green flowers and bracts; stamens 4 only. Similar habitats. June-August. N Scandinavia and Spitsbergen.

11* **Opposite-leaved Golden Saxifrage** *Chrysosplenium oppositifolium* L. Low, slightly hairy, creeping perennial; stems rooting readily, square. Leaves opposite, rounded or oblong, blunt-toothed. Flowers greenish, 3-4mm, surrounded by bright yellowish-green leafy bracts; anthers bright yellow; stamens 8. Damp and shaded habitats, streamsides, woodland flushes, wet rocks, generally on acid soils. March-July. Throughout, except Iceland, the Faeroes, C & N Scandinavia. **P**: Various small insects or self-pollinated. **B**: Throughout, scarce in the east.

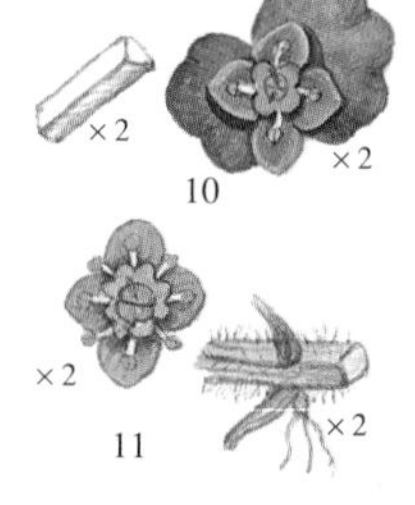

*Saxifraga biflora*

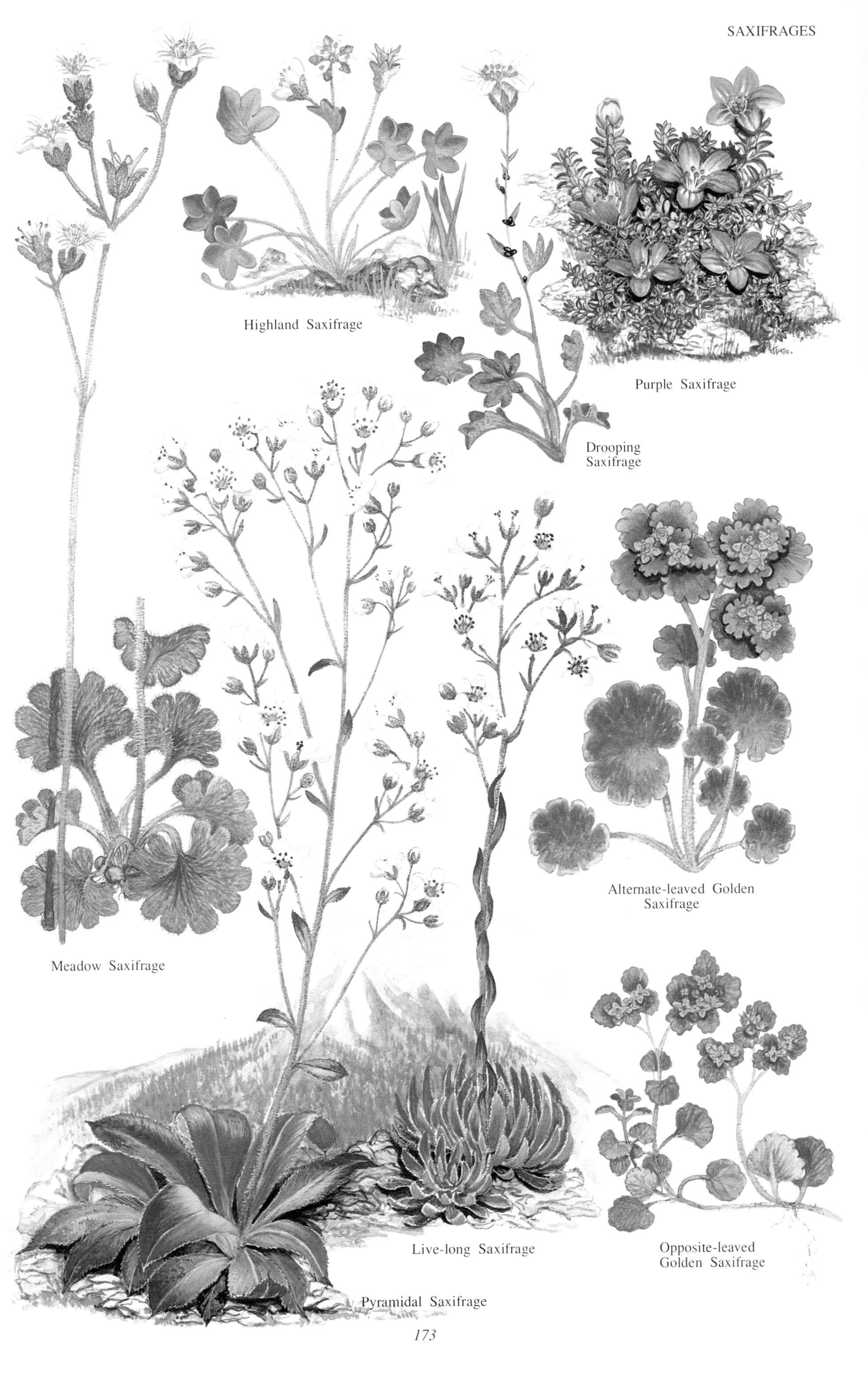
Highland Saxifrage
Purple Saxifrage
Drooping Saxifrage
Meadow Saxifrage
Alternate-leaved Golden Saxifrage
Live-long Saxifrage
Opposite-leaved Golden Saxifrage
Pyramidal Saxifrage

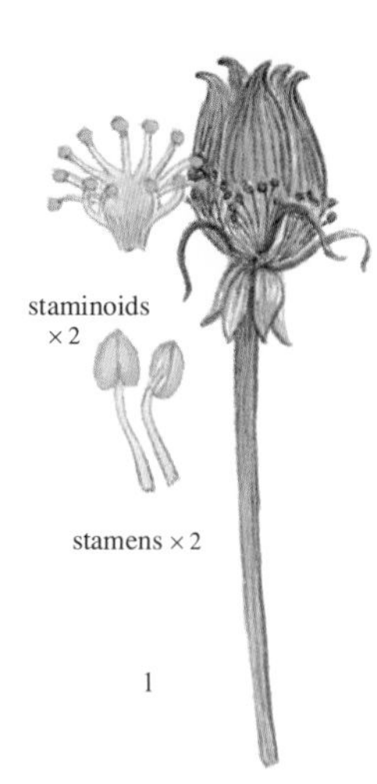


## GRASS OF PARNASSUS FAMILY Parnassiaceae

Perennial herbs with simple untoothed leaves, mostly basal. Flowers regular, with 5 sepals and 5 petals; stamens 5, with alternating staminodes. 50 species in N temperate regions. The staminodes are opposite the petals and have a solid nectar-secreting base, glistening like a drop of honey.

1* **Grass of Parnassus** *Parnassia palustris* L. Short to medium, hairless, tufted perennial. Leaves heart-shaped, the basal long-stalked; stem-leaf solitary, near centre of stem, clasping. Flowers solitary white, 15-30mm, with 5 oval petals, branched staminodes and 5 stamens; petals with translucent veins. Fruit a small capsule. Moist habitats, moors and marshes, wet pastures, fens, streamside flushes, generally on base-rich soils, to 2500m. June-September. Throughout except the extreme north, often local, sometimes forming extensive colonies. **P**: Various insects. Widespread in the temperate northern hemisphere, from Europe to Asia and the Himalaya. In lowland areas of Europe it has declined due to land-drainage. **B**: Widespread but decreasing; scarce in much of the SW and parts of E England.

subsp. *obtusiflora* (Rupr.) D.A. Webb. Like the type, but stem leaf absent or near the base of the stem, not clasping. Similar habitats. Arctic and sub-Arctic Europe.

## GOOSEBERRY FAMILY Grossulariaceae

Deciduous shrubs with alternate, 3-5-palmately lobed leaves. Flowers regular, 5-parted, with the ovary below (inferior), borne in small lateral clusters or long slender racemes; petals small, erect. Styles 2. Fruit a fleshy berry. 150 species in 2 genera, widely scattered in temperate regions of the northern hemisphere and in South America. Several are cultivated for their edible berries.

2* **Gooseberry** *Ribes uva-crispa* L. [*R. grossularia*]. Densely branched spiny shrub to 1.5m, spines often in threes. Leaves rather small, deeply-lobed. Flowers pale pinkish-green, often edged with purple, in small clusters of 1-3; petals white, tiny, sepals reflexed. Berry greenish-yellow or purplish-red, 10-20mm, bristly. Woodland, scrub, hedgerows, streamsides and beneath old walls, on waste ground, to 1800m. March-March. Throughout W & C Europe, including Britain, but widely naturalized in Ireland, Iceland and much of N Europe. **P**: and wasps. The wild plant from which the culinary forms have been bred. **B**: Throughout Britain except the north; naturalized in Ireland.

3* **Mountain Currant** *Ribes alpinum* L. Deciduous shrub 1-2m tall. Leaves 3-lobed, generally longer than wide. Flowers yellowish-green, 4-6mm, male and female on separate plants, borne in upright racemes, each flower with a conspicuous bract below. Berry scarlet, hairless, insipid. Rocky woodland, cliffs and rocky stream margins, generally on limestone, to 1900m. April-May. Much of Europe, though confined to the mountains in the south of its range. **P**: Bees and wasps. The racemes are longer in the male than the female plant. Cultivated and sometimes, like other species of *Ribes*, escaping and becoming locally naturalized. **B**: Confined to N England and N Wales; rarely naturalized elsewhere.

4* **Blackcurrant** *Ribes nigrum* L. Shrub to 2m. Leaves generally 5-lobed, covered with small brownish glands beneath, strong smelling. Flowers reddish-green or brownish-green, bell-shaped, 7-8mm, in long drooping racemes; sepals recurved, hairy. Berry black when ripe, 12-15mm, sweetish and aromatic, with persistent sepals. Wet woodland, hedgerows, fen carr and stream banks, to 1500m. April-May. Throughout, but widely naturalized; possibly not truly native in much of NW Europe, but long established there. **P**: Bees, but often self-pollinated. **B**: Throughout Britain and Ireland, local, but most common in the south and east.

5 **Rock Redcurrant** *Ribes petraeum* Wulfen. Laxly branched shrub, 1-3m. Leaves generally larger than *R. rubrum*, often rather wrinkled. Flowers pinkish, bell-shaped, 4-5mm, in long drooping racemes of 20 or more flowers, sepals spreading, hairy on margin. Berries dark purplish-red, acid tasting. Mountain woods, rocky places and stream margins, to 2450m. April-June. C & E France and S Germany.

6* **Redcurrant** *Ribes rubrum* L. [*R. sylvestre*]. Rather erect shrub to 1.5m. Leaves generally 5-lobed, heart-shaped at base, not aromatic, scarcely hairy beneath. Flowers pale green, often tinged with purple, 5mm, in drooping racemes of up to 20 flowers, sepals spreading, hairless; anther-lobes diverging. Berries bright shiny red when ripe, 6-10mm, hairless, acid tasting. Wet woods and hedgerows, fen carrs, shaded rocks and streamsides. April-May. Probably native to France, Belgium and Holland, but widely naturalized throughout, except for the extreme north. **P**: bees and wasps.

7* *Ribes spicatum* Robson. Like *R. rubrum*, but leaves with a truncated or slightly heart-shaped base with the basal lobes far apart (with a narrow gap between them in *R. rubrum*). Flower somewhat larger, the anther lobes touching one another. Local, in woods and rocky places, on calcareous soils. April-May. Europe. The cultivated Redcurrant is believed to be of hybrid origin between *R. rubrum* and *R. spicatum*. **B**: Scattered localities in N England and Scotland; declining.

Grass of Parnassus

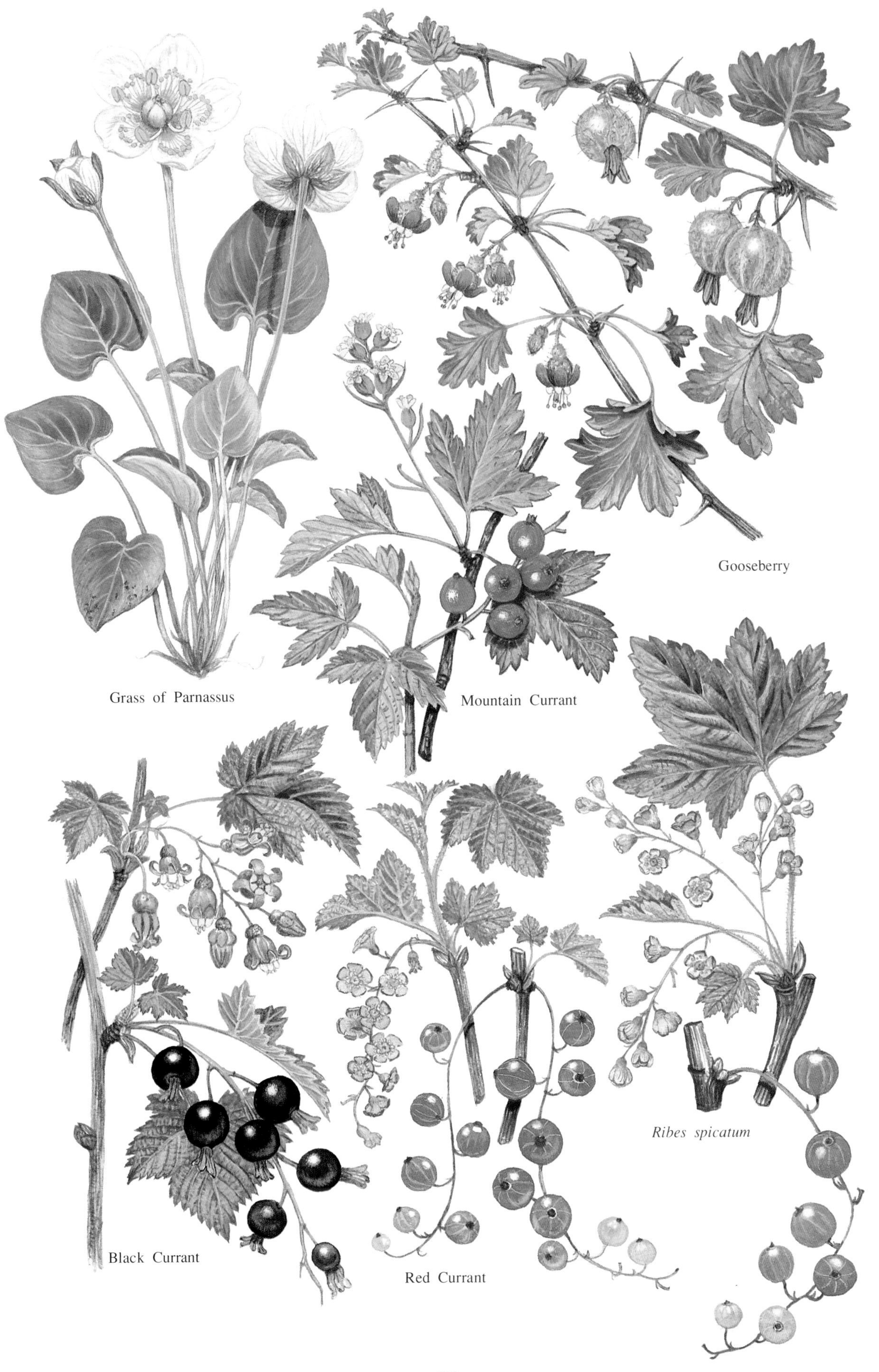
Gooseberry
Grass of Parnassus
Mountain Currant
*Ribes spicatum*
Black Currant
Red Currant

# ROSE FAMILY Rosaceae

A diverse family of trees, shrubs and herbs with alternate simple or compound leaves; stipules normally present. Flowers terminal, solitary or in racemes, cymes or panicles, often 5-parted, but generally with numerous stamens and few to many carpels; receptacle generally hollowed, but very variable, often with all the flower organs (except the carpels) attached to the rim. Fruit extremely variable from a capsule, to a collection of achenes, a drupe (plum and cherries) or a pome (apples and pears). Over 2000 species in 100 genera, widespread, but especially in temperate regions.

1* **Goatsbeard Spiraea** *Aruncus dioica* (Walter) Fernald. [*Actaea dioicus, A. vulgaris*]. Tall, stout, rhizomatous perennial; stems erect, up to 2m. Leaves large, 2-pinnate, with oval, toothed, leaflets. Flower small, 5mm, white or pale cream, in large pyramidal panicles, the branches slender and spike-like; dioecious. Damp shaded habitats, mountain woods, ditches, and marshy areas close to woods, generally on acid soils, to 1700m. May-August. Belgium, France and Germany; naturalised in Britain.

2* **Dropwort** *Filipendula vulgaris* Moench [*F. hexapetala, Spiraea filipendula*]. Medium to tall, tufted perennial; stems erect. Leaves pinnate with 8-25 pairs of oblong toothed leaflets, with smaller leaflets in between. Flowers pale cream, purplish beneath, 8-16mm, in dense flat-headed clusters; petals usually 6. Dry grassy habitats, meadows, pastures, roadsides and woodland margins, on calcareous soils, to 1500m. June-September. Throughout, except the far north. **B**: Widespread N to Yorkshire, but also on the east coast of Scotland as far as Angus; Ireland – only in Co. Clare.

3* **Meadowsweet** *Filipendula ulmaria* (L.) Maxim. [*Spiraea ulmaria*]. Similar to *F. vulgaris*, but a taller plant to 1.5m. Leaves with up to 5 pairs of large leaflets. Flowers fragrant, smaller, cream, 4-8mm; petals often 5. Damp habitats, marshes, fens and swamps, wet woodland, river and stream margins, on calcareous, neutral or somewhat acid soils, to 1500 m. June-September. Throughout, except Spitsbergen. **P**: The sickly-sweet scent of the flowers is attractive to flies. **B**: Throughout.

**Brambles, Blackberries** *Rubus*. Perennial herbs or shrubs, often prickly. Leaves simple, pinnate or digitate. Flowers 5-parted, white, pink or red with numerous stamens. Fruit a close head of fleshy segments or drupelets, (blackberry and raspberry). A very complex genus with some 250 species, but numerous segregates and microspecies. Many, like species of *Alchemilla*, reproduce apomictically, so that local variants are often maintained and can in time form extensive colonies.

4* **Cloudberry** *Rubus chamaemorus* L. Low to short creeping, hairy perennial; stems annual, not prickly. Leaves simple, rough, kidney-shaped, lobed and toothed. Flowers white, 15-20mm, solitary; male and female on separate plants. Fruit orange when ripe, edible, with up to 20 segments. Upland bogs, damp moors and tundra, generally in acid peaty places, to 1400m. June-August. N Britain, Ireland, Scandinavia and Germany. **B**: N Wales and Yorkshire north to Sutherland; Ireland – Tyrone, rare.

5 *Rubus humulifolius* C.A. Meyer. Like *R. chamaemorus*, but the stems often bristly and the leaves simple or ternate; flowers hermaphrodite. Fruit red when ripe, acid-tasting. Similar habitats and flowering time. Confined to Finland and the neighbouring regions of the USSR.

6* **Arctic Bramble** *Rubus arcticus* L. Short, creeping perennial; stems annual, without prickles, all flowering. Leaves 3-lobed or ternate. Flowers bright reddish-pink, 15-25mm, 1-3 on long stalks, with 5-7 petals, often slightly toothed. Fruit dark red when ripe, with many divisions. Grassy places, thickets and moors, to 1150m. June-July. Scandinavia; extinct in Britain.

7* **Stone Bramble** *Rubus saxatilis* L. Short to medium perennial with creeping vegetative, prickly stems, and upright annual flowering stems. Leaves ternate, with oval, toothed, leaflets. Flowers white, 8-10mm, in small clusters; petals narrow upright, sepals deflexed. Fruit red when ripe, with few large segments, edible. Shaded rocky and wooded habitats, screes, generally in damp places on calcareous or base-rich soils, to 2400m. June-August. Throughout, except Spitsbergen. **B**: Widespread in uplands but generally declining; rare or absent in much of E & SE England.

8* **Raspberry** *Rubus idaeus* L. Tall suckering perennial to 1.5m; stems biennial, unbranched, with weak prickles and a whitish bloom when young. Leaves pinnate, leaflets 5-7, oval, toothed, whitish beneath. Flowers white, 9-11mm, nodding, in small clusters. Fruit red or orange when ripe, edible. Open woodland, heaths, moors, commons, railway embankments, waste place, on light often sandy soils, to 2300 m. May-August. Throughout, except the Faeroes, Iceland and Spitsbergen, but only on the mountains in the south of its range. **B**: Widespread.

*Rubus spectabilis* Pursh. Deciduous shrub to 2m with biennial prickly stems. Leaves trifoliate, with oval, toothed leaflets. Flowers large, bright purple, 22-30mm, solitary. **I**: North America. Naturalized in Britain, Ireland, France, Holland and Germany. June-July. **B**: Extensively naturalised in N England & Scotland.

9* **Blackberry, Bramble.** *Rubus fruticosus* L. Complex large aggregate with hundreds of described species, mostly having arisen by apomixis and polyploidy. A very variable scrambling shrub with long arching thorny, angled stems, often rooting down at the tip and forming dense tangles; stems biennial. Leaves trifoliate or digitate, prickly; leaflets round to elliptical, toothed. Flowers white or pink, 20-32mm. Fruit red at first, purplish-black when ripe, edible. Many habitats from woodland to waste ground. May-November. Throughout. **B**: Throughout.

10* **Dewberry** *Rubus caesius* L. Weak sprawling subshrub; stems round, slightly prickly, unridged, with a whitish bloom. Leaves trifoliate, white, occasionally pink, 20-25mm, in small clusters; sepals pressed close to fruit. Fruit bluish-black, with a waxy bloom, edible. Damp habitats, usually over limestone, rough grassland and scrub, to 1200m. May-September. Throughout, except the Faeroes, Iceland and Spitsbergen. Hybridises with *R. fruticosus* aggregate – the progeny generally have large waxy fruits. **B**: Throughout except the extreme north, but scarcer in W and Ireland.

Goatsbeard Spiraea
× 1/10
Dropwort
Meadowsweet
Cloudberry
Arctic
Bramble
Stone Bramble
Raspberry
Blackberry
Dewberry

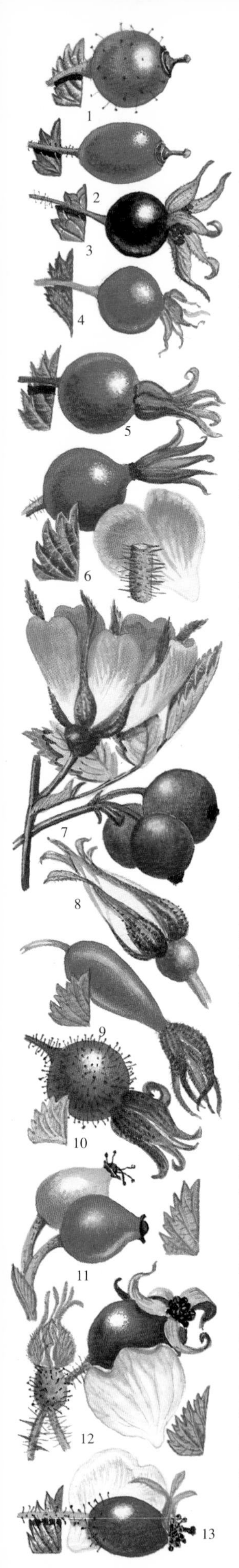


**Roses, Briars** *Rosa*. Prickly shrubs, usually deciduous. Leaves pinnate, leaflets often 5-7, toothed; stipules present, fused to the base of the leaf-stalk. Flowers terminal, solitary or clustered. Sepals 8. Petals usually 5. Stamens numerous. Styles protruding slightly, sometimes fused into a short column. Fruit a rounded to pear-shaped hip, with or without sepals attached at the top. 250 species in temperate N Hemisphere, also a few on tropical mountains.

1* *Rosa sempervirens* L. Evergreen, with long creeping stems covered in sparse curved prickles. Leaflets, thick, hairless and shiny. Flowers white, 25-35mm, in clusters of 3-7; sepals long-pointed, unlobed, falling; styles united into a column. Hip rounded, red when ripe. Scrub and dry banks. June-July. S France. Widespread in S Europe.

2* **Field Rose** *Rosa arvensis* Hudson. Deciduous scrambling shrub to 3m, usually with weak trailing green stems; prickles sparse, hooked. Leaflets oval to elliptical, dull green above, usually hairless. Flowers always white, 30-50mm, solitary or 2-3; sepals deflexed, soon falling; styles joined into a long column. Hip oval, red, smooth. Hedgerows, woodland margins, scrub, on rather heavy soils, to 2000m. June-August. W Europe. Generally scrambling among other shrubs and small trees. **B**: Widespread in England, Wales and Ireland; occasionally naturalized in Scotland.

3* **Burnet Rose** *Rosa pimpinellifolia* L. [*R. spinosissima*]. Suckering deciduous shrub to 1m; stems erect, with straight prickles mixed with bristles. Leaflets 5-11, small, hairless. Flowers white, rarely pink, 20-40mm, solitary; sepals unlobed. Hip small, globose, black when ripe. Dry open habitats, limestone pavement, stabilized sand-dunes and dune slacks, heaths, often on calcareous soils, to 2000m. May-July. Throughout, except for parts of N Europe, extinct in Sweden, absent from Finland. Hybrids with *R. canina* and its allies have a mixture of hooked prickles and stiff bristles along the stems. **B**: Throughout, except for the Shetland Is.

4* **Yellow Briar** *Rosa foetida* J. Herrmann. Rather like *R. pimpinellifolia*, but leaves with only 5-7 leaflets and yellow flowers. Hip red. Naturalized in dry sunny habitats in rocky places and scrub. June-July. **I**: SW Asia. Naturalized in France and Germany.

5* **Cinnamon Rose** *Rosa majalis* J. Hermann [*R. cinnamonea*]. Deciduous shrub to 2m, forming large patches; bark reddish-brown with slender prickles in pairs at the nodes. Leaflets bluish-green, elliptical, hairy. Flowers purplish-pink, 30-50mm, solitary, on hairless stalks. Hips rounded, red, smooth. Scrub, woodland margins, hedgerows and banks, to 2200m. June-July. Continental Europe, but naturalized in Belgium, Holland and Denmark; not in Sweden.

6 *Rosa acicularis* Lindley. Like *R. majalis*, but stems with long straight prickles mixed with stiff bristles. Similar habitats and flowering time. Finland and Sweden.

7 *Rosa glauca* Pourret [*R. rubrifolia*]. Like *R. majalis*, but young stems with a whitish bloom and flowers usually in clusters of 2-5. Hips reddish-brown when ripe. Woodland margins and scrub in mountains to 2100m. June-July. C & S France and Germany; naturalized in Finland and Sweden.

8* *Rosa rugosa* Thunb. Like *R. acicularis*, but stem densely bristly and prickly as well as downy. Leaflets, rough, deep shiny green. Flowers larger, 40-70mm, bright purplish-pink, occasionally white. Hip large and pendent, pumpkin-shaped. Hedgerows, scrub. June-August. **I**: N China and Japan. Widely naturalized. Commonly planted for hedging, and on Continental motorways.

9 **Alpine Rose** *Rosa pendulina* L. Rather like *R. majalis*, but stems smooth and leaves double-toothed (biserrate). Hips oval or pear-shaped, pendent. Mountain habitats, scrub, banks and rocky places, occasionally in hedgerows, to 2600m. June-August. Belgium, France and Germany.

10* **Provence Rose** *Rosa gallica* L. Deciduous shrub to 0.8m, forming large patches; stems with straight prickles and glandular-bristles. Leaflets 3-7, leathery, dull bluish-green, hairy beneath. Flowers deep pink, large, 60-90cm, fragrant. Hip rounded to spindle-shaped, bristly, bright red. Dry open habitats, to 2200m. May-August. Belgium to C France and Germany.

ROSA CANINA group. Deciduous shrubs; stems armed with stout curved or hooked prickles; leaves not leathery; hairy or hairless, scarcely glandular and not markedly scented.

11* **Dog Rose** *Rosa canina* L. Stems stout, arching, green, to 5m. Leaflets hairless, oval to elliptical, dark green or bluish-green, shiny or dull above. Flowers pink or white, 45-50mm, in cluster of 2-5, on hairless stalks; styles not joined into a column. Hip rounded to elliptical, smooth, red, without sepals when ripe. Hedgerows, woodland margins, roadsides, rough grassy places, to 2200m. June-July. Throughout, except the far north. Often the commonest wild rose. **B**: Throughout.

12 *Rosa jundzillii* Besser. Rather erect deciduous shrub to 2m; stems with slender straight or slightly curved prickles. Leaves rather leathery, with 5-7 elliptic or oval leaflets, hairless, but with glands along the margin and on the veins beneath. Flowers pale to deep pink, 30-50mm, usually solitary, faintly scented. Hip rounded to egg-shaped, red, with a few bristles, without sepals when ripe. Dry open and rough grassy places. June-July. C & E France and S Germany. C & E Europe.

*Rosa squarrosa* (Rau) Boreau. Like *R. canina*, but leaflets glandular beneath; glands also present on the leaf-stalks and stipules. Similar habitats and flowering time. Britain, Belgium, Holland, France and Germany southwards.

*Rosa andegavensis* Bast. Like *R. squarrosa* and *R. canina*, but flower-stalks with stalked glands and the hips often bristly. Similar habitats, distribution and flowering time.

*Rosa nitidula* Besser. Similar, but leaflets double-toothed (biserrate), the teeth glandular. Habitat and distribution similar.

13* *Rosa stylosa* Desv. Stout deciduous shrub to 3m tall; stems with hooked prickles. Leaves with 5-7 elliptical, sharply toothed leaflets, hairy at least on the veins beneath, but not glandular. Flowers white, occasionally pink, 30-60mm, solitary or in clusters, the styles united into a distinct column. Hips egg-shaped, red, smooth, without sepals. Hedgerows, woodland margins, rough grassy banks, to 2000m, sometimes clambering into low trees. June-July. C & S Britain, Ireland, France and S Germany.

*R. arvensis* has similar flowers but is a slighter bush with weak trailing stems and slender prickles, and its hips bear persistent sepals. **B**: Denbigh and Leicestershire. southwards; in Ireland mostly in W & S.

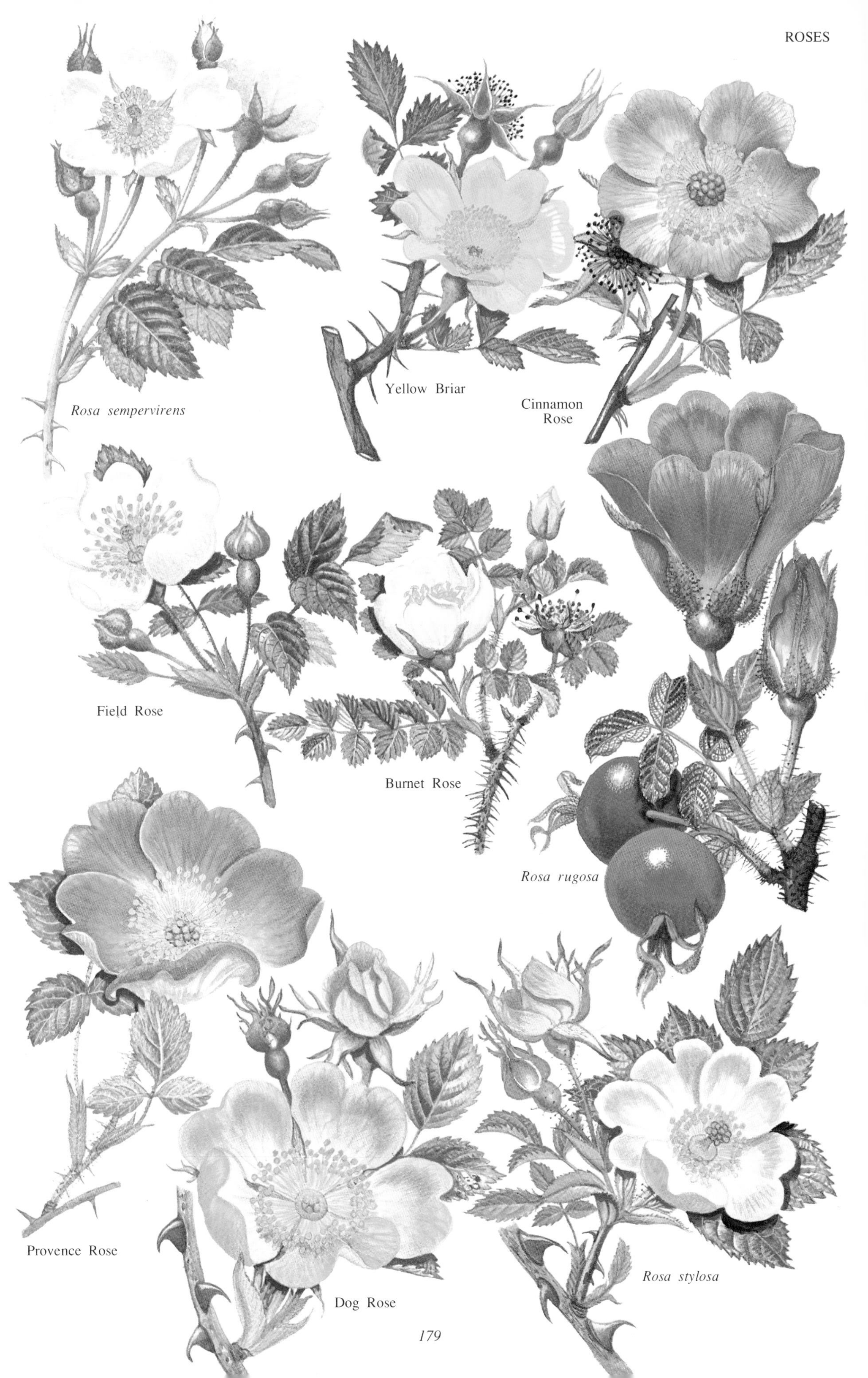

*Rosa sempervirens*
Yellow Briar
Cinnamon Rose
Field Rose
Burnet Rose
*Rosa rugosa*
Provence Rose
Dog Rose
*Rosa stylosa*

1 **Whitish-stemmed Briar** *Rosa vosagiaca* Desportes [*R. afzeliana* subsp. *vosagiaca*]. Like *R. canina* (p. 178), but stems reddish, with a whitish bloom when young. Leaflets hairless, usually bluish-green. Sepals erect or spreading, persistent on the ripe hips. Scrub, banks and lightly wooded areas, to 2200m. June-August. Continental Europe, except the extreme north.

*Rosa subcanina* (Christ) Dalla Torre & Sarnth. [*R. glauca* subsp. *subcanina*]. Like *R. vosagiaca*, but flowerstalks longer, 20-30mm (not 2-20mm); sepals often deflexed on the ripe hips. Scrub, hedgerows and woodland margins. June-July. Continental Europe N to S Norway.

2* *Rosa obtusifolia* Desv. [*R. tomentella*, *R. klukii*]. Like *R. canina*, but stems up to 2m and leaflets softly hairy on both surfaces, or just beneath. Flowers white or pale pink, 20-35mm, on hairless stalks; sepals reflexed at first. Hip rounded to oval, smooth, red, without sepals when ripe. Scrub, banks and rough grassy habitats. June-August. Britain, Ireland, Belgium, Germany, Denmark and S Sweden. **B**: N to Northumberland and Ireland.

3 *Rosa corymbifera* Borkh. [*R. dumetorum*]. Like *R. obtusifolia*, but leaflets broader, simply (not double) toothed. Similar habitats and flowering time. Throughout, except for the extreme north.

*Rosa deseglisei* Boreau. Like *R. obtusifolia*, but flowerstalks bristly-glandular. Similar habitats and flowering time. Britain, France, Germany, S Finland.

ROSA TOMENTOSA group. Like *R. canina* group, but leaves always densely downy and with a resinous smell; prickles straight or slightly curved.

4* **Downy Rose** *Rosa tomentosa* Sm. Compact shrub to 2m, with arching stems; young stems pale green. Leaflets soft, oval to lanceolate. Flowers pink or white, 30-40mm, usually in groups of 2-5, borne on bristly-glandular stalks. Hip rounded to pear-shaped, red, usually bristly, without sepal when ripe. Hedgerows, scrub, woodland margins and railway embankments, generally in hilly regions, to 2200m. June-July. Much of Europe, except the far north. **B**: Throughout, but rather scarce in Scotland.

*Rosa scabriuscula* Sm. Like *R. tomentosa*, but taller, to 3m. Leaflets rough. Hips with persistent sepals. Similar habitats and flowering time. Throughout but absent from Ireland.

5* *Rosa sherardii* Davies [*R. omissa*]. Like *R. scabriuscula*, but stems with a whitish bloom when young; leaves bluish-green. Flowers always pink. Hedgerows, scrub and open woodland. June-July. Europe N to S Sweden. **B**: Mainly confined to N England and Scotland, scarce elsewhere.

6* **Apple Rose** *Rosa villosa* L. [*R. pomifera*]. Like *R. tomentosa*, but shrub rarely exceeding 1.5m; stems erect, not arching, armed with slender straight prickles. Leaflets with a resinous-apple scent. Flowers pink, 30-45mm, solitary or clustered. Hip rounded or pear-shaped, bristly, with persistent erect sepals when ripe. Scrub, hedgerows, woodland margins, banks and rough grassy habitats, to 2200m. June-July. Belgium, Holland, France and Germany southwards; naturalized in Scandinavia. Sometimes cultivated in gardens.

7* *Rosa mollis* Sm. Like *R. villosa*, but stems with a whitish bloom when young. Flowers deep pinkish-red. Hips smooth or sparsely bristly. Hedgerows, woodland margins and scrub on mildly acid or calcareous soils. June-July. Europe except the extreme north. **B**: Mainly in N and Wales, rather scarce elsewhere.

ROSA RUBIGINOSA group. Leaflets thin, distinctly sticky with brown glands beneath, smelling of apples. Sepals pinnately lobed, erect or spreading.

8* **Sweet Briar** *Rosa rubiginosa* L. [*R. eglanteria*]. Deciduous shrub to 3m, with erect stems, armed with stout curved or hooked prickles mixed with bristles. Leaflets 5-7, rounded to oval, yellowish-green, often tinged with brown. Flowers fairly small, deep pink, 18-28mm, solitary or 2-3, on bristly stalks. Hip rounded to elliptic, bright red, smooth or bristly, with persistent sepals. Rough grassy habitats, scrub, bank and coastal habitats, especially stabilized shingle, but rather rare in hedgerows, to 2100m. Throughout, except the extreme north and Iceland. Differs from both the Canina and Tomentosa Groups in its densely glandular leaves which have a strong apple smell when crushed. **B**: Widespread S of the Wash; scattered localities elsewhere.

9 *Rosa elliptica* Tausch [*R. graveolens*]. Like *R. rubiginosa*, however, stems prickly, but without bristles and with smooth flower-stalks. Scrub, hedgerows and woodland margins. June-July. Belgium, France and Germany.

10* **Field Briar** *Rosa agrestis* Savi [*R. sepium*]. Like *R. rubiginosa*, but rarely exceeding 2m. Leaflets dull green, narrowed (not rounded) at the base. Flowers white, 22-38mm, on hairless stalks usually. Hips smooth, rounded to elliptic, red and without sepals when ripe. Hedgerows, woodland margins and banks, scrub and thickets, on calcareous soils, to 2200m. June-July. Throughout, except Iceland, the Faeroes and much of Scandinavia. **B**: Very rare; scattered localities in England and C Ireland.

11* *Rosa micrantha* Borrer ex Sm. Like *R. agrestis*, but leaflets rounded at the base. Hips smooth or bristly, on bristly stalks. Similar habitats and flowering time, also on heaths and in rough pasture on lighter, non-calcareous soils. Britain, Ireland, Belgium, Holland, France and Germany southwards. **B**: Widespread in England and Wales, scarce elsewhere.

Downy Rose

Rosa obtusifolia
Downy Rose
Rose sherardii
Apple Rose
Rosa mollis
Sweet Briar
Field Briar
Rosa micrantha

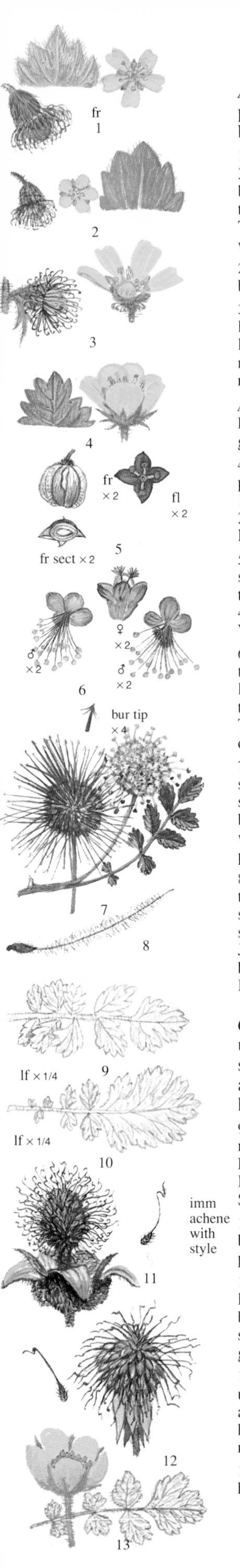


*Agrimonia.* Perennial rhizomatous herbs with erect, glandular-hairy stems. Leaves irregularly pinnate. Flowers in terminal spikes, 5-parted; stamens 5-20. Fruit small, 2-parted, with hooked bristles. 15 species in N temperate regions.

1* **Agrimony** *Agrimonia eupatoria* L. Short to medium, hairy perennial. Basal leaves in a rosette, with 3-6 pairs of main leaflets, with smaller leaflets inbetween, dark green above, whitish- or greyish-downy beneath. Flowers golden-yellow, 5-8mm, in slender spikes. Fruit 7-10mm, grooved, with erect hooks at the top. Dry grassy habitats and field borders, on mildly acid to calcareous soils, to 1800m. June-August. Throughout, except the far north. The fruits, as in the other species, are readily detached from the plant when ripe and easily attach themselves to animals fur, clothing, etc. **B**: Throughout, scarcer in the north.

2* *Agrimonia pilosa* Ledeb. [*A. dahurica*]. Like *A. eupatoria*, but taller, to 1.5m; leaves green above and beneath. Flowers pale yellow, 3-5mm. Similar habitats and flowering time. S Finland.

3 **Fragrant Agrimony** *Agrimonia procera* Wallr. [*A. odorata*]. Like *A. eupatoria*, but often taller, the leaves stickily-hairy beneath, the lower not in basal rosettes and flowers fragrant. Fruit grooved only in the lower part, the lower bristles bent backwards. Scrub, woodland margins and rough grassy habitats, on moist, generally heavy, but rarely calcareous soils, to 1800m. June-August. Throughout, except the far north. **B**: Throughout, except N Scotland.

*Agrimonia repens* L. [*D. anglica*]. Like *A. eupatoria*, but flowers larger, 8-10mm, the lower bracts unlobed. Fruit like *A. procera*. **I**: Turkey. Naturalized in grassy places and scrub, Belgium and Germany. June-August.

4* **Bastard Agrimony** *Aremonia agrimonoides* (L.) DC. [*Agrimonia agrimonoides*]. Short, rather slender perennial. Leaves mostly trifoliate, the lower in a basal rosette; leaflets oval, toothed. Flowers yellow, 7-10mm, in lax, branched, clusters; stamens 5-10. Flowers sometimes cleistogamous and not opening. Fruit 2-parted, not bristly. Mountain woods and scrub, to 1900m. May-July. S Germany; naturalized in N Britain. **B**: Locally naturalized in C Scotland.

5* **Great Burnet** *Sanguisorba officinalis* L. [*S. polygama*]. Medium to tall, tufted, hairless perennial; stems erect, branched. Leaves pinnate, the lower with 3-7 pairs of oval leaflets, greyish beneath. Flowers tiny, dull crimson, in dense oval heads, 10-30mm long, with both stamens and styles, but no petals; stamens 4. Damp habitats on base-rich soils, to 2300m. June-September. Throughout, but local in the north. **B**: Widespread in England and Wales, but local in the SW; rare elsewhere.

6* **Salad Burnet** *Sanguisorba minor* Scop. [*S. gaillardotii*, *Poterium sanguisorba*]. Short, rather greyish, tufted perennial, slightly hairy or hairless. Leaves mostly basal, pinnate, with 3-12 pairs of rounded or elliptical leaflets. Flowerheads globose, 10-20mm, upper flowers with reddish styles, lower with yellow anthers. Dry grassy and rocky habitats, especially over chalk and limestone, to 2200m. May-September. Throughout, except the far north. The young leaves are edible.**B**: Widespread in England and Wales; local or rare in Scotland and Ireland.

7 **Pirri-pirri Bur.** *Acaena novae-zelandiae* Kirk [*A. anserinifolia*]. Low creeping, hairy subshrub with short erect leafy stems. Leaves pinnate, with 3-5 deeply toothed oblong leaflets. Flowers greenish-white, small, but in dense, long-stalked, globose heads, 5-10mm, each flower with 4 long soft spines, reddish-brown and barbed at the tip. June-July. **I**: SE Australia and New Zealand, where in places it is a serious weed. Naturalized from gardens in dry open places, paths and coastal sand-dunes. Britain and Ireland, especially along the coast. Plants form tough mats which are remarkably resistant to trampling.

8* **Mountain Avens** *Dryas octopetala* L. Low carpeting, downy evergreen subshrub. Leaves oblong, toothed, leathery, deep green above, but white-felted beneath. Flowers white, 20-40mm, solitary, long-stalked; petals often 8, occasionally more. Fruit a head of achenes, each with a long persistent feathery style. Rocks and rocky ledges, sea cliffs and sand-dunes, on neutral or calcareous rocks, to 2500m. May-July. N Britain, Ireland, Scandinavia and the mountains of France and Germany. An Arctic-alpine species, but descending to sea-level in the northW of its range. **B**: Widespread but fairly local in Scotland and W Ireland (the Burren); rare in N England and Wales.

*Geum.* Perennial herbs with pinnate leaves. Flowers solitary or in branched clusters, 5-parted usually. Petals clawed; stamens and carpels numerous. Fruit a head of achenes each with a persistent feathery, often hooked, style. 40 species widespread in the north and south temperate areas and in the Arctic. The feathered fruits are dispersed by the wind as in *Dryas*, but several have hooked or barbed achenes which attach themselves to fur and clothing.

9* **Creeping Avens** *Geum reptans* L. [*Sieversia reptans*]. Low to short, spreading, hairy perennial, with reddish rooting runners. Patch-forming. Leaves pinnate, mostly in basal rosettes, leaflets deeply toothed. Flowers bright yellow, 25-40mm, solitary, usually 6-petalled. Achenes with feathery styles, not hooked. Mountain habitats, meadows, rocks, gravelly places and moraines, 1500-2800m. July-August. E France and S Germany. **P**: Bees and butterflies.

10* **Alpine Avens** *Geum montanum* L. [*Sieversia montana*]. Like *G. reptans*, but without runners and basal leaves with a large end leaflet. Flowers golden-yellow, petals often slightly notched. Rocky mountain pastures. June-August. E France and S Germany; mountains of C & S Europe.

11* **Water Avens** *Geum rivale* L. Short, hairy, tufted perennial. Leaves with 3-6 pairs of rounded, deeply-lobed leaflets; stem leaves usually trifoliate. Flowers, nodding bells, cream to dull pink with a purplish-brown calyx, 8-15mm, borne in lax, branched, clusters. Achenes with a hooked, feathery style. Moist, often shaded habitats, on calcareous or base-rich soils, to 2100m. April-September. Throughout, except Spitsbergen. **B**: Throughout; commonest in the north.

12* **Herb Bennet** *Geum urbanum* L. Medium, hairy perennial; stems erect. Basal leaves with 1-5 pairs of unequal, deeply-lobed, toothed leaflets; stem leaves 3-lobed or pinnate with a pair of large leaf-like stipules at the base of each. Flowers pale yellow, 8-15mm, erect, in branched clusters. Achenes hairy, with feathery, hooked styles. Woods, and other shaded habitats, to 1850m. May-September. Throughout, except the far north. Hybridizes with *G. rivale*: *G. × intermedium* Ehrh. **B**: Throughout, except parts of the north.

13 *Geum hispidum* Fries. Like *G. urbanum*, but rarely as tall, with smaller stipules (less than 10mm) and hairless achenes. Damp grassy and wooded habitats. June-July. SE Sweden. Very local.

Bastard
Agrimony
Salad Burnet
Great Burnet
Agrimony
*Agrimonia pilosa*
Water Avens
Alpine Avens
Mountain Avens
Creeping Avens
Herb Bennet

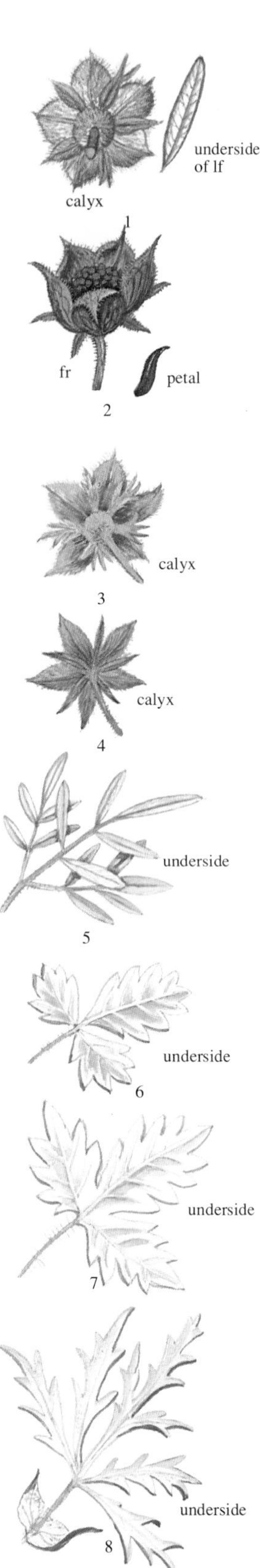


**Cinquefoils** *Potentilla.* Herbs, usually perennial, or small shrubs. Leaves digitate, pinnate or trifoliate. Flowers solitary or in branched clusters, 4-6-parted, usually 5-parted; epicalyx present. Stamens 10 to many. Fruit a head of single-seeded achenes. A complex genus with about 500 species scattered in temperate regions, especially in the northern hemisphere.

1* **Shrubby Cinquefoil** *Potentilla fruticosa* L. [*Dasiphora fruticosa*]. Much-branched deciduous shrub to 1m, with slender, often rather contorted branches, covered in the remains of old leaf-bases. Leaves greyish, pinnate, generally with 5 elliptical, untoothed, leaflets. Flowers yellow, 20mm, solitary or several together; petals rounded, not notched. Rocky habitats, damp hollows and river banks, moraines and scrub, on calcareous soils, to 2600m. May-July. N Britain, Ireland, Norway, Sweden and S France. **P**: Bees and butterflies. Widespread from Europe to Asia, including the Himalaya. In some parts of its range, including Britain, plants are dioecious, with male and female flowers on separate plants. Many forms are grown in gardens. **B**: Rare, confined to scattered localities in Teesdale, the Lake District and W Ireland.

2* **Marsh Cinquefoil** *Potentilla palustris* (L.) Scop. [*Comarum palustre*]. Short to medium, slightly hairy, rhizomatous perennial. Leaves pinnate, with 5-7 oblong, toothed, leaflets. Flowers maroon or purplish, 20-30mm, star-shaped; sepals much larger than the linear petals. Wet habitats, meadows, marshes, fens, bogs, ditches and dykes, to 2100m. May-July. Throughout. Locally common. **P**: Various insects.The stipules are conspicuous and papery. **B**: Widespread in N Britain and Ireland, local elsewhere.

3* **Silverweed** *Potentilla anserina* L. Short hairy perennial with long, creeping and rooting, runners. Leaves in tufts, pinnate, with 15-25 oblong, sharply toothed leaflets, silvery-hairy beneath. Flowers yellow, 15-20mm, solitary, the petals twice the length of the sepals. Damp and grassy habitats, meadows, pastures, pathways, roadsides, open woodland, farmtracks, waste and disturbed ground, sand-dunes, to 2400m. May-August. Throughout. Tolerant of trampling and often a weed of recently abandoned cultivated land and gardens. **P**: Various insects or self-pollinated. **B**: Common throughout.

subsp. *egedii* (Wormsk.) Hiitonen [*Potentilla egedii*]. Like the type, but leaves with 7-15 pairs of leaflets, not silvery beneath. Coastal habitats. June-September. Sweden and Finland.

4* **Rock Cinquefoil** *Potentilla rupestris* L. [*P. corsica*]. Medium, hairy perennial, without runners. Basal leaves pinnate with 5-7 oval to rounded, toothed leaflets; stem leaves usually trifoliate. Flowers white, 16-28mm, solitary or laxly clustered. Rocky habitats, slopes and ledges, open woodland, often on basic rocks, to 2200m. May-June. Europe north to Norway and S Sweden. **P**: Normally self-pollinated. Distinctive on account of its white flowers and pinnate leaves. **B**: Very rare, confined to isolated localities in Wales and Scotland.

5* **Cut-leaved Potentilla** *Potentilla multifida* L. (incl. *P. lapponica* (F. Nyl.) Juz.). Short, hairy perennial. Leaves 2-pinnate, with 5-9 deeply lobed leaflets, the lobes linear, green, but grey-silky beneath. Flowers yellow, 10-14mm, numerous; petals slightly longer than the sepals. Meadows and rocky habitats, generally on acid soils, 1000-3000m. Sweden, Spitsbergen, E & S France; naturalized in Finland.

6* **Snowy Cinquefoil** *Potentilla nivea* L. Low to short tufted perennial, usually with downy-white stems. Leaves trifoliate, the leaflets oval, toothed, downy-white beneath. Flowers yellow, 12-18mm, in lax-branched, clusters; petals slightly longer than the sepals. Rocky places, cliffs, screes and moraines, usually on limestone, 1600-2600m. June-August. Scandinavia and Spitsbergen.

7 *Potentilla chamissonis* Hultén. Like *P. nivea*, but a taller plant, the leaves sometimes with 5 leaflets, each leaflet pinnately lobed. July-August. Arctic Europe.

8* **Hoary Cinquefoil** *Potentilla argentea* L. Short to medium, downy perennial, with spreading to ascending stems. Leaves digitate, with 5 narrow oval, deeply toothed or lobed leaflets, deep green above, silvery-white, felted beneath. Flowers yellow, 10-12mm, numerous in branched clusters; carpels pale yellow. Rocky stony and grassy habitats, road verges, tracksides, abandoned cultivation and coastal habitats, to 1950m. June-September. Throughout, except the extreme north, Ireland and Iceland; mainly on the mountains in the south of its range. Throughout most of Europe. **B**: Local in England and S Scotland, scarce in SW England and Wales.

9 *Potentilla neglecta* Baumg. Like *P. argentea*, but taller and usually erect, the leaves grey-green above; flowers larger, 12-16mm, with dark yellow or orange-yellow carpels. Similar habitats and flowering time. Throughout, except most of the smaller islands, but because of confusion with *P. argentea* the precise distribution is not known.

Marsh Cinquefoil

Shrubby Cinquefoil
Marsh Cinquefoil
Silverweed
Rock Cinquefoil
Cut-leaved Potentilla
Snowy Cinquefoil
Hoary Cinquefoil

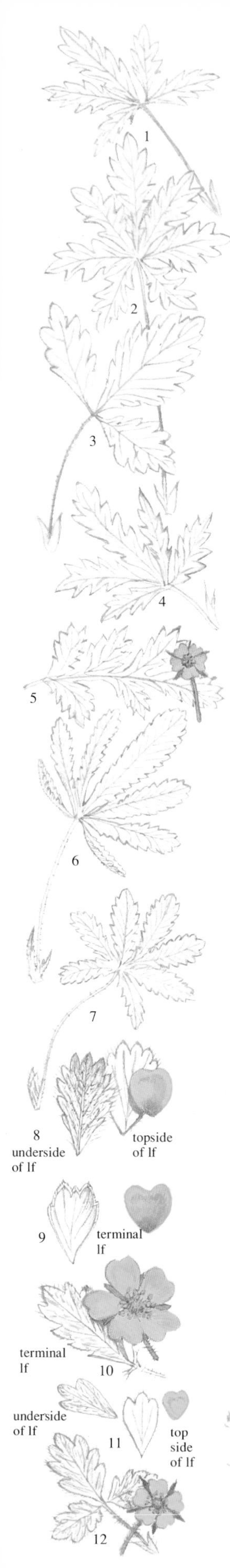


1* *Potentilla inclinata* Vill. [*P. canescens*]. Short to medium, more or less erect, hairy perennial. Leaves digitate, with 5-7 oblong, sharply toothed or pinnately lobed, leaflets, grey-hairy beneath. Flowers yellow, 11-15mm, numerous in branched clusters, petals slightly longer than the sepals. Rocky and grassy habitats. June-August. C & E France and S Germany. Central and southern Europe. **B**: Naturalised in a few places.

2* *Potentilla collina* Wibel. Like *P. inclinata*, but less tall and with more pronounced basal leaf rosettes. Leaves grey or silvery-silky beneath. Flowers 8-12mm, the petals equalling or slightly longer than the sepals. Similar habitats and flowering time. France, Denmark, Germany and S Sweden.

3* **Norwegian Cinquefoil** *Potentilla norvegica* L. Short to medium, hairy annual or short-lived perennial. Leaves green, mostly trifoliate; leaflets elliptical, toothed or lobed. Flowers rather bright yellow, 10-15mm, in branched clusters; petals equal to, or shorter than, the sepals which enlarge in fruit. Grassy, rocky and waste places, roadsides. June-September. Scandinavia and Germany; naturalized in Britain, Belgium, France and Holland. Mainly in N Europe and the neighbouring parts of the USSR. **B**: Scattered localities, except for N Scotland.

4 *Potentilla intermedia* L. [*P. heidenreichii*]. Like *P. norvegica*, but biennial or perennial, with 5 leaflets per leaf; petals equalling or slightly longer than the sepals. **I**: N and C USSR. Naturalized or casual in grassy places and waste ground in much of Europe.

5 *Potentilla supina* L. Like *P. norvegica*, but a lower plant with pinnate leaves, each with 5-11 oblong, sharply toothed leaflets. Flowers small, 6-8mm, the petals shorter than the sepals. Dry grassy habitats, meadows, banks and roadsides. June-August. Belgium, Holland, France and Germany southwards.

6* **Sulphur Cinquefoil** *Potentilla recta* L. Short to medium, stiffly erect, hairy perennial. Leaves digitate, with 5-7 oblong, coarsely-toothed to pinnately lobed leaflets, grey or greyish. Flowers pale yellow, 15-25mm, in lax, branched clusters; petals equalling or slightly longer than the sepals. Dry grassy and waste habitats, meadows, banks, woodland margins, roadsides and field boundaries. June-September. C & E France and Germany southward; naturalized in Britain and parts of N Europe. Cultivated in gardens and occasionally escaping. **B**: Scattered localities in England N to Humberside, rare elsewhere.

7* **Thuringian Potentilla** *Potentilla chrysantha* Trev. [*P. thuringiaca*, *P. goldbachii*, *P. nestlerana*]. Short to medium, very hairy perennial. Leaves digitate, with 5-9 oblong, coarsely-toothed leaflets. Flowers yellow, 15-22mm, in lax, branched clusters; petals longer than the sepals. Meadows and rocky habitats, to 2500m. June-August. E France and S Germany; naturalized in several places in Scandinavia.

8* **Alpine Cinquefoil** *Potentilla crantzii* (Crantz) G. Beck ex Fritsch [*P. alpestris*, *P. salisburgensis*]. Very variable, low to short, hairy perennial with a woody stock, not mat-forming. Leaves digitate, with usually 5, rounded to oblong, toothed leaflets, the end tooth equalling those adjacent to it; leaflets hairy beneath and along margins. Flowers yellow, 10-25mm, the petals often with an orange spot near the base, few in lax clusters. Open rocky habitats, rock ledges and crevices, occasionally on rocky grassland, generally on calcareous rocks, to 3000 m. N Britain, Iceland, Scandinavia and the mountains of France and Germany. An Arctic-alpine species, occasionally cultivated in gardens. The inflorescence stands up well above the foliage. **B**: Very local in N England and Scotland; rare in N Wales.

9* *Potentilla aurea* L. Like *P. crantzii*, but more mat-forming, the leaflets with a silky margin and a small end tooth; petals often with an orange blotch near the base. Rocky and grassy mountain habitats, on acid soils, 1400-2600m. June-August. Mountains of France and Germany.

10 *Potentilla hyparctica* Malte [*P. emarginata*]. Like *P. crantzii*, but leaves often glandular as well as hairy; no orange on the petals. Arctic habitats. July-August. Sweden and Spitsbergen.

11* **Dwarf Cinquefoil** *Potentilla brauniana* Hoppe [*P. dubia*, *P. minima*]. Low tufted, slightly hairy, perennial. Leaves small, trifoliate, the leaflets oval, slightly toothed near the tip, hairless above. Flowers yellow, small, 7-11mm, solitary of 2-3 together; petals equalling or longer than the sepals. Grassy and rocky mountain habitats, often by late lying snow patches, 1800-3150m. E France and S Germany – Alps and the Jura Mts.

12 *Potentilla frigida* Vill. Like *P. brauniana*, but a taller, hairier plant, the leaflets glandular, hairy above; petals shorter than or equalling the sepals. Acid rocks, 2400-3700m. E & S France and S Germany.

Dwarf Potentilla

*Potentilla inclinata*
*Potentilla collina*
Norwegian Cinquefoil
Sulphur Cinquefoil
Dwarf Cinquefoil
Thuringian Potentilla
Alpine Cinquefoil
*Potentilla aurea*

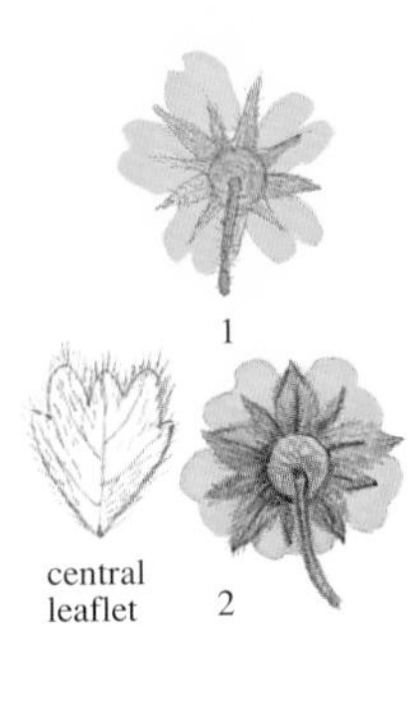


1* *Potentilla heptaphylla* L. [*P. opaca*, *P. rubens*]. Short, softly-hairy perennial, not mat-forming, with erect stems; hairs often reddish. Leaves digitate, with 5-7 oval, toothed leaflets. Flowers yellow, 10-15mm, usually in small lax clusters. Dry grassland, meadows, banks, roadsides, generally at low altitudes and on calcareous soils. April-July. France, Germany, Denmark and S Sweden.

2* **Spring Cinquefoil** *Potentilla tabernaemontani* Ascherson [*P. verna* subsp. *vulgaris*]. Variable, low to short, hairy, mat-forming perennial; stems rooting freely at the nodes, becoming woody near the base. Leaves digitate, with 5-7 wedge-shaped, toothed leaflets. Flowers yellow, 10-20mm, in small lax clusters; petals longer than the sepals. Dry grassy and rocky habitats, often of south-facing aspect, generally on calcareous soils, from low altitudes, but up to 3000m in the south of its range. April-June. Europe, except the extreme north. Sometimes confused with *P. crantzii*, but differs by forming spreading mats with the inflorescence scarcely longer than the leaves and with narrow, rather than broad, stipules. **B**: Very local from Wiltshire north to S Scotland.

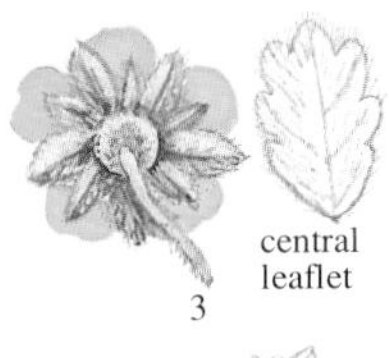


3* **Grey Cinquefoil** *Potentilla cinerea* Chaix ex Vill. [*P. arenaria*, *P. glaucescens*, *P. incana*]. Low mat-forming, densely grey starry-hairy perennial; stems often rooting at the nodes. Leaves trifoliate or digitate with 3-5 oblong, toothed leaflets. Flowers yellow, 10-16mm, solitary or in clusters of 2-6. Dry meadows and rocky habitats, to 1600m. April-July. Continental Europe north to Denmark and S Sweden.

*Potentilla pusilla* Host. Like *P. cinerea*, but leaves with 5-7 leaflets, with star-shaped hairs mixed with simple hairs. Flowers 10-14mm. Subalpine grassland. June-July. E France and S Germany. Thought to be of hybrid origin between *P. cinerea* and *P. tabernaemontani*.

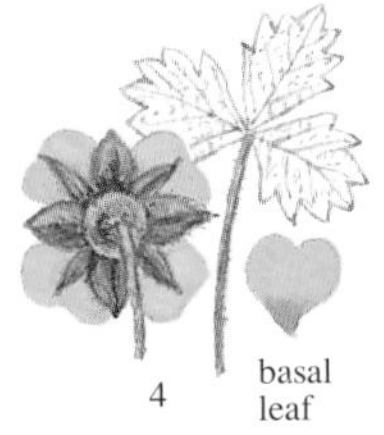


4* **Tormentil** *Potentilla erecta* (L.) Räuschel [*P. tormentilla*]. Low creeping, patch-forming downy perennial, not rooting at the nodes. Leaves usually trifoliate, occasionally with 4-5 leaflets, toothed towards the tip, silvery below, the stem leaves unstalked; basal rosette leaves withered by flowering time. Flowers yellow, 7-11mm, mostly 4-parted, few borne in lax clusters. Grassy habitats, heaths, moors, woodland rides and pathways, cultivated land, lawns, commons, on a wide range or soils, but avoiding highly calcareous soils; to 2500m. May-September. Throughout, except Spitsbergen. **B**: Common Throughout. Hybridises with *P. anglica* and *P. reptans*, *P.* × *italica* Lehm.

5* **Trailing Tormentil** *Potentilla anglica* Laicharding. Like *P. erecta*, but less sturdy and with persistent basal leaf-rosettes, the stems rooting at least at some of the nodes. Leaves often with 5 leaflets; flowers 14-18mm, with 4-5 petals, solitary or several together. Woodland margins, grassland, banks, hedgerows, roadsides and field boundaries, generally at rather low altitudes, usually on slightly acid soils. June-September. Europe north to S Scandinavia. Often confused with *P. erecta* and *P. reptans* and almost intermediate between these 2, however, the flower size and leaflet and petal numbers are generally sufficient for recognition purposes. Natural hybrids between *P. erecta* and *P. reptans* may look superficially like *P. anglica*, but are sterile. **B**: Throughout, though rare in Scotland.

6* **Creeping Cinquefoil** *Potentilla reptans* L. Low creeping invasive perennial, with long trailing shoots rooting at the nodes. Leaves digitate, with 5-7 leaflets, oblong and toothed; basal rosette leaves persisting. Flowers yellow, 17-25mm, solitary, with 5 petals. Waste ground and bare places, grassland, cultivated land, road-verges, hedgebanks and tracksides, on slightly acid, neutral or calcareous soils, to 1700m. June-September. Throughout, except the extreme north, the Faeroes and Iceland. An invasive weed in some areas, often difficult to eradicate. **B**: Common throughout.

7* **White Cinquefoil** *Potentilla alba* L. Low to short, hairy perennial. Leaves digitate, with 5 narrow-oblong, scarcely toothed leaflets, green and hairless above, silvery-hairy beneath. Flowers white, 15-20mm, in small lax clusters; petals longer than sepals. Rocky and grassy mountain habitats. April-June. C & E France and S Germany.

8 **Lax Potentilla** *Potentilla caulescens* L. Like *P. alba*, but leaflets green on both surfaces. Flowers more numerous, with rather narrow petals. Limestone rock fissures, to 2400m. May-July. E & S France and S Germany.

9* **Mountain Cinquefoil** *Potentilla montana* Brot. [*P. splendens*]. Low to short, hairy perennial, with long runners. Leaves trifoliate, sometimes digitate, with oblong leaflets toothed only towards the apex, green above, silky-grey beneath. Flowers white, 13-20mm, solitary or 3-4 together. Woodland habitats, rocky places, banks, from sea level to 1600m. May-June. W & C France.

10* **Barren Strawberry** *Potentilla sterilis* (L.) Garcke [*Potentilla fragariastrum*, *Fragaria sterilis*]. Low, hairy perennial with long rooting runners. Leaves bluish-green, trifoliate, with broad-oval toothed leaflets, grey-silky beneath. Flowers white, 10-15mm, with gaps between the notched petals. Dry grassy habitats and open woodland, scrub road-verges, hedgebanks and old walls, on well-drained sandy and calcareous soils. February-May, occasionally later. Europe, except the north. Often confused in flower with the Wild Strawberry, but differing in the bluish green, rather than deep green, leaves which have a small end tooth. The fruit is small and dry, not expanded and fleshy as in the strawberry. **P**: Various insects, but may also be self-pollinated. **B**: Common throughout, except N Scotland.

11 **Pink Barren Strawberry** *Potentilla micrantha* Ramond ex DC. Like *P. sterilis*, but without runners. Flowers smaller, pink, occasionally white, 7-9mm, with hairy (not hairless) filaments. Wooded and rocky habitats, mostly in the mountains, to 1600m. May-July. C & E France and SW Germany. Widespread in the mountains of C & S Europe.

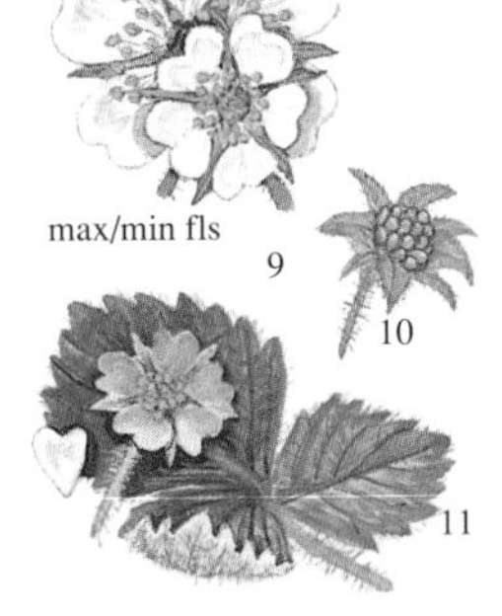

Pink Barren Strawberry

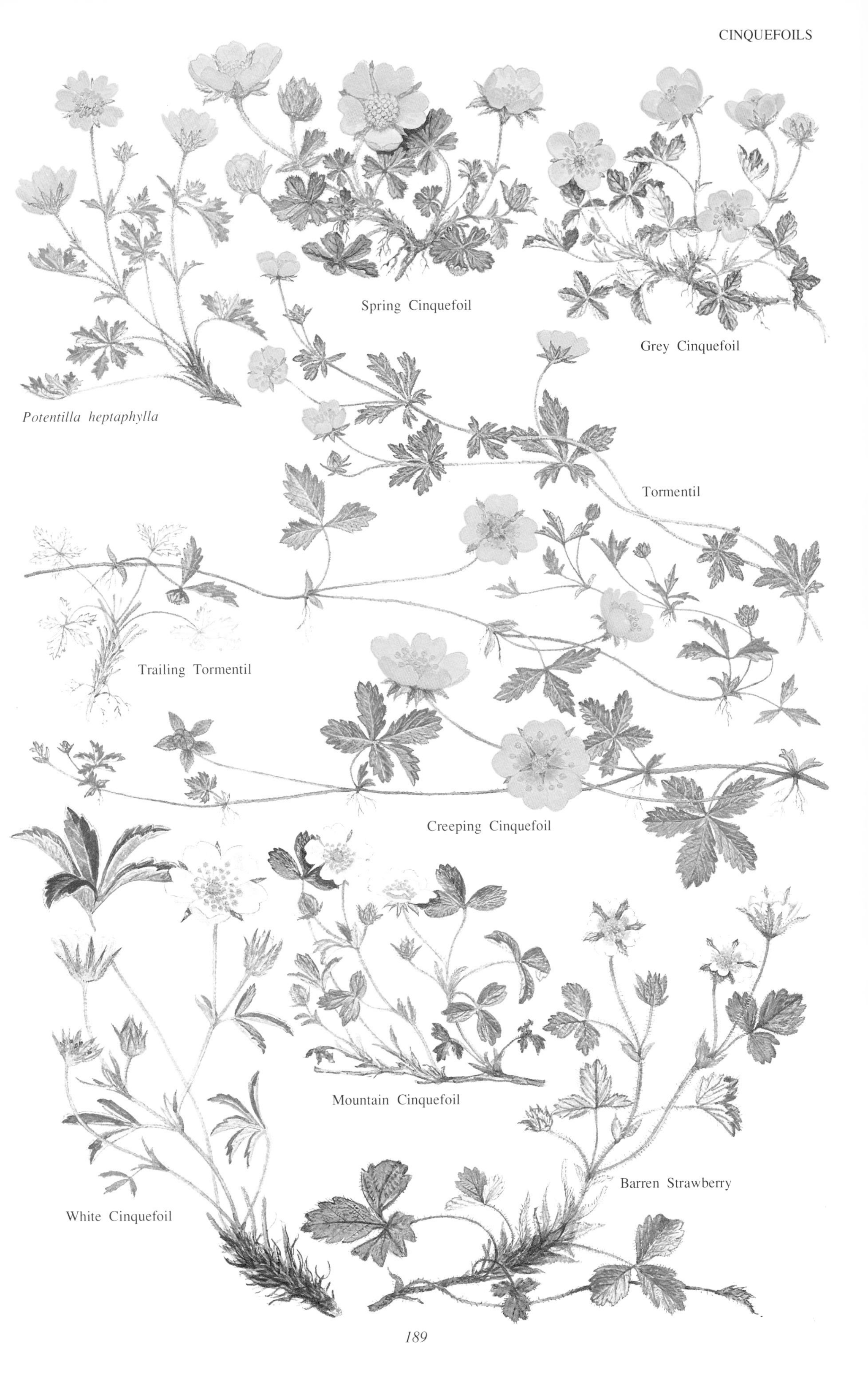
Potentilla heptaphylla
Spring Cinquefoil
Grey Cinquefoil
Tormentil
Trailing Tormentil
Creeping Cinquefoil
Mountain Cinquefoil
Barren Strawberry
White Cinquefoil

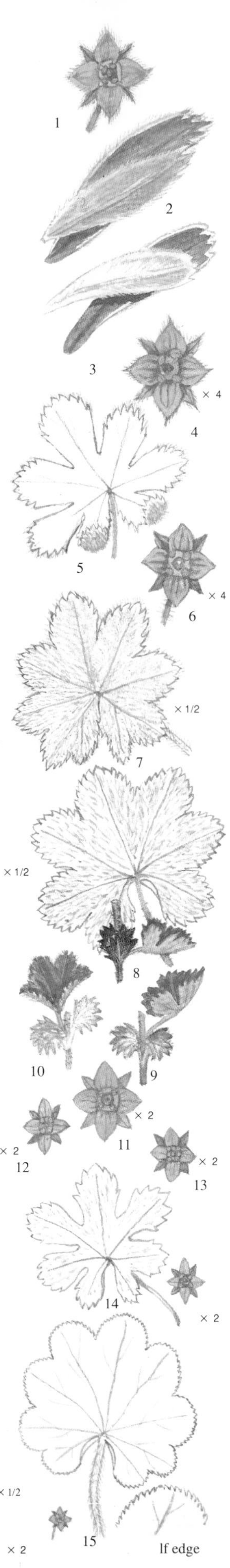


**Lady's Mantles** *Alchemilla.* Tufted perennial herbs. Leaves mostly basal, palmate or palmately lobed, toothed. Inflorescence branched, cymose. Flowers small, green or yellowish-green with 4 fused sepals, but no petals; epicalyx present. Stamens usually 4. Fruit a simple achene. This is an outline of species types in an extremely large and complex genus.

1* **Alpine Lady's-mantle** *Alchemilla alpina* L. [*A. alpina* subsp. *glomerata*]. Low to short, rhizomatous perennial. Leaves longer than the flower-stems, divided to the base into 5-7 leaflets, green or yellowish-green, hairless above, silvery-silky beneath, the leaflets lanceolate, sharply toothed; stem leaves few, 3-lobed. Flowers pale greenish-yellow, 3mm, in dense clusters. Mountain habitats, mainly on acid soils, to 2600m. June-August. Throughout, except Belgium, Holland and Denmark, but only on the mountains in the south of its range. **B**: N England, Scotland; very rare in Ireland.

2 *Alchemilla pallens* Buser. Like *A. alpina*, but leaves pale bluish-green, slightly hairy beneath, flat, not folded, generally with 7 elliptical, pointed segments, slightly joined at the base. Similar habitats and flowering time. E & SC France and S Germany.

3* **Hoppe's Lady's-mantle** *Alchemilla hoppeana* (Reichenb.) Dalla Torre. Low to short, hairy perennial. Leaves rounded, deeply divide into 7-9 linear-oblong, lobes, toothed at the blunt apex, hairless above, but hairy and greenish or silvery beneath. Flowers pale green, 3-4mm, in whorled clusters. Mountain meadows, rocky and stony places, generally on limestone, to 2600m. June-August. E France and SW Germany, from the Jura Mountains southwards.

4* *Alchemilla conjuncta* Bab. Short to medium perennial, to 40cm. Leaves rounded, thick, divided for about two thirds into 7-9 elliptical lobes, dull bluish-green above, shiny and silvery-hairy beneath, the lobes toothed at the apex. Flowers pale greenish-yellow, 3-4mm, in branched clusters. Rocky mountain habitats and rocky grassland, streamsides, to 2500m. June-July. E France – Alps and Jura. **B**: Cultivated in gardens and naturalized in N Britain from garden escapes. Established at Glen Cova and Glen Sannox, Scotland; very rare.

5 **Faeroes Lady's-mantle** *Alchemilla faeroensis* (Lange) Buser. Low to short perennial, not more than 15cm. Leaves kidney-shaped, divided for up to two-thirds into 7 oblong lobes, green above, but silvery-hairy and shiny beneath; lobes coarsely toothed. Flowers pale green, 3mm, in compact clusters. Rocky habitats and streamsides. June-July. The Faeroes and Iceland.

6* **Small Lady's-mantle** *Alchemilla glaucescens* Wallr. [*A. hybrida*, *A. minor*]. Low to short compact, silkily-hairy plant, to 20cm. Leaves rounded, with 7-9 shallow, rounded, few-toothed lobes, hairy on both surfaces; gap (sinus) at the leaf-base open; stalks and stem bases brownish. Flowers greenish, 3mm, in dense, branched clusters, silvery-hairy outside; flower-stalks hairy. Meadows, rocky and stony habitats, stream banks, generally on limestone, to 2500m. June-September. Throughout, except the far north. **B**: Yorkshire northward, but elsewhere as a garden escape; in Ireland only in Antrim.

7 *Alchemilla monticola* Opiz [*A. pastoralis*]. Medium, hairy, tufted perennial. Leaves rounded in outline, divided for about one third into 9-11 rounded, sharply toothed lobes, the gap (sinus) at the leaf-base closed, densely hairy on both surfaces. Flowers greenish-yellow, 2-3mm, in dense clusters; flower-stalks hairless. Grassy habitats, especially hay meadows, to 3100m. June-September. Throughout, except the far north. **B**: Locally abundant in Yorkshire and Durham; casual in Surrey and Buckinghamshire.

8 *Alchemilla filicaulis* Buser subsp. *vestita* (Buser) M.E. Bradshaw. Like *A. monticola*, but basal gap of leaves (sinus) wide and open, the leaves only thinly hairy and the stipules generally tinged with purplish-red. Flowers with a slightly hairy receptacle. Throughout, though mainly on the mountains in the south of its range, to 2200m. June-September. **B**: The most widespread *Alchemilla* in Britain, especially in the south.

9 *Alchemilla subcrenata* Buser. Like *A. filicaulis*, but the stipules brownish and the leaves more wavy. Receptacle hairless. Grassy meadows and banks. June-September. Britain, Scandinavia and the mountains of France and Germany. **B**: Only in Teesdale.

10* *Alchemilla minima* Walters. Very small plant with leaves not more than 3cm across, kidney-shaped, hairy mainly along the edges and along the folds, with a wide basal gap; stipules brownish. Flowers greenish-yellow, 2mm, in small dense clusters. Grassland over limestone. July-September. **B**: Only in Britain – N Yorkshire.

11* *Alchemilla xanthochlora* Rothm. Medium tufted, rather robust yellowish-green perennial, hairy in part. Leaves broadly kidney-shaped, occasionally rounded, with a wide basal gap (sinus) , divided for one third into 9-11 rounded, coarsely toothed lobes, hairless above. Flowers greenish-yellow, 2.5-3mm, in dense clusters; flowerstalks hairless. Grassy habitats, at low altitudes. May-July. North to C Scandinavia; naturalized in Finland. **B**: Throughout, except SE Ireland and the Channel Islands.

12* *Alchemilla glomerulans* Buser. Medium, rather robust perennial, the stems scarcely exceeding the leaves, silkily-hairy. Leaves rounded to kidney-shaped, divided for one third into usually 9 rounded, coarsely toothed lobes, undulate (folding when flattened and dried), hairy above. Flowers greenish, 4mm, in dense clusters on short stems; flower-stalks hairless. Wet habitats, grassy meadows, rock-ledges, river and streamsides, mostly in the mountains, sometimes close to snow patches, generally on acid soils, to c 3000m. May-September. N Britain and N Europe, and the mountains of France and Germany. The previous 4 species belong to the '*A. vulgaris*' aggregate, the members of which are widespread in Europe, N & W Asia, North America and Greenland. **B**: Scattered localities in N England – Teesdale, and the Scottish Highlands.

13* *Alchemilla glabra* Neyg. Like *A. glomerulans*, but leaves hairless above and stems hairless, or hairy only at the base. Flower clusters very lax. Grassy habitats, open woods and rocky places, to 2800m. June-September. N Europe, except the extreme north. **B**: Locally common in N England and Scotland only.

14 *Alchemilla incisa* Buser. Low to short, delicate, hairy perennial, not exceeding 15cm. Leaves kidney-shaped to almost rounded, small, divided for up to half-way into 7-9 narrow, toothed, lobes. Flowers yellowish-green, 3-4, in small lax clusters. Moist rocky and grassy mountain habitats, to 3100m. June-September. Mountains of E France and Germany from the Vosges and Jura southwards.

15 *Alchemilla tytthantha* Juz. [*A. multiflora*]. Plants short to medium with the leaves hairy on both surfaces, the stems and leaf-stalks with the hairs pointing downwards. Flowers very small, 1.5-2mm, hairless. Grassy moist habitats. June-September. **I**: USSR – Crimea. Naturalized locally in S Scotland.

Alpine Lady's-mantle
Hoppe's Lady's-mantle
*Alchemilla conjuncta*
Small Lady's-mantle
*Alchemilla minima*
*Alchemilla xanthochlora*
*Alchemilla glomerulans*
*Alchemilla glabra*

1* **Sibbaldia** *Sibbaldia procumbens* L. Low, tufted, hairy perennial with a branched woody stock. Leaves green, mostly in basal tufts, trifoliate, the oval leaflets 3-toothed at the apex. Flowers small, yellow or greenish, 4-5mm, in rather dense clusters; petals shorter than sepals, sometimes absent. Mountain grassland and rocky detritus, often by snow patches, on various soil types, 1100-3200m. July-August. Throughout, except Ireland, but only on the mountains in the south of its range. Distinguished from *Potentilla* by its small flowers, usually with 5 stamens, and the 3-toothed apex to the leaflets. **B**: Widespread, but local in the Scottish Highlands, especially on mountain tops.

**Strawberries** *Fragaria*. Like *Potentilla*, but fruit (typical strawberry) fleshy with seeds pitted over the surface. 15 species in the north temperate zone and in South America. The cultivated strawberry, *F.* × *ananassa*, is of hybrid origin, *F. chiloensis* × *F. virginiana*.

2* **Wild Strawberry** *Fragaria vesca* L. Low to short perennial, with long rooting runners. Leaves in basal tufts, trifoliate, with bright green oval or rhombic, toothed leaflets. Flowers white, 12-18mm, in lax clusters scarcely exceeding the leaves. Strawberry small, with deflexed sepals, hairless. Dry grassy habitats, woodland, hedgebanks, embankments and roadsides, generally on calcareous or base rich soils, to 2400m. April-July. Throughout, except the Faeroes and Spitsbergen. **B**: Throughout.

3 **Hautbois Strawberry** *Fragaria moschata* Duchesne. Like *F. vesca*, but a larger plant with few or no runners. Flowers held above the leaves, 15-25mm, often either male or female; fruit without seeds near the base. Similar habitats and flowering time. Belgium, France and Germany; widely naturalized from gardens in Britain and parts of N Europe.

4 *Fragaria viridis* Duchesne. Like *F. vesca*, but with short runners and the hairs on the flower-stalks spreading (not closely pressed to the stalks). Flowers creamy-white, occasionally unisexual. Fruit like *F. moschata*. To 1900m. Throughout, except the far north. **B**: Throughout.

5 **Garden Strawberry** *Fragaria* × *ananassa* Duchesne. Coarse plant with large leaves and flowers 15-30mm and large fruits up to 40mm, sometimes more. Widely cultivated and sometimes escaping and becoming naturalized. Plants generally produce long vigorous runners.

6* **Parsley-piert** *Aphanes arvensis* L. [*Alchemilla arvensis*]. Low to short hairy, slender, sometimes prostrate, annual, branching from the base. Leaves pale grey-green, 3-lobed, the lobes oblong and toothed. Flowers tiny, 1.5-2mm, petalless, in tight leaf-opposed clusters, each cluster surrounded by a leaf-like stipule with short triangular teeth. Bare ground and waste places, open patches on grassy heaths, roadsides and arable land, on well-drained light soils. April-October. Throughout, except the Faeroes, Iceland and much of Scandinavia. Tolerant of a variety of soils from acid to calcareous, but intolerant of competition from coarser herbs. **B**: Widespread, except the extreme NW.

7 **Slender Parsley-piert** *Aphanes inexspectata* Lippert [*A. microcarpa*]. Like *A. arvensis*, but a slighter greener plant; sepals close together, not spreading, and stipules with long oblong lobes. Open grassland, heaths, bare places, commons, woodland margins and tracks, an occasional weed of arable land, often on sandy and acid soils. April-October. Throughout north to S Sweden. **B**: Throughout.

**Pears** *Pyrus*. Deciduous trees or shrubs, often spiny, with simple leaves. Flowers 5-petalled, in rounded clusters, usually white; stamens 15-30 with reddish anthers. Fruit rounded or pear-shaped, with hard stone cells. 30 species in the E Mediterranean region and temperate Asia.

8* **Plymouth Pear** *Pyrus cordata* Desv. Shrub or small tree to 8m, with spiny branches and purplish twigs. Leaves oval to heart-shaped, finely-toothed, hairy when young. Flowers white, sometimes pinkish outside, 12-18mm. Fruits rounded, small, 8-15mm, without sepals at apex. Woodland and hedgerows, sometimes in scrub. April-May. SW England and W France. Very local. **B**: Very rare, restricted to SW Devon – Plymouth; extinct in Cornwall.

**Willow-leaved Pear** *Pyrus salicifolia* DC. Small, often semi-weeping tree to 5m with grey-downy, narrow-elliptical leaves. Fruit with persistent sepals. Widely cultivated; naturalized in Belgium and France. May-June. Frequently planted in parks and along road-verges and motorways. **B**: Occasionally along road-margins.

9* **Wild Pear** *Pyrus pyraster* Burgsd. Tree to 20m, usually with spiny branches; twigs grey or brown. Leaves variable, oval to heart-shaped, rather thin, slightly toothed, hairless when mature. Flowers white, 22-34mm. Fruit rounded to pear-shaped, 13-35mm, hard, not sweet, with persistent sepals. Open woods, thickets, scrub and hedgerows, often as isolated trees, to 1700m. April. Britain, Ireland, Belgium, France and Germany. **B**: Widespread in S England and Wales, scarce elsewhere.

10* **Cultivated Pear** *Pyrus communis* L. Like *P. pyraster*, but without spines and twigs stouter, reddish-brown, becoming shiny. Leaves oval to elliptical, slightly toothed or not. Fruit oblong to pear-shaped, 60-160mm, fleshy and sweet tasting, when ripe. Origins complex. Naturalized along hedgerows, woodland margin and on abandoned cultivated land, especially near old farm buildings, to 1600m. April-May. Europe, except the north. **B**: Isolated trees throughout, but rare in Scotland and Ireland.

11* **Crab Apple** *Malus sylvestris* Miller [*M. communis* subsp. *sylvestris*, *M. acerba*]. Deciduous shrub or tree to 10m with spiny branches, bark grey-brown, fissured and scaly. Leaves oval to elliptical, toothed, hairless when mature. Flowers pink or white, 30-40mm, in rounded clusters, fragrant. Fruit the familiar apple, 25-30mm, yellowish-green, sometimes reddening when ripe. Woods, especially of oak, hedgerows and scrub, to 1600m. May-June. Throughout Europe, north to C Finland, but not in the Faeroes and Iceland. Often used as a rootstock for cultivated varieties of apple. The sour fruits are used for making cider. **B**: Throughout, except for C & N Scotland.

12* **Cultivated Apple** *Malus sylvestris* subsp. *mitis* (Wallr.) Mansf. [*M. domestica*]. Like *M. sylvestris*, but branches not spiny; twigs downy and leaves hairy beneath at maturity. Naturalized from cultivation, in hedges, woodland and abandoned orchards. April-May. Throughout, except the extreme north.

Sibbaldia
Parsley-piert
Wild Strawberry
Plymouth Pear
Wild Pear
Cultivated Pear
Crab Apple
Cultivated Apple

*Sorbus*. Deciduous shrubs or trees, without spines. Leaves simple, lobed or pinnate. Flowers in condensed rounded or flat-topped clusters, white or occasionally pink, generally with 5 petals and 15-25 stamens; styles 2-5. Fruit a berry. 100 species in temperate regions of Europe, Asia and North America. A complex genus in which apomixis is frequent, giving rise to numerous local variants – some of these have been given species status. Many are cultivated in parks and gardens.

**True Service Tree** *Sorbus domestica* L. Deciduous tree to 20m; bark orange and dark brown, finely fissured. Leaves pinnate, with 6-8 pairs of oblong, toothed leaflets, hairy beneath when young. Flowers white, 16-18mm, in domed clusters. Berry large, 20-30mm, green tinged with reddish-brown. Woods and rocky habitats, sometimes in hedgerows. May. C & E France and S Germany southwards. Sometimes grown in parks and gardens.

**Mountain Ash, Rowan** *Sorbus aucuparia* L. Small, rather slender tree, with smooth silvery-grey branches. Leaves pinnate, with 5-7 pairs of oblong, toothed leaflets, green, hairy beneath. Flowers 8-10mm, in domed clusters. Berry 6-9mm. Woodland, hedgerows, moors, mountains to 2400m, mainly on light soils. May-June. Throughout, except the far north. Its profuse display of autumn berries attracts birds. **B**: Widespread, mainly in uplands. *S. a.* subsp. *glabrata* (Wimmer & Grab.) Cajander is less hairy, with flowering shoots hairless and berry longer than wide. N & C Europe.

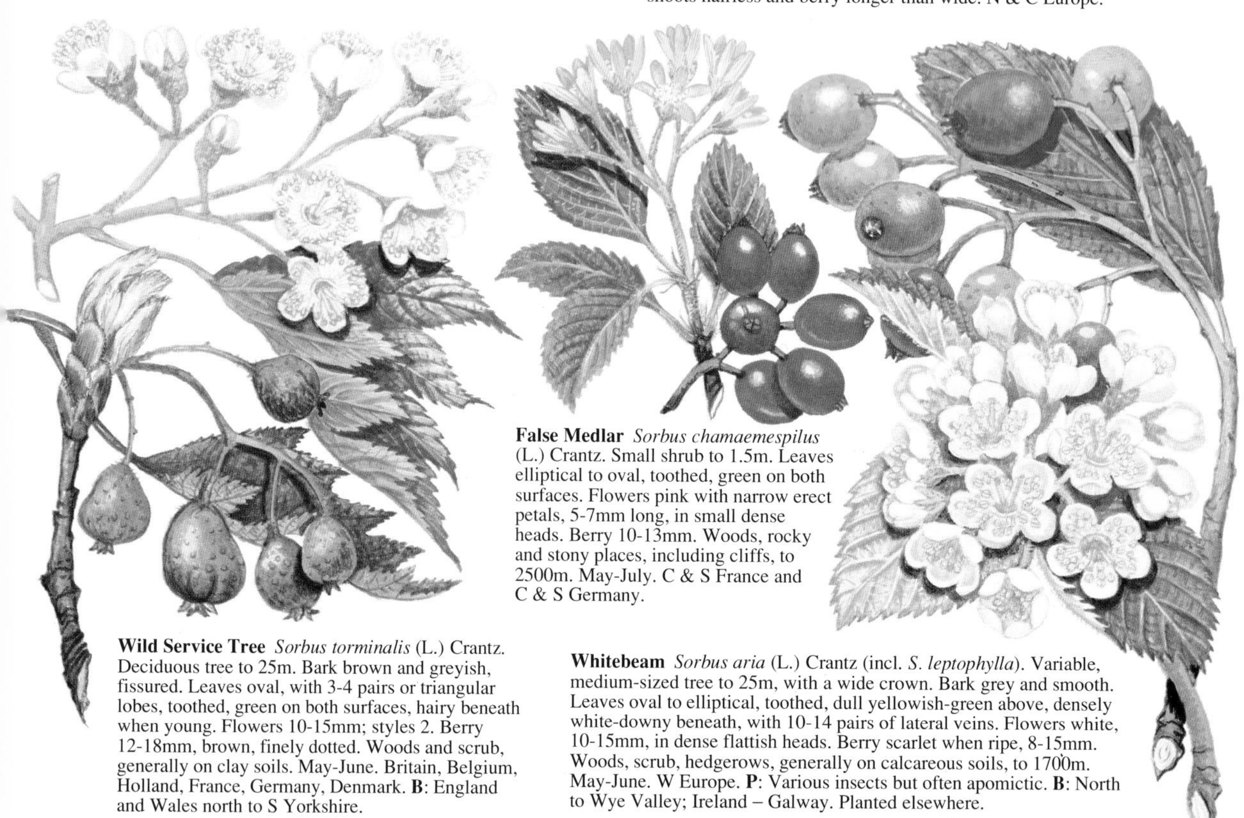

**False Medlar** *Sorbus chamaemespilus* (L.) Crantz. Small shrub to 1.5m. Leaves elliptical to oval, toothed, green on both surfaces. Flowers pink with narrow erect petals, 5-7mm long, in small dense heads. Berry 10-13mm. Woods, rocky and stony places, including cliffs, to 2500m. May-July. C & S France and C & S Germany.

**Wild Service Tree** *Sorbus torminalis* (L.) Crantz. Deciduous tree to 25m. Bark brown and greyish, fissured. Leaves oval, with 3-4 pairs or triangular lobes, toothed, green on both surfaces, hairy beneath when young. Flowers 10-15mm; styles 2. Berry 12-18mm, brown, finely dotted. Woods and scrub, generally on clay soils. May-June. Britain, Belgium, Holland, France, Germany, Denmark. **B**: England and Wales north to S Yorkshire.

**Whitebeam** *Sorbus aria* (L.) Crantz (incl. *S. leptophylla*). Variable, medium-sized tree to 25m, with a wide crown. Bark grey and smooth. Leaves oval to elliptical, toothed, dull yellowish-green above, densely white-downy beneath, with 10-14 pairs of lateral veins. Flowers white, 10-15mm, in dense flattish heads. Berry scarlet when ripe, 8-15mm. Woods, scrub, hedgerows, generally on calcareous soils, to 1700m. May-June. W Europe. **P**: Various insects but often apomictic. **B**: North to Wye Valley; Ireland – Galway. Planted elsewhere.

*Sorbus rupicola* (Syme) Hedl. (incl. *S. vexans*, *S. salicifolia*). Like *S. aria*, but a shrub to 2m, the leaves with only 7-9 pairs of lateral veins. Berries carmine, 7-15mm, longer than wide. Exposed upland and mountain limestone, to 500m. June-July. England, Ireland, Norway and S Sweden. **B**: Very local in N England; Wales, S Devon; N & W Ireland. *Sorbus lancastriensis* E.F. Warburg has leaves only 1½ times as long as broad (not twice), with 8-10 pairs of lateral veins. **B**: Rare: NW England – Morecambe Bay.

*Sorbus minima* (A. Ley) Hedl. (incl. *S. arranensis*, *S. lancifolia*, *S. leyana*, *S. neglecta*, *S. subpinnata*). Deciduous shrub to 3m. Twigs rather slender. Leaves with 8-9 pairs of lateral veins, grey-downy beneath. Berries 6-8mm. Rocky places, usually on limestone. W Britain, Norway. Probably of hybrid origin between *S. aucuparia* and *S. rupicola*. **B**: Local in S Wales – Brecon.

*Sorbus hybrida* L. [*S. fennica*]. Medium tree to 15m. Leaves oval, lobed, with 8-10 pairs of lateral veins, the basal 2 pairs of lobes free, coarsely toothed, grey-downy beneath. Flowers white, 9-12mm. Berry red when ripe, 10-12mm. Rocky habitats and open woodland. May-June. Scandinavia.

*Sorbus meinichii* (Lindeb.) Hedl. Like *S. hybrida*, but leaves with 4-5 pairs of free leaflets. Berry slightly larger. Rocky habitats. June-July. S & W Norway.

**Swedish Whitebeam** *Sorbus intermedia* (Ehrh.) Pers. [*S. suecica*, *S. scandica*]. Medium-sized deciduous tree to 15m generally intermediate between *S. aria* and *S. aucuparia*. Bark dull purplish-grey. Leaves elliptical, pinnately lobed to halfway, with 7-9 pairs of lateral veins, downy yellowish-grey beneath. Flowers dull white, 12-20mm. Berry 12-15mm, longer than wide. Rocky habitats. May-June. N Germany and Scandinavia; introduced in Britain. **B**: Similar plants with leaves whitish-grey beneath occurring in SW England, Wales and Ireland have been assigned various names including *S. bristoliensis*, *S. devoniensis*, *S. franconica* and *S. subcuneata*.

**Broad-leaved Whitebeam** *Sorbus* × *latifolia* (Lam.) Pers. Small tree. Leaves broadly elliptical, grey-downy beneath. Berry 12-14mm, yellowish-brown or orange when ripe. Woods and limestone rocks. May-June. Britain, Ireland, France, SW Germany. Apomictic; of hybrid origin between S. aria and *S. torminalis*. **B**: W Britain – Wye Valley.

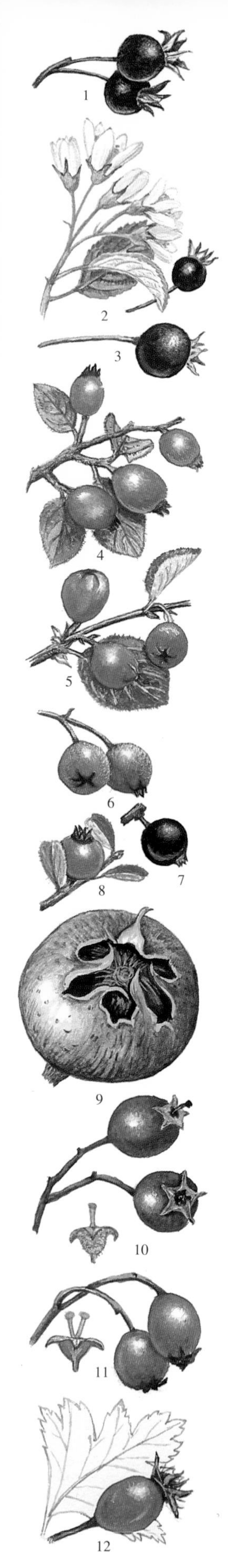

*Amelanchier*. Deciduous trees or small shrubs, without spines. Leaves simple, toothed; stipules soon falling. Flowers usually in terminal racemes, with narrow petals, white occasionally pink; stamens 10-20. Fruit like a tiny apple, with persistent sepals at the top. 25 species.

1* **Snowy Mespilus** *Amelanchier ovalis* Medicus [*A. vulgaris, A. rotundifolia*]. An erect or spreading shrub to 3m tall, with blackish bark and downy young shoots. Leaves oval, rounded or notched at the apex, coarsely toothed, downy beneath when young only. Flowers creamy-white, 10-13mm long, 3-8 in erect racemes; styles 5, separate. Fruit bluish-black 8-10mm. Rocky habitats, open woodland and thickets, mainly in mountains and on limestone, to 2400m. April-May. Belgium, France and Germany, north as far as Luxembourg. The small fruits are edible and sweet. The leaves turn deep red in the autumn.

2 *Amelanchier spicata* (Lam.) C. Koch [*A. humilis*]. Similar to *A. ovalis*, but leaves more finely toothed and densely white-downy beneath, especially when young. Flowers smaller, 4-10mm long; styles fused together at the base. Locally naturalized in open woodland, scrub and hedgerows in France, Germany and parts of Scandinavia. April-Mach. Provenance uncertain, possibly E North America.

3* **June Berry** *Amelanchier lamarckii* Schroeder [*A. confusa, A. grandiflora*]. Shrub or small tree to 9m with hairy young twigs. Leaves broad, elliptical-heart-shaped, finely toothed, purplish and downy when young, but soon grey and hairless. Flowers large, white, 35-40mm, in nodding racemes, with wide-spreading petals. Fruit blackish-purple. Garden origin. Sometimes naturalized in W Europe. April-May.

**Cotoneasters**. Deciduous or evergreen shrubs or small trees, without spines, and with simple untoothed leaves. Flowers in rounded clusters on short lateral shoots, or sometimes solitary; petals white or pinkish; sepals persistent. Fruit a fleshy red or black 'berry'. 50 species.

4* **Himalayan Cotoneaster** *Cotoneaster simonsii* Hort. ex Baker. An erect shrub to 4m, deciduous or semi-evergreen; twigs rough hairy. Leaves broad-oblong, dark green above, paler beneath, hairy when young. Flowers white with red markings, 7-8mm, in clusters of 2-4. Berry scarlet, elliptical, 8-10mm long. **I**: NE India – Khasia Hills. Occasionally naturalized in scrub and hedgerows and by old walls and buildings. June. Britain, Ireland, France and Norway. **B**: Scattered localities.

5* **Wild Cotoneaster** *Cotoneaster integerrimus* Medicus [*C. vulgaris, Mespilus cotoneaster*]. An erect deciduous shrub, with rather twisted branches, to 1m, occasionally taller; young twigs downy at first. Leaves subrounded to oval, more pointed on long shoots, grey-downy beneath. Flowers white tinged with pink, 6-7mm, in nodding clusters of 2-3. Berries red, 6-8mm, pendulous. Dry rocky and stony habitats, screes, in hilly and mountainous districts, often on calcareous soils, to 2800m. April-June. Throughout, except Ireland and the extreme north. **B**: Only known from one place in N Wales – Great Orme's Head.

6* *Cotoneaster nebrodensis* (Guss.) C. Koch [*C. tomentosus*]. Deciduous shrubn to 3m with downy young shoots. Leaves subrounded to elliptical, whitish-downy at first, but soon becoming green above. Flowers reddish outside, pinkish-white inside,5-6mm, 3-12 in nodding clusters. Berries red, subglobose, 7-8mm, downy white. Dry rocky habitats, mainly in mountains, usually on calcareous soils, to 2400m. May-June. France, Germany; naturalized in Denmark, Norway and Sweden. Rare.

7 **Black Cotoneaster** *Cotoneaster niger* Fries [*C. orientalis, C. melanocarpus*]. Deciduous shrub to 2.5m; young shoots slightly hairy, but becoming shiny red-brown. Leaves oval, 2-5cm long, dark green, but downy-white beneath. Flowers reddish to white flushed with red, 3-8 in nodding clusters. Ripe berry black with a bluish bloom, 6-9mm. Rocky habitats and scrub. June-July. Scandinavia except Finland.

8* **Small-leaved Cotoneaster** *Cotoneaster microphyllus* Wallich ex Lindley. Low spreading, sometimes prostrate, densely branched evergreen shrub, to 1m; twigs roughly-hairy. Leaves oblong, small, 5-8mm, blunt to notched at the apex, deep shiny green, grey-hairy beneath. Flowers white, 8-9mm, solitary or 2-3 together. Berry crimson-scarlet, globose, 6-7mm. **I**: Himalaya and SW China. Naturalized on limestone rocks near the sea, but also occasionally inland. June-July. Britain and Ireland. **B**: Scattered localities.

9* **Medlar** *Mespilus germanica* L. Small deciduous, spiny tree to 6m, with stiff branches; twigs roughly-hairy. Leaves oblong, finely toothed or untoothed, shiny dark green above, grey-hairy beneath. Flowers white, or slightly pink, 25-36mm, solitary on short leafy shoots; sepals slender, spreading between the petals. Fruit apple-shaped, 24-26mm, crowded by persistent sepals. Hedges, thickets and open woodland, sometimes by buildings or in old orchards. May-June. **I**: S Europe and W Asia. Naturalized in W Europe. The fruits are edible only when half-rotten. In cultivated forms, branches non-spiny and fruits larger. The flexible wood is used for whip handles. **B**: Occasionally naturalized.

**Hawthorns** *Crataegus*. Spiny deciduous trees with lobed, toothed, leaves. Flowers in rather flat-topped clusters borne on short lateral shoots (spurs), with 5 petals, numerous stamens and 1-3 styles. Fruit an edible fleshy 'berry' or haw, containing 1-3 hard stones. 200 species.

10* **Hawthorn, May Blossom** *Crataegus monogyna* Jacq. Shrub or small tree, 2-10m tall, branchlets usually spiny. Leaves wedge-shaped, deeply 3-7-lobed, slightly toothed or untoothed, somewhat hairy beneath, especially at the vein axils. Flowers white or sometimes pinkish, 8-15mm, fragrant with rounded petals; styles generally 1. Berry red with a mealy exterior, 8-10mm, oval in outline, containing a single stone. Woodland, scrub, hedgerows, thickets, rough grassy habitats, heaths and moors, lowland and upland, to 1700m. May-June. Throughout, except the extreme north. **B**: Throughout; often abundant.

11* **Midland Hawthorn** *Crataegus laevigata* (Poiret) DC. [*C. oxyacanthoides*]. Like *C. monogyna*, but leaves less deeply 3-5-lobed (the lobes rarely reaching halfway to the midrib) and without hair tufts at the vein-axils beneath. Flowers slightly larger, 9-12mm, always white, and usually with 2-3 styles. Berry generally with 2-3 stones. Woodland, scrub, hedgerows, on clay and loamy soils, lowland or upland, to 1700m. May-June. Throughout, except the extreme north. Tolerant of more shade than *C. monogyna*, though less abundant and starting to flower somewhat earlier. Sometimes considered a subspecies of *C. monogyna*, when the correct name is subsp. *nordica* Franco. **B**: Local in England, especially in the east, and in Wales; not in Scotland or Ireland. Hybridises with *C. monogyna*: *C.* × *media* Bechst.

12 *Crataegus calycina* Peterm. Spiny shrub or small tree to 10m with hairless, purplish or cinnamon-brown twigs. Leaves oval, 3-5-lobed usually, the lobes finely toothed. Flowers white, 15-20mm, with 1 style only. Berry pale red, oblong, with erect sepals, containing a single stone. Woods, scrub and hedgerows. May-June. Continental Europe, except W France, Holland and the extreme north. Hybridises with *C. monogyna, C.* × *kyrtostyla* Fingerh. in lowland NW Europe. subsp. *curvisepala* (Lindman) Franco [*C. curvisepala*]. Fruits dark red or blackish, with reflexed sepals. S Sweden eastwards.

Snowy Mespilus
June Berry
Himalayan Cotoneaster
Small-leaved Cotoneaster
*Cotoneaster nebrodensis*
Wild Cotoneaster
Midland Hawthorn
Medlar
Hawthorn

**Cherries, Plums** *Prunus*. Deciduous or evergreen trees or shrubs, spiny or not. Leaves simple, toothed, often with a pair of glands at the top of the stalk. Flowers 5-parted, solitary, clustered or in slender racemes; stamens numerous; style solitary. Fruit a fleshy or hard drupe. 200 species in the temperate northern hemisphere, but also in South America, with the largest concentration in China and Japan. Many are cultivated for their flowers and autumn colours.

1* **Cherry Plum** *Prunus cerasifera* Ehrh. [*P. divaricata*]. Deciduous shrub or tree to 8m, generally without spines; young twigs shiny and hairless. Leaves oblong, shiny and hairless above, hairy beneath. Flowers white, 15-20mm, mostly solitary, appearing with or slightly before the young leaves. Fruit rounded, 20-30mm, smooth, red or yellow. **I**: W Asia. Widely planted for fruit or for hedging and locally naturalized. February-March. Britain, France, Germany and Denmark. Used as a stock for grafting selected varieties of plums and gages. **B**: Scattered localities.

2* **Blackthorn, Sloe** *Prunus spinosa* L. Dense deciduous shrub, 1-4m, suckering freely, with spiny branches; bark dark brown or blackish, the young twigs downy. Leaves oval, broadest above the middle, dull green, toothed. Flowers white, 10-15mm, solitary, but dense on branches, appearing before the leaves. Fruit rounded, 10-15mm, black with a bluish bloom, acid-tasting. Woodland, scrub and hedgerows, often forming dense and extensive thickets, to 1600m. March-May. Throughout, except the Faeroes, Iceland and Spitsbergen. The fruits when ripe can be used for making Sloe Gin, but are too harsh-tasting to eat. **P**: Mainly bees. **B**: Throughout, except for some of the northern isles.

3* **Wild Plum** *Prunus domestica* L. Deciduous shrub or tree to 10m, not spiny, often suckering; young twigs dull, slightly hairy at first. Leaves oval to elliptical with a wedge-shaped base, dull green above, hairy and paler beneath, toothed. Flowers white tinged with green, 15-25mm, in small clusters with the emerging leaves. Fruit egg-shaped, 40-75mm, bluish-black, purple or red, edible. Widely planted and frequently naturalized, especially along hedgerows and woodland margins and close to habitations. April-May. **I**: Caucasus Mts. Throughout, except the north. Cultivated for many centuries, probably first in the region of ancient Persia. Various forms have become naturalized. **B** Scattered localities

4 subsp. *insititia* (L.) C.K. Schneider [*P. insititia*]. Like the type, but with downy young twigs, the branches often spiny and the flowers pure white. Fruit smaller, 20-50mm, purple, red, yellow or green. Widely naturalized in woods and hedgerows. Similar flowering time and distribution.

5* **Wild Cherry, Gean** *Prunus avium* L. [*Cerasus avium*]. Medium, spreading, deciduous tree 10-25m; young twigs hairless, bark reddish-brown, peeling in thin strips and paper-like. Leaves oblong, dull green above, toothed, often reddish, somewhat hairy beneath. Flowers white, 15-25mm, in clusters of 2-6 with the leaves. Fruit rounded and fleshy, dark red, sometimes creamy-yellow, bright red or black, 9-12mm, sweet or bitter. Woods, hedgerows and close to habitation on fertile, well-drained soils, to 1700m. April-May. Widely cultivated and naturalized throughout, except the Faeroes, Iceland and Spitsbergen. **P**: Mainly bees and hoverflies. A parent of some of the cultivated cherries. **B**: Widespread in England, Wales and Ireland, scarcer in Scotland.

6* **Dwarf or Sour Cherry** *Prunus cerasus* L. [*Cerasus vulgaris*]. Rather like *P. avium*, but usually a shrub, to 5m, generally suckering freely. Leaves glossy above, oblong-elliptical, becoming hairless beneath. Flowers white or pinkish, 12-18mm, in small clusters. Fruit bright red, rarely black, acid-tasting. Hedgerows and scrub. April-May. **I**: SW Asia.Widely naturalized in Europe, except for the extreme north. Long cultivated and a parent of the edible Cherries. **B**: S England, Wales and Ireland, scarce elsewhere.

7 *Prunus fruticosa* Pallas [*Cerasus fruticosa*]. Like *P. cerasus*, but a low spreading shrub, 0.3-1.5m, hairless. Fruit dark red, 7-10mm. Scrub and dry grassy habitats. March-April. W Germany. C Europe to Siberia. Cultivated in gardens for over 350 years – the common form is pendulous.

8* **St Lucie's Cherry** *Prunus mahaleb* L. [*Cerasus mahaleb*]. Deciduous shrub or sometimes a small tree, to 10m; young twigs glandular-hairy. Leaves broadly-oval to subrounded, toothed, glandular along margin. Flowers white, 8-12mm, in short rounded racemes on short leafy shoots, fragrant. Fruit fleshy, black, bitter. Dry hillslopes, thickets and open woodland, to 1700m. April-May. S Belgium, France and Germany; naturalized in Britain, Norway and Sweden. The flowers are fragrant. **P**: Mainly bees. C & S Europe. **B**: Occasionally naturalized.

9* **Bird Cherry** *Prunus padus* L. [*Cerasus padus*]. Deciduous tree or shrub to 17m; bark foetid, brown, peeling. Leaves elliptical, pointed and toothed, dull green above, paler and hairless beneath. Flowers white, 10-16mm, in long slender, pendent racemes, heavy-scented. Fruit small, shiny black, 6-8mm, bitter and astringent. Woodland, hedgerows, scrub and on moors, to 2200m. May-June. Throughout, except the Faeroes, Iceland and the extreme north. Widespread in Europe and Asia. Many forms are cultivated in gardens and are sometimes seen along roadsides and motorways. **B**: Throughout, except for S England.

subsp. *borealis* Cajander [subsp. *petraea*] Like the type, but a shrub to 3m, the young shoots and undersurface of the leaves hairy. Flowers scarcely scented. Similar habitats and flowering time. N & W Finland and S Germany.

10 **Rum Cherry** *Prunus serotina* Ehrh. Like *P. padus*, but with aromatic bark and the leaves dark shiny green above. Flowers 5-10mm, creamy-white. Fruit purplish-black, with a persistent calyx. Planted for timber and as an ornamental tree, particularly in C Europe. May-June. Naturalized in Britain, Holland, France, Denmark, Germany and S Sweden. The foliage turns yellow in the autumn. **B**: Occasionally planted in open native woodland.

11* **Cherry Laurel** *Prunus laurocerasus* L. [*Cerasus laurocerasus*, *Laurocerasus officinalis*]. An evergreen shrub or small tree to 8m; young twigs pale green. Leaves elliptical, bright mid-green, leathery, scarcely toothed. Flowers white, 7-9mm, in slender erect, rather stiff racemes. Fruit purplish-black, 10-12mm. **I**: E Europe and Asia Minor. Naturalized in open woodland and hedgerows, especially in coastal districts in Britain and France. April-May. **B**: W & S Britain.

Cherry Plum
Blackthorn
Wild Plum
Wild Cherry
Dwarf Cherry
St Lucie's Cherry
Bird Cherry
Cherry Laurel

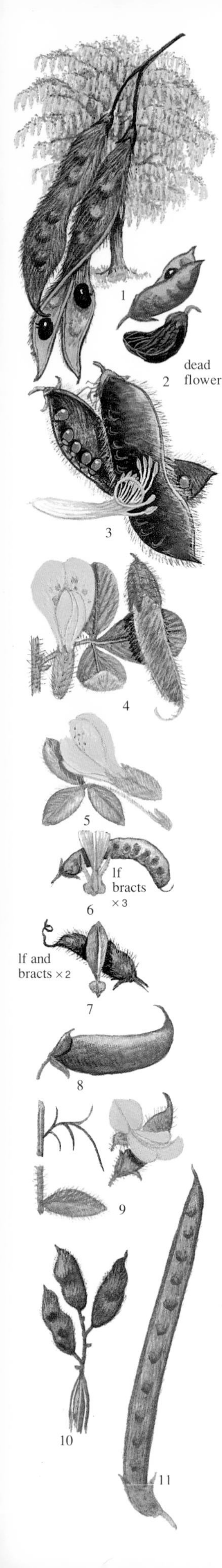


# PEA FAMILY Leguminosae

Trees, shrubs or herbs, mostly with alternate leaves. Leaves simple trifoliate or variously pinnate, generally with stipules at the base. Flowers typically pea-shaped, solitary, in spikes, racemes or panicles. Calyx tubular to bell-shaped, with 5 teeth. Petals 5, differing; upper the 'standard', the lateral 2 the 'wings' and the lower 2 fused together to form a 'keel' which encloses the 10 stamens and ovary. Fruit a pod, usually splitting lengthwise, but sometimes breaking up into 1-seeded segments. The third largest family of flowering plants (exceeded only by the Compositae and the Orchidaceae). Around 16,000 species, worldwide but particularly in tropics and subtropics. Many bear root-nodules with bacteria, *Rhizobium*, able to fix atmospheric nitrogen.

1* **Laburnum** *Laburnum anagyroides* Medicus [*L. vulgare*, *Cytisus laburnum*]. Deciduous tree to 7m with smooth pale brown bark; twigs grey-downy. Leaves trifoliate, grey-green; leaflets elliptical, untoothed. Flowers golden-yellow, 18-20mm, crowded in pendent racemes. Pod narrow-oblong, 40-60mm, hairy when young, blackish-brown when mature. Woods and scrub, very occasionally in hedgerows, also along roads and motorways, to 2000 m. May-June. S Germany and E France. Widely planted. A very poisonous plant, particularly the seeds. **P**: Bees. **B**: Naturalized in scattered localities throughout much of Britain and Ireland. *L. alpinum* (Miller) Bercht. & J.S. Presl [*Cytisus alpinus*], with hairless twigs and pods, is locally naturalised from gardens in Scotland.

2* **Black Broom** *Lembotropis nigricans* (L.) Griseb. [*Cytisus nigricans*]. An erect deciduous shrub to 1m, without spines; twigs silvery-hairy. Leaves trifoliate, dark green, paler beneath; leaflets linear to elliptical. Flowers yellow, black or brownish on ageing, 7-10mm, in terminal leafless racemes; wings shorter than keel. Pods linear-oblong, 20-25mm appressed-hairy. Dry habitats, woodland margins, roadsides, to 1800m. June-July. C & S Germany.

3* **Broom** *Cytisus scoparius* (L.) Link [*Sarothamnus scoparius*]. Much-branched, erect deciduous shrub, 1-2, m with angled slender green branches, no spines, usually hairless. Leaves small, simple, lanceolate or trifoliate. Flowers golden-yellow, 16-18mm, in leafy spikes, scented. Pod oblong, 25-40mm, black when ripe. Dry sunny habitats, generally at low altitudes. April-June. Throughout Europe except the far north. Poisonous to livestock. The bark is used in rope-making and for tanning. **P**: Bees. **B**: Throughout.

subsp. *maritimus* (Rouy) Heywood. Like the type, but a spreading or prostrate shrub not more than 40cm high; young twigs silkily-hairy. Maritime cliffs. May-June. Britain and parts of NW Europe. Very local.

4 **Clustered Broom** *Chamaecytisus supinus* (L.) Link [*Cytisus supinus*]. An erect or ascending deciduous shrub to 60cm, hairy. Leaves trifoliate, the leaflets elliptical, hairy or hairless above. Flowers yellow, the standard often brown-spotted, in small terminal clusters with a leafy ruff beneath. Pod 20-35mm, hairy. Scrub and rocky slopes, on calcareous soils. May-June. C & S France and S Germany.

5 *Chamaecytisus ratisbonensis* (Schaeffer) Rothm. [*Cytisus ratisbonensis*, *C. biflorus*]. Small procumbent or ascending, deciduous shrub to 45cm; young twigs silkily-hairy, no spines. Leaves trifoliate, the leaflets lanceolate, silkily-hairy beneath at first. Flower yellow, the standard spotted with orange red, 16-22mm, in leafy racemes. Pod 20-30mm, appressed-hairy. Scrub, heaths and other dry open habitats. May-June. Germany. Local.

**Greenweeds** *Genista*. Spiny or non-spiny shrubs with simple leaves. Flowers yellow, in racemes or heads, with a 2-lipped calyx. Pod usually splitting open. 75 species in Europe, W Asia and N Africa.

6* **Dyer's Greenweed** *Genista tinctoria* L. Prostrate to erect, deciduous, spineless subshrub to 1.5m, often less, very variable. Leaves oval to linear-lanceolate, hairy or not. Flowers 8-15mm, in leafy, stalked spikes; standard equalling keel. Pod narrow-oblong, hairless. Grassy habitats, meadows, road-verges and banks, heaths and open woods, to 1800m. Much of Europe, except the extreme north. Long used as a dye plant – yielding both yellow and green dyes. The seeds are purgative. **B**: Throughout, except for N Scotland; rather rare in Ireland.

7* **Hairy Greenweed** *Genista pilosa* L. Semi-evergreen prostrate or suberect shrub, to 1.5m, much-branched. Leaves usually oval, dark green, short-stalked, silvery-hairy beneath like the calyx. Flowers 8-10mm, in long racemes; standard and keel silvery-hairy. Pod narrow-oblong, 15-25mm, appressed-hairy. Rocky habitats, heaths and open woods, to 1300m. May-June. Britain, Belgium, Holland, France, Denmark and Germany.

8* **Petty Whin, Needle Furze** *Genista anglica* L. Rather slender, spreading to erect, usually spiny, subshrub, practically hairless. Leaves oval, pointed. Flowers 6-8mm, in short leafy racemes; standard shorter than keel. Pod narrow, curved and rather inflated. Heaths and moors. April-June. Northwards to S Sweden. **B**: Scattered localities throughout.

9 **German Greenweed** *Genista germanica* L. Much-branched subshrub to 60cm, with branched spines. Leaves lanceolate, hairy below and on margins. Flowers hairy, 10-12mm, in lax racemes, with very small bracts; standard pointed, much shorter than the keel. Pod oval-pointed. Rocky habitats and heaths, to 2300m. May-June. Continental Europe north to S Sweden.

10* **Winged Broom** *Chamaespartium sagittale* (L.) P. Gibbs [*Genista sagittalis*, *Genistella sagittalis*, *Pterospartium sagittale*]. Prostrate mat-forming subshrub with erect broadly-winged stems 10-50cm, constricted at the nodes. Leaves simple, elliptical. Flowers yellow, 10-12mm, in small terminal clusters. Meadows, rocky habitats and open woods, roadsides, to 1950m. May-July. SE Belgium, France and Germany.

11* **Spanish Broom** *Spartium junceum* L. An erect shrub to 3m with unspiny, grey-green, rush-like, hairless stems. Leaves small linear to lanceolate, sparsely scattered along stems, soon falling. Flowers rich yellow, 20-25mm, in lax racemes, scented. Pod narrow-oblong, silkily-hairy at first, 50-80mm, flattened. Rocky habitats, scrub, embankments, roadsides. May-July. C & S France; widely naturalized in other parts of Europe except the north. Much planted. The seeds are poisonous. **B**: Naturalized in the south.

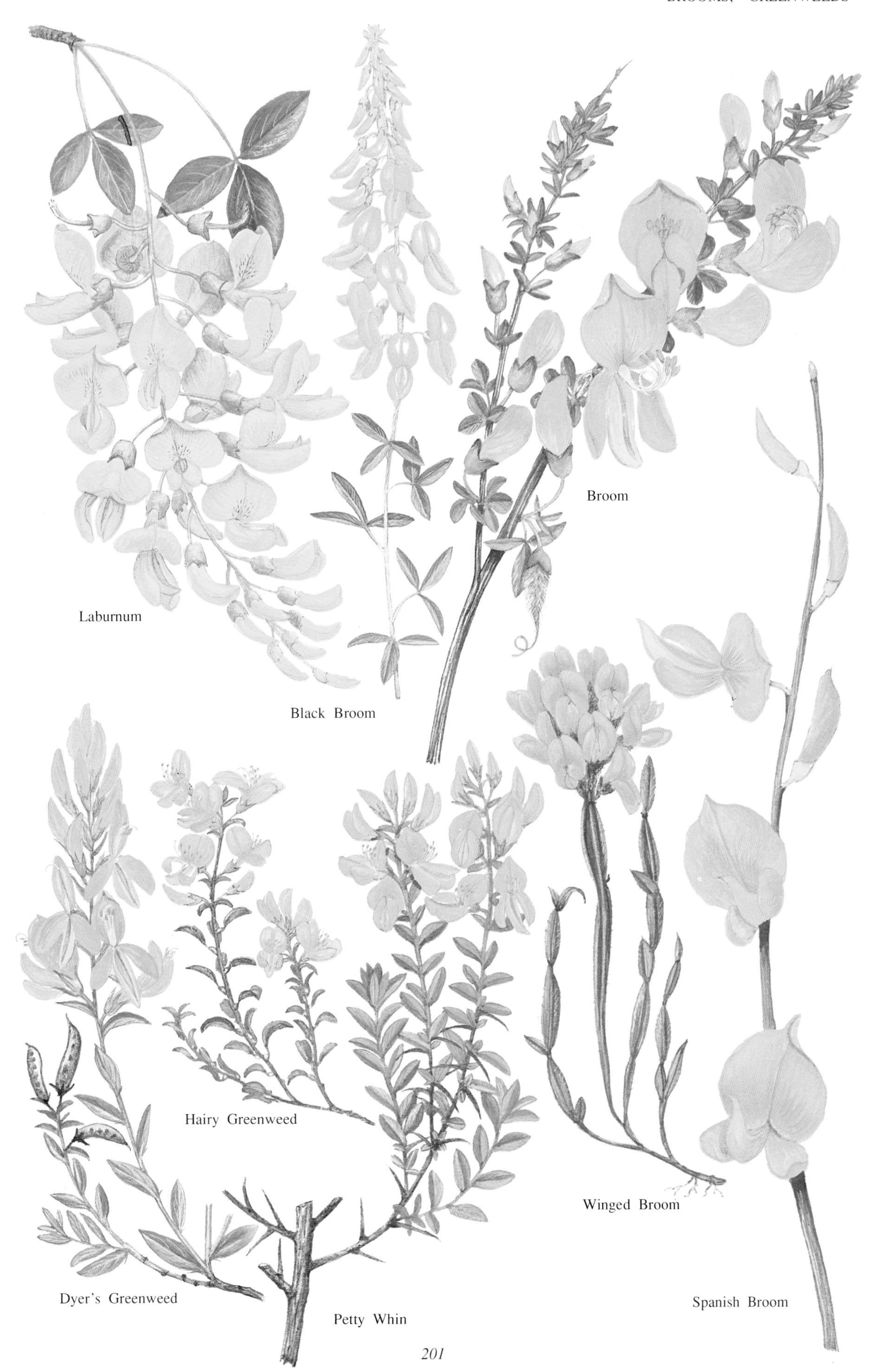
Laburnum
Black Broom
Broom
Hairy Greenweed
Winged Broom
Dyer's Greenweed
Petty Whin
Spanish Broom

**Gorses** *Ulex*. Very spiny evergreen shrubs with small alternate, scale-like, leaves (trifoliate in seedlings). Flowers yellow, solitary or in small clusters among spines; calyx greenish-yellow, 2-lipped; corolla persisting. Pod hairy, scarcely protruding beyond the calyx. 20 species in W Europe and N Africa. The flowers explode, as in *Genista*, when visited by insects, showering the pollinator with pollen. The fruit pod also explodes when ripe, flinging the seeds some distance from the plant.

1* **Gorse, Furse** *Ulex europaeus* L. Densely-branched shrub to 2m, often forming large colonies; young twigs with greyish- or reddish-brown hairs; main spines stout 12-25mm, furrowed. Flowers golden-yellow, 15-20mm long; sepals almost as long as petals. Pod 11-20mm, densely hairy. Grassland, heaths, scrub and hedgerows, woodland margins, banks and roadsides, on neutral or slightly acid soils, to 1200m. Flowering throughout the year, but particularly in late winter and spring. Britain, Ireland, Holland, France and Germany; naturalized in Belgium and Scandinavia. The seeds are poisonous. Sometimes cultivated in gardens or used as a hedging plant. After fire, plants are normally able to sprout again from the base. **B**: Widespread.

2* **Dwarf Gorse** *Ulex minor* Roth [*U. nanus*]. Like *U. europaeus*, but a much slighter shrub, 10-100cm, the twigs with reddish-brown hairs and the main spines rather weak, 8-15mm. Flowers smaller, clear yellow, 8-10mm long, the wings equalling the keel (not longer). Pod 7-8mm, very hairy. Heaths, grassy places, banks and roadsides. July-November. Britain and W & C France. Sometimes confused with *U. gallii*, but with weaker spines and rather smaller flowers. **B**: Scattered localities in England N to Cumbria, including East Anglia.

3 **Western Gorse** *Ulex gallii* Planchon. Like *U. minor*, but more sturdy with longer spines and deep golden-yellow flowers, 10-13mm. Heath, grassy places and banks, on acid soils. July-November. Britain, Ireland and W France. **B**: Throughout, except for C & N Scotland, especially in the West

**Lupinus** *Lupinus*. Annual or perennial herbs, rarely shrubby. Leaves digitate, long-stalked. Flowers in terminal racemes; calyx 2-lipped, divided almost to the base; keel-petal beaked. Pod narrow-oblong, splitting, often somewhat constricted between the seeds. 200 species in America and the Mediterranean region. The seeds are poisonous when fresh.

4 **Sweet or Yellow Lupin** *Lupinus luteus* L. Medium to tall, hairy annual. Leaflets oblong, pointed, sparsely hairy. Flowers bright yellow, 13-16mm long, sweetly-scented, in regular whorls in the raceme. Pod 40-50mm, densely hairy, black when ripe. Grassy, waste and disturbed places, widely cultivated for fodder and as a green manure. June-August. **I**: North America. Occasionally naturalized in Belgium, Holland, France and Germany.

5* **Garden Lupin** *Lupinus polyphyllus* Lindley. Tall, usually unbranched, minutely-hairy perennial. Leaflets 9-17, lanceolate, sparsely silkily-hairy beneath. Flowers blue, purple, pink or white, rarely yellow, but often bicoloured, 12-14mm long, in dense whorls in the raceme, scented. Pod 25-40mm, sparsely hairy. Grassy habitats, banks and roadsides, abandoned cultivation. June-August. **I**: W North America. Naturalized in Europe except the extreme north; grown as a fodder crop. Most garden Lupins grown in gardens are hybrids between this and *L. arboreus* and *L. polyphyllus*; they sometimes become naturalized, especially along roadways. **B**: Throughout, except the north; local.

6* **Nootka or Wild Lupin** *Lupinus nootkatensis* Donn ex Sims. Medium to tall hairy perennial. Like *L. polyphyllus*, but leaves with only 6-8 leaflets. Flowers blue or purple, occasionally bicoloured; scented. Pod 30-50mm. **I**: NW North America and NE Asia. Naturalized on moors and along river shingles. June-August. Scotland and Norway. **B**: Scattered localities in Scotland and Orkney.

7* **Tree Lupin** *Lupinus arboreus* Sims. An evergreen subshrub, 0.5-2.5m. Leaflets 5-12, narrow-oblong, stiffly hairy beneath. Flowers yellow or white, sometimes flushed with blue or purple, 14-17mm long, whorled in lax racemes, sweetly scented. Pod 40-80mm, stiffly hairy. Naturalized in coastal habitats, cliffs and shingle, sometimes on roadway embankments or waste places inland. May-August. **I**: California. Naturalized in scattered localities in Ireland, S England, East Anglia, Channel Islands.

8* **False Acacia** *Robinia pseudacacia* L. Deciduous tree to 25m, often suckering; bark grey-brown, deeply furrowed; shoots, especially of suckers, spiny. Leaves pinnate with 3-10 pairs of oval leaflets. Flowers white, 15-20mm, in pendent racemes, scented. Pod linear-oblong, 5-10cm, hairless. Dry sandy and rocky habitats, scrub and woodland margins, roadsides. June-July. **I**: C & E North America. Naturalized in Britain and parts of W Europe north to Holland. Widely cultivated for soil stabilization, particularly on light dry sandy soils. The wood is extremely rot-resistent and is used for fence posts, wheels and floors, while the fragrant flowers are used in perfumery. Trees are sometimes heavily infested with mistletoe, *Viscum album*, particularly in France. **B**: Scattered localities, but absent from much of the north.

9* **Goat's-rue** *Galega officinalis* L. Medium to tall erect, hairless, perennial. Leaves pinnate with 4-8 pairs of oblong or lanceolate leaflets; stipules green. Flowers white to pale purplish-lilac, 10-15mm long, in lateral racemes; calyx with bristle-like lobes. Pod 20-50mm, cylindrical, constricted between the seeds. Damp meadows, river and stream banks, ditches and waste ground. July-September. France and Germany; naturalized in Britain and Belgium. Widespread in C and S Europe. Sometimes confused with *Onobrychis*, but a taller plant with green (not brown membranous) stipules. **B**: Locally naturalized.

10* **Bladder Senna** *Colutea arborescens* L. Much-branched, deciduous shrub 2-6 m, with pale brown shoots. Leaves pinnate, with 4-5 pairs of oval leaflets. Flowers yellow, sometimes marked with red, 16-20mm long, in short lateral racemes. Pods very inflated-bladders, 5-7cm, brown and papery when ripe. Dry slopes, scrub and open woods, roadsides. June-August. Germany to E France southwards; naturalized in Britain and Belgium. The leaves and seeds have purgative properties and are sometimes used instead of senna. The dried fruit pods persist well into winter after the leaves have fallen. **B**: Scattered localities, especially in the south.

subsp. *gallica* Browicz. Like the type but, the ovary sparsely hairy. NC & C France.

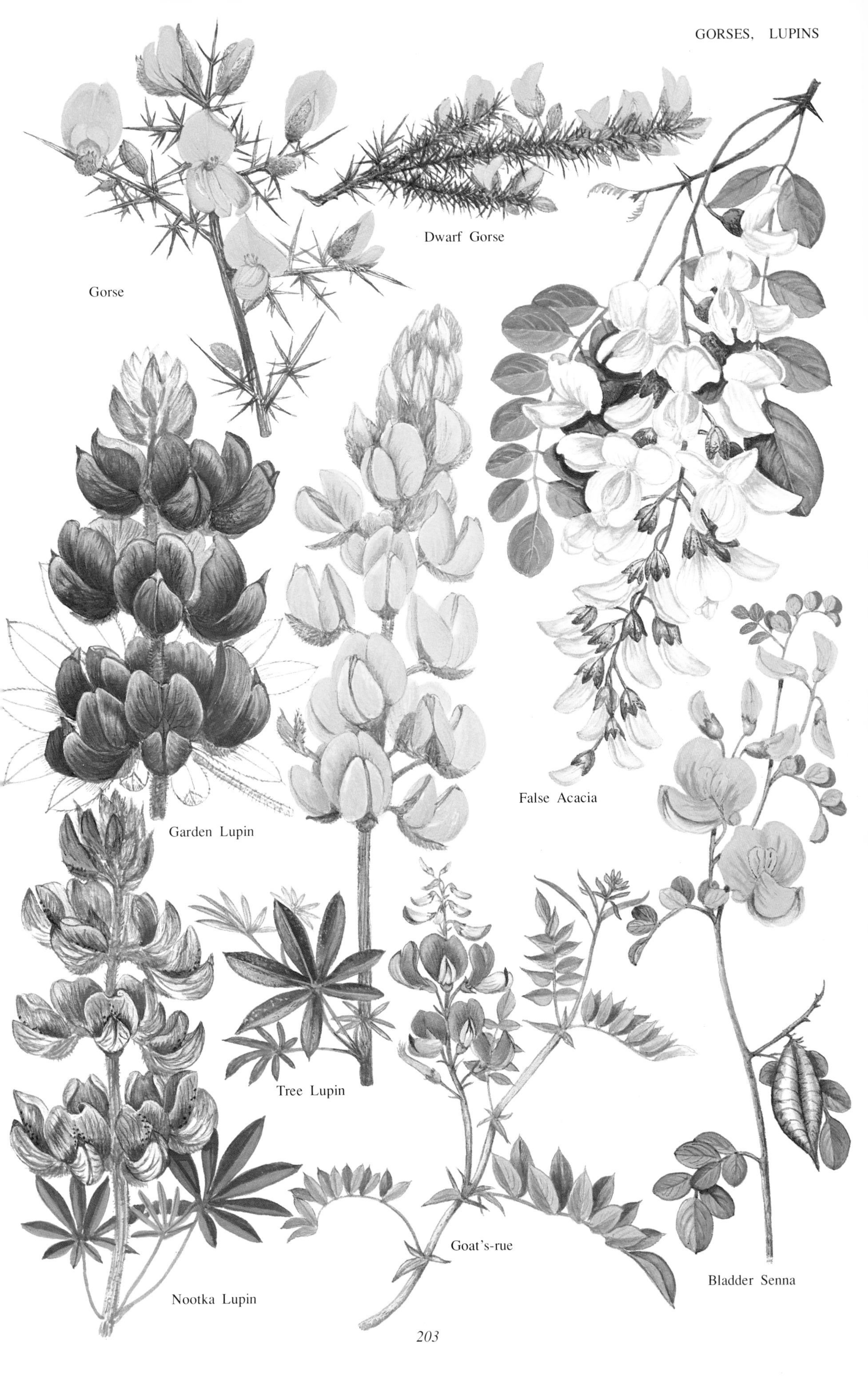
Gorse
Dwarf Gorse
False Acacia
Garden Lupin
Tree Lupin
Goat's-rue
Bladder Senna
Nootka Lupin

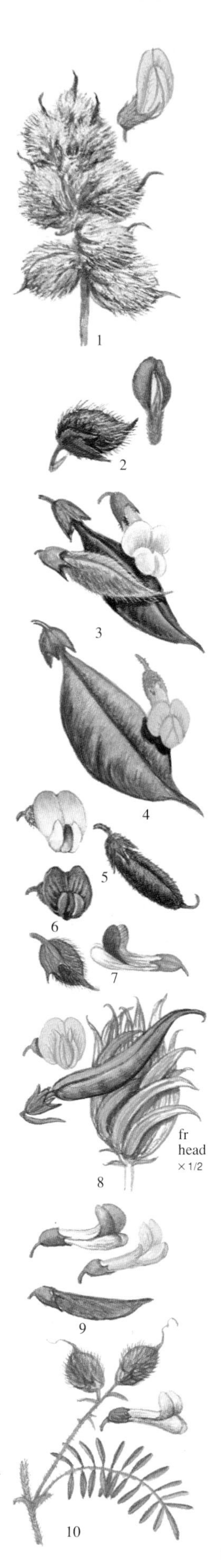


**Astragalus** *Astragalus*. Annual or perennial herbs or small shrubs. Leaves pinnate, with an end leaflet (in our species). Flowers in lateral racemes or clusters. Calyx tubular or bell-shaped, 5-toothed. Pod very variable. In excess of 2000 species scattered across the northern hemisphere, but particularly in C Asia and the Middle East. Sometimes confused with *Oxytropis*, but differing primarily in the blunt (not sharply pointed) keel.

1* **Wild Lentil** *Astragalus cicer* L. Medium to tall, rather straggly, erect, hairy perennial. Leaflets 10-15 pairs, lanceolate to oval. Flowers pale yellow, 14-16mm long, in dense oblong, long-stalked racemes. Pod rounded, pointed, inflated and membranous, 10-15mm, with black and white hairs. Meadows and open scrub, roadsides and woodland margins, to 1800m. June-July. Belgium, France and Germany; occasionally naturalized further north. Sometimes cultivated in gardens.

2* **Purple Milk-vetch** *Astragalus danicus* Retz. [*A. hypoglottis*]. Low to short hairy perennial with slender ascending stems. Leaflets 6-13 pairs, oblong, blunt. Flowers purplish or bluish-violet, 15-18mm long, in rounded, long-stalked clusters. Pods oval, inflated, 7-8mm, white-hairy. Meadows and short-grassy places, usually over limestones, sometimes on coastal sand-dunes, to 2400m. May-July. Europe north to S Sweden, but primarily on the mountains in the south; absent from Belgium and Holland. **B**: Scattered localities N of the Thames, absent N of Sutherland; Ireland – only on the Aran Islands.

3* **Yellow Alpine Milk-vetch** *Astragalus frigidus* (L.) A. Gray [*Phaca frigidus*]. Short greyish perennial, almost hairless; stems unbranched. Leaflets 4-8 pairs, elliptical. Flowers yellowish-white, 12-14mm long, in lax, long-stalked racemes. Calyx often reddish, the teeth with black hairs. Pod elliptical, flattened above, 20-30mm, brown, hairy. Mountain meadows, rocky places and moraines, usually on limestone, to 2800m. July-August. Faeroes, Norway, France and Germany.

4* **Mountain Lentil** *Astragalus penduliflorus* Lam. [*Phaca alpina* p.p.]. Short to medium hairy perennial; stems erect, usually branched. Leaflets 7-15 pairs, elliptical. Flowers yellow, 10-12mm long, in lax, long-stalked clusters; calyx green. Pod oval, inflated, black-hairy at first. Mountain meadows, rocky and stony habitats and open woodland, to 2850 m. July-August. C & E France, S Germany and S Sweden. Local The flowers are a deeper yellow than the closely related *A. frigidus*.

5* **Alpine Milk-vetch** *Astragalus alpinus* L. [*Phaca alpina* p.p., *P. astragalina*]. Short, somewhat hairy, perennial, with prostrate or ascending slender stems. Leaflets 7-12 pairs, elliptical, often pointed. Flowers whitish with a bluish or violet keel, 10-15mm long, in long-stalked racemes. Pod oblong, 10-15mm, brown, scarcely inflated, black-hairy at first. Mountain grasslands, rocky and stony habitats, cliffs, on base-rich or calcareous soils, 700-3100m. July-August. N Britain, Scandinavia, France and Germany. Sometimes confused with *A. danicus*, but a more slender plant with horizontal or deflexed, rather than vertical, flowers. **B**: Very rare, confined to a few scattered localities in C & N Scotland.

6 *A. a.* subsp. *arcticus* Lindman [*A. subpolaris*]. Like the type, but flowers purplish-violet; pod 8-11mm, oval. Arctic Scandinavia.

7* **Norwegian Milk-vetch** *Astragalus norvegicus* Weber [*A. oroboides*]. Short, rather stout, erect, hairless or slightly hairy perennial. Leaflets 6-7 pairs, oblong, usually notched. Flowers pale violet, 10-12mm long, in dense, long-stalked racemes. Pod egg-shaped, 9-10mm, black-hairy when young. Dry sunny limestone rocks, 1900-2500m. July-August. Arctic Scandinavia.

8* **Wild Liquorice** *Astragalus glycyphyllos* L. Medium to tall, straggling, slightly hairy or hairless, perennial. Leaflets 4-6 pairs only, oval to elliptical, blunt, but often with a tiny point. Flowers pale cream or yellowish-green, 11-15mm long, in dense short-stalked racemes. Pod narrow-oblong, 30-40mm, slightly curved, hairless. Rough, grassy habitats, open woodland and scrub, roadside embankments, to 2000m. June-August. Throughout, except the extreme north, Ireland, Iceland and the Faeroes, but only on the mountains in the south. Sometimes cultivated in gardens. **B**: Generally local but absent from Wales and Ireland.

9* *Astragalus arenarius* L. Short, hairy, perennial; stems slender, spreading or ascending, rather woody at the base. Leaflets 2-9 paris, linear to lanceolate. Flowers purple or lilac, occasionally whitish or yellowish, 13-17mm long, few in a lax raceme. Pod oblong, pointed, 12-20mm, hairy. Grassy and rocky habitats. May-July. C Germany north to S Sweden and Finland. Local.

10 *Astragalus baionensis* Loisel. Like *A. arenarius*, but leaflets 2-6 pairs, oblong-linear and flowers pale blue, 12-14mm long. Pod 8-10mm. Maritime sands. May-July. W France. Very local.

*Astragalus baionensis*

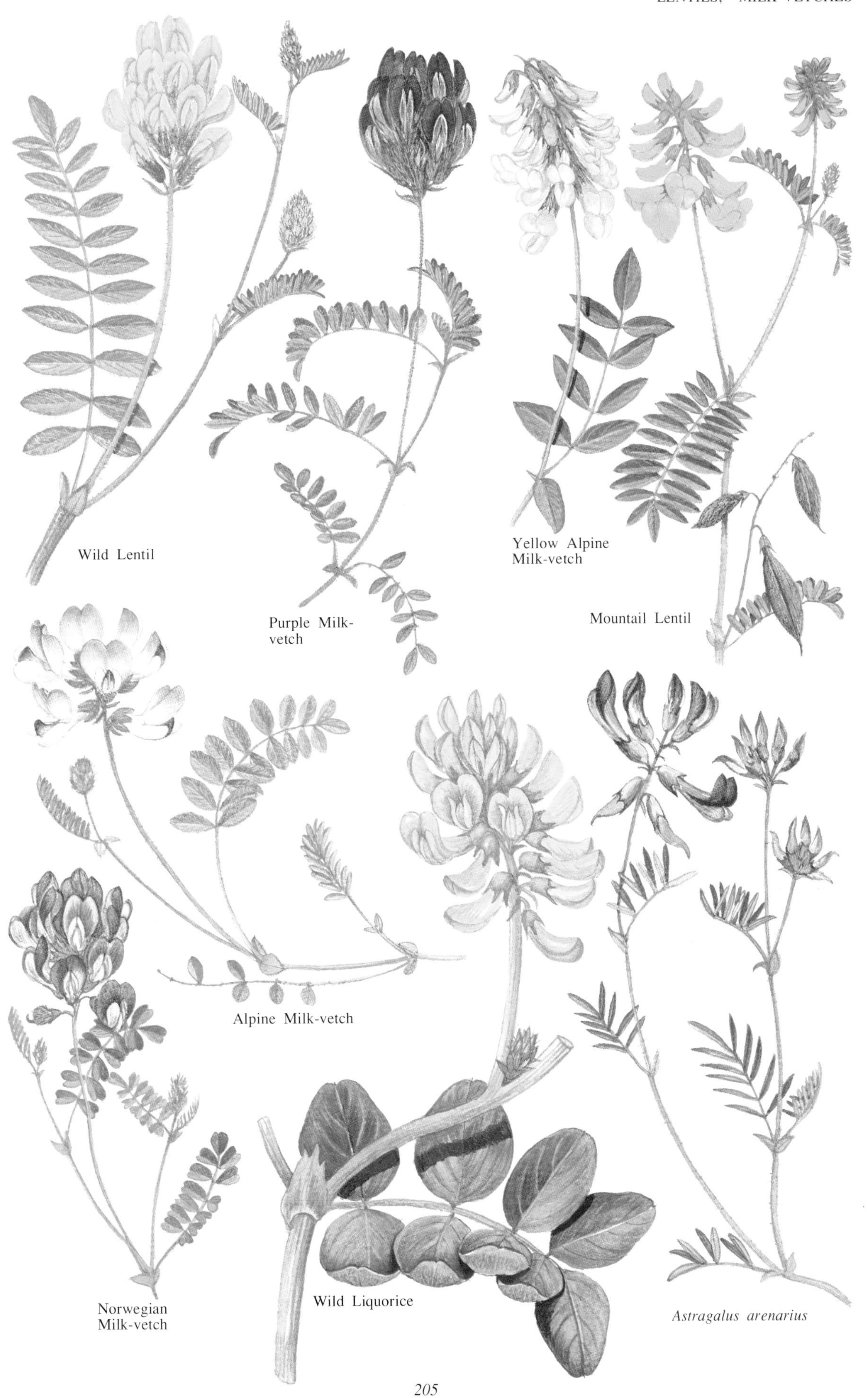
Wild Lentil
Purple Milk-vetch
Yellow Alpine Milk-vetch
Mountail Lentil
Alpine Milk-vetch
Norwegian Milk-vetch
Wild Liquorice
*Astragalus arenarius*

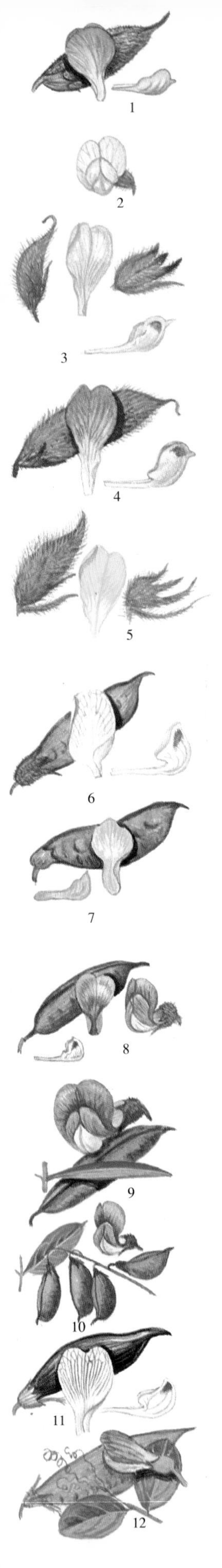


*Oxytropis*. Perennial herbs rather like *Astragalus* and often confused with it, but the keel petals always ending in a small tooth-like point. Flowers in dense rounded, long-stalked racemes. Leaflets appressed-hairy. 300 species in N temperate regions of the world. The small point or tooth on the keel-petal is the best means of distinguishing species of *Oxytropis* from *Astragalus*.

1* **Northern Milk-vetch** *Oxytropis lapponica* (Wahlenb.) Gay [*Astragalus lapponicus*]. Low tufted, hairy perennial, with stems; stipules joined together. Leaflets 8-14 pairs, lanceolate. Flowers violet-blue, 8-12mm. Pod narrow oblong, pendent, 8-15mm, with short hairs. Meadows, stony places and screes, 1800-3050m. July-August. Scandinavia and C France. Local.

2 *Oxytropis deflexa* (Pallas) DC. Like *O. lapponica*, but leaves with 11-16 pairs of leaflets and flowers whitish. Similar habitats. Arctic Norway. Rare.

3* **Yellow Oxytropis** *Oxytropis campestris* (L.) DC. [*Astragalus campestris*]. Short tufted, silkily-hairy, perennial, stemless. Leaflets 8-12 pairs, elliptical. Flowers pale yellow to whitish, sometimes violet or purple tinged, 15-20mm. Pod oval, 14-18mm, erect, appressed hairy. Mountain rocks and grassland, on base-rich or calcareous rocks, to 3000m, occasionally close to sea-level. June-July. N Britain, C & E France and Scandinavia. **B**: Rare; confined to a few localities in the Scottish Highlands.

subsp. *sordida* (Willd.) Hartman f. [*O. sordida*]. Like the type, but flowers yellowish or pale violet; pod oblong-cylindrical, slightly curved. Similar habitats and flowering time. N & E Norway and N Finland.

4* **Purple Oxytropis** *Oxytropis halleri* Bunge ex Koch [*O. sericea*, *Astragalus sericeus*]. Low silkily-hairy tufted perennial, stemless. Leaflets 10-14 pairs usually, lanceolate. Flowers bluish-purple, 15-20mm. Pod oval to elliptical, pointed, 15-20mm, appressed-hairy. Dry grassy habitats, banks and rock cliffs, generally on acid soils, to 2950m, but sometimes near sea level. July-August. N Britain, C & E France. **B**: Very local and very rare; scattered localities in Scotland, near sea-level in Perthshire.

5* **Woolly Milk-vetch** *Oxytropis pilosa* (L.) DC. [*Astragalus pilosus*]. Short to medium downy, tufted perennial, with leafy stems. Leaflets 9-13 pairs, oblong to linear-oblong. Flowers pale yellow, 12-14mm. Pod narrow-oblong, pointed, 15-20mm, densely hairy. Mountain habitats, grassy and stony places, moraines and gravels, to 2600m. June-August. S Sweden, E France and S Germany.

**Vetches** *Vicia*. Annual or perennial herbs, often climbing with the aid of leaf-tendrils; stems often ridged, but not winged. Leaves pinnate, terminating in a tendril or sometimes a point, but not a leaflet; stipules small and green. Flowers solitary, clustered or in racemes; calyx even or 2-lipped. Legume oblong, often rather flattened, curved and pointed. 150 species in N temperate regions and South America. Several are cultivated in gardens. The Broad Bean or Horse Bean, *Vicia faba*, cultivated since prehistoric times, is widely cultivated today and is sometimes casual.

6* **Wood Bitter-vetch** *Vicia orobus* DC. Medium, erect, hairy perennial; stems not ridged. Leaves ending in a short point, not a tendril; leaflets 6-15 pairs, oblong to elliptical. Flowers white with lilac or purple veins, 12-15mm, in lax, long-stalked, racemes. Pod yellowish-brown, 20-30mm, hairless. Rocky and shaded habitats, woodland and scrub, sometimes on cliffs, on mildly acid or calcareous soils, at low altitudes. May-June. Europe, except for Holland, Finland and Sweden. Local and decreasing in some areas -particularly Britain. **B**: Primarily in W England, Wales and W Scotland; coastal in N Scotland; local in Ireland.

7* *Vicia pisiformis* L. Clambering, hairless perennial to 2m. Leaflets 3-5 pairs, oval. Flowers yellow, 13-20mm, 8-30 in a rather one-sided raceme. Pod pale brown, 25-40mm, hairless. Open woodland, scrub and hedgerows. June-August. E France, Germany, C & S Sweden and SE Norway.

8* **Tufted Vetch** *Vicia cracca* L. Clambering perennial herb to 2m, often less; stem hairy or hairless. Leaflets 6-15 pairs, linear to oblong. Flowers bluish-violet, 10-12mm, in a one-sided, long-stalked raceme. Pods brown, 10-25mm, hairless. Rough grassy habitats, pastures and meadows, scrub, hedgerows, embankments and roadsides, occasionally along woodland margins, in open woods, or on coastal shingle, to 2200 m. June-August. Throughout, except Spitsbergen. One of the commonest vetches in our area. **P**: Chiefly bees. **B**: Throughout most of area.

9 **Fine-leaved Vetch** *Vicia tenuifolia* Roth (incl. *V. boissieri* Freyn, *V. elegans* Guss.). Like *V. cracca*, but flowers larger, 12-18mm, pale lilac or bluish-lilac, and the leaflets linear-lanceolate. Pod 20-35mm long. Similar habitats and flowering time. Continental Europe north to S Sweden; locally naturalized in Britain and Holland. **B**: Casual on disturbed ground in Britain.

10 **Danzig Vetch** *Vicia cassubica* L. Like *V. cracca*, but a smaller plant to only 60cm, with purple or pinkish flowers with a whitish keel, only 4-15 to a raceme (not 10-30). Pod yellowish. Grassy and rocky habitats. June-August. France, Germany and S Scandinavia.

11* **Wood Vetch** *Vicia sylvatica* L. Clambering perennial to 2m, though often less, usually hairless. Leaflets 5-12 pairs, oblong. Flowers white or pale lilac with purple veins, 12-20mm, in rather one-sided, long-stalked racemes. Pod black, 25-30mm, hairless. Open rocky woodland, scrub and woodland margins, sometimes on cliffs and in maritime habitats, to 2000m. June-August. Europe, except for Belgium and Holland. In coastal habitats, especially shingle beaches, a dwarf leafy, few-flowered form is sometimes found. **B**: Local throughout much of the area.

12 *Vicia dumetorum* L. Like *V. sylvatica*, but with only 3-5 pairs of leaflets and blue or purple flowers. Pods brown, 25-60mm. Rocky and grassy habitats. E France, Germany, Denmark and SC Sweden.

leaf of *Vicia fumetorum*

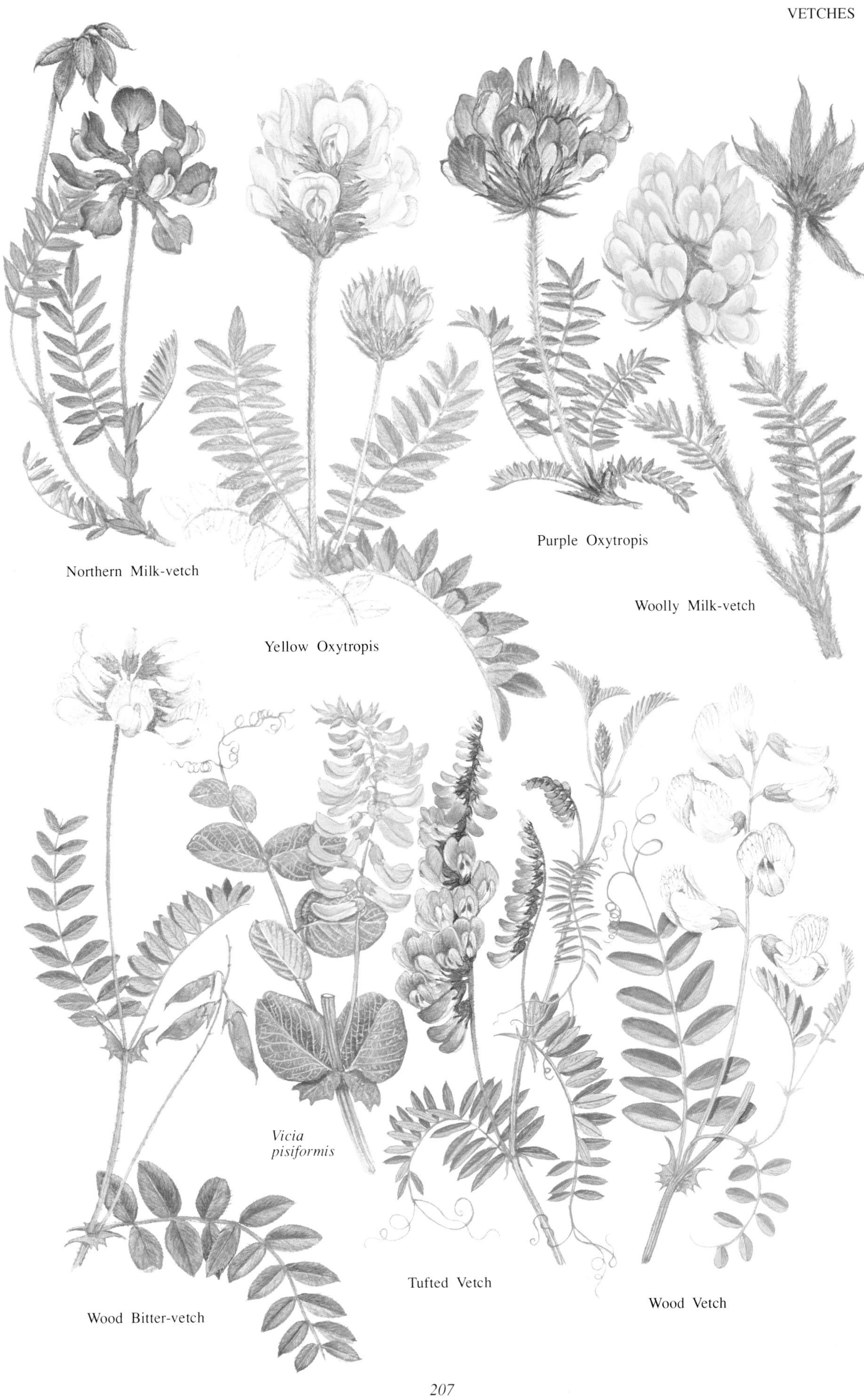
Northern Milk-vetch
Yellow Oxytropis
Purple Oxytropis
Woolly Milk-vetch
Vicia pisiformis
Wood Bitter-vetch
Tufted Vetch
Wood Vetch

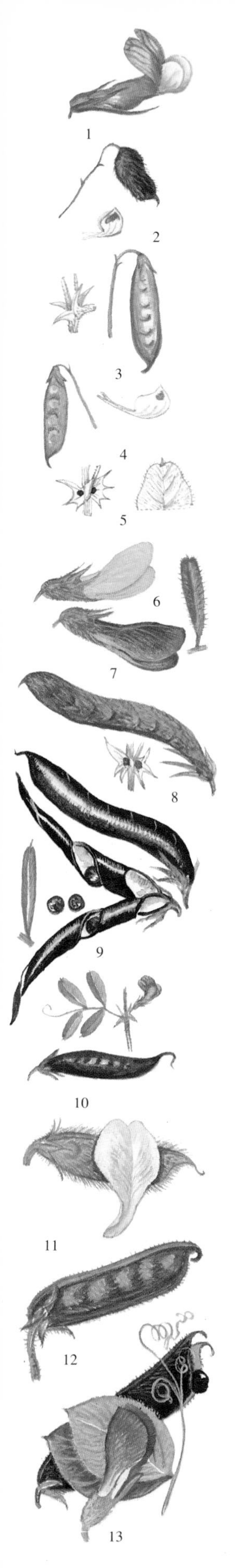


1* **Fodder Vetch** *Vicia villosa* Roth. Medium to tall, variable, hairy annual. Leaflets 4-12 pairs, linear to narrow-elliptical. Flowers violet, purple or blue, sometimes with white or yellowish wings, 10-20mm, in long-stalked racemes; calyx tube swollen at the base. Pod brown, 20-40mm, hairless. Cultivated, bare, waste and disturbed ground. June-November. France and Germany; naturalized in Britain and parts of N Europe. **B**: Scattered localities in Britain.

subsp. *varia* (Host) Corb. [*V. varia*]. Like the type, but hairless or with adpressed (not spreading) hairs and purple or white flowers with violet wings. Similar habitats and distribution.

2* **Hairy Tare** *Vicia hirsuta* (L.) S.F. Gray. Short to medium hairy annual. Leaflets 4-10 pairs, linear to oblong. Flowers whitish with a pale lilac tinge, small, 2-4mm, 1-8 in a short raceme. Pod black 6-11mm, usually hairy. Dry habitats, rough grassy meadows, woodland margins, roadsides and scrub, cultivated land, on mildly acid to calcareous soils, to 1800m. May-August. Throughout, except the Faeroes and Spitsbergen; naturalized in Iceland. A widespread weed of cultivation established almost throughout the world. **B**: Throughout, except for parts of N Scotland.

3* **Slender Tare** *Vicia laxiflora* Brot. [*V.tenuissima*]. Short to medium almost hairless annual. Leaflets 2-5 pairs, linear. Flowers pale purple, 6-9mm, 2-5 in a long-stalked raceme. Pod brown, 12-17mm, hairy or hairless, with 5 or more seeds. Rough grassy habitats, scrub, banks, hedgerows, tracks and roadsides, on neutral or slightly acid soils usually. June-August. C & S Britain, Belgium, Holland and France. **B**: Confined to C & S England, north as far as Bedfordshire; local and declining.

4* **Smooth Tare** *Vicia tetrasperma* (L.) Schreber. Similar to *V. tenuissima*, but leaflets 3-6 pairs, often wider and racemes only 1-2 flowered; flowers pale purple, 4-8mm. Pod brown, often 4-seeded, hairless. Rough grassy habitats, pastures, open scrub, hedgerows and roadsides on moist calcareous to slightly acid soils, to 1800m. May-August. Throughout, except the Faeroes, Iceland and Spitsbergen. **B**: North to the Humber; rare and casual in Scotland and Ireland.

5* **Bush Vetch** *Vicia sepium* L. Medium to tall, usually hairy, clambering perennial. Leaflets 3-9 pairs, oval to oblong; stipules with a dark spot near the base. Flowers purplish-blue, 12-15mm, in clusters of 2-6. Flower clusters with a very short stalk. Pod black, 20-35mm, hairless. Rough grassy habitats, meadows, woodland margins, scrub, track and roadsides, hedgebanks on slightly acid to calcareous soils, to 2150m. May-November. Throughout, except the Faeroes and Spitsbergen. **P**: Bumble-bees. Throughout Britain and Ireland; scarce in parts of the east.

6 *Vicia pannonica* Crantz. Like *V. sepium*, but a short to medium plant with narrower leaflets. Flowers pale yellow, 14-22mm, in clusters of 2-4. Pod yellowish, with adpressed hairs. Grassy and rocky habitats. June-August. C & S France; naturalized in Holland and Germany, casual in Britain.

7 *V. p.* subsp. *striata* (Bieb.) Nyman [subsp. *purpurascens*, *Vicia purpurascens*]. Like the type, but flowrs dirty-purple or brownish-purple. Similar distribution.

8* **Common Vetch** *Vicia sativa* L. Variable medium to tall clambering hairy annual; tendrils sometimes unbranched; stipules toothed, with a dark spot near the base. Flowers pink to dark-reddish purple with darker wings, 18-30mm long, solitary or two together. Pod yellowish-brown to brown, 25-70mm, usually hairy, breaking calyx when mature. Grassy habitats, meadows and pastures, hedgebanks, roadsides, cultivated and bare ground, to 2200m. April-September. Throughout; naturalized in much of the north and in Iceland. Often persisting as a relict of former cultivation – long cultivated as a fodder crop and for green manure. **B**: Naturalized or casual over much of the area.

9 *V. s.* subsp. *nigra* (L.) Ehrh. [*V. angustifolia*, *V. cuneata*, *V. heterophylla*, *V. pilosa*]. Like the type but flowers only 10-18mm long and pod dark brown or black. Grassy, rocky and bushy habitats. May-July. Similar distribution.

10 **Spring Vetch** *Vicia lathyroides* L. [incl. *V. olbiensis*]. A short spreading, often prostrate, hairy annual. Leaflets 2-4 pairs, linear to elliptical. Flowers purple, 5-8mm, solitary. Pod black, 15-30mm, hairless. Grassy habitats, arable land, heaths, commons, road verges and waste places, generally on light well-drained soils. May-June. Throughout, except the Faeroes, Iceland and Spitsbergen. **B**: Local throughout Britain and E Ireland though most commonly near the coast.

11* **Yellow Vetch** *Vicia lutea* L. A low tufted, prostrate, densely hairy to almost hairless annual with branched or unbranched tendrils. Leaflets 3-10 pairs, linear to oblong. Flowers pale yellow, often tinged with purple, 20-35mm, solitary or 2-3 together. Pod yellowish-brown to black, 20-40mm, usually hairy. Rough grassy habitats near or along the coast, stabilized shingle and sand-dunes, cliff tops and cliff ledges. June-September. Britain and France; naturalized locally in Germany. Readily distinguished by its yellow flowers. **B**: Rare; scattered coastal localities north to Angus.

12* **Bithynian Vetch** *Vicia bithynica* (L.) L. A medium clambering, almost hairless annual; stems 4-angled; stipules large and toothed. Leaflets 2-3 pairs, oval to lanceolate. Flowers purple with creamy-white wings and keel, 16-20mm, solitary or 2-3 together on a short stalk. Pod brown or yellowish, 25-50mm, hairy. Grassy habitats, hedgerows and scrub, field boundaries and cliff tops, generally close to the sea; local. May-June. England and France. **B**: Rare and declining; confined to scattered localities in England and Wales north to Wigtownshire.

13 *Vicia narbonensis* L. [incl. *V. serratifolia*]. A short to medium erect hairy annual like *V. bithynica* with broader leaflets; lower leaves often without tendrils; stipules toothed or not. Flowers dark purple, 10-30mm, 1-6 to a raceme. Pod black or dark brown, 30-70mm, hairless with a toothed margin. Grassy, rocky and bushy habitats. May-July. France; naturalized in Germany.

Hairy Tare
Fodder Vetch
Smooth Tare
Slender Tare
Bithynian Vetch
Common Vetch
Bush Vetch
Yellow Vetch

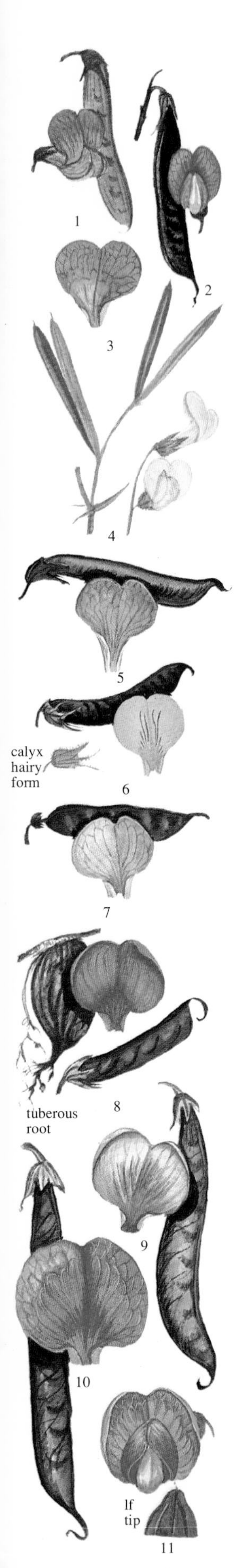


**Peas** *Lathyrus*. Annual or perennial herbs, often with leaf tendrils. Like *Vicia* but usually with winged stems and fewer leaflets. Leaves sometimes reduced to grass-like phyllodes. Flowers in axillary racemes, sometimes solitary. Pod usually narrowly oblong and compressed. 110 species concentrated primarily in N temperate regions but also in the mountains of tropical Africa. A number are cultivated in gardens, including the fragrant Sweet Pea, *Lathyrus odoratus*, from Crete.

1* **Spring Vetch** *Lathyrus vernus* (L.) Bernh. Short to medium, hairless or slightly hairy, tufted perennial; stems, not winged. Leaves terminating in a point, not a tendril; leaflets 2-4 pairs, oval to lanceolate, pointed. Flower reddish-purple, becoming blue, 13-20mm long, 3-10 in long racemes. Pod brown, 40-60mm, hairless. Woodland, scrub and hedgebanks, generally on calcareous soils, to 1900m. April-June. Continental Europe, except the extreme north; naturalized in Belgium and Holland. Widely cultivated in gardens where white, pink and blue forms exist as well as the normal-coloured form. **P**: bees.

2* **Black Pea** *Lathyrus niger* (L.) Bernh. Medium to tall, erect, almost hairless perennial. Like *L. vernus*, but with 3-6 pairs of blunt leaflets. Flowers smaller, 10-15mm. Rocky wooded habitats. May-July. Continental Europe north to C Scandinavia, but absent from Holland. **B**: Formerly native in Scotland, but now thought to be extinct there.

3* **Sea Pea** *Lathyrus japonicus* Willd. Low prostrate rather greyish-green perennial, often hairless; stems angled, not winged. Leaflets 2-5 pairs, elliptical, pinnately-veined; leaves occasionally without a tendril. Flowers purple, becoming blue, 18-22mm long, 2-7 to a raceme. Pods brown, 30-50mm, hairless. Coastal habitats, shingle and stabilized sand-dunes, occasionally along stony lake margins inland. June-August. Coasts of W & N Europe including Britain and Ireland. Widespread from Europe to China and Sakhalin, as well as North America. An early coloniser of shingle, often forming extensive colonies. The seeds have a long viability (4-5 years) in sea water. **B**: Very local along coasts, mainly in the S and E, but also in S Wales; Ireland, found only in Kerry.

subsp. *maritimus* (L.) P.W. Ball [*Lathyrus maritimus*, *Pisum maritimum*]. Like the type, but flowers smaller, 14-18mm long, in racemes of 5-12. Similar habitats. Coasts of W Europe and the Baltic.

4 *Lathyrus pannonicus* (Jacq.) Garcke [*L. albus* subsp. *asphodeloides*]. Short to medium, hairless or slightly hairy perennial; stems narrowly winged or ridged. Leaflets 1-4 pairs, linear to narrowly-lanceolate. Flowers pale cream, tinged reddish-purple, 12-20mm, in long-stalked racemes. Pods pale brown, 30-65mm, hairless. Grassy habitats, banks, woodland margins, scrub, to 1200m. May-June. NW France and Germany. The roots are tuberous and fleshy.

5* **Bitter Vetch** *Lathyrus linifolius* (Reichard) Bassler var. *montanus* (Reichard) Basler [*L. montanus*, *L. macrorrhizus*]. Short to medium erect, hairless, perennial with winged stems. Leaves ending in a point, not a tendril; leaflets 2-4 pairs, linear to elliptical. Flowers crimson, becoming bluish, 10-16mm, in long-stalked racemes. Pod reddish-brown, 25-45mm, hairless. Grassy habitats, road verges, banks, shrub, woodland margins and open woodland, on acid soils in hilly and mountainous regions, to 2200m. April-July. Throughout, except the Faeroes and Iceland. Scattered throughout most of Europe except for the SE. **B**: Scattered localities throughout, but rare in East Anglia. The fleshy root-tubers were formerly gathered for food and in Scotland were used to flavour some types of whisky.

6* **Meadow Vetchling** *Lathyrus pratensis* L. Variable, clambering perennial to 120cm, hairy or hairless, with winged stems. Leaves with tendrils; leaflets 1 pair, linear-lanceolate to elliptical. Flowers yellow 10-16mm, in long-stalked racemes. Pods black, sometimes hairy. Rough grassy habitats, meadows, pastures, hedgerows, scrub and woodland margins, roadsides, sometimes close to the sea, on calcareous to slightly acid soils, to 2150m. May-August. Throughout, except Spitsbergen. Widespread from Europe and N Africa to Asia, Siberia and the Himalaya, but rare in the Mediterranean region. **B**: Common throughout.

7* **Marsh Pea** *Lathyrus palustris* L. [*L. pilosus*]. Medium to tall clambering, slightly hairy perennial, to 120cm; stems narrowly winged. Leaves with branched tendrils; leaflets 2-5 pairs, linear to lanceolate. Flowers purplish-blue, 12-20mm, in long-stalked racemes. Pods brown, 25-60mm, hairless. Damp or wet grassy habitats among tall herbaceous vegetation, sometimes among bushes, on calcareous soils, generally at low altitudes. June-July. Throughout, except Spitsbergen. Widespread from Europe to E Asia, Japan and E North America. **B**: Local in scattered localities in England, Wales and Ireland.

8* **Tuberous Pea, Fyfield Pea** *Lathyrus tuberosus* L. Medium to tall, usually hairless, clambering perennial, to 120cm; stems angled but not winged. Leaves with branched tendrils; leaflets 1 pair, elliptical to oblong, with weak parallel veins. Flowers bright crimson, 12-20mm, in long-stalked racemes, slightly fragrant. Pod brown, 20-40mm, hairless. Grassy habitats, rough scrub, hedgerows and arable land, on neutral or calcareous soils. June-July. Belgium, Holland, France and Germany; naturalized in Britain, Denmark and Sweden. Sometimes cultivated in gardens, more widely so in the last century. The tuberous roots, which give the plant its name, are fleshy and edible. **B**: Scattered localities north to S Scotland.

9* **Narrow-leaved Everlasting-pea** *Lathyrus sylvestris* L. Medium to tall clambering, hairy or hairless, perennial to 2m; stipules narrower than the winged stem. Leaves with branched tendrils; leaflets 1 pair, linear to lanceolate. Flowers purplish-pink, 13-20mm, in long-stalked racemes. Pods brown, 40-70mm, hairless. Woodland and woodland margins, scrub, hedgebanks, roadsides. June-August. Europe, except the extreme north. **B**: Local; scattered localities north to N Wales, scarce further north.

10* **Broad-leaved Everlasting-pea** *Lathyrus latifolius* L. [*L. megalanthus*]. Very variable, tall clambering perennial to 3m, with widely winged stems, hairy or hairless. Leaflets 1 pair, linear to elliptical, with prominent parallel veins; tendrils branched. Flowers magenta-purple or pink, rarely white, 20-30mm, in long-stalked racemes; petals thick. Pods brown, 50-110mm, hairless. Rough grassy habitats, scrub, hedgerows and railway embankments, to 1500m. France; widely naturalized in S Britain, Belgium and Germany. Widely cultivated in gardens including forms with pink or white flowers. Plants may sprawl on the ground or clamber over bushes and fences. **B**: Scattered localities.

11 *Lathyrus heterophyllus* L. Similar to *L. latifolius*, but leaves with 2-3 pairs of leaflets and flowers smaller, 12-22mm. Similar habitats and flowering time. France, Germany and S Sweden. Occasionally cultivated in gardens.

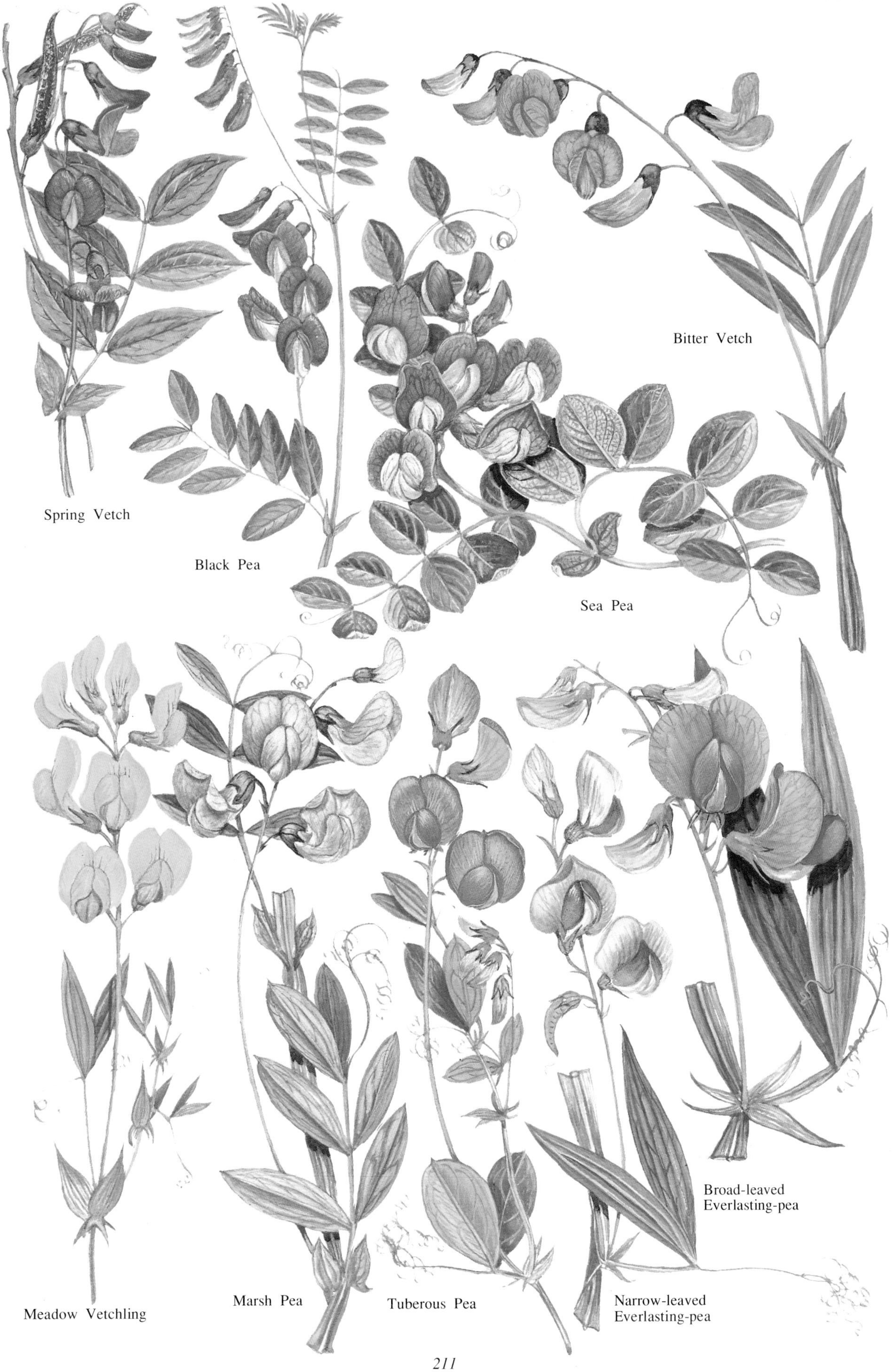

Spring Vetch

Black Pea

Bitter Vetch

Sea Pea

Meadow Vetchling

Marsh Pea

Tuberous Pea

Narrow-leaved Everlasting-pea

Broad-leaved Everlasting-pea

1* *Lathyrus sphaericus* Retz. Short to medium, hairless or somewhat hairy annual; stems angled, but not winged. Leaves ending in a branched tendril; leaflets 1 pair, linear to linear-lanceolate. Flowers orange-red, 6-13mm, solitary on slender stems. Pod brown, 30-70mm, hairless. Grassy and rocky habitats, scrub, hedgebanks and roadsides, generally at rather low altitudes. June-August. N & W France, Denmark and S Sweden.

*Lathyrus angulatus* L. Like *L. sphaericus*, but flowers purple or pale blue, occasionally 2 together, borne on longer stalks, 20-70mm. Similar habitats and flowering time. France.

2* **Hairy Vetchling** *Lathyrus hirsutus* L. Medium to tall clambering, slightly hairy, annual, to 120cm; stems winged. Leaves ending in a branched tendril; leaflets 1 pair, linear to oblong. Flowers reddish with pale blue wings and a whitish keel, 7-15mm, solitary or 2-3 together. Pod brown, 20-50mm, silky-hairy. Grassy habitats, bare and waste places, occasionally on cultivated land, especially arable fields. June-August. France, Belgium and Germany. **B**: Naturalized but rare and generally casual N to S Scotland.

3* **Grass Vetchling** *Lathyrus nissolia* L. Short to tall, hairless or slightly hairy, annual; stems not winged. Leaves simple, grass-like, pointed; stipules small. Flowers crimson, 8-18mm, solitary or 2 together, on a long slender stalk. Pod pale brown, 30-60mm, hairy or not. Grassy habitats, scrub and banks, sometimes along field and woodland boundaries, on mildy acid or neutral soils. May-July. Britain, Belgium, Holland, France and Germany. Difficult plant to observe or find ,especially when not flowering; the leaves are reduced to a blade-like midrib (phyllode) without a tendril. An autumn germinating annual like *L. aphaca*. **B**: Local in England and Wales, rare in Scotland.

4* **Yellow Vetchling** *Lathyrus aphaca* L. Medium to tall, hairless, clambering annual with winged stems. Tendrils present; stipules large and leaf-like, paired, arrowhead-shaped, grey-green. Flowers yellow, 16-18mm, usually solitary, on long stalks. Pod brown, 20-35mm, hairless. Dry grassy habitats, edge of scrub and woodland, waysides and field boundaries, generally on light well-drained calcareous soils. May-August. C & E France; naturalized in Britain, Belgium, Holland and Germany. The leaf is reduced to a tendril, the function of the leaf-blade being taken over by the large leaf-like stipules. Although an autumn-germinating annual, many seedlings may perish during severe weather. **B**: Local and declining; England N to Cambridgeshire, casual elsewhere.

**Restharrows** *Ononis*. Herbs or dwarf shrubs, usually glandular hairy. Leaves trifoliate or simple, toothed. Flowers in panicles or spikes with a beaked keel. Fruit pod oblong or oval, splitting. 70 species primarily in Europe and W Asia. **P**: Bees.

5* **Large Yellow Rest-harrow** *Ononis natrix* L. Variable short to medium sticky subshrub, often forming dense mounds. Leaves trifoliate, with linear to oval leaflets. Flowers deep yellow, veined red or violet on the outside and in bud, 12-20mm, in leafy panicles. Pod 12-25mm. Dry open habitats, rocky places and open scrub, often on limestone, to 2100 m. May-August. France and Germany. Occasionally cultivated in gardens. Variable – mountain forms tend to be more compact with smaller leaves.

6 *Ononis pusilla* L. [*O. columnae*] Like *O. natrix*, but a short perennial with long-stalked leaves. Flowers plain yellow, smaller than *O. natrix*, only 5-12mm, and in lax leafy spikes. Pod 6-8mm. Rocky and grassy habitats. June-August. France; naturalized in Belgium.

7* **Small Rest-harrow** *Ononis reclinata* L. Low, half-prostrate annual, sticky and hairy; the stem and leaves covered with shiny gland dots. Leaves trifoliate, with oval leaflets. Flowers purplish or pink, 5-10mm, in leafy racemes. Pod 8-14mm, pendent when ripe. Rough grassland, scrub and cliffs, near or on the coast, generally on sandy soils. May-June. SW Britain and France. W Europe to the Mediterranean region, the Middle East, Arabia and NE Africa. **B**: Very rare; isolated localities in SW England, S Wales and the Channel Islands.

8* **Common Rest-harrow** *Ononis repens* L. [*Ononis spinosa* subsp. *procurrens*]. Short to medium erect to spreading subshrub; stems often rooting near base, sometimes with soft, weak, spines. Leaves simple or trifoliate; leaflets oval, usually notched at the tip. Flowers pink or purplish, 15-20mm, in lax leafy racemes; wings equalling keel. Pod 5-7mm. Dry grassy habitats, meadows and pastures, scrub, banks and road-verges, on well-drained calcareous or basic soils. June-September. Throughout, except the extreme north, the Faeroes and Iceland. Probably hybridises with *O. spinosa* where the two grow in close proximity. **B**: Throughout much of region but rare in C & N Scotland.

9* **Spiny Rest-harrow** *Ononis campestris* G. Koch [*O. spinosa*]. Short to tall subshrub with stiffly-spiny erect stems, not rooting below, with 2 lines of hairs. Leaves mostly trifoliate; leaflets linear to oval, pointed or blunt but not notched. Flowers pink or reddish-purple, 15-20mm, in lax racemes. Rough grassy habitats, pastures, meadows, banks and roadsides, as well as rocky places, sometimes in coastal habitats, on well-drained calcareous soils usually, to 1800m. April-September. Throughout, except Ireland, the Faeroes, Iceland, Finland and Spitsbergen. Europe to Asia and N Africa. **B**: N to the Humber, rare further N and absent from SW England.

10 *Ononis arvensis* L. [*O. hircina*, *O. intermedia*]. Like *O. spinosa*, but not spiny and generally with 2 flowers at each node (not 1). Similar habitats. May-August. C & S Scandinavia and Germany.

Common Rest-harrow

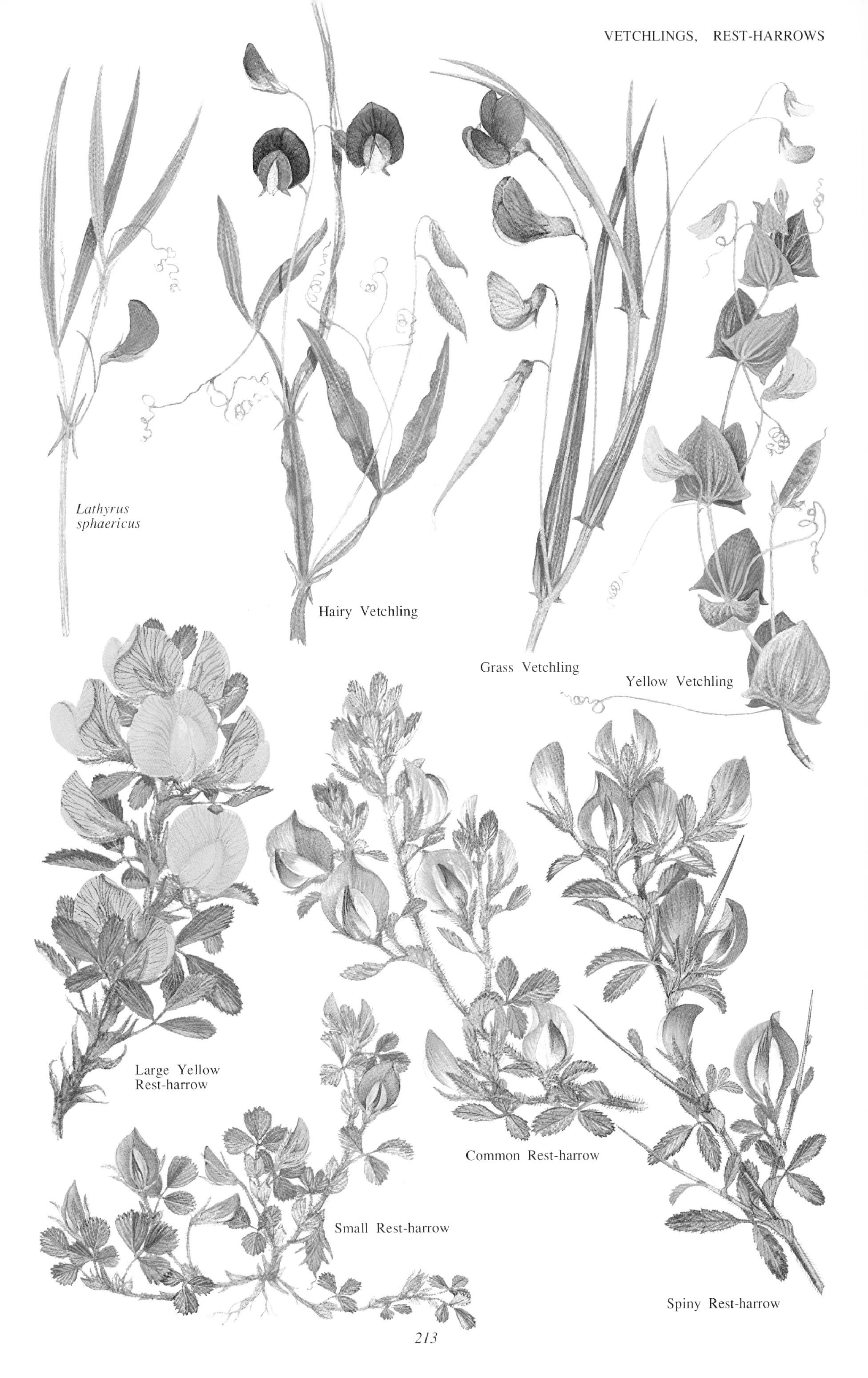
*Lathyrus sphaericus*
Hairy Vetchling
Grass Vetchling
Yellow Vetchling
Large Yellow Rest-harrow
Common Rest-harrow
Small Rest-harrow
Spiny Rest-harrow

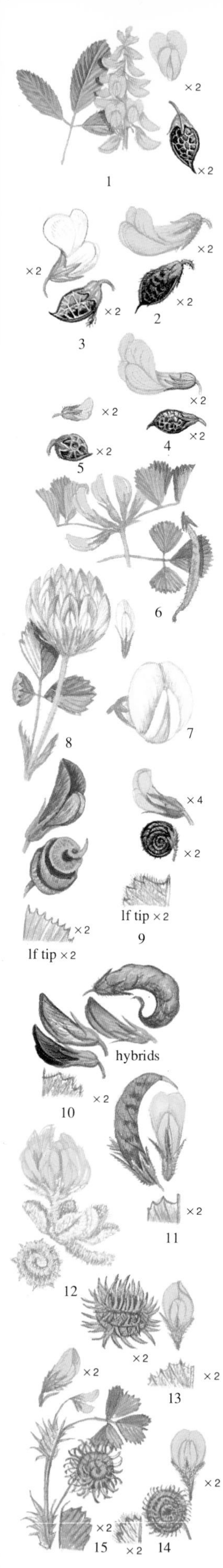


**Melilots** *Melilotus*. Annual or short-lived perennial herbs. Leaves trifoliate, leaflets usually toothed. Calyx has slightly uneven teeth. Corolla yellow or white, deciduous. Fruit pod, rounded to oblong, 1-2-seeded, usually indehiscent. 12 species. Leaves smell of new mown hay.

1 **Small-flowered Melilot** *Melilotus dentata* (Waldst. & Kit.) Pers. Medium to tall branched biennial to 1.5m tall, generally with erect stems. Leaflets elliptical, sharply toothed; stipules toothed at the base. Flowers bright yellow, small, 3-3.5mm, in many-flowered racemes, the wing petals longer than the keel. Pod oval, 4.5-5.5mm, hairless, slightly net-veined on the surface, blackish-brown when ripe. Saline meadows, salt steppes, occasionally on river banks. June-August. Denmark, Germany and S Sweden.

2* **Tall Melilot** *Melilotus altissima* Thuill. Medium to tall hairless biennial or short-lived perennial to 1.5m, stems erect and branched. Leaflets oblong, sharply toothed; stipules slender, untoothed. Flowers yellow, 5-7mm long, in many-flowered racemes, the wings equalling the keel. Pod oval, 5-6mm, hairy and net-veined, black when ripe, with a persistent style. Damp and saline habitats, roadsides, woodland margins, waste ground and coastal habitats, to 2000m. June-August. North to S Scandinavia; naturalized in S Britain, S and E Ireland. **B**: South of the Tyne; rarer in the SW and Wales; mainly S and E Ireland.

3* **White Melilot** *Melilotus alba* Medicus. A spring-germinating annual rather like *M. altissima*, but the flowers white, slightly smaller, 4-5mm, the standard petal longer than the wings and keel. Pod 3-5mm, hairless, greyish-brown when ripe. Open habitats, especially on arable land and fields, in waste places and along roadsides, a fairly frequent weed or ruderal, to 1850m. July-September. Widespread, more local or rare in Holland and Scandinavia; naturalized in Britain and Belgium. **B**: Widely naturalized north to S Scotland, though primarily in C and S England and Wales.

4* **Ribbed Melilot** *Melilotus officinalis* (L.) Pallas (*M. arvensis* Wallr.). Medium to tall hairless annual or perennial to 2.5m tall, with spreading to erect stems. Leaflets of lower leaves oval, sharply toothed, those of the upper leaves narrower; stipules mostly untoothed. Flowers yellow, 4-7mm, in slender lax racemes, the wings longer then the keel. Pod oval, 3-5mm, rough but hairless, brown when ripe, usually one-seeded. A common weed, generally on rather heavy or saline soils, to 2000m. July-September. France and Germany, rarer in Belgium and Holland; naturalized in Britain, Ireland and Scandinavia. **B**: Much of England south of the Tyne, but rarer in the W and in Ireland, casual elsewhere.

5 **Small Melilot** *Melilotus indica* (L.) All. [*M. parviflora*]. Like *M. officinalis*, but always a short to medium annual with narrower leaflets. Flowers much smaller, 2-3mm. Pod 1.5-3mm, olive-green when ripe. Cultivated land, waste places, generally at low altitudes. June-October. C and S France; naturalized in W Europe. **B**: Naturalized north to S Scotland, commonest in S England and Wales.

6 **Trigonella** *Trigonella monspeliaca* L. Short, hairy annual. Leaves trifoliate, the leaflets oval, untoothed or finely toothed. Flowers yellow, 4mm, in lateral umbel-like clusters, short-stalked. Pod linear, slightly curved, 7-17mm, pendent, usually hairy and with oblique veins. Waste ground, bare places and cultivated land. June-July. Belgium, W and C France.

7* **Classical Fenugreek** *Trigonella foenum-graecum*. Short to medium, erect, slightly hairy annual. Leaves trifoliate, the leaflets oblong, toothed near the apex. Flowers cream flushed with purple at the base, 10-15mm long, solitary or paired, in the axils of the upper leaves. Pod erect, linear, 70-100mm, hairless. April-June. Continental Europe except much of the north.

8 *Trigonella caerulea* (L.) Ser. Taller and less hairy, with toothed leaflets notched at the end. Flowers blue or white, 5-6mm, in stalked clusters. Pod oval, 4-5mm, erect or spreading. Arable land, waste and bare places. June-August. Cultivated for fodder and widely naturalized. Unknown origin.

**Medicks** *Medicago*. Annual or perennial herbs or small shrubs with trifoliate leaves. Flowers in axillary, stalked racemes. Calyx with more or less equal teeth. Corolla deciduous. Fruit-pod generally longer than the calyx, coiled or curved, often spiny. 120 species.

9* **Black Medick** *Medicago lupulina* L. Low, often prostrate, usually hairy annual. Leaflets rounded to rhombic or oblong, toothed or not. Flowers yellow, 2-3mm, many to a raceme. Pod kidney-shaped, 1.5-3mm, without spines, black when ripe. Short permanent grassland, often over limestone, on well drained soils, sometimes on mown verges, to 2300m. April-September. Throughout, except Spitsbergen. **B**: Throughout, mostly coastal in the north; scarce in Scotland and N Ireland.

10* **Lucerne** *Medicago sativa* L. Medium to tall hairy perennial, erect to spreading. Leaflets oval to linear, toothed at the apex; stipules lanceolate, untoothed, or toothed at the apex. Flowers blue to violet, 7-11mm, in rather short racemes. Pod spiralled, with 1-3 turns, leaving a hole in the centre, 4-6mm diameter, hairless to hairy. Widely cultivated as Alfalfa, and naturalized, to 1800m. June-July. Throughout, except the far north. **B**: Throughout, but scarce in NW Britain and Ireland. Hybridizes with subsp. *falcata*.

11 **Sickle Medick** *M. s.* subsp. *falcata* (L.) Arcangeli [*M borealis*, *M. falcata*]. Like the type, but flowers yellow, 5-8mm; pods almost straight or kidney-shaped. Grassland on calcareous or slightly acid soils; very sensitive to grazing. **B**: Restricted to East Anglia.

12 **Coastal Medick** *Medicago marina* L. Like *M. sativa*, but a white-hairy, densely leafy plant with flowers in stalked heads. Pod coiled, 5-7mm diameter, white-hairy and with small spines. Maritime habitats, sandy and rocky places. May-July. W France.

13* **Spotted Medick** *Medicago arabica* (L.) Hudson. Short to medium, slightly hairy or hairless spring-germinating annual. Leaflets heart-shaped, toothed near the apex, generally with dark blotches in the centre.Flowers yellow, 5-7mm, solitary or 2-4 together. Pod spiralled with 4-7 turns, 5-6mm diameter, hairless, usually spiny. Grassy habitats, waste ground, on well-drained soils. April-August. Britain, Belgium, Holland, France and Germany; naturalized in Ireland, and S Sweden. **B**: S and E England, especially by the coast, scattered localities further north and in Wales; naturalized but rather rare in Ireland.

14* **Bur Medick** *Medicago minima* (L.) Bartal. Low to short, hairy, erect to spreading autumn-germinating annual. Leaflets oval to heart-shaped, toothed towards the apex; stipules lanceolate, usually untoothed. Flowers bright yellow, 4-4.5mm, in stalked clusters of 1-6. Pod laxly spiralled with 3-5 turns, 3-5mm diameter, spiny and slightly hairy. Disturbed ground, tracks, sand-dunes, May-July. North to S Sweden. **B**: Breckland and coasts of SE England.

15 **Toothed Medick** *Medicago polymorpha* L. [*M. denticulata*, *M. hispida*, *M. nigra*, *M. polycarpa*]. Like *M. minima*, but stipules toothed and pods hairless, with 1 1/2-4 turns, usually spiny. Open habitats, short turf or sandy places, often close to the sea. May-August. Britain, France and Germany; naturalized in Belgium and Holland. B: S and E England, naturalized or casual elsewhere, including S Scotland.

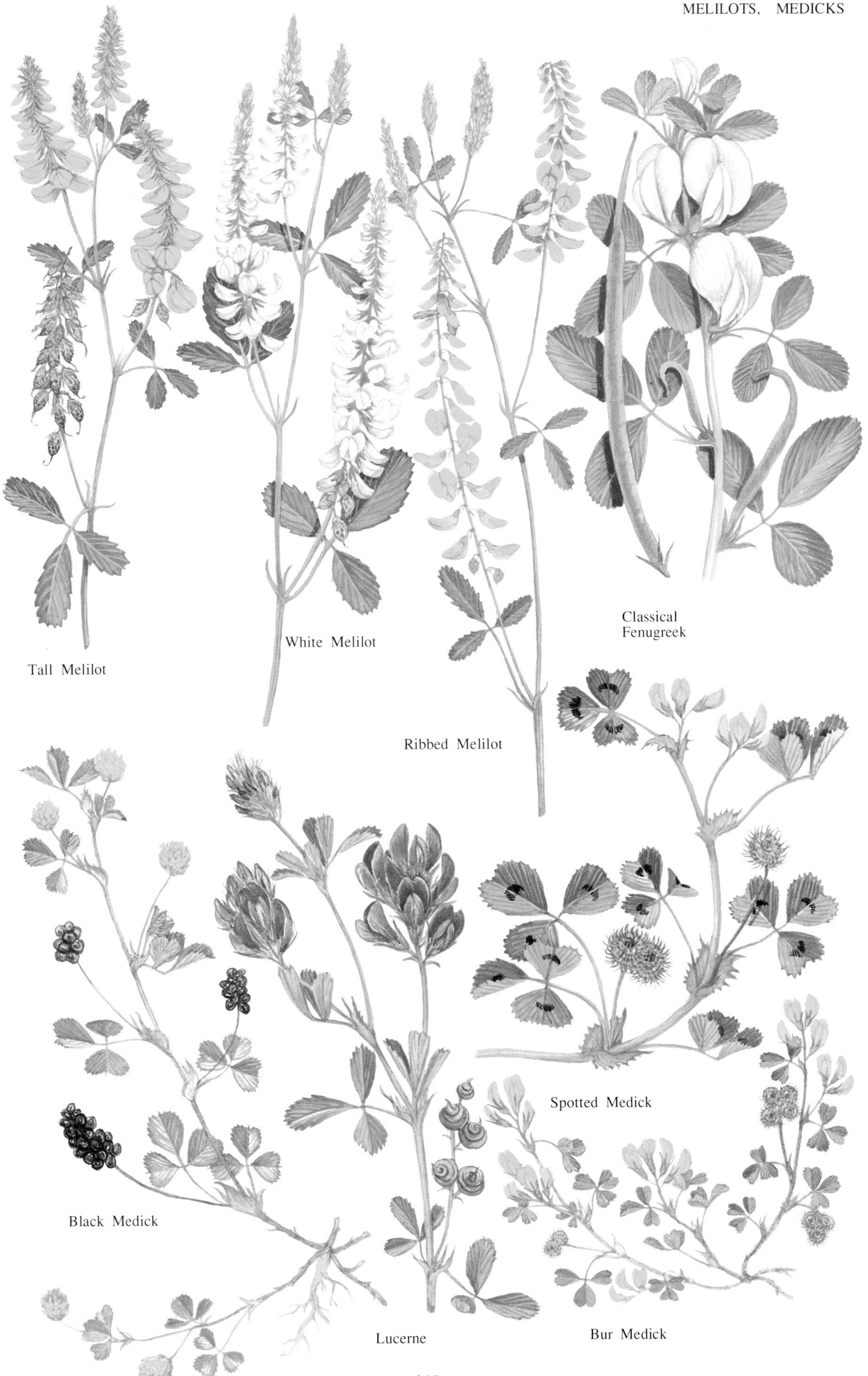

Tall Melilot

White Melilot

Ribbed Melilot

Classical
Fenugreek

Black Medick

Lucerne

Spotted Medick

Bur Medick

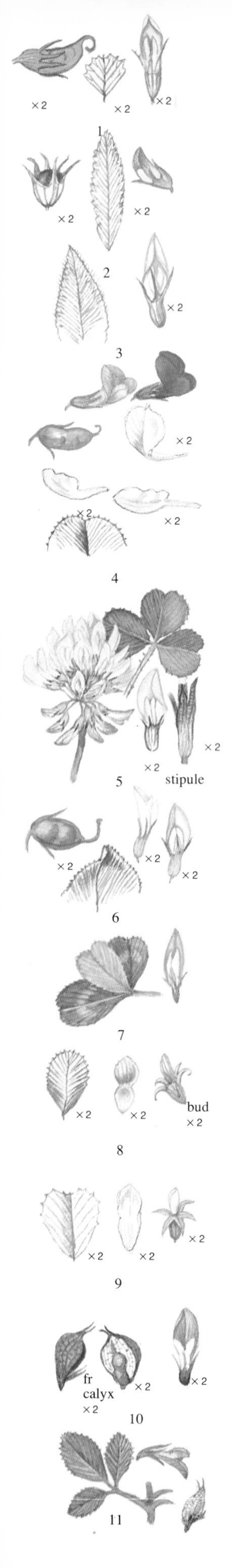


**Clovers** *Trifolium*. Annual, biennial or perennial herbs with trifoliate leaves; leaflets usually toothed. Flowers in heads or short spikes. Calyx with equal or subequal teeth. Corolla usually persistent, the wings longer than the keel. Pods usually rather small and enclosed within the calyx, usually containing 1-4 seeds. 300 species in the temperate subtropical regions.

1* **Fenugreek** *Trifolium ornithopodioides* L. [*Trigonella ornithopodioides*]. Low spreading hairless annual, scented of new mown hay or coumarin. Leaflets oval, narrowed at the base, finely toothed, with stalks longer than the leaf-blade; stipules lanceolate, long-pointed. Flowers white or pink, 6-8mm, in small clusters of 2-4. Pod oblong, 6-8mm long, projecting beyond the calyx, containing 5-9 seeds. Open habitats, on dry sandy or gravelly soils, but often wet in winter, often close to the sea on short turf or cliffs. May-September. Britain, Ireland, France, Holland and Germany. **B**: Local north to the Wash and Anglesey, sometimes further north, rare in S and E Ireland; mostly coastal.

2* **Upright Clover** *Trifolium strictum* L. [*T. laevigatum*]. Low to short, rather stiff, erect or ascending, hairless annual. Leaflets elliptical to linear, the veins ending in stalked glands; stipules conspicuous, oval to rhombic, glandular toothed like the leaves. Flowers pink, 7-10mm, in long-stalked heads, either lateral or terminal. Pod rounded, 2.5mm, 1-2 seeded. Short open grassland on acid soils, at low altitudes. May-July. SW Britain and W France; naturalized in Germany. Often succeeding best where the habitat is disturbed by grazing, trampling or fire. **B**: Very rare, only a few localities in W Cornwall and one on Jersey.

3* **Mountain Clover** *Trifolium montanum* L. Short-medium, tufted hairy perennial, with a rather woody base; stems mostly unbranched. Basal leaves with oval or elliptical leaflets, upper leaves with narrower leaflets, all hairless above but usually silkily-hairy beneath. Flowers white, yellowish or sometimes pink, becoming yellowish-brown on ageing, 7-9mm, borne in dense globose heads which are frequently paired. Pods usually 1-seeded. Mostly a plant of hills and mountains. Dry grassy habitats, open woods, often on calcareous soils, to 2600m. May-July. Belgium, France, Germany northwards, except for the far north.

4* **White Clover, Dutch Clover** *Trifolium repens* L. Very variable low to short, more or less hairless perennial, with creeping stems rooting at the nodes. Leaflets bright green, usually with a pale or dark mark in the centre, elliptical, with translucent veins; stipules large, sheathing the stem, membranous. Flowers white or pale pink, rarely deep red, 7-10mm, sweetly scented, in dense globose heads. Pod linear, constricted between the seeds. Grassy habitats, meadows and pastures, lawns, banks and tracksides, on a wide variety of soils, but more especially well-drained fertile calcareous soils, to 2750m. June-September. Throughout, except the extreme north and Spitsbergen. Plants from the higher mountains are usually dwarfer with short creeping stems, smaller leaves and fewer flowers. **B**: Throughout, often abundant.

5 **Western Clover** *Trifolium occidentale* D.E.Coombe. Similar to *T. repens*, but leaflets thicker and more rounded, deep bluish-green and without markings, the lateral veins not translucent; stipules often tinged with red. Flowers scentless. Windy exposed places, especially close to the sea, sand-dunes, rocky places and grassy habitats, often over granite or serpentine. April-July. SW England and NW France. **B**: Confined to Devon, W Cornwall, the Isles of Scilly and the Channel Islands.

6* **Alsike Clover** *Trifolium hybridum* L. [*T. fistulosum*]. Short, sometimes medium, hairless, erect perennial, not rooting at the nodes; stipules green with a slender tip. Leaflets oval to heart-shaped. Flowers purple or whitish, becoming pink, then later brown, 7-10mm, in long-stalked, globose heads. Pod 2-4 seeded. Meadows and pastures, roadside verges, to 2150m. June-September. Naturalized throughout Europe, except the extreme north. The native distribution of this species is uncertain, but it has been long cultivated as a forage crop. Superficially similar to *T. repens*, but lacking the creeping and rooting stems. **B**: Throughout, but less common in the north and west and in Ireland; widely cultivated since the 18th century.

subsp. *elegans* (Savi) Syme. Like the type, but differing in its much-branched stems and smaller flowerheads, 16-19mm (not 22-25mm). Mainly in continental Europe; rather uncommon in Britain and Ireland.

7* *Trifolium michelianum* Savi. Rather like *T. hybridum*, but an annual, often constricted at the nodes. Leaflets oval, pointed, often with a pale patch in the centre. Flowers pink, 8-11mm, deflexed shortly after they open. Pod 2-seeded, slightly hairy. Wet meadows and by standing water. June-July. France. Primarily from S Europe.

8* **Clustered Clover** *Trifolium glomeratum* L. Like *T. michelianum*, but shorter, rarely exceeding 20cm. Flowerheads smaller, only 8-12mm across (not 20-25mm) and flowers 4-5mm, only slightly longer than the calyx; calyx teeth reflexed. Dry places, short turf, sand-dunes, old quarries, wall tops, sandy or gravelly areas near the sea, sometimes along paths. June-August. S Britain, SE Ireland and W France. **B**: Coastal England from Norfolk to Kent, S Devon, Cornwall, Channel Islands, Isles of Scilly; Ireland – Wexford and Wicklow; doubtfully native in Wales.

9* **Suffocated Clover** *Trifolium suffocatum* L. Very low prostrate, tufted, hairless annual. Leaflets oval, narrowed towards the base, notched at the tip. Flowerheads unstalked, at the leaf-axils, the flowers white, 3-4mm, the petals shorter than the calyx. Pods 2-seeded. Dry grassy habitats near or on the coast, dune turf, pathways, sandy or gravelly ground, occasionally inland on light sandy soils. March-May. S Britain and W France. Mediterranean. **B**: Local and declining in England and Wales, from S Yorkshire and Cheshire southwards, including the Channel Islands and the Isles of Scilly.

10* **Strawberry Clover** *Trifolium fragiferum* L. Low prostrate and creeping, almost hairless perennial; stems rooting at the nodes.Leaflets oval to elliptical; stipules lanceolate, membranous. Flowers pale pink, 6-7mm, in dense stalked heads which expand to 10-20mm in fruit due to the upper lip of the calyx which becomes swollen. Pod 1-2-seeded. Rather local on short rough grassland, commons and grazed pastures, sometimes coastal, generally on heavy clay, sometimes saline, soils. July-September. Europe north to S Scandinavia. **B**: Local throughout except the extreme north, rarer and mostly coastal in Ireland.

subsp. *bonannii* (C.Presl) Soják [*T. neglectum*]. Like the type, but with larger fruiting heads, 15-25mm, the calyx 8-10mm (not 4-6mm). S England, France and S Germany. Sometimes cultivated.

11 **Reversed Clover** *Trifolium resupinatum* L. Rather like T. fragiferum, but a very variable short hairless annual with spreading to erect stems and flowers pink to reddish-purple, 2-8mm, twisted upside down with the standard petal below. Calyx becoming somewhat swollen in fruit, making heads 8-20mm. Grassy habitats, disturbed and waste ground, generally on rather damp soils. May-July. **I**: C & E Europe and W Asia. Naturalized or casual in Britain, France, Belgium, Holland and Germany, local or rather rare. **B**: Casual and usually coastal in Britain and generally close to docks.

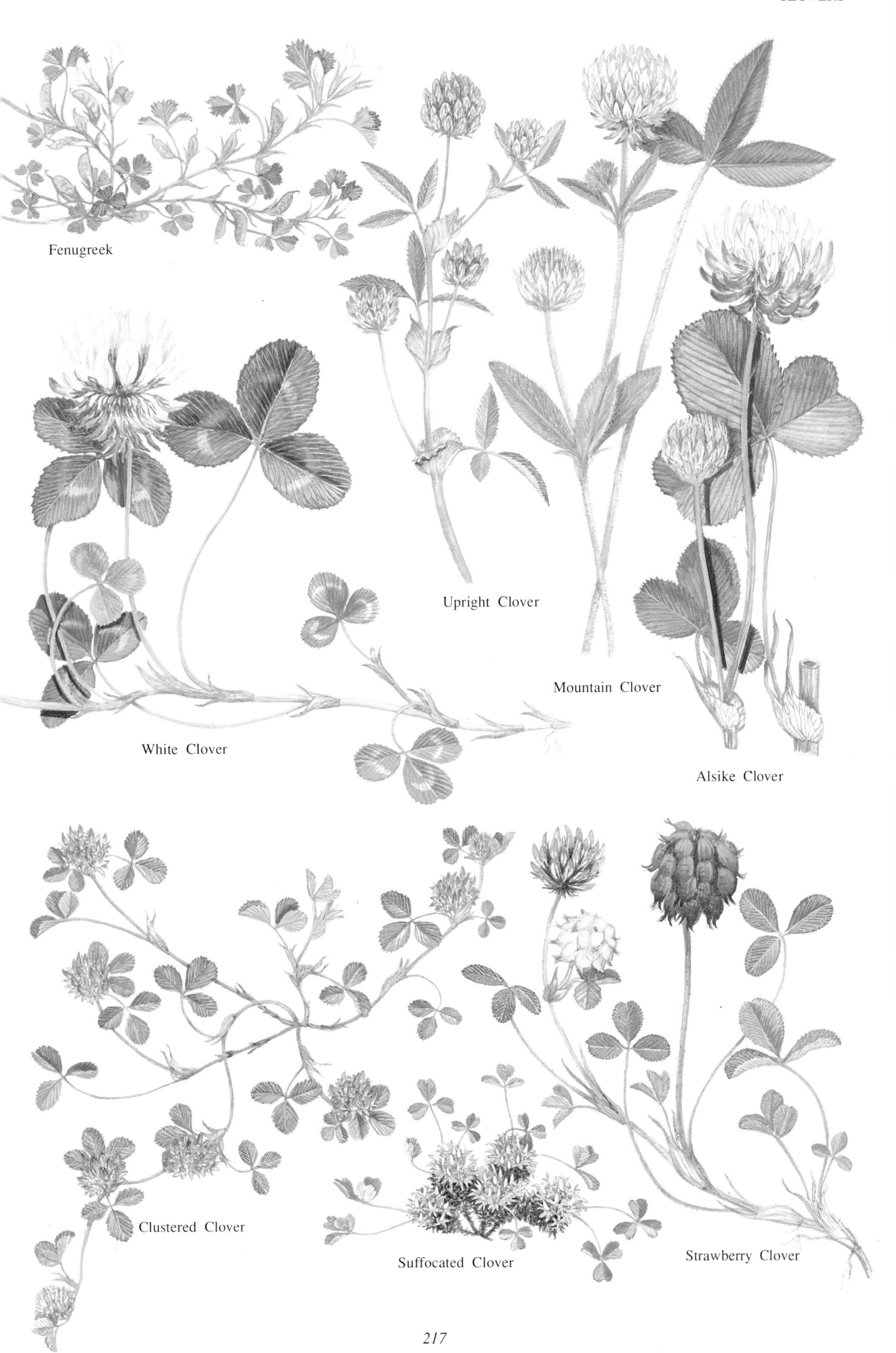
Fenugreek
Upright Clover
Mountain Clover
Alsike Clover
White Clover
Clustered Clover
Suffocated Clover
Strawberry Clover

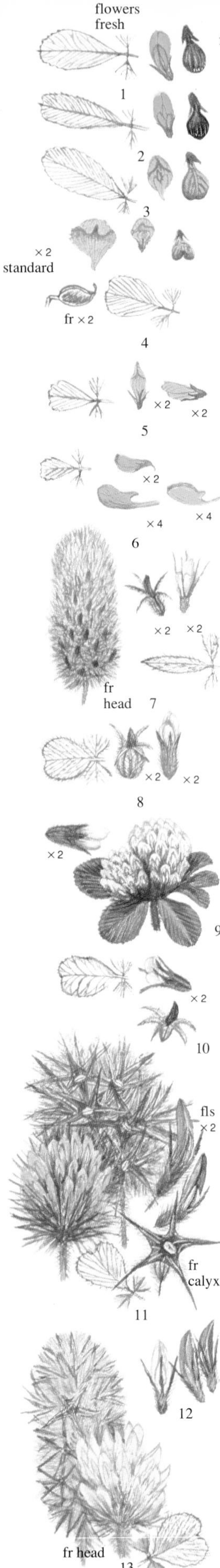


1* **Brown Clover** *Trifolium badium* Schreber. Short, tufted, generally hairy perennial with a stout taproot; hairs closely pressed to the stem. Upper leaves opposite, lower alternate; leaflets elliptical to rhombic. Flowers golden-yellow, becoming chestnut-brown as they age, 7-9mm long, in dense, stalked, globose heads. Meadows and damp stony habitats, generally on calcareous soils, 1400-28oom. July-August. C & E France, S Germany.

2* **Large Brown Clover** *Trifolium spadiceum* L. Short, tufted, almost hairless perennial, with erect mostly unbranched stems. Uppermost leaves opposite, the lower alternate; leaflets elliptical to oblong. Flowers golden-yellow, becoming very dark brown on ageing, 5-6mm long, in often paired, oblong, stalked heads. Grassy habitats on acid soils, to 2200m. June-August. France, Germany and Scandinavia, except Denmark and the extreme north.

3* **Large Hop Trefoil** *Trifolium aureum* Pollich [*T. agrarium*]. Robust, short, tufted, hairy annual or biennial, with erect stems. Leaflets oblong to rhombic, the terminal one stalkless; stipules linear-lanceolate. Flowers golden-yellow, 6-7mm long, in dense globose, stalked heads. Pod oblong, usually 2-seeded. Rough fields, scrub, wood margins and woodland clearings, waste places. July-August. Continental Europe except the far north; naturalized in Britain. Spring or autumn germinating. **B**: Throughout, but especially in the SE of England; generally rather uncommon.

4* **Hop Trefoil** *Trifolium campestre* Schreber (*T. procumbens* L. p.p.). Short, hairy, erect annual. Leaflets oval, narrowed towards the base, the central one short-stalked; stipules oval, pointed, rather swollen and rounded at the base. Flowers yellow, becoming pale brown eventually, 4-5mm long, in small globose, stalked heads, to 15mm across. pod oval, generally 1-seeded. Dry grassy habitats, short rough grassland, road verges, tracks and sand-dunes, generally at low altitudes. June-September. Throughout, except the extreme north, Iceland and Spitsbergen; naturalized in the Faeroes and Finland. Spring and autumn germinating. **B**: Throughout, though scarcer in the north; local in Ireland.

5* **Lesser Trefoil, Suckling Clover** *Trifolium dubium* Sibth.[*T. minus*]. Low to short, usually hairy annual, with spreading to ascending stems. Leaflets oval to heart-shaped, the terminal leaflet short-stalked; stipules broad-oval, exceeding the leaf-stalks. Flowers yellow, becoming yellowish-brown on ageing, 3-3.5mm long, few in small globose, stalked heads, not exceeding 9mm across. Pod oval, usually 1-seeded. Dry grassy habitats, short grassland, road verges, commons and lawns, sometimes on arable land, usually at low altitudes. May-October. Throughout, except the extreme north, the Faeroes, Iceland and Spitsbergen. **B**: Throughout, except for parts of the Scottish Highlands. This is the Shamrock of Ireland.

6* **Slender Trefoil** *Trifolium micranthum* Viv. Low, almost hairless annual with spreading or ascending stems.Leaves short-stalked; leaflets heart-shaped. Flowers yellow, becoming yellowish-brown on ageing, small, 2-3mm long, in small heads with up to 6 flowers. Pod oval, usually 2-seeded. Grassy habitats on dry soils, road verges, grassy tracks, waste ground, often on poor sandy or gravelly soils at low altitudes. June-August. Britain, Ireland, Belgium, Holland, France, Germany, Denmark and S Norway. Small and spring germinating, from Europe to C Asia. **B**: Throughout, but rather rare in N England and local in Scotland.

7* **Hare's-foot Clover** *Trifolium arvense* L. Low to short, hairy annual or biennial; hairs whitish or reddish. Lower leaves stalked, upper unstalked, the leaflets linear-oblong; stipules lanceolate, pointed. Flowers whitish or pinkish, 4mm long, numerous, in dense oblong or egg-shaped, stalked, silkily-hairy heads; the petals much shorter than the calyx. Pod 1-seeded. Dry grassy habitats, often on slightly acid soils, field borders, grassy heaths, road verges, woodland pathways, waste places, sometimes on sand-dunes, at low altitudes. June-September. Throughout, except the extreme north, the Faeroes, Iceland and Spitsbergen. **B**: Scattered throughout, except N Scotland; local and mostly on the coast of Ireland.

8* **Knotted Clover** *Trifolium striatum* L. Like *T. arvense*, but leaflets oval and flowerheads unstalked, usually solitary; petals equalling or slightly longer thyan the calyx. Habitats on well-drained soils, grassy heaths and banks, open grassy dunes. May-July. Europe north to Sweden, not Norway. Very similar to *T. bocconei* and *T. scabrum*, but flowers pink, not white. **B**: Scattered localities throughout, except the extreme north, often coastal; Ireland, local and coastal.

9 **Twin-headed Clover** *Trifolium bocconei* Savi. Like *T. striatum*, but leaflets hairless (not hairy) above, flowerheads often paired and flowers white, at least to begin with. Dry grassy habitats close to the sea. May-June. SW England and W France; naturalized in N Germany. Primarily Mediterranean. **B**: Very rare, confined to the extreme W of Cornwall and Jersey.

10* **Rough Clover** *Trifolium scabrum* L. Low to short, usually hairy annual, with spreading to ascending stems. Leaflets oval, narrowed towards the base, thick, finely toothed; stipules oval with a slender. pointed apex. Flowers whitish, occasionally with a pink tinge, 4-5mm, in unstalked, globose clusters, 5-12mm across; petals usually shorter than the calyx. Pod oblong, usually 1-seeded. Dry well-drained soils, open turf, fixed dunes, tracksides, on thin sandy or gravelly soils, calcareous or slightly acid; at low altitudes. May-July. Britain, Belgium, Holland, France and Germany. **B**: Local throughout, except N Scotland, mostly coastal but present in the East Anglian Brecklands; Ireland, confined to the SE coast.

11 **Starry Clover** *Trifolium stellatum* L. Short, erect, hairy annual. Leaflets ellitical. Flowers pink, sometimes yellowish, 4-8mm, in terminal, stalked, globose or oblong heads; calyx densely white-hairy, with slender spreading lobes, star-like, often reddish. Fields, roadsides, stony places and waste ground, at low altitudes. June-July. C France southwards; naturalised in SE Britain. Easily identified by its persistent starry calyces. **B**: Naturalized in a small area near Shoreham, Sussex.

12* **Crimson Clover** *Trifolium incarnatum* L. Robust short to medium, erect, hairy annual; stems unbrached or branched near the base. Leaflets rounded to oval, finely toothed towards the apex; stipules green, occasionally reddish, oval, blunt. Flowers blood-red, rarely white, 10-12mm, in dense solitary, oblong or cylindrical, heads, terminal; petals equalling or exceeding the calyx. Widely cultivated as a forage crop and widely naturalized in fields and waste places, sometimes coastal, to 1500m. May-September. Naturalized throughout, except the extreme north and most of the smaller islands. The origin of this distinctive species is uncertain. **B** Casual in various parts of Britain, but more especially in S England.

13 **Long-headed Clover** *T. i.* subsp. *molinerii* (Balbis ex Hornem.) Syme. Less robust then the type, but with yellowish-white or occasionally pink flowerheads; petals considerably longer than the calyx. Short turf on cliff-tops, exposed to sea spray. May-June. France and SW England. Sometimes confused with *T. ochroleucon* and *T. squamosum*, but the calyx much hairier. **B**: Very rare, confined to the Lizard Peninsular and to Jersey.

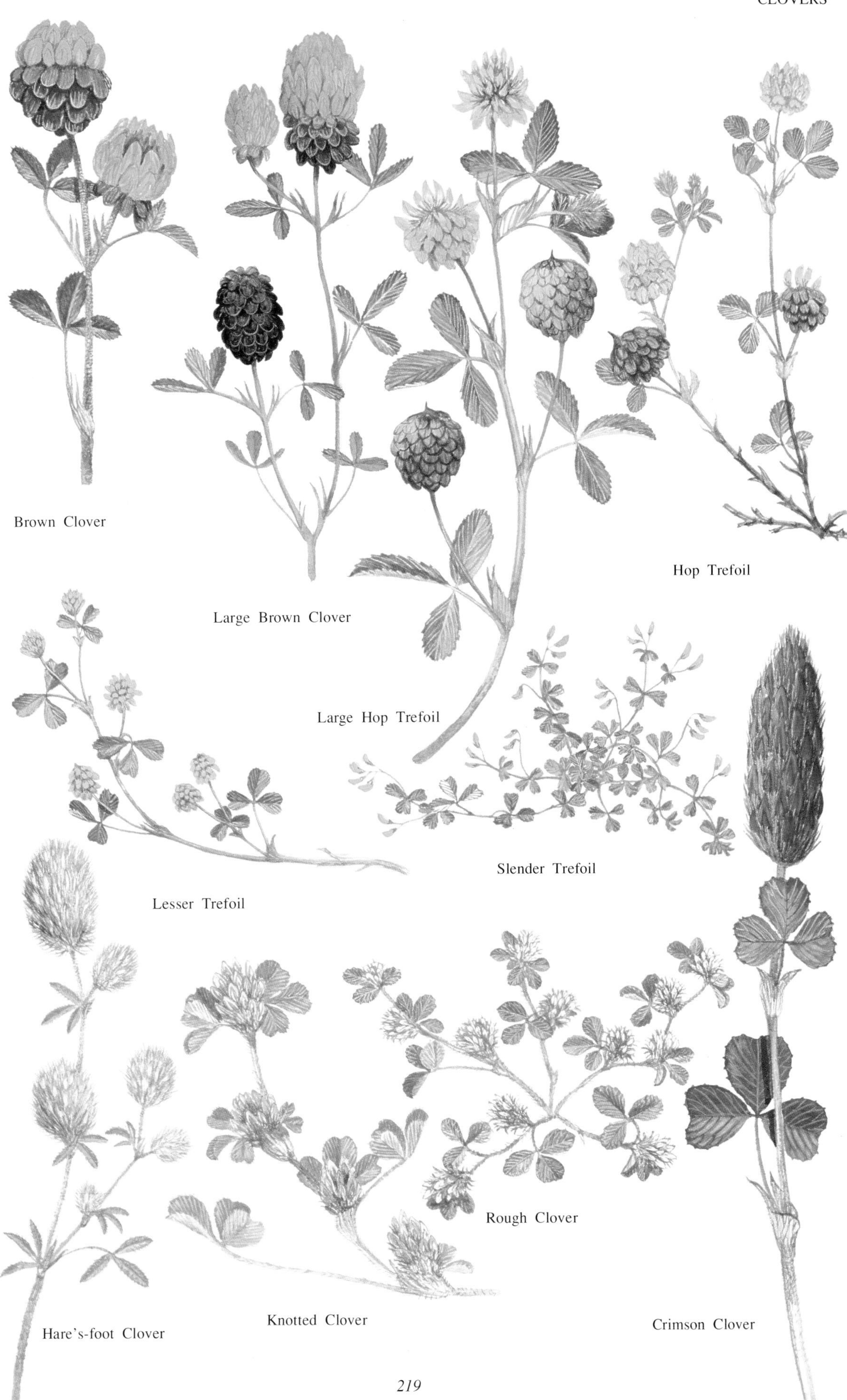
Brown Clover
Large Brown Clover
Large Hop Trefoil
Hop Trefoil
Lesser Trefoil
Slender Trefoil
Hare's-foot Clover
Knotted Clover
Rough Clover
Crimson Clover

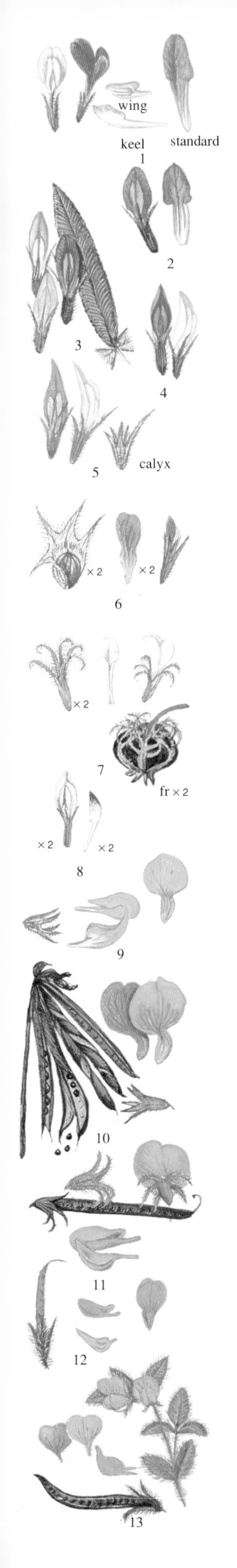


1* **Red Clover** *Trifolium pratense* L. Very variable, short to tall, tufted, hairy perennial. Leaflets oval to elliptical, often with a white crescent in the centre, hairy beneath; stipules triangular, the upper with a very wide base. Flowers reddish-purple or pink, 12-15mm, rarely cream or white, in dense globose heads, usually solitary and unstalked. Pod oval with a thickened tip. Grassy places, cultivated land; well-drained but moist, fertile, calcareous or weakly acid soils, to 3150m. May-September. All except parts of far N; naturalized Faeroes, Iceland. Several forage varieties are grown, especially var. *sativum* Sturm [*T. sativum*], scarcely hairy and with pink, often paired, flower-heads. **B**: Common.

2* **Zig-zag Clover** *Trifolium medium* L. Variable, medium, hairy, rhizomatous perennial, with flexuous, rather zig-zagged stems. Leaflets elliptical to oval, scarcely toothed; stipules lanceolate, green, with a hairy margin at apex. Flowers 25-35mm, in dense globose or egg-shaped, short-stalked heads. Pod splitting lengthwise. Grassy habitats, open woodland, scrub; usually on poor soils, to 2100m. May-July. Throughout, except the far north. **B**: Throughout; scarce in East Anglia, N Scotland and most of Ireland.

3 *Trifolium alpestre* L. Like *T. medium*, but flower-heads unstalked, smaller, 15-25mm; flowers purple, rarely pink or white. Same habitats but generally on more fertile well-drained soils, to 2300m. May-July. Belgium, France, Denmark, Germany.

4* **Red Trefoil** *Trifolium rubens* L. Medium, usually hairless, rhizomatous perennial with erect stems. Leaflets lanceolate, toothed, with curved veins; stipules oval to lanceolate, fused together for half their length to form a conical sheath below. Flowers reddish-purple, occasionally white, 14-15mm, in long cylindrical heads, solitary or paired. Dry open woodland, scrub and stony places, to 2050m. June-August. Mountains of C and E France, S Germany, C and S Europe.

5* **Sulphur Clover** *Trifolium ochroleucon* Hudson. Short to medium, rather hairy perennial, with short rhizomes and ascending stems. Leaflets oblong to lanceolate; stipules oblong with a slender apex. Flowers whitish-yellow, rarely pink, 15-20mm, in short-stalked or unstalked globose or oblong heads, terminal. Grassy places, usually shaded or damp; heavy clay soils, to 1800m. June-July. W, C and S Europe, east to W Asia. **B**: Local in E England up to Northants; occasional elsewhere.

6* **Sea Clover** *Trifolium squamosum* L [*T. maritimum*]. Short to medium, rather hairy annual with erect or ascending stems. Leaflets oblong, narrowed at the base, often with a notch-tip; stipules linear, broader towards the base. Flowers pale pink, 5-7mm, in egg-shaped heads, short-stalked, with a pair of leaves immediately below. Short open turf near the sea, saltmarshes, tidal estuaries. June-July. Britain, France, S Europe. **B**: Rare and declining; coastal England north to Lancs and Lincs.

7* **Subterranean Clover** *Trifolium subterraneum* L. Low hairy prostrate annual. Leaflets broadly heart-shaped, variable in size; stipules oval, pointed. Flowers whitish, 8-14mm, in dense clusters at base of leaves. Each head a mixture of a few fertile flowers and many infertile ones, the former reflexed after flowering. Pod 1-seeded, on recurved stalks which push the fruit into the soil surface. Dry grassy habitats, pastures, cliff and wall tops, often close to the sea; generally sandy or fine gravelly soils. May-June. Britain, Belgium, France, Holland. **B**: Local on coasts N to Cheshire and Lincs. Wicklow, Ireland.

8* **Dorycnium** *Dorycnium pentaphyllum* Scop. subsp. *germanicum* (Gremli) Gams. [*D. germanicum*]. Short to medium, hairy perennial, slightly woody at the base. Leaves pinnate, with 5 oblong leaflets, untoothed. Flowers white with a dark reddish-black keel, in axillary stalked clusters. Pod oval, 3-5mm. Grassy habitats and scrub; generally light, well-drained soils. June-July. W and C France, S Germany. The lower pair of leaflets resemble stipules, but minute true stipules are present at very base of leaf-stalk.

**Trefoils** *Lotus*. Annual or perennial herbs, often rather woody below. Leaves pinnate with 5 leaflets, the lower pair closely resembling stipules; the real stipules minute. Flowers solitary or in small heads. Calyx bell-shaped. Corolla generally yellow, with a beaked keel. Fruit-pod cylindrical, splitting lengthwise, many-seeded. The annual species have narrower pods (1-2mm across) than the perennials (2-3mm). 70 species in Europe, Asia, Africa and Australia.

9* **Narrow-leaved Bird's-foot Trefoil** *Lotus tenuis* Waldst. & Kit. ex Willd. [*L. tenuifolius*]. Low to short, often prostrate, hairless or slightly hairy perennial. Leaflets linear to lanceolate. Flowers bright yellow, 6-12m, solitary or in clusters of 2-4. Pod 15-30mm. Lowland grassy places, stabilised dunes, often near sea but sometimes inland; calcareous soils. June-August. All except N Scandinavia; naturalized in Finland and Norway. **B**: Local E and SE England south of the Humber, rarer elsewhere.

10* **Common Bird's-foot Trefoil** *Lotus corniculatus* L. Very variable, low to short, rather sprawling, hairy or hairless perennial, with a woody base; stems solid. Leaflets 4-18mm, rounded to lanceolate. Flowers yellow to orange-yellow, 10-16mm, in clusters of 2-7. Pod 15-30mm, straight. Short grassy habitats, heathland, road-verges, sometimes along coast, to 1600m. June-September. All except Spitsbergen; naturalized in Iceland. Once much grown for forage. Coastal forms in N and W Europe are almost hairless, with small fleshy leaflets. **P**: Bumble bees. **B**: Widespread.

subsp. *alpinus* (DC.) Schlechter ex Ramond [*L. alpinus*]. Smaller leaflets, only 2-6mm, and flowers often marked with red, usually in clusters of 2-3 or solitary. Meadows and rocky habitats, 2000-3100m. May-August. E France and S Germany.

11* **Greater Bird's-foot Trefoil** *Lotus uliginosus* Schkuhr [*L. pedunculatus*]. Like *L. corniculatus*, but a short to medium perennial with hollow, more or less erect stems. Leaflets bluish-green. Flowers 10-18mm, in larger heads of 5-12. Pod often slightly longer. Damp pastures, wet ditches, woodland pathways, freshwater margins; mildly acid soils, generally lowland. June-August. North to S Sweden; naturalized Faeroes, Finland, Norway. **B**: Throughout, except N Scotland and parts of C Ireland.

12* **Hairy Bird's-foot Trefoil** *Lotus suaveolens* Pers. [*L. hispidus, L. subbiflorus*]. Low to short, often prostrate annual, densely hairy, especially above, not woody at base. Leaflets lanceolate to oblong. Flowers yellow, 5-10mm, in clusters of 2-4, stalk longer than the subtending leaf. Pod 6-16mm, the valves contorted as they split apart. Grassy places, dunes, walls, fields; generally dry sandy soils. July-August. Britain, W France, S Europe. **B**: Mostly coastal. S Ireland, SW England, S Wales, Scilly, Channel Is.

13 **Slender Bird's-foot Trefoil** *Lotus angustissimus* L. Like *L. suaveolens*, but flowers sometimes with purple veins and keel petal with a right-angle along the lower edge, short-beaked. Pod 15-30mm. Dry grassy habitats, generally near sea. July-August. S Britain, W France to N Africa. **B**: Rare, a few localities along the south coast of England and Channel Is.

Red Clover
Zig-zag Clover
Red Trefoil
Sulphur Clover
Sea Clover
Subterranean Clover
Dorycnium
Narrow-leaved Bird's-foot Trefoil
Common Bird's-foot Trefoil
Greater Bird's-foot Trefoil
Hairy Bird's-foot Trefoil

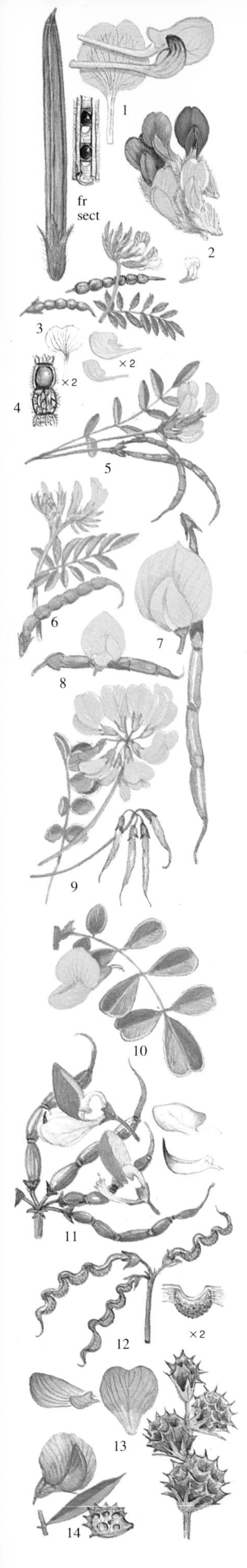


1* **Dragon's-teeth** *Tetragonolobus maritimus* (L.) Roth [*Lotus siliquosus*]. Low, hairy or hairless, prostrate perennial. Leaves trifoliate; leaflets oval, broadest above middle; stipules like the leaflets, pointed. Flowers pale yellow, 25-30mm, solitary on long stalks with a bract-like leaf just below the flower; calyx with equal teeth. Pod 30-60mm, square in section, winged on the angles. Dry grassy habitats and waste ground; calcareous soils, to 1200m. May-September. Denmark, France, Germany and S Sweden. **B**: Naturalized in a few localities in S England – Buckinghamshire, Essex and Kent; casual elsewhere.

2* **Kidney Vetch** *Anthyllis vulneraria* L. [*A. linnaei*]. Very variable, low to medium, silkily-hairy perennial. Leaves pinnate, lowermost with 5-7 leaflets, upper with 9-15; leaflets oblong to elliptical. Flowers normally yellow but may be red, purple, orange or white.; 12-15mm, in dense heads with a leaf-like bract immediately beneath; calyx inflated, white-woolly, the teeth often red-tipped. Dry coastal grassland, sometimes inland; often on calcareous soils, cliff-tops and rock ledges, to 3000m. June-September. Throughout, except the far north. **B**: Common.

Kidney Vetch is an extremely variable species, with many European subspecies recognized. A full treatment of them would be beyond the scope of this book, but for some of the more important variants, see p. 486.

3 **Cultivated Bird's-foot** *Ornithopus sativus* Brot. Short to medium, hairy annual, with spreading to erect stems. Leaflets 9-18 pairs, lanceolate to oval. Flowers white or pink, 6-9mm, in heads of 2-5. Pods 12-25mm, with 3-7 oblong segments, the terminal one beaked. SW Europe; cultivated for fodder and naturalized in parts of France and Germany. May-August.

4* **Bird's-foot** *Ornithopus perpusillus* L. Tiny, low, prostrate, hairy annual. Leaflets 7-13 pairs, elliptical to oblong. Flowers white or pink, 3-5mm, in heads of 3-8; bracts with 5-9 leaflets, longer than the flowers. Pods 10-18mm, with a hooked beak and 4-9 segments, in clusters resembling a bird's-foot. Open patches in short turf, waste and arable ground; generally on rather dry, acid, sandy or gravelly soils. May-August. North to S Sweden, not Norway, west to Russia. **P**: Self-pollinated. **B**: Widespread in England and Wales, scarcer or rare elsewhere, including much of Ireland.

5 **Orange Bird's-foot** *Ornithopus pinnatus* (Miller) Druce [*O. ebracteatus*]. Slender, short to medium, hairless or slightly hairy annual, with ascending stems. Leaflets 3-7 pairs, linear to oblong. Flowers orange-yellow, 6-8mm long, solitary or in heads of 2-5; bracts tiny or absent. Pods 20-35mm, with 8-12 segments, but not noticeably constricted. Grassy and open places. April-October. **B**: Very rare, only on Scilly and Channel Is.

6 *Ornithopus compressus* L. Like *O. pinnatus*, but a hairier plant, the leaves with 7-18 pairs of leaflets. Bracts present, leaf-like. Flowers yellow, 5-8mm. Pods 20-50mm, with 5-8 segments. Grassy and rocky habitats, especially on rather light sandy soils. May-August. C & S France.

7* **False Senna** *Coronilla emerus* L. Small, loosely-branched shrub to 1m, with green twigs. Leaves with 2-4 pairs of oval leaflets, broadest above the middle, bluish or greyish green; stipules tiny, membranous. Flowers large, pale yellow, often red-tinged, 14-20mm, solitary or in clusters of 2-5. Pods 50-110mm, with 3-12 bluntly-angled divisions. Woodland clearings and rocky habitats, to 1800m. April-May. France and Germany, S. Norway and S Sweden; naturalized Denmark.

8* **Small Scorpion Vetch** *Coronilla vaginalis* Lam. Small deciduous shrub to 50cm. Leaves with 2-6 pairs of oblong to rounded green leaflets with whitish margins, short-stalked; stipules leaflet-like, 3-8mm, fused together. Flowers yellow, 6-10mm, in heads of 4-10. Pods 15-35mm, linear-oblong, constricted between the seeds. Scrub, open woods and dry grassy habitats; calcareous soils in mountainous places, to 2250m. June-August. E France, S Germany.

9 *Coronilla minima* L. Like *C. vaginalis*, but leaflets unstalked and stipules only 1mm; flowers 10 or more in a head. Dry open scrub and grassland. Late June-September. W and C France.

10 **Scorpion Vetch** *Coronilla coronata* L. Short to medium perennial, not woody, with spreading or ascending stems. Leaves with 3-6 pairs of elliptical leaflets, stalked, with narrow membranous margins; stipules membranous, fused together, soon falling. Flowers yellow, 7-11mm, in heads of 12-20. Pods 15-30mm, with 1-5 bluntly-angled segments. Grassland, woodland and scrub; dry calcareous soils. June-August. C and E France, S Germany.

11* **Crown Vetch** *Coronilla varia* L. Short to tall perennial with spreading to erect stems. Leaves with 7-12 pairs of elliptical or oblong leaflets, each with a narrow membranous margin; stipules membranous, small, separate. Flowers pink, lilac-purple or bicoloured, 10-15mm, in heads of 10-20. Pods 20-60mm, with 3-8, 4-angled segments. Grassy and waste places, once grown for fodder and sometimes in gardens. France and Germany, rare in Holland; naturalized north to S Sweden. **B**: Naturalized in scattered localities N to Aberdeen, more frequent in the south.

12* **Horseshoe Vetch** *Hippocrepis comosa* L. Very variable, low, usually prostrate, hairy perennial, woody at the base. Leaves pinnate with 3-8 pairs of linear to oval leaflets. Flowers yellow, 6-10mm, in stalked heads of 5-12, the standard petal with a distinct claw. Pods 15-30mm, twisting with typical horseshoe-shaped segments. Short dry turf on chalk and limestone, to 2800m. April-July. Britain, Belgium, France, Holland, Germany. **P**: Bumble-bees. **B**: N to Kincardine in Scotland.

13* **Sainfoin** *Onobrychis viciifolia* Scop. [*O. sativa*]. Short to tall erect, hairless or somewhat hairy perennial. Leaves pinnate with 6-14 pairs of oblong, occasionally linear, leaflets. Flowers pink with purple veins, 10-14mm, in long-stalked axillary racemes; calyx with long teeth. Pods small, 5-8mm, with toothed sides. Grassy habitats, arable land, waste places, especially on calcareous soils. June-September. Probably a native of C and E Europe. Widely naturalized north to Sweden. A relict of former cultivation for fodder. **B**: Naturalized England, local in SW and Wales, rare elsewhere.

14 **Small Sainfoin** *Onobrychis arenaria* (Kit.) DC. Like *O. viciifolia*, but leaves with 5-12 pairs of leaflets and smaller flowers, 8-10mm; calyx-teeth with appressed, not spreading, hairs. Pastures and stony habitats, to 2500m. June-September. C and E France and C and S Germany. Very variable.

Dragon's-teeth
Kidney Vetch
Bird's-foot
False Senna
Small Scorpion Vetch
Crown Vetch
Horseshoe Vetch
Sainfoin

# WOOD SORREL FAMILY Oxalidaceae

Herbs, sometimes with a bulbous stock. Leaves usually trifoliate, the leaflets notched at the top, otherwise untoothed. Flowers solitary or in cymose or umbel-like clusters, 5-parted with separate sepals and petals, regular and hermaphrodite; stamens 10. Ovary superior. Fruit a capsule, splitting lengthwise. Some 8 genera and 900 species (only *Oxalis* in NW Europe), in tropical and temperate regions. *Oxalis* alone has almost 800 species; several introduced as garden plants and now weeds. Some exhibit sleep movements with the leaflets nodding at night, but spreading during daylight hours.

1* **Procumbent Yellow Sorrel** *Oxalis corniculata* L. Low to short rather hairy perennial with creeping stems rooting at the nodes. Leaves alternate, long-stalked; leaflets heart-shaped. Flowers yellow, petals 4-7mm, in small 1-7-flowered umbels. Capsules hairy, on deflexed stalks. Dry open habitats, cultivated land, waste places, generally low, but up to 1000m occasionally. May-October. Naturalized in Britain, much of Europe, except Denmark (country of origin uncertain). Now a garden weed. **B**: Most of England and Wales, local and often rare in Scotland and Ireland.

var. *atropurpurea* van Houtte. Purple leaves. Sometimes naturalized garden escape.

2* *Oxalis exilis* A. Cunn. [*O. corniculata* var. *microphylla*]. Like *O. corniculata*, but very dwarf with creeping, thread-like stems and leaflets only 2-6mm (not 8-15mm). Flowers usually solitary. Capsule smaller, 5-7mm (not 10-25mm). Cultivated land and waste places; sometimes a persistent weed of gardens. May-August. **I**: Australia; naturalized a few places in S England and Channel Is., occasionally elsewhere.

3 *Oxalis stricta* L. [*O. dillenii, O. navieri*]. Like *O. corniculata*, but a more tufted, short-lived perennial, the stems not rooting at the nodes. Leaves subopposite or in groups. Cultivated ground and waste places. July-October. **I**: E and C North America; naturalized in S Britain, France, Denmark, Germany. **B**: Very local W Sussex and Channel Is (Sark).

4 **Upright Yellow Sorrel** *Oxalis fontana* Bunge [*O. europaea*]. Rather like *O. corniculata* but a short to medium perennial, sometimes apparently annual, with slender underground stolons. Leaves opposite or in whorls; leaflets heart-shaped, rather broader than long. Flowers yellow, petals 4-10mm, in small, 2-5-flowered umbels. Capsule not hairy, 8-12mm, the stalks not deflexed. Cultivated land and waste places; often on sandy soils. June-September. Naturalized throughout, except the far N, Faeroes, Iceland and Spitsbergen. **I**: North America and E Asia. **P**: Probably self-pollinated. **B**: Mainly C and S England, local elsewhere.

5* **Pink Oxalis** *Oxalis articulata* Savigny. Short, tufted, hairy perennial, with swollen rhizomes. Leaves in terminal rosettes, leaflets heart-shaped, covered in orange or brownish dots (tubercles). Flowers pink, petals 12-20mm, in a broad, umbel-like cluster. Capsule 9-10mm. **B**: Waste ground, disturbed habitats, road verges and seashores, sometimes among native vegetation. May-October. **I**: E temperate South America; naturalized parts of Britain and France. The commonest pink-flowered weedy *Oxalis*. **B**: Frequent in SW Britain and S Ireland, rarer elsewhere.

6* **Wood Sorrel** *Oxalis acetosella* L. Low, creeping, slightly hairy perennial, with slender rhizomes. Leaves scattered, leaflets heart-shaped, pale green. Flowers white or pale lilac with lilac or purplish veins, solitary, bell-shaped, half-nodding, petals 8-15mm; followed later by petalless cleistogamous flowers. Capsule 3-4mm, oval and angled, hairless. Woods and shaded places; generally on humus-rich soils, sometimes among shaded rocks, hedge banks or limestone grikes, to 2100m. April-June. All except Spitsbergen. Europe to China and Japan. Often in ancient woodland. Most of the seed is produced by the cleistogamous flowers through the summer months. Formerly a salad vegetable. **B**: Widespread except for parts of E England.

7* *Oxalis debilis* Kunth [*O. corymbosa, O. martiana*]. Short, stemless, hairy perennial with an underground bulb. Leaves long-stalked, leaflets broadly to narrowly heart-shaped, with translucent dots beneath, especially near the margin. Flowers purplish-pink, petals 15-20mm, in a broad, umbel-like cluster. Fruit rarely produced. Cultivated land, waste places, pathways. July-September. **I**: South America; naturalized S Britain and France. A garden weed, spreading by small bulblets produced at base of the parent bulb. **B**: Mainly gardens in S and SW England.

8 *Oxalis latifolia* Kunth. Like *O. debilis*, but almost hairless and leaflets not dotted beneath. Flowers pale to deep pink, the petals 8-13mm. Cultivated ground, gardens, waste ground, sometimes fields. May-September. **I**: tropical South America; naturalized S Britain, France. Now a garden weed. **B**: Mainly S Ireland, SW England, Channel Is (Jersey).

9* **Bermuda Buttercup** *Oxalis pes-caprae* L. [*O. cernua*]. Low to short, slightly hairy, tufted perennial, often carpeting, with an underground bulb; no aerial stem. Leaves in small rosettes at the soil surface, long-stalked, leaflets heart-shaped. Flowers yellow, petals 20-25mm, in broad umbel-like clusters. Capsule rarely formed. Cultivated land, waste places. March-June. **I**: South Africa; naturalized S Britain, France. Once a glasshouse plant, occasionally in gardens, but not frost-hardy so only naturalized in mild coastal districts – extensively in the Mediterranean region. **B**: Bulb fields of Cornwall and the Isles of Scilly; casual S Devon, Channel Is.

10* *Oxalis incarnata* L. Low to short, hairless perennial with an underground bulb and erect aerial stem bearing small bulbils. Leaves subopposite, fairly crowded, leaflets rather delicate, heart-shaped. Flowers pale lilac with darker veins, petals 12-20mm, solitary on long stalks. Capsule not produced in Europe. Hedgerows and old walls, sometimes on cultivated land. May-July. **I**: South Africa; naturalized SW Britain. A garden plant. **B**: Naturalized locally in Devon, Cornwall, Channel Is; sometimes casual elsewhere.

Procumbent Yellow Sorrel

*Oxalis exilis*

Pink Oxalis

*Oxalis debilis*

Bermuda Buttercup

*Oxalis incarnata*

Wood Sorrel

# GERANIUM FAMILY Geraniaceae

Herbs, mostly with alternate leaves, palmately-lobed in *Geranium*, pinnately-lobed in *Erodium*; stipules present. Flowers in cymes or umbels, sometimes solitary, 5-parted, regular or slightly irregular; sepals separate. Stamens in two whorls of 5. Fruit with 5, one-seeded portions, united together into a prominent long beak, springing apart from the base when ripe, but usually remaining attached at the apex of the beak. 700 species from temperate and subtropical regions, including also the genus *Pelargonium*, the pot-plant 'Geranium'. The Crane's-bills (*Geranium*) are a primarily temperate genus with some 300 species.

1* **Bloody Crane's-bill** *Geranium sanguineum* L. Short to medium, hairy perennial with a stout horizontal rhizome; stems much-branched, erect to spreading. Leaves rounded in outline, cut into 5-7 pinnately-divided lobes. Flowers bright reddish-purple, rarely deep pink, 25-30mm, usually solitary, the stalks with a pair of small bracts near the centre; petals slightly notched. Grassland and woodland; light well-drained soils, limestone rocks and screes, sometimes fixed calcareous dunes, to 1900m. July-August. Throughout, except Holland and the far N, Faeroes, Iceland and Spitsbergen. **P**: Various insects. From W Europe to the Urals and the Caucasus. **B**: Scattered and local throughout, except for SE England and S Ireland. Two coastal forms growing in maritime sands can be recognised in Britain. var. *prostratum* (Cav.) Bab. has prostrate stems and flowers like the type. var. *lancastriense* (Miller) Gray (2) is like var. *prostratum*, but with pale pink flowers.

3* **Meadow Crane's-bill** *Geranium pratense* L. Medium to tall tufted, hairy perennial with erect stems; stems with deflexed hairs below, glandular hairs above. Leaves rounded in outline, divided almost to the base into 5-7 oval, deeply pinnately-divided lobes. Flowers bright violet-blue, 25-30mm, cup-shaped, in pairs; petals rounded. Fruit hairy. Meadows, hedgebanks and road verges; base-rich or calcareous soils, to 1900m. June-September. Throughout, except the far north. Naturalized Denmark. Frequently cultivated and often becoming naturalized from gardens. From W Europe across temperate Asia to Japan. **P**: Usually bees. The flower-stalks are deflexed after flowering but become erect again as the fruits mature. **B**: Widespread, but often local, except N Scotland; Ireland, confined to Antrim and Londonderry.

4* **Wood Crane's-bill** *Geranium sylvaticum* L. Medium, tufted perennial rather like *G. pratense*, but leaves less deeply cut. Flowers variable in colour from reddish-purple to pinkish-lilac or bluish, often with a whitish centre, 22-26mm; flower-stalks remaining erect after flowering. Meadows, pastures, damp woodland, hedgebanks; generally on base-rich or calcareous soils, sometimes in rocky habitats, to 2400m. June-July. Throughout, except Spitsbergen, but only on mountains in the south; naturalized Holland. **B**: Widespread and common south to Yorks, rarer elsewhere including most of Ireland.

5 *G. s.* subsp. *rivulare* (Vill.) Rouy [*G. rivulare*] is a shorter plant, rarely more than 30cm tall. Flowers white with red veins, rather smaller. Rocky and grassy habitats. E France – confined to the Alps.

6* **French Crane's-bill** *Geranium endressii* Gay. Medium to tall hairy perennial with a slender creeping rhizome; stems erect or ascending. Leaves rounded or pentagonal in outline, divided for over halfway into 5 broadly oval, irregularly-toothed lobes. Flowers silky-pink, 24-28mm, without darker veins, cup-shaped, generally in pairs; petals not or only slightly notched. Fruit hairy. Cultivated ground and road verges. June-July. SW France – Pyrenees. Naturalized W Europe as a garden escape. **P**: Various insects including bees and hoverflies. **B**: Local, mainly in the south.

7 **Pencilled Crane's-bill** *Geranium versicolor* L. [*G. striatum*]. Similar to *G. endressii*, but flowers white or pale lilac with conspicuous violet veins; petals more deeply notched. Shaded habitats, especially hedgerows and banks. May-September. **I**: S Italy, Balkans and the Caucasus. Naturalized from gardens in S Britain. Garden hybrids of *versicolor* and *endressii*, *G. × oxonianum* (8), are occasionally naturalized in Britain and Ireland.

9* **Knotted Crane's-bill** *Geranium nodosum* L. Short to medium, slightly hairy perennial with a slender creeping rhizome; stems erect. Leaves divided almost to the base into 3-5 oval, toothed lobes. Flowers bright pink or violet, with darker veins, 20-28mm, in pairs on densely hairy stalks, forming small clusters; petals deeply notched. Fruits hairy. Open woodland, to 1600m. May-September. Native of the mountains of C and S Europe, C and E France; naturalized Britain, Belgium, N, France, Holland and Germany. **B**: Occasional garden escape, mainly in England.

10 **Iberian Crane's-bill** *Geranium ibericum* Cav. Medium, hairy, tufted perennial, with erect stems branched in the upper part. Leaves divided almost to the base into 5-7 rhombic, pinnately-divided, lobes. Flowers large, deep purple, 36-46mm, in compact clusters; petals notched, often with a tooth in the base of the notch. Fruits hairy. Grassy habitats and hedgebanks. June-July. **I**: Caucasus. Naturalized from gardens in NW France. Widely grown in gardens.

11* **Dusky Crane's-bill** *Geranium phaeum* L. Medium, hairy, tufted perennial; stems erect, branched towards the top. Leaves polygonal in outline, divided for just over halfway into 5-7 broadly oval, deeply-toothed lobes. Flowers a sombre very dark purple, 15-20mm, in pairs, forming lax terminal clusters; petals slightly reflexed, often with a short point. Fruit hairy. Damp or shady places, open woodland and hedgebanks. May-July. C and E France and S Germany; naturalized north to S Sweden. **P**: Bees and probably also hoverflies. **B**: Naturalized throughout, but rare in much of Scotland and Ireland. subsp. *lividum* (L.'Her.) DC. flowers lilac. Occasionally naturalized from gardens in NW Europe.

12* **Marsh Crane's-bill** *Geranium palustre* L. Medium, tufted perennial with short rhizomes; stems erect or spreading. Leaves cut beyond the middle into wedge-shaped, sharply-toothed lobes. Flowers purple or lilac, 22-30mm, cup-shaped, in pairs forming lax clusters; petals not or slightly notched. Fruits hairy, on deflexed stalks. Damp meadows, to 1500m. June-August. North to S Sweden, but not Holland.

13* **Spreading Crane's-bill** *Geranium divaricatum* Ehrh. Medium, hairy annual; stems branched, erect to ascending. Leaves divided beyond the middle into 3-5 oval, deeply toothed lobes. Flowers pink, 8-12mm, in small clusters; petals notched, equalling the pointed sepals. Fruits hairy, on deflexed stalks. Woods, hedgerows and stony habitats, to 2000m. May-August. C and E France, C and S Germany.

Bloody Crane's-bill
Meadow Crane's-bill
Wood
Crane's-bill
French
Crane's-bill
Marsh
Crane's-bill
Dusky Crane's-bill
Spreading Crane's-bill
Knotted Crane's-bill

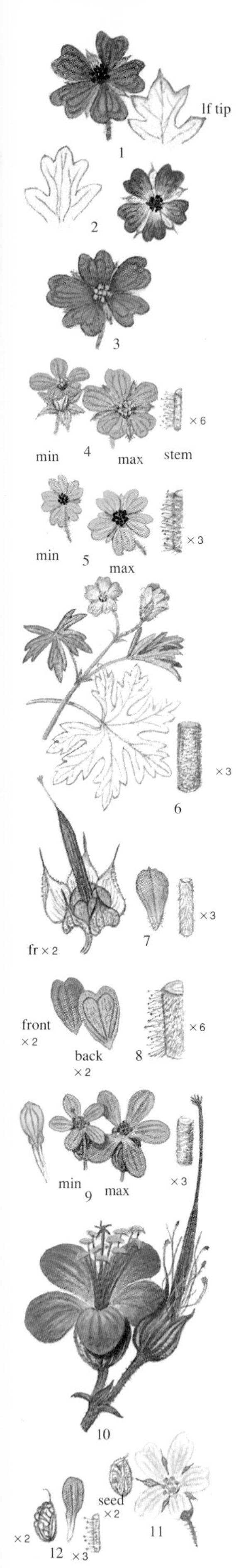


1 *Geranium bohemicum* L. Short to medium, hairy, annual or biennial, with erect stems. Leaves divided beyond half-way into 5-7, pinnately-divided, oval or rhombic lobes. Flowers bright violet-blue with darker veins, 15-18mm, in lax clusters, the petals shallowly notched; sepals with a short point, shorter than the petals. Fruit hairy, on erect stems; seeds yellowish-grey with brown patches. Grassy habitats, scrub and open woodland. May-September. E France, Germany, S Norway, S Sweden and S Finland. Confined to C and E Europe.

2 *Geranium lanuginosum* Lam. Like *G. bohemicum*, but the leaf-lobes blunter and flowers rather smaller, often with a whitish centre. Seeds uniformly brown. Grassy habitats and scrub. June-August. S Sweden – rare. Primarily Mediterranean.

3* **Hedgerow Crane's-bill** *Geranium pyrenaicum* Burm. f. Medium, hairy perennial with erect or ascending stems; hairs both glandular and non-glandular. Leaves rounded in outline, divided just over halfway into 5-7 wedge-shaped lobes, toothed at the top. Flowers purple to mauve-pink, 14-18mm, in pairs forming lax clusters, the petals deeply notched; sepals pointed, half the length of the petals. Fruit hairy, on densely hairy, deflexed stalks. Meadows, field margins, hedgerows, waste and cultivated ground; well-drained soils, to 1900m. June-August. Britain and France; naturalized Belgium, Denmark, S Sweden, S Finland, rare in Germany. **P**: Various insects, also self-pollinated. Widespread in SW and S Europe. **B**: Local in S and E England; scattered localities elsewhere, including Ireland.

4* **Round-leaved Crane's-bill** *Geranium rotundifolium* L. Short, hairy annual with erect or ascending stems; hairs glandular and non-glandular. Basal leaves divided under halfway into 5-7 rounded lobes, toothed at the top; upper leaves more deeply divided and with pointed segments. Flowers pink, 10-12mm, in lax clusters, the petals not or only slightly notched. Fruits hairy, not ridged, on spreading stalks turned up at the end. Dry hedgebanks, walls and occasionally on arable land, on sandy or calcareous soils, to 1600m. June-August. Britain, Belgium, SE Holland, France and Germany. **P**: Generally self-pollinated. Widespread from W Europe to C Asia. **B**: C and S England, S Wales and S Ireland; local or casual elsewhere.

5* **Dove's-foot Crane's-bill** *Geranium molle* L. Short, densely hairy, semi-prostrate annual with spreading or ascending branches; stems with long white hairs. Leaves grey-green, rounded or kidney-shaped in outline, divided beyond halfway into 5-7 wedge-shaped lobes, each 3-lobed at the top; upper leaves more deeply lobed and short-stalked or unstalked, alternate. Flowers pinkish-purple, 6-10mm, in lax clusters, the petals deeply notched, scarcely longer than the sepals. Fruit hairless, with transverse ridges. Dry grassland, lawns, cultivated land, especially among arable crops, waste ground, sand-dunes, on both sandy and calcareous soils, to 2000m. Throughout, except the far north; naturalized in the Faeroes. **P**: Possibly insects, but probably often self-pollinated. Widespread from Europe to Asia. **B**: Throughout Britain but scarcer in the north.

6 **Small-flowered Crane's-bill** *Geranium pusillum* L. Like *G. molle*, but the stems with short hairs and the leaves more narrowly cut. Flowers pale lilac, smaller, 4-6mm, half the stamens without anthers. Fruits hairy. Short grassland, cultivated land and waste places, generally on well-drained soils, to 2000m. June-September. Throughout, except the far north; naturalized in Ireland. **P**: Probably self-pollinated. **B**: Widespread in England and Wales, local and often casual elsewhere, including Ireland.

7* **Long-stalked Crane's-bill** *Geranium columbinum* L. Short to medium, hairy annual with erect or ascending stems. Lower leaves divided almost to the base into oblong, pinnately-divided lobes; upper leaves opposite and less lobed. Flowers purplish-pink, 12-18mm, long-stalked, the petals not notched. Fruit hairless or with a few hairs, not ridged. Open grassland, scrub, occasionally on arable land, mainly on dry calcareous or base-rich soils. June-August. Throughout, except the far north. **B**: Local in England, scarcer and in scattered localities in Wales, Scotland and Ireland.

8* **Cut-leaved Crane's-bill** *Geranium dissectum* L. Like *G. columbinum*, but basal leaves larger and the flower-stalks shorter, not more than 15mm, and sepals densely hairy. Fruit hairy. Waste and cultivated ground, grassland, hedgebanks and roadsides. May-August. Throughout, except the far north. **B**: Widespread throughout, but scarce in N Scotland.

9* **Shiny Crane's-bill** *Geranium lucidum* L. Short, almost hairless annual; stems erect to ascending. Leaves shiny green, often flushed with red, rounded in outline, divided just over halfway into 5 oval lobes, blunt-toothed, the uppermost leaves short-stalked. Flowers pink, 10-14mm, often paired, the petals oval, not notched, with a well marked claw. Fruit hairless. Shaded rocks, walls and hedgebanks, mainly on calcareous soils, an occasional weed of cultivated land. May-August. Throughout, except the far north. The flowers are similar in shape to *G. robertianum*, but rather smaller and the leaves less deeply divided. Widespread in upland England, Wales and W Ireland, scarcer elsewhere.

10 **Balkan Crane's-bill** *Geranium macrorrhizum* L. Short spreading, sticky and pungent, hairy perennial with horizontal rhizomes. Leaves rounded in outline, deeply divided into 5-7 deeply toothed lobes. Flowers dull purple, 24-28mm, in pairs or small clusters, the petals rounded, not notched. Fruits hairless. Shaded rocks and walls, hedgebanks and cultivated ground. June-August. **I**: S & SE Europe. Naturalized S Britain, Belgium and Germany. Widely grown in gardens. **P**: Bees and other insects. **B**: Naturalized in a few places in S England.

11* **Herb Robert** *Geranium robertianum* L. Short to medium, hairy annual or biennial, often flushed with red and with a strong and disagreeable smell. Leaves palmate, the lower with 5 pinnately-divided lobes, the upper with 3 lobes usually. Flowers bright pink, sometimes white, 14-18mm, the petals scarcely notched, with a distinct claw; pollen orange. Fruit usually hairy, ridged. Shaded habitats, banks, woods, coastal shingle and rocky places, to 2000m. May-September. Throughout, except the far north. White-flowered forms are locally frequent – these generally originating in gardens. Coastal forms growing in shingle are usually smaller and prostrate; they are sometimes distinguished as subsp. *maritimum* (Bab.) H.G. Baker. **B**: Throughout, but scarcer in the Scottish Isles.

12* **Little Robin** *Geranium purpureum* Vill. Rather like *G. robertianum*, but a less reddish plant, with smaller flowers, 7-14mm, and the pollen yellow. Open rocky habitats and hedgebanks on fertile calcareous soils, generally close to the sea. May-September. S Britain, S Ireland and W France. Restricted to W and SW Europe and Madeira. Coastal forms growing in shingle are prostrate and are sometimes confused with coastal forms of *G. robertianum*; these are sometimes distinguished as subsp. *fosteri* (Wilmott) H.G. Baker. **B**: Local in S England from Cornwall to Sussex, Channel Is and S Ireland, Cork.

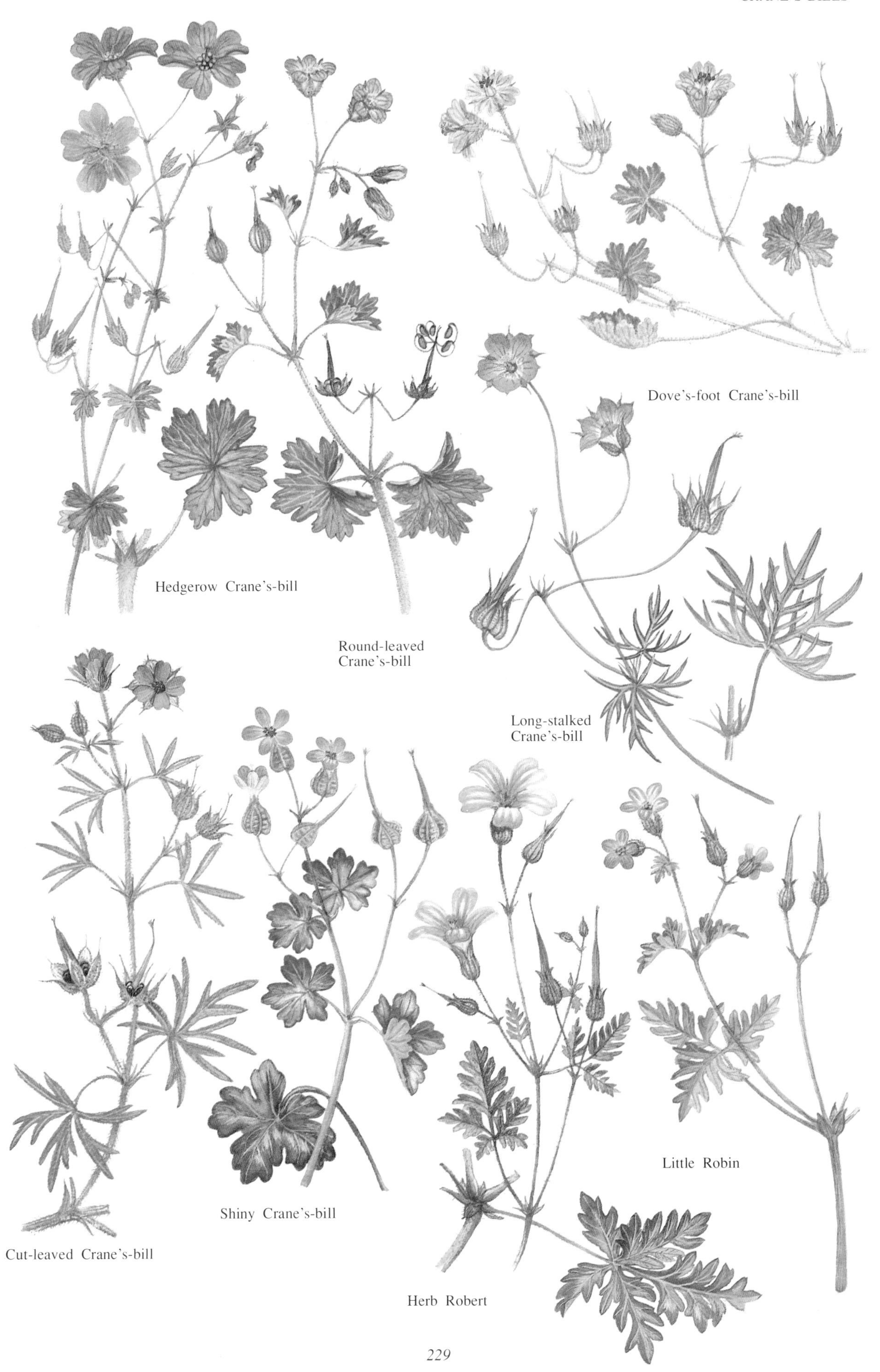
Dove's-foot Crane's-bill
Hedgerow Crane's-bill
Round-leaved Crane's-bill
Long-stalked Crane's-bill
Little Robin
Shiny Crane's-bill
Cut-leaved Crane's-bill
Herb Robert

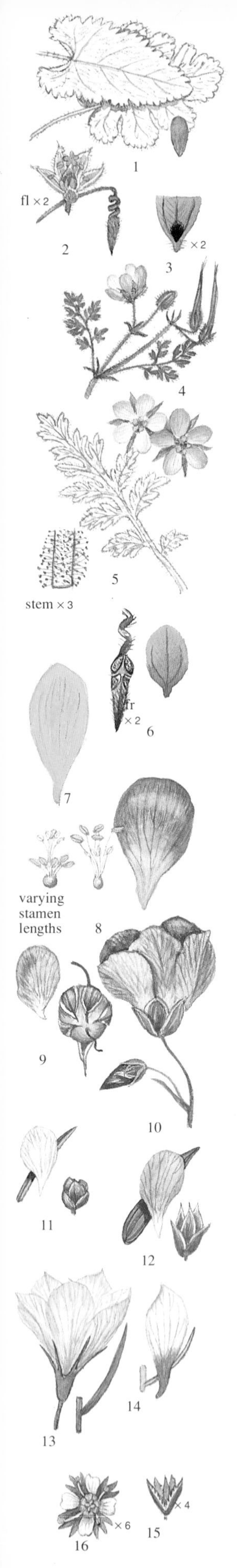


**Stork's-bills** *Erodium*. Distinguished from *Geranium* by bearing pinnately rather than palmately-divided leaves and by often having somewhat zygomorphic flowers. Leaves mostly in basal rosettes. 60 species mainly from the Mediterranean and W Asia.

1* **Mallow-leaved Stork's-bill** *Erodium malacoides* (L.) L'Her. Short to medium, hairy annual or biennial, often rather glandular. Leaves oval, toothed or shallowly-lobed, sometimes 3-lobed, with a heart-shaped base. Flowers purplish, 10-16mm, in lax umbels of 3-7, the petals not notched. Fruit beak 18-35mm. Dry open habitats. June-August. W France. Restricted mainly to S Europe.

2* **Sea Stork's-bill** *Erodium maritimum* (L.) L'Her. Low, often prostrate, white-hairy annual or biennial, with or without a stem. Leaves broadly oval, toothed to pinnately-lobed. Flowers pink or white, 4-6mm, but the petals often absent or soon falling, solitary or sometimes in pairs, on stalks equalling the subtending lf. Fruit beak 8-10mm. Short dry grassland, fixed dunes, generally near the sea. May-September. Britain and NW France. **B**: South of the Solway Firth, occasionally inland, Channel Is and S Ireland.

3* **Common Stork's-bill** *Erodium cicutarium* (L.) L'Her. Very variable low to medium, hairy annual or biennial, generally rather foetid, at first stemless but later branching from the base. Leaves 2-pinnately-lobed, or pinnate. Flowers usually purplish-pink, 7-18mm, up to 12 in an umbel with brownish bracts, the upper two petals often slightly larger and with a blackish blotch at the base. Fruits hairy, the beak 10-40mm long. Disturbed ground, dry grassland. June-September. Throughout, except the far north **B**: Widespread but often local; Ireland, mostly inland.

4 *E. c.* subsp. *bipinnatum* (Cav.) Tourlet [*E. bipinnatum*]. Less robust; usually densely glandular-hairy. Lilac or white petals equal and unmarked. Sea sands. May-September. Coasts of Britain and W Europe.

5 *E. glutinosum* Dumort (incl. *E. g.* subsp. *dunense* (Andreas) Rothm., *E. cicutarium* subsp. *dunense* Andreas) is intermediate between *cicutarium* and subsp. *bipinnatum*, possibly of hybrid origin.

6* **Musk Stork's-bill** *Erodium moschatum* (L.) L'Her. Short to medium, stickily-hairy annual or biennial, smelling of musk, almost stemless or with a distinct stem. Leaves pinnate, with oval, toothed leaflets. Flowers purple or violet, large, 16-24mm, in umbels of 5-12. Fruit hairy, the beak 20-25mm long. Cultivated and waste ground, often near the coast. May-July. Britain, France, Holland; naturalized Ireland, Belgium, Germany. **B**: Local, often coastal, from Cumbria and Lancs to Channel Is and coastal Ireland.

## FLAX FAMILY Linaceae

Herbs with slender, fibrous stems and simple untoothed leaves. Flowers 4 or 5-parted, sepals separate, petals contorted in bud; stamens 5. Fruit an 8 or 10-valved capsule. 200 species.

**Flaxes** *Linum*. Flowers 5-parted. Fruit capsule 10-valved, containing flat, shiny seeds. 200 species, mostly in the temperate and subtropical northern hemisphere.

7* **Yellow Flax** *Linum flavum* L. Variable, medium, hairless tufted perennial. Leaves alternate, spoon-shaped to lanceolate, 3-veined. Flowers yellow, 25-30mm, in branched clusters. Sepals scarcely longer than the fruit. Dry grassy habitats, to 1600m. May-August. C Europe. C & S France.

8* **Perennial Flax** *Linum perenne* L. Medium, erect, hairless perennial. Leaves linear to lanceolate, rather crowded, mostly 1-veined. Flowers dark blue, 20-26mm, numerous in raceme-like cymes; outer pointed sepals narrower than the blunt inner ones. Capsule subglobose, with or without a short beak. Dry permanent grassy habitats on calcareous soils. May-July. E France and S Germany. subsp. *anglicum* (Miller) Ockendon [*L. anglicum*]. Stems ascending and curved at the base; flowers sky blue. June-July. Britain. Rare, decreasing. **B**: Essex to Durham and Cumbria. subsp. *alpinum* (Jacq.) Ockendon. Shorter, to only 30cm (not 30-60cm). Mountain grassland, to 1700m. E France and S Germany.

9* *Linum austriacum* L. Middle leaves 1-3-veined and the flower-stalks downturned (not straight and erect). Grassy habitats on calcareous soils. June-July. France and Germany; naturalized Denmark.

10 *Linum leoni* F.W. Schultz. Low to short, not exceeding 30cm tall and the cymes with only 1-6 flowers; flower-stalks deflexed or ascending. Dry grassy habitats. June-July. France and W Germany.

11* **Pale Flax** *Linum bienne* Miller [*L. angustifolium*]. Low to medium, hairless biennial or perennial; stems erect or ascending, usually branched, slender. Leaves linear-lanceolate, 1-3-veined. Flowers pale blue or lilac-blue, 16-24mm, in lax cymes; sepals unequal but all pointed, the inner with a hairy margin. Capsule subglobose with a short beak. Dry permanent grassland on neutral or calcareous soils, generally close to the sea. May-September. Britain and France. The flowers are homostylous and the petals generally drop rather quickly after flowers open. W and S Europe, Madeira and Canary Is. **B**: Lancs. and N Wales south and in SE Ireland; always local.

12 **Flax** *Linum usitatissimum* L. [*L. humile*]. Like *L. bienne*, but annual and more robust; stem normally solitary, not clustered. Leaves 3-veined, 1.5-3mm wide (not 0.5-1.5mm). Sepals 6-9mm (not 4-5.5mm). June-August. Origins uncertain. Long grown throughout Europe for fibre (linen) and linseed.

13 *Linum tenuifolium* L. Short to medium, hairless tufted perennial with erect flowering shoots as well as short non-flowering shoots near the base. Leaves alternate, linear, 1-veined, with rough margins, sometimes slightly inrolled. Flowers pink or whitish, 12-18mm, in lax raceme-like cymes; sepals all pointed, 1-veined, with glandular-hairy margins. Dry grassy and rocky places. Late May-July. Belgium, France and Germany. Mainly in C and S Europe.

14 *Linum suffruticosum* L. Like *L. tenuifolium*, but with numerous non-flowering shoots and leaves with markedly inrolled margins. Flowers white with a violet or pink centre; sepals 3-veined and styles erect. Dry grassy and rocky places. Late May-July. NC France south to Spain and Italy.

15* **Fairy Flax** *Linum catharticum* L. Slender, low to short hairless annual, generally with a solitary stem. Leaves opposite, oblong to lanceolate, 1-veined. Flowers white with a yellowish centre, small, 4-6mm, in lax, branched clusters, nodding in bud; sepals with glandular-hairy margins. Calcareous grassland, coastal dunes, grassy heaths, cliffs, to 2350m. Late May-September. Throughout, except Spitsbergen.

16* **Allseed** *Radiola linoides* Roth. Low, hairless annual, rarely exceeding 8cm tall and often tiny, with thread-like stems and paired branching. Leaves opposite, elliptical, 1-veined. Flowers white, tiny, 1-2mm, in branched clusters, short-stalked; sepals 3-toothed at the apex. Capsule globose, tiny. Seasonally damp sandy or peaty ground on grassland and heaths, on acid soils. July-August. North to S Norway, but not Sweden. **B**: Local and declining; rare in most of Ireland except the W.

Sea Stork's-bill
Mallow Leaved Stork's-bill
Common Stork's-bill
Musk Stork's-bill
Fairy Flax
*Linum austriacum*
Allseed
Perennial Flax
Yellow Flax
Pale Flax

# SPURGE FAMILY Euphorbiaceae

Herbs or shrubs, often with milky latex and alternate, simple leaves. Flowers regular, usually without petals, male and female separate, on the same or on different plants; male flowers with 1 to many stamens, female with a superior, 3-locular ovary and 3 styles. Fruit a capsule, often splitting explosively; seeds with a lump (caruncle) on one end. 7000 species.

**Mercury** *Mercurialis*. Herbs with watery juice and opposite leaves. Male and female flowers on separate plants, the male in axillary spikes, with 8-15 stamens, the female solitary or in small clusters, with 2 styles. Fruit 2-lobed. 8 species.

1* **Annual Mercury** *Mercurialis annua* L. Short to medium, hairless or slightly hairy, annual; stems branched from the base. Leaves oval to lanceolate, toothed, short-stalked. Female flowers few, 3-4mm, in small axillary clusters. Fruit 2-3mm, bristly. Cultivated ground and waste places, especially on lighter soils, to 1800m. July-October. W Europe; naturalized Ireland and parts of Scandinavia. **B**: S And SE England, south of the Wash, rare and local elsewhere; Ireland except the north.

2* **Dog's Mercury** *Mercurialis perennis* L. Low to short rhizomatous, hairy perennial, becoming black on drying; stems unbranched. Leaves lanceolate, stalked, crowded towards the stem tops, the lowermost leaves reduced and scale-like. Flowers greenish, 4-5mm, the female solitary or 2-3 together, the male in long tassel-like spikes. Fruit 5-8mm, hairy. Shaded habitats, limestone grikes, to 1800m. February-April. Throughout, except the far N, Faeroes and Iceland. **B**: Throughout, but rare in N Scotland and Ireland.

3 *Mercurialis ovata* Sternb. & Hoppe. Like *M. perennis*, but the plant yellowish-green when dry. Leaves all foliage-like, not crowded or scale-like; leaf-stalks 1-2mm (not 5-18mm). Woods and other shaded habitats, to 1800m. March-May. S Germany.

**Spurges** *Euphorbia*. Herbs or subshrubs with milky latex. Male and female flowers separate but in discrete groups (cyathia) with several male and a solitary female set in a cup-shaped involucre, the male with a single stamen, the female with an ovary and 3 styles. Fruit a 3-lobed capsule. Over 1500 species in both tropical and temperate regions.

4* **Purple Spurge** *Euphorbia peplis* L. Low hairless, prostrate annual with purple stems, generally with 4 branches from the base. Leaves opposite, grey-green, oblong, with a rounded lobe on one side at the base, short-stalked, untoothed. Flowers greenish with semicircular, reddish-brown, glands. Capsule 3.5-4mm, nearly smooth and purplish. Sandy or shingly seashores, above high-water mark. July-September. S Britain and France. Rather rare and decreasing, fluctuating. **B**: Very rare, formerly on Lundy Island, Guernsey, Alderney and occasional in S England; possibly now extinct.

5 **Hairy Spurge** *Euphorbia villosa* Waldst. & Kit. ex Willd. [*E. pilosa*]. Medium to tall, rather stout, usually hairy, rhizomatous perennial; stems numerous, often with non-flowering branches, scaly below. Leaves oblong to elliptical, untoothed or slightly toothed near the apex. Flowers in umbel-like clusters with 5 or more main rays and yellowish-green leaf-like bracts; glands kidney-shaped, yellowish. Capsule 3-6mm, finely warted, hairy or not. Damp meadows, open woods, coppices and riverbanks, generally at low altitudes. May-July. France and Germany; extinct in Britain this century.

6 **Coral Spurge** *Euphorbia coralloides* L. Like *E. villosa*, but a medium, more densely tufted perennial, with slender stems, not scaly below. Capsule 3-4mm, slightly granular and densely hairy. **B**: Shaded habitats. June-July. **I**: S Italy & Sicily. Naturalized and local in S England – Sussex.

7* **Irish Spurge** *Euphorbia hyberna* L. Medium, practically hairless, rhizomatous perennial, with erect unbranched stems. Leaves oblong, often broadest above the middle, untoothed, slightly hairy beneath, eventually turning pinkish-red. Flowers in broad umbel-like clusters, with yellowish-green, leaf-like bracts; glands yellowish or brownish, kidney-shaped, untoothed. Capsule 5-6mm, covered by elongated warts. Damp or shaded places, on acid soils, to 2000m. April-July. S Britain, SW Ireland and France. Only in N Cornwall, N Devon, Somerset and localities in SW Ireland.

8* **Marsh Spurge** *Euphorbia palustris* L. Like *E. hyberna*, but taller, to 1.5m, with numerous non-flowering branches and greyish leaves, hairless on both surfaces. Capsule with short warts. Damp habitats, especially in swampy woodland, by rivers or close to the sea. May-July. N France, Holland, Germany to S Scandinavia; naturalized locally in Belgium.

9* **Sweet Spurge** *Euphorbia dulcis* L. Short to medium, rhizomatous perennial, with slender, unbranched, erect stems, scaly below. Leaves elliptical to oblong, untoothed, usually hairless. Umbel usually 5-rayed, with green bracts. Flowers with rounded glands, green at first, but soon becoming dark purple. Capsule 3-4mm, grooved, usually hairless, with prominent elongated warts. Damp or shaded habitats, generally at low altitudes. May-July. France, Belgium, Holland and Germany; naturalized Britain and Denmark. **B**: Naturalized in a few scattered localities throughout, except the far N.

10 *Euphorbia brittingeri* Opiz ex Samp. Short to medium, tufted, usually somewhat hairy perennial with thin erect stems, slightly woody but not scaly below, often with a few non-flowering branches. Leaves oblong-elliptical, finely toothed. Umbel often 5-rayed with yellowish-green bracts, becoming gradually green or purplish. Capsule with crowded warts. Wooded and grassy habitats. Late May-July. Belgium, France and Germany.

11* **Broad-leaved Spurge** *Euphorbia platyphyllos* L. Short to tall hairy or hairless annual; stems simple and erect with many lateral umbels as well as a terminal one. Leaves oval, broadest above the middle, finely toothed at the pointed tip, slightly heart-shaped the base. Main umbel with 5 main rays usually and greenish-yellow bracts which are mostly triangular in outline, becoming purplish later; glands suborbicular, not toothed. Capsule 2-3mm, with crowded warts; seeds olive-brown. Arable land and waste places, generally on heavy soils. June-October. Belgium, France and Germany, rare in Britain and Holland. **B**: Local in S & E England from the Humber south.

12 **Upright Spurge** *Euphorbia serrulata* Thuill. Like *E. paltyphyllos*, but a smaller and more slender, hairless plant. Umbels with 4-5 rays. Capsule with elongated warts; seeds reddish-brown. Woodland clearings, over limestone. Always local, but sometimes abundant. June-September. W Europe. **B**: Very rare; confined to a few scattered localities in Monmouthshire and Gloucs, casual elsewhere.

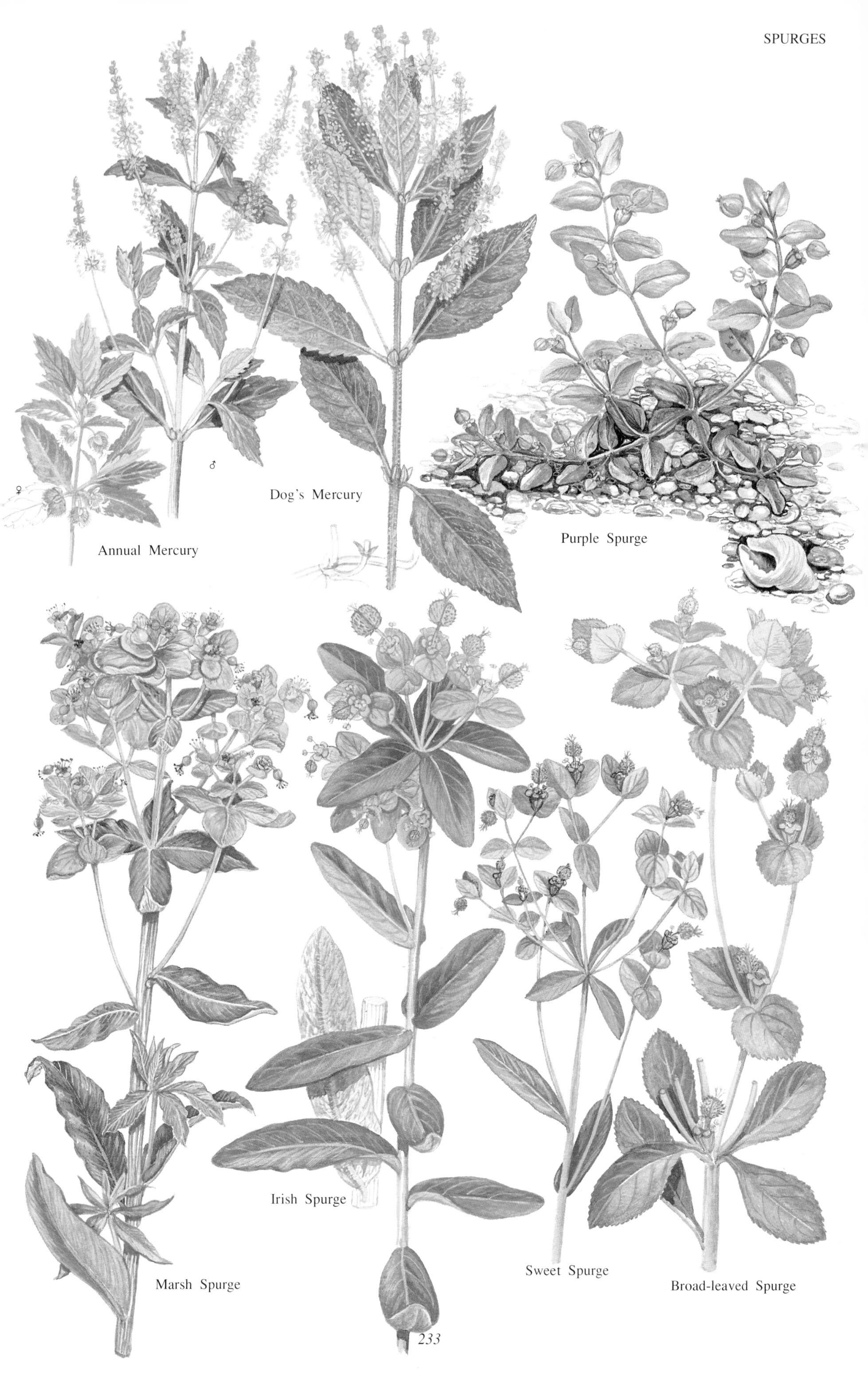
♀
♂
Annual Mercury
Dog's Mercury
Purple Spurge
Marsh Spurge
Irish Spurge
Sweet Spurge
Broad-leaved Spurge

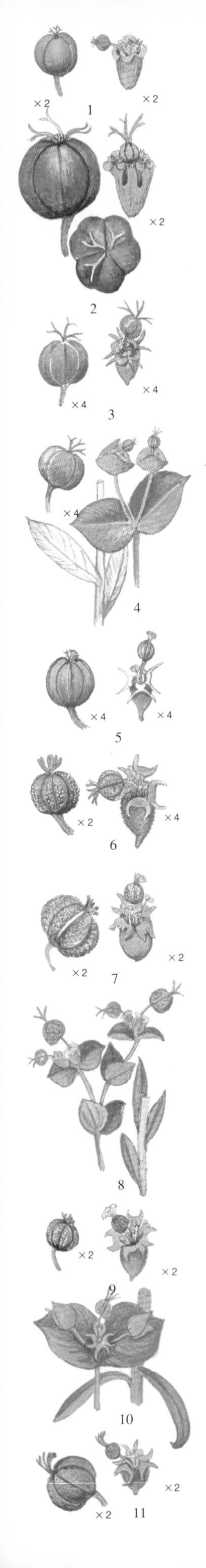


1* **Sun Spurge** *Euphorbia helioscopia* L. Short to medium, more or less hairless annual; stem usually solitary, erect. Leaves oval, broadest above the middle, blunt, finely toothed in the upper half. Umbel with 5 main rays usually and yellowish bracts similar in shape to the leaves. Capsule 2.5-3.5mm, smooth and unwinged. Seeds brown. Disturbed and waste ground, a frequent weed of arable land and gardens, sometimes along roadsides, generally at low altitudes, but up to 800m. May-August. Throughout, except the far north. Variable – plants on poor dry soils may be very dwarf. A quick growing summer annual. The seeds are distributed by ants. **B**: Throughout, scarcer and more coastal in the north.

2* **Caper Spurge** *Euphorbia lathyrus* L. Medium to tall, erect, hairless bluish-green biennial, to 1.5m. Leaves opposite, linear to oblong, untoothed and unstalked. Umbels with 2-6 main rays; bracts oval-triangular, pointed, bright green; glands kidney-shaped, with blunt horns. Capsule 13-17mm, smooth; seeds brown or grey. Disturbed ground and waste places, a frequent weed of gardens; sometimes found in open woodland. June-July. **I**: C & S Europe. Widely naturalized in W Europe. The very poisonous fruits are sometimes mistaken for the edible flower-buds of the true **Caper**, *Capparis spinosa*. **B**: Perhaps native in Sussex. Sometimes established, but mostly casual north to S Scotland.

3* **Dwarf Spurge** *Euphorbia exigua* L. Low to short, greyish, hairless annual, often much-branched from the base. Leaves linear to narrow-oblong, untoothed. Umbels with 3-5 main rays, the bracts triangular-oval with a heart-shaped base; glands kidney-shaped, with long slender horns. Capsule 1.6-2mm, smooth; seeds pale grey. Disturbed and waste ground, arable land, on base-rich or calcareous soils. June-October. Throughout, north to C Scandinavia; not in Faeroes, Iceland, Finland or Spitsbergen. **B**: Lowland, but scarcer in the north and west; Ireland, mainly in the east.

4 *Euphorbia falcata* L. Like *E. exigua*, but stem often unbranched or with very few branches at the base and with 4 or more lateral umbels (not 0-3). Disturbed and waste ground, or a weed of cultivated ground. June-August. NC France and S Germany. From C and S Europe.

5* **Petty Spurge** *Euphorbia peplus* L. Medium, hairless, green annual; stem erect, with 2 or more branches from the base. Leaves oval to almost rounded, untoothed, short-stalked. Umbels with 3 main rays usually and triangular-oval green bracts; glands kidney-shaped with long, slender horns. Capsule 2.8-3mm, smooth and shallowly keeled on the back of each valve; seeds pale grey. An often abundant weed of cultivated land, especially of arable crops and gardens, disturbed and waste ground. April-October. Throughout, north to C Scandinavia; not in Faeroes, Iceland and Spitsbergen. A spring and summer annual widespread in Europe east to W Asia and S Siberia. **B**: Common throughout England, Wales and Ireland, but scarcer in Scotland.

6* **Portland Spurge** *Euphorbia portlandica* L. Short greyish, hairless perennial, often becoming red; stems usually branched from the base. Leaves oval, often broadest above the middle, blunt, usually untoothed, with a prominent midrib beneath. Umbels generally with 4-5 main rays; bracts triangular or rhombic, yellowish-green; glands yellowish, kidney-shaped, with a pair of long horns. Capsule 2.8-3mm, granular along the back of the valves; seeds pale grey. Maritime sands, especially young dunes, sandy limestone cliffs. May-September. Britain and W France. **B**: S and W north to Wigtown, and coastal Ireland.

7* **Sea Spurge** *Euphorbia paralias* L. Short to medium, rather fleshy, greyish, hairless tufted perennial. Leaves oblong, the uppermost oval, overlapping, untoothed; midrib not prominent beneath. Umbels with 3-6 main rays usually with bracts oval; glands kidney-shaped with long horns. Capsule 3-5mm, granular on the back of the valves; seeds pale grey. Coastal habitats, sandy places and dunes, fine shingle, locally abundant. July-October. Britain, Belgium, Holland and France. From W Europe to most of the coastal Mediterranean. Sometimes confused with *E. portlandica*, but the latter has more spreading leaves with a pronounced midrib beneath. **B**: N to the Wash and the Solway Firth on the W side; coastal Ireland.

8 *Euphorbia seguierana* Necker [*E. gerardiana*]. Like *E. paralias*, but leaves less fleshy, linear to narrow-oblong, pointed; rays 5-15; glands truncated, without horns. Dry habitats. June-August. Belgium, Holland, France and Germany. Mainly in C and S Europe.

9* **Cypress Spurge** *Euphorbia cyparissias* L. Short to medium, hairless, rhizomatous perennial, often forming tufts; stems usually branched from the base. Leaves linear, crowded, untoothed, rather dull green but turning yellow or reddish later. Umbels with 9-18 main rays and rounded or kidney-shaped greenish-yellow bracts; glands kidney-shaped, with short horns. Capsule 3mm, granular on the backs of the valves; seeds grey, slightly shiny. Grassy and rocky habitats, scrub and waste places, often on calcareous soils, to 2650m. April-June. Belgium, Holland, France and Germany; naturalized Britain, and Scandinavia. Native in much of C and S Europe. Also a frequent garden escape. **P**: Mostly flies. **B**: Naturalized and local throughout except most of Scotland; absent from Ireland.

10 **Leafy Spurge** *Euphorbia esula* L. Like *E. cyparissias*, but sometimes hairy, the stems generally not branched at the base. Leaves less crowded, linear to oval, those of the flowering stems not more than 2mm (not 4mm or more) broad; glands slightly notched or with short horns. Grassy and waste places, open woods and streamsides. May-July. Same distribution as *E. cyparissias*. Often forming large patches. In most of C and S Europe. **B**: Local in England and Scotland.

subsp. *Euphorbia waldsteinii* (Sojak) A.R. Smith [*E. virgata*]. Like *E. esula*, but the leaves broadest at or just below the middle (not near the apex) and the umbel bracts 12-35mm (not 5-15mm). Grassy and waste places. May-July. **I**: E Europe & W Asia.

11* **Wood Spurge** *Euphorbia amygdaloides* L. Medium to tall, hairy, rhizomatous perennial often forming large patches, evergreen, often red-tinged; stems erect, unbranched usually. Leaves oblong, often broadest above the middle, untoothed, deep green, crowded towards the stem tops. Umbels with 5-10 main rays and oval or rounded bracts fused together to encircle the stem, yellowish-green; glands kidney-shaped, with converging horns. Capsule 3-4mm, granular; seeds blackish. Open woods, woodland clearings, coppices, generally on damp neutral or mildly acid soils. March-May. Britain, Belgium, France and C and S Germany, rare in Ireland. Forming extensive colonies, especially in recently coppiced woodland. Often associated with ancient woodland. Frequently grown in gardens, including a purple-leaved form. The shoots are biennial, flowering early in their second season. **B**: England S of the Wash and in Channel Is, rarer in Wales and scarce elsewhere.

Sun Spurge
Caper Spurge
Dwarf Spurge
Petty Spurge
× 1/4
Sea Spurge
Portland Spurge
Cypress Spurge
Wood Spurge

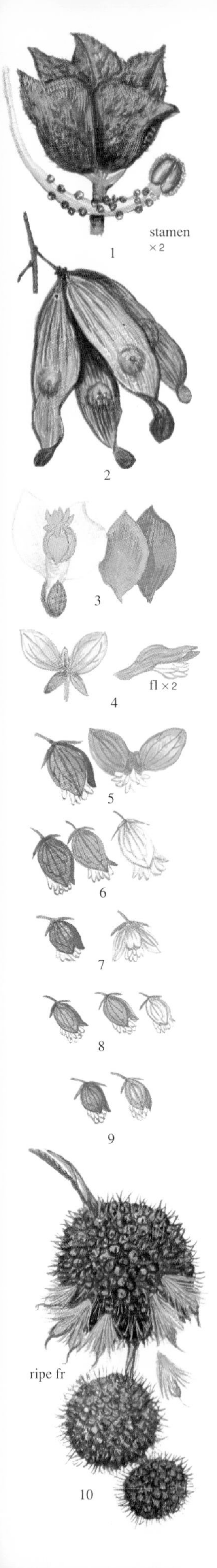


1* **Burning Bush** *Dictamnus albus* L. (Rue Family – Rutaceae – 900 species including the Orange, Lemon and Grapefruit). Medium to tall tufted, pungent and somewhat sticky, slightly hairy perennial. Leaves alternate, pinnate, with an end leaflet; leaflets 3-6 pairs, oval to lanceolate, untoothed. Flowers white to pink or bluish with prominent purple stamens, large, 40-50mm, 5-parted, in rather dense racemes, stamens 10. Fruit erect, deeply 5-lobed, very glandular. Dry grassy habitats and scrub. June-August. S Germany. Widespread in SC and S Europe. On hot still days the plant gives off volatile oils which can be ignited without apparently harming the plant. Flowers irregular with four petals pointing up and one down. Widely grown in gardens.

2* **Tree of Heaven** *Ailanthus altissima* (Miller) Swingle. (Tree of Heaven Family – Simaroubaceae – 150 species of trees or shrubs, usually with bitter tasting bark). Large deciduous tree to 20m, often producing numerous suckers; bark smooth and grey. Leaves pinnate, more or less hairless; leaflets oval-lanceolate, pointed, generally with a few coarse teeth near the base. Flowers greenish or creamish, 7-8mm, in large terminal panicles, mostly 5-parted, male and female often on separate trees. Fruit a group of reddish-brown samaras. Widely planted for ornament and frequently naturalized. June-July. **I**: China. Naturalized locally in Britain, Belgium, France and Germany.

## MILKWORT FAMILY Polygalaceae

Herbs and small shrubs with simple, generally alternate, leaves; no stipules. Flowers hermaphrodite, irregular, in slender racemes or spikes. Sepals 5, separate, the inner 2 (wings) much larger and petal-like. Corolla (in *Polygala*) with 3 fused petals, fringed at the apex. Stamens 8. Fruit a compressed, usually 2-lobed capsule. 70 species in 10 genera.

3* **Shrubby Milkwort** *Polygala chamaebuxus* L. Dwarf evergreen shrub, 5-15cm tall, often mat-forming. Leaves leathery, deep shiny green, linear to lanceolate. Flowers solitary or in pairs with a yellow keel and white, yellow or purple wings, 10-18mm. Capsule 6-8mm, with narrow wings. Woods, pastures and rocky slopes, mainly in the mountains, to 2500m. April-September. E France and C and S Germany.

4* **Tufted Milkwort** *Polygala comosa* Schkuhr. Low to short perennial herb with erect or ascending stems. Leaves oval, broadest above the middle, the lower ones usually fallen by flowering time, the upper narrower. Flowers usually lilac-pink, 4-6mm, in dense, rather conical racemes of 15-50, the lip equalling the wings; bracts linear, exceeding the flower-buds. Capsule shorter than the persistent wings. Dry grassy habitats and open woods, to 2200m. May-July. North to S Sweden, but not Denmark or Norway.

5 **Nice Milkwort** *Polygala nicaeensis* Risso ex Koch. Like *P. comosa*, but flowers larger, 8-11mm, the wings 3-5-veined (not 1-3-veined). Flowers pink, sometimes blue or white. Dry grassy and stony habitats, to 1700m. April-July. Mountains of E France, S Germany, C and S Europe.

6* **Common Milkwort** *Polygala vulgaris* L. Variable low to short, hairless or slightly hairy, perennial herb, with a branched woody stock and erect to ascending stems. Leaves all alternate, not bitter-tasting, oval to elliptical, the uppermost narrower. Flowers blue, pink or white, 5-8mm, in rather dense racemes of 10-40; bracts membranous, shorter than the flower-buds. Capsule equal in length to the persistent sepals. Short turf over chalk and limestone, heaths, commons and sand-dunes, at low altitudes. May-September. Throughout, except Iceland and Spitsbergen. **B**: Common.

7* **Thyme-leaved Milkwort** *Polygala serpyllifolia* Hose [*P. serpyllacea*]. Low to short, rather slender perennial, with spreading or ascending stems, not woody at the base. Leaves elliptical to oval, often broadest above the middle, lower leaves opposite, upper alternate. Flowers usually blue, 5-6mm, in small racemes of 3-10; corolla slightly longer than the wings. Capsule shorter and wider than the persistent wings. Grassy habitats and heaths, usually on acid soils, to 1800m. May-August. North to SW Norway, but not Sweden or Finland. **B**: The commonest and most widespread species.

8* **Chalk Milkwort** *Polygala calcarea* F.W. Schultz. Low to short spreading, hairless or slightly hairy perennial herb, with stolons terminating in leafy rosettes, each producing several erect flowering stems. Rosette-leaves oval, broadest above the middle, not bitter-tasting; leaves of flowering stems linear-lanceolate. Flowers blue or white, 5-6mm, in racemes of 6-20, the corolla slightly longer than the wings. Capsule slightly longer than the persistent wings. Calcareous grassland, especially when closely grazed, generally at rather low altitudes. May-June. Britain, Belgium, France and W Germany. Confined to W Europe, east to the Jura mountains. **B**: Confined to S and SE England north to Rutland; local.

9* **Dwarf Milkwort** *Polygala amarella* Crantz [*P. amara* subsp. *amarella*, *P. austriaca*]. Low to short perennial herb with numerous stems arising from basal leaf rosettes. Leaves bitter-tasting, hairless, elliptical to oval, the upper narrower. Flowers blue, violet, pink or white, small, 2-4.5mm, in racemes of 8-25. Capsule 3-4mm, much wider than the persistent wings. Damp meadows and open grassland, over chalk or limestone, to 2600m. May-August. North to S Scandinavia, but not Holland. **B**: Confined to two areas, one in N England, the other in SE England. S British population sometimes called *P. austriaca*.

## PLANE FAMILY Platanaceae

Deciduous trees with alternate leaves; buds hidden in the base of the leaf-stalks. Flowers unisexual, but on the same tree, in dense rounded heads arranged in a raceme; female with 3 sepals forming a cup, the petals often absent, carpels 5-9; male flowers with or without a cup-like calyx, petals present and stamens 3-5. Fruit a cluster of nutlets. A single genus and some 6 species. **P**: By wind.

10* **London Plane** *Platanus × hybrida* Brot. [*P. occidentalis × P. orientalis*]. Large deciduous tree to 35m. Trunk with bark peeling away in large flakes, giving a patchwork appearance. Leaves palmate, with 3-5 lobes, practically hairless, untoothed or slightly toothed. Flowers greenish, female flowerheads often 2, occasionally 3 or solitary; male generally more. Fruitheads pendent, 22-26mm. Widely planted in towns and cities or in large parks, very occasionally self-sown, although most of the seeds are infertile. May-June. Throughout except the far north. It is remarkably tolerant of polluted atmospheres. The leaves rot down very slowly. **B**: Planted in many parts.

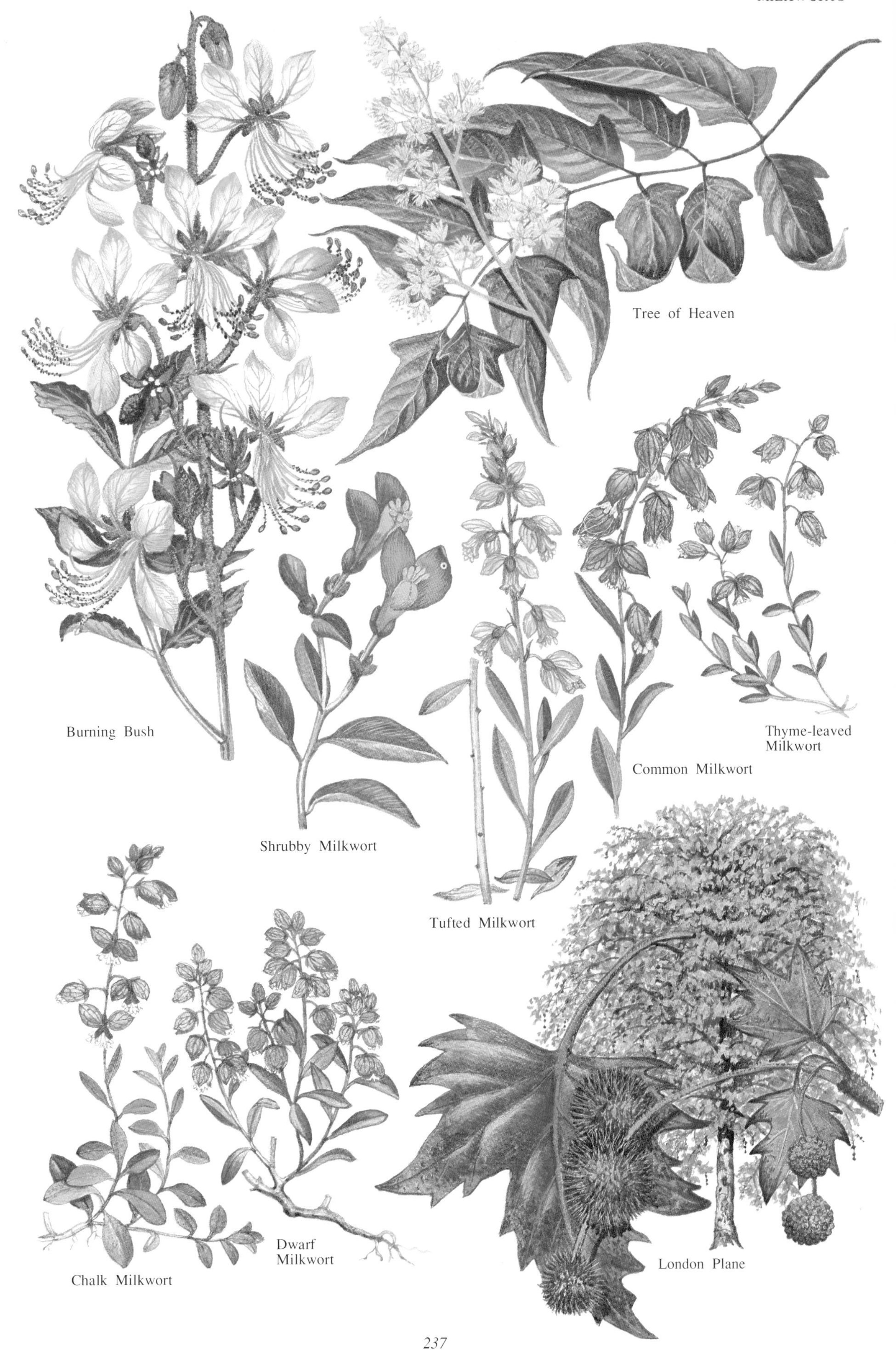
Tree of Heaven
Burning Bush
Shrubby Milkwort
Tufted Milkwort
Common Milkwort
Thyme-leaved Milkwort
Chalk Milkwort
Dwarf Milkwort
London Plane

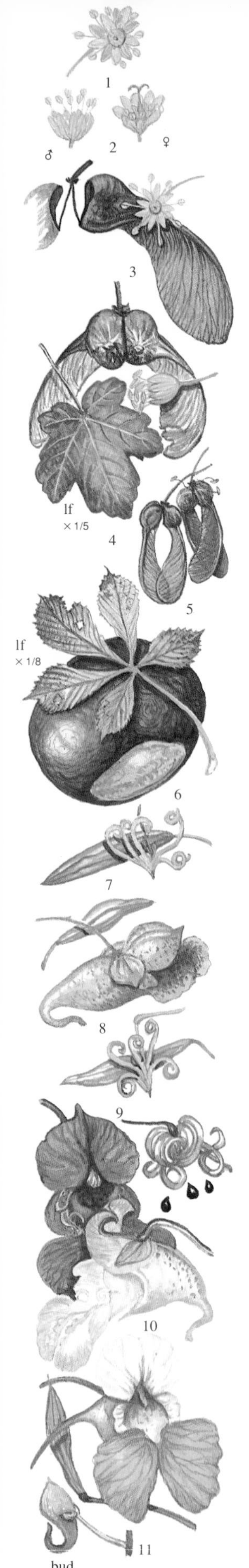


## MAPLE FAMILY Aceraceae

Trees or shrubs with opposite leaves, no stipules. Flowers often greenish or yellowish, regular, in racemes or panicles. Sepals and petals each 5, all separate; stamens 8 and ovary superior, of 2 carpels. Fruit composed of 2 winged samaras. 000 species in 2 genera.

1* **Norway Maple** *Acer platanoides* L. Large, spreading, deciduous tree to 30m; bark pale grey, smooth or shallowly ridged. Leaves large, 10-15cm across, palmately 5-7-lobed, the lobes long-pointed, with a few coarse teeth. Flowers an acid yellowish-green, 7-8mm, before the leaves in broad erect panicles. Fruits with widely separating horizontal wings. Woods and hedgerows, fields, to 1600m, but widely planted for shelter and for ornament. April-May. Belgium, France, Germany and Scandinavia north to S Sweden and S Norway; naturalized B and Holland. **B**: North to S Scotland, local and mostly planted.

2* **Field Maple** *Acer campestre* L. A variable deciduous bush or tree to 20m, with pale grey, fissured bark and downy twigs. Leaves rather small, 4-7cm across, 3-5-lobed, the lobes oblong, blunt, generally untoothed and with a hairy margin. Flowers greenish-yellow, 5-6mm, in small erect hairy clusters, with the leaves. Fruits usually hairy, with horizontal wings. Open woods, hedgerows, coppices, occasionally as an isolated old specimen in a field, to 1500m. April-May. North to S Sweden; naturalized Ireland. **P**: Various small insects. **B**: North to Cumbria and Durham.

3* **Sycamore** *Acer pseudoplatanus* L. Large spreading deciduous tree to 30m, with smooth grey bark, flaking when old; twigs hairless. Leaves large, 10-25cm across, palmately 5-lobed, the lobes pointed and coarsely toothed. Flowers yellowish-green, 6-7mm, in narrow pendent panicles usually with the leaves. Fruits hairless, with wings at right-angles. Woods, coppices, fields and hedgerows, along streamsides and also in coastal habitats, to 2000m. April-May. Belgium, Holland, France and Germany; widely naturalized Britain, Denmark and S Sweden. Probably introduced by the Romans. **B**: Naturalized almost throughout.

4 **Italian Maple** *Acer opalus* Miller [*A. opulifolium*]. Small deciduous tree or shrub, to 15m with pinkish-or orange-grey flaking bark; twigs hairless. Leaves 4-10cm across, palmately 5-lobed, the lobes broad, toothed, hairy beneath. Flowers pale yellow, 5-6mm, in pendent clusters, before or with the very young leaves. Fruits hairless with wings diverging at an acute angle. Woods and hedgerows, to 1900m. April. C and E France and SW Germany.

5* **Montpelier Maple** *Acer monspessulanus* L. Small deciduous tree or shrub, to 12m with dark greyish-black, finely cracked bark. Leaves 3-8cm across, leathery, 3-lobed, the lobes rather blunt and untoothed, shiny deep green above.Flowers greenish-yellow, 4-5mm, in broad clusters, erect at first but becoming pendent. Fruits hairless with the wings almost parallel. Open woods, hedgerows and fields. June. W and S France, C and S Germany. Mainly in S Europe and W Asia. Rather uncommon in cultivation.

6* **Horse-chestnut** *Aesculus hippocastanum* L. (Horse-chestnut Family – Hippocastanaceae). Large deciduous tree to 25m with dark greyish-brown bark, smooth at first but eventually flaking; buds large, deep brown and sticky. Leaves palmate with 5-7 separate elliptical leaflets, coarse-toothed. Flowers white, each of the 4 petals with a yellow or pink blotch at the base, 9-11mm, in large erect conical panicles. Fruit globose, up to 6cm, yellowish-green when ripe, spiny, splitting to reveal 1-2 large deep shiny-brown seeds (conkers). May-June. **I**: Balkan Peninsula. Britain, France, Germany. **B**: Especially in the C and S.

## BALSAM FAMILY Balsaminaceae

Succulent annual and perennial herbs. Flowers lateral, solitary or in racemes, basically 5-parted, but sepals usually only 3, the lower one elaborated into a distinctive spur. Lateral pairs of petals united near the base, the upper one free. Stamens united together around the ovary. Fruit an explosive capsule. Only 2 genera. *Impatiens* has 1200 species in the world.

7* **Touch-me-not** *Impatiens noli-tangere* L. Medium to tall, hairless annual to 1.8m; stems usually branched. Leaves alternate, oval to elliptical, the margin with a few coarse teeth. Flowers yellow with small brownish spots, 20-35mm, in lax racemes of 3-6; lower sepal sac-like, with a curved spur. Capsule linear. Moist habitats, stream and river banks, dyke-sides, wet woods and other damp shady places, to 1500m. July-September. All except the far north. **B**: Local and native in N England, Lake District, N Wales; casual elsewhere.

8 **Orange Balsam** *Impatiens capensis* Meerb. Medium to tall, hairless annual, to 1.2m. Leaves oval to lanceolate, the margin with a few coarse teeth and often rather wavy. Flowers orange with large reddish-brown spots, 20-30mm, in lax racemes of 2-5; lower sepal sac-like, with an incurved spur, shorter than in the previous species, 5-9mm (not 6-12mm). Capsule linear. Moist habitats, especially along rivers and canals. June-August. **I**: North America. Naturalized in France, England and Wales.

9* **Small Balsam** *Impatiens parviflora* DC. Low to tall, hairless annual; stem unbranched or branched. Leaves alternate, elliptical to oval, pointed, toothed, generally with glands towards the base; upper leaves larger than the lower. Flowers small, pale yellow, unspotted, small, 6-18mm, in racemes of 3-10, held above the foliage; spur straight or slightly curved. Capsule narrow club-shaped. Moist ground, waste and disturbed places, woodland and other shaded habitats, cultivated land. July-November. **I**: C Asia. Widely naturalized except in the far north. **B**: Widespread north to S Scotland, but still spreading.

10* **Policeman's Helmet, Himalayan** or **Indian Balsam** *Impatiens glandulifera* Royle [*I. roylei*]. Tall, stout, hairless annual to 2m tall, sometimes more. Leaves opposite or in whorls of 3-5, lanceolate to elliptical, sharply toothed. Flowers purple or pink, sometimes red or white, mostly with various spots and marks inside, 25-40mm, with a sweet somewhat sickly scent, in racemes of 5 or more; lower sepal sac-like, with a short incurved spur. Capsule club-shaped. Moist habitats, river and stream banks, lake margins, waste and cultivated land. July-October. **I**: W and C Himalaya. Naturalized most of Europe north to C and S Scandinavia. **B**: Widespread throughout, but avoiding dry habitats.

11 **Kashmir Balsam** *Impatiens balfourii* Hook. f. Medium to tall hairless annual, generally branched. Leaves alternate, oval, pointed, toothed, the upper rather larger than the lower, all stalked. Flowers pinkish-purple with a white helmet, 25-40mm, in racemes of 3-8, held just above the foliage; spur slender, straight or curved. Capsule narrow club-shaped. Moist habitats, often shaded, waste, disturbed and cultivated land, woodland margins. July-October. **I**: W Himalaya. Naturalized France and SE England.

Norway Maple
Sycamore
Field Maple
Montpelier Maple
Touch-me-not
Policeman's Helmet
Horse Chestnut
Small Balsam

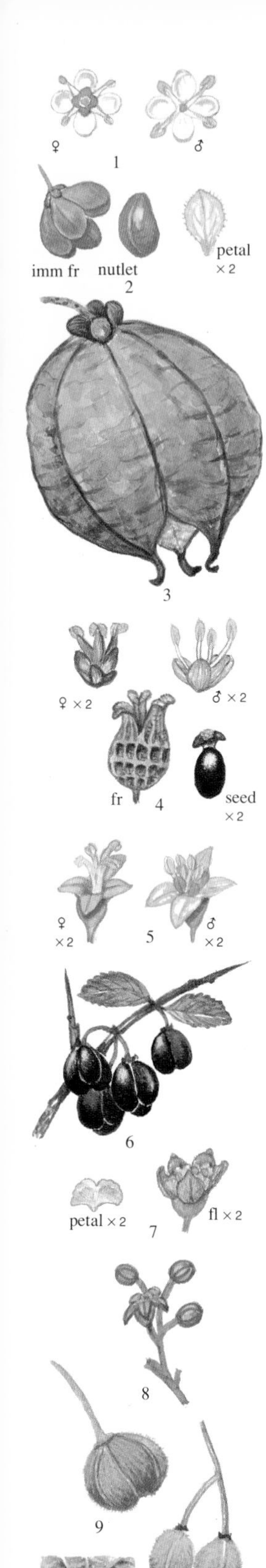


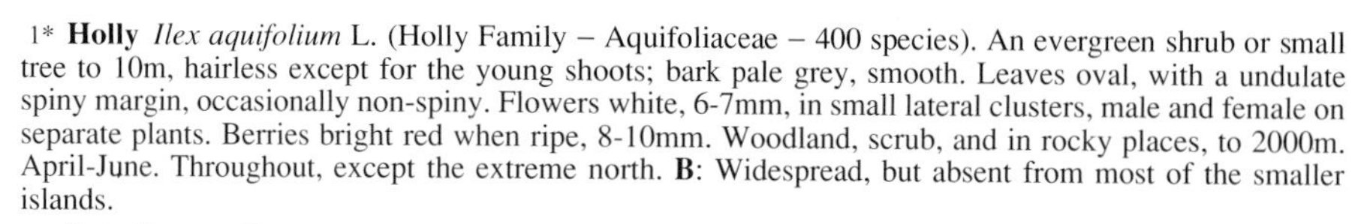

1* **Holly** *Ilex aquifolium* L. (Holly Family – Aquifoliaceae – 400 species). An evergreen shrub or small tree to 10m, hairless except for the young shoots; bark pale grey, smooth. Leaves oval, with a undulate spiny margin, occasionally non-spiny. Flowers white, 6-7mm, in small lateral clusters, male and female on separate plants. Berries bright red when ripe, 8-10mm. Woodland, scrub, and in rocky places, to 2000m. April-June. Throughout, except the extreme north. **B**: Widespread, but absent from most of the smaller islands.

2* **Spindle-tree** *Euonymus europaeus* L. [*E. vulgaris*]. (Spindle-tree Family – Celastraceae – 850 species in 55 genera). Much-branched shrub or small tree, 2-6m, hairless; twigs green, square in section. Leaves opposite, lanceolate to elliptical, pointed, finely toothed. Flowers greenish, 8-10mm, usually 4-parted, several borne in branched cymes. Capsules bright pink, 4-lobed, 10-15mm, pendent, splitting to reveal orange seeds. Woods and scrub mostly on calcareous soils, to 1600m. May-June. Throughout, except the far north; on mountains in the south. **B**: Throughout but not in N Scotland.

3* **Bladder-nut** *Staphylea pinnata* L. (Bladder-nut Family – Staphyleaceae – 50 species in 5 genera). Deciduous shrub to 5m. Leaves opposite, pinnate, with 5-7 oval-oblong leaflets, hairless, toothed. Flowers whitish, 6-10mm long, in large terminal panicles; petals oblong. Fruit much inflated, subglobose, 2-3-lobed, yellowish-brown when ripe, not splitting. Woods and scrub on calcareous soils. May-June. C and E France to S Europe. **P**: Various flies. **B**: Naturalized in a few localities in C and S.

4* **Box** *Buxus sempervirens* L. (Box Family – Buxaceae – 100 species in 6 genera). Evergreen shrub or small tree to 5m with a rather foetid smell. Leaves opposite, oval to elliptical. Flowers greenish-yellow, tiny, 2mm, in small lateral clusters, the male and female borne on the same plant. Fruit a small, 3-horned capsule, containing black seeds. Dry hillslopes, open woods, often of beech, *Fagus sylvatica*, scrub and rocky habitats, often on calcareous soils, to 1600m. April-May. S Britain, Belgium, France, C and S Germany, to S Europe and N Africa. **B**: Local in beechwoods in Buckinghamshire, Gloucestershire, Kent and Surrey.

## BUCKTHORN FAMILY Rhamnaceae

Trees or shrubs with simple leaves; stipules present. Flowers in cymes, 4-5-parted, the petals often small, sometimes absent, often hooded over the stamens. Ovary 2-4-celled. Fruit a fleshy berry. 500 species in 40 genera, cosmopolitan.

5* **Buckthorn** *Rhamnus catharticus* L. Deciduous shrub or small tree, to 6m; old branches ending in a spine-tip. Leaves oval, opposite or clustered, long-stalked, toothed. Flowers greenish, 3-4mm, in dense clusters arising from the older branches, or solitary, the male and female on separate plants usually. Berry green at first but ripening black, 6-8mm. Woodland, scrub and hedgerows, usually on calcareous soils, to 1500m. May-June. Throughout north to S Sweden. **P**: Various insects. **B**: England, except the SW, Wales and Ireland; doubtfully native in Scotland.

6 **Rock Buckthorn** *Rhamnus saxatilis* Jacq. Like *R. catharticus*, but a much-branched, very spiny, often prostrate shrub, sometimes to 2m. Leaves lanceolate to oval, hairless beneath, short-stalked, 13cm long (not 4-7cm). Rocky habitats and scrub, to 1500m. May-July. C and E France and S Germany.

7* **Alder Buckthorn** *Frangula alnus* Miller. [*Rhamnus frangula*]. An erect deciduous shrub or small tree to 4-5m, not spiny. Leaves oval, broadest above the middle, shiny-green, hairy beneath when young, with 7-9 pairs of veins. Flowers greenish, 3mm, solitary or clustered on the younger stems. Berry green at first, but turning red then finally black, 6-10mm. Damp woods, hedgerows and heaths, to 1800m. May-June. Throughout, except the extreme north. Like *Rhamnus*, but winter buds naked and flowers 5 not 4-parted. **B**: Local, often abundant in England and Wales, not in Scotland; rare in Ireland.

8* **Virginia Creeper** *Parthenocissus quinquefolia* (L.) Planchon. (Vine Family – Vitaceae – 700 species in 12 genera). Vigorous deciduous climber to 30m. Leaves digitate with 5 elliptical, coarsely toothed leaflets; tendrils branched. Flowers greenish, the petals 3-4mm long, with deflexed petals. Berry globose, 6mm, turning red then finally black with a whitish bloom. Old walls and buildings, waste ground. June-July. **I**: E USA. Naturalized occasionally in Britain, Holland and Germany. **B**: Occasionally naturalized in C and S.

## LIME TREE FAMILY Tiliaceae

Deciduous trees with simple alternate leaves, often asymmetric at the base and with 3 main veins. Flowers hermaphrodite, 5-parted with separate sepals and petals and numerous stamens, fragrant, in small pendulous clusters with a single large bract towards the base. Fruit a globular nut. About 300 species in 35 genera, widespread in both temperate and tropical regions.

9* **Large-leaved Lime** *Tilia platyphyllos* Scop. [*T. officinarum* subsp. *platyphyllos*]. Large deciduous tree to 40m with wide-spreading branches; bark dark brown, smooth, without bosses. Leaves broadly oval, abruptly pointed at the apex, obliquely heart-shaped at the base, toothed, hairy on the veins beneath. Flowers yellowish-white, fragrant, in clusters of 2-5, often 3. Nut subglobose to pear-shaped 8-10mm, hairy, 5-ribbed. Woodland on calcareous or base-rich soils, occasionally on limestone cliffs, to 1800m. June-July. E France, Germany and Denmark. **P**: Bees. subsp. *cordifolia* (Besser) C.K. Schneider [*T. cordifolia*]. Differs in the leaves being hairy on both surfaces; young twigs hairy. Similar habitats and flowering time. Belgium, Holland, France, Denmark and SW Sweden; widely planted in Britain. **B**: England and Wales north to Yorkshire; possibly native in the Wye Valley.

10* **Small-leaved Lime** *Tilia cordata* Miller [*T. parvifolia*]. Large deciduous tree to 30m with a wide, spreading crown; trunk with smooth grey bark, eventually with narrow brown cracks; young twigs usually hairless. Leaves almost rounded with a heart-shaped base, toothed, hairless except for tufts of reddish-brown hairs at the vein angles beneath. Flowers yellowish-white, fragrant, in clusters of 4-15. Nut globose, 5-6mm, scarcely ribbed. Woodland, but widely planted, on a wide range of soils, more especially over limestone, and on limestone cliffs, to 1500m. June-July. **P**: Bees. Most of Europe, except the extreme north, east to Siberia and the Caucasus. **B**: England and Wales north to Yorkshire.

11 **Common Lime** *Tilia × vulgaris* Hayne [*T. europaea*, p.p., *T. cordata × T. platyphyllos*]. Like *T. cordata*, but the flower cymes pendent (not obliquely erect) and the leaves with white hairs in the vein angles beneath. Nut 7-8mm, slightly ribbed. France, Germany and S Sweden. **B**: Naturalized throughout except in the extreme north.

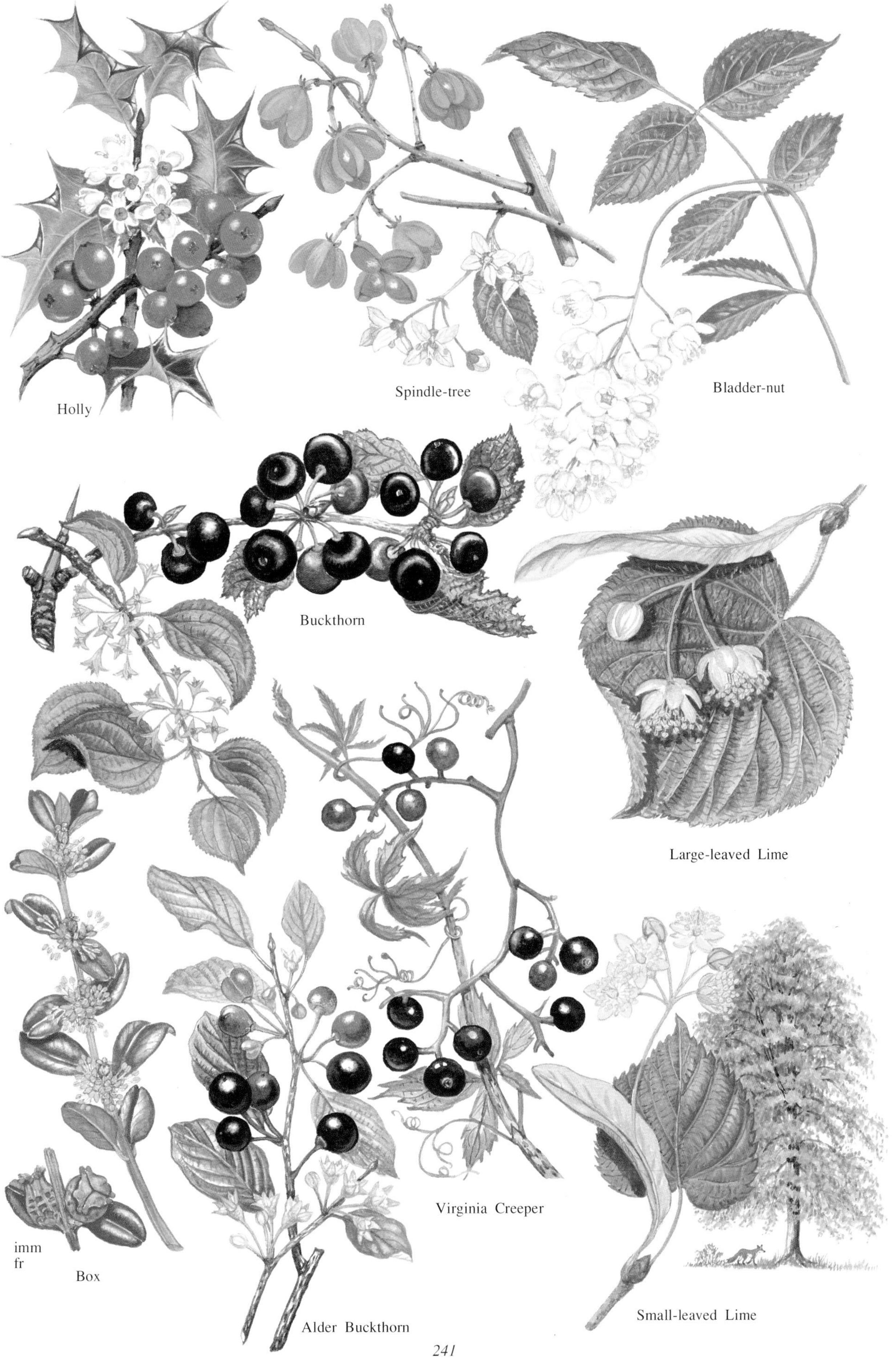

Holly

Spindle-tree

Bladder-nut

Buckthorn

Large-leaved Lime

imm
fr

Box

Alder Buckthorn

Virginia Creeper

Small-leaved Lime

all seeds × 2

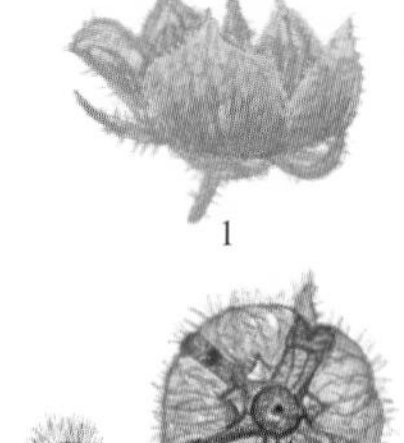

# MALLOW FAMILY Malvaceae

Herbs or shrubs, often with stellate hairs; stipules present. Leaves alternate, usually palmately-lobed. Flowers regular, hermaphrodite. Sepals and petals 5, separate; epicalyx often present. Stamens many, fused together below into a column to form a tube around the ovary and styles. Fruit with a ring of closely packed seeds (mericarps). Over 1000 species in 80 genera.

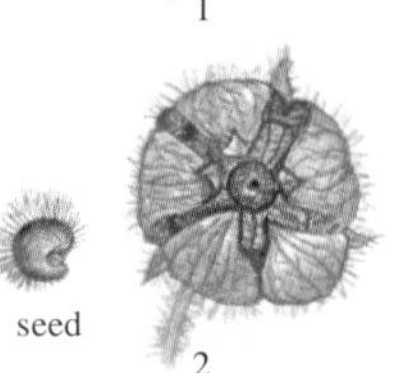

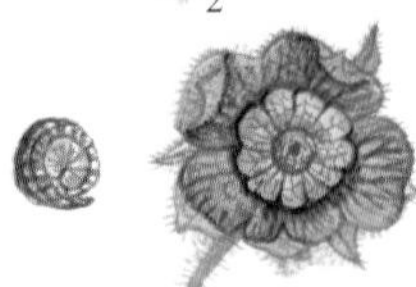

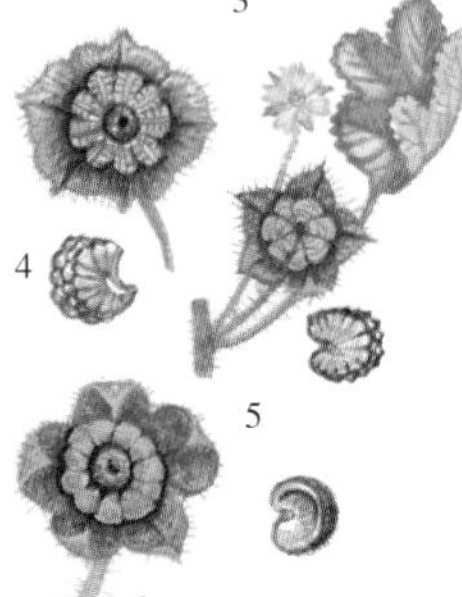



**Mallows** *Malva*. Herbs. Epicalyx consisting of 2-3 segments and petals notched, purple, pink or white. 30 species from northern temperate regions.

1* **Cut-leaved Mallow** *Malva alcea* L. Variable erect, medium to tall, hairy perennial, to 1.2m. Leaves rounded-heart-shaped in outline, palmately 5-lobed, toothed, the upper leaves more deeply cut, with narrower, often pinnately-divided lobes. Flowers bright pink, 35-60mm, solitary, at the upper leaf-axils; epicalyx of 3 oval, hairy, segments. Mericarps hairy or hairless, faintly ribbed. Grassy places and scrub, open woods, to 2000m. June-September. Belgium, Holland, France, Denmark, Germany and S Sweden; naturalized in S Norway and S Finland. Sometimes hybridises with *M. moschata*.

2* **Musk Mallow** *Malva moschata* L. Like *M. alcea*, but the leaves with 5-7 more pointed lobes which are all pinnately-divided. Epicalyx segments linear to narrowly oval, hairless or slightly hairy. Mericarps with long white hairs. Pastures and field margins, roadsides and hedgebanks, generally on dry fertile soils, sometimes in rocky habitats, to 1500m. July-August. Britain, Belgium, Holland, France and Germany; naturalized in Scandinavia. The flowers have a musk-like fragrance, especially when taken indoors. **B**: Throughout, but scarcer in the north and N Wales and rather rare in Ireland.

3* **Common Mallow** *Malva sylvestris* L. Medium to tall hairy biennial, to 1.5m, but often with sprawling stems. Leaves kidney-shaped to rounded-heart-shaped in outline, with 3-7 shallow, rather rounded, toothed lobes. Flowers pink to purple with darker veins, 20-50mm, in clusters of 2 or more at the leaf axils; epicalyx segments elliptical, hairy. Mericarps hairy or hairless, netted. Meadows and waste places, road-verges and hedgebanks, generally on dry well-drained soils, to 1800m. June-September. Throughout, except the extreme north, the Faeroes and Iceland; to W Asia. **P**: Usually bees. The form with hairy mericarps is sometimes assigned to var. *lasiocarpa* Druce. **B**: Throughout, but scarcer in the north and mainly on the east coast there.

4* **Least Mallow** *Malva parviflora* L. Short to medium, hairy or hairless annual with erect or ascending branches. Leaves rounded-heart-shaped, long-stalked, with 5-7 shallow triangular, toothed lobes. Flowers pale lilac-blue or mauve, small, 7-9mm, in clusters of 2-4, short-stalked, the petals slightly longer than the sepals; epicalyx segments linear to lanceolate. Fruit with much enlarged and spreading sepals; mericarps hairy or hairless, strongly netted, winged. Waste places and field margins, generally at low altitudes. June-September. France; naturalized in Britain and often casual in other parts of NW Europe. **P**: Bees and other insects. **B**: Occasionally naturalized, especially in the south but often casual.

5 **Small Mallow** *Malva pusilla* Sm. [*M. rotundifolia* p.p.]. Like *M. parviflora*, but the stems more spreading and flowers pale pink, in clusters of up to 10, 5-6mm, the sepals edged with hairs, green (not membranous) and not much enlarging in fruit; mericarps not winged. Waste places, field margins, cultivated land, rubbish tips and coastal habitats. June-September. Belgium, Holland, Germany and C and S Scandinavia, except Finland; naturalized in Britain, France and Finland. **P**: Self-pollinated. **B**: Local north to Aberdeen, often coastal.

6* **Dwarf Mallow** *Malva neglecta* Wallr. Low to short, generally sprawling, hairy annual. Leaves kidney-shaped to rounded-heart-shaped in outline, with 5-7 shallow, toothed lobes. Flowers pale lilac to whitish, 15-25mm, in clusters of 3-6, the petals with bearded claws, at least twice as long as the sepals; epicalyx segments linear to oval, shorter than the sepals. Mericarps hairy, scarcely ridged. Fields and waste places, grassy banks, roadsides and coastal habitats, usually on rather dry soils, to 1900m. May-September. Throughout, except the far north. Hybrids have been reported between this species and *M. pusilla* as well as *M. sylvestris*. **B**: Throughout but less frequent in the north, often coastal.

7 **Chinese Mallow** *Malva verticillata* L. Like *M. neglecta*, but the flowers smaller, 10-15mm, the petals only twice (not three times) as long as the sepals. Fruit stalks mostly less (not more) than 10mm long. Cultivated and waste ground. July-September. Widely naturalized in W Europe. Var. *crispa* L. is sometimes cultivated as a salad plant. **B**: Casual or locally naturalized.

**Tree Mallows** *Lavatera*. Very like *Malva*, but often more woody and the epicalyx segments fused together at the base, particularly noticeable in bud. About 25 species.

8* **Small Tree Mallow** *Lavatera cretica* L. Medium to tall, hairy annual or biennial to 1.5m. Leaves rounded-heart-shaped in outline, with 5-7 shallow, toothed lobes, stalked. Flowers lilac, 18-38mm, in clusters of 2-8 on unequal stalks; epicalyx segments oval, joined at the base, not much enlarging in fruit. Mericarps smooth or slightly ridged. Disturbed and waste ground, roadsides, old quarries and sometimes on cultivated ground. April-July. SW Britain, W France. Sometimes confused with *Malva sylvestris*, but epicalyx segments broader and joined at the base, and mericarps with rounded, not sharp, angles. **B**: Very rare; confined to Scilly, Jersey and Guernsey.

9 *Lavatera thuringiaca* L. Similar to *L. cretica*, but a hairier perennial, the leaves generally 5-lobed and the flowers usually solitary, larger, purplish-pink; epicalyx segments fused to halfway. Rocky and grassy habitats. June-August. S Germany; naturalized in France and Finland.

10* **Tree Mallow** *Lavatera arborea* L. Tall biennial to 3m, stems woody below, hairy when young. Leaves large, to 20cm, rounded in outline, 5-7-lobed, toothed. Flowers lilac with purple veins, 30-40mm, in clusters of 2-7; epicalyx segments rounded to oblong, longer than the calyx, fused to halfway and much enlarging in fruit. Mericarps sharply angled, hairy or hairless. Rocky habitats, cliffs and stony ground, especially close to the sea, but also along hedgerows and waste places. June-September. W and SW Britain, W Ireland and W France; **B**: Coastal SW England and Wales, Isle of Man and W Ireland; casual or naturalized in S England and W Scotland.

Cut-leaved Mallow
Musk Mallow
Common Mallow
Dwarf Mallow
Least Mallow
Small Tree Mallow
Tree Mallow

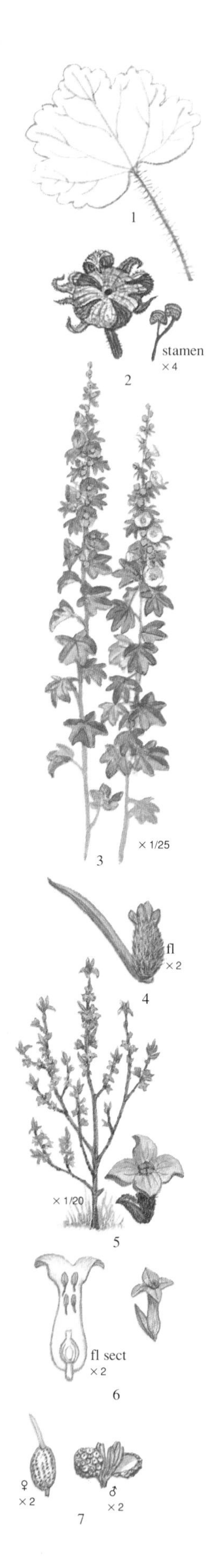


1* **Rough Marsh Mallow** *Althaea hirsuta* L. Short to medium, occasionally semi-prostrate annual; stems with simple bristles as well as starry hairs. Leaves rounded-heart-shaped in outline, toothed, the lower very shallowly lobed, the upper more deeply 3-5-lobed. Flowers pinkish-lilac, 24-28mm, solitary, but forming leafy racemes, with yellow anthers; epicalyx segments 6-9, narrowly triangular, bristly, almost equalling the sepals. Mericarps hairless, ridged. Field borders, scrub and woodland margins, usually on rather dry, somewhat calcareous soils. July-September. France and S Germany; naturalized or casual in Britain and Belgium, occasionally elsewhere; to C and S Europe and W Asia; *Althaea* is superficially very similar to *Malva*, but is distinguished by its epicalyx of 6-9, rather than 3, segments. Native of C and S Europe and W Asia. **B**: Very rare in Somerset and Kent, sometimes casual elsewhere.

2* **Marsh Mallow** *Althaea officinalis* L. Tall, soft grey-woolly, tufted perennial to 2m. Leaves triangular-oval in outline, toothed, mostly slightly 3-5-lobed. Flowers pale lilac-pink, 25-40mm, solitary or in clusters forming leafy racemes or panicles, anthers purplish-red; epicalyx segments 8-9 linear-lanceolate, velvety like the sepals and leaves. Mericarps hairy. Upper parts of salt marshes, brackish marshes, ditches and stream margins, close to the sea generally and at low altitudes. August-September. Britain, Belgium, France, Holland, Denmark and Germany. **P**: Bees or self-pollinated. Edible marsh-mallow was originally prepared from mucilage obtained from the roots and the plant was formerly used as an ingredient of soothing ointments. Occasionally cultivated today. **B**: Local coastal species in England S of the Wash and in S Wales, scattered localities further north; rare in SW Ireland.

3* **Hollyhock** *Alcea rosea* L. [*Althaea rosea*]. Tall, rather bristly perennial to 3m. Leaves large, to 30cm, rounded-heart-shaped in outline, usually shallowly 5-7-lobed, coarsely toothed, long-stalked. Flowers generally pink or white, sometimes red, violet or yellowish, large, 60-80mm, in long leafy, slender racemes; epicalyx segments 6, triangular, shorter than the sepals. Waste places, field boundaries, along walls and in rocky places. July-September. Widely naturalized from garden escapes in Britain, France and Germany; often casual elsewhere. Garden forms often double-flowered or deep-coloured. **P**: Mainly various bees. Origins unknown, possibly arose as a hybrid between *A. setosa* (Boiss.) Alef. and *A. pallida* (Willd.) Waldst. & Kit., both natives to E Europe and W Asia. **B**: Frequent.

## DAPHNE FAMILY Thymelaeaceae

Small shrubs, occasionally herbs, with simple, untoothed leaves. Flowers in clusters, racemes or umbels, hermaphrodite; calyx with a tube and 4 spreading lobes, petal-like, true petals absent. Stamens 8, fused to the side of the calyx-tube; style solitary, often very short. Fruit a small berry or nut. Cosmopolitan; 500 species, particularly well represented in Africa, Asia and the Pacific islands. A number of shrubs, particularly of the genus *Daphne*, are cultivated in gardens. Most produce stem fibres, used especially in the Himalaya and Tibet for paper-making.

4* **Annual Thymelaea** *Thymelaea passerina* (L.) Cosson & Germ. [*Lygia passerina*, *Passerina annua*]. Short to medium, erect, more or less hairless annual. Leaves linear-lanceolate, alternate, pointed. Flowers greenish, tiny, 2-3mm, solitary, or 2-3 together at the leaf-axils, bell-shaped, the tube hairy outside. Fruit a hairy nut. Dry rocky and waste places, to 1200m. July-September. Belgium, France and C and S Germany, to S Europe.

5* **Mezereon** *Daphne mezereum* L. Deciduous shrub up to 1.5m tall, though often less, with erect to spreading, greyish-brown branches; young shoots hairy. Leaves oblong-lanceolate, alternate, crowded towards the shoot tips, often rather pale green, short-stalked. Flowers pinkish-purple, 7-10mm long, in lateral clusters of 2-4, very fragrant, borne before the leaves appear, hairy outside. Fruit a bright shiny red berry. Woods, pastures and scrub, on calcareous soils, to 2600m. February-May (occasionally later at high altitudes). Britain and much of Europe except the extreme north, but generally scarcer than formerly due to land clearance and collecting. The berries are exceedingly poisonous. **P**: Early-flying butterflies, bumble-bees. Much cultivated in gardens including a white-flowered form, var. *alba* Aiton, which has yellow berries. **B**: Local and rare from Sussex to Yorkshire; a protected species.

6* **Spurge Laurel** *Daphne laureola* L. An evergreen shrub to 1m tall, though often less, with suberect branches; young shoots greenish and hairless. Leaves alternate, oblong, broadest above the middle, leathery, deep shiny green, crowded towards the shoot tips. Flowers greenish-yellow, 8-12mm long, in short, congested lateral racemes among the leaves, half-nodding, honey-scented. Berry black when ripe. Local in woods, woodland clearings, hedgerows, sometimes in rocky habitats, chiefly on dry calcareous soils, to 1600m. January-April. Britain, Belgium, France and Germany; naturalized in Denmark. Native W and S Europe N. Africa, the Azores, N Africa. **P**: Early-flying butterflies as well as bumble-bees. Mountain forms often have spreading or almost prostrate branches. **B**: Widespread, local in England, but scarce in the SW and in Wales; absent or alien in Scotland and Ireland.

## ELEAEAGNUS FAMILY Elaeagnaceae

Trees or shrubs often with scale-like hairs on the leaves and young stems. Leaves alternate, untoothed. Flowers without petals, the male and female borne on separate plants (in *Hippophae*), the male with 2 large sepals and 4 stamens, the female with 2 minute sepals and a large ovary with a single style. Some 50 species in 3 genera, from northern temperate and subtropical regions and E Australia. A number, especially *Elaeagnus*, are grown for ornament or shelter.

7* **Sea Buckthorn** *Hippophae rhamnoides* L. Large deciduous, *spiny* shrub to 4-5m tall, occasionaly more, suckering freely; twigs covered with silvery scales, later dark brown. Leaves linear to linear-lanceolate, alternate, covered in silvery scales when young, crowded on the lateral branches – spurs. Flowers greenish, very small, appearing before the leaves. Berry almost globose, 6-8mm, orange when ripe. Coastal cliffs, stabilized sand-dunes, river gravels and alluvium in mountain regions, to 2000m. March-May. Most of Europe, except the extreme north. The berries are edible, though rather astringent. Often planted in coastal regions for shelter or sand-binding, where the dense growth provides excellent cover for nesting birds. The roots bear nodules which harbour soil bacteria capable of fixing atmospheric nitrogen. **P**: Wind. **B**: S And E coastal areas, widely planted elsewhere; naturalized in Ireland.

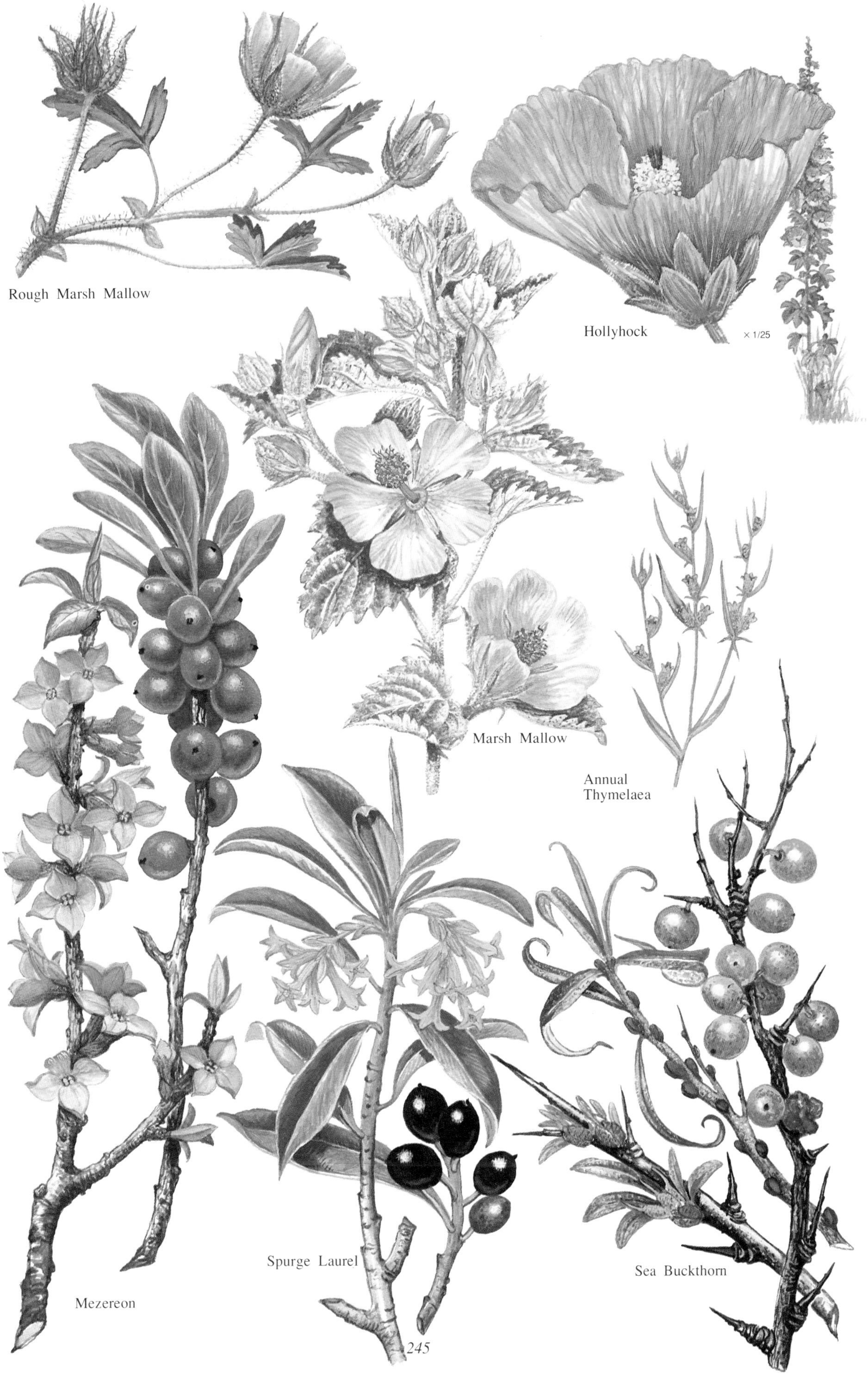

Rough Marsh Mallow

Hollyhock × 1/25

Marsh Mallow

Annual Thymelaea

Mezereon

Spurge Laurel

Sea Buckthorn

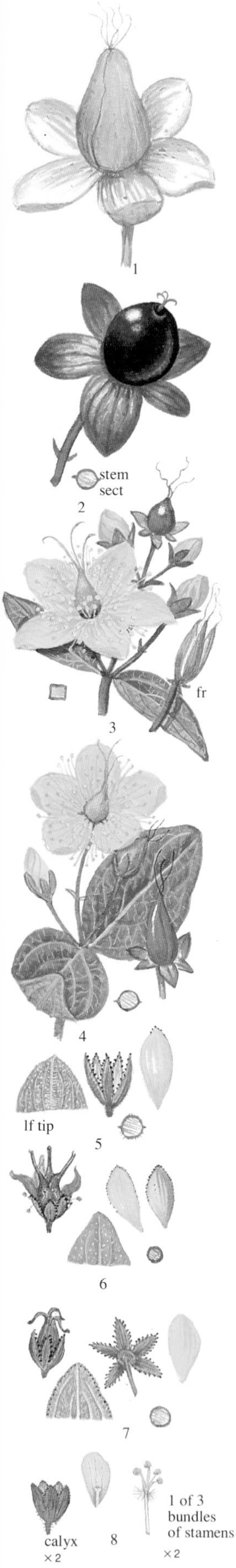

# HYPERICUM FAMILY Guttiferae

Shrubs or herbs with simple, entire, opposite or whorled leaves, often with translucent glands containing essential oils, or red or black gland-dots containing Hypericin. Flowers regular, generally with 4-5 sepals and petals, all free, the sepals overlapping in bud and the petals contorted in bud. Stamens numerous, often in 5, more or less, distinct bundles. Ovary superior, with 3-5 styles. Fruit a capsule. Cosmopolitan; 1000 species in 40 genera, primarily in the tropics. The only genus represented in our region is *Hypericum*, which is widespread in temperate and subtropical regions – a number are cultivated in gardens.

1* **Rose-of-Sharon** *Hypericum calycinum* L. Evergreen subshrub to 60cm, with a creeeping rhizomatous rootstock, often forming extensive patches. Leaves opposite, oblong to elliptical, scarcely stalked. Flowers yellow, 70-80mm, terminal, solitary or 2-3 together, with reddish anthers, the petals asymmetric; sepals unequal, persistent; styles 5. Shady habitats, parks, banks, roadsides, railway embankments. June-September. **I**: E Europe and N Turkey. Naturalized in Britain and France. Widely cultivated, very shade-tolerant. **P**: Various bees. **B**: Locally naturalized in many places.

2* **Tutsan** *Hypericum androsaemum* L. [*Androsaemum officinale*]. Semi-evergreen, hairless shrub to 70cm; stems spreading, with 2 raised lines. Leaves opposite, oval to oblong, unstalked, sometimes clasping the stem, slightly aromatic when crushed. Flowers pale yellow, 18-22mm, in small terminal clusters, the petals shorter than the sepals; sepals unequal, enlarging and deflexed in fruit; styles 3. Fruit fleshy, berry-like, 7-10mm long, reddish, purple-black when ripe, not splitting. Damp or shaded habitats, deciduous woodland, hedge-banks, generally on base-rich soils, also in limestone grikes. June-August. Britain, Ireland, Belgium and France. Native of W Europe, parts of S Europe and W Asia. Widely cultivated in gardens. In severe winters plants may loose all their leaves. **B**: Throughout W and S, scarce elsewhere.

3 **Stinking Tutsan** *Hypericum hircinum* L. subsp. *majus* (Aiton) N. Robson [*Androsaemum hircinum*]. Rather like *H. androsaemum*, but taller, to 1m, with square stems and foliage smelling of goat. Flowers larger, 25-30mm, the petals longer than the sepals; sepals shrivelling and falling before the fruits ripen. Capsule red or green, 8-13mm long, splitting when ripe. Damp habitats, often along river and stream banks. May-August. **I**: Mediterranean region and W Asia. Naturalized from gardens in Britain and France. Occasionally cultivated in parks and gardens, sometimes planted in woods. **B**: Naturalized in scattered localities in England, Wales and Ireland.

4 **Tall Tutsan** *Hypericum × inodorum* Miller [*H. elatum*]. Intermediate between *H. androsaemum* and *H. hircinum*. Stems slightly 2-edged and leaves aromatic when crushed, but not goat-scented. Flowers 20-30mm, the petals longer than the persistent sepals. Capsule red or green, splitting when ripe. Hedges, scrub, thickets and woodland margins. July-August. **I**: Madeira and Canary Is. Naturalized from gardens in Britain and France. Widely cultivated in gardens. **B**: Scattered localities in England, particularly the SW, Wales, W Scotland and Ireland.

5* **Hairy St. John's-wort** *Hypericum hirsutum* L. Medium to tall, erect, *hairy perennial* to 110cm; stems rounded in section, with 2 raised lines and with a creeping and rooting base. Leaves opposite, oblong to elliptical, rough-hairy, with translucent dots, but without marginal black glands. Flowers pale yellow, sometimes red-veined, 14-15mm, borne in a lax, many-flowered panicle; sepal-margin with black glands; anthers yellow; styles 3. Capsule surrounded by persistent petals and stamens. Woodland, scrub, rough grassland, road-verges and riverbanks, generally on calcareous soils. July-August. Most of Europe except the extreme north. **P**: Various insects. **B**: Throughout England except the SW; scattered or scarce elsewhere, including Ireland.

6* **Slender St. John's-wort** *Hypericum pulchrum* L. Short to tall, hairless, erect or ascending perennial; stems smooth and rounded, often reddish. Leaves opposite, ovate to oblong, clasping the main stems, smaller and short-stalked on lateral shoots, dotted with translucent glands. Flowers yellow with a red tinge, 14-15mm, in lax narrow panicles; sepal-margins with black glands; anthers orange to reddish-pink. Woods, heaths and dry grassy places, on acid, well-drained soils. June-August. The Faeroes and Europe north to S Scandinavia, not in Finland. In exposed places, dwarf few-flowered plants may be found, sometimes referred to forma *procumbens* Rostrup. **B**: Widespread but scarce in the E Midlands and East Anglia.

7* **Mountain St. John's-wort** *Hypericum montanum* L. Short to tall, almost hairless perennial; stems erect, unbranched, rounded. Leaves opposite, oval to lanceolate, unstalked, rough beneath, with a marginal row of black glands beneath. Flowers pale yellow, 10-15mm, in a fairly dense flat-topped clusters, fragrant; sepals with black-glandular teeth. Styles 3. Woods, scrub, hedgebanks, thickets and rough grassy places, usually on calcareous soils, to 1900m. July-September. Most of Europe, except the extreme north, to W Asia and N Africa. **B**: Scattered localities in England, Wales and SW Scotland; local and declining.

8* **Marsh St. John's-wort** *Hypericum elodes* L. [*H. palustre, Elodes palustris*]. Short, hairy perennial; stems round, erect from a creeping and rooting base, often with swollen internodes. Leaves rounded to oval, greyish with soft hairs, half-clasping the stem. Flowers pale yellow, 12-15mm, bell-shaped, not opening widely as in other species, borne in small clusters; sepals erect with marginal reddish glands. Capsule with persistent petals and stamens. Damp muddy habitats and shallow water, heaths, bogs, pond and pool margins, on acid soils. June-September. Britain, Belgium, Holland, France and Germany. The dense hairs on the leaves and stems prevent the plant from being wetted when submerged; plants often form dense floating mats with erect flower stems. Easily identified by its half-open flowers. **B**: Scattered localities throughout, but scarcer or absent in much of the east.

1 stem sect 2 3 fr 4 lf tip 5 6 7 calyx ×2 8 1 of 3 bundles of stamens ×2

Rose-of-Sharon
Tutsan
Hairy St. John's-wort
Slender St. John's-wort
Mountain St. John's-wort
Marsh St. John's-wort

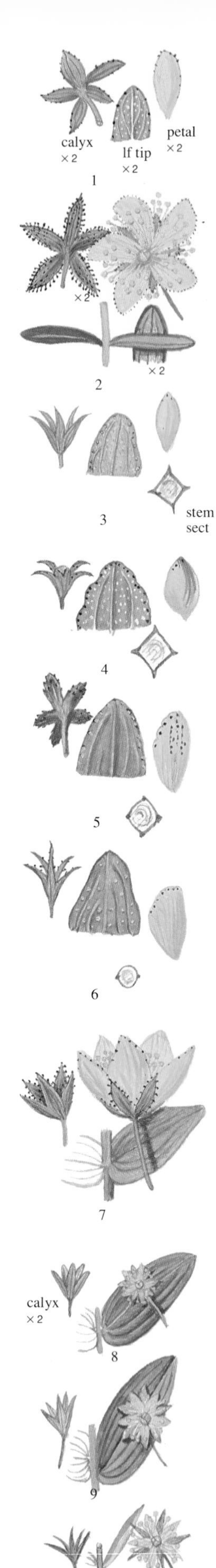


1* **Trailing St. John's-wort** *Hypericum humifusum* L. Slender, low, generally prostrate, hairless perennial; stems branching and rooting at the base, round. Leaves opposite, oblong to lanceolate, generally with translucent dots and some marginal black glands, unstalked. Flowers pale yellow, small, 8-10mm, in few-flowered lax clusters; sepals unequal, toothed or not, sometimes with marginal black glands. Capsule equalling the sepals. Open woods, scrub, heaths, moors, short turf, generally on acid or peaty soils, to 1800m. July-October. Europe north to S Scandinavia; to C Europe, the Azores and Madeira. **B**: Throughout but scarcer in the north.

2 **Toadflax-leaved St. John's-wort** *Hypericum linariifolium* Vahl. Rather like *H. humifusum*, but stems erect to spreading and leaves with margins rolled under and without translucent dots. Flowers often red-tinged beneath; petals twice as long (not slightly longer than) the sepals; sepals dotted all over with black glands. Dry sunny habitats, especially rocky habitats on thin acid soils, often growing among mosses and lichens. June-July. W and S Britain, W France; to Spain, Portugal, Madeira. **P**: Insects or self-pollinated. **B**: Very rare; Cornwall, S Devon, N Wales and the Channel Islands.

3* **Square-stalked St. John's-wort** *Hypericum tetrapterum* Fries [*H. quadrangulum, H. acutum*]. Short to tall, erect, hairless perennial; stems square with 4 narrow wings, spreading and rooting at the base. Leaves opposite, rounded to oval or elliptical, often half-clasping the stem, with translucent dots only. Flowers pale yellow, not red-tinged, 9-10mm, many in a spreading panicle; petals equalling sepals or slightly longer. Damp places, grassland, meadows, woodland clearings, road-verges, pond and river margins. June-September. Widespread throughout, except the Faeroes, Iceland, Norway, Finland and Spitsbergen; to N Africa and Madeira. Occasionally forms with slightly red-veined petals may be observed. **B**: Throughout except N Scotland and the Outer Hebrides.

4* **Wavy St. John's-wort** *Hypericum undulatum* Schousboe ex Willd. A short to tall, hairless, erect perennial; stems square with 4 narrow wings, spreading and rooting at the base. Leaves opposite, oval to elliptical, unstalked and half-clasping the stem, with a wavy margin, with translucent dots. Flowers mid-yellow, generally red-tinged beneath, 18-20mm, in lax panicles; sepals untoothed, with black glands, much shorter than the petals. Boggy and marshy habitats, stream and pond margins, on acid soils. August-October. S Britain and W France; to W Spain, Portugal and the Azores. Very attractive, easily recognised by its wavy-margined leaves. **B**: Very local and declining, W Wales, Devon and Cornwall.

5* **Imperforate St. John's-wort** *Hypericum maculatum* Crantz [*H. quadrangulum*]. Short to tall, hairless, erect perennial; stems square, 4-lined, but not winged, spreading and rooting at the base. Leaves opposite, oval to elliptical, unstalked, net-veined, without translucent dots, or with a few on the upper leaves. Flowers golden-yellow, 18-20mm, in broad panicles; sepals with black glands, untoothed. Damp habitats, meadows, wood margins, road-verges and stream-margins, to 2650m. June-September. Scandinavia, France and Germany east to Siberia. Rather local and generally found in hilly or mountain habitats. **B**: Local, W Scotland.

subsp. *obtusiusculum* (Tourlet) Hayek. Like the type, but the inflorescence branches making an angle of 50° (not 30°) with the main axis and the sepals with toothed margins; petals with black streaks in the middle (not black dots). Similar habitats and flowering time. Lowland NW Europe, Britain, the Faeroes, Belgium, Holland and W Germany. **B**: Widespread.

6* **Perforate St. John's-wort** *Hypericum perforatum* L. [*H. noeanum*]. Short to tall, hairless, erect perennial; stems round with 2 raised lines, spreading and rooting at the base. Leaves opposite, linear to oval, unstalked, with large translucent dots. Flowers yellow, 18-22mm, in broad panicles; sepals narrow, with or without black glands, much shorter than the petals. Dry fields, woods and scrub, rough grassland, road-verges and hedgebanks, on neutral or calcareous soils, to 2000m. May-September. Throughout, except the far north. The flowers do not produce nectar and are sometimes apomictic. **B**: Throughout, but rare in C and N Scotland.

7 *Hypericum elegans* Stephan ex Willd. Like *H. perforatum*, but always a short to medium plant; sepals edged by black-glandular teeth and petals with only marginal black glands (none in the centre). Dry habitats, generally on calcareous soils. June-August. Germany. restricted to C Europe. The following 3 North American species are all naturalized in damp habitats. They are probably introduced by wading birds or as a contaminant of agricultural seed or fodder. All have small flowers and usually 5 stamens.

8* *Hypericum mutilum* L. Rather like *H. perforatum*, but a short plant to 40cm, the stems branched from above the middle, 3-5-veined; sepals 1.5-3mm long (not 5-9mm), blunt. Marshy habitats. July-September. **I**: North America. Rare in France and Germany.

9 *Hypericum majus* (A. Gray) Britton. Like *H. mutilinum*, but stems unbranched or branched above the middle and leaves 5-7-veined; sepals 5-7mm long. Stream and pond margins. July-September. **I**: North America. E France and C and S Germany.

10* **Irish St. John's-wort** *Hypericum canadense* L. Low to short annual, or occasionally perennial, hairless, but often red-tinged; stems square, winged on the angles. Leaves opposite, linear to oblong, narrowed at the base, 1-3-veined. Flowers golden-yellow, 6-8mm, in few-flowered clusters, starry with widely spaced petals; sepals with pale or reddish streaks, but no black glands, 2-4.5mm. Wet heaths, grassland, often grazed, and stream margins, generally on acid peaty soils. August-October. **I**: North America. Naturalized, or possibly native in Ireland (Mayo) and in one locality in Holland. **B**: Rare in Ireland.

Trailing St. John's-wort
Square-stalked St. John's-wort
Wavy St. John's-wort
Imperforate St. John's-wort
Perforate St. John's-wort
*Hypericum mutilum*
Irish St. John's-wort

## VIOLET FAMILY Violaceae

Small herbs or shrubs with alternate leaves, stalked, often in basal tufts; stipules present, often prominent. Flowers zygomorphic, solitary; sepals 5 separate, extended behind into a short appendage; petals 5, separate, the lower forming a small lip and extended behind into a short or long spur. Stamens 5, held in a close ring around the ovary. Fruit a 3-valved capsule. 900 species in 22 genera, including trees, shrubs and herbs.

1* **Sweet Violet** *Viola odorata* L. Low, somewhat hairy perennial, with long, rooting runners. Leaves oval with a deeply heart-shaped base, in basal rosettes, blunt; stipules oval, usually fringed, hairless or with a slightly hairy margin. Flowers dark violet, or white with a violet spur, 13-15mm, fragrant; short spur, 6mm, the sepals blunt; flower-stalks with bracts in the middle. Capsule hairy. Woods, coppices, scrub, hedgerows and plantations, on neutral or calcareous soils, to 1400m. February-May, occasionally in August-September. Throughout, except the extreme north. During the summer small apomictic flowers are produced. **B**: Widespread, though scarcer in the north.

2 *Viola suavis* Bieb. Similar, but with short stout runners and more oblong leaves; stipules lanceolate, deeply fringed. Flowers violet with a white throat, 15-20mm, fragrant; bracts below the middle of the flowerstalk. Grassy and rocky habitats, scrub, to 1400m. March-May. France; naturalized in Germany.

3* **White Violet** *Viola alba* Besser. Low, somewhat hairy perennial with basal leaf-rosettes and slender, non-rooting runners. Leaves dark green, oval or somewhat triangular, with a deep heart-shaped base, pointed, long-stalked; stipules linear-lanceolate, deeply fringed. Flowers white, very rarely violet, 15-20mm, fragrant, the lateral petals bearded, the spur yellowish-green; bract at middle or above middle of the flower-stalk. Unlike *V. odorata*, the runners do not root down. Woodland, scrub, coppices and hedgerows, to 1200m. March-June. C and E France, Germany and S Sweden – Öland.

4* **Hairy Violet** *Viola hirta* L. Low, hairy perennial, without runners, but with a short rhizome and a dense leaf-rosette. Leaves pale green, oblong-oval with a heart-shaped base; stipules triangular to lanceolate, shortly fringed. Flowers violet, 10-15mm, not fragrant, with a dark violet spur; bracts below the middle of the flower-stalk. Capsule hairy. Open woods, woodland margins, coppices, scrub, grassland and pastures, on calcareous soils, to 2000m. March-June. Throughout, except the far north. Plants with smaller flowers and leaves folded upwards along the midrib, and with a very short spur are subsp. *calcarea* (Bab.) E.F. Warb. – C and S Britain. **B**: Throughout England, scarcer in Wales, Scotland and Ireland.

5 **Hill Violet** *Viola collina* Besser. Like *V. hirta*, but leaves paler green and stipules narrower, deeply fringed. Flowers pale blue with a whitish spur, fragrant. Open woods and scrub, to 2000m. March-April. Scattered localities in Belgium, France, Germany and C and S Scandinavia.

6* **Early Dog-violet** *Viola reichenbachiana* Jordan ex Boreau [*V. sylvestris*]. Low, somewhat hairy perennial, with *basal leaf rosettes* amd lateral leafy flowering stems. Leaves heart-shaped, as long as wide; stipules narrow lanceolate, fringed. Flowers violet, generally darker in the centre, 12-18mm, not fragrant, with a slender, straight, dark violet-blue spur, not furrowed or notched. Capsule hairless, with enlarged calyx appendages. Woods and shady habitats, coppices, hedgebanks, rarely in unshaded habitats, to 1800m. March-June. Throughout Europe except the extreme north. **B**: Locally abundant in England, scarce in Wales and Ireland, very rare in Scotland.

7 *Viola mirabilis* L. Stipules entire, not fringed, eventually brown and stems with a single line of hairs. Flowers pale violet, fragrant, 18-20mm, with a whitish spur. Woodland and scrub, mainly on base-rich soils. April-June. Belgium, France, Germany and Scandinavia, except for the extreme north.

8* **Teesdale Violet** *Viola rupestris* F. W. Schmidt. [*V. arenaria*]. Like *V. reichenbachiana*, but a lower more tufted plant with more rounded leaves with hairy leaf-stalks. Flowers reddish-violet, pale blue or white, 10-15mm, with a short pale violet, slightly furrowed spur. Capsule hairy. Open habitats, dry meadows, gravels and heaths over limestone, to 3100m. March-July. N Britain, Belgium, France, Germany, Holland, Norway, Sweden and Finland. Hairless forms occur sometimes on the Continent, where it is primarily a mountain species. **B**: Very rare; confined to Upper Teesdale and one place in Cumbria.

9* **Common Dog-violet** *Viola riviniana* Reichenb. Very variable low to short plant not unlike some forms of *V. reichenbachiana*, hairless or slightly hairy. Leaves heart-shaped, long-stalked. Flowers deep bluish-violet, 14-25mm, not scented; spur rather stout, whitish or pale purple, often upcurved, notched or furrowed towards the tip. Deciduous woodland, grassy heaths, old pastures, chalk downs, to 1800m. April-June, occasionally later. **B**: Throughout.

10* **Heath Dog-violet** *Viola canina* L. Variable low to short, hairless or slightly hairy perennial, without a basal rosette of leaves; stems spreading to erect. Leaves oval to triangular or almost lanceolate, usually with a heart-shaped base, bluntly toothed; stipules lanceolate, untoothed or toothed towards the tip. Flowers bright blue or violet, 10-18mm, not fragrant, with a whitish or greenish-yellow spur; sepals pointed. Capsule hairless. Open woods, grassy heaths, sandy commons, fens and coastal dunes, often on acid soils, to 2500m. April-July. Throughout. Generally easy to recognise by its absence of a leaf rosette and by its flower colour. **B**: Scattered localities throughout. subsp. *montana* (L.) Hartman. has erect stems and the leaves often asymmetric; stipules half as long or equalling the leaf-stalk. Flowers larger, 15-22mm, pale blue, often with a greenish spur. Heaths, open woods and fens. Throughout, but often local. **B**: Only known from the fens – Huntingdon and Cambridgeshire.

11* **Pale Dog-violet** *Viola lactea* Sm. Low to short, generally hairless perennial, without a basal leaf-rosette; stems ascending, often solitary. Leaves often purple-tinged, lanceolate, to oval-lanceolate, with a rounded or wedge-shaped base, broadest above the base; stipules lanceolate, coarsely toothed. Flowers pale bluish-white or greyish-violet, 15-20mm, with a short yellowish or greenish spur. Capsule hairless. Dry acid heaths. May-June. From Ireland, S England, W and C France south to N Spain and Portugal. Local and generally declining. **B**: S and SW England on short-grassy heaths and in S Ireland.

12* **Fen Violet** *Viola persicifolia* Schreber [*V. stagnina*]. Low to short, generally hairless perennial, with a creeping underground rhizome, without a basal leaf-rosette; stems erect. Leaves triangular-lanceolate with a truncated or slightly heart-shaped base, broadest at the base; stipules lanceolate, toothed or not. Flowers white or bluish-white, generally with violet veins, 10-15m, almost circular in outline; spur short, greenish. Capsule hairless. Marshes, fens and fresh peat cuttings, generally at low altitudes. May-June. Throughout except the far north. Declining. **B**: Very rare, confined to isolated sites in Huntingdon, East Anglia and scattered localities in Ireland.

Sweet Violet
White Violet
Hairy Violet
Early Dog-violet
Teesdale Violet
Common Dog-violet
Heath Dog-violet
Pale Dog-violet
Fen Violet

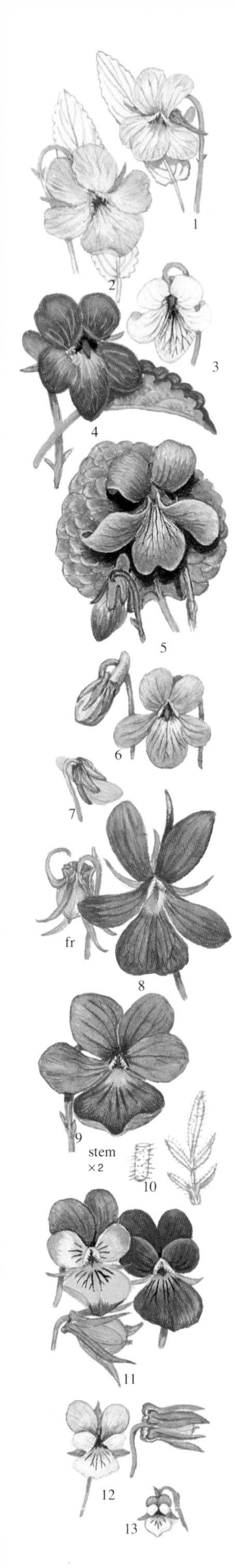


1* **Meadow Violet** *Viola pumila* Chaix. Low to short, hairless perennial without a basal leaf-rosette; stems ascending to erect. Leaves narrow, lanceolate with a wedge-shaped base; stipules large untoothed to coarsely toothed. Flowers pale blue, 15mm, rounded in outline, with a short greenish spur. Capsule hairless. Grassy habitats, at low altitudes. May-June. France, Germany and S Sweden, C and E Europe.

2 *Viola elatior* Fries. Like *V. pumila*, but a taller plant to 50cm, the leaves somewhat heart-shaped at the base and the stipules large, at least equalling the leaf-stalks. Flowers pale blue, 20-25mm. Damp grassland and scrub. May-July. Germany to C France and S Sweden; naturalized in Belgium.

3* **Marsh Violet** *Viola palustris* L. Low generally hairless perennial, *stemless*, with a creeping underground rhizome. Leaves kidney-shaped, obscurely toothed.; stipules oval-lanceolate, untoothed or finely toothed. Flowers pale violet with darker veins, rarely white, 10-15mm, not scented, with a blunt pale lilac spur. Capsule hairless. Easily told from having no aerial stem. Acid bogs, marshes, wet heaths, wet woodland and woodland flushes, often associated with *Sphagnum*, 1200-2600m. April-July. Throughout, except Spitsbergen. Europe and W Asia to Greenland, North America and the mountains of N Africa. **B**: Widespread in upland N and W Britain. subsp. *juressi* (Link ex K. Wein) P. Fourn. Differs primarily in its hairy leaf-stalks and bract in the middle (not in the lower half) of the flower-stalk. Britain, Belgium, Holland and France.

4 *Viola uliginosa* Besser. Like *V. palustris*, but stipules half-attached to the leaf-stalks and leaves heart-shaped, pointed. Flowers violet, 20-30mm, with a stout violet spur. Moorlands and marshes. April-July. Denmark, S Sweden and Finland, to W and C USSR. Generally declining from land drainage.

5 *Viola epipsila* Ledeb. Like *V. palustris*, but larger in all its parts. Leaves kidney-shaped to rounded heart-shaped, with scattered hairs beneath. Flowers 15-20mm. Marshy habitats. May-August. N Germany, Scandinavia and Iceland.

6 **Northern Violet** *Viola selkirkii* Pursh. ex Goldie. Very similar to *V. uliginosa*, but plant without a creeping rhizome and flowers smaller, pale violet, 13-15mm, with a longer spur, 5-7mm (not 3-4mm). Coniferous woods and other damp habitats. April-May. Norway, Sweden and Finland.

7* **Yellow Wood Violet** *Viola biflora* L. Low, rather fragile, slightly hairy perennial with a slender, creeping rhizome. Leaves in a basal tuft, but also borne on short leafy flowering stems, kidney-shaped, pale green, bluntly toothed; stipules small, oval to lanceolate, usually with a hairy margin. Flowers bright yellow, 15mm, solitary or in pairs, not scented, the upper 4 petals directed upwards. Capsule hairless. Damp or shaded habitats, woodland, banks and rocky habitats, primarily in mountainous regions, to 3000m. May-August. Norway, Sweden, Finland, C and E France, C and S Germany; confined to the mountains in the south. Widespread in the mountains of Europe and Asia. Sometimes cultivated.

8* **Horned Pansy** *Viola cornuta* L. Short, somewhat hairy perennial, with slender creeping rhizomes; stems erect or ascending. Leaves oval, pointed, bluntly-toothed, hairy beneath; stipules equalling, not exceeding, the leaf-stalks, deeply and palmately-lobed. Flowers violet or lilac, 20-30mm, fragrant; spur long, 10-15mm, slightly curved. Capsule hairless. Grassy meadows and banks. June-July. **I**: Pyrenees. Naturalized from gardens in Britain. Widely cultivated, including violet-purple and white-flowered forms. In its native habitats *V. cornuta* is a plant of meadows and other grassy habitats. **B**: Naturalized in scattered localities in England and Wales, rarely elsewhere.

9* **Mountain Pansy** *Viola lutea* Hudson. Low to short, hairless perennial with a slender, branched underground rhizome, but without leafy stolons. Leaves oval to lanceolate, bluntly toothed, the upper leaves markedly narrower than the lower; stipules palmately or pinnately lobed. Flowers usually yellow, sometimes violet or bicoloured, the lower petal with a dense network of veins, 20-25mm; spur short snd slender, 3-6mm. Capsule hairless. Grassy habitats, particularly short turf, rocky habitats, on acid or calcareous soils, to 2000m. April-July. Britain, Belgium, Holland, C and E France, W and S Germany; mostly in C and W Europe. Cultivated. **B**: Hilly districts of Wales, N England and S Scotland, very rare in Ireland.

10* **Rouen Pansy** *Viola hispida* Lam. [*V. rothomagensis*]. Short perennial, the whole plant covered in spreading hairs. Leaves oval with a somewhat heart-shaped base, blunt-toothed, the upper leaves narrower; stipules palmately-lobed with slender segments. Flowers violet or yellowish, 18-20mm; spur 4mm. Closely related to *V. lutea*, but much hairier with rather smaller flowers. Calcareous cliffs. June-July. NW France – near Rouen. Not known elsewhere.

11* **Heartsease, Wild Pansy** *Viola tricolor* L. Variable low to short, hairless or slightly hairy annual or short-lived perennial; stems erect or ascending. Leaves oval with a heart-shaped base, the upper leaves narrower, generally lanceolate, all bluntly toothed; stipules deeply pinnately-lobed. Flowers violet-blue, or a mixture of violet-blue, yellow and white, occasionally entirely yellow, 10-25mm, the spur 3-5mm. Rough grassland, waste places, cultivated land and a weed of arable land, on acid or neutral soils, to 2700m. April-October. Throughout, except Spitsbergen. **B**: Throughout, but less frequent in the south-east of England. subsp. *curtisii* (E. Forster) Syme [incl. *V. litoralis*]. is usually perennial and low growing, rarely exceeding 15cm, with fleshy leaves. Flowers rather variable in colour. Dunes and dry rough grassland close to the sea. Shores of the North Sea, English Channel and the Baltic. **B**: Coastal SW and W, Ireland; also inland on the East Anglian brecklands.

12* **Field Pansy** *Viola arvensis* Murray. Short, hairy annual, usually with erect stems. Leaves oblong, generally broadest above the middle, bluntly-toothed; stipules coarsely pinnately-lobed with a large endleaflet. Flowers cream or yellow with cream or bluish-violet upper petals, small, 10-15mm, the sepals equalling or exceeding the petals; spur short. Common weed of arable land, sometimes in waste places or on disturbed ground, on neutral or calcareous soils. April-October. Throughout, except the Faeroes, Iceland and Spitsbergen. **P**: Variable: various insects, also self-pollinated. Probably hybridises with *V. tricolor*. **B**: Throughout England and Wales, though scarce in the north and north-west; also Scotland and Ireland.

13* **Dwarf Pansy** *Viola kitaibeliana* Schultes. Low, densely grey-hairy annual. Leaves mostly oblong, broadest above the middle, coarsely toothed, the lowermost leaves generally rounded; stipules pinnately-lobed with an oblong terminal lobe. Flowers cream, often with a yellow centre, sometimes pale violet-blue, very small, 4-8mm, the sepals longer than the petals. Sometimes confused with *V. arvensis*, but the flowers are smaller and rather concave, not flat, and the plant small and little-branched. Dry open habitats, short-cropped turf, sandy arable fields, disturbed ground, sand-dunes. March-July. SW Britain and France; naturalized in Germany. **B**: Very rare; Channel Islands and the Isles of Scilly.

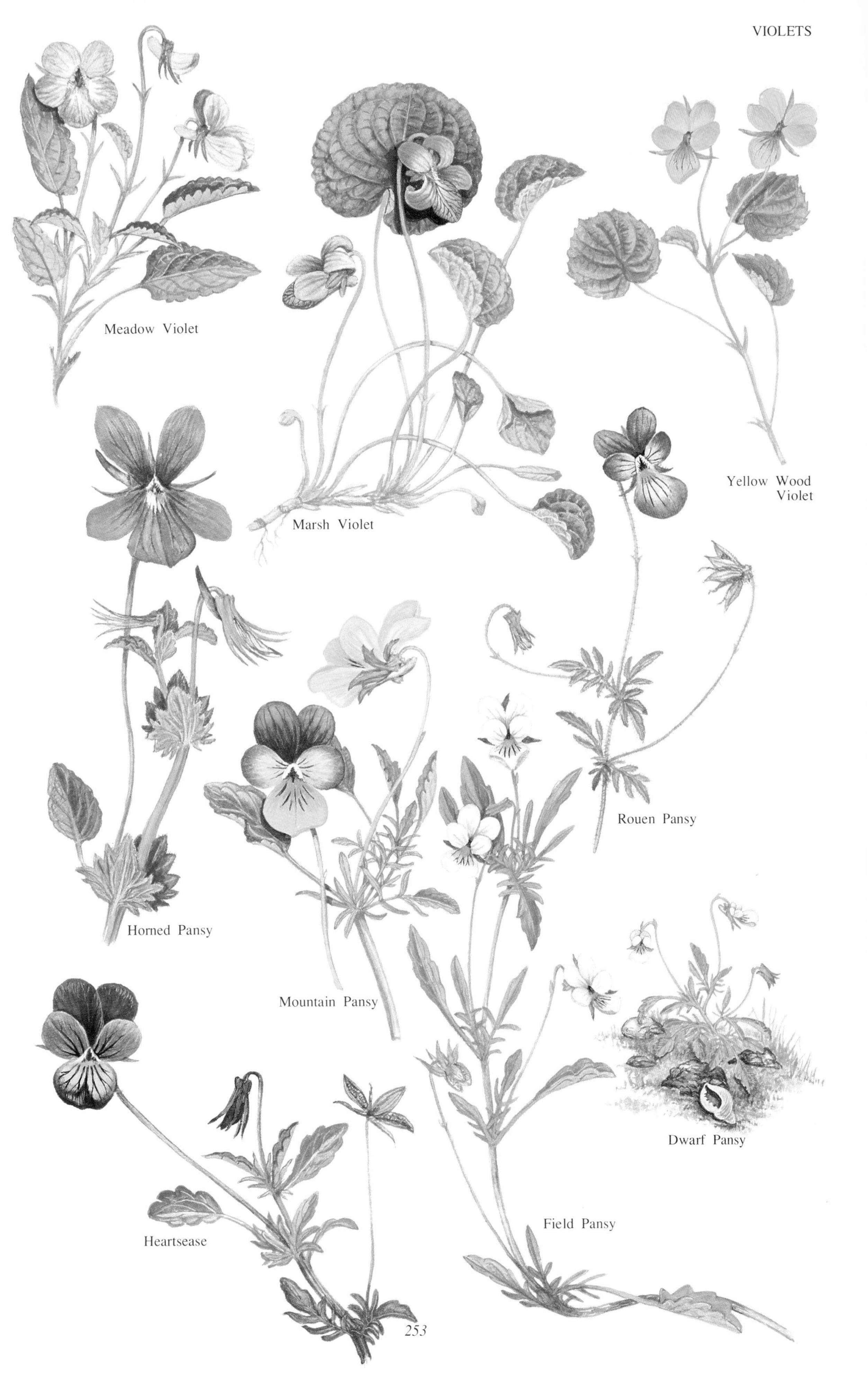
Meadow Violet
Marsh Violet
Yellow Wood Violet
Horned Pansy
Rouen Pansy
Mountain Pansy
Dwarf Pansy
Heartsease
Field Pansy

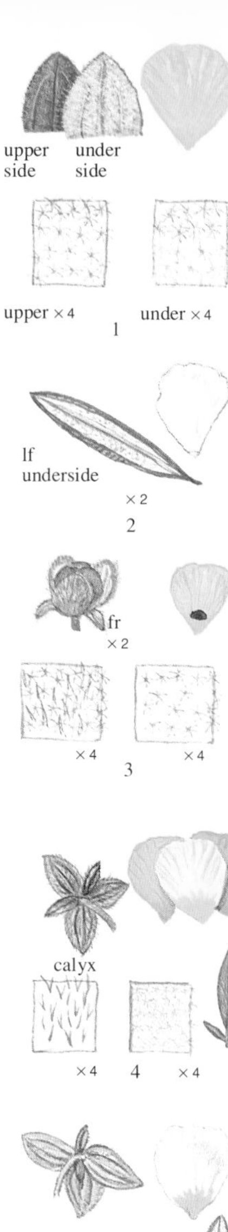


# ROCKROSE FAMILY Cistaceae

Shrubs and herbs with simple, often opposite leaves, furry-hairy; stipules often present. Flowers hermaphrodite, regular, solitary or in cymes, often raceme-like. sepals 3 or 5. Petals 5. Stamens numerous. Style solitary. Fruit a small capsule, splitting usually with 3 or 5 valves. Under 200 species in 8 genera in Europe, W Asia, North Africa, as well as parts of North and South America. A number are cultivated, particularly *Cistus* and *Helianthemum.*

1* **Halimium** *Halimium alyssoides* (Lam.) C. Koch [*H. occidentale, Helianthemum alyssoides*]. Small evergreen shrub to 1m tall, though often much less; stems erect or spreading, grey-hairy. Leaves opposite, oblong to lanceolate, blunt, dark green above, white-felted beneath, short-stalked or unstalked; stipules absent. Flowers yellow, 15-25mm, in short terminal cymes, the petals unspotted; sepals 3, hairy. Woods, scrub and sandy heaths, usually on dry siliceous soils and at low altitudes. April-June. W and C France to NW Spain and Portugal. Differs from *Helianthemum* in its style which is short or absent and in the presence of 3, not 5, sepals. The fruit capsule in both is 3-valved.

2* **Umbellate Halimium** *Halimium umbellatum* (L.) Spach [*Helianthemum umbellatum*]. Small, sticky shrub, with erect or spreading, often rather twisted, branches. Leaves crowded towards the branch tips, linear to linear-lanceolate, the margin rolled under, dark green above, white-felted beneath. Flowers white, 20-25mm, in small umbel-like clusters; sepals long-hairy. Capsule hairy. Bushy places and pine woods on light dry soils, at low altitudes. March-May. SW and C France extending south to N Spain and Portugal. **P**: Bees and other insects.

3* **Spotted Rockrose** *Tuberaria guttata* (L.) Fourr. [*T. variabilis, Helianthemum guttatum*]. Short, usually erect, hairy annual with a basal rosette of leaves. Leaves elliptical to oval, 3-veined, the uppermost leaves linear-lanceolate and often alternate, with stipules at the base. Flowers yellow, the petals with a conspicuous chocolate-brown or purplish basal blotch, 10-20mm, borne in terminal racemes; sepals 5, the outer 2 smaller. Capsule 3-valved. Bare patches among heather and gorse, exposed rocks and maritime cliffs, generally on thin soils over acid rocks, at low altitudes. May-August. SW Britain, France, Holland, C and S Germany. **P**: Insects or self-pollinated. The basal rosette of leaves has generally withered by flowering time. **B**: Very rare; found only in the Channel Islands, Jersey and Alderney.

subsp. *breweri* (Planch.) E. F. Warb. *Helianthemum breweri* Planch.). Like the type, but differing in being a more spreading plant, branched from the base, generally less than 10cm tall. Upper leaves without stipules, scarcely narrower than the lower leaves. Exposed rocky moorland, usually close to the sea. **B**: Restricted to a few localities in N Wales and W Ireland.

**Rockroses** *Helianthemum*.. Dwarf shrubs and subshrubs with opposite leaves; stipules usually present. Flowers yellow, sometimes white or pink, in raceme-like cymes. Sepals 5, the outer 2 smaller, often linear. Capsule 3-valved. About 80 species, mostly Mediterranean but some widespread in Europe, W & C Asia and the Cape Verde Islands. Several are grown in gardens.

4* **Common Rockrose** *Helianthemum nummularium* (L.) Miller [*H. chamaecistus, H. vulgare, H. arcticum*]. Low to short evergreen subshrub, often prostrate, but up to 30cm. Leaves oblong to lanceolate, usually green above, grey- or whitish-hairy beneath, the margins sometimes rolled under. Flowers golden-yellow, occasionally cream or orange, 12-20mm, in racemes of 1-12. Some wild yellow forms have a distinctive orange blotch at the base of the petals. Dry meadows, often short-grazed, banks and rocky habitats, usually on calcareous soils, sea level to 2800m. June-September. Throughout, except the extreme north, the Faeroes, Iceland and Norway. **P**: Bees and other insects, but also self-pollinated. Widely cultivated. Widespread from W Europe to C Asia. **B**: Throughout, except in SW and NW; rare in Ireland.

subsp. *obscurum* (Celak) J.Holub. Like the type but leaves green on both surfaces. C Europe north to S Sweden.

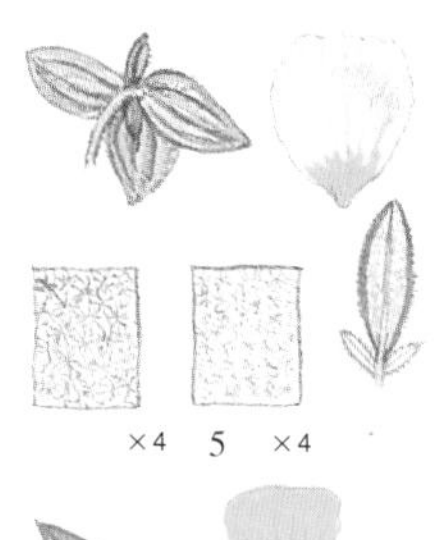


5* **White Rockrose** *Helianthemum apenninum* (L.) Miller. Like *H. nummularium* but flowers white with a yellow centre. Similar habitats and flowering time. SW England, Belgium, France and Germany southwards. **B**: SW England, local.

6* **Alpine Rockrose** *Helianthemum oelandicum* (L.) DC. Dwarf, tufted shrub to 20cm. Leaves elliptical to linear-lanceolate, green above and beneath, hairless. Flowers yellow, small, 9-12mm. Grassy and rocky habitats. May-July. Sweden – Öland.

subsp. *alpestre*. Like the type but leaves and sepals hairy. Mountain habitats. C & S France and S Germany.

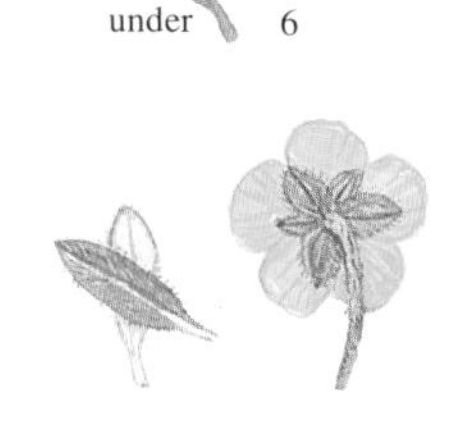

7* **Hoary Rockrose** *Helianthemum canum* (L.) Baumg. Low to short hairy perennial. Leaves elliptical to linear, usually greyish with branched hairs. Flowers small, 8-15mm, 3-5 borne in small racemes of 3-5. Dry meadows and rocky places, often on thin dry turf, also among and on cliffs, generally on limestone, to 1650m. May-July. W Britain, W Ireland, France, Germany and S Sweden. **P**: Bees. The stamens, like those of the previous species, are irritable, moving apart to brush visiting insects. **B**: Very local in N Wales, Cumbria and in the Burren in W Ireland.

subsp. *canescens* (Hartman) M.C.F. Proctor. Like the type, but leaves lanceolate to linear. Inflorescence borne on lateral branches of the current years growth (not at the apex of the previous year's growth). S Sweden – Öland. subsp. *levigatum* M.C.F. Proctor has darker green leaves, hairless or only slightly hairy above. Flowers only 1-3 usually. N England – Upper Teesdale. subsp. *piloselloides* (Lapeyr.) M.C.F. Proctor [*Cistus piloselloides*]. has leaves persistent on the lower part of vegetative stems, not clustered towards the tips. W Ireland. Also known from N Spain.

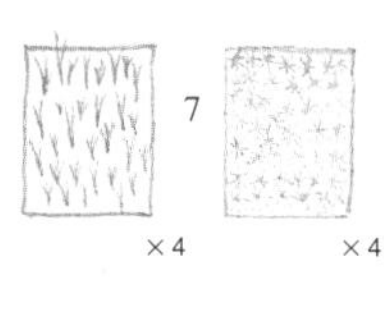


8* **Fumana** *Fumana procumbens* (Dunal) Gren. & Godron [*Cistus fumana, Helianthemum procumbens, Fumana nudifolia, F. vulgaris*]. Spreading subshrub to 40cm. Leaves alternate, linear, finely pointed, with a hairy margin; stipules absent. Flowers yellow, 8-12mm, in small lateral clusters or solitary. Fruit capsule 3-valved, on strongly hooked stalks. Rocky habitats and dry, bare places, at low altitudes. May-June. Belgium, France, Germany and S Sweden – Öland and Gotland; mainly C and S Europe. *Fumana* is superficially similar to *Helianthemum*, but differs in the outer stamens being sterile and in at least some of the leaves being alternate.

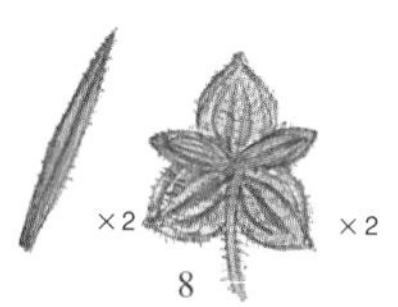

Spotted Rockrose
Umbellate Halimium
Halimium
Common Rockrose
White Rockrose
Alpine Rockrose
Hoary Rockrose
Fumana

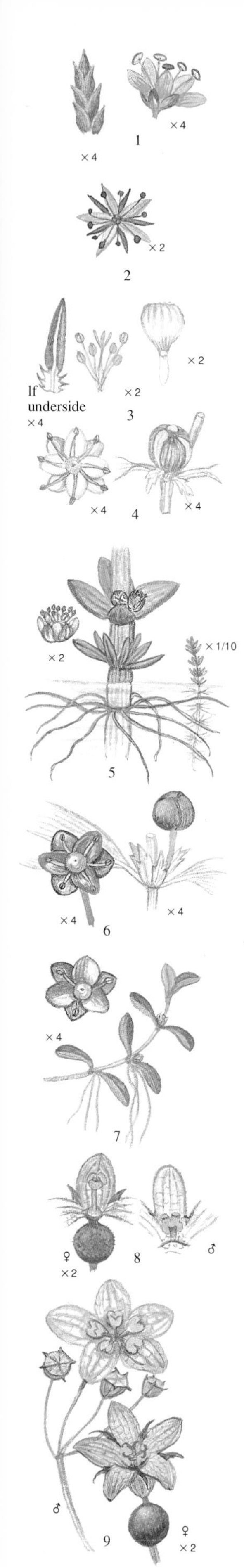


## TAMARISK FAMILY Tamaricaceae

Shrubs or small trees with small, simple, ericoid leaves. Flowers in spike-like racemes, hermaphrodite and regular. Sepals and petals 4-5, generally all separate. Stamens 4-15, free in *Tamarix*. Styles 3-4, but stigmas without styles in *Myricaria*. Fruit a capsule with fluffy seeds. About 120 species in 4 genera, in temperate and subtropical Europe and Asia, N, NE and SW Africa. Most grow in dry, sandy or rocky habitats. A number yield dyes and medicine.

1* **Tamarisk** *Tamarix gallica* L. [incl. *T. anglica*]. Hairless deciduous shrub or bushy tree, 1-3m tall, of feathery appearance, with dark brown or dark purple bark. Leaves green, occasionally grey-green, scale-like, only 1-3mm long, pointed, closely overlapping. Flowers pink, 2-3mm, 5-parted, in dense racemes which in turn form large terminal panicles; stamens 5. Capsule 3-sided. Coastal habitats. July-September. From NW France to NW Spain, Portugal, N Africa and Mediterranean Europe. Plants with grey-green leaves were formerly distinguished as *T. anglica*. **B**: Naturalized in a few localities on the S and E coasts of England from Suffolk southwards, also in the Channel Islands.

2* **Myricaria** *Myricaria germanica* (L.) Desv. Hairless grey-green shrub 0.5-2.5m tall, with erect branches. Leaves bluish-green, scale-like, linear-lanceolate, 2-5mm long, closely overlapping. Flowers pink, 5-6mm, in long, usually terminal, catkin-like spikes, 5-parted; stamens 10. River gravels, rocky places and other open habitats, often close to rivers in mountains, to 2400m. May-August. C and E France, Germany, Norway, Sweden and Finland.

## SEA HEATH FAMILY Frankeniaceae

Herbs with opposite ericoid leaves, untoothed. Flowers usually hermaphrodite. Sepals and petals 4-6, the petals clawed. Stamens usually 6. Fruit a small capsule. 90 species.

3* **Sea Heath** *Frankenia laevis* L. Low prostrate, finely hairy, much-branched perennial, often forming dense mats. Leaves linear-lanceolate, 2-5mm long, with the margin rolled under, often with a whitish crust. Flowers purplish to whitish, 9-11mm, solitary or in small clusters; petals 5 rounded or oval. Sandy and muddy habitats at the upper end of salt marshes, stabilized shingle and dune slacks. June-September. S Britain, W France, Madeira and the Mediterranean to W Asia. **B**: Local along coast on S and E England north to Norfolk and Anglesey; also in the Channel Islands.

## WATERWORT FAMILY Elatinaceae

Aquatic and marsh plants. Leaves simple, opposite or whorled; stipules present. Flowers hermaphrodite, regular, 3-4-parted, solitary or in small clusters. Stamens equal to or twice as many as petals. Styles 2-5. Fruit a small capsule, splitting lengthwise. Cosmopolitan; 33 species in 2 genera. In *Bergia* the flowers have 5 sepals: in *Elatine* there are either 2 or 4.

4* **Eight-stamened Waterwort** *Elatine hydropiper* L. [*E. oederi*]. Low to short hairless annual. Leaves opposite, elliptical, broadest above the middle. Flowers red, tiny, 4-parted, the petals equalling or longer than the sepals; stamens 8. Capsule 4-sided. Ponds, small lakes, canals. July-August. Throughout, except the extreme north, the Faeroes and Iceland. Declining from land drainage and the clearance of ditches, but most species of *Elatine* are adaptable to changing water levels and can survive in ditches and pools which dry out during the summer. **B**: Very rare; scattered localities in W England, Scotland and Ireland.

5 *Elatine alsinastrum* L. Like *E. hydropiper* but a larger plant with leaves in whorls of 3-18, linear in aquatic habitats, but lanceolate or oval in terrestrial forms. Pools, ditches and muddy habitats. July-September. N and C France and SW Finland.

6* **Six-stamened Waterwort** *Elatine hexandra* (Lapierre) DC. Slender low to short, hairless annual or short-lived perennial. Leaves opposite, narrowly oval, broadest above the middle. Flowers pinkish-white, tiny, stalked, solitary at the leaf-axils, 3-parted, the petals longer than the sepals; stamens 6. Capsule globose. Shallow ponds and lake margins and wet mud, at low altitudes. July-September. Throughout, except the extreme north, Finland, the Faeroes and Iceland.

7 **Three-stamened Waterwort** *Elatine triandra* Schkuhr. [*E. callictrichoides*]. Like *E. hexandra*, but always annual. Flowers white or red, unstalked with only 3 stamens. Similar habitats and flowering time. Belgium, Holland, France, Germany, Norway, Sweden and Finland.

## GOURD FAMILY Cucurbitaceae

Herbs, often climbers with alternate leaves and stem tendrils. Flowers unisexual; calyx 5-lobed; corolla deeply 5-6-lobed; stamens 3. Ovary inferior, conspicuous at the base of the female flowers. Fruit fleshy, berry-like. Over 700 species in some 90 genera, primarily in the tropics and subtropics, with a few in temperate parts of the world. Includes many food plants: cucumbers, marrows, melons, gherkins etc.

8* **White Bryony** *Bryonia dioica* Jacq. *B. sicula* (Jan.) Guss.). Tall climbing, hairy perennial to 4m, with coiled tendrils. Leaves palmately 5-lobed, the lobes untoothed or with a few blunt teeth, the central lobe scarcely larger than the adjacent ones. Flowers greenish-white with darker veins, 10-18mm, the female in small lateral clusters, the male on a separate plant, in drooping racemes. Berry green at first, red when ripe, 6-10mm. Hedgerows, woodland margins, scrub, on calcareous and base-rich soils, generally at low altitudes. May-September. S Britain, Belgium, Holland, France and Germany. Poisonous. Once grown medicinally and widely naturalized. **P**: Bees and other small insects. **B**: Widespread in lowlands from Northumberland southwards, occasionally naturalized elsewhere.

9* *Bryonia alba* L. Similar to *B. dioica*, but central lobe of leaves larger than the adjacent ones, male and female flowers separate, but on the same plant; stigma hairless (not hairy). Berry black when ripe, 7-8mm. Hedgerows, scrub, woodland margins and rocky places. May-September. Germany; widely naturalized in Belgium, France and S Scandinavia.

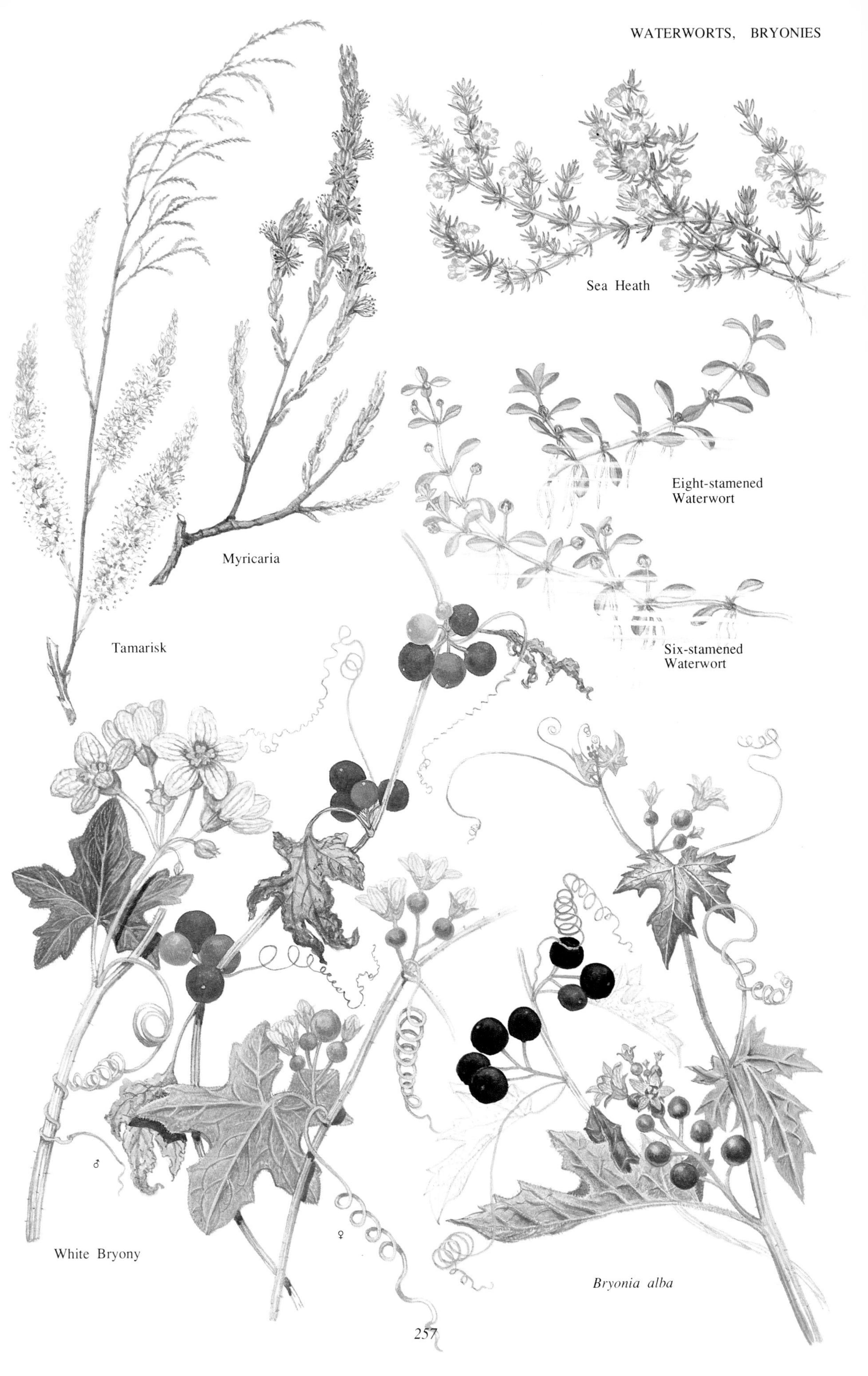
Sea Heath
Myricaria
Eight-stamened Waterwort
Tamarisk
Six-stamened Waterwort
♂
♀
White Bryony
Bryonia alba

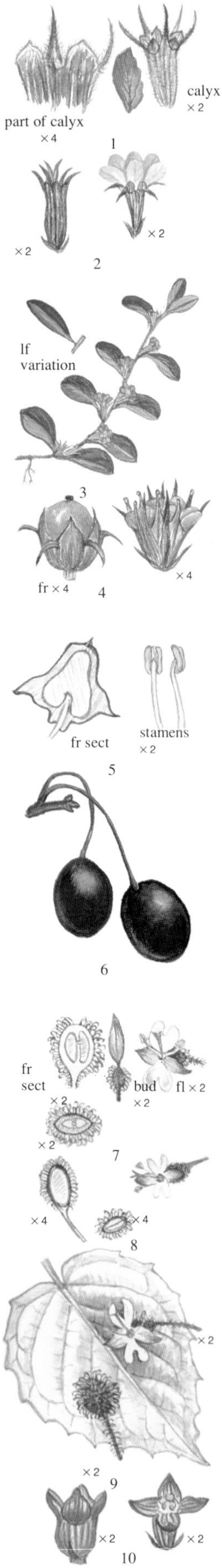

## LOOSESTRIFE FAMILY Lythraceae

Herbs with simple leaves; stipules tiny or absent. Flowers hermaphrodite, regular, 4-6-parted, solitary or in small cymes, or in terminal racemes. Epicalyx often present. Petals separate, pink or purple, inserted on the rim of a cup-like hypanthium. Stamens 2-12. Fruit a many-seeded capsule. 450 species in 22 genera. Most are tropical or subtropical.

1* **Purple Loosestrife** *Lythrum salicaria* L. Stout tall, grey-hairy, tufted perennial to 1.5m.; stems with 4, sometimes more, raised lines. Leaves opposite or in whorls of 3, the uppermost sometimes alternate, oval to lanceolate, untoothed, unstalked. Flowers reddish-purple, 10-15mm, in whorls forming long spikes; petals usually 6 and stamens 12. Capsule oval, 3-4mm. Freshwater margins, usually avoiding acid soils. June-August. Throughout, except the extreme north, the Faeroes and Iceland. Flowers trimorphic: 3 types of flower with the stamens and style set at different levels. **P**: Long-tongued insects, bees and butterflies. **B**: Widespread throughout, but scarce in Scotland except for the west coast.

2* **Grass-poly** *Lythrum hyssopifolia* L. Low to short, more or less hairless annual, with erect or ascending stems. Leaves alternate, linear to oblong, untoothed, suberect. Flowers pink, 5mm, solitary or paired at the base of the upper leaves, forming slender spikes, with 6 petals; stamens 4-6, not protruding. Disturbed or seasonally flooded ground, at low altitudes. June-September. Britain, Belgium, France and Germany; naturalized or casual elsewhere, especially in Norway and Sweden. **B**: Very rare and erratic in appearance; E England and Jersey, occasionally casual elsewhere.

3 *Lythrum borysthenicum* (Schrank) Litv. [*Middendorfia borysthenicum*]. Like *L. hyssopifolia*, but a low, rather rough annual with alternate or opposite leaves. Flowers with tiny purplish petals, sometimes absent; stamens 6. Seasonally wet places. June-August. NW and C Europe.

4* **Water Purslane** *Lythrum portula* (L.) D.A. Webb [*Peplis portula*]. Low, more or less prostrate, creeping, hairless annual; stems rooting at the nodes. Leaves opposite, rather fleshy, sometimes reddish, oval, tapered to a short stalk. Flowers purple, small, 1-2mm, solitary at the leaf bases, with 6 sepals and petals, the latter sometimes absent. Wet places in open communities, bare ground and the muddy margins of pools, generally on acid soils. June-October. Throughout, except the far north. **B**: Throughout, but often local, scarce in N and NW Scotland.

5* **Water Chestnut** *Trapa natans* L. (Water Chestnut Family – Trapaceae). An aquatic annual or short-lived perennial, hairless, unbranched; stems rooting into the mud, the roots greenish and paired at the lower nodes. Floating leaves forming a wide flat rosette, rhombic, coarsely toothed, leafstalk fleshy, swollen in the centre; submerged leaves linear, untoothed. Flowers white, 10-20mm, in the axils of floating leaves; sepals triangular, persisting to become woody and part of the fruit. Fruit shaped like a top, 25-30mm across. Lakes, ponds and canals, generally in nutrient rich, though not particularly calcareous, water, at low altitudes. June-July. C and E France and S Germany, extinct in Sweden; sometimes cultivated elsewhere. Formerly far more widespread in N Europe. Very variable from population to population. The fruits are eaten as a vegetable, particularly in Asia.

## WILLOWHERB FAMILY Onagraceae

Herbs and shrubs with simple, alternate or opposite leaves. Flowers hermaphrodite, regular or somewhat irregular. Sepals and petals 2, 4 or 5. Stamens 2, 4, 8 or 10. Style solitary. Ovary inferior, often conspicuous below the calyx. Fruit a capsule or a berry. 650 species.

6* **Fuchsia** *Fuchsia magellanica* Lam. Much branched shrub to 3m tall; stems thin, generally arching; bark pale brown and peeling in long strips. Leaves opposite or in threes, occasionally alternate, elliptical, pointed, toothed, short-stalked. Flowers purple with red sepals, occasionally white, 12-15mm, pendent, on long slender stalks, solitary or paired at the leaf axils, 4-parted; stamens and style long-exserted. Fruit an oblong berry, black and very juicy when ripe. Locally naturalized by old walls. July-October. **I**: Temperate S America. SW Britain and W Ireland. **B**: Local, SW England and W Ireland.

**Enchanter's Nightshades** *Circaea*. Perennial herbs with 2-parted flowers, the petals deeply notched and appearing to be 4. Fruit with hooked bristles, not splitting. 10 species.

7* **Enchanter's Nightshade** *Circaea lutetiana* L. Short to medium, somewhat hairy, perennial with long slender stolons. Leaves oval, slightly heart-shaped at the base, pointed, opposite, stalked, toothed. Flowers white or pinkish, small, 4-7mm, in lax racemes with the open flowers well spaced. Fruit club-shaped, equally 2-celled, with whitish bristles. Woodland, coppices, plantations and other shady habitats, on base-rich or calcareous soils, generally at low altitudes. June-August. Throughout, except the Faeroes, Finland, Iceland and Spitsbergen. **P**: Probably small flies. **B**: Widespread, except for N Scotland and many of the northern isles.

8* **Alpine Enchanter's Nightshade** *Circaea alpina* L. Low to short stoloniferous perennial. Leaves heart-shaped, toothed, hairless. Flowers white, tiny, 1-2.5mm, in short racemes which elongate after the petals drop. Fruit oblong, with only a single compartment, with soft bristles. Damp upland woods, rocky places, stream banks, generally in shaded habitats and on acid soils, to 2200m. June-August. Most of Europe except the far north. **P**: Small flies. **B**: Very rare; scattered localities in N Wales, Lake District, Nothumberland and Scotland – Argyll and Arran.

9 **Upland Enchanter's Nightshade** *Circaea* × *intermedia* Ehrh. *C. alpina* × *C lutetiana*. Like *C. lutetiana*, but plants sterile with a few hairs on the leaves and flowers 3-7mm. Generally growing with one or both parents, but often growing in the absence of *C. alpina*. The hybrid very occasionally sets a fruit, but usually the flowers fall off at maturity, plants persisting because of their vigorous stoloniferous habit. Throughout the range of *C. lutetiana*, but not in N Scandinavia. **B**: From S Wales northwards and in Ireland.

10* **Hampshire Purslane** *Ludwigia palustris* (L.) Elliott [*Isnardia palustris*]. Low creeping, hairless perennial; stems thin, deep purplish-red, rooting below. Leaves opposite, oval, broadest above the middle, red-veined. Flowers green, often red-edged, 3mm, solitary at the leaf axils, 4-parted but without petals. Fruit a small capsule, angled, with persistent sepals. Often confused with *Lythrum portula*, but leaves pointed (not blunt) and flowers without petals. Wet habitats, shallow pools and stream margins, generally in acid fen, at low altitudes. June-July. Belgium, Holland, France and Germany. **P**: Self-pollinated. **B**: Very rare; scattered localities in forests in Hampshire and Essex.

Water Purslane
Grass-poly
Water Chestnut
× 1/15
Purple Loosestrife
Alpine
Enchanter's Nightshade
Fuchsia
Enchanter's Nightshade
Hampshire Purslane

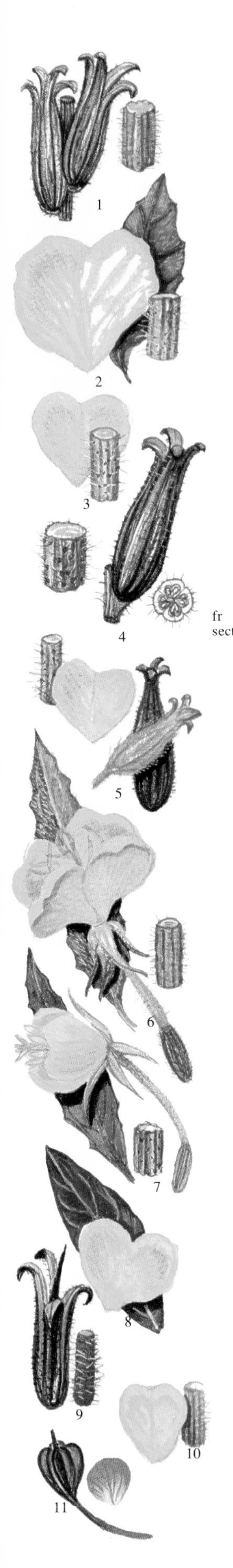


**Evening-primroses** *Oenothera*. Annual, biennial or perennial herbs with alternate leaves. Flowers usually large, 4-parted, in leafy spikes, sometimes solitary. Sepals usually strongly reflexed, generally falling. Petals often yellow, sometimes white or pink. Stamens 8, in two whorls. Stigma club-shaped or 4-lobed. Fruit a capsule, splitting into 4 valves. Almost 100 species, American in origin, but a number have arisen through hybridization in Europe. The flowers of most open in the evening, or remain open during dull weather. Most are fragrant. Pollinated by moths, though some are apparently self-pollinated. A number are garden plants and recently several are being grown for their oil-bearing seeds.

1* **Common Evening-primrose** *Oenothera biennis* L. An erect, short to tall, somewhat hairy annual or biennial; stem without red spots. Leaves lanceolate, green or bluish-green, finely toothed, the basal ones in a broad rosette at first, gradually decreasing in size up the stem; older leaves with red veins. Flowers primrose yellow, 40-50mm, in erect leafy spikes; stigmas equalling the anthers. Waste ground and open habitats, road verges, railway embankments and sand-dunes. June-September. **I**: North America. Naturalized throughout, except the extreme north, the Faeroes, Iceland and Finland. Widely established in many parts of Europe. A spring germinating species often grown in gardens, but more often appearing there as a weed. **B**: Scattered localities, except the extreme north, but often casual.

2 *O. rubricaulis* Klebahn. Like *O. biennis*, but up to 2m tall with well-branched stems, which are red-spotted. Flowers slightly smaller, the petals 22-23mm long. Similar habitats and flowering times. France and Germany.

3 *Oenothera novae-scotiae* Gates [*O. cambrica*]. Like *O. biennis*, but stem hairs with a red base and leaves rather crinkled. Flowers smaller, the petals 13-16mm long (not 24-30mm). Sand-dunes, rocky habitats and waste places inland. July-September. W Britain – coasts of the Bristol Channel; casual elsewhere in Britain. Described as recently as 1977. Many records of *O. parviflora* are in fact of this species.

4* **Large-flowered Evening-primrose** *Oenothera glazioviana* Micheli ex Martius [*O. erythrosepala*, *O. lamarkiana*]. Robust, medium to tall, hairy biennial, to 1.5m; stems erect, hairs with expanded red bases. Leaves broadly lanceolate, with crinkled margins, the basal ones in a broad rosette, the upper leaves narrower and smaller, usually with a white midrib. Flowers pale yellow, large, 50-80mm, in large leafy spikes; sepals usually reddish or red-striped. Capsule with red-based hairs. Waste places, railway embankments, road-verges, banks and sand-dunes. June-September. Britain, Belgium, Holland, France and Germany. Widely cultivated; probably arose spontaneously in gardens. Generally easy to identify by its large flowers and red-based stem hairs. Together with *O. biennis*, probably the commonest species found. Often casual, or at least not persisting for very long. **B**: England and Wales north to Yorkshire; casual elsewhere including Ireland.

*Oenothera suaveolens* Pers. Like *O. glazioviana*, but leaves not crinkled and stems without red-based hairs. Stigmas equalling anthers (not longer). Occasionally naturalized in France and Germany. Probably European in origin. Sometimes mistaken for *O. biennis*, but leaves not red-veined and flowers larger.

*Oenothera strigosa* (Rydb.) Mackenzie. Like *O. suaveolens*, but leaves greyish-hairy, sepals reddish (not greenish) and flowers rather smaller, the petals 18-24mm long (not 30-50mm). **I**: Temperate North America. Naturalized, but often local in France and Germany.

5* **Small-flowered Evening-primrose** *Oenothera parviflora* L. Short to tall, erect, hairy annual or biennial, to 2m; stems usually unbranched, red below, and with red-based hairs above. Leaves elliptical, the lower broadest above the middle and in a rosette, the stem leaves mostly lanceolate, but decreasing in size up the stem, the uppermost bract-like. Flowers yellow, small, 14-18mm, in a dense leafy spike whose tip nods somewhat when young; sepals green at first, but becoming streaked with red. Open habitats, waste and disturbed ground. June-September. **I**: Temperate North America. Naturalized in Holland, France, Germany and Norway. Long-established in Europe, though many of its records, especially those for Britain, are *O. novae-scotiae*. Unlike the preceding species the sepal tips (in bud) are spreading and free from one another and not held closely together.

6 *Oenothera ammophila* Focke. Like *O. parviflora*, but a shorter plant not exceeding 1m tall, the stems spreading somewhat at the base, without red spots. The leaves white-hairy, toothed. Tip of inflorescence drooping for a considerable distance and capsule red-striped. Similar habitats and flowering times. **I**: North America. Naturalized in Holland, France, Denmark and Germany.

7 *Oenothera syrticola* Bartlett. Like *O. ammophila*, but stems erect (not spreading at the base). Leaf-margin reddish and flowers smaller, the petals about 13mm long (not 16mm). **I**: E North America. Occurs sporadically in Belgium, Holland, France and Germany.

8 *Oenothera rubricuspis* Renner ex Rost. Like *O. ammophila*, but stems erect, up to 2m, and leaves dark green, untoothed. Casual or perhaps naturalized in SE Belgium and Germany – Hessen.

9 **Fragrant Evening-primrose** *Oenothera stricta* Ledeb. Medium to tall, densely hairy annual or biennial. Basal leaves linear to elliptical, broadest above the middle; stem leaves lanceolate, finely toothed, with wavy margins. Flowers yellow, soon turning red, 35-60mm, in leafy spikes, very fragrant. Capsule large, 20-25mm, enlarged in the upper half. Waste and disturbed habitats, sand-dunes. June-September. **I**: Temperate South America. Locally naturalized in Britain, France and Germany. Occasionally grown in gardens and sometimes casual in NW Europe. **B**: Naturalized in SW England and the Channel Islands and scattered localities north to S Scotland.

10* *Oenothera laciniata* Hill [*O. sinuata*]. Rather like *O. stricta*, but leaves pinnately-lobed and conspicuously wavy. Flowers smaller, 10-25mm. **I**: North America. Very rare, casual in Britain, France, Belgium and Holland.

11* **Pink Evening-primrose** *Oenothera rosea* L'Hér. ex Aiton. An erect, rather stiffly hairy, medium to tall perennial, occasionally an annual. Basal leaves coarsely pinnately-lobed, in a rosette, short-stalked; stem leaves elliptical, often with a few lobes towards the base. Flowers pink, small 8-18mm, in slender spikes. Capsule 8-10mm, with 8 angles. Dry rocky habitats, waste places and sand-dunes. July-September. **I**: Warmer regions of North and South America. Naturalized in W and C France; casual in other parts of NW Europe, including S Britain. Sometimes cultivated in gardens and very distinctive on account of its small pink flowers and divided leaves.

Common
Evening-primrose

Small-flowered
Evening-primrose

Large-flowered
Evening-primrose

*Oenothera laciniata*

Pink
Evening-primrose

Fragrant Evening-primrose

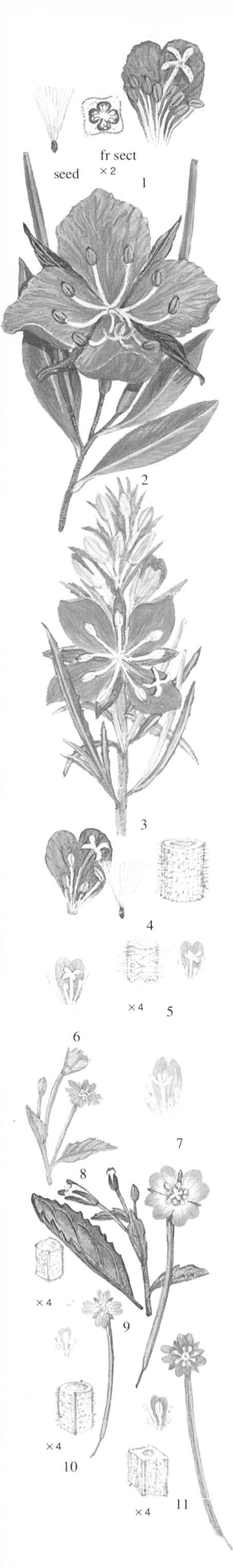


**Willowherbs** *Chamerion* and *Epilobium*. Perennial herbs, but often flowering in the first year, with stolons or overwintering leaf-rosettes. Leaves alternate, opposite or in whorls. Flowers in leafy racemes or spikes, occasionally solitary. Sepals and petals 4. Stamens 8. Stigma club-shaped or 4-lobed. Fruit a slender capsule, splitting lengthwise and containing numerous fluffy seeds. About 160 species in temperate and Arctic regions of both the northern and southern hemispheres. *Chamerion* differs from *Epilobium* in all the leaves being alternate and in the flowers being somewhat irregular (zygomorphic) rather than regular, with a deflexed style.

1* **Rosebay Willowherb** *Chamerion angustifolium* (L.) Holub [*Chamaenerion angustifolium, Epilobium angustifolium, E. spicatum*]. Robust, almost hairless, tall patch-forming perennial, to 2.5m. Leaves all alternate, lanceolate, slightly toothed, with a vein running along close to the margin. Flowers violet or rose-purple, 20-30mm, in long tapered racemes, the flowerbuds sharply deflexed; petals slightly notched. Rocky habitats, screes, woodland margins, disturbed sites, waste places, felled woodland, riverbanks, scrub, derelict building sites, on a variety of soils from peat to sand, to 2500m. June-September. Throughout, except Spitsbergen. Widespread and common, sometimes becoming a serious weed, especially in gardens. Very variable. Precise native distribution is uncertain. **B**: Widespread throughout, less common in the north.

2 **River Beauty** *Chamerion latifolium* (L.) Holub [*Chamaenerion latifolium, Epilobium latifolium*]. Low to medium, more or less hairless perennial with ascending stems. Leaves elliptical, untoothed, alternate, with obscure veins. Flowers rose-pink or purplish, large, 35-50mm, up to 7 in a lax one-sided raceme, the petals not notched. Rocky habitats, especially close to rivers and streams, moraines and screes, scrub. July-September. Iceland. Widespread in Arctic and sub-Arctic regions from North America to N Asia.

3 *Chamerion dodonaei* (Vill.) Holub [*Chamaenerion dodonaei, C. angustissimum, Epilobium dodonaei, E. rosmarinifolium*]. Rather like *C. angustifolium*, but leaves linear, slightly hairy and with rather obsure veins. Flowers pink or purplish, 20-25mm, nodding in bud; style at first deflexed, but later erect. Rocky habitats, banks, screes, riverbanks, to 1500m. July-September. E France and S Germany. Confined mainly to mountain regions of C and S Europe.

4* **Greater Willowherb** *Epilobium hirsutum* L. Robust, softly hairy, tall perennial, to 2m, often forming extensive patches; stems erect. Leaves opposite, sometimes whorled, oblong to lanceolate, unstalked and half-clasping the stem, coarsely toothed. Flowers bright purplish-pink, 15-25mm, in a leafy raceme, sometimes branched, petals notched; stigma 4-lobed. Damp and waste places, dykes, ditches, river and pond margins, to 2500m. Throughout, except the far north, Faeroes and Iceland. June-September. Often abundant, forming large colonies, but intolerant of shade. **P**: Various bees and flies. Widespread in Europe, Asia and Africa. **B**: Widespread in England, Wales and Ireland; Scotland, confined mainly to the east coast.

5* **Hoary Willowherb** *Epilobium parviflorum* Schreber. Short to medium, hairy perennial, overwintering by leafy rosettes; stems erect, rather robust. Leaves mostly opposite or whorled, oblong to linear-lanceolate, scarcely stalked, slightly toothed, not clasping the stem. Flowers pale purplish-pink, small, 7-12mm, in a lax raceme; stigma 4-lobed. Variable species – some forms are scarcely hairy. Waste ground on damp soils, streams banks, marshes and fens, to 2500m. July-August. Throughout except the extreme north, the Faeroes and Iceland. **P**: Bees, or may be self-pollinated. From Europe to Asia and N Africa. **B**: Throughout, though scarcer in the north.

6* **Western Willowherb** *Epilobium duraei* Gay ex Godron. Short to medium, slightly hairy perennial with long fleshy stolons; stem hairs bristly and close-pressed to the stem. Leaves opposite, oval with a rounded base, scarcely stalked, coarsely toothed; uppermost leaves alternate. Flowers pink, 8-14mm, in a lax often one-sided raceme; petals notched; stigma 4-lobed. Rocky habitats and banks in the mountains, on acid rocks, to 2500m. June-September. C and E France and SW Germany. Confined to the mountains of W and SW Europe from the Vosges south-westwards.

7* **Broad-leaved Willowherb** *Epilobium montanum* L. [*E. hypericifolium*]. Short to medium, slightly hairy perennial with a rounded stem. Leaves all opposite, the uppermost alternate, oval, toothed, sometimes untoothed, very short-stalked. Flowers purplish-pink, 6-12mm, in lax racemes, the petals notched; stigma 4-lobed. Woodland, hedgebanks, ditches, waste and cultivated ground, old walls, to 2600m. May-August. Throughout, except Iceland and Spitsbergen. **P**: Insects or self-pollinated. Very widespread from W Europe east across Asia to Japan. Sometimes a troublesome garden weed. **B**: Widespread throughout.

8 *Epilobium collinum* C.C. Gmelin. Like *E. montanum*, but with smaller leaves, 1-5cm long (not 3.5-8cm); flowers smaller, the petals only 3-6mm long (not 6-10mm). Similar habitats and flowering time. Continental Europe.

9 **Spear-leaved Willowherb** *Epilobium lanceolatum* Sebastiani & Mauri. Short to medium, somewhat hairy perennial, overwintering by leafy rosettes; stems slightly 4-angled. Leaves mostly alternate, though the lowermost usually opposite, oblong, narrowed at the base, stalked, toothed. Flowers white, gradually turning pink, 8-12mm, in a lax raceme; stigma 4-lobed. Waste ground, road-verges, railway embankments, rocky places and quarries, walls and cultivated ground, to 2500m. June-September. Belgium, Holland, France and Germany; naturalized in Britain, occasionally casual elsewhere. **P**: Self-pollinated. Native from W and S Europe to the Caucasus and N Africa. Sometimes a weed of gardens. **B**: Naturalized in the north to Norfolk and mid-Wales, but still spreading.

10* **Short-fruited Willowherb** *Epilobium obscurum* Schreber. Medium to tall, somewhat hairy perennial, with leafy above-ground stolons produced during the autumn, but not terminating in a rosette; stem round, with raised lines. Leaves opposite, but the upper few alternate, oval to lanceolate, toothed, unstalked. Flowers purplish-pink, 7-10mm, the petals notched; stigma club-shaped. Moist woodland, marshes, stream margins and ditches, on neutral or mildly acid soils, generally at low altitudes. July-August. Throughout, except the extreme north, the Faeroes and Iceland. **P**: Self-pollinated. Europe east to the Caucasus, N Africa and Madeira. **B**: Throughout, but absent from the Shetland Is.

11* **Square-stalked Willowherb** *Epilobium tetragonum* L.[*E. adnatum*]. Like *E. obscurum*, but without stolons. It has square stems which produce rosettes at the base in the autumn. Sepals without glandular hairs and fruit capsule longer, 7-10cm (not 4-6cm). Damp woodland clearings, road-verges, streamsides, ditches, waste and cultivated land, to 2400m. July-August. Throughout, except the extreme north, Ireland, the Faeroes and Iceland. **P**: Self-pollinated. Variable. **B**: Lowland England and Wales north to Yorkshire.

Greater Willowherb

Hoary Willowherb

Western Willowherb

Broad-leaved Willowherb

Short-fruited Willowherb

Rosebay Willowherb

× 1/16

Square-stalked Willowherb

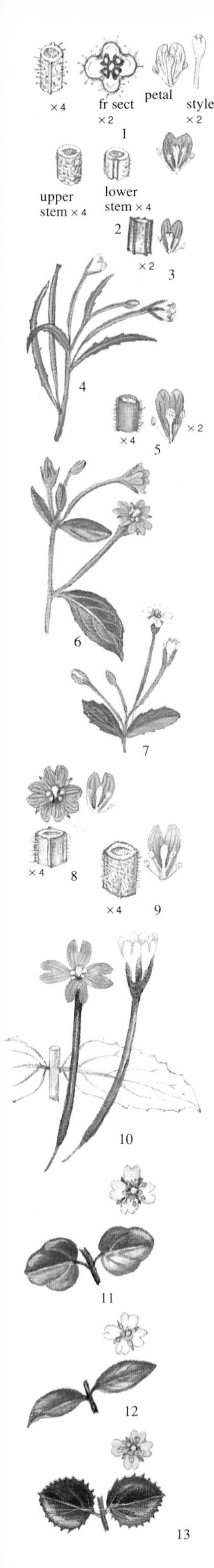


1* **Pale Willowherb** *Epilobium roseum* Schreber. Short to tall, rather slender, somewhat hairy perennial; stem rather brittle, with raised lines and almost stemless rosettes produced at the base of the stem during the autumn. Leaves mostly opposite, lanceolate to elliptical, narrowed at the base, toothed and stalked. Flowers white, becoming pinkish-streaked, 8-10mm, in a lax raceme; inflorescence whitish-hairy and glandular. Damp habitats, waste ground, woodland, copses, hedgerows and cultivated land, to 2500m. June-August. Throughout, except the extreme north, the Faeroes and Iceland. W Europe to W Asia. Distinguished from most other species by its relatively long-stalked leaves. A common weed of gardens in some areas. **B**: Throughout lowland England and Wales, but rare in Ireland and most of Scotland.

2* **Marsh Willowherb** *Epilobium palustre* L. Short to medium, hairy or almost hairless perennial with slender underground stolons; stems erect, unridged. Leaves mostly opposite, lanceolate, pointed, untoothed and unstalked. Flowers pale pink or white, 8-12mm, in a lax raceme; inflorescence often somewhat bristly, but rarely glandular; stigma club-shaped. Wet habitats, marshes, fens, valley bogs, woodland flushes, on mildly acid soils, to 2300m. July-August. Throughout, except Spitsbergen. Locally common, widespread in Europe, temperate Asia, Greenland and North America. **B**: Widespread.

3* **Nodding Willowherb** *Epilobium nutans* F.W. Schmidt. Low to short perennial, somewhat hairy above, with leafy overwintering stolons; stem ridged. Leaves mostly opposite, elliptical, unstalked, scarcely toothed. Flowers pale violet, small, 6-8mm, in a lax raceme, nodding at the tip; stigma club-shaped. Moors and waysides, to 2500m. June-August. C & E France and C & S Germany.

4 *Epilobium davuricum* Fischer ex Hornem. Like *E. nutans*, but plants overwintering by leafy-rosettes and stems glandular-hairy. Leaves linear to elliptical. Flowers white, rarely pale pink. Moist habitats. July-September. Norway, Sweden and Finland, except for parts of the south.

5* **Alpine** or **Pimpernel-leaved Willowherb** *Epilobium anagallidifolium* Lam. [*E. alpinum*]. Low creeping, almost hairless perennial with leafy stolons; flowering stems ascending, somewhat ridged. Leaves mostly opposite, the uppermost alternate, yellowish-green, elliptical, short-stalked, more or less untoothed. Flowers pale purple, 5-7mm, in few-flowered racemes nodding at the tip in bud; stigma club-shaped. Moist habitats, mossy mountain slopes, stream margins, springs, flushes, on acid or base-rich soils, to 3000m. July-early September. N Britain, the Faeroes, Iceland, Scandinavia, Spitsbergen, France and Germany, primarily on the mountains, especially in the south of its range. N and C Europe to Asia, Greenland and North America. **P**: Probably self-pollinated.

6 *Epilobium hornemannii* Reichenb. Like *E. anagallidifolium*, but a taller plant with stolons short and below ground. Flowers pale violet, slightly larger. Wet habitats in the Arctic and sub-Arctic. July-September. Iceland and Scandinavia, extending east into N Russia.

7 *Epilobium lactiflorum* Hausskn. Like *E. hornemannii*, but with inconspicuous leafy stolons above ground. Leaves weakly toothed. Flowers white, the petals 2.5-4mm (not 4-6mm). Similar habitats and flowering times. The Faeroes, Iceland and Scandinavia, extending east into N Russia.

8* **Chickweed Willowherb** *Epilobium alsinifolium* Vill. Low to short, slightly hairy perennial, with long spreading underground stolons. Leaves bluish-green, mostly opposite, oval to elliptical, toothed, short-stalked. Flowers purplish-pink, 8-9mm, in few-flowered racemes nodding at the tip in bud, the petals notched; stigma club-shaped. Moist and wet habitats, stream margins, springs and mossy flushes, generally on base rich soils, mostly in mountain habitats, to 2000m, perhaps higher. June-August. Britain, the Faeroes, Iceland, Scandinavia, France and Germany. Often confused with *E. anagallidifolium*, but with bluish-rather than yellowish-green leaves and thicker stems, 2-3mm diameter (not 1-2mm). **B**: Pennines northwards; rare in Ireland.

9* **American Willowherb** *Epilobium ciliatum* Rafin. [*E. adenocaulon*]. Medium to tall, glandular-hairy, often reddish perennial, without stolons; stems erect, with 4 raised lines, producing leafy rosettes at the base during the autumn. Leaves mostly opposite, oblong-lanceolate, toothed, with a slightly heart-shaped base and a short stalk. Flowers purplish-pink or white, 8-10mm, erect in bud, the petals deeply lobed; stigma club-shaped. Waste ground, roadsides, streamsides, banks, walls, damp woodland and cultivated land, generally at low altitudes, but ascending to about 1000m. June-August. Most of Europe; probably naturalized in the north of its range. **P**: Self-pollinated. Generally distinguished by its glandular hairs and rounded leaf bases. **B**: Doubtfully native. Rapidly spreading in lowland Britain, north to Angus.

10 *Epilobium glandulosum* Lehm. Like *E. ciliatum*, but a more robust plant with leaves scarcely stalked. Flowers slightly larger. Similar habitats and flowering times. **I**: N North America. Naturalized in Norway, Sweden and Finland.

11* **New Zealand Willowherb** *Epilobium brunnescens* (Cockayne) Raven [*E. nerterioides*]. Low prostrate, almost hairless, often matted perennial; stems slender, rooting at the nodes. Leaves nodes. Leaves mostly opposite, oval, short-stalked, slightly toothed. Flowers very pale pink, often with reddish sepals, 3-4mm, solitary, long-stalked; petals notched. Moist habitats, rocks, gritty and sandy ground, stream beds, cliffs, generally in rather bare places, occasionally on damp walls or in quarries. June-July. **I**: New Zealand. Naturalized in Britain. It was an escape from rock gardens, but has become so well established in some mountain regions as to appear to be a true native. **B**: Widely naturalized in the wetter and higher regions from N Wales and N England northwards, and in Ireland.

12 *Epilobium komarovianum* A. Léveillé [*E.inornatum*]. Like *E. brunnescens*, but leaves more elliptical. **I**: New Zealand. Naturalized on damp soils in various parks and gardens in Holland and Denmark; casual in Ireland.

13* *Epilobium pedunculare* A. Cunn. [*E. linnaeoides*]. Like *E. brunnescens*, but leaves sharply toothed and fruit-stalks 5-10cm (not 2-6cm.). **I**: New Zealand. Naturalized in 2 localities in W Ireland. Like the previous species, widely cultivated on rock gardens.

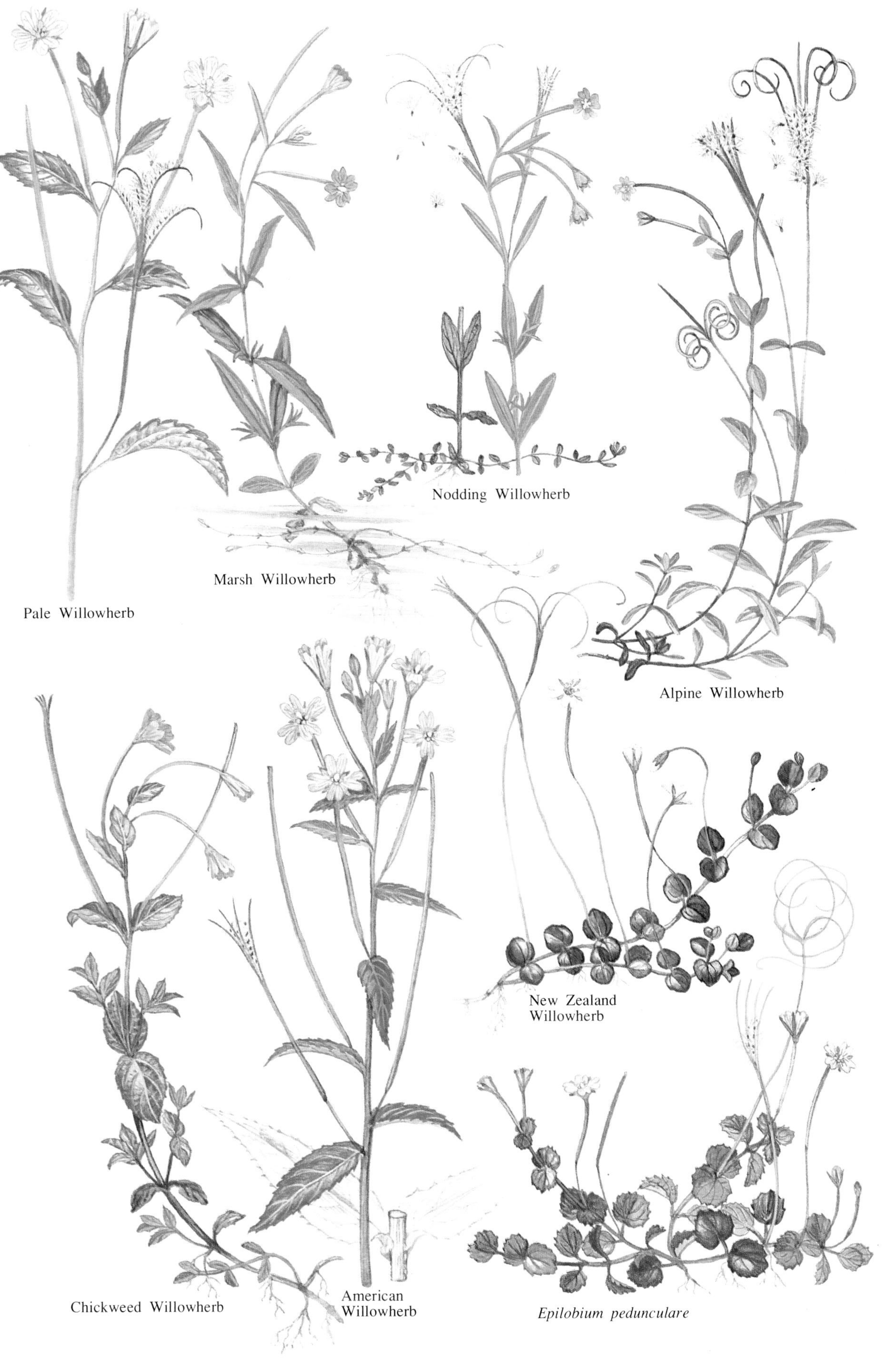
Pale Willowherb
Marsh Willowherb
Nodding Willowherb
Alpine Willowherb
New Zealand Willowherb
Chickweed Willowherb
American Willowherb
*Epilobium peduncular*e

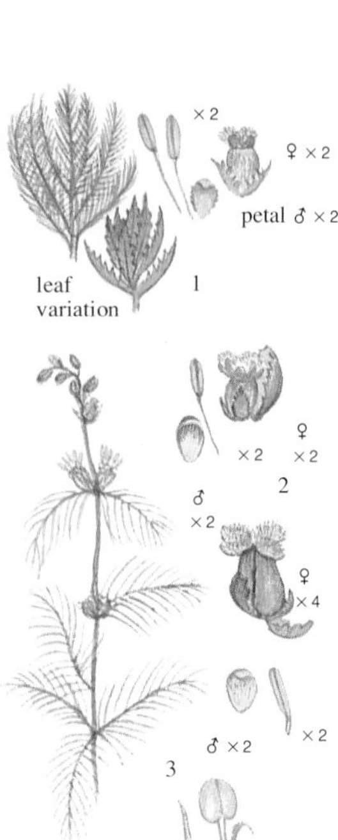


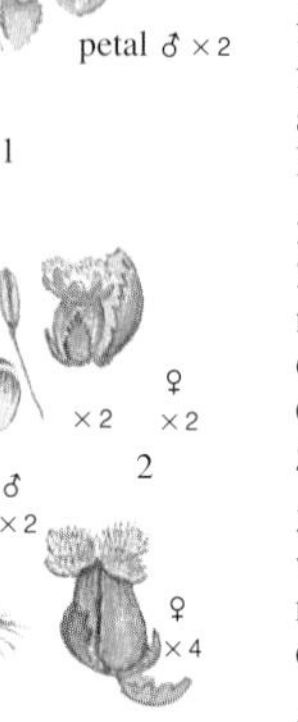

1* **Whorled Water-milfoil** *Myriophyllum verticillatum* L. (Water-milfoil Family – Haloragaceae – 150 species in 7 genera). An aquatic perennial with long slender, branching stems up to 3m long. Leaves in whorls of 5 usually, pinnately-lobed, with slender unbranched segments. Bracts pinnate, longer than the flowers. Flower-spike 7-25cm long; flowers tiny, greenish-yellow or reddish, the male with 4 petals, but the female petalless. Fruit rounded, 3mm, smooth. Freshwater habitats, especially in base-rich waters, at low altitudes. July-August. Throughout except the far north; often local. **B**: Uncommon; scattered localities in England, Wales and Ireland.

2* **Spiked Water-milfoil** *Myriophyllum spicatum* L. An aquatic perennial with long slender stems to 2.5m long, without detachable vegetative buds. Leaves in whorls of 4, pinnately-lobed, with slender segments. Bracts mostly unlobed and shorter than the flowers. Flower-spike 5-15cm long; flowers tiny, reddish, both the male and female with petals. Fruit rounded, finely warted. Freshwater habitats, generally in clear, calcium-rich waters, mostly at low altitudes, but to 750m. June-July. Throughout, except Spitsbergen; locally common. **P**:Wind. **B**: Throughout.

3 **Alternate-flowered Water-milfoil** *Myriophyllum alterniflorum* DC. Like *M. spicatum*, but leaves with fewer segments. Flower-spike much shorter, up to 3cm, with a drooping tip; flowers yellowish with red streaks, solitary or in groups of 2-4. Fruit more elongated, finely warted. Similar habitats, but generally over base-poor peaty soils, to 720m. Throughout, except N Scandinavia and Spitsbergen. **B**: Throughout.

4* **Mare's-tail** *Hippuris vulgaris* L. (Mare's-tail Family – Hippuridaceae). An aquatic, hairless perennial herb with a far-creeping rhizome and erect shoots coming above the water surface. Leaves in close whorls of 6-12, linear-lanceolate, pointed, untoothed. Flowers tiny, greenish with reddish anthers, borne in the axils of the upper leaves. Achene 2-3mm, smooth. Fresh-water habitats, especially in base-rich waters, to 600m. June-July. Throughout, except Spitsbergen. Local. **B**: Throughout.

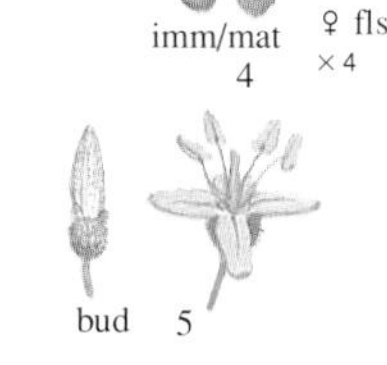


## DOGWOOD FAMILY Cornaceae

Trees or shrubs, occasionally herbs, with simple leaves; no stipules. Flowers 4-or occasionally 5-parted; bracts conspicuous, sepals small or absent; petals conspicuous, though small. Stamens 4, alternating with the petals. Fruit a berry-like drupe. 100 species in 15 genera.

5* **Dogwood** *Cornus sanguinea* L. [*Thelycrania sanguinea*]. Deciduous shrub to 4m, with dark red twigs. Leaves opposite, elliptical to oval, pointed, untoothed, hairy, with 3-4 pairs of main veins. Flowers dull white, 8-10mm, in umbel-like clusters. Fruit almost globose, 5-8mm, black when ripe. Woods, scrub and hedgerows, on calcareous soils, sometimes locally dominant, to 1550m. June-July. Throughout, except the extreme north; naturalized in Finland. The leaves often turn a rich purplish-red in the autumn and the bare red shoots are very conspicuous in the winter. **B**: North to Cumbria and Durham, but more or less absent in the SW and much of Wales; introduced further N; local in Ireland.

6* *Cornus sericea* L. [*C. stolonifera*, *Thelycrania stolonifera*]. Like *C. sanguinea*, but with blood-red twigs, suckering freely. Leaves more pointed. Flowers yellowish-white, 6-8mm. Fruit elliptical, in outline, 7-8mm, white when ripe. **I**: North America. Cultivated as an ornamental shrub and locally naturalized in Britain and Finland. June-July. **B**: Locally naturalized.

7* **Cornelian Cherry** *Cornus mas* L. Deciduous shrub or small tree to 8m, with greyish-yellow twigs. Leaves opposite, oval to elliptical, untoothed, dull green beneath; main veins 3-5 pairs. Flowers yellow, small, 4-5mm, borne in small clusters before the leaves emerge. Fruit oval in outline, 12-15mm, shiny-scarlet when ripe. Open woodland, scrub and occasionally along hedgerows. February-March. Belgium, France and S Germany; naturalized in Britain. The fruits are edible, being sometimes sold in markets in E Europe and W Asia. **B**: Rarely naturalized.

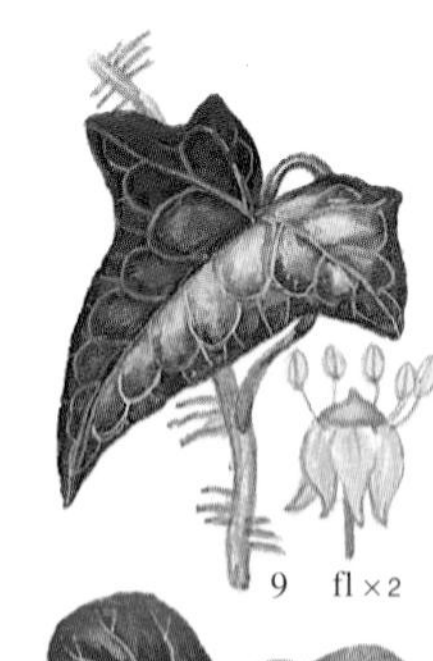


8* **Dwarf Cornel** *Chamaepericlymenum suecicum* (L.) Aschers. & Graebn. [*Cornus suecica*]. Low to short rhizomatous perennial with erect flowering shoots, generally unbranched. Leaves opposite, elliptical or rounded, untoothed, unstalked, sometimes hairy above. Flowers dark purple, 2-3mm, in small tight heads surrounded by 4 large, oval, white bracts. Fruit egg-shaped, 5mm, red when ripe. Mountain moors, on acid soils, to 1300m. July-September. N Europe. Plants are often shy flowering. Widely cultivated in gardens. **B**: Very local in N England north of Lancashire and Yorkshire, but more frequent in the Scottish Islands, especially in the east.

9* **Ivy** *Hedera helix* L. (Ivy Family – Araliaceae – 700 species in 60 genera). An evergreen woody climber to 30m, though often less, sometimes scrambling over the ground. Leaves deep shiny green and leathery, often with paler veins; those of immature plants often 3-5-lobed, those of mature flowering branches heart-shaped to elliptical – all untoothed. Flowers yellowish-green, with yellow anthers, 7-9mm, borne in small rather dense umbels, petals eventually reflexed. Fruit globose, 6-8mm, green at first, then brown, but finally black. Woodland, hedgerows, walls and old buildings, often in dense shade, to 1000m. September-November. Throughout, except the far north and Finland. **B**: Common throughout.

10 **Irish Ivy** *Hedera hibernica* hort. ex (Kirchner) Bean. non-flowering shoots with large deep green, heart-shaped leaves with 5-7 lobes. SW Ireland; occasionally naturalized elsewhere. Possibly of hybrid origin.

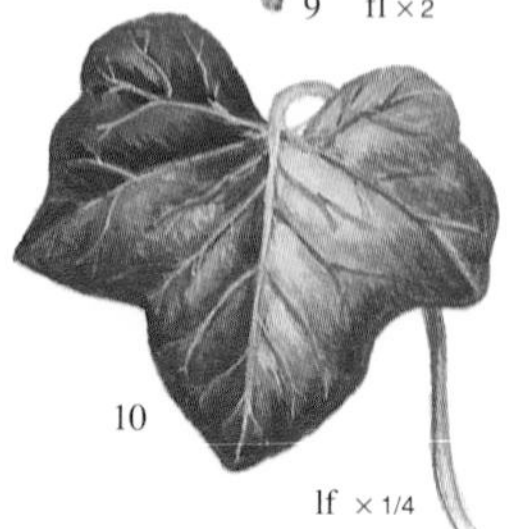

Whorled
Water-milfoil
Spiked
Water-milfoil
Mare's Tail
Dogwood
*Cornus sericea*
Cornelian Cherry
Dwarf Cornel
Ivy

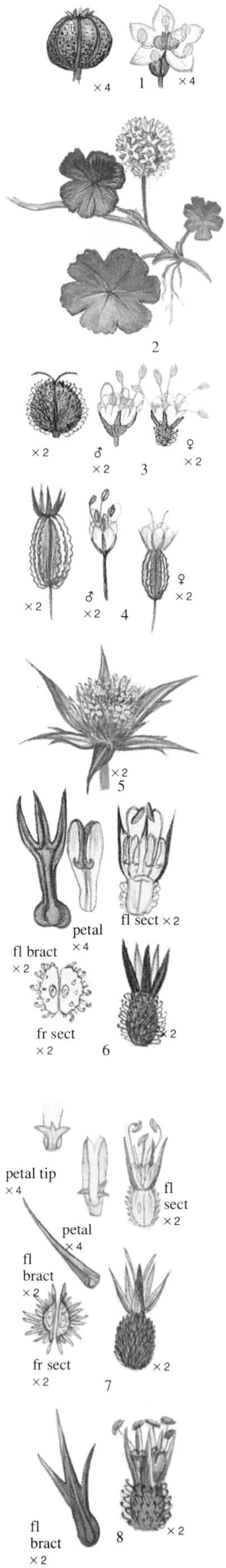

# CARROT FAMILY Umbelliferae

Annual, biennial or perennial herbs, occasionally shrubby, with alternate leaves. Leaves large generally pinnately-divided, but occasionally simple, often with inflated and sheathing bases. Inflorescence an umbel, usually compound, the primary umbel with or without bracts, the main branches (rays) supporting the secondary umbels. Flowers small, 5-parted, usually hermaphrodite. Calyx with 5 small teeth, or absent. Petals separate, usually notched and with an incurved point. Carpels 2, joined along a central axis, each terminated by a short style. Fruit dry, 2-parted, flattened or rounded in section, usually ribbed and with 4 resin canals (vittae) between the primary ridges. 3000 species in about 300 genera in most corners of the world, but particularly in temperate regions. Often difficult to identify, particularly in early flower. Most can be accurately named in fruit, whose characters are visible to the naked eye or with the help of a x10 hand-lens. Plants can normally be found with both flowers and fruit at the same time.

1* **Marsh Pennywort** *Hydrocotyle vulgaris* L. Low prostrate, somewhat hairy perennial with creeping stems rooting at the nodes. Leaves circular in outline, blunt-toothed, erect on long stalks. Flowers pinkish-green, tiny, borne in small whorls. sometimes one above another, hidden below the leaves. Fruit rounded, 2mm, ridged, on short stalks. Moist habitats, marshes, bogs, fens, grassy places, stream and pond margins, acid woodland flushes, generally at low altitudes. June-August. Europe, north to S Scandinavia. **P**: Usually self-pollinated. Often a difficult plant to detect among other vegetation. Widespread from W Europe east to C Asia and south to N Africa. **B**: Throughout.

2 *Hydrocotyle moschata* G. Forster. Like *H. vulgaris*, but leaves with a deep basal sinus and inflorescence with 10-20 flowers (not 2-5). **I**: New Zealand. Naturalized in SW Ireland – Valentia Is. *Hydrocotyle* is sometimes placed in a family of its own, the *Hydrocotylaceae.*

3* **Sanicle** *Sanicula europaea* L. Short to medium hairless perennial. Leaves shiny-green, palmately-lobed, with 3-5 wedge-shaped, toothed lobes, occasionally bilobed; teeth ending in a short bristle. Flowers pinkish or greenish-white, the outer male, the inner female, borne in tight uneven umbels; bracts small. linear. Fruit oval, 4-5mm, covered in hooked bristles. Deciduous woodland, especially of Ash, Beech or Oak, generally on calcareous or base-rich soils, to 1600m. May-August. Throughout, except the extreme north, the Faeroes and Iceland. A very shade tolerant plant, often forming colonies. **P**: Small insects or self-pollinated. The ripe fruits are distributed on the coats of animals or clothing. **B**: Throughout, but often local; rather rare in N Scotland.

4* **Great Masterwort** *Astrantia major* L. Robust, medium to tall tufted perennial; stems generally unbranched, except at the top. Leaves palmately-lobed, with 3-5 lanceolate to elliptical, toothed lobes, the basal ones long-stalked, the upper short-stalked or unstalked. Flowers pinkish, whitish or greenish, in dense pin-cushion-like umbels surrounded by a ruff of lanceolate, pointed bracts, scented; bracts whitish, often with a pink or purplish apex. Fruit short-cylindrical, 6-8mm, ridged. Woods and woodland clearings, meadows, generally in mountain districts, to 2000m. May-September. C and E France, C and S Germany; naturalized in Britain, Denmark and Finland. Widely cultivated in gardens. **P**: Beetles, sometimes also flies. Native to the mountains of C and S Europe and W Asia. **B**: Naturalized in a few localities in the north and west.

**Eryngos** *Eryngium*. Hairless, often prickly, biennial or perennial herbs with simple or pinnately-divided leaves. Inflorescence often branched. Flowers in unstalked dense, rounded or egg-shaped cone-like heads, surrounded by conspicuous spiny, generally coloured bracts. Fruit rounded to egg-shaped, slightly ridged and often scaly. Over 200 species from temperate and subtropical regions, particularly of South America. **P**: Bees, beetles, sometimes butterflies, for the flowers' nectar.

5* *Eryngium viviparum* Gay. Low biennial with spreading stems. Basal leaves 1-3, persistent, linear-oblong, toothed, the blade running down the stalk; upper leaves much smaller. Flowers greenish, in a spreading inflorescence with up to 50 small heads, each 4-6mm across and consisting of only 5-8 flowers; bracts lanceolate, with a few spines. Habitats liable to winter flooding. June-August. W and NW France. Local, with a narrow distribution from NW France to NW Spain and Portugal.

6* **Sea Holly** *Eryngium maritimum* L. Short to medium rather stiff tufted perennial, bluish-green. Basal leaves leathery, rounded or heart-shaped in outline, but 3-5-lobed with an undulate, densely spiny margin; the veins and margins whitish; stalk unwinged. Flowers powder-blue, in dense rounded heads 15-30mm, forming a spreading inflorescence; bracts oval, coarsely spiny. Fruit egg-shaped, densely scaly. Maritime sands and fine shingle, occasionally among rocks, sometimes forming extensive colonies. June-September. Coasts of Europe north to S Scandinavia. Local from NW Europe to the Mediterranean and the Black Sea. Sometimes cultivated in gardens. **B**: Coasts north to Flamborough Head and on the W coast of Ireland.

7* **Field Eryngo** *Eryngium campestre* L. An erect, short to medium perennial. Basal leaves leathery and usually persistent, oval in outline, 3-lobed, the terminal lobe pinnately-divided and with opposite lobes, spiny-toothed; stem leaves unstalked, clasping the stem. Flowers pale greenish-white, in dense rounded heads 10-15mm, the whole forming a flat-topped inflorescence; bracts linear-lanceolate, with or without one pair of spines. Fruit with dense, overlapping scales. Dry grassy habitats, especially rough grassland near the coast. July-August. S Britain, Belgium, Holland, France and Germany; naturalized in Denmark. Local from W Europe to C Asia and N Africa. **P**: Bees. **B**: Very rare; confined to a few localities in Hampshire, Kent and Guernsey.

8* *Eryngium planum* L. Medium to tall, stiffly branched, hairless perennial. Basal leaves oblong-heart-shaped, toothed, occasionally 3-lobed; stem leaves short-stalked or clasping. Flowerheads bluish, numerous, in branched oblong clusters, each egg-shaped, 10-20mm, with up to 8 narrow, spiny bracts. Dry sunny habitats. C & S Germany. Mainly in C and SE Europe.

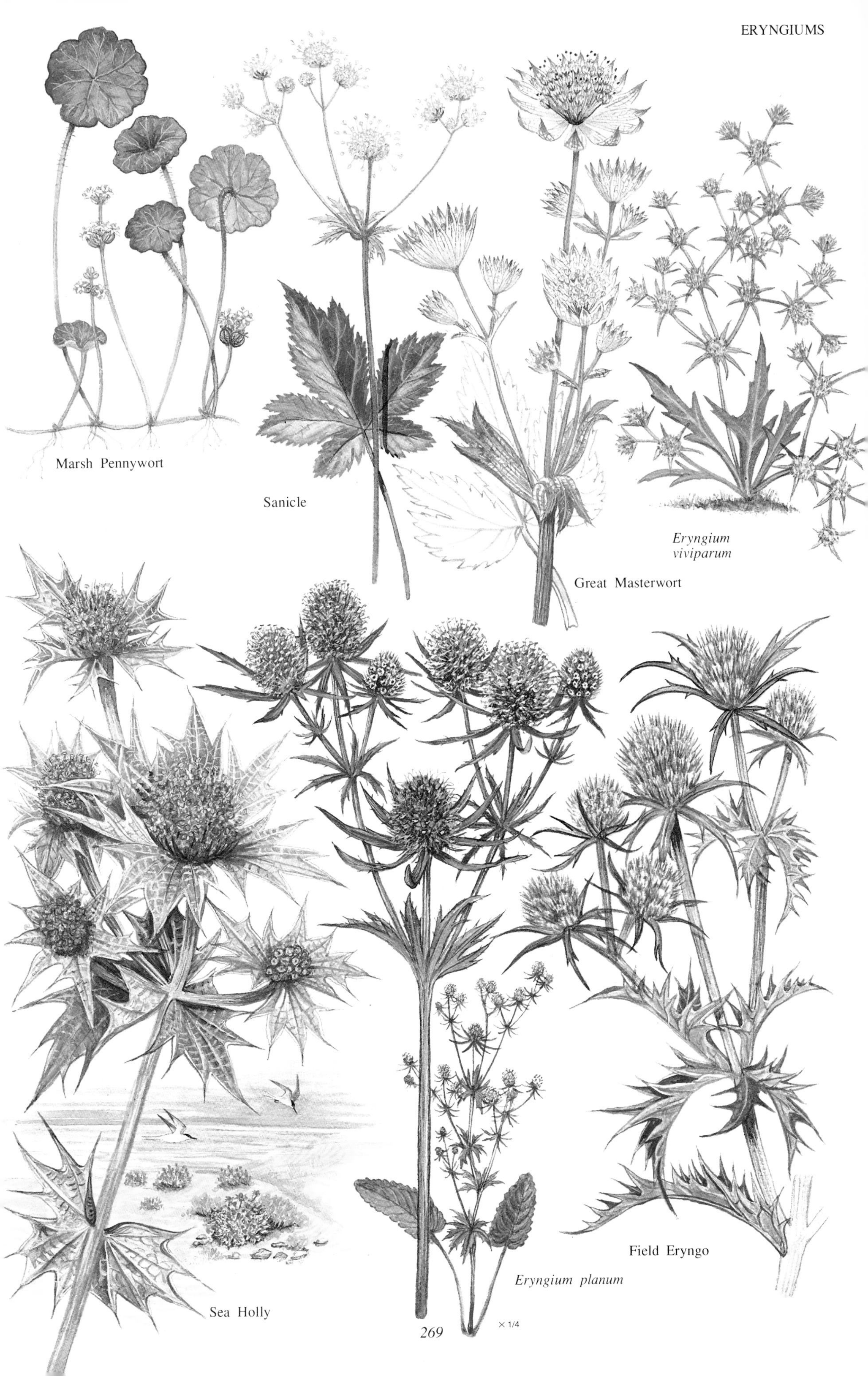
Marsh Pennywort
Sanicle
Great Masterwort
*Eryngium viviparum*
Sea Holly
*Eryngium planum*
Field Eryngo
× 1/4

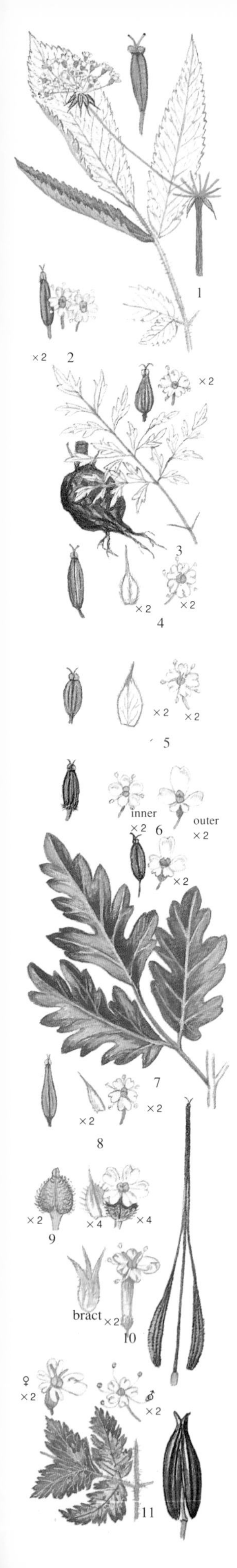


**Chervils** *Chaerophyllum.* Biennial or perennial herbs with 1-3-pinnate leaves. Flowers without sepals, the petals white or pink, in compound umbels; bracts few or absent. Fruits narrow oblong, narrowed to the beaked apex, with rounded ridges. 36 species.

1 *Chaerophyllum aromaticum* L. Tall robust, slightly hairy perennial, to 2m, with a far-creeping rhizome. Leaves 2-trifoliate, occasionally pinnate, the lobes undivided, lanceolate to oval, sharply toothed, the uppermost on the plant pointed. Flowers white. Fruit egg-shaped or oblong, often rather narrow, 8-15mm, with rather wavy styles. Scrub, hedgebanks and woodland margins. Late June-August. S Germany – rather rare.

2* **Hairy Chervil** *Chaerophyllum hirsutum* L. Robust, tall, softly hairy perennial; stems erect, green. Leaves deep green, 2-3-pinnate with relatively broad, little divided wedge-shaped segments which tend to overlap. Flowers white or occasionally pinkish, the petals hairy. Bracts usually absent, bracteoles unequal. Fruit narrow-oblong, 8-12mm, with more or less straight styles. Damp meadows, open woodland and other shaded habitats, to 2500m. July-August. C and E France, S Germany.

*Chaerophyllum villarsii* Koch. Like *C. hirsutum*, but hairs stiffer and leaf-segments narrower and not overlapping and more pointed. Fruit 8-20mm. E France and S Germany.

3 *Chaerophyllum bulbosum* L. Generally taller than *C. temulem* (below), the stems hairless above and bracteoles hairless. Grassy habitats. E France and C and S Germany, rare in Holland; naturalized in Belgium. The rootstock is short and tuberous, hence the common name. subsp. *prescottii* (DC.) Nyman [*C. prescottii*] is generally only 50cm tall (not 100-200cm) and with less finely divided leaves. Styles erect (not spreading) in fruit. Confined to C and S Sweden, Finland and the neighbouring part of the USSR.

4* *Chaerophyllum aureum* L. [*C. maculatum*]. Rather like *C. temulem*, but more robust and with yellowish-green, aromatic foliage, the leaflets with slightly hairy margins. Umbel with 12-18 rays (not 6-12). Fruit 8-12mm. Rough grassy habitats, meadows and banks. June-July. C and E France and S Germany; naturalized in N Britain. Stem solid and with some purple spots. **B**: Locally naturalized in S Scotland.

5* **Rough Chervil** *Chaerophyllum temulem* L. [*C. temulentum*]. Medium to tall, rather bristly biennial; stem erect, purple or purple-spotted. Leaves 2-3-pinnate, dark green, but eventually turning purple; leaflets oval, toothed. Flowers white, 2mm, in compound umbels, which are nodding in bud, the petals hairless; bracts usually absent, bracteoles hairy. Fruit oblong, tapered towards the apex, 4-7mm, often purple. Rough grassland, semi-shaded habitats, on well-drained soils, generally at low altitudes. May-July. Absent from the Faeroes, Iceland, Norway, Finland and Spitsbergen. **B**: Throughout England, Wales and E Scotland; very local in Ireland.

**Cow Parsleys** *Anthriscus.* Like *Chaerophyllum*, but stem hollow in the centre. Leaves 2-3-pinnate. Sepals tiny or absent; petals white. Fruit narrowly oblong, generally with a well developed beak, the ridges confined to the beak; vittae solitary, or absent. 12 species.

6* **Cow Parsley** *Anthriscus sylvestris* (L.) Hoffm. [*A. torquata, Chaerophyllum sylvestre, Cerefolium sylvestre*]. Medium to tall, rather robust, slightly hairy biennial or perennial, rarely an annual, to 1.5m. Leaves dull green, 3-pinnate. Flowers white, 3-4mm, the umbels with 4-15 rays, without lower bracts. Fruit 7-10mm, short beaked, bristly at the base, brown or black when ripe. Rough grassy habitats, generally at low altitudes. April-June. Throughout, except the far north. Common; one of the earliest umbels to come into flower. **B**: Common throughout, though scarcer in the north.

7 *Anthriscus nitida* (Wahlenb.) Garcke [*A. sylvestris* subsp. *alpestris*]. Similar to *A. sylvestris*, but the leaves dark green and rather glossy. Fruit 5-7mm, without a ring of bristles at the base. Rough grassy habitats and scrub. May-July. C and E France, C and S Germany.

8* **Garden Chervil** *Anthriscus cerefolium* (L.) Hoffm. [*A. longirostris, Cerefolium cerefolium*]. Short to medium, somewhat hairy annual; stem wiry, lined. Leaves 3-pinnate, with lobed segments, hairy beneath. Flowers white, 2mm, the umbels with 2-6 rays, borne opposite the leaves; flowerstalks shorter than the subtending bracteoles. Fruit almost linear, 7-10mm, with a slender long beak, hairless; styles erect. Grassy places, hedgebanks and waste ground, at low altitudes. May-June. **I**: E Europe to C Asia. Naturalized north to Denmark.

9* **Bur Chervil** *Anthriscus caucalis* Bieb. [*A. scandicina, A. vulgaris, Chaerophyllum anthriscus, Cerefolium anthriscus*]. Medium to tall annual; stems often purplish towards the base, hairless, branched and spreading. Leaves 2-3-pinnate, the segments toothed to finely divided, slightly hairy beneath. Flowers white, 2mm, the umbels with 2-6 rays, opposite the leaves; bracteoles finely pointed. Fruit egg-shaped, 3mm, covered with hooked bristles, short beaked. Dry habitats, rough grassland, waste ground, hedgebanks, on well-drained, often sandy soils. May-June. North to S Sweden. **B**: Scattered throughout, except for the north-west. var. *neglecta* (Boiss. & Reuter) P. Silva & Franco has hairless, non-bristly fruits. E. France and S Germany.

10* **Shepherd's Needle** *Scandix pecten-veneris* L. Short, almost hairless annual, rather variable. Leaves 2-3-pinnate. with linear lobes, toothed. Flowers white, 3-5mm, in simple umbels opposite the leaves, each with 1-3 rays; petals oblong, those of the outer flowers generally larger. Fruit subcylindrical, 20-80mm, ridged, with a very long beak; styles short and erect. A weed of arable land and waste places. April-August. North to S Sweden. Declining. **B**: Throughout, but local and rare in Wales and Scotland. *S. p.* subsp. *brachycarpa* (Guss.) Thell. [*S. brachycarpa*] has smaller fruits, 15mm, short-beaked. **I**: SE Europe. Naturalized in France and Germany.

11* **Sweet Cicely** *Myrrhis odorata* (L.) Scop. Tall, stout, hairy aromatic perennial, to 2m. Leaves 2-3-pinnate, with lanceolate, deeply toothed lobes; basal sheaths conspicuous. Flowers white, 2-4mm, in large umbels with 4-20 rays; petals unequal, sepals minute; bracts absent. Fruit linear-oblong, 15-25mm, deeply ridged, beaked, deep shiny-brown when ripe. Grassy habitats, to 2000m. May-July. France and Germany; naturalized elsewhere. Primarily a mountain plant in its native habitats. The whole plant is very aromatic and the roots and young fruits in particular are used as a low calorie sweetener. **B**: N Wales, N England, S Scotland, rare or casual elsewhere.

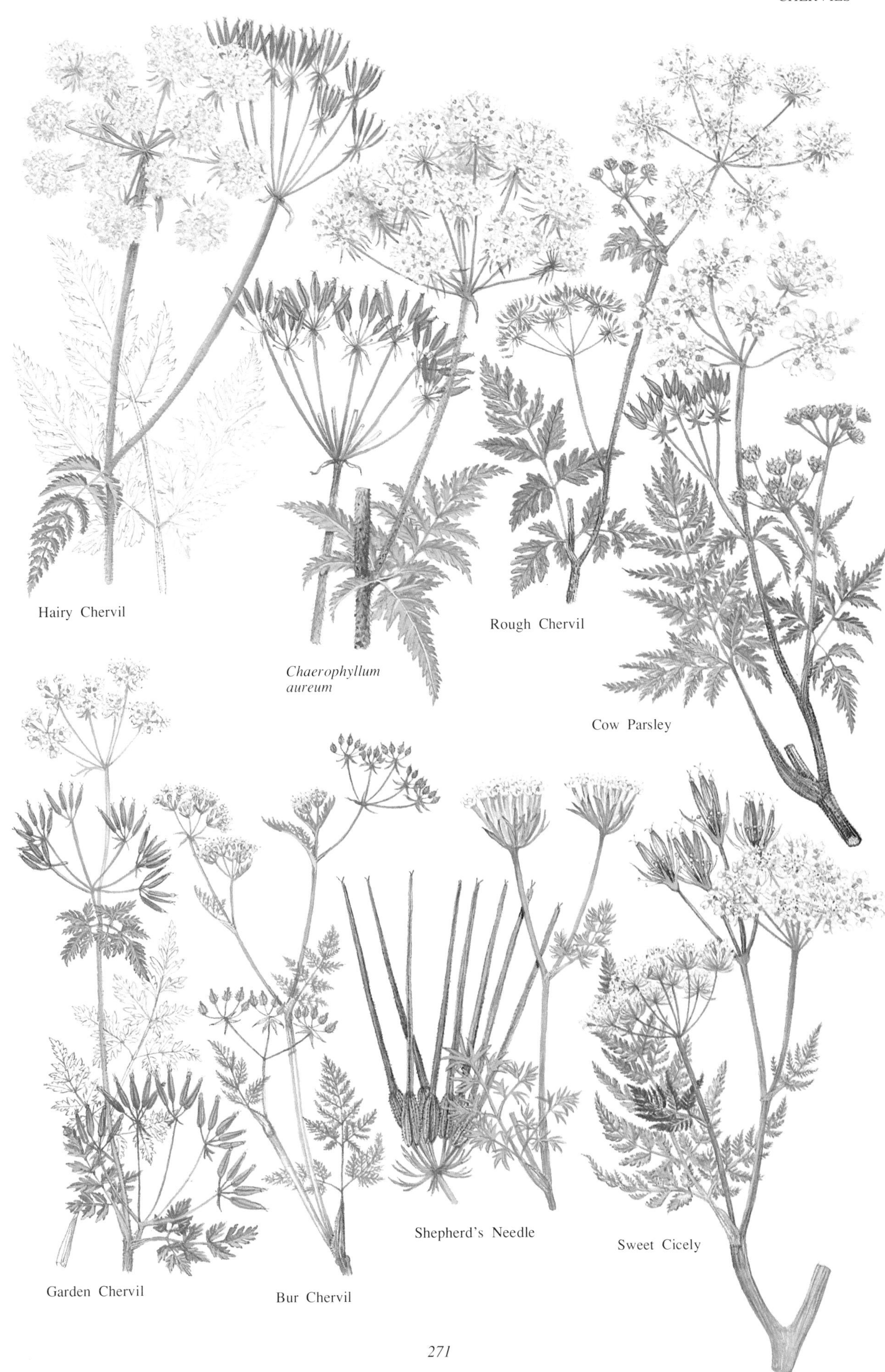
Hairy Chervil
Chaerophyllum aureum
Rough Chervil
Cow Parsley
Garden Chervil
Bur Chervil
Shepherd's Needle
Sweet Cicely

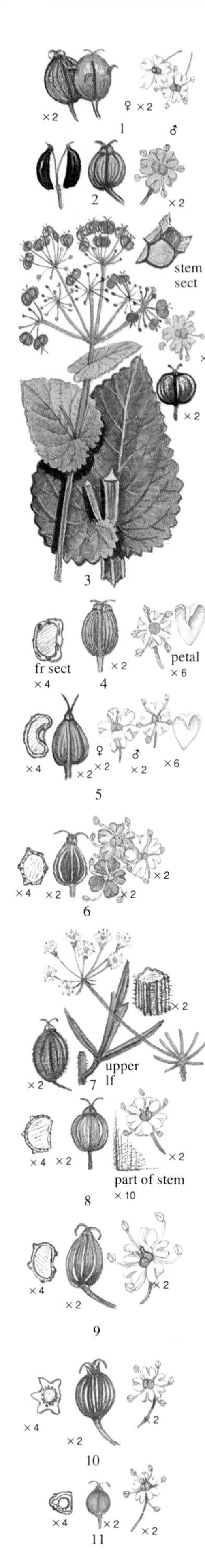


1* **Bifora** *Bifora radians* Bieb. Medium, non foetid, hairless annual. Leaves 1-2-pinnate with linear lobes. Flowers white, often tinged pink, the outer petals of the marginal flowers much larger than the others, sepals small or absent; umbels with 3-8 rays. Fruit globose, rough, not beaked, clearly 2-lobed. Waste and grassy habitats. June-August. **I**: E Europe. Naturalized in France and Germany.

2* **Alexanders** *Smyrnium olusatrum* L. Tall, pungent, hairless biennial; stems stout, becoming hollow when old, the upper branches usually opposite. Leaves dark green and shiny, the basal ones triangular in outline, 3-ternate, the segments toothed or lobed; stem leaves smaller and less divided, the uppermost chrome yellow, not clasping the stem. Flowers yellow, 3mm, in dense umbels, without sepals; umbels with 7-15 rays. Fruit egg-shaped, 7-8mm, with slender ridges, black when ripe. Hedgebanks, woodland margins, waste ground, roadsides and cliffs, often near the sea. April-June. NW France southwards; extensively naturalized in Britain and Holland. **B**: Mostly coastal from the Isle of Wight, N Wales and Norfolk south as well as the E coast of Ireland.

3 **Perfoliate Alexanders** *Smyrnium perfoliatum* L. Like *S. olusatrum*, but stems angled and narrowly winged, hairy at the nodes; upper branches alternate; upper leaves heart-shaped, clasping the stem, yellowish-green, toothed. Fruits 3-3.5mm, brownish-black when ripe. Woodland and rocky habitats, at low altitudes. May-July. **I**: S Europe to W Asia. Naturalized in Britain, France, Denmark and Germany. **B**: Established in a few places.

4* **Great Pignut** *Bunium bulbocastanum* L. Medium to tall, slightly hairy perennial; stems erect, ridged, solid. Leaves 2-3-pinnate, with linear-lanceolate lobes, the lower soon withering. Flowers white, 2mm, the umbels 3-8cm, with 10-20 rays; umbels with 5-10 lanceolate bracts, ridged. Fruit oblong, 3-5mm, with slender ridges; styles down-turned. Dry rough grassland and arable land, generally at low altitudes and on calcareous soils. May-July. S Britain, Belgium, Holland, France and C & S Germany; naturalized in Denmark. Root tuber black, up to 2.5cm across, edible. **B**: Very local; confined to Bedfordshire, Buckinghamshire, Cambridgeshire and Hertfordshire.

5* **Pignut** *Conopodium majus* (Gouan) Loret [*C. denudatum*]. Short to medium, almost hairless perennial; stem erect, slightly ridged, hollow after flowering. Leaves 2-3-pinnate, the lower with elliptical lobes, the upper with linear lobes and prominent sheathing bases. Flowers white, often brown-veined on the back, 1-3mm, usually unisexual, the umbels 3-7cm, with 6-12 rays; bracts absent or only 1-2, membranous. Fruit oblong, 3-4mm, with slender obscure ridges and erect styles. Open woodland, rough grassland, scrub, commons, grassy heaths, generally on dry slightly acid soils, at low altitudes. May-July. Britain, France and Norway; naturalized in the Faeroes. Root tuber dark brown, edible with a pleasant, mildly nutty flavour when cooked. The basal leaves have usually withered by flowering time. **B**: Throughout, but scarce in E Anglia, E Ireland and the Outer Hebrides.

**Burnet Saxifrages** *Pimpinella*. Perennial herbs, rarely annual. Basal leaves entire or 3-lobed; other leaves pinnate with narrow lobes. Flowers white or yellow, rarely pink or purplish, with minute sepals; petals not or slightly lobed. Fruit oblong to subglobose, with slender ridges and usually 3 vittae. 90 species, primarily Mediterranean.

6* **Great Burnet Saxifrage** *Pimpinella major* (L.) Hudson [*P. magna*]. Very variable, medium to tall, usually hairless perennial, without non-flowering leaf-rosettes; stem hollow, deeply ridged. Lower leaves pinnate, with 3-9 oval, toothed or lobed leaflets; stem leaves smaller and with prominent inflated bases. Flowers white to deep pink, 3mm, the umbels flat-topped, 3-6cm, with 10-15 rays. Fruits oblong, 2.5-3.5mm, with prominent whitish ridges, hairless. Rough grassland, woodland margins, road verges, hedgebanks, on dry calcareous soils, to 2300m. June-August. Most of Europe, except the far north. **B**: North to C Scotland; Ireland, confined mostly to the SW.

7 **Anise** *Pimpinella anisum* L. [*Anisum vulgare*]. Like *P. major*, but a strongly aromatic hairy annual, the basal leaves kidney-shaped, the upper 2-3-pinnate. Flowers white. Fruit 3-5mm, bristly. Waste and bare places, cultivated ground. June-August. **I**: Asia. Naturalized in France, Germany and Norway. Widely cultivated in S and E Europe, for medicine and flavouring.

8* **Burnet Saxifrage** *Pimpinella saxifraga* L. Like *P. major*, but a medium, short-hairy perennial; stem scarcely ridged, tough. Flowers 2mm. Fruit oval, 2-2.5mm, with rather inconspicuous ridges. Rough or grazed grassland, rocky places, on calcareous and base-rich soils, to 2300m. June-September. Similar distribution. Very variable: can look very like *P. major* and only really be separated in mature fruit. **B**: Throughout, except the Channel Islands; rare in N Scotland and N Ireland.

9* **Ground Elder** *Aegopodium podagraria* L. Vigorous, medium to tall, hairless rhizomatous perenial, often patch-forming. Basal leaves diamond-shaped in outline, 1-2-ternate, with lanceolate or oval, toothed segments; upper leaves smaller, with short inflated stalks. Flowers white, occasionally pink, 2-3mm, in umbel 2-6cm, with 10-20 rays; bracts usually absent. Fruit egg-shaped, with slender ridges; styles reflexed. Shady places, cultivation, generally at low altitudes. May-August. Throughout, except the far north, but naturalized in Britain, the Faeroes and Iceland. An often invasive and tenacious weed. **B**: Throughout, but local in W Scotland.

10* **Greater Water-parsnip** *Sium latifolium* L. Semi-aquatic, tall, hairless perennial, to 1.5m; stems strongly ridged, hollow. Submerged leaves 2-3-pinnate, with lanceolate, toothed leaflets, uneven at the base; aerial leaves usually pinnate, often with only 5 pairs of leaflets. Flowers white, 4mm, in umbels 6-10cm across, with 20-30 rays; bracts usually 2-6. Fruit oval, 3-4mm, with thick ridges and short styles. Shallow fresh water. June-September. Throughout, except the far north; probably naturalized in Norway. The submerged leaves are only present during the spring. Declining, from land drainage. **B**: Local in lowlands north to S Scotland; rare in Ireland.

11* **Lesser Water-parsnip** *Berula erecta* (Hudson) Coville [*Sium angustifolium*, *S. erectum*]. Hairless, medium to tall, stoloniferous semi-aquatic perennial to 1m, but often rather sprawling; stem hollow, lined. Submerged leaves 3-4-pinnate, with linear lobes; aerial leaves pinnate, with 7-14 pairs of lanceolate, toothed leaflets, the uppermost leaves small. Flowers white, 2mm, in leaf-opposed umbels, 3-6cm across, each usually with 10-20 rays; bracts leaf-like, often pinnate. Fruit globose, 1.5-2mm, faintly ridged, with long styles. Growing in shallow fresh water, occasionally along lake margins, usually on calcareous soils, at low altitudes. July-September. Throughout, except the far north. **B**: Local in lowland England and C Ireland, scarce elsewhere.

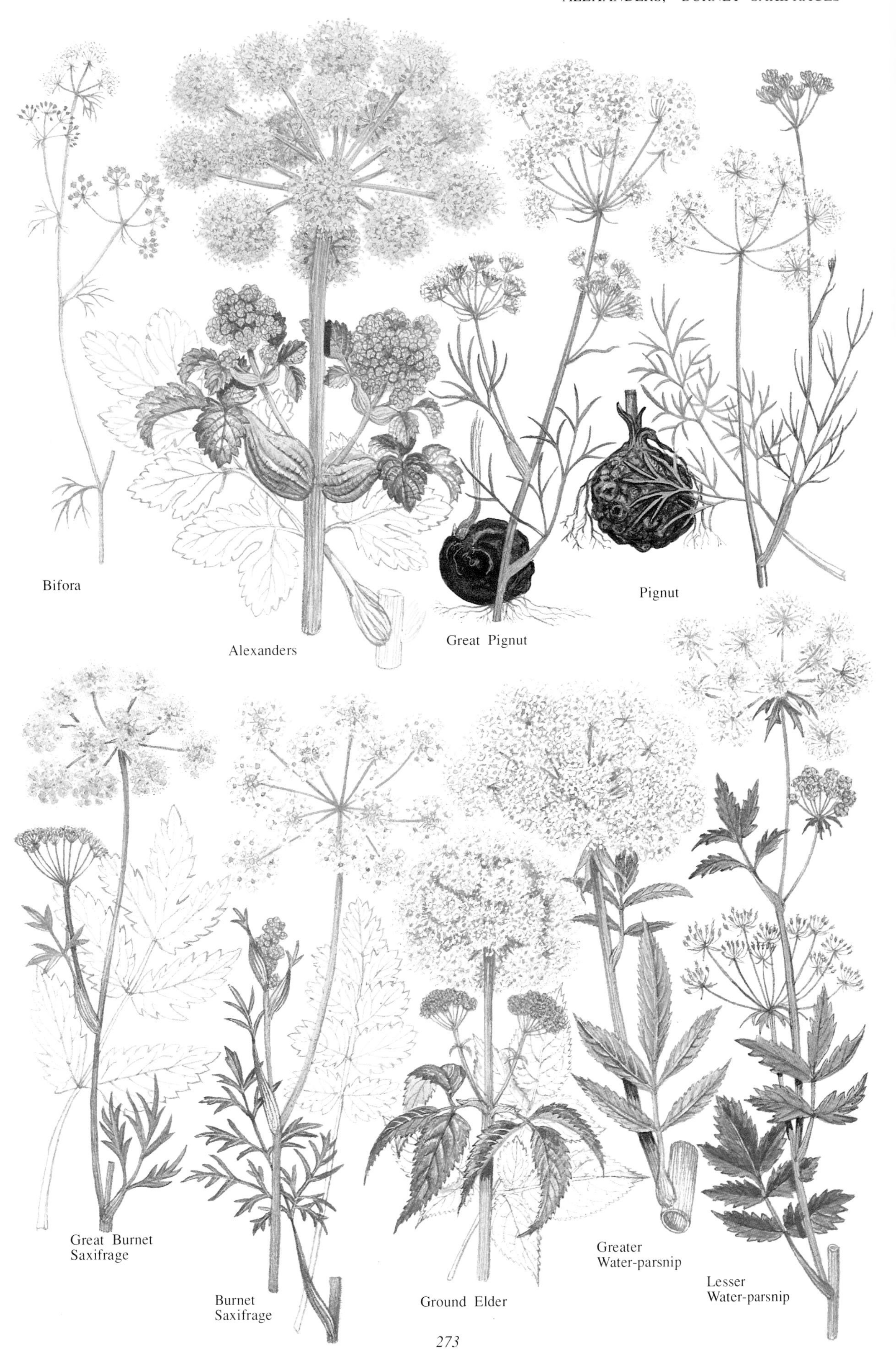
Bifora
Alexanders
Great Pignut
Pignut
Great Burnet
Saxifrage
Burnet
Saxifrage
Ground Elder
Greater
Water-parsnip
Lesser
Water-parsnip

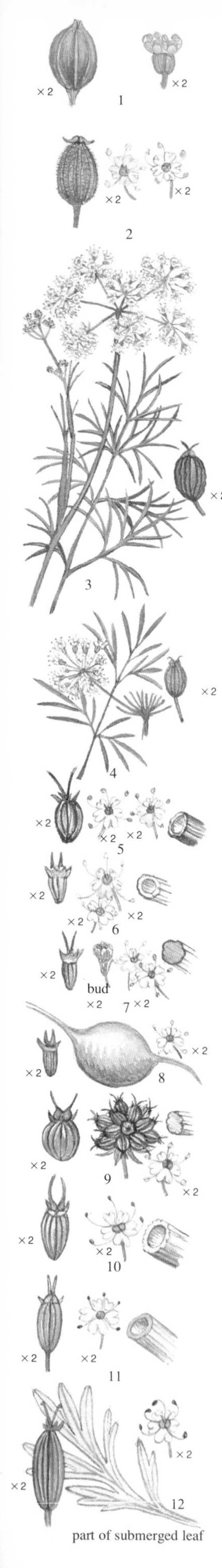


1* **Rock Samphire** *Crithmum maritimum* L. Short to medium, greyish, hairless, branched perennial, often rather woody at the base. Leaves diamond-shaped in outline, 1-2-pinnate, with slender fleshy, untoothed segments, the base membranous and sheathing the stem. Flowers yellowish-green, 2mm, in umbels 3-6cm across, with 8-36 rays; bracts present, lanceolate. Fruit oblong, 5-6mm, ridged, yellowish to purplish. Maritime habitats, rocky places including cliffs, sand and shingle. June-October. Atlantic coasts of Europe north to France and Holland. Edible. **B**: Local along the coasts of Suffolk, the south, the west and Ireland.

**Moon Carrots** *Seseli*. Biennial or perennial herbs, generally with a fibrous stock. Leaves pinnately or ternately divided. Flowers white or pink, with or without sepals. Fruit oblong to elliptical, scarcely compressed, with prominent ridges and 1-3 vittae. 60 species.

2* **Moon Carrot** *Seseli libanotis* (L.) Koch [*Libanotis montana*]. Variable medium to tall, hairy or hairless biennial or perennial to 1.2m; stems ridged. Leaves 1-3-pinnate with linear to oblong, often sickle-shaped leaflets, pointed; bracts numerous, linear. Flowers white or pink, 1-1.5mm, the petals hairy on the back, borne in umbels with 20-60 rays; sepals linear; bracts many, linear or lobed. Fruit elliptical, 2.5-4mm, usually rounded. Rough grassland and open scrub on dry calcareous soils. July-September. Throughout, except Ireland, Holland, and the far north. **B**: Very rare, restricted to a few localities in Beds., Cambs., Heref. and E Sussex.

3 *Seseli montanum* L. [*S. glaucum*]. Like *S. libanotis*, but a short to medium hairless perennial. Leaves 2-3-pinnate with linear or linear-lanceolate segments. Umbels without bracts and petals hairless. Fruit oblong, 2.5-4.5mm, with sharp (not blunt) ridges. Grassy and rocky habitats. June-August. N and C France; extinct in Germany. Native mainly of S Europe.

4 *Seseli annuum* L. Like *S. libanotis*, but umbels with 12-40 rays and fruits smaller, 1.5-3mm, hairless. Dry grassy habitats. June-July. N & C France and C and S Germany.

**Water-dropworts** *Oenanthe*. Perennial herbs with pinnate or pinnately-divided leaves. Flowers white or pale pink, the outer flowers with some larger petals; sepals persistent. Fruit egg-shaped to cylindrical, with thickened lateral ridges and a solitary vittae. Marshy or aquatic habitats. 35 species confined to temperate regions of the Old World.

5* **Tubular Water-dropwort** *Oenanthe fistulosa* L. Medium to tall greyish, hairless, stoloniferous perennial; stems little-branched, ridged, hollow, often constricted at the nodes. Basal leaves 1-2-pinnate, with oval, lobed leaflets; upper leaves pinnate, with narrower leaflets, untoothed, with sheathing long, hollow stalks. Flowers white or pink, 3mm, in small umbels with only 2-4 thickened rays, generally without bracts. Fruit cylindrical, 3-4mm, unstalked; styles spreading, as long as the fruit. Shallow water and marshy habitats on fertile soils, generally at low altitudes. June-September. Europe north to S Sweden. **B**: Throughout, though scarcer in the SW and Ireland.

6* **Narrow-leaved Water-dropwort** *Oenanthe silaifolia* Bieb. [*O. media*]. Tall erect, sparsely branched perennial; stems hollow, grooved. Basal leaves 2-4-pinnate, with linear-lanceolate leaflets, soon withering; stem leaves 1-2-pinnate. Flowers white, occasionally pink, 3mm, the umbels terminal with 4-10 rays, which become markedly thickened in fruit. Fruit subcylindrical, 2.5-4mm; styles spreading, almost as long as the fruit; fruit-stalks constricted at the top. Wet places, old fertile water-meadows. June-July. Britain, Belgium, France and Germany. **B**: Rare and decreasing; mainly in C and SE England.

7* **Corky-fruited Water-dropwort** *Oenanthe pimpinelloides* L. Like *O. silaifolia*, but stems solid, the basal leaves with wedge-shaped leaflets. Fruits with thickened ridges, short-styled, the stalks thickened (not constricted) at the top. Similar habitats and flowering time. S Britain, S Ireland, Belgium, Holland and France. **B**: Scattered localities in S England from Gloucestershire southwards; Ireland, restricted to Cork.

8 *Oenanthe peucedanifolia* Pollich. Like *O. silaifolia*, but root-tubers egg-shaped (not spindle-shaped) and rays and fruit-stalks not thickened in fruit. Fruit with short styles. Wet grassland, especially near rivers, at low altitudes. June-July. Belgium, Holland, France and Germany.

9* **Parsley Water-dropwort** *Oenanthe lachenalii* C.C. Gmelin [incl. *O. jordanii*, *O. marginata*]. Medium to tall perennial with cylindrical root-tubers; stems erect, solid, sometimes with a slight hollow when older. Basal leaves 2-pinnate, with linear or spatular-shaped leaflets; stem leaves 1-2-pinnate, with linear or linear-lanceolate leaflets. Flowers white, 2-3mm, in terminal umbels with 5-15 rays, the rays not thickening in fruit nor constricted at the top. Fruit egg-shaped, 2-3mm, the styles shorter than the fruit. Wet grassland, often on brackish soils near the coast. June-September. Europe north to S Sweden. **B**: Scattered localities, but mainly coastal in Ireland and SW Britain.

10* **Hemlock Water-dropwort** *Oenanthe crocata* L. Tall hairless, parsley-scented perennial, to 1.5m; stem stout, hollow and grooved. Basal leaves 3-4-pinnate with broad oval or rounded leaflets, lobed and toothed; stem leaves 2-3-pinnate, with narrower leaflets than the basal. Flowers white, 2mm, in umbels with 10-40 rays, not thickened in fruit; bracts linear. Fruit cylindrical, 4-6mm, the styles half the length of the fruit. Wet habitats, grassy places, woodland edges, freshwater margins, ditches, on acid soils at low altitudes. June-August. Britain, Belgium and France. Extremely poisonous in all parts. **B**: Widespread in S and W Britain, N and S Ireland, rather scarce elsewhere.

11* **Fine-leaved Water-dropwort** *Oenanthe aquatica* (L.) Poiret. Medium to tall, pale green, hairless biennial or perennial, to 1.5m, stoloniferous usually; stem stout, hollow and grooved. Basal leaves submerged, 3-4-pinnate, with linear or thread-like divisions; aerial leaves 3-pinnate, with oval, deeply-lobed leaflets, pointed. Flowers white, 2mm, the umbels terminal or opposite the leaves and with 5-15 rays, not thickened. Fruit oblong, 3.5-4.5mm, often curved, with very short styles. Shallow fresh water, on fertile soils, at low altitudes. June-September. Throughout, except the extreme north and Iceland. **B**: Throughout, but absent from most of Wales and Scotland and rather local in Ireland.

12 **River Water-dropwort** *Oenanthe fluviatilis* (Bab.) Coleman. Similar, but the submerged leaves 2-pinnate and with broader wedge-shaped segments, cut at the ends. Umbels always opposite the leaves. Fruit elliptical, 5-6.5mm. C & S Britain, Ireland, Belgium, France, Denmark and Germany. **B**: Scattered localities in England from the Wash southwards; rare elsewhere including Ireland.

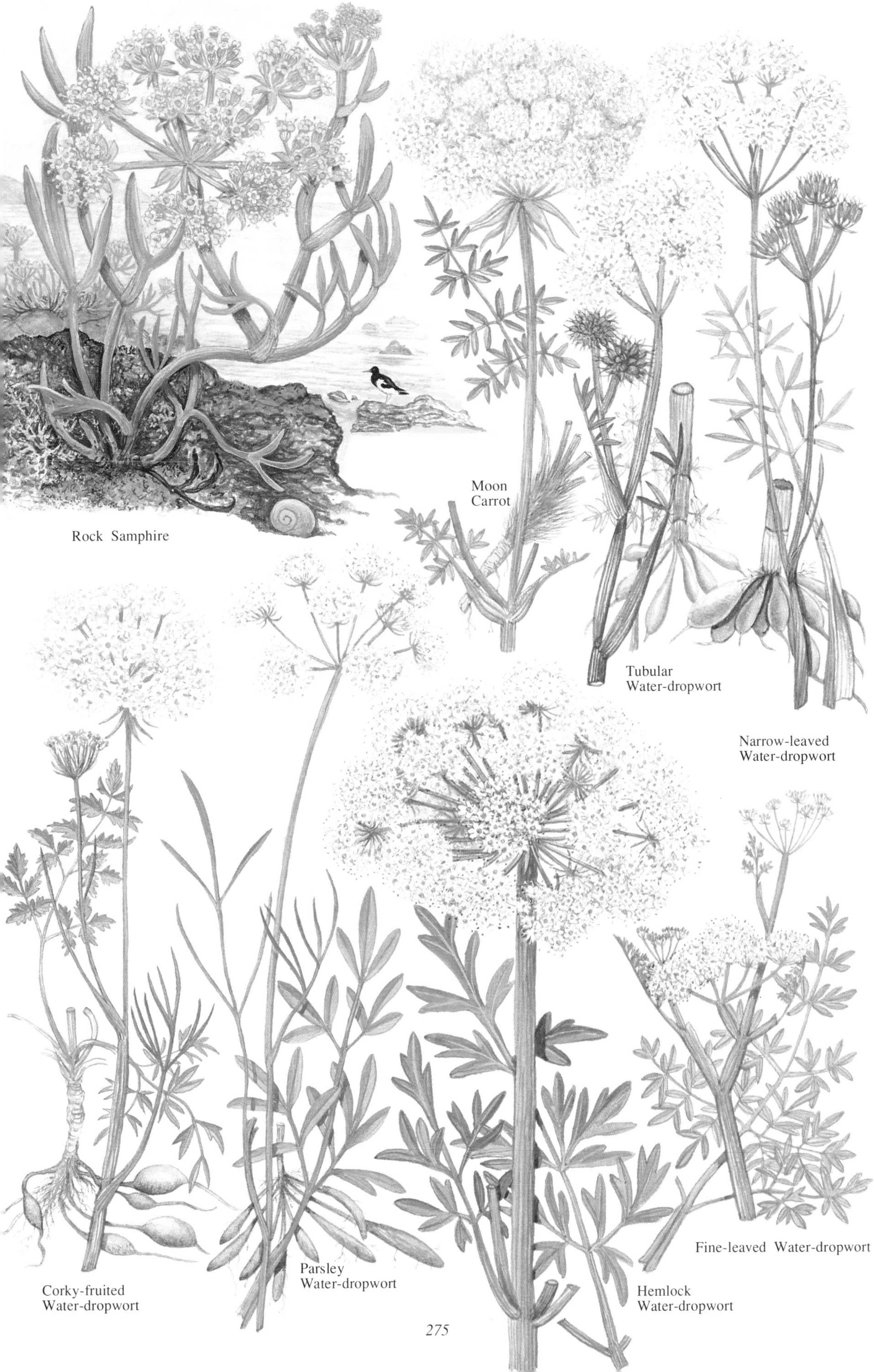

Rock Samphire

Moon Carrot

Tubular Water-dropwort

Narrow-leaved Water-dropwort

Corky-fruited Water-dropwort

Parsley Water-dropwort

Fine-leaved Water-dropwort

Hemlock Water-dropwort

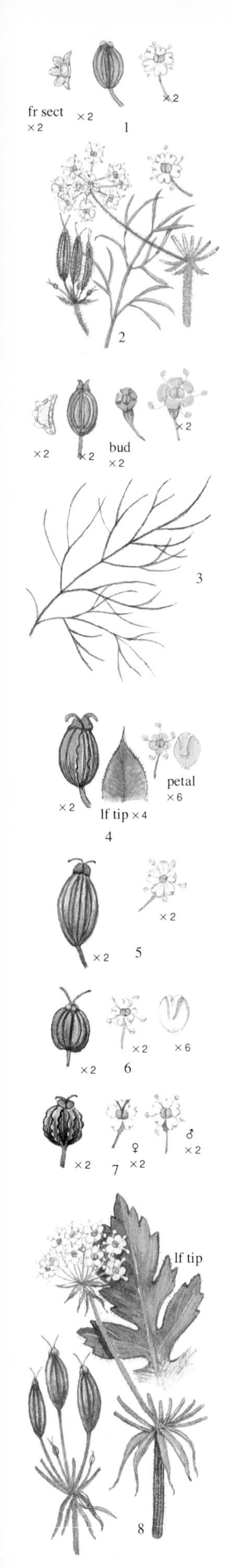


1* **Fool's Parsley** *Aethusa cynapium* L. Low to medium, hairless biennial, occasionally biennial; stem green, hollow, finely ridged. Leaves diamond-shaped in outline, 2-pinnate, with dark green, oval, lobed leaflets. Flowers white, 2mm, in umbels with 10-20 rays, the petals unequal; bracts absent, but bracteoles 3-4, pendent, giving a bearded appearance. Fruit egg-shaped, 3-4mm, with thick ridges, narrowly winged on the sides. Waste places, arable land, farmyards, gardens and occasionally in open woods, generally at low altitudes. June-October. Throughout, except the extreme north, the Faeroes and Iceland. A very poisonous plant, containing Coniine and Cynopine. However, these disappear when the plant is dried, so that the plant is harmless in a crop of hay. Widespread from Europe to the Caucasus and N Africa.

subsp. *agrestis* (Wallr.) Dostál [*A. cynapium* var. *agrestis*]. Like the type, but a low to short plant, not exceeding 20cm, with an angled stem. Mainly a weed of arable land. June-September. Throughout the range of the species. **B**: Rare and decreasing.

subsp. *cynapioides* (Bieb.) Nyman. Like the type, but always biennial, usually 1-2m tall; stem with a whitish bloom. Leaflets oblong to linear. Woodland. June-September. Confined to E France, Germany and S Europe.

2 **Athamanta** *Athamanta cretensis* L. Short to medium, hairy, erect perennial. Leaves 3-5-pinnate, rather feathery, with linear to linear-oblong lobes. Flowers white, 2-3mm, in umbels with 5-15 rays; bracts often present, linear, sometimes pinnately-lobed. Fruit oblong, 6-8mm, ridged, finely hairy, with a short beak and diverging styles. Mountain habitats, rocky places and screes, to 2700m. June-August. E France and S Germany. Confined to the mountains of C and S Europe. Some times confused with *Meum athamanticum*, a more strongly aromatic plant.

3* **Fennel** *Foeniculum vulgare* Miller [*F. officinale*]. Tall, greyish, hairless, strongly pungent perennial forming large tufts; stems tough, erect, to 2.5m, though often less, shiny, hollow when old. Leaves feathery, triangular in outline, 3-4-pinnate, with slender thread-like divisions and sheathing bases. Flowers yellow, 2-3mm, in umbels with 4-30 rays, without bracts. Fruit oblong, 4-10mm, ridged. Waste ground, roadsides, cliffs and rocky habitats, often close to the sea and at low altitudes. July-October. Britain and France; naturalized in Belgium, Holland and Germany. Widely cultivated, grown as an ornamental as well as a pot herb, the leaves being used for flavouring. Cultivated forms include those with bronze or purplish foliage and these sometimes become naturalized. The form widely grown as a vegetable, the so called Florence Fennel, which has widely expanded and overlapping leaf-bases, is var. *azoricum* (Miller) Mell. **B**: Coastal England, Wales, Isle of Man, local in Ireland, occasionally along motorways, rare elsewhere.

**Dill** *Anethum graveolens* L. (now *Peucedanum anisum*) [*Anisum vulgare*] Strongly aromatic annual, medium to tall, finely hairy. Flowers white. Fruit 3-5mm (see below), covered in short adpressed bristles, winged. **I**: Asia. A casual garden escape, occasionally naturalized. France and S Germany.

4* **Pepper Saxifrage** *Silaum silaus* (L.) Schinz & Thell. [*Silaus pratensis*]. Very variable short to tall, hairless perennial; stem solid, ridged. Leaves triangular in outline, 2-4-pinnate, the segments long-stalked, often with a reddish tip; leaflets lanceolate to linear, finely toothed; uppermost leaves smaller, often only 1-pinnate. Flowers yellowish, 1.5mm, in umbels with 5-15 rays; bracts present, few, or absent. Fruits oblong, 4-5mm, with slender ridges. Rough grassland, grassy commons, old meadows, road verges, hedgebanks, usually on clay soils, at low altitudes. June-August. S Britain, Belgium, Holland, France, Germany and S Sweden. W Europe, through C and S Europe to C Asia. Very variable in both height and in the shape of the leaf-lobes. **B**: Primarily in C and SE England, local elsewhere, but absent from Scotland and most of Wales.

5* **Baldmoney, Spignel** *Meum athamanticum* Jacq. Strongly aromatic, short to medium, hairless perennial, the base surrounded by fibrous leaf remains. Leaves mostly basal, 3-4-pinnate, with crowded linear segments. Flowers white or yellowish, sometimes tinged with purple, 2-3mm, the umbels with 3-15 rays; bracts 1-2, or absent. Fruit oblong, 4-10mm, stoutly ribbed. Mountain habitats, rough grassland, rocks and screes, generally on limestone, to 2700m. June-July. N Britain, Belgium, France and C and S Germany; naturalized in S Norway. Rather local. In the mountains of NW, C and parts of S Europe. The roots were formerly used as a spice when dried. **B**: Scattered localities in C and S Scotland and N England.

6* **Bladderseed** *Physospermum cornubiense* (L.) DC. [*P. aquilegifolium*, *Danaa nudicaulis*, *D. cornubiensis*]. Medium to tall, almost hairless perennial; stems solid, ridged. Basal leaves long-stalked, dark green, 2-ternate, with lobed leaflets; stem leaves small or reduced to a stalk. Flowers white, 2mm, in umbels with 6-20 rays; bracts present, lanceolate. Fruit egg-shaped, 3-4mm, narrowly ridged, dark brown when ripe. Open woodland, scrub, hedgebanks and arable land, often forming large colonies. July-August. S Britain and France. W and S Europe, the Mediterranean region and W Asia, east to the Caucasus. **B**: Very local, confined to E Cornwall, S Devon and Buckinghamshire.

7* **Hemlock** *Conium maculatum* L. Medium to tall hairless annual or biennial to 2.5m; stems erect, hollow, purple-brown spotted and with a whitish bloom. Leaves 2-4-pinnate, rather soft, finely cut into lanceolate segments. Flowers white, 2mm, in umbels with 10-20 rays, sepals absent; bracts present, lanceolate, deflexed. Fruits subglobose, 2.5-3.5mm, with wavy ridges. Damp habitats, meadows, open woods, scrub, river, stream, canal and lake margins, ditches, road-verges, sea walls, generally on rather heavy soils. June-July. lThroughout, except the extreme north, the Faeroes and Iceland. An extremely poisonous plant; all parts are toxic due to the presence of several polyacetylenes, especially Coniine, which paralyse the respiratory system and cause suffocation. Widespread from Europe to C Asia and N and NE Africa, as well as the Canary Isles. **B**: Throughout, but mostly coastal in the north and in Wales; more local in Ireland.

8 **Pleurospermum** *Pleurospermum austriacum* (L.) Hoffm. Tall, slightly hairy biennial or short-lived perennial, to 2m; stems ridged, fibrous at the base. Lower leaves triangular in outline, 2-3-ternate, with oval, toothed segments and hairy margins. Flowers white, 2-3mm, in umbels with 12-20 rays, petals rounded; bracts and bracteoles numerous, deflexed. Fruit egg-shaped, 9-10mm, strongly ridged. Mountain habitats, damp grassy and rocky places, open woods, to 2000m. June-September. E France, C & S Germany and S Sweden. Conspicuous on account of its relatively large downturned bracts. Primarily restricted to the mountain of C and S Europe.

Fennel, fr × 2

Dill, fr × 2

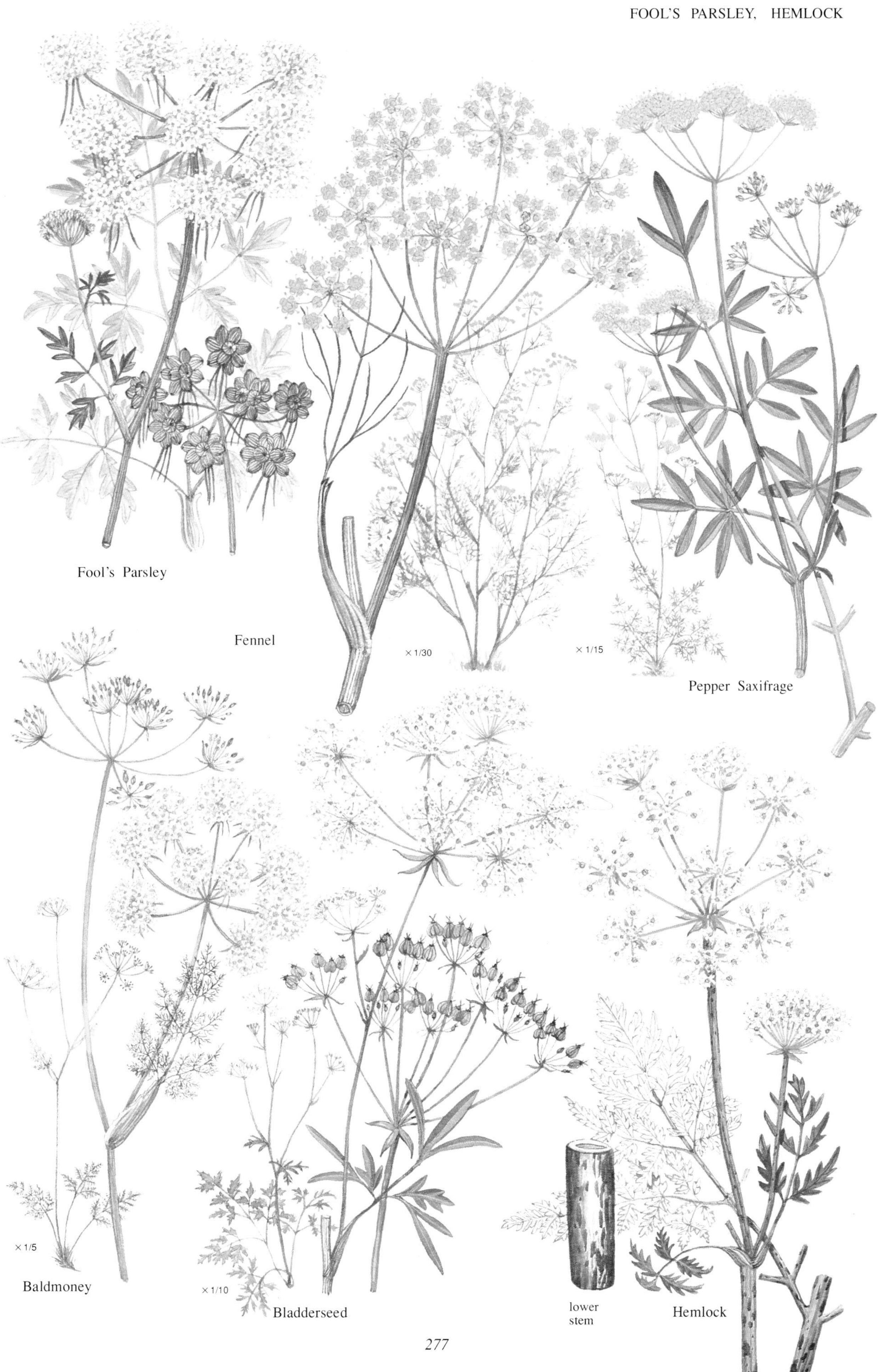
Fool's Parsley
Fennel
× 1/30
× 1/15
Pepper Saxifrage
× 1/5
Baldmoney
× 1/10
Bladderseed
lower
stem
Hemlock

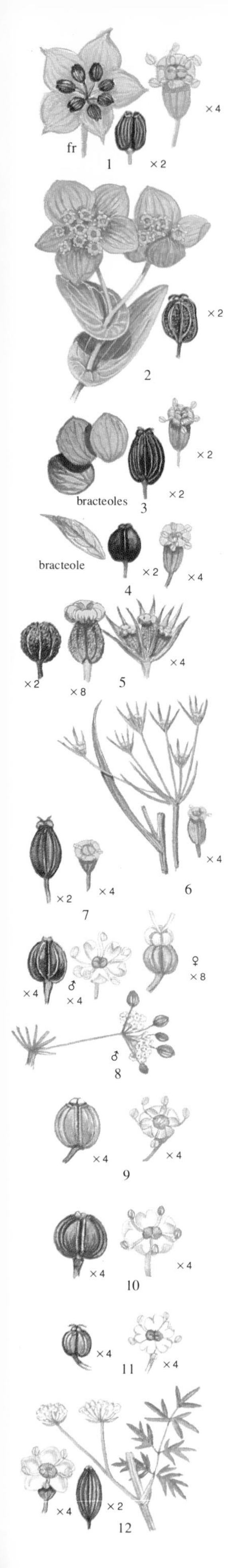


**Thorow-waxes** *Bupleurum*. Hairless annual or perennial herbs with simple, undivided leaves. Flowers usually green or yellow, in small umbels, without sepals; petals not notched. Fruit generally oblong or egg-shaped, with prominent ridges and 1-5 vittae. 100 species.

1* **Thorow-wax** *Bupleurum rotundifolium* L. Short to medium, greyish, often purple-tinged annual, erect. Leaves elliptical to almost rounded, blunt, the upper joining around the stem, the lower short-stalked. Flowers yellow, 1.5mm, in small umbels with 5-10 rays; bracteoles 5-6, oval, yellowish-green, forming a conspicuous cup around the flower clusters, becoming whitish in fruit. Fruit oblong, 3-3.25mm, smooth, blackish-brown when ripe. Arable land, waste places and open dry habitats on light slightly acid or calcareous soils, at low altitudes. June-July. Belgium, France and Germany; naturalized in Britain and Holland. Generally a weed of arable land. **B**: Formerly in scattered localities in England from Yorkshire southwards, now extinct.

2 **False Thorow-wax** *Bupleurum subovatum* Link [*B. lancifolium, B. protractum*]. Like *B. rotundifolium*, but leaves generally oval to oblong. Umbels with only 2-3 rays and rounded bracteoles. Fruit egg-shaped, 3-5mm, warted. Arable land and other dry open habitats, including gardens. June-October. **I**: S Europe and W Asia. Naturalized or casual in Britain, Belgium, France. Often introduced in wild bird food. **B**: Scattered localities throughout, as a casual.

3* **Long-leaved Hare's-ear** *Bupleurum longifolium* L. Medium to tall perennial, yellow or purple-tinged; stems stout, erect and hollow. Lower leaves elliptical-spatulate, with long broadly winged stalks, and sheathing at the base; stem leaves oval to heart-shaped, clasping the stem. Flowers yellowish, often red tinged, 2mm, in umbels with 5-12 rays; bracts 2-4, bracteoles 5-8 similar but smaller, all oval, green, occasionally yellowish or purplish. Fruit elliptic-oblong, 4-5.5mm, dark brown or black. Meadows, open woods and stony ground, to 2000m. July-August. France, Germany.

4* **Small Hare's-ear** *Bupleurum baldense* Turra. Low to short, greyish, little or much-branched annual, often under 8cm tall. All leaves linear-lanceolate to narrowly oblong, 3-5-veined, only the lowermost stalked. Flowers yellow, 2m, in small umbels with only 1-4 rays; bracts lanceolate, pointed, greyish, yellowish or brownish, half-enclosing the flower clusters. Fruit egg-shaped, 2mm. Dry open habitats, often close to the sea, generally on calcareous soils. June-July. S Britain and France. **B**: Very rare, south coast between Beachy Head and Berry Head; Channel Islands.

5* **Slender Hare's-ear** *Bupleurum tenuissimum* L. Short to medium, somewhat greyish, generally much-branched annual; stem wiry, solid. Leaves linear to linear-lanceolate, the lowermost short-stalked, the others unstalked, all with 5-7 veins. Flowers yellowish, the umbels with only 1-3 rays; bracts linear, 3-veined, often slightly toothed. Fruit subglobose, 1.5-2.5mm, granular. Rough grassland, upper levels of salt marshes, behind coastal shingle ridges and on dry saline soils. July-September. North to S Sweden. **B**: Local, S and E coasts of England south of the Tees.

6 *Bupleurum gerardii* All. [*B. affine, B. australe*]. Like *B. tenuissima*, but the leaves often sickle-shaped, pointed. Umbels mostly with 3-7 rays. Dry habitats. July-September. NW and C France; naturalized in Germany.

7* **Sickle-leaved Hare's-ear** *Bupleurum falcatum* L. Very variable, medium to tall perennial; stem hollow. Basal leaves oblong to elliptical, stalked, with 5-7 veins; stem leaves lanceolate to linear, often sickle-shaped, unstalked and half-clasping the stem. Flowers yellow, 1mm, in umbels with 3-15 rays; bracts 2-5, lanceolate, very unequal, 3-5-veined. Fruits oblong, 3-4mm. Grassy and waste places, hedgebanks, to 1600m. July-October. **I**: C & S Europe, Asia. Naturalized in SE Britain, Belgium, France and Germany. Declining. **B**: Very rare, a single site in Essex.

8* **Honeywort** *Trinia glauca* (L.) Dumort [*T. stakovii, T. vulgaris*]. Low to short, greyish, hairless perennial, with angled much-branched solid stems; stem base with old fibrous leaf remnants. Leaves 2-3-pinnate, with slender divisions. Flowers white or yellowish, minute, monoecious; umbels with 4-7 rays, without bracts. Fruit egg-shaped, 2-3mm, ridged. Short dry grassland over limestone. May-June. S Britain, France and SW Germany. The male often has longer leaf-lobes. **B**: Very rare, restricted to a few localities in S Devon, N Somerset and Gloucestershire.

**Marshworts** *Apium*. Annual, biennial or perennial herbs with pinnate or ternate leaves. Flowers with whitish petals and with minute or no sepals; petals usually not notched. Fruit egg-shaped to oblong, somewhat flattened, stoutly ridged; vittae solitary. 45 species.

9* **Wild Celery** *Apium graveolens* L. Medium to tall, yellowish-green, rather stout hairless biennial, with a strong celery smell; stem solid, grooved. Leaves shiny 1-2-pinnate, with rhombic or lanceolate, lobed and toothed segments. Flowers greenish-white, minute, in short-stalked or unstalked umbels with 4-12 rays, generally subtended by a small ternate leaf; bracts absent. Fruit egg-shaped, 1.5mm. Damp habitats, river margins, ditches, on fertile, often brackish soils, sometimes saline meadows inland. June-August. North to Denmark; naturalized in parts of C and S Scandinavia. Var. *dulce* (Miller) DC is edible celery and var. *rapaceum* (Miller) DC celeriac. **B**: Mainly coastal, scarce in E England.

10* **Fool's Water-cress** *Apium nodiflorum* (L.) Lag. [*Helosciadium nodiflorum*]. Low to medium, prostrate to ascending perennial, the stems finely grooved, rooting at the lower nodes. Leaves 1-pinnate, with 5-13 oval or lanceolate, toothed segments. Flowers white, minute, the umbels leaf-opposed, short-stalked or unstalked, with 3-12 rays; bracts usually absent. Fruit oval, longer than wide, 1.5-2mm. Wet habitats, on nutrient-rich and calcareous soils. July-August. W Europe. Locally abundant, Confused with *Berula erecta* (p. 272). **B**: Widespread throughout, but scarce in N England and Scotland.

11* **Creeping Marshwort** *Apium repens* (Jacq.) Lag. [*Helosciadium repens*]. Like *A. nodiflorum*, but stem prostrate and rooting at every node. Umbels with 3-6 rays and 3-7 bracts. Fruit wider than long. Old wet meadows, ditches, shallow pools. July-August. North to Denmark. Apparently hybridises with *A. nodiflorum*. **B**: Very rare, restricted to Oxfordshire, possibly in N England.

12 **Lesser Marshwort** *Apium inundatum* (L.) Reichenb. f. [*Helosciadium inundatum*]. An often aquatic, partly or wholly submerged perennial; stems slender, smooth. Leaves pinnate, the submerged ones thread-like, hollow, the floating or aerial ones with oval, often 3-lobed, segments. Flowers white, 2mm, in leaf-opposed short-stalked umbels, with 2-3 rays; bracts absent. Fruit elliptic-oblong, 2-3.5mm. Still and slow-moving water, generally on acidic or nutrient-poor soils. June-August. North to SE Sweden. **B**: Local and decreasing throughout, though rare in Scotland.

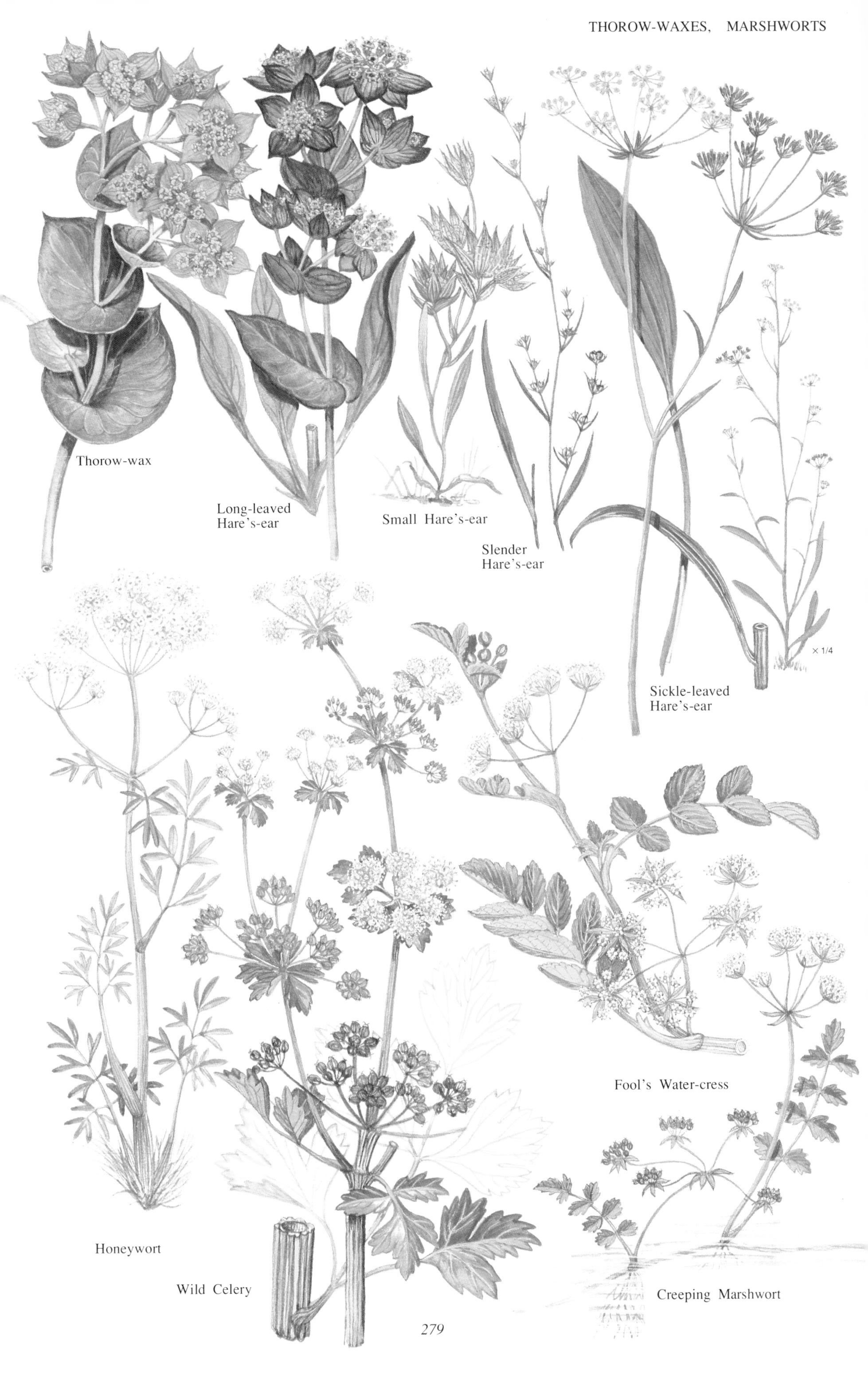
Thorow-wax
Long-leaved
Hare's-ear
Small Hare's-ear
Slender
Hare's-ear
× 1/4
Sickle-leaved
Hare's-ear
Fool's Water-cress
Honeywort
Wild Celery
Creeping Marshwort

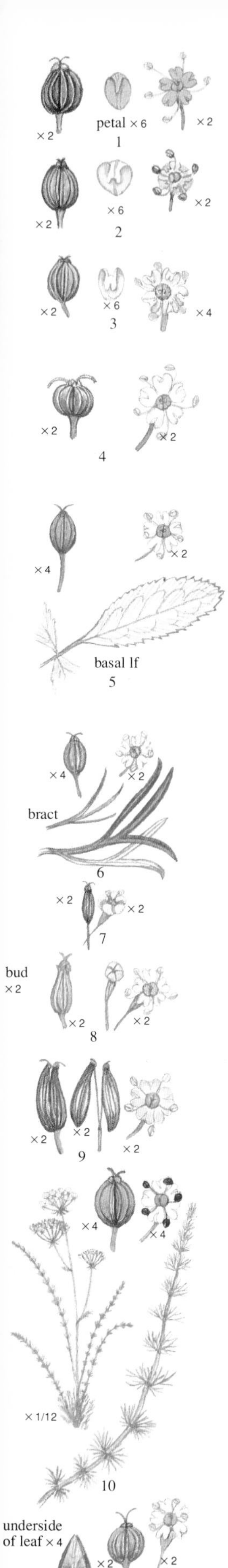


1* **Garden Parsley** *Petroselinum crispum* (Miller) A. W. Hill [*P. sativum*] Medium to tall, hairless, parsley-scented biennial; stems erect, solid, shallowly ridged. Leaves 3-pinnate, bright shiny green, with small wedge-shaped segments, lobed and often somewhat crisped. Flowers yellowish, 2mm, in flat-topped umbels with 8-20 rays; bracts 1-3, sometimes lobed. Fruit egg-shaped, 2.5-3mm, with slender ridges. Grassy and waste places, walls, rocks, cultivated land, at low altitudes. June-August. Origins uncertain. Naturalized throughout, except Holland and the far north. Cultivated forms include some with very curled leaflets, which may be found locally naturalized. **B**: Throughout, but rare in N Scotland.

2* **Corn Parsley** *Petroselinum segetum* (L.) Koch. Medium to tall, dark greyish-green, hairless parsley-scented biennial, occasionally annual; stem solid, finely ridged. Leaves oblong in outline, pinnate, with oval, toothed leaflets; margin thick with forward pointing teeth; smell of Parsley when crushed. Flowers white, sometimes pink or lilac, 2mm, in irregular umbels with 2-5 rays; bracts subulate. Fruit egg- shaped, 2-4mm. Arable fields, brackish grassland, hedgerows, river and stream banks, road-verges, on fertile, often calcareous soils, at low altitudes. August-October. C and S Britain, Belgium, Holland and France. **B**: Local and decreasing, C and S England, S Wales, N to Yorkshire.

3* **Stone Parsley** *Sison amomum* L. An erect hairless biennial, with a nauseating smell when crushed; stems solid, finely ridged. Leaves pinnate, with 5-9 pairs of oblong, toothed leaflets, often lobed. Flowers white, 1-2mm, in umbels with 3-6 slender, uneven rays; bracts linear, short. Fruit almost globose, 1.5-3mm, with slender ridges. Grassy banks, hedgerows, road-verges, occasionally on waste ground, generally on heavy soils. July-September. C and S Britain, France. Like *Petroselinum* the fruit have a solitary, rather short vittae. The smell of the crushed leaves has been well described as a mixture of petrol and nutmeg. **B**: Local, from Yorkshire southwards, including coastal N Wales.

4* **Cowbane** *Cicuta virosa* L. Tall stout, hairless perennial, to 1.2m. Leaves 2-3-pinnate, triangular in outline; leaflets linear-lanceolate to linear, sharply-toothed; leaves with long hollow stalks. Flowers white to pink, 3mm, in terminal umbels with 10-20 rays; bracts absent, but bracteoles present, linear. Fruit subglobose, with wide ridges and a solitary conspicuous vittae. Shallow freshwater and damp muddy habitats, at low altitudes. July-August. Throughout, except the far north and absent from much of S Europe. Extremely poisonous. **B**: Very local throughout, though absent from much of NE & NC Ireland.

5* **False Bishop's Weed** *Ammi majus* L. Medium to tall, hairless annual or occasionally a biennial; stems erect, branched. Leaves 1-2-pinnate, with crowded elliptical to oval, toothed segments. Flowers white, 3mm, in terminal umbels with 15-60 rays; bracts several, often more than half the length of the rays. Fruit oval, 1.5-2mm, pale, with slender ridges and a solitary vittae. Waste, rocky and bare places, at low altitudes. June-October. Primarily Mediterranean, to N and E France; a casual further north. **B**: Rare casual.

6 *Ammi visnaga* (L.) Lam. Rather more robust than *A. majus*, but the leaves with tangled, linear segments. Umbels with up to 150 rays; bracts usually lobed. Flowers white or yellowish. A weed of cultivated land and waste places. June-August. W and C France. The umbel rays become erect and finally infolded in fruit. Sometimes confused with *Peucedanum officinale*, which always has yellow flowers.

7* **Ptychotis** *Ptychotis saxifraga* (L.) Loret & Barrandon [*P. heterophylla*]. Medium, hairless biennial with a basal rosette of leaves in the first year. Rosette and lower leaves pinnate, with 3-9 oval, few-toothed leaflets, sometimes slightly lobed; upper leaves 1-2-pinnately divided with pronounced sheathing bases. Flowers whitish, 2mm, in umbels with 6-12 unequal rays, sepals conspicuous; bracts few, soon falling. Fruit oblong, 2-3mm, with slightly winged ridges. Dry habitats, banks, waste and rocky places, at low altitudes. June-August. NE and C France. Local. Primarily W Mediterranean.

8* **Longleaf** *Falcaria vulgaris* Bernh. [*F. rivinii*, *F. sioides*]. Medium to tall, greyish-green, hairless, much-branched perennial, occasionally annual; stems solid, often forming a low tangled mass. Leaves 1-2-ternate, with linear-lanceolate or linear, somewhat sickle-shaped segments, toothed. Flowers whitish, in umbels with 12-18 rays, sepals conspicuous; bracts present. Fruit oblong, 3-4mm, with low ridges. Grassy and waste habitats, at low altitudes. July-September. N France southwards; naturalized in SE Britain, Belgium, France, Holland and S Sweden. **B**: Naturalized or casual in East Anglia, SE England and the Channel Islands.

9* **Caraway** *Carum carvi* L. Medium to tall, erect, hairless biennial to 1.5m, though often less; stem hollow, ridged. Leaves 2-3-pinnate, with linear-lanceolate or linear lobes. Flowers whitish or pink, 2-3mm, in rather irregular umbels with 5-16 rays; bracts usually absent. Fruit oval, 3-6mm, with slender ridges, aromatic when crushed. Waste places, meadows and other grassy habitats, to 2000m. June-July. Native or widely naturalized or casual throughout, though local and rather rare in the north. **B**: Local and declining throughout, but perhaps only truly native in SE England.

10 **Whorled Caraway** *Carum verticillatum* (L.) Koch. An erect medium to tall, hairless annual; stems solid, scarcely branched, with few leaves. Leaves mostly basal, slender, with numerous whorled thread-like segments. Flowers white or pinkish, 2-3mm, in flat-topped umbels with 8-12 rays; bracts linear, pointed. Fruit elliptical, 2.5-4mm, with prominent, almost winged ridges. Marshy and damp meadows, stream margins, on acid soils. July-August. Britain, Belgium, Holland and France, extinct in Germany. Unusual looking, the leaves with distinctive whorled leaf segments, the lowermost smaller than the upper. Native to W and SW Europe. **B**: Local; primarily in W Britain and Ireland.

11* **Cnidium** *Cnidium dubium* (Schkuhr) Thell. Medium to tall, slightly hairy perennial; stems hollow, rounded below, grooved above. Leaves oblong in outline, 2-3-pinnate, with narrowly oblong, toothed segments, with recurved margins and a whitish apex. Flowers white, in umbels with 20-30 slightly winged rays; bracts few or absent, bracteoles numerous. Fruit almost globose, 2-3mm, with prominent ridges. Damp woods and disturbed ground, often close to the sea. July-October. Germany, E Denmark and S Sweden. Central European.

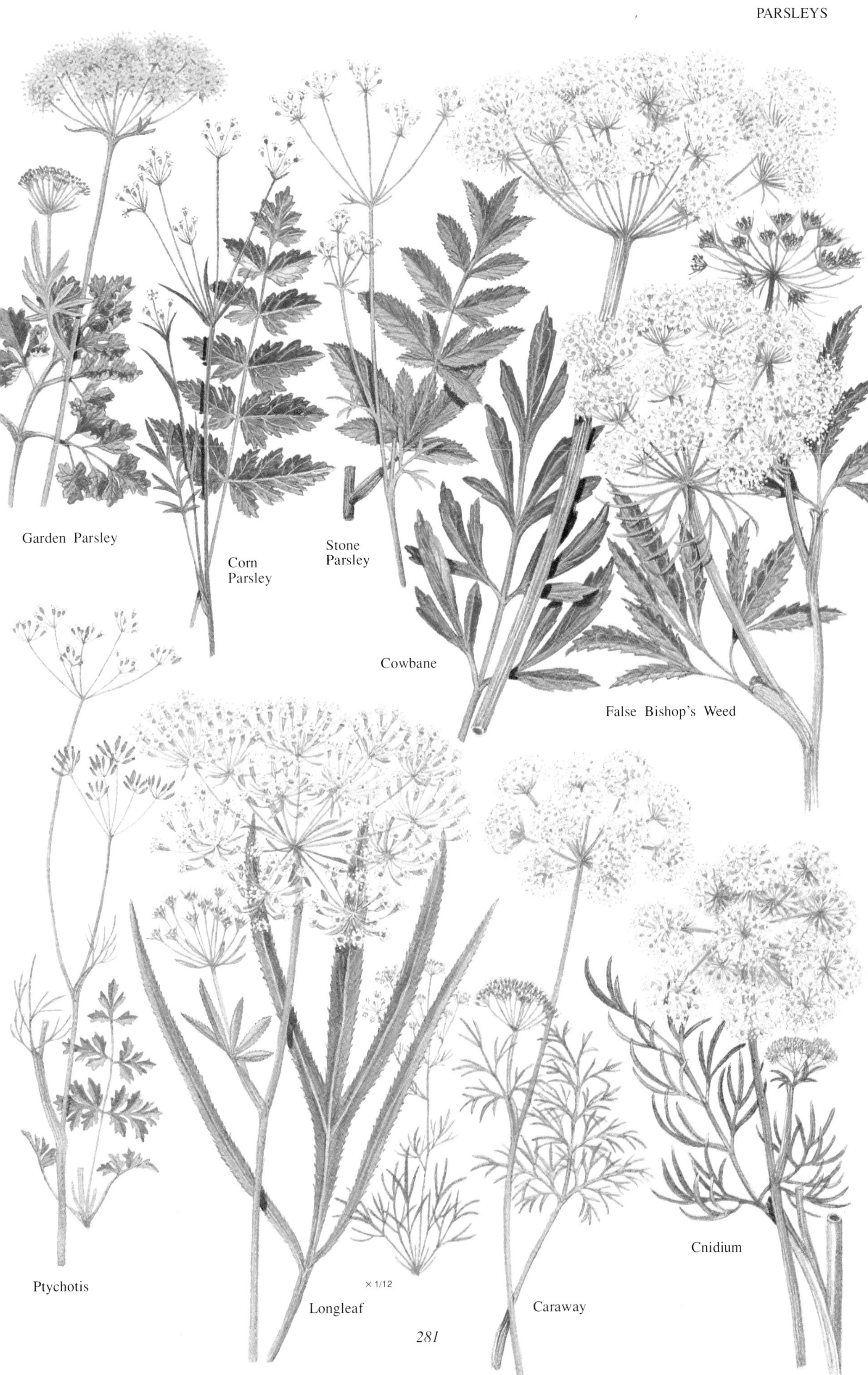
Garden Parsley
Corn Parsley
Stone Parsley
Cowbane
False Bishop's Weed
Ptychotis
× 1/12
Longleaf
Caraway
Cnidium

1* **Coriander** *Coriandrum sativum* L. Short, hairless annual, foetid when fresh. Lower leaves lobed, others 1-3-pinnate, the segments oval to linear, often toothed. Flowers white, occasionally pink, in small umbels with only 3-5 rays, the petals of the outer flowers larger, sepals conspicuous; bracts absent; bracteoles usually 3, linear. Fruit globose, 2-6mm, with narrow ridges, aromatic when crushed. Waste and bare places, at low altitudes. June-August. **I**: W Asia and N Africa. Naturalized in C & E France and Germany; casual in Britain. The fruits do not split into two mericarps when ripe. Widely cultivated as a culinary herb, especially in the USSR, the fruits being used for flavouring meat and other dishes. Oil is also extracted from the fruit for use in the pharmaceutical and perfumery industries.

2* **Cambridge Milk-parsley** *Selinum carvifolia* (L.) L. Medium to tall, almost hairless perennial, parsley-scented; stem solid, angled, branched and leafy. Leaves 2-3-pinnate, the segments oval to linear, toothed, occasionally lobed. Flowers white, 2mm, in umbels up to 7cm across, with 5-33 rays; bracts 1-2 or absent, soon falling. Fruit oblong, 3-4mm, winged on the ridges. Damp habitats, damp meadows and fens, usually at low altitudes. July-October. Throughout, except the extreme north, Ireland, the Faeroes and Iceland. Europe to C Asia, but absent from most of S Europe. **B**: Now confined to Cambridgeshire, formerly more widespread.

3 **Pyrenean Angelica** *Selinum pyrenaeum* (L.) Gouan [*Angelica pyrenaea*]. Like *S. carvifolia*, but a short to medium plant, stems ridged and stem leaves few; leaf-lobes linear-lanceolate to linear. Flowers pale yellowish, 2mm, the umbels with 3-9 unequal rays. Fruit 3.5-4mm. Mountain habitats, pastures and rocky places, to 2300m. June-August. C and E France north to the Vosges.

4 **Alpine Lovage** *Ligusticum mutellina* (L.) Crantz [*Meum mutellina*]. Short to medium, hairless perennial; stem erect, ridged, the base with numerous coarse fibres. Leaves triangular in outline, 2-3-pinnate, the segments linear-lanceolate. Flowers red or purple, 2-3mm, in umbels with 3-9 very unequal rays; bracts 1-2 or absent. Fruit egg-shaped, 4-6mm, the ridges narrowly winged, with numerous vittae. Damp meadows and open woods in mountain habitats, to 3000m. July-August. C & S France and Germany. C and S Europe east to W Russia.

5* **Scots Lovage** *Ligusticum scoticum* L. Like *L. mutellina*, but leaves 2-ternate, bright shiny green, the base of the stem without fibres. Flowers white, 2mm, the umbels with 8-20 rays. Fruit oval, 5-8mm. Rocky coastal habitats. July-August. N Britain, W Ireland, the Faeroes, Iceland, Denmark, Norway and E Sweden. Restricted to NW Europe. **B**: From Northumberland northwards and in W Ireland.

6 **Conioselinum** *Conioselinum tataricum* Hoffm. [*C. vaginatum*]. Tall, hairless perennial to 1.5m. Lower leaves long-stalked, triangular or rhombic in outline, 2-3-pinnate, with lanceolate to linear lobes; upper leaves with greatly inflated stalks and a small leaf-blade. Flowers white, 2-3mm, in umbels with 15-30, sepals absent; bracts several or absent. Fruit elliptical, 5mm, flattened, with winged ridges. Rocky and stony habitats. July-September. Norway, particularly in the Arctic regions, and Finland. Restricted to northern and north-eastern parts of Europe and N Russia.

**Angelicas** *Angelica*. Large perennial herbs with 2-3-pinnate or ternate leaves. Flowers white or pinkish, rarely yellowish, sepals inconspicuous usually. Fruit oval to oblong, flattened, the lateral ridges with wide wings, often wavy-edged; vittae variable in number. 70 species in the northern hemisphere as well as New Zealand.

7* **Wild Angelica** *Angelica sylvestris* L. [*A. illyrica, A. elata, A. brachyradia*]. Robust tall, almost hairless perennial, to 2m; stems hollow, ridged, generally tinged with purple. Leaves 2-3-pinnate, with oblong, sharply toothed segments; upper leaves reduced to large inflated sheaths and partially enclosing the developing umbels. Flowers white or pinkish, 2mm, in umbels 3-15cm across, with numerous rays; bracts few and soon falling, or absent. Fruit oval, 4-5mm, with membranous wings. Damp habitats, meadows, fens and woods, to 1800m. July-October. Throughout, except for parts of the extreme north and Spitsbergen. Europe, except for much of the south, east to C Asia. Sometimes cultivated as an ornamental plant. **B**: Fairly common throughout much of Britain.

8 *Angelica palustris* (Besser) Hoffm. [*Ostericum palustre*]. Like *A. sylvestris*, but a shorter plant, rarely more than 1.2m. Flowers white, with well-developed petals. Fruit oval-elliptical, 5-6mm. Wet habitats. July-September. C and S Germany.

9* **Garden Angelica** *Angelica archangelica* L. [*Archangelica officinalis*]. Stout tall, pleasantly aromatic perennial, to 2m; stem green, sometimes purple tinged, hollow and ridged. Leaves triangular or diamond-shape in outline, 2-3-pinnate, with large rather irregular divisions, coarsely toothed; upper leaves reduced to large inflated sheaths. Flowers greenish or cream, 3-4mm, in large umbels up to 25cm across, with numerous rays; bracts few or absent. Fruit oblong, 6-8mm, with corky wings. Damp habitats, riverbanks, ditches, damp meadows, thickets and waste places, locally abundant, to 1800m. June-September. The Faeroes, Iceland, Holland, Germany and much of Scandinavia; naturalized in Britain, Belgium and France. Long cultivated as a drug plant – the roots and fruits are used for various stomach disorders – but also used as a confection, for seasoning and for flavouring liqueurs. Often naturalized from gardens. **B**: Widely naturalized, especially in the north.

10* **Lovage** *Levisticum officinale* Koch. Tall strong smelling, branched perennial, to 2.5m; stems stout, the lower branches alternate, but the upper opposite or whorled. Leaves triangular or diamond-shape in outline, deep shiny green, 2-3-pinnate, the segments long-wedge-shaped, irregularly toothed. Flowers greenish-yellow, 2-3mm, in umbels up to 10cm across, with 12-20 rays, without sepals; bracts numerous, downturned. Fruit oblong, 5-7mm, winged on edges, yellow or brown, with a solitary vittae. Meadows, hedgebanks and other grassy places, to 1200m. June-August. **I**: Iran. Naturalized throughout Europe, except for the extreme north, particularly in mountain or hilly regions and near to habitations; occasionally naturalized or casual in Britain. Formerly cultivated as a medicinal plant, but now grown and used for flavouring and for salads, or as a decorative plant.

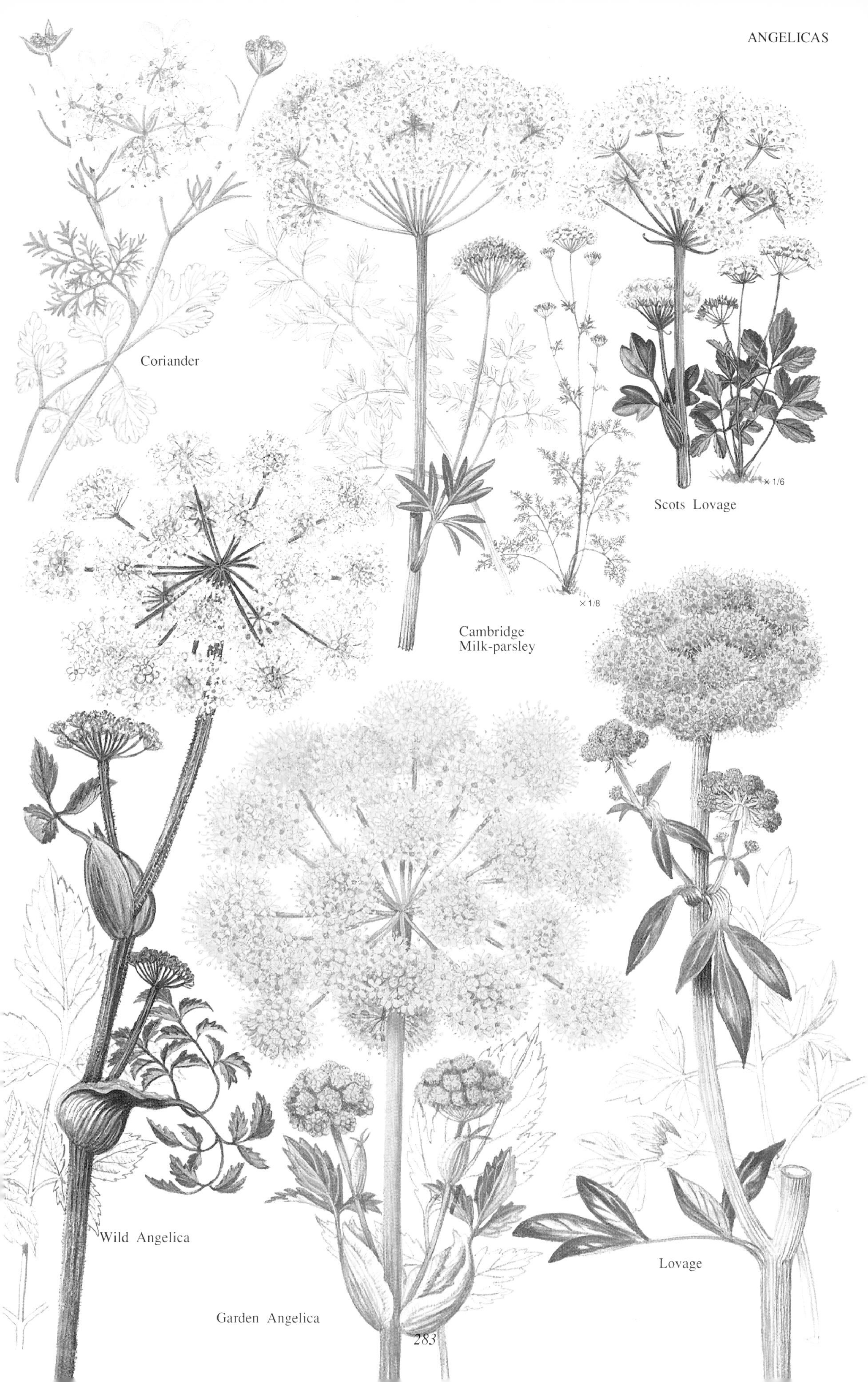
Coriander
Cambridge
Milk-parsley
× 1/8
Scots Lovage
× 1/6
Wild Angelica
Garden Angelica
Lovage

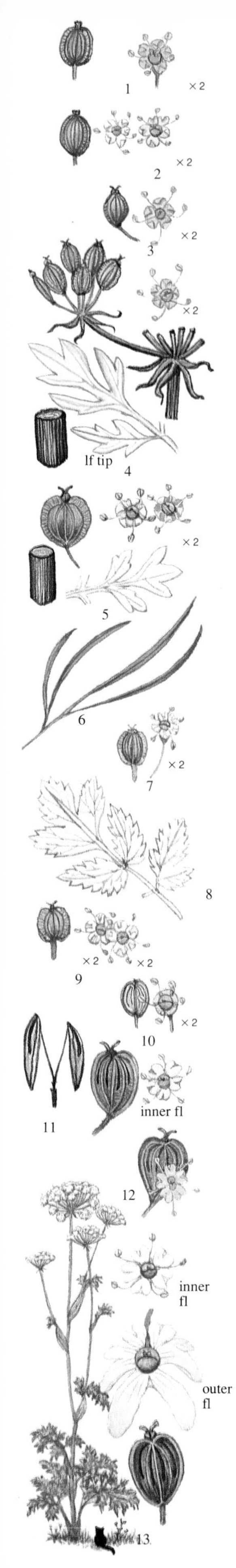


**Hog's Fennels** *Peucedanum*. Perennial, sometimes annual or biennial herbs, with pinnately or ternately divided leaves. Flowers white, yellow or pink, with broad oval petals; sepals sometimes present. Fruit flattened, oval or round in outline, narrowly or broadly winged on the margin, with 1-3 vittae. 150 species, widespread in the world.

1* **Hog's Fennel** *Peucedanum officinale* L. Medium to tall, robust, hairless perennial to 2m; stem rounded or slightly angled, with numerous fibres around the base, solid, branched above. Leaves 2-6-ternate, with linear lobes, rather tangled, midrib prominent. Flowers yellow, 2mm, in umbels 5-15cm across, nodding in bud, rays 10-40; bracts few or absent. Fruit elliptical to oblong, 5-10mm. Rough grassland, cliffs close to the sea. July-September. Britain, France, W & S Germany. **B**: Rare and local in N Essex and N Kent.

2 *Peucedanum gallicum* Latourr. Like *P. officinale*, but rays 10-20 and flowers white or pink. Fruit elliptical, 4-6mm. Grassy and waste habitats. July-August. W and C France.

3* *Peucedanum carvifolia* Vill. [*P. chabraei*]. Medium to tall, more or less hairless perennial; stem solid, ridged above, branched, with fibres around the base. Leaves shiny green above and beneath, oblong in outline, 1-pinnate, with lobed, untoothed divisions, net-veined. Flowers yellowish or greenish, 2mm, in umbels with 6-18 very unequal rays; bracts absent. Fruit broadly elliptical, 4-5mm, with transparent wings. Grassy and waste places. June-August. Belgium, Holland, France and Germany.

4 *Peucedanum alsaticum* L. [*Johrenia pichleri*]. Like *P. carvifolia*, but to 1.8m tall, often purplish and with whorled branches. Leaves 2-4-pinnate. Umbels with several persistent bracts; flowers dull yellow. Similar habitats and flowering time. France and Germany.

5 *Peucedanum oroselinum* (L.) Moench. Medium to tall, almost hairless, often reddish perennial; stem finely ridged, solid. Lower leaves 2-3-pinnate, with rather leathery, oval, lobed segments and narrowly winged stalks; upper leaves less divided and with inflated stalks. Flowers white or pinkish, 2mm, in umbels with 15-30 rays; bracts and bracteoles numerous, downturned. Fruit broadly oval, 5-8mm, with thick wings. Grassy habitats, scrub and rocky places. July-September. France, Germany and S Sweden, extinct in Denmark.

6 *Peucedanum lancifolium* Lange. Like *P. oroselinum*, but the base of the stem without fibres and leaf divisions linear, untoothed. Wet meadows and marshes. July-September. NW and W France.

7* **Milk-parsley** *Peucedanum palustre* (L.) Moench. Tall hairless biennial to 1.5m; stem hollow, strongly ridged, often purplish, with fibres at the base. Leaves mostly 2-4-pinnate, with oval, lobed segments, finely toothed, with a hard apex. Flowers greenish-white, 2mm, in umbels with 20-30 rays; bracts 4 or more, often forked and unequal. Fruit elliptical, 4-5mm, with thick wings. Wet habitats, generally on calcareous soils and growing among lush vegetation. July-September. Britain and most of Europe except the extreme north. All parts of the plant produce a thin milky juice when young. **B**: Very local in England north to Lancashire, but particularly in East Anglia.

8 *Peucedanum cervaria* (L.) Lapeyr. Like *P. palustre*, but the leaflets very sharply toothed and the bracts numerous, the lowermost pinnately-divided. Fruit elliptical to almost rounded, 4-9mm, with thick wings. Grassy, rocky and wooded habitats. July-September. Belgium, France and Germany.

9* **Masterwort** *Peucedanum ostruthium* (L.) Koch. Medium to tall, hairy biennial; stem usually hollow, ridged. Leaves 1-2-ternate, with oval, irregularly toothed lobes, hairy beneath; stalks of stem leaves greatly inflated. Flowers white or pinkish, 2mm, in umbels with 30-60 rays; bracts absent or 1-2 only. Fruit almost round, 4-5mm, widely winged. Mountain habitats, 1400-2800m. June-August. France and Germany; naturalized in Britain, Belgium and parts of C and S Scandinavia. **B**: Naturalized from Staffordshire northwards and in NE Ireland.

10* **Wild Parsnip** *Pastinaca sativa* L. An erect, medium to tall, strong-smelling, hairy biennial; stem hollow or solid, angled or ridged. Leaves pinnate, with 5-11 oval, lobed and toothed segments. Flowers yellow, 1.5mm, in umbels with 9-20 unequal rays; bracts 0-2, soon falling. Fruit elliptical, 5-7mm, narrowly winged, flattened. Rough grassy habitats, generally on dry calcareous soils, at low altitudes. July-August. Britain, Belgium, Holland, France and Germany; naturalized in Ireland and parts of Scandinavia. subsp. *sylvestris* (Miller) Rouy & Camus [*P. sylvestris*] has hairs softer and curled (not straight). Similar distribution and habitat, though absent from parts of the north; often the commonest form encountered. Can cause serious blistering if handled in strong sunlight. **B**: Widespread in C and S England, local and often coastal elsewhere.

11* **Hogweed** *Heracleum sphondylium* L. Stout medium to tall, rather bristly biennial or short-lived perennial, to 2.5m; stem hollow, ridged. Leaves pinnate with often 5 broad, lobed and toothed segments, bristly; upper leaves with large inflated bases. Flowers white, rarely pink, 5-10mm, in large umbels up to 15cm across with 12-25 rays; petals of outer flowers very unequal; bracts few or absent. Fruit elliptical to rounded, 7-10mm, flattened and broadly winged. Open woodland, banks, and rough grassland, to 1700m. April-September. Throughout, except the extreme north. Generally the commonest umbellifer flowering during the late summer. Variable. **B**: Throughout much of Britain.

subsp. *alpinum* (L.) Bonnier & Layens: has lowermost leaves simple, not pinnate, rounded or oval with 3-5 shallow lobes; flowers always white. Similar habitats. July-September. E France – Jura.

subsp. *pyrenaicum* (Lam.) Bonnier & Layens [*H. pyrenaicum*]: has simpleleaves with 5-7 pointed main lobes and flowers always white; umbels with 12-45 rays. Similar habitats. June-September. C and E France and S Germany.

12 subsp. *sibiricum* (L.) Simonkai [*H. sibiricum*]: has greenish flowers, the petals of the outer flowers in the umbel not markedly dissimilar in size. Similar habitats and flowering time. Britain – East Anglia, France, Belgium, Holland, Germany and S Sweden.

13 **Giant Hogweed** *Heracleum mantegazzianum* Sommier & Levier. Very robust biennial to 5m; stems ridged, red-spotted, up to 10cm diameter. Leaves often very large, pinnately-divided, with broad, toothed and lobed divisions. Flowers white, 8-20mm, in very large umbels up to 50cm across, with 50-150 rays, the petals of the outer flowers very unequal. Fruit oval-elliptical, 9-11mm, broadly winged, hairless or hairy. Grassy and waste places and stream margins. June-August. **I**: Caucasus and SW Asia. Originally introduced as a garden plant, but now naturalized in much of Europe and still spreading. The juice is photo-sensitive in sunlight and can cause serious skin blisters. **B**: Naturalized in many parts.

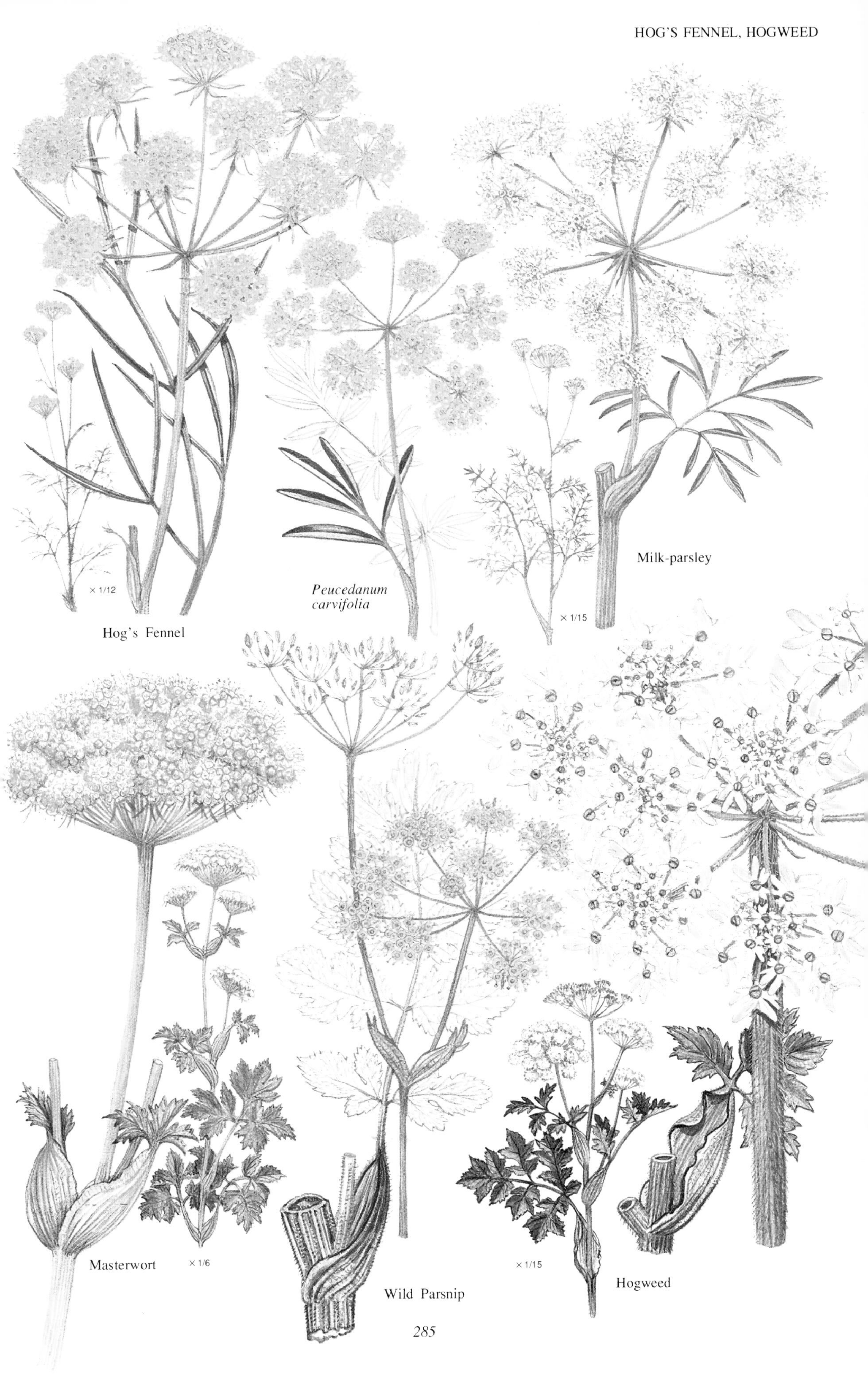
× 1/12
Hog's Fennel
*Peucedanum carvifolia*
Milk-parsley
× 1/15
Masterwort
× 1/6
Wild Parsnip
× 1/15
Hogweed

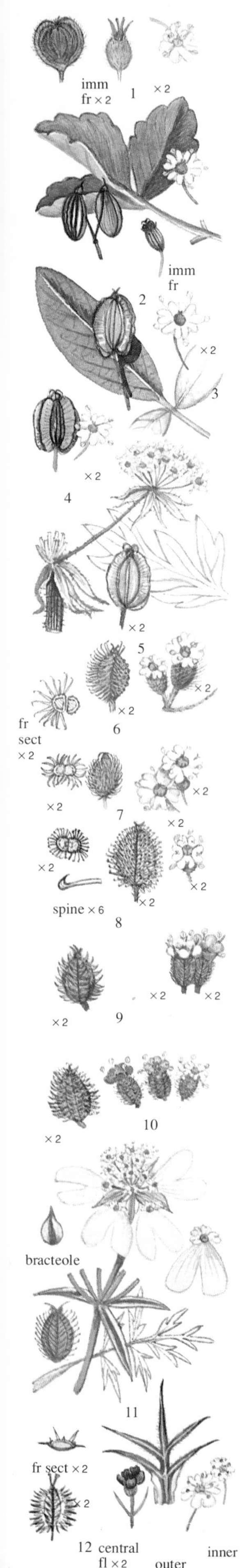


1* **Hartwort** *Tordylium maximum* L. Medium to tall, rather bristly annual or biennial, to 1.3m; stems stout, much branched, hollow, ridged. Leaves pinnate, the lower with oval, toothed segments, the upper with lanceolate or linear-lanceolate, toothed segments. Flowers white, occasionally pinkish, in flat umbels with 5-15 rays, the petals of the outer flowers markedly unequal; rather conspicuous sepals about half the length of the petals; bracts numerous. Fruit oblong, 5-8mm, flattened, with thickened whitish margins, bristly. Grassy habitats, hedgebanks and riverbanks, at low altitudes. June-July. S Britain and France; naturalized in Belgium and Germany. **B**: Rare and confined to Essex.

2 **Laser** *Laser trilobum* (L.) Borkh. [*Siler trilobum*]. Tall hairless perennial to 1.2m; stem base with fibrous leaf remains. Leaves green, 2-3-ternate, with oval or heart-shaped, toothed and lobed segments; leaf-stalks not inflated. Flowers white, in umbels with 11-20 rays; sepals conspicuous; bracts few or absent. Fruit rounded or oblong, 5-10mm, with prominent thickened ridges. Grassy habitats on calcareous soils, often local and rather rare. June-August. C & S France and Germany.

**Sermountains** *Laserpitium*. Perennial herbs with 2-4-pinnate or ternate leaves. Flowers white, sometimes pinkish, yellowish or yellowish-green; sepals conspicuous. Fruit elliptical to rectangular, with wavy ridges, winged.

3 *Laserpitium siler* L. [*Siler montanum*]. Medium to tall, almost hairless perennial; stem ridged, the base with numerous fibrous leaf remains. Leaves triangular in outline, the lower 3-4-pinnate, with oblong or lanceolate, greyish-green, untoothed segments, with a prominent whitish midrib and parallel veins. Flowers white, the umbels with 20-50 rays; bracts numerous. Fruit oblong, 6-12mm, narrowly winged. Open woodland and stony habitats, on calcareous soils, to 2000m. July-September. C & E France, C & S Germany.

4* **Broad-leaved Sermountain** *Laserpitium latifolium* L. Medium to tall, greyish, almost hairless perennial, to 1.5mm; stems solid, ridged, with fibrous leaf remains at the base. Leaves triangular in outline, 2-pinnate, with oval-heart-shaped segments, toothed or with a hairy thickened margin; upper leafstalks markedly inflated. Flowers white, in umbels with 25-40 rays; bracts numerous, downturned. Fruit oval, 5-10mm, with broad wavy wings. Mountain habitats, open woods and rocky places, to 2000m. July-August. France, Germany, C and S Scandinavia; naturalized in Belgium.

5 *Laserpitium prutenicum* L. Like *L. latifolium*, but a hairier plant with slender angled stems; bracts hairy-edged. Flowers white, sometimes tinged yellow; umbels with 12-20 rays. Fruit 3.5-4.5mm. Mountain habitats, woods, grassy places. July-September. France, Germany.

6* **Knotted Hedge-parsley** *Torilis nodosa* (L.) Gaertner. Low to short, usually prostrate, rough hairy annual; stem solid. Leaves 1-2-pinnnate, with deeply lobed segments. Flowers pinkish-white, 1mm, in unstalked or short-stalked, leaf-opposed umbels, with 2-3 very short rays, giving the flowers a clustered appearance. Fruit egg-shaped, with warts and straight spines. Sparsely grassy habitats, open and bare places, on dry soils generally at low altitudes. May-July. Britain, Belgium, Holland and France; naturalized in Germany. **B**: Local in S and E England, scattered localities elsewhere.

7* **Upright Hedge-parsley** *Torilis japonica* (Houtt.) DC. [*T. anthriscus*]. Medium to tall hairy annual, occasionaly a biennial, to 1.25m; stems stiffly erect, solid, ridged. Leaves 1-3-pinnate, the segments oval to lanceolate, coarsely toothed. Flowers white to pinkish or pinkish-purple, 2-3mm, in long-stalked umbels with 5-12 rays, the petals somewhat unequal; bracts present. Fruit egg-shaped, 3-4mm, with purple hooked spines. Grassy habitats, especially on dry soils, at low altitudes. July-September. Throughout, except the far north. **B**: Throughout, except N Scotland.

8* **Spreading Hedge-parsley** *Torilis arvensis* (Hudson) Link [*T. arvensis* subsp. *divaricata*, *T. helvetica*]. Rather like *T. japonica*, but a shorter plant, the umbels with only 1-2 or no bracts and the fruits with straight spines. A weed of cultivated land, particularly arable fields, usually on dry calcareous soils. July-September. W Europe. **B**: Local and decreasing, mainly confined to C and S England.

9* **Small Bur-parsley** *Caucalis platycarpos* L. [*C. daucoides*, *C. lappula*]. Short to medium, hairy, pale green annual; stems erect, branched, solid and angular. Leaves 2-3-pinnate, the segments deeply lobed. Flowers white or pink, 2mm, in long-stalked umbels with 2-5 rays, sepals conspicuous, greenish; bracts usually absent. Fruit elliptical, 6-13mm, with hooked bristles. Arable fields, waste and bare places, especially on calcareous soils, at low altitudes. June-July. France, Holland and Germany. Closely related to *Torilis*. **B**: Formerly a local casual in C and S Britain.

10 **Greater Bur-parsley** *Caucalis latifolia* L. [*Turgenia latifolia*]. Medium to tall, rather bristly, erect annual; stem hollow, ridged. Leaves pinnate, with oblong or lanceolate, toothed or lobed segments. Flowers white, pink or purplish, 4-5mm, in long-stalked umbels with 2-5 rays, petals unequal; bracts 3-5. Fruit elliptical, 6-9mm, with hooked bristles. Cultivated or bare ground, local. July-August. Belgium, France and Germany. The inner long-stalked flowers are male, the others hermaphrodite. In *C. platycarpos* this is reversed, with the long-stalked male flowers outermost. **B**: Casual.

11 **Orlaya** *Orlaya grandiflora* (L.) Hoffm. [*Caucalis grandiflora*]. Short to medium, erect annual; stem simple or branched, somewhat hairy at the base. Leaves 2-3-pinnate, with oval, toothed segments. Flowers white or pink, 3-4mm, in long-stalked umbels with 5-12 rays, the petals of the outer flowers very unequal. Fruit egg-shaped, 7-8mm, the ridges with 1-2 rows of hooked bristles. Dry grassy places, at low altitudes. June-August. Local. W, C and S Europe.

12* **Wild Carrot** *Daucus carota* L. Very variable, short to tall, hairy or hairless annual or biennial; stem solid, often ridged. leaves 2-3-pinnate, feathery, with linear or lanceolate segments; uppermost leaves often bract like. Flowers white, the central one of the umbel sometimes purple, 2mm, the umbels with numerous rays which become markedly contracted and concave in fruit; bracts conspicuous, usually 3-lobed. Fruit oblong, 2-4mm, shortly spiny. Rough grassland, on well-drained soils, usually at low altitudes. June-August. Throughout, except the far north. The taproot is typically white and the umbel rays become erect and incurved as the fruit matures. Very variable. **B**: The commonest form, particularly in coastal regions, but also in C and S England. Subsp. *gummifer* Hook.f. [*D. gummifer*] is a short plant, stem densely hairy; leaves shiny; umbels convex, the rays only slightly contracting in fruit. Mainly on cliffs and sand-dunes. W and NW Europe. **B**: Mainly confined to the S coast of England. Subsp. *gadecaei* (Rouy & Camus) Heywood has stems spreading, semi-prostrate, not erect. Coastal habitats. NW France. The **Cultivated Carrot** (*D. c.* subsp. *sativus* (Hoffm.) Schuebler & Martens) has a more swollen and fleshy rootstock, usually orange. Sometimes locally naturalized or casual.

Hartwort
Broad-leaved Sermountain
Knotted Hedge-parsley
Upright Hedge-parsley
Spreading Hedge-parsley
Small Bur-parsley
Wild Carrot
×1/3

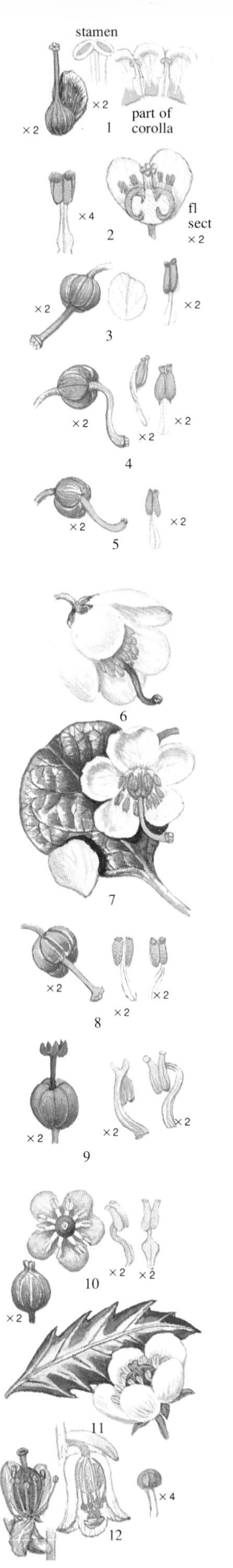


1* **Diapensia** *Diapensia lapponica* L. (Diapensia Family – Diapensaceae). Low evergreen subshrub forming dense, rather flattish cushions. Leaves small, oval-spoon-shaped, leathery, shiny deep green, crowded. Flowers white, 12-16mm, solitary, with 5 rounded petals; sepals reddening in fruit; style 3-lobed. Fruit a dry capsule, with many seeds. Rocky mountain habitats and tundra, in exposed places, to 1600m. May-June. N Britain, Iceland and Arctic Europe, to Asia and N America. Only discovered in Britain in 1951. **B**: Very rare; on a single mountain top in W Inverness-shire. Protected.

## WINTERGREEN FAMILY Pyrolaceae

Small perennial herbs. Leaves often in basal tufts, alternate or opposite. Flowers hermaphrodite, 4-5 parted, with free petals. Stamens usually twice as many as petals. Stigma solitary 5-lobed. Fruit a dry capsule containing numerous seeds. 4 genera and 35 species in N temperate and Arctic regions.

**Wintergreens** *Pyrola*. Hairless rhizomatous herbs with rather leathery leaves in basal tufts. Stem leaves small and scale-like. Flowers in terminal, long-stalked, racemes; anthers with pores at the end of short tubes. 25 species in N temperate and Arctic regions.

2* **Common Wintergreen** *Pyrola minor* L. Low to short perennial. Leaves broadly-elliptical, toothed, stalk shorter than leaf-blade. Flowers whitish or pale lilac-pink, 5-7mm, globose; style straight not protruding. Woodland, moors and damp rocky ledges, generally in mountains, rarely on sand-dunes, generally on calcareous soils, to 2700m. June-August. Throughout, except Spitsbergen. **B**: Local and declining; not in most of the Scottish Isles, most of Wales, SW England and East Anglia.

3* **Intermediate Wintergreen** *Pyrola media* Swartz. Rather taller than *P. minor* and with oval to rounded leaves. Flowers larger, white or pale pink, 8-10mm; style straight, protruding beyond the petals. Woodland, especially of Pine, and moors, heaths, on humus-rich soils, to 2200m. June-August. Throughout, except Belgium, Holland, the Faeroes, Iceland and Spitsbergen. Local. From W Europe east to the Caucasus. **B**: Rare and declining; mostly in Scotland and N Ireland – very rare elsewhere.

4* **Round-leaved Wintergreen** *Pyrola rotundifolia* L. Short perennial. Leaves oval, toothed, the stalk longer than leaf-blade. Flowers pure white, 8-12mm, opening widely; style S-shaped, protruding, 6-10mm. Woodland, bogs, fens, rock ledges and short turf, often on base-rich soils, to 1450m. June-September. Throughout, except the Faeroes and Iceland. **P**: Various insects or self-pollinated. Sometimes cultivated in gardens. **B**: Very local; confined to scattered localities; absent from SW England.

subsp. *maritima* (Kenyon) E.F. Warburg. Like the type, has smaller leaves and flower-stalks only 2-5mm long (not 4-8mm); style 4-6mm. Maritime habitats – dune slacks. Coasts of NW France to Denmark. **B**: Flint and Lancashire.

5* **Yellow Wintergreen** *Pyrola chlorantha* Swartz [*P. virens*, *P. virescens*]. Short, rather pale green perennial. Leaves oblong to rounded, toothed, the stalk longer than the leaf-blade. Flowers yellowish-green or yellowish-white, bell-shaped, 8-12mm; style curved, protruding. Coniferous woodland, rocky and grassy habitats, to 2200m. June-July. Continental Europe, except Holland, rare in the west.

6* **Norwegian Wintergreen** *Pyrola norvegica* Knaben. Short perennial. Leaves oval to elliptical, toothed, the stalk equal or shorter than the leaf-blade. Flowers large, white to lilac, 10-25mm, broad bell-shaped; style curved, protruding. Dry mountain slopes, on calcareous soils, to 1450m. June-August. W & N Scandinavia. Local.

7 *Pyrola grandiflora* Radius. Like *P. norvegica*, but with rounded, shiny leaves, the stalk longer than the leaf-blade. Flowers white, greenish-white or pink, 15-20mm. Arctic tundra. July-September. Very local. Confined to Iceland and Arctic Russia.

8* **Nodding** or **Toothed Wintergreen** *Orthilia secunda* (L.) House [*Pyrola secunda*, *Ramischia secunda*]. Low to short, rather pale green, rhizomatous perennial. Leaves small, in basal rosettes, oval, toothed, short-stalked. Flowers greenish-white, 5-6mm, in a nodding one-sided raceme; style straight, protruding. Mountain woodland and damp rocky ledges, to 2200m. June-August. Throughout, except the far north. **B**: Scattered localities from C Wales and Cumbria to the Scottish Highlands; confined to Fermanagh in Ireland.

9* **One-flowered Wintergreen** *Moneses uniflora* (L.) A. Gray [*Pyrola uniflora*]. Low to short rhizomatous perennial. Leaves rounded to oval, toothed, opposite, mainly basal. Flowers white, 13-20mm, solitary, nodding, opening widely; style straight. Native woodland, mainly coniferous, on acid soils, to 2100m. May-August. Most of Europe, but often very local. One of two *Moneses* species. **B**: Very rare and declining; confined to Scotland – mainly in the east.

10* **Umbellate Wintergreen** *Chimaphila umbellata* (L.) W. Barton [*Pyrola umbellata*]. Short, hairless subshrub. Leaves narrowly oval, leathery, dark green, opposite or in whorls. Flowers pinkish, 7-12mm, globose, in umbel-like clusters; style not protruding. Coniferous woods and damp rocky habitats, to 500m only. June-July. E France, Germany and Scandinavia.

11 *Chimaphila maculata* (L.) Pursh. Like *P. umbellata*, but leaves with white veins and flowers white, 18-20mm. **I**: E North America. Naturalized in woodland near Paris.

12* **Dutchman's Pipe, Yellow Bird's Nest** *Monotropa hypopitys* L. [*Hypopitys monotropa*] (Dutchman's Pipe Family – Monotropaceae). Low to short, downy or hairless, saprophytic perennial, whole plant creamy-white or yellowish, browning with age. Stems erect, covered with alternating, oval scale-leaves. Flowers narrow tubular-bells, 9-12mm long, borne in long nodding spikes; petals 4-5. Fruit a rounded capsule. Damp woodland, often coniferous or beech, on deep decaying leaf-litter, to 1800m. June-September. Throughout, except the far north. **B**: Very local in S England; rare elsewhere in Wales, Scotland and Ireland. subsp. *hypophegea* Holmboe [*Monotropa hypophegea*] has flowers shorter, 8-10mm, the filaments, styles and inside of the petals hairless (not stiffly hairy). but also in coastal habitats, especially dune slacks among dwarf willow scrub, *Salix repens*.

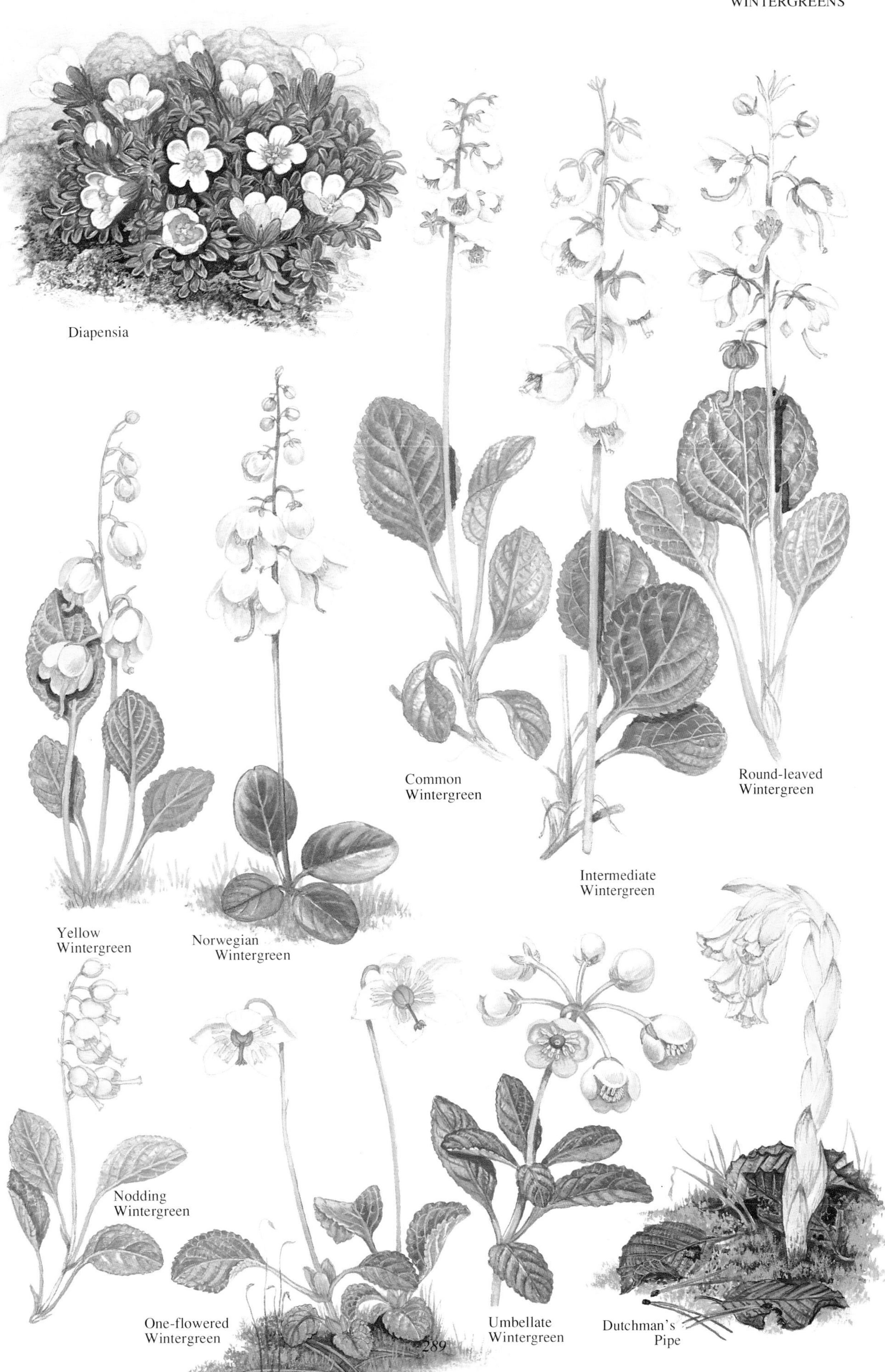
Diapensia
Yellow Wintergreen
Norwegian Wintergreen
Common Wintergreen
Intermediate Wintergreen
Round-leaved Wintergreen
Nodding Wintergreen
One-flowered Wintergreen
Umbellate Wintergreen
Dutchman's Pipe

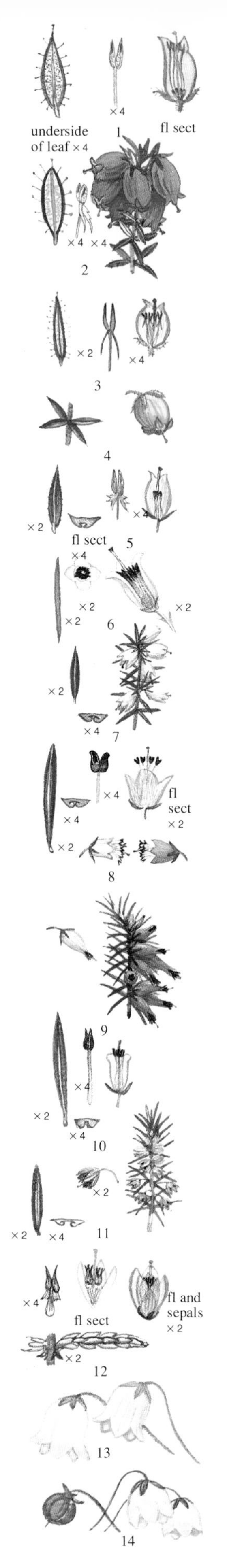


# HEATHER FAMILY Ericaceae

Dwarf or large shrubs or undershrubs with alternate, opposite or whorled leaves, untoothed usually. Flowers in racemes, often spike-like, or clusters, sometimes solitary. Petals 4-5, joined. Stamens usually twice as many as petal-lobes. Anthers opening by pores, not slits. Fruit a dry, many-seeded capsule or a berry. Over 1200 species in 50 genera.

**Heathers** *Erica*. Dwarf or medium-sized evergreen shrubs. Leaves narrow, whorled, the margin often rolled under. Flowers globular to bell-shaped, the corolla persisting in fruit, generally 4-lobed. Fruit a dry capsule. 500 species in the world.

1* **Dorset Heath** *Erica ciliaris* L. Dwarf shrub 30-80cm; young twigs hairy. Leaves oval to lanceolate in whorls of 3, usually hairy above when young, whitish beneath. Flowers bright reddish-pink, 8-12mm long, urn-shaped, somewhat curved, in terminal racemes; anthers not protruding. Fruit hairless. Wet heaths, bogs, scrub and open woods, on acid soils. May-October. S Britain, W Ireland, W & C France. **B**: Very local and decreasing; Dorset, Devon, Cornwall, Galway in Ireland.

2 **Mackay's Heath** *Erica mackaiana* Bab. [*E. mackaii*]. Like *E. ciliaris*, but with darker green leaves, borne in whorls of 4 and flowers smaller, 5-7 mm long, purplish-pink, in terminal clusters. Blanket bogs. August-September. Very local in W Donegal and W Galway. Also known from NW Spain. The Irish populations apparently do not set seed.

3* **Cross-leaved Heath** *Erica tetralix* L. Straggly, greyish dwarf shrub, 20-70cm; young twigs hairy. Leaves linear-lanceolate, in whorls of 4, hairy at least when young. Flowers pale pink, 6-9mm long; globular, in compact terminal clusters; anthers not protruding. Fruit downy. Bogs, moors, wet heaths, pine woods, occasionally on drier heaths, on acid soils, to 2200m. June-October. Throughout, except the far north. Hybridises with *E. mackaiana*, *E.* x *watsonii* Benth., in Ireland. **B**: Throughout, but absent from most of the English Midlands.

4 *Erica terminalis* Salisb. [*E. stricta*]. Like *E. tetralix*, but a much larger shrub, 1-2.5m, the leaves mostly in whorls of 4. Flowers bright pink, 5-7mm long, the corolla with recurved lobes. **I**: Spain, Italy, Morocco. Naturalized on sand-dunes in Derry, Ireland.

5* **Bell Heather** *Erica cinerea* L. Dwarf shrub 15-75mm; young twigs hairy, otherwise plant hairless. Leaves linear, in whorls of 3, dark green, often bronzed. Flowers bright magenta-purple, 4-7mm, bell-shaped, in racemes or compact clusters. Fruit hairless. Sub- alpine and maritime heaths, moors and open woodland on rather dry acid soils, to 1500m. July-September. Throughout, except much of the north and Iceland. Often growing in association with other dwarf ericaceous shrubs.

6* **Portuguese Heather** *Erica lusitanica* Rudolphi. Large shrub, 1-3.5m, with erect branches and smooth stems. Leaves hairless, linear to oblong, bright green in whorls of 3-5. Flowers white tinged with pink, 4-5mm long, in lateral racemes forming large heads; stigma red. Damp heaths and woodland margins, railway embankments. March-May. **I**: France, Spain and Portugal. Naturalized in Dorset and Cornwall.

7 **Tree Heather** *Erica arborea* L. Like *E. lusitanica*, but with rough hairs and smaller, pure white flowers, 2.5-4mm long; stigma white. **I**: Mediterranean. Widely cultivated, occasionally semi-naturalized in SW Britain and SW France. The roots and old knotted stems are used for making briar pipes.

8* **Cornish Heath** *Erica vagans* L. Short to medium, hairless shrub, to 60cm. Leaves linear, in whorls of 4-5. Flowers lilac, pink or white, 2.5-3.5mm, bell-shaped, in dense leafy racemes; anthers chocolate-brown, fully protruding. Fruit hairless. Dry heaths and open woods over acid rocks, often on ultra-basic and magnesium-rich soils, to 1800m. July-September. SW Britain, NW Ireland and W & C France. Also found in N Spain. Garden forms are occasionally naturalized. **B**: Very local; confined to W Cornwall and Fermanagh in Ireland.

9 **Spring Heath** *Erica herbacea* L. [*E. carnea*]. Dwarf shrub, spreading, only to 25cm tall; young twigs practically hairless. Leaves linear, in whorls of 4, congested. Flowers bright red, or flesh-pink, 5-6mm long, narrow bells borne in dense one-sided racemes; anthers dark purple, protruding. Mountain woods and stony slopes, over calcareous rocks, to 2700m. March-June. N & EC Germany.

10* **Irish Heath** *Erica erigena* R. Ross. [*E. mediterranea*]. Like *E. herbacea*, but an erect shrub to 2m with racemes often aggregated into a panicle; anthers reddish, only half-protruding. Damp habitats, moors and bogs. March-May. SW France and W Ireland – rare.

11 *Erica scoparia* L. Slender erect shrub to 1.5m, with whitish hairless twigs. Leaves in whorls of 3-4. Flowers green tinged with red, small, 2.5-3mm long, in narrow racemes. Woods, often of pine, and among heather, on acid soils. December to June. NC France southwards; naturalized in Holland.

12* **Heather, Ling** *Calluna vulgaris* (L.) Hull. Short to medium carpeting subshrub, to 80cm, scarcely hairy. Leaves opposite, scale-like, congested. Flowers pale purple to pinkish-lilac, 3-4mm, in slender racemes; sepals free and petal like, longer than the true petals; anthers not protruding. Open woodland, moors, bogs, banks, roadsides and stabilized dunes, on acid soils, to 2700m. July-September. Throughout, except Spitsbergen. Often forming extensive colonies.

13* **Cassiope** *Cassiope tetragona* (L.) D. Don. Dwarf shrub, 10-30cm, erect, much-branched. Leaves opposite, scale-like, lanceolate, closely overlapping along stems. Flowers creamy-white, often flushed with pink, 6-8mm, bell-shaped, nodding on slender downy stalks, 5-parted. Dry stony or sandy heaths and tundra, usually on slightly calcareous soils, to 1650m. Arctic Europe.

14* **Matted Cassiope** *Cassiope hypnoides* (L.) D. Don [*Harrimanella hypnoides*]. Prostrate, mat-forming subshrub. Leaves alternate, linear-oval, pointed, crowded but not closely pressed to the stem. Flowers white, often flushed with pink, and with pink sepals, 4-5mm, solitary, nodding, rounded-bells borne on slender stalks. Damp mossy tundra, streamsides and snow hollows, to 1900m. June-August. Scandinavian mountains, Arctic Europe and Iceland.

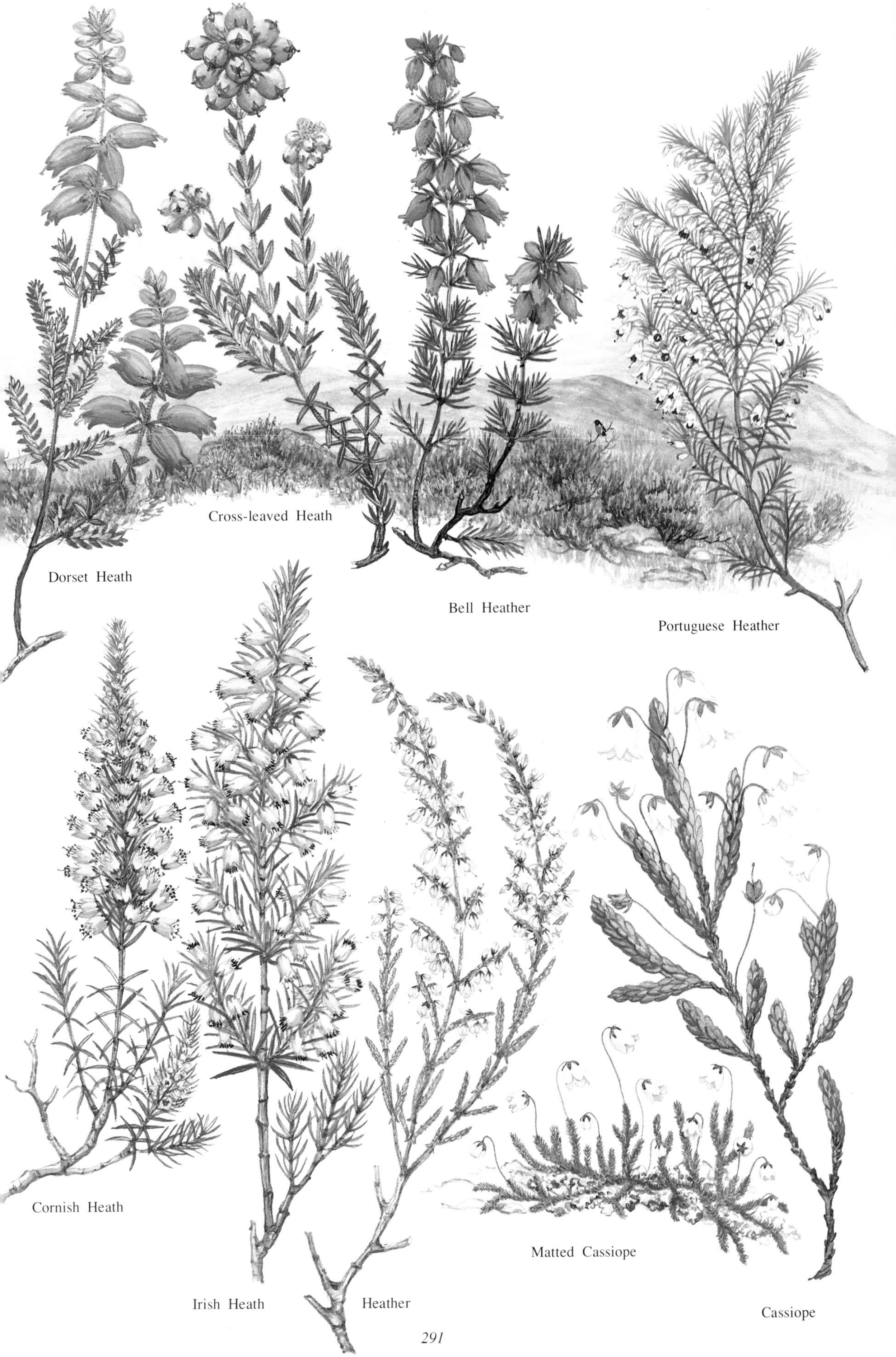

Cross-leaved Heath

Dorset Heath

Bell Heather

Portuguese Heather

Cornish Heath

Matted Cassiope

Irish Heath

Heather

Cassiope

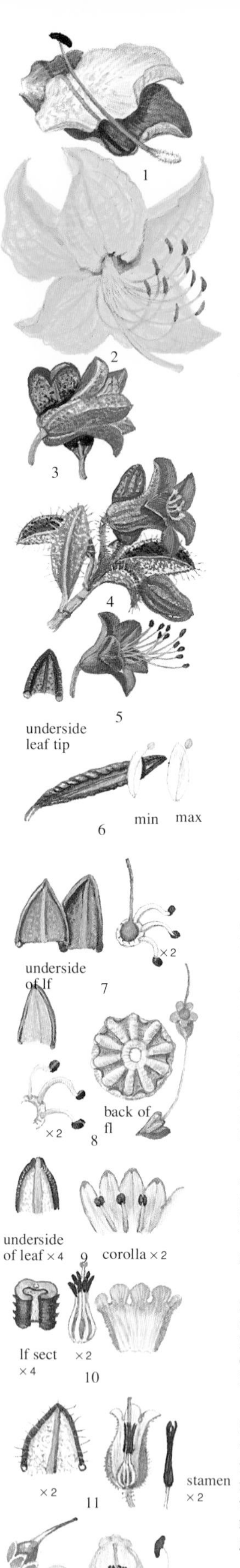


**Rhododendrons** *Rhododendron*. Shrubs, mostly evergreen, with alternate leathery leaves. Flowers in terminal clusters, trumpet or funnel-shaped with the petal partly joined in a tube. Stamens 10 usually. Fruit a dry capsule containing many tiny seeds. 600 species scattered across the northern hemisphere and in SE Asia, with a concentration of species in the E Himalaya, SW China and the islands of SE Asia. The nectar of some is poisonous.

1* *Rhododendron ponticum* L. Vigorous evergreen shrub to 5m, with spreading branches, hairless, often forming large patches. Leaves large, elliptical, deep shiny green, paler beneath. Flowers violet-purple, mauve, pinkish or whitish, generally spotted within, trumpet-shaped, 40-60mm long, in clusters of 8-15. Woodland and woodland margins, scrub, roadsides, railway embankments, rocky places, river and stream margins. May-August. **I**: W Asia. Widely naturalized in Britain, Belgium and France. Cultivated in large gardens, parks and estates, but well established in many places, often becoming a menace and difficult to control. **P**: Bees. **B**: Widely naturalized.

2 *Rhododendron luteum* Sweet [*Azalea pontica*]. Deciduous shrub, 2-4m, with smaller flowers than *R. ponticum*, yellow and richly fragrant. Woodland and scrub. May-June. **I**: Asia Minor. Occasionally naturalized in Britain. Widely cultivated in gardens including forms with salmon, apricot and orange, as well as semi-double flowers.

3* **Alpenrose** *Rhododendron ferrugineum* L. An evergreen shrub to 1.2m, often forming dense thickets. Leaves small, elliptic-oblong, shiny deep green above, reddish-scaly beneath. Flowers pale to deep reddish-pink, trumpet shaped, 13-16mm long, borne in small clusters. Stamens 10. Mountain slopes, open woods, scrub and roadsides, on neutral or slightly acid soils usually, to 3200m. May-August. E France and S Germany. Widespread in the mountains of C and S Europe. **P**: Mainly bees.

4 **Hairy Alpenrose** *Rhododendron hirsutum* L. Like *R. ferrugineum*, but leaves bright green and with a hairy margin; flowers bright pink. Open mountain woods, scrub and screes, on limestone, to 2600m. May-July. E France and S Germany. Unusual because it grows on calcareous soils whereas most *Rhododendron* species are adapted to acid soils and rocks.

5* **Lapland Rhododendron** *Rhododendron lapponicum* (L.) Wahlenb. Spreading evergreen shrub to 0.5m high, forming dense mats. Leaves oblong, dark green above, rusty with scales beneath, margins rolled under. Flowers violet-purple, bell-shaped, 8-15mm long, borne in small clusters; stamens only, 5-8. Dry heaths, tundra, stony places, on calcareous soils, to 1350m. May-June. Mountains of Scandinavia. Rather rare in cultivation.

6* **Labrador Tea** *Ledum palustre* L. An evergreen shrub to 1.2m, erect or spreading, young twigs rusty-hairy. Leaves narrow-oblong, leathery, dark green, rusty-hairy beneath, margins rolled under. Flowers creamy-white, 10-15mm, in dense umbels, with 5 separate petals; stamens 10 usually. Fruit a dry capsule. Bogs, heaths, coniferous woodland and scrub. May-July. Scandinavia and Germany.

subsp. *groenlandicum* (Oeder) Hultén [*L. groenlandicum*]. Like the type, but with narrower, more elliptic leaves, with the midrib concealed beneath. Flowers with 8 stamens. Similar habitats. C Scotland and Germany. Very local.

7* **Sheep Laurel** *Kalmia angustifolia* L. An evergreen shrub up to 1.5m, with hairless rounded twigs. Leaves opposite or in threes, elliptical to oblong, stalked, untoothed, slightly hairy, green or rusty beneath when young. Flowers bright reddish-pink, saucer-shaped, 8-9mm, in lateral clusters; stamens 10. Fruit a dry capsule. Moist peaty soils, scrub and open woodland. May-July. **I**: E North America. Naturalized in NW England and Germany. **B**: Very locally – naturalized in Cumbria.

8* *Kalmia polifolia* Wangenh. Like *K. angustifolia*, but young twigs ridged and hairy and leaves narrower, mostly opposite. Flowers in terminal clusters, rosy-lilac, 8-18mm. **I**: North America. Naturalized on bogs in SE England.

9* **Creeping or Mountain Azalea** *Loiseleuria procumbens* (L.) Desv. [*Azalea procumbens*]. Prostrate, mat-forming, hairless evergreen subshrub. Leaves small, oblong, shiny deep green, the margins rolled under. Flowers pale pink, tiny, 4-6mm, cup-shaped, solitary or in small terminal clusters; stamens 5. Dry bare stony and rocky ground, moors and peaty habitats, mainly in the mountains, to 3000m, but to 400m in N Britain. May-July. N Britain, the Faeroes, Iceland, Scandinavia, E & S France and S Germany. **P**: Various small insects, also self-pollinated. **B**: Local in the Scottish Highlands and Orkney.

10* **Mountain Heath** *Phyllodoce caerulea* (L.) Bab. Dwarf, heath-like, evergreen shrub, 10-35cm, stems often rooting at the base. Leaves alternate, narrow-oblong, rough-edged. Flowers lilac to purplish-pink, pendent bells, 7-12mm long, in clusters of 2-6. Fruit a dry, glandular, capsule. Mountain heaths, rocky and bushy moorland, on acid soils, to 2600m. June-August. N Britain, Scandinavia, Iceland and SW France. **P**: Various insects or self-pollinated. **B**: Very rare; restricted to a few localities in the Scottish Highlands but much depleted by over-collecting.

11* **St Dabeoc's Heath** *Daboecia cantabrica* (Hudson) C. Koch. Dwarf evergreen shrub to 70cm, with rather straggling stems. Leaves lanceolate to oblong, dark green, glandular-hairy above, white-downy beneath, with inrolled margins. Flower purplish-pink, bell-shaped, 9-14mm long, 4-lobed, in lax clusters. Fruit a dry capsule. Heaths, open woods, rocky habitats, on acid soils. May-October. W Ireland and France. Garden forms may be white, pink and purple-flowered. Also known from NW Spain. **B**: Local in W Ireland.

12* **Strawberry Tree** *Arbutus unedo* L. Small evergreen tree or shrub to 8m; bark reddish-brown, peeling in small flakes. Leaves elliptical, alternate, dark green, leathery, toothed, almost hairless. Flowers pearly-white, tinged with pink or green, bell-shaped, 7-9mm, 5-lobed, borne in dense drooping panicles, honey-fragrant. Fruit a rough-warty berry, 18-20mm, ripening through yellow to scarlet and crimson. Woodland margins, scrub and rocky habitats. August-December. W Ireland, W & S France. A characteristic plant of the Mediterranean maquis. **P**: Bees; the flowers, which have a strong honey-fragrance, produce abundant nectar. The fruits are edible and can be made into jams and preserves. Commonly cultivated in parks and gardens. **B**: Locally abundant in W Ireland.

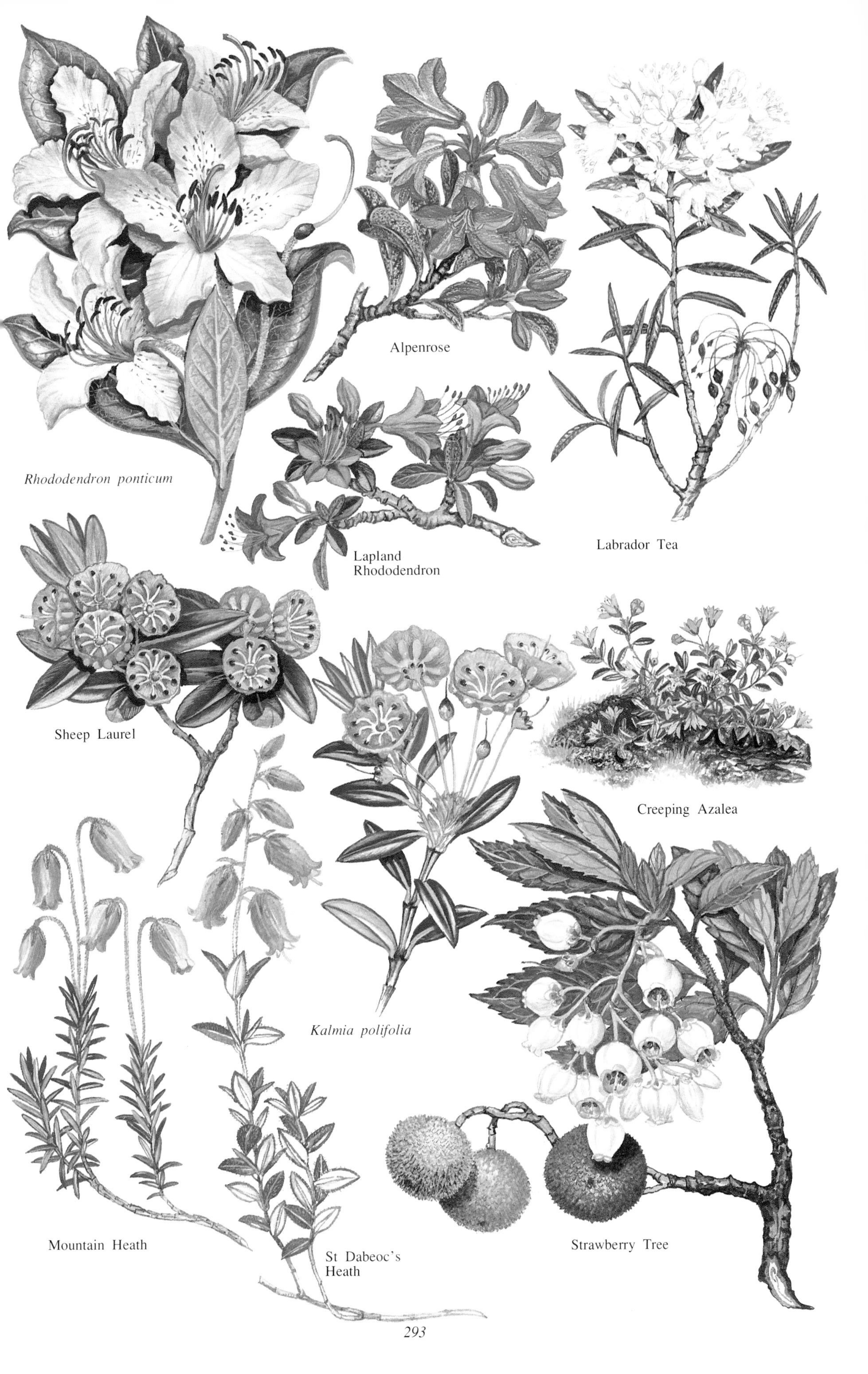
Rhododendron ponticum
Alpenrose
Labrador Tea
Lapland Rhododendron
Sheep Laurel
Creeping Azalea
Kalmia polifolia
Mountain Heath
St Dabeoc's Heath
Strawberry Tree

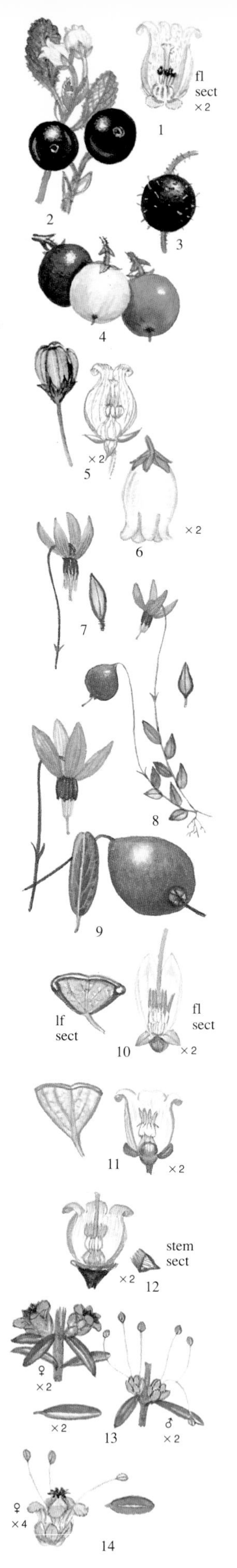


1* **Bearberry** *Arctostaphylos uva-ursi* (L.) Sprengel [*Arbutus uva-ursi*]. Prostrate, evergreen shrub with long rooting branches to 1.5m. Leaves alternate deep green and leathery, oval, broadest towards the tip, untoothed and with flat margins. Flowers greenish-white to pink, bell-shaped, 5-6mm, in small terminal clusters. Berry shiny-red, 6-8mm, edible. Rocky moors and heaths, open woodland, often in exposed places and on thin peaty soils, to 2800m. July-September. Throughout, except the far north. **B**: Scottish Highlands and the Scottish Islands; local in N England – Pennines and Cumbria, and in NW Ireland.

2 **Alpine Bearberry** *Arctostaphylos alpina* (L.) Sprengel [*Arbutus alpina, Arctous alpinus*]. Like *A. uva-ursi*, but smaller, stems to 60cm. Leaves not leathery, toothed, withered by late autumn but persisting until the following spring. Berry black when ripe, 6-10mm. Mountain moors and heaths, on acid rocks, sometimes together with *Calluna*, to 2700m. May-August. Throughout, except the Holland and the far north. **B**: Local in N & NW Scotland including Orkney and Shetland.

3* **Shallon** *Gaultheria shallon* Pursh. An erect evergreen shrub, 0.5-1.75m, suckering to form dense thickets, glandular hairy. Leaves alternate, elliptical leathery, toothed, short-stalked. Flowers pink, bell-shaped, 8-10mm, in terminal clusters. Berry dark purple-black, succulent, hairy. **I**: W North America. Naturalized on sandy and peaty soils, especially in open woodland. May-June. Britain. **B**: Scattered localities.

4* **Prickly Heath** *Pernettya mucronata* (L. f.) Gaud. ex Spreng. An erect or spreading, prickly, evergreen shrub, to 1m, suckering freely to form dense thickets, almost hairless. Leaves alternate, elliptical, spine-tipped, leathery and untoothed. Flowers white, 5-6mm, bell-shaped, borne in terminal drooping clusters. Fruit a berry, purple, pink or white, 10-12mm. **I**: S Chile. Naturalized on well-drained sandy soils in Britain. May-June. **B**: Occasional garden escape.

5* **Bog Rosemary** *Andromeda polifolia* L. Dwarf evergreen shrub, to 35cm, spreading to erect. Leaves alternate, linear to oblong, untoothed, whitish-grey beneath and with the margin rolled under. Flowers bright pink, bell-shaped, 5-8mm, nodding in terminal clusters. Fruit an erect greyish capsule. *Sphagnum* bogs and other moist acid habitats, to 1150m. May-September. Throughout, except Holland and the far north.

6* *Chamaedaphne calyculata* (L.) Moench [*Cassandra calyculata*]. An erect evergreen shrub, to 50cm, slightly hairy when young. Leaves, alternate, elliptical, sometimes slightly toothed, scaly-brown beneath. Flowers white, bell-shaped, 5-6mm long, in rather horizontal leafy racemes. Fruit a rounded capsule. Bogs, marshy habitats and wet woodland, on acid soils. June-July. Finland and Sweden.

*Vaccinium*. Deciduous or evergreen shrubs with alternate leaves. Flowers 4-5-parted, in lateral or terminal clusters or racemes, or solitary; petals fused in the lower part. Stamens 8-10. Fruit a berry. 100 species in Arctic and temperate regions. A number have edible berries.

7* **Cranberry** *Vaccinium oxycoccus* L. [*Oxycoccus quadripetalus, O. palustris*]. Low creeping evergreen subshrub. Leaves oval, untoothed, deep green above, whitish beneath, 3-6mm wide. Flowers pinkish-red, 6-10mm, with 4 narrow spreading or reflexed petal lobes, long-stalked. Berry rounded to pear-shaped, red or brownish, often speckled, 8-10mm, edible. Peat bogs, wet heaths and open wet woodland, generally creeping over *Sphagnum*, on acid soils, to 2000m. June-August. Throughout, except the far north. **B**: N England, Wales, Scotland and parts of Ireland.

8 **Small Cranberry** *Vaccinium microcarpum* (Turcz. ex Rupr.) Schmalh. [*Oxycoccus microcarpus*]. Like *V. oxycoccus*, but a smaller plant with leaves not more than 2.5mm wide, more triangular with hairless (not hairy) stalks. Berries red, 5-8mm, occasionally speckled. Drier parts of peat bogs. July-August. N Britain, Scandinavia, Iceland and S Germany. **B**: Very local; confined to E Scottish Highlands.

9 **American Cranberry** *Vaccinium macrocarpum* Aiton. Like *V. oxycoccus*, but a more robust plant with oblong leaves, 3.5-5mm wide. Flowers larger, 8-16mm. Berry red, 10-20mm. **I**: E North America. Locally naturalized in N Britain, Holland and Germany. Cultivated for its edible fruits.

10* **Cowberry** *Vaccinium vitis-idaea* L. More-or-less prostrate, creeping evergreen, subshrub. Leaves elliptical to oblong, leathery, dark green, the margins rolled under, untoothed. Flowers white or pale pink, bell-shaped, 5-8mm, in crowded racemes. Berry globose, red, 5-10mm, acid-tasting. Moors, heaths and coniferous woods, tundra, on poor acid soils, to 3050m. June-August. Throughout. Hybridises with *V. myrtillus*: *V. x intermedium* Ruthe. **B**: Throughout much of Scotland, more local in N England and Wales; scarce in Ireland.

11* **Bog Bilberry** *Vaccinium uliginosum* L. Deciduous creeping shrub with erect stems to 75cm, twigs rounded in section. Leaves oval, bluish-green, with netted veins, untoothed. Flowers white, often pink tinged, bell-shaped, 4-6mm long, in clusters of 1-3. Berry bluish-black, 7-10mm, sweet and edible. Moors, heaths and coniferous woodland, tundra, on poor damp acid soils, often in association with *V. myrtillus*, to 3000m. May-June. Throughout. **B**: From Cumbria and Durham northwards, and in Ireland. subsp. *microphyllum* Lange. has smaller leaves 6-15mm long (not 10-25mm). Arctic and sub-Arctic Europe.

12* **Bilberry** *Vaccinium myrtillus* L. Hairless creeping deciduous shrub to 60cm with numerous erect stems; twigs 3-angled, green. Leaves oval, bright green, toothed, flat. Flowers pale green tinged with pink, lantern-shaped, 4-6mm, usually solitary. Berry globose, bluish-black and sweet when ripe, edible. Heaths, moors and open woods (mainly birch, oak or pine) on dry acid soils, to 2800m. April-June. Throughout, except Spitsbergen. Often dominant on high moors and forming extensive colonies. **P**: Mainly bees.

## CROWBERRY FAMILY Empetraceae

Small evergreen shrubs with a heather-like appearance. Leaves alternate, untoothed. Flowers 4-6-parted. 3 genera and about 10 species in N temperate regions, the Andes and Falkland Is.

13* **Crowberry** *Empetrum nigrum* L. Low mat-forming shrub, 15-45cm; young twigs reddish. Leaves crowded, linear-oblong, deep shiny green, margins rolled under. Flowers pink or purplish, tiny, 1-2mm, at base of leaves, 6-petalled, dioecious. Fruit a rounded black berry, green at first, 5-6mm. Moors and bogs, birch and pine woodland, on rather dry peaty soils, sea level to 3050m. May-June. Throughout. **B**: Throughout N England, Wales and Scotland, N & W Ireland; more local in the south.

14 **Mountain Crowberry** *E. n.* subsp. *hermaphroditum* (Hagerup) Böcher [*Empetrum hermaphroditum*] stems not rooting down and greenish; leaves often slightly grooved beneath; flowers normally hermaphrodite. Mountain tops, high moors. Throughout. **B**: only in C & N Scotland.

Bearberry
Shallon
Prickly Heath
Bog Rosemary
*Chamaedaphne calyculata*
Cranberry
Cowberry
Bog Bilberry
Bilberry
Crowberry

# PRIMROSE FAMILY Primulaceae

Herbs, usually with simple leaves, often in basal rosettes. Flowers regular, usually 5-parted, the petals sometimes partially joined to form a tube; stamens opposite the petal-lobes. Ovary with a single compartment (loculus) and a solitary style. Fruit a small capsule. 20 genera and 1000 species almost throughout the world, but particularly in N temperate regions. Many of the genera have species which are cultivated in gardens – particularly *Anagallis*, *Androsace*, *Cyclamen*, *Dodecatheon*, *Lysimachia* and *Primula*.

**Primulas** *Primula*. Perennial herbs with leaves in basal rosettes. Flowers solitary or in long-stalked umbels; calyx bell-shaped; corolla with a distinct tube and spreading notched petal-lobes. Fruit capsule splitting with 5 valves, containing many seeds. About 500 species in N temperate regions and temperate South America. Most exhibit heterostyly, plants bearing either long-(pin-eyed) or short- (thrum-eyed) styles, the flowers with the stamens placed in an opposing position in the same flower. By this device cross-pollination between different plants is assured.

1* **Primrose** *Primula vulgaris* Hudson [*P. acaulis*]. Low hairy perennial. Leaves oval, tapered to the stalk, finely toothed, bright green above, paler and hairy beneath. Flowers pale yellow, 20-40mm, solitary, usually with orange markings in the centre, fragrant. Moist shaded habitats, woods, coppices, thickets, scrub, grassy banks, ditches, hedgebanks and sea cliffs, generally on heavy soils, to 1500m. Throughout, except Finland and Iceland. Widely cultivated in gardens, including white and pink as well as semi-double-flowered forms – these sometimes become naturalized. Formerly an important medicinal plant – the roots are said to be a strong, but safe, emetic. **B**: Throughout, but much diminished in some areas due to modern agricultural practices and over-collecting.
subsp. *sibthorpii* (Hoffmanns.) W.W. Smith & Forrest. Like the type, but flowers pink, red or purple. **I**: E Europe and W Asia. Naturalized occasionally from gardens, especially in C Europe.

2* **Oxlip** *Primula elatior* (L.) L. Low to short hairy perennial. Leaves rounded to elliptical, narrowed abruptly into the winged stalk. Flowers pale yellow, 15-25mm, several to many in a nodding one-sided umbel on a long stalk, not fragrant. Moist habitats, woods, coppices, ditches, moist grassy meadows and streamsides, on heavy soils, to 2700m. April-May. S Britain and Continental Europe, north to Denmark and S Sweden. Cultivated in gardens. Mountain forms are generally far dwarfer and with smaller leaves. **P**: Bees and butterflies. Hybrids with *P. vulgaris*, *P.* x *digenea* A. Kerner (3), may occur when the parent species grow in close proximity. **B**: Locally common, especially in ancient coppice woodland in E England.

4* **Cowslip** *Primula veris* L. [P. officinalis]. Low to short hairy perennial. Leaves oblong, broadest near the base and abruptly narrowed into the stalk. Flowers deep yellow, 9-15mm, with orange markings in the centre, sweetly fragrant, up to 30 in a nodding, one-sided cluster. Grassy habitats, meadows and pastures, scrub and open woodland, banks and roadsides, on drier calcareous soils than the previous 2 species, to 2200m. April-May. Throughout, except the extreme north and Iceland. Locally abundant. Cultivated in gardens. A food plant of the Duke of Burgundy Fritillary, *Hamearis lucina*. Hybrids with *P. vulgaris*, *P. tommasinii* Gren. & Godron (5), the False Oxlip, sometimes occur, generally close to one or both parents and these may be mistaken for *P. elatior*. **B**: Widespread in much of England, C & S Scotland and C Ireland; scarce elsewhere.
subsp. *canescens* Hayek. Like the type, but the leaves grey-hairy beneath; flowers 8-20mm; calyx larger 16-20mm (not 8-15mm). Similar habitats. C Europe to S France.

6* **Birdseye Primrose** *Primula farinosa* L. Low perennial, white mealy on the leaves beneath and on the stems. Leaves spoon-shaped or elliptical, finely toothed or not. Flowers lilac-pink or purple with a yellow 'eye', rarely all white, 8-16mm, 2 or more in a long-stalked umbel; sepals often tinged with black or purple, mealy. Damp habitats, grazed grassland, peaty places and the drier parts of mires, on basic or calcareous soils, to 3000m. May-August. N Britain and Continental Europe north to Denmark and S Sweden. Usually local, sometimes abundant. **B**: Confined to N England; local.

7* **Scottish Primrose** *Primula scotica* Hooker. Low perennial, mealy-white on leaves and stems. Leaves oblong to elliptical, slightly toothed or untoothed. Flowers dark purple with a yellow 'eye', rarely white, 5-8mm, 1-several on a short or long common stalk. Coastal turf and dunes, grassy heaths, on poorly-drained, base-rich sands or clay, at low altitudes. May-June, occasionally later. N Scotland. Unusual in that the flowers are all of the same type, not heterostylous. **B**: Rare; confined to the north coast of Scotland and Orkney.

8* *Primula scandinavica* Bruun. Like *P. scotica*, but a taller plants, the flowers 9-12mm. Calcareous mountain rocks. June-August. Norway and NW Sweden. Local.

9* *Primula stricta* Hornem. Like *P. scotica*, but leaves not mealy and flowers violet or lilac. Meadows and cliffs, to 750m. June-August. Scandinavia and Iceland.

10 *Primula nutans* Georgi [*P. sibirica*, *P. finmarchica*]. Like *P. stricta*, but flowers much larger, 10-20mm (not 4-9mm), 'pin'-or 'thrum-eyed'; leaves rather fleshy, hairless. Damp meadows, usually close to the sea. Scandinavia.

11 *Primula egaliksensis* Wormsk. Like *P. stricta*, but leaves elliptical to oblong (not spoon-shaped); flowers white or lilac. Moist pastures close to the sea. N Iceland. Also found in Greenland and Canada. Very local.

12* **Auricula, Bear's-ear** *Primula auricula* L. Low to short perennial. Leaves lance-shaped to rounded, rather fleshy, toothed or not, green or mealy-white. Flowers deep yellow, mealy in the centre, 15-25mm, in a mealy-stalked umbel. Damp grassy places and rock crevices, mountain habitats to 2900m. May-July. E France and S Germany. Widely cultivated in gardens where a wide range of flower colours are known, some with banded or speckled flowers.

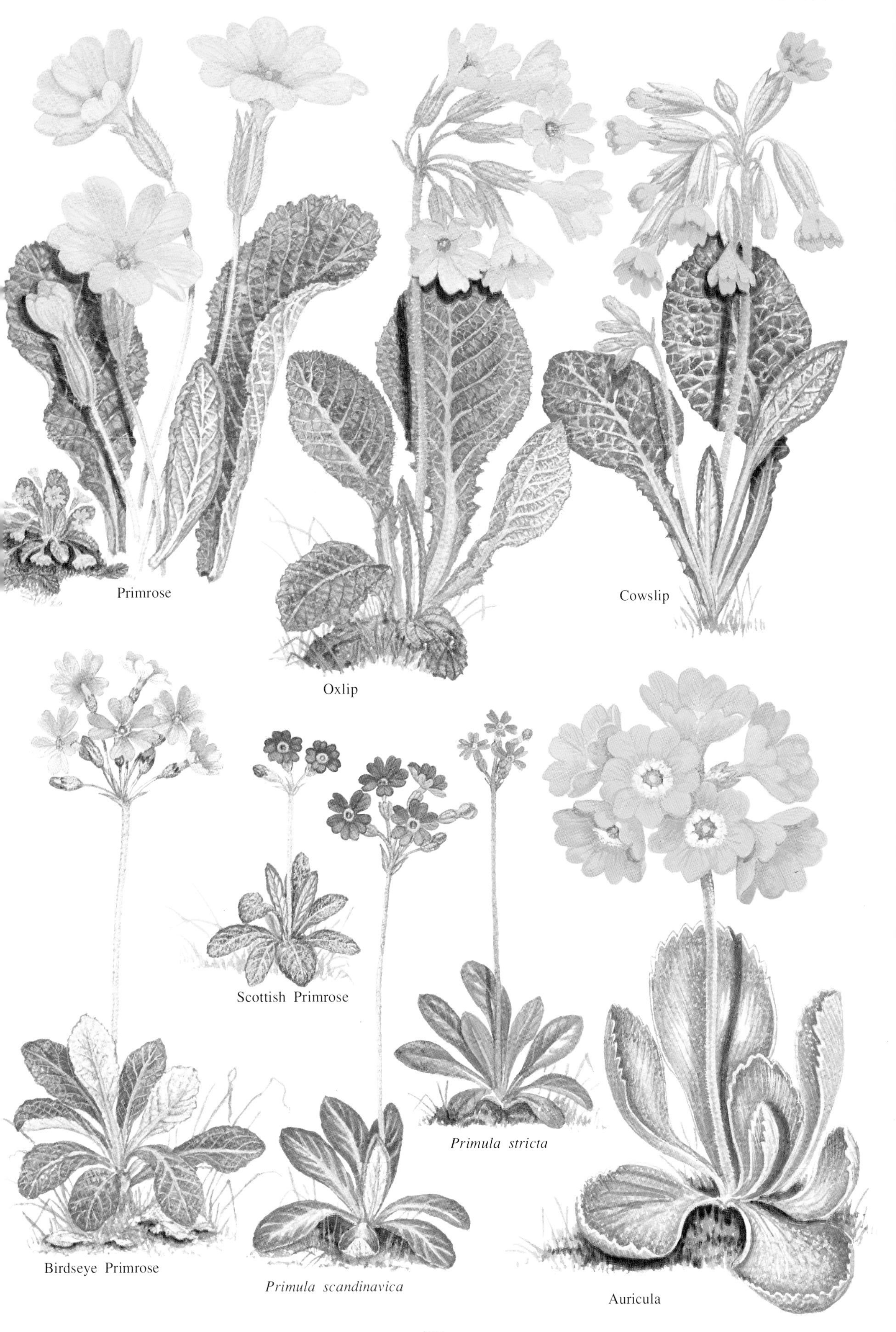
Primrose
Oxlip
Cowslip
Birdseye Primrose
Scottish Primrose
*Primula scandinavica*
*Primula stricta*
Auricula

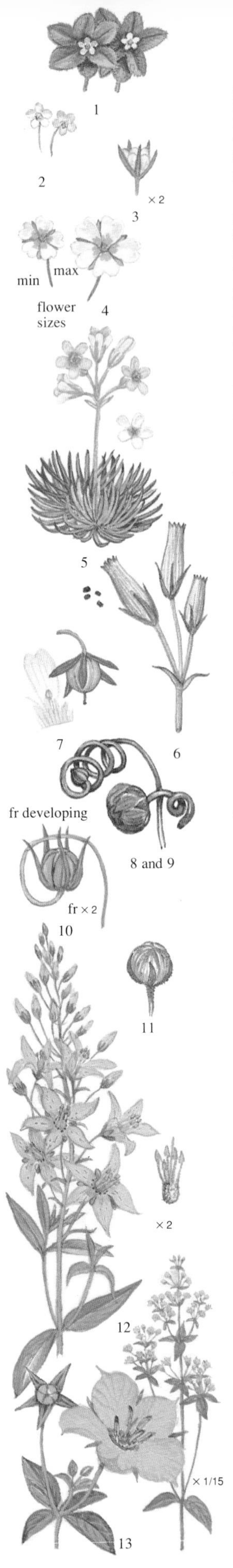


**Rock-jasmines** *Androsace*. Annual or perennial herbs, often forming low cushions. Leaves in basal rosettes. Flowers solitary or in small umbels, all similar (homostylous). Calyx bell-shaped. Petals with a short tube and spreading rounded or notched lobes, restricted in the throat. Fruit capsule splitting almost to the base. About 100 species in N temperate regions, especially Europe and the Himalaya. Many of the alpine species form dense cushions with leaf-rosettes.

1* **Annual Androsace** *Androsace maxima* L. [*A. turczaninowii*]. Low hairy annual. Leaves green, rounded or oblong, toothed, scarcely stalked, in a lax rosette. Flowers white or pink, small, 2-3mm, but surrounded by larger green sepals, in a long-stalked umbel. Dry fields, rocky and waste places, generally on slightly calcareous soils, to 1600m. April-May. C & S France and S Germany. Rather insignificant plant with very small flowers; widespread from C Europe to C Asia.

2* **Northern Androsace** *Androsace septentrionalis* L. Low, partly downy, annual or biennial. Leaves oblong to elliptical, toothed, in a lax basal rosette. Flowers white or pale pink, 4-5mm, 5-30 in long-stalked umbels; petals longer than sepals. Dry meadows and sandy habitats, to 2200m. Scandinavia, France and Germany. Local, generally found only on the mountains in the south of its range.

*Androsace chaixii* Gren. & Godron. Like *A. septentrionalis*, but with laxer, 5-8-flowered umbels; flowers larger 5-7mm. Open woodland, short turf and rocky ground, to 1800m. April-June. E & S France.

3* **Elongate Androsace** *Androsace elongata* L. Rather like *A. septentrionalis*, but almost hairless and leaves often untoothed. Flowers white, 3mm, the sepals slightly longer than the petals, many in long-stalked umbels. Dry open grassy habitats, to 2200m. May-July. C & S France and C & S Germany.

4* **Milkwhite Rock-jasmine** *Androsace lactea* L. Low, slightly hairy perennial, forming loose mats of small rosettes. Leaves linear, pointed, untoothed. Flowers milk-white with a yellow eye, 8-12mm, solitary or up to 6 in a long-stalked umbel; petal-lobes notched. Calcareous rocks, scree and rocky turf, to 2400m. E France – including the Jura Mts, S Germany. Cultivated – mainly in alpine houses and frames.

5 **Pink Rock-jasmine** *Androsace carnea* L. Low, tufted, slightly hairy perennial, with closely packed, bright green leaf-rosettes. Leaves linear, pointed, with bristly margin. Flowers pink or white, 6-9mm, 2-8 in a short or long-stemmed umbel; petals rounded. Acid rocks and screes, or short turf, in the mountains, 1400-3100m. June- August. C & S France. **P**: Bees and butterflies. Widely cultivated in alpine houses and frames. Several subspecies are recognised in C & SW Europe.

6* **Alpine Snowbell** *Soldanella alpina* L. Low hairless, mat-forming perennial. Leaves deep green and leathery, kidney-shaped, untoothed, long-stalked. Flowers violet or violet-blue, deeply fringed nodding bells, 8-13mm long, in clusters of up to 4, long-stalked. Fruit capsule oblong, erect. Mountain habitats, wet pastures, stony places, flushes and moraines, most frequent around snow patches and snow hollows, generally on calcareous soils, 450-3000m. May-August. C & E France – including the mountains of the Auvergne, S Germany – including the Schwarzwald. The stems greatly elongate as the fruits swell. Occasionally cultivated in gardens – primarily in alpine houses.

7* **Water Violet** *Hottonia palustris* L. Pale green aquatic, hairless perennial with submerged or floating, pinnate leaves. Flowers pale lilac with a yellow 'eye', in whorled racemes held above the water, with 5 notched lobes. Still fresh water, lakes, ponds and ditches, to 1500m. May-July. Europe, north to S Sweden. Cultivated in water gardens. **B**: Local in lowland England; scarce in Wales and Ireland.

8* **Sowbread** *Cyclamen purpurascens* Miller [*C. europaeum*]. Low tuberous-rooted perennial, evergreen and practically hairless. Leaves all basal rounded-heart-shaped, shiny deep green, often marbled, purplish beneath. Flowers carmine-pink, 15-20mm, solitary, with reflexed petal lobes, sweetly fragrant. Fruit capsules with coiling stalks. Rocky woodland and scrub, on moist loamy calcareous soils, to 1800m. June-October. E France. C & E Europe. Cultivated in gardens. The sweet, sticky seeds of this and other species of *Cyclamen* are much sought by ants which aid in their dispersal.

9* **Ivy-leaved Sowbread** *Cyclamen hederifolium* Aiton [*C. neapolitanum*]. Like *C. purpurascens*, but leaves lobed, pointed, ivy-like, strongly variegated with green and grey. Flowers pale pink with magenta streaks near the mouth, with an angled 5-sided throat, appearing before or with the young leaves. Fruit capsules with coiling stalks. Rocky and stony habitats, woodland and scrub, to 1200m. August-November. S France; naturalized locally in Britain. The most commonly cultivated *Cyclamen* in the open garden, often seeding itself around freely and sometimes becoming naturalized. In time the tubers reach a large size, 20cm or more across.

**Lysimachias** *Lysimachia*. Erect or prostrate herbs with opposite or whorled leaves. Flowers 5-parted, often rather starry, solitary, clustered or in racemes; petal-lobes joined near the base. Fruit-capsule 5-valved. 200 species in Europe, Asia and North America. Many grow in damp or semi-shaded habitats and a number are cultivated in gardens for their ornamental value.

10* **Yellow Pimpernel** *Lysimachia nemorum* L. Low evergreen creeping hairless perennial. Leaves opposite, pale green, rounded to lanceolate, pointed. Flowers bright yellow, saucer-shaped, 10-15mm, solitary, long-stalked, with slender sepal-teeth. Damp and shady places. May-July. Throughout, except the extreme North and Iceland. **B**: Throughout much of Britain.

11* **Yellow Loosestrife** *Lysimachia vulgaris* L. Medium to tall softly-hairy, stoloniferous perennial. Leaves opposite or in whorls of 3-4, oval to lanceolate, dotted with black or orange glands, stalked. Flowers yellow, 15-20mm, in pyramidal panicles, leafy below; sepal margins orange. Moist habitats, marshes, fens, river, stream and lake margins, on neutral or calcareous soils. Late June-August. Throughout, except the Faeroes, Iceland and Spitsbergen. Sometimes cultivated in gardens. **B**: North to Aberdeen.

12 *Lysimachia terrestris* (L.) Britton [*L. stricta*]. Medium to tall hairless perennial; stems erect. Leaves opposite, lanceolate, dotted with black glands, scarcely stalked. Flowers yellow, streaked and dotted with red or black, 9-13mm, in terminal racemes. **I**: North America. Naturalized along river and lake margins. June-August. NW England. Very local. Cultivated in gardens.

13 *Lysimachia ciliata* L. Medium to tall rhizomatous perennial, with erect hairless stems. Leaves opposite or in 4's, oval to lanceolate, with a hairy margins. Flowers yellow, 17-24mm, the petal-lobes with red basal blotches, solitary or paired. **I**: North America. Locally naturalized in damp, generally semi-shaded habitats. June-August. Britain and Belgium. Cultivated in gardens. **P**: Bees and other insects. **B**: Local throughout much of Britain, generally close to habitation.

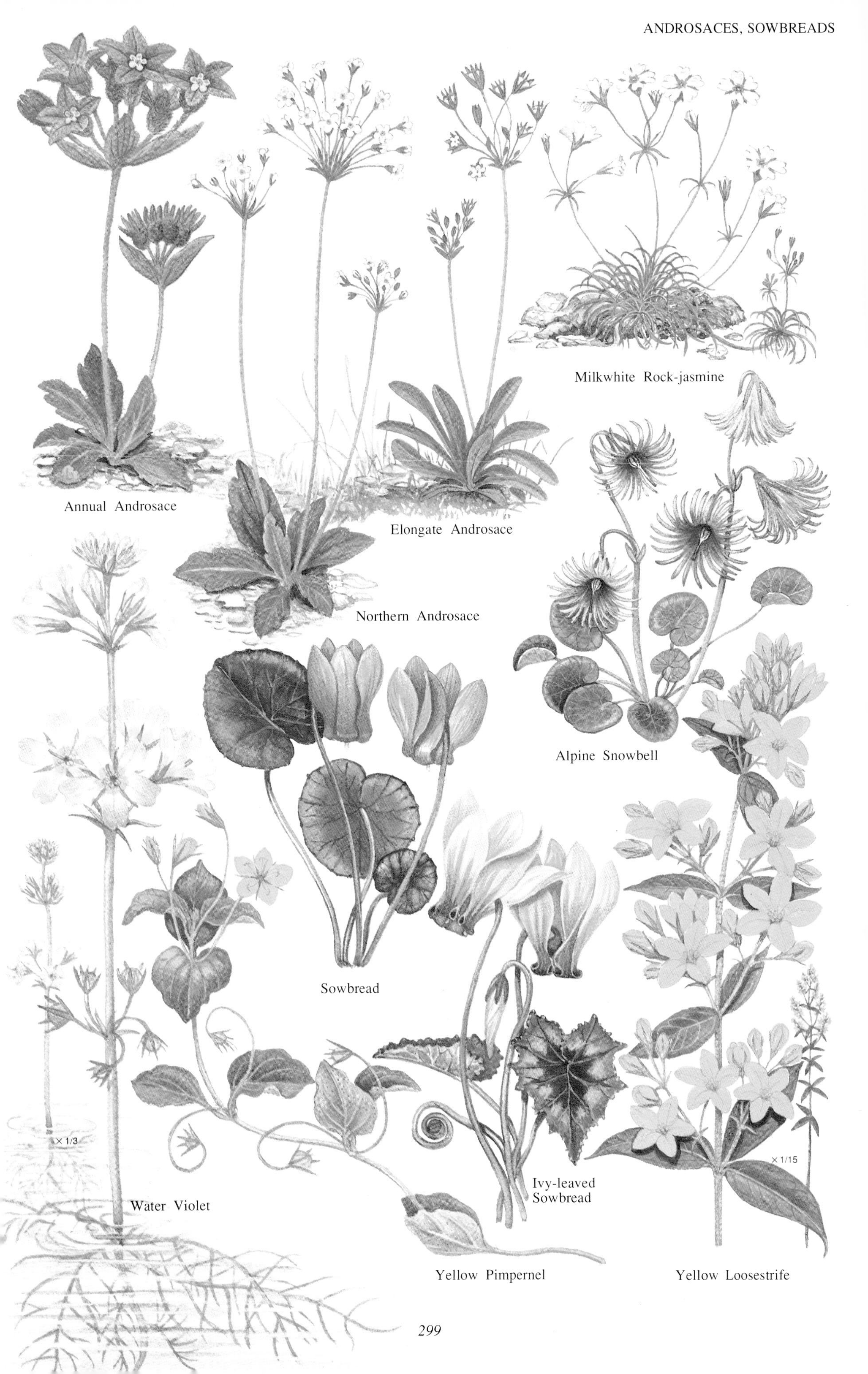
Annual Androsace
Elongate Androsace
Milkwhite Rock-jasmine
Northern Androsace
Alpine Snowbell
Sowbread
Water Violet
× 1/3
Yellow Pimpernel
Ivy-leaved
Sowbread
× 1/15
Yellow Loosestrife

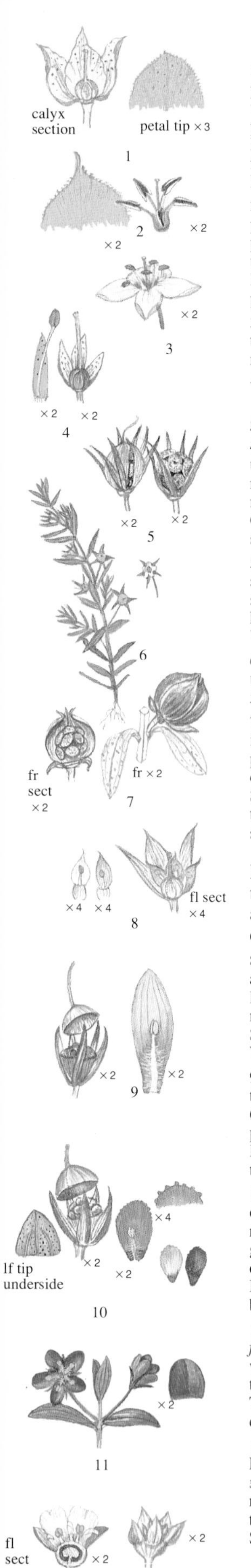


1* **Creeping Jenny** *Lysimachia nummularia* L. Low evergreen, hairless, creeping perennial. Leaves opposite, oval to rounded, short-stalked, gland-dotted. Flowers yellow, 12-18mm, solitary, cup-shaped with wide sepal-lobes (not narrow as in *L. nemorum*), the petals dotted with fine black glands. Capsule rarely produced. Wet habitats, streams margins, ditches, open woodland, damp grassy hedgerows and lake-shores. May-July. Throughout, except the extreme north and Iceland. Widely cultivated in gardens including a form with yellow foliage. **B**: England and Wales north to the Humber, rare further north and in much of Ireland.

2* **Dotted Loosestrife** *Lysimachia punctata* L. (incl. *L. verticillaris* Sprengel). Medium to tall erect patch-forming perennial, minutely-downy. Leaves in whorls of 3-4 or opposite, elliptical to lanceolate, pointed, with a hairy margin, glandular beneath. Flowers yellow, 18-26mm, borne in clusters at the base of upper leaf whorls; petal-lobes broad. Fringed with glandular hairs. Marshes, riverbanks, moist open woodland and other wet habitats, at low altitudes. June-August. **I**: SE Europe. Naturalized in Britain and many parts of Continental Europe. Widely cultivated in gardens. **P**: Mainly bees. **B**: Widely naturalized.

3* *Lysimachia clethroides* Duby. Medium to tall erect, slightly hairy, stoloniferous perennial. Leaves alternate, lanceolate, scarcely stalked, covered with black gland-dots. Flowers white, 5-7.5mm, in terminal racemes which are often bent to one side. **I**: E Asia. Naturalized in shaded moist habitats in Holland. July-August. Widely cultivated in gardens. **P**: Bees and beetles.

*Lysimachia ephemerum* L. Like *L. clethroides*, but leaves opposite and flowers larger, 8-11mm. Damp grassy habitats, often close to springs in SW France. Cultivated in gardens.

4* **Tufted Loosestrife** *Lysimachia thyrsiflora* L. [*Naumburgia thyrsiflora*]. Medium, erect perennial, usually hairless. Leaves opposite, lanceolate, unstalked, with black gland-dots. Flowers yellow, 4-6mm, in rounded, long-stalked clusters at base of middle stem leaves; stamens exceeding the petals. Wet habitats, marshes, bogs and fens, rivers and canals, generally growing in shallow water. June-July. N Britain and Continental Europe, except the extreme north. Easily recognised by its long-stalked flower clusters. Occasionally cultivated in gardens. **B**: Rare, confined to E Yorkshire and the Scottish lowlands.

5* **Chickweed Wintergreen** *Trientalis europaea* L. Low to short hairless perennial with a creeping rootstock. Leaves mostly in a single whorl at the stem top, lanceolate to elliptical, shiny green. Flowers white, sometimes tinged with pink, starry, 11-19mm with 5-9 petals, solitary or several. Damp grassy and mossy habitats, coniferous woods, on acidic soils, to 2000m. June-July. Much of Europe, especially the north. **P**: Various insects. **B**: From Derbyshire N, but also isolated localities in Suffolk.

6 *Asterolinon linum-stellatum* (L.) Duby [*A. stellatum*]. Low hairless annual. Leaves opposite, lanceolate, unstalked. Flowers very small, 0.5-2mm, white, solitary at the base of upper leaves, with 5 petals. Fruit a 5-valved capsule. Dry open habitats at low altitudes. May-July. C & S France.

7* **Sea Milkwort** *Glaux maritima* L. Low, more or less prostrate, rather fleshy, hairless, perennial, rooting at some leaf-nodes. Leaves elliptical, mostly opposite (except the uppermost), unstalked. Flowers pale pink, purplish or white, 3-6mm, petalless, solitary at the base of leaves. Capsule 5-valved. Coastal habitats, dry salt marshes, coastal sward, rock crevices, stabilized shingle, sometime inland on saline soils. May-September. Coastal Europe including Ireland, occasionally inland. **P**: Small flies or self-pollinated. Sometimes confused with small members of the *Caryophyllaceae*, but with the petals fused below, a solitary style and only 5 stamens. **B**: Coastal; inland in Worcestershire and Staffordshire.

**Pimpernels** *Anagallis*. Hairless herbs with opposite or alternate leaves. Flowers 5-parted, solitary and axillary; petals often brightly coloured. Fruit capsule rounded, the upper half splitting away like a small cap. 28 species, mainly in W Europe, Africa and Madagascar. Several are cultivated in gardens.

8* **Chaffweed** *Anagallis minima* (L.) E.H.L. Krause [*Centunculus minimus*]. Low erect annual. Leaves alternate, oval, scarcely stalked. Flowers white or pale pink, tiny, 1-2mm, almost hidden at the base of leaves; petals shorter than the pointed sepals. Damp open habitats, especially on sandy soils, heaths, commons, woodland pathways, coastal dune slacks. June-August. Throughout, except the Faeroes, Iceland and Spitsbergen. Often grows with the superficially similar *Radiola linoides* which has opposite leaves. **B**: Widespread in S England, coastal elsewhere, including W Ireland.

9* **Bog Pimpernel** *Anagallis tenella* (L.) L. Low, mat-forming, rather delicate perennial, rooting freely at the nodes. Leaves mostly opposite, rounded to elliptical, short-stalked. Flowers pink, occasionally white, 6-10mm, rather bell-like, borne on long slender stalks. Damp turf, bogs, marshy ground, spring flushes and pool margins, woodland rides, on acid soils, to 1200m. May-September. Britain and the Faeroes, generally local. Cultivated in gardens. The flowers open widely only in sunshine. **B**: Local throughout much of Britain, more especially in the west.

10* **Scarlet Pimpernel** *Anagallis arvensis* L. [*A. phoenicea*] *A. platyphylla*, *A. parviflora*]. Low prostrate or ascending annual with square stems. Leaves mostly opposite, oval, pointed, with black gland-dots beneath, unstalked. Flowers scarlet, occasionally pink or blue (especially in S), 4-7mm, petals hairy-margined, often slightly toothed at apex. Cultivated, disturbed and waste ground, pathways, coastal dunes and other sandy habitats, generally on light, well-drained soils. May-October. Throughout, except the Faeroes, Iceland and Spitsbergen. Grown in gardens – particularly the blue form which is commonest in the south but rare in the north. The flowers open widely in bright light. **B**: Throughout, mostly coastal in the north.

11 **Blue Pimpernel** *Anagallis foemina* Miller [*A. caerulea*, *A. arvensis* subsp. *caerulea*, *A. arvensis* subsp. *foemina*]. Rather like *A. arvensis*, but upper leaves narrow, lanceolate and flowers blue; petals narrower and without a hairy margin. Similar habitats to *A. arvensis*, but particularly arable fields and gardens. May-October. Often confused with blue forms of *A. arvensis* and distribution uncertain because of this. Throughout, except Ireland, Iceland, the Faeroes and Spitsbergen – probably casual or naturalized in much of the north and north-west. **B**: Scattered localities throughout.

12* **Brookweed** *Samolus valerandi* L. Low to short creeping perennial; stem branched or unbranched. Leaves oval or spoon-shaped, pale green, in a basal rosette and alternate up the stem. Flowers white, cup-shaped, 2-3mm, in lax racemes, petals joined to half-way. Damp or shaded habitats, wet grassland, pool, river and stream margins, dykes and ditches, rocky places, generally on calcareous sandy or brackish soils, to 1200m, but often near the coast. June-August. Throughout, except the Faeroes, Iceland, Norway and Spitsbergen. **B**: Scattered coastal localities, occasionally inland, especially in East Anglia.

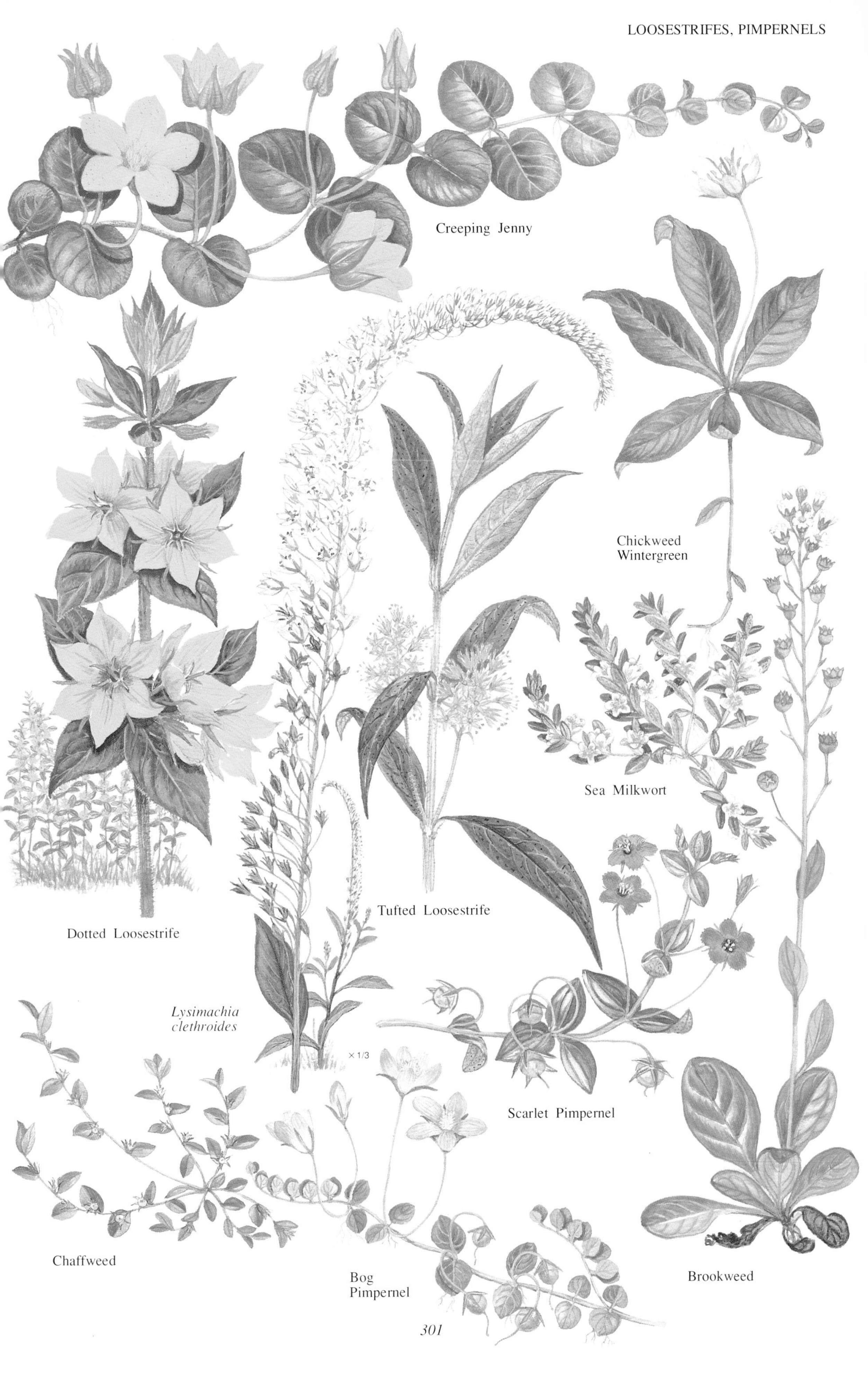
Creeping Jenny
Chickweed Wintergreen
Sea Milkwort
Tufted Loosestrife
Dotted Loosestrife
Lysimachia clethroides
×1/3
Scarlet Pimpernel
Chaffweed
Bog Pimpernel
Brookweed

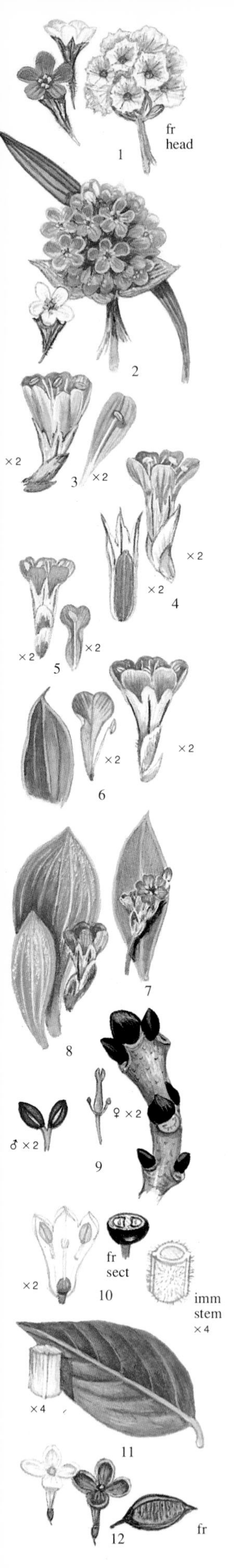


## THRIFT FAMILY Plumbaginaceae

Perennial herbs with basal rosettes of leaves; alternate if present on the stem, untoothed. Flowers in lax cymes or tight heads, 5-parted; calyx lobes persisting in fruit, petals fused together close to the base; styles 5. Fruit dry, with a papery wall, 1-seeded. 775 species in 19 genera.

1* **Thrift, Sea Pink** *Armeria maritima* (Miller) Willd. [*A. vulgaris, Statice armeria*]. Low to short, hairy, cushion-forming perennial with a branching stock. Leaves deep green, linear, usually 1-veined. Flowers pink, red or white, in dense heads 15-25mm across, borne on long leafless stalks, fragrant. Maritime cliffs and meadows, salt marshes and mountain rocks. April-August. Throughout, except Spitsbergen. **P**: Various insects including bees and butterflies. Very variable, forms in different habitats showing some ecological variations. **B**: Throughout; only on mountains inland. subsp. *sibirica*: (Turcz. ex Boiss.) Nyman:the leaves only 30-50mm long (not 40-120mm) and flowers pink or whitish. Dry stony habitats and heaths. N Finland. subsp. *elongata* (Hoffm.) Bonnier: leaves hairy-edged and flowerhead stalks longer and hairless; flowers pale pink. Sandy and rocky habitats. E England – Lincolnshire, S Denmark, Holland, NW Germany, S Finland and S Sweden. subsp. *halleri* (Wallr.) Rothm.: flowerheads smaller, 10-15mm, bright pink or red. Dry pastures and gravelly soils. Holland, Denmark and Germany. subsp. *purpurea* (Koch) Á & D Löve: like subsp. *elongata*, but flowers purplish. Marshy lake shores and wet meadows. S Germany.

2 **Jersey Thrift** *Armeria alliacea* (Cav.) Hoffmanns & Link [*A. allioides, A. arenaria, A. bupleuroides, A. plantaginea*]. Short to medium tufted perennial like *A. armeria*, but leaves fewer, 3-7-veined, broader, linear-lanceolate, flat. Flowers purplish to white, in dense heads 10-20mm across, on erect hairless stems. Dry grassland, mainly in mountains, to 3100m. May-September. France to SW Germany. **B**: Confined to sand-dunes in Jersey.

**Sea-lavenders** *Limonium*. Leaves leathery, simple, sometimes absent by flowering time, mostly in basal rosettes. Flowers in branched lax cymes, the branches like a one-sided spike. 120 species. Some are heterostylous; plants bearing either long or short styles and correspondingly placed anthers. Such plants are self-incompatible.

3* **Common Sea-lavender** *Limonium vulgare* Miller [ *Statice limonium*]. Short, hairless perennial, often carpeting the ground. Leaves elliptical to spoon-shaped, pinnately-veined, long-stalked. Flowers crowded, reddish or lavender-lilac, 6-8mm on stems branched in the upper half; branches spreading, short and crowded. Muddy salt marshes, often abundant. July-September. Europe north to SW Sweden. **P**: Bees, beetles and flies. **B**: North to Dumfries.

4* **Lax-flowered Sea-lavender** *Limonium humile* Miller [*Statice bahusiensis*]. Like *L. vulgare*, but often taller and with angular (not rounded) stems and erect branches. Maritime, muddy salt-marshes. July-August. Britain, NW France, Germany, Scandinavia. Hybridises with *L. vulgare*: *L. x neumanii* C.E. Salmon.

5* **Matted Sea-lavender** *Limonium bellidifolium* (Gouan) Dumort. [*Statice bellidifolia*]. Low to short tufted plant. Leaves oval, broadest above the middle, 3-veined, generally absent by flowering time. Flowers pale lilac, 4-5.5mm; stems rough, with numerous non-flowering branches. Maritime sandy salt-marshes. July-August. E Britain and W France to W Asia. Similar plants in NW Europe are generally referred to *L. dubyi* (Gren. & Godron.) O. Kuntze. **B**: Coasts of East Anglia.

6* **Rock Sea-lavender** *Limonium binervosum* (G.E. Sm.) Salmon [*Statice occidentalis, S. dodartii*]. Short plant with smooth stems. Leaves linear-lanceolate, 1-3-veined. Flowers violet-blue or lilac, 7-8mm, in pyramidal masses without non-flowering branches; petals wide and overlapping. Maritime cliffs, rocks and stabilized shingle. July-September. S Britain, Ireland and France. Restricted to W and SW Europe. Probably hybridises with *L. bellidifolium*. Highly variable. **B**: North to Lincolnshire; local.

7 *Limonium paradoxum* Pugsley. Like *L. binervosum*, but leaves shorter, to 4.5cm long (not 4-12.5cm), 1-veined. Maritime cliffs; basic igneous rocks. July-September. Only SW Wales and NW Ireland.

8 **Alderney Sea-lavender** *Limonium auriculae-ursifolium* (Pourret) Druce [ *Statice lychidifolia*]. Short to medium plant with smooth stems. Leaves like *L. binervosum*, but broader and greyish, 3-7-veined. Flowers violet-blue, 7-8mm; no non-flowering branches present. Maritime rocks. June-September. W France and the Channel Islands to N Africa. **B**: Confined to Alderney and Jersey.

## OLIVE FAMILY Oleaceae

Trees or shrubs with hard wood and opposite, untoothed leaves. Flowers 4-parted, usually regular; calyx bell-shaped; petals joined below. Fruit variable, winged, or a capsule or berry. 29 genera and 600 species; the majority occur in Asia.

9* **Ash** *Fraxinus excelsior* L. Large deciduous tree to 40m with smooth greyish bark, fissured with age; twigs flattened at the nodes, with black winter buds. Leaves pinnate, opposite, with 7-13 oval, pointed and toothed leaflets. Flowers with tufts of dark brownish-purple stamens, becoming greenish, without sepals or petals, borne in terminal or lateral, clusters, appearing before the leaves. Fruit a winged nut or samara. Woodland and scrub, mainly on calcareous soils, to 1600m. April-May. Throughout, except the far north. **P**: Wind. The leaves and twigs are poisonous to livestock. **B**: Throughout.

10* **Privet** *Ligustrum vulgare* L. Semi-deciduous shrub, 1-3m tall, densely branched with smooth greyish bark. Leaves lanceolate, thin untoothed, short-stalked. Flowers white, 4-6mm, in dense pyramidal panicles, fragrant. Fruit a berry, 6-8mm, shiny black when ripe. Wood margins, scrub and hedgerows, embankments, roadsides and abandoned cultivation, generally at low altitudes. May-June. Europe north to S Scandinavia. The berries are poisonous. **B**: North to Durham, but widely naturalized elsewhere.

11 **Garden Privet** *Ligustrum ovalifolium* Hassk. Like *L. vulgare*, but young shoots hairless and leaves broader, usually fully evergreen. June-July. **I**: Japan. Occasionally naturalized in Britain and W Europe, generally close to habitation. May lose all leaves in severe winters.

12* **Lilac** *Syringa vulgaris* L. Deciduous tree to 7m, suckering freely; bark greyish-brown, fibrous. Leaves broadly oval, opposite, short-stalked. Flowers lilac, occasionally white, 6-8mm, fragrant, in dense panicles, often paired. Fruit a 4-valved capsule. **I**: S E Europe. Naturalized from gardens; hedgerows and thickets, railway embankments. May-June. Britain, Belgium, France and Germany.

Thrift
dwarf clifftop form
tall form
Common Sea-lavender
Lax-flowered Sea-lavender
Matted Sea-lavender
Rock Sea-lavender
♀
♂
Ash
Privet
Lilac

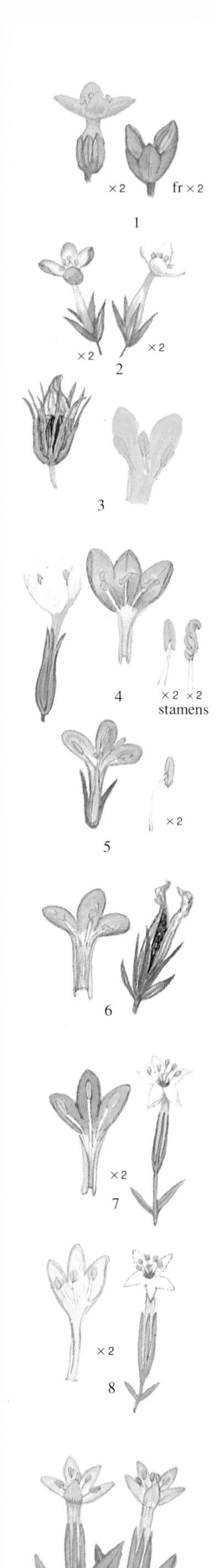


# GENTIAN FAMILY Gentianaceae

Hairless, bitter-tasting herbs with opposite, untoothed leaves. Flowers 4-5-parted; calyx deeply lobed; petals contorted in bud, united into a short or long tube. Fruit a 2-parted capsule. 80 genera and some 800 species throughout the world, though primarily in temperate regions. The flowers often only open widely in sunshine or bright light. Many are cultivated in gardens.

1* **Yellow Centaury** *Cicendia filiformis* (L.) Delarbre [*Microcala filiformis*]. Low to short, slender annual herb, sometimes no more than 2cm tall; stem single or few-branched. Leaves linear. Flowers yellow, 3-6mm, 4-petalled, solitary on long stalks. Damp sandy and peaty habitats, often close to the sea; local. June-October. S Britain, Ireland, Holland, Belgium, France and Germany. W & S Europe to W Asia and N Africa. **B**: Britain, north to Caernavon and Norfolk, also in Ireland and the Channel Islands; very local.

2* **Guernsey Centaury** *Exaculum pusillum* (Lam.) Caruel [*Cicendia pusilla*]. Low, often minute, annual, branched from the base or scarcely branched. Leaves linear-lanceolate. Flowers pink or creamish, 3-4mm, 4-petalled, with long sepal-lobes. Capsule slender. Damp sandy and grassy habitats, commons, at low altitudes. July-September. W & C France and the Channel Islands. Local, often hard to detect in grass. W and S Europe, east to Italy and in N Africa. **B**: Confined to Guernsey – rare.

3* **Yellow-wort** *Blackstonia perfoliata* (L.) Hudson [*Chlora perfoliata*]. Short to medium, greyish, erect annual. Leaves oval to triangular, the upper fused around the stem. Flowers yellow, 8-15mm, with 6-8 petal-lobes, in lax-branched clusters; calyx-lobes 1-veined. Grassland, dunes and rocky places, on calcareous soils, to 1300m. June-October. Britain, S Ireland, Belgium, Holland, France and SW Germany. W, C & S Europe, SW Asia and NW Africa. The plant is the source of a yellow dye. Formerly used as a medicinal plant – bitter tasting. **P**: Self-pollinated. **B**: North to Kirkcudbright and in Jersey and S Ireland.

subsp. *serotina* (Koch ex Reichenb.) Vollmann [*Blackstonia serotina*, *Chlora serotina*]. Like the type, but flowers smaller, 8-10mm, the calyx lobes often 3-veined below. Grassy inland habitats. Holland and Germany.

**Centauries** *Centaurium*. Biennial or perennial herbs with flowers in broad, often flat-topped, branched clusters. Petals pinkish or purple; calyx with keeled lobes. A difficult cosmopolitan group with about 30 species, though not in tropical and S Africa.

4* **Perennial Centaury** *Centaurium scilloides* (L.f.) Samp. [*Erythraea diffusa*, *E. portensis*]. Low to short, spreading perennial with flowering and non-flowering shoots. Leaves lanceolate to suborbicular, the lower short-stalked. Flowers pink, rarely white, 15-20mm, solitary or only a few together, stalked, 5-petalled. Grassy sea-cliffs. June-August. SW Britain, NW France. Also known from Portugal and NW Spain. Very local Atlantic seaboard species readily recognised by its perennial habit and relatively large flowers. **B**: Confined to W Cornwall and Pembrokeshire.

5* **Common Centaury** *Centaurium erythraea* Rafn. Variable low to short biennial, often with a solitary erect stem, branched above. Leaves elliptical to oval, mostly 3-7-veined, the lower in a distinct rosette, the upper much smaller. Flowers pink to purplish, 9-15mm, scarcely stalked, borne in flat-topped, branched clusters. Grassy habitats, woodland margins, scrub and mountains slopes, generally on dry, well-drained soils, to 1400m. June-September. Throughout, except the Faeroes, Iceland, Norway, Finland and Spitsbergen. Hybridises occasionally with *C. littorale* and *C. pulchellum*. Plants yield a yellowish-green dye and since early times has been used for controlling fevers. **B**: Common in England and Ireland, scarcer elsewhere.

6* **Seaside Centaury** *Centaurium littorale* (D. Turner) Gilmour [*C. vulgare*, *Erythraea littoralis*]. Low to short biennial with a basal rosette of rather leathery leaves; stem solitary or several, erect. Leaves narrow oval, parallel-sided, 1-3-veined only. Flowers pink, 11-14mm, unstalked, in dense rather flat-topped clusters. Dry grassy habitats, scrub and sand-dunes. June-September. Throughout, except the Faeroes, Iceland, and the extreme north. Mainly coastal in W Europe, but extending inland in parts of N and C Europe. **B**: Mainly confined to Wales, N England and Scotland; very local elsewhere.

subsp. *uliginosum* (Waldst. & Kit.) Melderis [*Centaurium uliginosum*]. Like the type, but stems and leaf-margins rough (not smooth) and flowers smaller, 5-11mm. Saline habitats, especially inland. Germany.

7* **Lesser Centaury** *Centaurium pulchellum* (Swartz) Druce [*Erythraea pulchella*]. Low to short slender annual without a basal rosette of leaves; stems branched or not. Leaves oval to lanceolate, pointed, 3-7-veined, branched from below the middle. Flowers pinkish-purple, rarely white, 5-9mm, in lax wide-spreading clusters, occasionally solitary; petals narrow. Open habitats, damp grassy places, especially close to the sea. June-September. Throughout, except the extreme north, the Faeroes and Iceland. Widespread in Europe, W & C Asia east to the Tian Shan Mts; also in the Azores. **P**: Various insects, but may also be self-pollinated. **B**: Widespread in C & S England, especially near the coast; very local elsewhere including Wales, Scotland and Ireland.

8* **Slender Centaury** *Centaurium tenuiflorum* (Hoffmanns. & Link) Fritsch [*Erythraea latifolia*]. Similar to *C. pulchellum*, but stems only branched above the middle, branches rather erect. Flowers deep pink, 6-9mm, short-stalked, in dense, flat-topped, clusters. Damp grassy habitats, especially along the coast. June-September. S Britain and France. W Europe to the Mediterranean. Sometimes confused with *C. pulchellum*, but with more erect (not wide-spreading) branches and a denser inflorescence. **B**: Confined to S England – Dorset and Isle of Wight.

9 **Yellow Centaury** *Centaurium maritimum* (L.) Fritsch [*Erythraea maritima*]. Like *C. tenuiflorum*, but an annual or biennial with a solitary stem. Basal leaves withered by flowering time. Flowers yellowish, sometimes tinged with pink or purple, 8-10mm, stalked. Sandy and grassy coastal habitats, heaths. April-June. N & NW France. Local.

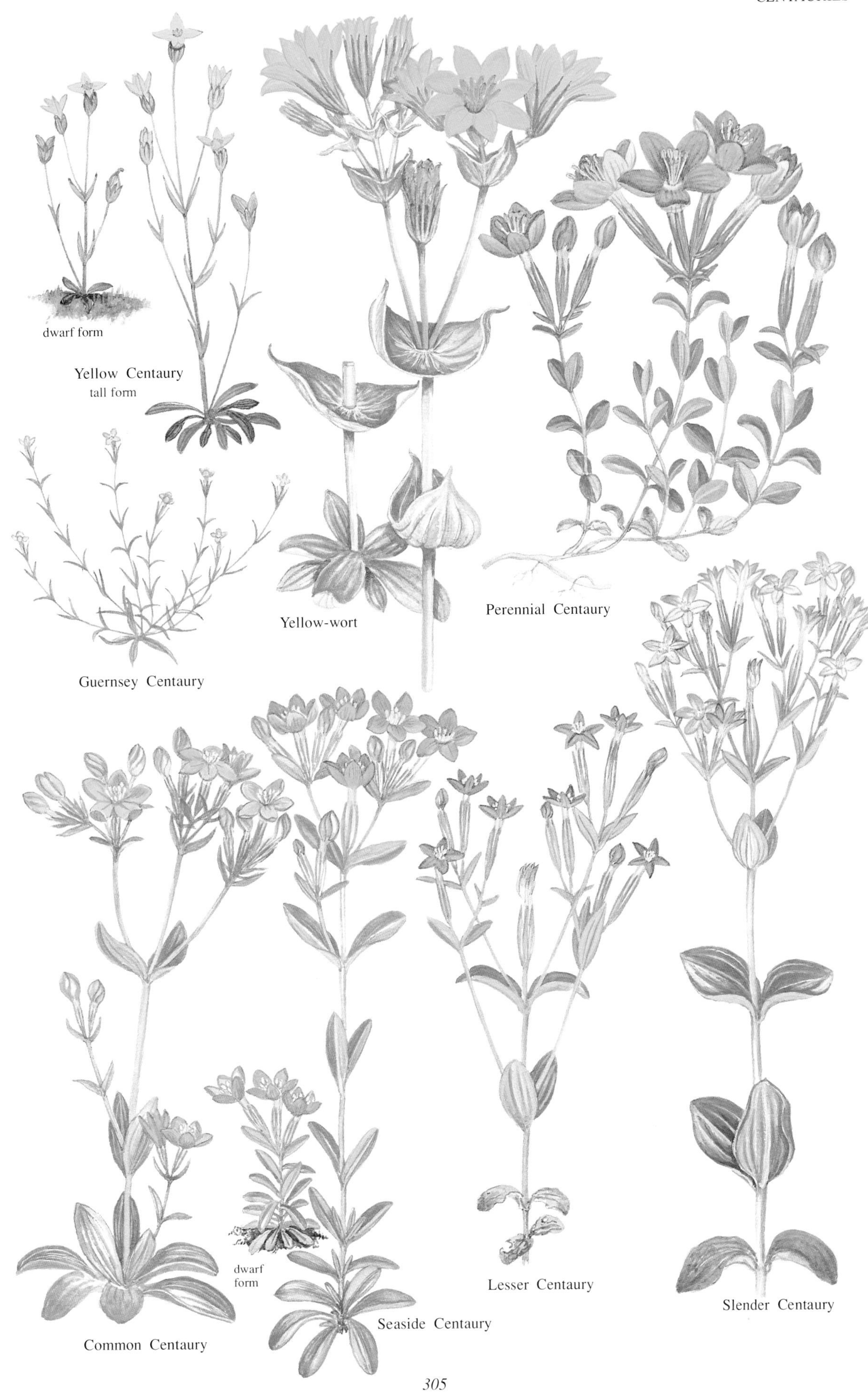
dwarf form
Yellow Centaury
tall form
Yellow-wort
Perennial Centaury
Guernsey Centaury
Common Centaury
dwarf form
Seaside Centaury
Lesser Centaury
Slender Centaury

**Gentians** *Gentiana*. Hairless annual or perennial herbs with opposite leaves, the lower often in basal rosettes. Flowers solitary or clustered. Calyx with 5 teeth, sometimes more, or untoothed and split along one side. Corolla bell, funnel or trumpet-shaped, with 5, or occasionally up to 9, lobes, the main lobes with lesser lobes (plicae) in between, these often scale-like. 400 species scattered almost throughout the world, though particularly in mountain regions. Many are cultivated in gardens, being especially noted for their brilliant blue flowers, but with a wide range of colour, including deep red in some Andean species.

* **Great Yellow Gentian** *Gentiana lutea* L. Stout, medium to tall perennial; stems erect, unbranched, solitary or several together. Leaves bluish-green, lanceolate to elliptic, the lower stalked, the upper clasping the stem, with prominent parallel main veins. Flowers yellow, 18-24mm long, in dense whorled clusters at the base of the upper leaves; petal lobes separated almost to the base, linear-elliptical and spreading; calyx membranous. Grassy mountain habitats, meadows, banks, woodland margins and clearings, roadsides, marshy ground and rocky slopes, to 2500m. June-August. C & E France and S Germany. Characteristic plant of mid-altitude alpine meadows – unpalatable to grazing animals. The roots are the source of Gentian of commerce.

* **Spotted Gentian** *Gentiana punctata* L. Short to medium, rather coarse perennial; stems erect, often with a metallic tinge above, unbranched. Leaves greyish, lanceolate to oval, ribbed and pointed, the lower stalked, the upper half-clasping the stem. Flowers pale greenish-yellow with purple spots, upright, bell-shaped, 14-35mm long, in whorls or small clusters at the base of the uppermost leaves, each flower with 5-8 lobes; calyx green, with short, erect teeth. Mountain meadows and rocky habitats, open woods and scrub, to 3050m. July-September. C & E France and S Germany. Distinctive on account of its purple spotted flowers.

* **Purple Gentian** *Gentiana purpurea* L. Like *G. punctata*, but flowers dull reddish-purple with deeper purple spots, upright, 15-25mm long. Calyx membranous, split down one side. Mountain habitats, grassy and rocky places, woodland margins and open woods, 1600-2750m. July-October. C & E France, S Germany and S Norway, local.

1* **Willow Gentian** *Gentiana asclepiadea* L. Short to medium, tufted perennial with slender erect, or arching stems. Leaves green, lanceolate to oval, pointed, with 3-5 longitudinal veins. Flowers mid to deep blue, rarely white, 35-50mm long, trumpet-shaped, with pointed lobes, usually in pairs at the nodes. Damp habitats, mountain meadows, rocks and woods, often on limestone, to 2200m. August-October. C & E France and S Germany. Widespread from the mountains of C & S Europe to W Asia. Widely cultivated in gardens where a number of different forms exist.

* **Marsh Gentian** *Gentiana pneumonanthe* L. Low to short perennial with slender, more or less erect, stems. Leaves linear to oblong, 1-veined, unstalked. Flowers blue, green-striped on the outside, trumpet-shaped, 25-45mm long, solitary or several together at the stem tops; calyx green, with 5 slender teeth. Marshy habitats, wet heaths, on thin peaty acid soils, to 1500m. July-October. Throughout except the extreme north and most of the islands; local in Britain. Declining in many areas due to land drainage and reclamation. **B**: Local; scattered localities in England from Sussex to Cumbria and Anglesey.

* **Cross Gentian** *Gentiana cruciata* L. Low to short tufted perennial; stems usually several, erect, arising from a basal leaf-rosette. Leaves, oblong to lanceolate, shiny green. Flowers dull blue, paler or greenish on the outside, 20-25mm long, trumpet- shaped, 4-lobed, in clusters at the base of the upper leaves; calyx membranous, often splitting along one side. Dry grassy habitats, banks, scrub and open woodland, to 2000m. July-October. Belgium, Holland, France and Germany.

* **Trumpet Gentian** *Gentiana clusii* Pers. & Song. Low tufted perennial. Leaves mostly basal, bright green, elliptical to lanceolate, rather leathery, the lowermost with a short broad stalk; stem leaves small, generally 1-2 pairs. Flowers mid to vivid deep blue, plain or with a few green spots in the throat, 40-60mm long, trumpet-shaped, solitary; calyx green occasionally purplish, with short triangular lobes. Mountain meadows and pastures and stony places, generally on limestone, to 2800m. April-August. E France and S Germany. A very conspicuous plant in flower, the flowers only opening in bright or sunny, warm weather. **P**: Various bees.

2 *Gentiana acaulis* L. [*G. kochiana*]. Like *G. clusii*, but flowers heavily spotted with green in the throat and calyx teeth narrowed at the base. Similar habitats, but on acid soils, 1400-3000m. May-August. S Germany.

3* **Spring Gentian** *Gentiana verna* L. Low tufted or mat-forming perennial. Leaves mostly basal, bright green, oval to lanceolate. Flowers deep blue, greenish-blue outside, with a white stigma, rarely all white, 12-18mm, with a narrow tube and widespreading lobes; calyx winged on the angles, with slender pointed teeth. Short, often stony turf, meadows, heaths, wet flushes and glacial deposits, fixed calcareous dunes, on mildly acid to calcareous soils, to 3000m. April-June, occasionally later. N England, W Ireland, C & E France and S Germany. Locally common but declining in some areas due to over-collecting. **P**: Bees and butterflies. Widely cultivated in gardens. **B**: Very rare, confined to Upper Teesdale in N England and the Burren in Galway, Ireland; protected.

* **Bavarian Gentian** *Gentiana bavarica* L. Like *G. verna*, but leaves yellowish-green, not in a basal rosette, broadest above the middle. Flowers deep blue with a pale blue tube, 16-20mm; calyx wings very narrow. Damp mountain meadows and marshy places, above 1800m. July-September. E France and S Germany.

* **Snow Gentian** *Gentiana nivalis* L. Low to short, slender annual herb; stems branched or unbranched, erect. Leaves oval to elliptical, unstalked. Flowers intense blue, small 6-8mm, with a narrow tube and spreading pointed lobes, solitary or in branched clusters. Meadows, marshes and rocky habitats, Arctic heaths, only on the mountains in the S of its range. June-August. Throughout, but often very local. **P**: The flowers open only in bright light and are self-pollinated. Has declined in some areas due to collecting and overgrazing. **B**: Very rare; confined to a few localities in the C Scottish Highlands.

* **Bladder Gentian** *Gentiana utriculosa* L. Low to short slender annual herb; stems erect, branched above. Leaves lanceolate to elliptical, the lower in a basal rosette, pale green. Flowers intense blue, 12-16mm, the narrow petal-tube with 5 pointed lobes; calyx inflated, bladder-like, winged on the angles. Damp meadows, marshes, bogs and stony places, to 2500m. May-August. E France and S Germany. Local. Easily recognised by its inflated calyces, which persist until the fruits are ripe.

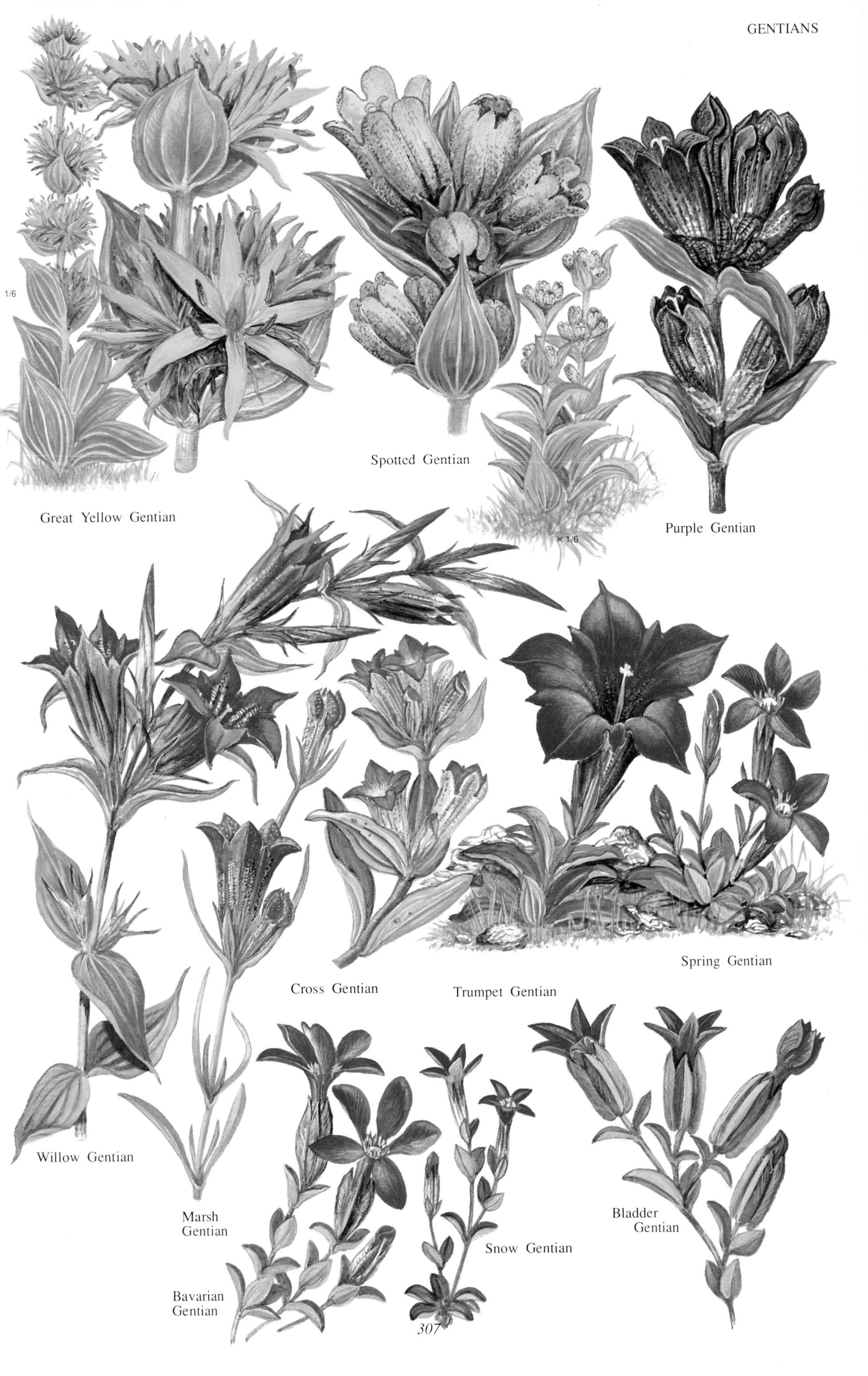
1/6
Great Yellow Gentian
Spotted Gentian
× 1/6
Purple Gentian
Willow Gentian
Marsh Gentian
Cross Gentian
Trumpet Gentian
Spring Gentian
Bavarian Gentian
Snow Gentian
Bladder Gentian

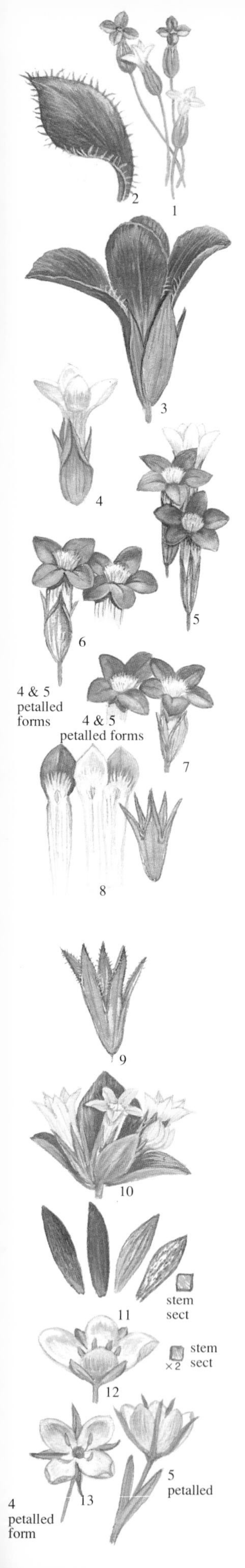


*Gentianella*. Annual or biennial herbs generally branched from above the basal leaf-rosette. Flower solitary or in branched clusters; calyx divided to at least halfway. The corolla funnel-shaped with 4-5 spreading lobes without smaller lobes (plicae) between, generally with a fringe of hairs in the throat. 140 species widespread in the temperate northern hemisphere.

1* **Slender Gentian** *Gentianella tenella* (Rottb.) Börner [*Gentiana tenella*]. Low unbranched annual or biennial, sometimes only 2-3cm tall. Basal leaves spoon-shaped to elliptical; stem leaves 1-4 pairs. Flowers sky blue or violet, rarely whitish or yellowish, small, 4-6mm, 4-parted, borne on long slender stalks. Damp pastures, stony habitats and screes, river gravels, usually on acid rocks, to 3100m. July-September. Iceland, Scandinavia and the mountains of France and Germany. The small 4-lobed flowers are very distinctive.

2* **Fringed Gentian** *Gentianella ciliata* (L.) Borkh. [*Gentiana ciliata*]. Low to short, erect biennial, unbranched. Lower leaves spoon-shaped to elliptical, not in a basal rosette, the upper linear-lanceolate, pointed. Flowers blue, large, 35-50mm across, long-stalked, the 4 lobes fringed with fine bluish hairs. Dry rocky habitats, meadows and open woods. August-October. Scandinavia, except the far north. France.

3 *Gentianella detonsa* (Rottb.) G. Don fil. [*Gentiana detonsa*]. Like *G. ciliata*, but often taller and with uneven (not even) calyx lobes. Flowers smaller, 35-40mm long, dark blue, the lobes only slightly fringed near the base. Damp habitats, especially near the sea. July-September. Iceland and Norway. Extends into Central Asia as far as the Pamir Mountains.

4* **Field Gentian** *Gentianella campestris* (L.) Börner [*Gentiana campestris*]. Low to short, erect biennial with a simple or branched stem. Leaves oval to lanceolate, generally increasing in size up the stem, the lowermost withered and brown by flowering time. Flowers bluish-lilac or white, 15-30mm long, readily distinguished from closely related species by their 4-lobed rather than 5 petal-lobes; calyx with 2 broad and 2 narrow lobes. Grassland, pastures, heaths and dunes, on acid or neutral soils, to 2750m. June-September. Throughout, except the extreme north of Scandinavia. **B**: Scotland, N England, Wales and Ireland; local in S England. subsp. *baltica* (Murb.) Tutin [*Gentiana baltica*] is an annual with a few basal green leaves. S Sweden to NW France and Germany.

5* **Felwort, Autumn Gentian** *Gentianella amarella* (L.) Börner [*Gentiana amarella*]. Low to short annual or biennial with an erect, simple or branched, stem; branches ascending, above the base of the plant. Leaves oval to lanceolate, the basal ones in a distinct rosette usually. Flowers dull purple, blue, pink or whitish, 14-22mm long, usually 5-parted; calyx-lobes narrow and equal. Dry hill pastures, cliffs and dunes, often on calcareous soils, to 1800m. June-October. Throughout, except Ireland, Iceland and N Scotland. Hybridises with *G. germanica*, *G.* x *pamplinii* (Druce) E. F. Warburg, and *G. uliginosa*, where the parent species grow in close proximity. subsp. *septentrionalis* (Druce) Pritchard has smaller flowers, 10-16mm long, creamy-white with reddish-purple on the outside, 4- or 5-lobed. Dunes and limestone pastures. **B**: NW Scotland and Outer Hebrides. subsp. *hibernica* Pritchard has leaves narrower and the flowers 19-22mm long. Similar habitats and flowering time. Ireland.

6 **Dune Gentian** *Gentianella uliginosa* (Willd.) Börner [*Gentiana uliginosa*]. Like *G. amarella*, but usually a smaller plant, 8-19cm tall, if branched then branched from the base. Flowers dull purple, 10-20mm long, 4-5-parted, the terminal flower borne on a long stalk; calyx-lobes very unequal. Damp meadows and dune slacks. W Britain and N Europe, including Holland and Germany. July-November. Annual forms, usually mixed with larger biennial ones, generally bear only 1-2 flowers. **B**: Local in Glamorgan, Carmarthen and Pembroke.

7* **Early Gentian** *Gentianella anglica* (Pugsley) E. F. Warburg [*Gentiana anglica*]. Low to short biennial with long branches from the base of the plant. Leaves lanceolate to linear. Flowers dull purple, 13-16mm long, 4 or 5-parted; calyx-lobes somewhat unequal. Grassland over chalk, rarely on sand-dunes. April-June. **B**: S England from Lincolnshire southwards, not Cornwall, local. subsp. *cornubiensis* Pritchard is branched from the middle, with broad basal leaves and flowers usually 17-20mm long. Cliff habitats. **B**: W Cornwall. Readily identified by its early flowering habit; spring and early summer.

8* **Chiltern Gentian** *Gentianella germanica* (Willd.) Börner [*Gentiana germanica*]. Low to short biennial, often branched from the base. Basal leaves generally withered by flowering time; stem leaves oval to lanceolate, minutely hairy on the margin. Flowers violet, pink or whitish, 20-40mm long, 5-parted; calyx-lobes equal and spreading. Chalk grassland and scrub, often hidden among tall grasses, sometimes in more open habitats, to 2750m. August-October. S England to Holland and Germany southwards. Annual plants, smaller in all their parts, are often found mixed in populations of the typical plant. **B**: Local from Bedfordshire to Hampshire and Middlesex.

9 *Gentianella aspera* (Hegetschw. & Heer) Dostál ex Skalicky, Chrtek & Vill. Like *G. germanica*, but with hairy-edged calyx-lobes. Mountain habitats. S Germany.

10 *Gentianella aurea* (L.) H. Sm. Like *G. germanica*, but flowers pale yellow, rarely bluish, less than 12mm long, crowded at the shoot tips. Sea and lake shores. August-October. Arctic Scandinavia and Iceland.

11* **Marsh Felwort** *Swertia perennis* L. Variable short to medium, erect, unbranched perennial, forming tufts; stems square in section, reddish or purplish. Leaves oval to elliptical, yellowish-green, the lower crowded, the upper clasping the stem, sometimes alternate. Flowers blue or violet-red, occasionally white or yellowish-green, 14-30mm, star-shaped with 4-5-pointed petals separated almost to the base, in branched clusters. Marshy habitats, particularly wet meadows in the mountains, to 2500m. From NC France to Germany. An often local plant, particularly variable in flower colour; those with greenish-yellowish petals are sometimes placed in a separate species, *S. punctata* Baumg., but they do not appear to merit even subspecific status. The same also applies to plants with alternate stem leaves and dark violet flowers previously referred to *S. alpestris* Baumg. ex Fuss.

12* **Lomatogonium** *Lomatogonium carinthiacum* (Wulfen) Reichenb. Low to short rather thin-stemmed, hairless annual; stems square. Leaves mostly basal, oval to oblong, pale green, the lowermost short-stalked. Flowers pale blue or white, 12-20mm, saucer-shaped, the 5 blunt petals joined near the base, solitary on long stalks; sepals distinctly shorter than the petals. Dry grassy places and meadows, often on banks above streams and rivers, in mountain habitats, to 2700m. August-October.

13 *Lomatogonium rotatum* (L.) Fries ex Fernald. Like *L. carinthiacum* but taller, the sepal-lobes narrower, equalling or longer than the petals. Iceland, but also native in N & C Asia and North America.

Slender Gentian
Fringed Gentian
Field Gentian
Felwort
Early Gentian
Chiltern Gentian
Marsh Felwort
Lomatogonium

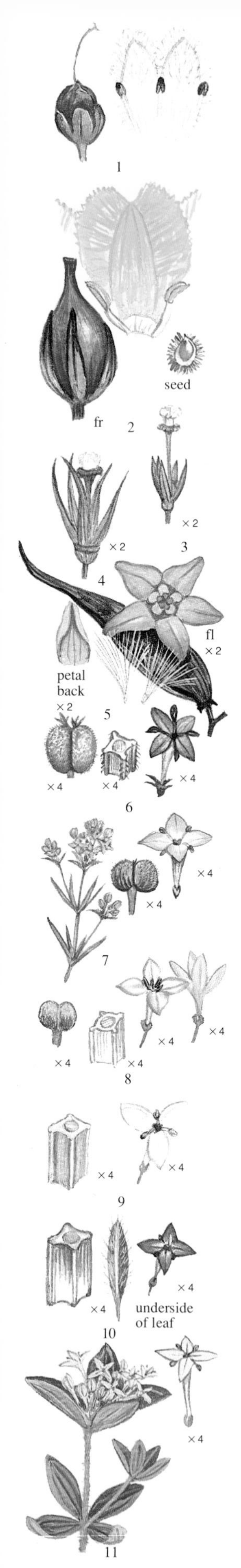


## BOGBEAN FAMILY Menyanthaceae

Aquatic or semi-aquatic perennial herbs with alternate leaves. Flowers 5-parted with a deeply divided calyx and petals fused below, often hairy or crested inside; stamens 5. Fruit a rounded or egg-shaped capsule; ovary superior, carpels 2. Cosmopolitan; 40 species in 5 genera.

1* **Bogbean** *Menyanthes trifoliata* L. Short, almost hairless aquatic or semi-aquatic perennial with stout creeping runners. Leaves trifoliate, with oval or diamond-shaped, untoothed leaflets, held above the water on long stalks. Flowers pink outside, whitish inside, 14-16mm, starry, the petals fringed with long whitish hairs, borne in lax erect racemes. Shallow water, or semi-aquatic in fens and bogs, to 1800m. April-June. Throughout, except parts of the far north. **B**: Throughout, sometimes locally abundant.

2* **Fringed Water-lily** *Nymphoides peltata* (S. G. Gmelin) O. Kuntze [*N. flava, Limnanthemum peltatum*]. Low aquatic perennial with long creeping stems. Leaves floating, rounded, with a slit to the long stalk, untoothed, plain green or purple blotched above, purplish beneath. Flowers yellow, 30-40mm, the petals with a fringed margin, borne on long stalks in small clusters, just above the water surface. Slow-moving or still fresh water, ponds, lakes or ditches, to 1200m. July-September. Britain, Belgium Holland, France and Germany; naturalized in Denmark and S Sweden. Widely cultivated. **B**: Local, E, C and SE England north to S Yorkshire; occasionally naturalized elsewhere.

3* **Lesser Periwinkle** *Vinca minor* L. (Periwinkle Family – Apocynaceae – 1500 species in 100 genera. Most produce a poisonous whitish latex when cut). Low to short, practically hairless, evergreen subshrub with trailing stems, often rooting at some of the nodes. Leaves leathery, lanceolate. Flowers reddish-purple to pink, occasionally white, 25- 30mm, solitary; calyx-lobes not hairy on margins. Fruits forked, up to 25mm, but rarely formed in NW Europe. Woods, coppices, hedgerows, banks and rocky ground, to 1320m. February-May. Europe north to Denmark. Doubtfully native in much of NW Europe. **P**: Long-tongued bees and bee-flies. **B**: Throughout, except N Scotland.

4* **Greater Periwinkle** *Vinca major* L. Short to medium spreading, evergreen subshrub, with trailing and arching stems, often rooting down at the tip. Leaves shiny bright green, oval, with a hairy margin. Flowers purplish-blue, 30-50mm, solitary, the calyx-lobes with a hairy margin. Fruits forked, 40-50mm, rarely formed in NW Europe. Woodland margins, copses, scrub, banks and hedgerows, at low altitudes. March-May. **I**: C & S Europe, N Africa. Naturalized in S Britain, Ireland and parts of France. **P**: Long-tongued bees. **B**: Local in S England, occasionally in S and W Ireland.

5* **Swallow-wort** *Vincetoxicum hirundinaria* Medicus [*V. officinale, Cynanchum vincetoxicum*]. (Swallow-wort Family – Asclepiadaceae – 2000 species in 250 genera). Medium to tall tufted perennial, hairless or slightly hairy; stems erect, unbranched. Leaves opposite, heart-shaped to lanceolate, short-stalked. Flowers greenish-yellow or yellowish-white, 5-10mm, in clusters of 6-8 at the base of the upper leaves. Fruit 5-6cm, hairless. Woods, rocky habitats, bare ground, generally on calcareous soils, to 1800m. June-September. Continental Europe north to S Scandinavia. Very poisonous plant. The fruits split along one side to release the seeds, which each have a tuft of hairs at one end for dispersal by wind.

## BEDSTRAW FAMILY Rubiaceae

Herbs or dwarf shrubs. Leaves opposite or in distinctive whorls, simple, untoothed; stipules often very like leaves. Flowers funnel-shaped, with a short or long tube, in dense heads, branched cymes or panicles, 4-5-parted; ovary below the corolla and calyx – inferior. Fruit fleshy or dry, 1-2-parted. 7000 species in 500 genera, mainly tropical.

6* **Field Madder** *Sherardia arvensis* L. Low, often prostrate annual, branched from the base, bristly. Leaves lanceolate, mostly in whorls of 6, pointed, unstalked. Flowers violet, pink or pale purple, 3mm, in terminal clusters surrounded by a ruff of leaf-like bracts; corolla 4-lobed. Fruit globose, 2-7mm, bristly, surrounded by the 4-6 enlarged sepal teeth. Cornfields, occasionally on bare ground. May-September. Throughout, except the far north. **B**: Widespread.

**Woodruffs** *Asperula*. Small herbs with square stems and leaves in whorls of 4-8. Flowers in branched clusters or heads. Calyx absent or of 4 minute teeth. Flowers hermaphrodite, 4-parted. Fruit 2-lobed (a pair of nutlets) without hooks or prickles. 90 species.

7 **Western Woodruff** *Asperula occidentalis* Rouy. Low creeping, losely tufted green perennial, with orange subterranean stolons, hairy below. Basal leaves oval, the others linear-lanceolate, pointed, rather fleshy. Flowers pink, 3mm, in egg-shaped clusters, unstalked, the corolla-tube equalling the lobes, rough outside. Fruit finely hairy. June-July. SW Britain, S & W Ireland and W France. **B**: Local, coastal.

8* **Squinancywort** *Asperula cynanchica* L. Very variable, slender, more or less prostrate, hairless perennial, green or greyish; stems much-branched. Leaves narrow-lanceolate to linear, in whorls of 4. Flowers pale pink to purplish outside, whitish inside, 2.5-3.5mm, in much-branched clusters, rough outside, vanilla-scented. Fruit 2- 3mm, finely warted. Dry pastures and other grassy habitats, dunes, on calcareous soils, to 2100m. June-September. W Europe east to the Caucasus. **B**: SW Ireland and Britain south from S Yorkshire and Cumbria.

9* **Dyer's Woodruff** *Asperula tinctoria* L.[*Galium triandrum*]. Very variable short to medium, often sprawling glabrous perennial blackening on drying, with orange, horizontal stolons. Leaves lanceolate to linear, in whorls of 4-6, rough-edged. Flowers white, 3-4mm, with a rather short tube, 3-parted. Fruit hairless, 1.5-2mm, granular. Grassy, sandy habitats, scrub, generally at low altitudes. June-September. France, N Germany, Scandinavia, rare in Denmark and Sweden. The roots yield a red dye.

10* **Blue Woodruff** *Asperula arvensis* L. Short, slender, hairless annual. Leaves lanceolate to linear-lanceolate, in whorls of 6-8, 1-veined. Flowers bright blue or bluish-violet, 4mm, in clusters surrounded by a ruff of leaf-like bracts, 4-parted. Fruit smooth, 2-3mm, brown. Fields and waste places, to 1500m. April-July. France, Holland and S Germany; naturalized in Britain and Scandinavia. From C and S Europe to W Asia and N Africa. **B**: A rare casual in England.

11 **Pink Woodruff** *Asperula taurina* L. Like *A. arvensis*, but the flowers white tinged with yellowish-pink, the flower clusters surrounded by a ruff of 4 large, leaf-like bracts. Waste places. May-June. **I**: S Europe. Naturalized in a few localities in England.

Fringed Water-lily
Bogbean
Lesser Periwinkle
Greater Periwinkle
Field Madder
Swallow-wort
Squinancywort
Dyer's Woodruff
Blue Woodruff

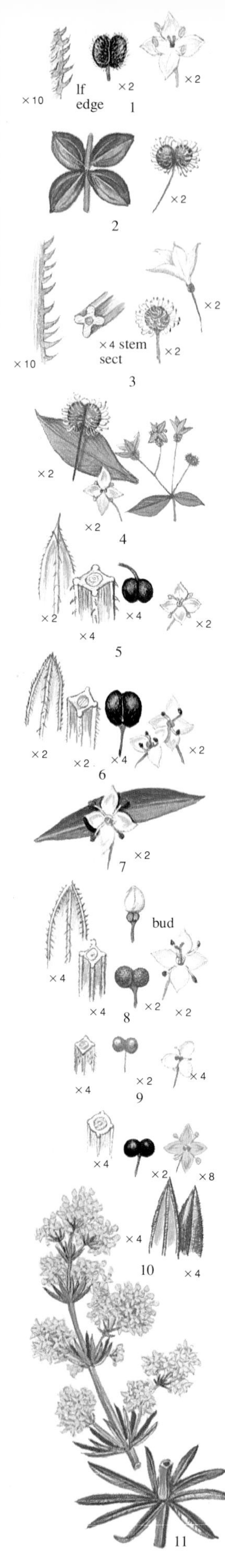


**Bedstraws** *Galium*. Like *Asperula*, but stems occasionally rounded. Leaves mostly in whorls of 4 or more. Flowers usually 4-parted, the corolla with a very short tube; stamens protruding. Fruit dry, usually 2-lobed, often with hooked bristles. A cosmopolitan genus with about 300 species. Many are extremely diificult to distinguish in flower and the fine details of the fruits are often very useful in separating similar looking species. Plants often turn deep brown or black on drying.

1* **Northern Bedstraw** *Galium boreale* L. A short, rather stiff, erect perennial, sometimes slightly hairy; stems rather stout, square. Leaves dark green, long-lanceolate, in 4's, thick and with a rough edge, 3-veined. Flowers white to yellowish, 3-4mm, in branched oblong clusters. Fruits 1.5-2mm, olive-brown with short, hooked, non-spreading bristles, sometimes hairless. Grassy, rocky and bushy places, streamsides, screes and shingle, to 2200m. June-August. Britain, Iceland and much of Continental Europe. The flowers are visited by various small insects. Distributed from N & C Europe to W Asia. **B**: From Lancashire northwards; N & W Ireland.

2 *Galium rotundifolium* L. [*G. scabrum*]. Like *G. boreale*, but a shorter plant with rather weak slender stems and oval or rounded leaves. Inflorescence often few-flowered. Fruit greenish, with spreading hooked bristles. Mountain woods. July-September. C & E France, Germany and S Sweden – Gotland; naturalized in Holland.

3* **Woodruff** *Galium odoratum* (L.) Scop. A spreading, mat-forming, short, almost hairless perennial, with numerous stolons; stems erect, square, with a ring of hairs at each node, leaves elliptical in whorls of 6-9, edged with tiny forward-pointing prickles. Flowers pure white, 4-7mm, fragrant, in lax heads; corolla lobes slightly recurved. Fruit 2-3mm, with hooked, black-tipped bristles. Deciduous woodland on calcareous soils, to 1600m. May-June. Throughout, except the Faroes, iceland and Spitsbergen. The entire plant has a sweet, hay-like scent. Shade-tolerant, soon dying out once the shade cover is removed. It can sometimes be confused with *Asperula cynanchica*, which, however, grows in dry pastures and sandy habitats and has leaves in fours. **P**: Bees and flies. The ripe fruits cling readily to animals and clothing. **B**: Throughout, except the Outer Hebrides and Orkneys.

4 **Coniferous Bedstraw** *Galium triflorum* Michx. Like *G. odoratum*, but flowers greenish to whitish, 1.5-3mm, in a more oblong inflorescence. Rocky coniferous woods to 1200m. June-August. Norway, Sweden, Finland, extending into NW USSR.

5* **Fen Bedstraw** *Galium uliginosum* L. Short to tall hairless perennial, with rooting rhizomes; stems spreading or ascending, square, with down-turned prickles on the angles. Leaves lanceolate to linear-lanceolate, in whorls of 6-10, with a prickly margin, 1-veined. Flowers white with yellow anthers, 2.5-3mm, in a narrow panicle. Fruit 1mm, eventually dark brown, on deflexed stalks. Wet habitats, particularly marshes and fens, locally common, to 2100m. June-August. Throughout, except the extreme north and most of the smaller islands. The plant is hay-scented (coumarin). **P**: Various small insects. Sometimes confused with *G. palustre*, but leaves in larger whorls and plant not blackening on drying. **B**: Throughout, particularly near the coast.

6* **Common Marsh Bedstraw** *Galium palustre* L. Very variable, straggly, short to medium, hairless perennial; stems stout or weak, square, usually rough on the whitish angles. Leaves elliptical to oblong, mostly in whorls of 4-6, with a rough margin, 1-veined. Flowers white, occasionally greenish, with red anthers, 2-3mm, usually 4 but sometimes 3-parted, in a lax pyramidal panicle. Fruit 2-3mm, black when ripe, smooth. Moist or wet habitats, sometimes growing in water, to 2100m. June-August. Throughout, except Spitsbergen, from Europe to W Asia. The leaves turn black on drying. **P**: Mainly bees. A very variable species with several different races, sometimes regarded as distinct species, including the following. **B**: Common throughout.

7 **Great Marsh Bedstraw** *Galium elongatum* C. Presl [*G. palustre* subsp. *elongatum*]. Like *G. palustre* but a more vigorous plant, to 1m tall; leaves 20-35mm long (not less than 20mm); flowers 3-4mm. Wet habitats, reed swamps, river or lake margins. June-July. Throughout except the Faeroes, Iceland and Spitsbergen. **P**: Mainly bees. Intermediates between *G. elongatum* and *G. palustre* occasionally occur. **B**: Throughout most of Britain.

8* **Slender Marsh Bedstraw** *Galium debile* Desv. [*G. constrictum*]. Like *G. palutstre*, but stems usually smooth and leaves linear, pointed. Inflorescence with erect, not spreading branches. Flowers pinkish-white, occasionally greenish, 2.5mm. Fruit granular, not on deflexed stalks. Marshes and pond-margins, local. May-July. Only S Britain and France. **B**: New Forest, Hampshire.

9* *Galium trifidum* L. A low, delicate perennial; stems slender, angled, with backward-pointing prickles. Leaves mostly in whorls of 4, linear, with a rough margin and midrib. Flowers white, 1.5mm, 3-parted, few to an inflorescence. Fruit smooth, 1.5-2.5mm, black when ripe. Wet habitats, particularly marshes and bogs, to 2100m. June-August. Mountains of Scanduiinavia, isolated stations in the Alps and Pyrenees.

10* **Lady's Bedstraw** *Galium verum* L. Low to short, sprawling perennial, stoloniferous; stems rounded with 4 raised lines of hairs. Leaves dark green, shiny,in whorls of 8-12, hairy beneath. Flowers golden-yellow, 2-3.5mm, fragrant, in dense oval panicle. Fruit smooth, 1.5mm, black when ripe. Grassland, open woodland, hadgebanks, road verges, sand-dunes. June-September. Throughout, except Faeroes and Spitsbergen. A common wayside flower, blooming throughout the summer months. The flowers have a strong coumarin scent, and attract a great variety of insects. Formerly used to stuff mattresses; the pleasant smell of the dried plant was said to discourage fleas. The flowers were also used to also to coagulate milk. Hybridises with *G. album, G. glaucum* and *G. mollugo*. **B**: Throughout.

11 *Galium* × *pomeranicum* Retz [*G. ochroleucum*]. A natural hybrid between *G. album* and *G. verum*, rather like *verum*, but stems 4-angled, leaves not darkening on drying. Flowers very pale yellow; inflorescence less hairy. Meadows and roadsides. June-August. Through, except the Faeroes and Spitsbergen. Often forming large and variable populations, intermediate between and often growing with the parent species.

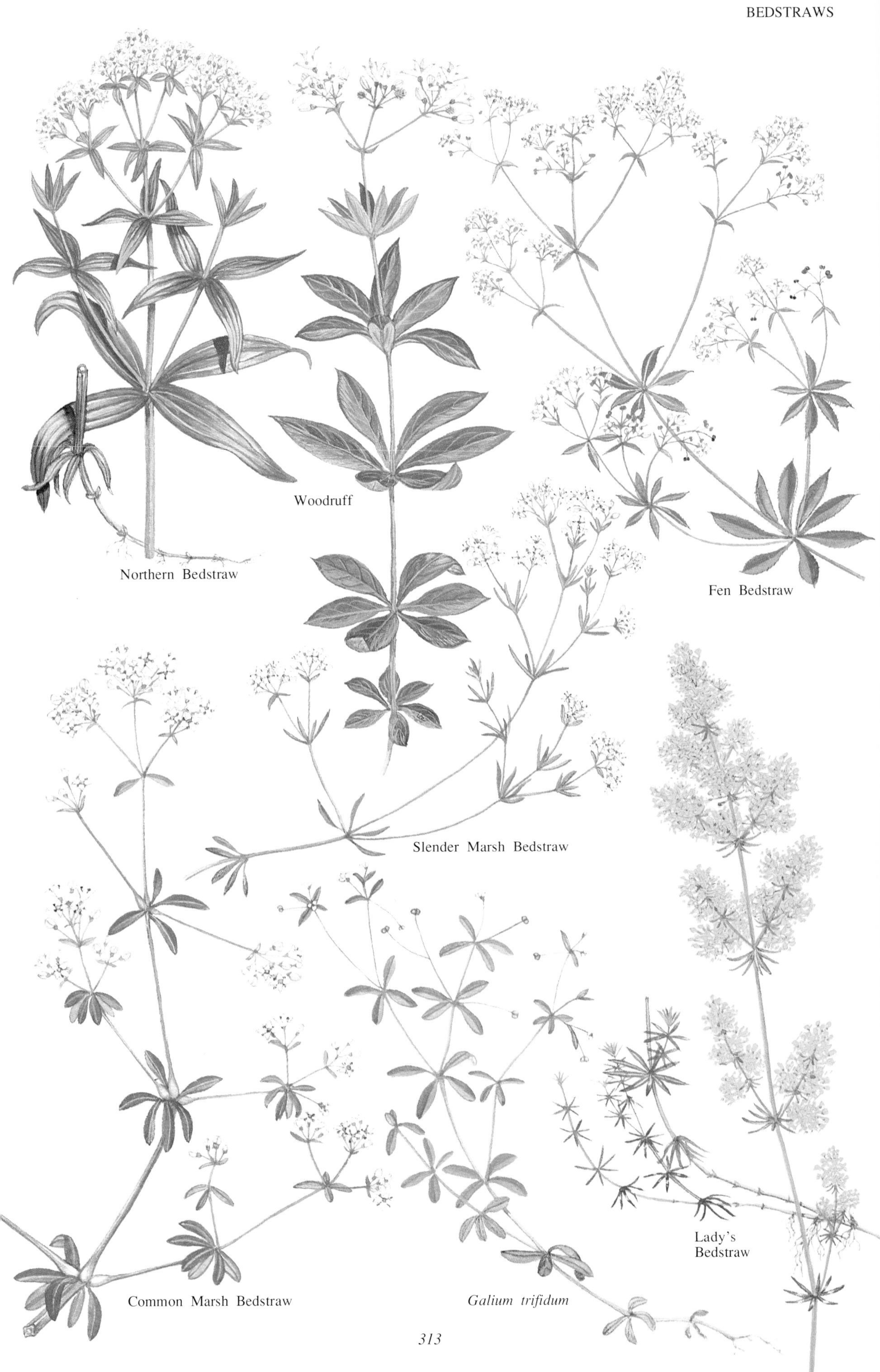
Northern Bedstraw
Woodruff
Fen Bedstraw
Slender Marsh Bedstraw
Lady's
Bedstraw
Common Marsh Bedstraw
Galium trifidum

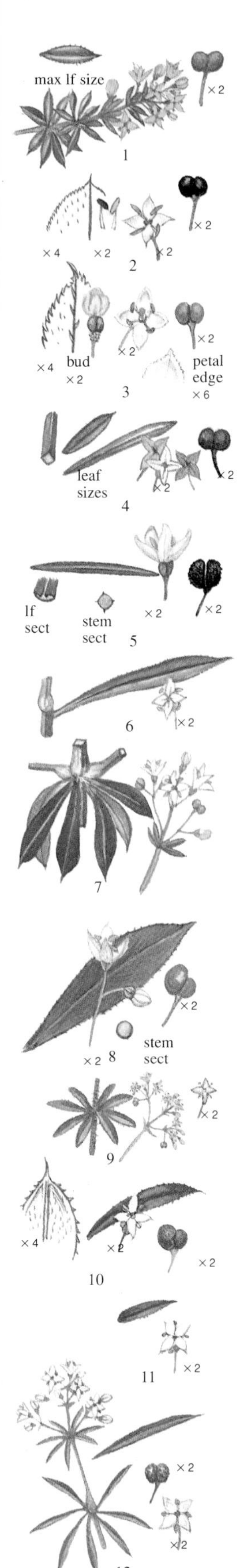


1 *Galium arenarium* Loisel. Low to short hairless or occasionally slightly hairy perennial, with underground stolons and numerous non-flowering shoots; stems 4-angled. Leaves broadly lanceolate, with a short translucent apex, rather fleshy and shiny, slightly rough at the edges. Flowers yellow, 3-4mm, in a rather narrow inflorescence. Fruit 3mm, hairless and smooth, but roughened when dry. Coastal habitats – maritime sands. Late May-July. W France. Very local. Also found in the extreme northern coastal part of Spain.

2* **Hedge Bedstraw** *Galium mollugo* L. Very variable medium to tall, sometimes scrambling, usually hairless perennial; stems square and smooth, the rootstock usually reddish, with long underground stolons. Leaves oblong to elliptical, in whorls of 6-8, thin and rather pale green, 1-veined. Flowers white, 2-3mm, in loose, branched clusters. Fruit smooth, black when ripe. Open woods, scrub, rough grassland and meadows, on base-rich soils, to 2100m. June-September. Throughout, except Iceland and much of Scandinavia. Easily recognised by its smooth stems, quite unlike the rough stems of species like *G. aparine*. **P**: Small flies; the anthers ripen before the stigmas and bend away as the latter ripen, thus ensuring cross-pollination. Widespread from Europe to C and N Asia. Hybridises with *G. verum*. **B**: Throughout, but rather scarce in Wales, Scotland and Ireland.

3* **Upright Bedstraw** *Galium album* Miller [*G. erectum*, *G. mollugo* subsp. *erectum*]. Like *G. mollugo*, but the stems erect (not spreading) and the leaves rather thicker. Inflorescence with ascending (not spreading branches); flowers 3-5mm. Pastures and grassy habitats, waste ground, generally in rather open dry habitats, to 2100m. June-September. Britain, Belgium, Holland, Denmark, France and Germany; naturalized in Ireland, Iceland and parts of Scandinavia. **P**: Small flies. **B**: Lowland Britain; rare and introduced in Ireland.

4 *Galium lucidum* All. [*G. rigidum*]. Medium to tall, hairless or slightly hairy, stoloniferous perennial; stems erect or ascending, square. Leaves linear-lanceolate, often in whorls of 6, remaining green when dried, with a rough margin, with a transparent tip. Flowers white, rarely yellowish or greenish, 3-5mm, in rather dense oblong panicles. Fruit dark brown when ripe, smooth. Calcareous rocks and screes, and other dry habitats, to 1600m. May-August. C & E France and S Germany. Widespread in the mountains of S & SC Europe.

5* **Glaucous Bedstraw** *Galium glaucum* L. [*Aspreula glauca*, *A. galioides*]. Medium to tall, bluish-green, stoloniferous perennial; stems whitish, often rooting at the base, rounded, with 4 ridges, swollen at the nodes, hairy or hairless. Leaves mostly in whorls of 8-10, linear, the margin rolled under almost to the midrib. Flowers white, 4-6mm, cup-shaped, in branched oval clusters. Fruit rough, but without bristles. Forest margins, dry grassland, stony and rocky habitats, often on sandy soils. May-July. Belgium, France and Germany; naturalized in S Norway and S Sweden. Local. Sometimes hybridises with *G. verum*.

6 *Galium aristatum* L. Like *G. glaucum*, but non-stoloniferous, the stems square and not rooting down at the base. Leaves bright green (not bluish-green) and with a rough margin. Flowers white 2-3mm. Open deciduous woods and scrub, to 1600m. June-September. C and E France and S Germany. Primarily central and southern European. The leaves are often more than 25mm long, while less than 25mm in the preceding species.

7 *Galium schultesii* Vest. Like *G. glaucum*, but the young shoots markedly bluish-green and leaves elliptical, deep green above but bluish-green beneath, turning black on drying. Open woodland. June-August. S Germany. Native to C and SE Europe.

8* **Wood Bedstraw** *Galium sylvaticum* L. Variable medium to tall, rather stout and bushy, without stolons; stems round with faint ridges, hairless, the young shoots greyish-green. Leaves elliptical, often broadest above the middle, mostly in whorls of 6-8, membranous, greyish beneath, not generally turning black on drying. Flowers white, 2-3mm, often nodding in bud. Ovary and fruit smooth, bluish-green. Woodland, coppices and scrub, to 1600m. June-September. Belgium, Holland, France and Germany. Primarily in C & E Europe. Generally easy to recognise by its bluish-green, non-bristly fruits and rounded stems.

9 *Galium obliquum* Vill. Like *G. sylvaticum*, but a shorter plant, the stems hairy at least at the base. Leaves in whorls of 7-10, 9-20mm long (not 20-40mm). Flowers yellow, greenish or occasionally purplish, 1-2mm. Dry grassy and rocky habitats, to 2000m. C & E France. Confined to the mountains of C Europe, especially the Alps, but extending west to the Cevennes and including the Jura.

10* **Slender Bedstraw** *Galium pumilum* Murray [*G. asperum*, *G. laeve*, *G. sylvestre*]. Variable, low to short, generally prostrate, tufted perennial, hairy or hairless; stems slender. Leaves in whorls of 8-10, narrow-oblong to sickle-shaped, the margin with a few backward directed bristles, drying pale green. Flowers creamy-white, 2-3mm, in a long lax panicle. Fruit hairless, often finely warted. Open woodland and grassland, usually on calcareous soils and at low altitudes. June-July. C & S Britain, Belgium, Holland, France, Germany and Denmark; naturalized in Finland and Sweden. Native to W and C Europe. **P**: Flies and other small insects. Intermediates between *G. pumilum* and *G. fleurotii* occur where the two grow in close proximity to one another. **B**: Scattered localities in SW Britain north to Lincolnshire and the Humber.

11* *Galium fleurotii* Jordan. Like *G. pumilum*, but a more densely tufted plant drying to dark brown or black and leaves linear-lanceolate, straight, mostly 6-10mm (not 10-16mm), usually bristly. Calcareous cliffs and screes, at low altitudes. June-July. S Britain and NW & C France – not found elsewhere. A rather rare and little known species. **B**: Very rare, a single locality in Cheddar.

12 *Galium valdepilosum* H.Braun. Like *G. fleurotii*, but leaves longer, mostly 11-18mm, and the plant greenish on drying. Leaves mostly in whorls of 7-8, rather thin. Flowers 2-3.5mm. Dry grassland and open woods. June-August. Denmark and Germany. The Danish plant is sometimes distinguished as subsp. *slesvicense* (Sterner) Ehrend.

13 **Swedish Bedstraw** *Galium suecicum* (Sterner) Ehrend. [*G. pumilum* subsp. *suecicum*]. Like *G. pumilum*, but leaves mostly in whorls of 7-8, shorter, 7-12mm long; plants with few (not many), vegetative shoots at flowering time. Flowers slightly smaller and fruits with more pointed warts. Dry grassland and scrub. July- September. C & S Sweden and NE Germany.

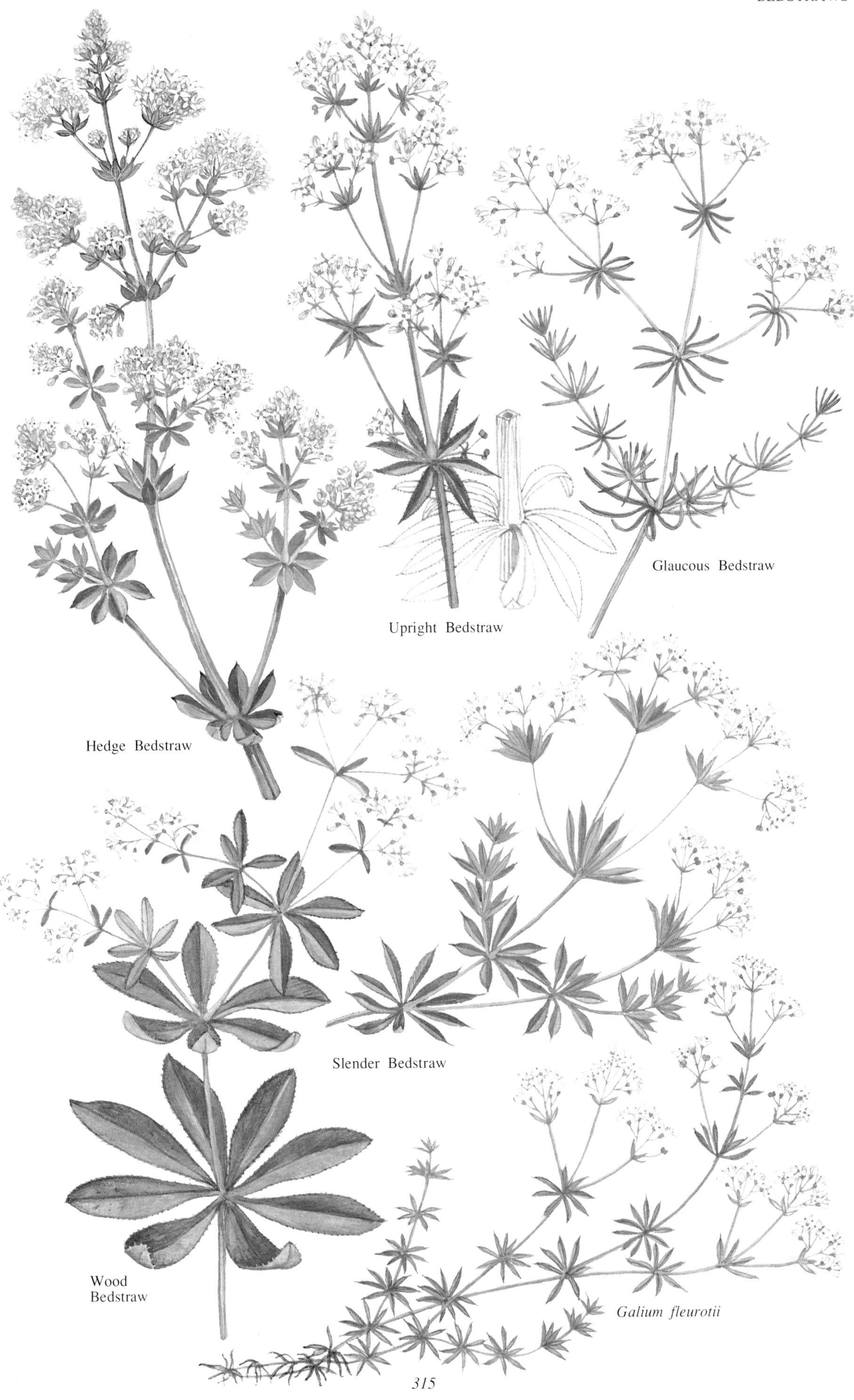
Upright Bedstraw
Glaucous Bedstraw
Hedge Bedstraw
Slender Bedstraw
Wood
Bedstraw
*Galium fleurotii*

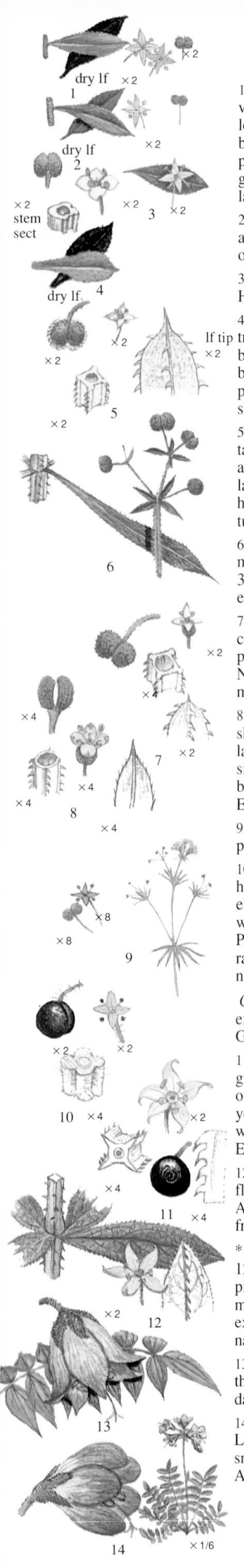


1* **Limestone Bedstraw** *Galium sterneri* Ehrend. Variable low to short, prostrate, perennial with a mat of vegetative shoots and ascending or erect flowering stems. Leaves in whorls of 7-8 usually, narrowly oblong, widest above the middle, dark green to blackish on drying; margin with many backward pointing bristles. Flowers creamy-white, occasionally greenish or yellowish, 2.3-3.3mm, in broad pyramidal panicles. Fruit hairless, but covered in small warts. Dry grassy and rocky habitats, often abundant on grazed grassland, on calcareous or igneous rocks, to 750m. June-July. NW Europe to S Finland. **B**: Wales, N England, Scotland, NE Ireland.

2 *Galium oelandicum* (Sterner & Hyl.) Ehrend. [*G. pumilum* subsp. *oelandicum*]. Like both *G. pumilum* and *G. sterneri*, turns metallic green when dried, stems often red at the base. Leaves mostly in whorls of 9, only 4-10mm long. Rocky and grassy habitats. July-August. S Sweden – Öland.

3 *Galium normanii* O.C. Dahl. More compact with broader leaves. Flowers yellowish-white, 3-4mm. Heaths and dry grassland. July-September. Iceland and two isolated localities in W Norway.

4* **Heath Bedstraw** *Galium saxatile* L. [*G. harcynicum*]. Low to short, hairless perennial, with a lax prostrate mat of vegetative shoots and spreading or ascending flowering shoots; stems smooth, square, much-branched. Leaves in whorls of 6-8, elliptical, broadest above the middle, the margin with forward directed bristles, turning black on drying. Flowers white, 2.5-4mm, rather sickly-scented, in lax, short-branched panicles. Fruit hairless, covered in small pointed warts. Scrub, pastures on dry acid soils, to 1300m, occasionally higher. June-August. Throughout. **B**: Widespread throughout.

5* **Cleavers, Goosegrass** *Galium aparine* L. Very variable, often scrambling, rather brittle, medium to tall bristly annual; stems 4-square, hairy at the nodes. Leaves in whorls of 6-9, narrowly elliptical, broadest above the middle, pointed, the margin with backward pointing bristles. Flowers dull white, 1.5-1.7mm, in lax, stalked clusters at the base of the upper leaves. Fruit green or purplish, 3-5mm, covered in dense hooked bristles, stalks bent sharply below the fruit. Woods, scrub, maritime shingle, generally at low altitudes. May-September. Throughout, except the far north. A widespread weed. **B**: Widespread.

6 **False Cleavers** *Galium spurium* L. [*G. vaillantii*]. Stems less hairy at the nodes and leaves narrower, mostly 30-35mm long (not 30-60mm). Flowers greenish-yellow, 0.8-1.3mm. Fruit blackish when ripe, 2-3mm, occasionally with reduced or no bristles. Hedgerows, scrub, waste ground. June-July. Throughout, except the extreme north. **B**: Locally established in Essex and a few other localities in C & S England.

7* **Corn Cleavers** *Galium tricornutum* Dandy. Like *G. aparine*, but leaves hairless above and flowers creamy-white, the clusters on stalks shorter (not longer) than the subtending leaves. Fruit rough, but not prickly, on downcurved stalks. Cultivated and waste ground, on dry calcareous soils. May-September. North to S Norway and S Sweden. The flowers are often borne in threes, the outer ones are sometimes male. **B**: Locally naturalized in England and Wales.

8* **Wall Bedstraw** *Galium parisiense* L. (incl. *G. parisiense* subsp. *anglicum* (Huds.) Clapham). Low to short, prostrate or scrambling annual; stems square, weak, hairless, bristly. Leaves in whorls of 5-7, lanceolate, downturned, the margin with backward pointing bristles. Flowers tiny, greenish inside, reddish outside, 0.5-1mm, in clusters forming a lax panicle. Fruit small, 1mm, hairless or with a few curved hairs, blackish. Sandy habitats and old walls. June-July. W C and S Europe. **B**: Rare and local; SE England and East Anglia.

9 *Galium divaricatum* Pourret ex Lam. Like *G. parisiense* but leaves in whorls of 6-8, often linear, and peduncles 5-20mm long (not 3-7mm). Dry open habitats. France, except NE.

10* **Crosswort** *Cruciata laevipes* Opiz. [*Galium cruciata*, *Valantia cruciata*] Short to medium, softly hairy, perennial, with a creeping stock; stems square, branched near the base. Leaves in whorls, of 4, oval-elliptical, 3-veined, yellowish-green. Flowers pale yellow, honey-scented, 2-2.5mm, in clusters forming whorls at the base of the leaves. Fruit globose, smooth and hairless, blackish, on recurved stalks when ripe. Pastures, road-verges, often on calcareous soils. April-June. Europe, north to Holland and Germany; naturalized in Ireland. The roots yield a red dye. **B**: Throughout, except the extreme north of Scotland; Ireland, naturalized in Co. Down.

*Cruciata glabra* (L.) Ehrend. [*Galium glabra*, *G. vernum*]. Like *C. laevipes*, but a slenderer plant, hairless except for the young leaves. Flowerstalks without bracts. Similar habitats and flowering time. Holland and Germany southwards.

11* **Wild Madder** *Rubia peregrina* L. Medium to tall, trailing or scrambling, hairless, rather rampant evergreen perennial, with a creeping rootstock; stems square, rough with downturned prickles. Leaves in whorls of 4-6, oval to elliptical, leathery, deep shiny green, 1-veined, rough on the midrib beneath. Flowers pale yellowish-green, 4-5mm, forming a leafy panicle, 5-lobed. Fruit subglobose, 4-6mm, black and fleshy when ripe. Scrub, woods. June-August. S Britain, W & S France. **B**: Mostly coastal S & W Wales and S England.

12 **Madder** *Rubia tinctorum* L. Like *R. peregrina*, but leaves softer, distinctly net-veined beneath, the flowers bright yellow, with less pointed lobes. Fruit reddish-brown. June-July. **I**: Mediterranean and W Asia. Naturalized in France and Germany; formerly in Britain. Formerly grown for the red dye, madder, from the roots.

* **Jacob's Ladder** *Polemonium caeruleum* L. (Jacob's Ladder Family – Polemoniaceae – 270 species in 12 genera). Medium to tall perennial forming tufts with erect angled stems, hairy above. Leaves alternate, pinnate, with 8-12 pairs of lanceolate leaflets. Flowers blue, occasionally white, 20-30mm, in a dense terminal cluster. Damp meadows, rocky habitats, often on limestone, to 2300m. May-August. Throughout, except the far north; naturalized in Belgium, Denmark, Holland. **B**: N England from Stafford northwards, naturalized elsewhere.

13 *Polemonium acutifolium* Willd. Like *P. caeruleum*, but the leaves with 8 or fewer pairs of leaflets and the flowers in smaller clusters, more bell-shaped and with pointed petal-lobes. River banks and thickets on damp soils. July-August. Arctic and sub-Arctic Europe.

14 *Polemonium boreale* Adams. Low to short, foetid perennial; stems leafless or with a solitary leaf. Leaves nearly all basal, pinnate, with 6-8 pairs of lanceolate leaflets, hairy. Flowers blue, 14-16mm, in small terminal clusters, bell-shaped, the petal-lobes rounded. Gravelly and sandy habitats. July-August. Arctic Europe – Norway and Spitsbergen extending into Arctic Russia.

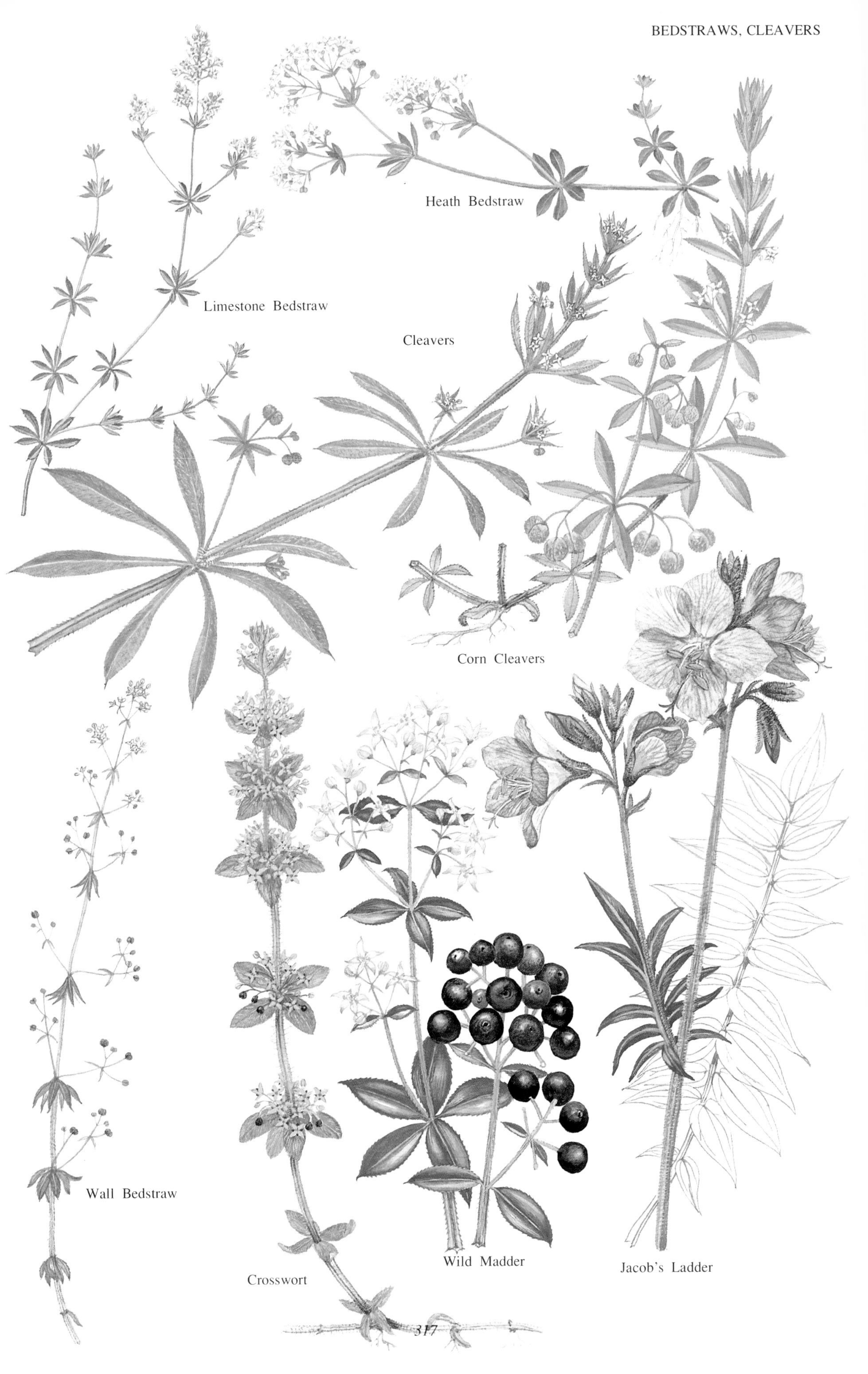
Heath Bedstraw
Limestone Bedstraw
Cleavers
Corn Cleavers
Wall Bedstraw
Crosswort
Wild Madder
Jacob's Ladder

## BINDWEED FAMILY Convolvulaceae

Herbs, often with twining or scrambling stems and alternate leaves; stems twining in an anti-clockwise direction. Flowers regular, hermaphrodite, funnel or bell-shaped, often large (not in *Cuscuta*), 4-5-parted. Fruit a 2-4-valved capsule. 1000 species in 40 genera. *Cuscuta* (100 species) are twining annual parasitic herbs with thread-like stems; without chlorophyll. Leaves reduced to tiny scales. Flowers in small tight clusters at the nodes (4-5-parted, bell-shaped, the petals reduced to tiny triangular scales; style 2. Parasitic on a variety of plants.

1* **Field Dodder** *Cuscuta campestris* Yuncker. Stems yellowish. Flowers yellowish-green, 2-3mm, in tight rounded clusters, 5-parted; flower-stalks generally shorter than the flowers; stamens protruding. Parasitic primarily on cultivated *Trifolium* and *Medicago* species. June-September. **I**: North America. Naturalized or casual in W Europe. **B**: Rare and generally casual.

2 *Cuscuta gronovii* Willd. Like *C. campestris*, but flowers in a longer, laxer cluster, their stalks often longer than the flowers themselves. Corolla attached to the top of the fruit capsule like a tiny cap, not persisting around the fruit base. Parasitic primarily on *Populus*, *Salix* and other trees and shrubs, particularly along river and stream banks. June-October. **I**: C & E North America. Naturalized in Belgium, France, Germany and Holland.

3* **Greater Dodder** *Cuscuta europaea* L. Stems stouter than *C. campestris*, often reddish, branching and often forming a dense entanglement. Flowers pinkish, 2mm, in dense rounded heads, often 4-parted, the stamens not protruding; flower-stalks very short. Parasitic primarily on *Urtica dioica* or *Humulus lupulus*, more rarely on other herbs. July-October. Throughout, except the far north and Ireland. **B**: Most of Britain, but primarily in S England, decreasing.

4 **Flax Dodder** *Cuscuta epilinum* Weihe. Like *C. europaea*, but stems scarcely branched or simple; flowers yellowish, 5-parted, the petal-lobes incurved (not spreading). Parasitic on flax, *Linum usitatissimum*, and other species of *Linum*, sometimes also on *Camelina sativa*. July-September. Europe, except Belgium and Denmark. **B**: Formerly in E England.

5* **Common Dodder** *Cuscuta epithymum* (L.) L. Forms a mass of slender, much-branched reddish or purplish stems. Flowers pale pink, 3-4mm, scented, 5-parted, with pointed spreading petals borne in tight rounded clusters; stamens protruding. Parasitic on various shrubs and herbs, but particularly *Calluna vulgaris*, *Ulex europaeus* and *Trifolium* species; heaths and grassy habitats, to 2200m. June-October. Throughout, except the far north.

6 *Cuscuta lupuliformis* Krocker. Like *C. epithymum*, but flowers with a solitary (not 2) style and stamens not protruding. Parasitic on species of *Salix* and other trees. Germany and Holland.

*Calystegia*. Perennial herbs with climbing or prostrate stems, producing a white latex when cut. Flowers large and funnel-shaped. Calyx enclosed by large bract-like scales. 10 species.

7* **Sea Bindweed** *Calystegia soldanella* (L.) R. Br. [*Convolvulus soldanella*]. Low prostrate, spreading, hairless perennial; stems sometimes twisting weakly. Leaves kidney-shaped, rather fleshy, long-stalked. Flowers pink or purplish, often with whitish stripes, 30-50mm, solitary, 'bracts' generally somewhat shorter than the calyx. Coastal dunes, sand and shingle. June-August. Coasts of Europe north as far as Denmark. **P**: Mainly bumble-bees. From Europe to Asia, N Africa and the Americas. **B**: S Britain, rare in the north and apparently decreasing.

8* **Larger Bindweed, Bellvine** *Calystegia sepium* (L.) R. Br. [*Convolvulus sepium*]. Vigorous climbing and twining perennial to 3m, hairy or hairless. Leaves arrow-shaped, bright green. Flowers white, very rarely pale pink, 30-50mm; 'bracts' longer than the calyx, scarcely overlapping. Woodland and river margins, scrub, hedges, fences and fens, waste places. June-September. Throughout, except the far north. **B**: Throughout, rarer in the north. subsp. *roseata* Brummitt has stems and leaf-stalks hairy and flowers pink, 40-55mm. Coastal salt-marshes, sands and waste places. W Europe, including Britain. subsp. *spectabilis* Brummitt is sometimes hairless and leaves with a rounded gap at the base (not sharply angled). Flowers 50-60mm. Naturalized in Finland (probably a native of Siberia).

9* **Great Bindweed** *Calystegia silvatica* (Kit. ex Schrader) Griseb. [*Convolvulus silvaticus*, *Calystegia sylvestris*]. Like *C. sepium*, but flower larger, 50-90mm, white, sometimes striped with pale pink; 'bracts' widely overlapping, pouched at the base. Hedgerows, around buildings and waste places. July-September. **I**: S & SE Europe. Naturalized in Britain. Probably hybridising with *C. sepium*. **B**: Commonest from the Midlands to SE England, scarce elsewhere.

10* **Hairy Bindweed** *Calystegia pulchra* Brummitt & Heywood. Like *C. silvatica*, but stems and leaf-stalks hairy, at least at first, the leaves often parallel-sided. Flowers always pink, 50-75mm. Hedges, scrub and waste places. July-September. Naturalized from gardens. Britain to Denmark and Germany southwards. Origin uncertain, possibly a garden hybrid or from NE Asia.

11* **Bindweed** *Convolvulus arvensis* L. Climbing or creeping perennial, hairy or hairless, to 2m, though often far less. Leaves arrow-shaped to oblong, stalked. Flowers white, or pink with white stripes, 10-25mm, solitary, weakly scented. Fruit a rounded capsule. Cultivated and waste land, pathways, short turf and coastal habitats, to 1850m. June-September. Throughout, except the far north. The flowers are short-lived and are open only during the day. **B**: Throughout.

## NEMOPHILA FAMILY Hydrophyllaceae

Herbs with alternate leaves and 5-parted, partly tubular, or funnel-shaped flowers. Fruit a capsule. Primarily from North America.

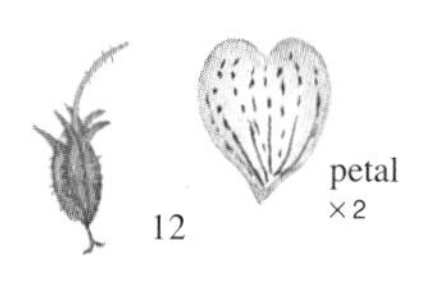


12* **Nemophila** *Nemophila menziesii* Hooker & Arnott. Low to short hairy annual herb. Leaves opposite, pinnately-lobed. Flowers blue or white, 15-30mm, solitary; calyx with sepal-like segments between the main lobes. **I**: W N America. Naturalized from gardens in Norway and Sweden. July-August.

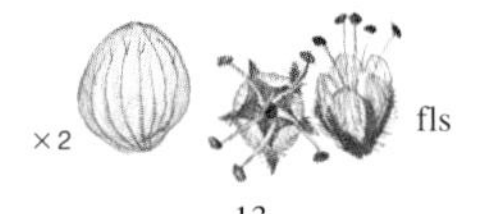


13* **Phacelia** *Phacelia tanacetifolia* Bentham. Medium to tall hairy annual with erect, branched stems. Leaves alternate oblong, lobed near the base. Flowers blue, 6-9mm, in branched, coiled spikes like a *Myosotis*; stamens protruding prominently. **I**: W N America. Naturalized from gardens in Scandinavia, Belgium, France and Holland. July-August.

Field Dodder
Common Dodder
Greater Dodder
Sea Bindweed
Larger Bindweed
Great Bindweed
Hairy Bindweed
Bindweed
Nemophila
Phacelia

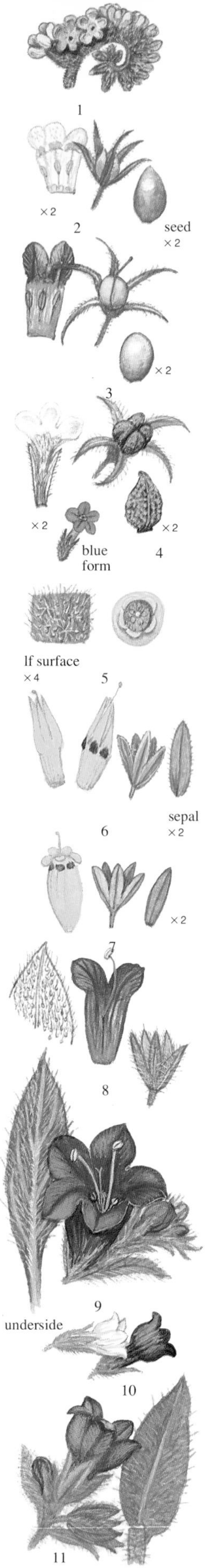

# BORAGE FAMILY Boraginaceae

Herbs or small shrubs, often with bristly stems and leaves, the bristles often with a swollen base. Leaves simple and alternate. Flowers in spiralled clusters (scorpioid cymes), short-stalked, 5-parted, the corolla funnel-shaped or constricted at the mouth. Stamens 5, joined to the petal-tube. Ovary superior, with a solitary style. Fruit consisting of 4, occasionally 2, nutlets, often ornamented. 2000 species in about 100 genera, throughout temperate and subtropical regions of the world, but particularly in the Mediterranean region.

1* **Heliotrope** *Heliotropium europaeum* L. Variable low to short, erect or spreading, hairy annual. Leaves oval to elliptical, grey-green, stalked. Flowers white with a yellow 'eye', 2-4.5mm, in one-sided, leafless, forked spikes; sepal-lobes linear-oblong. Fruit splitting into 4 nutlets, separating from the calyx. Waste places, waysides and cultivated land. May-July. France and S Germany. Widespread in S & SW Europe.

2* **Common Gromwell** *Lithospermum officinale* L. Medium to tall, often tufted, hairy perennial; stems erect, much-branched above. Leaves lanceolate to oval, pointed, with prominent veins, the upper unstalked. Flowers creamy- or greenish-white, funnel-shaped, 4-6mm long, in leafy spiralled clusters. Fruit with 4 shiny-white nutlets. Woodland margins, scrub, thickets and hedgerows, to 1600m. May-August. Throughout, except the far north, though rare in parts of the north and west. **B**: Widespread in lowland Britain, but scarce in Scotland and Ireland.

3* **Purple Gromwell** *Buglossoides purpurocaerulea* (L.) I.M. Johnston [*Lithospermum purpurocaerulea*]. Short to medium, tufted, hairy, rhizomatous perennial; stems erect, usually unbranched. Leaves lanceolate to narrowly elliptical, dark green, pointed. Flowers reddish-purple but soon turning bright blue, 14-19mm long, the terminal leafy cymes elongating into fruit. Nutlets 4, shiny-white. Woodland margins and scrub, generally on calcareous soils, to 1200m. April-June. S Britain, Belgium, France and C & S Germany. **B**: Rare, confined to a few localities in S Wales and SW England.

4* **Corn Gromwell** *Buglossoides arvensis* (L.) I.M. Johnston [*Lithospermum arvensis*]. Short to medium, somewhat bristly annual; stems solitary, erect, little branched. Leaves oblong to linear, with indistinct side veins, the upper unstalked. Flowers creamy-white, the tube usually violet, sometimes blue, 6-9mm long, borne in solitary terminal cymes. Nutlets hard, greyish-brown, warted. Cultivated ground, waste ground, to 2300m. April-September. Throughout, except the far north. A widespread weed. The tuft of hairs above the anthers keep rain out of the flowers. **B**: Common in England, rarer elsewhere.

5 *B. a.* subsp. *permixta* (Jordan ex F.W. Schultz) R. Fernando [*Lithospermum permixtum*]. Like the type, but the flowers blue, 5-7mm long, the calyx generally shorter (not equalling or longer) than the corolla-tube. Similar flowering time and habitats. France and S Germany.

6* **Golden Drop** *Onosma arenaria* Waldst. & Kit. Short to medium, tufted, bristly perennial, rarely a biennial, with some non-flowering leaf-rosettes. Leaves oblong, often broadest above the middle, sparsely to densely bristly, yellowish-green. Flowers pale yellow, tubular, with 5 short reflexed lobes, 12-17mm long, borne in a branched pyramidal cluster; calyx expanding in fruit. Rocky habitats, to 1700m. May-July. S Germany. Throughout C & SE Europe. Often very local. The root of the related *O. echioides* L. from S Europe, yields a red dye, 'orcanette jaune', used for colouring food.

7* **Lesser Honeywort** *Cerinthe minor* L. Very variable, scarcely hairy, short to medium annual or biennial. Leaves grey-green, often white spotted, the lower oval, stalked, the upper unstalked and clasping the stem. Flowers yellow, occasionally with 5 violet spots in the throat, 10-12mm long, narrow bell-shaped with pointed lobes, not recurved. Fruit-stalks spreading or reflexed. Pastures, woodland, rocky and cultivated ground, often over limestone and primarily in the mountains, to 2200m. May-September. C & E France and S Germany, occasionally further north. Mainly in C & E Europe.

8* **Smooth Honeywort** *Cerinthe glabra* Miller [*C. alpina*]. Short to medium, hairless biennial or perennial. Leaves grey-green, unspotted, the lower oblong, stalked, the upper heart-shaped, clasping the stem. Flowers yellow, generally with 5 red spots in the throat, 8-13mm long, tubular-bells with blunt recurved lobes, borne in leafy cymes. Fruit stalks reflexed. Meadows and damp woods, often in shaded habitats, generally on calcareous soils, to 2600m. May-July. C & S France and S Germany. C & S Europe.

9* **Viper's Bugloss** *Echium vulgare* L. Short to medium, variable, erect bristly biennial, rarely perennial; stems solitary or several. Leaves elliptical to lanceolate, stalked, with obscure lateral veins, the uppermost narrower and unstalked. Flowers pale to bright blue or blue-violet, pink in bud, 15-20mm long, with an oblique mouth, borne in branched coiled cymes; stamens long-protruding. Fruit hidden by the calyx-lobes. Dry open habitats, to 1800m. June-September. Throughout, except the far north. Some plants bear only smaller, solely female flowers. **B**: North to S Scotland and in E Ireland, scarce elsewhere.

10 **Purple Viper's Bugloss** *Echium plantagineum* L. [*E. lycopsis, E. maritimum*]. Like *E. vulgare*, but a lower annual or biennial, the leaves much softer, with prominent lateral veins, the uppermost with a heart-shaped base. Flowers blue, becoming pink, 18-30mm long, only 2 of the stamens protruding. Dry sandy habitats, often close to the sea, June-August. S Britain and France, occasionally casual elsewhere. **B**: Very rare; a few coastal localities in Cornwall, Scilly and Jersey.

11* **Nonea** *Nonea pulla* (L.) DC. Short to medium, greyish, bristly annual or perennial; stems erect, branched above. Leaves lanceolate to linear-lanceolate, pointed, untoothed, the upper clasping the stem. Flowers dark reddish-brown or blackish-purple, 10-14mm long, with short spreading lobes, borne in leafy one-sided clusters; sepals enlarging in fruit. Nutlets oval with a short lateral beak. Dry grassy habitats and stony places, at low altitudes. April-August. Germany; naturalized in France and S Finland. E & EC Europe.

12 *Nonea versicolor* (Steven) Sweet. Like *N. pulla*, but generally a shorter plant, the stems simple or somewhat branched above. Flowers purple to violet, 12-17mm long. Similar habitats and flowering time. **I**: Caucasus & E Turkey. Naturalized in Denmark, S Norway and S Sweden.

*Nonea rosea* (Bieb.) Link. Like *N. pulla*, but flowers purplish or brownish, 15-18mm. Cultivated land, waste and grassy places. May-August. **I**: Caucasus. Naturalized in France, Holland, Belgium, Germany and parts of S Scandinavia.

Heliotrope
Common Gromwell
Purple Gromwell
Corn Gromwell
Golden Drop
Lesser Honeywort
Smooth Honeywort
× 1/9
Viper's Bugloss
Nonea

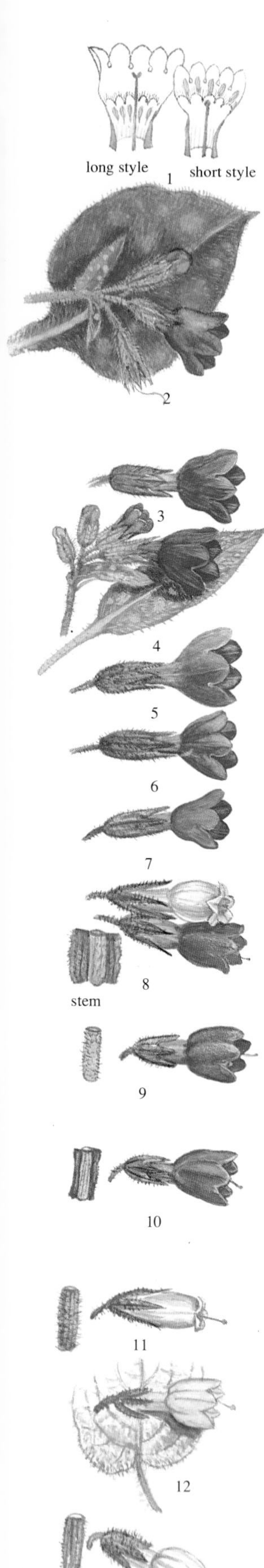


**Lungworts** *Pulmonaria*. Tufted perennial herbs with large basal leaves and smaller stem leaves; stems simple, rough and bristly like the leaves. Flowers in terminal cymes, with bracts, heterostylous, the corolla purple or blue, but often pink in bud, funnel-shaped, with 5 tufts of hairs alternating with the stamens. Nutlets with a raised ring around the base. 10 species, often variable. The summer leaves enlarge after flowering.

1* **Common Lungwort** *Pulmonaria officinalis* L. [*P. officinalis* subsp. *maculosa*]. Low to short, hairy, tufted, very rough perennial. Basal leaves heart-shaped, green with distinctive white blotches, stalked; stem leaves oval, unstalked, half-clasping the stem. Flowers reddish- to bluish-violet, pink in bud, 13-18mm long; calyx cylindrical, the teeth triangular, one third the length of the calyx. Shaded and semi-shaded habitats on rich deep, humus soils, generally over limestone, to 1900m. March-May. Northwards to S Sweden; naturalized in Britain. **B**: Naturalized in scattered localities throughout England, Wales and S Scotland.

2 *Pulmonaria obscura* Dumort. [*P. officinalis* subsp. *obscura*]. Like *P. officinalis*, but the leaves unblotched or with pale green blotches, with uneven (not even) bristles. Similar habitats and flowering time. Belgium, France, Germany, Denmark, S Sweden and Finland; naturalized in Britain.

3* *Pulmonaria mollis* Wulfen ex Hornem. Very variable, low to short, stickily-hairy tufted perennial. Basal leaves oval, gradually narrowed into a short, winged stalk, unspotted; stem leaves oval, half-clasping the stem. Flowers violet to violet-blue, 12-18mm long; inflorescence very sticky-glandular. Shaded habitats, open woodland, scrub and rocky places. March-May. E France and S Germany.

4 *Pulmonaria affinis* Jordan. Like *P. mollis*, but leaf-blade white-spotted, abruptly narrowed into a long, winged stalk. Flowers blue or violet-blue. C & W France to N Spain.

5* **Mountain Lungwort** *Pulmonaria montana* Lej. Short to medium, tufted perennial, sticky-glandular. Basal leaves long-lanceolate, shining above and with soft hairs, unspotted, or occasionally with rather indistinct green blotches, narrowed into the stalk; upper leaves heart-shaped, half-clasping the stem. Flowers pink, but turning bright blue, 15-20mm long, in small clusters. Meadows and damp mountain woods, to 1900m. April-May. Belgium, France and C & S Germany; extinct in Holland.

6* **Narrow-leaved Lungwort** *Pulmonaria longifolia* (Bast.) Boreau. Short to medium, rough tufted perennial. Basal leaves narrow lanceolate, gradually narrowed into the stalk, generally with white spots, but occasionally unspotted or with pale green spots; stem leaves similar to the basal, but unstalked. Flowers red at first, but soon turning to violet or violet-blue, 8-12mm long, in rather dense clusters. Semi-shaded habitats, sometimes among rocks, usually on rather heavy clay soils, to 2000m. April-June. S Britain and France. Forms with unspotted leaves are common in W and C France. **B**: Local in the New Forest, Dorset and the Isle of Wight.

7* *Pulmonaria angustifolia* L. [*P. azurea*, *P. angustifolia* subsp. *azurea*]. Short to medium, rough, tufted perennial. Basal leaves narrowly lanceolate, gradually tapered to a short stalk, unspotted and without glands; stem leaves similar but unstalked. Flowers bright blue, reddish in bud, 12-20mm long; sepal-tube narrow in fruit. Woods and meadows on acid soils, to 2600m. April-July. Denmark, Germany and S Sweden, isolated localities in France. Mainly in NE & E Europe.

**Comfreys** *Symphytum*. Tufted perennial herbs, generally with creeping rhizomes and rough, rather brittle stems and leaves. Flowers in dense spiralled cymes; corolla tubular-bell-shaped with 5 short triangular lobes, the throat with 5 scales alternating with the stamens. Style protruding. Nutlets egg-shaped, smooth or granular. 25 species in Europe and W Asia north to Siberia. Plants often of damp habitats.

8* **Common Comfrey** *Symphytum officinalis* L. Very variable medium to tall, stout perennial, to 1.2m; stems erect, widely winged. Leaves large and coarse, oval-lanceolate to lanceolate, untoothed, the base running onto the stem, only the lowermost stalked. Flowers purple-violet, dirty pink or whitish, 12-18mm, bell-shaped with reflexed lobes, borne in forked clusters. Nutlets black and shiny. Damp grassland, river, stream and canal banks, ditches and fens, to 1600m. May-July. Naturalized in Ireland and much of northern Europe. Often forming extensive colonies. **B**: Throughout, though scarcer in the north; naturalized in Ireland.

9* **Rough Comfrey** *Symphytum asperum* Lepechin. Tall stout, rough-hairy perennial herb, to 1.8m, though often less; stems not winged. Leaves oval to oblong, narrowed or almost heart-shaped at the base, the lower stalked, the upper unstalked, not clasping the stem. Flowers pink in bud, but turning blue, 10-17mm long. Nutlets granular. Waste places and margins of cultivated land. June-July. **I**: Iran & the Caucasus. Widely naturalized in Britain, Belgium, France and parts of C & S Scandinavia. Formerly widely grown as a cattle food, **B**: Scattered throughout.

10 **Russian Comfrey** *Symphytum* x *uplandicum* Nyman. Hybrid between *S. asperum* and *S. officinalis*, growing to 2m tall, the stems narrowly winged below the leaves and the flowers pink at first, but soon becoming blue or violet. Road-verges, banks, woodlands and waste places, usually away from the waterside. June-August. The commonest roadside comfrey. **B**: Naturalized throughout much of Britain.

11* **Tuberous Comfrey** *Symphytum tuberosum* L. Short to medium, rough, perennial, with stout tuberous rhizomes; stem simple or only slightly branched, somewhat winged. Leaves elliptical to lanceolate, the basal ones generally disappeared by flowering time, the upper sessile. Flowers pale creamy-yellow, 13-19mm long, with reflexed lobes. Nutlets finely granular. Damp and shaded places,to 1600m. May-July. Britain, France and Germany; naturalized in Ireland. Widely cultivated in gardens. The rhizomes have alternating thin and thick portions. **B**: Scattered localities throughout.

12 *Symphytum grandiflorum* DC. [*S. ibiricum*]. Like *S. tuberosum*, but rhizomes slender and far-reaching. Leaves oval to elliptical, rounded to heart-shaped at the base, the upper short-stalked. Flowers pale yellow on stems rarely more than 20cm tall. **I**: Caucasus. Naturalized in grassland and along hedgerows. May-June. S Britain. **B**: England from the Midlands southwards.

13* **White Comfrey** *Symphytum orientale* L. [*S. tauricum*]. Medium to tall, softly hairy biennial, or short-lived perennial, without creeping rhizomes; stems erect, much-branched. Leaves rather pale green, oval, rounded or slightly heart-shaped at the base, short-stalked, the uppermost usually unstalked. Flowers white, 14-18mm, the lobes not recurved. Nutlets dark brown, granular. Damp shaded habitats, woodland, hedgebanks and grassy places. April-May. **I**: Turkey. Locally naturalized in Britain and France. **B**: Scattered localities in E England and S Scotland, rare elsewhere.

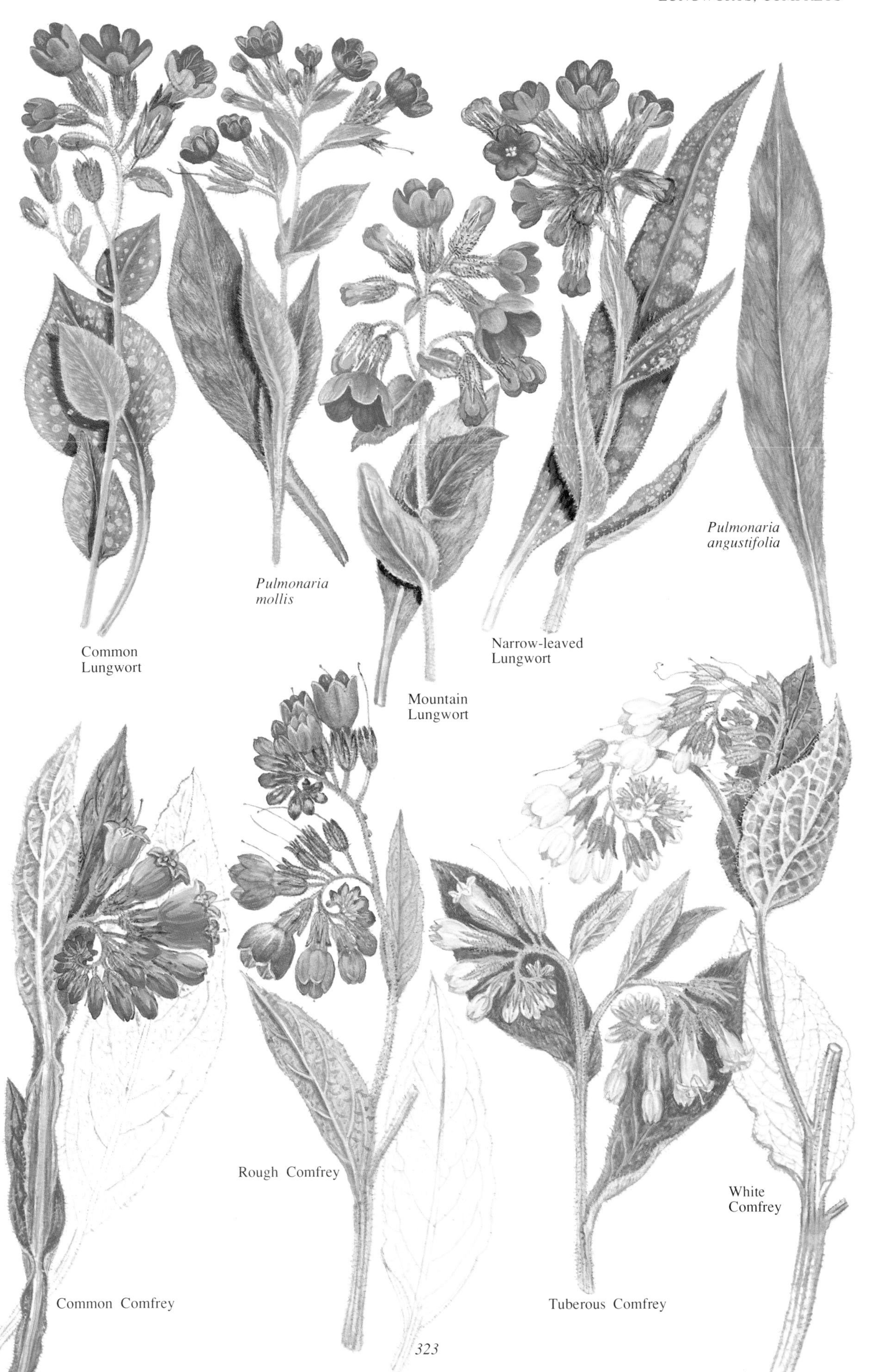
Common Lungwort
*Pulmonaria mollis*
Mountain Lungwort
Narrow-leaved Lungwort
*Pulmonaria angustifolia*
Common Comfrey
Rough Comfrey
Tuberous Comfrey
White Comfrey

**Anchusas** *Anchusa.* Annual or perennial herbs, generally rather bristly. Flowers in lateral or terminal cymes, the corolla bell-shaped, with a straight or curved tube and 5 spreading, rounded lobes; calyx deeply divided, usually enlarging in fruit. Stamens and style not protruding from the mouth of the corolla, but the mouth closed by scales or a tuft of hairs. 30 species in Europe and Asia. Several are grown in gardens.

1* **Yellow Alkanet** *Anchusa ochroleuca* Bieb. Medium to tall, rough-hairy perennial, occasionally a biennial; stems erect. Leaves lanceolate to linear, mostly less than 10mm wide. Flowers pale yellow, 7-10mm, in short crowded cymes; calyx divided for less than half-way, the teeth white-edged Waste places and cultivated land. June-August. **I**: SE & EC Europe. Naturalized in Holland; casual elsewhere in W & NW Europe; only species in our area with yellow flowers.

2* **Alkanet** *Anchusa officinalis* L. (incl. *A. procera* Besser ex Link). Short to tall, rough-bristly perennial, occasionally biennial; stems erect, unbranched. Leaves long-lanceolate, 10-20mm wide, the lower stalked. Flowers bluish-red or violet, rarely white or yellowish, 7-15mm, borne in elongating coiled cymes; calyx divided to the middle. Nutlets conical, unstalked. Meadows, banks and rocky habitats, waste places, often over limestone, to 1800m. May-September. France, Holland, Germany, Denmark, S Norway and S Sweden; casual or naturalized, but often rare in Britain, Belgium and Finland. Over much of Europe, except parts of the north. Also present in W Asia. **B**: Casual in waste place in S Britain.

3* **Large Blue Alkanet** *Anchusa azurea* Miller [*A. italica*]. Like *A. officinalis*, but often taller, the leaves 15-50mm wide. Flowers violet or deep blue, larger, 15-25mm with a tuft of white hairs in the centre; calyx divided almost to the base. Fields, waysides and waste places, at low altitudes. May-August. C France; naturalized or casual in Britain, N France and Germany. Primarily from Mediterranean Europe. **B**: Casual in waste places, especially in the south.

4* **Bugloss** *Anchusa arvensis* (L.) Bieb. [*Lycopsis arvensis*]. Short to medium bristly annual with ascending stems. Leaves lanceolate to linear-lanceolate, irregularly toothed, wavy-edged, the lower stalked, the upper unstalked and half-clasping the stem. Flowers bright blue, occasionally whitish, 4-6mm, the tube curved in the middle, borne in forked cymes, the sepals enlarging slightly in fruit; bracts leaf-like. Nutlets netted. Arable land, sandy heaths, waste and bare places, generally on light sandy or chalky soils, often close to the sea and at low altitudes. April-September. Throughout, except for the far north. **B**: Locally common.

5* **Green Alkanet** *Pentaglottis sempervirens* (L.) Tausch ex L.H. Bailey [*Anchusa sempervirens*, *Caryolopha sempervirens*]. Medium to tall, bristly perennial; stems erect or ascending, branched. Basal leaves oval to oblong, pointed, narrowed into a long stalk; stem leaves unstalked. Flowers bright blue with a white scaly throat, 8-10mm, in small, long-stalked, leafy cymes. Nutlets netted. Damp or shaded habitats, woodland margins, hedgebanks and cultivated land, often close to buildings, at low altitudes. April-July. **I**: SW Europe. Locally naturalized in much of Britain and Belgium. **B**: Scattered localities in the south.

6* **Borage** *Borago officinalis* L. Medium, bristly annual, occasionally overwintering; stems often rather robust, usually branched. Basal leaves oval to lanceolate, stalked, in a rosette to begin with, wavy-margined; stem leaves smaller, the uppermost clasping the stem. Flowers bright blue with a whitish centre, 20-25mm, half-nodding and rather star-shaped, with spreading pointed lobes and a prominent cone of purple-black stamens, borne in broad branched cymes. Waste ground, waysides and cultivated land, often in rather dry sunny places, to 1800m. May-September. C & S France; naturalized in Britain, Holland and France. **B**: A garden escape, usually near habitation.

7 **Abraham, Isaac & Jacob** *Trachystemon orientalis* (L.) G. Don f. Short to medium tufted, rhizomatous, sparsely bristly perennial; stems erect, branched. Basal leaves large, oval-heart-shaped, long-stalked; stem leaves rather few, oval to lanceolate, unstalked. Flowers bluish-violet, 14-18mm, starry, with spreading narrow pointed, often reflexed lobes, borne in branched cymes; stamens protruding, hairy, forming a cone. Damp woods and other shady habitats. April-May. **I**: E Mediterranean region. Naturalized in S Britain. **B**: Locally naturalized in Devon, Kent and Yorkshire.

8* **Oyster Plant** *Mertensia maritima* (L.) S.F. Gray. Low to short, prostrate, mat-forming, rather fleshy, greyish, hairless perennial. Leaves lanceolate or spoon-shaped, the lower stalked, the upper unstalked. Flowers pink, becoming pink and blue, 6mm, funnel-shaped, constricted in the throat, borne in branched leafy cymes. Fruit on recurved stalks, the nutlets flattened and fleshy, the outer coat rather inflated and papery. Coastal habitats, maritime sands and shingle. June-August. Britain, France, Denmark, Scandinavia, except for the Baltic region. Often germinates where there is an accumulation of vegetation, especially seaweed, along the shoreline. An exceptional plant in the genus, being quite smooth and hairless. **B**: Rare and decreasing; confined mainly to Britain from Norfolk and Anglesey northwards;also in E & N Ireland.

9* **Amsinckia** *Amsinckia micrantha* Suksd. [*A. lycopodioides*]. Erect, short to medium, bristly-hairy annual, often branched. Leaves linear to elliptical, often widest above the middle. Flowers deep orange-yellow, 5-8mm long, funnel-shaped, hairy inside, in coiled cymes without bracts. Nutlets wrinkled. Bare and disturbed ground, at low altitudes. April-August. **I**: E North America. Naturalized in NE England – Farne Islands; a frequent casual in France and parts of Scandinavia.

10 *Amsinckia calycina* (Moris) Chater [*Lithospermum calycinum*, *A. hispida*, *A. angustifolia*]. Like *A. lycopsoides*, but leaves linear-lanceolate and flowers hairless inside the throat. Waste places. May-August. **I**: Southern North America & South America. Locally naturalized in France but often casual in parts of NW Europe.

11* **Madwort** *Asperugo procumbens* L. Low spreading, often prostrate, bristly annual, sometimes clambering over other plants; stems angled. Leaves lanceolate, mostly opposite, untoothed or slightly toothed. Flowers violet to purplish with a white throat, 2-3mm, solitary or paired at the base of the leaves. Fruits surrounded by the enlarged leaf-like sepals. Cultivated fields, waste places, farmyards, often on nitrogen-rich soils, to 2600m. May-November. Germany and Scandinavia; naturalized in Britain, Holland and France. **B**: Scattered localities, especially close to sea-ports; usually casual.

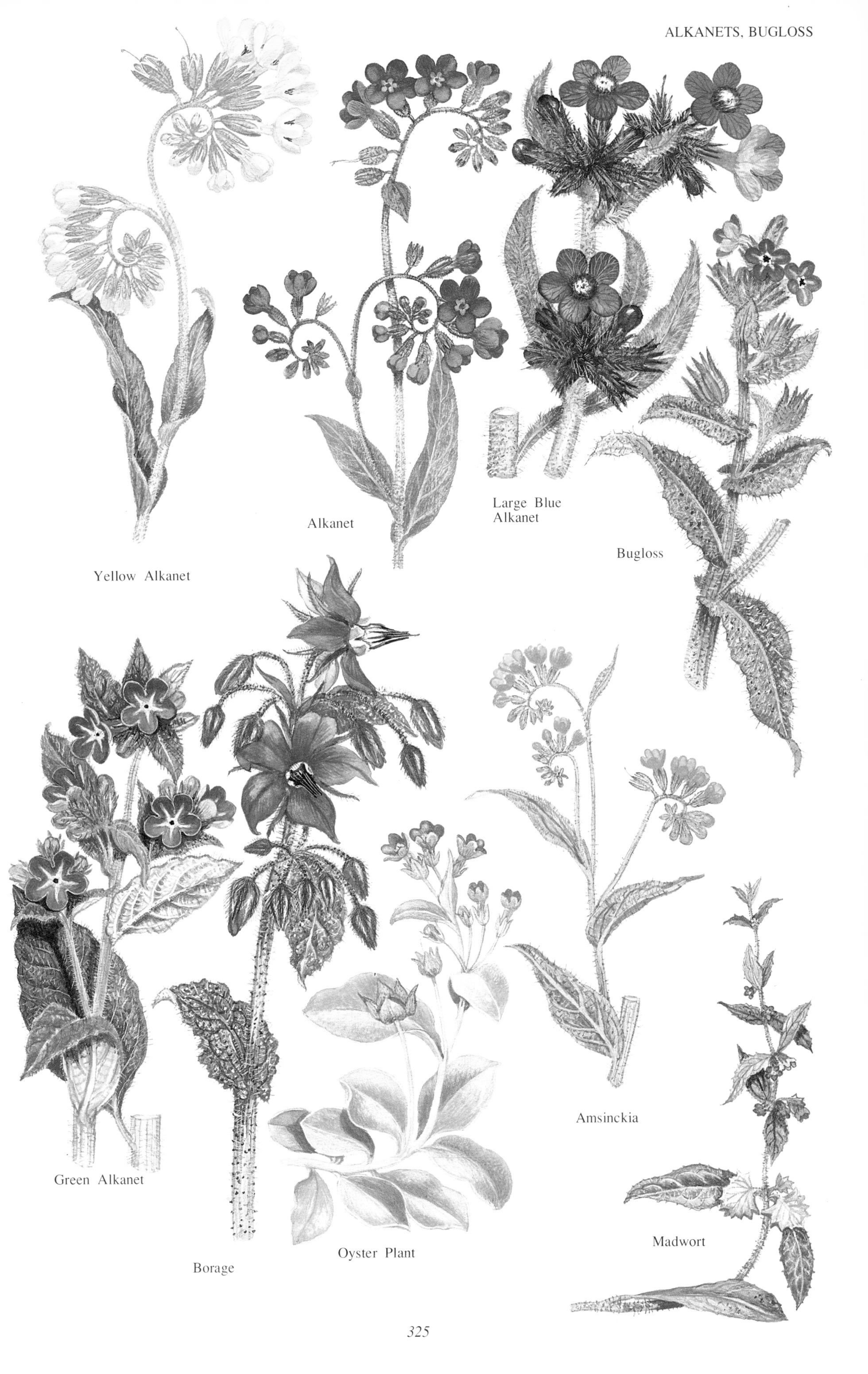
Yellow Alkanet
Alkanet
Large Blue Alkanet
Bugloss
Green Alkanet
Borage
Oyster Plant
Amsinckia
Madwort

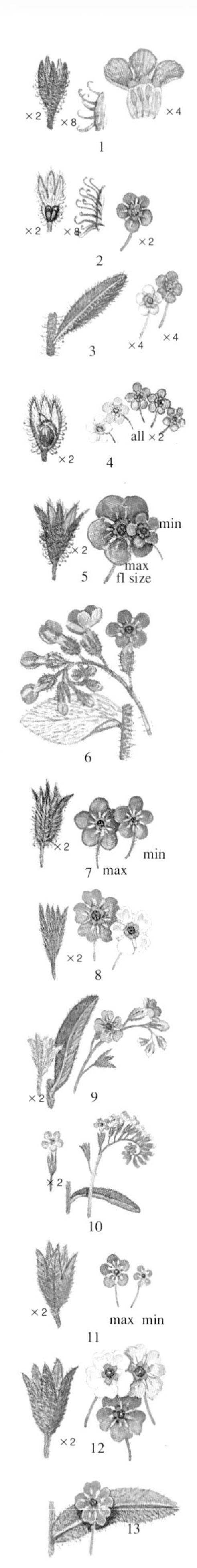


**Forget-me-nots** *Myosotis.* Annual or perennial herbs with narrow, alternate leaves. Flowers small, often blue, generally in paired cymes, the corolla with a short tube and a flat limb with 5 rounded lobes, with 5 scales in the throat. Nutlets small and shiny, enclosed in the persistent calyx. 50 species in temperate regions.

1* **Field Forget-me-not** *Myosotis arvensis* (L.) Hill. [*M. intermedia*]. Very variable, low to short, softly hairy annual or biennial, branched at the base. Basal leaves in a lax rosette, elliptical, broadest above the middle, scarcely stalked; stem leaves lanceolate, pointed. Flowers bright grey-blue, saucer-shaped, 3-5mm; calyx closed in fruit, the tube with numerous hooked hairs. Nutlets dark brownish-black. Dry habitats, arable soils, dunes, to 2000m. April-October. Throughout, except for Spitsbergen. Widespread from Europe to W Asia and Siberia. subsp. *umbrata* (Rouy) O.Schwarz from W Europe has a longer calyx, 7mm (not up to 5mm). **B**: Common throughout, except for the extreme north.

2* **Early Forget-me-not** *Myosotis ramosissima* Rochel [*M. gracillima, M. hispida*]. Slender, low to short, softly-hairy annual, often only 2-5cm tall. Basal leaves in a rosette, lanceolate; stem leaves oblong, unstalked. Flowers bright blue, 2-3mm, the petals scarcely exceeding the calyx; calyx open in fruit, with hooked hairs in the lower part. Nutlets pale brown. Dry open habitats, to 2000m. April-June. Throughout, except the far north. **B**: Throughout, scarce in W Britain and much of Ireland.

subsp. *globularis* (Samp.) Grau [*M. globularis*]. Like the type, but flowers extending to the base of the stem, not confined to the upper half; fruiting calyx 2mm (not 3-4mm) long. Sandy habitats, especially close to the sea. April-May. W France.

3* *Myosotis stricta* Link ex Roemer & Schultes [*M. vestita*]. Like *M. ramosissima*, but the lower stems and leaf-midrib beneath with hooked hairs; flowers 1-2mm, pale to bright blue, the stems of the inflorescence with spreading (not appressed) hairs. Dry sandy habitats, to 2000m. May-July. Throughout, except Britain, the Faeroes and Spitsbergen.

4* **Changing Forget-me-not** *Myosotis discolor* Pers. [*M. collina, M. versicolor*]. Low to short, hairy, rather slender annual; stems often branched at the base. Basal leaves lanceolate, blunt; stem leaves narrower and more pointed, often with at least one pair opposite. Flowers pale yellow or cream at first, becoming pink, violet or blue, 2mm; calyx 4-5mm, the teeth incurved in fruit. Nutlets dark brown. Bare open grassy places, open woods, generally on light soils, to 2000m. May-September. Throughout, north to S Scandinavia. Widespread in Europe; also present in the Azores. subsp. *dubia* (Arrondeau) Blaise. Like the type, but the stems without any opposite leaves and flowers cream at first. Confined to W Europe.

5* **Wood Forget-me-not** *Myosotis sylvatica* Hoffm. Short to medium, softly hairy biennial or perennial; stems much-branched, with spreading hairs. Basal leaves in a rosette, oval, mostly broadest above the middle; stem leaves lanceolate, pointed, unstalked. Flowers bright sky-blue, relatively large, 6-10mm, on slender stalks in lax cymes; calyx with a few spreading hooked hairs in the lower half, open in fruit. Damp woodland and mountain grassland, locally abundant, to 2000m. April-July. Europe north to S Scandinavia; probably naturalized in Finland. **B**: Rather local, but generally absent from the extreme N and SW.

6 *Myosotis decumbens* Host. Like *M. sylvatica*, but always perennial with a creeping rhizome and the calyx shorter (not equalling or longer) than the corolla-tube. Mountain habitats, grassy and rocky places, to 2000m, perhaps higher. May-August. C & E France, Germany and Scandinavia.

7* **Alpine Forget-me-not** *Myosotis alpestris* F.W.Schmidt [*M. sylvatica* subsp. *alpestris*]. Low to short, tufted, rather rough-hairy perennial. Basal leaves in a rosette, oblong-lanceolate, the lowermost stalked; stem leaves oval to linear, unstalked. Flowers bright to deep blue, relatively large, 6-9mm, in rather short cymes, fragrant; calyx silvery-hairy, with some hooked hairs below, the teeth erect or spreading in fruit. Nuts black. Damp woods and meadows, on basic mountain rocks, 700 to 2800m. May-September. N Britain, C & E France, C & S Germany. **B**: Very rare with a few localities in N England and Scotland.

8* **Creeping Forget-me-not** *Myosotis secunda* A. Murray. Short, hairy, rather slender annual or biennial with rooting runners, produced from the axils of the lowermost leaves; stems erect, with spreading hairs. Lower leaves oval, broadest above the middle, somewhat hairy; upper leaves elliptical, pointed. Flowers bright blue, rarely white, 4-8mm, petals slightly notched, the lower flowers with bracts; calyx bell-shaped in fruit, with straight hairs only. Nutlets dark brownish-black. Wet habitats, on acid, often peaty soils, to 800m. May-August. Britain, the Faeroes and France, primarily on the mountains in the south of its range. **B**: Scattered localities throughout.

9 **Pale Forget-me-not** *Myosotis stolonifera* (DC.) Gay ex Leresche & Levier. Like *M. secunda*, but a smaller plant producing numerous runners; leaves bluish-green and stem hairs not spreading. Calyx 3mm long in fruit (not 5mm). Flowers pale blue, 4-5mm. Wet hilly habitats, stream-sides and flushes. June-July. N Britain. Also found in Spain and Portugal. **B**: Very local, N England and S Scotland.

10 **Jersey Forget-me-not** *Myosotis sicula* Guss. Like *M. secunda*, but plants without stolons and leaves linear-oblong. Flowers small, pale blue, 2-3mm. Damp habitats, particularly coastal dune slacks. May-June. Channel Islands and NW France. Confined mainly to SW and S Europe from Portugal to Greece. **B**: Very rare, confined to Jersey.

11* **Tufted Forget-me-not** *Myosotis laxa* Lehm. subsp. *caespitosa* (K.F.Schultz) Hyl. [*M. caespitosa*, M. *lingulata*]. Short to medium erect, hairy annual or biennial, branched from the base; stem hairs not spreading. Leaves lanceolate, blunt. Flowers bright blue, 2-5mm, the petals not notched; calyx to 5mm in fruit, long-stalked, with straight hairs only. Nutlets dark brown. Wet habitats, marshes, stream and pond margins, ditches and dykes, stony and grassy ground, to 1600m. May-September. Throughout, except Iceland and Spitsbergen. Less frequent than *M. scorpioides*. **B**: Throughout.

subsp.*baltica* (Sam.) Hyl. ex Nordh. [*M. baltica, M. scorpioides* subsp. *laxa*] has stems ascending rather than erect and the fruiting calyx larger, to 8mm. Finland and Sweden.

12* **Water Forget-me-not** *Myosotis scorpioides* L. [*M. palustris, M. scorpioides* subsp. *palustris*]. Low to medium rhizomatous perennial, with runners; stems ascending to erect, hairless or with hairs pressed closely to the stem, occasionally spreading. Leaves oblong to lanceolate. Flowers sky-blue, occasionally pinkish or white, 4-10mm, flat, the petals slightly notched; calyx with short slender teeth, with straight hairs only. Nutlets black. Wet habitats, to 2000m. May-September. Throughout, except Spitsbergen. **B**: Throughout, except parts of N Scotland and S Ireland.

13 *Myosotis nemorosa* Besser. Lower leaves distinctly hairy beneath, with long, downward-pointing hairs. Flowers 5-6mm. Wet meadows, open woods. May-September. Throughout, except the far north.

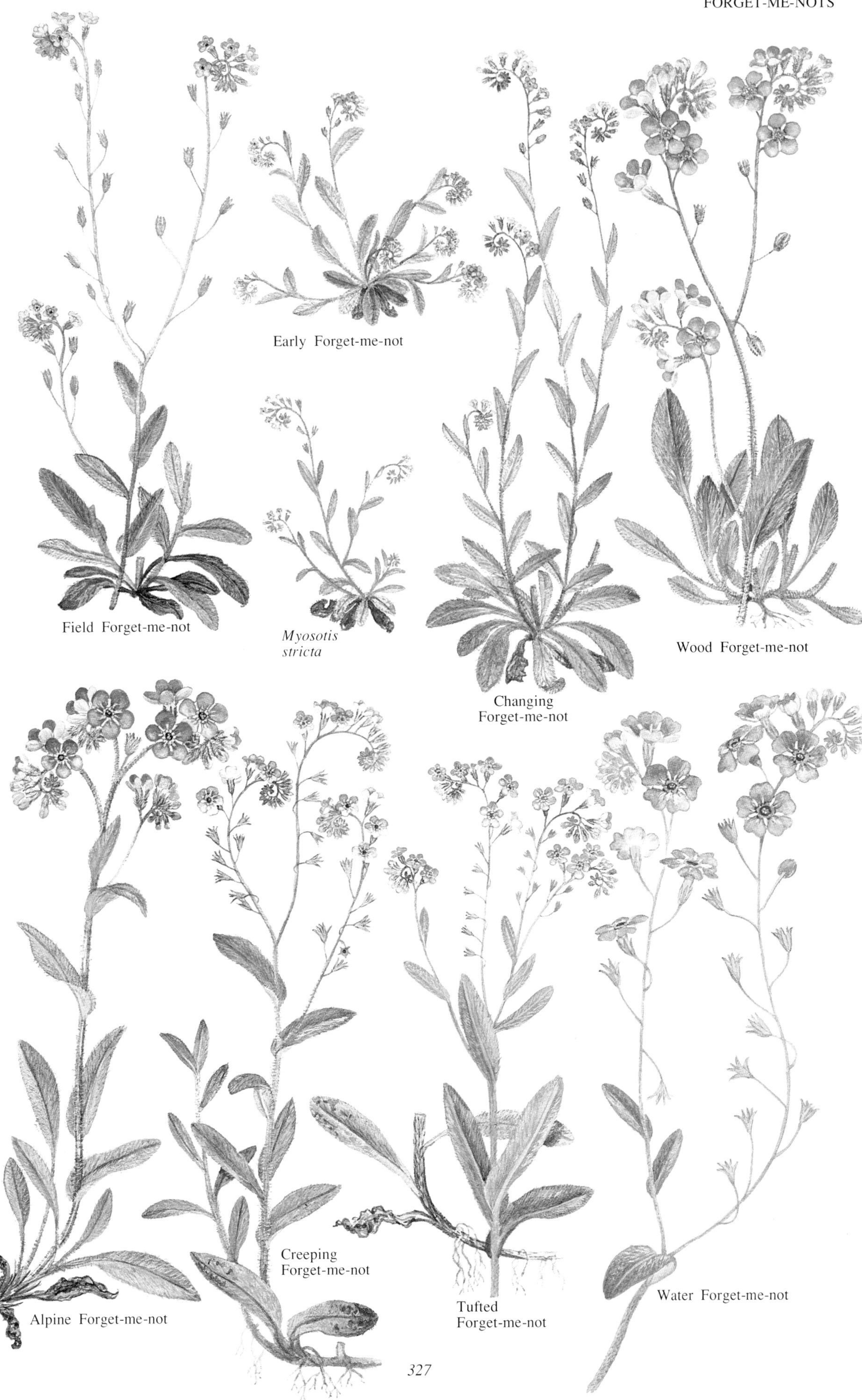
Early Forget-me-not
Field Forget-me-not
*Myosotis stricta*
Changing Forget-me-not
Wood Forget-me-not
Alpine Forget-me-not
Creeping Forget-me-not
Tufted Forget-me-not
Water Forget-me-not

1* **Deflexed Bur Forget-me-not** *Lappula deflexa* (Wahlenb.) Gürcke. [*Echinospermum deflexum, Hackelia deflexa*]. Short to tall, softly hairy annual or biennial; stem ascending to erect, branched, with spreading hairs. Leaves oblong-lanceolate to linear, untoothed, the lower stalked. Flowers pale blue or white, 5-7mm, bell-shaped, in branched clusters. Fruits with downturned stalks and an expanded calyx; nutlets with hooked spines. Rocky habitats, mainly in the mountains in the south of its range, to 2200m. June-August. C & S Scandinavia, France and Germany. The fruits, as in *L. squarrosa*, attach themselves to animal fur and clothing and are so dispersed to other suitable habitats.

2* **Bur Forget-me-not** *Lappula squarrosa* (Retz.) Dumort [*L. myosotis, Echinonspermum lappula*]. Short to medium, rough-haired, greyish annual or biennial, stems erect, usually branched, with adpressed hairs. Leaves oblong to linear-lanceolate, untoothed, mostly unstalked. Flowers pale blue, 4mm, forget-me-not-like, in leafy clusters. Fruit on erect stalks, with an expanded calyx; nutlets covered with hooked spines, in 2 rows, mitre-like. Dry habitats, disturbed and waste ground, waysides, cultivated fields, vineyards and sand-dunes, to 2500m. June-August. France and S Sweden; widely naturalized in other parts of Scandinavia, Denmark, Germany and locally so in Iceland, casual in Britain.

3* *Omphalodes scorpioides* (Haenke) Schrank. Short, hairless or slightly hairy, biennial; stems branched. Leaves lanceolate to elliptical, sometimes broadest above the middle, untoothed, the lower opposite and stalked, the upper alternate and unstalked. Flowers blue with a yellow centre, 3-4mm, solitary in the axils of the upper leaves. Fruit with an expanded calyx and hairy nutlets. Damp shaded habitats, woodland and rocky places, in the mountains, to 1400m. May-July. S Germany.

4* **Blue-eyed Mary** *Omphalodes verna* Moench. Low to short, mat-forming stoloniferous perennial, creeping with long runners which root down occasionally. Leaves bright green, oval to heart-shaped, untoothed, slightly hairy, the basal long-stalked. Flowers sky blue with yellowish 'folds' 8-10mm, Forget-me-not-like, in lax terminal racemes without bracts. Nutlets hairy. Damp mountains woods, but often naturalized in lowland woods and near habitation, to 1200m. February-May. **I**: C and SE Europe. Naturalized in Britain, Belgium, Holland, France and Germany (C & SE Europe). Widely cultivated in gardens, including a form with white flowers. **B**: A garden escape, locally naturalized.

**Hound's-tongues** *Cynoglossum*. Annual, biennial or perennial herbs. Flowers borne in cymes, without bracts. Calyx 5-lobed, divided almost to the base. Corolla with a short tube and widely spreading petal-lobes; with 5 scales in the throat. Fruit with an expanded calyx, the nutlets covered with hooked spines. 60 species, in temperate and tropical regions of the northern hemisphere. The bur-fruits attach themselves to fur and clothing and are dispersed by this means to other suitable habitats.

5* **Hound's-tongue** *Cynoglossum officinale* L. Medium greyish, softly-hairy biennial. Leaves lanceolate to oblong, untoothed, the lower short-stalked, the upper clasping the stem. Flowers dull purple, 5-6mm, funnel-shaped. Fruit flattened, the nutlets with hooked spines and a thickened flange. Dry grassy habitats, woodland margins, gravelly ground and stabilized sand-dunes, often near the sea, to 2400m. May- August. Throughout, except the extreme N, the Faeroes and Iceland. The plant smells distinctly of mice and was formerly used as a medicinal plant. **B**: Widespread in England and Wales north to C Scotland and in S & E Ireland, mainly coastal in the north.

6* **Green Hound's-tongue** *Cynoglossum germanicum* Jacq. Medium, rough-hairy biennial. Leaves lanceolate, slightly stalked to unstalked and clasping the stem, shiny and hairless or sparsely hairy above, hairy beneath. Flowers reddish-violet, 5-6mm, in lax cymes. Nutlets covered by dense hooked spines, but without a thickened border. Woodland margins on dry calcareous soils. May-July. Belgium, S Britain, France and Germany. **P**: Bees or self-pollinated. **B**: Very rare and declining; confined to scattered localities in the Chilterns, Cotswolds and on the North Downs.

7* **Blue Hound's-tongue** *Cynoglossum creticum* Miller. Like *C. germanicum*, but a softly-hairy plant, the leaves densely hairy on both surfaces. Flowers deep blue, 7-9mm, with netted veins. Open dry habitats, at low altitudes. May-July. NC France southwards. Widespread in S Europe.

## VERVAIN FAMILY Verbenaceae

Herbs or shrubs with opposite leaves. Flowers 4-5-parted. Calyx small. Corolla with a short tube and a flat limb (rotate), rarely 2-lipped. Fruit 4-parted, often fleshy. 80 genera and 800 species, mainly in tropical and subtropical regions of the world.

8* **Vervain** *Verbena officinalis* L. Medium, rough-hairy perennial; stems slender, stiffly erect, square. Leaves diamond-shaped to lanceolate, the lower stalked, deeply lobed to 1-2-pinnately-lobed, the upper leaves smaller and often unlobed. Flowers pale pink, 2-5mm, weakly 2-lipped, in terminal clusters to begin with but soon elongating into slender, branched spikes, leafless. Waste and rocky ground, rough grassland, waysides, generally on well-drained soils, to 1500m. June-October. Throughout, except the Faeroes, Iceland and Scandinavia, but probably naturalized in much of the north and north-west of its range. **P**: Bees, butterflies and hoverflies, or may be self-pollinated. Sometimes mistaken for a Labiate but distinguished by its leafless flower spikes. **B**: Throughout England and Wales; rare in Scotland – Fife.

Blue-eyed Mary

Deflexed Bur Forget-me-not

Bur Forget-me-not

*Omphalodes scorpioides*

Blue-eyed Mary

× 1/10

Hound's-tongue

Green Hound's-tongue

Blue Hound's-tongue

Vervain

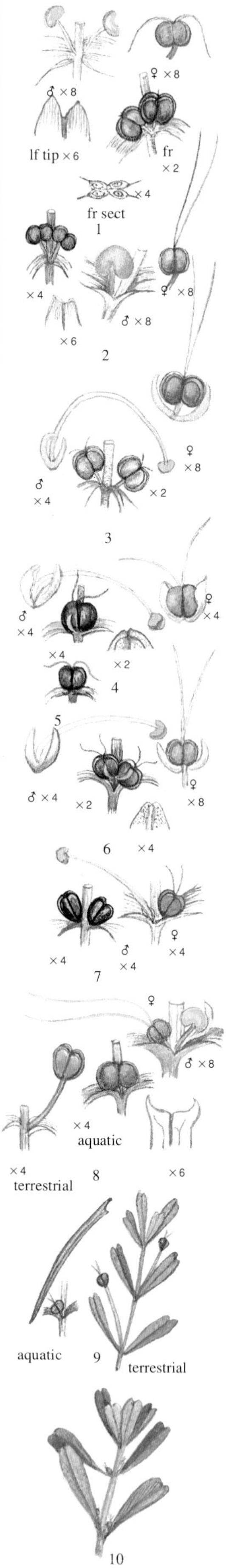

# STARWORT FAMILY Callitrichaceae

Aquatic or semiterrestrial annual or perennial herbs with opposite simple leaves. Flowers solitary, tiny and usually greenish, male and female separate, but borne on the same plant, sometimes a male and a female at the same leaf-axil; sepals and petals absent. Stamen 1. Styles 2; ovary 4-locular. Fruit 4-lobed, the lobes often keeled or winged, separating at maturity into 4, 1-seeded parts. A particularly complicated group with variable species which often prove difficult to identify accurately. Species can be amphibious, the land forms looking markedly different from the aquatic ones. Ripe fruits are often essential for reliable identification. Cosmopolitan; 17 species.

1* **Autumnal Water-starwort** *Callitriche hermaphroditum* L. [*C. autumnalis*]. Submerged aquatic herb. Leaves yellowish-green, transparent, linear, widest at the base, 1-veined, the apex deeply notched. Flowers solitary; styles spreading to deflexed, generally longer than the ovary, soon falling; bracts absent. Fruit subrounded, variable in size from 1-3.3mm, the lobes easily separating, with broad sharp wings. Lakes, rivers, streams, canals, to 450m. May-September. Throughout, except Spitsbergen. Widespread in Europe and NE Russia. **P**: Under water. The seeds of most species of *Callictriche* are probably dispersed by birds. **B**: Throughout Ireland; Britain from Snowdonia northwards.

2* **Short-leaved Water-starwort** *Callitriche truncata* Guss. subsp. *occidentalis* (Rouy) Schotsman [*C. truncata* subsp. *occidentalis*]. Like *C. hermaphroditum*, but the leaves deep bluish-green, translucent, elliptical. Fruit 1.4-1.6mm, clearly wider than long, usually slightly stalked. Ponds and ditches, in fresh or slightly brackish water and on fertile soils. May-September. Britain, Belgium and France. W Europe to S & SW Europe. The stems often have a reddish colour. **B**: Very rare; confined to a few localities in Kent and Nottinghamshire.

3* **Common Water-starwort** *Callitriche stagnalis* Scop. Submerged or terrestrial annual or perennial, with stems up to 60cm long. Submerged leaves narrowly elliptical; floating leaves often 6, in a rosette, broadly elliptical to rounded, pale green, usually with 5 veins; terrestrial forms with smaller and thicker leaves. Flowers in the axils of rosette leaves only, solitary or 1 male and 1 female together; styles recurved; bracts sickle-shaped. Fruit subrounded, pale, 1.6-1.8mm, deeply grooved between the lobes which are broadly winged. Still or slow-moving water, occasionally fast-flowing fresh water, often calcium-rich, streams, ditches and ponds, to 2000m. May-September. Throughout, except Spitsbergen.. **P**: Usually wind, occasionally water-pollinated. Terrestrial plants grow in muddy places adjacent to streams and ponds; forms with small thyme-like leaves are sometimes referred to var. *serpyllifolia* Lönnr. **B**: Throughout.

4* **Blunt-fruited Water-starwort** *Callitriche obtusangula* Le Gall. An aquatic or terrestrial plant. Submerged leaves linear, deeply notched; floating leaves 12 or more, in a rosette, diamond-shaped, short-stalked; leaves of terrestrial plants fleshy, yellowish-green, elliptical or narrow diamond-shaped. Flowers solitary, only in the axils of floating leaves; bracts sickle-shaped. Fruit elliptical, 1.5mm, brown, slightly keeled, but not winged. Slow-moving or stagnant fresh or brackish water habitats, streams, dykes, ditches and ponds, often calcium-rich, at low altitudes. May-September. Britain, Belgium, Holland, France and W Germany. Sometimes confused with *C. platycarpa* and *C. stagnalis*, but distinguished by its diamond-shaped floating leaves and unwinged fruits. **P**: Wind. W, C and S Europe as well as N Africa. **B**: Widespread in lowlands, rare in the north and absent from Scotland.

5* *Callitriche cophocarpa* Sendtner [*C. polymorpha*]. Like *C. obtusangula*, but rosette leaves oblong to subrounded. Fruit smaller, 0.8-1.2mm, oblong to rounded, often with persistent styles, less grooved. Slow-moving shallow water, generally alkaline, streams and dykes. May-September. Throughout, except Iceland. Local. Most records referring to this species in fact belong to *C. platycarpa*. **B**: Very rare.

6* **Various-leaved Water-starwort** *Callitriche platycarpa* Kutz. Aquatic to terrestrial, perennial plant. Submerged leaves linear, notched at the tip; floating leaves often deep green, few to a rosette, mainly 3-veined, stalked; leaves of terrestrial plants smaller and thicker. Flowers only at the axils of rosette leaves; stigmas erect; bracts sickle-shaped. Fruit subrounded, 1.5mm, brown, with keeled lobes. Fresh, occasionally brackish, water, flowing or still, streams, ditches, ponds and mudflats, usually at low altitudes. April-September. Throughout, except Iceland, Finland, N Scandinavia and Spitsbergen. **P**: Wind. **B**: Throughout lowlands.

7* *Callitriche palustris* L. Like *C. platycarpa*, but floating leaves elliptical to rounded and flowers generally paired, 1 male and 1 female; styles erect soon falling. Fruit oval, 1mm, blackish, winged only at the apex. Male flowers reduced or absent in terrestrial plants. Shallow still water, to 2600m. May-September. Iceland and Continental Europe, except for the extreme north; doubtfully recorded from Britain. Sometimes confused with *C. stagnalis*, but fruits smaller and winged only at the top.

8* **Intermediate Water-starwort** *Callictriche hamulata* Kutz ex Koch [*C. intermedia*]. Rather robust perennial, aquatic or terrestrial. Submerged leaves linear with an expanded notched apex, spanner-like; floating rosette leaves elliptical; terrestrial leaves smaller, elliptical, usually deep green. Flowers solitary, submerged, sometimes aerial; styles deflexed, bracts sickle-shaped. Fruit subrounded, 1.2-1.5mm, black, the styles or style-bases pressed closed to the side of the fruit, lobes narrowly winged. Lakes, streams, ponds, ditches, in still or slow-moving water, generally acid and nutrient poor, to 2000m. April-September. Throughout. The flowers are mostly water-pollinated. Sometimes confused with *C. platycarpa*, but the leaf-rosetes concave (not convex) above, and the submerged leaves characteristically expanded and spanner-like at the apex. **B**: Throughout; most common in SE England.

9 *Callitriche brutia* Petagna [*C. intermedia* subsp. *pedunculata*, *C. pedunculata*]. Like *C. hamulata*, but tip of submerged leaves asymmetrical, less claw-like, and fruit broadly winged. Lakes, pools, ditches, dykes and slow streams. April-September. Throughout to S Scandinavia, but not in Holland and Germany. Sometimes growing in pools that dry up in the summer. **B**: Distribution uncertain, but probably throughout; apparently decreasing.

10 **American Water-starwort** *Callitriche peploides* Nutt. Low, always terrestrial herb. Leaves elliptical, often widest above the middle. Flowers generally paired, 1 male and 1 female at the upper leaf-axils. Fruit subrounded, 0.6-0.8mm, scarcely stalked, the lobes narrowly winged. Muddy habitats. April-July, occasionally later. **I**: North America. N France – Seine-et-Oise. Distinguished from terrestrial forms of the preceding species by its smaller fruits which are distinctly wider than long.

Autumnal
Water-starwort
Short-leaved
Water-starwort
Common Water-starwort
Blunt-fruited
Water-starwort
*Callitriche*
*cophocarpa*
*Callitriche*
*palustris*
Various-leaved
Water-starwort
aquatic form
terrestrial form
Intermediate Water-starwort

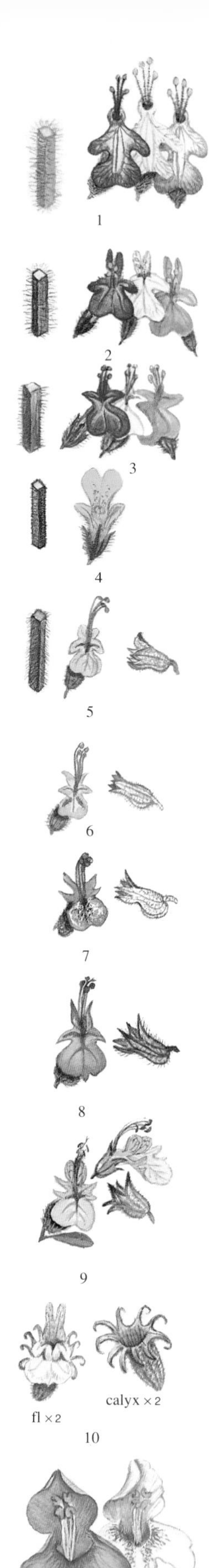


# MINT FAMILY Labiatae

Herbs or shrubs, generally with square stems, often glandular and aromatic. Leaves opposite, usually simple. Flowers irregular (zygomorphic), in distinct lateral clusters or verticillasters that often form a whorl around the stem, several of these making a leafy 'spike' or panicle. Calyx with 5 teeth, sometime 2-lipped. Corolla 2-lipped, except in *Ajuga* and *Teucrium*, the lower lip 3-lobed, the upper lip 2-lobed. Stamens 4, rarely 2. Fruit a cluster of 4, 1-seeded, nutlets. Over 3000 species in 170 genera.

**Bugles** *Ajuga*. Annual or perennial herbs. Calyx with 5 equal teeth. Corolla with a very reduced, inconspicuous, upper lip and a 3-lobed lower lip. Stamens protruding, 50 species.

1* **Blue Bugle** *Ajuga genevensis* L. Low to short, hairy perennial, without runners; stems hairy all round, often rather woolly. Leaves oblong, toothed, sometimes shallowly lobed, the lower stalked and generally withered by flowering time. Bracts leaf-like, tinged with blue or violet, the uppermost shorter than the flowers. Flowers bright blue, sometimes pink or white, 12-20mm long, in leafy spikes. Stamens with hairy stalks. Woodland clearings, stony places, generally on rather dry calcareous soils, to 2000m. April-August. Belgium, Holland, France and Germany; naturalized in Finland.

2* **Pyramidal Bugle** *Ajuga pyramidalis* L. Low to short hairy perennial, without runners; stems hairy all round. Leaves oval, toothed or untoothed, the basal stalked and present at flowering time. Bracts all longer than the flowers, oval to rounded, occasionaly lobed, tinged with blue or violet. Flowers pale violet-blue, rarely pink or white, 10-18mm long, in a dense leafy spike. Stamen stalks hairless. Grassy and rocky habitats, generally on calcareous soils, to 2800m. April-August. Alps and N Europe. Hybridises with *A. reptans*, *A.* x *hampeana* Braun & Vatke. **B**: Cumbria northwards; Ireland – Clare and Galway.

3* **Bugle** *Ajuga reptans* L. Low to short, hairy perennial with long rooting runners; stems hairy only on two opposing sides. Leaves oval or slightly toothed, often flushed with bronze, the lower stalked. Bracts oval, usually tinged with blue, the upper shorter than the flowers. Flowers pale violet-blue, occasionally pink or white, 14-17mm long, in leafy spikes, dense at first. Woodland and waste places, on calcareous to slightly acid soils, to 2000m. April-June. Throughout, except the extreme north; naturalized in Finland. **B**: Ubiquitous throughout much of Britain, except the extreme N.

4* **Ground-pine** *Ajuga chamaepitys* (L.) Schreber. Low to short greyish, hairy annual with a faint smell of pine-resin. Leaves with 3 narrow lobes, the lobes often further lobed or toothed. Flowers yellow with reddish or purplish markings, 7-15mm long, 1-2 together, partly hidden among the upper leaves. Bare stony ground and sparsely grassy habitats, on calcareous soils, to 1600m. May-September. W Europe. **B**: Very local in S England, especially Kent, Hampshire and Surrey.

**Germanders** *Teucrium*. Herbs or small shrubs. Calyx usually 2-lipped. Corolla with a single, 5-lobed lip, rather like a small broad manikin. 100 species.

5* **Wood Sage** *Teucrium scorodonia* L. Short to medium hairy perennial, sometimes almost a subshrub; stems erect, branched. Leaves oval to heart-shaped, stalked; bracts oval, short. Flowers pale greenish-yellow, occasionaly whitish, 8-9mm long, hairy, stamens protruding with maroon anthers, in leafless spikes. Open woods, dunes, generally on dry, non-calcareous soils, at low altitudes. July-September. Europe N to S Norway; naturalized in Denmark and Sweden. **B**: Throughout, rare in E England and C Ireland.

6* **Water Germander** *Teucrium scordium* L. Low to short, sprawling, soft-hairy perennial, with a strong garlic smell when crushed. Leaves oval to oblong, coarsely toothed, unstalked or scarcely stalked. Bracts leaf-like, generally longer than the flowers. Flowers purplish, 7-10mm long, borne in whorls up the leafy stem; calyx downy. Damp habitats, on calcareous soils, generally at low altitudes. June-October. Throughout, except the far north. Declining. **B**: Very rare, restricted to N Devon and Cambridgeshire; more frequent in W Ireland.

7* **Cut-leaved Germander** *Teucrium botrys* L. Low to short downy annual or biennial; stems usually branched. Leaves pinnately-lobed, stalked, the lobes untoothed. Bracts similar to the leaves. Flowers pink or purplish-pink, 15-25mm long, borne in whorls along the leafy stem; calyx curved. Dry, often rocky places, open patches in permanent chalk grassland, on calcareous soils, generally at low altitudes. June-October. W Europe. **B**: Very rare, from Surrey to Hampshire and Gloucestershire.

8* **Wall Germander** *Teucrium chamaedrys* L. Very variable, short, slightly hairy, tufted perennial with a woody base. Leaves oblong, untoothed to sharply toothed, shiny dark green above and rather leathery. Bracts similar to the leaves, but smaller. Flowers pale to deep purplish-pink, rarely white, 9-16mm long, in leafy spikes. Dry bare habitats, open woods, to 1800m. May-September. **B**: E Sussex on chalk downland, but generally regarded as alien; occasionally naturalised elsewhere.

9* **Mountain Germander** *Teucrium montanum* L. Low to short, spreading, mat-forming subshrub. Leaves leathery, narrow-elliptical, untoothed and with the margin rolled under, white-hairy and with a prominent midrib beneath. Bracts similar to the leaves. Flowers cream or yellowish, 12-15mm long, in flattish heads. Dry rocky and stony habitats, screes and dry pastures, on calcareous soils in the mountains, to 2400m. May-August. W Europe.

10* **White Horehound** *Marrubium vulgare* L. Medium, white-downy perennial, thyme-scented; stems branched, often with many short non-flowering branches. Leaves rounded to oval, often with a slightly heart-shaped base and a wrinkled surface, stalked. Flowers white, 12-15mm long, in dense whorls up the leafy stem; calyx with 10 short hooked teeth. Dry bare habitats, grassy and waste places, generally on calcareous soils, usually at low altitudes. June-September. Throughout, except the far north; alien in Ireland. **B**: Mainly S & W coasts of England and Wales; rare.

11* **Bastard Balm** *Melittis melissophyllum* L. Short to medium, hairy, strong-smelling tufted perennial; stems ascending, generally unbranched. Leaves oblong to heart-shaped, coarsely toothed, short-stalked. Flowers white, pink, purple or bicoloured, 25-40mm long, fragrant, in clusters of 2-6 at the base of the upper leaves, the corolla-tube much longer than the calyx. Shady habitats, woodland margins, hedgebanks, roadsides and shaded rocks, to 1400m. May-July. W Europe. **B**: Local in S England from the New Forest to Devon and Cornwall; also in W Wales.

Blue Bugle
Pyramidal Bugle
Bugle
Ground-pine
Wood Sage
Mountain Germander
Cut-leaved Germander
Water Germander
Wall Germander
White Horehound
Bastard Balm

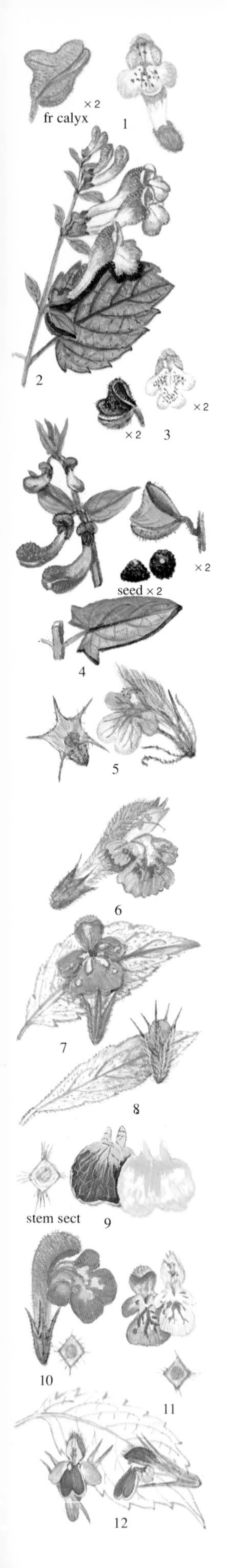


**Skullcaps** *Scutellaria*. Rhizomatous perennials. Flowers usually in pairs, often rather distant. Calyx 2-lipped, the upper lip with a distinctive erect, often rounded appendage. Corolla 2-lipped, with a long slender tube, the upper lip forming a small hood. the lower small, 3-lobed. 100 species, cosmopolitan except for southern Africa.

1* **Skullcap** *Scutellaria galericulata* L. Short to medium, usually hairy, shortly creeping perennial; stems erect, branched or unbranched. Leaves oval to lanceolate, slightly toothed, with a somewhat heart-shaped base, hairy or not, short-stalked; bracts similar to the leaves. Flowers bright violet-blue, rarely pink, with a whitish base, 10-18mm long, the corolla-tube abruptly upcurved. Wet grassy habitats, margins of rivers, streams, pools and marshes, on calcareous or neutral soils, at low altitudes. June-September. Throughout, except the far north. Hybridises with *S. minor*. **B**: Throughout, but scarce in N Scotland and in Ireland.

2 *Scutellaria altissima* L. Like *S. galericulata*, but taller, reaching to 1m. Flowers bluish with a white lower lip, 12-16mm long, the bracts shorter than the flowers. Grassy and waste habitats, waysides. June-August. **I**: S Europe. Naturalized in W Europe. Local. **B**: Naturalized in a few places in Somerset.

3* **Lesser Skullcap** *Scutellaria minor* Hudson. Short, rather slender perennial; stems branched or unbranched. Leaves oval to lanceolate, with a rounded or heart-shaped base, usually untoothed, mostly unstalked; bracts similar to the leaves. Flowers pink, 6-10mm long, with a more or less straight corolla-tube, spreading. Damp habitats, woodland rides, on peaty or mineral-rich soils. July-October. Throughout north to S Sweden. **B**: Throughout, though scarce or absent from much of the north.

4 **Spear-leaved Skullcap** *Scutellaria hastifolia* L. Like *S. minor* though often taller, the leaves arrow-shaped, untoothed and often purplish beneath. Flowers violet-blue, 15-20mm, the corolla-tube strongly curved; calyx glandular-hairy. Damp grassland, open woodland. July-September. France, Germany and Scandinavia; naturalized in E Britain and Belgium. **B**: Rare, confined to East Anglia.

5* **Phlomis** *Phlomis tuberosa* L. Stout, tall perennial, to 1.5m, with a tuberous rootstock. Basal leaves triangular to lanceolate with a heart-shaped base, stalked, coarsely toothed, often hairy beneath; upper leaves unstalked. Flowers pink or purple, 15-20mm long, in dense distant whorls at the base of the upper leaves; bracts linear, bristly. Dry grassy and rocky habitats, at low altitudes. June-August. C & E Germany. EC and SE Europe. **P**: Various bees.

**Hemp-nettles** *Galeopsis*. Annual herbs with dense whorls of flowers, the uppermost whorls crowded together. Calyx with spiny teeth. Corolla 2-lipped, the upper lip hooded and enclosing the stamens, the corolla tube with a ring of hairs inside. 10 species.

6* **Downy Hemp-nettle** *Galeopsis segetum* Necker [*G. dubia*]. Short to medium annual; stems with soft curly hairs. Leaves lanceolate to oval, narrowed at the base, toothed, stalked. Flowers pale yellow, rarely lilac with purple blotches, 25-30mm long, the corolla much longer than the very hairy calyx. A weed of cultivation, on acid soils, to 2200m. July-September. W Europe north to Denmark. **P**: Bees and other insects. Sometimes hybridises with *G. angustifolia*. Often erratic in appearance. **B**: Very rare, restricted to Caernarvonshire, formerly more widespread.

7 **Large Pink Hemp-nettle** *Galeopsis ladanum* L. [*G. intermedia*]. Short to medium annual; stems with curly hairs as well as some glands. Leaves lanceolate to oval, narrowed at the base, toothed, mostly stalked. Flowers deep pink with yellow blotches, 15-28mm long; calyx slightly hairy. Cultivated and waste land, generally on acid soils, to 2400m. July-October. Throughout, except the far north. Widespread in W and S Europe to Russia and C Asia. **P**: Insects or self-pollinated. **B**: Rare – naturalized in scattered localities, mainly in the south.

8* **Red Hemp-nettle** *Galeopsis angustifolia* Ehrh. ex Hoffm. [*G. ladanum* subsp. *angustifolia*]. Rather like *G. ladanum*, but leaves linear to lanceolate, usually stalked, few-toothed or untoothed. Flowers deep reddish-pink, with a yellow blotch in the throat, flecked with white, 14-24mm long; calyx white-hairy. A weed of arable land and bare places, occasionally on coastal shingle, to 2000m. July-October. W Europe. Naturalized in Ireland and Sweden. Local and rather erratic, sometimes forming large colonies but not permanent from one year to another. **B**: Local, S and E England, rare elsewhere.

9* **Large-flowered Hemp-nettle** *Galeopsis speciosa* Miller. Medium to tall, rough-hairy annual; stems well branched, swollen at the nodes, opposite sides hairy, the hairs yellowish. Leaves oval to lanceolate, toothed, stalked. Flowers pale yellow, the lower lip generally with a purple blotch, 27-34mm, in crowded whorls at the base of the uppermost leaves. Arable land, often in potato fields, often on dark peaty soils, to 1740m. July-September. Europe, except for the extreme north; naturalized in Ireland. Europe to Siberia. **B**: Rather local throughout, though absent from much of Ireland, SW England and S Wales.

10 **Hairy Hemp-nettle** *Galeopsis pubescens* Besser. Short to medium annual; stems usually branched, hairy on all 4 sides. Leaves heart-shaped or more or less square, toothed, stalked. Flowers bright pinkish-purple or reddish, generally with yellow blotches on the lower lip, 20-25mm long. Fields, banks, open woodland and hedgerows, sometimes on cultivated land, to 1600m. July-September. SC & E France, Germany; naturalized in Belgium and Holland.

11* **Common Hemp-nettle** *Galeopsis tetrahit* L. Short to medium, rough-hairy annual; stems well-branched, swollen at the nodes, hairy on opposite sides. Leaves lanceolate to oval, toothed, stalked. Flowers pinkish-purple with darker markings, rarely yellowish or whitish, 15-20mm long, the middle lobe of the lower lip not notched. Open woods, heaths, arable land, to 2400m. July-October. Throughout, except Spitsbergen. Europe to W Russia. A species probably of hybrid origin between *G. pubescens* and *G. speciosa*. **B**: Throughout.

12 *Galeopsis bifida* Boenn. Like *G. tetrahit*, but the flowers smaller, rarely exceeding 15mm long, the middle lobe of the lower lip deeply notched, the net-work of darker markings reaching the margin. Similar habitats and flowering time. Throughout, except the far north. Hybridises with *G. tetrahit*. **B**: Throughout.

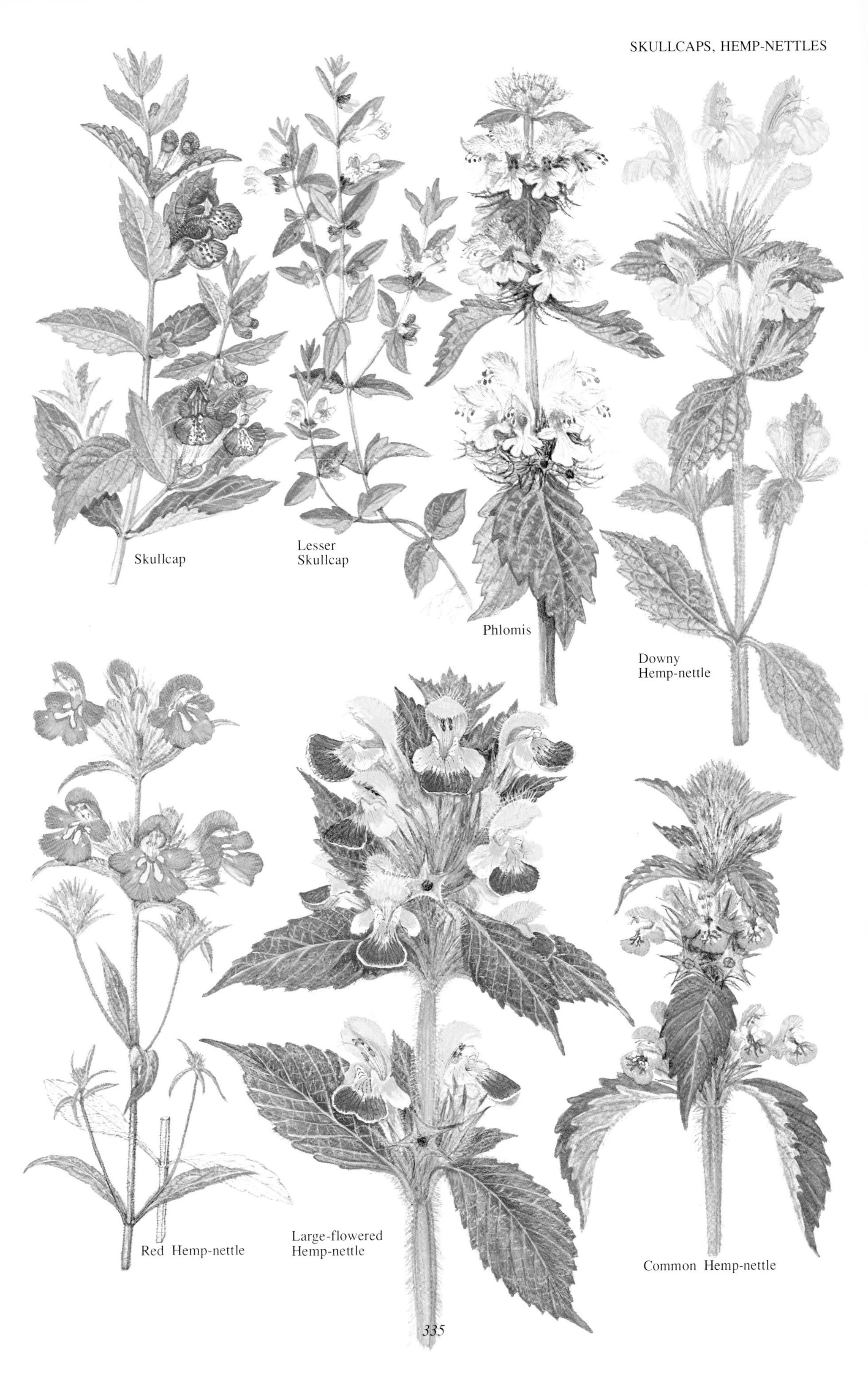
Skullcap
Lesser Skullcap
Phlomis
Downy Hemp-nettle
Red Hemp-nettle
Large-flowered Hemp-nettle
Common Hemp-nettle

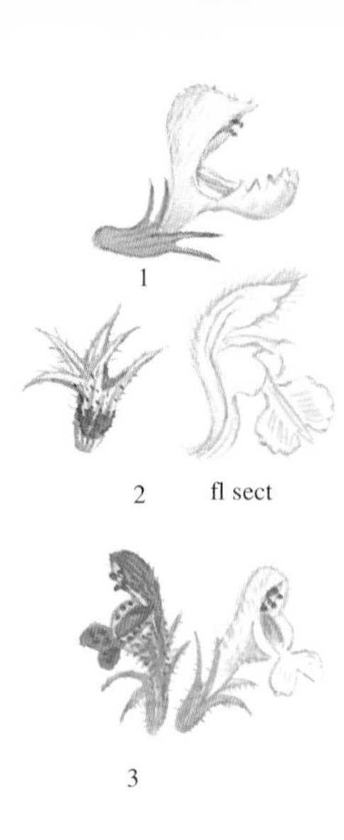


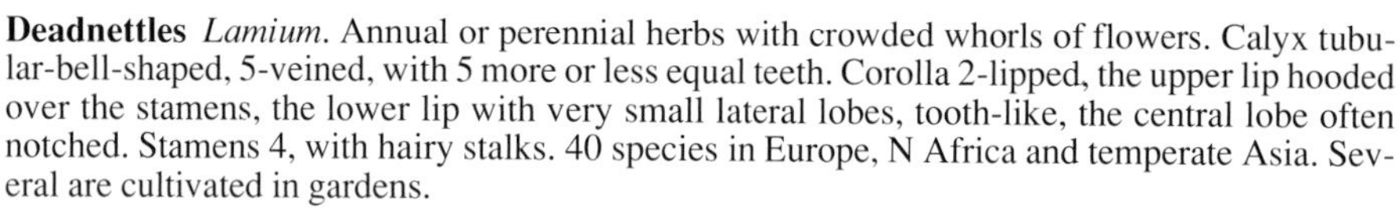

**Deadnettles** *Lamium*. Annual or perennial herbs with crowded whorls of flowers. Calyx tubular-bell-shaped, 5-veined, with 5 more or less equal teeth. Corolla 2-lipped, the upper lip hooded over the stamens, the lower lip with very small lateral lobes, tooth-like, the central lobe often notched. Stamens 4, with hairy stalks. 40 species in Europe, N Africa and temperate Asia. Several are cultivated in gardens.

1* **Spotted Deadnettle** *Lamium maculatum* L. Very variable, short to tall, hairy, aromatic perennial, patch-forming. Leaves triangular-ovate, coarsely toothed, often with a central whitish blotch, stalked. Flowers pinkish-purple, sometimes white or purplish-brown 20-35mm long, with a curved corolla-tube, the lower lip heart-shaped with tooth-like lateral lobes; calyx shorter than the corolla-tube. Grassy habitats, woodland margins, banks and hedgerows, to 2000m. April-October. Belgium, Holland, France and C & S Germany; naturalized in Britain and S Sweden. W and S Europe to W Asia. **P**: Bumble-bees. Widely cultivated, especially forms with white-blotched leaves. **B**: An occasional escape from cultivation.

2* **White Deadnettle** *Lamium album* L. Short to medium, faintly aromatic, patch-forming perennial, stoloniferous; stems spreading to erect. Leaves heart-shaped to oval, toothed, stalked. Flowers white, 20-25mm long, the corolla-tube curved near the base, the upper lip very hairy, the lower lip with 2-3 small teeth. Grassy habitats, hedgerows, banks, pathways, farmyards and waste places, generally on fertile deep soils, to 2300m. April-November. Europe except the extreme north; naturalized in Ireland and Iceland. Widespread from Europe across Asia to Japan. **P**: Various bees, especially bumble-bees, seeking nectar secreted at the base of the corolla-tube. **B**: Throughout, except the extreme north, scarcer in Wales and Ireland.

3* **Red Deadnettle** *Lamium purpureum* L. Low to short, often purplish, aromatic, hairy annual. Leaves oval, blunt-toothed, stalked; lower bracts longer than wide, stalked. Flowers pinkish-purple, 10-18mm long, with a straight corolla-tube; calyx teeth spreading in fruit. Cultivated and waste land, a frequent weed of gardens, to 2500m. March-December. Throughout, except Spitsbergen; naturalized in Iceland. Widespread from Europe to W Asia. **P**: Bees or self-pollinated. **B**: Throughout.

4 **Cut-leaved Deadnettle** *Lamium hybridum* Vill. [*L. hybridum* subsp. *dissectum*]. Like *L. purpureum*, but leaves deeply and irregularly toothed; bracts similar to the leaves. Flowers 10-15mm long, the corolla-tube without or with only a faint ring of hairs near the base (well marked in *L. purpureum*). A weed of arable land and other cultivated places, to 2500m. March-October. Throughout, except Iceland and Spitsbergen. **P**: Usually bees. Europe to W Russia and N Africa. **B**: Throughout, especially in E Britain.

5 **Northern Deadnettle** *Lamium molucellifolium* Fries [*L. hybridum* subsp. *intermedium*]. Similar to *L. hybridum*, but the bracts mostly unstalked and rather different from the leaves; flowers 14-20mm long. Cultivated ground and waste places, generally at low altitudes. May-September. Britain, the Faeroes, N & C Germany and Scandinavia; naturalized in Iceland. Primarily N and NW European. **B**: Mainly in C and N Britain and N Ireland.

6* **Henbit Deadnettle** *Lamium amplexicaule* L. Low to short, scarcely branched, hairy annual. Leaves rounded to oval, blunt-toothed or lobed, the lower stalked, but the upper unstalked and half-clasping the stem. Flowers pinkish-purple, 14-20mm long, with a straight corolla-tube; calyx softly hairy. A weed of cultivated land, particularly arable land, and waste places, to 2550m. March-December. Throughout, except the Faeroes and Spitsbergen; naturalized in Iceland. Europe to W Asia and N Africa; widely naturalized in North America. **P**: Mainly self-pollinated, but also by bees. **B**: Throughout, but rare north of the Humber and in Ireland.

7* **Yellow Archangel** *Lamiastrum galeobdolon* (L.) Ehrend. & Polatschek [*Galeobdolon luteum*, *Lamium galeobdolon*]. Short to medium, strong smelling, stoloniferous, hairy perennial, with long runners, often forming large patches; flowering stems erect. Leaves dark green, oval, coarsely toothed, mostly stalked; bracts similar to the leaves, but narrower. Flowers bright yellow, with greenish-brown markings, 17-21mm long, borne in whorls on the upper half of the stems; upper lip hooded, the lower lip 3-lobed. Shaded habitats, woods, coppices and scrub, generally on heavy calcareous soils, to 2000m. April-July. Britain and Continental Europe north to S Sweden; naturalized in Finland and Norway. Europe to W Russia and W Asia. **P**: Various bees. Cultivated in gardens, including forms with variegated leaves – invasive. **B**: Rare in England – confined to Lincolnshire.

subsp. *montanum* (Pers.) Ehrend. & Polatschek [*Lamium galeobdolon* subsp. *montanum*]. Flowers larger, 18-25mm long, in clusters of 9-15 (not 8 or less). Similar habitats and flowering time. Throughout the range of the species except for parts of the north. **B**: Throughout England and Wales, rare elsewhere.

8* **Motherwort** *Leonurus cardiaca* L. Medium to tall, hairy or hairless perennial, to 2m tall, though often less. Lower leaves palmately-lobed, with 5-7 segments toothed or shallowly lobed; upper leaves shallowly 3-lobed usually. Flowers white or pale pink, sometimes with purple spots, 8-12mm long, the upper lip very hairy on the back, borne in whorls up the leafy stem; calyx 5-veined. Hedgebanks, woodland margins, waysides and waste places, at low altitudes. July-September. Throughout, except the extreme north; naturalized in S Britain. Cultivated since medieval times as a medicinal plant to ease pain during childbirth. **P**: Mainly bees. W Europe to W Russia. **B**: Rare and usually casual in England and Wales.

9 **False Motherwort** *Leonurus marrubiastrum* L. [*Chaiturus leonuroides*, *C. marrubiastrum*]. Medium to tall, grey-hairy biennial, to 1.2m. Leaves oval to rounded, coarsely and irregularly toothed, the upper leaves lanceolate. Flowers pale pink, the corolla scarcely longer than the 10-veined calyx; upper lip of corolla hairy. Grassy and waste places. June-September. Germany; naturalized in several places in France, casual further north. Widespread in C and E Europe.

10* **Black Horehound** *Ballota nigra* L. Very variable medium to tall, almost hairless, often straggling, strongly aromatic perennial. Leaves heart-shaped to oval or oblong, short-stalked, toothed. Flowers lilac, 12-14mm, in dense whorls along the leafy stems; calyx funnel-shaped with triangular, finely-pointed, 4-6.5mm long, teeth. Woodland margins and rides, hedgebanks, road verges and waste ground, to 1530m. June-September. E France and Germany eastwards; naturalized or casual in Britain and parts of W and N Europe. Widespread from Europe to C Asia and N Africa. **P**: Various bees.

subsp. *foetida* Hayek [*B. borealis*]. Calyx with shorter teeth, only 1-3mm long; flowers lilac to white, often slightly larger. Similar habitats and flowering time. Throughout, except the Faeroes, Iceland, Finland, Norway, Spitsbergen and much of Germany. **B**: Throughout England south of the Humber, rare elsewhere.

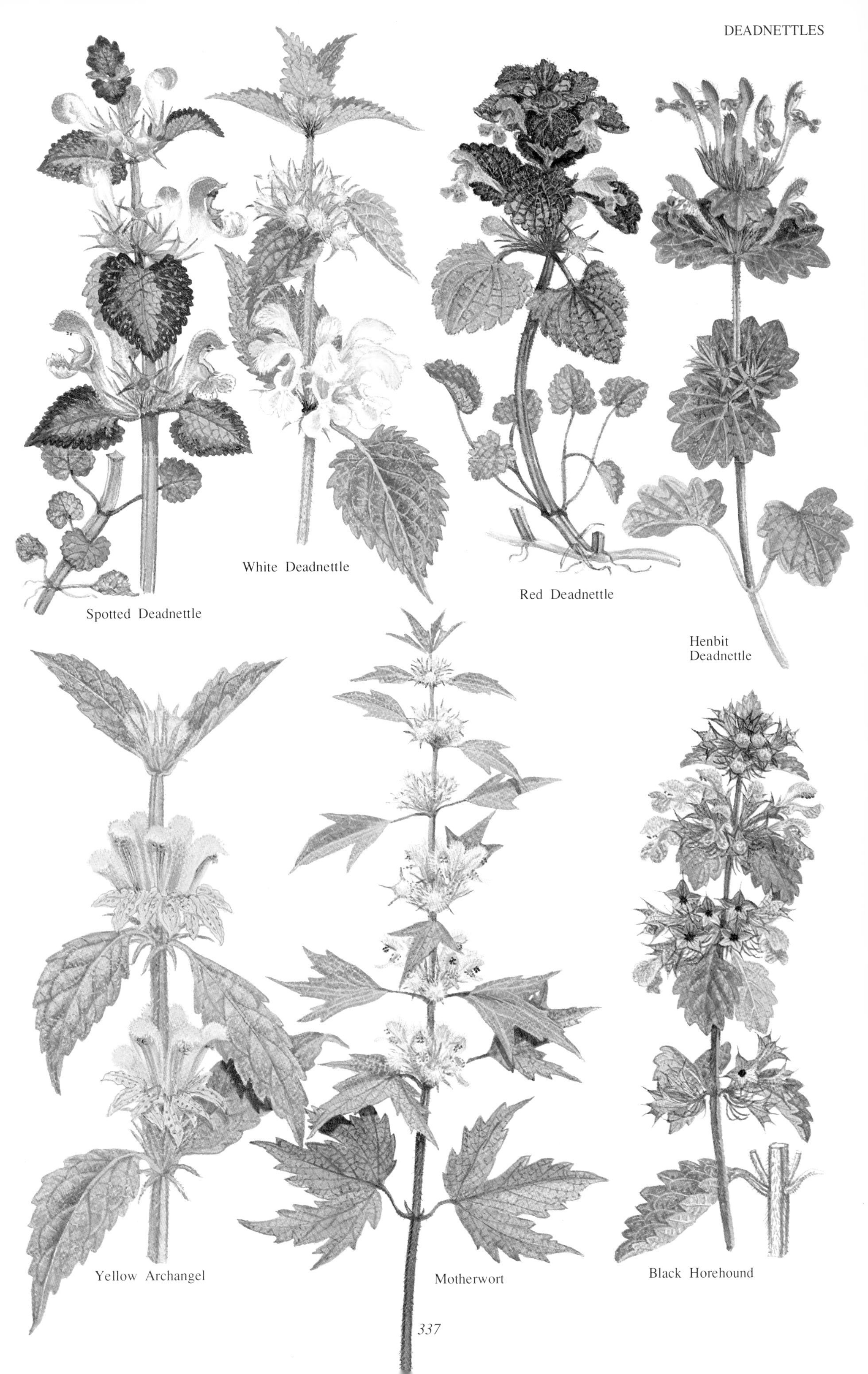
Spotted Deadnettle
White Deadnettle
Red Deadnettle
Henbit Deadnettle
Yellow Archangel
Motherwort
Black Horehound

**Betonys and Woundworts** *Stachys*. Annual or perennial herbs. Flowers in clusters of 2-8, forming leafy, spike-like inflorescences. Calyx tubular or bell-shaped, usually not 2-lipped, with 5 equal teeth, 5-10-veined. Corolla 2-lipped, the upper lip hooded or rather flat. The stamens not or only slightly protruding. 200 species, cosmopolitan except for Australia and New Zealand. Several are cultivated in gardens and are used as medicinal herbs.

1* **Betony** *Stachys officinalis* (L.) Trev. [*S. betonica, Betonica officinalis*]. Variable short to tall, softly-hairy perennial; stems erect, with persistent basal leaf-rosettes. Leaves oblong to oval with a heart-shaped base, coarsely toothed, the uppermost unstalked. Flowers bright reddish-purple, occasionally pink or white, 12-18mm long, the whorls forming a dense oblong spike, sometimes interrupted below. Permanent grassland, heaths, banks and open woodland, generally on rather light soils, to 1800m. June-October. Throughout north to S Scandinavia; naturalized in Finland and Norway. The plant was widely used by medieval herbalists who believed that it possessed magical properties to ward off various evils. The dried leaves were formerly used as a substitute for snuff. The plant can also be used as a herbal tea. **P**: Mainly bees. **B**: Throughout, though rare in Scotland, East Anglia and Ireland.

2* **Limestone or Alpine Woundwort** *Stachys alpina* L. Medium to tall, greyish, rather soft perennial; stems erect, glandular hairy. Leaves oblong to oval, with a heart-shaped base, blunt-toothed, stalked. Flowers dull purple, sometimes tinged with yellow, 15-22mm long, hairy, in a leafy spike; calyx tubular, glandular-hairy, with unequal teeth. Open woodland and damp rocky ground, generally over limestone, to 2000m. June-August. Britain, Belgium, France and S Germany. W Europe to C & S Europe, primarily on the mountains in the south of its range. **P**: Various bees. **B**: Very rare; confined to a few localities in Gloucestershire and Denbighshire.

3* **Downy Woundwort** *Stachys germanica* L. Medium to tall, grey-or white-downy, patch-forming biennial or perennial; stems ascending to erect, not glandular. Leaves oval to lanceolate, with a heart-shaped base, toothed, the lower long-stalked, the upper often unstalked, green above, but greyish or whitish beneath. Flowers pink or purple, 15-20mm long, hairy, the whorls forming a dense spike, interrupted below; calyx downy. Grassy habitats, road-verges, bare ground, sometimes along hedgerows, to 1750m. July-September. S Britain, Belgium, France and Germany. W Europe to S Europe, W Asia and N Africa. **P**: Bees. Widely cultivated in gardens **B**: Very rare; confined to two localities in Oxfordshire, but formerly more widespread.

4* **Hedge Woundwort** *Stachys sylvatica* L. Medium to tall, creeping perennial, to 1.2m, with a rather unpleasant smell; stems erect, glandular-hairy. Leaves heart-shaped, slightly hairy, all stalked. Flowers dull dark purple-red, with white markings, 13-18mm long, hairy, the whorls forming an interrupted spike; calyx glandular-hairy, with equal teeth. Woodland, hedgerows, banks and the margins of cultivated land, mostly in shaded habitats, to 1700m. June-September. Throughout, except the Faeroes, Iceland and Spitsbergen. Widespread and often abundant, from Europe to C Asia and the W Himalaya. **P**: Bees. Hybridises with *S. palustris*, *S.* x *ambigua* Sm., often forming extensive hybrid swarms. **B**: Throughout, except for parts of the Scottish Highlands.

5* **Marsh Woundwort** *Stachys palustris* L. Medium to tall, hairy, faintly aromatic perennial, to 2m; stems erect. Leaves oblong to lanceolate, with a heart-shaped base, blunt-toothed, only the lower stalked, but the upper unstalked. Flowers purple, 12-15mm long, slightly hairy, the whorls forming a dense spike, interrupted below, only the lower bracts leaf-like; calyx hairy, generally without glands. Damp habitats, the margins of rivers, streams, canals, ponds, dykes and ditches, or a weed of cultivated fields, especially arable land, to 1600m. June-October. Throughout, except the the Faeroes, Iceland and Spitsbergen. **P**: Mainly bees. The ring of hairs inside the corolla-tube helps to keep away unbidden insects such as flies. Often hybridises with the previous species. **B**: Throughout, though scarce in the Scottish Highlands.

6* **Yellow Woundwort** *Stachys recta* L. [*S. czernjajevii*]. Variable short to tall, sparsely hairy, aromatic perennial. Leaves oblong to oval, green, finely toothed, the lower stalked, the upper narrower and unstalked. Flowers pale yellow, with purplish streaks on the lower lip, 15-20mm long, borne in narrow, branched, more or less leafless spikes; stamens slightly protruding. Dry, rocky and waste habitats, waysides, to 2250m. June-September. Belgium, France and Germany; naturalized in W Britain. In much of W, C & S Europe and W Asia. **B**: Locally naturalized in Glamorgan.

7 **Annual Woundwort** *Stachys annua* (L.) L. Like *S. recta*, but usually annual. The flowers smaller, 10-16mm long, pale yellowish-white, occasionally with reddish markings. Cultivated land, especially arable land, open habitats and waste places, generally on calcareous soils, to 2000m. June-October. Belgium, Holland, France and Germany; naturalized or casual in Britain and Scandinavia. **P**: Bumble-bees. **B**: A casual in Britain, but very rarely locally naturalized.

8* **Field Woundwort** *Stachys arvensis* (L.) L. Short, erect, hairy annual; stem usually branched at the base. Leaves oval, often with a heart-shaped base, toothed, stalked. Flowers pale purple, 6-7mm long, in leafy spikes, the corolla scarcely longer than the calyx. Arable fields and sandy places, generally on acid soils and at low altitudes. April-November. Throughout, except the extreme north, the Faeroes and Iceland. A common weed in many parts of Europe, W Asia and N Africa. **P**: Usually self-pollinated, occasionally by insects. **B**: Throughout most of Britain, though scarcer in the east.

9* **Catmint** *Nepeta cataria* L. Medium to tall, branched, grey-woolly perennial, mint-scented. Leaves oval, often with a heart-shaped base, toothed, stalked. Flowers white with purple spots, 7-10mm long, in rather dense spikes, interrupted below; calyx-teeth straight. Hedgerows, roadsides and waysides, banks and rocky places, usually on calcareous soils, to 1500m. June-September. Britain, Belgium, Holland and France; widely naturalized in Ireland, Germany and Scandinavia. From W, C & S Europe to W & C Asia, east as far as the W Himalaya. **P**: Bees. Formerly widely cultivated as a medicinal herb. **B**: Local north to Cumbria and Northumberland; naturalized in scattered localities in Ireland.

10* **Hairless Catmint** *Nepeta nuda* L. Medium to tall, hairless or almost hairless perennial. Leaves oblong or oval, toothed, often slightly heart-shaped at the base, the lower short-stalked, the upper unstalked. Flowers pale violet, 6-8mm long, in branched spikes; calyx often tinged with blue. Open woodland, scrub, grassy and rocky places and waysides, generally at low altitudes. June-September. C & E France; naturalized in S Germany.

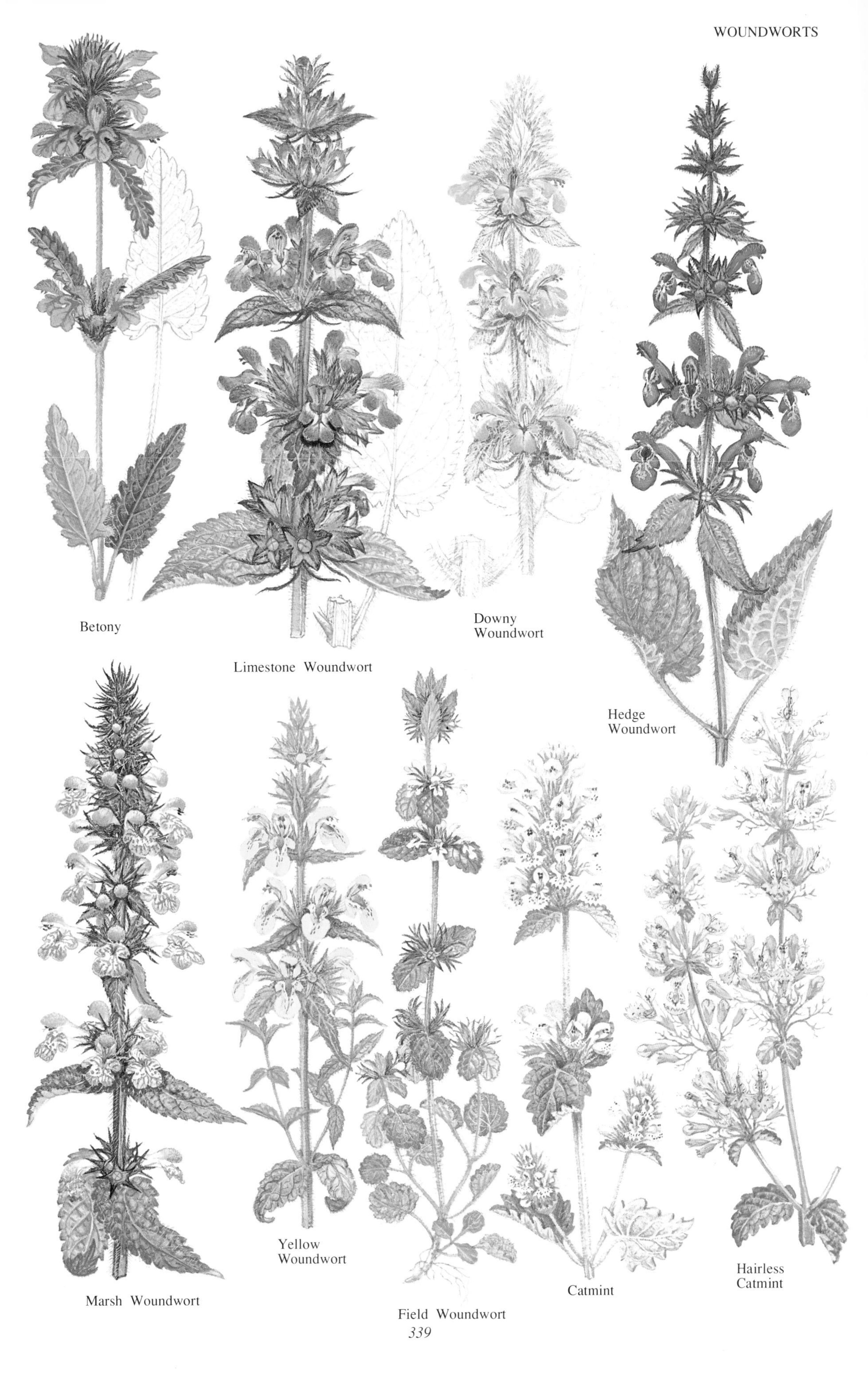
Betony
Limestone Woundwort
Downy Woundwort
Hedge Woundwort
Marsh Woundwort
Yellow Woundwort
Field Woundwort
Catmint
Hairless Catmint

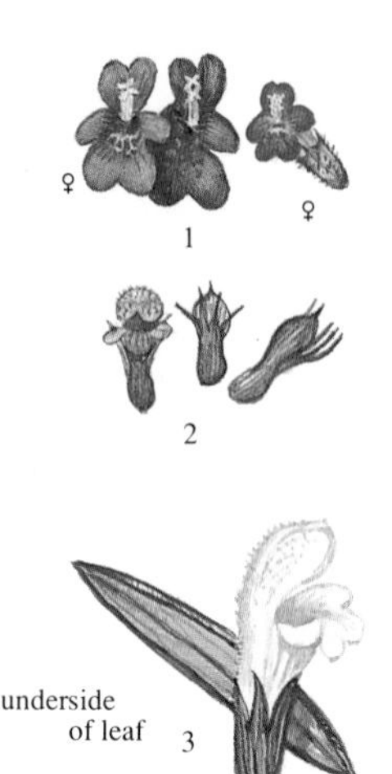


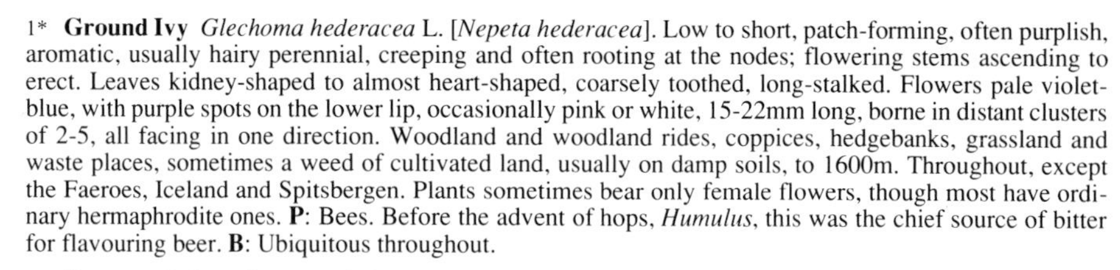

1* **Ground Ivy** *Glechoma hederacea* L. [*Nepeta hederacea*]. Low to short, patch-forming, often purplish, aromatic, usually hairy perennial, creeping and often rooting at the nodes; flowering stems ascending to erect. Leaves kidney-shaped to almost heart-shaped, coarsely toothed, long-stalked. Flowers pale violet-blue, with purple spots on the lower lip, occasionally pink or white, 15-22mm long, borne in distant clusters of 2-5, all facing in one direction. Woodland and woodland rides, coppices, hedgebanks, grassland and waste places, sometimes a weed of cultivated land, usually on damp soils, to 1600m. Throughout, except the Faeroes, Iceland and Spitsbergen. Plants sometimes bear only female flowers, though most have ordinary hermaphrodite ones. **P**: Bees. Before the advent of hops, *Humulus*, this was the chief source of bitter for flavouring beer. **B**: Ubiquitous throughout.

2* **Dracocephalum** *Dracocephalum thymifolium* L. Short to medium, slightly hairy annual. Basal and lower leaves oval to lanceolate with a heart-shaped base, toothed, long-stalked, the upper leaves narrower and short-stalked. Flowers small, bright lilac-blue, 7-9mm long, the clusters forming an interrupted spike; calyx 2-lipped, with 15 veins. Bare and waste places, sometimes on cultivated land. July-August. **I**: E Europe and W & C USSR. Naturalized in Denmark, Finland and Sweden.

3* **Northern Dragonhead** *Dracocephalum ruyschiana* L. Medium, usually hairless perennial; stems erect to ascending. Leaves linear-lanceolate, untoothed and with the margins rolled under, the lower short-stalked. Flowers blue or violet, occasionally white, 20-28mm long, in terminal clusters; bracts oval. Dry grassy habitats and open woods, always local, to 2200m. July-September. Mountains of France, Germany, Norway and Sweden. From C Europe to the USSR. **P**: Long-tongued bees. Occasionally cultivated in gardens.

**Self-heals** *Prunella*. Perennial herbs. Flowers generally in clusters of 6, forming dense rounded or oblong terminal heads; bracts different from the leaves. Calyx 2-lipped, closed in fruit. Corolla 2-lipped, straight, the upper lip hooded, the lower toothed. 5 species – mostly in Europe and the Mediterranean region, but one, *P. vulgaris*, cosmopolitan.

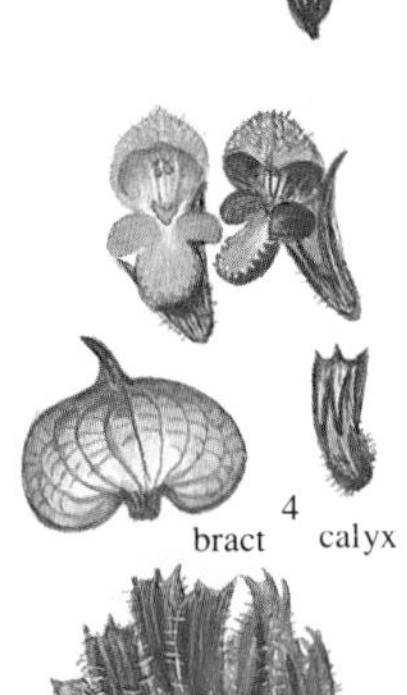


4* **Cut-leaved Self-heal** *Prunella laciniata* (L.) L. [*P. alba*]. Short, densely hairy, patch-forming perennial, with erect flowering stems. Leaves mostly pinnately-lobed, stalked. Flowers yellowish-white, occasionally rose-pink or purple, 15-17mm long, the flowerheads immediately subtended by a pair of leaves; calyx scarcely toothed. Dry grassy habitats and waste places, on dry calcareous soils, to 1320m. June- October. Belgium, France and Germany; naturalized in Britain and sometimes in N Europe. W Europe to W Asia and N Africa. **P**: Bees. Some plants bear small only female flowers. Hybridises with *P. vulgaris*, *P.* × *intermedia* Link. **B**: Naturalized, possibly native, in scattered localities in C and S England.

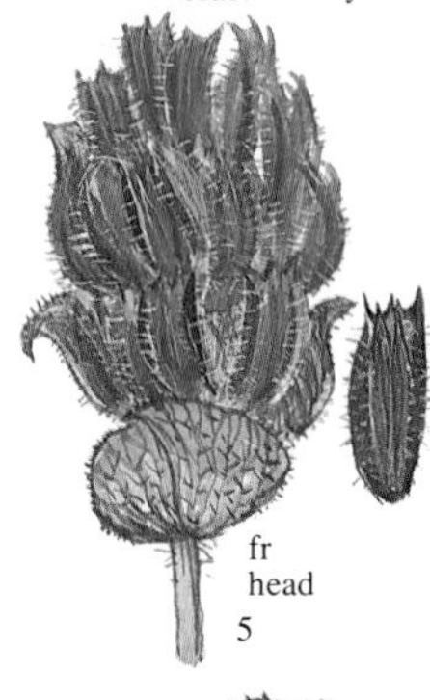


5* **Large Self-heal** *Prunella grandiflora* (L.) Scholler. Short to medium, slightly hairy, tufted perennial; stems erect. Leaves oval-lanceolate, slightly toothed or untoothed, stalked. Flowers deep violet-blue with a whitish tube, large, 25-30mm long, the flowerheads not immediately subtended by a pair of leaves; calyx teeth awned. Woodland and dry meadows, banks, to 2400m. June-October. Belgium, France, Germany, Denmark and S Sweden. Widely cultivated in gardens. **P**: Various bees.

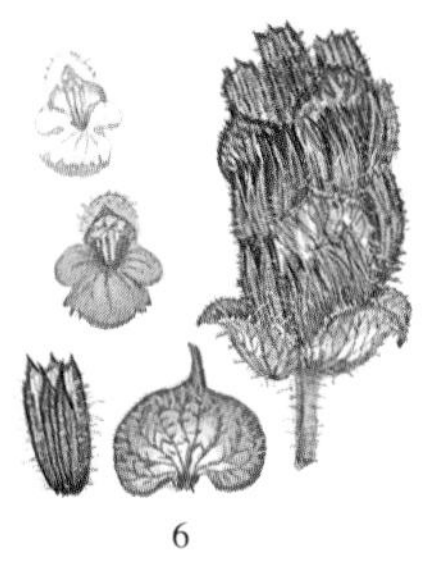


6* **Self-heal** *Prunella vulgaris* L. Low to short, patch-forming , usually hairy perennial; flowering stems ascending to erect. Leaves oval to diamond-shaped, slightly toothed or untoothed, stalked. Flowers deep violet-blue, occasionally pink or white, 13-15mm long, the flowerheads immediately subtended by a pair of leaves; calyx with fine-pointed teeth. Open woods, grassy habitats, meadows, pastures, road verges, lawns and cultivated land, generally on dry calcareous or neutral soils, to 2400m. June-November. **P**: Bees. Widespread from Europe to Asia, N Africa, Australia and North America. **B**: Common throughout.

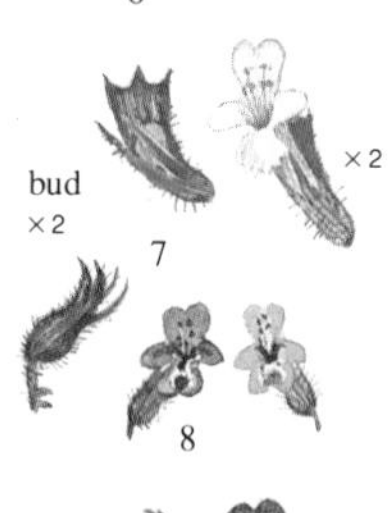


7* **Balm** *Melissa officinalis* L. Medium to tall, hairy, lemon-scented perennial, to 1.5m, though often less; stems erect and branched, glandular. Leaves yellowish-green, oval to diamond-shaped, deeply toothed, stalked. Flowers whitish or pale yellow, often becoming pinkish, 8-15mm long, in leafy whorled, interrupted spikes, the upper lip erect, the lower 3-lobed. Scrub and shaded habitats, cultivated land, at low altitudes. July-September. C & S France; widely naturalized in Britain, Belgium, Denmark, Germany and S Sweden. Widely cultivated and naturalized in many places as a culinary and medicinal herb. Today grown mostly for its pleasant aromatic foliage. **B**: Naturalized in C & S England, rare elsewhere.

8* **Basil-thyme** *Acinos arvensis* (Lam.) Dandy [*A. thymoides*, *Calamintha acinos*, *Satureja acinos*]. Short, hairy annual, rarely a short-lived perennial, branched from the base. Leaves lanceolate to oval, blunt or pointed, slightly toothed or untoothed, net-veined. Flowers violet with white markings, 7-10mm long, in loose terminal clusters; bracts similar to the leaves. Dry bare and sparsely grassy habitats, arable fields, rocky places, on calcareous soils, to 2000m. May-September. Throughout, except the extreme north, the Faeroes, Iceland and Spitsbergen. Europe to N Russia and W Asia. **P**: Bees. **B**: Local in C & S Britain, rare elsewhere.

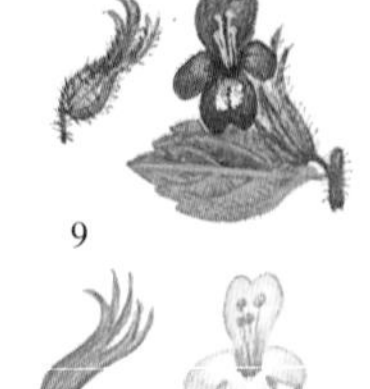


9 **Alpine Calamint** *Acinos alpinus* (L.) Moench [*Calamintha alpina*, *Satureja alpina*]. Short hairy perennial, forming small tufts, stems branched near the base. Leaves elliptical to almost rounded, slightly toothed or untoothed, with conspicuous lateral veins. Flowers violet with white markings on the lower lip,12-20mm long, borne in small whorls at the base of the uppermost leaves; calyx slightly curved, 2-lipped. Meadows, rocky places and screes, generally on calcareous soils, to 2500m. June-September. Mountains of C & E France and S Germany. **P**: Various insects including bees and butterflies. Sometimes cultivated in gardens. Primarily in the mountains of C and S Europe.

10* **Winter Savory** *Satureja montana* L. Short semi-evergreen, almost hairless subshrub, pleasantly aromatic. Leaves linear to elliptical, often broadest above the middle, rather leathery, hairy along the margin, unstalked, the opposite pairs joined together at the base, dotted with shiny glands. Flowers pale pink or white, 6-12mm long, with a straight tube, borne in loose leafy spikes; calyx tubular, 10-veined. Banks and old walls in sunny dry habitats. July-October. **I**: S Europe and N Africa. Naturalized in S Britain. Widely cultivated in gardens and used as a culinary herb for seasoning food. **B**: Naturalized on old walls and dry banks in S England.

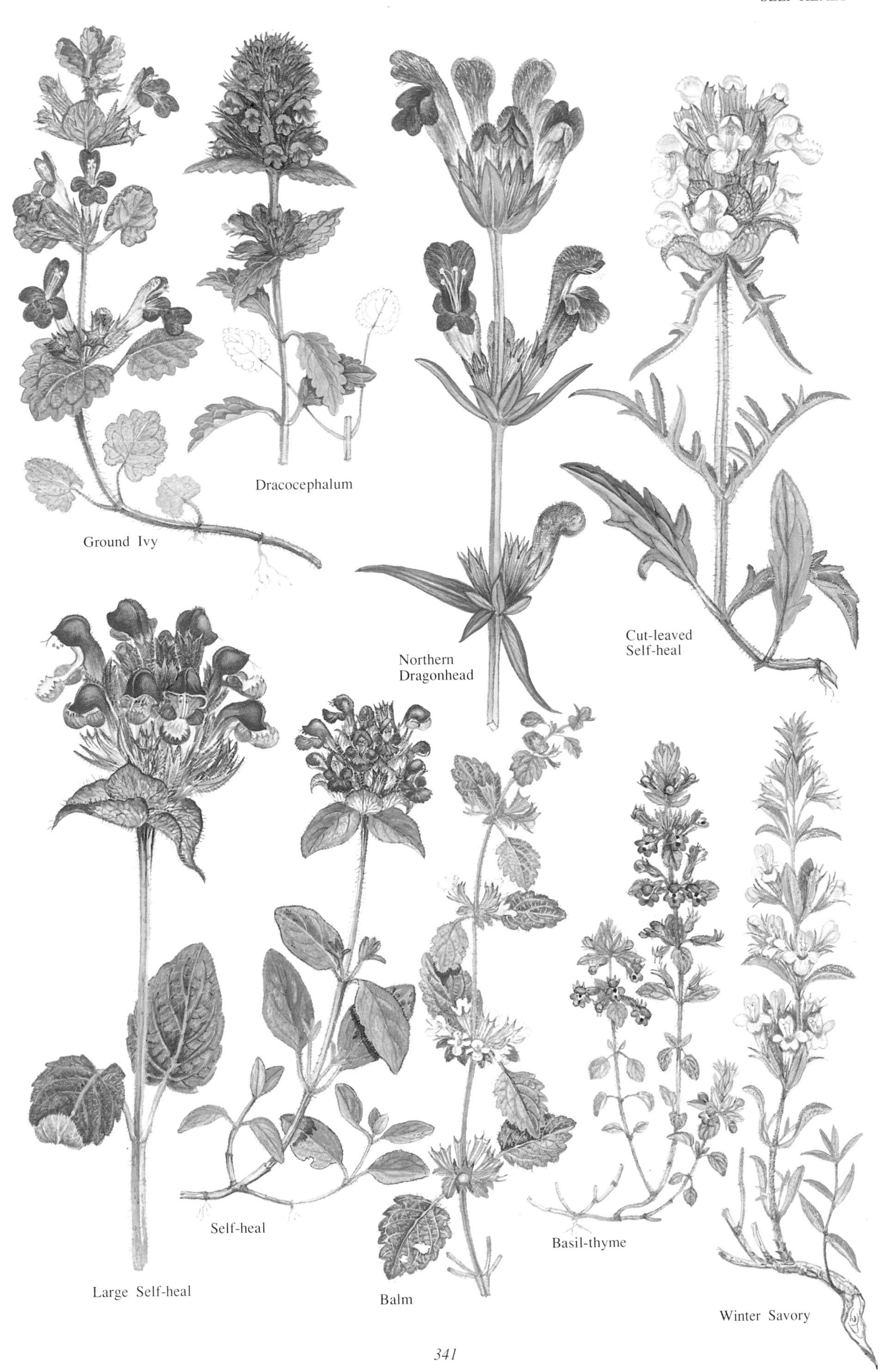
Dracocephalum
Ground Ivy
Northern Dragonhead
Cut-leaved Self-heal
Large Self-heal
Self-heal
Balm
Basil-thyme
Winter Savory

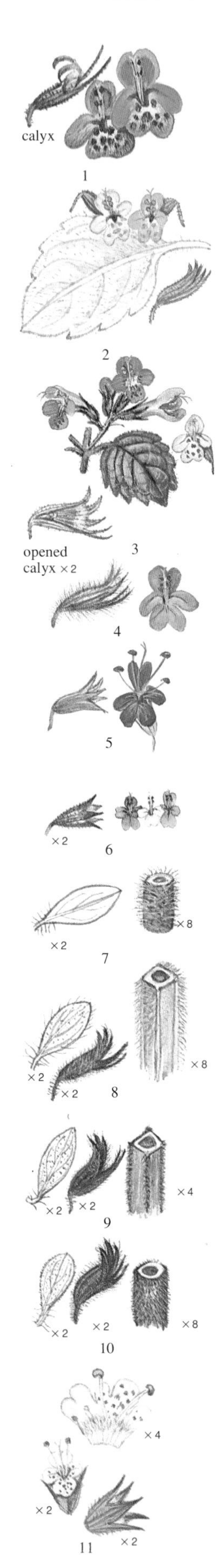


1* **Wood Calamint** *Calamintha sylvatica* Bromf. [*Satureja sylvatica*]. Short to medium, hairy, mint-scented, stoloniferous perennial; stems erect, scarcely branched, with long hairs. Leaves dark green, oval, coarsely toothed, stalked. Flowers pink to lilac with white spots on the lower lip. 15-22mm long, the lower lip 3-lobed, borne in lax whorls forming long leafy spikes; calyx tubular, 2-lipped, hairy in the mouth, 13-veined. Open woodland, banks and rocky ground, usually on calcareous soils, to 1600m. July-October. S Britain, France, W Germany. **B**: Very rare, confined to the Isle of Wight.

2 **Common Calamint** *C. s.* subsp. *ascendens* (Jordan) P.W. Ball [*C. ascendens*, *C. hirta*, *Satureja calamintha* subsp. *ascendens* and subsp. *menthifolia*]. Like the type, but generally a smaller plant, the leaves rarely more than 4cm long (not 3.5-6cm), slightly toothed or untoothed. Flowers 10-16mm long. Dry grassy habitats, hedgebanks and scrub, on calcareous soils. July-September. W Europe, often local. With a similar overall distribution in the wild to *C. sylvatica*. Forms with smaller, only female flowers occur. Sometimes confused with *C. nepeta*, but can be distinguished by examining the calyx; in *C. nepeta* a tuft of hairs protrudes from the mouth of the calyx.

3* **Lesser Calamint** *Calamintha nepeta* (L.) Savi subsp. *glandulosa* (Req.) P.W.Ball. [*Thymus glandulosus*, *Calamintha glandulosa*, *C. officinalis*, *Satureja calamintha* subsp. *glandulosa*]. Medium to tall, greyish, hairy perennial, with a long creeping rhizome; stems erect and much-branched. Leaves oval, shallowly toothed or almost untoothed, stalked. Flowers white or pale lilac, scarcely spotted, 10-15mm long, the whorls forming a lax leafy spike; calyx with a tuft of hairs protruding from the mouth. Dry habitats, scrub, hedgerows, banks and rough grassland, to 1300m. July-October. E Britain and France; naturalized in Germany. **B**: Very local; restricted to SE England, Norfolk and Suffolk.

4* **Wild Basil** *Clinopodium vulgare* L. [*Calamintha clinopodium*, *C. vulgaris*, *Satureja vulgaris*]. Short to medium, softly-hairy perennial, faintly aromatic; stems erect unbranched or slightly branched. Leaves oval-lanceolate, slightly toothed, short-stalked. Flowers pinkish-purple, 12-22mm long, in dense, rather distant whorls, the upper lip flat; calyx purplish, somewhat curved. Open woodland, scrub, hedgerows and dry grassy habitats and banks, generally on calcareous soils, at low altitudes. July-September. Throughout, except the far north. **B**: Throughout, though rare in N Scotland and SW Ireland.

5* **Hyssop** *Hyssopus officinalis* L. Medium, almost hairless, aromatic subshrub, generally with numerous erect stems. Leaves linear to linear-lanceolate, blunt, slightly shiny and glandular, untoothed. Flowers blue to violet, 7-12mm long, the upper lip erect, the lower 3-lobed, borne in small whorls forming leafy spikes; stamens protruding. Dry habitats, hill-slopes, banks and rocky ground, old walls, to 2000m. July-September. C & S France; naturalized in Britain, Belgium, Holland, N France and S Germany. **P**: Bees. **B**: Naturalized on old walls at Beaulieu.

6* **Marjoram** *Origanum vulgare* L. A very variable, medium to tall, tufted, hairy perennial, aromatic; stems erect, often purplish. Leaves oval, untoothed or slightly toothed, stalked, glandular. Flowers purplish red, pinkish or white, often darker in bud, 4-7mm, in broad, branched, flat-topped clusters; bracts dark purple usually, leaf- like; calyx 2-lipped. Rough dry grassland, scrub, hedgebanks, banks and road verges, usually on calcareous soils, to 2000m. Throughout, except the far north. A culinary herb. **B**: Throughout north to S Scotland, scarcer elsewhere, including Ireland.

**Thymes** *Thymus*. Small subshrubs, strongly aromatic, woody below. Flowers in crowded rounded or oblong clusters, sometimes interrupted below, small; female or hermaphrodite, borne on separate plants. Calyx bell-shaped, usually 2-lipped, hairy in the throat. Corolla with a straight tube and protruding stamens. 50 species in temperate regions of the old world. Several are used as culinary herbs.

7* **Hairy Thyme** *Thymus praecox* Opiz. Variable, low creeping woody plant, mat-forming, generally with numerous non-flowering shoots; stems hairy all round; flowering stems borne in rows, each with a basal cluster of small leaves. Leaves oval to rounded, untoothed, hairy on the margin at the base, with distinct lateral veins joining at the apex of the leaf. Flowers purple, 6mm long, in rounded clusters. Dry grassy habitats, heaths, grassland, pastures, banks, often closely grazed, sometimes on sand-dunes, screes and limestone pavement, to 3000m. May-September. Throughout, except Denmark, Finland and Sweden. **B**: Throughout, on a variety of soils.

8* **Wild Thyme** *T. p.* subsp. *brittanicus* (Ronnger) Holub [*T. pracox* subsp. *arcticus*, *T. drucei*]. Like the type, but stems hairy on two opposite sides only. Similar habitats and flowering time. Often associated with close-grazed pastures and meadows, pathways and ant hills. Britain, W & N France, the Faeroes, Iceland and W Norway – restricted to W & NW Europe. **B**: Throughout, but rare in East Anglia and C Ireland.

9* **Large Thyme** *Thymus pulegioides* L. [*T. alpestris*, *T. chamaedrys*, *T. montanus*]. Low to medium, tufted subshrub, with short creeping branches and ascending flower branches; stems square, hairy only on the angles. Leaves oval to lanceolate, untoothed, without marginal veins, short-stalked, hairy on the margin at the base. Flowers rose-purple, 6mm, in an oblong head, interrupted below. Dry grassy habitats, short turf, banks and hill-slopes, usually on calcareous soils, to 2000m. June-September. Europe north to S Scandinavia; naturalized in a few places in the Faeroes, Iceland and Finland. **B**: Widespread in C and S England and S Wales; scattered localities further north and in Ireland.

10* **Breckland Thyme** *Thymus serpyllum* L. [*T. serpyllum* subsp. *angustifolius*]. Low mat-forming, faintly aromatic plant; creeping stems rooting at the nodes, the flowering stems ascending, short, hairy all round. Leaves linear to elliptical, scarcely stalked, the lateral veins disappearing towards the apex. Flowers purple or pinkish, 6-7mm long, in rounded heads. Dry grassy habitats, often rabbit-grazed, bare ground, scrub, heaths and sand-dunes, to 3000m. May-September. Throughout, except Ireland, and much of the north. **B**: Very rare, confined to the Breckland soils of East Anglia and Cambridgeshire.

subsp. *tanaensis* (Hyl.) Jalas [*T. subarcticus*]. Like the type, but leaves larger, 7-13mm (not 5-10mm) and flowers 7-8mm long. Grassy habitats and rocky ground. Restricted to NE Sweden, Finland and the neighbouring part of the USSR.

11* **Gipsywort** *Lycopus europaeus* L. Very variable, medium to tall, slightly hairy perennial, not aromatic. Leaves oval to elliptical, pinnately-lobed at the base, toothed, mostly stalked. Flowers small, white, 4mm, in dense whorls at the base of the uppermost leaves; stamens 2, protruding; calyx with spine-like teeth. Wet habitats, wet woodlands, marshes, stream, pool and canal margins, dykes and ditches, mostly at low altitudes. July-September. Throughout, except the far north. Europe to C and N Asia. **B**: Common in England, Wales and Ireland, scarcer elsewhere.

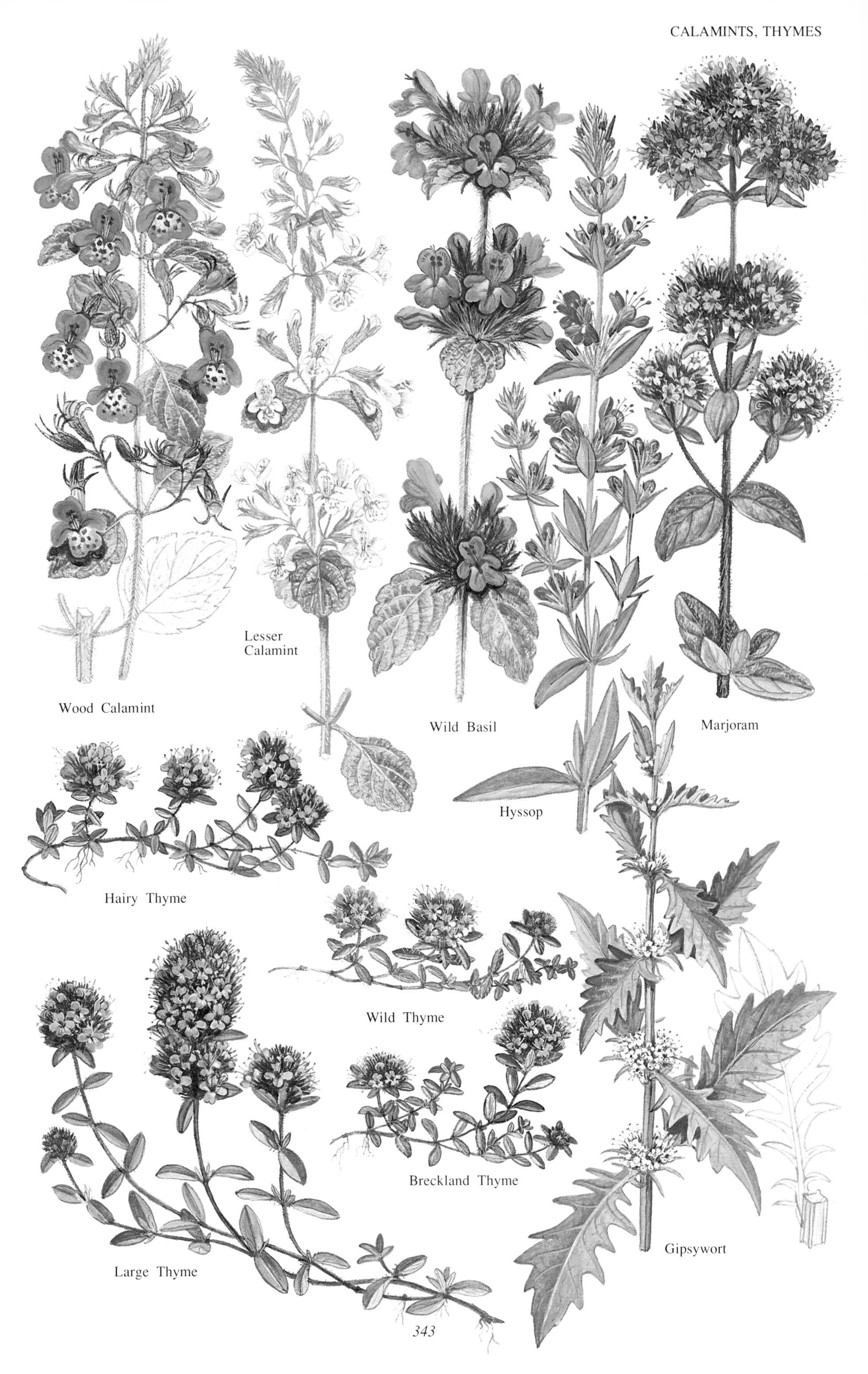
Wood Calamint
Lesser Calamint
Wild Basil
Marjoram
Hyssop
Hairy Thyme
Wild Thyme
Breckland Thyme
Large Thyme
Gipsywort

**Mints** *Mentha*. Stoloniferous perennial herbs with scented, often strongly aromatic foliage. Flowers hermaphrodite, or female on the same or on a separate plant, in dense whorls forming spikes or heads; calyx weakly 2-lipped; corolla also weakly 2-lipped, with 4 subequal lobes and short tube, generally shorter than the calyx. Stamens 4.

1* **Corsican Mint** *Mentha requienii* Bentham. Low creeping, hairless, pungent, carpetting perennial; stems thread-like, rooting at the nodes. Leaves small, oval or almost rounded, only 2-7mm long. Flowers pale lilac. 1.5-2mm long, in small clusters. Damp bare and shaded places. June-July. **I**: West Italy, Corsica and Sardinia; naturalized in Britain. **B**: Locally naturalized in S England, Wales and Ireland.

2* **Pennyroyal** *Mentha pulegium* L. [*M. vulgare*]. Short, usually hairy, prostrate perennial with erect flowering stems; strongly pungent. Leaves small, narrowly elliptical to oval, short-stalked, untoothed or with a few distant teeth, hairy at least below. Bracts leaf-like. Flowers lilac, 4.5-6mm long, in dense whorls, forming spikes without a terminal head of flowers, the stamens often protruding; calyx ribbed, hairy in the throat. Damp meadows and other damp habitats, particularly wet hollows, lake and pool margins, to 1800m. July-October. Throughout, except Scandinavia. **B**: Very rare; isolated localities from S Yorkshire southwards and in Ireland, diminishing overall.

3* **Corn Mint** *Mentha arvensis* L. Very variable, short to medium, hairy perennial, rarely behaving as an annual, with a rather sickly-sweet scent; stems ascending to erect. Leaves elliptical to oval, stalked, shallowly toothed. Flowers lilac or white, occasionally pink, 3-4mm long, in distant, dense whorls, forming spikes leafy at the apex, stamens protruding; calyx hairy. Damp habitats, open woodland, grassy roadsides, arable fields and ditches, to 1800m. July-September. throughout, except the extreme north. **B**: Throughout, scarcer in the north, absent from Orkney and Shetland.

4 **Whorled Mint** *Mentha × verticillata* L. [*M. sativa*], *M. aquatica × M. arvensis*. Like *M. arvensis*, but a hairier, more robust plant, with a rather sickly scent when crushed. Calyx 2.5-3.5mm long (not 1.5-2.5mm) and stamens not protruding. Damp habitats similar to the parent species. July-September. Throughout, except the Faeroes, Iceland and Spitsbergen. The commonest hybrid mint. **B**: Throughout, except for most of the Scottish Isles.

5 **Bushy Mint** *Mentha × gentilis* L. [*M. sativa* var. *gentilis*], *M. arvensis × M.spicata*. Very variable medium to tall plant, often red-tinged and sweetly scented, almost hairless. Leaves oval, often broadest above the middle. Flowers pink or lilac, in distant whorls subtended by leaf-like bracts, the stamens not protruding; calyx 2-3.5mm long. Damp habitats, especially ditches and the margins of rivers and ponds, but also waste ground with poor drainage. July-September. Britain, Belgium and France, widely naturalized except in the far north and most islands. **B**: Throughout, rare in the north.

6 **Tall Mint** *Mentha × smithiana* R. A. Graham [*M. rubra*], *M. aquatica × M. arvensis × M. spicata*. Tall perennial, sometimes reaching 1.5m tall, sweetly scented, often red-tinged. Leaves oval, almost hairless, stalked. Flowers lilac or pink, rarely white, in distant whorls with protruding stamens; calyx 3.5-4mm long. Ditches, river banks, waste ground, damp hedgebanks. July-September. Widely naturalized in W Europe. **B**: Widespread in England and Wales, rarer elsewhere, including Ireland.

7* **Water Mint** *Mentha aquatica* L. [*M. hirsuta*]. Variable short to tall perennial, hairy or almost hairless, often purplish, with a strong aromatic scent when crushed. Leaves oval to lanceolate, pointed, toothed and stalked. Flowers lilac-pink, 4-6mm long, in a dense oblong head, often with 1-2 distinct whorls of flowers below, stamens protruding; calyx 3-4mm long, hairy and distinctly veined. Swampy habitats. July-September. Throughout, except the extreme north; naturalized in Iceland. **B**: Common throughout, except for parts of the Scottish Highlands.

8 **Peppermint** *Mentha × piperita* L. [*M. nigricans*], *M. aquatica × M. spicata*. Medium to tall, pungent, often greyish perennial, usually hairy, often purple-tinged. Leaves lanceolate, toothed, long-stalked. Flowers lilac-pink, in long terminal spikes, interrupted below, the stamens not protruding. Ditches, damp road-verges, wet waste ground and other damp habitats. July-Spetember. Widely naturalized in Europe north to Denmark. **B**: Scattered localities throughout.

9* **Apple Mint** *Mentha suaveolens* Ehrh. [*M. rotundifolia*, *M. macrostachys*, *M. insularis*]. Medium to tall, sweetly-aromatic, hairy perennial; stem usually covered in white down. Leaves oval to almost rounded, grey-or white-downy beneath, unstalked or very short-stalked, toothed, the teeth bent downwards. Flowers whitish to pink, 2-2.5mm long, in a dense terminal, slender spike, interrupted below, often branched. Ditches, wet places along roadsides and in rough grassland. August-September. W Europe; naturalized in Ireland, Denmark and Sweden. **B**: Local S England, Wales, Ireland; rare further north.

10 **Apple-scented Mint** *Mentha × rotundifolia* (L.) Hudson [*M. niliaca*, *M. villosa*], *M. longifolia × M. suaveolens*. Very variable hybrid, more or less intermediate between the parent species in most characters. Ditches, roadsides, and other damp habitats. July-October. **B**: Throughout Britain, rare in Ireland.

11 **Large Apple Mint** *Mentha × villosa*. *M. spicata × M. suaveolens*. Very variable, some plants with hairy, others with hairless leaves. Many populations midway between parent species. July-October. Naturalized in N and NW Europe. **B**: Local.

12* **Horse Mint** *Mentha longifolia* (L.) Hudson [*M. sylvestris*, *M. incana*]. Very variable, hairy, medium to tall perennial, to 1.2m tall; stems usually white-downy. Leaves oblong-elliptical, 5-9cm long (to 4.5cm in *M. suaveolens*), green or grey, usually whitish beneath, short-stalked or unstalked, the margin with sharp spreading teeth. Flowers lilac or white, 3-4.5mm long, in dense terminal, branched spikes. Damp habitats, to 1900m. July-October. Belgium and France to Germany and S Sweden. The underside leaf hairs of *M. spicata* are branched, those of *M. longifolia* unbranched. **B**: Very occasionally naturalized.

*Mentha × villosonervata* auct., *M. longifolia × M. spicata*. A sterile hybrid, more or less intermediate in character. July-October. Locally naturalized in W & NW Europe. **B**: Locally naturalized.

13* **Spear Mint** *Mentha spicata* L. [*M. viridis*, *M. crispa*, *M. crispata*, *M. longifolia*, *M. sylvestris*]. Variable medium to tall, green or greyish perennial, strongly amd sweetly aromatic, sometimes rather musty. Leaves lanceolate to narrowly ovate, hairless or somewhat hairy, sharply toothed. Flowers pink or white, 2-3.5mm long, in a dense spike, sometimes branched. Damp habitats. July-October. Naturalized throughout, except the far north. **B**: Widely naturalized.

Corsican Mint
Pennyroyal
Corn Mint
Apple Mint
Water Mint
Spear Mint
Horse Mint

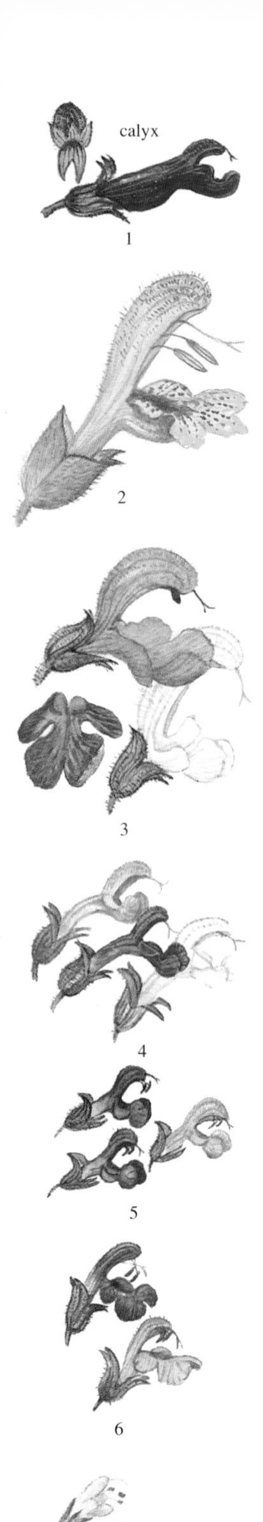


1* **Dragonmouth** *Horminium pyrenaicum* L. Low to medium, slightly hairy, tufted perennial, with erect or ascending flowering stems. Leaves mostly in basal rosettes, deep green, more or less shiny, oval to almost orbicular, stalked, hairy and blunt-toothed along the margin. Flowers dark bluish-violet, 17-21mm long, 2-lipped, in rather dense, one-sided spikes, the corolla tube curved upwards; stamens 4. Rocky habitats, grassy meadows and open woodland, in mountain habitats, to 2500m. June-August. E France & S Germany. Widespread species in the mountains of central and southern Europe. **P**: Various bees. Widely cultivated plant, generally grown on rock gardens. **B**: Cultivated, but not naturalized as far as it is known.

**Salvias** *Salvia.* Herbs with distinct whorls of flowers forming a lax, bracteate, spike or raceme. Both the calyx and the corolla are 2-lipped, the upper lip of the corolla forming a hood, the lower lip 3-lobed. Stamens only 2, hinged in the middle. Widespread in both the old and the new worlds. Many are very ornamental and are cultivated in gardens. One species, *S. officinalis*, is the source of the culinary herb, Sage, produced from the dried leaves.

2* **Jupiter's Distaff, Sticky Sage** *Salvia glutinosa* L. Medium to tall, stickily-hairy perennial, with erect, simple or branched stems. Leaves arrow-shaped or heart-shaped, stalked, toothed, rather a bright green. Flowers yellow, with reddish-brown markings, 30-40mm long, the lower lip more or less straight. Woods, copses and clearings, roadsides, mainly in mountain habitats on calcareous soils, to 1800m. June-September. C & E France and S Germany; naturalized in N Britain. Widespread in C & S Europe, extending east into Asia as far as the Himalaya. Plants often form large colonies, especially in woodland clearings. The sticky fruiting calyces attach themselves to passing animals and clothing, thus helping to spread the seeds to other suitable habitats. **B**: Naturalized in Scotland – Perthshire.

3* **Meadow Clary** *Salvia pratensis* L. [*S. tenorii, S. bertolonii*]. Very variable medium to tall, tufted, hairy perennial, slightly aromatic, with erect, branched stems, rather glandular above. Leaves oval to oblong with a heart-shaped base, somewhat wrinkled, deep green, the lowermost long-stalked, the upper smaller and unstalked. Flowers violet-blue, occasionally pink or white, large, 20-30mm long, in whorls of 4-6, the upper lip markedly curved; bracts green, half the length of the calyx. Meadows and grassy places on calcareous soils, occasionally in waste places, to 1900m. June-July. S Britain, Belgium, France, Holland, Germany and S Sweden. This species is often associated with old grassland over chalk or limestone. From Europe east into W Asia and south into N Africa. **P**: Long-tongued bumble-bees. *S. pratensis* is cultivated in gardens and is sometimes listed in catalogues under the name *S. haematodes*. **B**: Very rare and restricted to scattered localities in England, S of Lincoln, as well as S Wales.

4* **Wild Sage** *Salvia nemorosa* L. [*S. sylvestris*]. Medium, hairy perennial; stems without glands. Basal leaves simple, oblong with a heart-shaped base, blunt-toothed; stem-leaves similar, but unstalked. Flowers violet-blue, rarely pink or white, 8-12mm long, in loose whorls, each with 2-6 flowers; bracts violet, equalling or longer than the calyx. Meadows and grassy habitats, bare and waste places, to 1450m. June-August. C France and S Germany southwards; naturalized in Britain, France, Norway and Sweden. Widespread species in southern Europe. **P**: Various bees. Sometimes confused with *S. verbenaca*, but the leaves never with lobes at the base and the flowers in smaller whorls. **B**: Locally naturalized.

5* **Wild Clary** *Salvia verbenaca* L. [*S. clandestina, S. horminoides*]. Variable medium to tall hairy perennial; stems erect, simple or branched, glandular above. Leaves oblong, often pinnately-lobed, at least at the base, sharply toothed, the upper leaves and the stem often purplish. Flowers blue, violet or lilac, 6-10mm long, in lax or dense whorls of 6-10, generally with white markings in the throat; flowers either hermaphrodite or female, often cleistogamous; calyx enlarging in fruit. Dry grassy habiatats and roadsides, on well-drained calcareous or sandy soils. May-August. Britain and France. Primarily from southern Europe, particularly the Mediterranean region, but extending east into W Asia and also into N Africa. Plants often produce many small cleistogamous flowers which fail to open and are generally shorter than the calyces. The seeds produce a thick mucilage when soaked and this was used formerly as an eye-wash. **B**: Local in C & S England and Wales; rare in Scotland and Ireland.

6* **Whorled Clary** *Salvia verticillata* L. Medium to tall, rather foetid perennial, with erect, usually unbranched stems, often purplish, without glands. Leaves oval to lyre-shaped, with a square or heart-shaped base, the lower stalked and often with 1-2 pairs of basal lobes, toothed; upper leaves unstalked, generally purplish. Flowers lilac-blue to purplish, 8-15mm long, in dense whorls of 15-30, the upper lip more or less straight. Dry grassland, waste, bare and stony places, banks and paths. May-August. C & S France; widely naturalized in Britain, Belgium, N France, Holland, Germany and S Scandinavia. Widespread in S & E Europe and W Asia. **P**: Bees; smaller flowers on the plant are generally female without stamens, or with staminodes replacing the fertile stamens. **B**: Rare, casual or sometimes established in S Britain – scattered localities.

7* **Elsholtzia** *Elsholtzia ciliata* (Thunb.) Hyl. [*E. cristata, E. patrinii*]. Medium, erect annual, almost hairless. Leaves oval or elliptical, pointed, toothed, the lower stalked. Flowers lilac, small, 3-4mm long, in dense whorls forming a slender spike, the stamens slightly protruding; calyx hairy, 5-veined. Waste and bare ground on well-drained soils. June-August. **I**: C & E Asia. Naturalized or casual in Denmark, Germany and S Sweden. Sometimes cultivated as an ornamental plant. The genus is primarily Asian in origin and includes both small herbs as well as shrubs.

Meadow Clary

Dragonmouth
Meadow Clary
Jupiter's
Distaff
Wild Sage
Wild Clary
Whorled
Clary
Elsholtzia

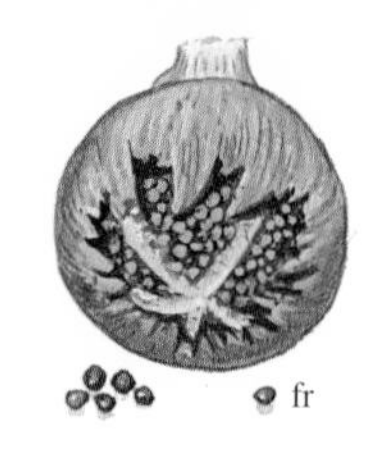


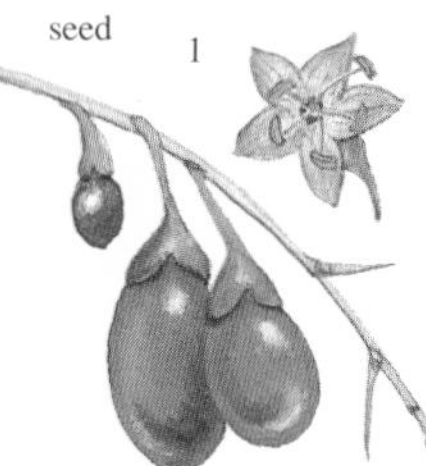

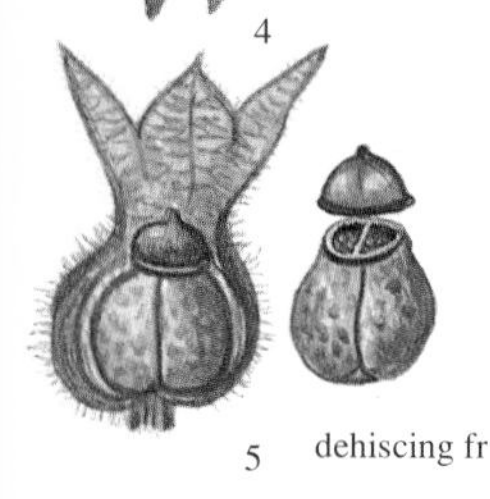


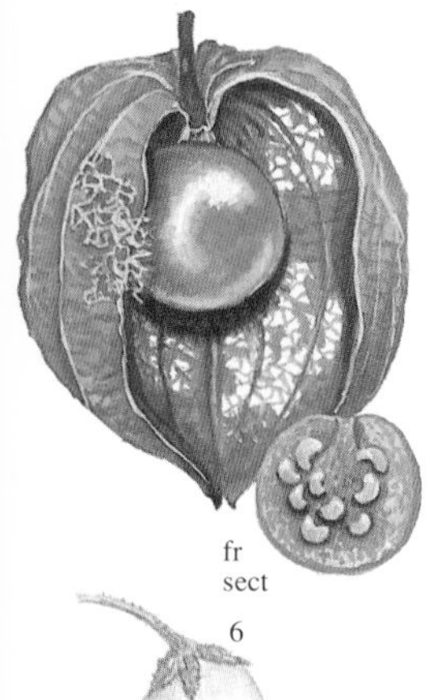


# POTATO FAMILY Solanaceae

Herbs or shrubs with simple or pinnate leaves. Flowers regular or irregular, hermaphrodite, generally 5-parted, the corolla bell-shaped or star-shaped with the petals fused together below. Stamens 5, attached to the corolla-tube. Ovary superior, usually with 2 compartments. Fruit a capsule of a berry.

1* **Apple of Peru** *Nicandra physalodes* (L.) Gaertner. Medium to tall hairless, rather foetid annual, sometimes reaching 2m in height. Leaves alternate, oval to elliptical, toothed and stalked, often lobed near the base. Flowers lilac to bluish, with a white centre, bell-shaped, 20-40mm long, solitary and half-nodding at the axils of the upper leaves; calyx with broad wings, greatly inflating in the fruiting stage. Fruit a brown berry encased in the membranous calyx. Bare and waste places, cultivated land. June-October. **I**: Peru. Locally naturalized in France and Germany, casual elsewhere. Very poisonous, though a garden plant; the fruit is very decorative. The flowers are short-lived and open for only a few hours each day. **B**: Casual in England and Wales.

2* **Duke of Argyll's Teaplant** *Lycium barbarum* L. [*L. halimifolium, L. vulgare*]. Deciduous shrub to 2.5m, with curved branches bearing a few slender spines. Leaves alternate or in small clusters, narrowly elliptical, widest at the middle, untoothed. Flowers purple, becoming brownish, trumpet-shaped, 8-9mm long, solitary or several together, with protruding stamens. Fruit a small red berry. Hedges and scrub, often growing over old walls. June-September. **I**: China. Naturalized over much of W Europe, including Britain, north to S Scandinavia. Cultivated in W Europe since the 18th century and although not a particularly decorative plant it is often used for hedging, especially in coastal districts. The berries are very attractive to a variety of birds. **B**: Widely naturalized in England, scarcer in Wales and Ireland.

3 **China Teaplant** *Lycium chinense* Miller [*L. rhombifolium*]. Like *L. barbarum*, but leaves oval to lanceolate, widest below the middle. Flowers larger, 10-15mm long, with the corolla-lobes 5-8mm (not 4mm). Hedges and scrub. June-September. Locally naturalized in Britain, France, Germany, Holland and S Sweden. Occasianally cultivated, but not nearly so frequently as *L. barbarum*. This species is sometimes regarded by botanists as an extreme form of the previous one.

4* **Deadly Nightshade** *Atropa bella-donna* L. Stout tall, hairless perennial, much branched, sometimes reaching 1.5m tall. Leaves alternate or opposite, oval, pointed, short-stalked, untoothed. Flowers brownish-violet or greenish, nodding bells, 25-30mm long, soliatary at the axils of the upper leaves. Fruit a succulent globose berry, shiny and black when ripe, surrounded by the starry persistent calyx. Damp or shaded habitats, woodland clearings, pathways, scrub and rocky places, often on calcareous soils and in mountain regions, to 1700m. Britain, Belgium, Holland, France and Germany; naturalized in Ireland, Denmark and S Sweden. A very poisonous plant; all parts of the plant are poisonous, especially the seeds and roots. The poisons are various alkaloids of which Hyoscyamine is the most important; Atropine is only present in small quantities. Long cultivated for its medicinal properties. **P**: Mainly bumble bees. **B**: Local in England south of the Tyne; scattered localities elsewhere, including Ireland.

5* **Henbane** *Hyoscyamus niger* L. Medium to tall, stickily-hairy, foetid annual or biennial, erect, branched or unbranched. Leaves oval to oblong, generally coarsely toothed or lobed, the basal stalked and forming a lax rosette, the stem leaves unstalked and clasping the stem. Flowers pale yellow, netted with purple veins, irregularly trumpet-shaped, 20-30mm long, with a well-defined lower lip, borne in one-sided branched spikes; calyx tubular, flared at the mouth with 5 short lobes. Fruit a capsule. Bare and disturbed ground, especially by the sea, near farm buildings, on light and nutrient-rich soils, to 1900m. May-September. Throughout, except the far north. All parts are poisonous and contain various alkaloids including Atropine, Hyoscyamine and Scopolamine. Its narcotic properties were used to alleviate toothache. Far less common than it used to be. **B**: Local throughout England, coastal in Wales and Scotland, very rare in Ireland.

6* **Cape Gooseberry** *Physalis alkekengi* L. Medium, slightly hairy, rhizomatous perennial, with erect, branched or unbranched stems. Leaves oval, stalked, with a few coarse teeth, or untoothed. Flowers dirty-white, more or less star-shaped, 15-25mm, solitary and half-nodding in the branch forks or in the axils of the upper leaves. Fruit an orange berry, completely surrounded by the greatly expanded orange-red calyx, like an inflated lantern, papery when ripe. Cultivated land, waste ground and scrub, sometimes forming large patches. July-August. Local in France and Germany; naturalized in Britain and Holland. Widely cultivated; inconspicuous in flower, but very striking in fruit. The fruits, the size of a cherry, are edible and the plant is cultivated for these in some countries, especially South Africa. **B**: Occasionally naturalized.

7* **Salpichroa** *Salpichroa origanifolia* (Lam.) Thell. Medium, much-branched, scrambling, hairy perennial, with flexuous stems becoming somewhat woody below. Leaves oval to almost rounded, short-stalked, untoothed. Flowers whitish, bell-shaped, 6-10mm long, solitary in the axils of the upper leaves. Fruit a small creamy-white berry. Coastal habitats, especially shingle, and waste places. August-October. **I**: E temperate South America. Locally naturalized in S Britain and W France. **B**: Scattered localities from Hampshire to Kent and in the Channel Islands.

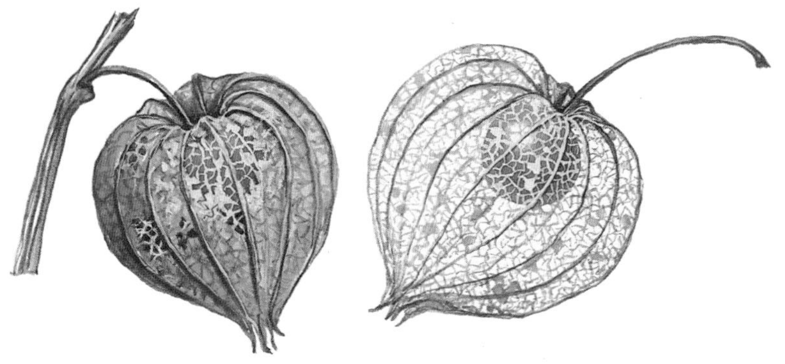

fruit of Cape Gooseberry in autumn

Duke of Argyll's Teaplant
Apple of Peru
Deadly Nightshade
× 1/25
Henbane
Cape Gooseberry
× 1/6
Salpichroa
×

**Nightshades** *Solanum*. Herbs or shrubs with alternate, sometimes opposite, leaves. Flowers in one-sided cymes or in umbels, at the leaf-axils or opposite the leaves (leaf-opposed). Corolla star-shaped, with spreading or reflexed lobes. Stamens protruding and forming a close cone around the stigma. Fruit a dryish or succulent berry. 1700 species widespread throughout the world. The fruits are generally poisonous.

1* **Black Nightshade** *Solanum nigrum* L. Variable low to medium, hairless or somewhat hairy annual; stems spreading to erect, often blackish. Leaves oval to lanceolate, untoothed to toothed or slightly lobed, stalked. Flowers white with yellow anthers, 10-14mm, in clusters of 5-10; petal-lobes spreading or reflexed. Berry dull black when ripe, green at first, globose, 6-10mm. Bare and disturbed ground, waste places, cultivated land and farmyards, often on nutrient-rich soils, to 1750m. July-October. Throughout, except the far north. Fruits poisonous with the alkaloid Solanine. **B**: Widespread in England south of the Humber; local elsewhere and absent from Scotland and much of Ireland.

2 **Green Nightshade** *Solanum physalifolium* Rusby [*S. sarrachoides*]. Like *S. nigrum*, but stems green and flowers in clusters of 3-8. Arable land, especially among rootcrops, bulbfields and waste ground, generally on light soils. July-September. **I**: Brazil. Naturalized in Britain, France and Germany. **B**: Established and increasing in S England; casual elsewhere.

3 *Solanum triflorum* Nutt. Rather similar to *S. nigrum*, but a foetid prostrate plant with deeply-lobed leaves. Berry white, marbled with green. Arable soils. August-October. **I**: W North America. Locally naturalized in Britain and Belgium. **B**: Rare: East Anglia and the neighbouring counties.

4* **Hairy Nightshade** *Solanum villosum* Miller [*S. luteum*]. Short to medium, hairy and glandular, annual; stems spreading to erect, with smooth, rounded ridges. Leaves diamond-shaped to lanceolate, untoothed to coarsely toothed. Flowers white with yellow anthers, 8-16mm, 3-5 in lax clusters. Berry orange-yellow or reddish-brown when ripe, subglobose, generally rather longer than wide. Bare and waste places, disturbed habitats. July-September. Britain, France and Germany; naturalized in Belgium, Holland, Denmark and S Sweden. subsp. *alatum*, is less hairy, with the hairs pressed close to the stems and leaves; the glands are absent and the stems more prominently ridged. Similar distribution.

5* **Bittersweet** *Solanum dulcamara* L. Medium to tall, scrambling, hairy or hairless perennial, the stems sometimes reaching 2m and becoming woody at the base. Leaves arrow-shaped or heart-shaped, pointed, generally with 1-4 small lobes at the base, stalked. Flowers purple with pale yellow anthers, very rarely white, 10-15mm, nodding, with reflexed petal-lobes which often have a pair of green spots near the base; flowers in lax clusters of 10-25. Berry shiny and red when ripe, egg-shaped. Damp woods, scrub, stream banks, sometimes on sea-shores. June-September. Throughout, except the far north. All parts of the plant are poisonous. A procumbent form found growing in coastal shingles is sometimes treated as a distinct species, *S. marinum* (Bab.) Pojark [*S. dulcamara* var. *marinum*] **B**: Throughout, except the extreme north of Scotland.

6 **Potato** *Solanum tuberosum* L. Medium, somewhat hairy, tuberous-rooted perennial with erect, winged stems. Leaves 1-2-pinnate, with rather uneven, oval, untoothed leaflets. Flowers white, pink or purplish, with yellow or orange anthers, 25-35mm, generally many in a broad cluster. Berry green or purplish, globose, 20-40mm. Waste places, rubbish tips, cultivated land. July-August. **I**: South America. Casual or partly established in parts of N & NW Europe, including Britain. Not fully hardy and often not surviving the winter unprotected. The edible tubers – potatoes – are developed from underground stems. The fruits, on the other hand, which resemble small unripe tomatoes, are very poisonous. **B**: Casual throughout.

7* **Tomato** *Lycopersicon esculentum* Miller. Medium to tall, bushy, hairy annual with rather sprawling stems; aromatic, especially when crushed. Leaves pinnate, with a mixture of alternating small and large, oval to lanceolate leaflets. Flowers yellow, 15-25mm, starry, with reflexed, pointed petal-lobes, the lobes generally 5-6 in number. Fruit a large fleshy berry, red when ripe, very occasionally yellow, 20-90mm, globose or pear-shaped, hairy when young. Arable land, waste places, rubbish-tips, often close to buildings. June-September. **I**: C and S America. Casual, very rarely naturalized in W & NW Europe. The seeds are generally killed by severe frosts. **B**: Well naturalized on sewage farms and rubbish tips in S Britain and Ireland; frequently casual.

8* **Thornapple** *Datura stramonium* L. [*D. tatula*]. Medium to tall erect, foetid annual, often reaching1.5m or more in height, generally hairless. Leaves oval to elliptical, usually lobed and with jagged teeth. Flowers white to purplish, erect funnel-shaped, 50-100mm long, solitary at the axils of the upper leaves, sometimes in the forks of branches; calyx large, half the length of the corolla, sharply angled. Fruit an erect, spiny, egg-shaped capsule, 35-70mm long. Bare, cultivated and waste ground, field margins. July-October. **I**: C & S America. Naturalized or casual throughout, except Ireland, Finland and the far north. An extremely poisonous plant. **B**: Irregular and spasmodic in appearance; mainly in S Britain.

9* **Tobacco** *Nicotiana tabacum* L. [*N. latissima*]. Very variable, tall, stickily-hairy, strong-smelling annual, up to 2m tall, sometimes more. Leaves large, elliptical to lanceolate, untoothed, narrowed and winged at the base, the wing running onto the stem. Flowers pale green or creamish, often with a pink tinge, trumpet-shaped, 35-55mm long, many borne in a broad terminal panicle; stamens uneven, sometimes slightly protruding. Fruit a green capsule. Cultivated land, waste places, occasionally on arable land. June-August. **I**: Argentina and Bolivia. Casual or naturalized in W Europe. Poisonous and narcotic, containing varying amounts of Nicotine. In the process of producing tobacco for smoking, the leaves are dried and cured. **B**: A rare casual.

10* **Small Tobacco** *Nicotiana rusticana* L. Like *N. tabacum*, but the leaves not winged onto the stem at the base. Flowers smaller, 12-17mm long, greenish-yellow. Cultivated land and waste places. July-September. **I**: North America. W Europe. Formerly cultivated for tobacco, but now largely replaced by the previous species.

Black Nightshade
Hairy Nightshade
Bittersweet
Tomato
Thorn Apple
Tobacco
Small Tobacco

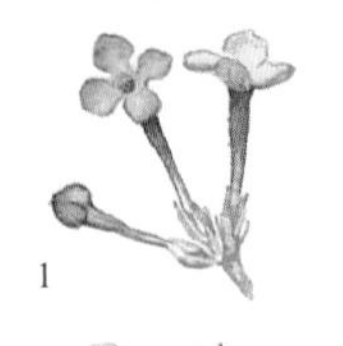


# BUDDLEJA FAMILY Buddlejaceae

Shrubs with simple opposite leaves. Flowers hermaphrodite, 4-parted, the stamens alternating with the petal-lobes. 10 genera and some 130 species, primarily in tropical and sub-tropical regions, but extending into parts of temperate Asia and South America.

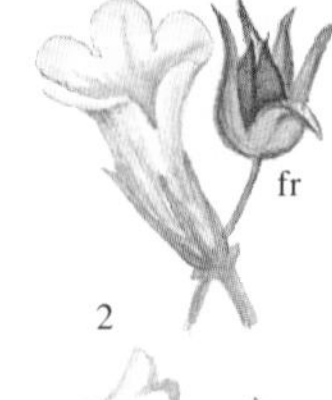


1* **Butterfly Bush** *Buddleja davidii* Franchet. Rather stiff deciduous shrub, 1-5m tall, with angled, downy stems. Leaves opposite, green or greyish, lanceolate, toothed, downy beneath, short-stalked. Flowers pale mauve or lilac to deep violet, with an orange 'eye', 9-11mm long, borne in a dense, pointed, spike-like panicle, with a strong sweet fragrance; corolla tubular, flared at the end into 4 broad lobes. Fruit a small capsule, splitting into two when ripe. Waste ground, walls, cliffs, old buildings, road and railway embankments. June-October. **I**: C & W China. A garden shrub, widely naturalized in W Europe. **B**: Naturalized in numerous places throughout Britain and S Ireland and still spreading – first noted at Harlech, N Wales, in c. 1921.

# FIGWORT FAMILY Scrophulariaceae

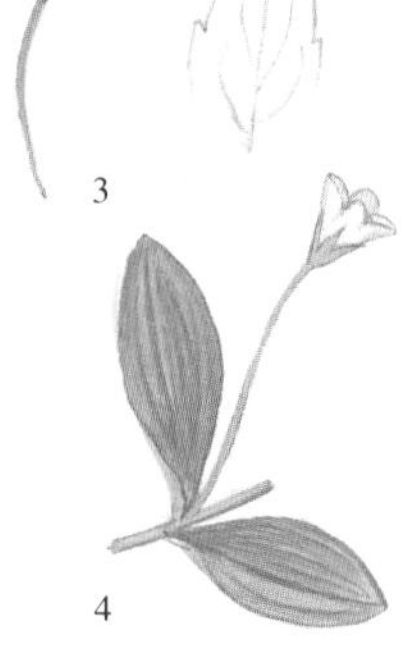


Herbs, rarely shrubs or trees. Leaves opposite or alternate. Flowers irregular (zygomorphic), in spikes or racemes, sometimes solitary; bracts present. Calyx 4-5-lobed, occasionally 2-lipped. Corolla 5-lobed, or clearly 2-lipped. Stamens 2 or 4, occasionally 5. Style solitary. Ovary superior, 2-celled. Fruit a capsule, generally 2-parted. 2500 species in 200 genera in most parts of the world. Distinguished from the *Labiatae* by their capsule fruits and by many species having alternate leaves.

2* **Gratiole** *Gratiola officinalis* L. Hairless, short to medium perennial; stems erect, square and hollow, with a creeping and rooting base. Leaves linear to lanceolate, opposite toothed or almost untoothed, half-clasping the stem; with translucent gland-dots. Flowers white, veined and tinged with purplish-red, 10-18mm, tubular 2-lipped, the lower lip 3-lobed, borne in leafy racemes. Wet habitats, meadows, marshes, river and stream banks, ditches. May-October. Belgium, Holland, France and Germany. From W and C Europe to W Asia.

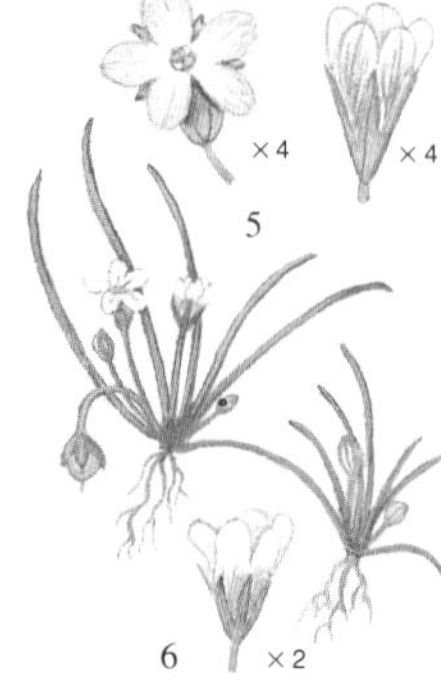


3* **Lindernia** *Lindernia dubia* (L.) Pennel. [*L. gratioloides*]. Short, hairless, erect annual. Leaves opposite, oblong, toothed. Flowers lilac, 7-8mm long, solitary at the upper leaf-axils, the corolla 2-lipped, spurred at the base; calyx deeply 5-lobed. stamens 4, but only 2 fertile. Muddy or sandy river banks. June-September. **I**: E North America. Naturalized in NW & C France and parts of SW Europe.

4 *Lindernia procumbens* (Krocker) Philcox [*L. gratioloides*]. Like *L. dubia*, but stems prostrate to ascending, the leaves untoothed or only very slightly toothed. Flowers pale pink, 5-6mm long; stamens 4, all fertile. Wet, muddy or sandy habitats. June-September. France and Germany. Plants sometimes produce small cleistogamous flowers. Widespread in C and SE Europe and W Russia.

5* **Mudwort** *Limosella aquatica* L. Low annual, sometimes perennial, with creeping and rooting runners, hairless. Leaves in basal clusters, linear-oblong to elliptical, often broadest above the middle, untoothed, rather fleshy. Flowers white, often marked with pink, 3mm, bell-shaped, 5-lobed, solitary; calyx 5-lobed. Muddy margins of pools, and places subjected to periodic flooding, often local, to 1800m. July-October. Throughout, except the far north. An erratically appearing plant. The runners are often erect at first before becoming horizontal and rooting down. **B**: Rare and declining in England and Wales; very few localities in Scotland and Ireland.

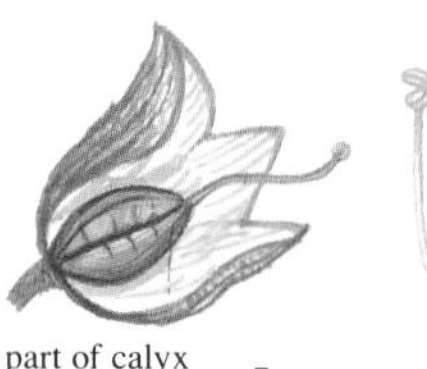


6 **Welsh Mudwort** *Limosella australis* R. Br. Like *L. aquatica*, but leaves all linear to awl-shaped, to 40mm long (not 20-120mm). Flowers white with an orange tube, 3.5-4mm long. Similar habitats and flowering time. June-October. Only W Britain, but widespread in E North America, Africa and Australia. Sometimes hybridises with the previous species. **B**: Very rare – only in Wales.

**Monkey-flowers** *Mimulus*. Perennial herbs, sometimes annual with opposite leaves. Flowers solitary at the upper leaves, forming leafy racemes. Calyx tubular-bell-shaped, 5-toothed. Corolla 2-lipped, the upper lip 2-lobed, the lower 3-lobed. Stamens 4. 100 species widespread in temperate America, E Asia, Australia and New Zealand.

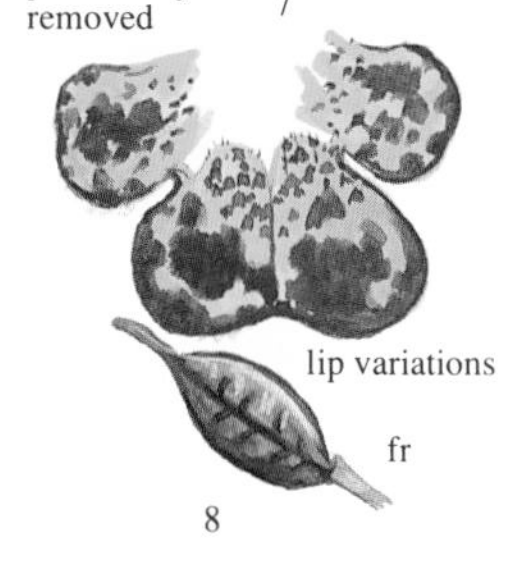


7* **Monkey-flower** *Mimulus guttatus* DC. Short to medium, almost hairless perennial; stem hollow, ascending to erect. Leaves oval, irregularly toothed, only the lower stalked. Flowers bright yellow, usually with reddish-brown spots in the mouth, 25-45mm long, the corolla mouth closed by 2 hairy ridges on the lower lip; flower-stem glandular-hairy. Wet marshy habitats, river and lake margins. July-September. **I**: W North America. Naturalized throughout, except the far north. **B**: Widespread throughout, except for the drier areas.

8* **Blood-drop Emlets** *Mimulus luteus* L. Short to medium hairless perennial, the stems often spreading at the base. Leaves narrowly oval, regularly toothed. Flowers yellow with small red spots in the throat and larger blotches on the lobes, 25-40mm long, borne on long slender stalks, the corolla with an open throat. Wet habitats, lake and stream margins, damp hollows. June-September. **I**: Chile. Naturalized in N Britain. Flower-stalks longer than in *M. guttatus*, 25-70mm long (not 12-25mm) and throat of flower open, not closed. **B**: Local in Scotland and a few localities in N England and Wales.

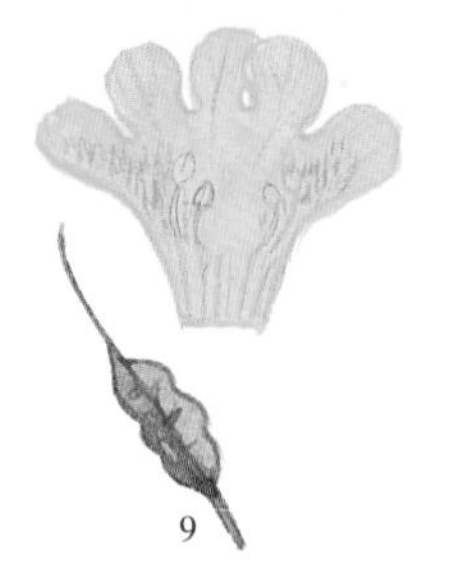


*Mimulus guttatus* × *M. luteus*. Complex hybrids, intermediate in most characters, are cultivated in gardens and may be locally naturalized. Some also involve another species, *M. cupreus*. Unlike *M. luteus* these hybrids have glandular-hairy flower-stems and yellow corollas, often suffused with pinkish-purple.

9* **Musk** *Mimulus moschatus* Douglas ex Lindley. Short, stickily-hairy perennial; stems erect with a spreading base. Leaves oval, toothed to untoothed, mostly short-stalked. Flowers pale yellow, sometimes striped with red in the throat, 10-20mm long, borne in leafy racemes; mouth of corolla open. Damp and

Butterfly Bush
Gratiole
Lindernia
Mudwort
Monkey-flower
Blood-drop Emlets
Musk

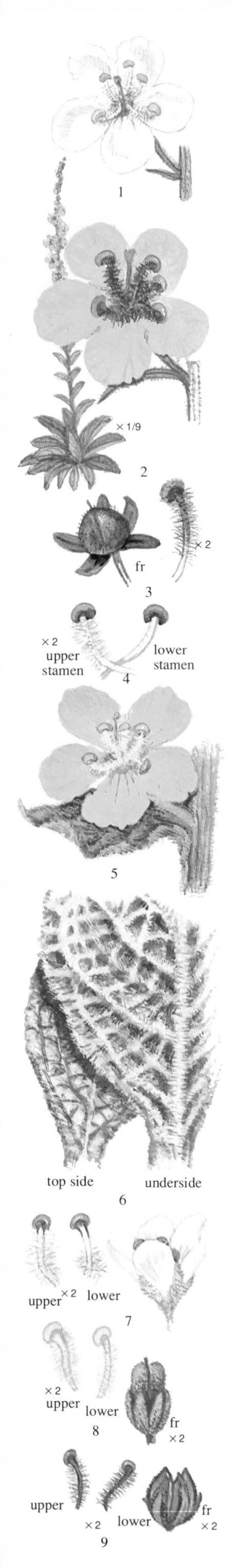


**Mulleins** *Verbascum*. Herbs, generally biennial or perennial, often with large basal leaf-rosettes. Leaves alternate, usually toothed or variously lobed. Flowers in racemes, spikes or panicles, often mealy. Calyx with 5 equal lobes. Corolla yellow, white, orange or violet, with a short tube and 5 widely spreading lobes. Stamens 4-5, with hairy stalks (filaments). Fruit a globose capsule. 250 or more species and numerous hybrids. Most are to be found in the Mediterranean region, but some are found as far east as N and C Asia. Some occur as casuals – *V. sinuatum*, *V. speciosum* and *V. densiflorum* for instance.

1* **Moth Mullein** *Verbascum blattaria* L. Medium to tall biennial, occasionally an annual; stems erect hairless below, but glandular-hairy above, angled. Basal leaves oblong to lanceolate, wavy, blunt-toothed to pinnately-lobed, shiny and hairless, short-stalked or unstalked, the upper leaves generally triangular and clasping the stem; bracts oval. Flowers yellow or rarely white, 20-30mm, in rather lax simple racemes, the flowers borne singly, long-stalked; stamens 5, the stalks of the lower 2 with purple hairs, the upper 3 smaller with white or purple hairs. Damp habitats, bare and waste places, waysides, at low altitudes. June-October. Belgium, Holland, France and Germany; naturalized in Britain. W and C Europe to S Europe, N Africa and W Asia. Easily recognised by the combination of lax inflorescence and solitarily spaced flowers. **P**: Various insects or self-pollinated. **B**: Rare, lowlands north to Durham, often not persisting for very long in any one place.

2 **Twiggy Mullein** *Verbascum virgatum* Stokes. Like *V. blattaria*, but plant more glandular-hairy throughout. Bracts 8-20mm long (not 7-8mm). Flowers 30-40mm, in small clusters of 2-5 along the raceme. Waste ground and waysides and the borders of cultivated land. June-October. S Britain and W & C France; sometimes casual or locally naturalized elsewhere. W and SW Europe. **B**: Very rare – restricted to Devon, Cornwall and the Isles of Scilly; casual elsewhere including Ireland.

3* **Purple Mullein** *Verbascum phoeniceum* L. Medium to tall perennial, hairy; stems erect hairy, glandular-hairy above. Basal leaves oval, slightly toothed or untoothed, rather wavy, the upper leaves unstalked. Flowers violet, 20-30mm, borne singly in lax racemes; stamens equal, the stalks with violet hairs. Grassy places, scrub, waste places and banks. June-August. W & C Germany; naturalized in Holland. C & E Europe to W Asia, including parts of Russia.

4* **Orange Mullein** *Verbascum phlomoides* L. Medium to tall, whitish or greyish woolly biennial, to 1.2m. Basal leaves oblong-elliptical, toothed or untoothed; stem leaves narrower and pointed, the base not or only shortly running down the stem. Flowers bright orange-yellow, 20-55mm, in dense spike-like woolly racemes, the corolla hairy outside; stamens 5, the lower 2 with hairless stalks, the upper 3 with whitish or yellowish hairs. Waste and bare ground, banks and waysides, dry stony habitats and scrub, to 1400m. July-September. France and Germany; naturalized or casual in Belgium, Holland and Britain. **B**: Casual or sometimes naturalized in C & S Britain.

5 *Verbascum densiflorum* Bertol. [*V. thapsiforme*]. Like *V. phlomoides*, but the base of stem leaves running down the stem almost to the leaf below. Bracts 15-40mm long (not 9-15mm). Similar habitats and flowering time, local. Belgium, Holland, France, Denmark, Germany and S Sweden. Native of W & C Europe. The spikes occasionally have one or two short branches towards the base.

6* **Great Mullein, Aaron's Rod** *Verbascum thapsus* L. Medium to tall soft, greyish or whitish woolly biennial, to 2m, though often less. Basal leaves elliptical to oblong, blunt, toothed or untoothed with a narrow winged stalk, the stem leaves smaller at the base running down the stem almost to the leaf below. Flowers yellow, 12-35mm, in a dense woolly, spike-like raceme, occasionally branched below; stamens 5, the stalks with white hairs, the lower 2 often hairless or almost so. Hedgebanks, scrub, road-verges, rough grassland and waste places, generally on dry stony gravelly or calcareous soils, to 1850m. June-August. Throughout, except the far north. Sometimes confused with *V. phlomoides*, but the leaf-bases running well down the stem and the stigma globose rather than spoon-shaped. **P**: Various insects or self-pollinated. **B**: Common in England and Wales, scarcer elsewhere.

7* **Hoary Mullein** *Verbascum pulverulentum* Vill. Tall mealy-white-woolly biennial, particularly when young; stem erect, ridged, to 1.2m. Basal leaves oblong, generally widest above the middle, toothed or untoothed, short-stalked or unstalked; stem leaves smaller, unstalked, the upper with a heart-shaped base. Flowers yellow, 18-25mm, in large pyramidal panicles; stamens 5, the stalks all with white hairs. Road-verges, banks, rough grassland, occasionally in waste places, to 1200m. July-August. E Britain, Belgium, France and Germany. Mainly in C and SE Europe. The thick mealy-wool covers the entire plant, although this wears off a little on the older leaves. **P**: Various insects. **B**: Very rare, restricted to a few localities in Norfolk and Suffolk, casual elsewhere.

8* **White Mullein** *Verbascum lychnitis* L. Medium to tall, grey-downy biennial, to 1.5m; stems angled, erect. Basal leaves oval to oblong, toothed or almost untoothed, green above, but white-mealy beneath, short-stalked, the upper leaves small, pointed and unstalked. Flowers white, occasionally yellow, 12-20mm, in a pyramidal panicle; stamens 5, all with whitish or yellowish hairs. Hedgebanks, waste places, scrub, rocky habitats, old quarries and railway embankments, often on calcareous soils, to 1200m. July-August. Britain, Belgium, Holland, France, Denmark and Germany; naturalized in parts of Scandinavia. Widespread from W Europe to the Caucasus and NW Africa. **B**: Rare, scattered localities in W Sussex, Somerset, Kent and occasionally further north and in Wales.

9* **Dark Mullein** *Verbascum nigrum* L. Tall, green, not mealy, hairy perennial; stems erect, hairy. Basal leaves oval to oblong, with a heart-shaped base, long-stalked, dark green above, paler beneath, thinly hairy; upper leaves smaller, almost unstalked. Flowers yellow, 18-25mm, in racemes, sometimes with one or two branches below; stamens 5, the stalks all with violet hairs. Rough grassland, road-verges, hedgebanks, rocky habitats, generally on dry calcareous soils, to 1800m. July-October. Europe north to S Scandinavia. Easily identified by its long-stalked leaves and by the violet hairs on the stamen filaments. W Europe to the Caucasus and Siberia. **B**: Fairly common in England and Wales north to Nottingham, scattered localities and probably naturalized further north.

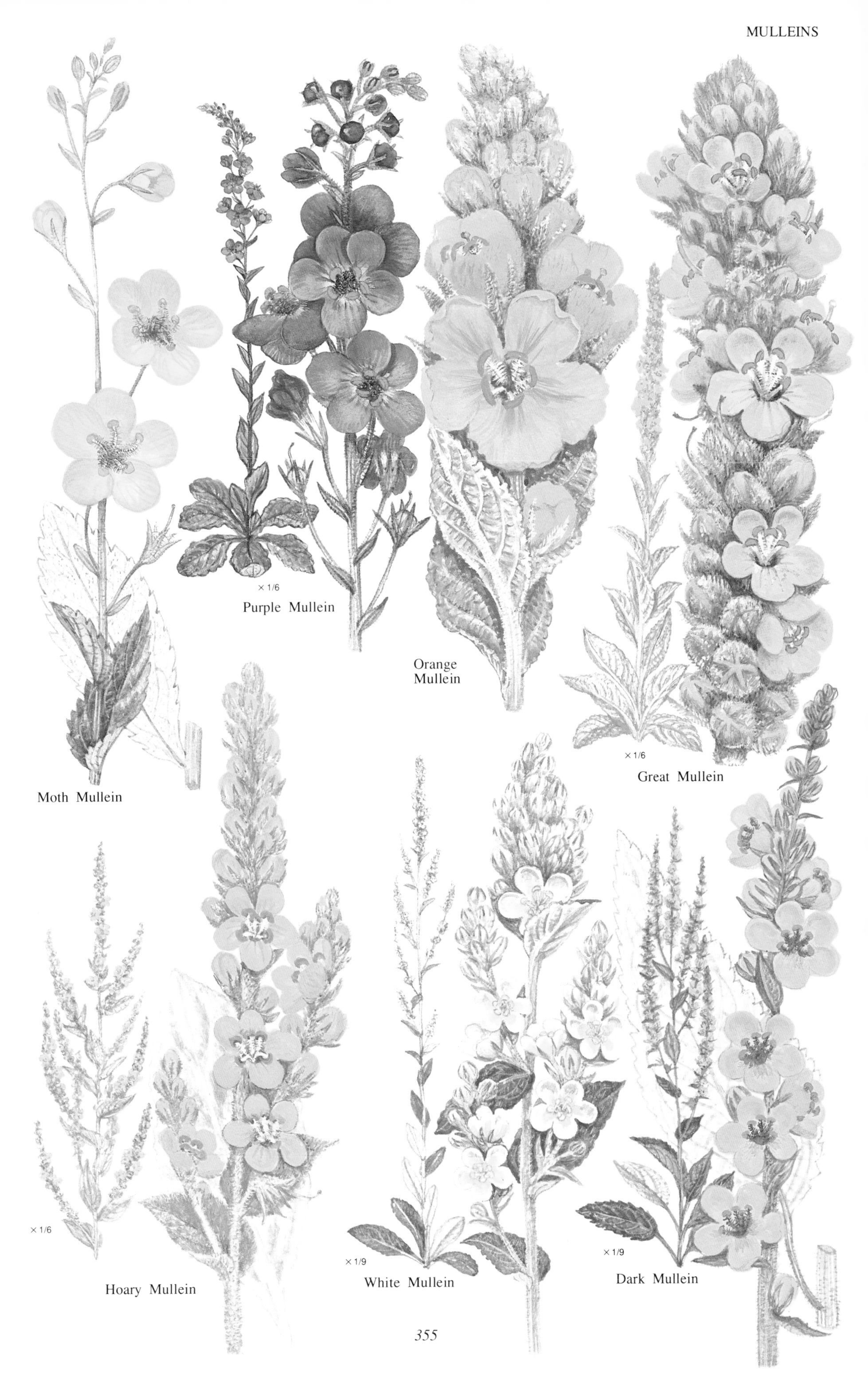
× 1/6
Purple Mullein
Orange Mullein
× 1/6
Great Mullein
Moth Mullein
× 1/6
Hoary Mullein
× 1/9
White Mullein
× 1/9
Dark Mullein

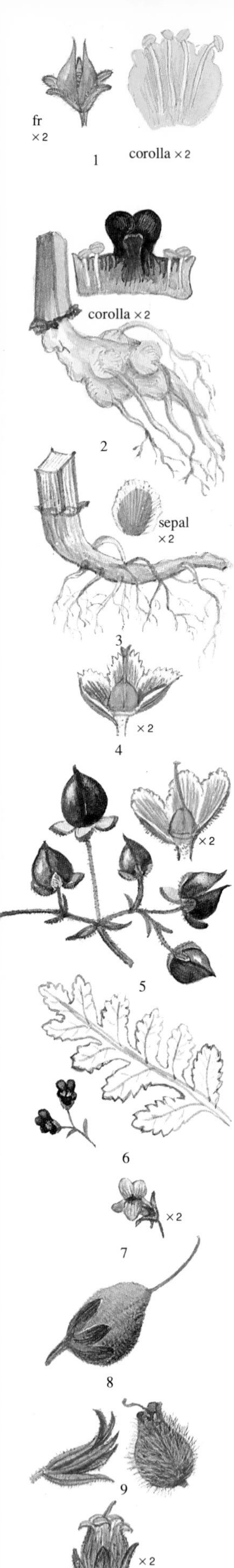


**Figworts** *Scrophularia*. Biennial or perennial herbs with erect or ascending, often square stems. Leaves usually opposite. Flowers in cymes, forming lax racemes or panicles. Calyx with 5 equal lobes. Corolla pouched, generally more or less 2-lipped, with 5 rounded lobes, the upper 2 joined together at the base. Stamens 4, the fifth stamen generally replaced by a scale-like staminoid. 120 species from the northern temperate hemisphere. The flowers are small and often rather sombre in coloration – most are pollinated by wasps.

1* **Yellow Figwort** *Scrophularia vernalis* L. Medium to tall, softly glandular-hairy biennial or perennial; stems obscurely square. Leaves oval, generally heart-shaped at the base, deeply toothed, stalked; bracts similar to the leaves. Flowers greenish-yellow, 6-8mm long, not 2-lipped; staminode absent. Shaded habitats, mountain woods, waste places and plantations, to 1800m. April-June. France and SC & S Germany; naturalized in Britain, Belgium, Holland, Denmark and S Sweden. **P**: Bees. **B**: Local; scattered localities north to Dundee.

2* **Common Figwort** *Scrophularia nodosa* L. Medium to tall hairless perennial; stems erect, square, the angles not winged. Leaves oval to lanceolate, generally with a heart-shaped base, double-toothed, stalked; bracts mostly linear, only the lowermost leaf-like. Flowers greenish with a purplish-brown upper lip, 7-9mm long; calyx-lobes with a narrow membranous margin. Hedgerows, woodland margins and woodland rides, river and stream margins and other damp habitats, generally on fertile soils, to 1850m. June-September. Throughout, except the extreme north. **P**: Wasps – the purplish-brown coloration of the flowers is typical of wasp-pollinated flowers. The rootstock bears nodular tubers, hence the scientific name of the species. **B**: Widespread throughout, except N Scotland.

3* **Water Figwort, Water Betony** *Scrophularia auriculata* L. [*S. aquatica*]. Like *S. nodosa*, but often taller, to 1m, the stems narrowly winged on the angles. Leaves often with 2 small lobes at the base of the blade. Calyx-lobes with broad membranous margins. Bracts nearly all linear. Wet habitats, river, stream, canal and lake margins, ditches, dykes, fens and marshes. June-September. Britain, Belgium, Holland, France and Germany. The rootstock bears swollen nodules. **B**: Widespread in England, local elsewhere but absent from most of Scotland.

4* **Green Figwort** *Scrophularia umbrosa* Dumort. [*S. aquatica*]. Rather like *S. nodosa*, but stems broadly winged and leaves narrowed at the base, sharply toothed. Bracts often leaf-like. Flowers olive-brown; calyx-lobes with membranous, slightly toothed margins. Staminode 2-lobed. Damp shaded habitats, woodland, riverbanks, marshes and fens, to 1850m. Britain, Belgium, Holland, France, Denmark and Germany. Sometimes confused with *S. auriculata*, but staminode 2-lobed and leaves blunt-toothed. **P**: Wasps. **B**: Very local, England and Scotland except the north; very rare in Ireland.

5* **Balm-leaved Figwort** *Scrophularia scorodonia* L. Medium to tall hairy perennial; stems square, but not winged. Leaves oval to lanceolate, with a heart-shaped base, sharply toothed, stalked, hairy on both surfaces; bracts mostly leaf-like. Flowers dull purple, 8-12mm long; calyx-lobes with broad membranous margins; staminode not lobed. Hedgerows, river and stream margins, meadows and dune slacks. June-August. S Britain and W France. Primarily coastal, sometimes locally abundant. **B**: Very local, restricted to Devon, Cornwall, the Isles of Scilly and the Channel Islands.

6 **French Figwort** *Scrophularia canina* L. Medium, hairless perennial; stems much-branched. Lower leaves pinnately-lobed, with rather narrow lobes, generally toothed; upper leaves elliptical to oblong, sometimes unlobed, the uppermost usually alternate; bracts small, not leaf-like. Flowers dark purplish-red, 4-5mm long, numerous; calyx-lobes with a broad membranous margin. Mainly in mountains, sandy and rocky habitats, river gravels, often on limestone, to 2200m. June-August. C & E France and S Germany. **P**: Wasps.

7* **Anarrhinum** *Anarrhinum bellidifolium* (L.) Willd. Short to tall, hairless biennial or perennial. Basal leaves in a rosette, oblong, broadest above the middle, irregularly toothed, the stem leaves numerous, crowded, alternate, palmately-lobed, with 3-5 narrow divisions. Flowers pale lilac or blue with a short spur, 4-5mm long, 2-lipped, borne in slender racemes, sometimes branched, the mouth of the corolla open; stamens 4. Dry habitats, waysides, walls, rocky habitats and pine woodland, at low altitudes. March-August. NC France and SC Germany southwards. C Europe to Italy and the Iberian Peninsula.

8* **Snapdragon** *Antirrhinum majus* L. Medium to tall perennial; stems branched, hairless below, but glandular-hairy above; stems erect or ascending. Leaves linear to elliptical, untoothed. the lower opposite and crowded, the upper alternate; bracts leaf-like, but smaller. Flowers pink, purple, yellow or bicoloured, 33-45mm long, fragrant, 2-lipped, pouched at the base, but not spurred, borne in slender spike-like racemes; mouth of the corolla closed. Capsule usually glandular-hairy, opening by 3 pores. Dry rocky habitats, arable fields, railway embankments, old walls, waste ground, to 1600m. July-September. **I**: Mediterranean region. Widely naturalized in Britain, Belgium, Holland, France and Germany. A common garden plant with many cultivars, including forms with open-mouthed flowers and double flowers. **P**: Bumblebees, or occasionally self-pollinated. **B**: Scattered localities in England and Wales; declining.

9* **Lesser Snapdragon** *Misopates orontium* (L.) Rafin. [*Antirrhinum orontium*]. Short, slightly branched annual, sometimes hairy below, glandular-hairy above. Leaves linear to elliptical, untoothed, the lower opposite, the upper alternate. Flowers pink, rarely white, 10-15mm, pouched at the base, scarcely stalked, borne in leafy racemes, the mouth of the corolla closed; calyx-lobes equalling the corolla. Capsule glandular-hairy, opening by 3 pores. Arable fields, roadsides, waste and bare places, cultivated ground, at low altitudes usually. July-October. S Britain, Ireland, Belgium, Holland and France; naturalized in Germany, Denmark and S Sweden. Often rare or local, declining in many areas due to modern farming practices, especially the use of herbicides. **P**: Bees or self-pollinated. **B**: Confined mainly to East Anglia and S England.

10* **Small Toadflax** *Chaenorhinum minus* (L.) Lange [*Linaria minor*]. Slender low to short, glandular-hairy annual; stems erect, branched. Leaves linear to lanceolate, blunt, untoothed, alternate. Flowers lilac with a yellow patch on the lower lip, spurred, 6-9mm long, solitary on slender stalks at the base of the upper leaves, the mouth of the corolla open. Arable land, disturbed and waste ground, cultivated land, short grassy places, railway lines, at low altitudes. May-October. Throughout, except the Faeroes, Iceland and Spitsbergen, but casual in much of the north of its range. **P**: Probably self-pollinated. **B**: Widespread, but absent from much of C and N Scotland.

Yellow Figwort
Common Figwort
Water Figwort
Green Figwort
Balm-leaved Figwort
Anarrhinum
Snapdragon
Lesser Snapdragon
Small Toadflax

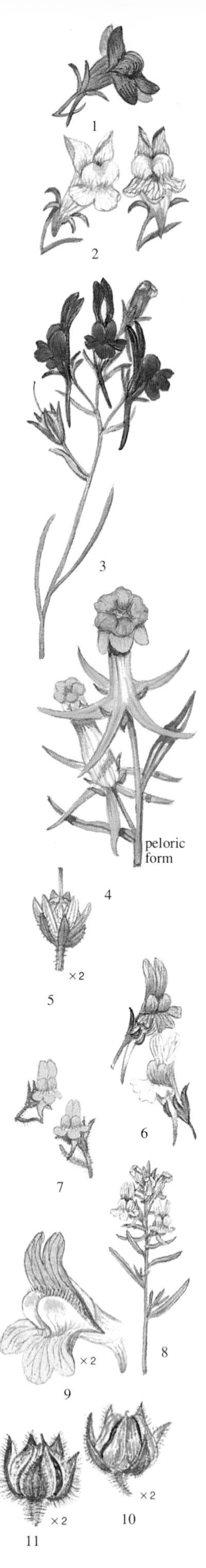


**Toadflaxes** *Linaria*. Annual or perennial herbs. Leaves simple, untoothed, unstalked, often whorled below, but alternate above. Flowers in spike-like racemes; bracts present. Calyx with 5 unequal lobes. Corolla 2-lipped, spurred. Stamens 4. Capsule globose. 150 species particularly in the Mediterranean region, but also present in temperate parts of Europe, Asia and North America. Distinguised from *Antirrhinum* by the presence of spurred flowers.

1* **Purple Toadflax** *Linaria purpurea* (L.) Miller. Medium, hairless tufted perennial; stems erect to ascending, often branched above. Leaves linear, numerous, the lower whorled. Flowers purplish-violet, 9-12mm long, in long, rather dense, tapered racemes, the spur 5mm long, curved; bracts equalling the flower-stalks. Waste places and cultivated land, sometimes growing on old walls, at low altitudes. June-August. **I**: C & S Italy and Sicily. Naturalized in Britain. Widely cultivated in gardens and from there sometimes becoming naturalized. **P**: Bees. **B**: Locally naturalized in England and Ireland.

2* **Pale Toadflax** *Linaria repens* (L.) Miller [*L. striata*]. Medium to tall, hairless, tufted perennial, to 1.2m, though often less, with a creeping rhizome; stems numerous, erect, usually branched above. Leaves linear, the lower whorled. Flowers white to pale lilac with violet veins, 8-15mm long, borne in long lax racemes, the spur 3-5mm long, straight; bracts equalling the flower-stalks. Dry habitats, cultivated and waste ground, stony fields and rocky places, to 2300m. June-September. Belgium, France and Germany; naturalized in Britain, Holland and Scandinavia. Native to W and SC Europe. **P**: The flowers are self-sterile; bees. Hybridises with *L. vulgaris*, *L.* x *sepium* Allman; hybrids have yellow flowers striped with red. Hybrids often occur where the parent species grow close to one another. **B**: Naturalized in England and Wales, north to Yorkshire, and in Ireland.

3 **Jersey Toadflax** *Linaria pelisseriana* (L.) Miller. Rather like *L. repens*, but a short to medium annual, the flowers purplish-violet with a whitish patch on the lower lip, 15-20mm long, the spur 7-9mm long. Cultivated land, waste places and heaths, at low altitudes. May-July. SW Britain, W & C France. **P**: Bees. **B**: Very local, restricted to Jersey.

4* **Common Toadflax** *Linaria vulgaris* Miller. Medium to tall, tufted perennial, with a creeping rhizomatous rootstock; stems numerous, erect, hairless or glandular-hairy above, usually branched in the upper half. Leaves linear to narrowly elliptical, 1-veined, crowded, mostly alternate. Flowers bright to pale yellow, 25-33mm long, in a long dense raceme, the spur 10-13mm long, straight; bracts equalling or longer than the flower-stalks. Grassy habitats and waste ground, road-verges, hedgebanks and cultivated land, to 1600m. July-October. Throughout, except the Faeroes, Iceland and Spitsbergen, perhaps naturalized in the north of its range. Cultivated in gardens – a 'peloric' form sometimes occurs with a regular 5-spurred corolla. **P**: Various bees. **B**: Britain north to S Scotland and in Ireland.

5* **Prostrate Toadflax** *Linaria supina* (L.) Chaz. Short, grey-green biennial or short-lived, tufted perennial; stems more or less prostrate with ascending ends, hairless below, but glandular hairy in the inflorescence. Leaves linear, the lower whorled. Flowers pale yellow, occasionally tinged with violet, 13-20mm long, in short dense racemes rarely with more than 10 flowers, the spur 10-15mm long, almost straight. Dry habitats, often sandy, open grassland, banks and waste and bare habitats, to 2000m. June-September. C & S France; naturalized in SW Britain and S Sweden. **P**: Bees. **B**: Naturalized in Cornwall, casual elsewhere.

6* **Alpine Toadflax** *Linaria alpina* (L.) Miller. Low, more or less prostrate, grey-green annual or short-lived perennial, hairless. Leaves linear to narrowly oblong, rather fleshy, mostly whorled. Flowers violet with a yellow patch on the lower lip, ocasionally pink or whitish, 13-22mm long, borne in short, few flowered racemes, the spur 8-10mm long, slightly curved. Mountain habitats, rocks, moraines, screes and river gravels on a variety of rocks, 1500-3800m. May-August. Local but widespread in the mountains of C and S Europe. **P**: Bees.

7* **Sand Toadflax** *Linaria arenaria* DC. Low to short, stickily-hairy, bushy annual, seldom more than 15cm tall; stems ascending to erect, branched from the base. Leaves elliptical, the lower whorled. Flowers small, yellowish, 4-7mm long, few in a lax raceme, the spur 2-3mm long, often violet; bracts much longer than the flower-stalks. Coastal habitats, particularly on sand-dunes. May-September. W France south of Dunkirk; naturalized in SW Britain. **B**: Naturalized in N Devon – Braunton.

8 *Linaria arvensis* (L.) Desf. Like *L. arenaria*, but only stickily-hairy in the inflorescence and the leaves linear. Flowers pale lilac-blue, 4-7mm long, the spur 1.5-3mm long, strongly curved. Waste places and margins of cultivated land, on dry soils. May-August. N & C France and Germany.

9* **Ivy-leaved Toadflax** *Cymbalaria muralis* P.Gaertner, B. Meyer & Scherb. [*Linaria cymbalaria*]. Trailing, often purplish,, tufted, hairless perennial; stems slender. Leaves alternate, kidney-shaped to almost rounded, 5-9-lobed, long-stalked. Flowers lilac to violet with a yellowish patch on the lower lip, 9-15mm long, solitary on long stalks at the base of the leaves; spur 1.5-3mm long. Shady rocks and woods, old walls, generally on calcareous soils, to 2000m. May-September. **I**: S Europe and W Asia. Widely naturalized in Europe, north to C Scandinavia. The flowers are very like those of *Linaria*, but the genus can be distinguished by its broad, lobed, rather than linear leaves. Locally common. **P**: Bees. **B**: Almost throughout.

10* **Sharp-leaved Fluellen** *Kickxia elatine* (L.) Dumort. [*Linaria elatine*]. Low to short, often prostrate, rather slender, hairy or stickily-hairy annual; stems branched from the base. Leaves oval to arrow-shaped, pointed, stalked. Flowers yellowish to bluish with a violet upper lip, 7-10mm long, solitary on long stalks at the base of the leaves; spur straight. Capsule globose. Arable land and other cultivated places, on light soils, generally at low altitudes. July-October. Throughout, except the Faeroes, Iceland, Norway and Spitsbergen; naturalized in Ireland, Denmark and Sweden. The flowers are similar to *Linaria*, but the species of *Kickxia* differ in their solitary flowers and broad leaves. **P**: Probably insects. **B**: Local north to Cumbria and in the Isle of Man; naturalized in S & W Ireland.

11* **Round-leaved Fluellen** *Kichxia spuria* (L.) Dumort. [*Linaria spuria*]. Low to short spreading, hairy or stickily-hairy annual; stems branched from the base. Leaves alternate, oval, the lower slightly toothed, the upper often with a heart-shaped base, short-stalked. Flowers yellow with a deep purple upper lip, 10-15mm long, solitary on slender stalks at the base of the leaves; spur curved. Cultivated fields and other open habitats, on light soils, generally at low altitudes. July-October. S Britain, Belgium, Holland, France and Germany; naturalized or casual further north. Local plant of cultivated land, mainly in C and S Europe. **P**: Bees; flowers are self-sterile but cleistogamous flowers sometimes occur. **B**: Local in England and Wales north to Nottingham, sometimes casual further north.

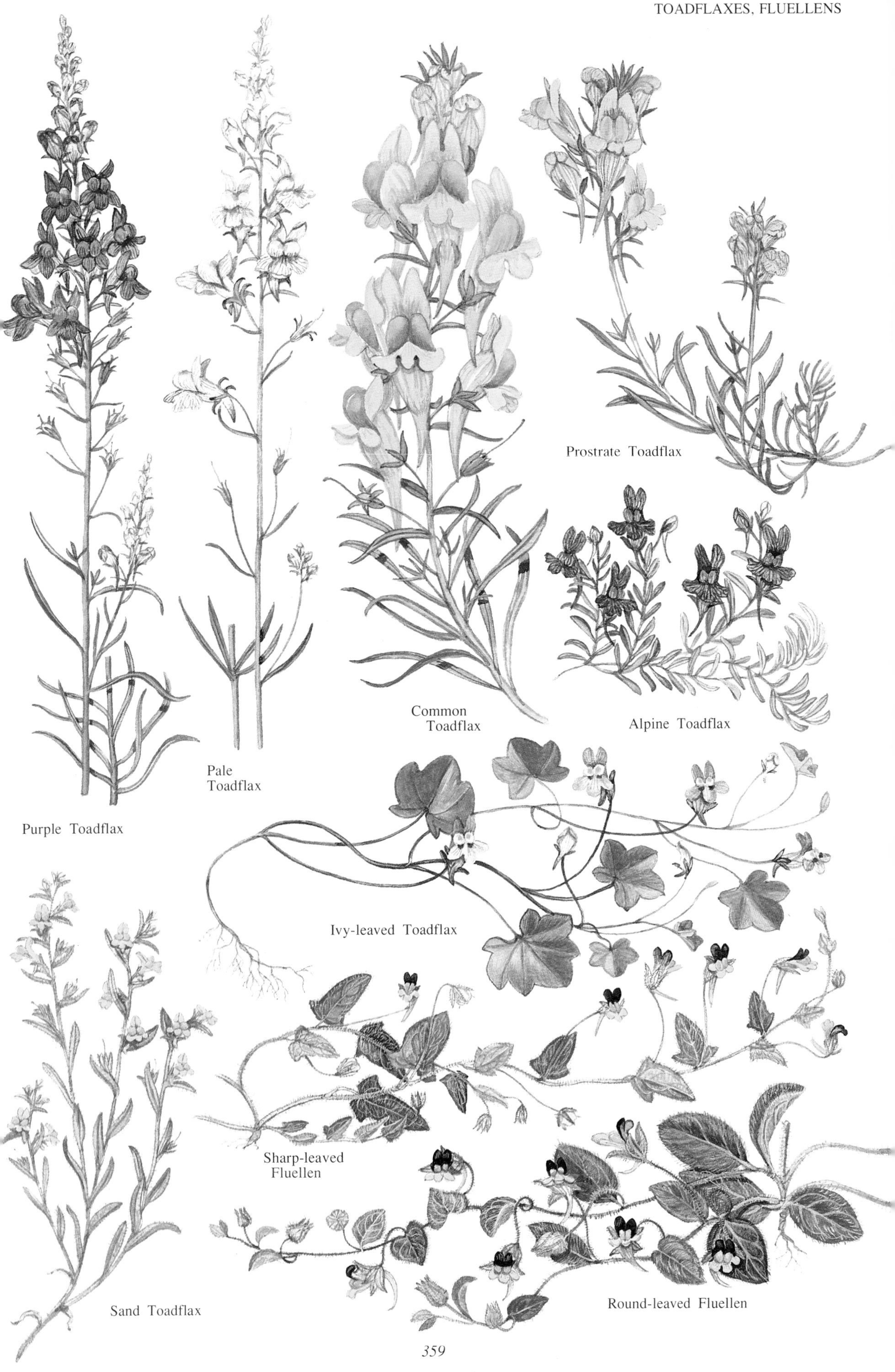

Prostrate Toadflax

Common Toadflax

Alpine Toadflax

Pale Toadflax

Purple Toadflax

Ivy-leaved Toadflax

Sharp-leaved Fluellen

Round-leaved Fluellen

Sand Toadflax

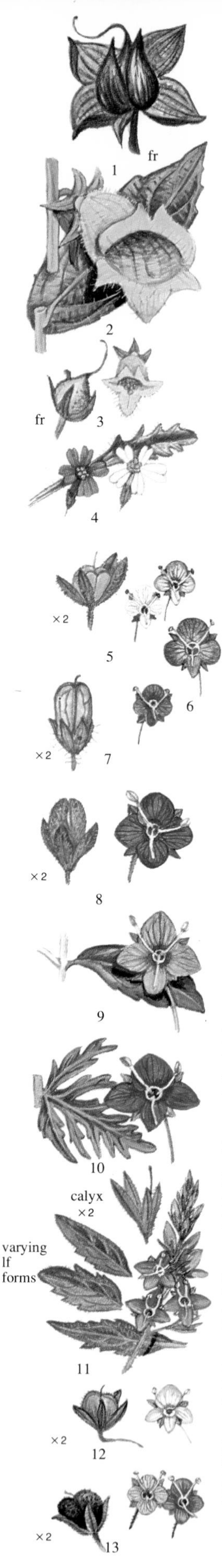


**Foxgloves** *Digitalis*. Biennial or perennial herbs with erect stems and simple, alternate leaves, the lower in a basal rosette. Flowers in long one-sided spikes; bracts present. Calyx 5-lobed. Corolla cylindrical, nodding, often rather inflated, but constricted at the base, 2-lipped. Stamens 4. Capsule conical, containing numerous small seeds. 20 species from Europe to W Asia.

1* **Foxglove** *Digitalis purpurea* L. Medium to tall, hairy, greyish biennial or short-lived perennial, to 1.5m tall; stems generally unbranched, hairless at the base. Basal leaves oval to lanceolate, tapered to the winged stalk, blunt toothed; upper leaves similar but smaller. Flowers purple, patterned with darker spots or rings inside, occasionally white, with a silky sheen, 40-55mm long, hairy on the lower lip inside and along the lip margin. Open woodlands, scrub, mainly on acid soils, to 1000m. June-September. North to Sweden; naturalized in Denmark and Holland. Garden cultivars are sometimes naturalized locally. Very poisonous, with Digitalin, still used in the treatment of heart complaints. **B**: Throughout.

2 **Large Yellow Foxglove** *Digitalis grandiflora* Miller [*D. ambigua*]. Tall, hairy, biennial or perennial. Leaves oval-lanceolate, finely toothed, shiny green above, slightly downy beneath, unstalked. Flowers pale yellow with faint reddish-brown netted markings inside, 40-50mm long, borne in loose spikes. Woods, scrub, rocky slopes and stabilized screes, to 2000m. June-August. Belgium, C & E France and C & S Germany. Poisonous.

3* **Small Yellow Foxglove** *Digitalis lutea* L. Like *D. grandiflora*, but a less hairy plant with narrower leaves, generally broadest above the middle. Flowers pale yellow to whitish, not netted inside, 15-22mm long, in rather dense spikes, the lateral lobes of the corolla recurved. Woodland margins, scrub, rocky habitats and stabilized screes, to 2000m. June-August. Belgium, France and C & S Germany. Poisonous. **B**: Naturalized in England.

4* **Fairy Foxglove** *Erinus alpinus* L. Low, tufted, hairy perennial herb; stems ascending to erect. Leaves mostly basal, oblong, broadest above the middle, toothed, the lower stalked. Flowers bright purplish-pink, rarely white, with 5 almost equal, notched, petal-lobes, borne in small clusters which elongate in fruit. Stamens 4, the lower 3 slightly larger than the upper. Capsule 4-valved. Rocky habitats, stony grassland, old walls, to 2400m. May-October. C & E France and S Germany; naturalized in Britain. **P**: Butterflies, but can be self-pollinated. **B**: Naturalized in scattered localities in N England, Scotland and Ireland.

**Speedwells** *Veronica*. Annual or perennial herbs. Leaves opposite or whorled, sometimes alternate above. Flowers in terminal clusters or racemes, or solitary in the leaf axils. Calyx deeply 5-lobed. Corolla flat or cup-shaped, the 4 lobes often uneven, fused together at the base. Stamens 2 only, protruding. Fruit flattened, heart-shaped or 2-lobed. 200 species.

5* **Thyme-leaved Speedwell** *Veronica serpyllifolia* L. Low creeping perennial, almost hairless. Leaves oval, untoothed or slightly toothed, short-stalked. Flowers white or pale blue with darker purplish-blue veins, 6-8mm, borne in leafy terminal spikes. Capsule heart-shaped, 4-5mm, hairy. Waste and bare places, sparsely grassy habitats and cultivated land, at low altitudes. March-October. Throughout, except Spitsbergen. **P**: Flies. **B**: Throughout.

6 *V. s.* subsp. *humifusa* (Dickson) Syme [*V. humifusa*]. Like the type, but racemes shorter with 8-15 flowers (not 20-40), the flowers larger 7-10mm, bright blue. Mountain habitats to 2500m. May-August. Throughout the range of the species.

7* *Veronica alpina* L. Low to short, hairy or hairless perennial. Leaves oval to elliptical, untoothed or slightly toothed, unstalked, bluish-green. Flowers deep dull blue, 7-8mm, in small terminal clusters, elongating somewhat in fruit. Fruit elliptical with a rounded apex, hairless. Damp mountain rocks, rocky meadows and stony places, 450- 2000m. July-August. N Britain, Scandinavia, the Faeroes, Iceland and the mountains of France and Germany. Often local. **B**: Local in the Scottish Highlands.

8* **Rock Speedwell** *Veronica fruticans* Jacq. [*V. saxatilis*]. Low to short tufted perennial; the stems branched, ascending, finely hairy above, woody below. Leaves narrow oblong to oval, often broadest above the middle, untoothed or slightly toothed, scarcely stalked. Flowers deep blue with a reddish centre, 11-15mm, in small terminal clusters, elongating in fruit. Fruit oval, scarcely notched at the apex. Grassland and rocky habitats, mainly in the mountains, to 3000m. July-September. N Britain, the Faeroes, Iceland, Scandinavia, C & E France and S Germany. Local. **B**: Local, Scottish Highlands.

9 *Veronica fruticulosa* L. Like *V. fruticans*, but flowers pink, 9-12mm. Capsule rounded, notched. Mountain rocks and screes on calcareous soils, to 2800m. C & E France and S Germany.

10* **Large Speedwell** *Veronica austriaca* L. subsp. *teucrium* (L.) D.A. Webb [*V. teucrium*]. Variable medium to tall perennial; stems erect or somewhat sprawling. leaves oval to oblong, often heart-shaped at the base, deeply toothed, unstalked. Flowers bright blue, 10-13mm, in long-stalked, paired racemes towards the stem tops. Capsule heart-shaped, hairy or hairless. Grassy mountain habitats, to 1800m. June-August. C & E France and S Germany. Rather local subspecies. *V. austriaca* is widespread in the mountains of C and S Europe. subsp. *vahlii* (Gaudin) D.A. Webb [*V. teucrium* subsp. *vahlii*]: a shorter plant, not exceeding 30cm, with shorter racemes. Grassy and rocky habitats. May-August. Belgium, Holland, France, Germany.

11 *Veronica prostrata* L. Like *V. austriaca*, but stems to only 25cm tall, the plant with prostrate vegetative shoots. Leaves oblong to oval, densely hairy, short-stalked, toothed or untoothed. Flowers pale blue, 6-8mm. Grassy and rocky mountain habitats, to 1800m. June-August. E France and C & S Germany. subsp. *scheereri* J.P. Brandt, leaves sparsely hairy, lanceolate to linear-oblong. Flowers deep blue, 8-11mm. Similar habitats and flowering time, but generally at lower altitudes. Belgium, Holland, France and Germany.

12* **Nettle-leaved Speedwell** *Veronica urticifolia* Jacq. Short to medium, sparsely hairy perennial; stems erect. Leaves triangular-oval, toothed, unstalked. Flowers lilac, 7mm, in lax, paired racemes. Capsule rounded, notched, hairy on the margin; fruit-stalks bent upwards just below the fruit. Shady habitats, woods, scrub and rocky places, to 2000m. May-July. C & E France and S Germany. Local.

13* **Leafless-stemmed Speedwell** *Veronica aphylla* L. Low stemless, or short-stemmed, matted, hairy perennial. Leaves elliptical-oblong, slightly toothed, short-stalked, in a lax rosette near the stem apex, often at soil level. Flowers deep blue or lilac, 6-8mm, in small clusters of 2-6. Capsule heart-shaped, purplish and hairy. Mountain rocks and stony alpine pastures, generally on limestone, 1200-3000m. July-September. E France and S Germany.

Fairy Foxglove
Thyme-leaved Speedwell
Veronica
alpina
Rock Speedwell
Large Speedwell
Foxglove
Small
Yellow
Foxglove
Nettle-
leaved
Speedwell
Leafless-stemmed
Speedwell

1* **Heath Speedwell** *Veronica officinalis* L. Low to short hairy perennial; stems prostrate to ascending. Leaves oval to elliptical, toothed, softly-hairy, stalked. Flowers lilac-blue, with darker veins, 6-8mm, borne in solitary or paired, dense pyramidal, long-stalked racemes; bracts linear. Capsule oval to slightly heart-shaped, longer than the calyx, hairy. Woods and heaths, grassland, generally on rather dry soils, to 2150m. May-August. Throughout, except Spitsbergen. Variable – some forms may have almost hairless leaves and more heart-shaped capsules. From Europe, excluding much of the Mediterranean region, to W Asia and the Azores. **P**: Various flies and bees. **B**: Common throughout.

2* **Germander Speedwell** *Veronica chamaedrys* L. Low to short, spreading, often sprawling, hairy perennial; stems with two opposite lines of hairs. Leaves oval to oblong-lanceolate, dark green, toothed, unstalked or short-stalked. Flowers bright blue with a white centre, 9-10m, borne in opposite, stalked racemes at the base of the upper leaves; bracts lanceolate. Capsule heart-shaped, with a hairy margin, shorter than the calyx. Grassland, thickets, woodland, hedgerows and banks, stony and waste places, to 2250m. March-July. Throughout, except the far north; naturalized in Iceland. **P**: Primarily bees and flies. **B**: Common throughout.

3* **Wood Speedwell** *Veronica montana* L. Low to short, often sprawling, softly-hairy perennial; stems hairy all round. Leaves oval, rather pale green, coarsely toothed, long-stalked. Flowers pale lilac-blue, 8-10mm, in lax, usually alternate racemes; bracts small, linear. Capsule kidney-shaped, longer than wide, glandular-hairy on the margin, longer than the calyx. Damp woodland, to 1400m. April-July. Britain, Belgium, Holland, France, Germany, Denmark and S Sweden. **P**: various bees and flies. **B**: Local throughout, except the extreme north of Scotland.

4* **Marsh Speedwell** *Veronica scutellata* L. Short to medium, usually hairless perennial; stems creeping below, but eventually ascending. Leaves linear-oblong, often reddish-brown, untoothed or slightly toothed, unstalked. Flowers pale pink or white, or pale blue with darker veins, 5-6mm, borne in alternate, lax, long-stalked racemes. Capsule kidney-shaped, hairless, much longer than the calyx. Marshes, bogs, wet meadows, pond and ditch margins and other wet habitats, generally on acid soils, to 1800m. June-August. Throughout, except the Faeroes and Spitsbergen. Europe, except most of the Mediterranean region, east across much of temperate Asia. **B**: Common throughout much of Britain.

A hairy form occurs throughout much of the range of the species and is generally distinguished as var. *villosa* Schumach

5* **Brooklime** *Veronica beccabunga* L. Short to medium, perennial; stems creeping and rooting at the base, then ascending, thick and fleshy and hairless. Leaves oval to oblong, blunt, with a rounded base, blunt-toothed, short-stalked. Flowers pale to dark blue, rarely pinkish, 5-8mm, in opposite, long-stalked racemes, at the base of the upper leaves. Capsule almost rounded, scarcely notched at the apex, hairless, shorter than the calyx. Wet habitats, river, stream, lake and pond margins, ditches and dykes, sometimes in wet meadows, to 2500m. May-September. Throughout, except Spitsbergen. **P**: Various bees and flies, but may often be self-pollinated. **B**: Common throughout.

6 *Veronica anagalloides* Guss. Like *V. beccabunga*, but an annual with erect stems and linear-lanceolate leaves, only the uppermost unstalked. Flowers whitish to lilac, 3-5mm. Capsule elliptical, not notched. Similar habitats. June-September. E France and C & S Germany.

7* **Blue Water Speedwell** *Veronica anagallis-aquatica* L. Medium perennial hairless or sometimes glandular-hairy in the inflorescence; stems erect, unbranched or much-branched. Leaves pale green, the lower oval, stalked, scarcely toothed, the upper lanceolate, half-clasping the stem, toothed towards the apex. Flowers blue with violet lines, 5-10mm, in slender. long-stalked racemes. Capsules rounded to elliptical, slightly notched, hairless. Wet habitats, river, stream, lake and pond margins, wet meadows and muddy habitats, to 1450m. June-August. Throughout, except the extreme north and the Faeroes. **P**: Flies, often self-pollinated. Sometime cultivated in gardens. **B**: Common throughout much of Britain.

8 **Pink Water Speedwell** *Veronica catenata* Pennell [*V. aquatica*]. Like *V. anagallis-aquatica*, but leaves dark green, linear to linear-lanceolate, all unstalked. Flowers pink with reddish veins, 3-5mm. In similar habitats to the previous species, though most frequently close to, or growing in, slow-moving water. June-August. Throughout, north to S Sweden. **B**: Throughout.

Hybridises with *V. anagallis-aquatica* where the two grow in close proximity; hybrids, *V. × lackschewitzii* J.Keller, are generally vigorous and sterile, with no capsules developing.

9 **French Speedwell** *Veronica acinifolia* L. Low to short, glandular-hairy, erect annual, often branched from the base. Leaves oval, untoothed or slightly toothed, short-stalked. Flowers blue, small, 2-3mm, in terminal racemes. Capsule kidney-shaped, almost 2-lobed, glandular-hairy. Damp habitats, grassland and cultivated ground. April-July. NC France and Germany, extinct in Belgium; casual or perhaps locally naturalized in England. **B**: Local in England, mostly casual.

10 **Breckland Speedwell** *Veronica praecox* All. Like *V. acinfolia*, but leaves more deeply toothed. Flowers in slender racemes occupying most of the stem, blue, 3mm. Capsule broader than long. Cultivated fields, waste and sparsely grassy places. March-June. Continental Europe north to S Sweden – Gotland. **B**: Very local in W Norfolk and Suffolk.

11* **Fingered Speedwell** *Veronica triphyllos* L. Low to short, glandular-hairy annual; stems more or less erect, branched or unbranched. Leaves digitately-lobed, with 3-7 oblong, sometimes toothed, lobes; lower bracts lobed like the leaves. Flowers deep blue, 3-4mm, in lax terminal racemes. Capsule kidney-shaped, glandular-hairy. Cultivated land, especially arable fields, waste places, generally on light sandy soils at low altitudes. April-July. Throughout, north to S Sweden. **P**: Small bees, often self-pollinated. **B**: Local and rare in East Anglia, sometimes casual elsewhere.

12* **Wall Speedwell** *Veronica arvensis* L. Low to short, hairy annual; stems unbranched to much-branched, erect to prostrate, sometimes glandular. Leaves triangular-oval, coarsely toothed, the lower short-stalked, the upper unstalked. Flowers blue, 2-3mm, in terminal racemes; bracts longer than the flower-stalks. Capsule heart-shaped, with a few hairs around the margin. Dry open habitats, grassy banks, heaths and old walls, to 2100m. March-October. Throughout, except the Faeroes and Spitsbergen; naturalized in Iceland. A common weed of cultivated land, including gardens. **P**: Small bees. **B**: Common throughout.

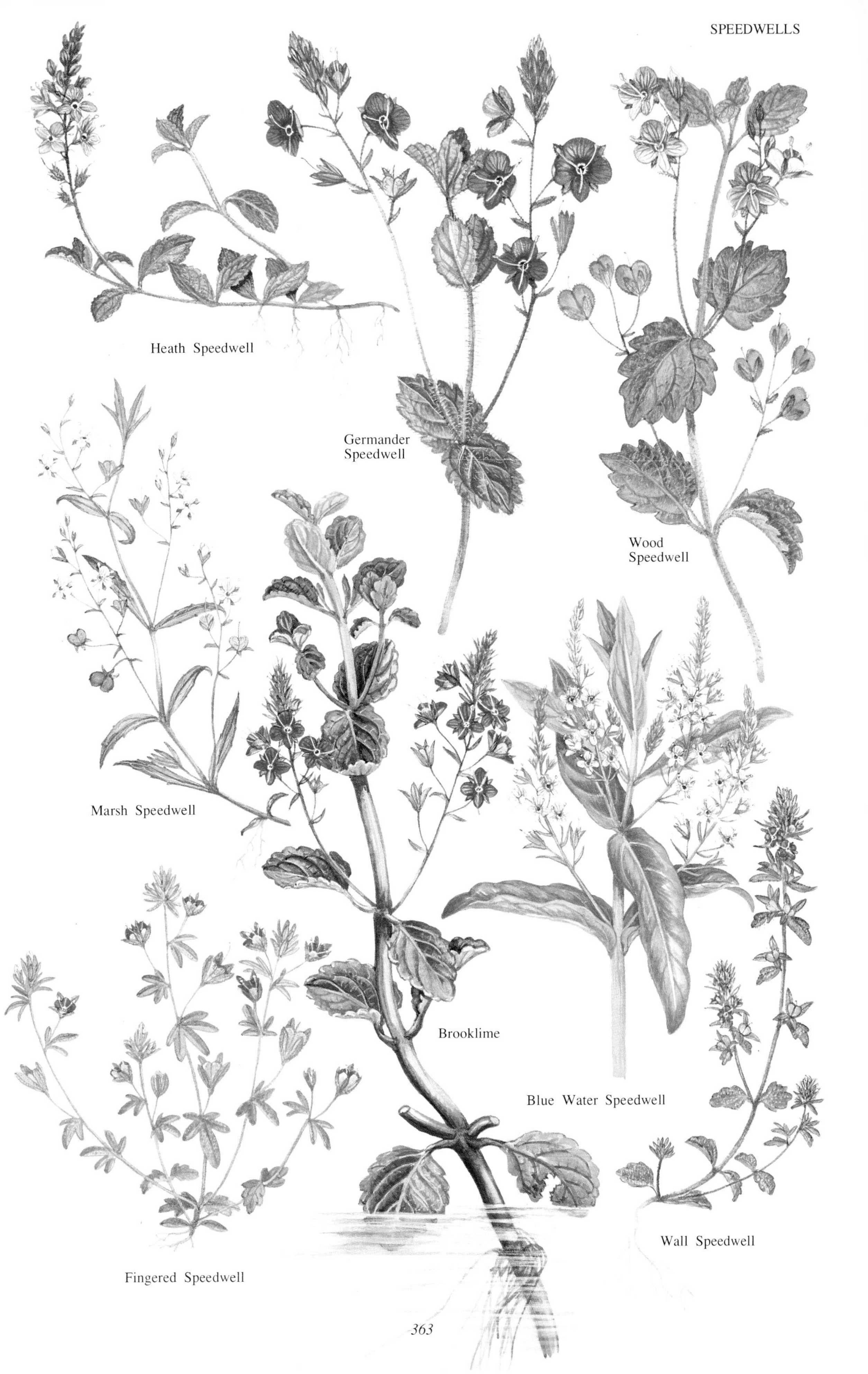
Heath Speedwell
Germander
Speedwell
Wood
Speedwell
Marsh Speedwell
Brooklime
Blue Water Speedwell
Wall Speedwell
Fingered Speedwell

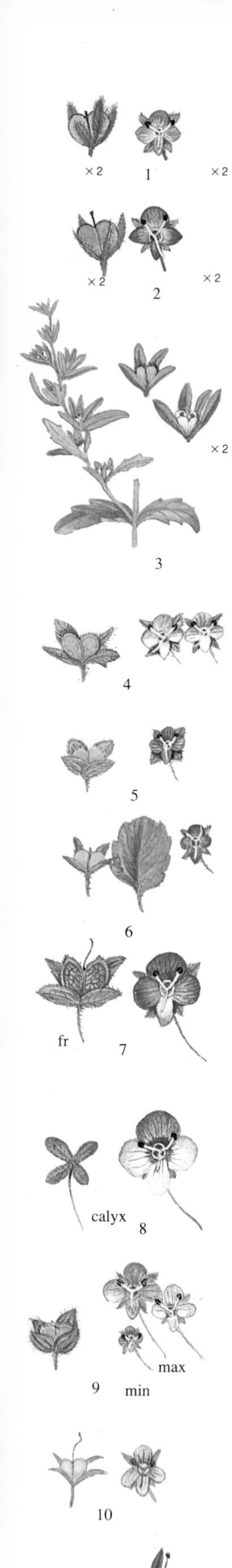


1* **Spring Speedwell** *Veronica verna* L. [*V. brevistyla*]. Low to short annual, sometimes only 3cm tall; stems erect, generally branched from the base, hairy below, glandular-hairy above. Leaves oval to lanceolate, coarsely toothed, the lower short-stalked, the upper unstalked and pinnately-lobed. Flowers blue, 3mm, in terminal, rather dense, glandular-hairy racemes. Capsule kidney-shaped, broader than long, glandular-hairy. Dry habitats, cultivated fields and grassland, to 2100m. May-July. Europe except the extreme north. Europe to W and C Asia and Morocco. Sometimes confused with *V. arvensis*, but easily separated by the deeply lobed upper leaves. **B**: Local; East Anglia – Breckland.

2 *Veronica dillenii* Crantz. Like *V. verna*, but taller and more robust and with rather fleshy leaves. Flowers deep blue, 4-5mm. Style on fruit 1.5mm (not 0.5mm). Dry habitats, grassy places and waste places, at low altitudes. April-July. France and Germany. Sometimes included within *V. verna.*

3 **American Speedwell** *Veronica peregrina* L. Like *V. arvensis*, but hairless and flowers whitish, 2-3mm. Capsule oval, slightly notched. Cultivated ground, damp waste places and streamsides. April-July. **I**: South America. Naturalized or casual in Europe north to S Scandinavia. **P**: Self-pollinated. A thoroughly naturalized weed in many parts of Europe. **B**: Locally naturalized, particularly in NW Ireland and parts of England.

4* **Green Field Speedwell** *Veronica agrestis* L. Low spreading, hairy annual. Leaves oval, pale green, mostly alternate except for the lowermost, toothed, short-stalked. Flowers whitish with a pink or blue upper petal, 3-6mm, solitary, borne on long slender stalks which are usually recurved in fruit. Capsule 2-lobed with parallel lobes, sparsely glandular-hairy. Cultivated ground, on a variety of soils, though often acid, to 1800m. March-November. Throughout, except the extreme north, the Faeroes and Iceland. Throughout much of Europe, though scarce in the Mediterranean region, extending to W Asia and N Africa. **P**: Various bees and flies, often self-pollinated. **B**: Throughout, though scarcer in the south of Britain.

5* **Grey Field-speedwell** *Veronica polita* Fries. Like *V. agrestis*, but leaves more evenly toothed and flowers uniformly bright blue. Capsule with broad overlapping calyx-lobes, borne on long glandular-hairy stalks. Cultivated ground. Flowering throughout the year. Throughout, except the Arctic. An often abundant weed of cultivated land. **P**: Usually self-pollinated. **B**: Throughout, but commonest in S England, local elsewhere.

6 *Veronica opaca* Fries. Like *V. agrestis*, but leaves very shallowly toothed and flowers deep blue. Capsule like *V. polita*, but calyx-lobes narrow and not overlapping at the base. Cultivated ground and waste places, often on calcareous soils. Continental Europe north to C Scandinavia.

7* **Common Field Speedwell** *Veronica persica* Poiret. [*V. buxbaumii*]. Low to short, generally sprawling hairy annual; stems branched. Leaves oval-triangular, mostly alternate except for the lowermost, pale green, coarsely toothed, short-stalked. Flowers bright blue, 8-12mm, the lowermost petal usually white, solitary on slender stalks at the base of the upper leaves. Capsule 2-lobed, borne on recurved stalks, the lobes diverging, keeled. Cultivated ground on a variety of soil types, locally abundant. Flowering throughout the year. **I**: SW Asia. Naturalized throughout, except the extreme north, the Faeroes and Iceland. The commonest species of *Veronica* on cultivated land, recorded in Europe since early in the nineteenth century. **P**: Various insects but usually self-pollinated. **B**: Common throughout.

8* **Slender Speedwell** *Veronica filiformis* Sm. Low creeping, mat-forming hairy perennial; stems slender, freely rooting at the nodes. Leaves kidney-shaped, coarsely toothed, stalked, mostly opposite on non-flowering stems, but alternate on the flowering stems. Flowers pale lilac-blue, 10-15mm, the lowermost petal often paler, solitary on long slender stalks from the base of the upper leaves. Capsule 2-lobed, the lobes almost parallel, slightly hairy, rarely produced. Damp grassland, river and stream margins, road-verges and lawns, often abundant. April-July. **I**: Turkey and the Caucasus. Naturalized in Europe north to Denmark and N Germany. An often troublesome weed of lawns and gardens in particular, forming extensive mats. Originally introduced to gardens as an ornamental subject. Plants are self-incompatible and rarely set seed. **B**: Throughout, particularly in the south.

9* **Ivy-leaved Speedwell** *Veronica hederifolia* L. Low sprawling, hairy annual; stems branched at the base. Leaves kidney-shaped, with 3-7 lobes, the end lobe the largest, pale green, 3-veined, all but the lowermost alternate, stalked. Flowers pale lilac or pale blue, 4-9mm, solitary, on stalks shorter than the leaves; sepals heart-shaped at the base. Capsule rounded, very shallowly notched, hairless. Cultivated land on a variety of soils, often abundant. March-August. Throughout, except the extreme north, the Faeroes and Iceland, probably introduced in much of N Europe. A common weed distinguished from the previous weedy species by its lobed rather than toothed leaves. **P**: Various insects, often self-pollinated. **B**: Common throughout.

subsp. *lucorum* (Klett & Richter) Hartl. Like the type, but the end lobe of the leaves longer than wide (not wider than long) and style on the capsule only 0.5mm long (not 0.7-1mm). Similar habitats as well as woods and hedgerows. April-July. Much of N and NW Europe including Britain and S Finland.

10* **Long-leaved Speedwell** *Veronica longifolia* L. [*Pseudolysimachium longifolium*, *Veronica maritima*]. Robust, medium to tall tufted, hairless or finely hairy perennial; stems erect, to 1.2m. Leaves opposite or in whorls of 3-4, lanceolate to linear-lanceolate, sharply toothed, short-stalked. Flowers lilac or pale blue, 6-8mm, in dense terminal spike-like racemes, often with 1-2 short branches below. Capsule heart-shaped, hairless. Moist habitats, river and stream banks and other wet habitats, to 1250m. June-July. Continental Europe north to C Scandinavia; occasionally naturalized in Britain. Cultivated in gardens. Easily recognised by its tall erect stems and slender pointed flower spikes.

11* **Spiked Speedwell** *Veronica spicata* L. [*Pseudolysimachium spicatum*]. Low to medium, tufted, often mat-forming, hairy perennial. Leaves opposite, linear-lanceolate to oval, blunt-toothed, mostly unstalked. Flowers violet-blue, 4-8mm, with a rather long tube, borne in dense terminal, spike-like racemes. Capsule rounded, notched, hairless. Dry grassy habitats, woodland margins and rocky slopes, on basic or calcareous rocks, to 2050m. July-October. Britain, France, Germany, Scandinavia; naturalized in Holland. Widely cultivated in gardens. **P**: Various insects. Hybridises occasionally with *V. longifolia*, especially in the Baltic region. **B**: Scattered localities in Avon, Wales, W Yorkshire and Cumbria; rare in East Anglia.

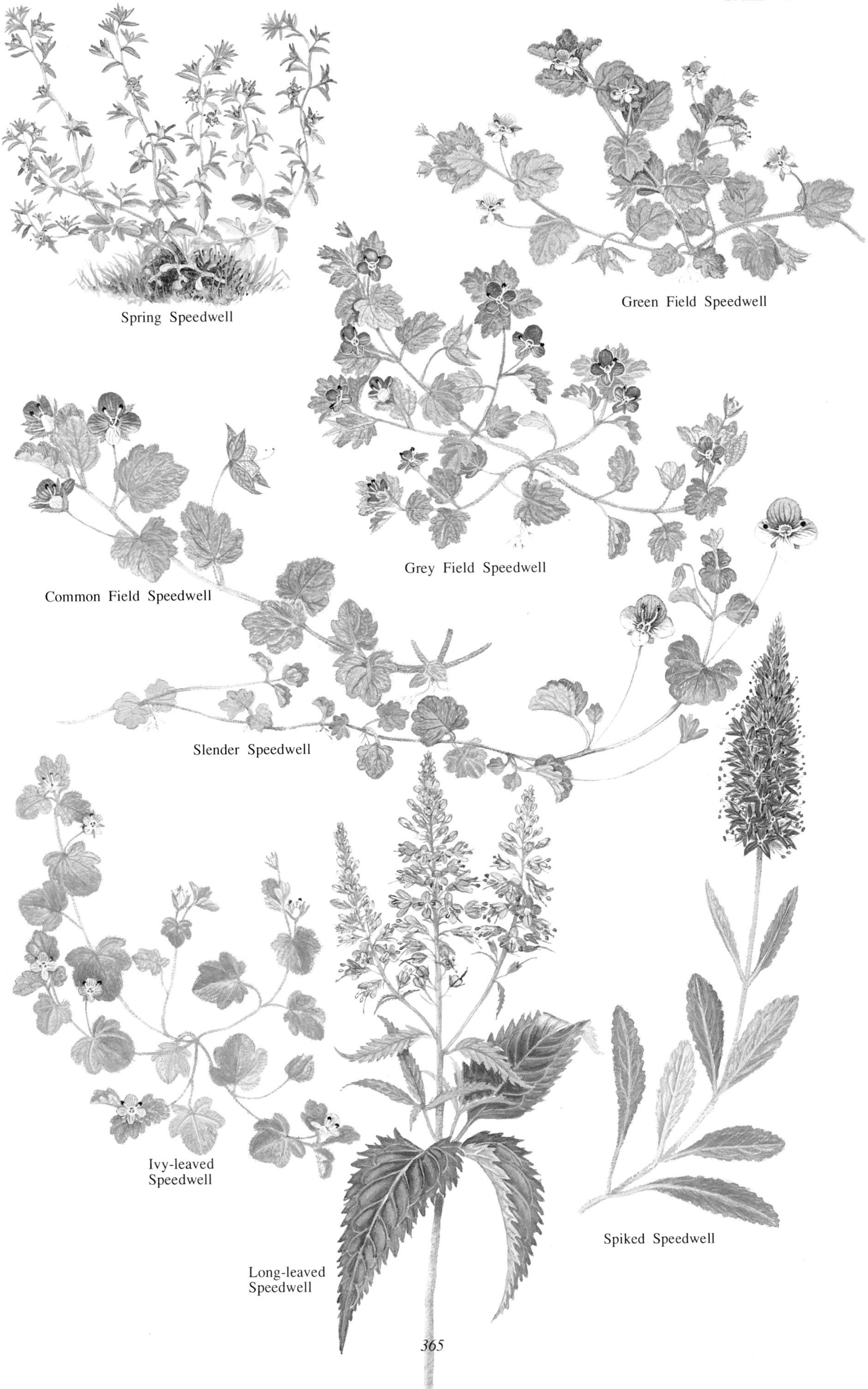

Spring Speedwell

Green Field Speedwell

Grey Field Speedwell

Common Field Speedwell

Slender Speedwell

Ivy-leaved Speedwell

Spiked Speedwell

Long-leaved Speedwell

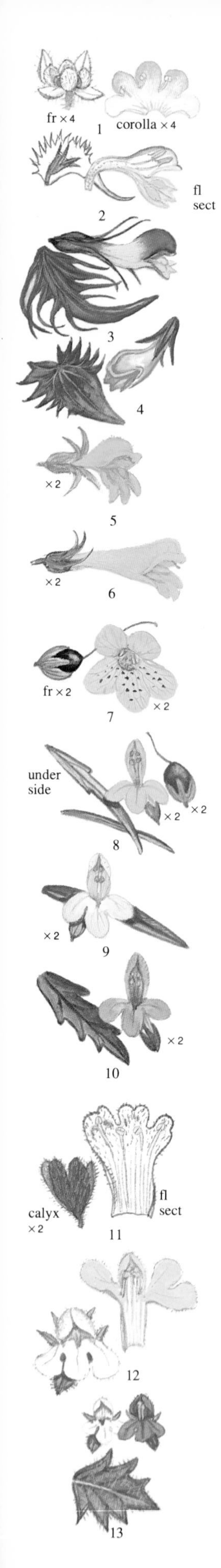


1* **Cornish Moneywort** *Sibthorpia europaea* L. Prostrate, far-creeping, slightly hairy perennial; stems thread-like slightly branched, rooting at the nodes. Leaves alternate, kidney-shaped to rounded, blunt-toothed, long-stalked, rather pale green. Flowers white or cream, often flushed with pink, especially on the lower lobes, small, 1.5-2.5mm, solitary, borne on slender stalks at the base of the leaves; corolla with 5 almost equal lobes. Capsule splitting into two. Damp shaded habitats, very local, to 500m. July-October. Britain and W France. **B**: Very local in S England, S Wales, the Channel Islands, Lewis and Kerry.

2* **Crested Cow-wheat** *Melampyrum cristatum* L. Short to medium, finely hairy, erect annual. Leaves linear-lanceolate, toothed or untoothed, unstalked; bracts narrow heart-shaped, folded along the middle, pointed, purplish-red and toothed in the lower half. Flowers pale yellow variegated with purple, 12-16mm long, in a dense 4-sided spike, throat of corolla closed. Capsule opening along only one margin. Dry grassy, often rocky habitats, woodland margins. June-September. W Europe. **B**: Very local, England north to Lincolnshire.

3* **Field Cow-wheat** *Melampyrum arvense* L. Short to medium, erect, hairy annual with spreading branches. Leaves lanceolate, untoothed or toothed at the base, unstalked; bracts oval-lanceolate, erect, not folded, sharply lobed in the lower half, green, whitish or pinkish-red. Flowers purplish-pink, with a yellow throat, 20-25mm long, the lower lip with an upturned margin, borne in cylindrical spikes, not 4-sided; corolla with a closed throat. Dry grassy and rocky habitats, to 1500m. June-September. W Europe. **B**: Rare, scattered localities in England north to Lincolnshire, including the Isle of Wight.

4 *Melampyrum nemorosum* L. Like *M. arvense*, but leaves lanceolate to oval-heart-shaped and bracts usually violet-blue, especially near the apex, occasionally greenish or whitish. Flowers bright yellow, 15-20mm; calyx hairy. Germany, Denmark, Finland and Sweden.

5* **Small Cow-wheat** *Melampyrum sylvaticum* L. Variable, short, rather slender, usually hairy annual with spreading to almost erect branches. Leaves lanceolate to elliptical, up to 12mm wide, untoothed; bracts lanceolate, green, untoothed or with 1-2 teeth at the base. Flowers golden yellow, rarely whitish, 8-10mm long, the lower lip often speckled with purple, borne in a lax, one-sided spike; corolla with an open throat. Capsule splitting along 2 margins. Open mountain woods, woodland clearings and scrub, 400-2500m. June-August. W Europe, Scandinavia; only on mountains in the south. **B**: Very local, N England, Scotland and Antrim and Derry in Ireland.

6* **Common Cow-wheat** *Melampyrum pratense* L. Very variable low to medium, hairless or slightly bristly annual, with spreading or almost erect branches. Leaves linear to oval, up to 35mm wide, untoothed and generally unstalked; bracts green, oval to linear-lanceolate, toothed at the base or untoothed. Flowers bright yellow to whitish, 10-18mm long, the upper lip often tinged with red or purple, in lax, one-sided spikes; throat of corolla usually closed. Capsule splitting along one margin only. Open woods, scrub and heaths, to 2250m. May-October. Not in the extreme north. Very variable. **B**: Widespread.

7* **Tozzia** *Tozzia alpina* L. Short to medium hairy, semi-parasitic perennial. Leaves opposite, oval, rather fleshy, coarsely toothed, unstalked. Flowers golden-yellow, 6-10mm long, purple-spotted in the throat, the upper lip erect, borne in leafy racemes; corolla funnel-shaped, 5-lobed, slightly 2-lipped; stamens 4. Capsule usually splitting. Damp habitats, generally among coarse herbs, meadows and streamsides, on calcareous soils, to 2250m. June-July. Mountains of C & E France and S Germany. Semiparasitic, generally locally common. **P**: Mainly bees.

8* **Yellow Odontites** *Odontites lutea* (L.) Clairv. [*Orthantha lutea*]. An erect, short to medium annual with numerous ascending branches, hairy. Leaves linear, blunt, untoothed or very slightly toothed, with inrolled margins; bracts similar to the leaves but somewhat smaller. Flowers bright yellow, 5-8mm long, hairy, the lower lip deflexed; anthers protruding. Capsule hairy. Dry grassland and scrub, often on calcareous soils, to 1800m. July-September. C & E France and S Germany.

9 *Odontites jaubertiana* (Boreau) D. Dietr. ex Walpers. Like *O. lutea*, but leaves linear to linear-lanceolate, untoothed. Flowers cream or yellow, sometimes tinged with red, the anthers not protruding, included under the upper lip of the corolla. Similar habitats and flowering time. W & C France. subsp. *chrysantha* (Boreau) P. Fourn. [*Odontites chrysantha*]: leaves generally with 1-2 pairs of teeth and flowers always bright yellow. Pastures and woodland margins. August-October. NC France.

10* **Red Bartsia** *Odontites verna* (Bellardi) Dumort [*O. rubra*]. Very variable, short, hairy, often purplish annual; stems erect to ascending, slightly square, generally branched below. Leaves lanceolate, usually toothed; bracts similar to the leaves, generally longer than the flowers. Flowers reddish-pink, 8-10mm long, hairy, the lower lip somewhat deflexed; anthers often slightly protruding. Capsule hairy. Meadows, pastures, scrub, roadsides, pathways, field boundaries, waste places, on a variety of soils, to 1800m. July-October. Throughout, except the far north. subsp. *pumila* (Nordst.) A. Pedersen, a dwarf form only in coastal grassland in N. Scotland and the Outer Hebrides. **B**: N Britain and Ireland, local or rare in the south. subsp. *litoralis* (Fries) Nyman [*Odontites litoralis*]: stems usually unbranched and leaves widest at or close to the base. Coastal meadows. The Baltic coast, S Norway, Denmark and Holland. subsp. *serotina* (Dumort.) Corb. [*Odontites serotina*]: taller, to 50cm, with long spreading branches; bracts shorter than the flowers. Not in the far north. **B**: Mainly in S Britain, rare in Scotland.

11* **Alpine Bartsia** *Bartsia alpina* L. Short, glandular-hairy perennial; stems erect, simple. Leaves oval tootthed, opposite, unstalked; bracts similar to the leaves, often purplish. Flowers dull dark purple, 15-20mm long, in rather dense spikes, the upper lip of the corolla slightly longer than the lower lip; anthers hairy. Capsule longer than the calyx. Damp mountain habitats, on basic soils, to 2700m. June-August. N Europe, France and Germany. Rather local. **B**: Very rare, a few places in N England and C & S Scotland.

12* **Yellow Bartsia** *Parentucellia viscosa* (L.) Caruel [*Bartsia viscosa*]. Short to medium, glandular-hairy annual; stems erect, generally unbranched. Leaves opposite, oblong to lanceolate, pointed, coarsely toothed, unstalked; bracts similar to the leaves, decreasing in size up the stem. Flowers yellow, occasionally white, 16-24mm long, in spikes, 2-lipped, open-mouthed, the lower lip 3-lobed, the upper hooded. Capsule hairy. Damp rough grassland, often close to the coast. June-September. Britain and France; naturalized in Belgium, Holland and Denmark. **B**: Mainly in SW England, S Wales, SW Ireland; scattered localities elsewhere.

13 *Parentucellia latifolia* (L.) Caruel [*Bartsia latifolia*]. Like *P. viscosa* but a shorter plant, 5-30cm tall with triangular-lanceolate, deeply toothed leaves. Flowers reddish-purple, occasionally white, smaller, 8-10mm long, the corolla persistent, not falling as the fruits develop. Sandy and stony habitats. June-September. NW France southwards. Mainly S Europe.

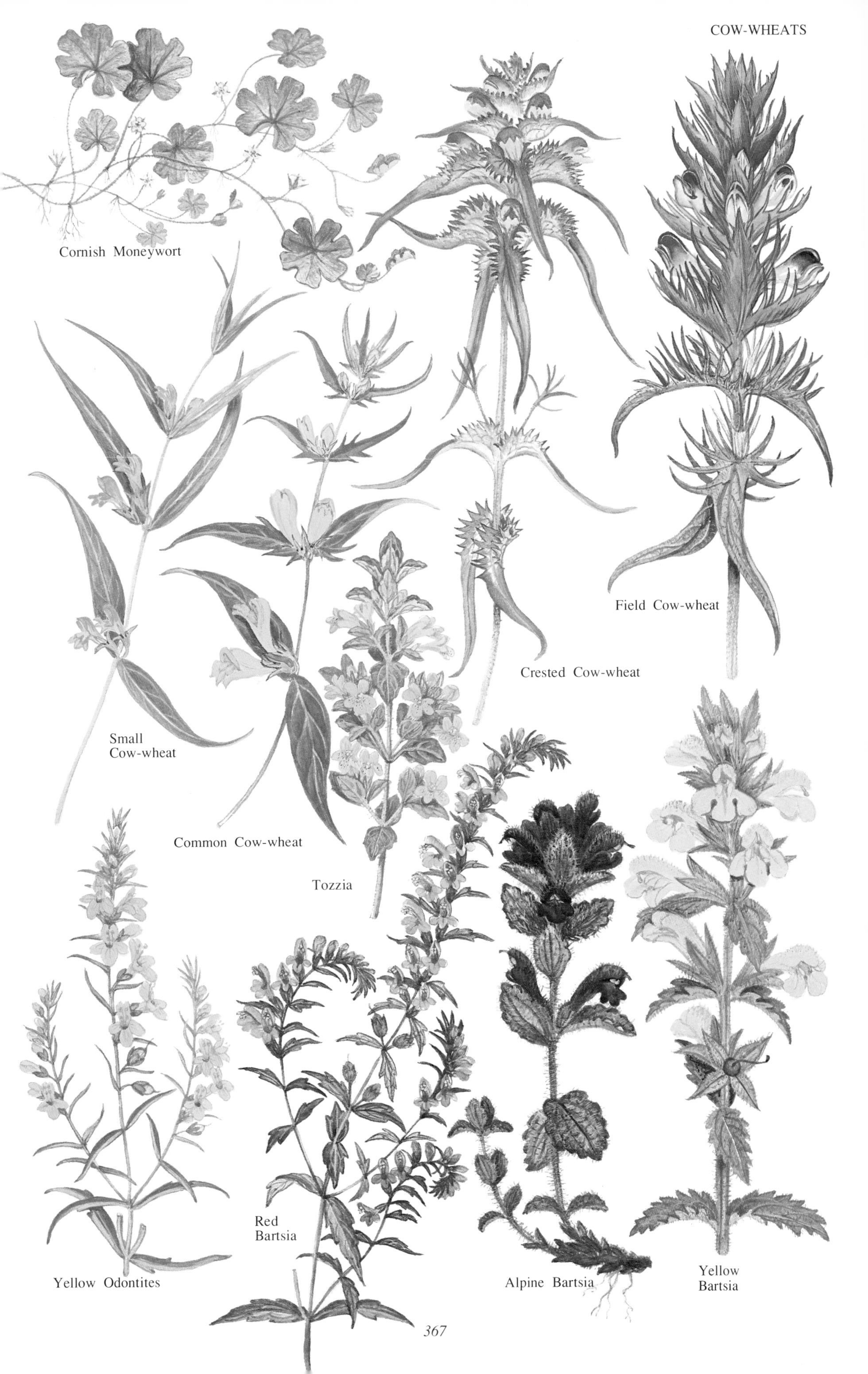
Cornish Moneywort
Field Cow-wheat
Crested Cow-wheat
Small
Cow-wheat
Common Cow-wheat
Tozzia
Red
Bartsia
Yellow Odontites
Alpine Bartsia
Yellow
Bartsia

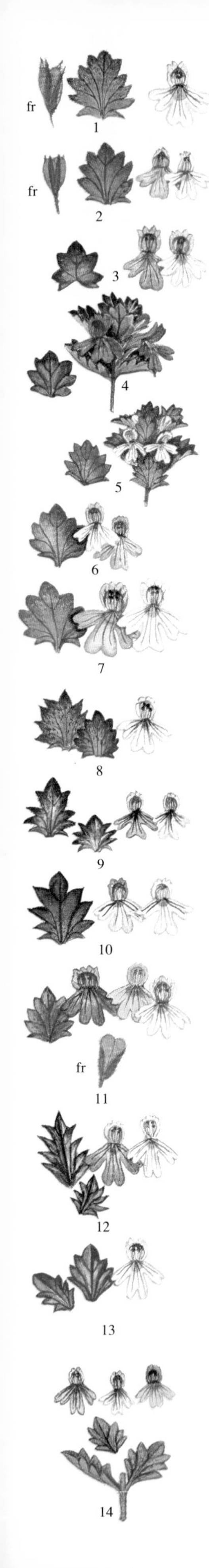

**Eyebrights** *Euphrasia*. Semiparasitic annual herbs with opposite toothed leaves; bracts large and leaf-like. Flowers borne in spikes. Calyx tubular to bell-shaped, 4-lobed. Corolla 2-lipped, white, purplish, pink or yellowish, the lower lip flat, 3-lobed. Stamens 4. Fruit capsule splitting lengthwise. Over 150 species in temperate regions of both hemispheres.

1* **Common Eyebright** *Euphrasia rostkoviana* Hayne. Short hairy, often red-tinged plant; stems branched, the main internodes much longer than the leaves. Leaves pale to dark green, oval to oblong, toothed; lower bracts triangular, to oval. Flowers white, yellow throated, the upper lip often lilac, 8-12mm long. Capsule not exceeding the calyx. Moist meadows and pastures, open woods and rocky places, generally lowland. July-October. Throughout, except Iceland and Spitsbergen. Local. **B**: Wales and neighbouring counties, N England and Ireland. subsp. *montana* (Jordan) Wettst. [*E. montana*]: has stems with up to 3 pairs (not up to 5) of branches and the lower bracts oval to rounded. Capsule often longer than the calyx. Meadows and pastures and other grassy habitats in the mountains, to 3000m. Throughout the range of the species. Plants are usually branched only in the upper half; often from the base in the type. **B**: Local, Shropshire and Brecon, N England and S Scotland. subsp. *campestris* (Jordan) P. Fourn.: 3-8 pairs of branches and bracts oval to trullate. Dry grassy habitats. Belgium and Germany.

2* *Euphrasia anglica* Pugsley. Like *E. rostkoviana*, but stems usually flexuous or curved, the internodes generally shorter than the leaves. Flowers white or pale lilac, smaller, 6.5-8mm long. Grassy pastures and heaths, generally on wet or rather heavy soils. May-September. **B**: England north to Lancashire, C & SE Ireland; locally common.

3 *Euphrasia rivularis* Pugsley. Slender low plant, unbranched or with up to 2 pairs of branches, some of the internodes longer than the leaves. Leaves elliptical to rounded, small, to 7mm long, often with a few glandular hairs, dark green, sometimes purple tinged; bracts broadly oval, sharply toothed. Flowers white with a lilac upper lip, or all lilac, 6-9mm long. Capsule elliptical, equalling the calyx lobes, hairy on the edge. Damp mountain pastures. May-July. N Wales and Cumbria, not known elsewhere.

4 *Euphrasia vigursii* Davey. Stems and branches erect and dark purplish. Leaves dull grey-green, often flushed with dull violet or blackish. Flowers purple to deep reddish-violet. Heaths. June-September. Told by its deep, generally purple, flowers. **B**: Only in Cornwall and S Devon, local.

5* *Euphrasia hirtella* Jordan ex Reuter. Like *E. rivularis*, but plants taller, to 20cm, unbranched or with 1-2 pairs of erect short branches. Flowers white, sometimes tinged with lilac on the upper lip, 5.5-7mm long. Grassland, mainly in mountain habitats, often on slightly acid soils, to 2300m. June-September. France and Germany.

6* **Eastern Eyebright** *Euphrasia picta* Wimmer. Short erect or flexuous plant with 0-3 pairs of erect branches. Leaves oval to oblong, the lower often short-stalked, hairy above; bracts like the leaves but often wider and more triangular. Flowers lilac or white with a lilac upper lip, strongly veined with violet, 7.5-10mm long. Capsule shorter than the calyx, with a hairy edge. Grassy places, 1500-2500m. July-September. Confined to the Alps. The lowest flowers are borne from the 2-6 stem node.

7* **Arctic Eyebright** *Euphrasia arctica* Lange ex Rostrup. Very variable, generally pale green, low to short plant, glandular-hairy in the upper half, with a few long suberect branches. Leaves oval to lanceolate, blunt-toothed; bracts like the upper leaves, oval, often slightly heart-shaped at the base. Flowers bluish-lilac to white, 6-11mm long. Capsule oblong, sometimes slightly notched at the apex, equalling the calyx. Meadows, pastures, roadsides and other grassy habitats. July-September. N Britain, Ireland, the Faeroes, Iceland and Scandinavia. Often rather local. **B**: Restricted to Orkney and Shetland. subsp. *borealis* (Townsend) Yeo [*E. borealis*, *E. rostkoviana* f. *borealis*]: capsule smaller, 4-5.5mm (not 5.5mm or more). Similar habitats and flowering time. Britain and W Norway. Very local plant. **B**: Throughout, but commonest in Scotland; scarce elsewhere, including Ireland.

8* *Euphrasia tetraquetra* (Bréb.) Arrondeau. Like *E. arctica*, but leaves rather fleshy, with shorter glandular hairs and with smaller flowers, 5-7mm. Short grassland in coastal habitats, coastal sands, rare inland. July-September. Britain and NW France. Commonly hybridises with *E. confusa*, where they grow near one another. **B**: Primarily in S Britain and S Ireland.

9* **Wind Eyebright** *Euphrasia nemorosa* (Pers.) Wallr. Short, erect, often purplish plant, with 1-9 pairs of long, slender, ascending branches. Leaves elliptical to oblong-triangular, dark green or purplish-green, often hairy, sharply toothed, with deeply impressed veins, the upper leaves generally larger; bracts oval, pointed, smaller than the uppermost leaves. Flowers white, sometimes with a bluish upper lip, sometimes lilac, 5-7.5mm long. Capsule oblong, slightly shorter than the calyx. Grassy habitats, woods, heaths, downs and pastures. July-September. Restricted to north and central Europe, not in the far north. Hybridises with *E. confusa* and *E. stricta*. **B**: Throughout, more local in Scotland and Ireland.

10 *Euphrasia pseudokerneri* Pugsley. Like *E. nemorosa*, but often shorter and leaves scarcely hairy; flowers 7-9mm long, sometimes more. Capsule much shorter than the calyx. Calcareous grassland. July-September. **B**: North to Lincolnshire, and in Ireland.

11* *Euphrasia confusa* Pugsley. Low to short plant with 2-8 pairs of long, slender spreading or ascending branches, generally profusely branched. Leaves oval to lanceolate, bluntly toothed, hairless or with a few hairs; lower bracts alternate, broader than the leaves. Flowers white, purple or yellow, 5-9mm, only at alternate nodes. Capsule broad-oblong, usually notched, longer than the calyx. Heaths, grassy cliffs, sand-dunes, July-September. Britain, the Faeroes and W Norway. Local. Mountain forms are generally less branched and sometimes only 2-3cm tall. **B**: Local in Britain; in Ireland, restricted to Wicklow.

12* **Glossy Eyebright** *Euphrasia stricta* D. Wolff ex J. F. Lehm. [*E. brevipila*, *E. condensata*]. Short, rather stiff plant, often strongly purple tinged, with 2-6 pairs of branches. Leaves glossy, narrowly oval to lanceolate, sharply toothed, hairless. Flowers white or lilac, veined with blue and with a yellow throat, 7.5-10mm long. Capsule oblong, sometimes slightly notched. Meadows, dry grassland and scrub, to 2600m. June-September. Throughout, except the extreme north; naturalized in Britain.

13 *Euphrasia hyperborea* Joerg. Like *E. stricta*, but often shorter and with a flexuous stem, with only 1-2 pairs of branches. Damp meadows and birch woodland. July-September. Arctic Norway, extending into W Russia. **B**: Locally naturalized in Scotland.

14* **Dwarf Eyebright** *Euphrasia minima* Jacq. ex DC. Like *E. stricta*, but plant not purple-tinged and stem leaves distinctly stalked, blunt-toothed, generally with some hairs. Flowers white with a lilac upper lip or all lilac, 4-6mm. Mountain habitats, grassy, often stony places, to 3250m. July-September. C & E France and S Germany.

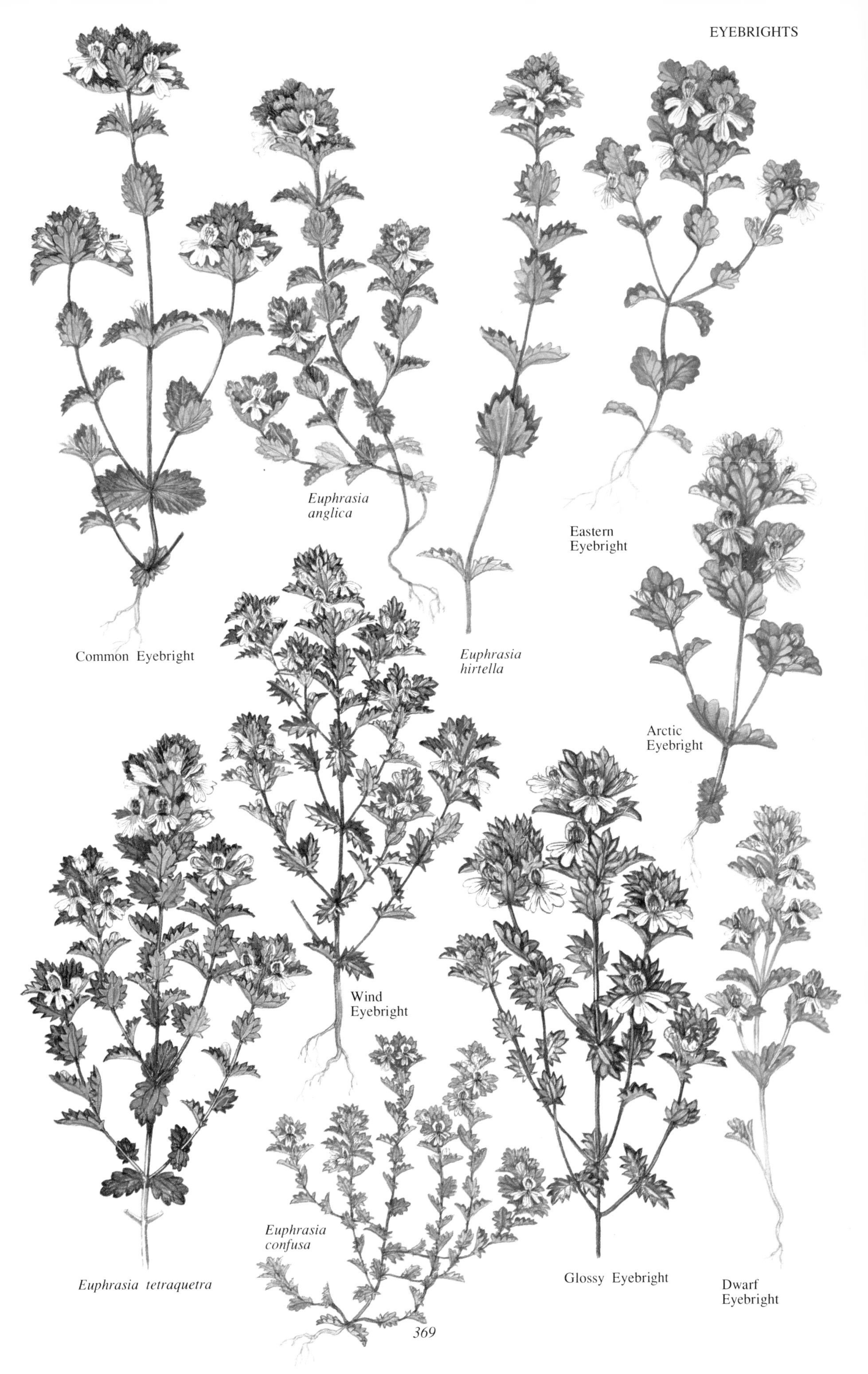
Common Eyebright
*Euphrasia anglica*
*Euphrasia hirtella*
Eastern Eyebright
Arctic Eyebright
Wind Eyebright
*Euphrasia tetraquetra*
*Euphrasia confusa*
Glossy Eyebright
Dwarf Eyebright

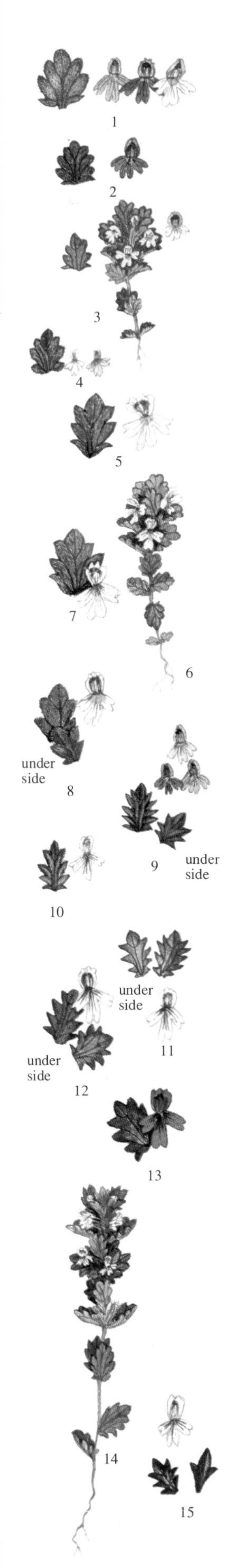


1* *Euphrasia frigida* Pugsley. Low to short plant, stem usually flexuous, green or reddish. Leaves oblong to rounded, with numerous short glandular-hairs, sometimes short-stalked, blunt-toothed; bracts usually rounded, usually larger then the upper leaves. Flowers white with a lilac upper lip, sometimes all lilac or sometimes purple, 4-7mm long. Capsule elliptical, deeply notched, equalling or longer than the calyx. Grassy places, rock ledges, mostly in the mountains and in damp places, to 2000m, sometimes higher. July-September. Britain, the Faeroes, Iceland, Scandinavia and Spitsbergen. Arctic and sub-Arctic regions from Europe to W Siberia and Labrador. The form from Iceland is generally hairier. **B**: Scottish Highlands and the mountains of Ireland.

2* *Euphrasia foulaensis* Townsend ex Wettst. Like *E. frigida*, but a more compact plant, the leaves smaller, dark green and blackening on drying, slightly hairy. Flowers usually violet, the lower lip not much longer than the upper. Clifftops and the margins of salt marshes, at low altitudes, generally close to the sea. July-August. **B**: N Scotland and the Faeroes.

3 *Euphrasia cambrica* Pugsley. Like *E. frigida*, but smaller, to 8cm, the stem with 0-2 pairs of flexuous branches. Flowers white or yellowish-white, sometimes with a lilac upper lip, 4-5.5mm long, the lower lip very small. Grassy slopes and rock ledges, locally common. **B**: Wales – mountains of Brecon and Caernarvon.

4* *Euphrasia ostenfeldii* (Pugsley) Yeo [*E. curta* var. *ostenfeldii*]. Low plant, the stem erect or flexuous with 0-4 pairs of long, slender, ascending branches. Leaves oblong to almost rounded, with few teeth, densely white-hairy; bracts broad, like the upper leaves. Flowers small, white or occasionally lilac, 3.5-6mm long. Capsule oblong, notched, equalling or longer than the calyx. Grassy, sandy or stony habitats, often close to the sea or in the mountains. June-September. N Britain, the Faeroes and Iceland. From Iceland to W Russia and Quebec. **B**: Locally common, N Wales to the Scottish Highland; coastal Ireland.

5* *Euphrasia marshallii* Pugsley. Like *E. ostendfeldii*, but leaves narrower, usually more than 8mm long (not 7mm or less). Flowers 5.5-7mm long, in rather dense clusters with overlapping bracts. Maritime grassland, especially sea-cliffs, very local. **B**: N Scotland, including the Inner Hebrides and Shetland; not known elsewhere.

6 *Euphrasia rotundifolia* Pugsley. Like *E. ostenfeldii*, but stem with 0-3 pairs of short branches and leaves more rounded. Sea-cliffs, very local. July. **B**: N Scotland – including Shetland and the Outer Hebrides.

7 *Euphrasia dunensis* Wiinst. Like *E. marshallii*, but the leaves with glandular hairs and stems with only 0-3 pairs of branches. Maritime grassland, on calcareous soils. Denmark – NW Jylland, not known elsewhere.

8 *Euphrasia campbelliae* Pugsley. Like *E. ostenfeldii*, but leaves hairy mainly in the upper half, white-bristly on the margin, sometimes purplish beneath, with very few teeth only – 1-3 pairs. Capsule shorter than the calyx. Heathy moors and grassland close to the sea. July. **B**: NW Scotland – Island of Lewis; not known elsewhere.

9* *Euphrasia micrantha* Reichenb. [*E. gracilis*]. Low to short, rather slender, strongly purple-tinged plant; stems with 2-7 pairs of erect branches. Leaves small, not more than 8mm long, oval, hairy or hairless, with a few sharp teeth; bracts similar to the leaves. Flowers lilac, violet or purple, rarely white, 4.5-6.5mm long. Capsule rounded, sometimes slightly notched, usually shorter than the calyx. Heaths and moors, generally associated with *Calluna vulgaris*, on damp light soils, to 2000m , sometimes higher. July-September. Throughout, except Iceland and Spitsbergen. Commonest in the north of its range, very local elsewhere. **B**: Throughout, scarcer in the south and east.

10* *Euphrasia scottica* Wettst. Like *E. micrantha*, but leaves only very slightly purple tinged and often darker beneath. Flowers usually white. Capsule equalling the calyx. Wet moorland, fens and flushes, to 1000m. July-August. Britain the Faeroes, Iceland and Scandinavia. **B**: N England, Scotland, C & N Wales and Ireland.

11 *Euphrasia calida* Yeo. Like *E. scottica*, but leaves not purple beneath and calyx whitish-membranous apart from the teeth. Capsule usually much shorter than the calyx. Associated with warm springs. July-August. Iceland, not known elsewhere.

12 *Euphrasia saamica* Juz. Like *E. scottica*, but purple colour of leaves not stronger below than above and flowers slightly larger. Moorland and tundra. July-August. Arctic Scandinavia.

13 *Euphrasia atropurpurea* (Rostrup) Ostenf. Like *E. scottica*, but often a lower plant with larger leaves; flowers deep purple, at least in part. Grassy and rocky habitats. July. Endemic to the Faeroes.

14 *Euphrasia bottnica* Kihlman. Like *E. scottica*, but flowers very small, 3-3.5mm. Capsule clearly notched, hairless. Sea-cliffs and salt marshes. July-August. Finland and Sweden – restricted to the region around the Gulf of Bothnia.

15* **Irish Eyebright** *Euphrasia salisburgensis* Funck. Low to short plant with 0-7 pairs of slender, erect or spreading branches. Leaves often strongly tinged purple or brown, oblong to oval, narrowed at the base, toothed mainly in the upper half, with only 1-3 pairs of sharp teeth; bracts lanceolate, as large as the leaves. Flowers white, 5-7.5mm long, generally with a rather small lower lip. Capsule oblong, occasionally slightly notched, hairless. Stony grassland, scrub, screes, rocky habitats and dunes, generally on calcareous soils, to 2600m. July-September. Britain, France, Germany and S Scandinavia. Only on the mountains in the south of its range. W Europe to W Russia. **P**: The small-flowered species of *Euphrasia* are generally self-pollinated, the larger-flowered ones by bees and other insects. **B**: Widespread in W Ireland, scarce elsewhere.

*Euphrasia campbelliae*

*Euphrasia saamica*

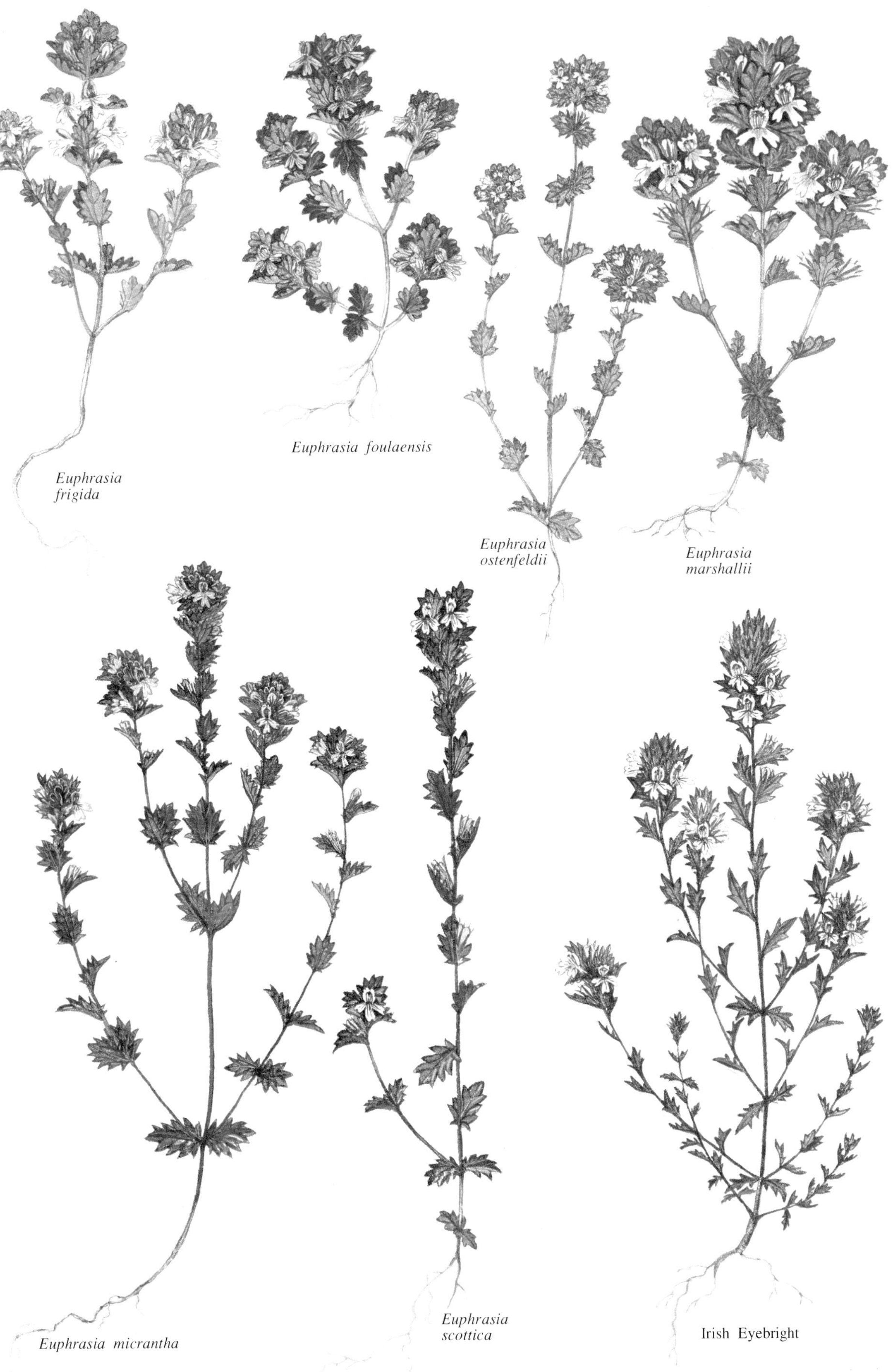
Euphrasia frigida
Euphrasia foulaensis
Euphrasia ostenfeldii
Euphrasia marshallii
Euphrasia micrantha
Euphrasia scottica
Irish Eyebright

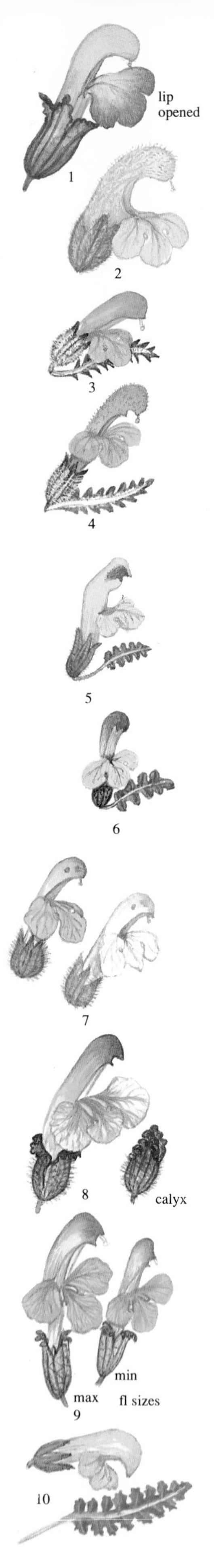


**Louseworts** *Pedicularis*. Semiparasitic perennial, occasionally annual, herbs, often tufted. Leaves usually alternate, but sometimes opposite or whorled, usually pinnately-lobed. Flowers in terminal, spike-like racemes; bracts present, usually leaf-like. Calyx tubular-bell-shaped, 5-toothed, often more or less 2-lipped. Corolla 2-lipped, the upper lip forming a hood or beak, the lower lip 3-lobed. Stamens 4, hidden within the upper lip of the corolla. 250 species in the northern temperate hemisphere and in the Andes of South America. Most are semiparasitic on various species of grass and turn black or dark brown on drying. **P**: Various bumble-bees. Flower shape and colour are important in accurate identification.

1* **Moor-king** *Pedicularis sceptrum-carolinum* L. Short to tall, often reddish perennial, stems erect hairless. Basal leaves in a rosette, lanceolate, pinnately-lobed, sometimes slightly hairy beneath; stem leaves few, alternate. Flowers pale yellow with a red margin to the lower lip, suberect, 22-32mm long, borne in a long lax, scarcely leafy spike; mouth of corolla usually closed. Fens, wet woodland and scrub, river and stream margins, generally at low altitudes. June-August. Scandinavia, local in S Germany. Readily recognised by its almost leafless stems and closed corollas in which the lower lip is partly wrapped around the upper.

2* **Leafy Lousewort** *Pedicularis foliosa* L. Rather stout, short to medium, erect hairy perennial. Leaves dull green, broadly lanceolate, 2-3-pinnately-lobed, hairy beneath. Flowers pale yellow, 15-25mm long, in dense leafy spikes; upper lip of corolla blunt, hairy; calyx membranous, hairy on the veins. Meadows, streamsides and scrub. often on calcareous soils, to 2500m. Late June-August. C & E France and Germany from the Vosges southwards. Readily identified by its pronouncedly leafy cone-shaped spikes and pale yellow flowers.

3* **Hairy Lousewort** *Pedicularis hirsuta* L. Low leafy perennial, woolly above. Leaves linear to lanceolate, pinnately-lobed, with sharply toothed segments; bracts similar to the leaves but shorter. Flowers bright pink, 10-15mm long, in dense heads; corolla hairless, the upper lip blunt. Damp habitats, stony tundra, seashores, river and stream banks, on calcareous soils, to 1400m. June-July. Arctic Europe, including Spitsbergen.

4 *Pedicularis dasyantha* Hadac. Like *P. hirsuta*, but flowers purplish-pink, 17-20mm long; corolla hairy, the upper lip straight, 2-toothed at the apex. Stony tundra. Spitsbergen, extending into Arctic Russia.

5* **Crimson-tipped Lousewort** *Pedicularis oederi* Vahl. Low to short, erect perennial, hairy above. Leaves alternate, lanceolate, pinnately-lobed, toothed, with hairless segments, the basal leaves shorter than the flower stems; bracts shorter than the flowers. Flowers yellow, the upper lip with a crimson apex, 12-20mm long, borne in dense spikes, but gradually elongating; upper lip of corolla straight and blunt. Damp grassland and rocky habitats, usually on calcareous soils, mainly in the mountains, to 1950m. July-August. C & E France, C & S Germany, C & S Norway and Sweden. Easily recognised by the red apex to the upper lip of the corolla; the red coloration often takes the form of 2 spots, one on each side. Rather local.

6 *Pedicularis flammea* L. Like *P. oederi*, but flowers smaller, 10-12mm long, the upper lip suffused with dark red, curved at the apex; calyx hairless (not hairy) often red-spotted. Damp habitats, to 1300m. July-August. Mountains of Norway and Sweden. Also found in Arctic and sub-Arctic North America.

7* **Verticillate Lousewort** *Pedicularis verticillata* L. Low to short, tufted, hairless or hairy, erect perennial. Basal leaves lanceolate, pinnately-lobed, long-stalked; stem leaves in whorls of 3-4, short-stalked; bracts similar to the leaves, but smaller and often purplish. Flowers purplish-red, 12-18mm long, rarely pinkish or white, in whorls forming a dense spike, the upper lip of the corolla blunt, almost straight. Damp mountain pastures and tundra, 1500-3100m. June-August. C & E France and S Germany. Restricted to the mountain areas of C and S Europe, but also present in parts of Arctic and sub-Arctic Russia. Easily distinguished by its distinct whorled stem leaves.

8* **Marsh Lousewort** *Pedicularis palustris* L. Short to medium, hairless or slightly hairy biennial, sometimes annual; stem single, branched, the lower branches longer than the upper giving the plant a pyramidal shape. Leaves triangular-lanceolate, finely pinnately-lobed, the stem leaves usually alternate but sometimes opposite or whorled; bracts similar to the leaves. Flowers reddish-pink, occasionally yellowish or white, 18-25mm long, in lax spikes, the upper lip of the corolla almost straight, with 4 teeth close to the apex; calyx strongly 2-liped, inflated in fruit. Damp meadows, fens, marshes and heaths, generally on acid soils, to 1800m. May-September. Throughout, except for Iceland, Spitsbergen and the extreme north of Scandinavia. N and C Europe, east to the Caucasus. The fruits are very characteristic with their markedly inflated calyces. **B**: Throughout much of Britain, but decreasing.

subsp. *opsiantha* (E. L. Ekman) Almquist. Like the type, but the flowers smaller, 14-18mm long; side-branches of even length giving the plant a cylindrical shape. Similar habitats and flowering time. Restricted to N and E Europe, primarily in N Germany and Scandinavia.

subsp. *borealis* (J.W. Zett.) Hyl. Similar to subsp. *opsiantha*, but stems unbranched and leaves narrow, scarcely lobed. Arctic and sub-Arctic habitats. Scandinavia and the Faeroes.

9* **Lousewort** *Pedicularis sylvatica* L. Low to short perennial or biennial; stems numerous, erect, unbranched, hairless or with 2 lines of hairs. Leaves lanceolate, 2-pinnately-lobed, hairless or slightly hairy; bracts similar to the leaves, though smaller. Flowers pink or red, 15-25mm long, in lax spikes, the upper lip of the corolla blunt, slightly curved, 2-toothed near the apex; calyx not 2-lipped, hairless, inflated in fruit. Bogs, marshes, damp heaths, moors and open woodland, on peaty soils, to 1800m. April-July. Throughout, except the Faeroes, Iceland and N Scandinavia. W and C Europe to W Russia. Variable plant – may sometimes be only 6-7cm tall. **B**: Common throughout much of Britain.

subsp. *hibernica* D.A.Webb. Like the type, but with hairy stems and calyces. Moors and bogs. Confined to Ireland and Norway.

10 **Lapland Lousewort** *Pedicularis lapponica* L. Short, hairless or slightly hairy perennial. Leaves linear-lanceolate, pinnately-lobed, toothed; bracts like the leaves but shorter. Flowers pale creamy-yellow, 14-16mm long, almost horizontal, a few in a lax head, the upper lip of the corolla short; calyx membranous, split to halfway into untoothed lobes. Heaths and dry tundra, to 1700m. June-August. Arctic and sub-Arctic Scandinavia, local. From Scandinavia to N Russia, parts of N Asia and North America.

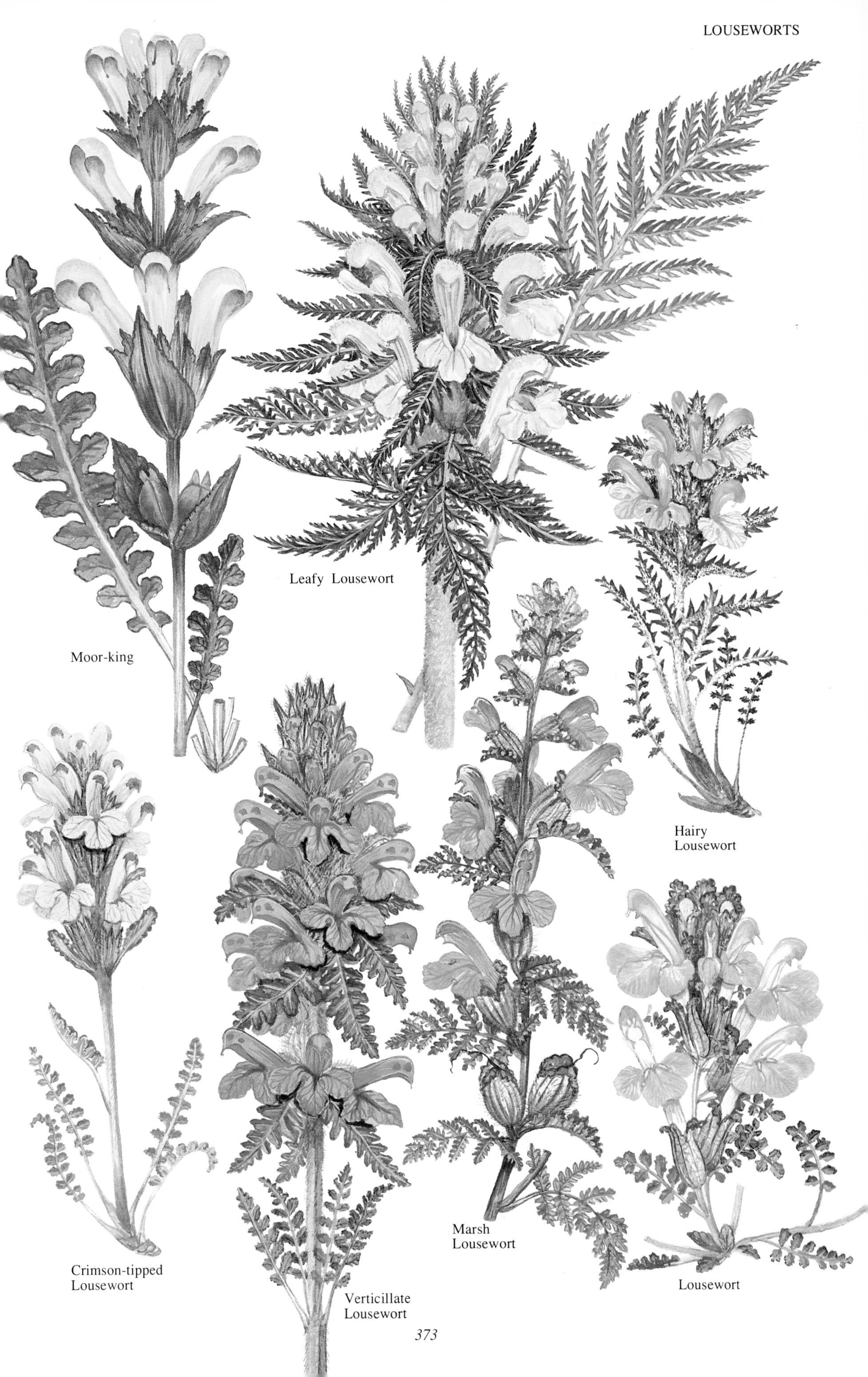
Leafy Lousewort
Moor-king
Hairy Lousewort
Crimson-tipped Lousewort
Verticillate Lousewort
Marsh Lousewort
Lousewort

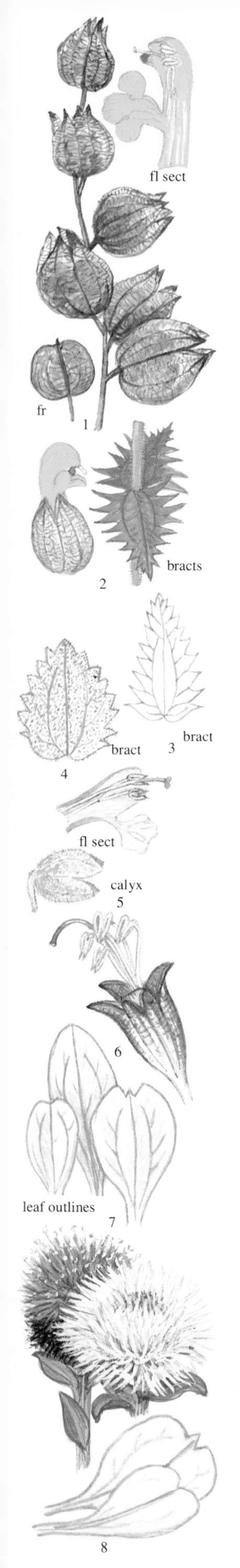


**Rattles** *Rhinanthus*. Annual semiparasitic herbs with opposite, unstalked leaves, toothed or untoothed. Flowers in terminal spike-like racemes; bracts present, leaf-like. Calyx flattened laterally, oval to rounded, 4-toothed, expanding and persistent in fruit. Corolla usually yellow, 2-lipped, with a long tube below, the upper lip with 2 teeth just below the apex. Stamens 4. 25 species in northern temperate regions. They are often difficult to determine accurately – details of the flowers and bracts are important in identification; the upper leaves and lower bracts are often transitional in form between the true leaves and the bracts.

1* **Yellow Rattle** *Rhinanthus minor* L. [*Alectorolophus minor*]. Low to medium plant; stem erect, branched or unbranched, often with black streaks or dots. Leaves oblong to linear-lanceolate, 5-15mm wide, dark green, toothed; bracts triangular, sharply toothed, equalling or longer than the calyx. Flowers yellow or brownish, 13-15mm long, with violet, or occasionally white teeth, mouth more or less open. Grassy habitats, meadows, pastures, fens, often on basic or calcareous soils, to 2000m. May-September. Throughout, except Spitsbergen. W and N Europe to Siberia, the Caucasus, S Greenland and Newfoundland. Very variable – a number of variants have been recognised in the past, but these generally overlap in gross morphological characters. However, high altitude variants found in northern parts of the range of the species, with finely hairy calyces, are sometimes referred to *R. borealis* (Sterneck) Druce. In typical *R. minor* the calyx is hairless or with a few hairs along the margin, but not hairy all over. **B**: Common throughout.

2 *Rhinanthus groenlandicus* (Ostenf.) Chab. Like *R. minor*, but stem hairy on 2 opposite sides and leaves bright yellowish-green; leaves and bracts deeply toothed, the teeth spreading sidewards (not pointing towards the apex). Grassy habitats. June-September. The Faeroes, Iceland and Scandinavia, except for S Sweden.

3* **Narrow-leaved Rattle** *Rhinanthus angustifolius* C.C. Gmelin [*R. serotinus*, *Alectorolophus major*]. Very variable, short to medium, plant; stem erect, branched or unbranched, hairy or hairless, often with black streaks. Leaves linear to elliptical, 2-5mm wide, toothed; bracts lanceolate, longer than the calyces, hairless, with large basal teeth. Flowers pale yellow, 16-20mm long, slightly curved, the mouth closed, calyx hairless. Meadows and other grassy habitats, especially cornfields, to 2500m. June-September. Throughout, except the extreme north, Ireland, the Faeroes and Iceland. Variable species in which many ecotypical variants have been recognised – the following subspecies is more or less distinct. **B**: Throughout, but rare in S Britain and most of Ireland.

subsp. *grandiflorus* (Wallr.) D.A. Webb [*R. grandiflorus*]. Leaves 8-15mm wide, lanceolate to almost oval. Throughout the range of the species, extending into N Scandinavia.

4* **Greater Yellow Rattle** *Rhinanthus alectorolophus* (Scop.) Pollich [*R. major*, *Alectorolophus hirsutus*]. Short to tall plant; stem erect, branched or unbranched, hairy, without black streaks. Leaves lanceolate to oval, toothed, pale green; bracts rhombic-triangular, evenly toothed, finely hairy. Flowers yellow, 18-20mm long, slightly curved, mouth closed; calyx with long whitish hairs, less marked in the fruiting stage. Meadows and other grassy habitats, to 2300m. May-September. Belgium, Holland and France. Rather local, occurring in both lowland and mountain regions, particularly in C Europe.

5* **Toothwort** *Lathraea squamaria* L. Low to short parasitic, rhizomatous, white or creamish perennial, often tinged with lilac-pink and slightly hairy above. Leaves scale-like, fleshy, alternate, rounded-heart-shaped; bracts similar but thinner. Flowers cylindrical, 2-lipped, 14-17mm long, half-nodding in a one-sided spike. Fruit capsule splitting along the midribs. Parasitic on the roots of various trees and shrubs, especially *Alnus, Corylus, Fagus* and *Ulmus*, in woodland, hedgerows and coppices, generally on fertile soils, to 1600m. March-May. Throughout, except the Faeroes, Iceland and the extreme north; extinct in Holland. Occasionally cultivated in gardens. Like other parasitic plants this species is without chlorophyll – plants turn blackish on drying. **P**: Bumble-bees. W Europe to Asia and the Himalaya. **B**: Throughout, except for N Scotland.

6* **Purple Toothwort** *Lathraea clandestina* L. Low parasitic perennial; stem subterranean, branched, yellowish, forming tufts at the soil surface. Scale-leaves kidney-shaped, clasping the stem, alternate or opposite; bracts similar to the scale-leaves. Flowers violet, with a reddish-purple lower lip, 40-50mm long, borne in small clusters; upper lip of the corolla hooded. Woodland, coppices and hedgerows, in damp shady places – parasitic on the roots of various trees, particularly *Alnus, Populus* and *Salix*. April-May. Belgium and France; naturalized in Britain. Widely cultivated in gardens. Native to W Europe, Spain and Italy. **B**: Local, naturalized in a few places in England.

## GLOBULARIA FAMILY Globulariaceae

Perennial herbs with alternate, untoothed leaves. Flowers irregular (zygomorphic), 5-parted, 2-lipped, borne in dense rounded heads surrounded by a ruff or involucre of bracts. Stamens 4. Fruit dry, surrounded by the persistent calyx.

7* **Globularia** *Globularia punctata* Lapeyr. [*G. aphyllanthes*, *G. willkommii*]. Short to medium, tufted, evergreen perennial; stems erect. Basal leaves in a rosette, oval to spoon-shaped, sometimes 3-lobed or notched at the tip, stalked, the lateral veins distinctly visible above; stem leaves lanceolate to oblong, unstalked. Flowerheads blue, 15mm; bracts lanceolate, pointed. Meadows, rocky habitats and open woods, often on limestone, to 1650m. May-June. Belgium, N France and Germany. Mainly in C and S Europe. Sometimes cultivated in gardens. Various bees and butterflies.

8* **Common Globularia** *Globularia vulgaris* L. Low to short perennial. Basal leaves in a rosette, oval, stalked, notched or 3-toothed at the apex, the lateral veins not or scarcely visible above; stem leaves lanceolate, pointed, unstalked. Flowerheads blue, occasionally lilac or white, 25mm; bracts lanceolate, pointed. Dry grassy and stony habitats. April-June. Mainly in NC & E France and S Germany; very local in S Sweden – Öland and Gotland.

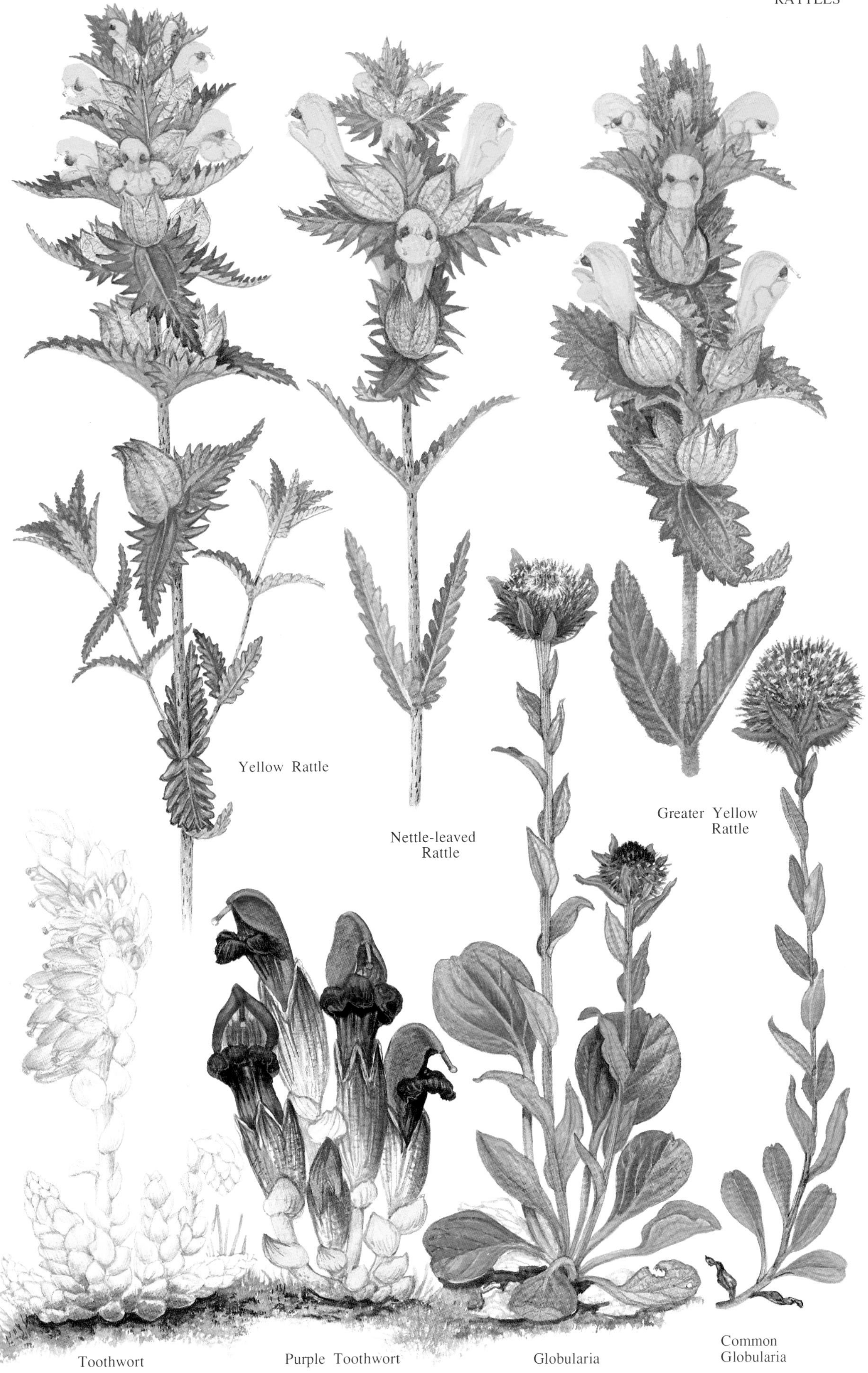
Yellow Rattle
Nettle-leaved Rattle
Greater Yellow Rattle
Toothwort
Purple Toothwort
Globularia
Common Globularia

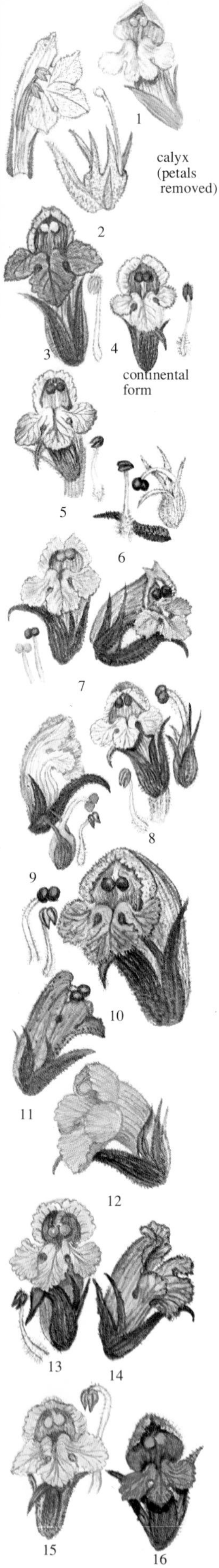

# BROOMRAPE FAMILY Orobanchaceae

Parasitic perennial herbs without chlorophyll. Stems erect, mostly unbranched. Leaves alternate, scale-like. Corolla 5-lobed, 2-lipped. Stamens 4. 130 species in 11 genera.

1* **Hemp Broomrape** *Orobanche ramosa* L. [*Phelypaea ramosa*]. Low to short; stem glandular-hairy, usually branched. Scale-leaves oval to lanceolate, pointed; bracts and bracteoles present, equalling the calyx. Flowers violet or pale bluish with a whitish base, 10-22mm long, glandular-hairy; anthers hairless or slightly hairy at the base; stigma white, cream or pale blue. On *Cannabis, Nicotiana, Solanum* etc. July-September. C & S France; naturalized elsewhere. **B**: Rather rare, S England, including East Anglia.

2* **Yarrow Broomrape** *Orobanche purpurea* Jacq. [*Phelypaea caerulea*]. Stems generally unbranched, greyish, finely glandular-hairy. Scale-leaves narrow-lanceolate; bracts and bracteoles present, appearing as 3 bracts to each flower. Flowers bluish-violet with deep violet veins, whitish below, 18-25mm long; anthers usually hairless; stigma white or pale blue. On *Achillea* etc. Rough grassland, to 1800m. June-July. North to S Sweden. **B**: Very rare in England S of Lincoln, S Wales and the Channel Is.

3 *Orobanche arenaria* Borkh. [*Phelypaea arenaria*]. Anthers densely hairy; flowers 25-35mm long with a white stigma. On *Artemisia*; often on alluvial soils, to 1800m. June-July. NE France and Germany.

4* **Thyme Broomrape** *Orobanche alba* Stephan ex Willd. [*O. epithymum*]. Short, reddish-purple; stems slightly swollen at the base, glandular-hairy. Scale-leaves lanceolate; bracts lanceolate, pointed. Flowers purplish-red, yellowish or whitish, 15-25mm long, fragrant, ascending, the lower lip with a glandular-hairy margin; stamens with densely hairy filaments; stigmas red or purple. On *Thymus* and other labiates; woods, rocky slopes, generally on calcareous soils, to 1800m. June-August. North to S Sweden. Local. **B**: Generally deeper coloured flowers. Very local, SW Britain, W Scotland, N & W Ireland.

5 **Thistle Broomrape** *Orobanche reticulata* Wallr. Like *O. alba*, but more robust, the flowers scarcely fragrant, the lower lip evenly lobed (without a larger central lobe) and not hairy margined. On species of *Carduus, Cirsium, Knautia* etc; fields and rough grassland, to 2500m. June-August. North to S. Sweden. **B**: Very rare and declining; found in a few localities in Yorkshire.

6* **Common Broomrape** *Orobanche minor* Sm. Stem yellowish tinged with red, unbranched, swollen at the base. Scale-leaves oval to lanceolate; bracts oval, pointed. Flowers pale yellow 10-16mm long, relatively small, curved, slightly hairy or glandular-hairy, the lower lip with even lobes without a hairy margin; filaments hairy below; stigma yellow, rarely purple. On *Trifolium* etc. Meadows, cultivated land, to 2200m. June-September. W Europe, rare in Holland; naturalized in Ireland, Denmark and Sweden. **B**: Widespread in England and Wales, rare in Scotland and Ireland.

**Carrot Broomrape** *Orobanche maritima* Pugsley. Stem purplish and lower lip of corolla unequally 3-lobed. On *Daucus carota, Ononis repens, Plantago maritima*. Rough maritime pastures. June-July. Very rare. W France, Channel Is., S England.

7 *Orobanche amethystea* Thuill. Scale-leaves linear-lanceolate, pointed. Flowers white or cream, tinged with violet throughout, 15-25mm long, the corolla sharply inflected near the base, the upper lip of the corolla deeply bilobed. On *Ballota nigra, Daucus carota* and *Eryngium* etc. Fields and stony places, to 2200m. June-August. S Britain, France and Germany. **B**: Isle of Wight and the Channel Islands.

8 **Oxtongue Broomrape** *Orobanche loricata* Reichenb. [*O. picridis*]. Corolla not sharply inflected near the base, pale yellow tinged and veined with violet; stigma always purple and filaments densely hairy near the base. On *Artemisia, Crepis, Picris, Daucus carota* etc. Rocky habitats and grassland, to 2200m. June-July. North to Denmark. **B**: Very rare, a few localities in Bucks, Kent and Somerset.

9* **Ivy Broomrape** *Orobanche hederae* Duby. Stem unbranched, reddish-purple, glandular-hairy, swollen at the base. Scale-leaves oblong to lanceolate, pointed; bracts lanceolate. Flowers dull cream tinged with reddish- purple towards the end, 12-22mm long, corolla narrowed in the throat; the upper lip slightly notched or un-notched, the lower lip unevenly lobed, not hairy on the margin; filaments hairless; stigmas yellow. On *Hedera helix*. Woodland, hedgebanks, old walls, at low altitudes. June-July. W Europe. **B**: Local, S England and Wales, rare in Ireland; particularly in coastal districts.

10* **Bedstraw Broomrape** *Orobanche caryophyllacea* Sm. [*O. vulgaris*]. Stems slightly swollen at the base, glandular-hairy, yellowish or purplish. Scale-leaves narrowly triangular-lanceolate, pointed; bracts lanceolate, short. Flowers pink or pale yellow tinged with dull purple, 20-32mm long, clove-scented, lower lip quite evenly lobed, hairy-margined; filaments hairy; stigma purple. On *Galium mollugo* etc. Open woodland, rough grassy places, to 1600m. June-July. North to S Norway. **B**: Only 2 places in Kent.

11 **Germander Broomrape** *Orobanche teucrii* Holandre. Calyx not more than 12mm long (not 10-17mm) and the bracts 12-20mm long (not 17-25mm). On *Teucrium*; dry pastures and rocky habitats, to 1600m. Belgium, France and Germany.

12 *Orobanche lutea* Baumg. Like *O. caryophyllacea*, but flowers yellowish or reddish-brown, the lower lip not or scarcely hairy-margined; stigma yellow or whitish. On *Medicago* and *Trifolium*, and other legumes; meadows and pastures, to 1600m. June-August. France, Holland and Germany.

13* **Knapweed Broomrape** *Orobanche elatior* Sutton. Stems slightly swollen at the base, usually glandular-hairy, reddish or yellowish-brown. Scale-leaves triangular-lanceolate; bracts lanceolate, as long as the flowers. Flowers yellow, often tinged with pink, 18-25mm long, curved, the upper lip toothed, the lower without a hairy margin; filaments hairy; stigma yellow; unscented. On *Centaurea, Thalictrum* etc. Dry grassy places, to 1600m. June-July. North to Sweden. **B**: Scattered, north to Cumbria and N Yorkshire.

14 **Alsace Broomrape** *Orobanche alsatica* Kirschleger. Similar but stem strongly swollen at the base and the flowers often tinged with purple or brown. On umbellifers. June-August. E France and S Germany.

15* **Greater Broomrape** *Orobanche rapum-genistae* Thuill. Short to tall; stems strongly swollen at the base, pale yellow, sometimes red-tinged. Scale-leaves oval to linear-lanceolate; bracts linear-lanceolate, longer than the flowers. Flowers yellow tinged with purple, 20-25mm long, slightly curved, fetid, the lower lip with a hairy margin; stigma yellow. On various shrubby leguminous species. Rough grassland on well-drained soils, mostly at low altitudes. May-July. W Europe. **B**: Throughout, except for most of Scotland, C & N Ireland; mostly on *Cytisus* and *Ulex* – declining.

16* **Slender Broomrape** *Orobanche gracilis* Sm. [*O. cruenta*]. Stem slender, glandular-hairy, yellowish or reddish. Scale-leaves oval to lanceolate; bracts triangular, pointed, shorter than the flowers. Flowers yellowish with a red-veined lip outside, shiny dark red inside, 15-25mm long; filaments hairy at least in the lower half; stigma yellow. On *Leguminosae*; grassy places, to 1350m. June-August. France, Germany.

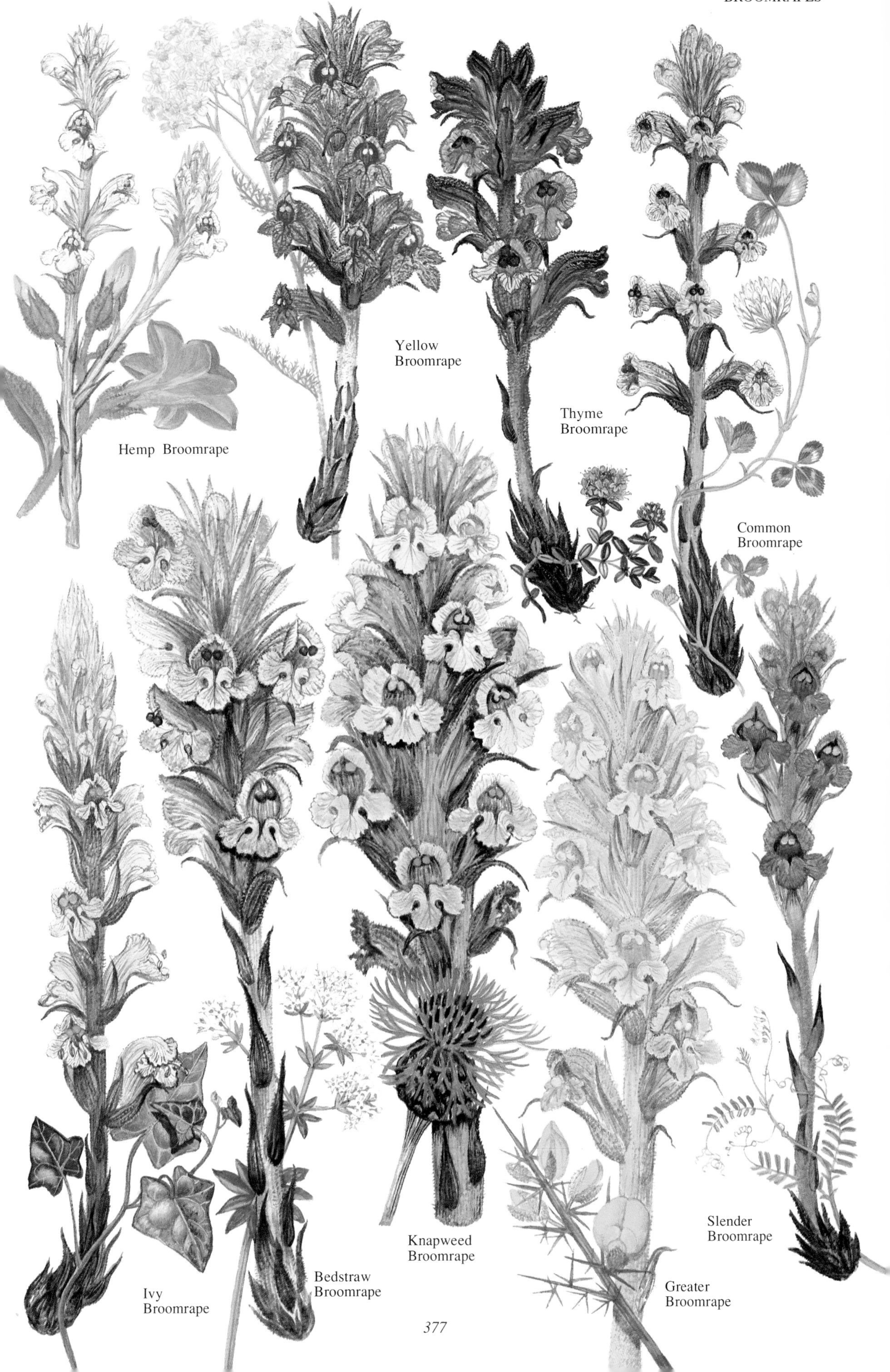

Yellow Broomrape

Thyme Broomrape

Hemp Broomrape

Common Broomrape

Slender Broomrape

Knapweed Broomrape

Bedstraw Broomrape

Greater Broomrape

Ivy Broomrape

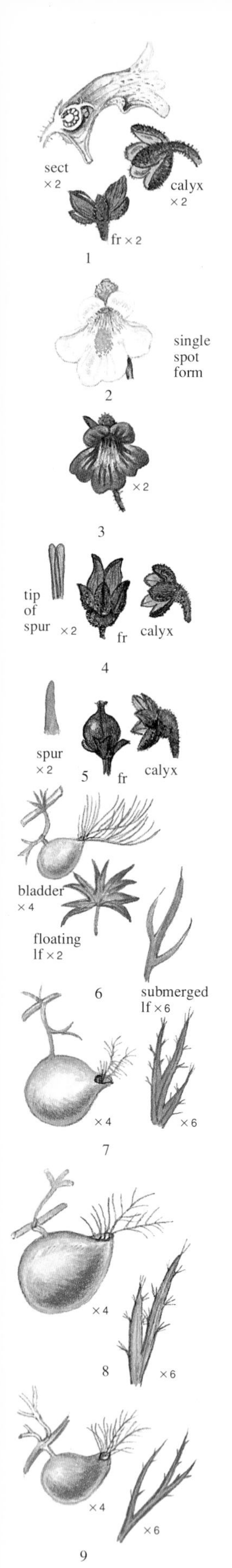


# BUTTERWORT FAMILY Lentibulariaceae

Small carnivorous plants, annual or perennial herbs, terrestrial or aquatic. Flowers solitary or in racemes. Calyx 2-5-lobed, often 2-lipped. Corolla 2-lipped, spurred, the upper lip 2-lobed usually, the lower lip 3-lobed. Stamens 2. Ovary of 2 carpels forming a single cell. Fruit a capsule. Plants trap small invertebrate animals by various means – in *Pinguicula* they are caught on the sticky surface of the leaf, while in *Utricularia* they are trapped in small sac-like bladders borne on the underwater leaf segments. Enzymes produced from glands on the leaves digest the soft parts of the animals which are then absorbed by the plant. As in other insectivorous groups, the species generally grow in acid habitats low in nutrients.

**Butterworts** *Pinguicula*. Perennials with sticky leaves in basal rosettes, generally rather pale yellowish-green, soft and fleshy, the margin inrolled. Flowers solitary, on long stalks, rather Violet-like, the lip of the corolla with a hairy patch or palette in the centre; spur pronounced. Capsule opening by 2 valves. 30 species characteristic of bogs and wet places, in the northern hemisphere and in South America.

1* **Pale Butterwort** *Pinguicula lusitanica* L. Plants overwintering as a rosette. Leaves greyish, oblong. Flowers pinkish to pale lilac, yellow in the throat, 6-7mm, borne on very slender, glandular-hairy stalks; corolla lobes rounded, notched; spur 2-4mm long, bent downwards. Bogs and wet heaths, acid flushes and wet moorland, ditches, to 500m. June-October. Britain and W France. **P**: Self-pollinated. **B**: Scattered localities in W & SW Britain, Isle of Wight, north to the Outer Hebrides and Orkney and most of Ireland.

2* **Alpine Butterwort** *Pinguicula alpina* L. Plants overwintering as a bud at ground level. Leaves elliptical to lanceolate, yellowish-green. Flowers white, with 1-2 yellow spots in the throat, 8-10mm; spur short and conical, 2-3mm, yellowish, bent downwards. Similar habitats, to 2600m. June-August. Arctic and sub-Arctic Europe, mountains of C & E France and S Germany; extinct in Britain.

3* **Villous Butterwort** *Pinguicula villosa* L. Plants overwintering as a bud at ground level. Leaves brownish, elliptical to almost rounded. Flowers pale violet, with 2 yellow spots or stripes in the throat, 7-8mm, borne on slender glandular-hairy stalks; spur short, 2-3mm. *Sphagnum* bogs, to 1100m. July. Norway, N & C Sweden and Finland. Local.

4* **Large-flowered Butterwort** *Pinguicula grandiflora* Lam. Low to short plant overwintering as a bud at ground level. Leaves oblong, pale green. Flowers deep violet, with a white patch in the throat, 15-20mm, borne on glandular-hairy stalks, the upper lip of the corolla cleft almost to the base; spur rather stout, 10-12mm, straight, directed backwards, sometimes slightly notched at the tip. Acid bogs and wet rocks and flushes, to 860m. May-July. SW Ireland & French Alps, Jura; naturalized in SW England. **P**: Various bees and butterflies. Hybridises with *P. vulgaris*; *P.* × *scullyi* Druce, **B**: Confined to SW Ireland, Kerry to Cork and Clare; naturalized in Cornwall.

5* **Common Butterwort** *Pinguicula vulgaris* L. Low plant overwintering as a bud at ground level. Leaves oblong, yellowish-green. Flowers violet, with a white patch in the throat, 10-15mm, borne on slender glandular-hairy stalks, the corolla lobes oblong; spur short, 3-6mm, straight or slightly curved, pointed. Bogs, wet heaths, moorland and wet rocks, ditches, to 980m (to 2300m in the Alps). May-July. Throughout, except Spitsbergen. The commonest species in our area, though often local. In C Europe hybridises with *P. alpina*, and in NW Europe with *P. grandiflora*. **B**: Throughout, but rare or absent in most of S England and S Ireland.

**Bladderworts** *Utricularia*. Rootless perennial, aquatic herbs overwintering by detachable winter buds. Stems horizontal with erect flowering stems. Leaves finely divided, bearing tiny sac-like bladders that trap minute water organisms. Flowers yellow, spurred, both the calyx and the corolla 2-lipped, the lower lip with a projecting palate, not lobed, larger than the upper lip. Capsule opening irregularly. 250 species, cosmopolitan.

6* **Lesser Bladderwort** *Utricularia minor* L. Aquatic perennial; stems of 2 kinds – bearing green thread-like leaves and few bladders or colourless and bearing bladders on very reduced leaves. Leaves rounded in outline, palmately divided; bladders 2mm. Flowers pale yellow, 6-8mm, 2-6 in a raceme held above the water surface; spur very short. Capsules borne on curved stalks. Ponds, ditches and bogs, in shallow, often peaty waters, to 1850m. June-September. Throughout, except Spitsbergen. **B**: Frequent in S & W Britain and Ireland.

7* **Intermediate Bladderwort** *Utricularia intermedia* Hayne. An aquatic or subaquatic perennial rather like *U. minor*, but the leaf-segments finely toothed, tipped by a small bristle and the bladders more or less confined to the colourless stems; bladders 3-4mm. Flowers bright yellow, generally with reddish-brown lines, 9-14mm, 2-4 borne in the raceme; spur conical. Ponds, shallow lakes and ditches, in acid peaty waters, to 1000m. July-September. Throughout, except the far north. **B**: Local from the Lake District northwards; rare elsewhere, including Ireland.

8* **Greater Bladderwort** *Utricularia vulgaris* L. Floating and submerged aquatic perennial with slender stems all of one kind. Leaves oval in outline, pinnately deivided into narrow, toothed segments, each tooth tipped by a bristle or several bristles; bladders 3mm. Flowers deep yellow, 12-18mm, 4-10 in the raceme, the lower lip of the corolla bent down; spur conical, pointed; fruiting freely. Capsules borne on strongly recurved stalks. Grows in up to 1m of water, to 800m or sometimes higher. July-August. Throughout, except Iceland and Spitsbergen. **B**: Throughout, usually local, rather rare in Wales.

9* *Utricularia australis* R. Br. [*U. neglecta*]. Similar to *U. vulgaris*, but a more slender plant with lemon yellow flowers, the lower lip of the corolla with a flat, slightly wavy margin, the spur tapered to a blunt tip; flower stalks 11-26mm (not 6-17mm). Rarely fruiting. Still, often rather infertile, acid waters. July-August. **P**: Bees. **B**: Throughout, but rare in E Britain and in N Scotland.

Pale Butterwort
Alpine Butterwort
Villous Butterwort
Large-flowered Butterwort
Common Butterwort
Intermediate Bladderwort
Lesser Bladderwort
Greater Bladderwort
*Utricularia australis*

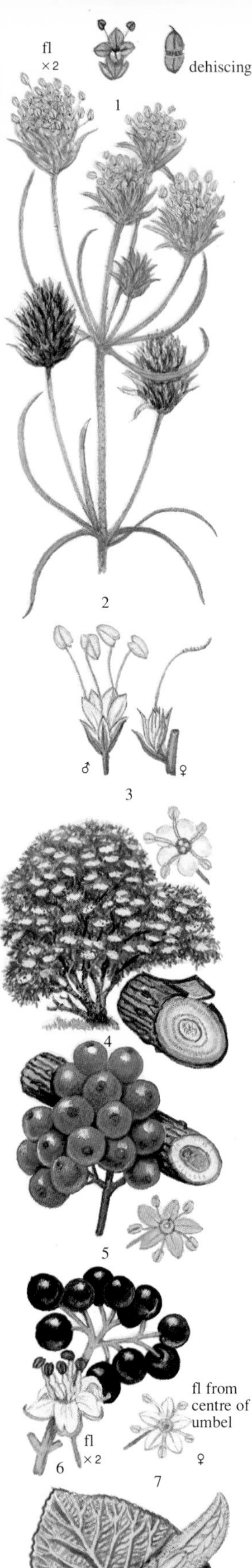


## PLANTAIN FAMILY Plantaginaceae

Terrestrial or occasionally aquatic herbs. Leaves either in a basal rosete or alternate, occasionally opposite. Flowers in clusters or slender spikes, 4-parted, regular, hermaphrodite in *Plantago*, unisexual in *Littorella.* Sepals fused together at the base, persistent. Petals membranous, fused together, inconspicuous. Stamens prominent, protruding, providing most of the colour. Fruit a capsule, dehiscing transversely in *Plantago*, indehiscent in *Littorella.* 200 species.

1* **Greater Plantain** *Plantago major* L. Low to short hairy or hairless perennial with a solitary or several leaf-rosettes. Leaves oval to elliptical, sometimes with a heart-shaped base, irregularly toothed or untoothed, 5-9-veined, thick and dark green. Flowers greenish-yellow, 3mm, in long spikes more or less equalling the unridged stalks; anthers pale purplish, turning yellowish-brown. Grassy and waste habitats, to 2800m. June-October. Throughout, except Spitsbergen; naturalized in Iceland. **B**: Common throughout. subsp. *intermedia* (DC.) Arcangeli [*P. intermedia*, *P. major* subsp. *pleiosperma*]. Leaves 3-5-veined, narrowed to the stalk, generally finely hairy (not hairless). Damp habitats, especially on saline soils. Almost throughout the species range, local.

* **Buck's-horn Plantain** *Plantago coronopus* L. Low biennial or perennial, sometimes annual, with a solitary or several rosettes. Leaves linear to lanceolate, often pinnately-lobed (occasionally unlobed), toothed, hairless or finely-hairy. Flowers yellowish-brown, 3mm, in long spikes on unridged stalks longer than the leaves; anthers pale yellow. Coastal habitats, sandy or gravelly soils, occasionally inland. May-July. Throughout, except for the far north. **B**: Coastal; inland in East Anglia and in S England.

* **Sea Plantain** *Plantago maritima* L. Low to short perennial with several or many leaf-rosettes. Leaves thick and rather fleshy, rigid, linear, untoothed, 3-5-veined. Flowers brownish, 3mm, in long greenish spikes on unridged stalks, generally exceeding the leaves; anthers yellow. Coastal habitats, generally on rather poor calcareous soils, occasionally inland besides mountain streams, to 2400m. June-August. Throughout, except Spitsbergen. Alpine forms are sometimes called *Plantago hudsoniana* Druce. **B**: On all coasts, inland in the Scottish Highland, the Pennines and Snowdonia.

* **Ribwort Plantain** *Plantago lanceolata* L. Very variable low to medium, hairy or hairless perennial, with several leaf-rosettes usually. Leaves linear-lanceolate to lanceolate, slightly toothed or untoothed, 3-5-veined, strongly ribbed, stalked. Flowers brown, 4mm, in short blackish spikes, on ridged stalks greatly exceeding the leaves; anthers pale yellow. Meadows, banks, waste and cultivated land, on neutral or basic soils, to 2300m. April-October. Throughout, except the extreme north. **B**: Ubiquitous.

* **Hoary Plantain** *Plantago media* L. Like *P. lanceolata*, but plants greyish, with a solitary or only a few leaf-rosettes. Leaves elliptic-oval, 7-9-veined, with curly hairs. Flowers whitish, fragrant, in spikes shorter than the long, ridged stalks; anthers lilac, occasionally whitish. Dry grassy habitats and waste places, on calcareous soils, to 2450m. May-August, occasionally later. Throughout, except for the far north, naturalized in Ireland. **B**: Widespread in England, rarer elsewhere.

2 **Branched Plantain** *Plantago arenaria* Waldst. & Kit. [*P. ramosa*, *P. psyllium*, *P. indica*]. Short to medium hairy annual; stems much-branched, leafy. Leaves linear to linear-lanceolate, opposite or whorled, not fleshy, generally untoothed. Flowers brownish-white, 4mm, in rounded or egg-shaped clusters; anthers pale yellow. Dry habitats, on sandy soils. May-August. C & E France and S Germany; naturalized in Belgium and Holland, frequently casual elsewhere. **B**: An infrequent casual.

3* **Shore-weed** *Littorella uniflora* (L.) Ascherson [*L. lacustris*]. Slender aquatic, stoloniferous, hairless perennial herb. Leaves in basal tufts, linear. Flowers whitish, 4-6mm, the male and female separate, but on the same plant; male solitary, 3-4-parted, long-stalked, female 2-4-parted, clustered, unstalked. By sea or shallow water. June-August. Throughout, except the far north. **B**: Commoner in the north.

## HONEYSUCKLE FAMILY Caprifoliaceae

Woody shrubs and climbers, occasionally herbs, with opposite leaves. Flowers 5-parted, hermaphrodite, solitary, paired, or borne in clusters or panicles. Calyx small usually. Corolla regular (actinomorphic) or 2-lipped, the lobes fused below into a short or long tube. Stamens usually 5, fused to the corolla-tube. Ovary inferior. Fruit a berry or nutlet. 400 species in 13 genera.

4* **Common Elder** *Sambucus nigra* L. Deciduous shrub or small tree to 10m; bark corky, grey-brown with a whitish pith; branches arching, rather brittle. Leaflets 5-7, oval to elliptical, pointed, sharply toothed, slightly hairy beneath; stipules absent or very small. Flowers white with yellowish-white anthers, in flat-topped clusters, 10-24cm across, scented. Berry turning red, then finally black, globose. Woods, old walls, near farm buildings, riverbanks, generally on calcareous and nitrogen-rich soils, to 1600m. June-July. Throughout, except the far north, **B**: Throughout, but introduced to many Scottish islands.

5* **Alpine Elder** *Sambucus racemosa* L. Deciduous shrub to 4m, with arching branches; bark grey, pith reddish-brown. Leaflets 3-7, oval to elliptical, pointed, sharply toothed, hairless, though often slightly hairy when young. Flowers creamy-white, in dense pyramidal panicles, 3-6cm across; anthers yellowish-white. Ripe berry shiny and scarlet-red, globose. Mountain woods and shady rocky places, to 2050m. April-June. W Europe; naturalized in Britain, especially Scotland, and many parts of Scandinavia.

6* **Dwarf Elder, Danewort** *Sambucus ebulus* L. Spreading, tall, hairless perennial herb, with creeping rhizomes; to 2m tall. Leaflets 5-13, oblong to lanceolate, sharply toothed; stipules conspicuous, oval. Flowers white, rarely pink outside, with purple anthers, borne in flat-topped clusters 5-16cm across. Ripe berry black, globose. Hedgerows, roadsides, local, to 1600m. July-August. W Europe; naturalized in Britain, Denmark and S Sweden. **B**: Scattered localities throughout, but rare in Scotland.

7* **Guelder-rose** *Viburnum opulus* L. Deciduous shrub to 4m, with hairless, greyish, angled young twigs; buds with scales. Leaves palmate with 3, occasionally 5, toothed lobes, usually hairy beneath. Flowers white, in broad flat-topped clusters, 4.5-10.5cm across; inner flowers fertile, 4-7mm, surrounded by a few large sterile flowers, 15-20mm. Ripe berry red. Woodland, fen carr, on wet soils, to 1600m. June-July. Throughout, except the far north. **B**: Throughout, but less common in Scotland.

8* **Wayfaring Tree** *Viburnum lantana* L. Downy deciduous shrub or small tree, to 6m; twigs grey, buds naked. Leaves oval-lanceolate, finely toothed, woolly-white beneath. Flowers creamy-white, 5-9mm, in dense clusters, 6-10cm across, all similar and fertile. Berry turning red, then finally black, oval. Open woods,scrub, on calcareous soils, to 1600m. April-June. W Europe; naturalized Norway, Sweden. **B**: S England, S Wales; absent from much of the SW, probably naturalized further north.

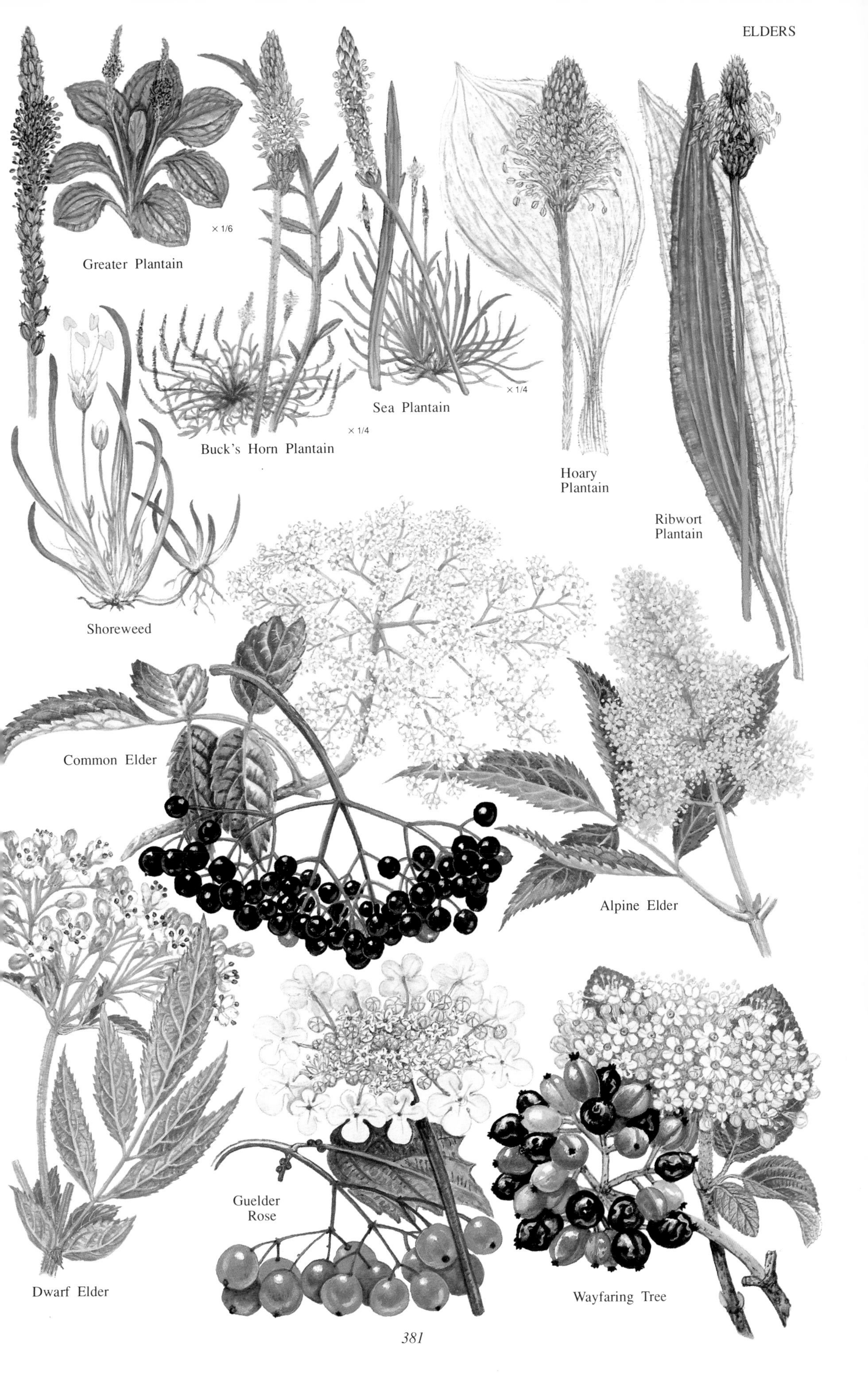
Greater Plantain
× 1/6
Buck's Horn Plantain
× 1/4
Sea Plantain
× 1/4
Hoary Plantain
Ribwort Plantain
Shoreweed
Common Elder
Alpine Elder
Dwarf Elder
Guelder Rose
Wayfaring Tree

1* **Snowberry** *S. racemosus* . A deciduous shrub, 1-3m; branches erect and arching, slender, the twigs hairless, yellowish-brown. Leaves oval, unlobed or lobed in the lower part, untoothed, greyish, usually slightly hairy beneath, stalked. Flowers pink, bell-shaped, 5-6mm long, in small spike-like racemes. Fruit a white pulpy berry, 10-15mm. Scrub and hedgerows, railway embankments, often forming dense thickets by a prolific growth of suckers. June-September. **I**: North America. Naturalized throughout, except the extreme north, Holland, the Faeroes, and Iceland. **B**: Widely naturalized but apparently rarely spreading by seeds. **B**: Locally naturalized close to cultivation.

2* **Twin Flower** *Linnaea borealis* L. Low creeping, evergreen subshrub, with slender trailing stems forming large mats. Leaves opposite, oval or rounded, blunt-toothed, short-stalked. Flowers pinkish-white, pendent bells, 5-9mm long, borne in pairs on long slender stalks, fragrant. Fruit an achene, glandular-hairy. Coniferous woods, heaths and mossy tundra, 1200-2200m. June-August. N Britain, Scandinavia, mountains of E France and S Germany. **P**: Various small insects, especially bees. A subspecies is found in North America. **B**: Rare, confined in E Scotland.

3 **Flowering Nutmeg** *Leycesteria formosa* Wall. Deciduous shrub, to 2m; stems hollow, swollen at the nodes, hairless. Leaves oval, untoothed or toothed, stalked. Flowers white, funnel-shaped, 15-20mm long, in pendent clusters at the shoot tips, with prominent reddish-purple bracts. Fruit a reddish-purple berry, 8-10mm. Scrub and waste places, sometimes planted as cover for game birds. July-September. **I**: India and China. A garden shrub naturalized in Britain and France. The stems rarely become very woody and generally die back during severe winter weather, sometimes to almost ground level.

**Honeysuckles** *Lonicera*. Deciduous or evergreen shrubs or climbers. Leaves untoothed, opposite. Flowers in lateral pairs or in terminal heads. Calyx 5-lobed, generally inconspicuous. Corolla regular or 2-lipped, in the latter case the upper lip 4-lobed and the lower lip 1-lobed. Fruit a berry. 180 species primarily in the northern temperate zone but extending southwards locally into Java and Mexico. Many species are cultivated in gardens.

4* **Blue-berried Honeysuckle** *Lonicera caerulea* L. Deciduous shrub to 2m, slightly hairy or hairless; bark yellowish-brown to reddish, flaking eventually. Leaves elliptical to oblong, pointed, hairless or with a few hairs on the midrib beneath. Flowers yellowish-white, bell-shaped with 5 equal lobes, 12-16mm long, borne in pairs, with 5 equal lobes. Berry bluish-black when ripe, succulent, the bracteoles forming a cup around the base of the fruit pairs. Local, mountain woods and scrub, on acid soils, 1300-2600m. May-July. France, Germany, Finland and Sweden; naturalized in Norway.

5* **Alpine Honeysuckle** *Lonicera alpigena* L. Deciduous shrub to 3m; twigs with a solid pith, occasionally slightly hairy when young. Leaves oblong to elliptical, pointed, hairy along the margins when young; bracts linear. Flowers yellowish or greenish-yellow, often tinged with reddish-brown, 2-lipped, 12-20mm long, borne in pairs. Berry scarlet when ripe, fused together in pairs. Mountain woods, scrub and rocky places, on limestone, to 2300m. May-July. C & E France and S Germany.

6* **Black-berried Honeysuckle** *Lonicera nigra* L. Like *L. alpigena*, but leaves bright green above, bluish-green beneath. Flowers pale pink, 6-10mm long, faintly fragrant. Berries black with a bluish bloom. Woods, scrub and rocky places, to 1800m. May-July. E France and S Germany. Mainly in the mountains of C Europe.

7* **Fly Honeysuckle** *Lonicera xylosteum* L. Deciduous shrub to 3m; young twigs grey-hairy usually, pith hollow. Leaves elliptical to almost rounded, pointed, hairy at least beneath, grey-green. Flowers yellowish-white, 2-lipped, 8-12mm long, hairy outside, borne in pairs. Berries bright red when ripe, globose, in pairs but not fused. Woods, scrub and hedgerows, generally on limestone, to 1800m. May-July. Throughout except for the far north. The most widespread of the shrubby honeysuckles in our area. Also in E Europe and N Asia. **P**: Bumble-bees. The berries have purgative and emetic properties. **B**: Very rare, confined to a few places in S England, particularly in W Sussex.

8* **Perfoliate Honeysuckle** *Lonicera caprifolium* L. Deciduous woody climber to 4m. Leaves elliptical, short-stalked, dark green above, but bluish-green beneath; leaves beneath the flowers fused together to form a cup. Flowers creamy-white or yellowish, tinged with purple, 30-50mm long, 2-lipped, borne in terminal clusters. Berry red or orange-red when ripe. Hedges and scrub. July-September. Naturalized in Britain, Belgium, France, Germany and parts of Scandinavia (C & S Europe). Widely cultivated in gardens, where it is generally grown on walls or fences. **B**: Scattered localities in S & N England and S Scotland.

9* **Japanese Honeysuckle** *Lonicera japonica* Thunb. Semi-evergreen climbing shrub to 4m, hairy. Leaves oval, hairy-margined. Flowers white at first, becoming yellowish, occasionally tinged with pink, 30-40mm long, 2-lipped, borne in pairs, sweetly scented. Berries red at first but black when ripe. Hedgerows and old walls. July-September. Locally naturalized in S Britain, France and Germany. There are various garden forms, of both this and the next species. **B**: Locally naturalized in SW England.

10* **Honeysuckle, Woodbine** *Lonicera periclymenum* L. A robust deciduous, twinning climber, to 6m, hairless or hairy. Leaves oblong to elliptical, dark green above, bluish-green beneath, the lower stalked, the upper unstalked but not fused together below the flowers. Flowers creamy-white, changing to yellowish, sometimes tinged with purple, 35-55mm long, 2-lipped, very fragrant, borne in terminal clusters. Berries globose, red when ripe. Woodland, scrub, hedgerows, rocks and cliffs, on acid or calcareous soils, to 1800m. June-October. North to C & S Norway and S Sweden. **P**: Night-flying moths, long-tongued bumble-bees by day. Various garden forms. **B**: Common throughout most of Britain and Ireland.

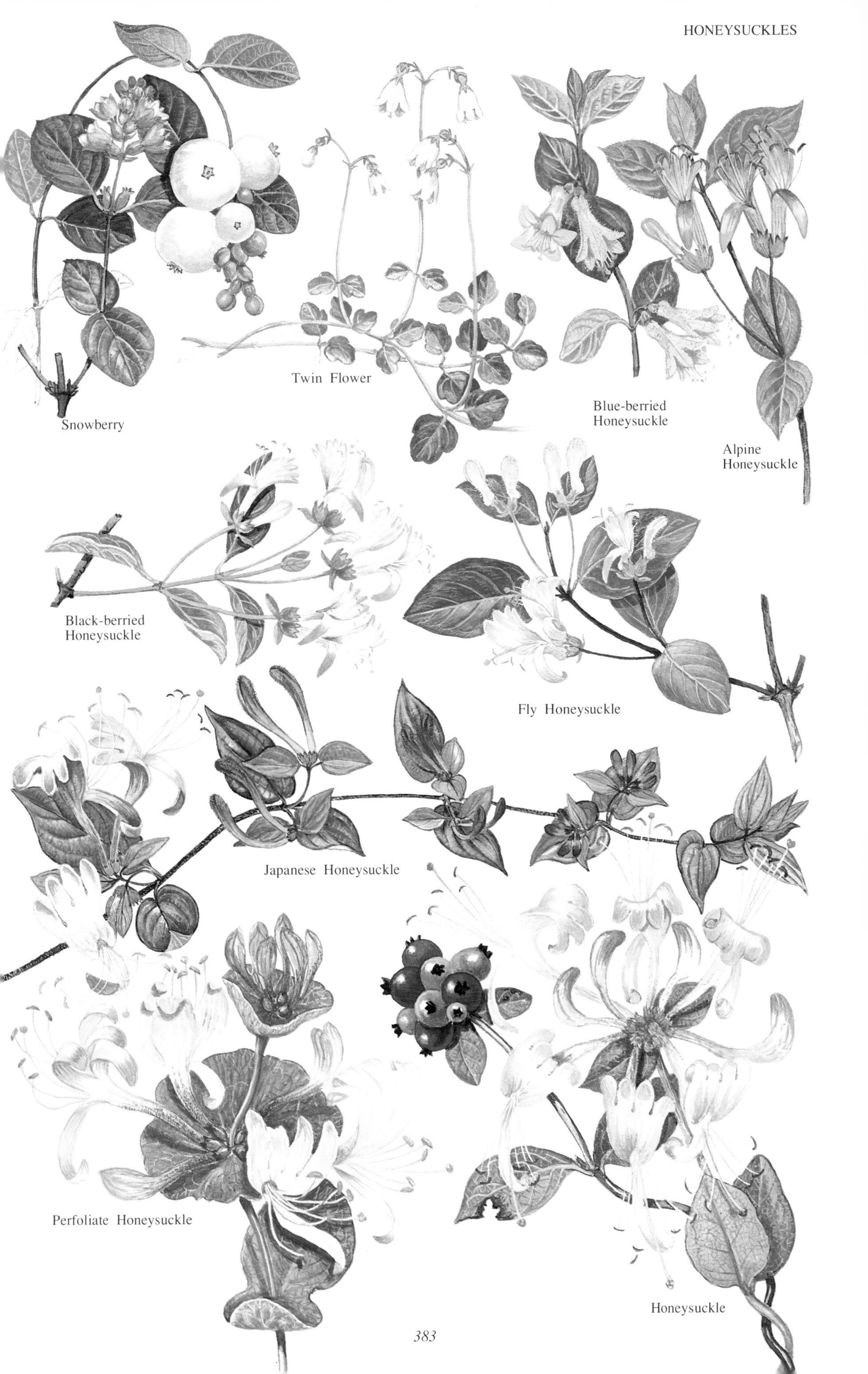
Snowberry
Twin Flower
Blue-berried Honeysuckle
Alpine Honeysuckle
Black-berried Honeysuckle
Fly Honeysuckle
Japanese Honeysuckle
Perfoliate Honeysuckle
Honeysuckle

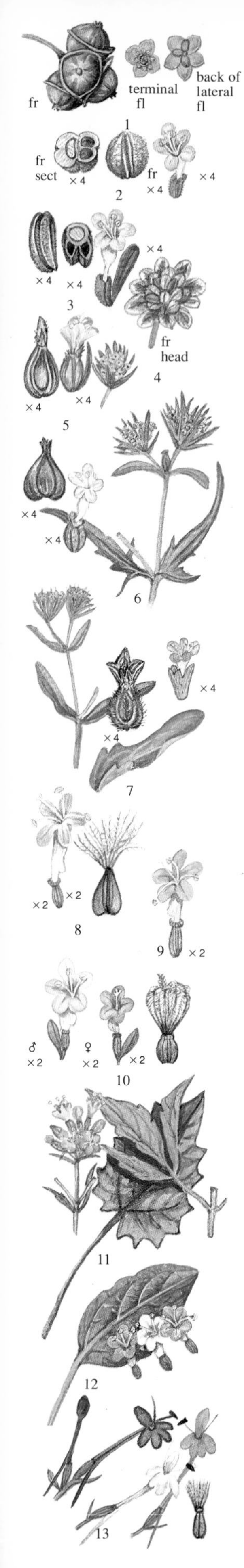


1* **Moschatel, Townhall Clock.** *Adoxa moschatellina* L. (Moschatel Family – Adoxaceae. A single species). A rather delicate, low, hairless, rhizomatous perennial forming carpets; stems slender, erect, unbranched. Basal leaves 2-ternate, with oval or oblong lobes, long-stalked; stem leaves one pair, ternate, the segments trilobed. Flowers greenish, in small clusters 6-8mm across, each normally with 5 flowers. Fruit greenish, but seldom produced. Shady habitats on moist soils, to 2400m. April-May. Throughout, except Holland and the far north, but only on the mountains in the south of its range. The uppermost flower in each cluster is generally 4-parted and faces upwards, the remaining 4 flowers are 5-parted and face sideways at 90° to each other. **B**: North to the Moray Firth; only Antrim, in Ireland.

## VALERIAN FAMILY Valerianaceae

Annual and perennial herbs. Leaves opposite or whorled, sometimes all basal. Flowers in cymose heads, often dense, sometimes paniculate, Calyx usually inconspicuous, toothed. Corolla funnel-shaped, sometimes spurred at the base, the limb generally 5-lobed. Stamens 1-4. Ovary inferior. Fruit dry and indehiscent with a persistent calyx. 350 species.

2* **Common Corn Salad, Lamb's Lettuce** *Valerianella locusta* (L.) Laterrade [*V. olitoria*]. Variable low to medium, hairless annual. Lower leaves broadly oval to spoon-shaped, blunt, untoothed; upper leaves oval to lanceolate, untoothed or slightly toothed; bracts green. Flowers pinkish or bluish, tiny, in flat, umbel-like clusters. Fruit almost rounded, 2mm, compressed in cross-section, smooth or with faint transverse ridges, the calyx reduced to tiny inconspicuous teeth; fruit clusters surrounded by a ruff of bracts. Arable and waste land, cliffs, dunes, on sandy or calcareous soils, generally at low altitudes. April-June. Throughout, except much of the north. **B**: Throughout; often local and primarily coastal in the north.

3 **Keeled-fruited Corn Salad** *Valerianella carinata* Loisel. fruits narrowly oblong, 0.75mm wide, squarish in cross-section, the fertile cell not corky on the back. A weed of crops, walls and rocks. April-June. W Europe. **B**: Very local in England north to Yorkshire, very rare elsewhere.

4 *Valerianella coronata* (L.) DC. Like *V. locusta*, but bracts broadly oval and membranous (not green and oblong, broadest above the middle). Fruit clusters falling as one piece, not separating into individual fruits. Cultivated and waste land. April-June. C & S France.

5* **Narrow-fruited Corn Salad** *Valerianella dentata* (L.) Pollich [*V. morisonii*]. Low to short hairless annual. Lower leaves oval, generally broadest above the middle, wavy; upper leaves oval to lanceolate, untoothed or slightly toothed; bracts green with a narrow membranous margin, slightly toothed. Flowers pale pink or lilac, tiny, in lax, flat-topped clusters. Fruit egg-shaped, flat on one surface, the calyx well developed and about half the width of the fruit itself, one tooth larger than the others; fruits falling separately. Arable land, usually on light soils. Local and decreasing. June-July. Throughout, except the far north and Finland. **B**: Local in England north to the Tyne, rare elsewhere.

6 **Broad-fruited Corn Salad** *Valerianella rimosa* Bast. [*V. auricula*] Fruits more rounded, bluntly 3-angled in cross-section, the calyx teeth minute. A weed of cereal crops, generally local. July-August. W Europe. Later flowering than previous species. **B**: Mainly in S England, south of Leicestershire; rare and declining.

7 **Hairy-fruited Corn Salad** *Valerianella eriocarpa* Desv. Like *V. dentata*, but the upper leaves more coarsely toothed and fruit generally hairy, the calyx as broad as the fruit. Fruits falling in clusters, not separately. Arable land, old walls. June-July. **I**: Probably Mediterranean. Naturalized in W Europe.

8* **Common Valerian** *Valeriana officinalis* L. Very variable medium to tall perennial, to 1.5m; stems hairless. Leaves pinnate or pinnately-lobed, toothed or untoothed, the lower long-stalked, the upper much smaller and unstalked. Flowers pink or white, hermaphrodite, 2.5-5mm long, in almost rounded clusters. Fruit hairy or hairless. Meadows, ditches, often on calcareous soils, to 2400m. June-August. Germany and S Sweden. subsp. *collina* (Wallr.) Nyman has stems densely hairy below, the middle leaves with 15 or more linear leaflets (not 6-13). Flowers often slightly larger, 3-6mm long. Dry, often calcareous meadows, scrub and woods. W Europe, Denmark, Iceland. **B**: Throughout, except the Shetlands. subsp. *sambucifolia* (Mikan f.) Celak [*V. sambucifolia*, *V. excelsa*] is stoloniferous, the middle leaves with 5-9 lanceolate, toothed, leaflets. Flowers larger, 4-8mm long. Damp shaded habitats. Scandinavia, N Germany, Holland.

9* **Pyrenean Valerian** *Valeriana pyrenaica* L. Tall perennial, to 1.1m; stems stout, solid, hairy at the nodes. Basal leaves oval to rounded, irregularly toothed, long-stalked; upper leaves with 1-2 pairs of small lateral leaflets. Flowers pink, 2.5-3mm long, in dense clustered heads, hermaphrodite. Fruit hairless. Damp habitats, to 2400m. June-August. **I**: Pyrenees. **B**: Naturalized in a few woods in W Britain and Ireland.

10* **Marsh Valerian** *Valeriana dioica* L. Short to medium, perennial; stems several usually, erect, slightly hairy at the nodes. Basal leaves oval to elliptical, untoothed, long-stalked, bright green; stem leaves pinnately-lobed, mostly unstalked. Flowers pink, occasionally white, 2-4mm long, in rounded clusters; monoecious, male flowers larger than females. Fruit hairless. Wet places on calcareous or slightly acid soils, to 1800m. May-July. Throughout north to S Scandinavia. **B**: North to the Firth of Forth; not in Ireland.

11 **Three-leaved Valerian** *Valeriana tripteris* L. Short perennial with both flowering and non-flowering shoots; stems erect, hairy at the nodes. Basal leaves oval, heart-shaped at the base, shallowly toothed, long-stalked; middle and upper leaves trifoliate, occasionally pinnate. Flowers pink or white, 3-5mm long, in lax rounded clusters. Fruit hairless. Woods, scrub, usually on calcareous soils, to 2600m. June-August. Mountains of France and Germany from the Vosges southwards.

12 *Valeriana montana* L. Basal leaves not heart-shaped, untoothed; upper leaves toothed,not lobed. Flowers lilac, pink or white, slightly larger. Scrub and rocky habitats, on calcareous soils, to 2600m. April-July. C & E France and S Germany.

13* **Red Valerian** *Centranthus ruber* (L.) DC. [*Valeriana ruber*]. Tufted, medium to tall, greyish, somewhat fleshy and waxy perennial. Leaves lanceolate to oval, pointed or blunt, the uppermost slightly toothed at the base, clasping the stem. Flowers red, pink or white, funnel-shaped, 8-12mm long, in large panicles, fragrant, spurred at the base, the spur 4-7mm long. Fruit a 1-seeded nut, with a feathery persistent calyx. Walls, banks, June-August. **I**: Mediterranean. Naturalized in W Europe. **B**: Naturalized north to S Scotland, especially in SW Britain and S Ireland.

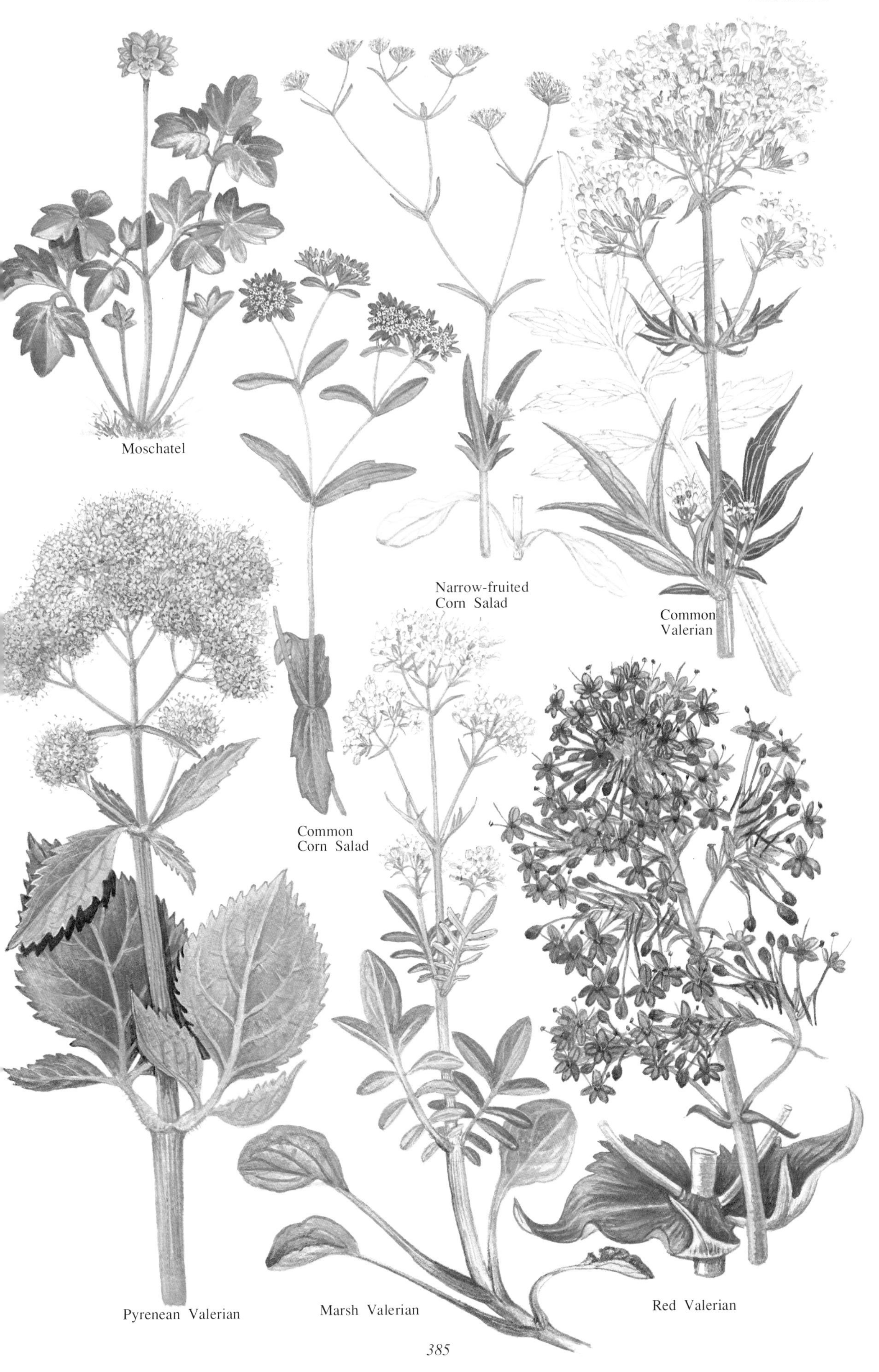
Moschatel
Narrow-fruited
Corn Salad
Common
Valerian
Common
Corn Salad
Pyrenean Valerian
Marsh Valerian
Red Valerian

# TEASEL FAMILY Dipsacaceae

Annual or perennial herbs with opposite or whorled leaves. Flowers small, in dense thistle-like, flowerheads, surrounded by a ruff of bracts, hermaphrodite or female. Calyx small, 4-5-toothed. Corolla 4-5-lobed or 2-lipped. Stamens 2 or 4. Fruit dry, often with a persistent, papery calyx. 115 species in 9 genera, mainly Mediterranean.

1* **Wild Teasel** *Dipsacus fullonum* L. [*D. sylvestris*]. Tall biennial, to 2m; stems erect, prickly on the angles. Basal leaves in a large rosette, oblong-elliptical, untoothed, prickly, withering early in the second season; stem-leaves linear-lanceolate, the pairs fused together around the stem at their bases. Flowers pinkish-purple, in large spiny oblong-cylindrical heads, 3-8cm long. Many habitats, especially on clay soils. July-August. Britain, Belgium, Holland, France and Germany; naturalized in Denmark. **B**: C & S England north to the Humber, S Wales; rare elsewhere including Ireland.

2 **Fuller's Teasel** *D. f.* subsp. *sativus* (L.) Thell [*Dipsacus sativus*, *D. fullonum* sensu Miller]. Like the type but the flowerheads 3-8cm long, with spreading bracts (not ascending and overtopping the flowerhead); corolla tube 13mm long (not 9-11mm). Cultivated ground and waste places. Similar flowering time. Naturalized in Britain, France and S Germany; origin uncertain. **B**: An occasional escape in S Britain, still cultivated in Somerset.

3* *Dipsacus laciniatus* L. Like *D. fullonum*, but stems with slender prickles and stem-leaves pinnately-lobed; flowerheads similar, the flowers pale pink. Meadows, streamsides and waste places. July-August. C & E France and Germany.

4* **Small Teasel** *Dipsacus pilosus* L. [*Cephalaria pilosa*]. Medium to tall biennial, to 1.2m; stems erect, sparsely prickly. Basal leaves in a rosette in the first year, oval, narrowed to a long stalk, hairy, toothed; stem leaves oval, short-stalked, with a basal pair of elliptical, unequal leaflets usually. Flowerheads globose, 1.5-2cm, the flowers whitish, 6-9mm long; bracts narrowly triangular with long silky hairs, spine-tipped. Damp or shady habitats, generally on calcareous soils. August-September. North to Denmark. **B**: England and E Wales north to Yorkshire.

5 *Dipsacus strigosus* Willd. Like *D. pilosus*, but a taller plant; flowerheads 2.5-4cm, the flowers pale yellow. Similar habitats and flowering time. **I**: W & S USSR. Naturalized in Britain, Denmark and S Sweden. **B**: Casual or locally naturalized.

6* **Devil's-bit Scabious** *Succisa pratensis* Moench. [*Scabiosa succisa*]. Medium to tall. hairy perennial; stems erect to ascending. Basal leaves elliptical, often broadest above the middle, generally untoothed, often blotched with purple, the upper leaves sometimes toothed. Flowers lilac to dark violet-blue, rarely pink or whitish, in rounded heads, 15-25mm; florets more or less equal in size, sometimes only female. Damp places, on calcareous to slightly acid soils, to 2400m. July-October. Throughout, except Spitsbergen. **B**: Common throughout.

7* **Wood Scabious** *Knautia dipsacifolia* Kreutzer [*Scabiosa dipsacifolia*, *Knautia sylvatica*]. Rather robust medium to tall, hairy perennial, to 1.5m; stems usually purplish. Leaves bright green, variable on the same plant, oblong-oval, tapered to the base, toothed, the lowermost stalked. Flowerheads bluish-violet to lilac, 25-40mm, the outermost florets somewhat larger than the inner. Shady habitats, woodland and woodland margins and scrub and tall herb communities, to 2000m. June-September. Mountains of France and Germany. Unlike *Scabiosa*, *Knautia* has 8 calyx teeth (not 4-5) and a hairy receptacle. subsp. *gracilis* (Szabo) Ehrend. [*K. gracilis*] is more slender, the peduncles not glandular and the upper leaves tapered at the base (not rounded). Similar habitats and flowering time. Belgium to C Germany and SC France.

8* **Field Scabious** *Knautia arvensis* (L.) Coulter. Very variable medium to tall, hairy perennial or biennial, often stoloniferous, with basal leaf-rosettes; stem usually with purple spots. Leaves pinnately-lobed, the lower stalked, the upper sometimes undivided. Flowerheads bluish-violet to lilac, occasionally purple, 20-40mm, hermaphrodite or female, the former rather larger. Meadows, open woods, hedgebanks, generally on dry calcareous soils, to 600m. July-September. Throughout, except the far north; naturalized in the Faeroes and Iceland. **B**: Throughout, except Outer Hebrides and Shetland Is.

**Scabious** *Scabiosa*. Annual or perennial herbs with opposite leaves, often in basal rosettes. Flowers in flattened or semi-spherical heads, long-stalked. Corolla with a short tube and 5 unequal lobes, the outer florets longer than the central ones. 80 species, mainly in Europe, Mediterranean and W Asia.

9* *Scabiosa canescens* Waldst. & Kit. [*S. suaveolens*]. Short to medium, leafy, hairy perennial; stems branched. Lower leaves lanceolate, pointed, untoothed; upper leaves pinnately-lobed with slender divisions. Flowerheads blue or lilac, 15-25mm, the marginal florets twice the size of the central ones; flower-bracts oval-lanceolate. Dry grassy habitats. July-September. Continental Europe north to S Sweden.

10 **Mournful Widow, Sweet Scabious** *Scabiosa atropurpurea* L. [*S. maritima*]. Like *S. canescens*, but a less hairy plant with dark purple to deep lilac flowerheads, 20-30mm, the outer florets only slightly longer than the inner; bracts narrow-lanceolate. Fruithead oblong (not rounded). Dry sunny habitats, banks and waste ground. June-August. **I**: Mediterranean. **B**: Locally naturalized in S England.

11* **Small Scabious** *Scabiosa columbaria* L. Short to medium, much-branched, hairy perennial. Basal leaves oval to lanceolate, unlobed or pinnately-lobed, long-stalked, the upper leaves 1-2-pinnately-lobed or pinnate. Flowerheads bluish-lilac, 20-40mm, the outer florets slightly longer than the central ones, with dark bristles at the base; bracts narrowly lanceolate, shorter than the florets. Grassy places, on dry calcareous soils, to 900m. July-August. Trhoughout north to S Sweden. **B**: Scattered throughout north to C Scotland.

12* **Shining Scabious** *Scabiosa lucida* Vill. Short, somewhat hairy, tufted perennial. Leaves rather glossy and almost hairless, the basal leaves oval-lanceolate with shallow blunt teeth, the upper pinnately-lobed with a large end leaflet. Flowerheads rose-pink to violet or deep mauve, 10-20mm, the outer florets distinctly longer than the inner ones, with dark bristles at the base. Dry meadows, stony and rocky places, to 2700m. June-September. French and German mountains.

13 **Yellow Scabious** *Scabiosa ochroleuca* L. Annual or biennial, the lower leaves less deeply cut than *S. columbaria*, the upper more deeply. Flowerheads yellow, 20-25mm. Grassy habitats, rough and uncultivated ground, scrub and waysides. July-August. C & E France and S Germany.

Wild Teasel
× 1/15
*Dipsacus laciniatus*
Small Teasel
Devil's-bit Scabious
Wood Scabious
Field Scabious
*Scabiosa canescens*
Small Scabious
Shining Scabious

# BELLFLOWER FAMILY Campanulaceae

Annual or perennial herbs nearly always with latex. Leaves usually alternate, simple; stipules absent. Flowers often large and showy, hermaphrodite, regular (actinomorphic) borne in heads, racemes or panicles, sometimes solitary. Calyx with 3-5 narrow teeth, closely fused to the ovary. Corolla shallowly to deeply lobed, often bell-shaped, with a short or a long tube, generally 5-lobed.Stamens 5, fused or free. Style solitary. Ovary inferior. Fruit a capsule, dehiscing by slits or pores, containing numerous seeds. Some 35 genera and 700, or more, species, almost throughout the world. Many are cultivated for their attractive flowers, especially in the genus *Campanula*. Certain genera like *Jasione* and *Phyteuma*, approach the *Compositae* in their compact flowerheads and fused anthers held closely around the style.

1* **Arctic Bellflower** *Campanula uniflora* L. Rather delicate, short, hairless perennial, rarely more than 14cm tall; stems erect, unbranched. Basal leaves oblong, often broadest above the middle, short-stalked, toothed or untoothed; upper leaves lanceolate to linear-lanceoalate, pointed. Flowers dark blue, black distally, 7-9mm long, solitary and pendent, borne on slender stalks. Capsule dark blue to blackish. Stony tundra, heaths and grassy places, on calcareous soils, to 1600m. July. Arctic and sub-Arctic Europe. Norway, Sweden, Finland, Iceland and Spitsbergen.

2* **Spreading Bellflower** *Campanula patula* L. Medium, hairy or hairless, rather rough biennial or perennial; stems erect or ascending, slender. Leaves slightly hairy, the basal ones oval, broadest above the middle, toothed, the upper leaves few, linear-lanceolate, unstalked. Flowers violet to pale blue, rarely white, 20-25mm long, erect wide bells, a few to many in a lax, long-branched inflorescence; buds pendent. Capsules erect, with 10 prominent veins. Grassy habitats, open woodland, scrub, hedgerows and waste places, to 1600m. July-September. Britain, Belgium, Holland, France, Germany and S Finland; naturalized in Denmark, Norway and Sweden. **P**: Various bees. **B**: Scattered localities in England and Wales, north to Durham; rare and rapidly declining.

3* **Rampion Bellflower** *Campanula rapunculus* L. Very variable medium to tall biennial; stems erect unbranched, hairless or slightly hairy. Basal leaves oval, broadest above the middle, toothed, stalked; upper leaves linear-lanceolate, unstalked. Flowers pale blue or white, 10-20mm long, broad erect bells, in a lax raceme, rarely branched; calyx teeth bristle-like, long.. Meadows, fields, forest and woodland margins, hedgebanks, waysides and waste places, often on gravelly soils, to 1600m. July-August. Belgium, Holland, France and Germany; naturalized in Britain, Denmark and Sweden. Formerly cultivated as a salad vegetable, both the carrot-like root and the shoots being edible. Sometimes confused with *C. patula*, but distinguished by its usually unbranched inflorescence and the bracteoles borne at the base of the flowerstalks (not in the centre). **B**: Scattered localities throughout England and S Scotland, rare.

4* **Peach-leaved Bellflower** *Campanula persicifolia* L. Medium to tall, hairless perennial; stems erect, unbranched. Basal leaves lanceolate to oval, bluntly toothed, stalked; upper leaves linear-lanceolate, toothed. Flowers blue or white, 30-40mm long, horizontal or ascending broad bells, not nodding, borne in a lax raceme; calyx teeth half the length of the corolla. Meadows and open woods, commons and waste places, to 2000m. June-August. Belgium, Holland, France, Germany and Scandinavia; naturalized in Britain. Widely cultivated in gardens including forms with double flowers. **P**: Bees or self-pollinated. **B**: Scattered localities especially in S England, but extending north to Scotland.

5* **Canterbury Bell** *Campanula medium* L. Medium, bristly biennial; stem erect, usually branched. Leaves in a basal rosette in the first year, oval, toothed, stalked; stem leaves lanceolate, unstalked, Flowers bluish-lilac, dark violet-blue, or white, 30-50mm long, deep suberect bells, in broad panicles, each branch usually terminated by a single flower; calyx-lobes heart-shaped, bristly. Dry open habitats. May-June. **I**: SE France & C Italy. Naturalized or casual in Britain and Germany. Some garden forms have an expanded and coloured calyx surrounding the corolla – the so-named 'cup and saucer' varieties. **B**: Naturalized on railway embankments in SE England, occasionally elsewhere.

6* **Bearded Bellflower** *Campanula barbata* L. Short tufted, bristly perennial, generally with non-flowering leaf-rosettes; stems usually unbranched. Basal leaves lanceolate to oblong, untoothed, wavy-edged, stalked; stem leaves few, linear-lanceolate. Flowers pale blue, with long whitish hairs inside, 20-30mm long, narrow pendent bells, in a lax one-sided raceme; calyx teeth 2 rows, narrow, much shorter than the corolla. Mountain habitats, woods, grassy places and scrub, stony places, to 3000m. June-August.C & E France and S Germany, rare in S Norway. Confined primarily to the mountains of C Europe. Sometimes cultivated in gardens.

7 *Campanula sibirica* L. Like *C. barbata*, but a more softly hairy biennial, to 50cm tall, without non-flowering leaf-rosettes; stems branched above. Grassy habitats. July-August. NE Germany.

8* **Clustered Bellflower** *Campanula glomerata* L. Short to medium, roughly hairy, stoloniferous perennial, with erect flowering stems. Basal leaves rounded or heart-shaped, toothed, long-stalked; stem leaves similar, though often narrower, and short-stalked or unstalked and half-clasping the stem. Flowers deep bright blue, occasionally paler blue or white, erect bells, 15-30mm long, in tight terminal clusters, sometimes with a few flowers below. Grassy habitats, open woodland, scrub, tracksides and waste places, rarely on sea cliffs, generally on calcareous soils, to 1700m. June-August. Throughout, except Ireland, Iceland, the Faeroes and the extreme north. Widely cultivated in gardens. P Various bees and butterflies. **B**: Locally common throughout north to Kincardine.

9 *Campanula cervicaria* L. Like *C. glomerata*, but a more bristly plant, with lanceolate leaves and winged leaf-stalks and pale blue flowers, 14-16mm long, with a protuding style. Similar habitats and flowering time. Continental Europe, except for the far north.

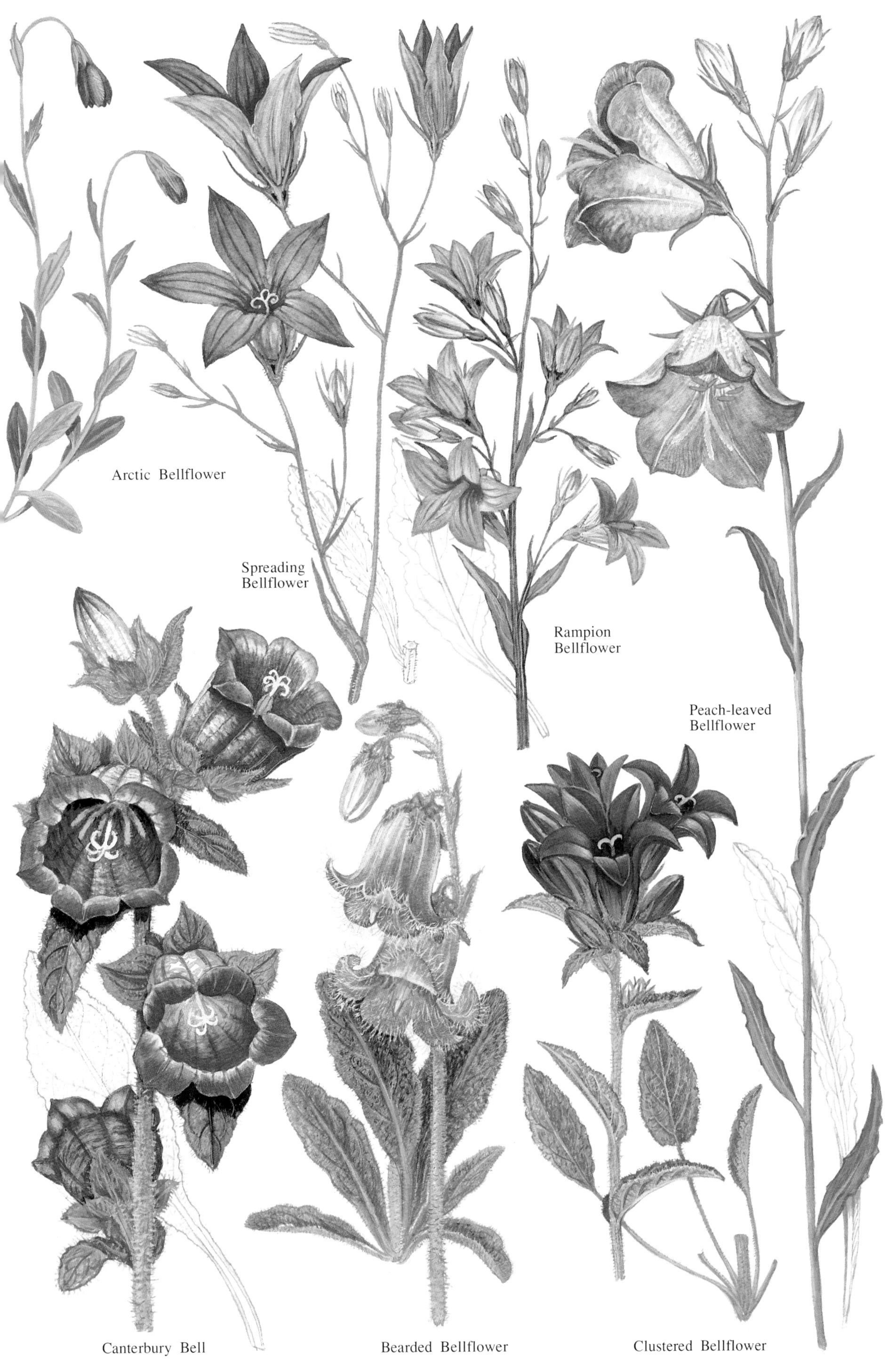
Arctic Bellflower
Spreading Bellflower
Rampion Bellflower
Peach-leaved Bellflower
Canterbury Bell
Bearded Bellflower
Clustered Bellflower

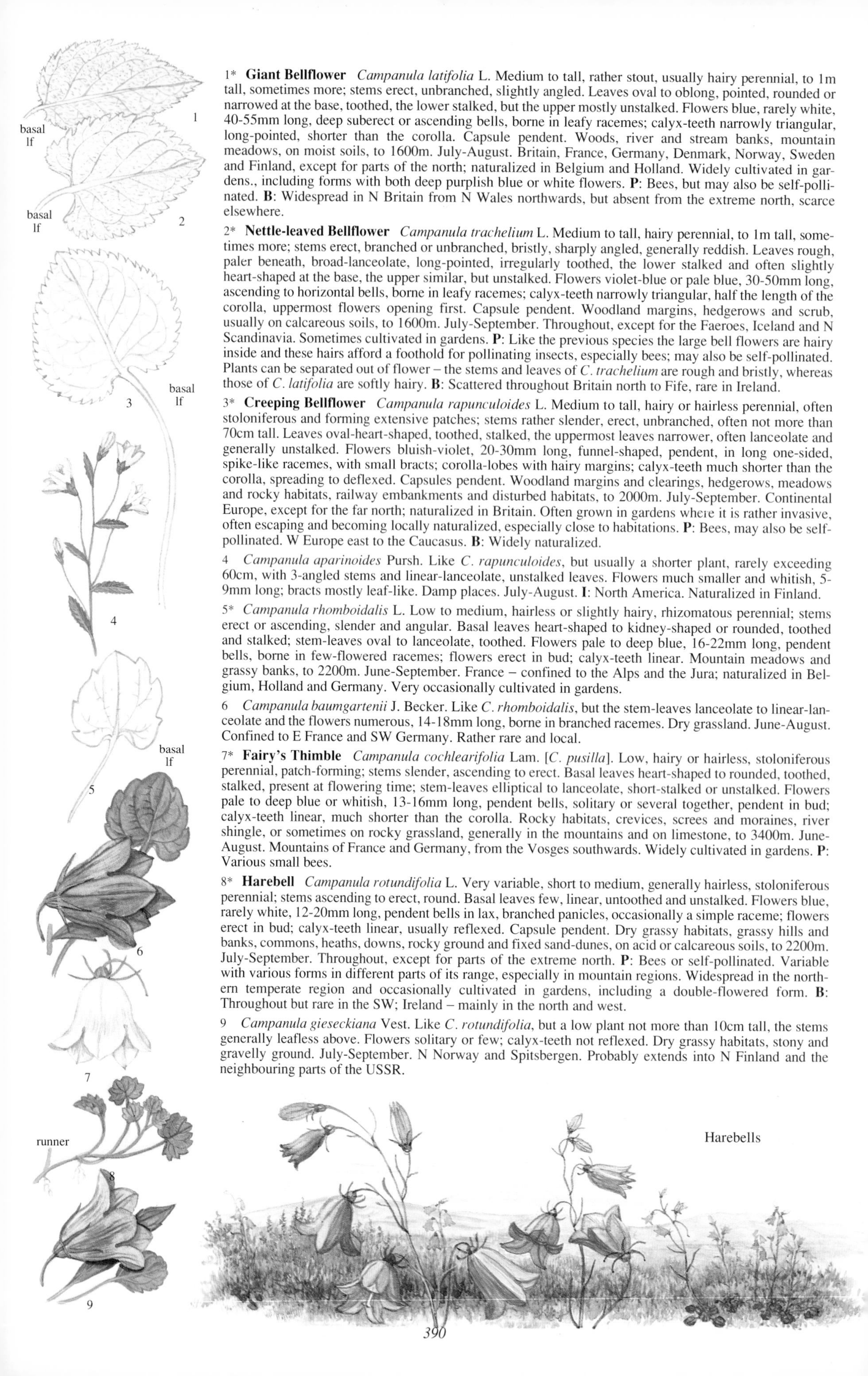


1* **Giant Bellflower** *Campanula latifolia* L. Medium to tall, rather stout, usually hairy perennial, to 1m tall, sometimes more; stems erect, unbranched, slightly angled. Leaves oval to oblong, pointed, rounded or narrowed at the base, toothed, the lower stalked, but the upper mostly unstalked. Flowers blue, rarely white, 40-55mm long, deep suberect or ascending bells, borne in leafy racemes; calyx-teeth narrowly triangular, long-pointed, shorter than the corolla. Capsule pendent. Woods, river and stream banks, mountain meadows, on moist soils, to 1600m. July-August. Britain, France, Germany, Denmark, Norway, Sweden and Finland, except for parts of the north; naturalized in Belgium and Holland. Widely cultivated in gardens., including forms with both deep purplish blue or white flowers. **P**: Bees, but may also be self-pollinated. **B**: Widespread in N Britain from N Wales northwards, but absent from the extreme north, scarce elsewhere.

2* **Nettle-leaved Bellflower** *Campanula trachelium* L. Medium to tall, hairy perennial, to 1m tall, sometimes more; stems erect, branched or unbranched, bristly, sharply angled, generally reddish. Leaves rough, paler beneath, broad-lanceolate, long-pointed, irregularly toothed, the lower stalked and often slightly heart-shaped at the base, the upper similar, but unstalked. Flowers violet-blue or pale blue, 30-50mm long, ascending to horizontal bells, borne in leafy racemes; calyx-teeth narrowly triangular, half the length of the corolla, uppermost flowers opening first. Capsule pendent. Woodland margins, hedgerows and scrub, usually on calcareous soils, to 1600m. July-September. Throughout, except for the Faeroes, Iceland and N Scandinavia. Sometimes cultivated in gardens. **P**: Like the previous species the large bell flowers are hairy inside and these hairs afford a foothold for pollinating insects, especially bees; may also be self-pollinated. Plants can be separated out of flower – the stems and leaves of *C. trachelium* are rough and bristly, whereas those of *C. latifolia* are softly hairy. **B**: Scattered throughout Britain north to Fife, rare in Ireland.

3* **Creeping Bellflower** *Campanula rapunculoides* L. Medium to tall, hairy or hairless perennial, often stoloniferous and forming extensive patches; stems rather slender, erect, unbranched, often not more than 70cm tall. Leaves oval-heart-shaped, toothed, stalked, the uppermost leaves narrower, often lanceolate and generally unstalked. Flowers bluish-violet, 20-30mm long, funnel-shaped, pendent, in long one-sided, spike-like racemes, with small bracts; corolla-lobes with hairy margins; calyx-teeth much shorter than the corolla, spreading to deflexed. Capsules pendent. Woodland margins and clearings, hedgerows, meadows and rocky habitats, railway embankments and disturbed habitats, to 2000m. July-September. Continental Europe, except for the far north; naturalized in Britain. Often grown in gardens where it is rather invasive, often escaping and becoming locally naturalized, especially close to habitations. **P**: Bees, may also be self-pollinated. W Europe east to the Caucasus. **B**: Widely naturalized.

4 *Campanula aparinoides* Pursh. Like *C. rapunculoides*, but usually a shorter plant, rarely exceeding 60cm, with 3-angled stems and linear-lanceolate, unstalked leaves. Flowers much smaller and whitish, 5-9mm long; bracts mostly leaf-like. Damp places. July-August. **I**: North America. Naturalized in Finland.

5* *Campanula rhomboidalis* L. Low to medium, hairless or slightly hairy, rhizomatous perennial; stems erect or ascending, slender and angular. Basal leaves heart-shaped to kidney-shaped or rounded, toothed and stalked; stem-leaves oval to lanceolate, toothed. Flowers pale to deep blue, 16-22mm long, pendent bells, borne in few-flowered racemes; flowers erect in bud; calyx-teeth linear. Mountain meadows and grassy banks, to 2200m. June-September. France – confined to the Alps and the Jura; naturalized in Belgium, Holland and Germany. Very occasionally cultivated in gardens.

6 *Campanula baumgartenii* J. Becker. Like *C. rhomboidalis*, but the stem-leaves lanceolate to linear-lanceolate and the flowers numerous, 14-18mm long, borne in branched racemes. Dry grassland. June-August. Confined to E France and SW Germany. Rather rare and local.

7* **Fairy's Thimble** *Campanula cochlearifolia* Lam. [*C. pusilla*]. Low, hairy or hairless, stoloniferous perennial, patch-forming; stems slender, ascending to erect. Basal leaves heart-shaped to rounded, toothed, stalked, present at flowering time; stem-leaves elliptical to lanceolate, short-stalked or unstalked. Flowers pale to deep blue or whitish, 13-16mm long, pendent bells, solitary or several together, pendent in bud; calyx-teeth linear, much shorter than the corolla. Rocky habitats, crevices, screes and moraines, river shingle, or sometimes on rocky grassland, generally in the mountains and on limestone, to 3400m. June-August. Mountains of France and Germany, from the Vosges southwards. Widely cultivated in gardens. **P**: Various small bees.

8* **Harebell** *Campanula rotundifolia* L. Very variable, short to medium, generally hairless, stoloniferous perennial; stems ascending to erect, round. Basal leaves few, linear, untoothed and unstalked. Flowers blue, rarely white, 12-20mm long, pendent bells in lax, branched panicles, occasionally a simple raceme; flowers erect in bud; calyx-teeth linear, usually reflexed. Capsule pendent. Dry grassy habitats, grassy hills and banks, commons, heaths, downs, rocky ground and fixed sand-dunes, on acid or calcareous soils, to 2200m. July-September. Throughout, except for parts of the extreme north. **P**: Bees or self-pollinated. Variable with various forms in different parts of its range, especially in mountain regions. Widespread in the northern temperate region and occasionally cultivated in gardens, including a double-flowered form. **B**: Throughout but rare in the SW; Ireland – mainly in the north and west.

9 *Campanula gieseckiana* Vest. Like *C. rotundifolia*, but a low plant not more than 10cm tall, the stems generally leafless above. Flowers solitary or few; calyx-teeth not reflexed. Dry grassy habitats, stony and gravelly ground. July-September. N Norway and Spitsbergen. Probably extends into N Finland and the neighbouring parts of the USSR.

Giant
Bellflower
Nettle-
leaved
Bellflower
Creeping
Bellflower
*Campanula
rhomboidalis*
Fairy's Thimble
Harebell

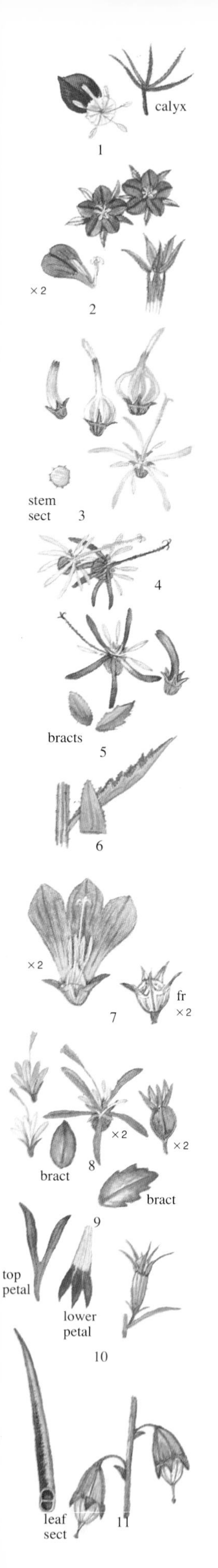


1* **Large Venus's Looking Glass** *Legousia speculum-veneris* (L.) Chaix [*Specularia speculum-veneris*]. Short to medium, often rather hairy annual, to 40cm; stems generally much branched. Leaves alternate, oblong, unstalked or occasionally the lowermost short-stalked, scarcely wavy. Flowers violet, 18-20mm, opening widely to form a blunt 5-pointed star, borne in large, often lax panicles; calyx-teeth shorter than or equalling the ovary and almost as long as the petals. Capsule 10-15mm long, with spreading calyx-teeth. Arable land and other cultivated land, bare and waste ground, at low altitudes. May-July. W Continental Europe, declining.

2* **Venus's Looking Glass** *Legousia hybrida* (L.) Delarbre [*Specularia hybrida*]. Like *L. speculum-veneris*, but a stiffly-hairy plant with markedly wavy leaves. Flowers fewer and smaller, 8-15mm, lilac to reddish-purple, the calyx-teeth longer than the petals. Capsule 15-30mm long, with erect calyx-teeth. Arable land, bare and stony places, generally on light sandy or calcareous soils, at low altitudes. May-August. Declining. **B**: In England from Cumbria southwards; absent elsewhere

**Rampions** *Phyteuma*. Perennial herbs with unbranched leafy stems. Flowers 5-parted, in dense heads or spikes, unstalked, with a ruff of bracts immediately below. Corolla with narrow lobes fused together at the top at first and forming a tube, the lobes later spreading. Stamens free. Stigma 2-3-lobed. Fruit a capsule with 2-3 pores. 45 species, Europe and W Asia.

3* **Spiked Rampion** *Phyteuma spicatum* L. Medium to tall, hairless perennial with erect stems. Basal and lower leaves oval, with a heart-shaped base, blunt, toothed, long-stalked, present at flowering time; upper stem-leaves lanceolate to linear, unstalked. Bracts linear, conspicuous. Flowers whitish to pale yellowish-green, with yellowish-brown styles, borne in dense cylindrical spikes up to 6cm long; stigmas 2-lobed. Meadows, woods, roadsides, often on acid soils, to 2100m. June-August. Throughout north to S Norway; naturalized in Finland and Sweden. **B**: Very rare, only in E Sussex.

4* **Black Rampion** *Phyteuma nigrum* F.W. Schmidt. Medium, hairless perennial; stems erect, leafless in the upper part. Basal leaves heart-shaped, stalked, blunt-toothed, present at flowering time; upper leaves very reduced, lanceolate to linear, not crowded. Bracts linear, longer than the flowers. Flowers blackish-violet, or occasionally blue, rarely white, in an egg-shaped or cylindrical spike, the corolla curved in bud; stigmas usually 3-lobed. Mountain meadows and open woods, to 1200m. June-August. Belgium, France, Germany.

5* **Round-headed Rampion** *Phyteuma orbiculare* L. Very variable short to medium, hairless perennial, occasionally very slightly hairy; stems erect. Basal leaves linear-lanceolate to elliptical, stalked, toothed; stem-leaves lanceolate to linear, often untoothed, the uppermost unstalked. Bracts oval, toothed or untoothed, variable in size. Flowers blue to violet-blue, in rounded heads, the corolla strongly incurved in bud; styles 2-3-lobed. Dry grassland, often closely grazed, rocky habitats, grassy banks and waysides, generally on calcareous soils, to 2600m. June-August. S Britain, France and Germany. Unlike *Jasione* flowerbuds almost hairless and curved. **B**: Very local; confined to south and south-east England from Dorset eastwards.

6 *Phyteuma tenerum* R. Schultz. Like *P. orbiculare*, but stems very leafy and sharply toothed; bracts triangular. Similar habitats and flowering time. France.

7* **Ivy-leaved Bellflower** *Wahlenbergia hederacea* (L.) Reichenb. Low and slender, hairless, trailing perennial. Leaves alternate, palmately-lobed to rounded or kidney-shaped and sharply angled, long-stalked. Flowers pale blue, 6-10mm long, bell-shaped, with 3-5 short lobes, solitary on long thin stalks at the base of the leaves. Capsule erect. Acid peaty places, generally at low altitudes. July-August. Britain, Belgium, France and Germany; extinct in Holland. A delicate plant with small ivy-like leaves. Closely related to *Campanula*, but with slender creeping stems and all the leaves similar and long-stalked. **B**: S & W Britain and S & W Ireland; local elsewhere.

8* **Sheep's-bit** *Jasione montana* L. Very variable, hairy, low to medium annual or biennial, occasionally perennial; stems erect to ascending., leafless above. Leaves linear-oblong to lanceolate, wavy, generally untoothed. Flowers blue, rarely pink or white, in terminal globular heads, each surrounded at the base by a ruff of oval or lanceolate, untoothed bracts; corolla 5-lobed almost to the base. Rough grassland, heaths, rocky hills, sea-cliffs, on light sandy soils, to, 1700m. May-August. Throughout, except the Faeroes, Iceland and parts of the far north. Locally abundant in some areas. **B**: Mainly in E & S Britain and coastal Ireland, local elsewhere.

9 *Jasione leavis* Lam. [*J. perennis*]. Like *J. montana*, but a perennial with numerous non-flowering shoots and linear-oblong leaves which persist at the stem base when dead; bracts toothed. Flowers always blue. Mountain habitats, dry meadows in particular, to 1900m. July-August. Belgium, France and Germany, north to Luxembourg; naturalized in Finland.

10* **Heath Lobellia** *Lobelia urens* L. Medium, almost hairless, erect perennial; stems solid, leafy, with juicy milk. Leaves alternate, oblong to linear-lanceolate, unstalked, slightly toothed. Bracts linear, equalling or longer than the flower-stalks. Flowers blue or purplish, 10-15mm, in lax spike-like racemes; corolla 2-lipped, the upper lip 2-lobed, the lower lip 3-lobed. capsule erect. Grassy places, woodland margins, generally at low altitudes. July-October. W and SW Europe, often rather local. **P**: Mainly flies. Very susceptible to being swamped by coarser vegetation, especially bracken. **B**: Very rare in S England from Sussex to Cornwall.

11* **Water Lobelia** *Lobelia dortmanna* L. Hairless, stoloniferous aquatic perennial with hollow flower-stems rising above the water surface. Leaves linear-oblong, in a basal tuft, untoothed and unstalked. Bracts oval, much smaller than the flower-stalks. Flowers pale lilac, 15-20mm, pendent, in a few-flowered raceme; corolla 2-lipped. Capsule pendent. Lakes, pools and tarns, in still acid waters up to 3m deep. July-August. Local in uplands of N and NW Europe, but absent from most of France except the north-west. **B**: Local in Wales, the Lake District, Scotland and W Ireland.

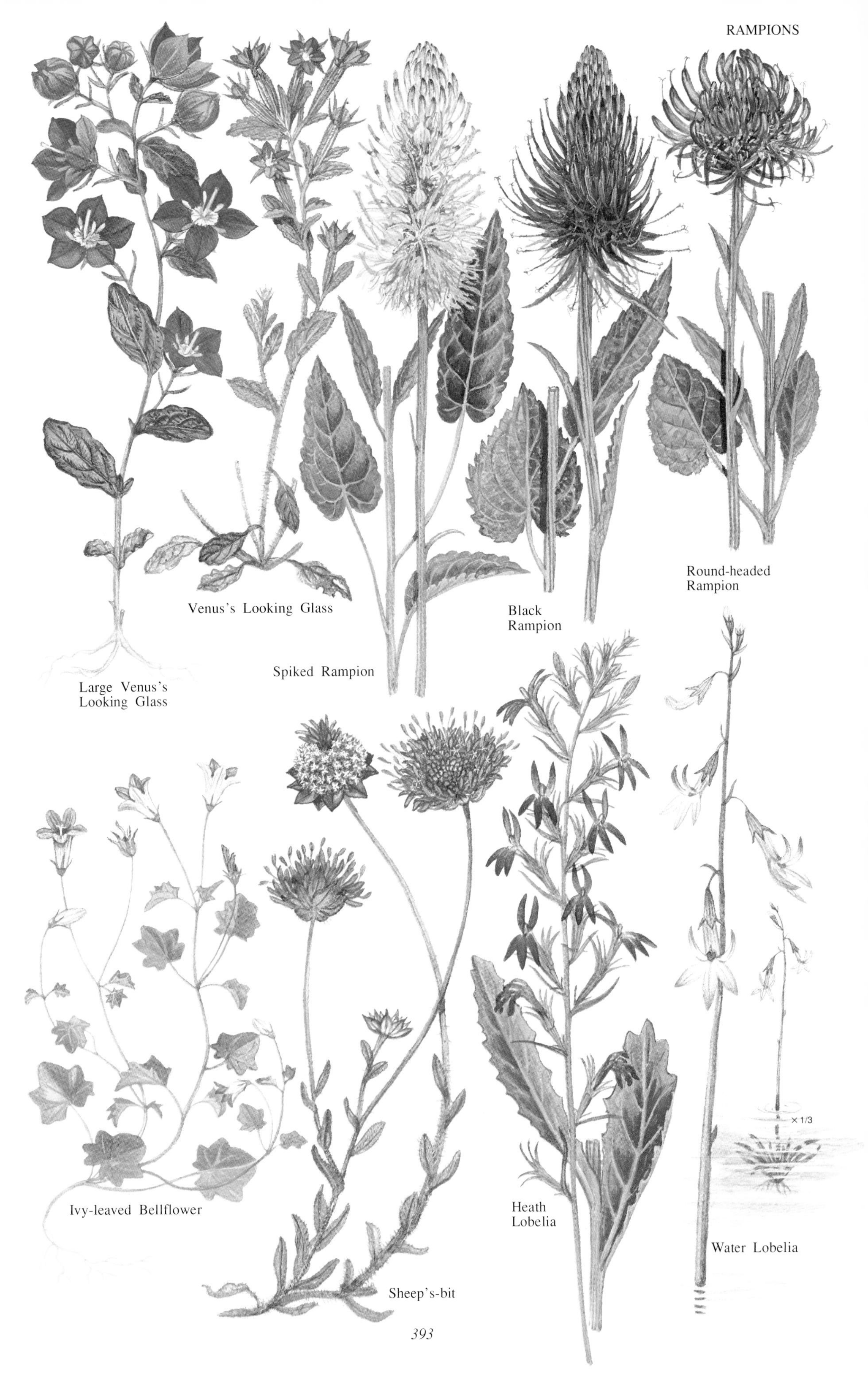
Large Venus's Looking Glass
Venus's Looking Glass
Spiked Rampion
Black Rampion
Round-headed Rampion
Ivy-leaved Bellflower
Sheep's-bit
Heath Lobelia
Water Lobelia
× 1/3

## DAISY FAMILY Compositae

Herbs or shrubs with alternate, opposite or rosetted leaves. Flowers small (florets) in congested heads (capitula), often with receptacle scales at the base of each floret. Flowerheads with an involucre of bracts around the base. Florets all similar in the capitulum or those in the centre (disc) different from the outer (rays) giving the typical daisy flowerhead; florets tubular and 5-toothed, tubular and 2-lipped or ligulate with a one-sided strap-like appendage. Stamens 5 fused together around the style. Ovary inferior. Fruit an achene, often with a feathery appendage (pappus). Over 14,000 species in 900 genera, worldwide. Mainly insect-pollinated.

1* **Hemp-agrimony** *Eupatorium cannabinum* L. Robust, medium to tall, hairy perennial; stems erect, often reddish, to 1.75m. Leaves opposite, palmately-lobed, leaflets coarsely toothed, short-stalked, the uppermost undivided. Flowerheads pink or purplish, occasionally white, 2-5mm, in dense, rather flat-topped clusters; florets all tubular. Damp habitats, at low altitudes. July-September. Throughout, except the far north. **B**: Common in England and Wales and parts of Ireland, scarcer elsewhere.

2* **Goldenrod** *Solidago virgaurea* L. Very variable, low to tall, hairy or hairless perennial. Basal leaves oblong, broadest above the middle, stalked; stem leaves linear-lanceolate, decreasing up the stem, unstalked. Flowerheads bright yellow, 15-18mm, in lax, rather narrow panicles. Dry places, on calcareous or acid soils, to 2800m. July-September. Throughout. Mountain forms with shorter stems and simpler inflorescences and hairless leaves are often called subsp. *minuta* (L.) Arcangeli. **B**: Throughout, scarce in C England, East Anglia and C Ireland.

3* **Canadian Goldenrod** *Solidago canadensis* L. A vigorous medium to tall, perennial; stems erect, to 1.5m tall, hairless below, but hairy above, densely leafy. Leaves lanceolate, pointed, toothed, with 2 prominent lateral veins, hairy at least on the margin and along the veins beneath. Flowerheads golden-yellow, 5-6mm, each with 10-17 short rays, in broad pyramidal panicles. August-October. **I**: North America. Widely naturalized. **B**: Naturalized or casual.

4 **Early Goldenrod** *Solidago gigantea* Aiton. Often taller, to 2.5m, with hairless, often bluish-green, stems, and hairless leaves. Similar habitats. July-September. **I**: North America. Locally naturalized north to S Sweden, but not in Norway. **B**: Locally naturalized in many areas.

5 *Solidago graminea* (L.) Salisb. Like *S. canadensis*, but stems often hairless and leaves narrow, crowded, untoothed, with rough margins, and 2-4 parallel veins. Flower panicle broader and flatter. Widely cultivated and sometimes naturalized on waste ground and waysides. July-October. **I**: North America. Naturalized in Britain, France and Germany. **B**: Occasionally naturalized.

6* **Daisy** *Bellis perennis* L. A variable, low, hairy perennial. Leaves in a basal rosette, bright green, oblong to spoon-shaped, toothed or untoothed. Flowerheads with a yellow disk and numerous narrow white rays, 15-30mm, solitary on long slender stalks, the rays often tipped with red or flushed with purplish-red underneath. Short grassy habitatssl on a wide variety of soils, to 2500m. Flowering almost all year round, except at altitude. Throughout, except Spitsbergen; naturalized in much of Scandinavia. **B**: Throughout.

**Asters** *Aster*. Herbs, mostly perennial, with alternate, simple leaves. Flowerheads solitary or in panicles or corymbs, each with a single row (1 whorl) of rays surrounding a central disk; blue, purple, pink or white, female or sterile, the disk-florets hermaphrodite. 500 species.

7* *Aster macrophyllus* L. A medium to tall perennial with non-flowering leafy rosettes; stems erect, green or purple-tinged, glandular-hairy. Lower and basal leaves heart-shaped, with slender unwinged stalks; upper leaves oval, short-stalked or unstalked. Flowerheads with pale violet rays, often fading almost to white, 25-30mm, and a yellowish disk, in broad, rather flat, branched clusters. Waste and grassy places. July-September. **I**: North America. Naturalized or casual in Holland and Germany.

8 *Aster schreberi* Nees. Like *A. macrophyllus*, but without glands and the lower leaves with a broad rectangular gap (sinus) at the base. Rays white. **I**: North America. **B**: Naturalized in Scotland – Renfrew.

9 *Aster divaricatus* L. Shorter exceeding 60cm tall, the stems flexuous above, blackish-purple, not glandular; upper leaves heart-shaped to narrowly triangular. Flowerheads with only 5-10 (not 9-20) white rays. **I**: North America. Locally naturalized in grassy and waste places in Holland. July-September.

10* *Aster laevis* L. A medium to tall perennial; stems erect, hairless, usually reddish-purple. Leaves oval-lanceolate to lanceolate, slightly toothed, with long winged stalks. Flowerheads violet-blue, 20-30mm, with a yellowish disk, borne in a long panicle; flower-bracts very unequal, whitish with a green tip. Scrub, damp woods and river and stream banks. July-October. Naturalized or casual throughout north to S Norway. **B**: Naturalized locally in S England and Ireland – Tyrone.

11 *Aster puniceus* L. [*A. hispidus*]. Like *A. laevis*, but plants stiffly-hairy and the leaves with a pair of small lobes at the base, smelling of crushed juniper berries; flower-bracts more or less equal, green. **I**: E North America. **B**: Locally naturalized in N England and S Scotland.

12* **Michaelmas Daisy** *Aster novi-belgii* L. [*A. brumalis*]. A medium to tall perennial; stems usually erect, often purplish, with lines of hairs. Leaves oval-lanceolate to linear-lanceolate, the stem-leaves mostly with broad clasping bases, slightly toothed. Flowerheads with violet-blue to purple rays and a yellow disk, 25-40mm, borne in broad to narrow, generally rather flattish clusters, the inflorescence with long ascending lower branches. Waste places, waysides, riverbanks. September-October. **I**: North America. Widely naturalized except in the far north. **B**: Widely naturalized.

13 *Aster × salignus* Willd. Like *A. novi-belgii*, but the stem-leaves narrowed at the base, without small basal lobes; flowerheads with pale violet-blue rays. Same habitats. August-October. **I**: North America. Widely naturalized or casual. **B**: Scattered, in Cambridgeshire, Oxfordshire, Cornwall, SE England, and parts of Scotland.

14 *Aster × versicolor* Willd. [*A. laevis × A. novi-belgii*]. Like *A. laevis*, but the leaves somewhat bluish-green beneath, the lower very short-stalked, the middle ones oval; flowerheads with white rays, becoming bluish-violet. Locally naturalized in Britain, Holland, France and Germany. August-October.

*15 *Aster novae-angliae* L. A medium to tall hairy perennial; stems erect, hairy, glandular above, to 2m tall. Leaves lanceolate to oval, with a pair of small lobes at the base, untoothed. Flowerheads with reddish-purple to deep violet rays and a yellow disk, 20-40mm, borne in rather flat-topped clusters. Waysides, stream banks. September-November. **I**: North America. Widely naturalized in W Europe. **B**: Local.

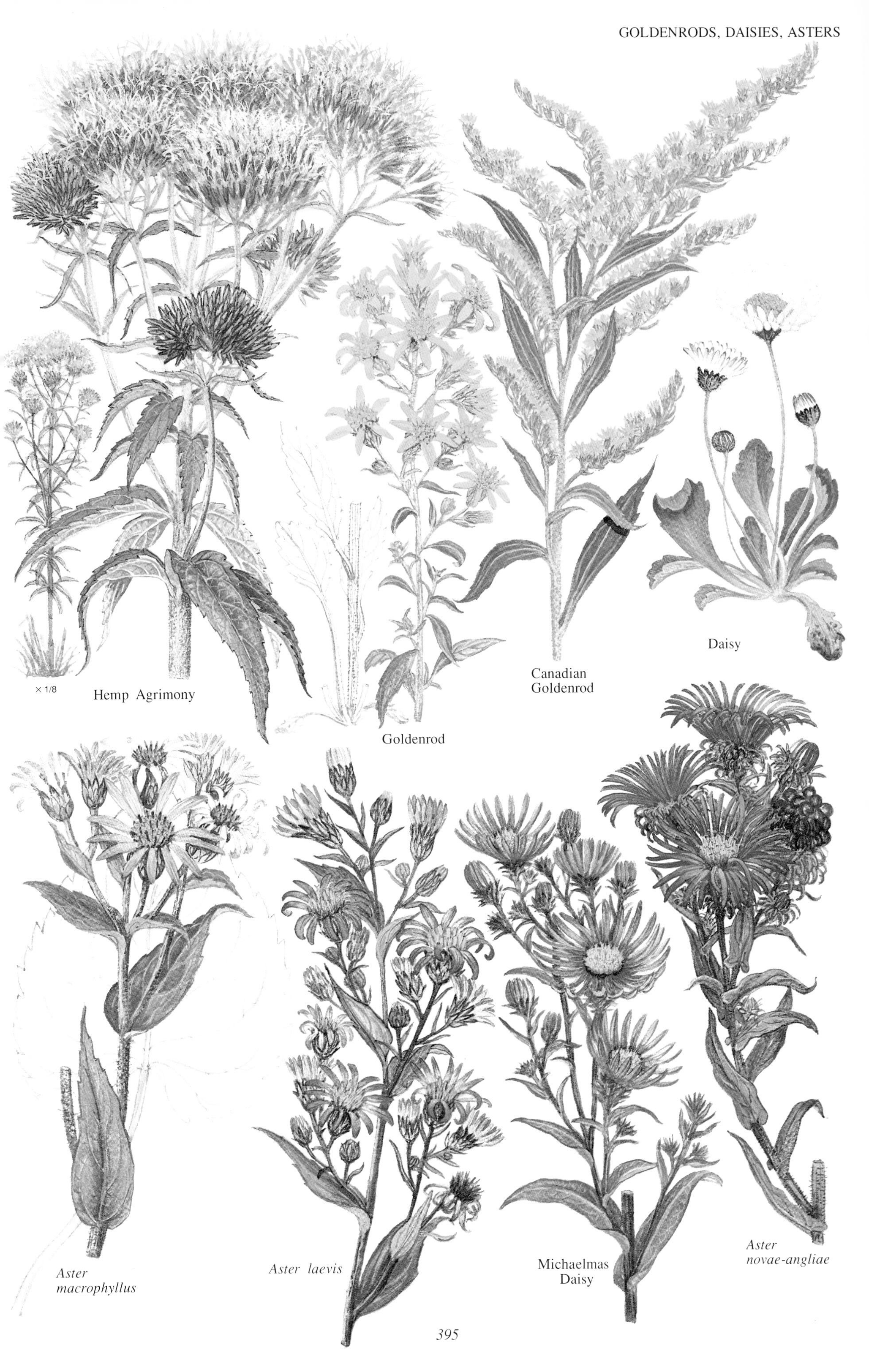
× 1/8
Hemp Agrimony
Goldenrod
Canadian Goldenrod
Daisy
*Aster macrophyllus*
*Aster laevis*
Michaelmas Daisy
*Aster novae-angliae*

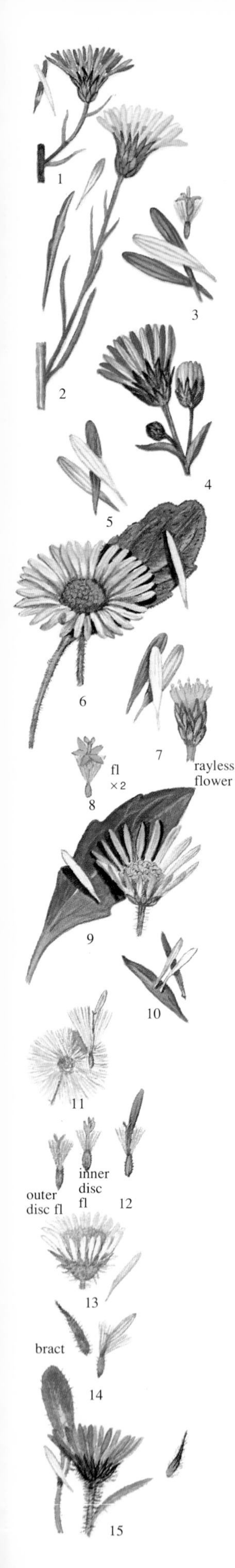


1* *Aster lanceolatus* Willd. Very variable tall perennial, to 1.3m; stems erect, green, sometimes tinged with purple. Leaves lanceolate to linear-lanceolate, narrowed at the base, hairless, untoothed or slightly toothed. Flowerheads with white, sometimes violet-blue, rays surrounding a yellow disk, 15-25mm, borne in narrow panicles; outer flower-bracts without a slender green tip. River banks, waste ground. August-October. **I**: North America. Naturalized north to Norway. **B**: Local in England.

2 *Aster pilosus* Willd. A medium perennial, not exceeding 60cm tall. Leaves linear to linear-lanceolate. Flowerheads in a wide panicle, the outer flower-bracts with a slender green tip. Riverbanks and waste ground. August-October. **I**: North America. Locally naturalized in Holland.

3* **European Michaelmas Daisy** *Aster amellus* L. Very variable short to medium, slightly hairy perennial; stems erect from a spreading base, not glandular. Basal and lower leaves broad-lanceolate to oval, narrowed to a short stalk, untoothed or blunt-toothed; upper leaves narrower, unstalked. Flowerheads few, up to 10, with blue, occasionally red or white rays, surrounding a yellow disk, 30-50mm, borne in lax flat-topped clusters. Scrub and woodland margins, open woodland meadows and rocky places, to 1400m. August-September. C & E France and Germany.

4 *Aster sibiricus* L. Shorter, the stems often purplish and the stem leaves half-clasping the stem at the base. Flowerheads with violet rays; outer flower bracts pointed (not blunt). August-September. One locality in Norway.

5* **Alpine Aster** *Aster alpinus* L. Low to short, tufted, hairy perennial; stem ascending to erect. Leaves untoothed, the lower elliptical, generally broadest above the middle, stalked, the upper narrower and unstalked. Flowerheads solitary, with violet-blue, rarely pink or white, rays, surrounding a yellow disk, 35-45mm. Dry meadows, stony and rocky habitats, in the mountains, to 3200m. July-September. C & E France and C & S Germany.

6 **False Aster** *Aster bellidiastrum* (L.) Scop. [*Bellidiastrum michelii*]. Like *A. alpinus*, though often taller, with leafless stems. Flowerheads with white or pink rays, 20-40mm. Similar habitats and flowering time, to 2800m. E France and S Germany, from the Jura Mountains southwards.

7* **Sea Aster** *Aster tripolium* L. [*Tripolium vulgare*]. Hairless, short to medium annual or short-lived perennial; stems often reddish, erect or ascending from a branched base. Leaves fleshy, linear to lanceolate, rounded in section, half-clasping the stem, the upper unstalked. Flowerheads with bright blue or purple rays surrounding a yellow disk, 8-20mm, borne in large, often flat-topped, panicles. Coastal habitats, sometimes in saline habitats inland. July-October. Throughout, except the far north. A form without ray florets occurs, var. *discoideus*, especially in the south. **B**: Coastal, most frequent in the south.

8* **Goldilocks Aster** *Aster linosyris* (L.) Bernh. [*Linosyris vulgaris*]. Hairless short to medium perennial; stems spreading to erect, slightly roughened, densely leafy. Leaves linear, pointed, unstalked, rough-edged, 1-veined. Flowerheads small, 12-18mm, bright yellow, without rays, borne in dense flat-topped clusters. Cliffs, open grassland, generally on calcareous soils at low altitudes. August-September. North to S Sweden; naturalized in Holland. Declining. **B**: Coasts of Devon, N Somerset, Wales, Lancashire.

## Fleabanes *Erigeron*. Annual or perennial herbs, similar to *Aster*. Stems often with long, non-glandular hairs. Flowerheads with linear ray-florets in 2 (not 1) rows. 200 species.

9 **Sweet Scabious** *Erigeron annuus* (L.) Pers. [*Stenactis annua*]. Medium to tall, somewhat hairy annual or short-lived perennial; stems slender, erect, branched above. Leaves oval, often broadest above the middle, mostly toothed, short-stalked. Flowerheads 15-20mm, with slender white or pale blue rays, borne in lax, rather flat-topped clusters. Fields and waysides. July-August. **I**: North America. Naturalized in W Europe. subsp. *strigosus* (Muhlenb. ex Willd.) Wagenitz. has stems with appressed hairs and subsp. *septentrionalis* (Fernald & Wieg.) Wagenitz & Hegi untoothed stem leaves and white-rayed flowers.

10* **Wall Daisy, Mexican Fleabane** *Erigeron karvinskianus* DC. [*E. mucronatus*]. Short to medium, loosely tufted, much branched perennial; stems slender, slightly hairy. Leaves small wedge-shaped or oval, often 3-lobed, short-stalked, the uppermost narrower. Flowerheads 12-15mm, with slender pale purple, pink or white rays, borne in lax leafy, branched clusters. Rocky places, dry habitats, pavements and walls. July-September. **I**: Mexico. Naturalized in S Britain and France. **B**: Naturalized in SW England, the Channel Islands and Scilly Islands.

11* **Blue Fleabane** *Erigeron acer* L. Very variable, densely grey-hairy, short to medium annual or biennial; stems erect. Basal leaves narrow elliptical to oval, toothed or untoothed, stalked; upper leaves lanceolate, unstalked. Flowerheads 10-15mm, with purple or lilac rays, borne in panicles, often flat-topped, occasionally solitary; rays scarcely longer than the disk florets; flower-bracts often purplish. Dry grassland, walls on well-drained calcareous soils, to 2200m. July-August. Throughout, except the far north. **B**: Common in England and Wales, scarce elsewhere. subsp. *droebachensis* (O.F. Mueller) Arcangeli: leaves hairless, the upper smaller than the basal; flower-bracts hairless with a lilac apex. Denmark, Germany, S Norway. subsp. *politus* (Fries) H. Lindb. fil.: flower-bracts uniformly purplish. Finland.

12* **Alpine Erigeron** *Erigeron alpinus* L. Variable, hairy, low to short perennial; stems erect, with basal leaf rosettes. Basal leaves narrow elliptical, often broadest above the middle, stalked, pointed, densely hairy, untoothed; upper leaves lanceolate, unstalked, half-clasping the stem. Flowerheads 20-30mm, with lilac rays, borne in clusters of 2-3 or solitary. Mountain meadows, stony places, open woods, 1500-3050m. July-September.

13* **One-flowered Fleabane** *Erigeron uniflorus* L. Low to short, slightly hairy perennial, rarely more than 15cm tall. Basal leaves spoon-shaped, stalked, untoothed, hairy-margined, when young at least; upper leaves few, narrower and unstalked. Flowerheads 10-15mm, held well above the leaf-rosettes, solitary, the rays white or pale lilac; flower-bracts lilac-tipped, with long white hairs. Mountain pastures and around permanent snow patches, 1200-3000m. July-September. Alps, mountains of N Europe.

14* **Alpine Fleabane** *Erigeron borealis* (Vierh.) Simmons. Short hairy or hairless, perennial; stems erect. Lower and basal leaves oval; broadest above the middle, stalked, untoothed; upper leaves narrower and unstalked. Flowerheads 10-16mm, with lilac rays, usually solitary or occasionally 2-3 together; flower-bracts hairy. Mainly in mountains on calcareous soils, to 1200m. July-August. N Britain, Scandinavia. **B**: Very rare, a few scattered localities in Scotland. *Erigeron neglectus* A. Kerner. Like *E. borealis*, but stems stiffer and basal leaves with a hairy margin. Limestone rocks, stony ground. 1800-2600m. July-September. E France and S Germany – Alps.

15 *Erigeron humilis* R.C. Graham. Flowerheads at or just above the leaf-rosettes; flower-bracts dark, with purplish hairs. Damp stony hillsides, tundra, to 1400m. Arctic and sub-Arctic Europe.

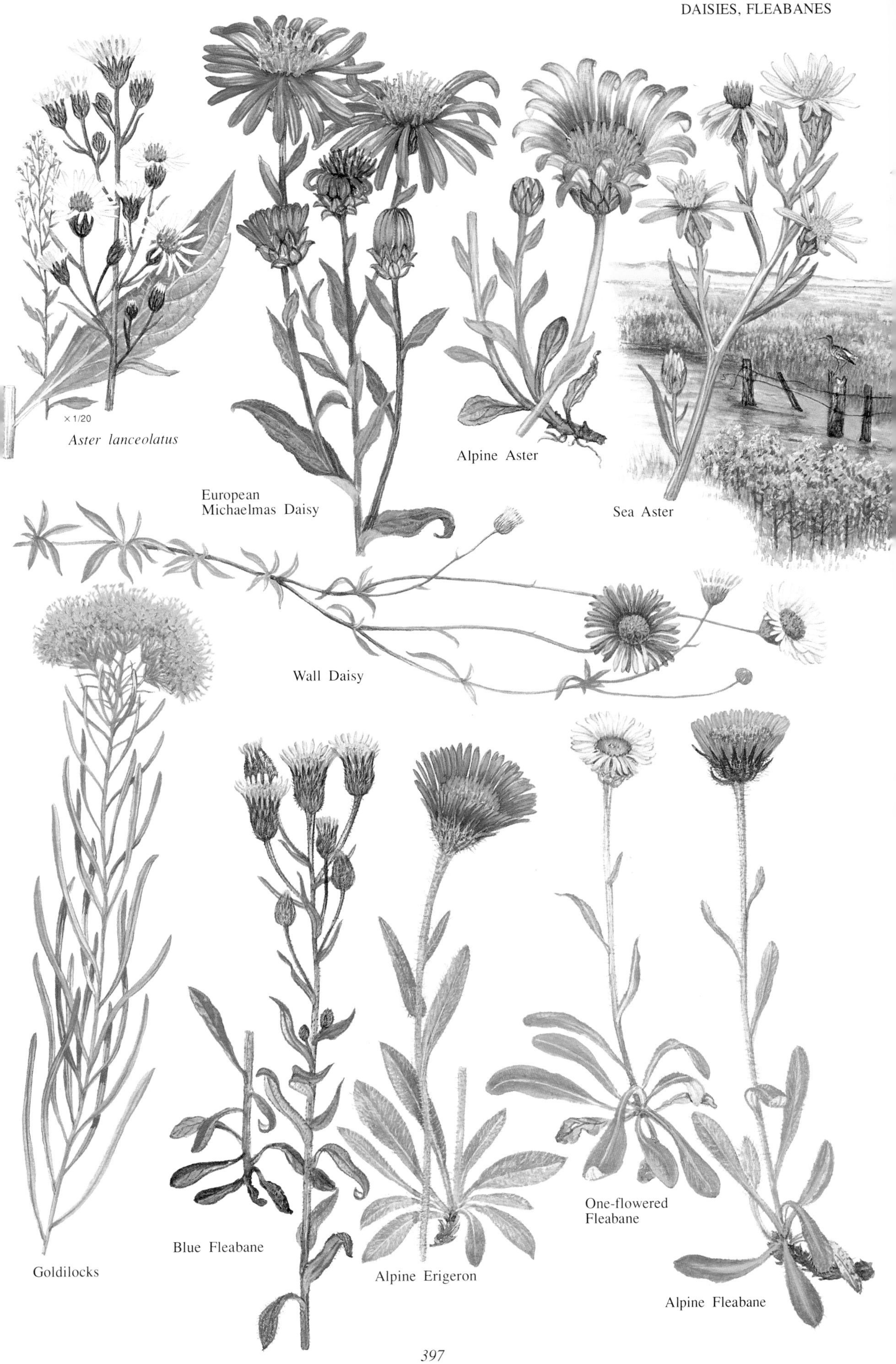

× 1/20

*Aster lanceolatus*

European Michaelmas Daisy

Alpine Aster

Sea Aster

Wall Daisy

Goldilocks

Blue Fleabane

Alpine Erigeron

One-flowered Fleabane

Alpine Fleabane

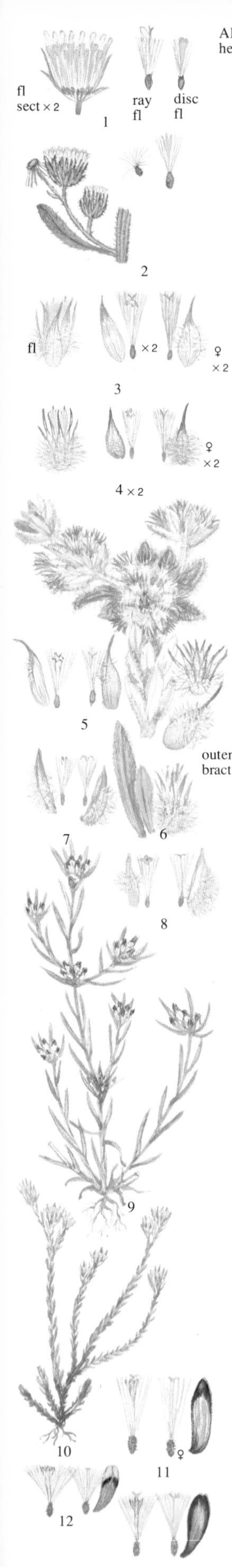


1* **Canadian Fleabane** *Conyza canadensis* (L.) Cronq. [*Erigeron canadensis*]. Short to tall, hairy annual, to 1.5m. Leaves alternate, narrow oblong, often broadest above the middle, stalked, the lower often deciduous before flowering. Flowerheads 2-5mm, with white, rarely pinkish ray florets scarcely longer than the yellow disk florets, borne in branched clusters; flower-bracts usually hairless. Cultivated and waste land, walls, dunes. July-September. **I**: North America. Naturalized throughout, except Ireland and the far north. **B**: Locally naturalized.

2 *Conyza bonariensis* (L.) Cronq. [*C. ambigua, Erigeron bonariensis, E. crispus*]. Taller, to 2.5m, more densely hairy. Inflorescence often with long branches overtopping the main axis. Flowerheads larger, 7-10mm. **I**: Tropical America. W & C France.

## **Cudweeds** *Filago*. Downy and woolly annuals with alternate, untoothed leaves. Flowerheads small, in lateral or terminal clusters, all the florets tubular, the outermost female, the others hermaphrodite. Pappus usually present. 12 species in Europe, North and South America.

3* **Common Cudweed** *Filago vulgaris* Lam. [*F. germanica, F. canescens*] Short to low, grey-white woolly annual; stems erect, regularly branched from the middle. Leaves linear-lanceolate to lanceolate, often rather wavy. Flowerheads yellow, in dense rounded clusters of 20 or more, 10-14mm across, not overtopped by the subtending leaves; flower-bracts yellowish with a red tinge. Heathy grassland, embankments, disturbed ground, on light acid sandy soils. July-August. Throughout, except Ireland and much of Scandinavia. **B**: Throughout England and Wales; rare and scattered elsewhere.

4* **Red-tipped Cudweed** *Filago lutescens* Jordan [*F. apiculata*]. Short yellowish-green plant covered in yellowish wool; stem erect, rather irregularly branched. Leaves oblong-lanceolate to spoon-shaped. Flowerheads yellowish, in small clusters of 10-25, overtopped by 1-2 narrow subtending leaves; flower-bracts red- or purple-tipped. Fields, gravel pits, waysides, on well-drained sandy soils. July-August. North to S Sweden. **B**: Very rare and decreasing; confined to Surrey, East Anglia and E Yorkshire.

5 **Broad-leaved Cudweed** *Filago pyramidata* L. Like *F. lutescens*, but a greyish-white, almost prostrate, much-branched plant, often with narrower leaves. Flowerheads in clusters of 5-20, sometimes overtopped by subtending leaves; flower-bracts curved, yellow-tipped. Borders of arable fields, road verges, disturbed habitats, waysides, on well-drained sandy or chalky soils. July-August. W Europe. **B**: Very rare and decreasing, confined to East Anglia, SE England, Hampshire, Berkshire and Hertfordshire.

## **Logfia**. Flowerheads solitary or in small clusters of 2-7 at the leaf-axils. Flower-bracts 15-20.

6* *Logfia arvensis* (L.) J. Holub [*Filago arvensis, F. montana*] White woolly, short annual; stem erect, with short lateral branches above. Leaves oblong to linear-lanceolate, 10-20mm long. Flowerheads yellowish, 2.5-6mm, in small clusters generally overtopped by the leaves at their base, in a raceme-like inflorescence or a panicle; flower-bracts with a transparent apex. Arable land, fields and sandy habitats. July-August. Continental Europe, except the far north. **B**: Casual.

7* **Small Cudweed** *Logfia minima* (Sm.) Dumort. [*Filago minima*]. Low to short, grey-silkily-hairy annual; stem slender, erect, branched above the middle. Leaves oblong-linear to linear, flat, 4-10mm long. Flowerheads 2.5-3.5mm, in clusters, not overtopped by the subtending leaves, terminal or in the stem forks; flower-bracts pale, hairless in the upper part. Fields, waste land, heaths, on sandy or gravelly soils. June-September. Throughout, except most of the north. **B**: Local; not in the Scottish Isles, rare in Ireland.

8* **Narrow-leaved Cudweed** *Logfia gallica* (L.) Cosson & Germ. [*Filago gallica*]. Very variable greyish-hairy, low to short annual; stems slender, much-branched. Leaves linear, often thread-like, 15-25mm long. Flowerheads 2.5-4mm, in clusters much overtopped by the subtending leaves; flower-bracts woolly and swollen below, with a thin yellowish tip. Dry gravelly grassland. July-September. W and S Europe. **B**: Declining, possibly alien; scattered localities in SE England and the Channel Islands.

9 *Logfia neglecta* (Soyer-Willemet) J. Holub [*Gnaphalium neglectum, Filago neglecta*]. Like *L. gallica*, but the stem generally branched from the base, flower-bracts brownish, terminal or at the branch forks. Similar habitats and flowering time. Belgium and France. Also found in Corsica.

10 *Bombycilaena erecta* (L.) Smolj. [*Micropus erectus*]. Low to short, greyish-white, woolly annual. Leaves alternate, linear to lanceolate, with wavy margins. Flowerheads small, yellowish, in small clusters of 2-3, generally overtopped by the subtending leaves. Dry sandy and stony habitats. June-September. NC & S France. Like *Logfia*, but the flowerheads without a pappus.

## *Omalotheca*. Hairy, often woolly perennials with non-flowering shoots. Leaves alternate. Flowerheads in spikes or racemes, the florets all tubular, the outer female, the inner hermaphrodite and usually reddish at the apex; flower-bracts mottled. 120 species.

11* **Highland Cudweed** *Omalotheca norvegica* (Gunn.) Schultz Bip. & F. W. Schultz., *Gnaphalium norvegica* Gunn.). Low to short, white-downy perennial, with short non-flowering leafy shoots and erect flowering stems. Leaves lanceolate, 3-veined, the lower stalked, mostly the same length but shorter close to the flower spike, white downy on both surfaces. Flowerheads 6-7mm, in compact leafy spikes, reddish or yellowish; flower-bracts with a dark brown papery margin. Meadows, open woods, stony places, generally on acid soils, to 2800m. July-August. N Europe, France and Germany. **B**: Very rare, only in the Scottish Highlands.

12* **Heath Cudweed** *Omalotheca sylvatica* (L.) Schultz Bip. & F. W. Schultz [*Gnaphalium sylvaticum*]. Like *O. norvegica* but leaves linear to lanceolate, 1-veined, gradually diminishing in size up the stem. Flowerheads in small clusters forming a long interrupted spike. Pappus reddish (not white). Grassy heaths, commons, open woods and woodland clearings, recently coppiced land and scrub, on acid sandy soils, to 2500m. July-September. Widespread throughout. **B**: Local throughout but rare in SW England, Wales and Ireland.

13* **Dwarf Cudweed** *Omalotheca supina* (L.) DC. [*Gnaphalium supinus*]. Low, tufted perennial, often less than 6cm tall, with numerous non-flowering shoots. Leaves linear-lanceolate to spoon-shaped, 1-veined, grey-woolly on both surfaces. Flowerheads few, up to 7, in a compact leafy head, reddish, partly hidden by the flower-bracts which have papery-brown margins. Rocky places, damp meadows, on acid and basic rocks, mostly between 1400-3400m. July-August. N Europe and the mountains of France and Germany. **B**: Widespread in the Scottish Highlands and on the island of Skye.

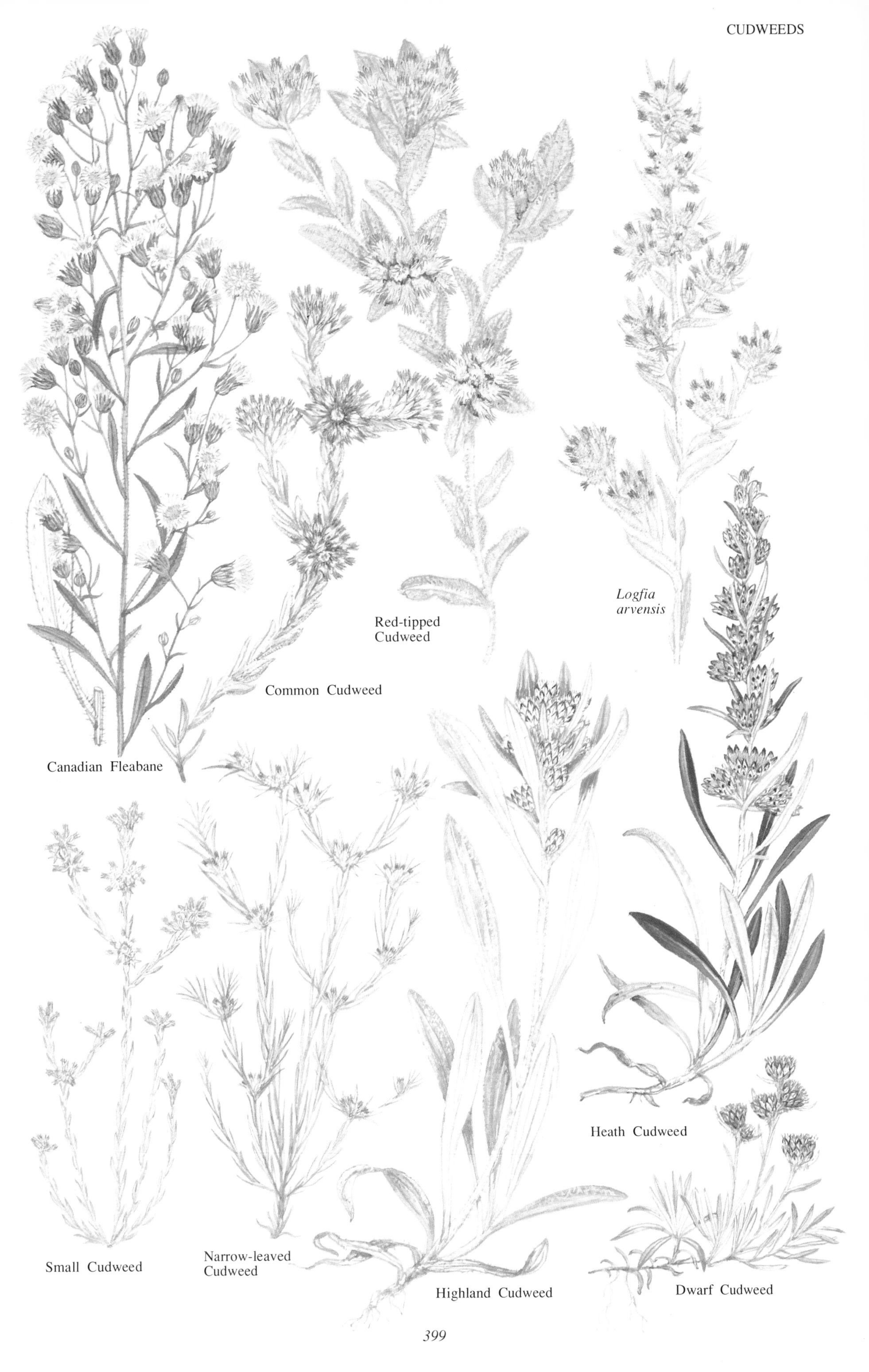

Canadian Fleabane
Common Cudweed
Red-tipped Cudweed
*Logfia arvensis*
Small Cudweed
Narrow-leaved Cudweed
Highland Cudweed
Heath Cudweed
Dwarf Cudweed

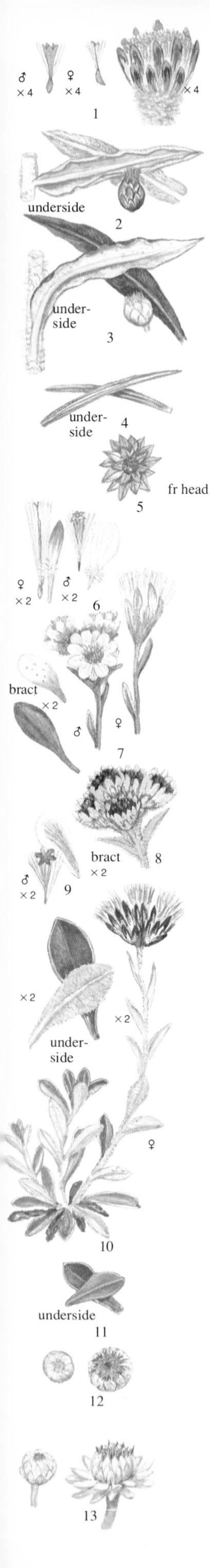


1* **Marsh Cudweed** *Filaginella uliginosa* (L.) Opiz [*Gnaphalium uliginosum*]. Much-branched, short to medium, silvery-grey-woolly annual. Leaves alternate, linear-lanceolate to oblong, green above. Flowerheads without rays, yellowish-brown, borne in clusters overtopped by the subtending leaves; flower-bracts brownish. Damp habitats, on clay or sandy soils, generally not calcareous. July-September. Throughout, except the far north. **B**: Widespread except N Scotland and C Ireland.

2* **Jersey Cudweed** *Gnaphalium luteoalbum* L. Short to medium white-woolly annual, without non-flowering shoots; stems branched near the base. Leaves alternate, linear to oblong, white-woolly on both surfaces, then margins inrolled. Flowerheads yellowish and reddish, without rays, borne in egg-shaped, leafless heads; flower-bracts yellowish, hairless. Damp, generally sandy but well-drained habitats. June-August. North to S Sweden. Differs from *Filaginella* by the absence of leaves immediately below the flower clusters. **B**: Rare and declining, scattered localities in East Anglia and the Channel Is., occasional casual elsewhere.

3 *Gnaphalium undulatum* L. Like *G. luteoalbum*, but taller and more robust, the leaves green above while white and woolly beneath. Flowerheads in lax flat-topped clusters; flower-bracts white, hairless. Roadsides, waste places. June-August. **I**: South Africa. Naturalized NW France, Channel Is.

4* *Helichrysum stoechas* (L.) Moench. Variable, short to medium subshrub; stems woody below, erect. Leaves linear to linear-elliptical, white-woolly, sometimes hairless above, the margins rolled under, strongly aromatic when crushed. Flowerheads rounded, in a lax cluster 15-30mm across; florets yellowish, all tubular, surrounded by conspicuous white papery bracts. Sandy habitats, dunes, cliffs and scrub. May-August. France.

5* *Helichrysum arenarium* (L.) Moench. Short, greyish-white, downy perennial, not aromatic when crushed; stems erect from a branched stock, with non-flowering leafy rosettes at the base. Leaves oblong, often broadest above the middle, flat, 1-veined, stalked; uppermost leaves lanceolate to thread-like. Flowerheads small, in dense clusters 20-50mm across; florets all tubular, yellowish, surrounded by yellowish or reddish-orange, shiny, flower-bracts. Sandy, often exposed habitats and rocks. July-August. North to S Sweden.

**Cat's-foot** *Antennaria*. Small tufted perennials with basal leaf-rosettes and erect, unbranched leafy stems; leaves untoothed. Flowerheads in terminal clusters, florets tubular, either male or female, borne on separate plants, surrounded by thin papery, persistent flower-bracts; pappus present, hairs in several rows. About 15 species, mainly N Arctic.

6* **Mountain Everlasting, Cat's-foot** *Antennaria dioica* (L.) Gaertner (incl. *A. hibernica* Br.-Bl.). Low to short, mat-forming perennial with slender, branched stolons. Basal and rosette leaves oval, to spoon-shaped, often notched at the apex, downy below but generally hairless above; upper leaves narrower and pointed. Flowerheads 6-12mm, in clusters of 2-8; flower-bracts white in male flowerheads, pink in female. Dry mountain grassland, on base-rich or calcareous soils, to 3000m. May-July. Throughout, except the far north. **B**: Widespread in upland Britain and W Ireland, rarer elsewhere. *A. d.* var. *hyperborea* (D. Don) DC. Like the type, but leaves broader, downy-white on both surfaces. **B**: Confined to the Hebrides.

7 *Antennaria nordhageniana* Rune & Ronning. Like *A. dioica*, but a low plant, rarely more than 6cm tall, of laxer habit. Flower stems and the base of the flower-bracts yellowish with brown spots; flowerheads in clusters of 1-3 only. June-July. Mountains of N Norway; possibly also in Finland.

8 **Carpathian Catsfoot** *Antennaria carpatica* (Wahlent.) Bluff & Fingerh. Low to short, white-downy perennial like *A. dioicia* but without stolons. Leaves oblong to linear, pointed. Flowerheads in small clusters, brown or blackish. Damp meadows and rocky places, usually to 1500m. July-August. Scandinavia and mountains of France and Germany.

9 *Antennaria villifera* Boriss. Like *A. dioica*, but flower-bracts purplish in the centre; male florets purple above, with yellow anthers. Close to snow patches and in other damp habitats, on calcareous soils. June-July. Arctic Norway, Sweden and Finland.

10 **Alpine Cat's-foot** *Antennaria alpina* (L.) Gaertner. Low to short, downy, mat-forming, stoloniferous perennial, seldom exceeding 15cm tall. Basal leaves narrow-oblong, broadest above the middle, pointed, downy beneath, usually hairless above; upper stem leaves narrow with a scarious papery apex. Flowerheads 4-8mm, unstalked, in clusters of 3-5; upper half of flower-bracts of female plants dark greenish-brown. Mountains rocks, heaths and tundra, on calcareous soils, to 2200m. Scandinavian mountains.

11 *Antennaria porsildii* Elis. Like *A. alpina*, but plants non-stoloniferous, rarely exceeding 10cm tall and basal leaves more or less hairless. Damp mountain heaths. N Scandinavia.

12* **Edelweiss** *Leontopodium alpinum* Cass. Low to short, tufted, grey-downy perennial. Basal leaves in a rosette, narrow-oblong, to spatulate, grey-green or whitish; stem leaves alternate, smaller, linear-lanceolate. Flowerheads small, yellowish-white, in dense clusters surrounded by a conspicuous ruff of white-woolly leaves forming a flattish head. Grassy and rocky mountain habitats, moraines, occasionally on cliffs and screes, generally on limestone, 1700-3400m. July-September. C & E France, S Germany. Mountains of Europe from the Jura southwards. Protected in some countries.

13* **Pearly Everlasting** *Anaphallis margaritacea* (L.) Bentham. Grey- or whitish-downy, medium to tall, tufted perennial. Leaves lanceolate to linear, untoothed, alternate, the margins often rolled under, becoming hairless above. Flowerhead small, 8-12mm, the yellowish tubular florets surrounded by numerous shiny, papery-white, persistent flower-bracts; flowerheads in crowded, rather flat-topped, clusters. Damp and waste places. August-September. **I**: N & E Asia and North America. Widely naturalized except the extreme north. **B**: Naturalized in scattered localities.

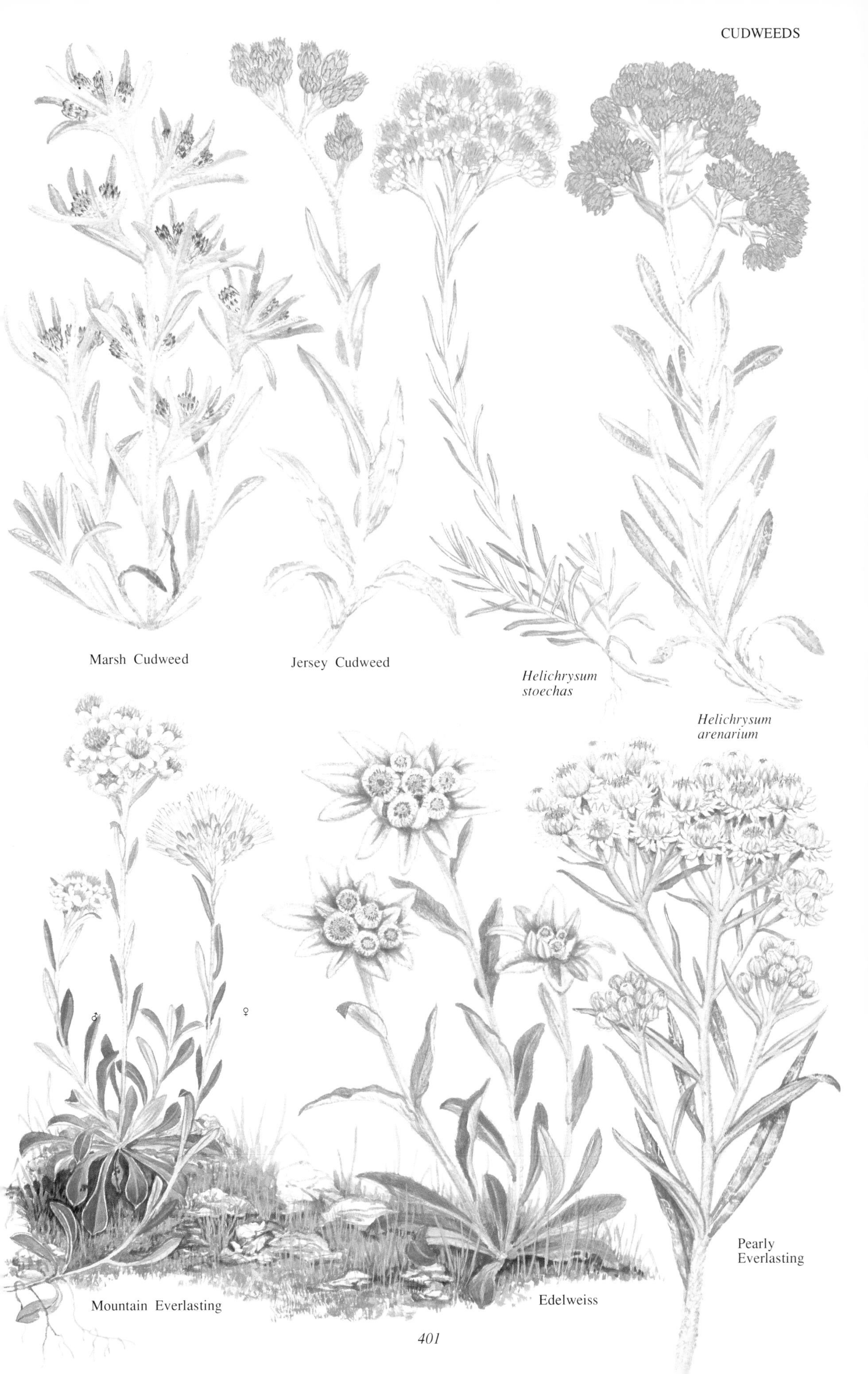
Marsh Cudweed
Jersey Cudweed
*Helichrysum stoechas*
*Helichrysum arenarium*
♂
♀
Mountain Everlasting
Edelweiss
Pearly Everlasting

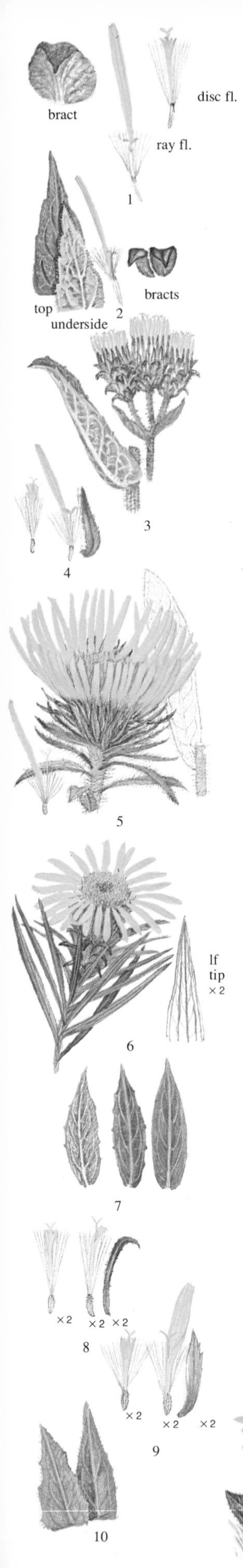


**Inulas** *Inula*. Perennial herbs, rarely biennial, with simple alternate leaves. Flowerheads solitary or in flat-topped clusters or panicles; florets yellow, the outer rayed, strap-like, female, the rays sometimes very short; disk florets all tubular, hermaphrodite; pappus present, of simple hairs. Over 100 species in temperate and subtropical regions of the Old World. They can generally be recognised by their simple leaves and flat, yellow flowerheads, with numerous narrow rays. Several are cultivated in gardens.

1* **Elecampane** *Inula helenium* L. Medium to tall robust, hairy perennial; stems erect, to 2.5m. Leaves oval to elliptical, the lower large and stalked, the upper unstalked and clasping the stem. Flowerheads large, 60-80mm, bright yellow, borne 2-3 together or solitary; rays long and narrow; outer flower-bracts recurved. Waste ground, old meadows, plantations, orchards, copses and roadsides. July-August. **I**: W & C Asia. Naturalized in Europe, except the extreme north. Formerly widely cultivated as a medicinal plant, the rootstock used as a tonic, diuretic and expectorant. Naturalized in many parts of Europe, W Asia and North America. **P**: Bees and hoverflies. **B**: Throughout.

2* *Inula helvetica* Weber [*I. vaillantii*]. Tall greyish-hairy perennial with erect stems to 1.5m. Leaves lanceolate to elliptical, toothed or untoothed, the lower stalked, the upper unstalked, narrowed at the base but not clasping the stem. Flowerheads 30-45mm, solitary or in small clusters, rays long and narrower; outer flower-bracts with recurved tips. Woods, river and streamsides. July-August. C & S France & SW Germany – Upper Rhine Valley. Mainly in SW Europe from Spain to NW Italy.

3 **German Inula** *Inula germanica* L. Medium, somewhat hairy perennial, with erect stems. Lower leaves oblong to oval, stalked, finely toothed, slightly hairy above; upper leaves heart-shaped, unstalked, clasping the stem. Flowerheads small, 7-11mm, in lax, branched clusters, the rays short; outer flower-bracts with recurved tips, downy. Grassy habitats. June-August. C & S Germany. In parts of C and SE Europe. The flower-bracts form a rather narrow cylinder with the flower rays scarcely longer.

4* **Irish Fleabane** *Inula salicina* L. Medium to tall, hairy or sparsely hairy perennial; stems erect, often slightly bristly at the base, to 75cm. Lower leaves linear-lanceolate to oval, stalked, stiff, hairless above; upper leaves narrow heart-shaped, unstalked, half-clasping the stem; all leaves prominently net-veined above. Flowerheads medium-sized, 25-40mm, with long slender rays; flower-bracts hairy only along the margin. Rocky and woody slopes, marshes, fens and stony limestone shores, at low altitudes. July-August. Ireland, Continental Europe, except the extreme north. Sometimes confused with *Buphthalmum salicifolium*, but easily distinguished by the more or less hairless leaves, the absence of receptacle scales and the narrower ray-florets.

5 *Inula hirta* L. Like *I. salicina*, but a shorter, more hairy plant, the upper leaves not or only slightly clasping the stem; flower-bracts hairy. Grassy habitats. July-August C & E France and S Germany. SC and E Europe.

6 **Narrow-leaved Inula** *Inula ensifolia* L. Like *I. salicina*, but leaves with 3-7 parallel veins, hairless except along the margin, untoothed, the upper leaves scarcely clasping the stem. Flower-bracts silkily-hairy at the base. Grassy habitats and scrub. July-August. S Sweden – Gotland. Mainly in E and EC Europe. Local, rare in our area.

7* *Inula britannica* L. Short to tall, hairy biennial or short-lived perennial; stems erect, to 75cm. Leaves lanceolate, untoothed to finely toothed, the lower stalked, the upper unstalked, slightly clasping the stem, all densely softly hairy beneath. Flowerheads medium to large, 20-50mm, solitary or 2-3 together, with long narrow rays; outer flower-bracts hairy, linear. Wet habitats, river and stream margins, marshes, ditches, wet grassland and wet woods, at low altitudes. July-August. Continental Europe, except the extreme north; naturalized in Finland; extinct in Britain. Seed is probably spread by water fowl.

8* **Ploughman's-spikenard** *Inula conyza* DC. [*I. squarrosa*, *Conyza squarrosa*]. Hairy, medium to tall perennial; stems erect, purplish, to 1.2m. Leaves elliptical to lanceolate, finely toothed, the uppermost unstalked, not clasping the stem. Flowerheads small, 9-11mm, dull yellow, the outer florets without or with a very short and inconspicuous ray, many flowerheads in a flat-topped cluster; inner flower-bracts purplish. Dry generally rocky habitats, open woods, scrub, rocky slopes and cliffs, and grassy places, generally on calcareous soils, at low altitudes. July-September. C & S Britain, Belgium, Holland, France, Denmark and Germany. The basal leaves are frequently mistaken for those of the foxglove, *Digitalis purpurea*. **B**: England and Wales N to the Tees.

9* **Golden Samphire** *Inula crithmoides* L. Hairless, short to tall, fleshy perennial or subshrub; stems branched and rather woody below. Leaves crowded, linear to linear-lanceolate, untoothed or with a 3-toothed apex. Flowerheads medium, 20-28mm, golden-yellow, rayed, borne in flat-topped clusters; outer flower-bracts linear, erect. Coastal habitats, salt marshes, maritime rocks and shingle, sea cliffs. July-September. Britain and France. W Europe and the Mediterranean, always local. Readily identified by its fleshy leaves and golden-rayed flowers. **B**: Very local, north to Kirkudbright; Ireland – confined to the S & E coasts.

10* **Stink Aster** *Dittrichia graveolens* (L.) W. Greuter [*Inula graveolens*]. Medium, densely glandular annual smelling of camphor; stems erect. Leaves lanceolate to oblong, untoothed or slightly toothed. Flowerheads small, 7-12mm, in spike-like panicles. Damp habitats. July-August. NC France southwards. In parts of W and in S Europe. Like *Inula* but flowerhead smaller with cylindrical, not angled, achenes in which the pappus-hairs are fused together close to the base.

*Inula hirta*

Narrow-leaved Inula

Elecampane
× 1/20
*Inula helvetica*
Irish Fleabane
*Inula britannica*
Ploughman's-spikenard
Golden Samphire
Stink Aster

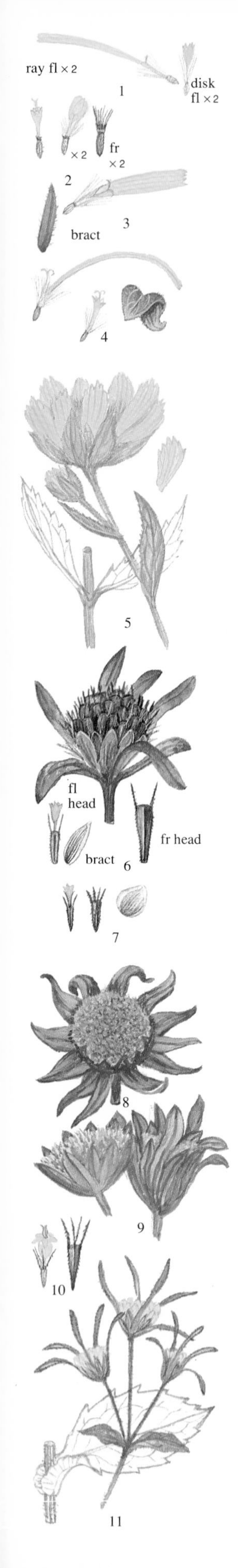


1* **Common Fleabane** *Pulicaria dysenterica* (L.) Bernh. Medium, hairy, stoloniferous perennial; stems erect, branched above. Basal leaves oblong, narrowed at the base, withered by flowering time; stem leaves heart- or arrow-shaped, clasping the stem with unstalked bases; all leaves green above, softly grey-downy beneath, toothed or untoothed. Flowerheads daisy-like, golden yellow, 15-30mm, in lax flat-topped clusters, rays numerous, linear; flower-bracts linear, downy. Damp habitats, river and stream banks, ditches, dykes, marshes and wet meadows. August-September. Throughout, except the far north. *Pulicaria* is similar to *Inula*, but differs in the pappus which has an outer row of closely fused scales. **B**: Throughout except for C & N Scotland.

2* **Small Fleabane** *Pulicaria vulgaris* Gaertner [*P. prostrata*]. Low to short, hairy, sometimes glandular annual; stems green or brownish, the side branches overtopping the main stem. Leaves alternate, wavy-edged, oblong to lanceolate, the basal stalked and withered by flowering time. Flowerheads numerous, 8-10mm, yellow, with very short rays; flower-bracts linear, downy. Seasonally wet habitats, moist sandy places and hollows, stream and pond margins, at low altitudes. August-September. North to S Sweden. Local and declining. **B**: Very rare, mostly confined to the New Forest.

3* **Yellow Ox-eye** *Buphthalmum salicifolium* L. Short to medium, somewhat hairy perennial; stems erect, branched or unbranched. Leaves alternate, the lower lanceolate, stalked, the upper oblong to linear-lanceolate, pointed, unstalked; all leaves untoothed or slightly toothed. Flowerheads bright yellow, large, 30-60mm, solitary, with long strap-like rays; flower-bracts lanceolate, slightly hairy. Damp stony and rocky habitats, open woodland, generally on calcareous soils, to 2050m. June-July. E France and S Germany. Local.

4* **Large Yellow Ox-eye** *Telekia speciosa* (Schreber) Baumg. Tall, somewhat hairy perennial to 2m; stem branched above. Leaves broadly oval-heart-shaped to diamond-shaped, coarsely toothed, slightly hairy beneath, the lower stalked, the upper with a rounded or narrow base. Flowerhead large, 50-60mm, orange-yellow with a brownish-yellow disk, borne in flat-topped clusters; rays long, strap-shaped. Damp habitats, river and stream banks and open woods, waysides. June-August. **I**: E & SE Europe. Naturalized in Britain, Belgium, France and Germany. A large, rather coarse plant. Like *Buphthalmum*, but flower-bracts with spreading tips and anthers bearded at the base. **B**: Scattered localities, particularly in the west.

5 **Guizotia** *Guizotia abyssinica* (L.fil.) Cass. Tall annual; stems erect, to 2m, branched, stickily-hairy above. Lower leaves opposite, the upper alternate, oblong-lanceolate, toothed to untoothed, clasping the stem with an unstalked base. Flowerheads yellow, 20-30mm, rayed. **I**: E Africa. Locally naturalized in rather dry habitats. July-August. France; casual elsewhere in many parts of Europe including Britain.

**Bur-marigolds** *Bidens*. Annual or perennial herbs with opposite leaves, simple to pinnately divided. Flowerheads solitary, stalked, generally without ray-florets; outer flower-bracts often leaf-like. About 200 species, mainly in America, but cosmopolitan in distribution. The achenes bear barbed bristles which attach themselves to animals, clothing etc.

6* **Trifid Bur-marigold** *Biden tripartita* L. Very variable, short to medium, hairy or almost hairless annual; stems erect. Leaves opposite, usually 3-lobed, occasionally 5-lobed with lanceolate, coarsely-toothed divisions; leaf-stalks short and winged. Flowerheads yellow, 10-25mm, erect, in branched clusters. Achenes with 3-4 barbed bristles. Damp habitats, on mildly acid or calcareous soils, mostly at low altitudes. July-September. Throughout, except the far north. Variable – habitat greatly affects the size of the plant, especially leaf-size. **B**: Throughout, but scarcer in Ireland, Wales and much of Scotland.

7* *Bidens connata* Muhl. ex Willd. Like *B. tripartita*, but leaves generally unlobed, the lowermost sometimes with 1-2 pairs of spreading lobes along the leaf-stalks. Achenes with 4-5 bristles. Similar habitats and flowering time. **I**: North America. Naturalized in Belgium, Holland, France and Britain – extensively established along canals in the London area.

8 *Bidens radiata* Thuill. Like *B. tripartita*, but with 10-12 outer, leaf-like bracts (not 5-8); achenes with 2 bristles only. Similar habitats and flowering time. NW & E France, Germany, Denmark, Sweden and Finland. Confined mostly to part of N & NE Europe.

9 **Beggarticks** *Bidens frondosa* L. [*B. melanocarpa*] Like *B. tripartita* but taller, to 1m, the leaves mostly pinnate with 3-5 separate leaflets. Flowerheads 10-20mm; outer flower-bracts shorter than the blackish inner ones, rarely leaf-like. Achenes with 2 bristles. Wet habitats and waste ground. July-October. **I**: North, Central and South America. Naturalized or casual throughout, except for Scandinavia and Iceland. **B**: Naturalised along canals in the Midlands and S Wales.

10* **Nodding Bur-marigold** *Bidens cernua* L. Short to tall annual; stems erect, hairless or slightly hairy. Leaves opposite linear-lanceolate to lanceolate, toothed, unstalked. Flowerheads yellow, 15-25mm, nodding, with or without rays; outer flower-bracts 5-8, leaf-like, the inner with dark streaks. Moist habitats. July-September. Throughout, except the far north. The form with ray-florets, which is common in some areas is generally referable to var. *radiata* DC. – the rays are short, oval and golden yellow. **B**: Throughout except the far north, but scarce in SW England, S Wales and Ireland.

11 *Sigesbeckia jorullensis* Kunth [*S. cordifolia*]. Medium to tall, hairy annual; stems erect to 1.2m. Leaves opposite, oval to heart-shaped, toothed, with broad, winged, clasping stalks. Flowerheads small, yellow, 5-8mm, in a lax panicle, rays short; outer flower-bracts linear. Pappus absent. Open habitats. July-September. **I**: Tropical America. Naturalized locally in Britain and Germany; casual elsewhere. **B**: Naturalised primarily in Lancashire.

Common Fleabane
Small Fleabane
Yellow Ox-eye
× 1/12
Large Yellow Ox-eye
Trifid Bur-marigold
*Bidens connata*
Nodding Bur-marigold

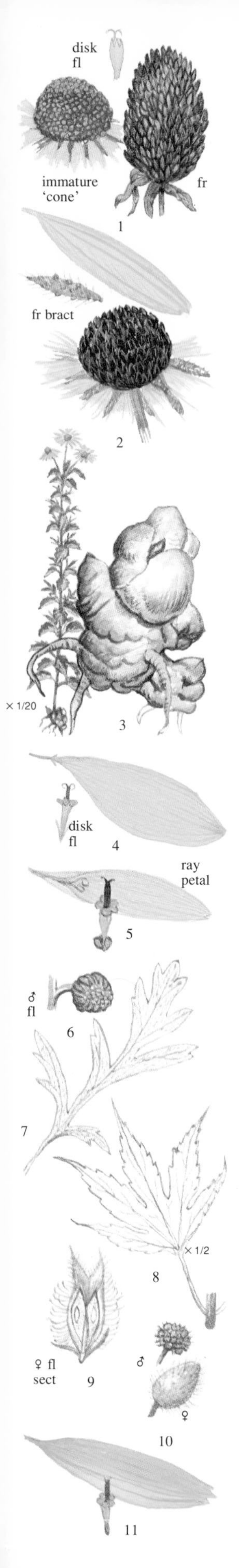


1* **Cone-flower** *Rudbeckia laciniata* L. Tall, almost hairless, rather greyish perennial, to 3m; stems erect, branched above. Leaves alternate, the lower pinnately-lobed, stalked, the upper oval, occasionally 3-lobed; all leaves coarsely toothed to untoothed. Flowerheads large, 60-120mm, yellow with a yellowish-green conical disk, solitary or several together, long-stalked. Damp habitats, open woods, stream and river banks. June-October. **I**: North America. Naturalized in Britain, Belgium, Holland, France and Germany. **B**: Locally naturalized from gardens.

2 **Hairy Rudbeckia** *Rudbeckia hirta* L. Medium to tall bristly or hairy biennial or short-lived perennial; stem erect, branched. Leaves alternate, 3-veined, the lower elliptical, often broadest above the middle, long-stalked, the upper leaves linear-lanceolate, scarcely-stalked; all leaves slightly toothed or untoothed. Flowerheads large 60-80mm, yellow with a purplish-brown conical disk, solitary or several together; rays large, elliptical, downturned. Pappus absent. Waste places, stream banks. July-October. **I**: North America. Naturalized in Belgium, France and Germany. **B**: Casual.

3 **Jerusalem Artichoke** *Helianthus tuberosus* L. Tall, rough, tuberous-rooted perennial; stems erect, branched above, to 2.8m, hairy above. Leaves mostly opposite, oval, coarsely toothed, narrowed into a winged stalk. Flowerheads yellow with a yellowish-green disk, 40-80mm, rays present, borne singly or several together, long-stalked; flower-bracts spreading, dark green. A frequent escape from cultivation. September-November. **I**: North America. Naturalized or casual in Britain, Holland and Germany, occasionally elsewhere. Widely cultivated for its edible tubers, which contain Inulin. Flowers rarely borne, except after long hot summers.

4* *Helianthus × laetiflorus* Pers. (*H. rigidus* (Cass.) Desf. × *H. tuberosus*). Like *H. tuberosus*, but rhizomes not tuberous and leaves broad-lanceolate, conspicuously 3-veined beneath, rough. Flowerheads 60-100mm, more freely borne. **I**: North America. Naturalized in waste places. August-October. Holland, France, Denmark and Germany. Widely cultivated in gardens and sometimes escaping. Both the parent species come from North America. *H. rigidus*, the Garden Sunflower, sometimes becomes naturalized.

5* *Silphium perfoliatum* L. Tall, hairless perennial herb to 2.5m; stems erect, square. Leaves opposite, triangular-oval, pointed, coarsely toothed, the lower long-stalked, the upper with winged stalks, the wings of each leaf pair joined around the stem to form a cup. Flowerheads yellow, rayed, in flat-topped clusters; disc flat. **I**: North America. A rare casual or locally naturalized along streams and rivers and in damp meadows. Germany.

**Ragweeds** *Ambrosia*. Annual or perennial herbs, mostly with opposite leaves. Flowerhead either male or female, inconspicuous, the male rounded, drooping in terminal racemes, the female borne in the axils of the uppermost leaves, erect with a single floret only; rays absent, all florets tubular. Flower-bracts with small spines or warts near the apex. About 20 species from America. The prickly fruit attaches itself to fur and clothing.

6* **Ragweed** *Ambrosia artemisiifolia* L. Medium to tall, hairy annual, not aromatic; stems branched and angled, often reddish, becoming somewhat woody below. Leaves mostly opposite, greyish beneath, usually pinnately-lobed, with lanceolate divisions. Male flowerheads greenish-yellow, 3-5mm, in slender spikes, female inconspicuous below at the base of the upper leaves; flower-bracts with spiny teeth. Disturbed ground. August-October. **I**: North America. Locally naturalized or casual in Britain, Belgium, France and Germany. A chief cause of hay fever, particularly in America.

7 *Ambrosia coronopifolia* Torrey & Gray. Like *A. artemisiifolia*, but a perennial with creeping rhizomes and more deeply divided leaves. Flower-bracts without or with very small teeth. **I**: North America. Locally naturalized in Britain, Belgium, Holland, France, Denmark and Germany.

8 **Great Ragweed** *Ambrosia trifida* L. Like *A. artemisiifolia*, but the leaves palmately-lobed with 3-5 lobes. Fruit 6-12mm (not 3-5mm). Cultivated ground and waste places. August-October. **I**: North America. Naturalized in France and Germany; casual in Britain and perhaps elsewhere.

9* **Rough Cocklebur** *Xanthium strumarium* L. Medium to tall, stiffly branched, annual, to 1.2m, not spiny. Leaves alternate, oval to triangular, with a heart-shaped base, unlobed or with 3-5 toothed lobes, bristly. Flowerheads greenish, male and female separate but on the same plant, 5-6mm, in lateral clusters, sometimes terminal, the globular male borne above the egg-shaped female; flower-bracts with straight or hooked spines, enlarging in fruit to 14-18mm. Damp habitats, lake, river and stream margins, pastures waste and disturbed ground, at low altitudes. July-October. **I**: Probably North America. Naturalized in France and Germany, frequently casual elsewhere. Poisonous, causing paralysis in livestock. **B**: A casual; locally established.

10* **Spiny Cocklebur** *Xanthium spinosum* L. Rather like *X. strumarium*, but plants with simple or 3-forked, orange-yellow spines at the base of each leaf-stalk. Leaves generally white-felted beneath. Fruiting heads 8-12mm, densely clothed with hooked spines. Waste ground, especially near sea-ports and mills where it has been introduced entangled in sheep's wool. July-October. **I**: North and South America. Casual or locally established in Britain, France and Germany.

**Stinking Cocklebur** *Xanthium echinatum* Murr. Like *X. spinosum*, but a strongly aromat.c bristly, yellow-green plant; leaves not white beneath. **I**: America. Casual on waste ground in Britain and C Europe, perhaps elsewhere.

11* *Heliopsis helianthoides* (L.) Sweet subsp. *scabra* (Dunal) Fisher. Medium to tall, rather rough perennial; stems erect, to 1.5m, simple or branched. Leaves opposite, lanceolate, stalked, coarsely toothed. Flowerheads yellow, rayed, 45-65mm, solitary on long stalks; disk conical; outer flower-bracts leaf-like usually. **I**: North America. Locally naturalized on waste ground and along canal and river banks in Denmark and Germany. July-September. The form called subsp. *occidentalis* is occasionally naturalized; it has more triangular, short-stalked, leaves and larger flowers.

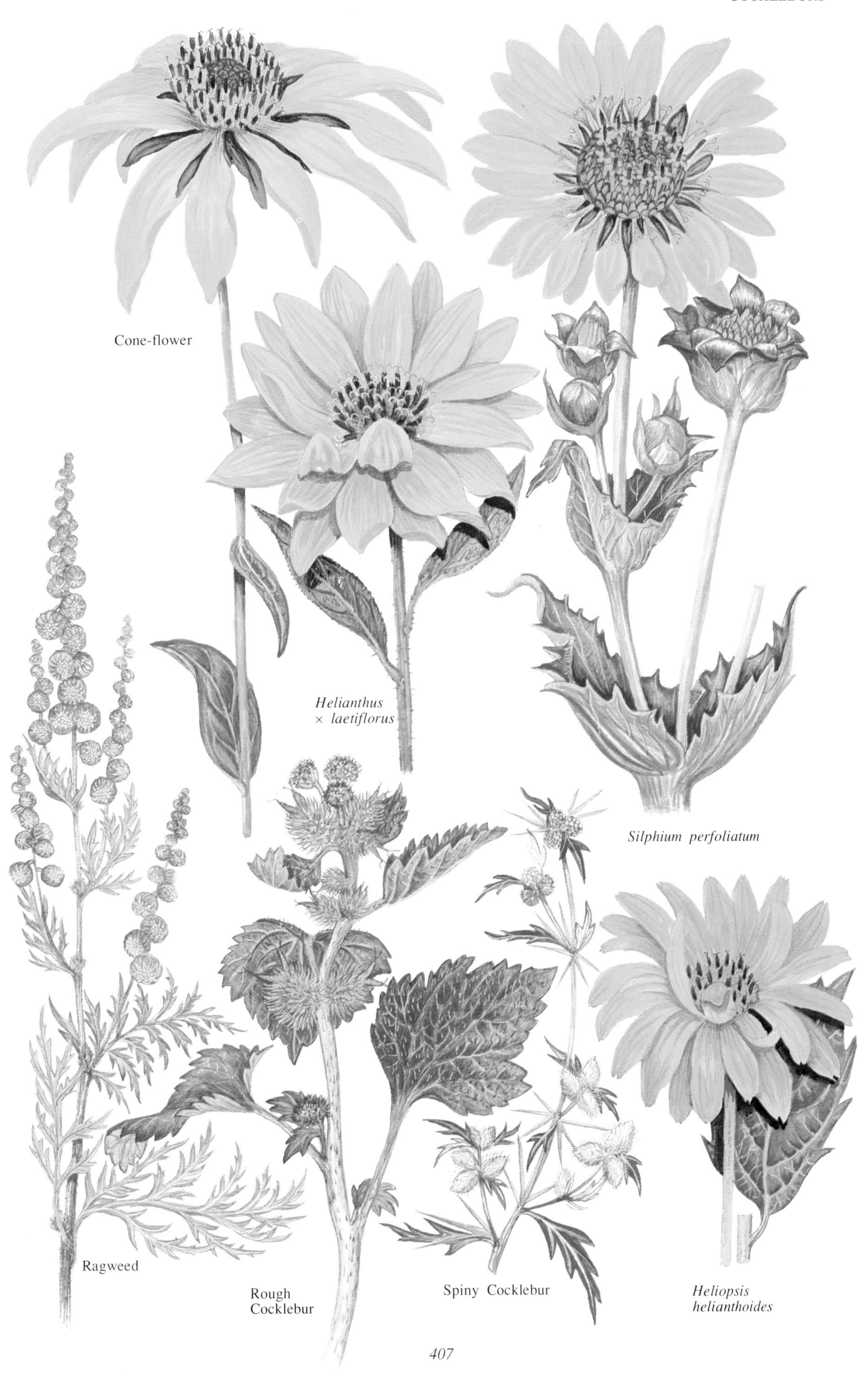
Cone-flower
*Helianthus × laetiflorus*
*Silphium perfoliatum*
Ragweed
Rough Cocklebur
Spiny Cocklebur
*Heliopsis helianthoides*

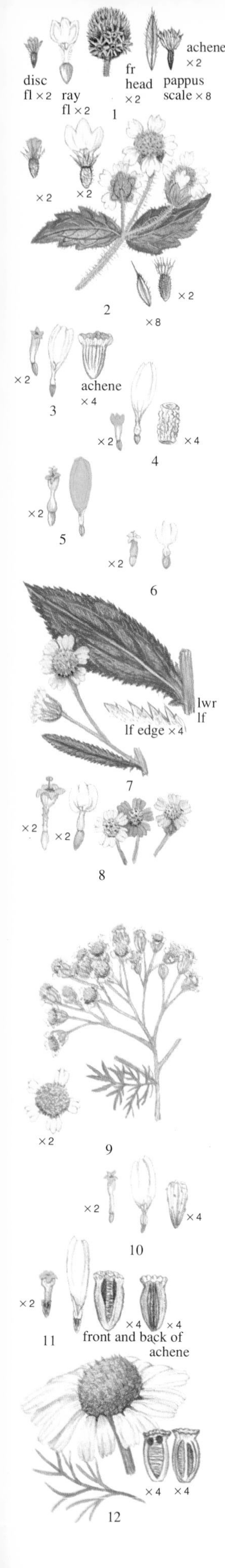


1* **Gallant Soldier, Kew-weed** *Galinsoga parviflora* Cav. Short to tall annual; stem erect, much branched, more or less hairless. Leaves opposite, oval, pointed, toothed, short-stalked, generally somewhat wavy. Flowerheads small, rayed, white with a yellow disk, 3-5mm, borne in branched clusters; rays usually 3-toothed. Disturbed and cultivated ground, arable land and gardens in particular. May-October. **I**: South America. Widely naturalized except for the far north. A cosmopolitan weed. Occasionally hybridises with *G. ciliata*. **B**: Widely naturalized, particularly in the SE.

2 **Shaggy Soldier** *Galinsoga ciliata* (Rafin) S.F. Blake [*G. quadriradiata*]. Like *G. parviflora*, but stem white-hairy below, the upper stem with long spreading hairs; pappus scales awned (unawned in *G. parviflora*). Similar habitats and flowering time. **I**: Central and South America. Naturalized in Continental Europe, except the far north. Very like the previous species, far less common and frequently overlooked. **B**: Confined mainly to S England and S Wales.

**Chamomiles** *Anthemis*. Herbs, occasionally dwarf shrubs with alternate, mostly divided leaves. Flowerheads solitary, terminal on long stalks; outer florets usually white and rayed, female; disk-florets yellow, on a prominent conical receptacle. 100 species.

3* **Corn Chamomile** *Anthemis arvensis* L. Short to medium hairy, aromatic annual, often overwintering; stems much-branched from the base, the branches longer than the main stem. Leaves oblong, pinnately-lobed, with linear, pointed segments, woolly beneath, especially when young. Flowerheads 20-30mm, white with a yellow disk. Cultivated ground, especially arable land, waste places, road and path sides, on calcareous, often sandy, soils, to 1950m. June-July. Throughout, except the Faeroes and Spitsbergen and N Scandinavia. Flowers fragrant. **B**: Scattered localities throughout, rarer in the north.

4* **Stinking Chamomile** *Anthemis cotula* L. Short to medium foetid, slightly hairy or hairless annual. Leaves 2-3-pinnately-lobed, with somewhat fleshy linear segments. Flowerheads 13-30mm, white with a yellow disk, the rays spreading at first but becoming reflexed, generally sterile. Cultivated land, particularly arable land, stubble fields, waste places, disturbed ground. Farmyards, generally on heavy base-rich clay soils. May-October. Widespread in Europe, W and N Asia. Plants germinate in the spring or in the autumn. **B**: Locally common in England and Wales, rare elsewhere.

5* **Yellow Chamomile** *Anthemis tinctoria* L. [*Cota tinctoria*]. Variable, slightly to densely hairy, medium perennial; stem usually branched. Leaves 2-pinnately-lobed, green above, white-woolly beneath. Flowerheads golden-yellow, 25-45mm, in branched, flat-topped clusters, rays generally present. Waste places, banks, waysides and cultivated land, generally at low altitudes. July-September. Belgium, Holland, France, Germany, Denmark and S Sweden; naturalized in Britain and Finland. **B**: Naturalized in scattered localities in England and Scotland.

**Sneezeworts** *Achillea*. Perennial herbs with pinnately-divided or undivided, alternate leaves. Flowerheads small, generally congested into dense, flat-topped clusters; ray-florets present, often short, female, often 3-toothed; disk-florets hermaphrodite. About 100 species, mainly in temperate regions of the old world.

6* **Sneezewort** *Achillea ptarmica* L. Medium to tall hairy perennial, with a creeping woody stock; stems erect, angular. Leaves lanceolate, sharply toothed, pointed, more or less hairless and unstalked. Flowerheads white with a greenish-white disk, 12-18mm, borne in lax, branched clusters. Damp grassy habitats, meadows, marshes, stream and river margins, on acid or neutral soils, to 1700m. July-September. Throughout except the extreme north; naturalized in Iceland. **B**: Common throughout.

7 *Achillea cartilaginea* Ledeb. ex Reichenb. Like *A. ptarmica*, but leaves wider, to 17mm (not 4-8mm) and generally double-toothed, densely hairy. Flowerheads with shorter rays. Similar habitats and flowering time. E Germany and Finland.

8* **Yarrow, Milfoil** *Achillea millefolium* L. Variable, short to medium, hairy, stoloniferous perennial, generally patch-forming, strong-smelling; stems erect, unbranched. Leaves feathery, lanceolate, finely 2-pinnately-lobed, with small, narrow segments. Flowerheads small, white, occasionally pink or reddish, with a white or cream disk, 4-6mm, borne in dense flat-topped heads; rays short. Grassy habitats, pastures, meadows, grassy heaths, road verges, hedgebanks, lawns and waste places, on calcareous or slightly acid soils, to 2800m. July-October. Throughout, except Spitsbergen. **B**: Throughout.

subsp. *sudetica* (Opiz.) Weiss. [*Achillea sudetica*]. Like *A. millefolium*, but leaves velvety-hairy. Flowerheads with pink rays. Mountain habitats. France and Germany.

9 *Achillea nobilis* L. Short to medium, aromatic, non-stoloniferous perennial; stem erect, hairy, unbranched. Leaves elliptical, 1-2-pinnately-lobed, flat, slightly hairy. Flowerheads numerous, yellowish-white, with a yellow disk, 3mm, borne in dense flat-topped clusters; rays very short, 1mm. Dry habitats, grassy and rocky places. July-August. C & E France and Germany; naturalized or casual in Britain and N Europe.

10* **Chamomile** *Chamaemelum nobile* (L.) All. [*Anthemis nobilis*]. Low to short, hairy, creeping and rooting perennial with spreading stems, aromatic. Leaves alternate, oblong, 2-3-pinnately-lobed, feathery, with slender segments, hairless beneath. Flowerheads white with a yellow disk, 18-25mm, solitary, long-stalked, rayed, or rays occasionally absent; flower-bracts thin, shiny, with white margins. Grassy heaths, commons, roadsides, banks and lawns, generally on light sandy soils. June-August. Britain and France; naturalized in Belgium and Germany. **B**: Local restricted to S England, S Wales and East Anglia – decreasing.

11* **Sea Mayweed** *Matricaria maritima* L. [*Tripleurospermum maritimum*]. Variable low to medium, scarcely scented biennial or perennial, occasionally annual; stems spreading to prostrate, hairless. Leaves alternate, irregularly 2-3-pinnately-lobed, with short fleshy segments. Flowerheads white witn a yellow disk, 15-50mm, solitary, long-stalked; rays spreading. Coastal habitats. July-September. Europe, except parts of the extreme north. **B**: Throughout, especially in coastal regions of the north, N Wales and Ireland.

12 **Scentless Mayweed** *Matricaria perforata* Merat [*M. inodora, Tripleurospermum inodorum*]. Like *M. maritima*, but the leaves not fleshy; achenes with well separated ribs (not close and touching). Cultivated land, waste and disturbed land, roadsides and waysides, on a wide range of soils, calcareous to slightly acid. July-September. Throughout, except the far north. **B**: Throughout, but scarcer in the north and C Ireland.

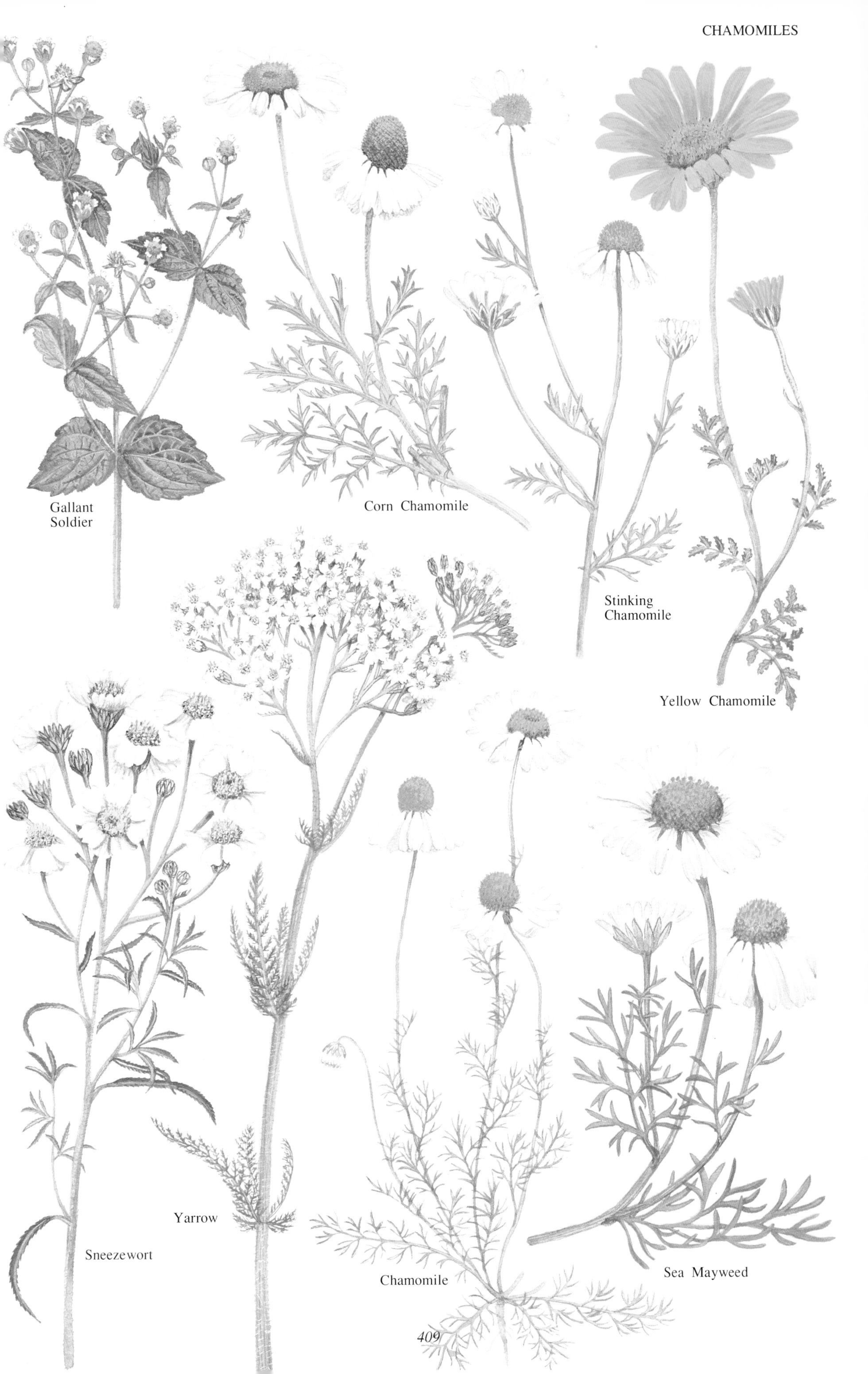
Gallant Soldier
Corn Chamomile
Stinking Chamomile
Yellow Chamomile
Yarrow
Sneezewort
Chamomile
Sea Mayweed

1* **Scented Mayweed** *Chamomilla recutita* (L.) Rauschert [*Matricaria recutita, M. chamomilla*]. An aromatic, short to medium, hairless annual; stems ascending to erect, branched. Leaves alternate, feathery, 2-3-pinnately-divided, with bristle-tipped divisions. Flowerheads white with a yellow, conical hollow disk, 10-25mm, solitary on long stalks; rays downturned soon after the flowers open. Cultivated land, especially arable fields, waste and disturbed ground, saline steppes, on sandy or loamy soils, generally at low altitudes. July-August. Throughout, except Ireland, the Faeroes, Iceland and Spitsbergen; probably naturalized in much of N Europe. **B**: Throughout England except much of the SW, scarce elsewhere.

2* **Pineapple Mayweed, Rayless Mayweed** *Chamomilla suaveolens* (Pursh.) Rydb. [*Matricaria matricarioides*]. Low to short hairless annual, pineapple-scented; stems well-branched, erect or ascending. Leaves feathery, bright green, rather fleshy, 2-3-pinnately-lobed, with thread-like segments. Flowerheads yellowish-green, rayless, 5-8mm, with a conical disk, solitary on short stalks. Cultivated, waste and disturbed ground, pathways, road verges, waysides and farmyards, generally at low altitudes, on a variety of soils. May-November. **I**: NE Asia. Widely naturalized in Europe. Tolerant of trampling. **P**: Rarely visited by insects. The seeds are rapidly spread on car tyres and shoes. **B**: Throughout.

3* **Cottonweed** *Otanthus maritimus* (L.) Hoffmanns & Link [*Diotis maritima*]. White-woolly, short perennial; stems ascending, stout, branched above. Leaves alternate, oblong to lanceolate, untoothed or blunt-toothed, fleshy, unstalked. Flowerheads yellow, 7-10mm, globose, button-like, rays absent, borne in close, flat-topped clusters; flower-bracts white-woolly. Maritime habitats, stabilized shingle and sand-dunes, August-October. SW Ireland and France. **B**: Very rare – confined to a single locality on the Wexford coast of Ireland.

4* **Corn Marigold** *Chrysanthemum segetum* L. Short to tall, greyish, hairless perennial, somewhat fleshy; stems erect, branched or unbranched. Leaves alternate, oblong, deeply and sharply toothed, the uppermost sometimes untoothed. Flowerheads yellow, 35-55mm, large and daisy-like with strap-shaped rays and a flat disk. Cultivated land, particularly arable fields, waste and disturbed ground, at low altitudes, generally on acid or neutral sandy soils. June-August. Throughout, except for N Scandinavia and Iceland. **B**: Locally common throughout; declining generally.

**Tansies** *Tanacetum*. Annual or perennial herbs, often aromatic, with alternate, pinnately divided leaves. Flowerheads rayed, in flat-topped clusters or solitary. Pappus generally forming a cup. About 70 species, formerly included in *Chrysanthemum*.

5* **Tansy** *Tanacetum vulgare* L. [*Chrysanthemum vulgare, C. tanacetum*]. Medium to tall, strongly aromatic, patching-forming, almost hairless perennial, to 1.5m. Leaves pinnately-lobed, with lanceolate, toothed segments, deep green; uppermost leaves unstalked. Flowerheads yellow, 7-12mm, button-like, rayless, in large, rather flat-topped, clusters. A variety of habitats and soils except very acid ones. July-September. Throughout, except the far north; naturalized in Ireland. **P**: Various insects, including bees and hoverflies. **B**: Throughout, but scarce in N Scotland and parts of Ireland.

6* **Feverfew** *Tanacetum parthenium* (L.) Schultz Bip. [*Chrysanthemum parthenium, Leucanthemum parthenium, Pyrethrum parthenium*]. Short to medium, strongly aromatic biennial or short-lived perennial, slightly hairy; stems erect to ascending, branched above. Leaves often rather yellowish-green, 1-2 pinnately-divided, mostly stalked. Flowerheads white with a yellow disk, daisy-like, 10-25mm, many in a lax flat-topped cluster, rays spreading. Scrub and rocky places, banks, old walls, waysides and cultivated land, often close to buildings, generally on calcareous or neutral soils. July-September. **I**: SE Europe. Naturalized in Europe north to S Sweden. **B**: Established in many parts.

7 *Tanacetum corymbosum* (L.) Schultz Bip. [*Chrysanthemum corymbosum, Leucanthemum corymbosum, Pyrethrum corymbosum*]. Like *T. parthenium*, but not aromatic and stem leaves unstalked. Flowerheads 25-40mm, generally fewer. Open woodland, scrub and meadows. June-August. NC, C & E France and Germany; naturalized in Denmark and S Sweden. S & E Europe and W Asia.

8 *Tanacetum macrophyllum* (Waldst. & Kit.) Schultz Bip. [*Chyrsanthemum macrophyllum, Pyrethrum macrophyllum*]. Like *T. parthenium*, but flowerheads very numerous and smaller, 8-15mm. **I**: SE Europe. Naturalized in scrub, grassy places and along woodland margins in Britain, Denmark and Germany.

9 **Alpine Moon-daisy** *Leucanthemopsis alpina* (L.) Heywood [*Chrysanthemum alpinum, Pyrethrum alpinum, Tanacetum alpinum*]. Low to short, slightly hairy, patch-forming perennial; stems ascending, rarely more than 15cm, few-leaved. Leaves grey-green, mostly basal, oval to spoon-shaped, pinnately-lobed. Flowerheads daisy-like, white with a yellow disk, 20-40mm, solitary; rays strap-shaped, often becoming pink on ageing. Mountain habitats, 1800-2800m. July-August. C & E France and S Germany. Like a dwarf form of *Leucanthemum vulgare*. **P**: Bees, flies and butterflies.

10* **Ox-eye Daisy** *Leucanthemum vulgare* Lam. [*Chrysanthemum leucanthemum*]. Short to tall, hairless or slightly hairy, patch-forming perennial with short leafy stolons; stems erect, ridged, often branched. Leaves dark green, oblong, toothed, the basal stalked, the stem leaves clasping and unstalked. Flowerheads white with a yellow disk, large daises, 25-50mm; rays long, strap-shaped. Rough grassy habitats, scrub, road verges, embankments, hayfields open woodland and woodland pathways, on calcareous or slightly acid soils, to 2700m. May-September. Throughout, except Spitsbergen; naturalized in the Faeroes and Iceland. **P**: Various insects.

11* **Shasta Daisy** *Leucanthemum maximum* DC., a more vigorous plant from the Pyrenees, with larger flowerhead, 70-90mm, is widely cultivated and sometimes found as a garden escape – including double-flowered forms. **B**: Throughout.

12* **Buttonweed** *Cotula coronopifolia* L. Low to short, hairless annual. Leaves fleshy, linear, untoothed or few-toothed, unstalked, sheathing at the base. Flowerheads button-like, yellow, 5-10mm, solitary on long stalks, rayless; outer florets without a corolla, long-stalked, inner with a corolla and short-stalked. Damp, often saline habitats, sandy places, often on or near the coast. July-October. **I**: South Africa. Naturalized in Britain, France, Germany, Denmark and S Norway. Widely established in both the north and south hemispheres. The flowerheads always face the sun.

Pineapple Mayweed
Cottonweed
Corn Marigold
Scented Mayweed
Tansy
Feverfew
Ox-eye Daisy
Shasta Daisy
Buttonweed

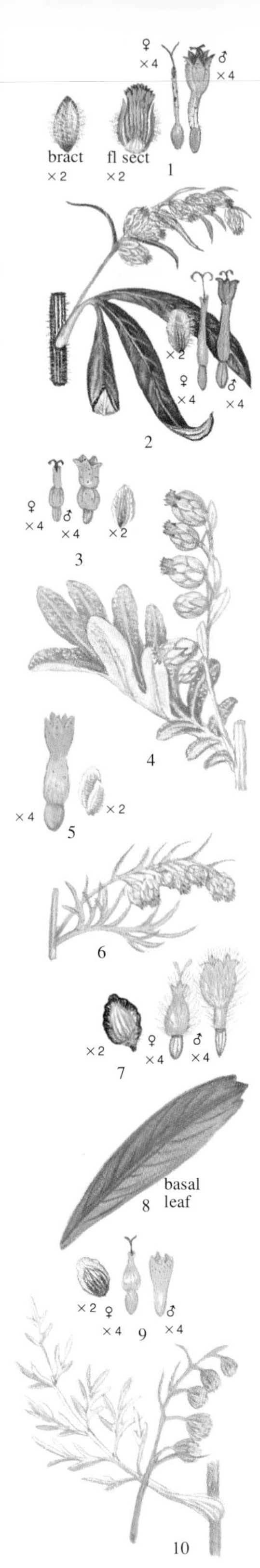


**Wormwoods and Mugworts** *Artemisia*. Herbs or small shrubs, often aromatic with alternate leaves. Flowerheads small, often pendent, in racemes or panicles, sometimes in terminal clusters; flower-bracts in few rows. Pappus usually present. Some 200 species, particularly in Europe, Asia and North America. Several have attractive silver or grey foliage and are cultivated in gardens.

1* **Mugwort** *Artemisia vulgaris* L. Medium to tall, somewhat hairy, tufted perennial, slightly aromatic; stems erect, often reddish or purplish, with a wide pith. Leaves 1-2-pinnately-lobed, the upper unstalked, dark green above, silvery-downy beneath. Flowerheads 3-4mm long, reddish-brown, yellowish or purplish, egg-shaped, unstalked, borne in panicles; lower bracts leaf-like. Roadsides, hedgerows, embankments, riverbanks, on a variety of soils. May-September. Throughout except the extreme north. **B**: Throughout, but mostly coastal in Scotland and scarcer in Ireland.

2 **Chinese Mugwort** *Artemisia verlotiorum* Lamotte. Like *A. vulgaris*, but a taller more aromatic plant with overwintering leaf-rosettes, long rhizomes and densely hairy stems with a narrow pith. Roadsides, banks and waste places. November-December. Naturalized in S Britain, Belgium, France and Germany. Later flowering than *A. vulgaris*. **B**: Naturalized in SE England.

3* **Wormwood** *Artemisia absinthium* L. Medium to tall, strongly aromatic, silkily-hairy, tufted perennial; stems grooved, somewhat woody below. Leaves 2-3-pinnately-lobed, stalked, silky-white on both surfaces. Flowerheads 3mm, yellowish, subglobose, nodding in long panicles. Roadsides, waste ground and coastal habitats. July-August. Throughout, except the far north; naturalized in Ireland. **B**: Throughout England and Wales, mostly coastal elsewhere and rare in Ireland.

4 **Hoary Mugwort** *Artemisia stellerana* Besser. Medium, densely white-felted, non-aromatic perennial. Lower leaves pinnately-lobed with blunt segments, white-felted above and beneath; upper leaves unstalked, sometimes unlobed. Flowerheads 5-9mm, yellow, numerous, in a raceme-like panicle, each surrounded by white-felted flower-bracts. Cultivated land, waste places and waysides. July-September. **I**: NE Asia. Naturalized in S Britain, E Ireland, Denmark and Sweden. **B**: Naturalized in S & SW England, Scotland and near Dublin in Ireland.

5* **Sea Wormwood** *Artemisia maritima* L. Short to medium, strongly aromatic downy perennial with long leafy non-flowering shoots; stems becoming woody below. Leaves 2-3-pinnately-lobed, stalked, woolly on both sides, the lower withered by flowering time. Flowerheads 10-20mm, yellow or orange-yellow, egg-shaped, nodding or erect, in leafy racemose-panicles. Coastal habitats, bare sandy places and sea walls, occasionally inland. August-October. Variable, with various subspecies, especially in S and E Europe. **B**: Coasts north to Kincardine and Cumbria; rare in Ireland.

subsp. *humifusa* (Fries & Hartman) K. Persson. Similar, but a lower plant, rarely more than 25cm, with numerous non-flowering shoots; inflorescence narrower. Baltic islands of Gotland and Öland.

6 *Artemisia oelandica* (Besser) Komarov. Like *A. maritima*, but not aromatic and leaves silkily-hairy, 1-2-pinnately-lobed. Flowerheads smaller, nodding in a one-sided panicle. Limestone pavement. July-September. SE Sweden – Öland.

7* **Norwegian Mugwort** *Artemisia norvegica* Fries. Low to short, silkily-hairy, tufted perennial, often very dwarf. Leaves mostly basal 2-pinnate or the lowermost more or less digitate, stalked; upper leaves 1-2-pinnate, unstalked. Flowerheads yellowish, 10-12mm, subglobose, solitary usually, nodding on long stalks. Rocky mountain habitats, often on sandstone in peaty places, 550-1900m. July-September. N Britain, C Norway. Very local, only discovered in Britain in 1950. Also found in the USSR – Ural Mountains. The British plant is often referred to var. *scotica* Hultén. **B**: Very rare – restricted to 2 mountains in NW Scotland – W Ross.

8* **Tarragon** *Artemisia dracunculus* L. Tall, aromatic, hairless perennial to 1.2m; stems erect, much-branched. Basal leaves 3-toothed at the apex; other leaves linear to lanceolate, untoothed or weakly toothed; all leaves green. Flowerheads globose, 5-7mm, yellowish-green, pendent, in raceme-like panicles. Scrub and waste places. July-September. **I**: S & E USSR. Naturalized in France and Germany. Widely grown as a culinary herb.

9* **Field Wormwood** *Artemisia campestris* L. Medium to tall, tufted, almost scentless, scarcely hairy perennial with a branched woody base; stems erect or ascending, reddish-brown. Leaves silkily-hairy, 2-3-pinnately-lobed, stalked, the upper leaves less divided and often unstalked. Flowerheads yellowish or reddish, 3-4mm, erect or spreading, borne in a wide panicle. Bare, often sandy habitats, margins of arable fields, road-verges, waste land, coniferous plantations, to 2000m. August-September. Throughout, except Ireland and the far north. **B**: Very rare and declining, confined to East Anglia.

subsp. *maritima* Arcangeli. Like the type, but leaves rather fleshy, the segments convex (not keeled) beneath. Flowerheads often recurved. Maritime sands, drier parts of salt marshes. August-September. Coasts of Belgium, Holland and France. Not to be confused with *A. maritima* which has woolly, not fleshy leaves.

10 *Artemisia pontica* L. Like *A. campestris*, but leaves downy-white on both surfaces, clasping the stem at the base; flowerheads nodding, pale yellow. Scrub and waste ground. July-September. C and S Germany; naturalized in parts of France, sometimes casual elsewhere. Occasionally cultivated.

11 *Artemisia rupestris* L. Small subshrub to 45cm tall, though often less, woody below and with numerous non-flowering shoots. Leaves 1-2-pinnately-lobed, green, hairy or hairless, unstalked. Flowerheads 5-7mm, yellowish, nodding, in raceme-like panicles. Scrub and rocky habitats. July-September. Baltic Sweden and the neighbouring parts of the USSR; extinct in Germany. East into Asia.

Mugwort

Wormwood

Sea Wormwood

Norwegian Mugwort

Tarragon

Field Wormwood

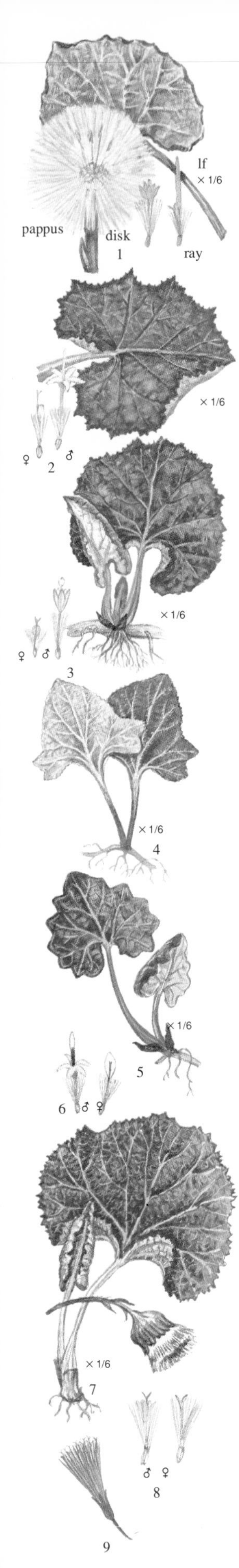


1* **Coltsfoot** *Tussilago farfara* L. Low to short, downy, creeping, perennial; stems erect, purplish-scaly, unbranched. Leaves all basal, rounded-heart-shaped, shallowly lobed or toothed, mealy white above when young, white felted beneath, appearing after the flowers. Flowers yellow, 15-35mm, with narrow rays, solitary, erect in bud, nodding as the flowers fade. Fruit a white 'clock', erect. Damp habitats particularly on calcareous clay soils, roadsides, embankments, hedgebanks, arable fields, sand-dunes, shingle, hillside flushes and woodland margins, to 2600m. February-April. Throughout, except the extreme north. **P**: Bees and flies. The leaves were formerly used in herbal remedies – dried and smoked they were employed in the treatment of coughs and asthma. **B**: Throughout.

**Butterburs** *Petasites*. Perennial herbs with large basal leaves and stems usually with scales. Flowerheads numerous, either male or female borne on different plants, in spike-like racemes or panicles which greatly elongate in fruit; florets of male flowers mostly tubular with a few rays; female with many rays. Fruit a 'clock'. About 15 species in Europe, N Asia and North America. Several are cultivated in gardens.

2* **White Butterbur** *Petasites albus* (L.) Gaertner. Low to medium, patch-forming perennial. Basal leaves rounded-heart-shaped, white-woolly beneath, the margin regularly-lobed and toothed. Flowerheads yellowish-white, or whitish, fragrant, up to 45, larger and more crowded in male plants, borne in a broad raceme or panicle, appearing before the leaves. Damp meadows, gullies, river and stream banks, woods and plantations, to 2200m. March-May. France, Germany, Denmark, S Norway and S Sweden; naturalized in Britain and the Faeroes. Often forming extensive colonies. **P**: Various insects. Occasionally cultivated in gardens. **B**: Naturalized in England and E Scotland.

3* **Butterbur** *Petasites hybridus* (L.) P. Gaertner, B. Meyer & Scherb. [*P. officinalis*]. Short to tall, hairy patch-forming perennial. Leaves rounded-heart-shaped, grey-hairy beneath, appearing after the flowers and often very large, irregularly toothed. Flowerheads pale reddish-violet, unscented, the male larger 7-12mm, the female 3-6mm, borne in cone-shaped panicles, more numerous than in the previous species. Damp habitats, river and stream-banks, roadsides, wet meadows, damp open woodland, to 1800m. March-May. Britain, Belgium, Holland, France, C & S Germany; naturalized in Scandinavia. Often forming extensive colonies, the mature leaves reaching up to 1 m across. The leaves were at one time used for wrapping butter hence the common English name – Butterbur. **B**: Throughout, except N & E Scotland.

4 *Petasites spurius* (Retz) Reichenb. Like *P. hybridus*, but leaves triangular-arrow-shaped, 2-3-lobed on each side at the base, hairy beneath; scale leaves generally without a rudimentary leaf-blade. Flowerheads yellowish, usually fewer; flowerbracts pale green, hairy, especially at the tip (not purplish and more or less hairless). Sandy sea-shores and riverbanks, often very local. March-April. Denmark, NW Germany, Finland (probably extinct) and Sweden.

5 *Petasites frigidus* (L.) Fries [*Nardosmia frigida*, *N. angulosa*]. Like *P. hybridus*, but flowerheads very few, 5-12 (not 30 or more), yellowish-white or reddish. Streamsides, bogs, wet woodland and other wet habitats, to 1760m. May-June. Scandinavia and Spitsbergen. East into the USSR to the Ural Mountains.

6* **Winter Heliotrope** *Petasites fragrans* (Vill.) C. Presl. Short to medium, patch-forming hairy perennial. Leaves appearing with the flowers, kidney-shaped or heart-shaped, slightly hairy beneath, hairless and rather shiny above, regularly toothed. Flowerheads pinkish-white, vanilla fragrant, few in a broad raceme; flower-bracts pale green or purplish, slightly hairy. Damp shaded habitats, hedgebanks, roadsides, river and stream margins, waste and cultivated land. November to March. **I**: C Mediterranean region. Widely naturalized in Britain, Belgium, France and Denmark. Cultivated in gardens since the nineteenth century. The male flowers have few or no rays, while the female have many broad rays. Often abundant. **B**: Throughout, especially in the south.

7 **Giant Butterbur** *Petasites japonicus* (Siebold & Zucc.) Maxim. [*Nardosmia japonica*]. Like *P. fragrans*, but a taller more robust plant, the leaves hairless beneath, the stems with 15 or more scale-leaves (not 2-7). Flowerheads creamy with pale green bracts. Stream banks and other wet habitats. May-June. **I**: Japan and Sakhalin. Naturalized in Britain, Holland and Denmark. Cultivated in gardens. The leaves can grow very large, up to 1 m across. **B**: Local garden escape.

8* **Purple Coltsfoot, Alpine Coltsfoot** *Homogyne alpina* (L.) Cass. Low to short, creeping hairy perennial; stems erect. Basal leaves rounded-heart-shaped, dark shiny green above, often purplish beneath, long-stalked, blunt-toothed; stem leaves few, lanceolate, unstalked. Flowerheads purple or purplish-red, 10-15mm, usually solitary; florets all tubular, spreading, the outer female, the inner hermaphrodite. Pappus a white 'clock'. Mountain habitats, damp meadows, open woods, rocky places and streamsides, to 3000m. May-September. C & E France and S Germany; naturalized in Britain. In the mountains of C Europe, the Pyrenees and the Balkans. **P**: Bees, flies and butterflies. **B**: Very rare – introduced to Scotland – Clova and S Uist in the Outer Hebrides.

9* **Adenostyles** *Adenostyles alliariae* (Gouan) A. Kerner [*A. albifrons*]. Medium to tall, hairy perennial; stems erect, to 2m, branched usually. Lower leaves triangular-heart-shaped to kidney-shaped, coarsely toothed, woolly beneath; upper leaves smaller, unstalked and half-clasping the stem. Flowerheads reddish-purple, rarely white, small and rayless, 6-8mm long, numerous, in dense, branched, flat-topped clusters; florets all tubular. Damp mountain habitats, woods, streamsides, wet meadows, scrub and damp rocky ground, to 2700m. July-August. C & E France from the Vosges southwards, S Germany. Widespread in the mountains of C and S Europe. **P**: Bees and butterflies.

*Adenostyles alpina* (L.) Bluff & Fingerh. Like *A. alliariae*, but upper stem leaves stalked, not clasping the stem. Similar habitats and flowering time. E France and S Germany from the Jura Mountains southwards.

Coltsfoot
♀
White Butterbur ♂
♂
Butterbur
♀
Winter Heliotrope
Purple Coltsfoot
Adenostyles

1* **Arnica** *Arnica montana* L. Short to medium, hairy, somewhat aromatic perennial. Basal and lower leaves oval to elliptical, often broadest above the middle, glandular-hairy above, almost unstalked; stem-leaves few, often only 2 and opposite, linear-lanceolate. Flowerhead yellow, daisy-like, 50-80mm, solitary on long stalks; rays strap-like. Mainly in mountains habitats on neutral or slightly acid soils, to 2850m. May-August. North to S Scandinavia. Locally common.

2 *Arnica angustifolia* Vahl subsp. *alpina* (L.) I.K.Ferguson [*A. alpina*]. Like *A. montana*, but basal leaves narrower, 0.5-2cm wide (not 2-4cm), elliptic-lanceolate, the stems with scattered leaves. Flowerheads smaller, 35-45mm. Grassy places, open woodland. June-August. N Scandinavia.

**Leopard's-banes** *Doronicum.* Perennial herbs with swollen stolons and alternate leaves. Flowerheads daisy-like, yellow, solitary on long slender stalks; ray florets long and narrow strap-like, female; disk florets tubular, hermaphrodite. About 34 species.

3* **Plantain Leopard's-bane** *Doronicum plantagineum* L. Medium to tall perennial, hairless below but hairy above. Basal leaves oval-elliptical, narrowed into a long stalk, untoothed or slightly toothed; stem leaves oval to lanceolate, unstalked, clasping the stem. Flowerheads bright yellow, 50-80mm, usually solitary; flower-bracts linear-lanceolate with a hairy margin. Woods, pastures and heaths, to 2000m. June-July. France; naturalized in Britain and Holland. Distributed from N France to SW Europe and Italy.

4 **Leopard's-bane** *Doronicum pardalianches* L. [*D. cordatum*]. Medium to tall, hairy, patch-forming perennial. Basal leaves oval-heart-shaped, long-stalked, toothed or untoothed; stem leaves oval to lanceolate, clasping the stem. Flowerheads bright yellow, 30-60mm, in branched clusters of 2-6; flower-bracts triangular-lanceolate, with a hairy margin. Woods, plantations, shrub, roadsides, waste gound on a variety of soils, especially calcareous ones, to 2200m. May-July. France and Germany, rare in Belgium and Holland; naturalized in Britain. **B**: Naturalized throughout, especially in E Scotland.

* **Austrian Leopard's-bane** *Doronicum austriacum* Jacq. [*D. orphanidis*]. Like *D. pardalianches*, but flowerheads generally 5-12 and all the stem leaves clasping. Shady woodland, woods and streamsides, to 2000m. July-August. C & E France and S Germany. In the mountains of C and S Europe.

**Ragworts** *Senecio*. Herbs or dwarf shrubs with alternate leaves. Flowerheads often numerous, in flat-topped clusters; outer florets generally rayed, female, the disk florets tubular, hermaphrodite; flower-bracts mostly in 1 row only. Pappus white or greyish. 2000 species.

5* **Silver Ragwort** *Senecio cineraria* DC. [*Cineraria maritima*]. A dwarf subshrub to 50cm, the stems and leaves densely white- or silvery-woolly; stems much-branched. Leaves mostly in loose basal rosettes, oval to lanceolate, deeply toothed to pinnately-lobed; upper leaves generally less deeply divided. Flowerheads bright yellow, 12-15mm, in dense flat-topped clusters, each flowerhead with 10-13 rays. Maritime cliffs and coastal rocks, occasionally inland. June-August. **I**: W & C Mediterranean. Naturalized in S W England, Wales and E Ireland. Hybrids with *S. jacobaea* have been recorded locally in Britain.

6 *Senecio inaequidens* DC. Like *S. cineraria* but leaves grey-green, rather fleshy but not silvery-woolly. Flowerheads golden-yellow. **I**: South Africa. Naturalized in rocky habitats in Belgium and France; occasionally established in Scotland.

7 **German Ivy** *Senecio mikanioides* Otto ex Walpers. Scrambling, hairless woody twiner to 3m, sometimes more; stems much-branched. Leaves fleshy, 5-7 lobed, triangular to kidney-shaped, long-stalked. Flowerheads yellow, rayless, 5-7mm, numerous, in dense terminal and lateral panicles. **I**: South Africa. Sometimes naturalized on trees and old walls in the Channel Islands and France. June-September.

8* **Broad-leaved Ragwort** *Senecio fluviatilis* Wallr. Medium to tall, stoloniferous perennial; stems erect, often branched, to 2m, slightly hairy above. Leaves numerous, elliptical to linear-lanceolate, pointed, unstalked but scarcely clasping the stem, toothed, hairless. Flowerheads yellow, 15-30mm, with 6-8 rays, in broad flat-topped clusters; flower-bracts black-tipped. Damp habitats, generally at low altitudes. July-September. France, Holland and Germany; naturalized in Britain and Denmark. **B**: Throughout, except for N Scotland.

9* **Wood Ragwort** *Senecio nemorensis* L. Medium to tall, occasionally shortly-stoloniferous; stems erect, branched above, to 2m, densely leafy. Leaves, hairy only beneath, greatly decreasing in size up the stem. Flowerheads yellow, 20-35mm, with 5-6 rays, numerous, borne in broad flat-topped clusters; flower-bracts black-tipped. Damp habitats to 2200m. July-September. Widespread in Belgium, Holland, France and Germany; naturalized in Sweden.

10* **Fen Ragwort** *Senecio paludosus* L. Tall, woolly, stems erect, generally branched above. Leaves linear-lanceolate to lanceolate, coarsely toothed short-stalked, shiny green above but white-woolly beneath; upper leaves unstalked and half-clasping the stem. Flowerheads bright yellow, 30-40mm, with 12-20 rays, in large flat-topped clusters or panicles; flower-bracts hairless or woolly. Damp places. May-July. North to S Sweden; extinct in Denmark. Declining. **B**: Rare, only in Cambridgeshire.

11* **Field Fleawort** *Senecio integrifolius* (L.) Clairv. Short to medium, downy; stems erect, unbranched. Basal leaves in a rosette, rounded to oblong, wrinkled, untoothed or slightly toothed, with a short, winged stalk, greyish or whitish-woolly at first; stem leaves few and small, bract-like. Flowerheads yellow or golden-yellow, 15-25mm, with 12-15 rays, few borne in a loose cluster; flower-bracts green, occasionally with a red tip. Dry, short, grassy habitats on well-drained calcareous soils, hills and banks, to 2200m. May-June. North to Sweden. **B**: Very local in S England north to Cambridge, on chalk or limestone. subsp. *maritimus* (Syme) Chater [*S. campestris* var. *maritimus*] has basal leaves coarsely toothed and more numerous stem leaves; flower-bracts 8-12mm long (not 6-8mm). Maritime cliffs of N and NE aspect. May-June. Only W Britain – near Holyhead Island.

*Senecio helenitis* (L.) Schinz & Thell. Basal leaves not pressed to the ground and generally withered by flowering time; stem leaves numerous and well developed. Damp grassy or stony mountain habitats, to 2200m. June-July. Belgium, France and Germany. subsp. *candidus* (Corb.) Brunerge. Shorter, to 45cm, the stems and undersurface of the leaves white-woolly. Flowerheads with 13-26 rays (not 12-13). Grassy and stony coastal habitats. June-July. N France.

*Senecio rivularis* (Waldst. & Kit.) DC. Taller, with erect basal and lower leaves; lower leaves with a broad, winged stalk and often a heart-shaped base. Flowers yellow or orange, 25-35mm, each with 15-21 rays. Damp grassy or rocky mountain habitats, to 2500m. June-August. C & S Germany.

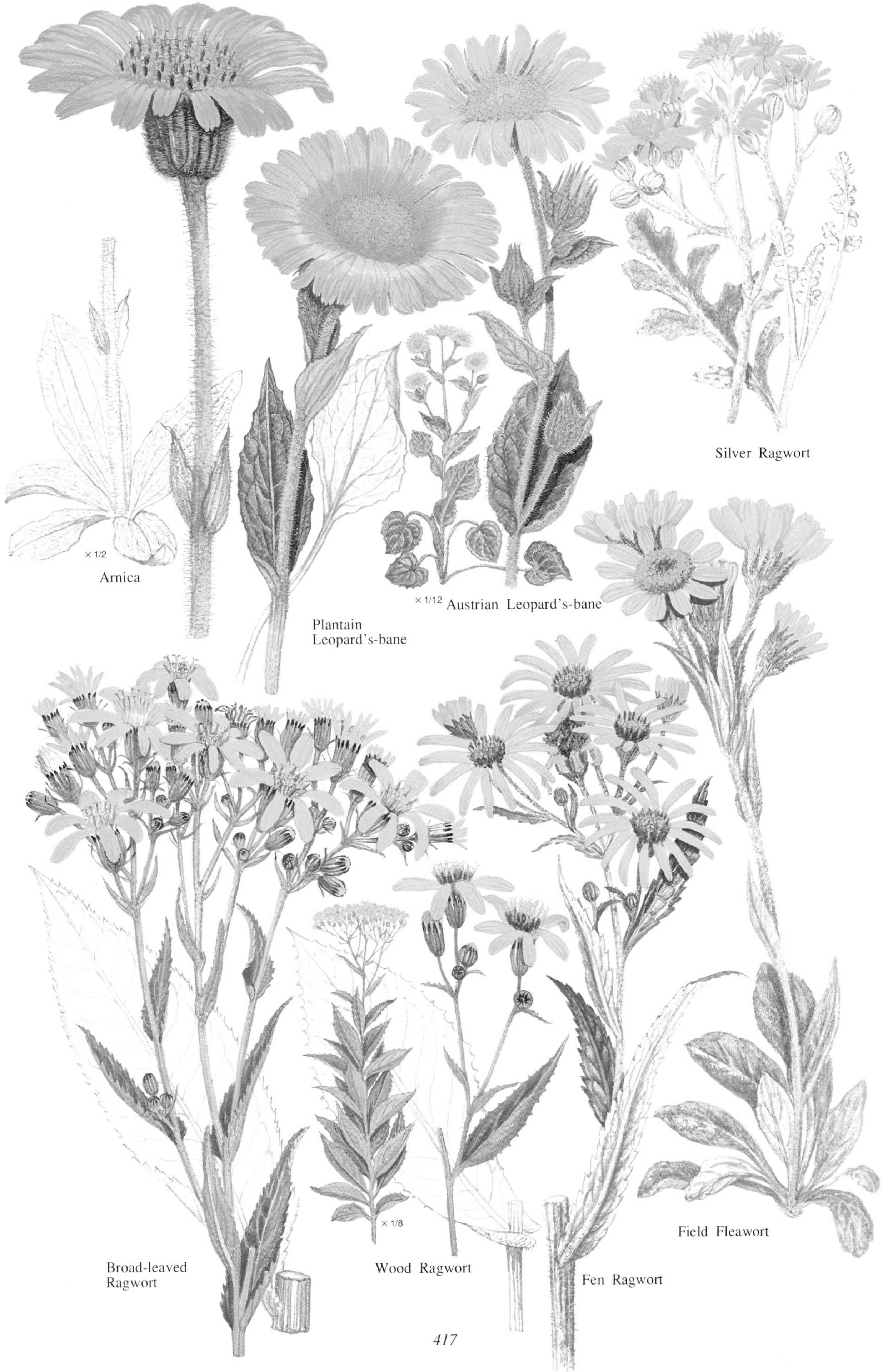

Silver Ragwort

× 1/2
Arnica

× 1/12 Austrian Leopard's-bane

Plantain
Leopard's-bane

× 1/8

Broad-leaved
Ragwort

Wood Ragwort

Fen Ragwort

Field Fleawort

1* **Common Ragwort** *Senecio jacobaea* L. Medium to tall, hairy or almost hairless biennial or perennial; stems erect, to 1.5m, branched above. Basal and lower leaves pinnately-lobed with a small end lobe, generally withered by flowering time; upper leaves 1-2-pinnately-lobed, half-clasping the stem. Flowerheads bright golden-yellow, 15-25mm, with 12-15 rays, borne in large flat-topped, branched clusters; inner flower-bracts dark-tipped. Waste and poor land, field boundaries, overgrazed meadows, woodland, roadsides, banks, sand-dunes, on a wide variety of soils, to 900m. June-November. Europe except the extreme north; naturalized in Finland. A widespread weed; extremely poisonous to livestock, cattle and horses but not to sheep. A widely naturalized weed in North America and New Zealand. Forms without rays are generally assigned to var. *flosculosus* DC. **B**: Throughout.

2 **Marsh Fleawort** *Senecio congestus* (R. Br.) DC. [*S. palustris*, *S. tubicaulis*]. Like *S. jacobaea*, but annual or perennial to 2m, though often less. Leaves oblong to linear-lanceolate, coarsely toothed or untoothed, rarely lobed, the lower generally present at flowering time. Flowerheads 20-30mm with 19-22 rays, numerous, borne in a broad panicle, rarely flat-topped. Damp meadows, marshes, fens and ditches. June-July. North to S Sweden. **B**: Extinct – formerly in East Anglia.

3* **Marsh Ragwort** *Senecio aquaticus* Hill. Like *S. jacobaea*, but usually a biennial with stems often branched in the lower part and basal leaves usually unlobed. Flowerheads 25-30mm with 12-15 rays, borne in loose, flat-topped clusters; flower-bracts not black-tipped. Wet habitats, meadows, marshes, ditches, dykes, river and stream margins, generally on heavy silt or alluvial soils, to 600m. July-August. Throughout, except the far north and Finland. **P**: Flies. **B**: Throughout. subsp. *barbareifolius* (Wimmer & Grab.) Walters [*S. erraticus*, *S. aquaticus* var. *barbareifolius*]: branches of the inflorescence spreading widely (not ascending) and flowerheads more numerous, 12-20mm. Roadsides, ditches and seasonally wet habitats. E & S France and Germany.

4* **Hoary Ragwort** *Senecio erucifolius* L. Medium to tall, grey-downy perennial with a shortly-creeping stock bearing terminal leaf-rosettes; stems erect, branched above the middle. Leaves pinnately-lobed, the lower stalked, usually present at flowering time; the upper leaves with narrower lobes; all leaves with somewhat down-rolled margins, woolly especially beneath. Flowerheads bright yellow, 12-15mm with 12-15 rays, borne in a narrow flat-topped cluster. Rough grassland, pastures, hedgebanks and roadsides, shingle beaches, on neutral or calcareous clay soils. July-September. Throughout except Norway and Finland. **B**: Locally common in England and Wales; rare in Scotland and Ireland.

5* **Oxford Ragwort** *Senecio squalidus* L. Very variable, hairy or almost hairless annual or short-lived, medium perennial; stems well branched. Leaves deep green, lanceolate to pinnately-lobed, the upper clasping the stem. Flowerheads bright yellow, 15-25mm, usually with 13 rays, borne in irregular flat-topped clusters; all flower-bracts black-tipped. Waste and disturbed ground, old walls, to 2300m. April-December. S Germany; widely naturalized in Britain, France and Denmark. C & S Europe. **B**: Widespread in England and Wales, local in S Scotland and Ireland. Has spread rapidly in Britain during the last 150 years, the seed being distributed along the railway network in particular

6 **Welsh Ragwort** *Senecio cambrensis* Rosser. Like *S. squalidus*, but the inflorescence more branched and leafy. Flowerheads numerous, 6-12mm, with 8-15 short rays which soon become recurved. Open habitats, roadsides and gardens. May-October. W Britain. Considered a natural hybrid between *S. squalidus* and *S. vulgaris*. Discovered in 1953. **B**: Very rare, restricted to Denbighshire, Flintshire, Shropshire and S Scotland.

7 *Senecio vernalis* Waldst. & Kit. Like *S. squalidus*, but young shoot white-woolly and branches of the inflorescence erect, (not spreading). Flowerheads larger, 20-25mm. Cultivated and waste ground, open sandy habitats. June-September. Throughout except the far north and Finland; naturalized in Britain. **B**: Naturalized in England, but rare.

8* **Heath or Wood Groundsel** *Senecio sylvaticus* L. Short to medium yellowish-green, hairy annual, not sticky; stems erect, slender and grooved. Basal and lower leaves elliptical pinnately-lobed, broadest above the middle, short-stalked; upper leaves similar, but broader and clasping the stem. Flowerheads bright yellow, 4-6mm, with 8-15 short rays which soon become recurved, borne in a larger terminal, flat-topped cluster. Open woodland and woodland margins, heaths, road verges, embankments, disturbed and waste ground, fired sites, generally on well-drained, acid, sandy soils. July-September. Throughout, except the far north. **B**: Throughout, but local and rather scarce in W Scotland and Ireland.

9 **Sticky Groundsel** *Senecio viscosus* L. Like *S. sylvaticus*, but a very sticky, foetid annual. Upper leaves unstalked but scarcely clasping the stem. Flowerheads pale yellow, 6-10mm; flower-bracts green (not dark tipped). Waste and disturbed ground, roadsides, railways, sand-dunes and shingle beaches on sandy or gravelly soils, to 2300m. July-October. Britain, Belgium, Holland, France and Germany; widely naturalized in Ireland, Holland, Denmark, Finland and Sweden. Common and widespread annual weed. **B**: Locally common north to S Scotland, rare elsewhere.

10* **Groundsel** *Senecio vulgaris* L. Low to short annual, occasionally hairless; stem simple or weakly-branched. Leaves bright shiny green, pinnately-lobed, the upper unstalked and clasping the stem. Flowerheads yellow, small and rayless, 4-5mm, in lax, branched clusters; flower-bracts often black-tipped, the outer very short. A common weed, to 2300m. Flowering throughout the year. Throughout, except Spitsbergen. **B**: Throughout much of Britain. subsp. *denticulatus* (O.F. Mueller) P.D. Sell. [var. *radiatus*]: flowerheads with rays; rays 6-12, recurving soon after the flowers open. Local throughout the range especially in coastal habitats.

11* **Pot Marigold** *Calendula officinalis* L. Rough-hairy, short to medium annual or short-lived perennial; stems branched and spreading, leafy to the top. Leaves oblong to elliptical, often widest towards the top, somewhat mealy, untoothed. Flowerheads large and daisy-like, orange or deep golden-yellow, 40-70mm, solitary; rays broad and strap-like. Achenes curved and beaked, boat-shaped. Waste, disturbed and cultivated land. May-October. Casual in many parts of W and NW Europe. **B**: Locally naturalized or casual.

**Field Marigold** *Calendula arvensis* L. Like *C. officinalis*, but leaves narrower and flowerheads smaller, 10-20mm. Widespread weed of arable land, disturbed and waste ground, waysides and vineyards. June-August. France and Germany.

Common Ragwort

Marsh Ragwort

Hoary Ragwort

Oxford Ragwort

Heath Groundsel

Groundsel

Pot Marigold

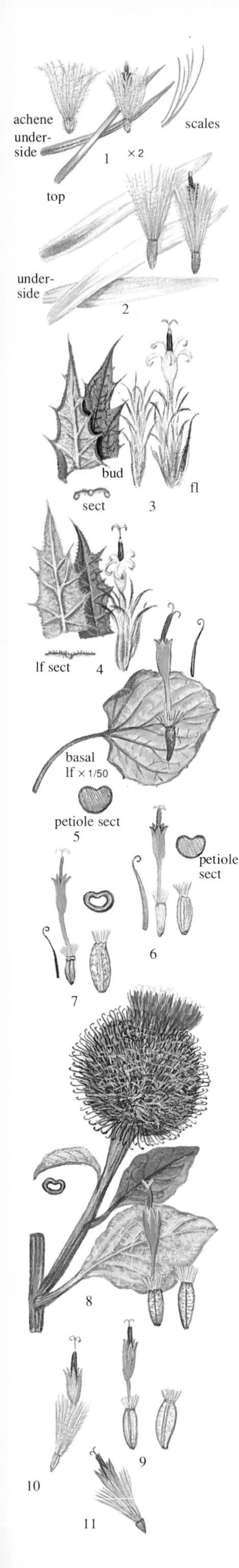


1* **Carline Thistle** *Carlina vulgaris* L. Low to short biennial with a branched or unbranched stem. Leaves alternate, not in a basal rosette, thistle-like, narrow-oblong, with a deep spiny, wavy margin, cottony, especially beneath. Flowerheads 15-40mm, solitary or in groups of 2-5, rayless but with numerous spreading, shiny yellowish bracts, linear, pointed; florets all tubular, yellowish-brown, hermaphrodite. Dry, often grazed grassland and sand-dunes on calcareous soils, to 1750m. July-September. Throughout, except the far north. During the first season plants form a leaf-rosette but this has withered by flowering time in the second year. Like *C. acaulis* the flowers open in dry weather. **B**: Locally common in lowlands. subsp. *intermedia* (Schur) Hayek [*C. biebersteinii*] is taller, 30-70cm (not 15-30cm). Lower leaves and upper half of the upper leaves weakly spiny. Germany and S Finland. subsp. *longifolia* Nyman [*C. longifolia*] has stems generally unbranched, the upper leaves flat and weakly spiny, the lateral veins distinctly parallel to the margin. Damp grassland. Mountains of Germany and S Finland.

2* **Stemless Carline Thistle** *Carlina acaulis* L. Low short-lived perennial, occasionally biennial, stemless. Leaves in a flat basal rosette, elliptic-oblong, pinnately-lobed, spiny, sometimes downy beneath. Flowerheads large 35-60mm, with conspicuous, wide-spreading, silvery or pinkish, linear, pointed bracts; disk florets whitish or purplish-brown. Mountain habitats, to 2800m. June-September C & E France and Germany. Plants die after flowering. subsp. *simplex* (Waldst & Kit.) Nyman. Like the type, but with a distinct stem, 15-60cm, carrying up to 6 flowerheads. **B**: Similar habitats and distribution.

3* **Globe Thistle** *Echinops sphaerocephalus* L. Tall, stout, stickily-hairy perennial; stems erect, ridged, branched or unbranched. Leaves thistle-like, oblong-elliptical, pinnately-lobed, with a fine spiny margin, white-downy beneath. Flowerheads globose, bluish-grey or whitish, 30-60mm, solitary, rayless; florets all tubular, hermaphrodite. Dry bare stony and grassy habitats, scrub. June-September. C France; naturalized in Belgium, Germany and Sweden. Throughout much of C and S Europe.

4 *Echinops exaltatus* Schrader [*E. commutatus*]. Like *E. sphaerocephalus*, but upper surface of leaves bristly and margin rough. Flowerheads whitish or greyish. Similar habitats and flowering time. **I**: Mediterranean. Naturalized in France, Germany and Denmark.

**Burdocks** *Arctium*. Erect leafy biennials with a stout tap-root. Leaves alternate, hairy, slightly toothed or untoothed. Flowerheads solitary or in lax flat-topped clusters or racemes, globose; florets tubular, hermaphrodite, reddish-purple or white; flower-bracts numerous, each terminating in a hooked bristle. 5 closely related and much confused species from Europe and Asia. Each shows a great deal of variation and all hybridize with one another when growing in close proximity. The hooked flower-bracts help in the dispersal of the seeds.

5* *Arctium tomentosum* Miller [*Lappa tomentosa*]. Tall biennial to 1.5m. Basal leaves large, up to 50cm, broadly heart-shaped, white cottony beneath, with solid stalks. Flowerheads small, 12-20mm (15-25mm in fruit) densely white-cottony. Woodland margins, hedgerows, scrub and waysides. July-September. Throughout except the far north. **B**: Occasionally naturalized.

6* **Greater Burdock** *Arctium lappa* L. [*A. majus*, *Lappa officinalis*, *L. major*]. Tall biennial to 1.5m; stems hairy to more or less hairless. Basal leaves large, up to 50cm, heart-shaped with solid stalks. Flowerheads globose, 20-25mm (35-42mm in fruit), shiny golden-green and more or less hairless on the outside. Open woodland, hedgerows, roadsides, rough grassy places, waysides and waste ground. July-September. Throughout, except the far north, the Faeroes, Iceland and Spitsbergen. **B**: Throughout lowlands, N to C Scotland, scarce in the SW, Wales and Ireland.

7* **Lesser Burdock** *Arctium minus* (Hill) Bernh. [*Lappa minor*]. Medium to tall biennial, to 1.5m; stems often reddish, with downcurved branches, hairy. Basal leaves large, to 50cm, broadly heart-shaped, with hollow stalks. Flowerheads egg-shaped, 15-18mm (15-25mm in fruit), green or purple-tinged and often white-cottony when young. On a variety of habitats and soils except very acid ones. July-September. Throughout, except the far north. The burs often remain on the plant through the winter. **B**: Widespread except the Scottish Highlands.

8 subsp. *pubens* (Bab.) J. Arenes [*A. pubens*]. Like the type, but flowerheads larger, 20-30mm, stalked, the stalks 15-40mm, (not short-stalked or unstalked); florets equalling the flower-bracts (not longer). Similar habitats and flowering time. Britain, Belgium, Holland, France and Denmark. Probably of hybrid origin between *A. lappa* and *A. minus*. **B**: Commonest subspecies in lowland England.

9 subsp. *nemorosum* (Lej.) Syme. Like the type, but plants up to 2.5 m with flowerheads 30-40mm; florets equalling the flower-bracts. Similar habitats and distribution. The typical subspecies can generally be recognized, however, intermediates frequently occur and should be borne in mind when attempting identification. **B**: Lowland England, Scotland and Ireland.

10* **Alpine Saw-wort** *Saussurea alpina* (L.) DC. [*Serratula alpina*]. Short to medium, stoloniferous, grey-hairy perennial. Leaves alternate, oval to lanceolate, toothed or untoothed, not spiny, white-cottony beneath, with a narrowed, winged stalk. Flowerheads purple, 15-20mm long, in a compact cluster, fragrant; florets all tubular and hermaphrodite. Mountain grassland, rocky places, alpine and maritime cliffs and screes on calcareous and base-rich soils, to 3000m. July-September. Britain, Scandinavia, mountains of France and Germany. **B**: Scattered localities in Scotland, Lake District, N Wales, N & W Ireland.

11* **Jurinea** *Jurinea cyanoides* (L.) Reichenb. Medium, leafy perennial. Basal leaves 1-2-pinnately-lobed, with narrow segments, occasionally unlobed, white-cottony beneath, with inrolled margins. Flowerheads pink or purplish, 15-18mm, subglobose, solitary or several together. Dry, generally sandy habitats. June-August. C & S Germany. Also native in W USSR.

Carline Thistle
Stemless Carline Thistle
Globe
Thistle
*Arctium*
*tomentosum*
Greater Burdock
Lesser Burdock
Alpine
Saw-wort
Jurinea

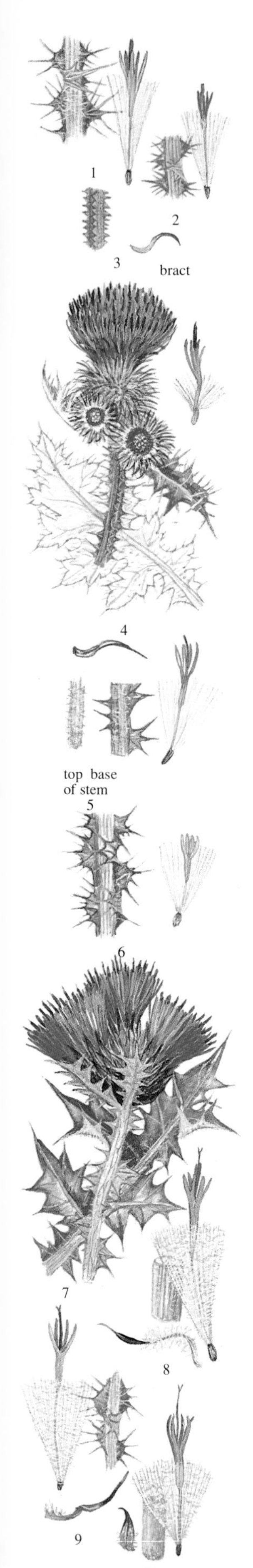


**Thistles** *Carduus.* Annual, biennial or perennial herbs with spiny-winged stems. Leaves alternate, often spine-toothed. Flowerheads globose to cylindrical, typically shaving-brush shaped; florets all tubular, often purple or pink; flower-bracts in many rows, spine-tipped. About 120 species primarily in Europe, Asia and N Africa. Pappus composed of simple hairs (feathery in *Cirsium.*

1* **Musk or Nodding Thistle** *Carduus nutans* L. Tall biennial to 1.5m; stem winged below, white-cottony. Leaves deeply pinnately-lobed, the lobes spine-toothed, woolly on the raised veins beneath. Flowerheads bright reddish-purple, large, 30-50mm, half-nodding, solitary or several together, borne on non-spiny stalks; outer flower-bracts strongly recurved. Pastures, arable land, roadsides and other grassy habitats, waste ground, to 2500m. May-September. Throughout Europe north to SE Norway. The flowers have a slightly musky or almond fragrance. **B**: Throughout north to the Firth of Forth, but scarce in Wales and rare in Ireland.

2* **Welted Thistle** *Carduus acanthoides* L. Tall hairy biennial to 2m; stem branched, spiny-winged except just below the flowerheads. Leaves oblong to lanceolate, pinnately-lobed, weak-spined, dull green slightly white-cottony beneath. Flowerheads reddish-purple, 10-25mm, erect, solitary or in small clusters; flower-bracts slightly spreading or erect, with a weak spine-tip. Rough grassy places, meadows and pastures, roadsides, hedgebanks, river and streamsides, waste ground, to 3000m. June-August. Europe north to S Sweden; naturalized in Norway. **B**: Lowlands north to the Moray Firth and in E Ireland; rare elsewhere.

3* **Great Marsh Thistle** *Carduus personata* (L.) Jacq. Tall perennial to 1.2m, sometimes taller; stems with narrow, spiny wings right to the top. Leaves deep green above, greyish-white-cottony beneath, the lower pinnately-lobed, the upper lanceolate, unlobed; all leaves with a soft-spiny margin. Flowerheads reddish-purple, 15-25mm, several in a tight erect head; flower-bracts mostly S-shaped, not spine-tipped. Damp meadows, woods, road-verges and stream-sides, to 2300m. Mountains of France and Germany from the Vosges southwards. Widespread in the mountains of C and S Europe.

4 *Carduus crispus* L. subsp. *multiflorus* (Gaudin) Franco [*C. multiflorus*]. Like *C. personata*, but all leaves lanceolate or elliptical, lobed or toothed, more strongly spined. Flower-bracts slightly curved at the apex, but not S-shaped. Grassy places, roadsides, waysides, waste places and streamsides, to 1900m. Throughout, except for N Scandinavia.

5* **Alpine Thistle** *Carduus defloratus* L. subsp. *glaucus* Nyman [*C. glaucus*]. Medium to tall perennial, to 1m; stems spiny-winged below, unwinged above. Leaves oblong-lanceolate, pinnately-lobed, spine-toothed, stalked, bluish-green beneath. Flowerheads purple to rose-purple, 25-40mm, solitary and slightly nodding; flower-bracts hairless, S-shaped, the inner erect. Meadows, open woods and stony places on calcareous soils, to 3000m. June-October. C & S Germany.

6* **Slender Thistle** *Carduus tenuiflorus* Curtis. Slender, short to tall, annual or biennial; stem wide, spiny-winged to the top, cottony-white. Leaves lanceolate, lobed and spiny-margined, white-cottony beneath. Flowerheads small, pinkish, 6-10mm, in compact clusters of 3 or more; flower-bracts oval-lanceolate with a curved spine-tip, hairless. Dry open habitats, roadsides, waste ground and grassy places close to the sea at low altitudes. May-August. Britain, Belgium, Holland and France; naturalized in Norway and Sweden. **B**: Local throughout, except N Scotland, often coastal.

7 *Carduus pycnocephalus* L. Like *C. tenuiflorus*, but flowerheads usually stalked, solitary or 2-3 together. Rough grassy and waste places. June-August. France, casual further north, including England.

**Thistles** *Cirsium.* Biennial or perennial herbs often with spiny-winged stems and spiny leaves. Leaves alternate, pinnately-lobed. Flowerheads purple, pink, yellow or white; flower-bracts usually spine-tipped. Pappus of feathery hairs. Ocer 100 species in the northern hemisphere.

8* **Woolly Thistle** *Cirsium eriophorum* (L.) Scop. Variable, medium to tall, stout biennial to 1.5m; stems unwinged, white-cottony, branched above. Leaves pinnately-lobed, with long rigid spines, white-cottony beneath. Flowerheads large cottony-hairy, reddish-purple, 25-50mm, usually solitary, erect; outer flower-bracts recurved, spine-tipped. Rough grassland, roadsides, scrub, railway embankments, on dry calcareous soils, to 2100m. July-September. Britain, Belgium, Holland, France and Germany. The British plant is generally referred to subsp. *britannicum* Petrak. **B**: Scattered localities in England and Wales N to NE Yorkshire.

9* **Spear Thistle** *Cirsium vulgare* (Savi) Ten. [*C. lanceolatum*]. Tall biennial to 1.5m; stem with sharp spiny wings to the top. Leaves lanceolate, pinnately-lobed, sharply spiny-margined, dull green, prickly-hairy on the upper surface. Flowerheads purple, 20-40mm, in a panicle or flat topped cluster; flower-bracts straight, with a yellow spine. Grassy and waste places, disturbed ground, often on fertile, base-rich soils. July-October. Throughout except the far north. A common and injurious weed in many places. **B**: Widespread throughout.

10* **Meadow Thistle** *Cirsium dissectum* Hill [*C. anglicum*]. Medium to tall, rhizomatous perennial; stem unwinged, white-cottony, generally unbranched. Leaves elliptical to lanceolate, with a soft-spiny margin, the lower sometimes pinnately-lobed, green above, white-cottony beneath. Flowerheads purple, 25-30mm, solitary usually; flower-bracts lanceolate, cottony, not spreading, the outer spine-tipped. Wet meadows, roadsides, on mildly acid or calcareous peaty soils. June-August. Britain, Belgium, Holland, France and Germany; naturalized in Norway. **B**: Locally common in England, Wales and Ireland.

Musk Thistle
Great Marsh Thistle
Alpine Thistle
Welted Thistle
Slender Thistle
Woolly Thistle
Spear Thistle
Meadow Thistle

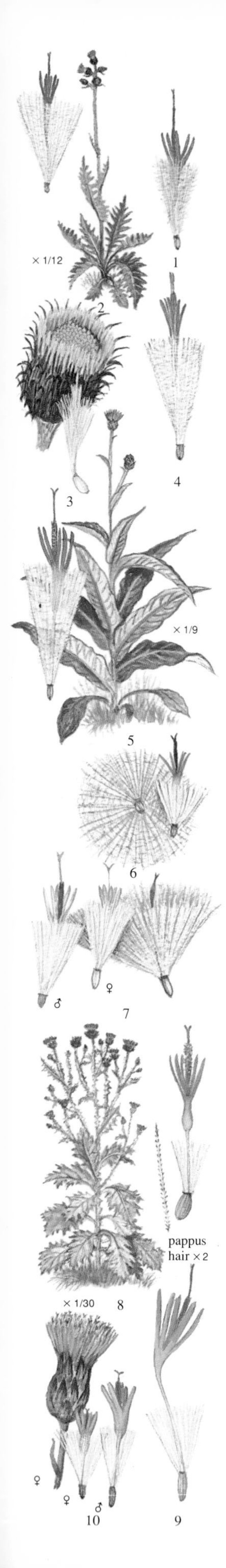


1* **Tuberous Thistle** *Cirsium tuberosum* (L.) All. [*Carduus tuberosus*]. Medium to tall perennial with tuberous roots; stems unwinged, cottony, leafless towards the top. Leaves deeply pinnately-lobed, the margin with soft bristles, green on both surfaces, slightly cottony beneath, the upper unstalked. Flowerheads purple, 20-30mm, solitary on long unwinged stalks; flower-bracts erect, cottony below. Grassy habitats, hills and downs, scrub, on calcareous soils, generally at rather low altitudes. June-August. S Britain, France and Germany; naturalized in Belgium. Declining. Hybridizes with *C. acaule*. Sometimes confused with *C. dissectum*, but leaves deeply lobed. **B**: Very rare, confined to a few places in C S England and S Wales.

2* **Brook Thistle** *Cirsium rivulare* (Jacq.) All. [*C. tricephalodes*]. Like *C. tuberosum*, but without tuberous roots; stem leafless above or with a few bract-like leaves. Flowerheads purple, occasionally white, smaller, 20-35mm, in tight terminal clusters of 2-8, stalkless; flower-bracts purple. Damp habitats, meadows, pastures, marshes, river and stream margins and ditches, on acid soils, to 1750m. June-September. France and Germany; naturalized in Sweden.

3* **Cabbage Thistle** *Cirsium oleraceum* (L.) Scop. Medium to tall perennial, to 1.5m; stems unbranched or slightly branched, unwinged, ridged, hairless. Leaves elliptical, pinnately-lobed to toothed, weakly spiny, green on both surfaces; upper leaves unlobed, clasping the stem with large basal lobes. Flowerheads pale yellow, 25-40mm, in terminal clusters of 2-6 with a cluster of leaves immediately below; flower-bracts erect. Wet habitats, wet woods, marshes, fens, river and stream margins, at low altitudes. July-September. Continental Europe north to S Scandinavia; naturalized in Britain, Ireland and Finland. **P**: Bees and butterflies. **B**: Naturalized in a few scattered localities in England, Scotland and Ireland.

4* **Dwarf Thistle** *Cirsium acaule* Scop. [*Carduus acaulos*]. Low stemless or almost stemless perennial, with a rosette of leaves, almost hairless. Leaves oblong, pinnately-lobed, spiny, deep green. Flowerheads purple, 20-40mm, generally solitary, borne directly on the leaf-rosette; flower-bracts erect, purplish, scarcely spiny. Dry grassy habitats, especially grazed pastures, on calcareous soils, to 2550m. June-September. Throughout, except N Scandinavia. Intolerant of heavy trampling despite often growing in grazed habitats. **P**: Bees. **B**: Local in England and S Wales, N to Yorkshire.

5 **Melancholy Thistle** *Cirsium helenoides* (L.) Hill [*C. heterophyllum*]. Medium to tall, stoloniferous perennial; stems generally unbranched, and usually leafless at the top, cottony. Leaves flat, oval to lanceolate, the lowest sometimes pinnately-lobed, green above, white-felted beneath, with a soft prickly margin. Flowerheads purple, rarely white, 35-50mm, solitary or in tight clusters of 2-4, borne on unwinged stems. Damp habitats, meadows, roadsides, banks, open woods and woodland margins, scrub, streamsides, calcareous or base-rich soils, to 2350m. July-September. Throughout, except Belgium and Holland; naturalized in Iceland. Only on mountains in C Europe. **P**: Bees. **B**: Widespread from mid-Wales and Staffordshire northwards, and in Ireland.

6* **Marsh Thistle** *Cirsium palustre* (L.) Scop. Medium to tall biennial, to 1.2m; stems spiny-winged to the top, usually branched above. Leaves linear-lanceolate, pinnately-lobed and very spiny, mostly unstalked, hairy above. Flowerheads purple, occasionally white, 10-20mm, in clusters of 2-8; flower-bracts purple tinged, erect, weakly spiny. Damp habitats, woodland clearings and margins, rough meadows, marshes, roadsides and hedgerows, on a wide variety of soils. July-September. Throughout, except the extreme north and Iceland. Distinguished from *C. vulgare* by its hairy, not prickly, upper leaf-surface and from *C. arvense* by its winged stems. The young shoots can be eaten as a salad. **P**: Bees, butterflies and flies. **B**: Abundant throughout.

7* **Creeping Thistle** *Cirsium arvense* (L.) Scop. Medium to tall stoloniferous perennial; stem usually branched, not winged or spiny. Leaves lanceolate to oblong, pinnately-lobed to unlobed, spiny, hairless above, sometimes cottony beneath, the upper leaves unstalked. Flowerheads pale purple or lilac, 15-25mm, fragrant, solitary or 2-5 together, stalked; flower-bracts erect with short spine-tips, generally purplish. Meadows, pastures, arable and cultivated land, roadsides, open woodland, waste places, on a wide variety of soils but often abundant on fertile ground. June-September. Throughout, except Spitsbergen; naturalized in the Faeroes and Iceland. An injurious and widespread weed of cultivation, spreading rapidly by vegetative means. Plants are normally dioecious, with male and female flowers on separate plants. **P**: Various insects. **B**: Abundant throughout.

8* **Scotch Thistle** *Onopordon acanthium* L. Very stout grey- or white-felted, tall biennial, to 3m; stem branched above, often yellowish, spiny-winged to the top. Leaves in a basal rosette in the first year, oval to lanceolate, with a spiny, undulate margin. Flowerheads thistle-like, purple or white, 30-50mm, usually solitary; flower-bracts numerous, spreading, ending in a yellowish spine. Hedgebanks, roadsides, waste and cultivated land on slightly acid, neutral or calcareous soils, to 1500m. July-September. France, Germany, Belgium and Holland; naturalized in Britain, Denmark and Sweden. Widely cultivated in gardens and often becoming naturalized. Introduced to Britain probably by the Romans. **P**: Long-tongued bees and butterflies. **B**: Widely naturalized in England and Wales, rare elsewhere.

9* **Milk Thistle** *Silybum marianum* (L.) Gaertner. Robust hairless or slightly hairy, medium to tall annual or biennial, to 1.5m; stems unwinged. Leaves oblong, pinnately-lobed, spiny-margined, pale shiny-green and variegated with white, especially along the veins above, wavy; upper leaves clasping the stem. Flowerheads purple, 40-50mm, solitary, thistle-like; flower-bracts large, spreading and spiny. Roadsides, waste places and cultivated ground. June-August. C & S France; naturalized in Britain, Belgium and Holland. **P**: Bees and butterflies. The leaves and stems are edible and can be used in salads. **B**: Scattered localities north to S Scotland, frequently casual.

10* **Saw-wort** *Serratula tinctoria* L. Very variable, hairless or slightly hairy, low to tall perennial, spineless; stems slender unwinged. Leaves lanceolate or oval, unlobed to pinnately-lobed, finely toothed, the lower stalked. Flowerheads purple, rarely white, 15-20mm, rayless, in branched clusters; flower-bracts purplish, erect; male and female flowers borne on separate plants. Rough permanent grassland, scrub, heaths, woodland margins, on slightly acid to calcareous soils, to 2400m. Throughout, except the Faeroes, Iceland, Finland and the extreme north. Europe to C Asia as well as Algeria. The thistle-like flowerheads are spineless. **P**: Bees and flies.

× 1/5
Tuberous Thistle
Brook Thistle
Dwarf Thistle
Cabbage Thistle
Marsh Thistle
× 1/15
Creeping Thistle
Scotch Thistle
Milk Thistle
Saw-wort
♂

**Cornflowers and Knapweeds** *Centaurea*. Annual or perennial herbs. Leaves alternate, divided or undivided. Flowerheads solitary or in groups of 2-3; florets tubular, the outer often larger, sterile and spreading, the inner hermaphrodite; involucre of flower-bracts globose or cylindrical, each bract with a characteristic cut or spiny appendage. About 450 species, 200 in Europe alone, in the northern hemisphere, South America and Australia. A number are cultivated in gardens. The characteristics of the flower-bracts are particularly important in accurate identification.

1* **Greater Knapweed** *Centaurea scabiosa* L. Medium to tall, robust, rough, somewhat bristly perennial, to 1.5m; stems erect, branched above. Leaves pinnately-lobed, occasionally unlobed, with oblong or linear segments, lower leaves stalked. Flowerheads purple, large, 30-50mm, the outer florets much longer than the inner; flower-bracts green with a brown or black, fringed horse-shoe shaped appendage. Rough grassy habitats, scrub, hedgebanks and roadsides, on dry calcareous soils, to 500m. July-September. Throughout, except Iceland and the far north; naturalized in the Faeroes. **P**: Bees and flies. **B**: Widespread, except the extreme north.

2* **Red Star-thistle** *Centaurea calcitrapa* L. Short to tall, much-branched perennial, almost hairless; stems grooved. Leaves pinnately-lobed, the lobes bristle-pointed, the lower withered by flowering time, the upper smaller and narrower. Flowerheads pale purple, 8-10mm, the florets all the same length; flower-bracts conspicuous, terminating in a stiff yellowish spine, often with short lateral spines. Rough grazed grassland, dry bare places, waysides and waste places, on chalky or sandy soils, at low altitudes. July-September. C & S France; naturalized in Britain, Belgium, Holland and Germany. Often introduced as an impurity in Lucerne and Clover seed for agricultural use. **P**: Bees and flies. **B**: Very rare, confined to the Sussex coast; casual elsewhere in S England and S Wales.

3* **Rough Star-thistle** *Centaurea aspera* L. [*C. heterophylla*]. Like *C. calcitrapa*, but a lower plant with cottony stems, the upper leaves often undivided. Flowerheads larger, 20-25mm; flower-bracts palmate, with 3-5 more or less equal, short spines. Stabilized sand-dunes and waste ground and other dry open habitats. July-September. W & S France and the Channel Islands – Guernsey and Jersey; naturalized in S Wales. **B**: Local in the Channel Islands; naturalized in Glamorgan.

4* **Yellow Star-thistle** *Centaurea solstitialis* L. Medium to tall, greyish-downy biennial; stems erect, branched from the lower half. Lower leaves pinnately-lobed, stalked, the upper linear-lanceolate, unstalked and untoothed. Flowerheads pale yellow, solitary, 10-12mm, the florets all the same length; flower-bracts with a stout narrow straw-coloured spine. Dry open habitats, grassy and rocky places, cultivated land and waste places. July-September. C & S France; naturalized in Britain and Germany. **P**: Mainly bees. **B**: Naturalized in the south.

5* **Brown Knapweed** *Centaurea jacea* L. [*C. amara*]. Medium to tall, hairy perennial, to 1.2m; stem slender but thickened below the flowerheads, unbranched or slightly branched. Leaves rough, oval to lanceolate, unlobed or sometimes pinnately-lobed; upper leaves unlobed, sometimes slightly toothed at the base, unstalked. Flowerheads purple, rarely white, 10-20mm, solitary, stalked, the outer florets longer than the inner; flower-bracts oval, with a rounded, pale brown, irregularly fringed appendage. Rough grassland, open woods, roadsides and waste ground, generally at low altitudes. August-September. Europe, except the far north. Hybridizes freely with *C. nigra* – forming extensive hybrid swarms in which pure *C. jacea* may be hard to distinguish. **P**: Bees, butterflies and flies. **B**: Formerly naturalized in S England; casual elsewhere.

6 *Centaurea decipiens* Thuill. Like *C. jacea*, but a shorter plant, the flowerheads slightly smaller, unstalked; flower-bract appendage more triangular. Pastures and other grassy places. July-September. Continental Europe north to S Norway.

7 *Centaurea microptilon* Gren. & Godron. Like *C. jacea*, but flowerheads smaller, purple or pink; flower-bracts with a triangular-lanceolate, fringed, black or reddish-brown appendage. Grassy habitats, pastures, roadsides and woodland margins. July-September. Belgium, Holland and France. W and SW Europe. Probably of hybrid origin between *C. jacea* and *C. nigra*.

8* **Doubtful Knapweed** *Centaurea nigrescens* Willd. Medium to tall perennial; stems erect, little-branched. Leaves rough, hairy, the lower oblong-lanceolate or lobed, stalked; upper leaves unlobed, occasionally slightly toothed. Flowerheads purple, 10-20mm, solitary; flower-bracts green with a fringed blackish-brown or pale-brown triangular appendage, not covering the adjacent bracts. Meadows, pastures and other grassy habitats, open woods, to 2200m. July-August. E France and S Germany. The French plant, which has grey-hairy, rather than green leaves, is generally assigned to subsp. *ramosa* Gugler.

9* **Common Knapweed, Hardheads** *Centaurea nigra* L. (incl. *C. debeauxii* Gren. & Godron). Medium to tall roughly hairy perennial; stems erect, branched or unbranched, thickened below the flowerheads. Leaves elliptical to oval, the lower somewhat toothed, the upper untoothed and unstalked. Flowerheads purple, 20-40mm, solitary or in branched clusters, the florets all equal in length; flower-bracts with a broad blackish, triangular, fringed appendage, concealing the adjacent bracts. Rough grassland, meadows, pastures, roadsides, railway embankments, waysides and cliffs, on a wide variety of soils, to 1000m. June-September. Europe north to C Scandinavia; naturalized in Denmark. The commonest species of *Centaurea* normally seen. **P**: Various insects.. **B**: Throughout most of Britain.

10 **Slender Knapweed** *C. n.* subsp. *nemoralis* (Jordan) Gugler [*C. nemoralis*]. Like the type, but a slenderer, more branched plant. Flowerheads pale purplish-red, the outer florets often longer than the inner; flower-bracts with brown appendages. Similar habitats and distribution but generally on lighter soils. Not always easy to separate from the type, especially in Britain. **B**: England and Wales from the Humber southwards.

× 1/15
Greater Knapweed
Red Star-thistle
Rough Star-thistle
Yellow Star-thistle
Brown Knapweed
Doubtful Knapweed
Common Knapweed

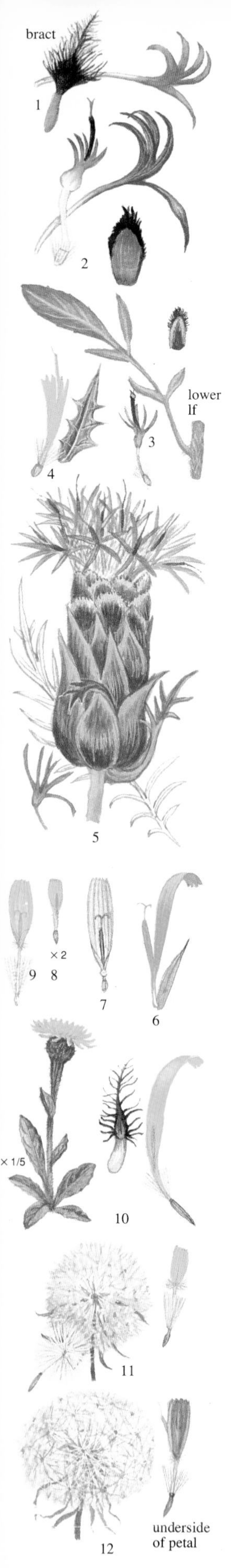


1* **Wig Knapweed** *Centaurea phrygia* L. [*C. austriaca*]. Medium to tall, hairy perennial; stems erect, branched or unbranched, thickened below the flowerheads. Leaves green, lanceolate to ovate, toothed or untoothed, the upper clasping the stem. Flowerheads pink or purple, 30-50mm, solitary, the outer florets spreading, greatly exceeding the inner; flower-bracts with prominent feathery, brown or black, recurved appendage. Meadows, woodland and scrub, to 2200m. July-September. Denmark, Germany, Norway and Finland; naturalized in Sweden.

subsp. *pseudophrygia* (C.A. Meyer) Gugler, leaves rough-hairy and the flowerheads often in clusters of 2-4. Similar habitats and flowering time. Denmark, S Norway and Germany.

2* **Perennial Cornflower** *Centaurea montana* L. Short to medium rhizomatous, cottony perennial; stems branched or unbranched, winged with the leaf-base. Leaves soft, oval to oblong, generally untoothed. Flowerheads violet or blue, 60-80mm, the outer florets spreading, much longer than the inner; flower-bracts with a blackish-brown fringed border. Dry open habitats, often on calcareous soils, to 2100m. June-August. Belgium, France and Germany; naturalized in Finland; casual elsewhere. **B**: Casual; common in gardens.

3* **Cornflower** *Centaurea cyanus* L. Medium to tall annual; stems erect, generally branched above, greyish-cottony. Leaves lanceolate, untoothed or slightly toothed, or the lower somewhat pinnately-lobed, cottony beneath. Flowerheads violet-blue or dark blue, rarely white or purple, 15-30mm, solitary, the outer florets spreading, much longer than the inner; flower-bracts narrowly fringed with brown or silver. Disturbed ground and other open habitats, to 1800m. June-August. Throughout except the far north, often casual. A widespread cornfield weed now declining. **B**: Rare and often casual.

4* **Carthamus** *Carthamus lanatus* L. [*Kentrophyllum lanatum*]. Medium, strong-smelling, cottony annual; stems branched above. Leaves pinnately-lobed, spiny-toothed, sticky-glandular, the upper lanceolate, spiny. Flowerheads thistle-like, golden yellow, 20-30mm, few, borne in flat-topped clusters; flower-bracts very spiny. Dry open habitats, at low altitudes. May-August. NC France southwards.

5 *Carduncellus mitissimus* (L.) DC. [*Carthamnus mitissimus*]. Low, scarcely hairy perennial, stemless or with a stem up to 10cm tall. Leaves in a basal rosette, pinnately-lobed, the lobes linear-lanceolate, untoothed or with a finely prickly margin. Flowerhead solitary, bluish-purple, 25-35mm, thistle-like; flower-bracts erect, held close together, pointed. Pappus feathery. Dry open habitats, grassy places, on calcareous soils, generally at low altitudes. Only in W & C France and Spain.

6* **Spanish Oyster Plant** *Scolymus hispanicus* L. Medium to tall, usually hairy biennial or perennial, with latex; stems with interrupted spiny wings. Lower leaves oblong, broadest above the middle, soft, pinnately-lobed, with a few spines, long-stalked; upper leaves narrower, more rigid and more spiny. Flowerheads yellow, thistle-like, 20-30mm, in a narrow panicle, florets all rayed; flower-bracts very spiny. Uncultivated ground, generally on sandy soils. May-August. NW France southwards.

7* **Chicory** *Cichorium intybus* L. Medium to tall, stiffly hairy or hairless perennial, with latex; stems erect, branched. Basal leaves pinnately-lobed to deeply toothed, short-stalked; upper leaves lanceolate, toothed or untoothed, clasping the stem. Flowerheads clear bright blue, rarely pink or white, 25-40mm, in leafy, branched spikes; all florets rayed, the rays strap-like with a toothed tip. Rough grassy habitats, roadsides, waste ground, fields and tracksides, usually on calcareous soils. July-October. Throughout, except the far north. The leaves have long been used as a salad, young shoots being forced and blanched, and the dried roots yield chicory. **B**: Frequent in lowlands, probably naturalized in the north.

8* **Lamb's Succory** *Arnoseris minima* (L.) Schweigger & Koerte. Low to short slightly hairy annual. Leaves all in a basal rosette, oblong, broadest above the middle, toothed, short-stalked. Flowerheads yellow 7-10mm, solitary, on leafless stems much swollen towards the top; all florets rayed. Pappus absent. Cultivated and waste ground, especially on sandy soils. June-August. Continental Europe north to S Sweden.

**Cat's-ears** *Hypochoeris*. Annual or perennial herbs, with latex. Leaves mostly in basal rosettes. Flowerhead solitary or few together; all florets rayed, yellow; flower-bracts numerous, lanceolate. Pappus feathery. 70 species in temperate northern hemisphere..

9* **Spotted Cat's-ear** *Hypochoeris maculata* L. [*Achyrophorus maculatus*]. Short to tall, bristly perennial. Stems simple or slightly branched, swollen below the flowerheads. Leaves elliptical to oval, unevenly toothed or untoothed, often blotched with dark purple; stem leaves few or absent. Flowerheads lemon-yellow, 30-45mm, often solitary. Grassy habitats, open woodland, quarries and ancient earthworks, on calcareous soils, to 1800m. June-July. Throughout except Arctic Europe. **P**: Bees and hoverflies. **B**: Rare and declining in England, Wales and the Channel Islands.

10 **Giant Cat's-ear** *Hypochoeris uniflora* Vill. [*Achyrophorus uniflorus*]. Short to medium, hairy perennial; stems usually unbranched; thickened in the upper half. Leaves oblong to elliptic, toothed, hairy-edged; stem leaves 1 or several, scale-like above. Flowerheads pale golden-yellow, 40-60mm; flower-bracts dark-hairy. Meadows and open woods, on acid soils, to 2600m. July-September. Mountains of C & E France and S Germany.

11* **Smooth Cat's-ear** *Hypochoeris glabra* L. Low to short, hairless or rarely rather prickly annual; stems slightly to strongly thickened above, usually branched. Leaves oblong, broadest above the middle, pale shiny-green, pinnately-lobed to toothed. Flowerheads pale or bright yellow, small, 10-15mm; flower-bracts with dark tips, unequal, the longest as long as the rays. Open habitats, grassy places, sand-dunes, on sandy soils, at low altitudes. June-October. Throughout, except the far north. **B**: Widespread in E and SE England, rare or absent elsewhere.

12* **Cat's-ear** *Hypochoeris radicata* L. Like *H. glabra*, but perennial, the flowerheads larger, 20-30mm and the leaves roughly hairy. Pastures, meadows, commons, lawns, road-verges and stabilized sand-dunes on mildly acid and sandy soils, to 1800m. June-September. Throughout, except Finland and the far north. Can be used as a salad plant during the winter. **B**: Common throughout.

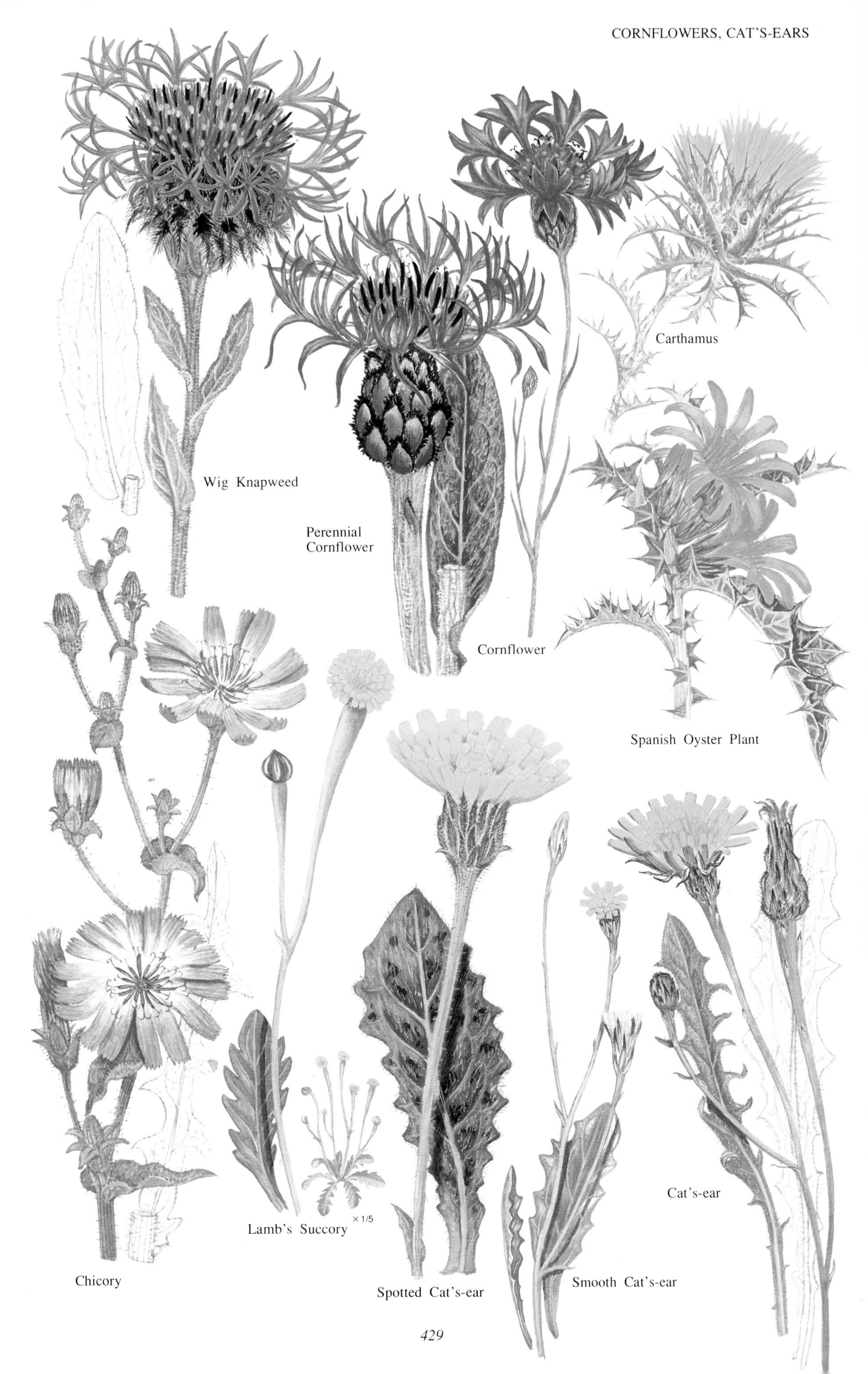
Carthamus
Wig Knapweed
Perennial
Cornflower
Cornflower
Spanish Oyster Plant
Cat's-ear
Lamb's Succory ×1/5
Chicory
Spotted Cat's-ear
Smooth Cat's-ear

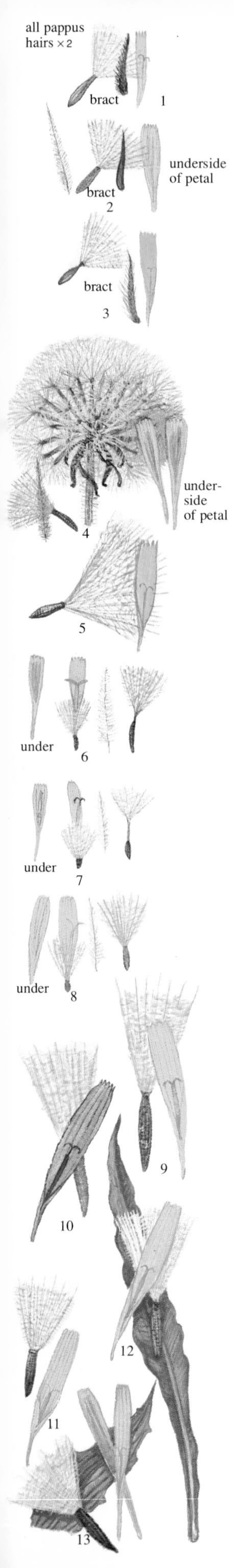


**Hawkbits** *Leontodon.* Annual or perennial herbs with latex and basal rosettes of leaves. Flowerheads solitary, or several on branched leafless stalks; florets yellow, rayed, the outer often striped with red or grey on the outside. Achenes cylindrical, ribbed. 45 species.

1* **Pyrenean Hawkbit** *Leontodon pyrenaicus* Gouan. subsp. *helveticus* (Merat) Finch & P.D. Sell. Low to medium perennial; stems unbranched, thickened towards the top and with a few dark hairs. Leaves linear to elliptical, toothed. Flowerhead plain yellow, 20-25mm; flower-bracts lanceolate, with dark hairs; stigmas yellow. Pappus pale brown. Mountain meadows, scrub and stony places, often on acid soils, to 3000m. June-August. C & E France and S Germany.

**Mountain Hawkbit** *Leontodon montanus* Lam. Like *L. pyrenaicus*, but a shorter plant, not more than 10cm tall, the leaves toothed or pinnately-lobed. Flowerheads 20-30mm, the flower-bracts with pale grey hairs. Pappus white or yellowish. Rocky and stony meadows, gravelly ground, generally on limestone, to 2900m. July-August. E France and S Germany. In the mountains of C & S Europe.

2* **Autumn Hawkbit** *Leontodon autumnalis* L. Very variable, low to medium perennial; stems usually branched, hairless or slightly hairy. Leaves narrow-oblong, deeply toothed to pinnately-lobed. Flowerheads 20-35mm yellow, the outer rays striped with red outside; flower-bracts linear-lanceolate, dark green, hairless to woolly. Grassy and rocky habitats, usually on calcareous soils, to 2600m. June-October. Throughout, except Spitsbergen. **B**: Throughout. subsp. *pratensis* (Koch) Arcangeli often shorter, the flower-bracts thickly covered with black woolly hairs. Primarily in grassy mountain habitats throughout W & NW Europe, including Britain.

3 *Leontodon keretinus* F. Nyl. Like *L. autumnalis*, but flower-bracts with long reddish hairs and rays orange, not striped beneath. Grassy habitats. Finland and N & C Russia.

4* **Rough Hawkbit** *Leontodon hispidus* L. Low to medium perennial; stems generally unbranched, white-hairy, often with 1-3 small leaf-like bracts. Leaves oblong, broadest above the middle, deeply toothed or pinnately-lobed, with a winged stalk. Flowerheads solitary, 25-40mm, bright yellow, the outer rays orange or reddish beneath; flower-bracts linear-lanceolate, white-bristly. Grassy habitats, usually on well-drained calcareous soils. June-October. Throughout, except the far north. Very variable. The dwarf form from the Alps is generally assigned to subsp. *alpinus*. **B**: Throughout much of Britain.

5* **Grey Hawkbit** *Leontodon incanus* (L.) Schrank. Low to short, hairy perennial. Leaves linear-oblong to narrow-oblong, untoothed to toothed, with a winged stalk, densely grey-hairy. Flowerheads deep yellow, 30-45mm, solitary; flower-bracts linear-lanceolate with a few to many greyish hairs. Mountain grassland. June-August. C France and S Germany.

6* **Lesser Hawkbit** *Leontodon taraxacoides* (Vill.) Merat [*L. saxatilis, Thrincia hirta*]. Low to short, slightly hairy, short-lived perennial; stems unbranched, scarcely thickened below the flowerheads, without leaf-like bracts. Leaves narrowly oblong, broadest above the middle, untoothed to pinnately-lobed. Flowerheads solitary, 12-20mm, deep yellow, the outer rays greyish-violet beneath; flower-bracts pale, narrow-lanceolate, with stiff hairs. Grassy habitats, stabilized sand-dunes, on sandy or calcareous soils. June-October. W Europe; naturalized in Denmark and Sweden. **B**: Throughout, except the extreme north.

7* **Bristly Oxtongue** *Picris echioides* L. [*Helmintia echioides*]. Medium to tall, bristly annual or biennial, each bristle arising from a pimple; stems well branched. Leaves elliptical to oblong, wavy-edged, pimply, with winged stalks; upper leaves unstalked, oval, clasping the stem with unstalked bases. Flowerheads yellow, 20-25mm, numerous, florets all rayed; flower-bracts with bristly margins, the outer heart-shaped. Pappus white. Stream sides, rough grassy and waste places, on heavy calcareous soils. June-November. A southern species, possibly native in W Europe. **B**: Locally common in lowland England, especially on heavy soils near the sea, and S & E Ireland, scarce elsewhere.

8* **Hawkweed Oxtongue** *Picris hieracioides* L. Short to tall, stiffly-hairy biennial, or short-lived perennial; stems branched. Leaves lanceolate to oblong, toothed or untoothed, stalked, the upper small and unstalked, clasping the stem. Flowerheads yellow, 20-35mm, stalked, in a cluster; flower-bracts lanceolate with blackish hairs, the outer shorter and spreading. Pappus creamy. Rough grassy places, on dry calcareous soils. July-October. Throughout, except the Faeroes, Iceland, Norway and Spitsbergen; naturalized in Ireland. Variable, especially as regards hairiness. **B**: Widespread in S England, scarce elsewhere; mostly coastal in Wales.

9* **Cut-leaved Viper's-grass** *Scorzonera laciniata* L. [*Podospermum laciniatum*]. Very variable short to medium annual or biennial. Basal leaves 1-2-pinnately-lobed, with slender segments; upper leaves less divided, clasping the stem. Flowerheads yellow or brownish, 15-25mm; flower-bracts slightly shorter than or equalling the rays. Grassy or rocky habitats. May-June. Belgium, France, S Germany. Rare.

10* **Purple Viper's-grass** *Scorzonera purpurea* L. Short to medium, hairless, tuberous-rooted perennial; stem solitary erect, slightly branched. Leaves grass-like, linear, channelled, erect; stem leaves smaller, clasping. Flowerheads pale lilac, 30-45mm, the florets longer than the sepal-like flower-bracts. Dry grassy and rocky habitats, sometimes shaded, to 2000m. May-July. SC & E France and S Germany. Local.

11* **Viper's-grass** *Scorzonera humilis* L. (incl. *S. candollei* Vis.). Low to medium perennial, scarcely hairy to cottony at leaf and stem bases; stems solitary or few, usually unbranched. Leaves linear-lanceolate, to lanceolate, untoothed, the uppermost small and scale-like. Flowerheads pale yellow, rarely whitish, 25-30mm, solitary, the florets much longer than the sepal-like flower-bracts. Marshy fields and other damp grassy habitats, on acid soils usually, to 1700m. May-July. Throughout, except Ireland and the far north. **B**: Very rare – restricted to a single locality in Dorset.

12 **Austrian Viper's-grass** *Scorzonera austriaca* Willd. Like *S. humilis*, but stems with several scale leaves immediately below the flowerhead; flowers brighter yellow. Dry meadows and rocky places, stony habitats, to 2800m. May-July. C & E France and S Germany. The stems generally have brown fibres at the base.

13 **Black Salsify** *Scorzonera hispanica* L. Like *S. humilis*, but taller, the stems usually branched and leafy; leaves slightly toothed or untoothed, the uppermost smaller than the lower, but not scale-like. Flowerheads yellow, sometimes purplish outside, 30-50mm. Grassy and rocky places, scrub. May-July. C & E France and S Germany; locally naturalized elsewhere. **B**: Casual in England.

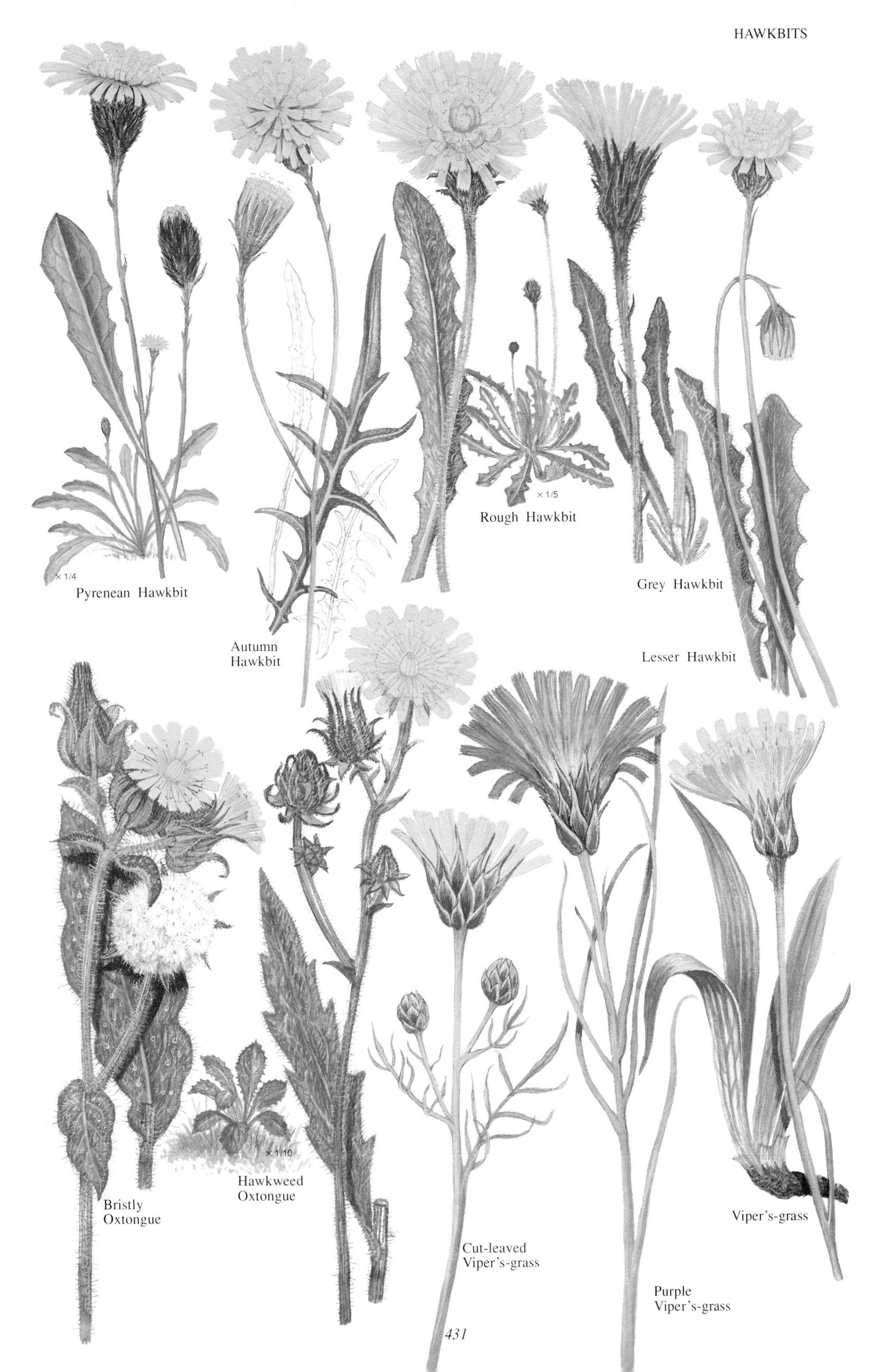
× 1/4
Pyrenean Hawkbit
Autumn
Hawkbit
× 1/5
Rough Hawkbit
Grey Hawkbit
Lesser Hawkbit
Bristly
Oxtongue
× 1/10
Hawkweed
Oxtongue
Cut-leaved
Viper's-grass
Purple
Viper's-grass
Viper's-grass

**Goat's-beard** *Tragopogon*. Annual to perennial herbs, with latex; stem usually solitary, branched or unbranched. Leaves linear-lanceolate to linear with parallel veins, those on the stem sheathing at the base. Flowerheads solitary or several, all the florets rayed; flower-bracts in a single row, long. Fruit a large 'clock', with beaked achenes. 45 species. The flowers generally open in the mornings only.

1* **Salsify** *Tragopogon porrifolius* L. Medium to tall, hairless biennial with a cylindrical rootstock; stem erect, usually branched, to 1.2m, broadening below the flowerheads. Leaves broad-linear, widened at the base. Flowerheads lilac to dull reddish-purple, 25-48mm; flower-bracts often 8, as long or slightly longer than the rays. Grassy habitats, waysides, on well drained soils. May-August. **I**: Mediterranean. Widely naturalized north to Denmark and S Sweden. The swollen fleshy rootstock, White Salsify, can be cooked and the young green shoots can be added to salads. **B**: Locally naturalized or casual, especially SE England, and Ireland.

2* **Goat's-beard, Jack-go-to-bed-at-noon** *Tragopogon pratensis* L. Variable, medium to tall annual or short-lived perennial; stem erect, unbranched or slightly branched, usually hairless. Leaves linear-lanceolate, channelled, the stem leaves half-clasping, tapered to a fine point. Flowerheads pale yellow, 18-40mm; flower-bracts generally 8-10, longer than the rays; flowerstalks not inflated. Fruit a large white clock. Rough grassy habitats, occasionally on sand-dunes, to 2600m. June-July. Throughout, except the far north. **B**: Scarce, the common Goat's-beard is subsp. *minor*.

3* *T. p.* subsp. *minor* (Miller) Wahlenb. [*T. minor*]. Like the type, but flowerhead bright yellow, generally smaller, the flower-bracts twice as long as the rays; the bracts usually have a pale reddish margin. Similar habitats and flowering time. Britain, France and Germany. **B**: Throughout England and Wales and S Scotland, scarce elsewhere.

4 *Tragopogon dubium* Scop. [*T. major*, *T. dubius* subsp. *campestris*]. Like *T. pratensis*, but a shorter plant, the flowerstalks strongly inflated below the flowerheads; flower-bracts 8-12. Dry woods and woodland margins, scrub. June-July. E France and Germany; naturalized in Belgium. C Europe to W Asia.

5 *Aetheorhiza bulbosa* (L.) Cass. [*Crepis bulbosa*]. Short to medium, bluish-green perennial, with whitish globose tubers. Leaves mostly basal, elliptical to oval, pointed, stalked, sometimes deeply lobed; flowerheads yellow, 18-32mm, the florets all rayed; flower-bracts linear-lanceolate with blackish hairs, in several overlapping rows. Cultivated fields, maritime sands and dry rocky ground. May-July. NW & W coastal France. Primarily a native of the Mediterranean region.

**Sow-thistles** *Sonchus*. Annual to perennial herbs, generally with solitary branched stems, producing copious latex. Leaves toothed to pinnately-lobed, often bristly; stem leaves clasping. Flowerheads yellow, the florets all rayed; flower-bracts numerous, in 3 rows. Fruit a clock; achenes not beaked. 70 species.

6* **Prickly Sow-thistle, Spiny Milk-thistle** *Sonchus asper* (L.) Hill. Short to tall, greyish annual, hairless except for a few glandular-hairs on the upper stem; stem angled, sometimes branched, to 1.2m. Leaves thin, pinnately-lobed or unlobed, with a soft spiny margin, the upper clasping the stem firmly with rounded basal lobes. Flowerheads golden-yellow, 20-25mm, in lax clusters; flower-bracts 35-45. Cultivated and waste land, field margins, waysides, on fertile mildly acid or calcareous soils, generally at fairly low altitudes, but to 700m. June-August. Throughout, except the far north. Plants occasionally overwinter. A weed of fields. **B**: Throughout, except parts of the Scottish Highlands.

subsp. *glaucescens* (Jordan) Ball [*Sonchus glaucescens*]. Like the type, but a biennial with thicker leaves, the basal in a distinct rosette. Similar habitats and flowering time. France.

7* **Smooth Sow-thistle, Milk Thistle** *Sonchus oleraceus* L. Like *S. asper*, but leaves generally more deeply lobed, clasping the stem with pointed, triangular, basal lobes. Flower-bracts up to 35 and achene rough (not smooth). Cultivated ground, waste places, burnt ground, waysides, on fertile, mildly acid to calcareous soils. June-August. Throughout, except the far north. Occasionally hybridises with *S. asper* – hybrids are apparently always sterile. **B**: Throughout, except for parts of the Scottish Highlands.

8* **Marsh Sow-thistle** *Sonchus palustris* L. Tall perennial to 2.5m, though often less; stem stout, erect, 4-angled, hollow, densely glandular-hairy above. Leaves greyish, the basal ones lanceolate, unlobed to pinnately-lobed with a soft-spiny margin; middle leaves arrowhead-shaped, clasping; uppermost leaves narrow and bract-like. Flowerheads pale yellow, 28-40mm, in tight clusters; flower-bracts with blackish-green glands. Achenes yellow. Wet habitats, by fresh or brackish water, on base-rich, peaty or alluvial soils. July-September. Throughout north to S Scandinavia. Decreasing in some areas. **B**: Very local and rare, confined to a few localities in East Anglia and Kent.

9 *Sonchus maritimus* L. Like *S. palustris*, but a short to medium rhizomatous perennial; upper leaves generally unlobed and clasping the stem with rounded (not pointed) basal lobes. Base of flower-stalks with whitish hairs. Damp saline soils, especially near the coast. June-August. W & NW France.

10* **Perennial or Field Sow-thistle** *Sonchus arvensis* L. Medium to tall, slightly greyish perennial, with far creeping stolons; stem erect, to 1.5m, furrowed, bristly above. Lower leaves hairless, unlobed to pinnately-lobed, with a soft-spiny margin; upper leaves larger and pinnately-lobed, clasping the stem with rounded basal lobes. Flowerheads golden-yellow, 40-50mm, in lax clusters; flower-bracts with sticky yellowish glands. Achenes brown. Cultivated and waste ground, banks, river and streamsides and brackish marshes, maritime sands and shingle along the drift line. July-October. Throughout, except the far north. Widespread weed, difficult to eradicate because of its invasive stolons. **B**: Throughout lowlands, scarcer in the north.

subsp. *uliginosus* (Bieb.) Nyman [*S. uliginosus*]. Like the type, but flowerheads and flower-stalks hairless. **B**: Similar distribution but scarcer in the north and NW.

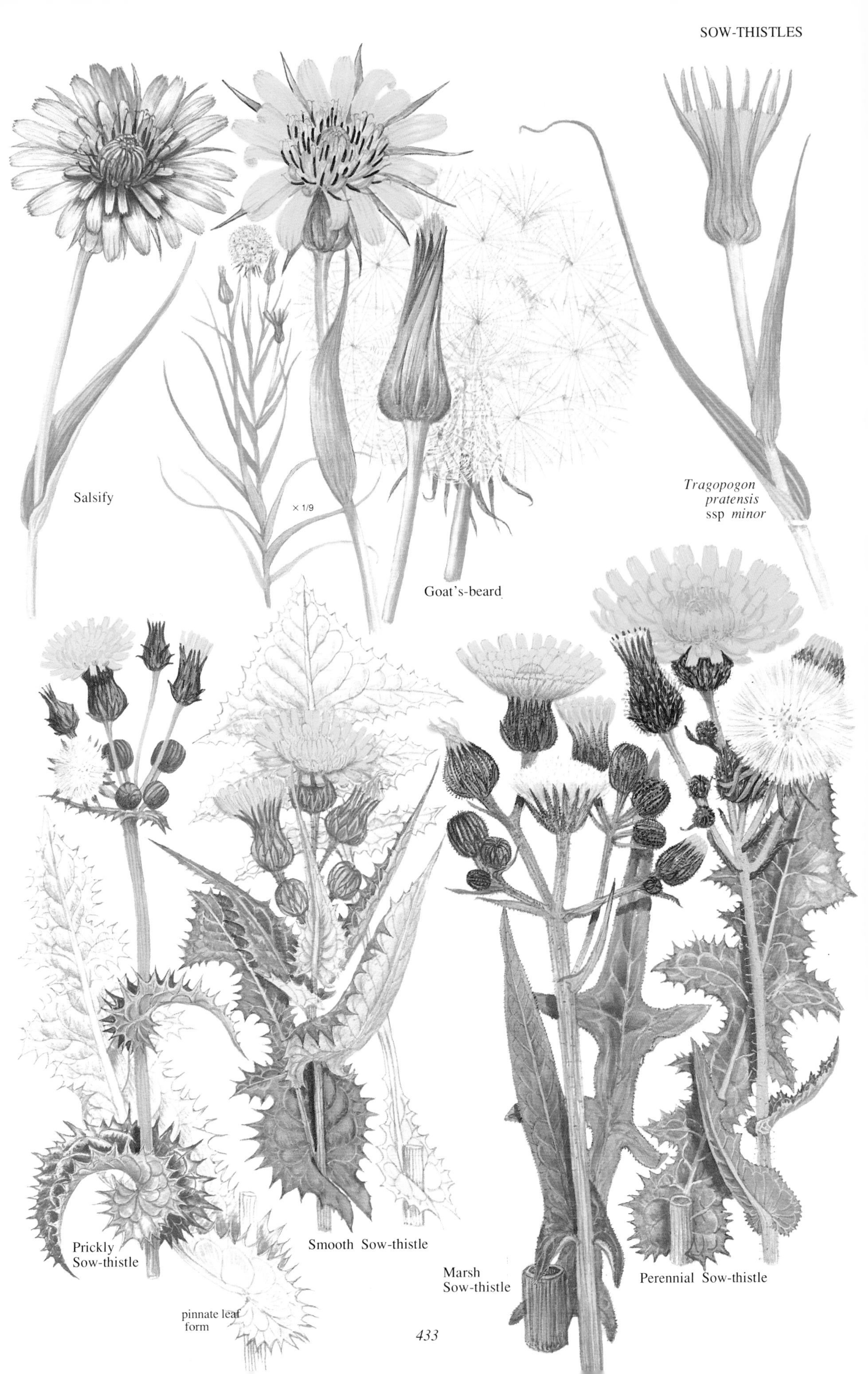
Salsify
× 1/9
Goat's-beard
*Tragopogon pratensis* ssp *minor*
Prickly Sow-thistle
pinnate leaf form
Smooth Sow-thistle
Marsh Sow-thistle
Perennial Sow-thistle

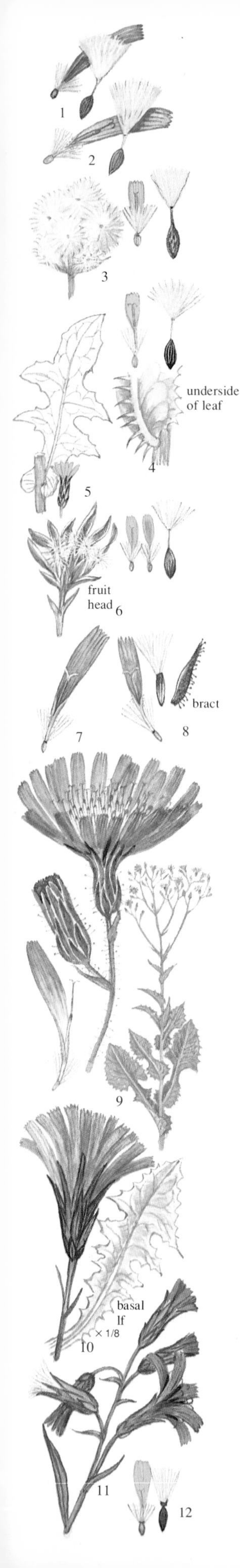


**Lettuces** *Lactuca*. Annual or perennial herbs with milky latex. Leaves unlobed to pinnately-lobed, often prickly. Flowerheads yellow or blue, all the florets rayed; flower-bracts narrow, in several rows. Pappus present; achenes beaked. About 70 species in the temperate northern hemisphere.

1* **Russian Lettuce** *Lactuca tatarica* (L.) C.A. Meyer. Medium to tall, scarcely hairy perennial with thin underground stolons; stem erect, branched above. Basal leaves pinnately-lobed, short-stalked with the lobes directed backwards; upper leaves lanceolate, clasping the stem with unstalked bases. Flowerheads lilac-blue, 18-26mm, in large panicles. Achenes yellowish to black with a white pappus. Disturbed and waste land, river-banks, seashores. June-August. **I**: W USSR. Naturalized or casual in Britain, Holland, Germany and Scandinavia.

2* *Lactuca sibirica* (L.) Maxim. Like *L. tatarica* but generally with a simple unbranched stem and the leaves lanceolate, untoothed to sharply toothed, clasping, only the lowermost stalked. Inflorescence rather flat-topped. Achenes yellowish-green with a greyish pappus. Wooded habitats, scrub, river sands and gravels, on moist but well-drained soils. July-September. Scandinavia. Also found in W USSR.

3* **Prickly Lettuce** *Lactuca serriola* L. [*L. scariola*]. Tall greyish annual or biennial; stem to 1.8m, stiff and erect, hairless or rather bristly. Leaves oblong, pinnately-lobed, prickly on margins and along the midrib beneath; upper leaves held in a vertical position, deeply-lobed, clasping the stem. Flowerheads pale yellow, 11-13mm, in a narrow or pyramidal panicle. Achenes greyish. Disturbed and waste ground, sand-dunes. July-September. Throughout, except Ireland and the far north. The erect upper leaves are held in a north-south plane. **B**: E & S England west to the Isle of Wight; scattered localities elsewhere in England.

4* **Great Lettuce** *Lactuca virosa* L. Like *L. saligna*, but taller, to 2m, the stems sometimes bristly below; upper leaves horizontal, finely prickly along the midrib beneath. Flowerheads in a pyramidal panicle. Achenes maroon to blackish. Disturbed ground, waste places, road-verges, river and canal banks, sandy and gravelly habitats, cliff ledges. July-September. Britain, Belgium, France and Germany. **B**: Widespread in England, except the SW; north to the Tyne.

5 *Lactuca quercina* L. [*L. stricta*, *L. quercina* subsp. *stricta*]. Like *L. serriola*, but stems and leaves thinner, lyre-shaped, pinnately-lobed, the uppermost lobed or unlobed. Flowerheads in a broad flat-topped cluster. Achenes with a blackish-beak. Similar habitats and flowering time. France, C & S Germany and S Sweden – Gotland.

6* **Least Lettuce** *Lactuca saligna* L. Medium to tall, hairless annual or biennial; stem erect, whitish, with stiff branches. Leaves greyish, the lower unlobed or pinnately-lobed, stalked, the upper oblong to linear, unlobed, vertical, clasping the stem with an arrow-shaped base. Flowerheads pale yellow, 9-11mm, in slender stalked spikes; flower-bracts green. Achenes black with a white beak. Bare and grassy habitats, waste places, poor pastures, coastal shingle, sea walls, and rocky places. June-August. W Europe. **B**: Very rare and decreasing, confined to a few localities in Essex, Kent and Sussex.

7* **Mountain or Blue Lettuce** *Lactuca perennis* L. Medium to tall hairless perennial; stems erect, branched above. Leaves greyish pinnately-lobed, short-stalked, the upper usually unstalked and half-clasping the stem. Flowerheads pale blue to lilac, large, 30-40mm, long-stalked, in broad, rather flat-topped panicles. Achenes black with a pale apex. Dry and rocky habitats, stony ground, grassy places, generally on limestone, to 2100m. May-August. Belgium, France and Germany. C & E Europe.

8* **Alpine Sow-thistle** *Cicerbita alpina* (L.) Wallr. [*Mulgedium alpinum*, *Sonchus alpinus*]. Tall stout perennial; stems erect to 2.5m, branched or unbranched and dense with reddish glandular-hairs in the upper half. Leaves hairless, large, bluish beneath, pinnately-lobed, with a large triangular end lobe, the upper clasping the stem, the lower with broad-winged stalks. Flowerheads pale blue-violet, about 20mm, in a long panicle; florets all rayed; flower-bracts linear, brownish glandular. Achenes whitish, not beaked. Moist habitats mainly in mountain regions, open woodland, grassy and rocky places, river and stream-banks, often among luxuriant vegetation, to 2200m. July-September. N Britain, mountains of France, Germany and Scandinavia. **B**: Very rare, a few localities in the Scottish Highlands.

9 **Blue Sow-thistle** *Cicerbita macrophylla* (Willd.) Wallr. Like *C. alpina*, but the leaves bristly and less lobed, the end segment heart-shaped. Flowerheads larger 28-32mm, in a broader and flatter inflorescence, lilac. Waysides and waste places on moist soils, often forming extensive patches. July-September. **I**: Caucasus Mts. Widely naturalized in Britain, France, Germany, Denmark and Scandinavia. The stock has far-reaching rhizomes, unlike *C. alpina*, eventually forming large colonies. The British as well as much of the northern European plants are probably referable to subsp. *uralensis* (Rouy) Sell. **B**: Established in scattered localities in N England, Scotland and Ireland.

10 *Cicerbita plumieri* (L.) Kirschleger [*C. orbelica*, *Mulgedium plumieri*, *Sonchus plumieri*]. Like *C. alpina*, but a hairless, non-glandular plant; flowerheads blue, in a rather flat-topped panicle. Mountain habitats, rocky places, streamsides and ravines, to 2200m. July-September. WC & E France and S Germany; occasionally naturalized in Britain. Cultivated in gardens.

11 **Purple Lettuce** *Prenanthes purpurea* L. Medium to tall, hairless perennial; stems to 1.5m. Leaves elliptical to oblong, sometimes linear, untoothed to toothed, bluish-green, often waisted, clasping the stem. Flowerheads half-nodding, violet or purplish, occasionally white, 18-20mm, in a lax, branched panicle; all the florets rayed, each flowerhead with only 3-6 rays. Pappus white. Woodland, river and stream-sides, rocky places and other moist shaded habitats, mainly in the mountains, to 2050m. July-September. C & E France and S Germany; naturalized in Denmark.

12* **Wall Lettuce** *Mycelis muralis* (L.) Dumort [*Lactuca muralis*]. Short to tall, hairless perennial; stem solitary, branched. Leaves soft, pinnately-lobed, toothed, with a large triangular end-lobe, the upper leaves clasping the stem, the lower stalked. Flowerheads small, pale yellow, 7-8mm, in a broad lax panicle, each with only 5 rayed-florets. Achenes blackish with a pale beak. Rocky woodland, hedgebanks, waste places, walls and other rocky habitats, on calcareous or base rich soils. July-September. Throughout except the far north. **B**: North to S Scotland, although scarce in SW England and East Anglia; naturalized in Ireland.

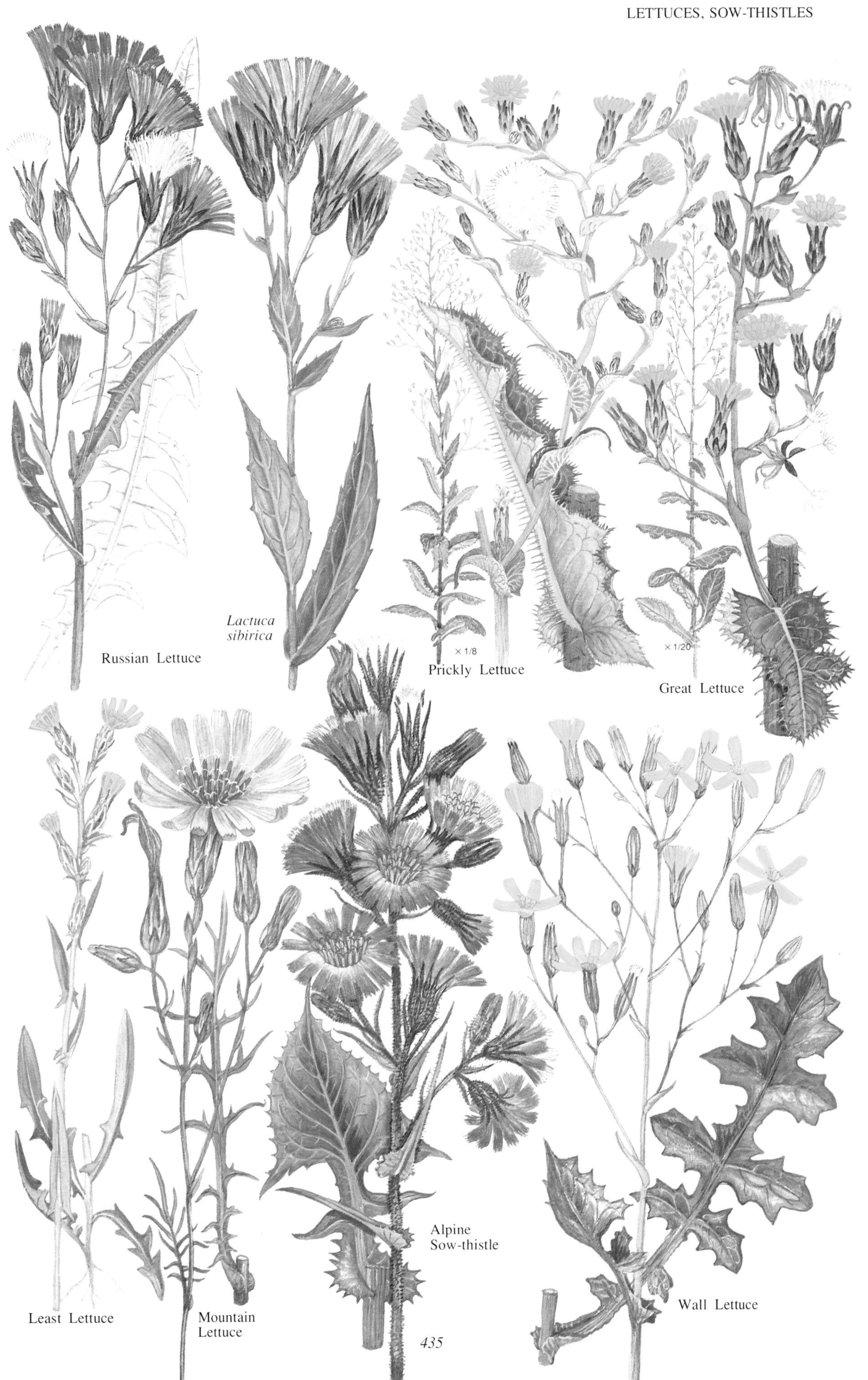
Russian Lettuce
Lactuca sibirica
× 1/8
Prickly Lettuce
× 1/20
Great Lettuce
Least Lettuce
Mountain Lettuce
Alpine Sow-thistle
Wall Lettuce

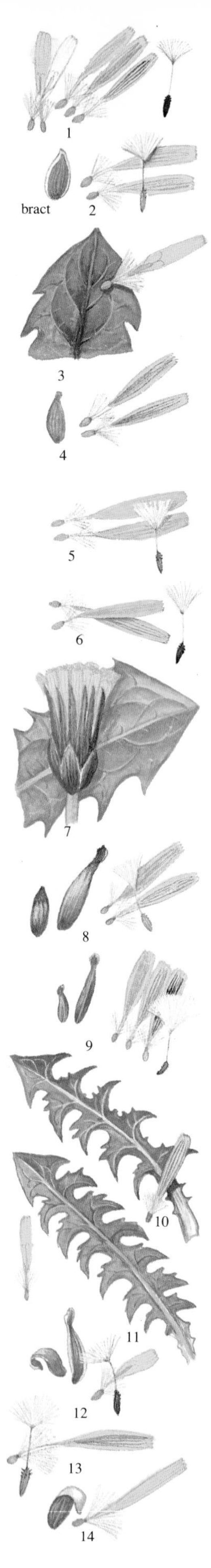


**Dandelions** *Taraxacum*. Perennial herbs with milky latex and rosettes of basal leaves sprouting from a tap root. Leaves lobed or unlobed. Flowerheads yellow or white, solitary, borne on hollow scapes, flat-topped when open; florets all rayed, the inner shorter than the outer; flower-bracts in 2 rows, the outer shorter and often recurved. Fruit a large and conspicuous 'clock'. A large and extremely complicated genus; some 200 microspecies are recognised in Britain alone. This is only an outline treatment.

1* *Taraxacum phymatocarpum* J. Vahl Group (incl. *T. arcticum* (Trautv.) Dahlst., *T. dovrense* (Dahlst.) Dahlst.). Low plant, 4-10cm. Leaves narrow-elliptical, usually broadest above the middle, untoothed or with shallow triangular lobes, bright green. Flowerheads 15-25mm, the rays yellow or white, grey, violet or purple striped beneath; outer flower-bracts grey-green to nearly black. Achenes blackish. Arctic habitats. June-September. Norway and Spitsbergen.

2* *Taraxacum praestans* H. Lindb. f. Group (incl. *T. euryphyllum* (Dahlst.) M.P. Christiansen, *T. lainzii* Van Soest, *T. landmarkii* Dahlst., *T. maculigerum* H. Lindb. fil., *T. naevosiforme* Dahlst., *T. naevosum* Dahlst., *T. stictophyllum* Dahlst.). Low to short plant, 8-25cm. Leaves dull green, lobed to cut, often dark-spotted and hairy, the stalk winged and toothed. Flowerheads pale yellow, 35-55mm, the rays with a grey or brown stripe beneath; outer flower-bracts oval-lanceolate with a pale margin, often bluish-green, erect to spreading. Moist and wet, generally grassy habitats, often in hills or mountains. April-August. Throughout, except Belgium and the far north. **B**: Throughout, especially the north.

3 **Red-veined Dandelion** *Taraxacum spectabile* Dahlst. (incl. *T. faeroense* (Dahlst.) Dahlst.). Like *T. praestans* but leaves often with a reddish midrib, outer flower-bracts always erect; flowerheads bright deep yellow. Wet habitats, grassland, river and stream banks, wet rocks, on calcareous and base rich soils. April-August. Throughout except Finland. **B**: Hilly districts in the north & west; occasionally elsewhere.

4 *Taraxacum ceratophorum* (Ledeb.) DC. Group (incl. *T. brachyceras* Dahlst., *T. melanostylum* T.C.E. Fries, *T. tornense* T.C.E. Fries, *T. lactucaceum* Dahlst.). Like *T. spectabile*, but leaves dark green, thin; outer flower-bracts horned, without a pale margin. Flowerheads deeper yellow, the rays red or purple striped beneath. Grassy habitats. Scandinavia, Iceland and the mountains of C & E France.

5 *Taraxacum unguilobum* Dahlst. Group (incl. *T. fulvicarpum* Dahlst.). Like *T. praestans*, but leaves never spotted and stalks more or less unwinged. Rays unstriped or with a pinkish stripe beneath. Achenes pinkish to reddish-brown. Similar habitats and flowering time. Britain and Norway. **B**: Mainly in the west and north.

6* *Taraxacum croceum* Dahlst. (incl. *T. ceratolobum* Dahlst., *T. craspedotum* Dahlst., *T. cymbifolium* H. Lindb. f. ex Dahlst., *T. pycnostictum* M.P. Christiansen). Like *T. praestans*, but the plant bright green and flowerheads deep yellow to golden-yellow, the rays with a grey or brown stripe beneath. Achenes pale brown. Grassy and rocky habitats. April-August. N Europe. **B**: Mainly in Scotland.

7 *Taraxacum adamii* Claire Group (incl. *T. litorale* Raunk., *T. nordstedtii* Dahlst.). Like *T. praestans*, but leaves bright green and unspotted; outer flower-bracts erect, not spreading or recurved. Wet grassy habitats. Throughout, except the Faeroes, Iceland and the extreme north. **B**: Throughout, especially Scotland.

8* **Narrow-leaved Marsh Dandelion** *Taraxacum palustre* (Lyons) Symons Group (incl. *T. anglicum* Dahlst., *T. austrinum* G. Hagl., *T. balticum* Dahlst., *T. germanicum* Van Soest, *T. hollandicum* Van Soest, *T. lividum* (Waldst. & Kit.) Peterm., *T. suecicum* G. Hagl.). Low to short plant, 6-15cm. *Leaves linear to linear-lanceolate* unlobed or lobed, erect to spreading, usually hairless, unspotted. Flowerheads pale yellow, 25-50mm, the rays with a grey or purplish stripe beneath, borne on reddish scapes; flower-bracts oval, often flushed with purple or violet, erect. Achenes pale brown or olive-green. Wet habitats, to 2200m. April-June. Europe. Declining in many areas. **B**: Local throughout much of Britain.

9* **Lesser Dandelion** *Taraxacum erythrospermum* Andrz. ex Besser Group. (incl. *T. brachyglossum* (Dahlst.) Dahlst., *T. commixtum* G. Hagl., *T. decipiens* Raunk., *T. dunense* Van Soest, *T. lacistophyllum* (Dahlst.) Raunk, *T. laetum* (Dahlst.) Dahlst., *T. proximum* (Dahlst.) Dahlst., *T. rubicundum* (Dahlst.) Dahlst., *T. scanicum* Dahlst., *T. silesiacum* Dahlst. ex G. Hayl.). Low to medium, rather delicate plant, 4-15cm. Leaves horizontal, oblong, with narrow deep lobes, with a slender, green, red or purplish stalk. Flowerheads pale yellow, 15-35mm, the rays striped with grey, brown or purple beneath, borne on slender scapes; flower-bracts erect, the outer generally horned. Achenes red, purple or violet. Dry habitats on well-drained, especially calcareous, soils. April-June. Throughout, except the Faeroes and Spitsbergen. A low delicate plant sometimes confused with *T. palustre*, but usually with more deeply lobed leaves and flower rays broad and short, rather than long and narrow. **B**: Throughout.

10 *Taraxacum obliquum* (Fries) Dahlst. Group (incl. *T. platyglossum* Raunk.). Like *T. erythrospermum*, but flowers golden-yellow, the rays red-purple striped beneath. Sand-dunes and dune slacks. May-July. **B**: W Scotland. Very local.

11 *Taraxacum simile* Raunk. Group (incl. *T. proximiforme* Van Soest, *T. pseudolacistophyllum* van Soest. Like *T. obliquum*, but leaves with 6-8 lobes (not 3-5) and flowerheads pale yellow. Dry habitats April-August. Britain and most of Continental Europe. **B**: Local.

12* *Taraxacum fulvum* Raunk Group (incl. *T. fulviforme* Dahlst., *T. glauciniforme* Dahlst., *T. oxoniense* Dahlst.). Low to short plant, 5-15cm. Leaves horizontal to almost erect, bright green, with narrow deep lobes and narrow unwinged, green or purple stalks. Flowerheads pale yellow, with short rays; flower-bracts green with pale margins, the outer recurved, usually horned. Achenes red to pinkish-brown. Dry grassy and rocky habitats. April-August. Throughout, except far N Scandinavia. **B**: Throughout.

13* **Common Dandelion** *Taraxacum officinalis* Weber Group. Very variable, low to medium plant, 5-40cm, often robust. Leaves lobed to unlobed, coarse, never spotted, with broad-winged, lobed stalks. Flowerheads mid-yellow, 25-50mm, often convex above, the rays usually with a brown or grey-violet stripe beneath, borne on rather stout scapes; flower-bracts usually dark bluish-green, the outer recurved, not horned. Grassy habitats, cultivated ground, to 2600m. March-October. Throughout, except Spitsbergen. Over 200 'species' have been included in this group alone in Europe. **B**: Throughout, often abundant.

14 *Taraxacum crassipes* H. Lindb. f. Like *T. officinalis*, but the outer flower-bracts very pale, often whitish on the inside and the leaves with an elongated end-lobe. Flowerheads pale yellow, the long narrow rays scarcely striped beneath. Meadows and clearings in coniferous woodland. Scandinavia.

*Taraxacum phymatocarpum*
*Taraxacum praestans*
varying leaf forms
*Taraxacum croceum*
varying leaf forms
Narrow-leaved Marsh Dandelion
Lesser Dandelion
*Taraxacum fulvum*
Common Dandelion

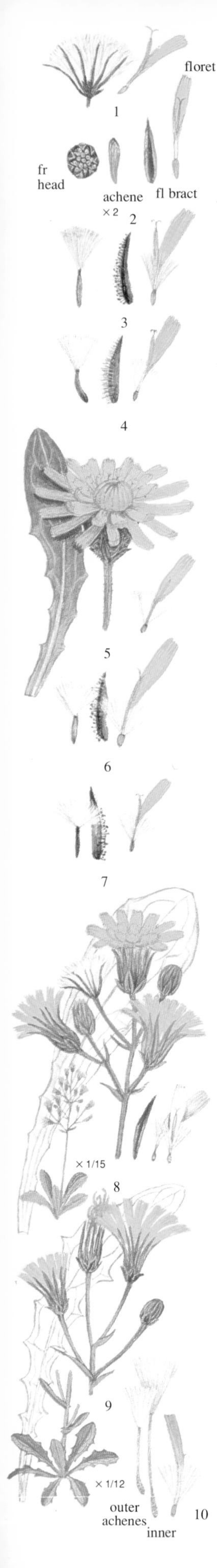


1* **Chondrilla** *Chondrilla juncea* L. [*C. canescens*, *C. latifolia*]. Medium to tall, greyish biennial or perennial, hairless or stiffly hairy, particularly below; stems stiff and broom-like, few-leaved. Basal leaves oblong, deeply-toothed to lobed, soon withering; upper leaves linear, unlobed, sometimes toothed. Flowerheads yellow, 10mm, in small clusters, unstalked, all the florets rayed; flower-bracts linear-lanceolate, erect, hairy or hairless. Pappus white, soft. Dry open habitats on sandy or stony ground. July-September. France and Germany. The seeds are poisonous.

2* **Nipplewort** *Lapsana communis* L. Short to tall, hairy annual, without milky latex; stems leafy, erect, to 1.2m, branched or unbranched. Leaves oval, toothed, often lobed at the base, the lower stalked, the upper sometimes unstalked. Flowerheads yellow, 10-20mm, in a lax-branched panicle, all the florets rayed; flower-bracts lanceolate, erect. Waste and disturbed ground, hedgerows, roadsides, woodland margins, walls, gardens, in sunny or shaded places, on slightly acid to calcareous soils. June-October. A fast-growing and often tiresome weed of cultivation. **B**: Throughout.

**Hawk's-beards** *Crepis*. Annual or perennial herbs with spirally arranged leaves, often lobed, with the lobes pointing backwards; stems normally branched. Flowerheads, mostly yellow, sometimes orange, pink or white; florets all rayed; flower-bracts in 2 rows, the outer shorter. Pappus usually white. About 200 species. The achenes of *Hieracium* are truncated at the apex and the pappus is normally brown and brittle.

3* **Marsh Hawk's-beard** *Crepis paludosa* (L.) Moench. Medium to tall, almost hairless, rhizomatous perennial. Leaves dark green, elliptical, sharply toothed, the basal with a short, winged stalk, the upper linear to lanceolate, unstalked, clasping the stem with pointed lobes. Flowerheads yellow or dull orange-yellow, 15-25mm, few in a lax cluster; flower-bracts linear-lanceolate, dark with sticky black glands. Pappus brownish. Damp and shady habitats to 2150m. July-September. Throughout, except the far north. **B**: Widespread in C & N England, Scotland, Wales and Ireland.

4* **Northern Hawk's-beard** *Crepis mollis* (Jacq.) Ascherson [*C. succisifolia*]. Very variable, rather slender, perennial, hairy or hairless. Basal leaves elliptical, untoothed or slightly toothed, with a winged stalk; upper leaves lanceolate or bract like, half-clasping the stem. Flowerheads yellow, 20-30mm, few in a lax cluster. Pappus soft, pure white. Damp or shady habitats, woods, river and streamsides, fens, on calcareous or base-rich soils, in hilly and mountain districts, to 1400m. July-August. N Britain, France and Germany. **B**: Rare and declining; confined to N England and Scotland.

5 **Alpine Hawk's-beard** *Crepis alpestris* (Jacq.) Tausch. Like *C. mollis*, but flowers usually solitary and the lowermost leaves often lobed. Achenes 10-12-ribbed (not 18-20). Meadows and stony places, generally on limestone, to 2650m. June-September. E France and S Germany, from the Jura southwards.

6* **Rough Hawk's-beard** *Crepis biennis* L. Short to tall, rough-hairy biennial; stem branched above. Basal leaves oblong to elliptical, toothed to pinnately-lobed, with a winged stalk; upper leaves lanceolate to linear, not clasping the stem, lobed or unlobed. Flowerheads golden-yellow, 20-35mm, in lax, flat-topped clusters; flower-bracts linear-lanceolate, downy with yellowish or black hairs. Pappus soft, white. Pastures, waste places, waysides, roadsides, woodland margins, arable fields, generally on calcareous soils, mostly at low altitudes. June-July. Throughout except the Faeroes, Iceland and the far north, but naturalized in Ireland, Finland and Norway. **B**: Scattered localities in England and C & S Scotland; rare in Wales and Ireland.

7* *Crepis tectorum* L. Short to tall, slightly hairy annual, with leafy stems. Basal leaves few, toothed to pinnately-lobed, sometimes waisted; upper leaves lanceolate to linear, unstalked. Flowerheads yellow, 15-25mm, many in branched clusters; flower-bracts linear-lanceolate, hairy. Pappus pure white. Dry sandy and grassy habitats. June-September. Continental Europe, except the far north. Variable species in which several subspecies are normally recognised.

subsp. *nigrescens* (Pohle) P.D. Sell [*Crepis nigrescens*]. Shorter plant than the type, not exceeding 30cm, with 3-6 stem leaves only; flower-bracts with grey, non-glandular, hairs. Rocky and sandy habitats along rivers and by the sea. Finland.

subsp. *pumila* (Liljeblad) Sterner. Like subsp. *nigrescens*, but rarely exceeding 7cm, with numerous basal leaves but only 2-4 stem leaves; flower-bracts with glandular-hairs. Shallow calcareous soils. Confined to S Sweden – Gotland & Öland.

8 *Crepis praemorsa* (L.) Tausch. Like *C. tectorum*, but leaves all basal, lanceolate, toothed or untoothed. Flowerheads yellow, rarely pink or white, 15-18mm; flower-bracts hairless. Achenes with about 20 ribs (not 10). Dry meadows and stony habitats, usually on limestone, to 1800m. May-July. France, Germany and much of Scandinavia and Denmark.

9 *Crepis pulchra* L. Like *C. tectorum*, but stems branched from the base and basal leaves in a well defined rosette, the upper leaves linear, often bract-like. Flower-bracts hairless, the outer very short. Outer achenes larger than the inner. Dry open sunny habitats. June-September. France and S Germany. Distributed mainly in the south and parts of C Europe.

10* **Stinking Hawksbeard** *Crepis foetida* L. Short to medium, unpleasant smelling, hairy annual; stem branched from the base or the middle. Basal leaves elliptical, broadest above the middle, toothed to 2-pinnately-lobed; stem leaves similar, but narrower and unstalked, clasping the stem with rounded basal lobes. Flowerheads yellow, reddish-purple beneath, 15-20mm, in lax, flat-topped clusters, nodding in bud. Outer achenes smaller than the inner; pappus white. Waste places, waysides, shingle, generally on calcareous soils. June-August. S Britain, Belgium, France and Germany. The plants smell of bitter almonds when bruised. **B**: Very rare, only known from Kent – Dungeness, formerly more widespread.

Chondrilla
Nipplewort
Marsh Hawksbeard
× 1/10
Northern Hawksbeard
× 1/10
Rough Hawksbeard
*Crepis tectorum*
Stinking Hawksbeard

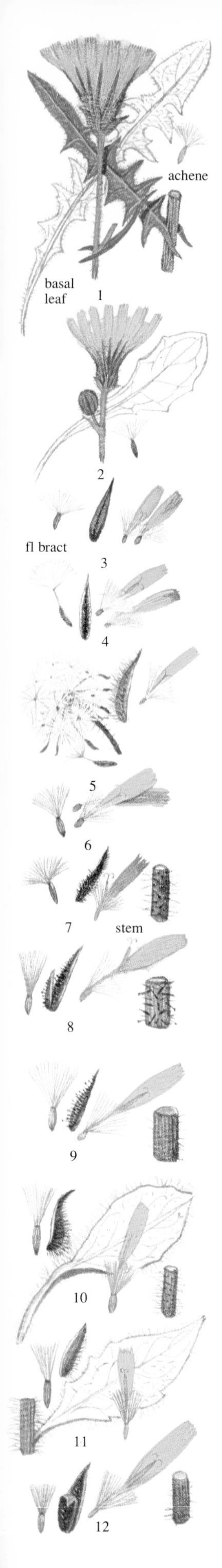


1 **French Hawksbeard** *Crepis nicaeensis* Balbis. Medium to tall, hairy biennial, sometimes annual; stem branched above, reddish and ribbed below. Leaves with yellowish hairs, the basal oblong, pinnately-lobed; upper leaves lanceolate, unstalked, clasping the stem with pointed basal lobes, arrow-like. Flowerheads yellow, the rays often red-tipped, 20-25mm, borne in flat-topped clusters; flower-bracts linear-lanceolate downy. Pappus white. Meadows, pastures, arable fields, waste places. June-July. **I**: Mediterranean. Naturalized north to Sweden. **B**: Rare casual in the south.

2 *Crepis multicaulis* Ledeb. Like *C. nicaeensis*, but uppermost leaves clasping the stem (not bract-like). Achenes reddish-brown (not golden), with 10-20 (not 10) ribs. Arctic and sub-Arctic habitats. Norway.

3* **Smooth Hawksbeard** *Crepis capillaris* (L.) Wallr. [*C. virens*]. Medium to tall, hairless or slightly hairy annual; stems slender, branched at base or above. Leaves shiny, the basal numerous, lanceolate, toothed to pinnately-lobed; stem leaves like basal, but smaller and clasping the stem with pointed basal lobes. Flowerheads yellow, often reddish beneath, 10-15mm in lax clusters; flower-bracts lanceolate, generally downy. Pappus soft, white. Grassland, roadsides, waste places, heaths and old walls. June-November. W Europe; naturalized in Denmark and Sweden. **B**: Throughout.

4* **Beaked Hawksbeard** *Crepis vesicaria* L. subsp. *haenseleri* (Boiss. ex DC.) P.D. Sell [*Crepis taraxacifolia, Barkhousia haensleri*]. Medium to tall, hairy annual or perennial; stems branched below. Leaves dandelion-like, oblong, pinnately-lobed, the lowermost stalked, the middle clasping the stem with pointed lobes; upper leaves lanceolate to linear, bract-like. Flowerheads yellow, the outer florets striped brown beneath, 15-25mm, borne in lax-branched flat-topped clusters, erect in bud; flower-bracts downy. Achenes long-beaked; pappus soft, white. Grassy and waste places, roadsides, cultivated land, railway embankments, generally on dry calcareous soils, at low altitudes. May-July. Continental Europe from Belgium southwards, including Germany; naturalized in Britain. The typical species, subsp. *vesicaria*, is a native of the Mediterranean region. **B**: Widespread north to Yorkshire, still spreading.

5* **Bristly Hawksbeard** *Crepis setosa* Hallerf. Like *C. vesicaria*, but upper part of stems and flower-bracts bristly; flowerheads pale yellow. Similar habitats. July-September. C & E France and Germany; occasionally casual elsewhere.

## Hawkweeds *Hieracium*.

Perennial herbs, sometimes stoloniferous. Leaves toothed but never pinnately-lobed, alternate. Flowers yellow, occasionally gold or red; flower-bracts linear-lanceolate. Achenes truncated at apex; pappus brown and brittle. A very large and vastly complex genus, mainly in temperate, alpine and Arctic regions of the northern hemisphere and South America. 260 species are recognized in Europe alone. Seed is produced apomictically, yielding numerous forms and local races.

6* **Mouse-ear Hawkweed** *Hieracium pilosella* L. [*Pilosella officinarum*]. Low to short, stoloniferous perennial, with woolly white hairs. Leaves all in a basal rosette, elliptical, often broadest above the middle, untoothed, white-felted beneath. Flowerheads lemon-yellow, 20-30mm, solitary, the outer florets often red striped beneath. Grassy, waste and bare places, pastures, commons, hill-slopes, sand-dunes, banks and walls, on dry acid or calcareous soils, to 3000m. Throughout, except parts of the extreme north. Sometimes placed in a separate genus, *Pilosella*. The stolons terminate in small leafy rosettes. **B**: Throughout; scarce in East Anglia and N Scotland.

7* **Orange Hawkweed, Fox-and-cubs** *Hieracium aurantiacum* L. Short to medium stoloniferous, blackish-hairy, perennial. Leaves mostly in a basal rosette, bluish-green, elliptical to lanceolate, untoothed, short-stalked; stem leaves 1-4, smaller. Flowerheads orange-brown or orange-red, 13-15mm, in fairly tight clusters. Grassy and waste habitats, meadows, waysides, cultivated land, to 2600m. June-August. France, Germany and Scandinavia; naturalized in Britain, Belgium, Holland, Denmark and Iceland. **B**: Widely naturalized, particularly in the north.

8* **Hawkweed** *Hieracium murorum* L. Group. Short to medium, hairy perennial. Leaves few, variable, lanceolate to oval, toothed or untoothed; stem leaves similar but unstalked. Flowerheads yellow, 20-30mm, few in a cluster; flower-bracts with glandular hairs. Grassy and rocky habitats and old walls. June-August. Throughout, except the Faeroes and Spitsbergen.

9* **Common Hawkweed** *Hieracium vulgatum* Fries [*H. levicaule*]. Short to tall hairy perennial. Basal leaves lanceolate to oval, toothed; stem leaves similar, the upper smaller and unstalked. Flowerheads yellow, in a spreading flat-topped cluster; flower-bracts hairy and glandular. Rocky grassland, banks, open woodland and heaths. June-August. Throughout, except the Faeroes, Iceland and Spitsbergen. **B**: N Britain, rarer in S England, Wales and Ireland.

10 **Alpine Hawkweed** *Hieracium alpinum* L. Low to short hairy perennial. Leaves mostly in a basal rosette, elliptical to spoon-shaped, toothed or untoothed, the outer smaller than the inner, with a winged stalk; stem leaves, when present, bract-like. Flowerheads yellow, 25-35mm, usually solitary; flower-bracts linear-lanceolate, blackish, incurved in bud. Mountain habitats, rocky and grassy places, screes, to 3000m. July-August. N Britain, Iceland, France, Germany and Scandinavia. **B**: Confined to the Scottish Highlands.

11 *Hieracium sabaudum* L. Group. (incl. *H. obliquum* Jordan, *H. vagum* Jordan, *H. virgultorum* Jordan). Medium to tall hairy perennial, without glandular hairs. All leaves borne on the stem, the lower crowded, oval to lanceolate, finely toothed, short-stalked; upper leaves with a rounded or heart-shaped base. Flowerheads 22-34mm, in broad, almost flat-topped clusters; flower-bracts olive or blackish-green, hairless. Styles usually dark. Woodland and woodland margins, copses, hedge-banks, scrub and rocky places, to 800m, occasionally higher. August-October. Throughout, except for the Faeroes and Iceland and most of Scandinavia, with the exception of Denmark. **B**: Throughout, except for parts of the far north.

12* **Leafy Hawkweed** *Hieracium umbellatum* L. Group. Medium to tall softly hairy perennial. All leaves borne on the stem, dark green, lanceolate to linear, untoothed or slightly toothed. Flowerheads yellow, 20-30mm, in a flat-topped cluster; flower-bracts blackish-green, hairless or almost so. Styles yellow. Woodland, grassy places and heaths. July-October. Throughout Europe, except the far north. **B**: Throughout.

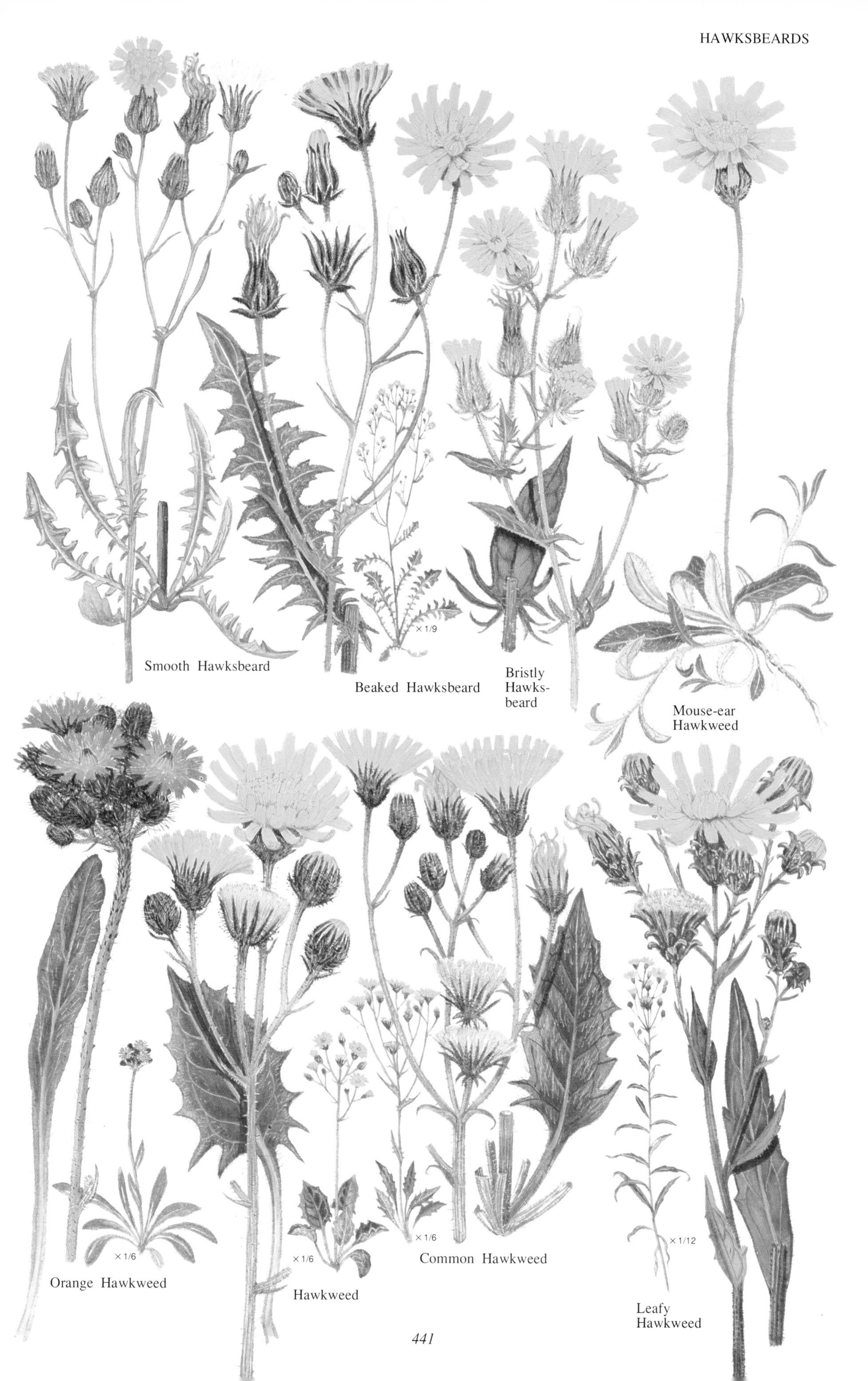
Smooth Hawksbeard
× 1/9
Beaked Hawksbeard
Bristly Hawks-beard
Mouse-ear Hawkweed
× 1/6
Orange Hawkweed
× 1/6
Hawkweed
× 1/6
Common Hawkweed
× 1/12
Leafy Hawkweed

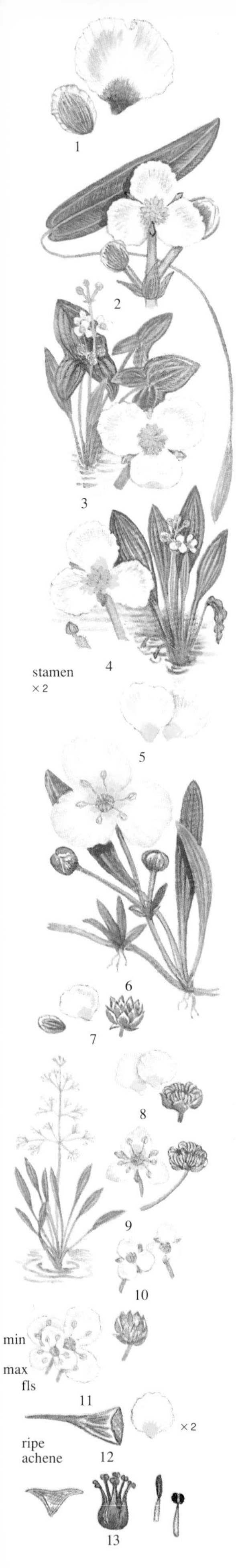


# Order – Monocotyledons

Embryo plants with 1 seed-leaf (cotyledon). Vessels of the stem in series of bundles or scattered. Leaves usually with parallel veins. Flowers typically 3-parted.

## WATER-PLANTAIN FAMILY Alismataceae

Perennial herbs of aquatic freshwater habitats. Leaves often basal and with sheath bases. Flowers buttercup-like, hermaphrodite or unisexual, borne in distinctive whorls composing a simple umbel, a raceme or panicle; sepals 3, petals 3, generally larger than the sepals; stamens 3 to many; carpels 3 to many. Fruit a group of achenes or follicles. 80 species in 13 genera.

1* **Arrowhead** *Sagittaria sagittifolia* L. Medium to tall perennial. Aerial leaves narrow to broadly arrow-shaped, with pointed basal lobes, long-stalked; floating leaves when present oval to lanceolate. Flowers white, 20-26mm, the petals with a basal purple blotch, borne in whorled racemes or panicles; male and female flowers on the same plant; stamens hairless with purplish-brown anthers. July-August. Throughout except the far north. **B**: Widespread; naturalized in Scotland.

2 *Sagittaria natans* Pallas with narrower floating leaves; aerial leaves, when present, with short rounded basal lobes, generally directed downwards. Flowers smaller, 14-18mm, the petals plain white, with yellow anthers, borne in a simple umbel or a whorled raceme. July-early September. Sweden and Finland.

3 **Broad-leaved Arrowhead** *Sagittaria latifolia* Willd. [*S. obtusa*]. Medium to tall perennial. Aerial leaves broadly to narrowly arrow-shaped, with long pointed basal lobes, but also with some linear or oval leaves without basal lobes. Flowers with plain white petals, 18-34mm, monoecious or dioecious, in whorled racemes or panicles, anthers yellow. **I**: North America. Naturalized in France and Germany.

4 **Canadian Arrowhead** *Sagittaria rigida* Pursh. Medium to tall perennial. Leaves all aerial, linear to elliptical or oval, pointed, occasionally with 2 short basal lobes. Flowers white, 14-24mm, the petals sometimes with a basal yellow blotch, borne in whorled racemes; the female often very short-stalked; stamens with dilated hairy stalks. **I**: North America. **B**: Locally naturalized around Exeter, Devon.

*Sagittaria subulata* (L.) Buchenau. Like *S. rigida*, but the filaments hairless and the leaves submerged or floating, linear to oval. Naturalized in a single locality in S England.

5* **Lesser Water-plantain** *Baldellia ranunculoides* (L.) Parl. [*Echinodorus ranunculoides, Alisma ranunculoides*]. Low to short, creeping perennial; stem short or elongated. Leaves aerial, lanceolate, tapered at both ends, stalked. Flowers white or pale pink, 10-16mm, in an umbel or in 2 whorls in a raceme, hermaphrodite; stamens 6 only. Fruit, many achenes in a rounded head. June-September. North to S Norway. **B**: Throughout, except the far north.

6 **Creeping Water-plantain** *Baldellia repens* (Lam.) van Ooststr. ex Lawalrée [*Alisma repens*]. Like *B. ranunculoides*, but stems long and creeping, rooting at the nodes. Flowers larger, 15-22mm, in whorls of 2-6 (not 10-12). Ponds and ditches. June-July. W & NW France.

7* **Floating Water-plantain** *Luronium natans* (L.) Rafin. [*Elisma natans*], *Alisma natans*]. A perennial with elongated stems floating or rising in the water, or submerged, rooting at the nodes. Leaves floating or aerial, elliptical to oval, blunt. Flowers white, 12-16mm, the petals with a yellow blotch at the base, hermaphrodite, borne on long stalks at the leaf-axils; stamens 6. Fruit a collection of up to 15 achenes in an irregular whorl. Still and slow-moving acid water. May-August. North to S Scandinavia, but not Finland. Local but apparently spreading. **B**: Wales, Shropshire, Cumbria to Ayr; introduced elsewhere.

8* **Common Water-plantain** *Alisma plantago-aquatica* L. Medium to tall perennial. Leaves lanceolate to oval, pointed, with a truncated or slightly heart-shaped base, stalked, mostly aerial. Flowers white, sometimes with a purplish tinge, small, 6-10mm; styles erect, equalling, or longer than the young achenes. Achenes 2-3mm long, with a long beak at or below the middle. June-August. Throughout, except the far north. **B**: Rather rare in the north, widespread elsewhere.

9 **Narrow-leaved Water-plantain** *Alisma lanceolatum* With. Like *A. plantago-aquatica*, but leaves lanceolate to elliptical with a narrowed base. Flowers purplish-pink, often slightly larger. Styles shorter and arising near the top of the achenes. Similar habitats and flowering time. Throughout, except the far north. **B**: Local north to Yorkshire; less common than *A. plantago-aquatica*.

10* **Ribbon-leaved Water-plantain** *Alisma gramineum* Lej. [*A. loeselii, A. plantago-aquatica* subsp. *graminifolium*]. Medium to tall perennial. Leaves submerged or aerial, the former ribbon-like, the latter linear, expanding into a narrow-elliptical blade. Flowers white, or white with a purplish tinge, smaller than previous 2 species, 4-7mm; the petals longer than the sepals, the inflorescence generally overtopping the leaves; styles recurved, or even coiled, shorter than the young achenes. Achenes 2-2.75mm. July-September. Denmark and Germany southwards. **B**: Rare, now only in one locality in Worcestershire.

*Alisma wahlenbergii* (Holmberg) Juz. [*A. gramineum* subsp. *wahlenbergii*] leaves usually all submerged. Flowers white, 3-5mm, the petals scarcely longer than the sepals. Sweden and Finland.

11* **Parnassus-leaved Water-plantain** *Caldesia parnassifolia* (L.) Parl. [*Alisma parnassifolium*]. Medium to tall perennial. Leaves all basal, floating or aerial, oval to elliptical, with a heart-shaped base, long-stalked. Flowers white, 8-12mm, hermaphrodite, borne in whorls forming racemes or panicles; stamens 6. Fruit consisting of a head of 6-10 fleshy 'achenes'. July-September. Germany to C & NC France.

12* **Star Fruit** *Damasonium alisma* Miller [*D. stellatum*]. Low to short perennial. Leaves all basal, floating or submerged (aerial in subterrestrial plants), oblong with a truncated base, blunt. Flowers white, 5-6mm, the petals with a basal yellow blotch, borne in whorls forming a raceme or panicle, or sometimes a simple umbel; stamens 6. Fruit a whorl of partly fused follicles forming a 6-pointed star, each follicle 5-12mm long. June-September. S Britain and France. **B**: One locality in Surrey, two in Buckinghamshire.

13* **Flowering Rush** *Butomus umbellatus* L. (Flowering Rush Family – Butomaceae). Rather stout, medium to tall, hairless perennial with short creeping rhizomes. Leaves basal, linear and rush-like, triangular below, sheathing at the base, somewhat twisted. Flowers pale to bright pink with darker pink veins, the sepals stained with green on the outside, cup-shaped, 16-26mm, hermaphrodite, borne in a long-stalked umbel, overtopping the leaves; sepals 3, slightly shorter than the 3 petals; stamens 9, red. Fruit 6 red-purple, partly-fused follicles, forming a small egg-shaped structure. July-August. Throughout, except the far north. **B**: Widespread, but rare in Wales; decreasing.

Arrowhead
× 1/8
Lesser
Water-plantain
× 1/9
Common
Water-plantain
Floating Water-plantain
× 1/8
× 1/2
Ribbon-leaved
Water-plantain
Parnassus-leaved
Water-plantain
Star Fruit
Flowering Rush

# FROGBIT FAMILY Hydrocharitaceae

Aquatic plants, submerged or floating. Leaves alternate or in whorls, often sheathing the stem at their base. Flowers unisexual or hermaphrodite, 1 or 2 enclosed in bract-like spathes; sepals and petals 3; stamens 2-15; styles 3-15. Fruit a capsule, splitting lengthwise. 70 species.

1* **Frogbit** *Hydrocharis morsus-ranae* L. Hairless floating plant with runners giving rise to tufts of leaves at intervals; roots with numerous long hairs near the tip. Leaves kidney- or heart-shaped, bronzey-green, untoothed, long-stalked. Flowers white, the petals with a yellow spot at the base, 18-20mm, monoecious; male in stalked clusters of 1-4, female solitary; stamens 12; styles 6, each 2-branched. July-August. Throughout, except the far north. **B**: Locally common north to Durham, diminishing overall.

2* **Water Soldier** *Stratiotes aloides* L. Stoloniferous submerged perennial, except at flowering time, looking like sunken pineapple tops. Leaves in coarse rosettes, linear-lanceolate, tapered, spine-toothed, rather brittle. Flowers white, 30-45mm, held just above the water surface, dioecious; male several to a cluster but the female solitary; stamens 12; styles 6, each 2-branched. June-August. Throughout, except the far north. but rarer in the west; naturalized in Ireland and France. Fruit is not always produced. **B**: Scattered localities throughout, but rare in W England; most British plants are female.

3 **Large-flowered Water-thyme** *Egeria densa* Planchon. [*Elodea densa*]. An aquatic perennial with submerged, elongated stems. Leaves linear, in dense whorls of 3-5, unstalked. Flowers white, 14-20mm, the male and female on separate plants, the male on long slender stalks; usually 9 stamens, styles 3, 2-branched. Freshwater ponds and lakes, ditches. June-August. **I**: South America – Argentine. Cultivated in aquaria and naturalized locally in Britain, France and Holland. **B**: Naturalized in Lancashire canals.

4* **Canadian Pondweed** *Elodea canadensis* Michx. Forming deep green masses at or below the water surface, much branched, the shoots not collapsing out of water. Leaves in whorls of 3, the lowermost opposite, linear-oblong, at least 2mm wide, curved, minutely toothed, with tiny scales at the base. Flowers white or pale purple, 4-5mm, the male borne on long thread-like stalks. Lakes and canals, mainly in base-rich waters. May-September. **I**: North America. Naturalized throughout, except the far north. Male plants absent or scarce. **B**: Widespread, but rare, declining.

5 **Greater Water-thyme** *Elodea callitrichoides* (L.C.M. Richard) Caspary [*E. ernstiae*]. A larger plant than *E. canadensis*, the leaves up to 25mm (not 17mm) long, and tapered to a point. Flowers white, the female, 6-7mm. May-September. **I**: South America. Naturalized in France. **B**: Casual.

6* **Nuttall's Water-thyme** *Elodea nuttallii* (Planchon) St John. A more delicate, pale green plant than the preceding two species, the shoots collapsing readily out of water, but forming invasive colonies. Leaves up to 15mm long, but not exceeding 1.8mm wide, in whorls of 3-4, the lowermost opposite. Flowers white, female 2-3.5mm. Male flowers breaking free of the parent plant and rising to the water surface. May-September. **I**: North America. Naturalized in W Europe. Still spreading and often replacing *E. canadensis*. **B**: Spreading rapidly throughout England except the SW; local in Wales, rare elsewhere.

7 **Esthwaite Waterweed** *Hydrilla verticillata* (L. fil.) Royle. Submerged aquatic perennial with elongated stems. Leaves linear, in whorls of 3-8, toothed. Flowers 3-5mm, the narrow petals transparent but with red streaks, sepals wider, dioecious; stamens and styles 3. Local in lakes and deep ponds. June-August. Britain and Germany. **B**: Cumbria, N Lancashire and W Galway.

8* **Curly Water-thyme** *Lagarosiphon major* (Ridley) Moss. Submerged aquatic with elongated stems, forming entangled masses. Leaves linear, alternate, spreading and recurved, crowded towards the shoot tips, minutely toothed. Flowers pinkish, 3-4mm, female spathes with 1-3 flowers, male usually with many, dioecious, stamens 3; styles 3, each 2-branched. July-August. **I**: South Africa. Naturalized in Britain and France.

9* **Tape Grass** *Vallisneria spiralis* L. Submerged, stoloniferous, aquatic perennial forming tufts rooting into the mud. Leaves pale green, linear, very long and ribbon-like, generally with reddish dots and streaks. Flowers pinkish-white, dioecious, the female 4-7mm, borne on long thread-like stalks, the male tiny, petals inconspicuous; stamens 2-3; styles 3, each 2-branched. Warm freshwater habitats, very local. June-October. Southern, naturalized in W Europe, often where effluent from factories warms the water. **B**: W England, naturalized in Lancs, S Yorkshire and round London.

10* **Rannoch Rush** *Scheuchzeria palustris* L. (Family Scheuchzeriaceae – the only species). Short grass-like herb with creeping stems and persistent leaf-bases. Leaves linear, with a pore at the tip, with sheathing, rather inflated base, equalling or longer than the erect flower stems. Flowers yellowish-green, 4-6mm, hermaphrodite, in lax, few-flowered racemes, tepals 6, all similar; stamens 6. Fruit of 3 inflated follicles fused at the base. Wet boggy habitats; usually in pools in *Sphagnum* bogs. June-August. Europe, except Holland. **B**: Scotland – Perth, rare; formerly in N England, Argyll and Offaly.

11* **Cape Pondweed** *Aponogeton distachyos* L. fil. (Cape Pondweed Family – Aponogetaceae – 25 species in one genus). An aquatic tuberous-rooted perennial; stem green, spongy. Leaves floating, elliptic-oblong, long-stalked, untoothed. Flowers white, in a Y-shaped cluster, the segments only 1-2, elliptical and petal-like, 10-20mm long; stamens 6 or more. Fruit a cluster of 3 follicles. Lakes and ponds. August-October. **I**: South Africa. Naturalized in Britain and France. **B**: Frequently planted, naturalized.

12* **Sea Arrow-grass** *Triglochin maritima* L. (Arrow-grass Family – Juncaginaceae – few species in 4 genera). Short to medium, tufted, erect perennial, hairless, not stoloniferous. Leaves linear, half-rounded in section, to 4mm wide, fleshy. Flowers green, 3-4mm, short-stalked, in a long spike. Fruit egg-shaped, 3-4mm long, all 6 compartments containing seeds. Saline marshes, generally close to the sea, often in short turf. May-September. Throughout, except Spitsbergen. **B**: Throughout coastally.

13 **Marsh Arrow-grass** *Triglochin palustris* L. Often taller and stoloniferous. Leaves deeply grooved, only 2mm wide and flower 2-3mm long, in a lax spike. Fruit larger, 7-10mm, club-shaped, but opening into a narrow arrow-shape, only 3 compartments fertile and containing seeds. Generally among tall grasses especially in coastal districts. Wet meadows, marshes, fens and stream banks. May-August. Throughout, except Spitsbergen. **B**: Throughout, but local in C & S England.

14 **Bulbous Arrow-grass** *Triglochin bulbosa* L. subsp. *barrelieri* (Loisel.) Rouy. Like *T. palustris*, but the stock a fibrous coated bulb, without stolons and leaves to 4mm wide. Flowers small, greenish, 1.5-2.5mm. Fruit elliptical, 6-12mm, spreading. Damp saline habitats. March-May. Mediterranean region reaching north to NW France. subsp. *laxiflora* (Guss.) Rouy [*Triglochin laxiflora*]. Fruits closely pressed to the flower axis. September-November. Similar distribution.

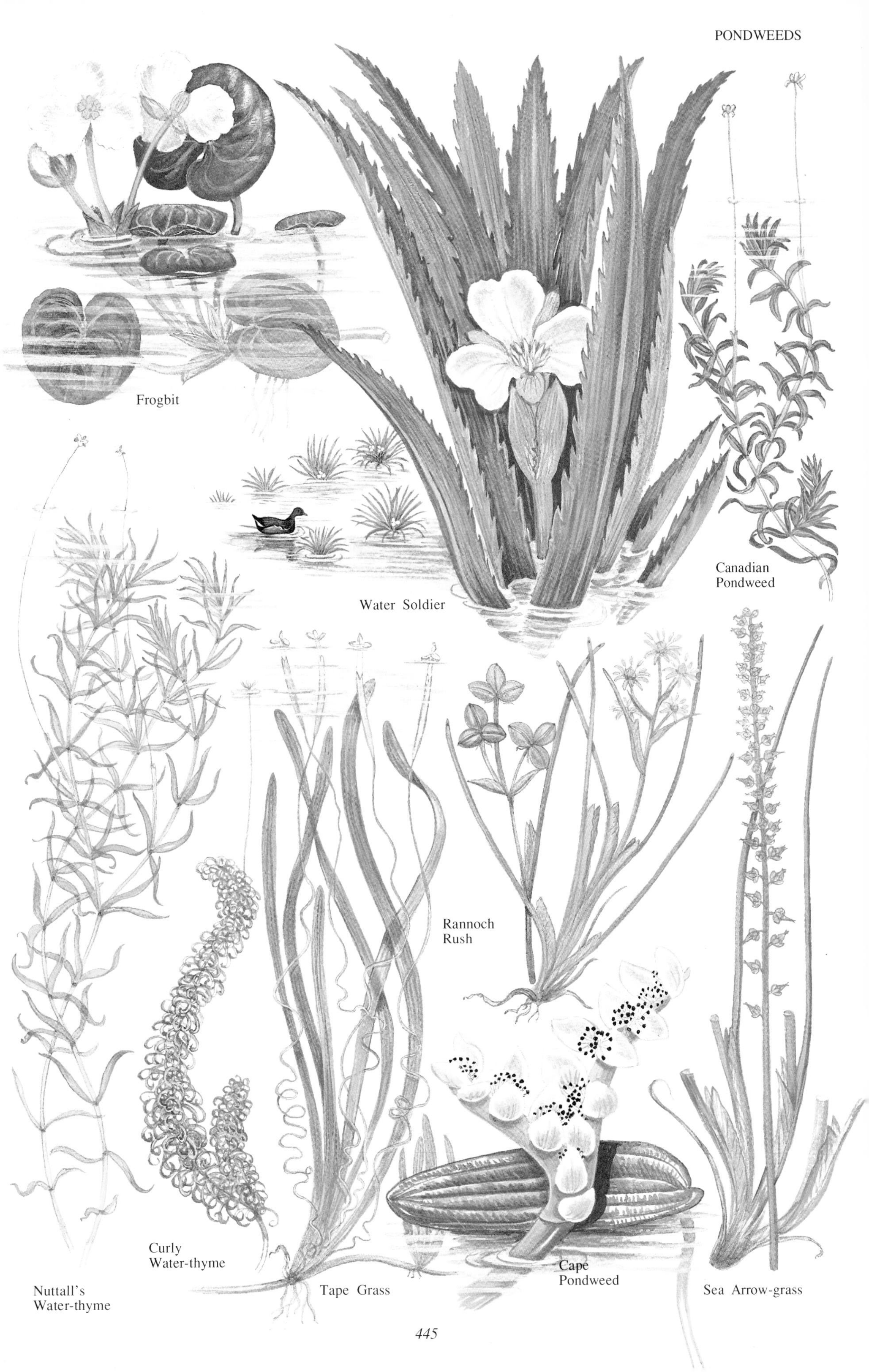
Frogbit
Water Soldier
Canadian Pondweed
Rannoch Rush
Curly Water-thyme
Nuttall's Water-thyme
Tape Grass
Cape Pondweed
Sea Arrow-grass

# PONDWEED FAMILY Potamogetonaceae

Herbs with floating or submerged leaves borne on long stalks, alternate or opposite and sheathing at the base. Flowers hermaphrodite in stalked spikes or clusters, no bracts, usually greenish, with 4 sepals but no petals; stamens 4. Fruit consisting of 4 nutlets, separate or partly fused. 2 genera, mainly in freshwater, occasionally in brackish waters, widespread in the world.

*Potamogeton.* Perennial herbs with alternate leaves. Fruitlets somewhat fleshy (dry and achene-like in *Groenlandia*). A difficult genus growing in a variety of freshwater habitats. Plants are occasionally marooned on muddy banks and may appear atypical with thickened leaves with short stalks. Identification is complicated by frequent hybridization. Hybrids are usually sterile and often form vigorous and persistent populations reproducing vegetatively.

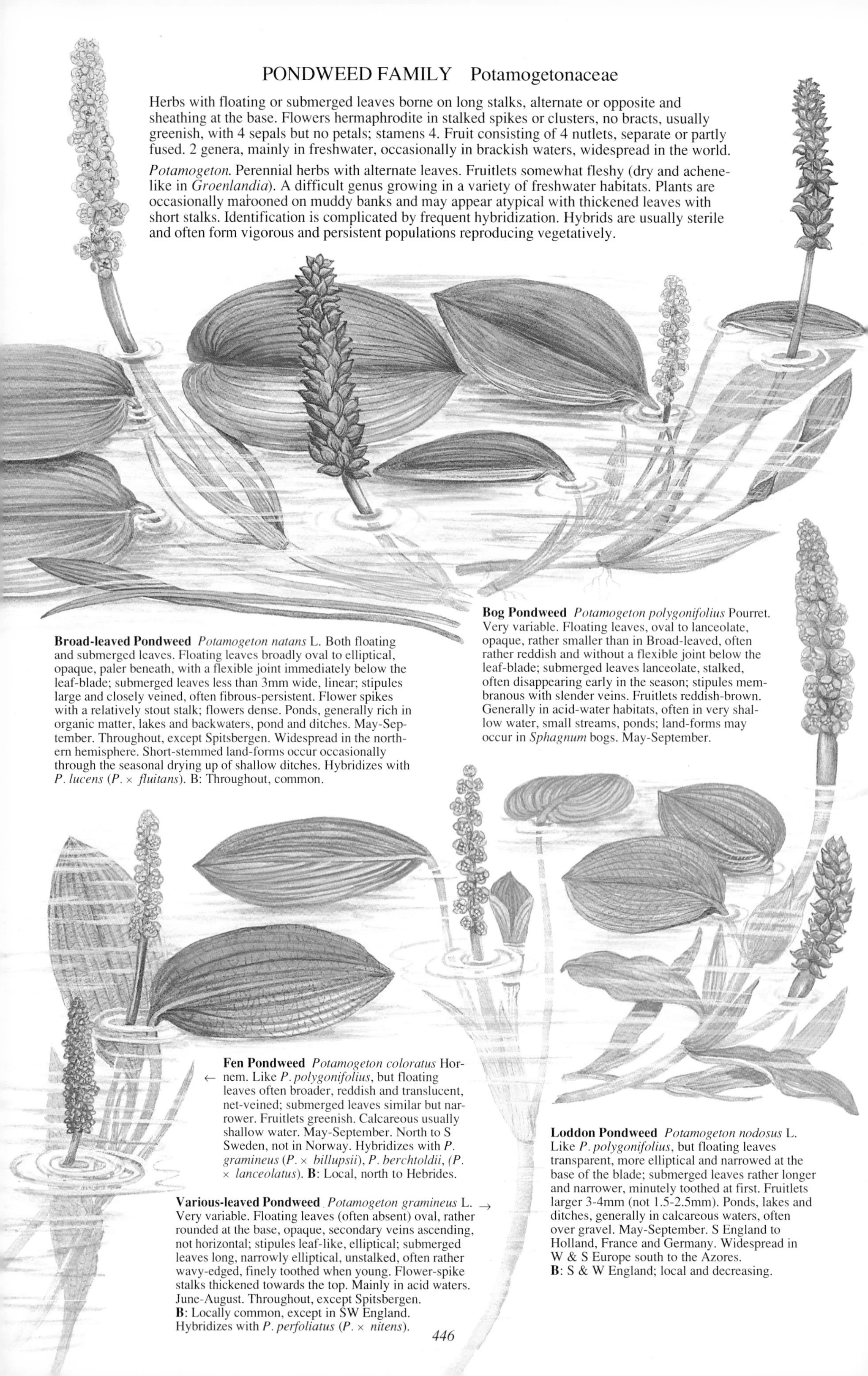

**Broad-leaved Pondweed** *Potamogeton natans* L. Both floating and submerged leaves. Floating leaves broadly oval to elliptical, opaque, paler beneath, with a flexible joint immediately below the leaf-blade; submerged leaves less than 3mm wide, linear; stipules large and closely veined, often fibrous-persistent. Flower spikes with a relatively stout stalk; flowers dense. Ponds, generally rich in organic matter, lakes and backwaters, pond and ditches. May-September. Throughout, except Spitsbergen. Widespread in the northern hemisphere. Short-stemmed land-forms occur occasionally through the seasonal drying up of shallow ditches. Hybridizes with *P. lucens* (*P.* × *fluitans*). B: Throughout, common.

**Bog Pondweed** *Potamogeton polygonifolius* Pourret. Very variable. Floating leaves, oval to lanceolate, opaque, rather smaller than in Broad-leaved, often rather reddish and without a flexible joint below the leaf-blade; submerged leaves lanceolate, stalked, often disappearing early in the season; stipules membranous with slender veins. Fruitlets reddish-brown. Generally in acid-water habitats, often in very shallow water, small streams, ponds; land-forms may occur in *Sphagnum* bogs. May-September.

**Fen Pondweed** *Potamogeton coloratus* Hornem. ← Like *P. polygonifolius*, but floating leaves often broader, reddish and translucent, net-veined; submerged leaves similar but narrower. Fruitlets greenish. Calcareous usually shallow water. May-September. North to S Sweden, not in Norway. Hybridizes with *P. gramineus* (*P.* × *billupsii*), *P. berchtoldii*, (*P.* × *lanceolatus*). **B**: Local, north to Hebrides.

**Various-leaved Pondweed** *Potamogeton gramineus* L. → Very variable. Floating leaves (often absent) oval, rather rounded at the base, opaque, secondary veins ascending, not horizontal; stipules leaf-like, elliptical; submerged leaves long, narrowly elliptical, unstalked, often rather wavy-edged, finely toothed when young. Flower-spike stalks thickened towards the top. Mainly in acid waters. June-August. Throughout, except Spitsbergen. **B**: Locally common, except in SW England. Hybridizes with *P. perfoliatus* (*P.* × *nitens*).

**Loddon Pondweed** *Potamogeton nodosus* L. Like *P. polygonifolius*, but floating leaves transparent, more elliptical and narrowed at the base of the blade; submerged leaves rather longer and narrower, minutely toothed at first. Fruitlets larger 3-4mm (not 1.5-2.5mm). Ponds, lakes and ditches, generally in calcareous waters, often over gravel. May-September. S England to Holland, France and Germany. Widespread in W & S Europe south to the Azores. **B**: S & W England; local and decreasing.

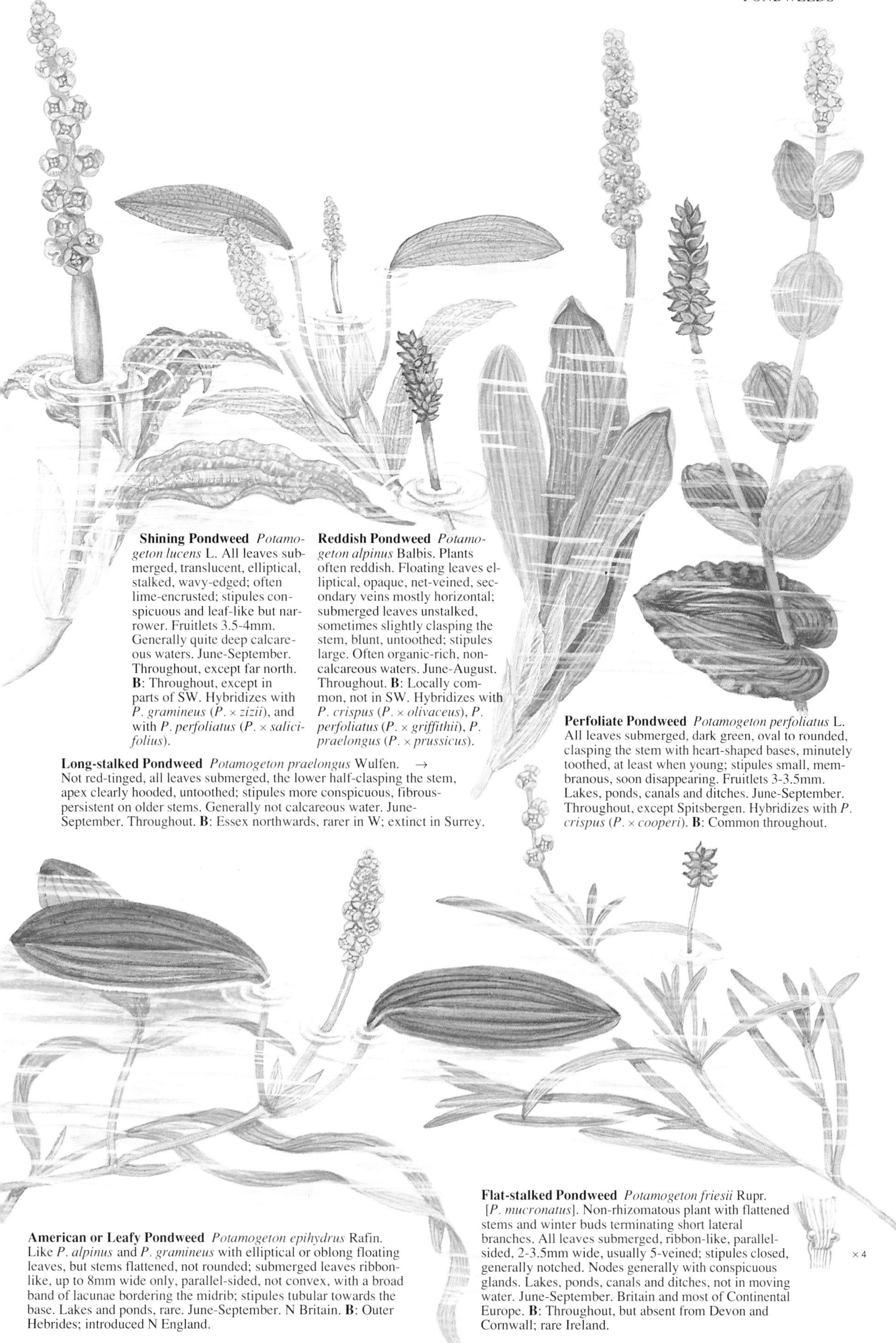

**Shining Pondweed** *Potamogeton lucens* L. All leaves submerged, translucent, elliptical, stalked, wavy-edged; often lime-encrusted; stipules conspicuous and leaf-like but narrower. Fruitlets 3.5-4mm. Generally quite deep calcareous waters. June-September. Throughout, except far north. **B**: Throughout, except in parts of SW. Hybridizes with *P. gramineus* (*P.* × *zizii*), and with *P. perfoliatus* (*P.* × *salicifolius*).

**Reddish Pondweed** *Potamogeton alpinus* Balbis. Plants often reddish. Floating leaves elliptical, opaque, net-veined, secondary veins mostly horizontal; submerged leaves unstalked, sometimes slightly clasping the stem, blunt, untoothed; stipules large. Often organic-rich, non-calcareous waters. June-August. Throughout. **B**: Locally common, not in SW. Hybridizes with *P. crispus* (*P.* × *olivaceus*), *P. perfoliatus* (*P.* × *griffithii*), *P. praelongus* (*P.* × *prussicus*).

**Long-stalked Pondweed** *Potamogeton praelongus* Wulfen. →
Not red-tinged, all leaves submerged, the lower half-clasping the stem, apex clearly hooded, untoothed; stipules more conspicuous, fibrous-persistent on older stems. Generally not calcareous water. June-September. Throughout. **B**: Essex northwards, rarer in W; extinct in Surrey.

**Perfoliate Pondweed** *Potamogeton perfoliatus* L. All leaves submerged, dark green, oval to rounded, clasping the stem with heart-shaped bases, minutely toothed, at least when young; stipules small, membranous, soon disappearing. Fruitlets 3-3.5mm. Lakes, ponds, canals and ditches. June-September. Throughout, except Spitsbergen. Hybridizes with *P. crispus* (*P.* × *cooperi*). **B**: Common throughout.

**American or Leafy Pondweed** *Potamogeton epihydrus* Rafin. Like *P. alpinus* and *P. gramineus* with elliptical or oblong floating leaves, but stems flattened, not rounded; submerged leaves ribbon-like, up to 8mm wide only, parallel-sided, not convex, with a broad band of lacunae bordering the midrib; stipules tubular towards the base. Lakes and ponds, rare. June-September. N Britain. **B**: Outer Hebrides; introduced N England.

**Flat-stalked Pondweed** *Potamogeton friesii* Rupr. [*P. mucronatus*]. Non-rhizomatous plant with flattened stems and winter buds terminating short lateral branches. All leaves submerged, ribbon-like, parallel-sided, 2-3.5mm wide, usually 5-veined; stipules closed, generally notched. Nodes generally with conspicuous glands. Lakes, ponds, canals and ditches, not in moving water. June-September. Britain and most of Continental Europe. **B**: Throughout, but absent from Devon and Cornwall; rare Ireland.

× 4

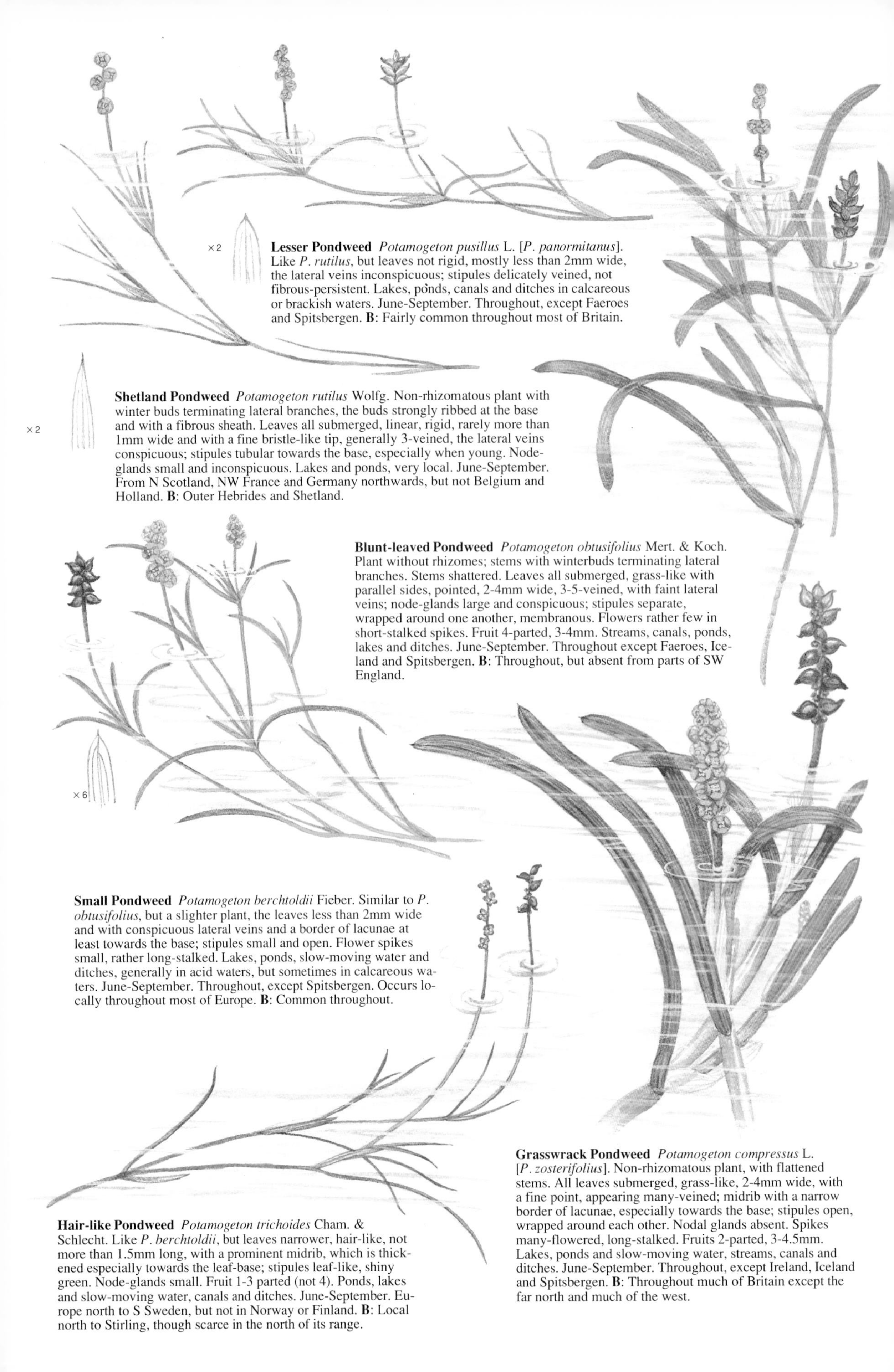


**Lesser Pondweed** *Potamogeton pusillus* L. [*P. panormitanus*]. Like *P. rutilus*, but leaves not rigid, mostly less than 2mm wide, the lateral veins inconspicuous; stipules delicately veined, not fibrous-persistent. Lakes, ponds, canals and ditches in calcareous or brackish waters. June-September. Throughout, except Faeroes and Spitsbergen. **B**: Fairly common throughout most of Britain.

**Shetland Pondweed** *Potamogeton rutilus* Wolfg. Non-rhizomatous plant with winter buds terminating lateral branches, the buds strongly ribbed at the base and with a fibrous sheath. Leaves all submerged, linear, rigid, rarely more than 1mm wide and with a fine bristle-like tip, generally 3-veined, the lateral veins conspicuous; stipules tubular towards the base, especially when young. Node-glands small and inconspicuous. Lakes and ponds, very local. June-September. From N Scotland, NW France and Germany northwards, but not Belgium and Holland. **B**: Outer Hebrides and Shetland.

**Blunt-leaved Pondweed** *Potamogeton obtusifolius* Mert. & Koch. Plant without rhizomes; stems with winterbuds terminating lateral branches. Stems shattered. Leaves all submerged, grass-like with parallel sides, pointed, 2-4mm wide, 3-5-veined, with faint lateral veins; node-glands large and conspicuous; stipules separate, wrapped around one another, membranous. Flowers rather few in short-stalked spikes. Fruit 4-parted, 3-4mm. Streams, canals, ponds, lakes and ditches. June-September. Throughout except Faeroes, Iceland and Spitsbergen. **B**: Throughout, but absent from parts of SW England.

**Small Pondweed** *Potamogeton berchtoldii* Fieber. Similar to *P. obtusifolius*, but a slighter plant, the leaves less than 2mm wide and with conspicuous lateral veins and a border of lacunae at least towards the base; stipules small and open. Flower spikes small, rather long-stalked. Lakes, ponds, slow-moving water and ditches, generally in acid waters, but sometimes in calcareous waters. June-September. Throughout, except Spitsbergen. Occurs locally throughout most of Europe. **B**: Common throughout.

**Grasswrack Pondweed** *Potamogeton compressus* L. [*P. zosterifolius*]. Non-rhizomatous plant, with flattened stems. All leaves submerged, grass-like, 2-4mm wide, with a fine point, appearing many-veined; midrib with a narrow border of lacunae, especially towards the base; stipules open, wrapped around each other. Nodal glands absent. Spikes many-flowered, long-stalked. Fruits 2-parted, 3-4.5mm. Lakes, ponds and slow-moving water, streams, canals and ditches. June-September. Throughout, except Ireland, Iceland and Spitsbergen. **B**: Throughout much of Britain except the far north and much of the west.

**Hair-like Pondweed** *Potamogeton trichoides* Cham. & Schlecht. Like *P. berchtoldii*, but leaves narrower, hair-like, not more than 1.5mm long, with a prominent midrib, which is thickened especially towards the leaf-base; stipules leaf-like, shiny green. Node-glands small. Fruit 1-3 parted (not 4). Ponds, lakes and slow-moving water, canals and ditches. June-September. Europe north to S Sweden, but not in Norway or Finland. **B**: Local north to Stirling, though scarce in the north of its range.

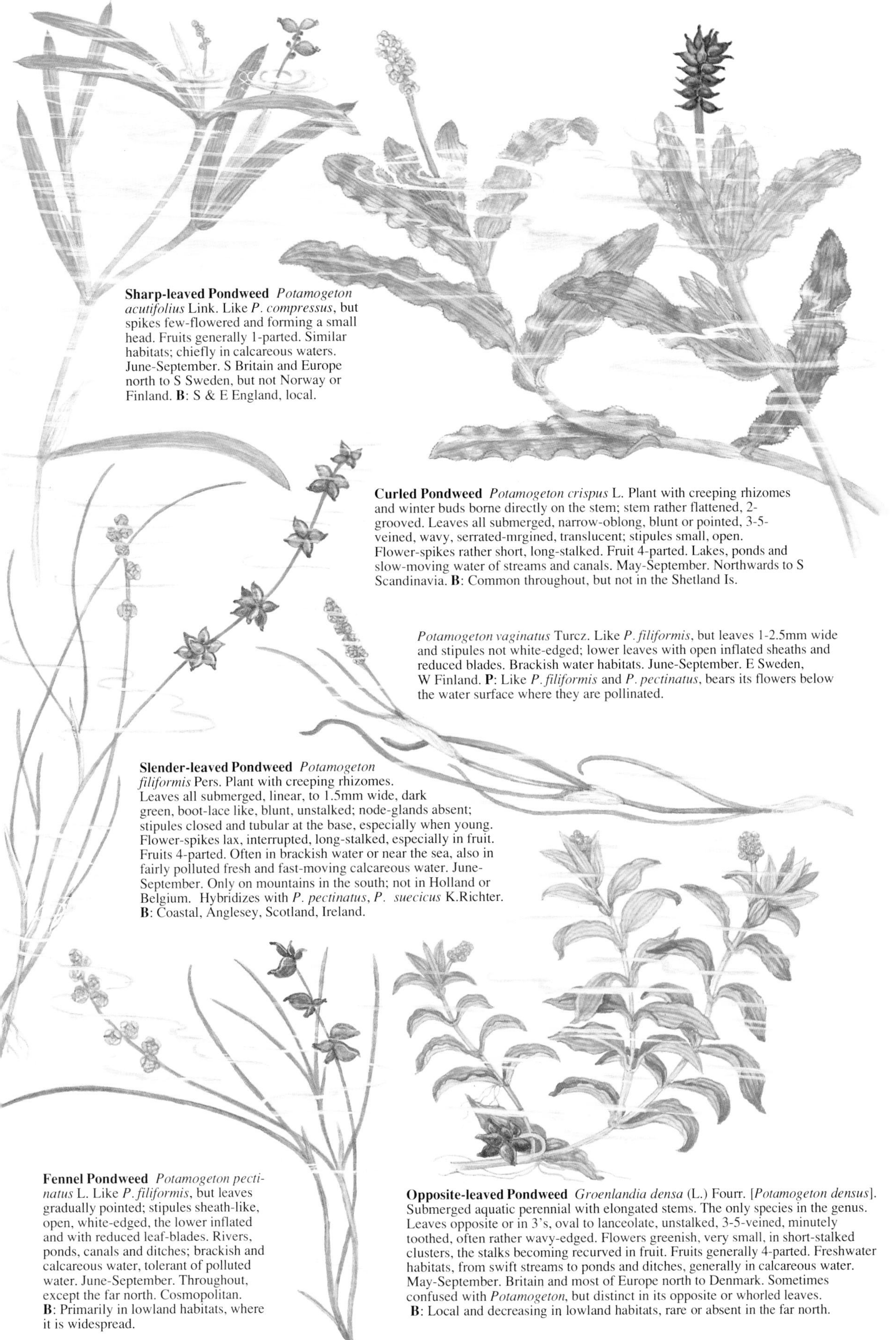

**Sharp-leaved Pondweed** *Potamogeton acutifolius* Link. Like *P. compressus*, but spikes few-flowered and forming a small head. Fruits generally 1-parted. Similar habitats; chiefly in calcareous waters. June-September. S Britain and Europe north to S Sweden, but not Norway or Finland. **B**: S & E England, local.

**Curled Pondweed** *Potamogeton crispus* L. Plant with creeping rhizomes and winter buds borne directly on the stem; stem rather flattened, 2-grooved. Leaves all submerged, narrow-oblong, blunt or pointed, 3-5-veined, wavy, serrated-mrgined, translucent; stipules small, open. Flower-spikes rather short, long-stalked. Fruit 4-parted. Lakes, ponds and slow-moving water of streams and canals. May-September. Northwards to S Scandinavia. **B**: Common throughout, but not in the Shetland Is.

*Potamogeton vaginatus* Turcz. Like *P. filiformis*, but leaves 1-2.5mm wide and stipules not white-edged; lower leaves with open inflated sheaths and reduced blades. Brackish water habitats. June-September. E Sweden, W Finland. **P**: Like *P. filiformis* and *P. pectinatus*, bears its flowers below the water surface where they are pollinated.

**Slender-leaved Pondweed** *Potamogeton filiformis* Pers. Plant with creeping rhizomes. Leaves all submerged, linear, to 1.5mm wide, dark green, boot-lace like, blunt, unstalked; node-glands absent; stipules closed and tubular at the base, especially when young. Flower-spikes lax, interrupted, long-stalked, especially in fruit. Fruits 4-parted. Often in brackish water or near the sea, also in fairly polluted fresh and fast-moving calcareous water. June-September. Only on mountains in the south; not in Holland or Belgium. Hybridizes with *P. pectinatus*, *P. suecicus* K.Richter. **B**: Coastal, Anglesey, Scotland, Ireland.

**Fennel Pondweed** *Potamogeton pectinatus* L. Like *P. filiformis*, but leaves gradually pointed; stipules sheath-like, open, white-edged, the lower inflated and with reduced leaf-blades. Rivers, ponds, canals and ditches; brackish and calcareous water, tolerant of polluted water. June-September. Throughout, except the far north. Cosmopolitan. **B**: Primarily in lowland habitats, where it is widespread.

**Opposite-leaved Pondweed** *Groenlandia densa* (L.) Fourr. [*Potamogeton densus*]. Submerged aquatic perennial with elongated stems. The only species in the genus. Leaves opposite or in 3's, oval to lanceolate, unstalked, 3-5-veined, minutely toothed, often rather wavy-edged. Flowers greenish, very small, in short-stalked clusters, the stalks becoming recurved in fruit. Fruits generally 4-parted. Freshwater habitats, from swift streams to ponds and ditches, generally in calcareous water. May-September. Britain and most of Europe north to Denmark. Sometimes confused with *Potamogeton*, but distinct in its opposite or whorled leaves. **B**: Local and decreasing in lowland habitats, rare or absent in the far north.

## TASSELWEED FAMILY Ruppiaceae

Submerged aquatic perennial herbs of saline waters with narrow alternate leaves, sheathing at the base. Flowers hermaphrodite, in pairs subtended by a pair of leaves, small, on long thread-like stems to the water surface, without sepals or petals. Stamens 2. Fruit generally of 4 carpels; fruitlets not splitting. 3 species in 1 genus.

1* **Beaked Tasselweed** *Ruppia maritima* L. [*R. rostellata*]. Slender plant with pale green linear leaves less than 1mm wide, untoothed, pointed, the basal sheaths only slightly inflated. Flower pairs on stalks up to 6cm long, recurved or flexuous in fruit; anthers 0.6-0.7mm long. Submerged herbs in saline or brackish-water habitats. July-September. Throughout except Spitsbergen. In this and Zosteraceae the flowers are reduced to just the essential male and female organs. **B**: Widespread, but often local.

2 **Spiral Tasselweed** *Ruppia cirrhosa* (Petagna) Grande [*R. spiralis*, *R. maritima* subsp. *spiralis*]. Similar to *R. maritima*, but leaves deep green and leaf-sheaths conspicuously inflated. Flower-stalks up to 8cm long, greatly elongating and coiling in fruit. Anthers 1.5-1.7mm long. Brackish or saline habitats near the sea. July-September. Throughout, except the far north. **B**: Rather rare.

## EEL-GRASS FAMILY Zosteraceae

Submerged marine herbs with creeping rhizomes and long internodes, generally with roots at each node. Flowers in branched spikes on lateral or terminal stems, enclosed in the sheathing base of a leaf-like spathes, simple, without sepals or petals, with 1 stamen and a 2-branched style. 15 species in 2 genera, widespread in temperate seas.

3* **Eel-grass or Grass-wrack** *Zostera marina* L. Tufted perennial. Leaves long and ribbon-like, dark green, often exceeding 200mm long, generally 5-10mm wide with a blunt or pointed tip, those on flowering stems usually narrower, all 3-9-veined. Flowers minute, greenish, in terminal much-branched spikes, male and female separate; stigma twice as long as the style. Fine gravelly, sandy or muddy shore lines, near and just below spring low water zone. June-September. Throughout, except Spitsbergen. A conspicuous plant at low tide. **B**: Throughout, though rare or absent in much of the north.

4 **Narrow-leaved Eel-grass** *Zostera angustifolia* (Hornem.) Reichenb. Like *Z. marina*, but leaves mostly thread-like, 100-300mm long, only 1-3mm wide, often notched at the apex, 1-3-veined. Flowering stems much shorter, only 10-30cm; stigma equalling the style. Often in estuaries, from half-tide to low-tide mark. June-September. Britain, Denmark and Sweden (distribution incompletely known). **B**: Scattered localities, north to Orkney.

5 **Dwarf Eel-grass** *Zostera noltii* Hornem. [*Z. minor*, *Z. nana*]. Like *Z. angustifolia*, but a smaller, slighter plant. Leaves mostly only 0.5-1.5mm wide, 1-veined, less than 200mm long. Flower stems up to 10cm only, simple or sparsely branched. Muddy, sheltered, marine habitats, often in estuaries, from half-tide to low-tide mark, but also in shallow, non-tidal saline pools and backwaters. June-October. North to S Scandinavia. Scattered localities, except for the far north; locally common.

## ZANNICHELLIA FAMILY Zannichelliaceae

Submerged aquatic herbs with creeping rhizomes. Leaves linear, sheathing at the base. Flowers small, unisexual, solitary or in small clusters, the male stalked, with 1-3 stamens, the female unstalked with 1-9 free carpels. Fruit a collection of achenes. 20 species in 6 genera, widespread.

6* **Horned Pondweed** *Zannichellia palustris* L. (incl. *Z. major* (Hartman) Boenn. ex Reichenb.). Variable slender perennial. Leaves alternate or sub-opposite, thread-like, up to 2mm wide, pointed, untoothed; basal sheaths membranous, tubular when young. Flowers tiny, greenish, petalless; male with 1-2 stamens, female with generally 4 carpels. Slow moving or still freshwater habitats, or brackish water pools. May-August. Throughout, except Faeroes and Spitsbergen. Very variable especially as regards the characters of the flowers and fruits. **B**: Throughout, often rather local.

## NAIAD FAMILY Najadaceae

Submerged aquatic herbs with elongated stems. Leaves opposite, sometimes in 3's, slender, sheathing at the base. Flowers small, unisexual, solitary or several together; male flower with a 2-lipped perianth and a single stamen, female with a perianth, with a single ovary and style. Fruit not splitting. One genus, with 35 species in the world..

7* **Holly-leaved Naiad** *Najas marina* L. [*N. major*]. Perennial herb with a smooth or rough stem. Leaves narrow-lanceolate, 1-6mm wide, spiny-edged; back of midrib also often spiny. Male and female flowers borne on separate plants; anthers 4-celled. Fruit 3-8mm. Fresh- or brackish-water habitats. July-September. Throughout north to S Finland. **B**: Norfolk Broads only, rare.

8* **Slender Naiad** *Najas flexilis* (Willd.) Rostk. & W.L.E. Schmidt. Slender perennial with a smooth stem. Leaves linear, 1-2.5mm wide, grass-like, opposite or in 3's, toothed, with 20-36 teeth on each side, translucent. Male and female flowers separate, but borne on the same stem; anthers 1-celled only. Fruit 2-3mm. Lakes and ponds, very local. August-September. Britain, Germany and Scandinavia. **B**: Lake District, Scotland and W Ireland; always local.

9 **Lesser Naiad** *Najas minor* All. Like *N. flexilis*, but leaves thread-like, less than 1mm wide and minutely toothed; leaf-sheaths with small rounded lobes at the base. Ponds and lakes. August-September. Belgium, France and Germany; extinct in Holland.

10 *Najas tenuissima* (A. Braun) Magnus. Like *N. minor*, but leaves not more than 1.2cm long and with only 6 teeth (not 5-17) on each side. Ponds and lakes. August-September. S Finland.

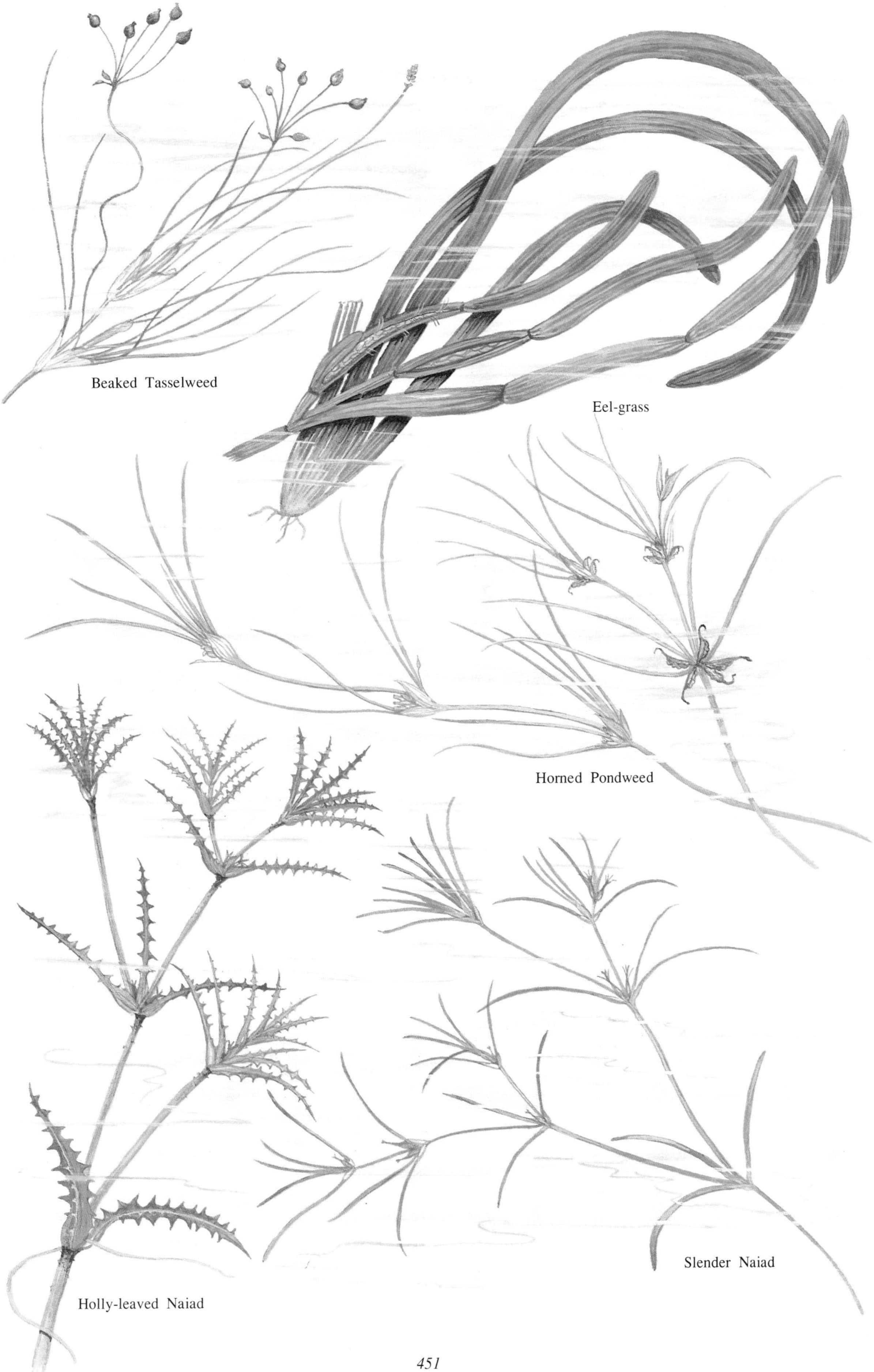

Beaked Tasselweed

Eel-grass

Horned Pondweed

Holly-leaved Naiad

Slender Naiad

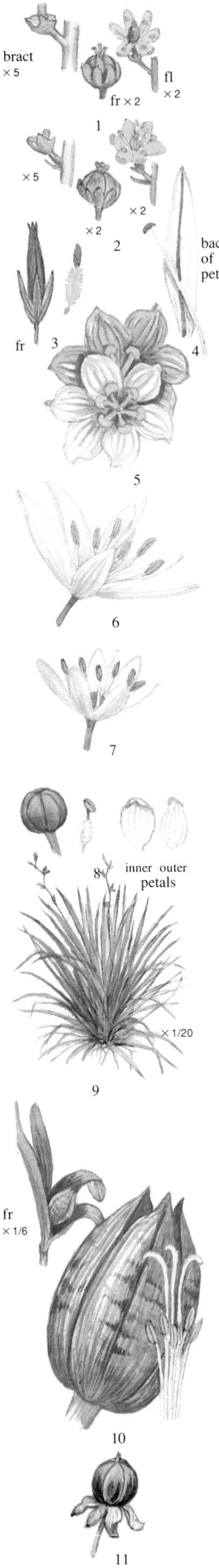

# LILY FAMILY *Liliaceae*

Perennial herbs, often with bulbous, rhizomatous or a tuberous stock. Leaves often linear or lanceolate with parallel veins, sometimes heart-shaped. Flowers solitary or in spikes, racemes or panicles, hermaphrodite. Perianth generally of 6 parts, often all similar and petal-like (tepals), separate or fused together. Stamens usually 6. Ovary 3-celled; styles generally 1, sometimes 3. Fruit a capsule or berry. Over 2500 species in 250 genera.

1* **Scottish Asphodel** *Tofieldia pusilla* (Michx.) Pers. [*T. palustris*]. Low to short densely tufted, hairless perennial with erect, slender flower-stems. Leaves forming small fans, sword-shaped, mostly basal, 3-4-veined, those on the stem small and often bract-like. Flowers white or greenish, small, 3-5mm, 5-10 borne in a short spike; bracts 3-lobed. Fruit a small sub-globose capsule. Wet habitats on mountain meadows and rocks and tundra, to 2500m. June-August. France and Germany, N Europe. Only found on high mountains in the south of its range. **B**: Local, N Yorkshire northwards.

2* **German Asphodel** *Tofieldia calyculata* (L.) Wahlenb. Like *T. pusilla*, but a larger plant with leaves 4-10-veined and flowering stems up to 35cm (not 12cm) tall. Flowers yellowish, rarely reddish, 4-7mm, up to 30 in a rather dense spike; bracts unlobed. Damp grassy habitats and bogs, usually on calcareous soils, to 2500m. From S Sweden (Gotland) to the mountains of France and Germany.

3* **Bog Asphodel** *Narthecium ossifragum* (L.) Hudson. Low to medium, rather variable, rhizomatous, hairless perennial with fans of fleshy, sword-shaped leaves, basal, often orange-tinged; stem leaves small and bract-like, the upper larger than the lower. Flowers greenish-yellow or orange-yellow, 10-16mm, starry, in a rather lax spike-like raceme; filaments of stamens densely hairy. Fruit a small narrow, elliptical capsule, to 12mm long. Bogs and wet acid heaths and moors, to 1200m. July-September. Throughout, except the far north. Generally regarded as poisonous, especially to livestock. The seeds have a slender bristle-like tail at each end. **B**: Generally widespread, but rarer and local in the east of England.

4* **Asphodel** *Asphodelus albus* L. Stout, medium to tall, hairless perennial with swollen roots and an erect flowering stem. Leaves in basal tufts, linear, pointed, grey-green. Flowers white, starry, 30-40mm, in dense, generally simple, spike-like racemes; bracts broad. Fruit a small subglobose capsule. Meadows, heaths and open woodland, generally in hilly and mountain habitats, to 1600m. June-August. France and S Europe. Often conspicuous in mountain pastures, also at low altitudes in NW France.

5* **White False Helleborine** *Veratrum album* L. Very variable, medium to tall, stout, rhizomatous perennial, often forming clumps; stems erect. Leaves alternate, oval to elliptical, deep green, pleated, untoothed, the lower closely overlapping, hairy at least on the veins beneath, the upper smaller and narrower. Flowers whitish or greenish, occasionally yellowish, starry, 14-26mm, borne in dense terminal panicles, branched in the lower half. Fruit a capsule, slightly hairy. Damp grassy meadows, hills and mountains, to 2700m. France, Germany, Norway, Finland.

6* **St Bernard's Lily** *Anthericum liliago* L. Rather slender short to medium, hairless perennial. Leaves linear, flat or slightly grooved, tapered to a sharp apex. Flowers glistening white, 20-35mm, starry, in simple or slightly branched lax racemes, the tepals much exceeding the stamens; style curved; bracts lanceolate, pointed, membranous. Open habitats in hills and mountains, generally over limestone, 300-1800m. May-July. France, Germany, Belgium, Denmark, S Sweden.

7* *Anthericum ramosum* L. Similar, but flowers generally in a well-branched panicle, larger, 22-40mm and with a straight style. Capsule smaller 5-6mm (not 8-10mm). Dry, sunny, grassy habitats, generally at low altitudes. June-July. France, Germany, Denmark, S Sweden.

8* **Kerry Lily** *Simethis planifolia* (L.) Gren. [*S. bicolor*, *Anthericum planifolium*]. Low to short hairless perennial. Leaves linear, all basal, grass-like, with fibrous, sheathing bases, greyish and often somewhat curved. Flowers white but purplish-violet beneath, 18-22mm, starry, borne in lax, branched, clusters on leafless stems. Stamens with conspicuously hairy filaments. Fruit a globose capsule, 5mm. Rocky habitats and heaths, particularly pine-woods. May-June. S England, Ireland and France. **B**: Very local plant found only in Dorset and in Kerry, Ireland.

9 **New Zealand Flax** *Phormium tenax* J.R. & G. Forster. Very large, tough, tufted perennial, up to 4m tall in flower. Leaves forming large fans, very leathery, sword-shaped, folded along the centre, deep green, bronzed or reddish. Flowers brownish-red and yellow, 3-5cm, in clusters forming a stiff, stout-stemmed panicle, the inner tepals recurved at the apex; stamens protruding on long reddish filaments. Fruit capsule dark brown, 7-9cm long, triangular in cross-section. **I**: New Zealand and Norfolk Island. Locally naturalized. July-August. **B**: W England and W Ireland.

10* **Autumn Crocus, Meadow Saffron** *Colchicum autumnale* L. Low to short cormous perennial flowering in the autumn before the leaves appear. Leaves 3-4 in a cluster, broad lanceolate, deep shiny green. Flowers pinkish or lilac-purple, rarely white, 4-6cm long, goblet-shaped and crocus-like, with yellow anthers and whitish styles, 1-6 borne from a tubular whitish spathe at ground level. Fruit an egg-shaped green capsule nestling in the middle of the leaves, below ground-level at first but pushed above ground-level with the developing leaves. Damp grassy meadows, woodland margins, on neutral or calcareous soils, to 2000m. August-September. Naturalized in Denmark and Sweden. Poisonous to livestock. Distinguish from the real autumn-flowering crocuses (*Crocus* in the *Iridaceae*) which have 3 stamens not 6, and linear leaves. **B**: Locally common north to Cumbria, and in SE Ireland.

11* **Snowdon Lily** *Lloydia serotina* (L.) Reichenb. Low, hairless, bulbous perennial with leafy stems, elongating in fruit. Basal leaves green and thread-like, generally only 2; stem leaves 2-4 similar but much shorter, the uppermost bract-like. Flowers solitary, white with reddish-purple veins, more or less bell-shaped, 18-22mm, half-nodding; tepals separate but all rather similar. Fruit a small rounded, 3-ribbed capsule, splitting at the apex. Stony slopes, rock ledges, alpine meadows and tundra, hilly or mountain habitats, to 3000m. May-July. W Britain, France and Germany. **B**: Rare and protected, Snowdonia.

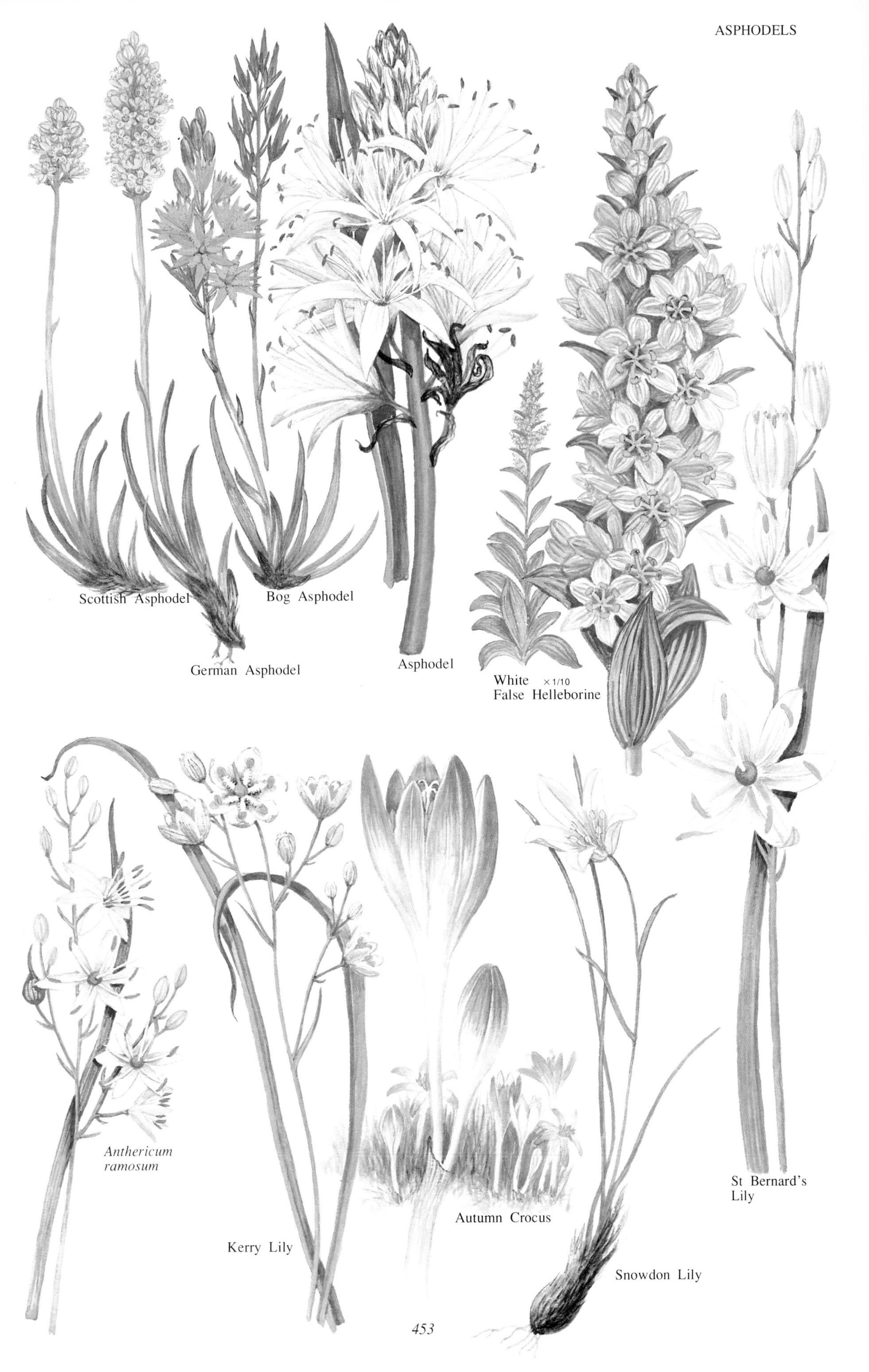

Scottish Asphodel
Bog Asphodel
German Asphodel
Asphodel
White ×1/10
False Helleborine
St Bernard's
Lily
*Anthericum*
*ramosum*
Kerry Lily
Autumn Crocus
Snowdon Lily

**Gageas** *Gagea*. Small bulbous perennials with erect stems. Leaves basal and on the unbranched stem. Flowers yellow, starry, with 6 separate tepals. Fruit a small capsule. 100 species. The leaf characters are important.

* **Meadow Gagea** *Gagea pratensis* (Pers.) Dumort. [*G. stenopetala*]. Low slender perennial with a hairless stem. Basal leaf solitary, broad, linear, flat; stem leaves 2 opposite, lanceolate; leaf-margin hairy. Flowers 2-6, yellow, slightly green-tinged, 20-30mm; petals rather blunt. Grassy habitats, often in or near fields, generally on calcareous soils, to 1000m. March-May. From France and Germany to Holland, Denmark and S Sweden, mainly on the mountains in the south.

* **Yellow Star of Bethlehem** *Gagea lutea* (L.) Ker-Gawler [*G. silvatica*]. Low to short, slender perennial with a hairless stem. Basal leaf solitary, linear-lanceolate, flat, 5-12mm broad, generally yellowish-green; stem leaves 2, opposite or subopposite, lanceolate, margin hairy. Flowers 7, yellow, 15-25mm, in an umbel-like cluster, each tepal with a band of green on the back; flower-stalks hairy or not. Damp grassland, scrub and open damp woods, especially on basic soils, to 1700m. March-May. Throughout, except Ireland, Faeroes, Iceland and Spitsbergen. Occasionally cultivated in gardens. **B**: Scattered localities in C & N England, S Scotland; very local in S England.

* **Least Gagea** *Gagea minima* (L.) Ker-Gawler. Low slender perennial with hairless stems. Basal leaf solitary linear, more or less flat; stem leaves 2, opposite, lanceolate. Flowers 1-7, yellow, 10-15mm, on long slender hairless or slightly hairy stalks; tepals sharply pointed, often reflexed. Meadows and open woods, on calcareous soils, to 1000m. March-May. Continental Europe, except France, Belgium and Holland.

* **Belgian Gagea** *Gagea spathacea* (Hayne) Salisb. Low to short slender perennial with a hairless stem. Basal leaves 2, narrow-linear, circular in section; stem leaves generally 3, flat, the lower oblong-lanceolate, the upper 2 opposite, smaller and bract-like. Flowers 2-4, yellowish-green, 18-20mm, borne on hairless stalks. Woodland in damp grassy places, scrub, to 1500m. April-June. Continental Europe north to Belgium, Holland, Germany, Denmark and S Sweden.

* *Gagea arvensis* (Pers.) Dumort. [*G. villosa*]. Low slender, greyish perennial with a minutely-hairy stem. Basal leaves 2, narrow-linear, grooved; stem leaves 2, lanceolate, hairy, often with tiny bulbils in their axils. Flowers 5-12, greenish-yellow, 15-20mm. Dry open habitats, meadows and waste ground, generally on acid soils, to 2200m, also occasionally on cultivated land. April-May. France and Germany north to Holland, Denmark and S Sweden; extinct in Belgium.

* **Bohemian Gagea** *Gagea bohemica* (Zauschner) Schultes & Schultes fil. [*G. bohemica* subsp. *zauschneri*]. Low slender perennial, often only 3-4cm tall, with a hairless stem. Basal leaves 2, thread-like, often wavy or coiled; stem leaves 2, alternate, lanceolate, pointed. Flowers 1-3, rather bright yellow, 10-20mm. Dry stony and rocky habitats, sometimes on thin sward. January-March. W Britain, C & S Germany. **B**: Only recently discovered in mid Wales; the Welsh plants are nearly always 1-flowered, very rare. subsp. *gallica* (Rouy) J.B.K. Richardson [*G. bohemica* var. *gallica*]. Like the type, but the basal leaves silky with whitish hairs. Similar habitat and flowering time. W France.

1* **Wild Tulip** *Tulipa sylvestris* L. Short to medium bulbous perennial, hairless except for the base of the stamens. Leaves 2-4, rather fleshy, alternate, linear-lanceolate, grooved, deep green outside, but grey-green on the upper surface. Flowers yellow, rarely cream, the 3 outer tepals tinged with green, pink or crimson on the back, 36-70mm long, the outer tepals narrow, often recurving; flowers erect, but nodding in bud, faintly fragrant. Fruit an oblong capsule containing many flattened seeds, but rarely produced. Meadows, grassy and rocky habitats, at low altitudes. April-May. Probably native in France; widely naturalized, except the far north and Finland. In the mountains of C & S Europe, however, a smaller version, subsp. *australis* (Link) Pamp. generally produces seed. **B**: Naturalized in woods and orchards, scattered localities mainly in C & S England and SE Scotland.

2* **Snakeshead Lily** *Fritillaria meleagris* L. Short bulbous perennial with a slender erect stem, sometimes growing in small groups. Leaves alternate, grey-green, linear, pointed, decreasing in size up the stem. Flowers solitary, nodding, lantern-shaped, purple or pink with conspicuous, purple chequering, or whitish with greenish markings, 30-45mm long. Fruit capsule erect, on elongated stems. Damp water meadows, open woodland at low altitudes, to 1200m. April-May. S Britain to Holland, France and Germany; naturalized in Scandinavia and extinct in Belgium. **B**: C & S England, introduced elsewhere; protected.

**Lilies** *Lilium*. Large genus of bulbous perennials right across the northern hemisphere. Plants have a fleshy, scaly, bulbous rootstock and whorled or alternate stem-leaves. The 6 tepals are all petal-like and very conspicuous as are the prominent stamens. The stigma is 3-lobed. Fruit a 3-lobed capsule, dehiscing to the base. 100 species primarily in northern temperate regions.

* **Martagon Lily** *Lilium martagon* L. Tall perennial with an erect stem dashed or spotted with red. Leaves mostly in whorls, except the uppermost, elliptical, dark green, mostly 7-9-veined, sometimes hairy on the veins beneath. Flowers 5-10 in a lax raceme, pink, spotted with purple, nodding, turk's cap-shaped with recurved tepals, 3-4cm; anthers orange-yellow or reddish-purple, prominent. Woods, scrub and mountain meadows, generally on calcareous soils, to 2800m. France, Germany, Belgium, Holland, and Scandinavia. Naturalized in Britain.**B**: Scattered localities in England and Wales north to Cumbria.

3* *Lilium bulbiferum* L. Medium erect perennial with green stems, with bulbils in the leaf axils. Leaves alternate, deep bright green, lanceolate to oval, 3-7-veined, hairy along the margin. Flowers 1-3, orange-red with small black spots, large erect broad trumpets, 8-10cm; anthers orange to reddish-brown. Woodland, scrub and rocky mountain slopes, to 2400m. June-July. E France & S Germany; naturalized in Norway and Sweden. A form without stem bulbils is called var. *croceum*.

4* **Pyrenean or Yellow Martagon Lily** *Lilium pyrenaicum* Gouan. Medium to tall erect perennial often forming small clumps. Leaves alternate, bright green, linear lanceolate, generally 3-veined, often rather crowded, especially below, the margin finely hairy. Flowers greenish-yellow with small purple lines and spots, small pendent turk's caps, 1-8 in a lax raceme, strong smelling; anthers prominent, reddish-brown. Woods, rocky places and occasionally along hedgerows. June-July. **I**: Pyrenees, where often on mountain meadows and rocky slopes. **B**: Widely naturalized in SW England, Wales, Ireland and C Scotland.

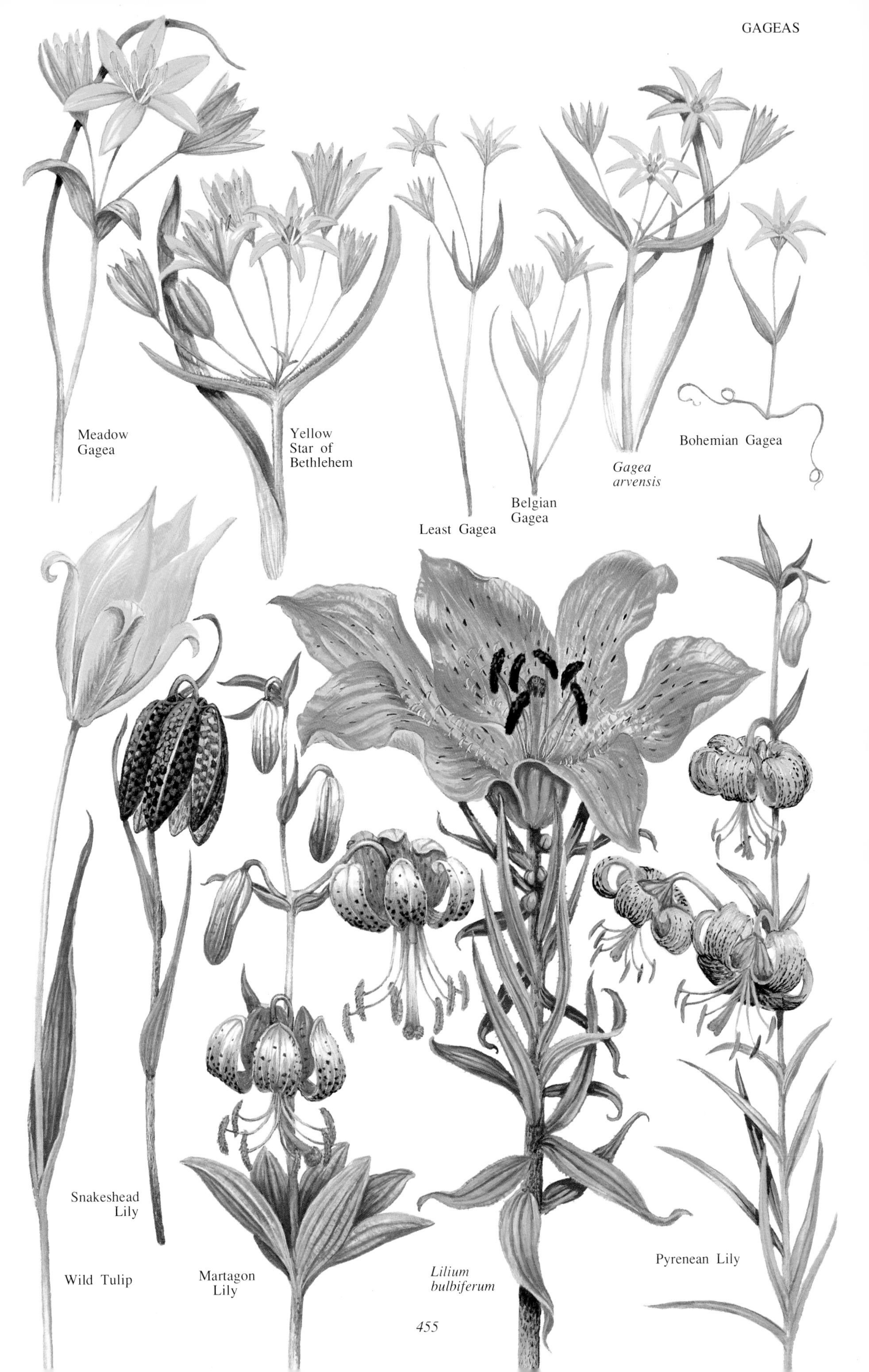
Meadow Gagea
Yellow Star of Bethlehem
Least Gagea
Belgian Gagea
*Gagea arvensis*
Bohemian Gagea
Snakeshead Lily
Wild Tulip
Martagon Lily
*Lilium bulbiferum*
Pyrenean Lily

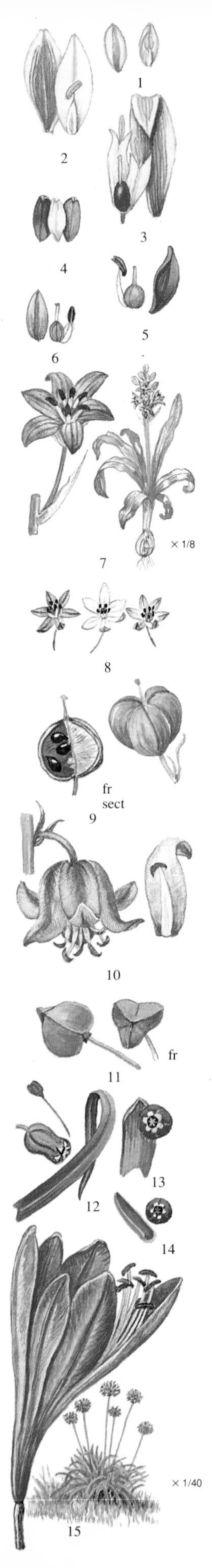


**Stars of Bethlehem** *Ornithogalum*. A genus of bulbous perennials with basal leaves and broad or spike-like racemes of starry flowers, white or greenish; flowers with bracts at base of the stalk. Tepals all similar 6, free. Stamens 6. Fruit a capsule. 100 species.

1* **Bath Asparagus** *Ornithogalum pyrenaicum* L. Medium to tall hairless perennial. Leaves linear, grey-green, often partly or wholly withered by flowering time. Flowers pale greenish-yellow or greenish-white, 18-24mm, many in a slender erect raceme, faintly fragrant; anthers pale yellow. Capsule on an erect stalk. Meadows, grassy banks, scrub and open woodland, to 1200m. May-July. S Britain, Belgium and France. The young flowering shoots are edible. **B**: Locally abundant in the south.

2* **Star of Bethlehem** *Ornithogalum umbellatum* L. Short hairless perennial. Leaves 6-9, linear, channelled and with a conspicuous white stripe along the groove, rather limp. Bracts whitish, up to the length of the flower-stalks. Flowers glistening white stars, 28-38mm, the tepals with a broad band of green on the outside, erect, borne in a pyramidal raceme; anthers yellow. Grassy and cultivated land, to 1600m. April-June. W Europe, naturalized in Ireland and Scandinavia. **B**: Naturalized; possibly only native in the east.

3* **Drooping Star of Bethlehem** *Ornithogalum nutans* L. Short to medium hairless perennial. Leaves grass-like, linear with a medium white stripe above. Bracts exceeding the flower-stalks. Flowers white, with a greenish stripe on the back of each tepal, 26-40mm, nodding bell-shaped, borne in a lax one-sided, spike-like raceme. Grassy habitats, banks and cultivated ground. April-May. **I**: E Europe. Widely naturalized except Finland and N Scandinavia. **B**: Often locally abundant.

4* **Alpine Squill** *Scilla bifolia* L. (incl. *S. nivalis* Boiss.). Low to short hairless perennial. Leaves usually 2, broadly linear, channelled, appearing with the flowers and sheathing the lower part of the scape. Flowers bright blue, occasionally lilac or white, 10-18mm, in lax short racemes; bracts generally absent. Meadows and woodland habitats and scrub, to 1500m. March-June. Belgium, France and Germany southwards; naturalized in Holland.

5* **Siberian Squill** *Scilla sibirica* Haw. Low hairless perennial. Leaves 2-4, oblong to broad-linear, wrapping around one another when young and only partly developed at flowering time. Flowers drooping, deep blue, bell-shaped, 9-14mm long, 1-3 on a common scape, but often several scapes per bulb. Woods and scrub. April-May. **I**: USSR, Turkey. Naturalized in Holland.

*Scilla amoena* L. Like *S. sibirica*, but leaves 3-7, well developed by flowering time. Flowers bright blue, erect, 9-12mm long, 3-6 to a scape. Cultivated, locally naturalized in France. Origin uncertain.

6* **Spring Squill** *Scilla verna* Hudson. Low to short hairless perennial. Leaves 2-7, narrow-linear, often rather curly, present before the flowers. Flowers violet-blue, rarely white, 10-16mm, in short racemes; bracts membranous, longer than the flower-stalks. Rocky and grassy places, to 2000m. April-June. Britain, Faeroes, France, Norway. **B**: W Britain and Northumberland northwards, E Ireland.

7 **Pyrenean Squill** *Scilla liliohyacinthus* L. Short hairless perennial. Leaves 6-10, broad, strap-shaped, shiny-green. Flowers bright blue, 16-20mm, relatively small in proportion to the leaves which are fully developed at flowering time, borne in short racemes, often more than one per bulb; bracts small and membranous. Woodland and damp grassland, to 2000m. May-June. C & S France.

8* **Autumnal Squill** *Scilla autumnalis* L. Low to short hairless perennial. Leaves 5-10, linear, grass-like, generally erect, appearing after the flowers. Flowers small, pinkish-blue to lilac, 6-10mm, in short racemes, elongating in fruit; bracts absent. Dry grassy and rocky habitats, in mountains to 2000m, or close to the sea. August-October. S England and France. **B**: S England & Channel Islands.

9* **Bluebell** *Scilla non-scripta* (L.) Hoffmanns & Link [*Hyacinthoides non-scripta*, *Endymion nutans*, *E. non-scriptus*]. Short hairless perennial. Leaves 3-6, linear to linear-lanceolate, keeled beneath, tip hooded. Flowers violet-blue, rarely pinkish or white, nodding tubular-bells, 14-20mm long, borne in a one-sided raceme, drooping towards the tip; tepals recurved near the apex; anthers cream; Each flower has a pair of bracts at the base of the flower-stalk. Woods, heaths, sometimes in mountains or on sea cliffs, to 1500m. April-June. W Europe; naturalized from gardens in Germany. **B**: Often dominant on lighter soils; not in Orkney or Shetland.

10 **Spanish or Garden Bluebell** *Scilla hispanica* Miller [*Hyacinthoides hispanica*, *Endymion hispanicus*]. A stouter, short to medium, plant than *S. non-scripta*, often forming large clumps. Leaves 4-8, linear-lanceolate to lanceolate, hooded at the tip. Flowers blue, broadly bell-shaped, nodding, 12-20mm long, borne in an erect raceme, not one-sided or drooping at the tip; anthers blue. Shady habitats, cultivated ground. May-June. **I**: Spain & Portugal. Naturalized from gardens in Britain and France.

11* **Tassel Hyacinth** *Muscari comosum* (L.) Miller [*Leopoldia comosa*]. Very variable short to medium perennial. Leaves 3-5, linear, channelled, rather floppy. Flowers pale brown with creamish or yellowish-brown teeth, 5-9.5mm long, the upper sterile flowers violet, smaller, often numerous in a conspicuous terminal tuft, occasionally white. Dry habitats, to 1300m. May-June. France and Germany, otherwise naturalized in W Europe. **B**: Naturalized in S Wales, occasionally casual elsewhere.

12 *Muscari tenuiflorum* Tausch [*Leopoldia tenuiflora*]. Like *M. comosum*, but with 3-7 narrower leaves. Flowers fewer generally, pale greyish-brown with pale cream and blackish recurved teeth; sterile flowers bright violet. Grassy habitats to 500m. June-July. S Germany.

13* **Small Grape Hyacinth** *Muscari botryoides* (L.) Miller. Short hairless perennial. Leaves 2-3, linear-oblong, the upper surface paler grey-green, broader at the hooded tip, often prominently ribbed. Flowers bright blue with white teeth, 3.5-5mm long, almost round, in a fairly dense spike-like raceme; sterile flowers smaller and paler. Grassy habitats, open woodland, to 2000m. March-May. France and Germany; naturalized in Belgium and Holland.

14* **Common Grape Hyacinth** *Muscari neglectum* Guss. ex Ten. [*Hyacinthus racemosus*, *Muscari racemosum*, *M. altanticum*]. Very variable short hairless perennial. Leaves 3-6, linear to linear-lanceolate, bright green, often reddish at the base, channelled. Flowers dark blackish-blue with white recurved teeth, 3.5-7.5mm long, borne in a fairly dense spike-like raceme; sterile flowers paler, bluish. Dry grassland and cultivated ground, to 1600m, occasionally rubbish tips. March-May. France and Germany. **B**: Rare, in East Anglia and Oxfordshire.

15 **Agapanthus** *Agapanthus praecox* Willd. Stout, medium to tall, tufted rhizomatous perennial, hairless. Leaves all basal, strap-shaped, 30-50mm wide, bright green. Flowers blue, occasionally white, flared trumpets, 40-45mm long, borne in a large umbel on a stout scape; spathe enclosing umbels 2-valved, soon falling. **I**: South Africa. Naturalized in rocky places in the Isles of Scilly. July-August.

× 1/5
Star of Bethlehem
Bath Asparagus
Drooping
Star of Bethlehem
Alpine Squill
Siberian
Squill
Spring
Squill
Autumnal
Squill
Bluebell
Tassel
Hyacinth
Small
Grape Hyacinth
Common Grape
Hyacinth

**Onions and Garlic** *Allium*. Bulbous perennials, the bulbs solitary or clustered, scaly. Leaves very variable from linear to oval, flat or circular in section, sheathing at the base. Flowers in distinctive umbels, enclosed in bud in a membranous spathe. Tepals 6, all similar and petal-like, free or fused together just at the base, spreading or not, persisting. Fruit a small capsule. Sometimes the flowers in the umbel are wholly, or partly, replaced by bulbils which detach themselves and form new plants. A large and complex genus with more than 500 species widespread in temperate regions of the northern hemisphere as well as in Ethiopia and Mexico. All parts of the plant have a distinctive strong smell of onion or garlic. Sometimes the flowers in the umbel are wholly or partly replaced by bulbils which detach themselves and form new plants. The genus includes a number of important crops including the onion, leek, shallot and garlic. Other are widely cultivated for their ornamental value.

1* *Allium angulosum* L. A short to medium tufted perennial, with bulbs aggregated on a short horizontal rhizome. Leaves 4-6, linear, prominently keeled beneath. Scape angled, 2-edged above. Spathes 2-5-lobed, persistent. Flowers pale purple, cup-shaped, 4-6mm long, numerous in a semi-rounded umbel; stamens slightly protruding. Damp meadows near rivers, generally subject to seasonal flooding. June-July. E France and S Germany. **P**: Bees and butterflies.

2* **German Garlic** *Allium senescens* L. subsp. *montanum* (Fries) J. Holub. Rather similar to *A. angulosum*, but leaves 4-9, almost flat, not keeled beneath, often twisted. Flowers slightly larger, lilac, with the stamens well protruding. Dry, usually rocky habitats. June-July. Central France, Germany and S Sweden; naturalized in Norway, extinct in Denmark. Looks very similar to *A. angulosum* but can be clearly separated by the leaf characters and their very different habitats.

3* *Allium suaveolens* Jacq. Medium tufted perennial with bulbs closely aggregated on a short rhizome. Leaves 2-5, linear, flat, keeled beneath, sheathing the lower third of the rounded scape. Spathe 2-valved, persistent, equalling the flowers. Flowers pale pink or white, generally with a pink midvein, cup-shaped, 4-5mm long, in a dense rounded umbel; stamens well protruding. Damp meadows and moors. June-July. France and Germany.

4* **Chives** *Allium schoenoprasum* L. Very variable, short to medium, tufted perennial. Leaves 1-2, linear-cylindrical, hollow, grey-green. Scape hollow, erect. Spathes papery, 2-3-lobed, equalling or shorter than the umbel, persistent. Flowers lilac or pale purple, with a deep mid-vein along each pointed tepal, narrow bell-shaped, 7-15mm long, in a small dense umbel; stamens not protruding. Rocky and grassy habitats, streamside, damp mountain rocks, primarily on the mountains in the south of its range, to 2600m. June-August. Throughout, except the Faeroes, Iceland and Spitsbergen. Cultivated as an edible herb, and thus frequently naturalized.

5 **Cultivated Onion** *Allium cepa* L. Very variable medium to tall biennial with bulbs up to 10cm in diameter, occasionally more. Leaves stout, hollow, tapering gradually from a somewhat swollen base, basal in the first year but borne on the lower part of the stem in the second. Scape stout, hollow, conspicuously swollen in the lower part. Flowers white, the tepals with a greenish midvein, starry, 5-8mm, forming large dense umbels with protruding stamens, ovary whitish. Widely cultivated and often naturalized throughout Europe. July-September. The origin of the cultivated onion is obscured by its long history in cultivation. Numerous forms and cultivars exist including both perennial forms and those in which the flowers in the umbel are replaced by prominent bulbils. It is possible that the cultivated onion was derived from a central Asian species, *Allium oschaninii* B. Fedtsch.

6* **Welsh Onion** *Allium fistulosum* L. Rather stout medium to tall perennial, often forming tufts. Leaves cylindrical, hollow, sheathing the lower third of the stem. Scape hollow, rounded, swollen near the middle. Spathe 1-2-valved, more or less equalling the umbel. Flowers creamy-white, cup-shaped, 6-7mm long, in a very dense rounded umbel with stamens well protruded; anthers yellow. Long cultivated: its wild status, like that of the *A. cepa*, uncertain. Widely naturalized, sometimes a casual garden escape persisting for only a few years. June-September. Probably throughout. In SC Norway it has become naturalized on the turf roofs of local dwellings.

7* *Allium victorialis* L. Medium rhizomatous perennial, solitary or clustered. Leaves 2-3, lanceolate to elliptical, short-stalked, deep green, sheathing the lower third of the stem. Scape 2-edged in the lower part. Spathes shorter than the umbels. Flowers greenish-white, often rather dull, starry, 8-10mm, in dense sub-rounded umbels with protruding stamens; anthers yellow. Grassy habitats and woodland margin in the mountains, 1400-2600m. Germany from the Vosges southwards, C & E France.

8* **Rosy Garlic** *Allium roseum* L. Short to medium perennial, solitary or several together; bulbs generally with numerous bulbils. Leaves 2-4, linear, slightly grooved, with a rough margin, sheathing the lower quarter of the stem. Scape rounded in section. Spathes 3-4-lobed, shorter than the umbels, persistent. Flowers rose-pink to whitish, bell- or cup-shaped, 10-12mm long, up to 30 in an umbel, the stamens not protruding. Dry grassy and rocky places, open habitats and cultivated ground, generally at low altitudes. June-July. C & S France; naturalized in Britain. In some forms some of the flowers in the umbel are replaced by small bulbils, subsp. *bulbiferum* (DC.) E.F. Warb. **B**: Naturalized in scattered localities.

9* **Triquetrous Leek** *Allium triquetrum* L. Short to medium tufted perennial. Leaves 2-3, floppy, triangular in section like the scape, mostly basal. Spathe 2-valved, equalling the umbel. Flowers white, bell-shaped, 10-18mm long, drooping in a one-sided umbel, the pointed tepals with a green midvein on the back; stamens not protruding. Woods, scrub, streambanks, hedgerows and beneath walls, generally in damp, rather shady habitats. April-June. **I**: W Mediterranean region. Naturalized from cultivation in SW Britain and Ireland. **B**: Naturalized in SW Britain, S Wales, S Ireland and the Channel Islands.

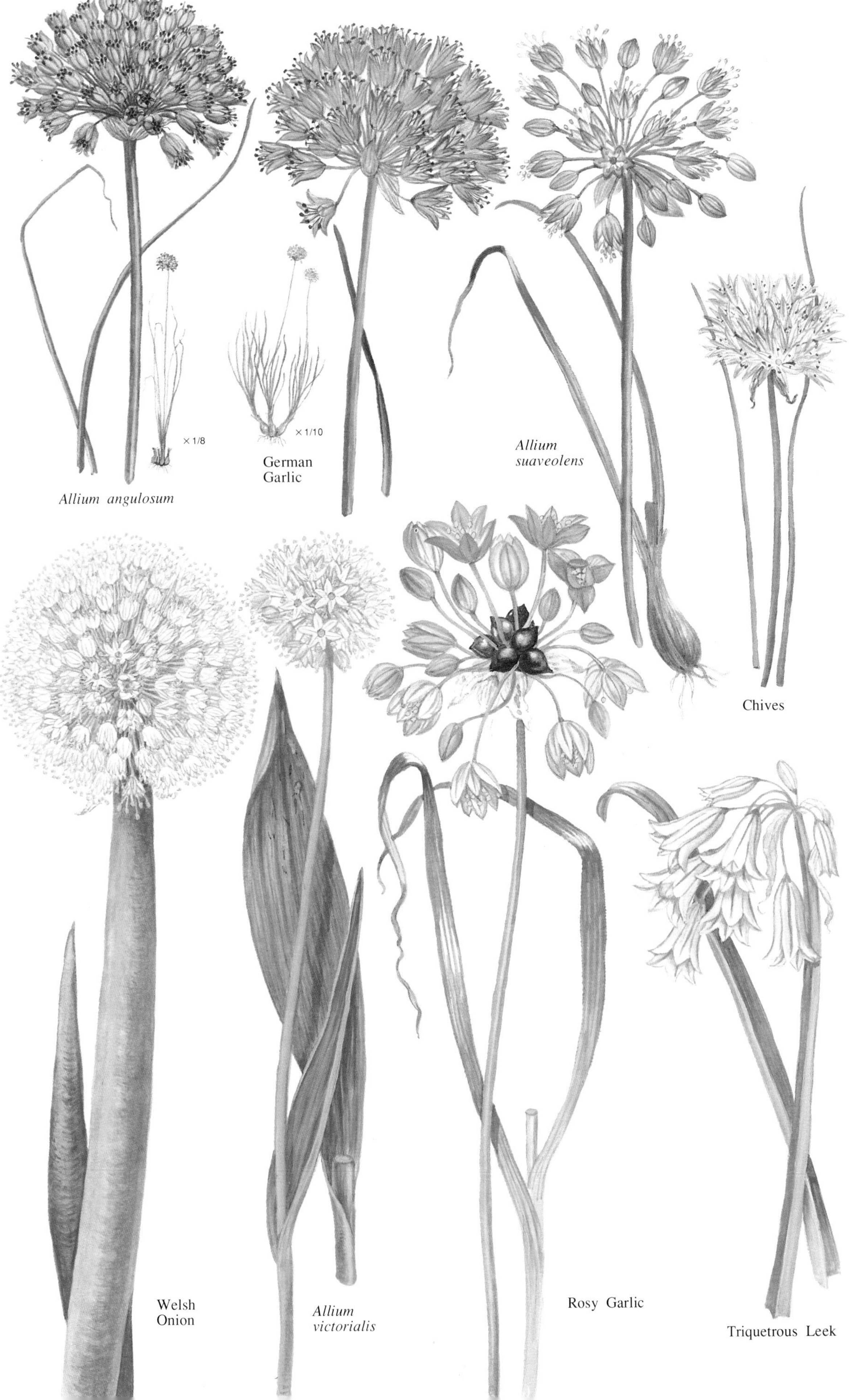

*Allium angulosum*

× 1/8

German Garlic

× 1/10

*Allium suaveolens*

Chives

Welsh Onion

*Allium victorialis*

Rosy Garlic

Triquetrous Leek

1* **Few-flowered Leek** *Allium paradoxum* (Bieb.) D. Don. Short perennial, not rhizomatous but often forming large colonies. Leaf solitary, strap-shaped, to 20mm wide. Flowers pearly-white, bell-shaped, 10-12mm long, relatively few in an umbel on stalks of varying lengths, often replaced by small rounded bulbils; tepals with a faint greenish stripe on the outside. Capsule rarely produced. Waste places and cultivated ground. April-May. **I**: Caucasus and N Iran. Naturalized in Britain, Holland, Denmark and Germany. Can become an invasive weed. **B**: Naturalized in a number of places.

2* **Ramsons or Wild Garlic** *Allium ursinum* L. Short to medium perennial, often forming extensive colonies. Leaves 2-3, flat, narrow to broadly elliptical, bright green, smelling powerfully of garlic when bruised. Scape 2-3-angled. Spathes shorter than the umbel. Flowers white, starry, 12-20mm, in a fairly dense rounded umbel; tepals pointed, longer than the stamens. Woods, scrub, hedgerows and shady banks, to 1900m. April-June. Europe north to S Scandinavia. **B**: Throughout, except some Scottish Islands.

3* *Allium paniculatum* L. Medium to tall perennial, solitary or clustered. Leaves 3-5, linear, about 2mm wide, bright green, sheathing the lower half of the stem, ribbed beneath. Spathes 2-valved, unequal, greatly exceeding the umbel. Flowers white or pale lilac, bell-shaped, 4.5-7mm long, in a fairly dense rounded umbel; stamens not protruding, anthers yellow. Rocky and grassy habitats. May-June. NC France southwards.

4* **Field Garlic** *Allium oleraceum* L. Medium to tall perennial with solitary or clustered bulbs. Leaves 3-4, linear to thread-like, hollow and circular below, but channelled above, prominently ribbed, sheathing at least the lower half of the stem. Spathe 2-valved, greatly exceeding the umbel, with long tails, up to 20cm. Flowers whitish, tinged with green, pink or brown, often rather dingy, bell-shaped, 6-7mm long, in an uneven umbel with the outer flowers drooping down on slender stalks; stamens not protruding. Rocky ground, scrub, roadsides and cultivated ground. June-August. Throughout, except Iceland, Faeroes and Spitsbergen; naturalized locally in Ireland. Cultivated in gardens but often becoming a weed. In some forms the flowers are partly or wholly replaced in the umbel by small bulbils. **B**: Throughout, north to Moray; Ireland – restricted to E.

5* **Keeled Garlic** *Allium carinatum* L. Medium perennial, bulb solitary or grouped. Leaves 2-4, linear, slightly channelled above, ribbed beneath, sheathing the lower half of the stem. Spathes 2-valved, uneven with tails much longer than the umbel. Flowers purple or pinkish-purple, cup-shaped, 4-6mm long, in an uneven lax umbel with the outer flowers drooping on slender stalks, the flowers often partly or wholly replaced by bulbils; stamens with purple anthers, protruding prominently. Capsule rarely produced. Meadows, open woods, heathland. July-August. France, Germany and Denmark; naturalized in Britain, Belgium and Holland. **B**: Scattered localities in Britain and NE Ireland.

6 **Garlic** *Allium sativum* L. Medium to tall perennial. Leaves 6-12, linear, flat but keeled beneath, sheathing the lower half of the stem. Spathe 1-valved with a very long beak, up to 25cm, falling early. Flowers white, pink, occasionally purplish, cup-shaped, 3-5mm long, few to an umbel and often failing to develop properly and quickly withering. Capsule rarely produced. Grassy and rocky habitats, cultivated land. July-August. Widely cultivated, except in the north, and sometimes an escape from cultivation, but often not persisting for very long. The cultivated garlic is probably derived from an Asian species *A. longicuspis* Regel.

7* **Wild Leek** *Allium ampeloprasum* L. Tall, often stout perennial, to 1.8m. Leaves 4-10, grey-green, linear, flat channelled above and keeled beneath, 5-40mm wide. Spathe papery, 1-valved, soon falling as the flowers emerge. Flowers whitish to pale purple, bell-shaped, 4-5.5mm long, very many, up to 500, in a large dense rounded umbel, 5-9cm across; stamens slightly protruding. Disturbed ground, rocky places, especially near the sea, and hedgerows. July-August. France, rare in Britain, often becoming naturalized or casual on rubbish heaps. **B**: Rather rare; scattered localities in S Wales, Somerset – Steep Holm, Cornwall and Guernsey. var. *babingtonii* (Borrer) Syme [*A. babingtonii*] has few flowers and bulbils, 8-15mm long, in the umbel. W Ireland and SW Britain, mostly near or along the coast. The flower-heads sometimes produce a secondary umbel. var. *bulbiferum* Syme. Similar but the bulbils smaller, 6-8mm. W France, Channel Islands.

8 **Cultivated Leek** *Allium porrum* L. [*A. ampeloprasum* var. *porrum*] The result of centuries of cultivation and selection, but was derived in the first instance from *A. ampeloprasum*. Widely cultivated and occasionally naturalized throughout much of the area.

9* **Sand Leek** *Allium scorodoprasum* L. Medium to tall perennial. Leaves 2-5, linear, 5-20mm wide, flat, sometimes channelled, rough-edged, narrowed at the base, sheathing the lower half of the stem. Spathe short-beaked, shorter than the umbel and falling early. Flowers pale lilac to deep purple, bell-shaped, 5-8mm, few in an umbel with stalks of varying lengths, stamens not protruding, often partly or wholly replaced by purplish bulbils. Apparently sterile and not producing a capsule. Sandy and grassy habitats, hedgebanks, waste and cultivated land. June-August. Most of Europe except the far north; naturalized in Ireland and Belgium. **B**: Local from Lincolnshire and Cheshire northwards; naturalized in scattered localities in Ireland.

10* subsp. *rotundum* (L.) Stearn [*Allium rotundum*]. Has leaves only 2-10mm wide and umbel many-flowered, without bulbils; flowers dark purple, the inner tepals paler with a whitish margin. C & S Germany. Widespread in S Europe.

11* **Round-headed Leek** *Allium sphaerocephalon* L. Variable short to tall slender perennial. Leaves 2-6, linear, hollow, grooved above, usually sheathing the lower half of the stem. Spathe usually 2-valved, short-beaked, persistent. Flowers reddish-purple or pink, bell-shaped, 3.5-5.5mm long, packed into a dense rounded umbel; stamens long-protruding. Grassy and rocky habitats, to 2650m. June-August. Belgium and Germany southwards. Widely cultivated. **B**: Rare near Bristol and on Jersey.

12* **Crow Garlic** *Allium vineale* L. Medium to tall perennial, to 1.2m. Leaves 2-4, subcylindrical, grooved, hollow, sheathing the lower half of the stem. Spathe 1-valved, beaked, shorter than the umbel, soon falling. Flowers pink to red or greenish-white, bell-shaped, 2-4.5mm long, long-stalked in a lax umbel, often mixed with bulbils; stamens generally more or less protruding. Dry grassy or rocky habitats, cultivated and waste ground or road-verges, to 1900m. June-August. Throughout, except the far north. Sometimes a serious weed. **B**: Widespread in England and Wales, local elsewhere. var. *compactum* (Thuill.) Syme. The tight umbels consist only of bulbils. Throughout the range of the species.

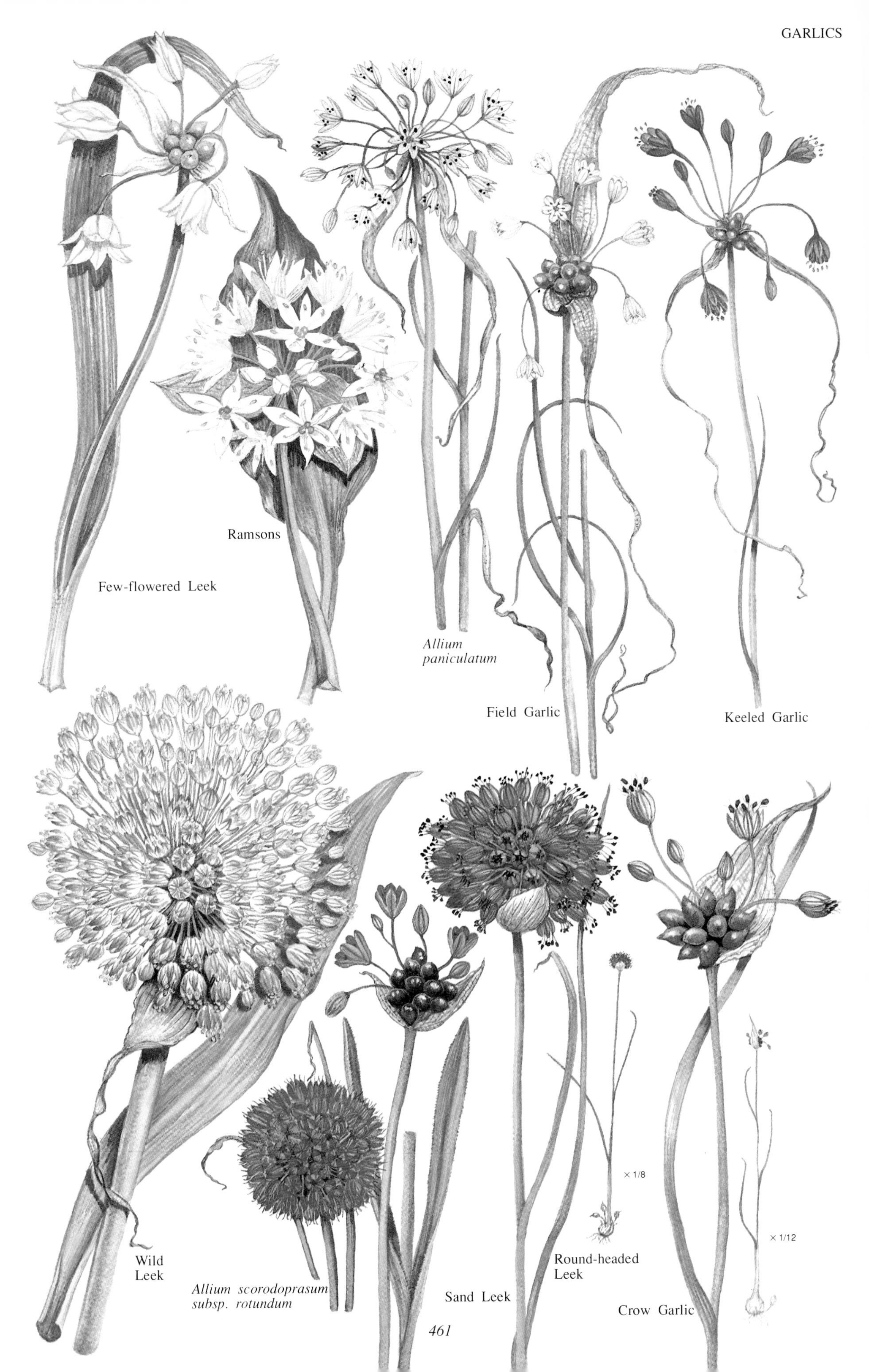
Few-flowered Leek
Ramsons
Allium paniculatum
Field Garlic
Keeled Garlic
Wild Leek
Allium scorodoprasum subsp. rotundum
Sand Leek
Round-headed Leek
× 1/8
Crow Garlic
× 1/12

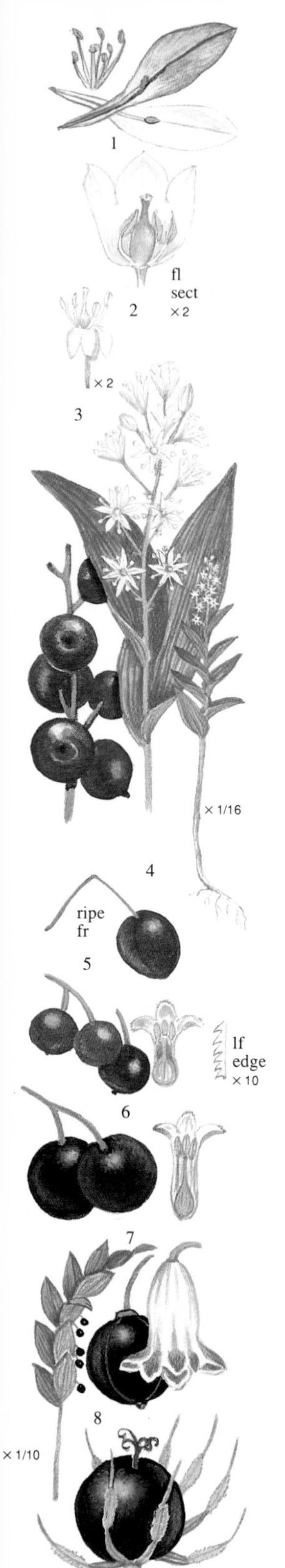


1* **Ipheion** *Ipheion uniflorum* (R.C. Graham) Rafin. Low to short, tufted, bulbous perennial smelling of garlic. Leaves linear grey-green, blunt and rather floppy. Spathes papery. Flowers solitary, on leafless stalks, white or violet-blue, 18-22mm, funnel-shaped, the tepals with a deeper midvein on the outside, united below into a short tube and containing the stamens and styles. Grassy banks and hedgerows. May-June, sometimes later. **I**: S. America. Naturalized in Britain and France. The leaves often begin to appear in the late autumn. **B**: Mainly in S, local.

2* **Lily of the Valley** *Convallaria majalis* L. Low to short rhizomatous, hairless perennial, the stem with green or violet sheathing scales at the base. Leaves deep dull green, 2 or occasionally 3, elliptical, untoothed, almost opposite, with close parallel veins. Flowers small rounded bells, white sometimes tinged with pink, 5-9mm, drooping in a slender one-sided raceme, very fragrant. Fruit a small red berry. Woods, scrub, mountain meadows to 2300m. May-June. Throughout, except the far north and Ireland. **B**: Widespread in England and Wales, especially in the east; local in Scotland.

3* **May Lily** *Maianthemum bifolium* (L.) F.W. Schmidt. Low to short creeping, rhizomatous, hairless perennial, carpeting the ground; stem with 2 scale leaves at the base. Leaves 2, alternate, heart-shaped, pointed, shiny-green, untoothed. Flowers white, starry, 4-6mm, in a short raceme, with 4 tepals and 4 stamens, fragrant. Fruit a small red berry, 4-5mm. Shady habitats in moist, slightly acid, humus-rich soils, to 2100m. May-July. Throughout except the far north. **B**: Rare in S England and S Wales.

4 **Smilacina** *Smilacina stellata* (L.) Desf. Medium rhizomatous perennial with erect leafy, unbranched stems. Leaves bright green, heart-shaped, unstalked and clasping the stem, alternate, with prominent parallel veins. Flowers white, small, 8-9mm, starry, in a lax terminal raceme; tepals 6, narrow, more or less separate. Fruit a dark red berry, with the remains of the style still attached. Shady rocky and wooded places. June-July. **I** North America. Naturalized locally in S Norway and S Sweden.

5* **Streptopus** *Streptopus amplexifolius* (L.) DC. Medium to tall rhizomatous perennial with leafy stems, branched. Leaves alternate, oblong-heart-shaped, unstalked and clasping the stem, decreasing in size up the stem. Flowers nodding, greenish-white, bell-shaped, 8-10mm long, solitary or 2 on slender stalks bent in the middle; tepals 6, pointed, fused near the base. Fruit a small red berry. Damp woods and rocks, sometimes in hilly and mountain habitats, to 2300m. June-July. France and Germany southwards.

**Solomon's Seals** *Polygonatum*. Rhizomatous perennials with erect stems bearing alternate or whorled leaves, usually unstalked and untoothed. Flowers nodding, solitary or several together at the leaf-axils, tubular with 6 short lobes, the stamens not protruding. Fruit a small red or blackish berry, several seeded. 20 species.

6* **Whorled Solomon's Seal** *Polygonatum verticillatum* (L.) All. Medium to tall patch-forming perennial; stem angled, hairless, Leaves mostly in whorls of 3-8, linear to lanceolate, minutely downy beneath. Flowers greenish-white, bell-shaped, 5-10mm long, solitary or 2-3 together, unscented. Berry red at first but becoming dark purple. Woods, scrub and rocky habitats, generally in hilly or mountain districts, to 2400m. Most of Europe. Extremely variable. **B**: Rather rare, from Northumberland northwards.

7* **Common Solomon's Seal** *Polygonatum multiflorum* (L.) All. Medium to tall hairless perennial, patch-forming; stems arching, rounded and smooth. Leaves elliptical to lanceolate. Flowers white with a greenish tip, 9-20mm long, the lower in clusters of 2-6, the uppermost paired or solitary, unscented, somewhat constricted in the middle. Berry bluish-black. Wooded and scrubby habitats, generally on calcareous soils, to 2200m. May-June. Throughout, except Ireland and the far north. **B**: Scattered localities; naturalized in Scotland. The garden plant a hybrid between *P. multiflorum* and *P. odoratum*.

8 **Angular Solomon's Seal** *Polygonatum odoratum* (Miller) Druce [*P. officinale*, *P. pruinosum*]. Rather like *P. multiflorum*, but generally a shorter plant, stems angled. Flowers solitary or paired, scented, not constricted in the middle. Woods and rocky habitats, generally on calcareous soils. May-June. Europe north to C Scandinavia. **B**: Local in W England and Wales; also on the Inner Hebrides.

9* **Herb Paris** *Paris quadrifolia* L. Low to short hairless, often patch-forming perennial with erect stems. Leaves 4 in a single whorl half-way up to stem, broadly-oval, 3-5-veined. Flowers solitary, erect yellowish-green, starry, 4-6-parted, the sepals lanceolate, pointed, the petals thread-like; stamens prominent. Fruit a 'berry-like' capsule, black when ripe, splitting to reveal bright red, fleshy, shiny seeds. Woodland and damp shady habitats on calcareous soils, to 2000m. May-June. Throughout, except Ireland and the far north. Berries poisonous. Flowers hermaphrodite, sometimes female. Often indicative of ancient woodland. **B**: Local north to Caithness, especially in the east; also Inner Hebrides.

10* **Wild Asparagus** *Asparagus officinalis* L. Medium to tall perennial, to 1.2m, hairless, with smooth, erect or ascending, much-branched stems. Leaves needle-like (actually reduced stems or cladodes) forming a feathery mass, bright green, linear, in clusters of 4-15, each slightly flattened and 10-25mm long. Flowers small, greenish-white, bell-shaped, 4.5-6.5mm long, solitary or paired at the nodes, but generally not mixed with the 'leaves'; tepals 6 fused, together in the lower half; male and female flowers on separate plants. Fruit a bright red berry, 6-10mm. Scrub, grassy and waste habitats, cultivated ground, at low altitudes. June-August. North to S Denmark and S Sweden; naturalized in Norway and Finland. Young shoots edible.

11 subsp. *prostratus* (Dumort) Corb. Like the type, but stems prostrate or spreading, to 30cm. Needle-leaves smaller, 5-10mm. Flowers often mixed with the 'leaf' clusters. Maritime sands and coastal rocks. SE Ireland, Britain and from France to NW Germany.

12 *Asparagus tenuifolius* Lam. Like *A. officinalis*, but 'leaves' in clusters of 15-40 and the flowers generally larger, 6-8mm long. Berry 10-16mm. Scrub and rocky places. June-July. C & NW France, very local.

13* **Butcher's Broom** *Ruscus aculeatus* L. Short to tall, evergreen, rhizomatous, tough bushy perennial with erect stems, often forming dense thickets. Leaves (actually flattened branches or cladodes) alternate, oval to lanceolate, spine-tipped, leathery, deep green but becoming brown eventually. Flowers small, dull green with purple spots, 4-5mm, solitary or paired and borne upon the upper surface of the 'leaves', male and female flowers usually on separate plants; tepals 6, starry. Fruit a red berry, 10-15mm. Woods, hedge-banks, sometimes sea-cliffs, calcareous soils, at low altitudes. January-April. Britain and France. **B**: S England N to Caernarvon and Norfolk, sometimes planted.

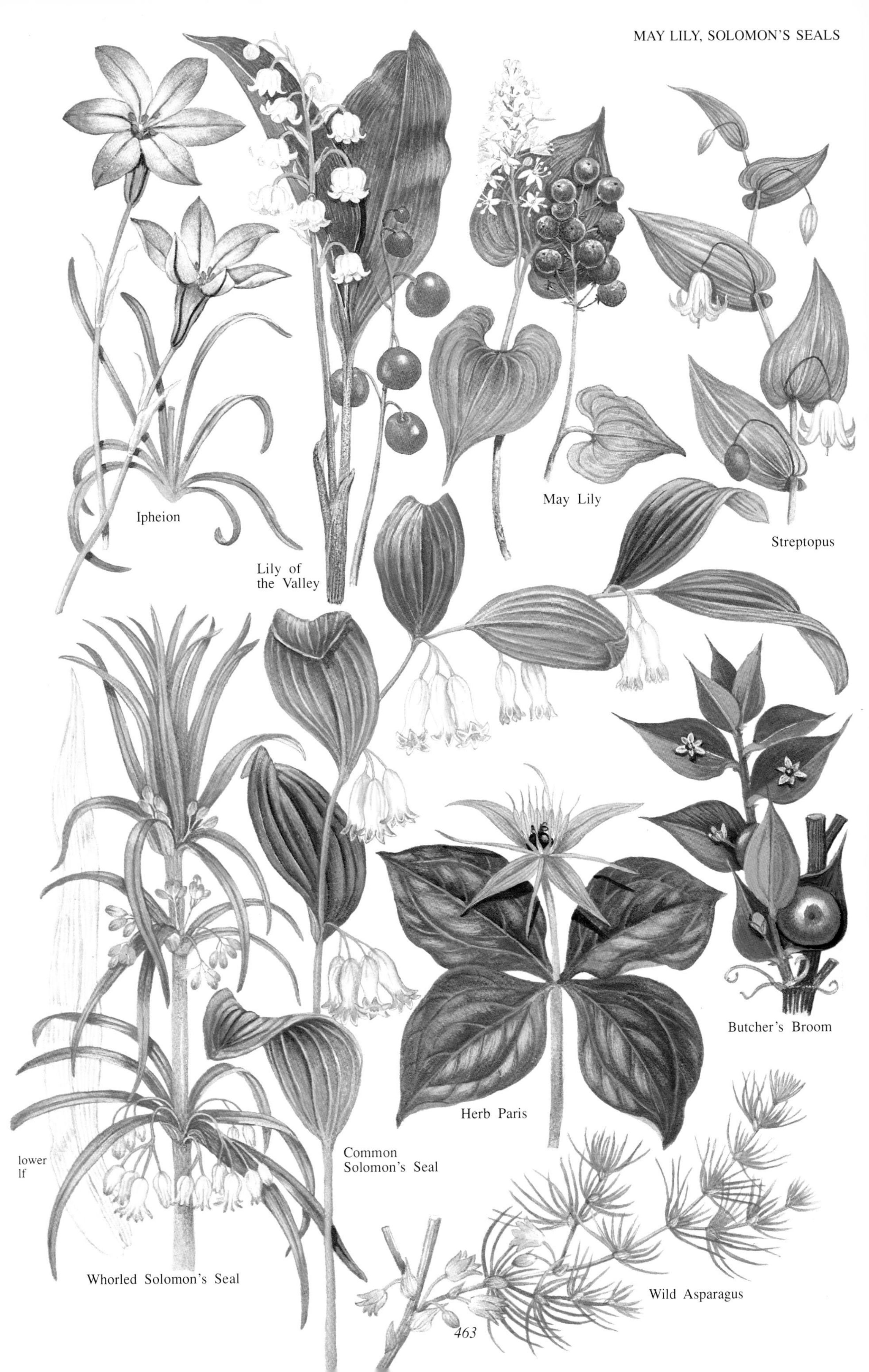
Ipheion
Lily of the Valley
May Lily
Streptopus
Butcher's Broom
Herb Paris
lower lf
Common Solomon's Seal
Whorled Solomon's Seal
Wild Asparagus

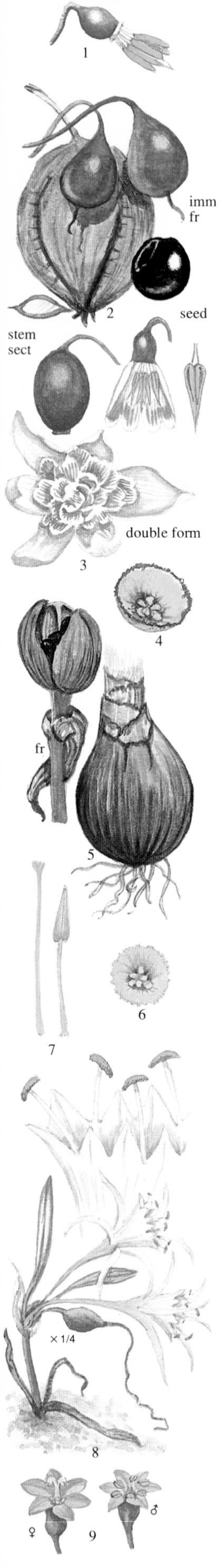

# DAFFODIL FAMILY Amaryllidaceae

Hairless, bulbous perennials with basal leaves and leafless flower-stems (scapes). Flowers subtended by 1-2-valved papery spathes enveloping the buds, solitary or a number in an umbel. Flowers hermaphrodite, regular or somewhat asymmetric, with 6 petaloid segments or tepals, the outer 3 similar or dissimilar to the inner 3; stamens 6; style solitary. The ovary is inferior (superior in Liliaceae). Fruit a 3-valved capsule. In *Narcissus* and *Pancratium* a conspicuous cup, the corona, is present between the tepals and the stamens. 500 species in 60 genera native to temperate and tropical regions of the world.

1* **Spring Snowflake** *Leucojum vernum* L. Short perennial. Leaves bright green, 2-3, strap-shaped, 5-15mm wide, generally half-developed at flowering time. Scape narrowly winged on each side, longer than the leaves. Flowers white, nodding bells, 15-25mm long, solitary or paired, the tepals all alike and with a green spot near the thickened tip; anthers orange. Damp woods, copses, meadows, occasionally in hedgebanks, to 1600m. February-March. Belgium, France and Germany; naturalized in Britain, Holland and Denmark. Widely cultivated, becoming locally naturalized. **B**: Rare, restricted to isolated localities in Dorset and Somerset.

2* **Summer Snowflake** *Leucojum aestivum* L. Rather robust medium perennial, often forming clumps. Leaves bright green, strap-shaped, 7-20mm wide, equally or slightly shorter than the scape, which is 2-winged. Flowers white, nodding, bell-shaped, 13-22mm long, in lax umbels of 2-6 on slender stalks of varying lengths, the tepals with a green spot near the swollen tip. Marshes and wet meadows, stream and riverbanks, to 1300m. April-June. Throughout except Scandinavia and Iceland; naturalized in Denmark. Widely cultivated. After flowering the stalks assume a more or less vertical position as the fruits develop. **B**: Native in the South, north to Oxford and in scattered localities in Ireland, sometimes a garden escape; often among willows.

3* **Snowdrop** *Galanthus nivalis* L. Variable low to short perennial. Leaves usually 2, grey-green, linear, 2.5-7mm wide, held flat against each other in bud and only partly developed at flowering time. Flowers white, solitary, nodding, 12-25mm long, the inner tepals quite different from the outer, shorter and with a greenish marking near the notched end; anthers green. Damp woodland habitats, copses, hedgerows, meadows and sometimes streambanks, to 1600m. January-March. France and Germany; widely naturalized in Britain, Holland, Norway and Sweden. The double-flowered form 'Flore Plena' is occasionally found naturalized. **B**: Scattered localities north to Moray, often locally abundant.

**Narcissi and Daffodils** *Narcissus*. Flowers solitary or 2-3 together on a common scape. Tepals all similar, petaloid. Corona present, cup- or trumpet-shaped, free from and with the stamens inside.

4* **Poet's Narcissus** *Narcissus poeticus* L. Medium perennial. Leaves grey-green, linear, 6-10mm wide, slightly grooved. Flowers solitary, half-nodding, white, 40-50mm, with a small cup-shaped corona, yellow with a crisped red or brownish rim, sweetly scented. Moist meadows, particularly in mountain regions, to 2300m. April-June. C & E France; naturalized in Britain, Belgium and Germany. Garden forms with flatter tepals are sometimes naturalized.

*Narcissus × incomparabilis* Mill. A hybrid between *N. pseudonarcissus* and *N. poeticus*. Grown in gardens and sometimes naturalized. Fields and waste places. April-May. C France southwards; locally naturalized elsewhere, including Britain.

5* **Wild Daffodil or Lent Lily** *Narcissus pseudonarcissus* L. Very variable short to medium perennial, often clump-forming. Leaves usually greyish-green, linear, 6-12mm wide. Flowers solitary, horizontal or nodding, cream to pale yellow, 20-35mm long, with a large trumpet-shaped, deep yellow corona, as long as the tepals. Meadows and open deciduous woodlands, waysides and waste ground, from lowland valleys to hills and mountains, to 2200m. Britain, France, Holland and Germany; widely cultivated and naturalized elsewhere. Cultivated varieties are sometimes naturalized; the true wild species is generally a slighter plant with half-nodding flowers.

6* **Tenby Daffodil** *N. obvallaris* Salisb. [*N. pseudonarcissus* subsp. *obvallaris*]. Probably a hybrid between *N. pseudonarcissus* subspecies and only known from SW Wales, though it is cultivated in gardens.

7* **Primrose Peerless** *Narcissus × medioluteus* Miller [*N. biflorus*] *N. poeticus × N. tazetta*. Medium, generally tufted perennial. Leaves grey-green, flat, 7-10mm wide. Flowers generally paired, creamy-white, 35-40mm, with a small bright yellow, cup-shaped corona, sweetly scented. Meadows and hedgerows. April-May. Widely naturalized in Britain and possibly elsewhere. Cultivated in gardens.

8* **Sea Daffodil** *Pancratium maritimum* L. Short to medium, stout-bulbous perennial. Leaves grey-green, 12-20mm wide, daffodil-like, appearing before the sweetly fragrant flowers, generally recurved. Scape thick but somewhat flattened. Flowers white, 100-120mm long with narrow-lanceolate, pointed tepals and a large trumpet-shaped white corona with the stamens attached to the toothed rim, about two thirds the length of the tepals. Maritime sands, occasionally rocks, local. July-October. W France.

# YAM FAMILY Dioscoreaceae

Tuberous-rooted perennnials; generally with alternate leaves. Flowers small in clusters or spikes, generally 6-lobed, male and female on separate plants. A primarily tropical and subtropical family with about 170 species in 9 genera, including the important tropical yams, whose fleshy rootstocks are a staple food in a numnber of countries.

9* **Black Bryony** *Tamus communis* L. Twining hairless perennial to 4m. Leaves heart-shaped, bright shiny-green, stalked. Flowers greenish-yellow, tiny, 3-6mm, the female in small clusters, the male in long slender racemes. Fruit a fleshy, bright shiny-red berry. Woods, scrub and hedgerows, fences, at low altitudes. May-August. Britain, Belgium, France and Germany; naturalized in Ireland. The poisonous berries often persist into late autumn and early winter.

Spring Snowflake
Snowdrop
Summer Snowflake
Poet's Narcissus
Wild Daffodil
♂
Tenby
Daffodil
Black Bryony
♀
Primrose
Peerless
Sea Daffodil

# IRIS FAMILY Iridaceae

Bulbous, cormous or rhizomatous plants. Leaves linear to sword-shaped, basal or alternate on the stem. Flowers solitary, clustered or in spikes. Perianth of 6 petaloid segments (tepals), often united below into a long tube; stamens 3 only; styles 3-parted. Fruit a 3-parted capsule. 1000 species in 70 genera.

**Irises** *Iris*. Rhizomatous or bulbous perennials, generally with a well developed stem. Leaves mostly basal, alternate. Flowers usually large, several together enclosed by leaf-like spathes. Outer tepals (falls) horizontal or drooping, generally with a large limb, inner tepals (standards) smaller and usually erect, all petaloid. Style arms petaloid, arching over the falls and each shielding a single stamen. About 200 species in temperate regions.

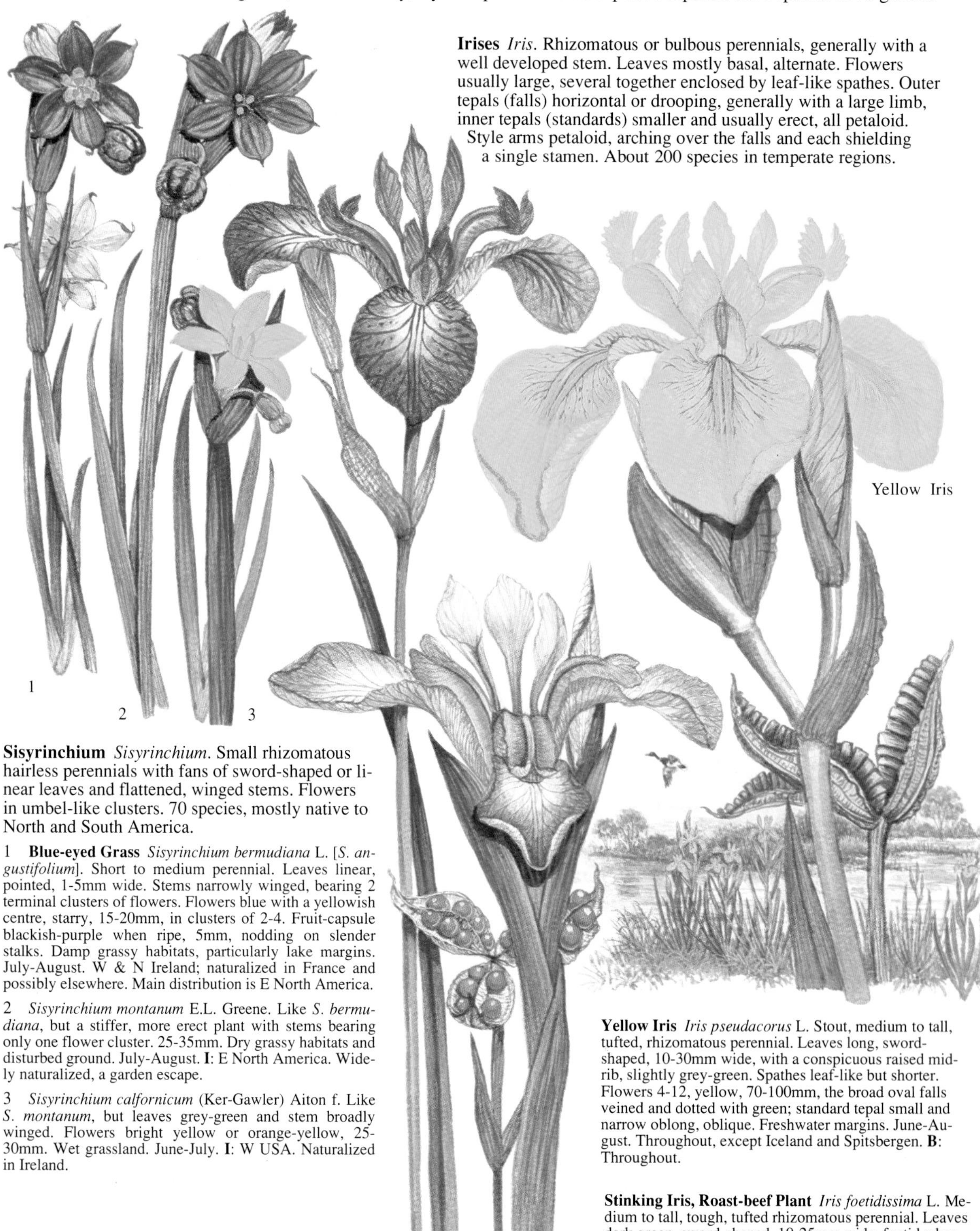


**Sisyrinchium** *Sisyrinchium*. Small rhizomatous hairless perennials with fans of sword-shaped or linear leaves and flattened, winged stems. Flowers in umbel-like clusters. 70 species, mostly native to North and South America.

1 **Blue-eyed Grass** *Sisyrinchium bermudiana* L. [*S. angustifolium*]. Short to medium perennial. Leaves linear, pointed, 1-5mm wide. Stems narrowly winged, bearing 2 terminal clusters of flowers. Flowers blue with a yellowish centre, starry, 15-20mm, in clusters of 2-4. Fruit-capsule blackish-purple when ripe, 5mm, nodding on slender stalks. Damp grassy habitats, particularly lake margins. July-August. W & N Ireland; naturalized in France and possibly elsewhere. Main distribution is E North America.

2 *Sisyrinchium montanum* E.L. Greene. Like *S. bermudiana*, but a stiffer, more erect plant with stems bearing only one flower cluster. 25-35mm. Dry grassy habitats and disturbed ground. July-August. **I**: E North America. Widely naturalized, a garden escape.

3 *Sisyrinchium calfornicum* (Ker-Gawler) Aiton f. Like *S. montanum*, but leaves grey-green and stem broadly winged. Flowers bright yellow or orange-yellow, 25-30mm. Wet grassland. June-July. **I**: W USA. Naturalized in Ireland.

**Siberian Iris** *Iris sibirica* L. Tall, tufted, stem slight branched. Leaves linear, grass-like, 4-10mm wide, bright green. Flowers 2-3, violet-blue, rarely white, the falls yellowish at the base, the blade heavily veined with deep violet-purple on a whitish centre, 55-65mm. Capsule dark brown when ripe, on stalks of uneven length. Damp grassy habitats, to 1100m. June-July. E

**Yellow Iris** *Iris pseudacorus* L. Stout, medium to tall, tufted, rhizomatous perennial. Leaves long, sword-shaped, 10-30mm wide, with a conspicuous raised mid-rib, slightly grey-green. Spathes leaf-like but shorter. Flowers 4-12, yellow, 70-100mm, the broad oval falls veined and dotted with green; standard tepal small and narrow oblong, oblique. Freshwater margins. June-August. Throughout, except Iceland and Spitsbergen. **B**: Throughout.

**Stinking Iris, Roast-beef Plant** *Iris foetidissima* L. Medium to tall, tough, tufted rhizomatous perennial. Leaves dark green, sword-shaped, 10-25mm wide, foetid when bruised. Spathes leaf-like but smaller. Flowers 1-5, dull violet tinged with dull yellow, the falls lightly veined, 55-80mm, the blade of the falls rather narrow, oblong. Capsule green, becoming brownish and splitting to reveal bright red berry-like seeds. Open woodland, rarely dunes, generally at low altitudes, but to 1200m. May-July. Britain and France. sist into the winter. **B**: Widespread in England and Wales; naturalized in Scotland and Ireland.

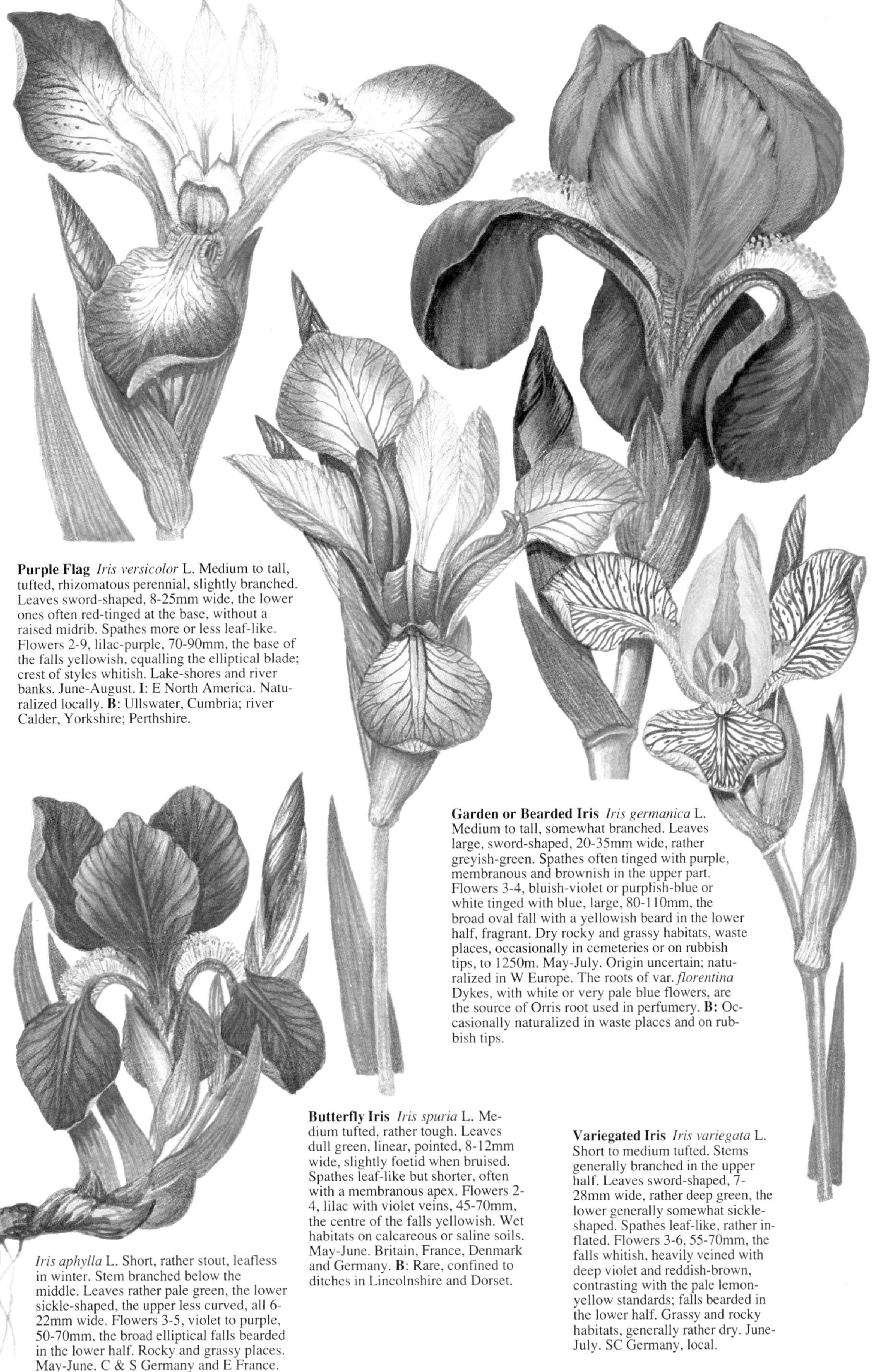

**Purple Flag** *Iris versicolor* L. Medium to tall, tufted, rhizomatous perennial, slightly branched. Leaves sword-shaped, 8-25mm wide, the lower ones often red-tinged at the base, without a raised midrib. Spathes more or less leaf-like. Flowers 2-9, lilac-purple, 70-90mm, the base of the falls yellowish, equalling the elliptical blade; crest of styles whitish. Lake-shores and river banks. June-August. **I**: E North America. Naturalized locally. **B**: Ullswater, Cumbria; river Calder, Yorkshire; Perthshire.

**Garden or Bearded Iris** *Iris germanica* L. Medium to tall, somewhat branched. Leaves large, sword-shaped, 20-35mm wide, rather greyish-green. Spathes often tinged with purple, membranous and brownish in the upper part. Flowers 3-4, bluish-violet or purplish-blue or white tinged with blue, large, 80-110mm, the broad oval fall with a yellowish beard in the lower half, fragrant. Dry rocky and grassy habitats, waste places, occasionally in cemeteries or on rubbish tips, to 1250m. May-July. Origin uncertain; naturalized in W Europe. The roots of var. *florentina* Dykes, with white or very pale blue flowers, are the source of Orris root used in perfumery. **B**: Occasionally naturalized in waste places and on rubbish tips.

*Iris aphylla* L. Short, rather stout, leafless in winter. Stem branched below the middle. Leaves rather pale green, the lower sickle-shaped, the upper less curved, all 6-22mm wide. Flowers 3-5, violet to purple, 50-70mm, the broad elliptical falls bearded in the lower half. Rocky and grassy places. May-June. C & S Germany and E France.

**Butterfly Iris** *Iris spuria* L. Medium tufted, rather tough. Leaves dull green, linear, pointed, 8-12mm wide, slightly foetid when bruised. Spathes leaf-like but shorter, often with a membranous apex. Flowers 2-4, lilac with violet veins, 45-70mm, the centre of the falls yellowish. Wet habitats on calcareous or saline soils. May-June. Britain, France, Denmark and Germany. **B**: Rare, confined to ditches in Lincolnshire and Dorset.

**Variegated Iris** *Iris variegata* L. Short to medium tufted. Stems generally branched in the upper half. Leaves sword-shaped, 7-28mm wide, rather deep green, the lower generally somewhat sickle-shaped. Spathes leaf-like, rather inflated. Flowers 3-6, 55-70mm, the falls whitish, heavily veined with deep violet and reddish-brown, contrasting with the pale lemon-yellow standards; falls bearded in the lower half. Grassy and rocky habitats, generally rather dry. June-July. SC Germany, local.

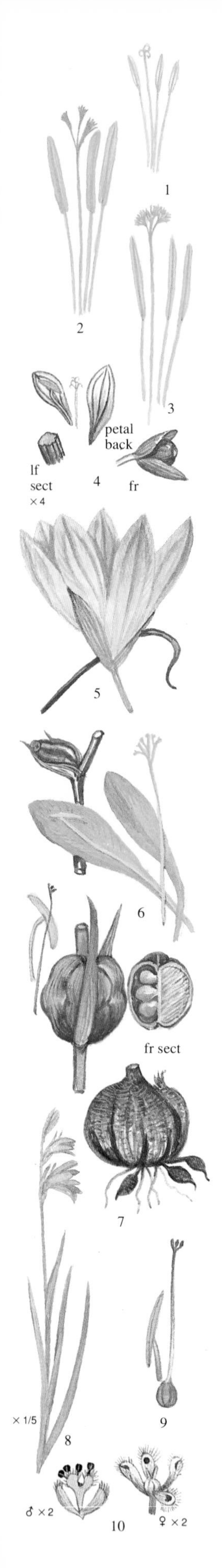


1* **Ixia** *Ixia paniculata* Delaroche. Medium to tall cormous perennial; stem usually branched. Basal leaves linear, pointed, 3-12mm wide, with prominent parallel veins; stem leaves few, similar to the basal, but smaller. Flowers pale to deep cream, salver-shaped, rather crocus-like, 22-38mm, in a narrow pointed spike; style protruding with 3 short, curved branches. **I**: South Africa – Cape of Good Hope. Cultivated as an ornamental and locally naturalized in Britain. June-July. **B**: S Britain.

2* **Spring Crocus** *Crocus vernus* (L.) Hill. [*C. purpureus*]. Very variable low, hairless, cormous perennial. Leaves 2-4, basal, linear, with a central white stripe, partly developed at flowering time, sometimes very short. Flower solitary, white or purple, or white striped with purple at the base, goblet-shaped, but expanding more widely in warm sunshine, 30-55mm long, with a long tube with whitish spathes at the base; anthers yellow; style branched, orange-red, longer than the stamens. Grassy meadows and open woodland in mountain localities, to 2700m. March-June. E France, from the Jura southwards, and S Germany; naturalized in Britain. This is the widespread spring-flowering crocus of the mountains of C & S Europe, often flowering around melting snow patches or in snow hollows. **B**: Naturalized in scattered localities in England, Wales and parts of Scotland.

subsp. *albiflorus* (Kit.) Ascherson & Graebner [*C. albiflorus*]. Flowers generally smaller and white, the styles shorter than the stamens. Similar habitat and distribution. Hybrids occur where subsp. *vernus* and subsp. *albiflorus* grow in close proximity.

3* *Crocus nudiflorus* Sm. Short, cormous, hairless, perennial. Leaves 3-4, basal linear, with a white stripe down the middle, appearing in the spring and withered by the time the flowers appear in the autumn. Flowers solitary, deep lilac-purple, goblet-shaped, 30-60mm, on a very long tube; anthers yellow; styles feathered, orange. Meadows and pastures. September-October. **I**: Pyrenees, N & C Spain. Naturalized in Britain. Distinguished from the so-called 'Autumn Crocus', *Colchicum autumnale*, p. 452, by its paler, pinker flowers, 6 (not 3) stamens and simple whitish 3-branched style. **B**: Rare; naturalized in scattered localities from Gloucestershire to Derby and SW Yorkshire.

4* **Sand Crocus** *Romulea columnae* Sebastini & Mauri [*Trichonema columnae*]. Low, hairless, cormous perennial. Leaves generally up to 6, basal, linear, to 1mm wide only, thread-like and curly, bright green, rounded in section. Flowers 1-3, pale lilac to pale violet with purple veins and a yellow throat, crocus-like, 10-20mm, with a short tube, borne on a short leafless stalk (scape). Dunes and sandy grassland, primarily near or on the coast. April-May. SW Britain and W & N France. Superficially like a *Crocus*, but the leaves without a white stripe and the flowers with a short tube. The flowers generally open widely only in bright sunshine. **B**: SW England and Channel Islands; rare and protected.

5 *Romulea rosea* (L.) Ecklon var. *australis* (Ewart) De Vos [*Trichonema purpurascens*]. Like *R. columnae*, but leaves 1-1.3mm wide and generally spreading. Flowers magenta-pink to lilac with a yellow throat, larger, 20-40mm. Rocky and grassy habitats. April. **I**: South Africa. Naturalized locally in the Channel Islands. Occasionally cultivated.

6* **Montbretia** *Crocosmia × crocosmiflora* (Lemoine ex Burbidge & Dean) N.E. Br. [*Tritonia × crocosmiflora*]. Vigorous medium, patch-forming, stoloniferous, hairless perennial; stems erect. Leaves rather pale green, narrow sword-shaped, 5-20mm wide, alternate. Flowers orange-red; 25-55mm, tubed below and with uneven spreading lobes, in a horizontal or ascending 2-sided spike, with the stamens and styles protruding; anthers yellow. Woods, sea cliffs, waste ground. July-September. Naturalized in Britain and France. This hybrid arose in cultivation from two South African species, *Crocosmia aurea* Pappe ex Hooker and *C. pottsii* (M'Nab ex Baker) N.E. Br. **B**: Widely naturalized, occasionally in Ireland.

**Gladioli** *Gladiolus*. Erect cormous perennials with stiff sword-shaped basal leaves enclosing the lower part of the stem and narrow spikes of flowers. Each flower is enclosed in bud in a pair of leaf-like spathes. Flowers irregular (zygomorphic) the lower part tubed, the upper with 6 lobes, the lower 3 forming a lip, often with a diamond-shaped pattern in the centre. Stamens and style curved below the upper tepals. About 120 species.

7* *Gladiolus illyricus* Koch. Medium perennial; basal sheaths (below the leaves) green, sometimes tinged with red on the veins. Leaves 4-10mm wide. Flowers reddish-purple, 25-45mm long, 3-10 in a spike, sometimes with a branch below, the lower tepals zoned with white and dark red; flowers in the spike alternating to left and right, the anthers equalling or shorter than the filaments. Open woodland, scrub and heaths, occasionally slightly marshy. June-July. S Britain and France. W Europe to W Asia. **B**: Only in the New Forest; rare and protected.

8 *Gladiolus palustris* Gaudin. Like *G. illyricus*, but flowers all facing in one direction and only 2-6 to a spike; spike without an axillary branch. Wet meadows. June-July. S Germany to SE France.

9* *Gladiolus italicus* Miller [*G. segetum*]. Medium to tall perennial with pale to dark red basal sheaths, often spotted with pale green or white. Leaves 5-16mm wide. Flowers bright purplish-red to pale pink, 25-45mm long, more or less pointing in one direction, 6-16 to a spike; anthers clearly longer than the filaments. Cultivated land. June-July. WC France southwards.

10* **Pipewort** *Eriocaulon aquaticum* (Hill) Druce. [*E. septangulare*]. (Pipewort Family – Eriocaulaceae). Low to short tufted, hairless aquatic perennial. Leaves linear, 3-5mm wide, forming dense mats just below the water. Flowerheads 6-11mm, raised above the water surface; female flowers greyish-black, male yellowish-grey, the 'head' surrounded by a ring of greyish bracts. Margins of peaty lakes and pools, shallow freshwater, to 320m. July-September. W Ireland and W Scotland. This is the only European representative of an otherwise entirely American or E Asian genus. **B**: Local in W Scotland – Skye and Coll, and in W Ireland – W Cork to W Donegal.

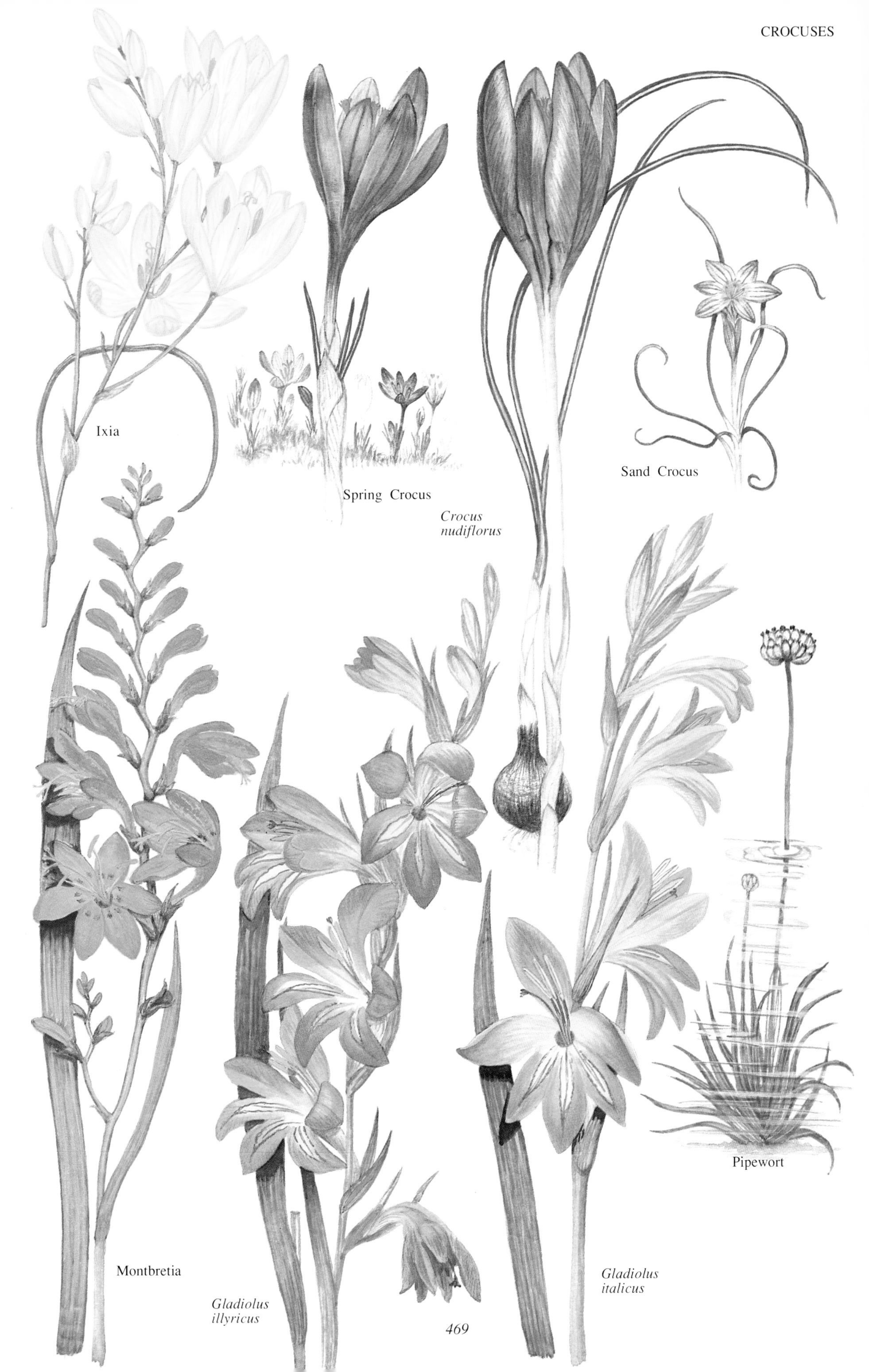
Ixia
Spring Crocus
*Crocus nudiflorus*
Sand Crocus
Montbretia
*Gladiolus illyricus*
*Gladiolus italicus*
Pipewort

## ARUM FAMILY Araceae

A large and primarily tropical or subtropical family with a few European representatives. Plants generally hairless, with underground rhizomes or tubers. Leaves (in European species) basal, stalked. Flowers small, unisexual, in a compact spike, the male and female on the same or on separate plants, in the former the female below the male; flower spike often terminating in a prominent appendage, the spadix, and wholly or partly enfolded in a large fleshy bract, the spathe. Fruit a fleshy or rather dry berry. 1200 species in 120 genera: primarily tropical.

1* **Sweet Flag** *Acorus calamus* L. Medium to tall, rather stout, tufted, rhizomatous perennial. Leaves linear, pointed, rather iris-like, 7-20mm wide, often with one or both margins crinkled, smelling sweetly when bruised. Flowers tiny, greenish-yellow, in a compact narrow ascending cone, 5-9cm long, the stem prolonged above the 'flower-cone' into a leaf-like spathe. Berry not formed in Europe. Shallow fresh-water habitats. June-July. **I**: Asia, North America. Naturalized throughout, except the far north. **B**: Local.

2* **Bog Arum** *Calla palustris* L. Short stoutly-rhizomatous, aquatic perennial, the green rhizomes bearing the persistent membranous bases of leaves. Leaves rounded to heart-shaped, plain green, stalked, held just above the water surface. Flowers yellowish-green, in a compact spadix, 1-3cm long, with a basal white spathe similar in shape to the leaves, not enfolding the spadix. Berries red, 5mm. Swampy places. June-August. Throughout except the extreme west. Widespread. **B**: Locally naturalized in Surrey.

3* **Skunk Cabbage** *Lysichiton americanus* Hultén & St John. Very robust medium to tall, clump-forming, rhizomatous perennial. Leaves very large (to 120cm) oval, rather spade-shaped, deep green, appearing after the flowers, spadix stout, greenish, 5-12cm long, loosely surrounded by a large yellow spathe, very arum-like, to 25cm long, short-stalked; with a disgusting smell. Berries greenish, half embedded in the spadix. Swampy ground. **I**: W North America. **B**: Occasionally naturalized in Ireland.

*Lysichiton camtschatcense* (L.) Schott. Like *L. americanus*, but a smaller plant with white spathes and white spadices. **I**: E Asia. Occasionally naturalized in boggy places in S England.

4* **Large Cuckoo Pint** *Arum italicum* Miller. Short to medium perennial with a horizontal tuber. Leaves triangular-heart-shaped with the basal lobes diverging, with conspicuous whitish veins. Spathe pale greenish-yellow, 15-40cm, drooping at the apex; appendix of spadix stout and yellowish. Fruiting spike large, 10-15cm. Hedgerows, disturbed land. May-June. S Britain and France; naturalized in Holland. **B**: Naturalized in a few localities in the South. subsp. *neglectum* (Townsend) Prime [*Arum neglectum*] has leaves evenly coloured or with dark spots or blotches, the basal lobes converging, sometimes even overlapping. **B**: Local from Cornwall to S Wales and the Channel Islands.

5* **Lords and Ladies, Cuckoo Pint** *Arum maculatum* L. Short perennial with a horizontal tuber, often patch-forming. Leaves appearing in the spring, blunt arrow-shaped, shiny bright green, often with small black blotches. Spathes flushed, spotted and streaked with purple, rarely entirely yellowish. Fruiting spike relatively small, 3-4cm, berries bright orange-red. Woodland, hedgerows, ditch-banks. April-May. Britain to Holland and Germany southwards; naturalized in Holland. **P**: Insects attracted by the smell are trapped by the downward-pointing hairs. At night the spathe loosens, allowing them to escape. **B**: Throughout much of Britain, rarer in Scotland.

*Arum orientalis* Bieb. subsp. *danicum* (Prime) Prime [*A. maculatum* subsp. *danicum*]. Very similar to *A. maculatum*, but tubers vertical and leaves always without dark blotches. Shady habitats. April-May. E Denmark and S Sweden.

## DUCKWEED FAMILY Lemnaceae

Very small freshwater, submerged or floating herbs, multiplying from buds produced by central or lateral pouches on the tiny fronds. The new fronds remain attached to the parent plant or quickly become detached. The flowers are minute, but consist of 1-2 stamens and a solitary ovary set in a cavity on the upper surface of the frond. 22 species in 3 genera.

6 **Rootless Duckweed** *Wolffia arrhiza* (L.) Horkel ex Wimmer [*W. michelii*, *Lemna arrhiza*]. Fronds floating rootless and veinless, not exceeding 1mm, egg-shaped, separate, pale green. Bud pouch solitary and basal with a circular opening. Flowers with a single stamen. Lakes, ponds and ditches. W Europe. The smallest European flowering plant, though probably never flowers with us.

7 **Ivy-leaved Duckweed** *Lemna trisulca* L. Submerged aquatic, translucent. Fronds oblong to narrowly oval, pointed, 5-15mm, linked together by conspicuous persistent stalks, each at right angles to the next and each with a single root. Flowering fronds smaller than the vegetative fronds. Lakes, ponds, dykes and ditches. May-July. Throughout, except the far north. The fronds rise to the surface at flowering time. **B**: Throughout, except for N Scotland; rare in SW England and SW Ireland.

8 **Fat Duckweed** *Lemna gibba* L. Fronds grey-green, floating, 2-5mm, opaque, oval to rounded, rather swollen beneath and with a single long root. Fronds linked by an inconspicuous stalk and soon parting, sometimes flushed reddish-brown. Fresh, brackish or somewhat polluted water. May-July. Throughout, except the north. **B**: Almost throughout.

9 **Common Duckweed** *Lemna minor* L. Fronds pale green, rounded to oval, 1.5-5mm opaque, floating, flat, 3-veined; linked by an inconspicuous stalk or separate and with a single root. Lakes, ponds, ditches and backwaters. May-July. Throughout, except the far north. Often forms a complete green layer on water. **B**: Absent only from parts of Scotland and the Shetland Is.

**Lesser Duckweed** *Lemna miniscula* Herter [*L. valdiviana*]. Fronds smaller, often narrower and oblong or curved, 1-veined. **I**: North and South America. Naturalized in Britain and France. Rapidly spreading. **B**: Scattered localities.

10 **Greater Duckweed** *Spirodela polyrhiza* (L.) Schleiden [*Lemna polyrhiza*]. Fronds floating, dark shiny-green, often purplish beneath, oval to rounded, 4-10mm, often somewhat asymmetrical, each with at least 2 roots and 3-15 veins, flat or somewhat inflated; separated or 2-5 together, each usually with 2 lateral bud pouches. Lakes, ponds, ditches. Rarely flowering. Throughout, except the far north. **B**: Absent from parts of the north, Wales and S Scotland; in Ireland confined mainly to the east.

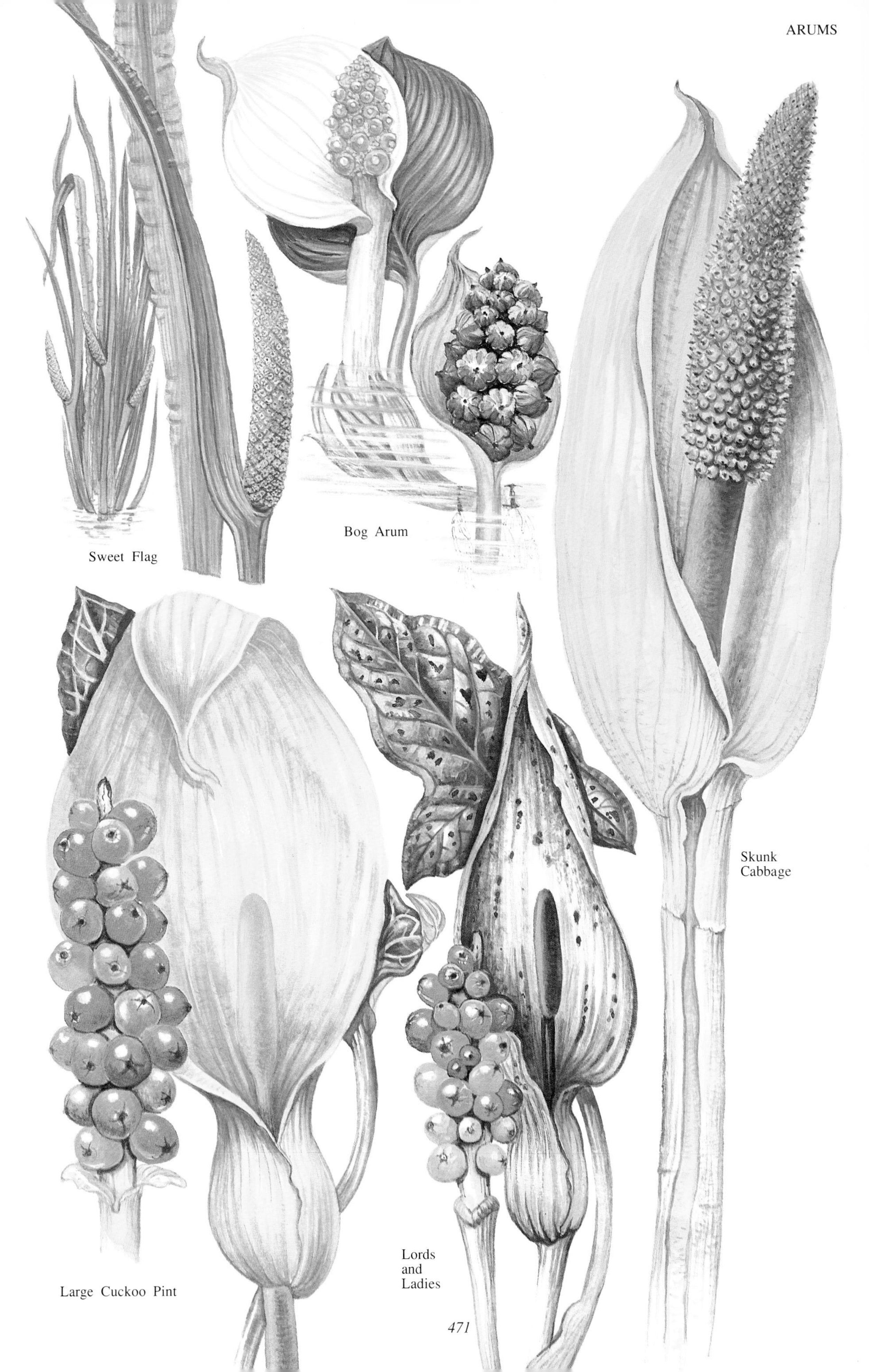
Sweet Flag
Bog Arum
Skunk Cabbage
Large Cuckoo Pint
Lords and Ladies

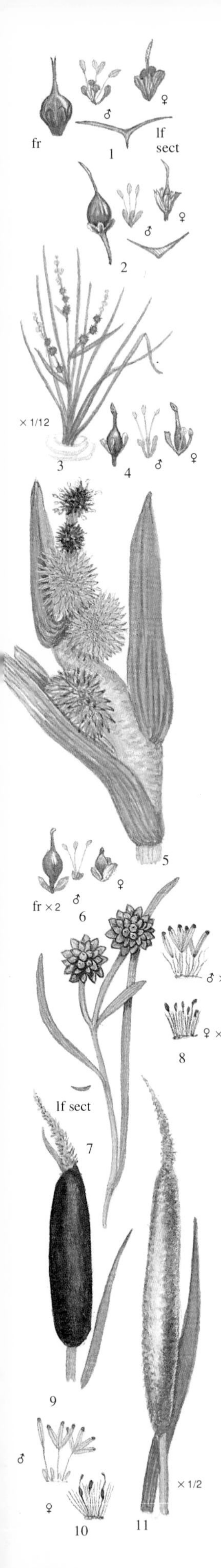


## BUR-REED FAMILY Sparganiaceae

Aquatic or semi-terrestrial perennials with leaves and inflorescences above water, or leaves floating. Leaves strap-shaped or ribbon-like. Flowers unisexual, in globose clusters on branched or unbranched stems, the female below, the male smaller and above; perianth consisting of 3-4 scale-like segments in female flowers, 1-6 in the male; stamens 1-8. Fruit dry, 1-seeded, not splitting. 15 species in a single genus. All are hairless, with creeping rhizomes and erect or floating stems.

1* **Branched Bur-reed** *Sparganium erectum* L. [*S. ramosum*]. Medium to tall robust perennial, to 1.5m, erect. Leaves usually erect, occasionally floating, strap-shaped, triangular in section near the base, keeled beneath. Inflorescence branched, but the individual flower clusters unstalked, the male clusters above the female on lateral branches, distinct from one another; perianth segments thick. Fruit shiny pale brown, with a dark brownish-black apex, angled, more or less triangular in section, 6-8mm long. Freshwater margins, marshland. July-August. Throughout, except the far north. **B**: S England north to the Wash. subsp. *microcarpum* (Neuman) Domin [*S. microcarpum*] has the fruits 6-7mm, generally with only 1 compartment (not 2) and domed above, not angular. **B**: Almost throughout. subsp. *neglectum* (Beeby) K. Richter [*S. neglectum*] has fruits larger, 7-9mm, shiny pale brown all over and rounded in section. North to Sweden. **B**: Commonest in S England, but north to Cumbria and Ireland.

2* **Unbranched Bur-reed** *Sparganium emersum* Rehmann [*S. diversifolium*, *S. simplex*]. Medium perennial with erect or floating stems. Leaves triangular in section, not keeled and not inflated at the sheathing base, erect or floating. Inflorescence simple, unbranched; female clusters 3-6, the lowermost often stalked, male clusters 3-10 separate from one another like the female; style straight. Freshwater margins. July-August. Throughout. **B**: Widespread.

3 *Sparganium gramineum* Georgi. Like *S. emersum*, but inflorescence generally with one or two branches below, each bearing 2-3 female flower clusters; style bent not straight. Scandinavia.

4* **Floating Bur-reed** *Sparganium angustifolium* Michx. [*S. affine*]. Medium to tall perennial to 1m with floating, rarely erect stems. Leaves flat in section, inflated and sheathing at the base. Inflorescence simple, unbranched, with 2-4 female flower clusters, the lowermost stalked, and generally only 2 male clusters which are so close as to appear one oblong cluster. Lake and river margins, peaty moor pools, only on the mountains in the south of its range. July-September. Throughout, except Spitsbergen. **B**: Britain, except for much of the south and west; Ireland, except for the centre.

5 *Sparganium glomeratum* Laest. ex Beurl. Like *S. angustifolium*, but stems always erect, to 40cm only and leaves distinctly keeled beneath, not inflated at the sheathing base. Female and male flower clusters all congested together. Similar habitats and flowering time. Scandinavia (not Denmark).

6* **Least Bur-reed** *Sparganium natans* L. [*S. minimum*]. Short perennial with a floating, rarely erect stem. Leaves floating, ribbon-like and translucent, slightly inflated at the sheathing base. Inflorescence simple, unbranched, with 2-3 female flower clusters and a solitary male cluster, all clearly separate from one another. Lakes, ponds and ditches on non-calcareous soils. June-July.

7 *Sparganium hyperboreum* Laest. ex Beurl. Like *S. natans*, but leaves elliptical in section, rather thick and not translucent. The female flower clusters irregular, at least the lowermost stalked. Similar habitats. July-August. Scandinavia (not Denmark).

## BULRUSH FAMILY Typhaceae

Marsh or aquatic perennials. Flowers unisexual but borne on the same inflorescence in dense cylindrical spikes, the male above the female; flowers small, surrounded by hairs or scales; stamens solitary or in clusters of 2-3; ovary 1-celled. Fruit dry and usually dehiscent: a quantity of downy seed is carried away by the wind. 10 species in 1 genus scattered throughout many parts of the world north to the Arctic Circle.

8* **Lesser Bulrush** *Typha angustifolia* L. Tall, tufted, hairless perennial to 2m; stem base corm-like. Leaves mostly basal, linear, dark greyish-green, flat, 3-6mm wide; leaf-sheaths closed at the throat. Flowering stems about two-thirds the length of the leaves, the female dark reddish-brown, becoming mottled with age, clearly separated from the male flowers by a gap of 3-8cm; pollen grains solitary. Freshwater margins, swamps, often forming extensive patches. July-August. North to C Scandinavia. Used in some countries for thatching roofs. Like other species of *Typha*, it likes plenty of decaying humus. Hybrids with *T. latifolia*, *T.* x *glauca* Godron, are known from S England. **B**: Local, rare in Ireland.

9 **Least Bulrush** *Typha minima* Funck. Slender, medium to tall, tufted perennial, to 75cm. Leaves stiff, linear, only 1-3mm wide, leaf-blades generally absent from the flowering stems; leaf-sheaths open at the throat. Spikes dark brown, rather short, the male and female zones continuous, sometimes with a very slight gap in between, often subtended by a small leaf-like bract; pollen grains in 4's. River and lake shore gravels, often on slightly calcareous soils. July-August. C & S Germany to SE France. Primarily from C Europe.

10* **Bulrush, Cattail** *Typha latifolia* L. Vigorous tall, clump forming perennial, to 2m, stout and creeping. Leaves long and strap-shaped, 8-20mm wide, a rather pale greyish-green; leaf-sheaths usually open at the throat. Flower stems slightly shorter than the leaves, dark brown blotched with white on ageing, to 3cm in diameter, the straw-coloured male flowers about the same length as, and continuous with, the female part, the whole up to 15cm in length; pollen grains in 4's. Lakes, ponds, rivers and marshes. July-August. Throughout, except the far north. Often completely fills ponds and ditches. **B**: Widespread.

11 *Typha shuttleworthii* Koch and Sonder. Like *T. latifolia*, but a shorter plant, not exceeding 1.5m and the female spike silvery-grey when mature and distinctly longer than the male part of the spike. **B**: Similar habitats but generally on acid soils. July-August. E France and S Germany.

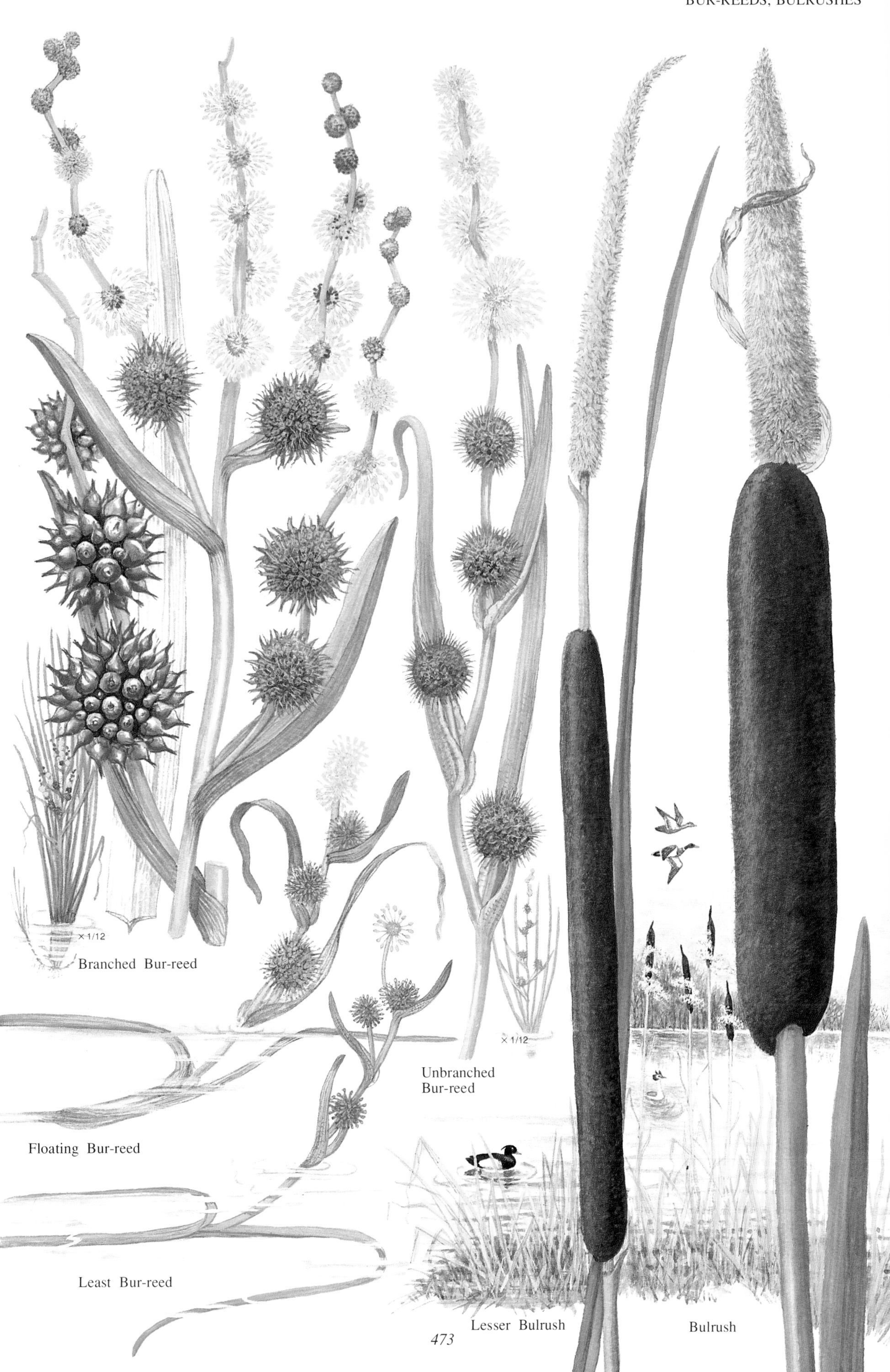


Branched Bur-reed

Unbranched Bur-reed

Floating Bur-reed

Least Bur-reed

Lesser Bulrush

Bulrush

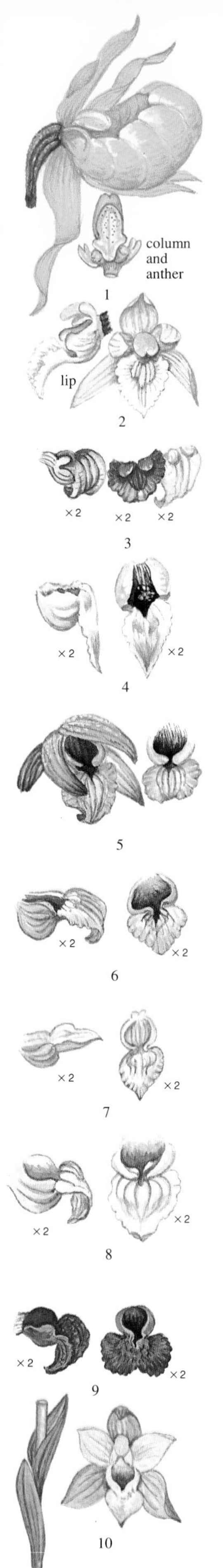


# ORCHID FAMILY Orchidaceae

Perennial herbs with a rhizomatous or tuberous stock, or the base of the stem swollen into a pseudobulb. Leaves spirally arranged, untoothed, sometimes reduced to scales. Flowers usually resupinate (rotated through 180o) in bud and appearing upside down. Perianth in 2 whorls of 3, the outer 3 (sepals) similar, the inner with the upper 2 (petals) alike but the lower one elaborated into a lip, often lobed or patterned. Anthers and stigma borne on a distinctive column; pollen grains aggregated into clusters or pollinia. Ovary inferior. Fruit a 3-parted capsule containing many minute seeds. 25,000 to 30,000 species in 800 genera. The majority are from the tropics and are epiphytic, growing on the branches and trunks of trees.Temperate species are mostly terrestrial. Most orchids have a mycorrhizal association with a fungus without which the orchid cannot grow and thrive.

1* **Lady's Slipper Orchid** *Cypripedium calceolus* L. Short to medium hairy. Leaves 3-4, pale green, elliptical, prominently ribbed. Flowers solitary, maroon-brown, rarely yellowish-green, with a large pouched yellow lip, spotted with red inside; petals narrower than the sepals and often twisted, lip 4-6.5cm. Fruit finely hairy. Hilly and mountainous woods and slopes, usually on calcareous soils, to 2000m. May-early July. N England and Continental Europe except the extreme north, Belgium and Holland. **B**: Extremely rare: a small colony in Yorkshire; protected.

**Helleborines** *Epipactis*. Flowers short-stalked, in slender racemes, often all more or less facing in the same direction. Sepals and petals held closely together or spreading. Lip folded or jointed near the middle, the basal part more or less cupped, the tip rounded to triangular. 25 species in northern temperate regions.

2* **Marsh Helleborine** *Epipactis palustris* (L.) Crantz [*Helleborine palustris*]. Short to medium; stem hairy above, purplish below. Leaves 4-8, oblong to lanceolate, pointed, folded lengthwise, decreasing in size upwards. Flowers 7-14; sepals greenish with faint violet or purplish-brown stripes; petals whitish with a pink base; lip white with a yellow blotch and purplish lines, the tip oval with a frilly margin. Occasionally the flowers may be pale yellowish-white with a whitish lip.Fruit hairy. Marshes, fens and other damp habitats, occasionally dune slacks, to 1800m. July-August. Throughout except the extreme north, the Faeroes and Iceland. **B**: Locally common, north to Perth; also found in the Inner Hebrides.

3* **Broad-leaved Helleborine** *Epipactis helleborine* (L.) Crantz [*Helleborine latifolia*]. Medium to tall; stem hairy above, often purplish beneath. Leaves 4-10, green, spirally arranged, oval-elliptical, pointed. Flowers 15-50, variable in colour but the sepals usually greenish or greenish-yellow and the petals pale pinkish-violet to purplish-red; lip 9-10mm, the lower part dark red inside, the tip oval to heart-shaped, greenish-white, pink or purplish, the tip curved under, giving the lip a rounded appearance. Fruit smooth or rough. Woodland, especially beech, woodland margins and scrub, on calcareous soils, sometimes on sand-dunes, to 1800m. July-September. Throughout, except the far north. **B**: Throughout much of Britain.

4* **Narrow-lipped Helleborine** *Epipactis leptochila* (Godfery) Godfery [*E. dunensis*]. Short to medium; stem slightly hairy above, the whole plant tinged with yellow. Leaves oval to lanceolate alternating in two opposing ranks. Flowers 5-25, green with a yellowish-green lip edged in white and with a pair of basal whitish or pinkish basal blotches, half-nodding and often appearing only half-open; lip 4-9mm, the tip heart-shaped, pointed, the basal part often mottled with red inside. Fruit warted and with sparse blackish hairs. Densely shaded places, especially in beech or conifer woods, on calcareous soils, to 1000m, sometimes on sand-dunes or in coastal pine plantations. July-August. From Britain to Denmark, Germany, Belgium, Holland and France. **B**: S England and S Wales, N to Shropshire.

5 *Epipactis muelleri* Godfery. Similar but leaves yellowish-green and flowers ascending to spreading; lip 7-9mm, pale pink or greenish, the tip heart-shaped with the apex curved under, not flat. Open woods and woodland clearings on calcareous soils, to 1800m. July-August. Belgium and Holland, France and Germany. Rather local.

6* **Dune Helleborine** *Epipactis dunensis* (T. & T.A. Stephenson) Godfrey. Like *E. muelleri* but the lip long-pointed, recurved, the apex rather flat. Maritime sand-dunes, coastal pine-plantations. June-July. Britain × N Wales and NW England.

7* **Green-flowered Helleborine** *Epipactis phyllanthes* G.E. Sm. Short to medium, the stem green, hairless or slightly hairy. Leaves 3-6, alternating in two opposing ranks, rounded to lanceolate, generally pointed, with slightly undulating margins. Flowers 15-35, pendent, only half opening, pale yellowish-green, the petals sometimes tinged with violet; lip 6-8mm with a whitish basal cup and an oval to heart-shaped tip, greenish-white to purplish. Fruit hairless. Open woodland, scrub on calcareous soils, and on coastal sand-dunes. July-September. Britain, N France, Denmark and S Sweden. Variable, often considered an aggregate. **B**: Local in England and Wales, especially in the south.

8* **Violet Helleborine** *Epipactis purpurata* J.E. Sm. Short to medium, stems hairy above, purplish beneath. Leaves spirally arranged, greyish or purplish, elliptical to lanceolate. Flowers many in a raceme, half-nodding, whitish, the sepals greenish or purplish on the outside, the petals sometimes tinged with pink; lip 8-10mm, the basal cup greenish mottled with violet inside, the tip triangular or heart-shaped, whitish, the apex curved under. Fruit rough. Woodland, often of beech, on calcareous soils, to 1400m. August-September. Britain, Belgium and France to Denmark and Germany, but not in Holland. Often growing in clumps unlike *E. helleborine*. **B**: England N to Cumbria; possibly recorded in S Scotland.

9* **Dark Red Helleborine** *Epipactis atrorubens* (Hoffm.) Besser [*E. atropurpurea, Helleborine atropurpurea*]. Short to medium; stem hairless or slightly hairy, violet tinged below. Leaves 5-10, oval to lanceolate, pointed, alternating in two opposing ranks. Flowers 8-18, spreading to half-nodding, deep purple-red, fragrant, opening widely; lip 5.5-6.5mm, with the basal cup green with a reddish margin and red-mottled inside, the tip heart-shaped, deep purple-red, the pointed apex curved under. Fruit densely hairy. Woodland, scrub, rocky places and sand-dunes, generally on calcareous soils, to 2200m. June-July. Throughout, except the far north. **B**: Local and often rather rare, from Brecon N to Sutherland.

10 **Small-leaved Helleborine** *Epipactis microphylla* (Ehrh.) Swartz [*Helleborine microphylla*] leaves small and only 3-6 in number. Flowers greumerous in a spike-like raceme, rather sickly scented, yellowish-brown like the rest of the plant, but occasionally yellowish or whitish, the sepals and petals similar, not spreading; lip 8-12mm, often rather greyish-brown, 2-lobed, spurless. Beech and other woods, to 1700m. May-July. Throughout, but rare or absent in the far north. **B**: Widespread but local, throughout, north to Inverness.

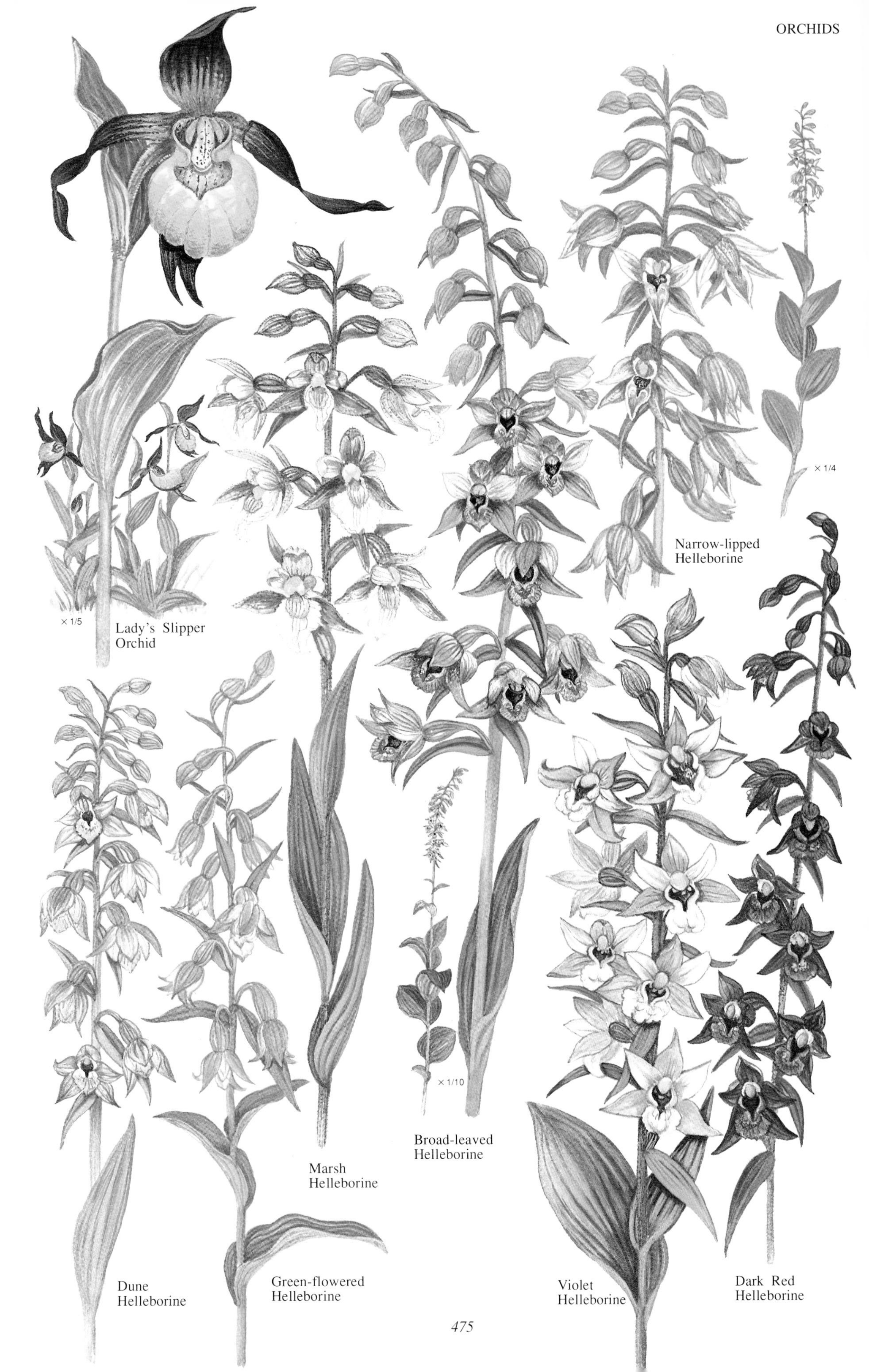

× 1/5
Lady's Slipper Orchid
Narrow-lipped Helleborine
× 1/4
× 1/10
Broad-leaved Helleborine
Marsh Helleborine
Dune Helleborine
Green-flowered Helleborine
Violet Helleborine
Dark Red Helleborine

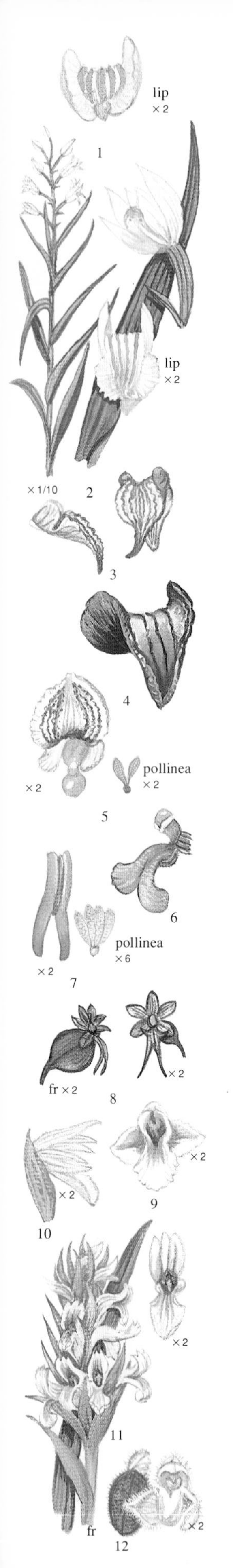


*Cephalanthera*. Flowers relatively large, in a lax spike, not scented, spurless; sepals and petals similar, often close together to form a bell-shape; lip constricted in the middle, forward projecting, the tip ridged above. Bracts present. Flowers unstalked, rather erect. 10 species.

1* **White Helleborine** *Cephalanthera damasonium* (Miller) Druce [*C. alba*]. Short to medium hairless; stem angled with 2-3 brown basal sheaths. Leaves 4-10, oblong to lanceolate, somewhat bluish-green. Flowers white or cream, all but the highest flowers much exceeded by the subtending bracts; sepals and petals blunt, the latter slightly shorter; lip 14-17mm, white, with 3-5 orange-yellow ridges above. Ovary hairless. Beechwoods and other shady habitats, generally on calcareous soils, to 1300m. May-July. North to S Sweden. **B**: Locally from S England north to Cumbria and Yorkshire.

2 **Narrow-leaved Helleborine** *Cephalanthera longifolia* (L.) Fritsch. Similar but stems slightly ridged above and with 2-4 whitish, often green-tipped basal sheaths. Leaves lanceolate, the uppermost linear, dark green. Only the lowest bracts longer than the pale white flowers; sepals and petals pointed. Shady habitats, to 1500m. May-June. Throughout, except the extreme north, extinct in Holland. The bracts are generally shorter than the ovary. **B**: Rare; north to Inverness; also in Ireland and the Inner Hebrides.

3* **Red Helleborine** *Cephalanthera rubra* (L.) L.C.M. Richard. Short to medium; stem tinged purple, hairy above, with several brownish sheaths below. Leaves 5-8, lanceolate, pointed, folded lengthwise. Flowers bright pink or purplish-pink, sepals and petals lanceolate, pointed, the lateral sepals often, though not always, spreading widely; lip 17-22mm, whitish with a purple and pink margin and with 7-9 narrow yellowish ridges above. Ovary hairy. Beechwoods, shady rocky slopes, to 1800m, usually on calcareous soils. June-July. North to C Scandinavia; not in Holland. **B**: Local, Gloucs, Bucks and Hants; protected.

4* **Violet Bird's Nest Orchid** *Limodorum abortiva* (L.) Swartz [*Ionorchis abortivum*]. Medium to tall, violet-tinged, saprophytic. Stem erect, dotted, with numerous violet-tinged scales – without green leaves. Flowers rather large, 38-46mm across, violet; sepals and petals spreading, pointed, the petals slightly shorter and narrower; lip 16-17mm long, yellowish stained with violet, with a wavy margin, unlobed and with a short spur at the base, 10-15mm long. Conifer woods, other shady places, sometimes grassland, generally calcareous soils, to 1200m. May-July. Belgium, France and Germany. Local, unpredictable.

5* **Ghost Orchid** *Epipogium aphyllum* Swartz. Low, saprophytic. Stem erect, pinkish with reddish streaks, somewhat swollen near the base, the lower part covered with 2-5 short brownish scale-like sheaths. Flowers solitary or 2-5, with a rather sickly smell, pale yellowish-brown, the petals slightly broader than the sepals, yellowish with pale violet lines; lip large, uppermost in the flower, pinkish or yellowish, oval-concave with a small basal lobe on each side; spur slender, 6-8mm, blunt. Mainly beechwoods. May-July. Throughout, except the far north and Holland, but only on the mountains in the south, to 1900m. Unpredictable. **B**: Very rare, only in Herefordshire, Shropshire, and Oxfordshire; protected.

6* **Bird's Nest Orchid** *Neottia nidus-avis* (L.) L.C.M. Richard. Short to medium, yellowish-brown, saprophytic. Stem erect without leaves but with scale-like, semi-membranous, sheaths overlapping one another. Flowers numerous in a spike-like raceme, rather sickly scented, yellowish-brown like the rest of the plant, but occasionally yellowish or whitish, the sepals and petals similar, not spreading; lip 8-12mm, often rather greyish-brown, 2-lobed, spurless. Beech and other woods,to 1700m. May-July. Throughout, but rare or absent in the far north. **B**: Widespread but local, throughout, north to Inverness.

7* **Common Twayblade** *Listera ovata* (L.) R. Br. Short to medium rhizomatous perennial. Stem erect, hairy above, with several brown sheathing-scales at the base. Leaves only 2, oval, low down on the stem, dull green, prominently ribbed. Flowers small, numerous in a slender spike-like raceme, with tiny bracts, stalked, yellowish-green, the sepals and petals not spreading widely but forming a loose hood; lip 7-15mm, pendent, notched at the end, spurless. Various habitats from woodland to marshy ground, often on calcareous soils, to 2100m. May-July. Throughout, except the far north. Often difficult to spot but probably the commonest European orchid. **B**: Throughout, except for the Shetland Is.

8* **Lesser Twayblade** *Listera cordata* (L.) R. Br. Low to short slender, creeping, rhizomatous perennial, often only 4-5cm tall. Stem erect, slightly hairy above and with 1-2 brownish sheathing-scales at the base. Leaves 2, subopposite, more or less heart-shaped, finely pointed, shiny green. Flowers up to 12 in a short spike-like raceme with minute bracts, reddish-green, the sepals and petals half-spreading and forming a rather open hood; lip 3.5-4.5mm, purplish, 2-lobed, spurless. Coniferous and other woods, moors and bogs, to 2300m. June-August. Throughout except Belgium and Spitsbergen. **B**: Throughout, rare in S England.

9* **Autumn Lady's Tresses** *Spiranthes spiralis* (L.) Chevall. [*S. autumnalis*]. Low to short glandular-hairy perennial with an erect stem; stem with several overlapping scale-leaves at the base. Leaves primarily in a basal rosette, oval to elliptical, bluish-green, withered by flowering time, but with next year's young rosette appearing to one side. Stem leaves small and bract-like. Flowers white, fragrant, 6-7mm long, in a single spiral along the flowering axis, the lip yellowish-green with an undulate apex; bracts shorter than the flowers. Dry grassy habitats 1000m. August-September. North to Denmark. **B**: Throughout England and Wales, especially in the west, and in S Ireland.

10* **Summer Lady's Tresses** *Spiranthes aestivalis* (Poiret) L.C.M. Richard. Often slightly taller with the basal rosette of leaves present at flowering time, and the flower-spike coming from the centre, not one side of the rosette. Leaves linear-lanceolate, pointed, suberect; stem leaves similar but smaller. Flowers all white, 7-8mm, less fragrant, the bracts as long as the flowers. Wet places to 1250m. July-August. W Europe. In *Spiranthes* the closely held sepals and petals form a tube. **B**: Rare and possibly extinct.

11 **Irish Lady's Tresses** *Spiranthes romanzoffiana* Cham. Low to short; stem slightly glandular-hairy. Leaves linear-lanceolate to narrow oblong, suberect, not in a pronounced basal rosette but decreasing in size up the stem, present at flowering time. Flowers tinged white with cream or green, 11-12mm long, fragrant, in a broad spiralled spike of 3 revolutions; sepals and petals united in the lower half; bracts as long as, or longer than, the flowers. Damp peaty habitats. August-September. Local and rather rare. **B**: W Scotland (S Hebrides) and Ireland only; formerly in Devon; protected.

12* **Creeping Lady's Tresses** *Goodyera repens* (L.) R. Br. Low to short, creeping rhizomatous perennial, the stolons creeping and branching overground; stem glandular-hairy above. Leaves 3-6 oval, pointed, in a loose basal rosette, conspicuously net-veined; stem leaves small and scale-like, half-sheathing. Flowers white, 3-4mm, in a slender, spiralled-spike, sickly-fragrant; lateral sepals spreading but the upper sepal and petals forming a close hood; lip triangular, flat, concave at the base, forward and projecting, spurless. Wooded habitats, especially coniferous, in hills and mountains to 2200m. July-August. Throughout. Rare in some areas, especially Belgium. **B**: N England, Sutherland and Orkney; probably introduced via pine plantations in Norfolk.

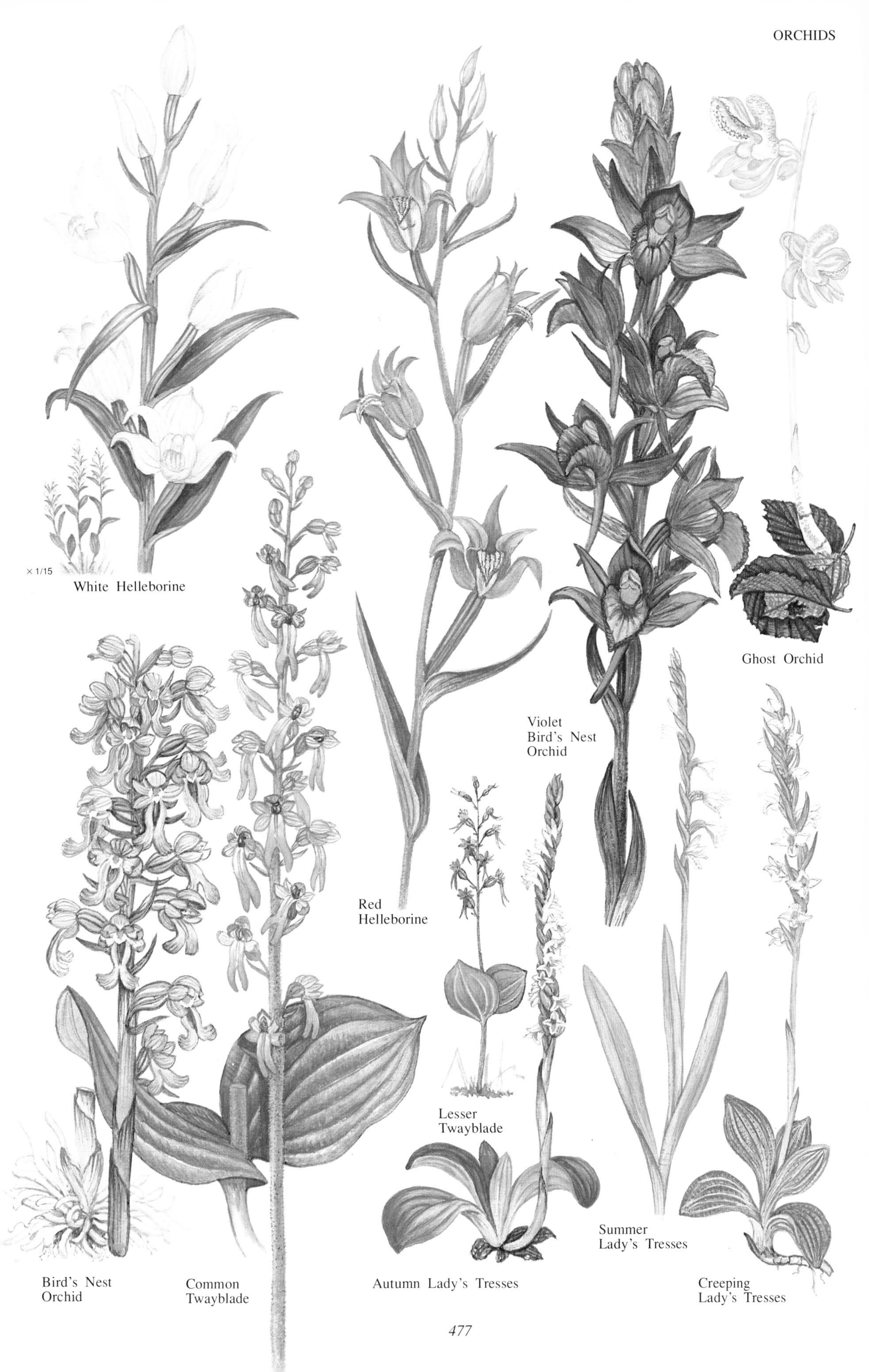
× 1/15
White Helleborine
Red
Helleborine
Violet
Bird's Nest
Orchid
Ghost Orchid
Lesser
Twayblade
Summer
Lady's Tresses
Bird's Nest
Orchid
Common
Twayblade
Autumn Lady's Tresses
Creeping
Lady's Tresses

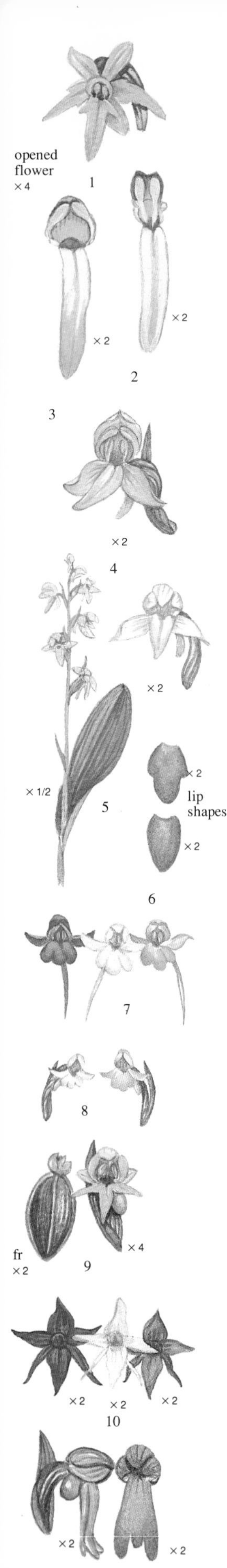


1* **Musk Orchid** *Herminium monorchis* (L.) R. Br. Low to short tuberous-rooted. Leaves 2-3, near the base, oval to lanceolate, rather yellowish-green; upper stem with 1-2 small bract-like leaves. Flowers small yellowish-green, in a lax spike, sometimes all facing in one direction, honey-scented; petals and sepals all spreading, the petals half the length of the sepals, the former often with a pair of short lobes near the base; lip 3.5-4mm, 3-lobed, the middle lobe longer than the lateral ones, spurless. Grassland on calcareous soils, to 1800m. June-July. Throughout, except the far north and Ireland. **B**: S England and S Wales, north to Norfolk and Oxfordshire; rare.

**Butterfly Orchids** *Platanthera*. Tuberous-rooted perennials. Leaves rather few, the basal large but the upper small and bract-like. Flowers white or green, in a broad erect spike; lateral sepals spreading, but the upper sepal and petals closed forming a hood. Lip narrow and unlobed with a cylindrical spur at the base. 80 species, north temperate and tropical.

2* **Lesser Butterfly Orchid** *Platanthera bifolia* (L.) L.C.M. Richard [*Orchis bifolia*]. Medium. Basal leaves 2, oblong to elliptical, subopposite; upper leaves 2-5, linear-lanceolate, much smaller than the basal. Flowers white with a slight green tinge, especially on the lip, sweetly fragrant; lip 8-12mm, linear-oblong, pendent; spur 25-30mm long, longer than the ovary; anther cells (containing the pollinia) parallel. Open woods, scrub, meadows, damp heaths, banks and marshy ground, especially on calcareous and base-rich soils, to 2300m. June-July. Throughout, except Iceland and Spitsbergen. **B**: Widespread, but local throughout, except for Orkney and Shetland.

3* **Greater Butterfly Orchid** *Platanthera chlorantha* (Custer) Reichenb. Flowers more strongly tinged with green, faintly smelling of vanilla; spur shorter, 18-27mm, slightly expanded near the tip; anther cells converging above, not parallel. Woods, scrub, meadows and damp heaths, particularly on calcareous and base-rich soils, to 1800m. June-July. Throughout, but rare or absent in the north. **B**: Throughout, except for Orkney and Shetland.

4 *Platanthera hyperborea* (L.) Lindley. Low to short with narrow cord-like roots. Leaves 4-7, oblong to lanceolate, diminishing in size up the stem. Flowers greenish, small, numerous, fragrant; lip 3-7mm, broad-lanceolate, horizontal; spur 3.5-4.5mm, curved, shorter than the ovary. Meadows and moorland. July-August. Iceland.

5 *Platanthera obtusata* (Pursh) Lindley subsp. *oligantha* (Turc.) Hultén. Similar, but the stem with only a solitary large, basal, elliptical leaf and one bract-like leaf on the stem above. Flowers greenish-white; lip 3-3.5mm, rather diamond-shaped, horizontal; spur 2.5-3mm, only half the length of the ovary. Calcareous mountain heaths. July-August. Arctic and sub-Arctic Norway and Sweden.

6* **False Musk Orchid** *Chamorchis alpina* (L.) L.C.M. Richard [*Herminium alpinum*]. Low, hairless, tuberous-rooted, sometimes only 3cm tall. Leaves 4-8, all basal, linear, grass-like, erect and as tall as the flower stem. Flowers few in a short lax spike, greenish-yellow tinged with purplish-brown, the sepals and petals closed to form a hood; lip 4mm long, oblong, slightly 3-lobed or unlobed, spurless; bracts linear, exceeding the flowers. Damp mountain grassland on calcareous soils, 1600-2700m. July-August. Scandinavia, not Denmark, and the Alps.

7* **Fragrant Orchid** *Gymnadenia conopsea* (L.) R. Br. Short to medium hairless, tuberous-rooted; stem with 2-3 basal, brownish, sheaths. Leaves 4-8, linear-lanceolate, plain green, decreasing in size up the stem, the uppermost small and bract-like. Flowers pink or reddish-lilac, sometimes white or purple, vanilla-scented, in a rather dense slender spike, the lateral sepals spreading, but the upper sepal and petals forming a hood; lip 3.5-5mm, 3-lobed with a longer, slender, slightly curved spur, 11-18mm, pointed, much longer than the ovary. In a variety of habitats from grassland and open scrub, to marshes and fens, but more especially on calcareous soils, to 2700m. June-July. Throughout, except the far north. **B**: Local, but sometimes abundant, throughout much of Britain, not Ireland.

8 *Gymnadenia odoratissima* (L.) L.C.M. Richard [*Orchis odoratissima*]. Like *G. conopsea*, but less tall, rarely exceeding 28cm, with linear, greyish leaves and smaller flowers, pale pink or whitish, vanilla-scented; lip 2.5-3mm long; spur short, only 4-5mm long, blunt, not longer than the ovary. Grassy habitats and scrub on calcareous soils, to 2700m. June-July. France, Germany, Belgium and S Sweden.

9* **Small White Orchid, White Frog Orchid** *Pseudorchis albida* (L.) Á. & D. Löve [*Leucorchis albida*, *Gymnadenia albida*]. Short erect, hairless, tuberous-rooted; stems with 2-3 basal sheaths. Leaves erect oblong, keeled, shiny green, the upper leaves narrower and pointed. Flowers small pale yellowish- or greenish-white, fragrant, in a dense cylindrical spike, the sepals and petals close together to form a hood; lip 2-3mm, 3-lobed, with a short 2mm long basal spur, much shorter than the ovary; bracts equalling the ovary. Grassy habitats, meadows and pastures in hills and mountains, to 2500m. May-June. North to C Scandinavia, extinct in Holland.

subsp. *straminea* (Fernald) Á. & D. Löve [*Leucorchis albida* subsp. *straminea*] has leaves broader, oblong, and less dense; flower-spikes shorter; bracts longer than the ovary. Similar habitats. June-July. Scandinavia south as far as S Norway.

10* **Black Vanilla Orchid** *Nigritella nigra* (L.) Reichenb. f. [*Gymnadenia nigra*]. Low to short tuberous-rooted, with an erect leafy stem, rather slender. Leaves linear-lanceolate to lanceolate, channelled, pointed, decreasing in size gradually up the stem. Flowers usually crimson-black, rarely reddish-purple, pink or whitish, vanilla-scented, in a tight conical head; sepals and petals all spreading, pointed; lip 5-10mm, triangular unlobed, scarcely differentiated from the sepals; spur short, violet or whitish, pouched. Mountains meadows, often on slightly calcareous soils, to 2800m. Alps, Norway, Sweden.

11* **Frog Orchid** *Coeloglossum viride* (L.) Hartman. Low to short, tuberous-rooted, sometimes only 5cm tall; stem with several brownish basal sheaths. Leaves 2-5, oval to lanceolate, decreasing in size up the stem, usually pointed. Flowers small and inconspicuous, yellowish-green, often tinged with reddish- or purplish-brown, faintly honey-scented; borne in a rather lax spike, the lateral sepals spreading, the upper sepal and petals forming a hood; lip 6-8mm, usually yellowish-brown, oblong, notched with a small tooth in the middle (so appearing 3-toothed); spur very short. Grassland and woodland margins, occasionally on rock ledges or dunes, generally in upland or mountains on calcareous soils, to 2500m. Throughout, except Spitsbergen. **B**: Throughout, especially in the north.

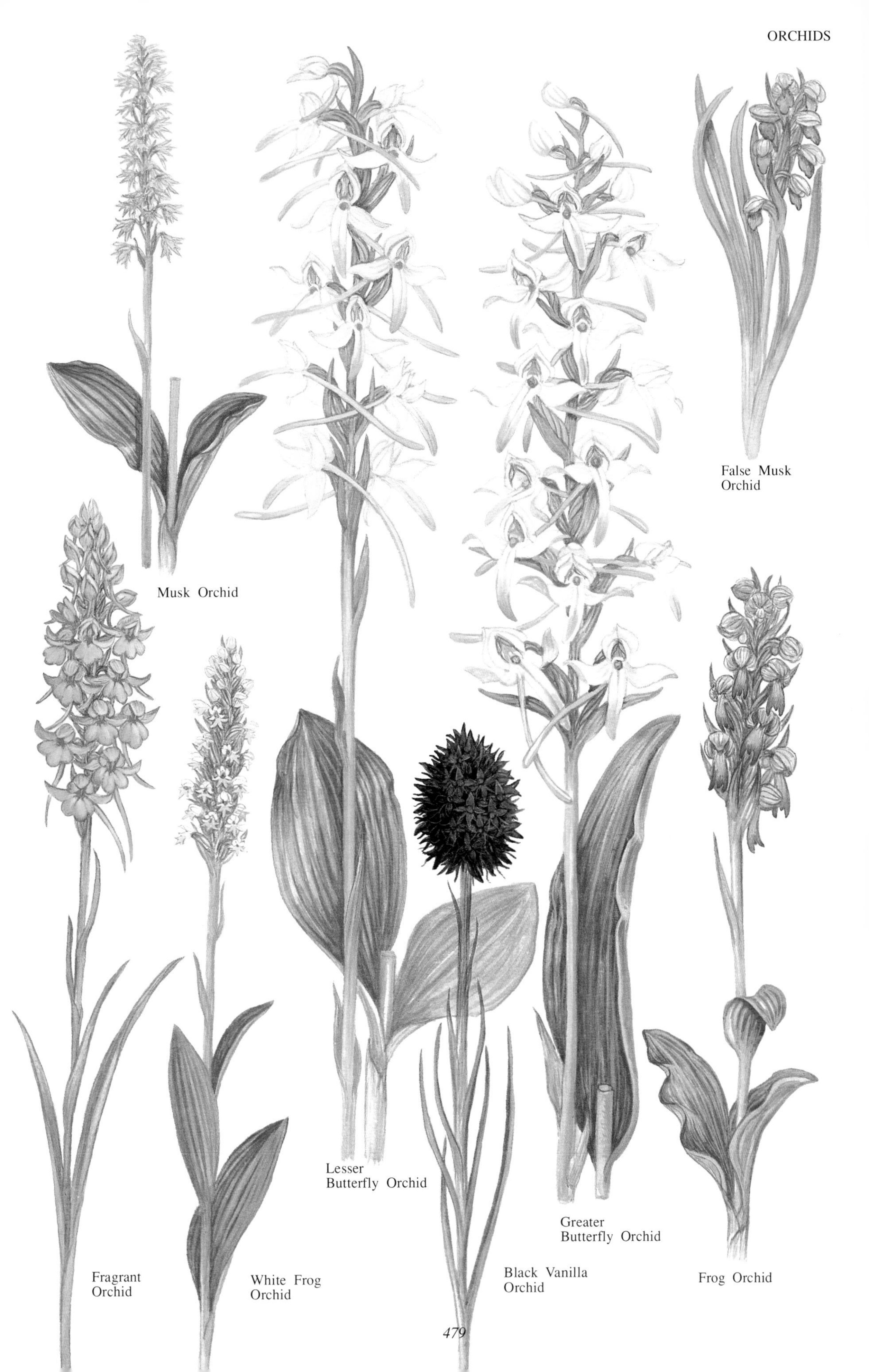
False Musk
Orchid
Musk Orchid
Lesser
Butterfly Orchid
Greater
Butterfly Orchid
Fragrant
Orchid
White Frog
Orchid
Black Vanilla
Orchid
Frog Orchid

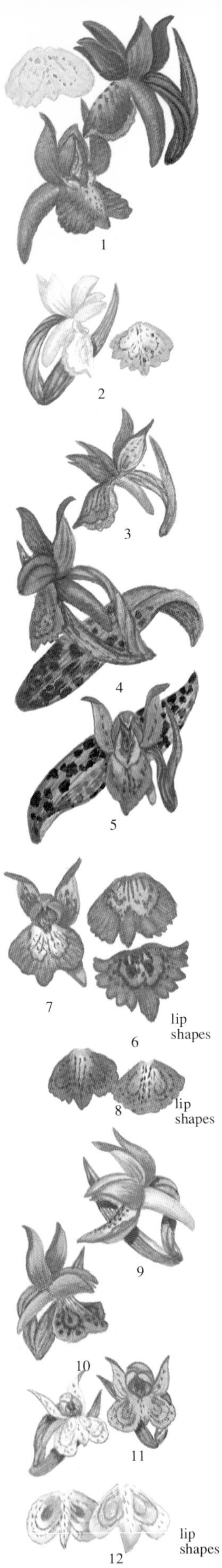


**Marsh Orchids** *Dactylorhiza* [*Dactylorchis*]. Tuberous-rooted, hairless; tubers hand-like, not a pair or rounded tubers as in *Orchis*. Leaves plain or dark-spotted. Lower bracts often leaf-like. Flowers in dense or lax spikes; sepals and petals hooded to spreading; lip 3-lobed, generally with an intricate pattern of lines and dots; spur present. 25 species.

1* **Elder-flowered Orchid** *Dactylorhiza sambucina* (L.) Soó [*Orchis sambucina*]. Short. Leaves 4-5, oblong to lanceolate, rather pale green and unspotted. Bracts longer than the ovaries, purple in purple-flowered forms, otherwise green. Flowers pale yellow, purple, sometimes bicoloured, often growing in mixed colonies; lip spotted purple; lateral sepals spreading, the upper sepals and petals forming a hood; lip 7-8mm, rounded, shallowly 3-lobed. Meadows, scrub and open woodland, in mountain habitats, to 2100m. April-July. Scandinavia, C & E France and S Germany. A dense egg-shaped flower-spike.

2* **Early Marsh Orchid** *Dactylorhiza incarnata* (L.) Soó [*Orchis incarnata*]. Short to medium, stem generally hollow. Leaves 4-5, erect, oblong to lanceolate, yellowish-green, keeled, pointed and often with a hooded apex, generally unspotted. Lower bracts at least equalling or longer than the flowers. Flowers pinkish to lilac, occasionally whitish, with a finely spotted lip, borne in a cylindrical spike, the lateral sepals spreading, the upper sepal and petals forming a hood; lip 5-7.5mm, unlobed to shallowly 3-lobed, the central lobes small and tooth-like, the side folded backwards giving the lip a very narrow appearance; spur pointing downwards, half the length of the ovary. Marshes, damp grassy habitats, occasionally on dune slacks, to 2100m. May-July. Throughout except the far north. Very variable: subsp. *coccinea* (Pugsley) Soó, distinctive deep crimson flowers; subsp. *pulchella* Druce (3), flowers purple, often streaked red, usually on acid peaty soils. **B**: Coastal Wales, N England, Scotland, Ireland.

4 *Dactylorhiza cruenta* (O.F. Mueller) Soó [*D. i.* subsp. *cruenta*]. Shorter than the type, rarely more than 25cm tall. Leaves with dark spots in the lower half, not hooded. Flowers purple, often streaked with red. Britain and N Europe, especially Scandinavia, but in the mountains in the S Alps.

5 *Dactylorhiza lapponica* (Hartman) Reichb. f. [*D. pseudocordigera, D. traunsteineri* subsp. *lapponica*]. Rather like *D. incarnata*, but less than 20cm tall, the leaves with dense dark spots. Flowers purplish-red, the lip 5.5-7mm; spur two thirds the length of the ovary Fens and damp calcareous grassland. June-July. Norway, Sweden and N Finland; possibly in Scotland.

6* **Broad-leaved Marsh Orchid** *Dactylorhiza majalis* (Reichenb.) P.F. Hunt & Summerhayes [*D. latifolia*]. Medium to tall. Leaves 4-8, the lower oblong to broad-lanceolate, the upper narrow-lanceolate, bluish-green, all with rather irregular dark brownish-black spots (sometimes absent). Bracts and upper stem often purplish, bracts generally longer than the flowers. Flowers purplish-lilac to deep reddish-purple, the lip with darker dotted or lined loop markings, borne in a broad egg-shaped or cylindrical spike; lateral sepals spreading upwards like a pair of dove wings, the upper sepal and petals forming a hood; lip 9-10mm, 3-lobed, the central lobe without markings; spur shorter than the ovary. Damp meadows, marshes and fens, on calcareous soils, to 2500m. May-July. North to S Sweden. Much more deeply-coloured flowers in mountain than in lowland forms. Continental forms generally have paler flowers and a less compact habit than those in Britain and Ireland. Plants from W Ireland with pale flowers and unspotted leaves are sometimes called *D. kerryensis*. *D.m.* subsp. *occidentalis* (Pugsley) P.D. Sell (7) is shorter, rarely much over 20cm, flowers deep violet-purple, 3-lobed, the central lobe marked like the lateral ones. **B**: W Scotland, Hebrides, Ireland.

8* **Northern Marsh Orchid** *Dactylorhiza purpurella* (T. & T.A. Stephenson) Soó [*Orchis purpurella, D. incarnata* subsp. *pulchella, D. majalis* subsp. *purpurella, Dactylorhiza purpurella*]. Like *D. majalis*, but stems 20-45cm tall (not up to 75cm); leaves lanceolate, unspotted or with a few small dark spots near the tip. Flowers bright purple-red; lip 5-9mm, unlobed or shallowly 3-lobed, marked with irregular lines and spots, the spur tapered. Similar habitats, but also on fens and dune slacks, especially on base-rich soils. Confined to NW Europe. **B**: N Wales and Derby northwards; Ireland – mainly in the C & N.

9* **Southern Marsh Orchid** *Dactylorhiza praetermissa* (Druce) Soó [*Orchis praetermissa, D. majalis* subsp. *praetermissa*]. Like *D. purpurella*, but stem to 75cm tall; lip 10-12mm, 3-lobed, with a central cluster of small spots; spur generally short and rather thick. NW Europe. Habitats as subsp. *purpurella*, on base-rich or calcareous soils. Throughout, except Ireland, much of Scandinavia, and Iceland. **B**: Confined mainly to the south and east and the Channel Is. – absent from Scotland and Ireland.

10 **Narrow-leaved Marsh Orchid,** *Dactylorhiza traunsteineri* (Sauter) Soó [*Orchis traunsteineri*]. Medium perennial with a rather flexuous stem and well spaced leaves. Leaves 3-5, lanceolate, pointed or blunt, generally spotted. Flowers deep rose-purple, the lip boldly marked with deep purple looped-lines and spots; lateral sepals spreading upwards, the upper sepals and petals forming a hood; lip 7-9mm, 3-lobed, the central lobe longer than the lateral ones which are half-folded backwards; spur pendent, half the length of, or equalling, the ovary. Marshy habitats and fens, to 2000m. June-July. Britain, N Germany and Scandinavia; probably extinct in Holland and Denmark.(See also p. 488). **B**: Scattered localities – most common in S & E England.

11* **Heath Spotted Orchid** *Dactylorhiza maculata* (L.) Soó [*Orchis maculata*]. To 60cm. Leaves 6-12 usually dark spotted, the lower oval to oblong, the upper lanceolate. Bracts mostly shorter than the flowers. Flowers pale pink, pale purple or whitish, the lip with darker looped-lines and spots; lateral sepals spreading, the upper sepal and petals forming a hood; lip 7-11mm, 3-lobed, wavy-edged, the central lobe small and tooth-like, often shorter than the adjacent lobes; spur up to three-quarters the length of the ovary. Moorland, damp grassy and wooded habitats, mainly on acid soils, to 2200m. June-early August. Throughout, except Spitsbergen. (See also p. 488). **B**: Throughout in suitable habitats.

12* **Common Spotted Orchid** *Dactylorhiza fuchsii* (Druce) Soó [*D. maculata* subsp. *fuchsii*]. Short to medium. Leaves 7-12, usually dark-spotted, keeled, often bract-like. Bracts usually shorter than flowers. Flowers white to reddish-purple, the lip with deeper looped-line markings and spots; lateral sepals spreading widely; lip 7-11mm, deeply 3-lobed; spur equalling the ovary. Grassy habitats, open woods, generally on calcareous soils, to 2200m. June-early August. Throughout, except the far north. Probably the commonest of the Marsh Orchids, often in large colonies. **B**: throughout. (See also p.488).

Elder-flowered Orchid
Early Marsh Orchid
Broad-leaved Marsh Orchid
Northern Marsh Orchid
Southern Marsh Orchid
Common Spotted Orchid
Heath Spotted Orchid

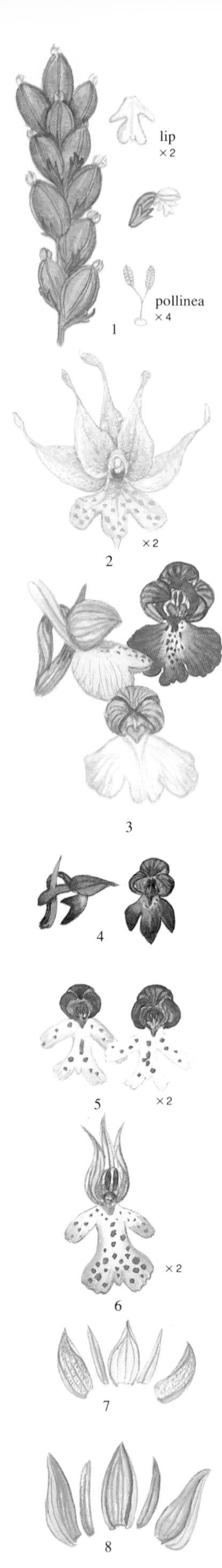


1* **Dense-flowered Orchid** *Neotinea maculata* (Desf.) Stearn [*N. intacta, Aceras densiflorum*]. Short, tuberous-rooted, hairless. Leaves 3-6, somewhat grey-green, oblong to lanceolate, the upper smaller and rather erect, generally with small purplish-brown spots in lines, occasionally unspotted. Flowers dull pink or greenish-white, in a short dense, slightly one-sided spike, vanilla-scented, the sepals and petals all forming a close hood; lip 3-4mm, horizontal, 3-lobed; spur very short, 1.5-2mm, blunt. Woods, scrub, grassland, dunes, often on calcareous soils, generaly at low altitudes. W Britain and France southwards. Rather inconspicuous. **B**: Very local, confined to Isle of Man and W Ireland – Galway to Offaly and Clare.

2* **Round-headed Orchid** *Traunsteinera globosa* (L.) Reichenb. Short to medium, tuberous-rooted, with a slender erect stem. Leaves 2-3, oblong to lanceolate, the stem with several pinkish bract-like leaves in the upper half. Flowers pinkish-lilac, in a rounded head, lengthening with age, the petals and lip with fine deep pink spots; sepals and petals spreading, all terminating in a small 'blob'; lip 4.5-5mm, squarish, 3-lobed, with a spur half the length of the ovary. Mountain meadows, sometimes in open grassy scrub and open woods, to 2600m. E France and S Germany, from the Vosges southwards.

*Orchis*. The largest European genus together with *Ophrys*, though the latter is more Mediterranean. Tuberous-rooted perennials, the tubers 2-3, oval or rounded, testicle-like. The sepals and petals often forming a hood or the lateral sepals spreading. Lip unlobed to 3-lobed, projecting behind into a short or long spur, often inclined downwards. 60 species native to Europe, North Africa, temperate Asia and the Canary Is.

3* **Green-winged Orchid** *Orchis morio* L. Short to medium. Leaves mostly in a basal rosette, oblong to broadly lanceolate, unspotted, the uppermost sheathing the lower stem. Bracts lanceolate equalling, or shorter than, the ovary. Flowers very variable in colour from purple to purplish-violet, reddish, pink or whitish, the lateral sepals (wings) often green-veined; sepals and petals all close together forming a hood; lip 8-10mm, shallowly 3-lobed, often with dark spots along the centre; spur shorter than the ovary, curved upwards. Grassy habitats, open scrub, sometimes in abandoned quarries or sand-pits, to 1800m. May-June. Throughout, except the far north. **B**: Locally common throughout north to S Scotland; rapidly decreasing.

4* **Bug Orchid** *Orchis coriophora* L. Short. Leaves 4-7, lanceolate to linear, the upper sheathing the stem. Bracts lanceolate, equalling or longer than the ovary. Flowers small, violet- or reddish-brown with a purplish-green, unspotted lip, with an unpleasant smell, borne in narrow spikes; sepals and petals forming a close pointed hood, the sepals faintly green-veined; lip 6-8mm, 3-lobed; spur pointing downwards. Damp grassy habitats, sometimes in open scrub, often on hill slopes, to 1800m. April-June. Belgium, France, C & S Germany; extinct in Holland.

subsp. *fragrans* (Pollini) Sudre from S Europe, has pleasantly vanilla-scented flowers. Rather a dingy-flowered orchid, often difficult to spot, but frequently growing with other species like *Anacamptis pyramidalis* and *Ophrys apifera*.

5* **Burnt-tip Orchid** *Orchis ustulata* L. Low to short with a rather slender stem. Leaves 2-3, oblong, pointed, basal, but the stem with several sheathing leaves above, all unspotted. Bracts oval-lanceolate, shorter than, or equalling the ovary. Flowers brownish-purple or maroonish, with a white or pale pink, purple-spotted lip, fragrant, borne in an oval spike, but lengthening with age; sepals and petals forming a close hood; lip 4-8mm, 3-lobed, but the middle lobe large and notched at the tip; spur half the length of the ovary, pointing downwards. Grassy habitats and open scrub, on calcareous soils, often in hilly or mountain regions, to 2100m. May-June. Europe north to S Sweden; extinct in Holland. The brownish-unopened flower-buds at the top of the spike give the orchid a burnt-tip appearance hence the common English name. Forms with yellowish or whitish sepals and petals do sometimes occur, but these are generally found in the southern parts of Europe. **B**: Widespread in England N to Cumbria – most frequent in the south.

6* **Toothed Orchid** *Orchis tridentata* Scop. Short to medium. Basal leaves 3-4, oblong, unspotted, the stem with several sheathing leaves in the lower half. Bracts lanceolate, pointed, shorter than, or equalling the ovary. Flowers pale violet-lilac, the lip spotted with purple, in a conical spike, elongating as successive flowers open, the sepals and petals forming a close hood; lip 6-9mm, 3-lobed, but the middle lobe broader and notched, with a small 'tooth' in the notch, all the lobes rather broad, not manikin-like; spur cylindrical, pointing downwards, half the length of the ovary. Grassy and rocky habitats, open woodland and scrub, to 1500m. May-June. E France and S Germany.

7* **Monkey Orchid** *Orchis simia* Lam. Short to medium, with an erect or somewhat flexuous stem. Basal leaves 3-5, oblong-lanceolate or oval, rather flat, shiny but unspotted, the stem with several sheathing leaves in the lower half. Flowers pale greyish-pink with a pinkish-purple lip with faint purplish spots, manikin-like, the sepals and petals forming a head-like hood, the lip 14-16mm long, the lateral lobes forming very narrow curved 'arms' while the central lobe is itself 2-lobed to form the very narrow curved 'legs' with a small tooth in between like a short tail; spur cylindrical, pointing downwards, shorter than the ovary. Grassy habitats and open scrub, woodland margins, primarily on calcareous soils, to 1500m. May-June. S Britain, Belgium, Holland, France and Germany. **B**: Oxfordshire and Kent only; rare and protected.

8* **Military Orchid** *Orchis militaris* L. Short to medium. Basal leaves 3-5, oval to lanceolate, flat and unspotted, the stem with several sheathing leaves in the lower half. Bracts much shorter than the ovary. Flowers manikin-like, pink or greyish-pink, with a pinkish-purple lip, the lip and inside of the hood purple spotted or marked; lip flat, 12-15mm, with narrow lateral lobes forming the 'arms'with the central lobe, itself 2-lobed, forming the 'legs' with a small tooth-like 'tail' in between; spur short, downcurved, half the length of the ovary. Grassland, open scrub and woodland margins or clearings, on calcareous soils, to 1800m. May-June. North to S Sweden. Often confused with *O. simia* but the lip is whitish at the base and the 'legs' and 'arms' broader and flat, not markedly curved. In *O. simia* the flowers at the top of the spike open first, in *O. militaris* the lowermost. **B**: Rare and found only in Suffolk and Buckinghamshire – formerly more widespread; protected.

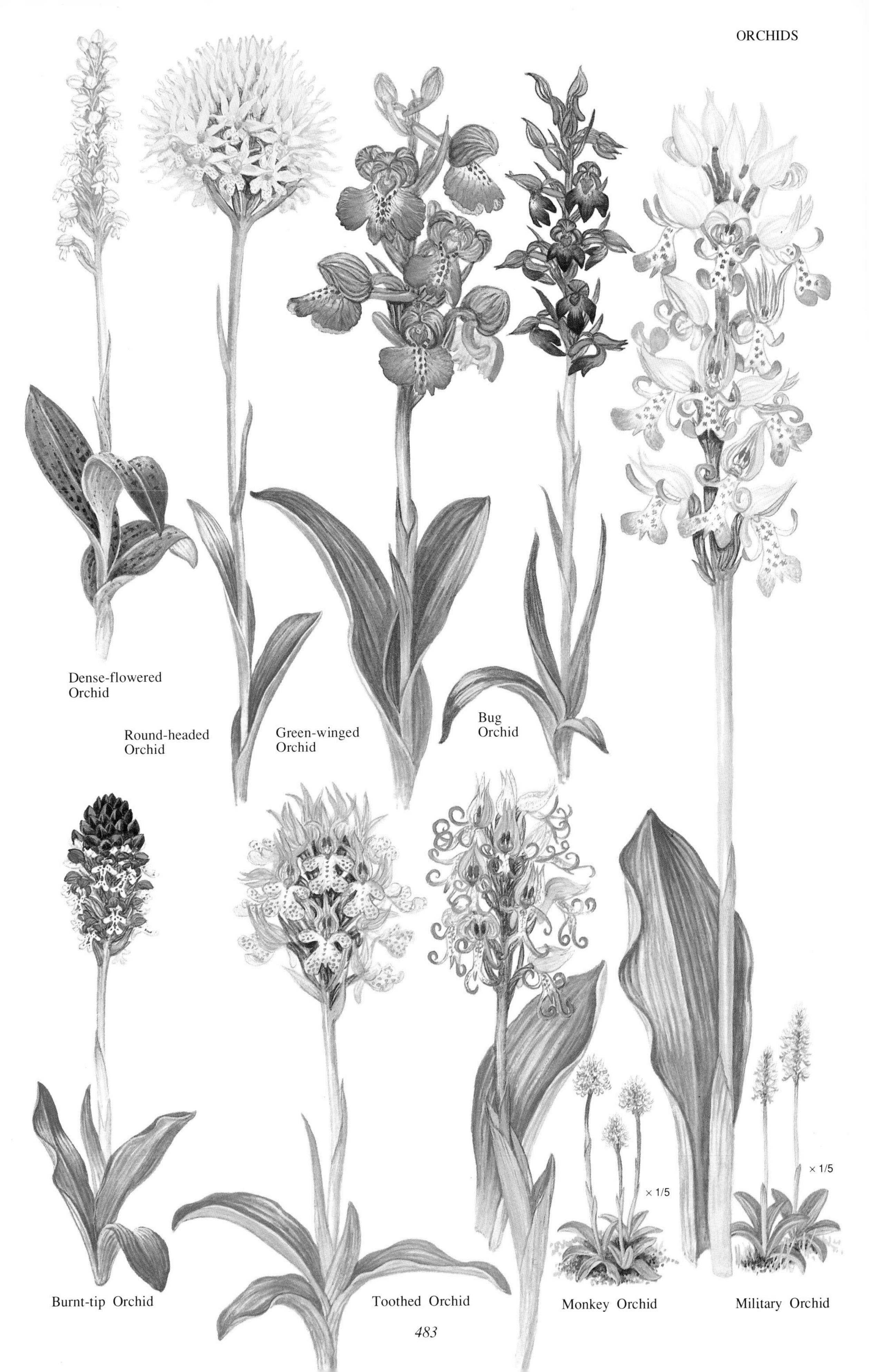
Dense-flowered Orchid
Round-headed Orchid
Green-winged Orchid
Bug Orchid
Burnt-tip Orchid
Toothed Orchid
Monkey Orchid
× 1/5
Military Orchid
× 1/5

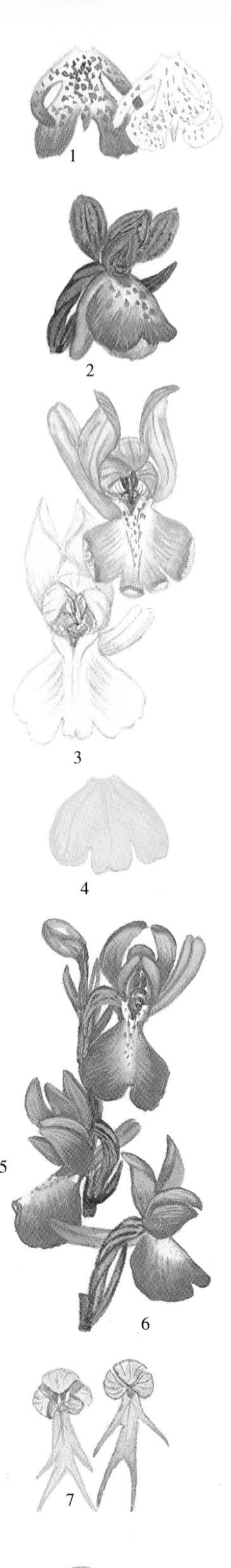

1* **Lady Orchid** *Orchis purpurea* Hudson. Medium to tall perennial. Basal leaves 3-6, oblong, shiny-green, unspotted, the stem with several sheathing leaves in the lower half. Bracts lanceolate, much shorter than the ovary. Flowers fragrant, brownish-purple with a white or pale pink, the lip dotted purple, lady-like, the sepals and petals forming a hood; lip 10-15mm, with the lateral lobes forming slender 'arms', but the central lobe much broader, 'skirt-like' with a notch at the end, with a small 'tooth' in the notch; spur short, only half the length of the ovary, pointing downwards. Woodland and scrub, occasionally on grassy slopes by woodland, or road verges, on calcareous soils, to 1500m. May-June. S Britain, Belgium, Holland and Denmark to France and Germany. **B**: Very local, now only in Kent, though formerly more widespread; protected.

2 *Orchis spitzelii* Sauter ex Koch. Like *O. purpurea*, but shorter, not exceeding 40cm tall, with dull green leaves and linear-lanceolate bracts slightly longer than the ovary. Flowers purple, the lip somewhat paler, but with darker dots, 7-8mm. Grassy scrub and woodland margins. May-June. S Sweden – Gotland; extinct in Germany. June-July.

3* **Early Purple Orchid** *Orchis mascula* (L.) L. Short to medium perennial. Basal leaves 3-5, narrow-oblong, often with purple blotches, especially in the north and west of its range; stem often purplish, with several sheathing leaves in the lower half. Bracts lanceolate, equalling the ovary. Flowers purple, occasionally pinkish, rarely white, with an unpleasant smell reminiscent of tom-cats, the lateral sepals spreading upwards, like 'wings', the upper sepal and petals forming a hood; lip 6-8mm, generally with a few small spots in the whitish middle, rather diamond-shaped, 3-lobed, the central lobe almost square, sometimes notched. Spur as long as the ovary, pointing upwards. Woodland, often of beech or oak, scrub, grassland, road-verges, to 2650m. April-June. Throughout, except Iceland and Spitsbergen. Mountain forms often have intensely deep purple-black flowers. In W Europe *O. mascula* frequently grows in association with *Scilla non-scripta*. B: Throughout much of Britain; locally common.

4* **Pale-flowered Orchid** *Orchis pallens* L. Short perennial. Basal leaves oblong, narrowed at the base, slightly shiny-green, unspotted; stem with several sheathing leaves in the lower half. Bracts lanceolate, equalling, or longer than, the ovary. Flowers pale yellow, fragrant, the sepals spreading, the petals forming a hood; lip 6-8mm, 3-lobed, the central lobe almost square, sometimes notched. Spur almost equalling the ovary, pointing downwards. Woodland and grassy habitats, on calcareous soils, to 2000m. May-June. E France and S Germany. Confined to the Alps and SE Europe. The smell of the flowers is similar to the Elderberry, *Sambucus nigra*.

5 **Lax-flowered Orchid** *Orchis laxiflora* Lam. Medium to tall perennial, with an erect or somewhat flexuous, purplish stem. Leaves 3-8, lanceolate to linear, spreading, unspotted. Bracts lanceolate, slightly shorter, to longer than the ovary, often purplish, 3-7 veined. Flowers purple, in a rather lax spike, the sepals spreading, the petals forming a hood; lip 7-9mm, oval, narrowed at the base, 3-lobed, but the central lobe small and shorter than the lateral lobes. Spur two-thirds the length of the ovary, widened at the tip. Damp meadows and marshes, generally at low altitudes. May-June. SW Britain, France, Belgium and Germany. W & S Europe, W Asia and North Africa, often forming extensive colonies. **B**: Only known on Jersey.

6* *Orchis palustris* Jacq. [*O. laxiflora* subsp. *palustris*]. Like *O. laxiflora*, but often with rather paler flowers, the lip more clearly 3-lobed, with the central lobe as long as, or longer than, the lateral lobes; spur narrowed at the tip. Similar habitats and flowering time. Germany and S Sweden.

7* **Man Orchid** *Aceras anthropophorum* (L.) Aiton f. Short perennial, the stem with several brown sheaths at the base. Leaves oblong, keeled, shiny-green, the upper leaves smaller and bract-like. Bracts membranous, shorter than the ovary. Flowers greenish-yellow, often with reddish margins and streaks, borne in a slender spike, often many-flowered, each flower manikin-like with the sepals and petals forming a close hood; lip 12-15mm, pendent, the lateral lobes forming short, narrow 'arms' and the central lobed divided into narrow 'legs'; spurless. Grassland, field boundaries, abandoned quarries, banks and open scrub, rarely along woodland margins, on calcareous soils, to 1500m. May-June. S & SE Britain, Belgium, Holland, France and Germany. Frequently hybridizes with *Orchis* when growing in close proximity; more especially with *O. militaris*, *O. purpurea* and *O. simia* in the southern part of their range. **B**: Local in England, except the SW, north to Derbyshire.

8* **Lizard Orchid** *Himantoglossum hircinum* (L.) Sprengel [*Loroglossum hircinum*, *Orchis hircina*]. Medium to tall, rather stout, tuberous-rooted perennial, the stem with faint purplish markings. Lower leaves 4-6, elliptic-oblong, dull green, unspotted, often partly withered by flowering time. Flowers pale grey-green, finely spotted with purple spots and streaks, borne in a long spike, many-flowered, rather foul-smelling; sepals and petals forming a small hood, the lip very long, often spiralling, 30-50mm, with 2 short arms near the top and a long central, tail-like central lobe notched at the tip; spur present, 7-12mm long, downward pointing. Grassland, scrub and open woods or woodland margins, road-verges or sand-dunes, generally at low altitudes, but up to 1800m in the south of its range. June-July. SE Britain, Belgium, Holland, France and Germany. Widespread in S Europe and W Asia with several subspecies, but often local in the north and west of its range and usually forming small colonies. It is fascinating to watch the long lip uncoil from the bud and lengthen into a slender 'lizard'. **B**: Scattered localities in England N to Yorkshire, but mostly in Kent and Sussex; generally rare.

Lax-flowered Orchid

*Orchis palustris*

Pale-flowered
Orchid
Lady
Orchid
Early Purple
Orchid
Man Orchid
× 1/12
Lizard Orchid

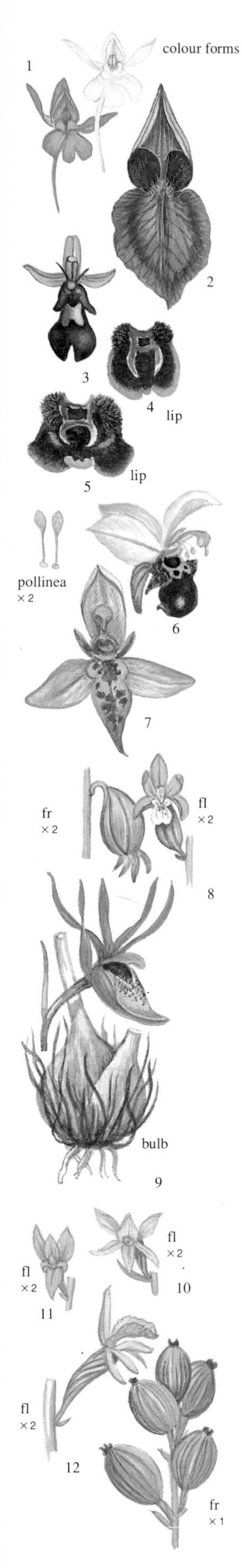


1* **Pyramidal Orchid** *Anacamptis pyramidalis* (L.) L.C.M. Richard. Short to medium hairless. Leaves narrow-oblong, pale green, unspotted. Flowers bright pink or purplish-red, rarely white, in a broad pyramidal spike, becoming more cylindrical with age; lip 3-lobed, 6-8mm, with a long, slender, curved spur. Grassy habitats, generally rough unimproved meadows and pastures, banks, roadsides, open woodland rides, coastal sand-dunes, generally on calcareous soils, to 1900m. June-August. Throughout, except the north, the Faeroes and Iceland. **P**: Butterflies and moths. One of the commonest orchids in NW Europe. **B**: Throughout much of Britain, but only coastal in western N Scotland; also local in the Hebrides.

2* **Heart-flowered Orchid** *Serapias cordigera* L. Short to medium. Leaves narrow lanceolate, the lowermost sheath-like and purple spotted. Flowers pale to deep reddish-purple with a deeper, often reddish veined, tongue-like lip, 20-35mm long, borne in a short oval spike, spurless. Grassy meadows and roadsides, scrub and open woods, stream margins. May-June. NW & C France southwards. **P**: Bumble-bees.

*Ophrys*. Perennials with rounded root-tubers. Leaves mostly basal with 1-2 usually sheathing the stem. Flowers in spikes with large spreading sepals, and 2 small rather insignificant petals. Lip elaborate, sometimes with side-lobes, often furry and with a distinctive shiny pattern-shield or 'speculum'. 40 species, mainly Mediterranean.

3* **Fly Orchid** *Ophrys insectifera* L. [*O. muscifera*]. Short to medium slender. Leaves narrow lanceolate, shiny plain green. Flowers in a narrow spike, deep violet-brown with green sepals; lip narrow, 9-10mm long, 3-lobed, resembling a fly, the larger central lobe forked at the tip, with a shiny pale violet-blue zone near the base, the lip occasionally greenish or whitish. Woods, scrub, coppices, fens and rough grassy, generally shaded habitats, usually on calcareous soils. May-June. Throughout, except the extreme N and Iceland. **P**: Male *Gorytes* wasps. **B**: England N to Cumbria; scarce in SW England and Wales; Ireland, confined more or less to the Burren.

4* **Early Spider Orchid** *Ophrys sphegodes* Miller [*O. aranifera*]. Short to medium. Leaves broad-lanceolate, plain green. Flowers 2-10 in a lax spike, pale to dark brown or blackish-brown with yellowish-green sepals; lip oval, 10-12mm, with an X- or H-shaped bluish-violet pattern, reminiscent of the body of a bumble-bee. Short turf, dry unimproved grassland and rocky places, roadsides, banks, in sunny often exposed habitats, on calcareous soils, to *c.* 1000m. April-June. SE Britain, Belgium, France and Germany. Often very local. **P**: Occasionally by bees. **B**: Rare and local; confined to S & SE England; protected.

5* **Late Spider Orchid** *Ophrys holoserica* (Burm. f.) W. Greuter [*O. fuciflora*]. Similar, but sepals pink, purple or whitish; lip 9-13mm, with a small pointed, heart-shaped tip, the pattern violet or blue with a yellowish margin – tip of lip forward pointing. Short turf on unimproved grassland, field borders, roadsides, on calcareous soils, to 1300m. June-July. SE Britain, Belgium, France and Germany. Widespread in C & S Europe, with several subspecies. **P**: In Europe by male *Eucera* bees, but in Britain pollination is rare, or self-pollination may occur. **B**: Very rare; confined to E Kent on chalk downs; protected.

6* **Bee Orchid** *Ophrys apifera* Hudson. Short to medium with oval or lanceolate, plain green leaves. Flowers 2-11, in a slender lax spike, brownish-purple to yellowish-green, with pink or purple sepals; lip 3-lobed, 10-13mm, with a yellowish pattern enclosing a reddish-brown shield-shaped zone, tip incurved. Short and rough grassy habitats, meadows and pastures, scrub, banks, and roadsides, railway cuttings, quarries, sand-dunes, on calcareous soils, to 1000m. Britain, Belgium, France and Germany. Widespread in C & S Europe as well as NW Africa where white-flowered forms are sometimes found. **P**: In Continental Europe mostly insect-pollinated, in Britain self-pollination is normal. Erratic in appearance. **B**: Most of England, the Channel Islands; rare in the SW and in Wales.

7 **Wasp Orchid** *O. a.* var. *trollii* (Hegelsch.) Reichenb. Like the type, but the lip narrower, pointed, brownish with yellowish markings, but no shield. C & S France.

8* **Coralroot Orchid** *Corallorhiza trifida* Chatel. Low to short hairless, yellowish-green, saprophytic. Stem without leaves, but with 2-4 overlapping scales. Flowers pale yellowish-green with slender sepals and petals; lip 5mm, slightly 3-lobed, whitish with red lines and blotches, spurless. Damp woodland, generally of pine or mixed pine and birch, on mossy peaty soils, sometimes in alder-willow scrub or on moist coastal sand-dunes. June-July. Throughout, except Ireland. The rootstock is a coral-like mass of pale yellowish, fleshy roots. **P**: Insects, or self-pollinated. **B**: Rare and local; confined to N England and S & C Scotland; protected.

9* **Calypso** *Calypso bulbosa* (L.) Oakes. Low to short, with a swollen basal pseudobulb. Leaf solitary, elliptical, ribbed. Flowers large, solitary, nodding, pinkish-purple, lip pouched, horned below, 10-20mm, whitish with pink or yellow blotches, sepals and petals narrow, similar, all pointing upwards. Wet coniferous woodland and marshes, on mossy acid soils. May-June. Finland and Sweden.

10* **One-leaved Bog Orchid** *Malaxis monophyllos* (L.) Swartz [*Microstylis monophyllos*]. Short, generally with a solitary elliptical leaf; pseudobulbs small, one above the other. Flowers many in a fairly dense spike, greenish-yellow, spurless, the sepals and petals spreading; lip 2-2.7mm, lanceolate, pointed, uppermost. Wet marshy ground, scrub and sphagnum bogs. June-July. Germany, Finland, Norway and Sweden.

11* **Fen Orchid** *Liparis loeselii* (L.) L.C.M. Richard [*Malaxis loeselii, Pseudorchis loeselii*]. Low to short hairless with a pseudobulb netted with old leaf bases. Leaves two, oblong. Flowers 3-8, in a lax spike, yellowish-green, spurless, petals and sepals spreading; lip 4-5mm, oblong, undulate, uppermost. Bogs, fens and wet places, dune slacks, generally on calcareous soils. June-July. S Britain and Continental Europe except the far north. Local, often rare. In fens it often flowers most prolifically for a year or two after the reeds are cut. Insect or self-pollinated, although this is by no means certain. **B**: Very rare and declining; a few localities in East Anglia and S Wales; protected.

12* **Bog Orchid** *Hammarbya paludosa* (L.) O. Kuntze [*Malaxis paludosa*]. Low hairless, often only 5cm tall, with pseudobulbs one above the other. Leaves 2-5, rounded, concave, with tiny bulbils along margin near the apex. Flowers many in a slender spike, yellowish-green, sepals and petals spreading, spurless; lip 2mm, lanceolate, uppermost. Wet acid *Sphagnum* bogs, generally in rather exposed places. July-September. Throughout, except Iceland and Spitsbergen. Very local plant, often difficult to detect. **P**: Small flies. **B**: Rare and local; scattered localities in SW England, W Wales, Cumbria, W Scotland and E Ireland.

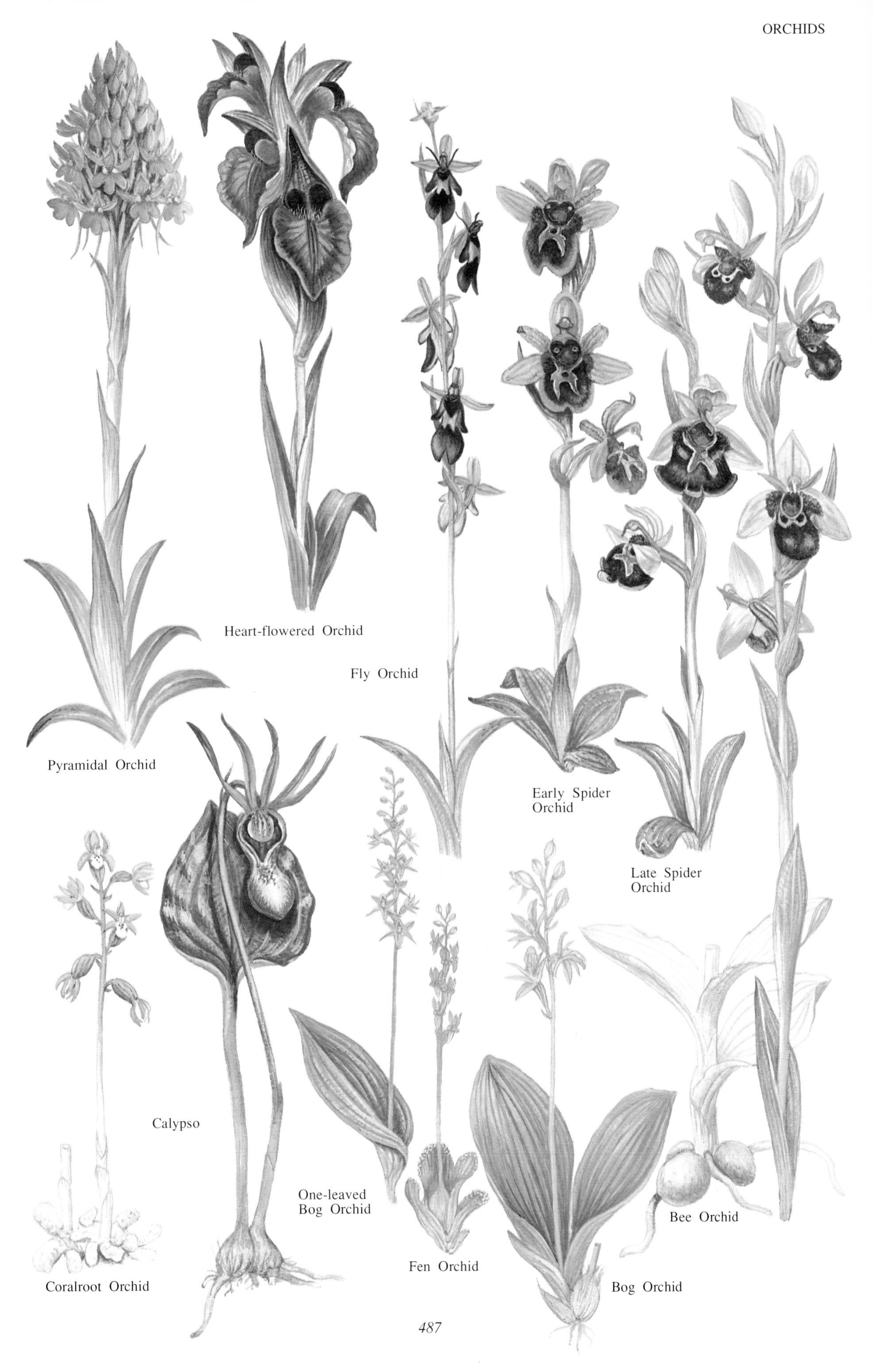

Heart-flowered Orchid
Fly Orchid
Pyramidal Orchid
Early Spider
Orchid
Late Spider
Orchid
Calypso
One-leaved
Bog Orchid
Fen Orchid
Bee Orchid
Coralroot Orchid
Bog Orchid

FURTHER RACES OF *Anthyllis vulneraria* L. (page 192)

*A. v.* subsp. *maritima* (Schweigger) Corb. [*A. maritima*] has stems white-hairy, spreading. Lower leaves with 1-3 leaflets, upper with 7-11. Flowers yellow with a plain greenish calyx. Denmark, N Germany and S Sweden.

*A. v.* subsp. *polyphylla* (DC.) Nyman [*A. arenaria, A. polyphylla*] is more erect, medium to tall, hairy below. Lower leaves with 3-7 leaflets, upper with 11-15. Flowers yellow, occasionally reddish, with plain greenish calyx. Denmark; naturalized Belgium and parts of Germany.

*A. v.* subsp. *corbierei* (Salmon & Travis) Cullen [*A. maritima* var. *corbieri*] has hairy stems and rather fleshy leaves, the lower with 1-3 leaflets, the upper with 9-11. Flowers yellow with plain greenish calyx. W Britain - Anglesey, Cornwall, Channel Islands – Sark.

*A. v.* subsp. *vulgaris* (Koch) Corbiere [subsp. *carpatica*] is short, slightly hairy. All leaves with 1-7 leaflets. Flowers pale yellow, occasionally deep yellow or reddish, with plain greenish calyx. Britain, Ireland, Belgium, France, Denmark, Germany.

*A. v.* subsp. *iberica* (W. Becker) Jalas is short and spreading, with branched stems. Leaves with 3-9 leaflets. Flowers red, the calyx with red-tipped teeth. Coastal France and Belgium.

*A. v.* subsp. *lapponica* (Hyl.) Jalas is short, the stems silkily-hairy. Leaves with 1-9 leaflets. Flowers yellow, calyx plain or teeth red-tipped.

*A. v.* subsp. *borealis* (Rouy) Jalas is hairier and has stems rarely more than 15cm. Iceland.

*A. v.* subsp. *alpestris* Ascherson & Graebner [*A. alpestris*] is similar, but the calyx has grey or blackish hairs. Mountain grassland and rocks. C and E France, S Germany.

FURTHER RACES OF MARSH ORCHIDS (page 480)

**Narrow-leaved Marsh Orchid** *Dactylorhiza traunsteineri* (Sauter) Soó

*D. t.* subsp. *curvifolia* (Nyl.) Soó has 2-4 leaves, strongly curved, lip shallowly 3-lobed or scarcely lobed at all, the central lobe, when present, scarcely longer than the adjacent ones. Sweden and Finland.

*D. t.* subsp. *lapponica* (Laest. ex Hartmann) Soó has an inflorescence with fewer and smaller flowers; lip only 5.5-7mm, the central lobe shorter than the adjacent lobes; spur two thirds the length of the ovary. Northern Finland.

**Heath Spotted Orchid** *Dactylorhiza maculata* (L.) Soó.

The British and NW European form is sometimes referred to subsp. *ericetorum* (E. F. Linton) Vermeul. **B**: Throughout in suitable habitats.

*D. t.* subsp. *elodes* (Griseb.) Soó is shorter, to only 30cm (not 60cm), all the leaves lanceolate to linear, with or without spots. Flowers pink, pale lilac or reddish-purple.

*D. t.* subsp. *islandica* (Á. & D. Löve) Soó is like subsp. *elodes*, but the plants do not exceed 18cm in height and have a hollow (not solid) stem. Only in Iceland.

# KEY TO FAMILIES

This is a key to the families of plants covered in this book. Nine of the larger families, although keyed out here, also have separate keys to the genera that they contain, on p. 498 onwards:

| | | |
|---|---|---|
| Caryophyllaceae, 498 | Rosaceae, 504 | Labiatae, 511 |
| Ranunculaceae, 500 | Leguminosae, 506 | Scrophulariaceae, 513 |
| Cruciferae, 501 | Umbelliferae, 508 | Compositae, 515 |

All the keys are of the type called dichotomous, where the reader is presented with successive sets of characters to choose from. Technical terms are explained in the Glossary (p. 31). A lens is useful for examining the smaller features.

**1** a Trees or bushes with green needle- or scale-like leaves. Flowers and seeds often borne in cones; ovules naked, solitary or on the upper surface of the cone-scales (Gymnosperms): **2**
**1** b Trees, bushes or herbs, not as above, if with scale-like leaves then also with showy and conspicuous flowers (e.g. Brooms, *Cytisus* and *Genista*); ovules enclosed within a carpel or carpels (Angiosperms): **5**

**2** a Leaves opposite or whorled; short shoots lacking: *Cupressaceae, 42*
**2** b Leaves alternate, or clustered on short lateral shoots: **3**

**3** a Seeds borne in woody cones; trees, generally with a solitary trunk: *Pinaceae, 40*
**3** b Seeds solitary, surrounded by a red fleshy aril when ripe: **4**

**4** a Evergreen trees or shrubs with flat needle-leaves: *Taxaceae, 42*
**4** b Low shrubs with rush-like stems and inconspicuous scale leaves: *Ephedraceae, 42*

**5** a Aquatic plants partly or wholly submerged, or floating; flowers often inconspicuous, green or brownish: GROUP A, *490*
**5** b Plants not as above: **6**

**6** a Flowers small (5mm or less), arranged in distinctive, simple or compound umbels, with main stalks or flower-stalks arising from one point like the spokes of an umbrella: GROUP B, *491*
**6** b Flowers not arranged as above, small or large: **7**

**7** a Flowers relatively small, massed into distinctive heads, shaving-brush-like, disc or daisy-like, rounded or button-like; flowerheads always with a ruff of bracts (flower-bracts) encircling the base: GROUP C, *491*
**7** b Flowers small to large, solitary, or arranged along a central axis in racemes, spikes, panicles or catkins: **8**

**8** a Leaves linear or strap-like to elliptical with parallel veins, or arrow-shaped (if inflorescence surrounded by a spathe – *Araceae, 470);* flowers usually with parts in 3's or 6's (Monocotyledons): GROUP D, *491*
**8** b Leaves very variable in shape, pinnately- or net-veined; flowers often with parts in 4's, 5's or more than 6 (Dicotyledons): **9**

**9** a Flowers tiny, in unbranched catkins or spikes: GROUP E, *492*
**9** b Flowers small or large, borne in racemes, branched clusters, cymes or panicles: **10**

**10** a Plants without green pigment (chlorophyll); leaves reduced to tiny scales: **11**
**10** b Plants with chlorophyll: **13**

**11** a Climbing plants with thread-like stems and flowers in tiny clusters: *Convolvulaceae, 318*
**11** b Erect herbs; flowers in spikes or racemes: **12**

**12** a Flowers regular, not lipped; whole plant cream or whitish: *Monotropaceae, 288*

**12** b Flowers lipped; ovary inferior, plants saprophytic: *Orchidaceae, 474*
**12** c Ovary superior, plants parasitic: *Orobanchaceae, 376*

**13** a Flowers irregular (zygomorphic), often 1- or 2-lipped, the petals or petal-lobes unequal: **14**
**13** b Flowers regular (actinomorphic), forming often a star-, cup- or bell-shape, with even segments – petals and/or sepals: **15**

**14** a Petals separate from one another: GROUP F, *492*
**14** b Petals fused together in various ways, often with a conspicuous tube: GROUP G, *493*

**15** a Leaves arranged in distinct whorls of 4 or more: *Rubiaceae, 310*
**15** b Leaves alternate or opposite: **16**

**16** a Flowers with a single perianth whorl or occasionally 2 whorls similar in size and colour: **17**
**16** b Flowers with a perianth consisting of 2 distinct whorls, petals and sepals, differing in both size and colour: **18**

**17** a Perianth segments petal-like, coloured or white: GROUP H, *493*
**17** b Perianth segments green: GROUP I, *494*

**18** a Petals fused below to form a tube: **19**
**18** b Petals separate from one another (free): **20**

**19** a Ovary superior: GROUP J, *494*
**19** b Ovary inferior: GROUP K, *495*

**20** a Ovary superior: GROUP L, *496*
**20** b Ovary inferior: GROUP M, *497*

GROUP A – *Aquatic plants partly or wholly submerged, or floating.*

**1** a Tiny plants without stems, floating, with round, oval or pointed scale-like segments; flowers minute: *Lemnaceae, 470*
**1** b Plants larger, with obvious stems: **2**

**2** a Leaves divided, toothed or not: **3**
**2** b Leaves undivided, toothed or not: **5**

**3** a Flowers tiny, often greenish, inconspicuous; leaves all similar: leaves pinnately-divided and flowers in whorled spikes: *Haloragaceae, 266*
**3** b Leaves repeatedly forked and flowers solitary: *Ceratophyllaceae, 104*
**3** c Flowers obvious, yellow, pink, lilac or white; leaves often of 2 types: **4**

**4** a Flowers yellow; leaves bearing tiny underwater bladders: *Lentibulariaceae (Utricularia), 378*
**4** b Flowers lilac, pink or white; leaves pinnate and flowers lilac or pink, in whorled racemes: *Primulaceae (Hottonia), 296*
**4** c Leaves palmately-divided and flowers white, solitary: *Ranunculaceae (Ranunculus), 106*

**5** a Leaves in whorls of 3-many: **6**
**5** b Leaves not in whorls, opposite or alternate, or all basal: **7**

**6** a Leaves in whorls of 8 or more; flowers in leaf-axils above the water surface: *Hippuridaceae, 266*
**6** b Leaves in whorls of 3-5: leaves toothed and flowers in the leaf-axils: *Najdaceae, 450*
**6** c Leaves untoothed, male flowers floating free, female borne on long stalks at water surface: *Hydrocharitaceae, 444*

**7** a Flowers small but in stout dense cylindrical spikes, males above females: *Typhaceae, 472*
**7** b Flowers small to large, not arranged as above: **8**

**8** a Flowers large and conspicuous, 3-5-petalled or lobed, white or pink: **9**
**8** b Flowers small and inconspicuous, greenish or brownish: **10**

**9** a Flowers 2-lipped: *Campanulaceae (Lobelia dortmanna), 388*
**9** b Flowers not 2-lipped: **10**

**10** a Leaves floating, rounded-heart-shaped to kidney-shaped; flowers white, 3-parted: *Hydrocharitaceae (Hydrocharis), 444*
**10** b Flowers yellow, 5-parted: *Menyanthaceae, 310*
**12** c Leaves not as above: **11**

**11** a Leaves in pineapple-top-like clusters, submerged except at flowering time: *Hydrocharitaceae (Stratiotes), 444*
**11 b** Leaves linear, strap-shaped to arrow- or heart-shaped, above the water; flowers rose-pink, in umbel-like clusters: *Butomaceae, 442*
**11** c Flowers white or pale pink, in whorled racemes or panicles: *Alismataceae, 442*

**12** a Flowers in distinctive spikes or globose stalked clusters: **13**
**12** b Flowers in basal clusters or solitary: **14**

**13** a Flowers in globose clusters, monoecious: *Sparganiaceae, 472*
**13** b Flowers in dense spikes: spikes stalked – plants of fresh water with floating or submerged leaves: *Potamogetaceae, 446*
**13** c Spikes unstalked – plants of saline water, all leaves submerged: *Zosteraceae, 450*

**14** a Leaves submerged and floating, some in terminal rosettes: *Callictriceae, 330*
**14** b Leaves submerged, alternate or opposite: **15**

**15** a Flowers in umbel-like clusters; leaves linear: *Ruppiaceae, 450*
**15** b Flowers in axils of leaves; solitary and with 3-4 sepals and petals, fruit a capsule: *Elatinaceae, 256*
**15** c Clustered and without obvious sepals and petals, fruit a nut: *Zannichelliaceae, 450*

GROUP B – *Flowers small (5mm or less), arranged in simple or compound umbels, main stalks or flower-stalks arising from one point.*

**1** a Leaves linear with parallel veins; flowers with 6 similar petal-like segments; stamens 6: *Liliaceae* (*Allium, Gagea*), *452*
**1 b** Leaves not as above; petals 4 or 5; stamens similar number or fewer: **2**

**2** a Leaves all in basal rosettes; sepals and petals fused below into a short or long tubes: *Primulaceae, 296*
**2** b Leaves mostly along stems; sepals and petals not fused below into tubes: **3**

**3** a Woody shrub or climbers with berry-like fruits: **4**
**3** b Herbs; fruits not berry-like, generally splitting in some manner, often capsular: **5**

**4** a Flowers 5-parted; evergreen climber with alternate leaves: *Araliaceae* (*Hedera*), *266*
**4** b Flowers 4-parted; deciduous shrubs or trees with opposite leaves, or creeping subshrubs (then flower clusters surrounded by 4 large white bracts): *Cornaceae, 266*

**5** a Plants with milky latex; leaves always undivided; flowers with a stalked 3-celled ovary and 3 styles, surrounded by 1-several 1-stamened male flowers in a toothed cup: *Euphorbiaceae* (*Euphorbia*), *232*
**5 b** Plants without milky latex; leaves often finely divided (except *Bupleurum*); flowers with 5, often notched, petals; fruits with 2 carpels and 2 styles: *Umbelliferae, 268* (and key, *508)*

GROUP C – *Flowers relatively small, massed into distinctive heads.*

**1** a Ovaries of florets (individual flowers in the flowerhead) superior: **2**
**1** b Ovaries of florets inferior: **3**

**2** a Leaves linear, all basal: flowerheads pink or white – plants of rocks and dry salt marshes: *Plumbaginaceae, 302*
**2** b Flowerheads greyish-white – plants of freshwater habitats, partly submerged: *Eriocaulaceae, 468*
**2** c Leaves oval, basal and on the stems; flowerheads blue: *Globulariaceae, 374*

**3** a Petals separate; leaves spine edged: *Umbelliferae* (*Eryngium*), *268*
**3** b Petals fused for part or most of their length into a tube; leaves sometimes spine-edged (in members of the *Compositae*): **4**

**4** a Flowers greenish, 5 per head forming a tiny 'box': *Adoxacaea, 384*
**4** b Flowers very variable in colour, but rarely greenish, few to many in flat or rounded heads, or side by side: **5**

**5** a Climbing shrubs with opposite leaves; florets 2-lipped: *Caprifoliaceae* (*Lonicera*), *380*
**5** b Herbs, leaves mostly alternate, sometimes opposite; florets not 2-lipped, the clusters of florets with a whorl or ruff of leaves or bracts immediately below, often closely overlapping: **6**

**6** a Stamens 5, often joined into a tube around the style, generally not protuding from the corolla-tube: **7**
**6** b Stamens separate (free), protruding from the corolla-tube: **8**

**7** a Florets all similar; calyx present: *Campanulacaeae* (*Campanula glomerata, Jasione*), *388*
**7** b Florets all tubular or all strap-shaped, or both in the flowerhead; calyx absent, but often replaced by hairs (pappus) or scales: *Compositae, 394* (and key, *515)*

**8** a Leaves alternate; corolla-lobes fused at first above the stamens: *Campanulaceae* (*Phyteuma*), *388*
**8** b Leaves opposite; corolla-lobes not fused together at first above the stamens: florets borne in daisy- or thistle-like flowerheads: *Dipsacaceae, 386*
**8** c Florets borne on a repeatedly forked axis: *Valerianaceae* (*Valerianella*), *384*

GROUP D – *Leaves linear or strap-like to elliptical with parallel veins, or arrow-shaped; flowers usually with parts in 3's or 6's (Monocotyledons).*

**1** a Flowers with a distinctive lip and an inferior ovary; pollen in clusters (pollinia): *Orchidaceae, 474*
**1** b Flowers not as above; ovaries superior or inferior: **2**

**2** a Ovary superior; stamens usually 6: **3**
**2** b Ovary inferior; stamens 3 or 6: **4**

**3** a Perianth green and calyx-like: *Juncaginaceae, 444*
**3** b Perianth, at least the inner whorl, coloured and petal-like: *Liliaceae, 452*

**4** a Stamens 3: *Iridaceae, 466*
**4** b Stamens 6: **5**

**5** a Leaves linear or strap-shaped; flowers hermaphrodite: *Amaryllidaceae, 464*
**5** b Leaves heart-shaped; male and female flowers borne on separate plants: *Dioscoreaceae, 464*

GROUP E – *Flowers tiny, in unbranched catkins or spikes.*

**1** a Trees or shrubs: **2**
**1** b Herbs: **8**

**2** a Flowers hermaphrodite, bright pink; stems with scale-like leaves: *Tamaricaceae, 256*
**2** b Flowers unisexual, male and female flowers separate, borne on the same or different plants, the male catkins often pendent; leaves not scale-like: **3**

**3** a Leaves covered in tiny scales; fruit a globose orange berry: *Elaeagnaceae, 244*
**3** b Leaves not covered in tiny scales; fruit dry, not a berry: **4**

**4** a Leaves resinous-aromatic; catkins erect: *Myricaceae, 52*
**4** b Leaves not resinous-aromatic; catkins often pendulous: **5**

**5** a Male and female catkins on separate plants, the male erect (*Salix*) or pendulous (*Populus*): *Salicaceae, 46*
**5** b Male and female catkins on the same plant, the male always pendulous: **6**

**6** a Fruiting cones erect, with numerous scales; fruits tiny winged nuts: *Betulaceae, 54*
**6** b Fruits not in cone-like structures, large and nut-like or borne in drooping clusters with large bracts: **7**

**7** a Nuts embedded in a woody cup or case (acorns, beechnuts, chestnuts): *Fagaceae, 56*
**7** b Nuts borne in a leafy husk, or small and winged, in drooping clusters: *Corylaceae, 54*

**8** a Stems green or reddish, fleshy and jointed, without obvious leaves: *Chenopodiaceae* (*Salicornia*), *72*
**8** b Stems not as above: **9**

**9** a Flower-spikes partially or wholly surrounded by a large coloured bract-like spathe: *Araceae, 470*
**9** b Flower-spikes or clusters not as above: **10**

**10** a Leaves strap-like with parallel veins; flowers tiny, in dense cylindrical spikes, male flowers above the female: *Typhaceae, 472*
**10** b Leaves broad, pinnately- or net-veined: **11**

**11** a Climbing herbs; fruit cone-like with papery scales: *Cannabaceae* (*Humulus*), *58*
**11** b Non-climbing plants: **12**

**12** a Plants with leaves in basal rossetes; flowers and fruits in slender rat's-tail-like spikes: *Plantaginaceae, 380*
**12** b Plants with leafy stems: **13**

**13** a Leaves all, or mostly alternate, with membranous stipules (ochrea) surrounding the stem above each leaf; flowers hermaphrodite: *Polygonaceae, 62*
**13** b Leaves opposite, without membranous stipules; flowers either male or female, borne on separate plants: **14**

**14** a Leaves and stems with stinging hairs: *Urticaceae, 60*
**14** b Plants without stinging hairs: *Euphorbiaceae* (*Mercurialis*), *232*

GROUP F – *Petals fused together in various ways, often with a conspicuous tube.*

**1** a Trees: **2**
**1** b Shrubs or herbs: **3**

**2** a Flowers pea-like, borne in pendulous racemes; fruit a pod: *Leguminosae* (*Laburnum* and *Sophora*), *200*
**2** b Flowers not pea-like, borne in erect candelabra-like panicles; fruit a conker: *Hippocastanaceae, 238*

**3** a Flowers pea-like, with standard, wing and keel petals: *Leguminosae, 200* (and key, *506*)
**3** b Flowers not pea-like: **4**

**4** a Leaves palmately-lobed: *Ranunculaceae* (*Aconitum*, *Consolida* and *Delphinium*), 106
**4** b Leaves pinnately-lobed or unlobed: **5**

**5** a Flowers spurred, the spur slender, pouch or sac-like: **6**
**5** b Flowers unspurred: **8**

**6** a Leaves deeply and pinnately-divided, ferny; flowers 2-lipped; fruit a 2-parted capsule or indehiscent (achene): *Papaveraceae* (*Corydalis* and *Fumaria*), *124*
**6** b Leaves heart-shaped to lanceolate, undivided: **7**

**7** a Fruit capsule 3-parted; lowest petal spurred: *Violaceae, 250*
**7** b Fruit capsule 5-parted, explosive; lowest sepal spurred: *Balsaminaceae, 238*

**8** a Petals 4, not lobed, 2 short and 2 long; stamens 6: *Cruciferae* (*Iberis* and *Teesdalia*), *132*
**8** b Petals 3, 4, 5 or 6, deeply lobed; stamens 8 to many: **9**

**9** a Flowers with 2 large, often petal-like, sepals and 3 tiny sepals; small herbs with petals joined into a tube; stamens 8: *Polygalaceae, 236*
**9** b Flowers with 5 more or less equal green sepals; tall herbs with long flower racemes; stamens many: *Resedaceae, 162*

## GROUP G – *Petals separate from one another.*

**1** a Ovary inferior: **2**
**1** b Ovary superior: **3**

**2** a Leaves opposite; lower lip of corolla 1-lobed; fruit a berry: *Caprifoliaceae, 380*
**2** b Leaves alternate, or all basal; lower lip of corolla 3-lobed; fruit a capsule: *Campanulaceae, 388*

**3** a Ovary 2-celled; fruit a 2-parted capsule usually with many seeds: *Scrophulariaceae, 352*
**3** b Ovary 4-lobed; fruit consisting of 4 nutlets: **4**

**4** a Leaves alternate, often bristly; inflorescence in one-sided, often spiralled (scorpoid) cymes; corolla generally rather indistinctly 2-lipped: *Boraginaceae, 320*
**4** b Leaves opposite, rarely bristly; inflorescence long spikes or whorls at the upper leaves: **5**

**5** a Style arising from the base of the ovary, from between the 4 ovules: *Labiatae, 332*
**5** b Style arising from the top of the ovary: *Verbenaceae, 328*

## GROUP H – *Perianth segments petal-like, coloured or white.*

**1** a Ovary inferior: **2**
**1** b Ovary superior: **5**

**2** a Leaves in whorls of 4 or more along the stems; ovary 2-celled: *Rubiaceae, 310*
**2** b Leaves opposite or alternate; ovary 4-5-celled: **3**

**3** a Leaves heart- or kidney-shaped: **4**
**3** b Leaves lanceolate or linear to pinnate; perianth 4-5-lobed: **6**

**4** a Low or erect herbs; perianth unlobed or with 3 short triangular lobes: *Aristolochiaceae, 62*
**4** b Climbing perennial herbs; perianth 6-lobed: *Dioscoreaceae* (*Tamus*), *464*

**5** a Leaves alternate, simple, lanceolate to linear; stamens 5: *Santalaceae, 60*
**5** b Leaves opposite, simple or pinnate; stamens 1-3: *Valerianaceae, 384*

**6** a Leaves with whitish, often translucent, stipules (ochrea) sheathing the stem above each leaf: *Polygonaceae, 62*
**6** b Leaves without stipules, or stipules free, often green and not sheathing the stem: **7**

**7** a Shrubs with berries: **8**
**7** b Herbs with capsular fruits or achenes, occasionally woody climbers with opposite leaves (*Clematis*): **10**

**8** a Perianth tubular below, 4-lobed: *Thymelaeaceae, 244*
**8** b Perianth segments free, with 6 or more segments: **9**

**9** a Heath-like shrub with tiny (1-2mm) flowers; leaves small, lanceolate, untoothed: *Empetraceae, 294*
**9** b Shrubs, sometimes with spiny stems; leaves prickly-margined: *Berberidaceae, 124*

**10** a Flowers with many stamens; fruit a head of free carpels: *Ranunculacaeae, 106*
**10** b Flowers with as many, or twice as many, stamens as perianth segments; fruit not as above, often capsular: **11**

**11** a Leaves pinnate; fruit a 1-seeded nutlet: *Rosaceae* (*Sanguisorba*), *176*
**11** b Leaves elliptical, small; fruit not as above: **12**

**12** a Fruit a capsule: *Primulaceae* (*Glaux*), *296*
**12** b Fruit a small berry: *Liliaceae* (*Maianthemum*), *452*

GROUP I – *Perianth segments green.*

| | | |
|---|---|---|
| **1** a | Parasitic on trees; with regularly forked branches and opposite leaves: | *Loranthaceae, 62* |
| **1** b | Plants not parasitic: | **2** |
| **2** a | Trees or shrubs: | **3** |
| **2** b | Herbs: | **8** |
| **3** a | Flowers borne on the surface of apparent leaves (cladodes), 6-petalled: | *Liliaceae* (*Ruscus*), *452* |
| **3** b | Flowers not borne on the 'leaf-blade': | **4** |
| **4** a | Shrubs of salt-marsh habitats with fleshy opposite or alternate leaves and tiny flowers: | *Chenopodiaceae* (*Halimione* and *Suaeda*), *72* |
| **4** b | Trees or shrubs, not of salt-marsh habitats: | **5** |
| **5** a | Leaves opposite and pinnate; fruit winged: | *Oleaceae* (*Fraxinus*), *302* |
| **5** b | Leaves simple, opposite or alternate; fruit a capsule or berry: | **6** |
| **6** a | Leaves grey or silvery, deciduous; perianth 2- lobed: | *Elaeagnaceae, 244* |
| **6** b | Leaves deep green, evergreen; perianth 4-lobed: | **7** |
| **7** a | Fruit a 3-horned capsule; flowers yellowish with 4 stamens: | *Buxaceae, 240* |
| **7** b | Fruit a black berry; flowers greenish with 8 stamens: | *Thymelaeaceae, 244* |
| **8** a | Plants with milky latex; flowers without a perianth, but each cluster with one 3-celled ovary and several stamens: | *Euphorbiaceae* (*Euphorbia*), *232* |
| **8** b | Plants not as above: | **9** |
| **9** a | Stamens many, more than twice as many as perianth segments: | *Ranunculaceae, 106* |
| **9** b | Stamens limited, at most twice as many as perianth segments, often fewer: | **10** |
| **10** a | Leaves pinnate: | *Cruciferae* (*Cardamine impatiens*), *132* |
| **10** b | Leaves palmately-lobed: | *Rosaceae* (*Alchemilla*), *176* |
| **10** c | Leaves simple, unlobed: | **11** |
| **11** a | Flowers solitary, small; fruit a long rat's-tail of achenes: | *Ranunculaceae* (*Myosurus*), *106* |
| **11** b | Flowers in groups, racemes or spikes: | **12** |
| **12** a | Flowers in flattened umbel-like heads, surrounded by green leafy bracts: | *Saxifragaceae* (*Chrysosplenium*), *168* |
| **12** b | Flowers in clusters at the leaf-axils, or in spikes or racemes, not surrounded by bracts: | **13** |
| **13** a | Plants with membranous stipules (ochrea) surrounding the stem above each leaf: | *Polygonaceae, 62* |
| **13** b | Plants without stipules, or if present then not as above: | **14** |
| **14** a | Flowers in irregular spikes or racemes, or axillary clusters; leaves often fleshy, usually alternate; perianth with 5 segments: | *Chenopodiaceae, 72* |
| **14** b | Flowers solitary or clustered; leaves opposite; perianth with 4 or 6 segments: | **15** |
| **15** a | Ovary inferior; perianth 4-parted: | *Onagraceae* (*Ludwigia*), *258* |
| **15** b | Ovary superior; perianth 6-parted: | *Lythraceae* (*Peplis*), *258* |

GROUP J – *Petals forming a tube, ovary superior.*

| | | |
|---|---|---|
| **1** a | Trees or shrubs with woody stems: | **2** |
| **1** b | Herbs, not woody, occasionally prostrate subshrubs: | **5** |
| **2** a | Corolla with a very short tube, the petals appearing almost free, but fused near the base; plants dioecious: | *Aquifoliaceae, 240* |
| **2** b | Corolla with a definite tube which is at least one third the length of the entire corolla, often more; plants with hermaphrodite flowers: | **3** |
| **3** a | Stamens 8 or 10, twice the number of the corolla lobes: | *Ericaceae, 290* |
| **3** b | Stamens equal to the number of corolla lobes or fewer: | **4** |
| **4** a | Corolla 4-lobed; stamens 2: | *Oleaceae, 302* |
| **4** b | Corolla 5-lobed; stamens 5: | *Solanaceae* (*Lycium*), *348* |
| **5** a | Corolla with a very short tube, the petals appearing almost free, but joined near the base: | *Portulacaceae, 80* |
| **5** b | Corolla with a definite tube, at least one third the length of the corolla, often more: | **6** |
| **6** a | Ovary 4-lobed or of several separate carpels: | **7** |
| **6** b | Ovary not as above: | **8** |

**7** a Ovary 4-lobed; bristly herbs usually; leaves not attached by their centre: *Boraginaceae, 320*
**7** b Ovary a cluster of free carpels; fleshy herbs; leaves circular in outline and attached by their centre: *Crassulaceae* (*Umbilicus*), *164*

**8** a Stamens opposite the corolla lobes (petals): **9**
**8** b Stamens alternating with the corolla lobes: **10**

**9** a Fruit a 1-seeded capsule; calyx papery: *Plumbaginaceae, 302*
**9** b Fruit a many-seeded capsule; calyx green: *Primulaceae, 296*

**10** a Floating or bog plants with fringed petals; leaves trifoliate or heart-shaped: *Menyanthaceae, 310*
**10** b Terrestrial plants; petals not fringed; leaves not as above: **11**

**11** a Leaves opposite, or all in basal rosettes: **12**
**11** b Leaves alternate, borne along a definite stem: **17**

**12** a Small prostrate subshrub with tiny leaves: *Ericaceae* (*Loiseleuria*), *290*
**12** b Plants not as above: **13**

**13** a Leaves alternate, all in basal rosettes: **14**
**13** b Leaves opposite, borne in basal rosettes or along stems, or both: **15**

**14** a Flowers 5-lobed, solitary, 8-15mm, borne close to the leaf cushions: *Diapensiaceae, 288*
**14** b Flowers 4-lobed, 2-4mm, borne in long slender spikes or in stalked clusters, generally above the foliage: *Plantaginaceae, 380*

**15** a Leaves unstalked; seeds without silky plumes: *Gentianaceae, 304*
**15** b Leaves stalked; seeds with silky plumes: **16**

**16** a Leaves and stems with milky juice when cut; stems erect with long-stalked flower-clusters in the leaf-axils: *Asclepiadaceae, 310*
**16** b Leaves and stems without milky juice when cut; stems sprawling and bearing solitary flowers: *Apocynaceae, 310*

**17** a Leaves pinnate; ovary 3-celled; stigmas 3: *Polemoniaceae, 316*
**17** b Leaves simple or, lobed, not pinnate; ovary 2-celled; stigmas 1-2: **18**

**18** a Herbs with twining stems; corolla funnel-shaped, scarcely lobed: *Convolvulaceae, 318*
**18** b Herbs without twining stems; corolla with obvious lobes: **19**

**19** a Flowers solitary or in arching cymes; fruit a berry or a 2-4-celled capsule: *Solanaceae, 348*
**19** b Flowers in spikes or racemes; fruit a 2-celled capsule: *Scrophulariaceae, 352* (and key, *513*)

GROUP K – *Petals forming a tube, ovary inferior.*

**1** a Flowers 5, forming a small box-like head; leaves 2-trifoliate: *Adoxaceae, 384*
**1** b Flowers many, borne in heads or spikes or racemes, or solitary; leaves not as above: **2**

**2** a Climbing by coiled tendrils; fruit a red berry; plants dioecious: *Cucurbitaceae, 256*
**2** b Plants not as above, rarely dioecious: **3**

**3** a Stamens 8-10 per flower, the anthers with terminal pores; heath-like plants: *Ericaceae, 290*
**3** b Stamens 1-5, the anthers opening by slits, not pores: **4**

**4** a Leaves in whorls of 4 or more along stems: *Rubiaceae, 310*
**4** b Leaves alternate or opposite: **5**

**5** a Anthers fused into a tube around the style: **6**
**5** b Anthers free, long-stalked: **7**

**6** a Calyx represented by a pappus of hairs or scales; flowers rarely blue: *Compositae, 394*
**6** b Calyx with 5 obvious green teeth; flowers often blue: *Campanulaceae* (*Jasione*), *388*

**7** a Leaves alternate; herbs: **8**
**7** b Leaves opposite; shrubs and herbs: **9**

**8** a Flowers cup-shaped; stamens opposite corolla lobes: *Primulaceae, 296*
**8** b Flowers bell-shaped; stamens alternating with the corolla lobes: *Campanulaceae, 388*

**9** a Stamens 1-3: *Valerianaceae, 384*
**9** b Stamens 4-5: *Caprifoliaceae, 380*

GROUP L – *Petals free, ovary superior.*

**1** a Ovary of 2 or more carpels, not or scarcely united, each carpel with its own style: **2**
**1** b Ovary of several carpels fused together to form a single structure, with a single style which is often lobed distally, or ovary of 1 carpel only: **4**

**2** a Plants with fleshy leaves, generally undivided: *Crassulaceae, 164*
**2** b Plants not fleshy: **3**

**3** a Stipules present; receptacle concave, or if convex then an epicalyx present: *Rosaceae, 176*
**3** b Stipules absent; receptacle always convex: *Ranunculaceae, 106*

**4** a Trees, shrubs or woody climbers: **5**
**4** b Herbs, not woody: **20**

**5** a Stamens and petals attached to the rim of a cup-like receptacle (hypanthium): *Rosaceae, 176*
**5** b Stamens and petals attached at the base, below the ovary: **6**

**6** a Trees: **7**
**6** b Shrubs, or occasionally woody climbers: **10**

**7** a Leaves opposite: **8**
**7** b Leaves alternate: **9**

**8** a Leaves unlobed: *Celastraceae, 240*
**8** b Leaves palmately-lobed: *Aceraceae, 238*

**9** a Leaves rounded, with a heart-shaped base: *Tiliaceae, 240*
**9** b Leaves oval, not heart-shaped at the base: *Rhamnaceae, 240*

**10** a Leaves opposite; stamens many more than the numberof petals (at least twice as many): **11**
**10** b Leaves alternate; stamens up to twice the number of petals: **12**

**11** a Styles 3 or 5; stamens united into 5 bundles: *Guttiferae, 246*
**11** b Styles 1-2; stamens all free: *Cistaceae, 254*

**12** a Deciduous shrubs: **13**
**12** b Evergreen shrubs: **16**

**13** a Leaves opposite: **14**
**13** b Leaves alternate: **15**

**14** a Leaves oval to elliptical; fruit a brightly coloured capsule: *Celastraceae, 240*
**14** b Leaves lobed; fruit a winged nut: *Aceraceae, 238*

**15** a Branches with or without simple, unbranched, terminal spines: *Rhamnaceae, 240*
**15** b Branches with 3-branched spines: *Berberidaceae, 124*

**16** a Leaves with spiny or prickly margins, or with a spine tip: **17**
**16** b Leaves without spines or prickles: **19**

**17** a Leaves spine-tipped; flowers borne on the apparent leaf-blade: *Liliaceae* (*Ruscus*), *452*
**17** b Leaves with spiny or prickly margins: **18**

**18** a Flowers yellow; leaves pinnate: *Berberidaceae* (*Mahonia*), *124*
**18** b Flowers white; leaves simple: *Aquifoliaceae, 240*

**19** a Flowers 3-parted, with 3 stamens (or stamens absent); heath-like shrubs with linear leaves up to 6mm long only: *Empetraceae, 294*
**19** b Flowers 5-parted, with 10 stamens; shrubs with oblong or linear leaves at least 10mm long: *Ericaceae* (*Ledum*), *290*

**20** a Sepals fused together into a long tube, lobed at the top: **21**
**20** b Sepals separate or fused together only at the very base: **23**

**21** a Calyx (sepal-)-tube 6-lobed; petals often 6: *Lythraceae, 258*
**21** b Calyx-tube 5-lobed; petals usually 5: **22**

**22** a Prostrate woody marsh plants; flowers with 6 stamens and 1 style: *Frankeniaceae, 256*
**22** b Plants not as above; stamens twice as many as the petals, styles 2-5: *Caryophyllaceae, 82* (and key, *498)*

**23** a Leaves opposite: **24**
**23** b Leaves alternate: **29**

**24** a Stamens many, united into 3 or 5 bundles; leaves and sepals often gland-dotted: *Guttiferae, 246*
**24** b Stamens few or many, not united into bundles; leaves and sepals rarely gland-dotted: **25**

**25** a Flowers solitary: **26**
**25** b Flowers in clusters or racemes: **27**

**26** a Leaves small, 2-6mm long; unstalked: *Saxifragaceae, 168*
**26** b Leaves larger, at least 10mm long, stalked: *Pyrolaceae, 288*

**27** a Sepals 2 only: *Portulaçaceae, 80*
**27** b Sepals 4 or 5: **28**

**28** a Stamens equal to the number of petals; ovary 5-celled: *Linaceae, 230*
**28** b Stamens twice as many as the number of petals; ovary 1-celled: *Saxifragaceae, 168*

**29** a Leaves covered with sticky, red-tipped, glandular hairs which trap insects: *Droseraceae, 162*
**29** b Leaves not as above: **30**

**30** a Leaves variously lobed, or trifoliate, or pinnate: **31**
**30** b Leaves unlobed: **37**

**31** a Stamens many more than the number of petals: **32**
**31** b Stamens equal to or up to twice as many as the number of petals: **34**

**32** a Sepals 2; petals 4; fruit a capsule: *Papaveraceae, 124*
**32** b Sepals and petals 5; fruit a berry or a ring of 1-seeded nutlets: **33**

**33** a Leaves 2-pinnate or 2-trifoliate; fruit a berry, black or red: *Ranunculaceae, 106*
**33** b Leaves palmately-lobed; fruit a ring of 1-seeded nutlets: *Malvaceae, 242*

**34** a Sepals and petals 4; stamens 6: *Cruciferae, 132* (and key, *510)*
**34** b Sepals and petals 5; stamens 10: **35**

**35** a Leaves trifoliate: *Oxalidaceae, 224*
**35** b Leaves palmately- or pinnately-lobed: **36**

**36** a Fruit a 2-parted capsule: *Saxifragaceae, 168*
**36** b Fruit 5-seeded with a long central beak: *Geraniaceae, 304*

**37** a Sepals 3 or 4: **38**
**37** b Sepals 5: **40**

**38** a Sepals and petals 3; fruit a triangular nutlet: *Polygonaceae* (*Rumex*), *62*
**38** b Sepals and petals 4, or petals absent; fruit a capsule or tiny, dry and 1-seeded: **39**

**39** a Stamens 6; fruit a 2-parted capsule splitting lengthwise: *Cruciferae, 132* (and key, *510)*
**39** b Stamens 4; fruit dry 1-seeded: *Rosaceae* (*Alchemilla* and *Aphanes*), *176*

**40** a Leaves all along stems, not in a basal rosette: *Linaceae, 230*
**40** b Leaves all or mostly in basal rosettes or tufts: **41**

**41** a Flowers borne in dense terminal clusters; leaves linear: *Plumbaginaceae, 302*
**41** b Flowers solitary or in branched clusters or racemes; leaves elliptical or rounded: **42**

**42** a Ovary with a solitary stigma or a solitary stout style: *Pyrolaceae, 288*
**42** b Ovary with 2 or 4 stigmas or styles: **43**

**43** a Stamens 10; staminodes absent; stems generally with several leaves: *Saxifragaceae, 168*
**43** b Stamens 5; staminodes 5; stem with a solitary leaf: *Parnassiaceae, 174*

GROUP M – *Petals free, ovary inferior.*

**1** a Shrubs and trees, or woody climbers: **2**
**1** b Herbs, not woody: **5**

**2** a Woody evergreen climbers, attaching to supports by stem roots: *Araliaceae, 266*
**2** b Trees or shrubs, generally deciduous: **3**

**3** a Leaves opposite, untoothed; flowers 4-parted: *Cornaceae, 266*
**3** b Leaves alternate, often toothed; flowers 5-parted: **4**

**4** a Flowers with 10 to many stamens; petals equal to or longer than the sepals: *Rosaceae, 176* (and key, *504)*
**4** b Flowers with only 5 stamens; petals shorter than the sepals: *Grossulariaceae, 174*

**5** a Petals 2 or 4, occasionally absent: *Onagraceae, 258*
**5** b Petals 5 or more, sometimes numerous: **6**

**6** a Petals numerous; plants with narrow opposite fleshy leaves, triangular in cross-section: *Aizoaceae* (*Carpobrotus*), *80*
**6** b Petals 5, rarely more; leaves not as above: **7**

**7** a Flowers in slender spike-like racemes; fruit with hooked spines: *Rosaceae* (*Agrimonia*), *176*
**7** b Flowers in dense to lax clusters, cymes or panicles; fruit without hooked spines: **8**

**8** a Stamens 10; fruit a dry capsule; flowers in cymes, panicles or solitary: *Saxifragaceae, 168*
**8** b Stamens 5; fruit composed of 2 dry, 1-seeded, nutlets pressed close together; flowers borne in heads or umbels: *Umbelliferae, 268* (and key, *508)*

# Subsidiary Keys

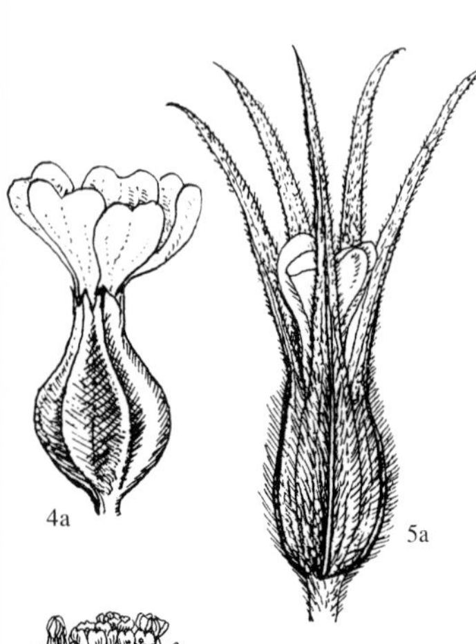


The following nine keys are to the genera (found in our region) in nine of the more important or complex Families. They are also covered in the preceding Key to Families, though in less detail.

## KEY TO CARYOPHYLLACEAE - pages 82-103

| | | |
|---|---|---|
| **1**a | Plants without stipules: | **2** |
| **1**b | Plants with stipules: | **26** |
| **2**a | Sepals joined into a tube, with 5 teeth: | **3** |
| **2**b | Sepals free (separate): | **11** |
| **3**a | Fruit a berry: | *Cucubalus, 100* |
| **3**b | Fruit a dehiscent capsule, occasionally indehiscent: | **4** |
| **4**a | Calyx winged: | *Vaccaria, 100* |
| **4**b | Calyx unwinged: | **5** |
| **5**a | Sepals leafy, much longer than the petals: | *Agrostemma, 96* |
| **5**b | Sepals neither leafy not longer than the petals: | **6** |
| **6**a | Styles 2: | **7** |
| **6**b | Styles 3-5: | **10** |
| **7**a | Calyx-tube with whitish membranous seams between the teeth: | **8** |
| **7**b | Calyx-tube without membranous seams: | **9** |
| **8**a | Flower clusters with a ruff (involucre) of bracts at the base: | *Petrorhagia, 100* |
| **8**b | Flowers separate, without a ruff of bracts: | *Gypsophila, 100* |
| **9**a | Epicalyx present; petal-limb without scales at the base (in the throat of the flower): | *Dianthus, 102* |
| **9**b | Epicalyx absent; petal-limb with scales at the base: | *Saponaria, 100* |
| **10**a | Capsule teeth twice as many as styles: | *Silene, 96* |
| **10**b | Capsule teeth equalling the number of styles: | *Lychnis, 96* |
| **11**a | Styles 2 only: | **12** |
| **11**b | Styles 3-5: | **14** |
| **12**a | Flowers without petals: | *Scleranthus, 92* |
| **12**b | Flowers with petals: | **13** |
| **13**a | Capsule dehiscing with 2 entire or notched teeth: | *Bufonia, 84* |
| **13**b | Capsule dehiscing with 4 equal teeth: | *Moehringia, 82* |
| **14**a | Capsule teeth equalling the number of styles: | **15** |
| **14**b | Capsule teeth twice as many as styles: | **18** |
| **15**a | Fleshy maritime plants: | *Honkenya, 84* |
| **15**b | Plants not as above: | **16** |
| **16**a | Styles fewer than sepals: | *Minuartia, 84* |
| **16**b | Styles as many as sepals: | **17** |
| **17**a | Leaves linear; capsule teeth not notched: | *Sagina, 92* |
| **17**b | Leaves oval; capsule teeth slightly notched: | *Myosoton, 90* |

9a 9b 10a 10b 12a 13a 13b 15a

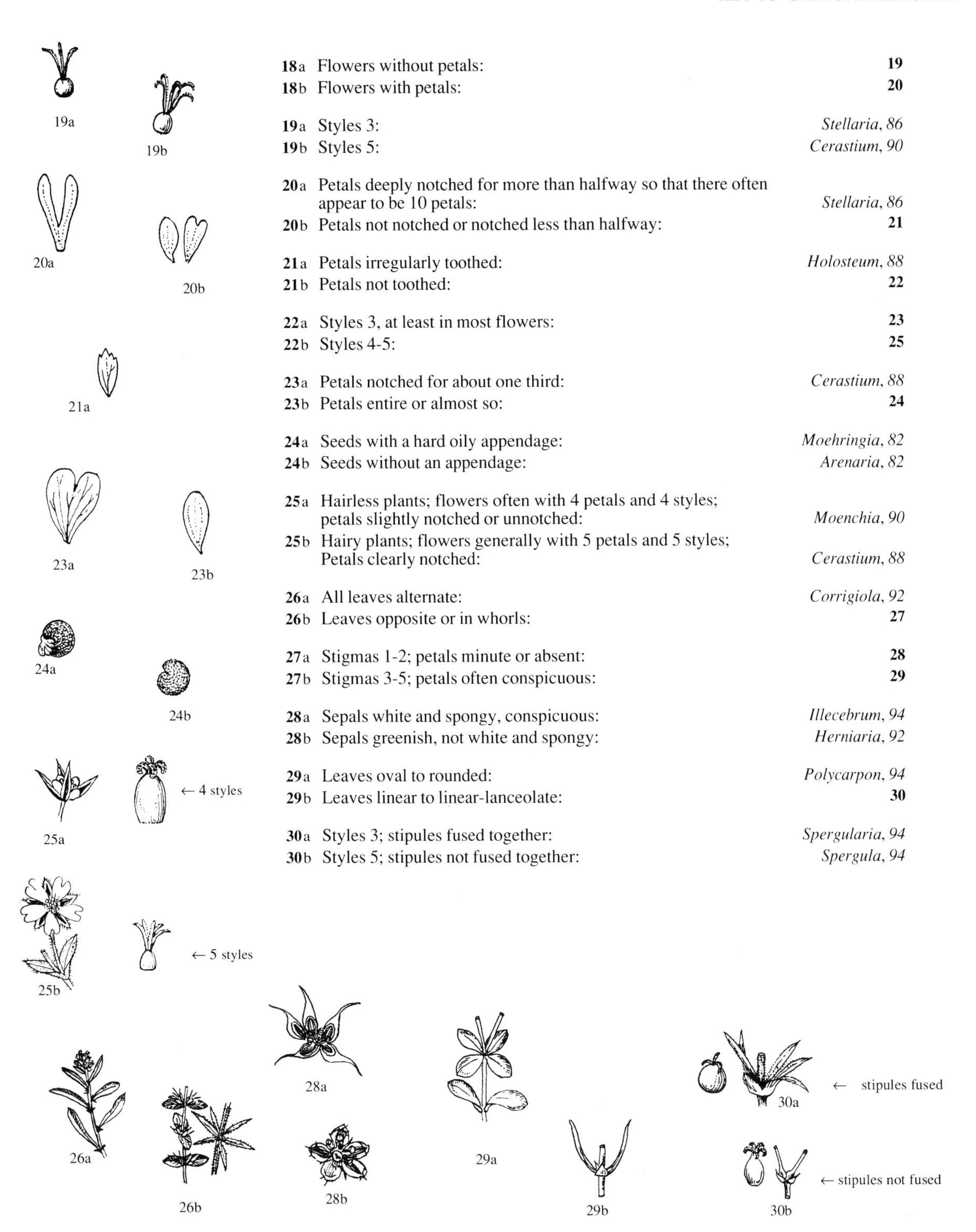


| | | |
|---|---|---|
| **18**a | Flowers without petals: | **19** |
| **18**b | Flowers with petals: | **20** |
| **19**a | Styles 3: | *Stellaria, 86* |
| **19**b | Styles 5: | *Cerastium, 90* |
| **20**a | Petals deeply notched for more than halfway so that there often appear to be 10 petals: | *Stellaria, 86* |
| **20**b | Petals not notched or notched less than halfway: | **21** |
| **21**a | Petals irregularly toothed: | *Holosteum, 88* |
| **21**b | Petals not toothed: | **22** |
| **22**a | Styles 3, at least in most flowers: | **23** |
| **22**b | Styles 4-5: | **25** |
| **23**a | Petals notched for about one third: | *Cerastium, 88* |
| **23**b | Petals entire or almost so: | **24** |
| **24**a | Seeds with a hard oily appendage: | *Moehringia, 82* |
| **24**b | Seeds without an appendage: | *Arenaria, 82* |
| **25**a | Hairless plants; flowers often with 4 petals and 4 styles; petals slightly notched or unnotched: | *Moenchia, 90* |
| **25**b | Hairy plants; flowers generally with 5 petals and 5 styles; Petals clearly notched: | *Cerastium, 88* |
| **26**a | All leaves alternate: | *Corrigiola, 92* |
| **26**b | Leaves opposite or in whorls: | **27** |
| **27**a | Stigmas 1-2; petals minute or absent: | **28** |
| **27**b | Stigmas 3-5; petals often conspicuous: | **29** |
| **28**a | Sepals white and spongy, conspicuous: | *Illecebrum, 94* |
| **28**b | Sepals greenish, not white and spongy: | *Herniaria, 92* |
| **29**a | Leaves oval to rounded: | *Polycarpon, 94* |
| **29**b | Leaves linear to linear-lanceolate: | **30** |
| **30**a | Styles 3; stipules fused together: | *Spergularia, 94* |
| **30**b | Styles 5; stipules not fused together: | *Spergula, 94* |

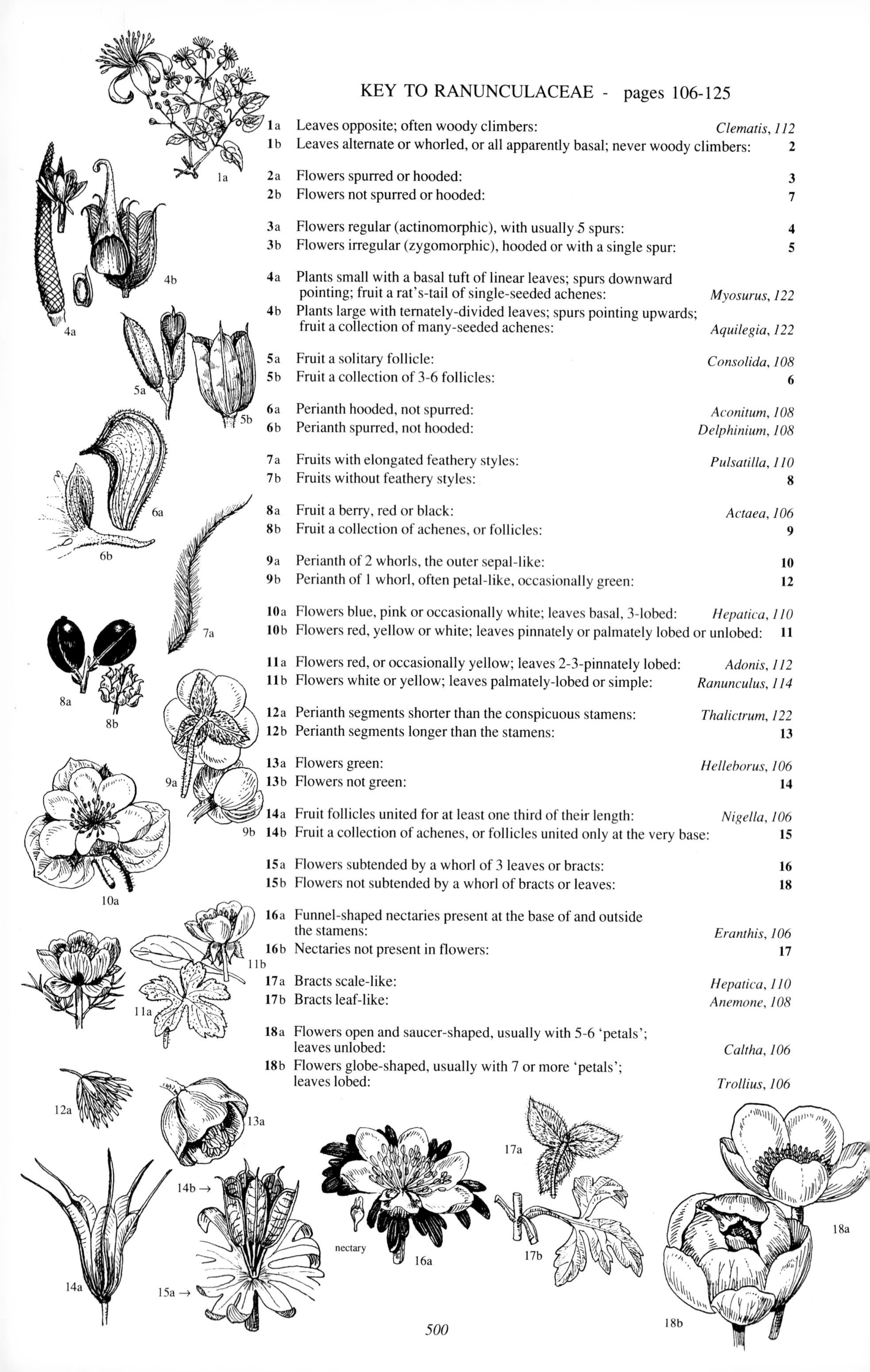


# KEY TO RANUNCULACEAE - pages 106-125

**1**a Leaves opposite; often woody climbers: *Clematis, 112*
**1**b Leaves alternate or whorled, or all apparently basal; never woody climbers: **2**

**2**a Flowers spurred or hooded: **3**
**2**b Flowers not spurred or hooded: **7**

**3**a Flowers regular (actinomorphic), with usually 5 spurs: **4**
**3**b Flowers irregular (zygomorphic), hooded or with a single spur: **5**

**4**a Plants small with a basal tuft of linear leaves; spurs downward pointing; fruit a rat's-tail of single-seeded achenes: *Myosurus, 122*
**4**b Plants large with ternately-divided leaves; spurs pointing upwards; fruit a collection of many-seeded achenes: *Aquilegia, 122*

**5**a Fruit a solitary follicle: *Consolida, 108*
**5**b Fruit a collection of 3-6 follicles: **6**

**6**a Perianth hooded, not spurred: *Aconitum, 108*
**6**b Perianth spurred, not hooded: *Delphinium, 108*

**7**a Fruits with elongated feathery styles: *Pulsatilla, 110*
**7**b Fruits without feathery styles: **8**

**8**a Fruit a berry, red or black: *Actaea, 106*
**8**b Fruit a collection of achenes, or follicles: **9**

**9**a Perianth of 2 whorls, the outer sepal-like: **10**
**9**b Perianth of 1 whorl, often petal-like, occasionally green: **12**

**10**a Flowers blue, pink or occasionally white; leaves basal, 3-lobed: *Hepatica, 110*
**10**b Flowers red, yellow or white; leaves pinnately or palmately lobed or unlobed: **11**

**11**a Flowers red, or occasionally yellow; leaves 2-3-pinnately lobed: *Adonis, 112*
**11**b Flowers white or yellow; leaves palmately-lobed or simple: *Ranunculus, 114*

**12**a Perianth segments shorter than the conspicuous stamens: *Thalictrum, 122*
**12**b Perianth segments longer than the stamens: **13**

**13**a Flowers green: *Helleborus, 106*
**13**b Flowers not green: **14**

**14**a Fruit follicles united for at least one third of their length: *Nigella, 106*
**14**b Fruit a collection of achenes, or follicles united only at the very base: **15**

**15**a Flowers subtended by a whorl of 3 leaves or bracts: **16**
**15**b Flowers not subtended by a whorl of bracts or leaves: **18**

**16**a Funnel-shaped nectaries present at the base of and outside the stamens: *Eranthis, 106*
**16**b Nectaries not present in flowers: **17**

**17**a Bracts scale-like: *Hepatica, 110*
**17**b Bracts leaf-like: *Anemone, 108*

**18**a Flowers open and saucer-shaped, usually with 5-6 'petals'; leaves unlobed: *Caltha, 106*
**18**b Flowers globe-shaped, usually with 7 or more 'petals'; leaves lobed: *Trollius, 106*

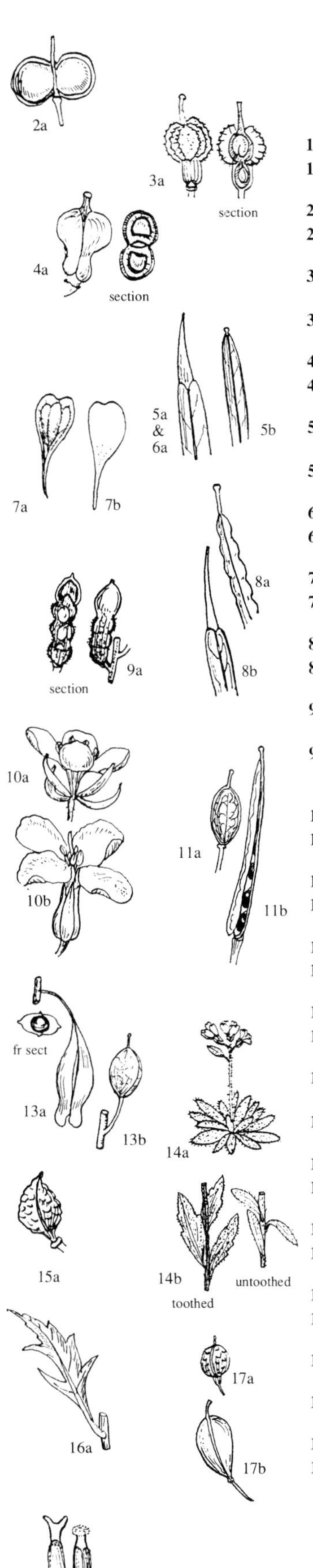


# KEY TO CRUCIFERAE - pages 132-161

**1**a Flowers yellow, occasionally with violet veins or markings: **2**
**1**b Flowers white, pink or purple: **27**

**2**a Fruit 2-lobed, with the lobes side by side: *Biscutella, 152*
**2**b Fruit not as above: **3**

**3**a Fruit composed of 2 segments, the upper globose and ribbed, the lower thickened and stalk-like: *Rapistrum, 160*
**3**b Fruit not as above: **4**

**4**a Fruit indehiscent, with 1 fertile and 2 sterile compartments: *Myagrum, 134*
**4**b Fruit dehiscent, with 2 fertile compartments: **5**

**5**a Fruit-capsule 2-valved below, but with a conspicuous beak above (the beak not splitting): **6**
**5**b Fruit-capsule not beaked, 2-valved almost to the top: **11**

**6**a Valves of fruit-capsule 1-veined: **7**
**6**b Valves of fruit-capsule 3-7-veined: **9**

**7**a Petals with violet veins: *Eruca, 158*
**7**b Petals plain yellow: **8**

**8**a Fruit-capsule distinctly beaded: *Erucastrum, 158*
**8**b Fruit-capsule not beaded: *Brassica, 158*

**9**a Fruits erect, closely pressed to the main axis; beak short, swollen below: *Hirshfeldia, 160*
**9**b Fruits spreading, not closely pressed to the main axis; beak long, not swollen below: **10**

**10**a Sepals spreading, separated from one another: *Sinapis, 158*
**10**b Sepals erect, closed: *Rhynchosinapis, 160*

**11**a Fruit a silicula, not more than 3 times longer than broad: **12**
**11**b Fruit a siliqua, at least 5 times longer than broad: **18**

**12**a Fruit-capsule flattened, elliptical in cross-section: **13**
**12**b Fruit-capsule globose or oblong, rounded or square in cross-section: **15**

**13**a Fruit-capsule oblong, pendent: *Isatis, 134*
**13**b Fruit-capsule elliptical, erect: **14**

**14**a Dwarf tufted plants, generally with untoothed leaves and with leafless flowering stems (scapes): *Draba, 146*
**14**b Plants larger, the leaves often toothed and the flowering stems leafy: *Alyssum, 146*

**15**a Fruit indehiscent (not splitting), warted and with crested wings: *Bunias, 134*
**15**b Fruit dehiscent, smooth, neither warted nor winged: **16**

**16**a Leaves toothed and or lobed; hairs simple: *Rorippa, 138*
**16**b Leaves neither lobed nor toothed; hairs branched, often star-shaped: **17**

**17**a Fruit-capsule netted; petals *c.* 2mm long: *Neslia, 152*
**17**b Fruit-capsule not netted; petals 3-5mm long: *Camelina, 150*

**18**a Fruit-capsule with 2 short horns at the apex: *Cheiranthus, 136*

**18**b Fruit-capsule without horns at the apex, the stigma disc-like or club-shaped: **19**

**19**a Fruit-capsule flattened, elliptical in cross-section; flowers cream: *Arabis, 142*
**19**b Fruit-capsule rounded or almost square in cross-section; flowers yellow to golden, rarely greenish: **20**

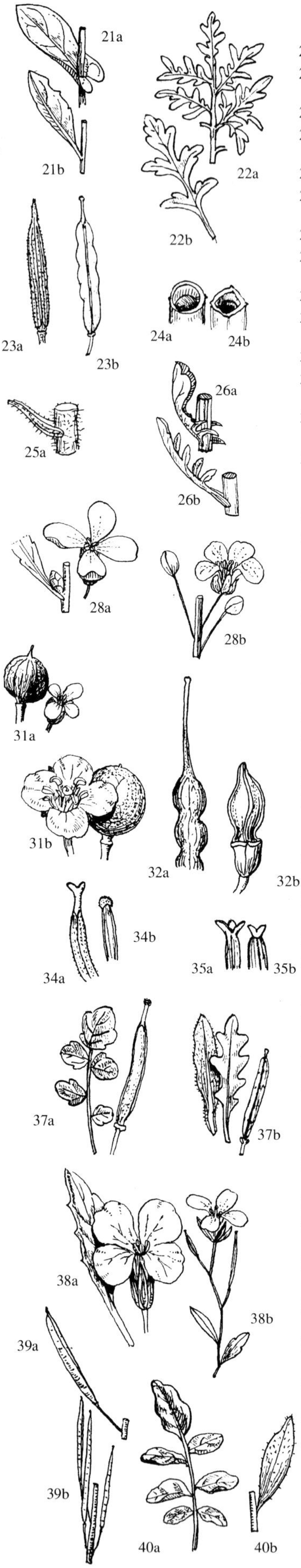

**20**a Leaves toothed or untoothed, not lobed: **21**
**20**b Leaves mostly pinnately-lobed: **22**

**21**a Stem-leaves clasping, unstalked: *Conringia, 156*
**21**b Stem-leaves not clasping, generally narrowed to the base: *Erysimum, 134*

**22**a Leaves 2-pinnately-lobed, ferny-looking: *Descurainia, 132*
**22**b Leaves 1-pinnately-lobed: **23**

**23**a Valves of fruit 3-7-veined: *Sisymbrium, 132*
**23**b Valves of fruit 1-veined: **24**

**24**a Fruit-capsule round in cross-section: *Brassica, 158*
**24**b Fruit-capsule square in cross-section: **25**

**25**a Stems hairy below; all leaves stalked: *Brassica, 158*
**25**b Stems not hairy below: **26**

**26**a Upper leaves clasping the stem, the end lobe generally larger than the adjacent lobes: *Barbarea, 136*
**26**b Upper leaves not clasping the stem, the end lobe not larger than the adjacent lobes: *Diplotaxis, 156*

**27**a Fruit not, or only rarely developed: **28**
**27**b Fruit always developed: **29**

**28**a Flowers usually pink or purple, borne in racemes; stems bearing bulbils in the leaf-axils: *Cardamine bulbifer, 140*
**28**b Flowers white, borne in broad panicles; stems without bulbils: *Armoracia, 138*

**29**a Fruit indehiscent, lacking 2 obvious parallel valves: **30**
**29**b Fruit dehiscent, splitting into 2 equal valves: **33**

**30**a Fruit small (pea-sized) and globose: **31**
**30**b Fruit larger, elongated: **32**

**31**a Flowers with unequal petals, 2-3mm long; fruit 1-parted: *Calepina, 160*
**31**b Flowers with equal petals, 6-10mm long; fruit 2-parted, with a lower seedless portion: *Crambe, 160*

**32**a Fruit-capsule tapered to the tip, sausage-like, or constricted between the seeds: *Raphanus, 160*
**32**b Fruit with a mitre-shaped end borne on a stout lower joint: *Cakile, 160*

**33**a Fruit a siliqua, at least 6 times longer than broad: **34**
**33**b Fruit a silicula, less than 4 times as long as broad: **44**

**34**a Style forked, particularly noticeable in fruit: **35**
**34**b Style not forked, the stigma disk or club-shaped: **36**

**35**a Stigmas horned on the back; maritime plants with grey leaves: *Matthiola, 136*
**35**b Stigmas not horned on the back; plants not normally maritime, with green leaves: *Hesperis, 136*

**36**a Fruit-capsule flattened, elliptical or linear in cross-section: **37**
**36**b Fruit-capsule round or square in cross-section: **40**

**37**a Leaves pinnate; fruit-valves without obvious veins: *Cardamine, 140*
**37**b Leaves untoothed to pinnately-lobed; fruit-valves with a strong mid-vein: **38**

**38**a Leaves somewhat fleshy; sepals erect, closed; flowering stems leafless: *Parrya, 136*
**38**b Leaves not fleshy; sepals spreading; flowering stems leafy: **39**

**39**a Fruit-capsules spreading, not closely pressed to the main stem axis: *Cardaminopsis, 142*
**39**b Fruit-capsules erect, closely pressed to the main stem axis: *Arabis, 142*

**40**a Leaves pinnate: *Nasturtium, 138*
**40**b Leaves not lobed or divided: **41**

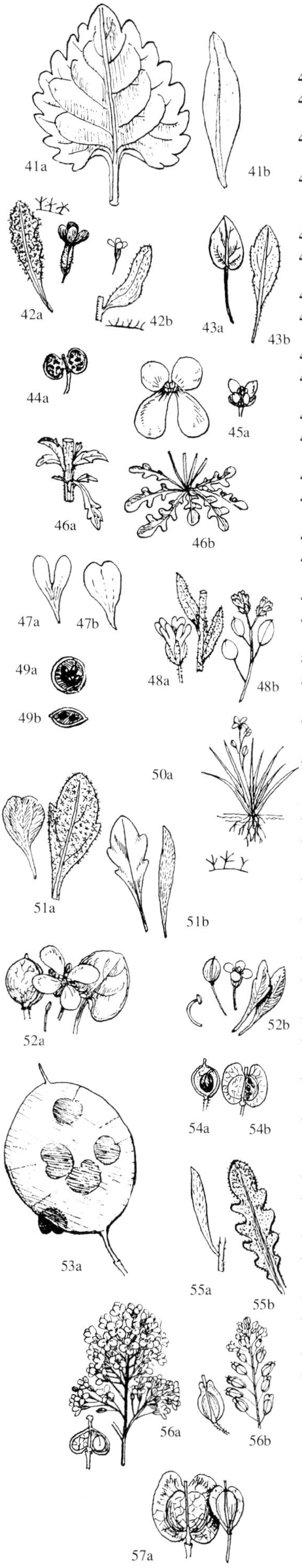

**41**a Leaves heart-shaped, smelling strongly of garlic when crushed: *Alliaria, 132*
**41**b Leaves elliptical to lanceolate, not smelling of garlic when crushed: **42**

**42**a Petals violet, 8-10mm long; plants with star-shaped hairs; stigma deeply 2-lobed: *Malcolmia, 136*
**42**b Petals white, 2-8mm long; plants with simple hairs, sometimes branched; stigma not 2-lobed: **43**

**43**a Basal leaves with a truncated or somewhat heart-shaped base: *Eutrema, 132*
**43**b Basal leaves narrowed to the base: *Arabidopsis, 134*

**44**a Fruit-capsule 2-lobed, ridged: *Coronopus, 156*
**44**b Fruit-capsule not as above: **45**

**45**a Petals unequal in length, 2 short and 2 long: **46**
**45**b Petals equal, or occasionally absent: **47**

**46**a Stems leafy: *Iberis, 152*
**46**b Stems leafless, the leaves all in a basal rosette: *Teesdalia, 152*

**47**a Petals deeply 2-lobed, lobed to at least halfway: **48**
**47**b Petals unlobed or very slightly notched: **49**

**48**a Stems leafy: *Berteroa, 146*
**48**b Stems leafless: *Erophila, 148*

**49**a Fruit-capsule round in cross-section: **50**
**49**b Fruit-capsule elliptical in cross-section: **53**

**50**a Leaves all basal; hairless aquatic plants: *Subularia, 156*
**50**b Leaves not all basal; plants not aquatic: **51**

**51**a Plants with star-shaped (stellate) hairs; petals reddish-purple or deep violet, 12mm long or more: *Aubrieta, 144*
**51**b Plants hairless or with simple unbranched hairs; petals white, pink or mauve, less than 10mm long: **52**

**52**a Leaves rather fleshy; stamens straight: *Cochlearia, 150*
**52**b Leaves not fleshy; stamens sharply curved: *Kernera, 150*

**53**a Fruit-capsule large, splitting to leave a large silvery membrane (septum): *Lunaria, 144*
**53**b Fruit-capsule not as above: **54**

**54**a Cells of fruit each with a single row of seeds: **55**
**54**b Cells of fruit each with 2 or more rows of seeds: **57**

**55**a Leaves linear, unlobed: *Lobularia, 146*
**55**b Leaves mostly lobed: **56**

**56**a Flowers borne in broad panicles; fruit-capsules with a heart-shaped base: *Cardaria, 156*
**56**b Flowers borne in spikes or racemes; fruit-capsules winged or flanged: *Lepidium, 154*

**57**a Fruit-capsules winged on the margins: *Thlaspi, 152*
**57**b Fruit-capsules not winged: **58**

**58**a Fruit-capsule triangular-heart-shaped: *Capsella, 152*
**58**b Fruit-capsule rounded or elliptical: **59**

**59**a Leaves unlobed: *Draba, 146*
**59**b Leaves pinnately-lobed: **60**

**60**a Fruits with 1-2 seeds to each compartment (loculus): *Hornungia, 152*
**60**b Fruits with 3-10 seeds to each compartment: *Hymenolobus, 152*

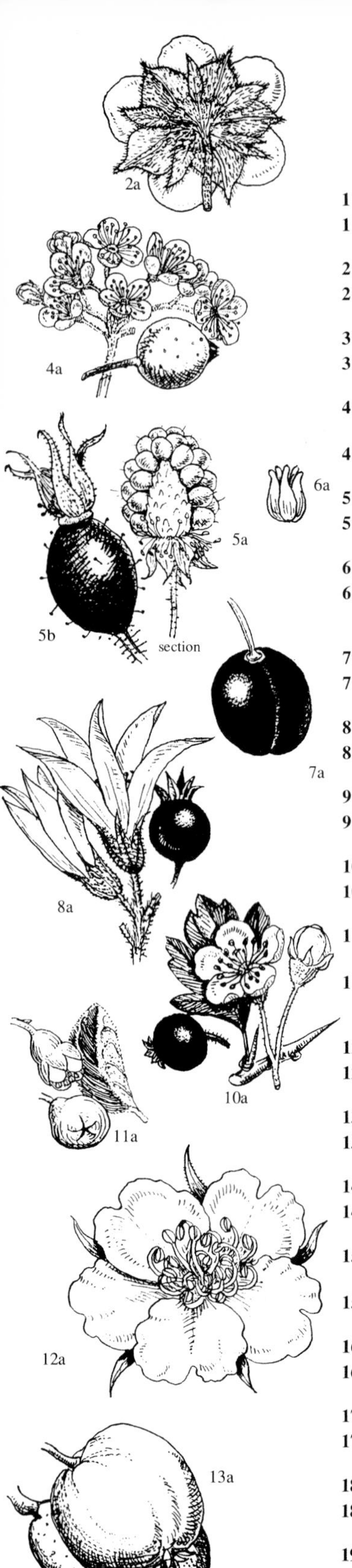


# KEY TO ROSACEAE - pages 176-199

| | | |
|---|---|---|
| **1**a | Trees or shrubs with erect, spreading or arching branches: | **2** |
| **1**b | Herb, or prostrate if woody: | **14** |
| **2**a | Flowers bright yellow; epicalyx present: | *Potentilla fruticosa, 184* |
| **2**b | Flowers white, pink or red; epicalyx absent: | **3** |
| **3**a | Leaves compound, pinnate or palmate: | **4** |
| **3**b | Leaves simple, lobed or unlobed: | **6** |
| **4**a | Flowers in large broad, rather flat, clusters; fruit berry-like; stems not prickly: | *Sorbus, 194* |
| **4**b | Flowers in panicles or small clusters; fruit raspberry-like or a hip; stems prickly: | **5** |
| **5**a | Fruit raspberry-like; receptacle convex: | *Rubus, 176* |
| **5**b | Fruit a hip; receptacle concave, forming a deep cup below the sepals: | *Rosa, 178* |
| **6**a | Fruit dry; receptacle somewhat convex: | *Spiraea, 176* |
| **6**b | Fruit fleshy, a pome, drupe or haw; receptacle concave, forming a cup below the sepals: | **7** |
| **7**a | Fruit a drupe (cherry), without sepal remains at the top: | *Prunus, 198* |
| **7**b | Fruit not as above, generally with sepal remains or marks at the top: | **8** |
| **8**a | Flowers in racemes: | *Amelanchier, 196* |
| **8**b | Flowers solitary or in clusters: | **9** |
| **9**a | Flowers in compound, umbel-like heads: | **10** |
| **9**b | Flowers solitary or 2-3 together, or in small clusters: | **11** |
| **10**a | Branches thorny; anthers purple: | *Crataegus, 196* |
| **10**b | Branches not thorny; anthers pink or cream: | *Sorbus, 194* |
| **11**a | Flowers and fruits less than 10mm; fruit red, berry-like; leaves generally untoothed: | *Cotoneaster, 196* |
| **11**b | Flowers and fruits more than 20mm; fruit green, yellowish or brown; leaves usually toothed: | **12** |
| **12**a | Flowers solitary; branches thorny: | *Mespilus, 196* |
| **12**b | Flowers in a cluster; branches not thorny: | **13** |
| **13**a | Fruit-flesh smooth – an apple; anthers yellow: | *Malus, 192* |
| **13**b | Fruit-flesh gritty – a pear; anthers pink or purple: | *Pyrus, 192* |
| **14**a | Epicalyx present: | **15** |
| **14**b | Epicalyx absent: | **21** |
| **15**a | Leaves with a large end leaflet; styles persistent, feathery and often hooked in fruit: | *Geum, 182* |
| **15**b | Leaves without a large end leaflet; styles not persistent or feathery: | **16** |
| **16**a | Flowers conspicuous, with prominent yellow, white or pink petals: | **17** |
| **16**b | Flowers tiny, greenish, with very small or no petals: | **19** |
| **17**a | Leaves pinnate or palmate: | *Potentilla, 184* |
| **17**b | Leaves trifoliate: | **18** |
| **18**a | Fruit fleshy – a strawberry; receptacle hairless: | *Fragaria, 192* |
| **18**b | Fruit dry; receptacle hairy: | *Potentilla sterilis, 188* |
| **19**a | Leaves trifoliate; sepals 5; petals tiny or absent: | *Sibbaldia procumbens, 192* |
| **19**b | Leaves palmately-lobed; sepals 4; petals absent: | **20** |

15a 17a 17b 18a 18b 19a

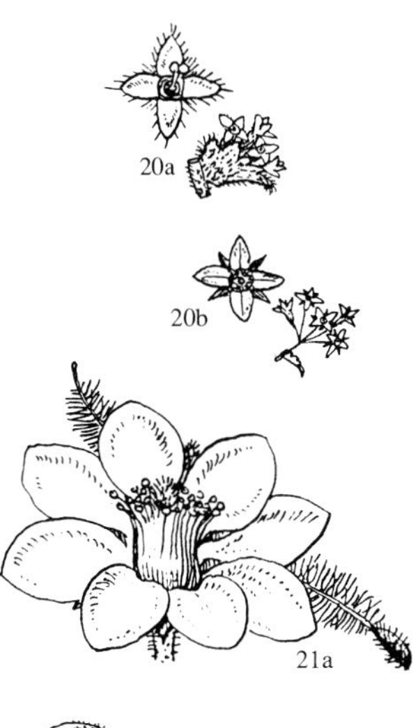

20a 21a 20b

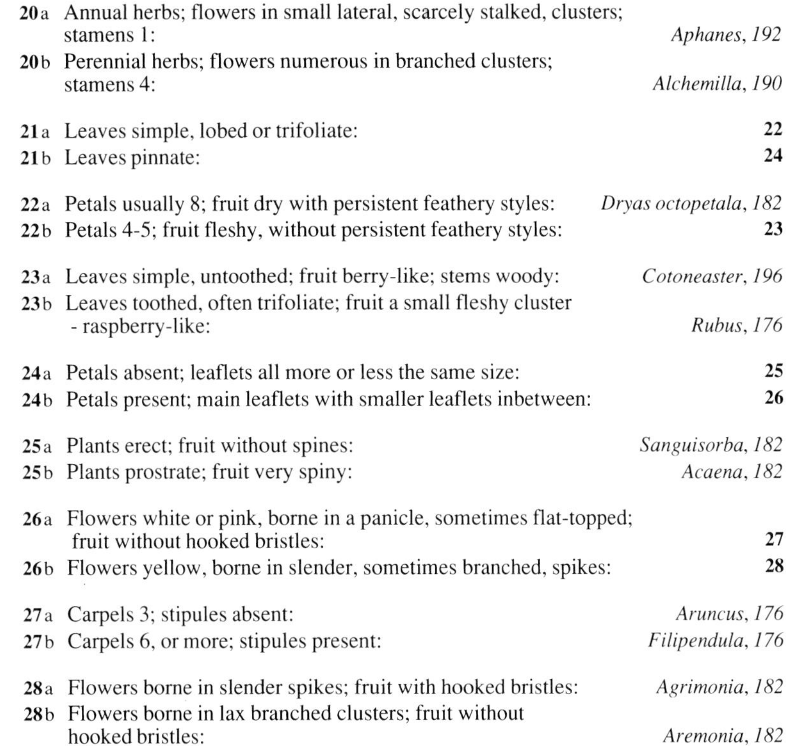

**20**a Annual herbs; flowers in small lateral, scarcely stalked, clusters; stamens 1: *Aphanes, 192*

**20**b Perennial herbs; flowers numerous in branched clusters; stamens 4: *Alchemilla, 190*

**21**a Leaves simple, lobed or trifoliate: **22**

**21**b Leaves pinnate: **24**

**22**a Petals usually 8; fruit dry with persistent feathery styles: *Dryas octopetala, 182*

**22**b Petals 4-5; fruit fleshy, without persistent feathery styles: **23**

**23**a Leaves simple, untoothed; fruit berry-like; stems woody: *Cotoneaster, 196*

**23**b Leaves toothed, often trifoliate; fruit a small fleshy cluster - raspberry-like: *Rubus, 176*

**24**a Petals absent; leaflets all more or less the same size: **25**

**24**b Petals present; main leaflets with smaller leaflets inbetween: **26**

**25**a Plants erect; fruit without spines: *Sanguisorba, 182*

**25**b Plants prostrate; fruit very spiny: *Acaena, 182*

**26**a Flowers white or pink, borne in a panicle, sometimes flat-topped; fruit without hooked bristles: **27**

**26**b Flowers yellow, borne in slender, sometimes branched, spikes: **28**

**27**a Carpels 3; stipules absent: *Aruncus, 176*

**27**b Carpels 6, or more; stipules present: *Filipendula, 176*

**28**a Flowers borne in slender spikes; fruit with hooked bristles: *Agrimonia, 182*

**28**b Flowers borne in lax branched clusters; fruit without hooked bristles: *Aremonia, 182*

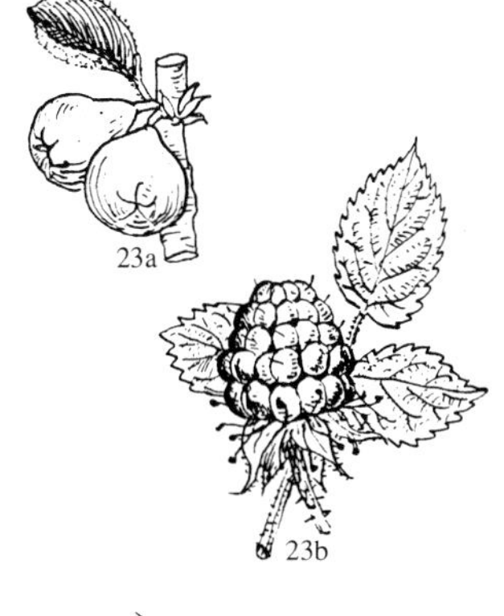

23a 23b

25a

27a

27b

25b

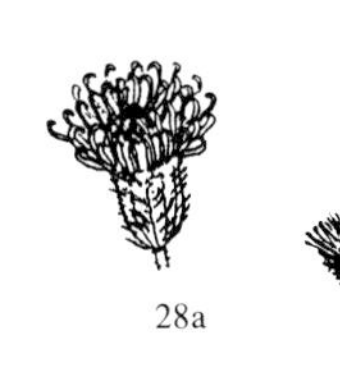

28a

28b

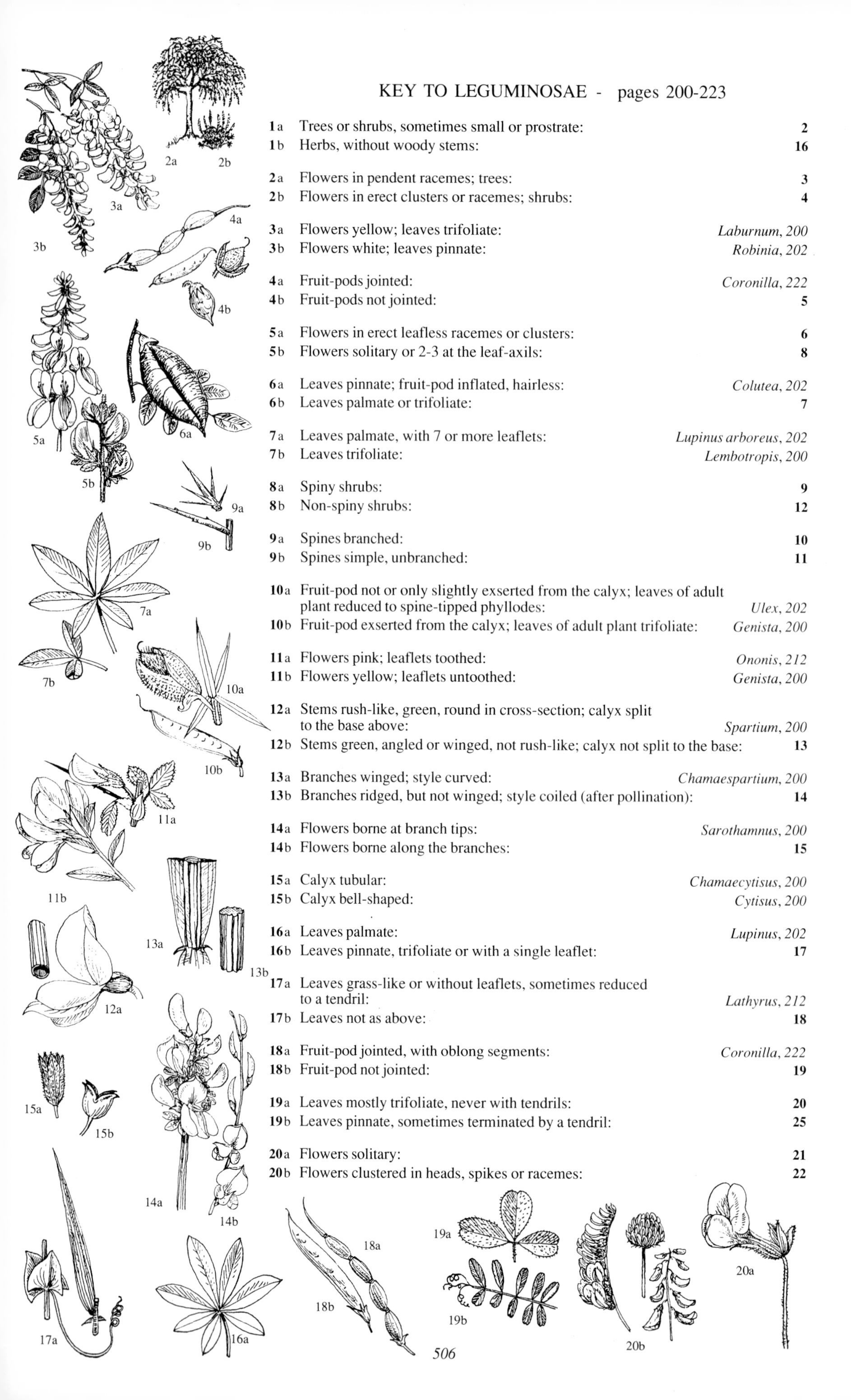


## KEY TO LEGUMINOSAE - pages 200-223

| | | |
|---|---|---|
| **1**a | Trees or shrubs, sometimes small or prostrate: | **2** |
| **1**b | Herbs, without woody stems: | **16** |
| **2**a | Flowers in pendent racemes; trees: | **3** |
| **2**b | Flowers in erect clusters or racemes; shrubs: | **4** |
| **3**a | Flowers yellow; leaves trifoliate: | *Laburnum, 200* |
| **3**b | Flowers white; leaves pinnate: | *Robinia, 202* |
| **4**a | Fruit-pods jointed: | *Coronilla, 222* |
| **4**b | Fruit-pods not jointed: | **5** |
| **5**a | Flowers in erect leafless racemes or clusters: | **6** |
| **5**b | Flowers solitary or 2-3 at the leaf-axils: | **8** |
| **6**a | Leaves pinnate; fruit-pod inflated, hairless: | *Colutea, 202* |
| **6**b | Leaves palmate or trifoliate: | **7** |
| **7**a | Leaves palmate, with 7 or more leaflets: | *Lupinus arboreus, 202* |
| **7**b | Leaves trifoliate: | *Lembotropis, 200* |
| **8**a | Spiny shrubs: | **9** |
| **8**b | Non-spiny shrubs: | **12** |
| **9**a | Spines branched: | **10** |
| **9**b | Spines simple, unbranched: | **11** |
| **10**a | Fruit-pod not or only slightly exserted from the calyx; leaves of adult plant reduced to spine-tipped phyllodes: | *Ulex, 202* |
| **10**b | Fruit-pod exserted from the calyx; leaves of adult plant trifoliate: | *Genista, 200* |
| **11**a | Flowers pink; leaflets toothed: | *Ononis, 212* |
| **11**b | Flowers yellow; leaflets untoothed: | *Genista, 200* |
| **12**a | Stems rush-like, green, round in cross-section; calyx split to the base above: | *Spartium, 200* |
| **12**b | Stems green, angled or winged, not rush-like; calyx not split to the base: | **13** |
| **13**a | Branches winged; style curved: | *Chamaespartium, 200* |
| **13**b | Branches ridged, but not winged; style coiled (after pollination): | **14** |
| **14**a | Flowers borne at branch tips: | *Sarothamnus, 200* |
| **14**b | Flowers borne along the branches: | **15** |
| **15**a | Calyx tubular: | *Chamaecytisus, 200* |
| **15**b | Calyx bell-shaped: | *Cytisus, 200* |
| **16**a | Leaves palmate: | *Lupinus, 202* |
| **16**b | Leaves pinnate, trifoliate or with a single leaflet: | **17** |
| **17**a | Leaves grass-like or without leaflets, sometimes reduced to a tendril: | *Lathyrus, 212* |
| **17**b | Leaves not as above: | **18** |
| **18**a | Fruit-pod jointed, with oblong segments: | *Coronilla, 222* |
| **18**b | Fruit-pod not jointed: | **19** |
| **19**a | Leaves mostly trifoliate, never with tendrils: | **20** |
| **19**b | Leaves pinnate, sometimes terminated by a tendril: | **25** |
| **20**a | Flowers solitary: | **21** |
| **20**b | Flowers clustered in heads, spikes or racemes: | **22** |

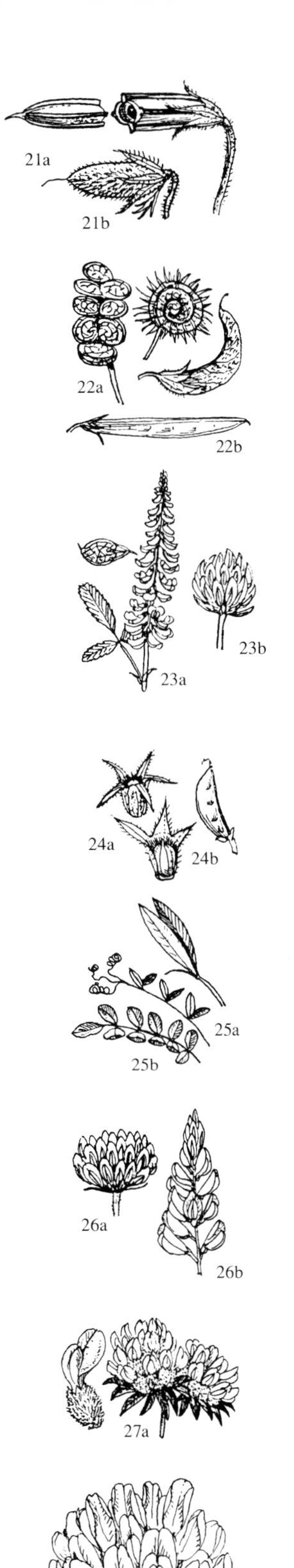


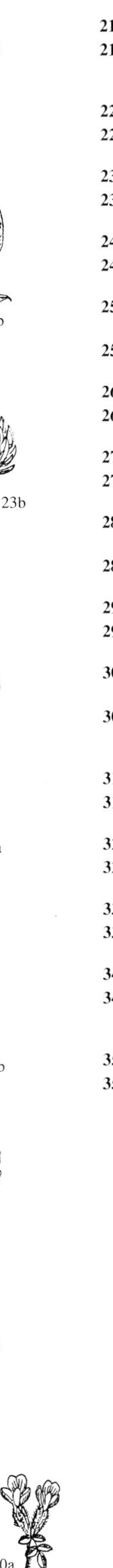

**21**a Flowers yellow, long-stalked; fruit-pod winged on angles: *Tetragonolobus, 222*
**21**b Flowers pink or yellow, short-stalked; fruit-pod not angled or winged: *Ononis, 212*

**22**a Fruit-pod spiralled or curved: *Medicago, 214*
**22**b Fruit-pod more or less straight: **23**

**23**a Flowers borne in long lax racemes: *Melilotus, 214*
**23**b Flowers borne in dense clustered heads: **24**

**24**a Fruit-pod hidden within the calyx: *Trifolium, 216*
**24**b Fruit-pod protruding well beyond the calyx: *Trigonella, 214*

**25**a Leaves terminating in a point or a tendril, sometimes with only one pair of leaflets: *Vicia* and *Lathyrus, 208, 210*
**25**b Leaves terminating in a leaflet: **26**

**26**a Flowers borne in umbel-like clusters: **27**
**26**b Flowers borne in racemes: **32**

**27**a Calyx woolly, inflated; inflorescences usually paired: *Anthyllis, 222*
**27**b Calyx not as above; inflorescences usually solitary: **28**

**28**a Flowers pink, occasionally white or purple; inflorescence with more than 10 flowers: *Coronilla varia, 222*
**28**b Flowers yellow or whitish; inflorescence with 2-10 flowers: **29**

**29**a Keel-petal dark red or blackish at the tip: *Dorycnium, 220*
**29**b Keel-petal not as above: **30**

**30**a Leaflets 5, the leaf-stalk very short or absent; fruit-pod straight, not jointed: *Lotus, 220*
**30**b Leaflets mostly more than 5 and leaves stalked; fuit-pod curved or jointed: **31**

**31**a Fruit-pods with horseshoe-shaped segments; plants hairless: *Hippocrepis, 222*
**31**b Fruit-pods with barrel-shaped segments; plants hairy: *Ornithopus, 222*

**32**a Stipules membranous, brown and papery; flowers pink or red: *Onobrychis, 222*
**32**b Stipules green; flowers yellow, blue or purple: **33**

**33**a Plants robust, hairless: **34**
**33**b Plants less robust, often low or spreading, usually hairy: **35**

**34**a Flowers white or mauve; plants erect: *Galega, 202*
**34**b Flowers greenish-yellow; plants ascending to spreading: *Astragalus glycyphyllos, 204*

**35**a Keel-petal blunt-tipped: *Astragalus, 204*
**35**b Keel-petal sharple pointed at the tip: *Oxytropis, 206*

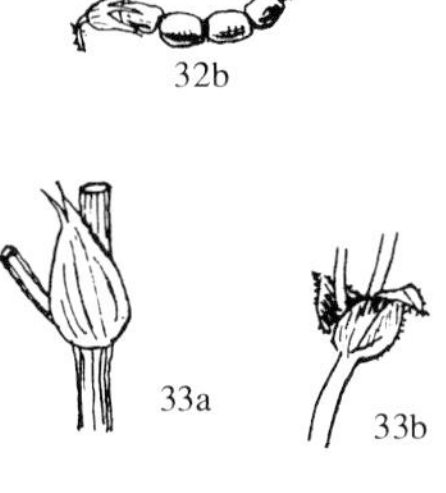


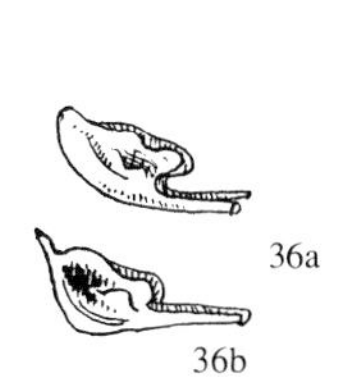

# KEY TO UMBELLIFERAE - pages 268-287

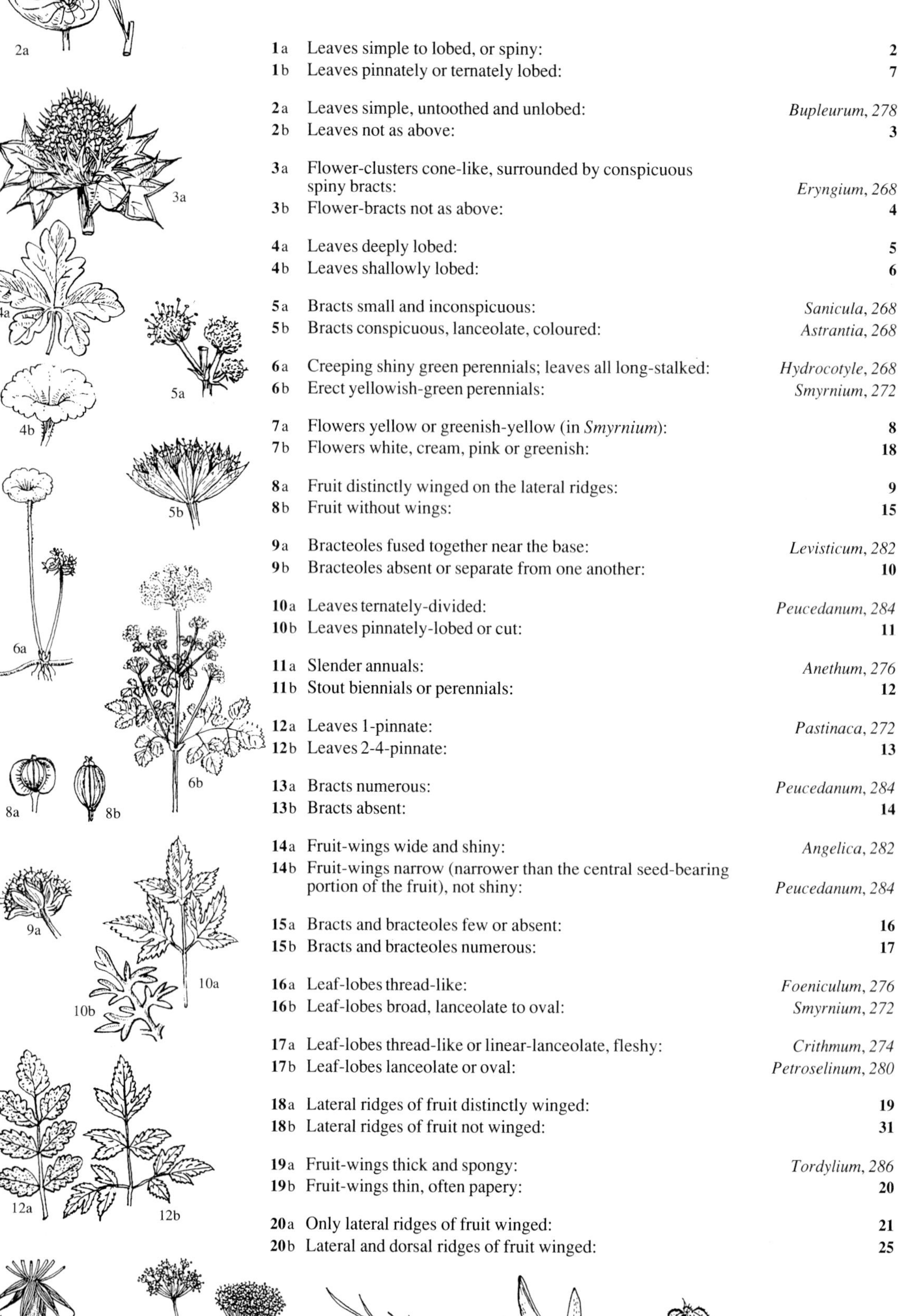


| | | |
|---|---|---|
| **1**a | Leaves simple to lobed, or spiny: | **2** |
| **1**b | Leaves pinnately or ternately lobed: | **7** |
| **2**a | Leaves simple, untoothed and unlobed: | *Bupleurum, 278* |
| **2**b | Leaves not as above: | **3** |
| **3**a | Flower-clusters cone-like, surrounded by conspicuous spiny bracts: | *Eryngium, 268* |
| **3**b | Flower-bracts not as above: | **4** |
| **4**a | Leaves deeply lobed: | **5** |
| **4**b | Leaves shallowly lobed: | **6** |
| **5**a | Bracts small and inconspicuous: | *Sanicula, 268* |
| **5**b | Bracts conspicuous, lanceolate, coloured: | *Astrantia, 268* |
| **6**a | Creeping shiny green perennials; leaves all long-stalked: | *Hydrocotyle, 268* |
| **6**b | Erect yellowish-green perennials: | *Smyrnium, 272* |
| **7**a | Flowers yellow or greenish-yellow (in *Smyrnium*): | **8** |
| **7**b | Flowers white, cream, pink or greenish: | **18** |
| **8**a | Fruit distinctly winged on the lateral ridges: | **9** |
| **8**b | Fruit without wings: | **15** |
| **9**a | Bracteoles fused together near the base: | *Levisticum, 282* |
| **9**b | Bracteoles absent or separate from one another: | **10** |
| **10**a | Leaves ternately-divided: | *Peucedanum, 284* |
| **10**b | Leaves pinnately-lobed or cut: | **11** |
| **11**a | Slender annuals: | *Anethum, 276* |
| **11**b | Stout biennials or perennials: | **12** |
| **12**a | Leaves 1-pinnate: | *Pastinaca, 272* |
| **12**b | Leaves 2-4-pinnate: | **13** |
| **13**a | Bracts numerous: | *Peucedanum, 284* |
| **13**b | Bracts absent: | **14** |
| **14**a | Fruit-wings wide and shiny: | *Angelica, 282* |
| **14**b | Fruit-wings narrow (narrower than the central seed-bearing portion of the fruit), not shiny: | *Peucedanum, 284* |
| **15**a | Bracts and bracteoles few or absent: | **16** |
| **15**b | Bracts and bracteoles numerous: | **17** |
| **16**a | Leaf-lobes thread-like: | *Foeniculum, 276* |
| **16**b | Leaf-lobes broad, lanceolate to oval: | *Smyrnium, 272* |
| **17**a | Leaf-lobes thread-like or linear-lanceolate, fleshy: | *Crithmum, 274* |
| **17**b | Leaf-lobes lanceolate or oval: | *Petroselinum, 280* |
| **18**a | Lateral ridges of fruit distinctly winged: | **19** |
| **18**b | Lateral ridges of fruit not winged: | **31** |
| **19**a | Fruit-wings thick and spongy: | *Tordylium, 286* |
| **19**b | Fruit-wings thin, often papery: | **20** |
| **20**a | Only lateral ridges of fruit winged: | **21** |
| **20**b | Lateral and dorsal ridges of fruit winged: | **25** |

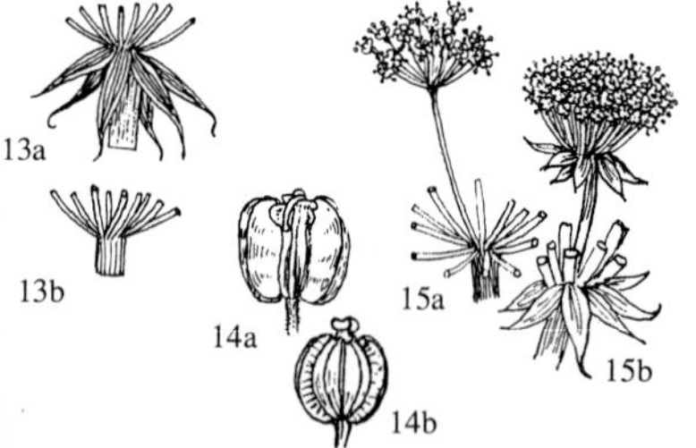


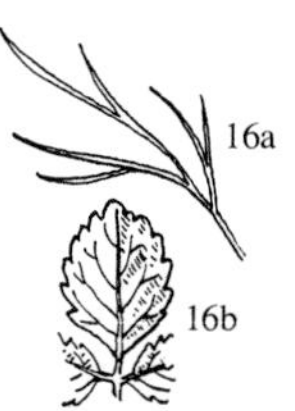


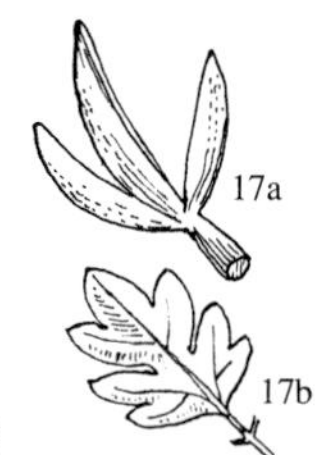


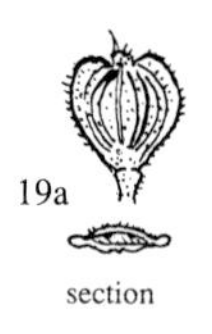


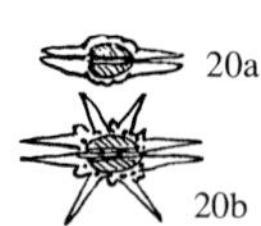

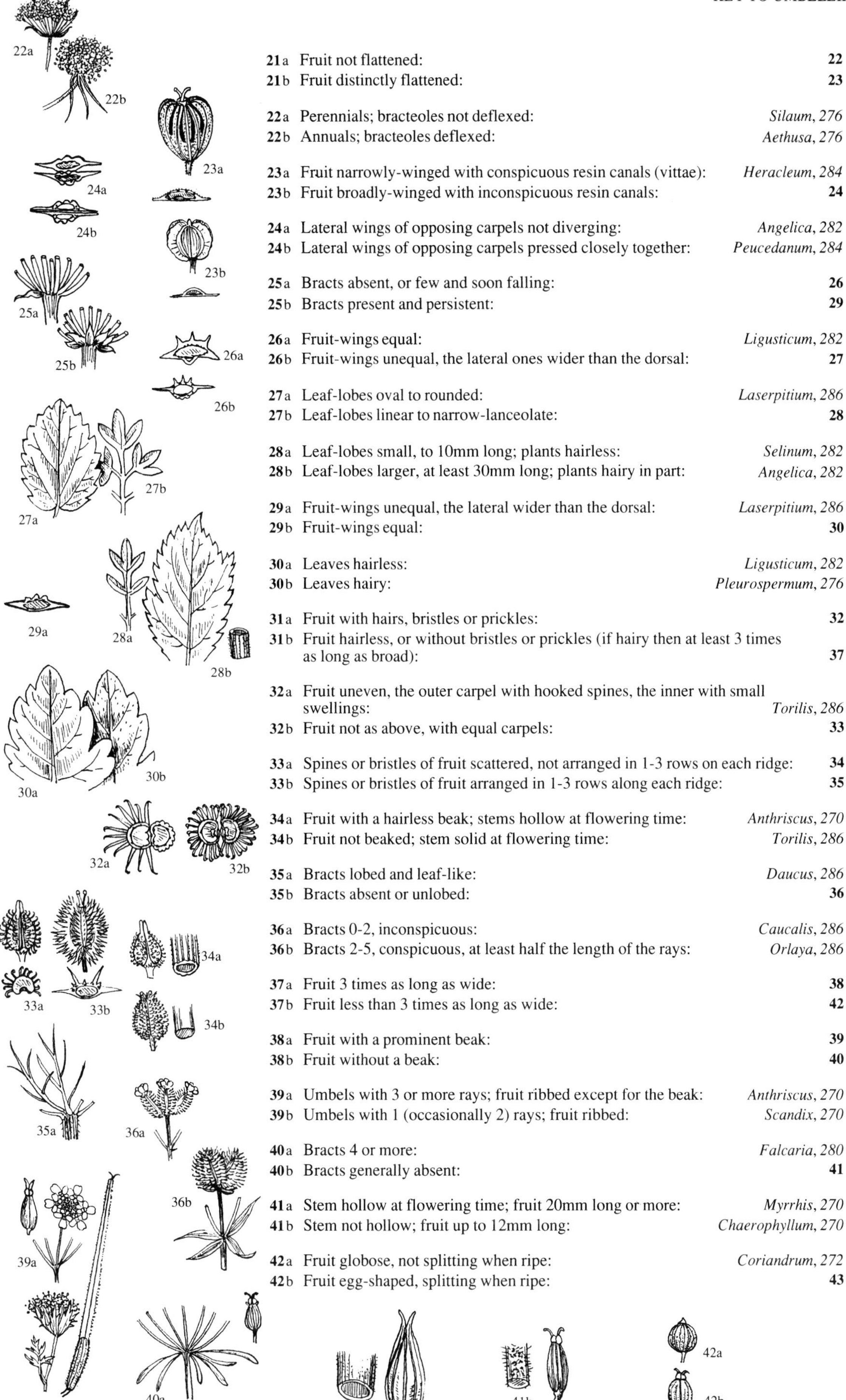


**21**a Fruit not flattened: **22**
**21**b Fruit distinctly flattened: **23**

**22**a Perennials; bracteoles not deflexed: *Silaum, 276*
**22**b Annuals; bracteoles deflexed: *Aethusa, 276*

**23**a Fruit narrowly-winged with conspicuous resin canals (vittae): *Heracleum, 284*
**23**b Fruit broadly-winged with inconspicuous resin canals: **24**

**24**a Lateral wings of opposing carpels not diverging: *Angelica, 282*
**24**b Lateral wings of opposing carpels pressed closely together: *Peucedanum, 284*

**25**a Bracts absent, or few and soon falling: **26**
**25**b Bracts present and persistent: **29**

**26**a Fruit-wings equal: *Ligusticum, 282*
**26**b Fruit-wings unequal, the lateral ones wider than the dorsal: **27**

**27**a Leaf-lobes oval to rounded: *Laserpitium, 286*
**27**b Leaf-lobes linear to narrow-lanceolate: **28**

**28**a Leaf-lobes small, to 10mm long; plants hairless: *Selinum, 282*
**28**b Leaf-lobes larger, at least 30mm long; plants hairy in part: *Angelica, 282*

**29**a Fruit-wings unequal, the lateral wider than the dorsal: *Laserpitium, 286*
**29**b Fruit-wings equal: **30**

**30**a Leaves hairless: *Ligusticum, 282*
**30**b Leaves hairy: *Pleurospermum, 276*

**31**a Fruit with hairs, bristles or prickles: **32**
**31**b Fruit hairless, or without bristles or prickles (if hairy then at least 3 times as long as broad): **37**

**32**a Fruit uneven, the outer carpel with hooked spines, the inner with small swellings: *Torilis, 286*
**32**b Fruit not as above, with equal carpels: **33**

**33**a Spines or bristles of fruit scattered, not arranged in 1-3 rows on each ridge: **34**
**33**b Spines or bristles of fruit arranged in 1-3 rows along each ridge: **35**

**34**a Fruit with a hairless beak; stems hollow at flowering time: *Anthriscus, 270*
**34**b Fruit not beaked; stem solid at flowering time: *Torilis, 286*

**35**a Bracts lobed and leaf-like: *Daucus, 286*
**35**b Bracts absent or unlobed: **36**

**36**a Bracts 0-2, inconspicuous: *Caucalis, 286*
**36**b Bracts 2-5, conspicuous, at least half the length of the rays: *Orlaya, 286*

**37**a Fruit 3 times as long as wide: **38**
**37**b Fruit less than 3 times as long as wide: **42**

**38**a Fruit with a prominent beak: **39**
**38**b Fruit without a beak: **40**

**39**a Umbels with 3 or more rays; fruit ribbed except for the beak: *Anthriscus, 270*
**39**b Umbels with 1 (occasionally 2) rays; fruit ribbed: *Scandix, 270*

**40**a Bracts 4 or more: *Falcaria, 280*
**40**b Bracts generally absent: **41**

**41**a Stem hollow at flowering time; fruit 20mm long or more: *Myrrhis, 270*
**41**b Stem not hollow; fruit up to 12mm long: *Chaerophyllum, 270*

**42**a Fruit globose, not splitting when ripe: *Coriandrum, 272*
**42**b Fruit egg-shaped, splitting when ripe: **43**

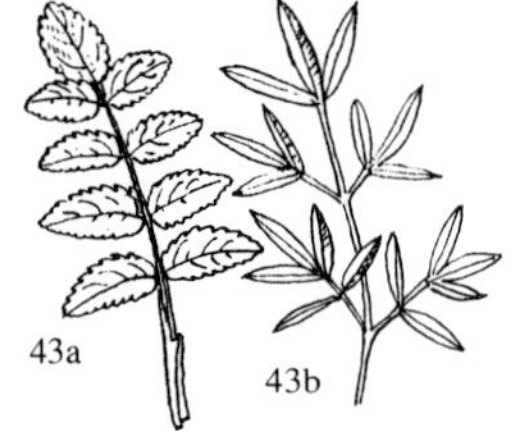


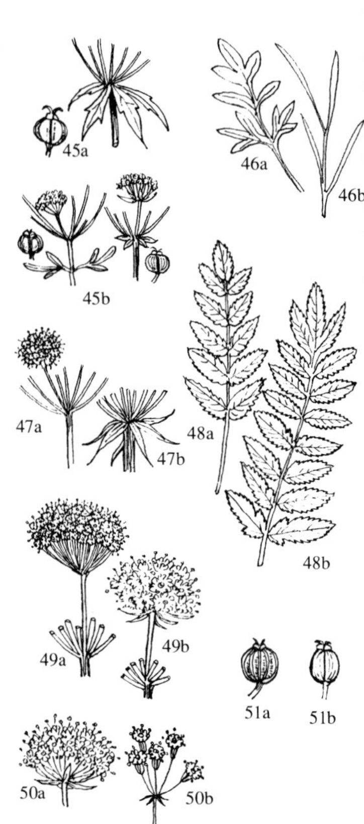


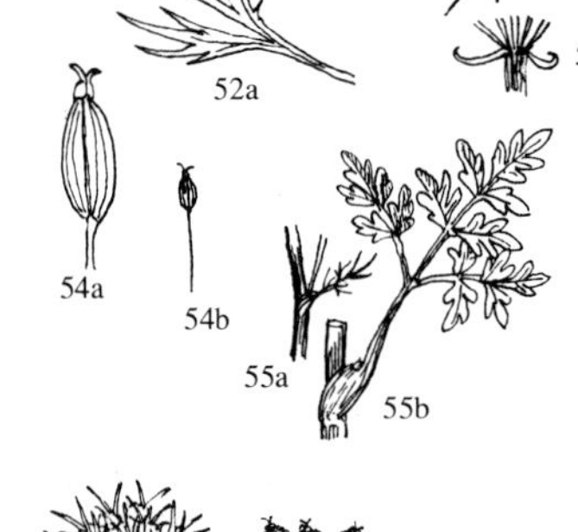


| | | |
|---|---|---|
| **43**a | Lower leaves simple or pinnate: | **44** |
| **43**b | Lower leaves 2-ternate or 2-pinnate or more divided: | **52** |
| **44**a | Plants with creeping stems, often rooting at the nodes: | **45** |
| **44**b | Plants erect, not rooting at the nodes: | **46** |
| **45**a | Bracts numerous, large and lobed: | *Berula, 272* |
| **45**b | Bracts few or none, if present unlobed: | *Apium, 278* |
| **46**a | Upper leaves with oval or lanceolate segments: | **47** |
| **46**b | Upper leaves with linear or thread-like segments: | **49** |
| **47**a | Bracts absent: | *Pimpinella, 272* |
| **47**b | Bracts numerous: | **48** |
| **48**a | Leaves with 4-6 pairs of segments: | *Sium, 272* |
| **48**b | Leaves with 7-10 pairs of segments: | *Berula, 272* |
| **49**a | Bracteoles absent: | *Pimpinella, 272* |
| **49**b | Bracteoles present: | **50** |
| **50**a | Bracteoles at least half the length of the flower-stalks: | *Oenanthe, 274* |
| **50**b | Bracteoles short, only one quarter the length of the flower-stalks: | **51** |
| **51**a | Plant foul-smelling; fruit globose: | *Sison, 280* |
| **51**b | Plant parsley-scented; fruit egg-shaped: | *Petroselinum, 280* |
| **52**a | Aquatic plants with finely divided underwater leaves: | *Oenanthe, 274* |
| **52**b | Plants not as above: | **53** |
| **53**a | Bracts large, at least half the length of the rays, the larger often divided: | **54** |
| **53**b | Bracts smaller, less than half the length of the rays, occasionally absent: | **55** |
| **54**a | Plants 1-2m tall; fruit at least 4mm long: | *Pleurospermum, 276* |
| **54**b | Plants to 1m tall; fruit to 2.5mm long: | *Ammi, 280* |
| **55**a | Stem-leaves very small or absent: | *Carum, 280* |
| **55**b | Stem-leaves conspicuous: | **56** |
| **56**a | Fruit-stalks thickened and mostly shorter than the fruit: | *Oenanthe, 274* |
| **56**b | Fruit-stalks not thickened, mostly longer than the fruit: | **57** |
| **57**a | Bracts several: | **58** |
| **57**b | Bracts absent: | **61** |
| **58**a | Leaf-lobes thread-like: | **59** |
| **58**b | Leaf-lobes lanceolate to oval: | **60** |
| **59**a | Leaves narrow-oblong in general outline: | *Carum, 280* |
| **59**b | Leaves oval to triangular in general outline: | *Meum, 276* |
| **60**a | Stems purple spotted: | *Conium, 276* |
| **60**b | Stems not purple spotted: | *Carum, 280* |
| **61**a | Basal leaves with linear to linear-lanceolate segments: | **62** |
| **61**b | Basal leaves with oval segments: | **63** |
| **62**a | Leaf-segments sharply toothed: | *Cicuta, 280* |
| **62**b | Leaf-segments untoothed or pinnately cut: | *Carum, 280* |
| **63**a | Rhizomatous perennials, patch-forming; lower leaves 2-ternate: | *Aegopodium, 272* |
| **63**b | Non-rhizomatous perennials; lower leaves 2-pinnate: | *Apium, 278* |

56a 56b 57a

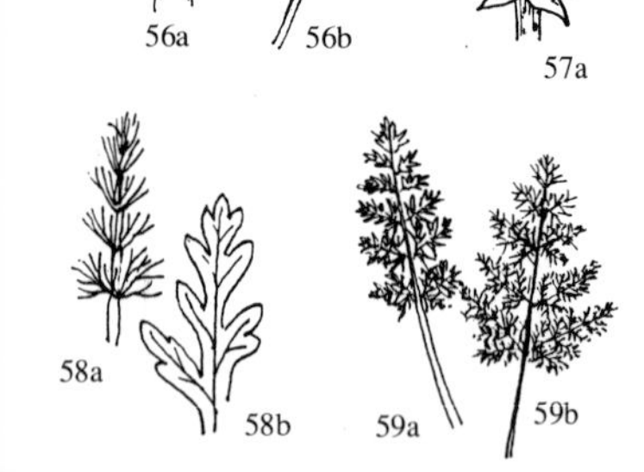

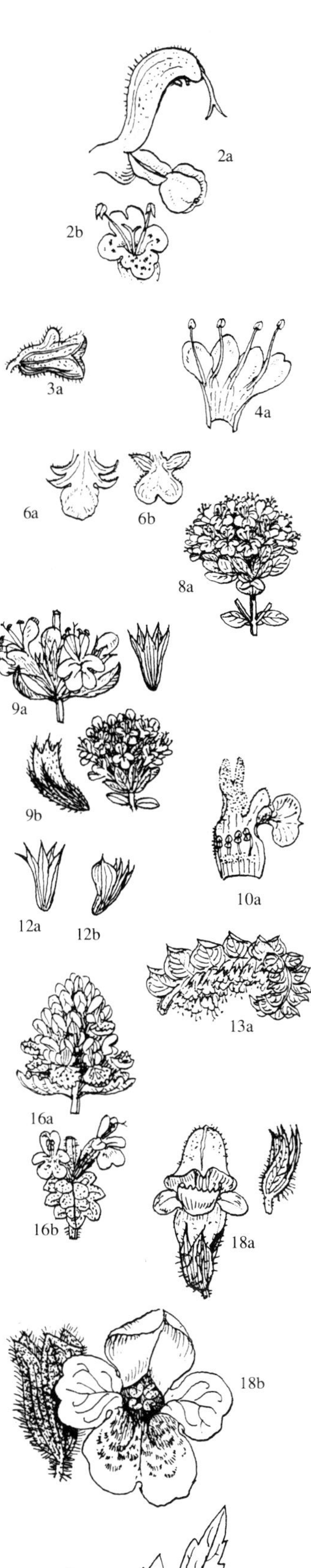


# KEY TO LABIATAE - pages 332-347

**1**a Stamens 2: **2**
**1**b Stamens 4: **3**

**2**a Corolla clearly 2-lipped: *Salvia, 346*
**2**b Corolla with 4 more or less equal lobes: *Lycopus, 342*

**3**a Calyx 2-lipped, the upper lip with a scale on the back (an erect dorsal projection): *Scutellaria, 334*
**3**b Calyx 5-toothed, if 2-lipped then uper lip clearly 2-toothed: **4**

**4**a Corolla with 4 more or less equal lobes: *Mentha, 344*
**4**b Corolla not as above: **5**

**5**a Corolla 1-lipped, the upper absent or reduced to 2 tiny teeth: **6**
**5**b Corolla clearly 2-lipped: **7**

**6**a Lower lip of corolla 5-lobed, the tube without a ring of hairs within: *Teucrium, 332*
**6**b Lower lip of corolla 3-lobed, the upper lip very small and 2-toothed, the tube with a ring of hairs within: *Ajuga, 332*

**7**a Stamens exserted beyond the upper lip of the corolla, diverging: **8**
**7**b Stamens included within the tube or the upper lip of the corolla: **10**

**8**a Flowers in branched panicles; bracts conspicuous: *Origanum, 342*
**8**b Flowers in whorls, often forming leafy spikes; bracts inconspicuous: **9**

**9**a Flowers violet-blue; calyx with 5 equal teeth: *Hyssopus, 342*
**9**b Flowers pink; calyx 2-lipped: *Thymus, 342*

**10**a Stamens included within the corolla-tube: *Marrubium, 332*
**10**b Stamens not as above - included under the hood of the corolla: **11**

**11**a Calyx with 11-15 veins: **12**
**11**b Calyx with 5-10 veins: **13**

**12**a Calyx with 5 equal teeth, 11-13-veined: *Satureja, 340*
**12**b Calyx 2-lipped, 15-veined: *Dracocephalum, 340*

**13**a Flowers in crowded one-sided spikes: *Elsholtzia, 346*
**13**b Flowers not as above: **14**

**14**a Stamens straight; upper lip of corolla concave: **15**
**14**b Stamens curved; upper lip of corolla flat (slightly concave in *Melissa*): **24**

**15**a Outer stamens shorter than the inner: **16**
**15**b Outer stamens longer than the inner: **17**

**16**a Flowers many in each whorl, borne in erect terminal inflorescences: *Nepeta, 338*
**16**b Flowers 2-6 in each whorl, borne in the leaf-axils of creeping stems: *Glechoma, 340*

**17**a Calyx 2-lipped: **18**
**17**b Calyx with 5 more or less equal teeth: **19**

**18**a Corolla up to 20mm long; calyx closed after flowering: *Prunella, 340*
**18**b Corolla 25mm long or more; calyx not closed after flowering: *Melittis, 332*

**19**a Lower leaves deeply lobed: *Leonurus, 336*
**19**b Lower leaves not deeply lobed: **20**

**20**a Calyx funnel-shaped, truncated and with 5 short teeth: *Ballota, 336*
**20**b Calyx bell-shaped to tubular, with long teeth: **21**

**21**a Lower lip of corolla with obscure lateral lobes: *Lamium, 336*
**21**b Lower lip of corolla with well developed lateral lobes: **22**

22a

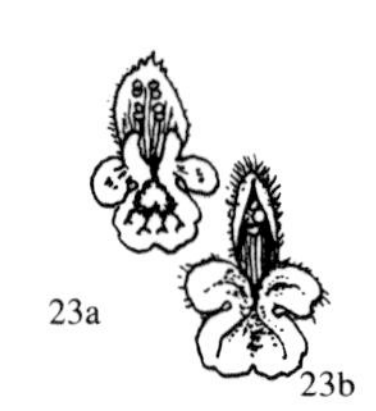
23a 23b

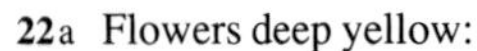

**22**a Flowers deep yellow: *Lamiastrum, 336*
**22**b Flowers purple or pink: **23**

**23**a Annuals; corolla with 2 bosses (lumps) at the base of the lower lip: *Galeopsis, 334*
**23**b Perennials, occasionally annuals, but then corolla without 2 bosses at the base of the lower lip: *Stachys, 338*

**24**a Corolla-tube sharply upcurved close to the base: *Melissa, 340*
**24**b Corolla-tube not as above: **25**

**25**a Style branches unequal; calyx with 5 equal teeth: *Phlomis, 334*
**25**b Style branches more or less equal; calyx 2-lipped: **26**

**26**a Flowers borne in opposite, stalked cymes; calyx straight: *Calamintha, 342*
**26**b Flowers not as above, borne in tight whorls; calyx curved: **27**

**27**a Erect plants with rosy-purple flowers; whorls with 3-8 flowers: *Clinopodium, 342*
**27**b Creeping or ascending plants with violet flowers; whorls many-flowered: *Acinos, 340*

24a

25a

26a

27a

27b

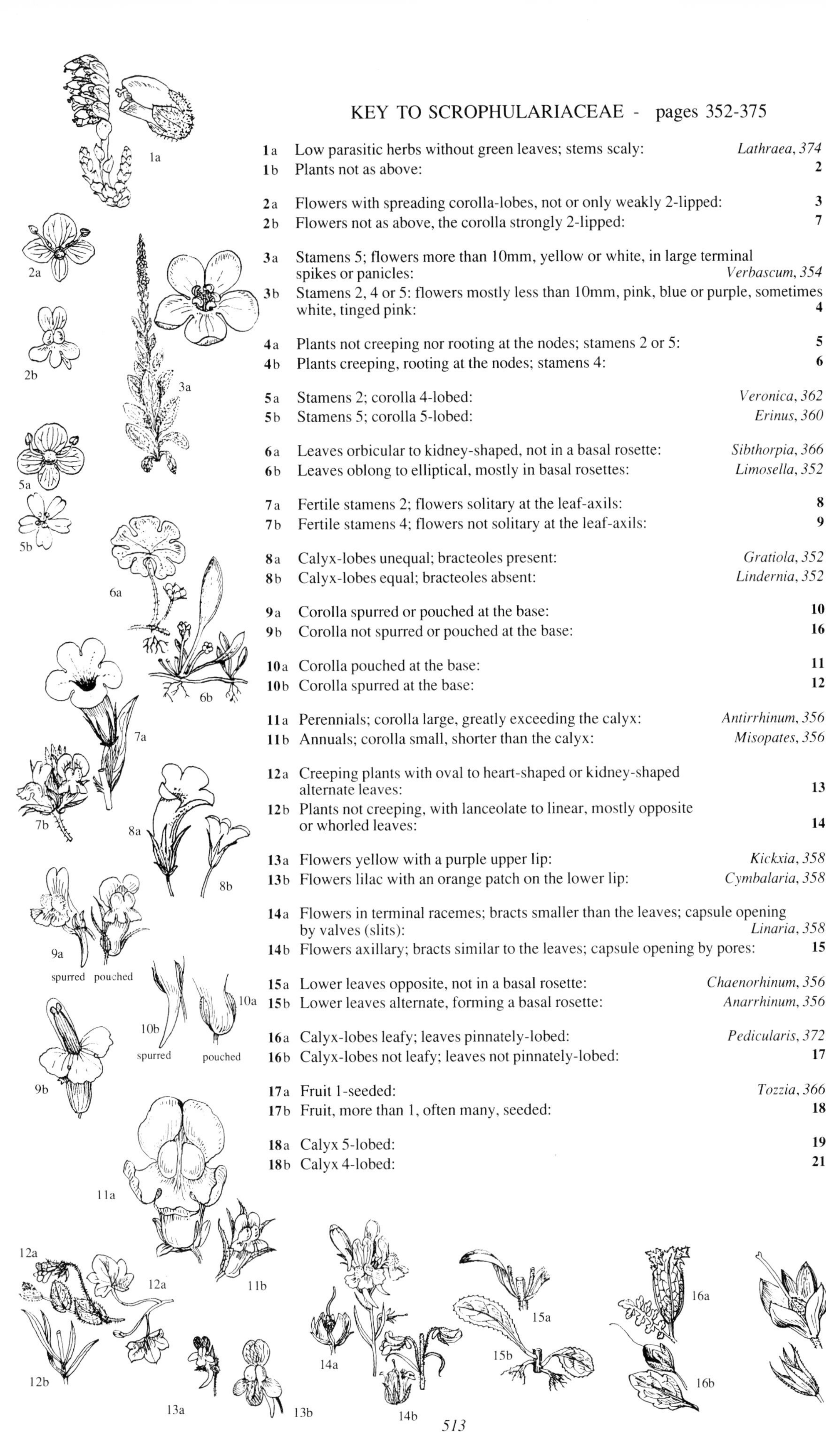


# KEY TO SCROPHULARIACEAE - pages 352-375

| | | |
|---|---|---|
| **1**a | Low parasitic herbs without green leaves; stems scaly: | *Lathraea, 374* |
| **1**b | Plants not as above: | **2** |
| **2**a | Flowers with spreading corolla-lobes, not or only weakly 2-lipped: | **3** |
| **2**b | Flowers not as above, the corolla strongly 2-lipped: | **7** |
| **3**a | Stamens 5; flowers more than 10mm, yellow or white, in large terminal spikes or panicles: | *Verbascum, 354* |
| **3**b | Stamens 2, 4 or 5: flowers mostly less than 10mm, pink, blue or purple, sometimes white, tinged pink: | **4** |
| **4**a | Plants not creeping nor rooting at the nodes; stamens 2 or 5: | **5** |
| **4**b | Plants creeping, rooting at the nodes; stamens 4: | **6** |
| **5**a | Stamens 2; corolla 4-lobed: | *Veronica, 362* |
| **5**b | Stamens 5; corolla 5-lobed: | *Erinus, 360* |
| **6**a | Leaves orbicular to kidney-shaped, not in a basal rosette: | *Sibthorpia, 366* |
| **6**b | Leaves oblong to elliptical, mostly in basal rosettes: | *Limosella, 352* |
| **7**a | Fertile stamens 2; flowers solitary at the leaf-axils: | **8** |
| **7**b | Fertile stamens 4; flowers not solitary at the leaf-axils: | **9** |
| **8**a | Calyx-lobes unequal; bracteoles present: | *Gratiola, 352* |
| **8**b | Calyx-lobes equal; bracteoles absent: | *Lindernia, 352* |
| **9**a | Corolla spurred or pouched at the base: | **10** |
| **9**b | Corolla not spurred or pouched at the base: | **16** |
| **10**a | Corolla pouched at the base: | **11** |
| **10**b | Corolla spurred at the base: | **12** |
| **11**a | Perennials; corolla large, greatly exceeding the calyx: | *Antirrhinum, 356* |
| **11**b | Annuals; corolla small, shorter than the calyx: | *Misopates, 356* |
| **12**a | Creeping plants with oval to heart-shaped or kidney-shaped alternate leaves: | **13** |
| **12**b | Plants not creeping, with lanceolate to linear, mostly opposite or whorled leaves: | **14** |
| **13**a | Flowers yellow with a purple upper lip: | *Kickxia, 358* |
| **13**b | Flowers lilac with an orange patch on the lower lip: | *Cymbalaria, 358* |
| **14**a | Flowers in terminal racemes; bracts smaller than the leaves; capsule opening by valves (slits): | *Linaria, 358* |
| **14**b | Flowers axillary; bracts similar to the leaves; capsule opening by pores: | **15** |
| **15**a | Lower leaves opposite, not in a basal rosette: | *Chaenorhinum, 356* |
| **15**b | Lower leaves alternate, forming a basal rosette: | *Anarrhinum, 356* |
| **16**a | Calyx-lobes leafy; leaves pinnately-lobed: | *Pedicularis, 372* |
| **16**b | Calyx-lobes not leafy; leaves not pinnately-lobed: | **17** |
| **17**a | Fruit 1-seeded: | *Tozzia, 366* |
| **17**b | Fruit, more than 1, often many, seeded: | **18** |
| **18**a | Calyx 5-lobed: | **19** |
| **18**b | Calyx 4-lobed: | **21** |

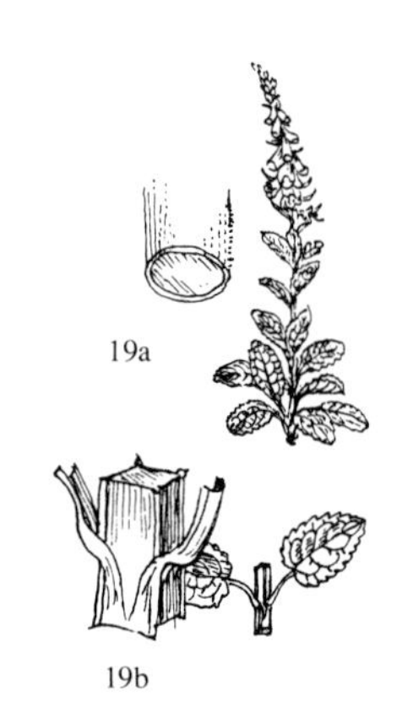


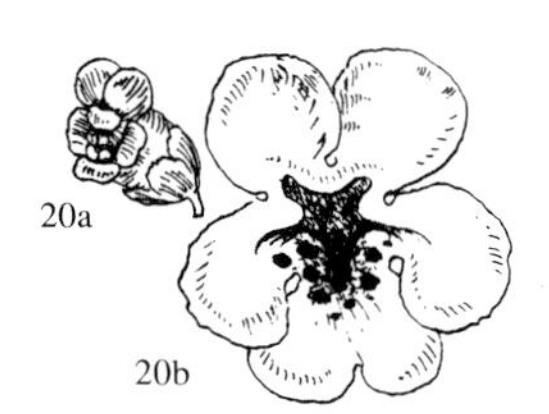


**19**a Leaves alternate; stems round: *Digitalis, 360*
**19**b Leaves opposite; stems square: **20**

**20**a Flowers small, globose, green, brown or reddish, with small corolla-lobes: *Scrophularia, 356*
**20**b Flowers large, tubular or trumpet-shaped, yellow: *Mimulus, 352*

**21**a Calyx flattened, expanding in fruit: *Rhinanthus, 374*
**21**b Calyx tubular or bell-shaped, not expanding in fruit: **22**

**22**a Corolla-lobes spreading, the upper lip 2-lobed: *Euphrasia, 368*
**22**b Corolla-lobes not spreading, the upper lip unlobed or notched: **23**

**23**a Upper lip of corolla laterally compressed, corolla mouth closed: *Melampyrum, 366*
**23**b Upper lip of corolla not laterally compressed, corolla mouth open: **24**

**24**a Flower-spikes 1-sided; corolla 4-8mm: *Odontites, 366*
**24**b Flower-spike not 1-sided; corolla 10mm long or more: **25**

**25**a Flowers purple; perennials: *Bartsia, 366*
**25**b Flowers yellow; sticky annuals: *Parentucellia, 366*

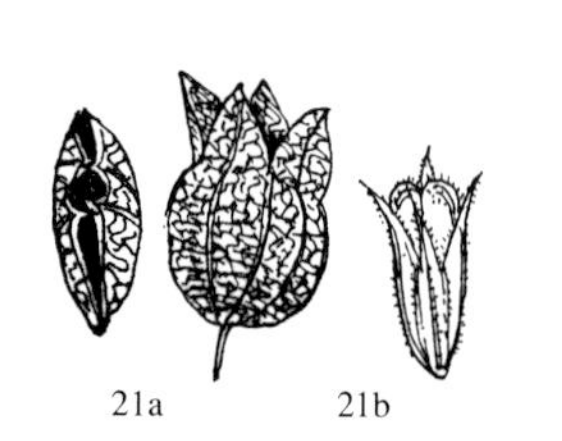


21a 21b

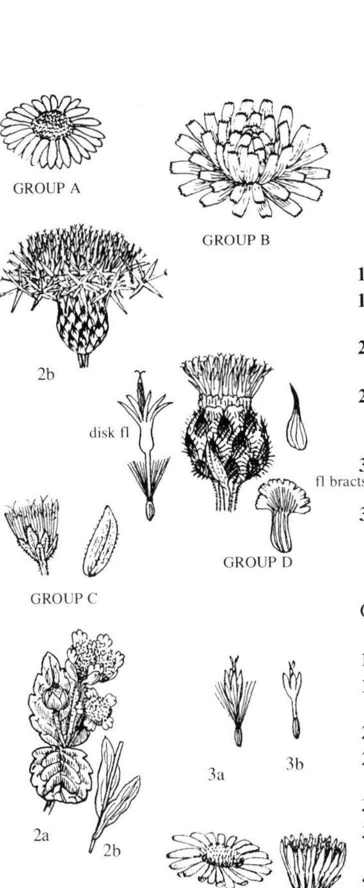


# KEY TO COMPOSITAE - pages 394-441

## Key to groups

| | | |
|---|---|---|
| **1**a | Flowerheads daisy-like, with both disk and ray florets: | *GROUP A* |
| **1**b | Flowerheads not as above: | **2** |
| **2**a | Flowerheads composed solely of strap-like ray-florets; plants with milky latex: | *GROUP B* |
| **2**b | Flowerheads composed solely of tubular disk-florets, the outer sometimes long and spreading: | **3** |
| **3**a | Flowerheads thistle-like; flower-bracts generally spine-tipped or with an appendage: | *GROUP D* |
| **3**b | Flowerheads not thistle-like; flower-bracts simple, without a spine or an appendage: | *GROUP C* |

GROUP A: Flowerheads daisy-like, with both disk and ray florets

| | | |
|---|---|---|
| **1**a | Ray-florets white, pink or blue; disk usually yellow: | **2** |
| **1**b | Ray-florets yellow; disk usually yellow: | **13** |
| **2**a | Leaves opposite: | *Galinsoga, 408* |
| **2**b | Leaves alternate or all basal: | **3** |
| **3**a | Florets with a pappus: | **4** |
| **3**b | Florets without a pappus: | **7** |
| **4**a | Ray-florets spreading: | *Aster, 396* |
| **4**b | Ray-florets erect: | **5** |
| **5**a | Ray-florets whitish, very short: | *Conyza, 398* |
| **5**b | Ray-florets pink, purplish or bluish, not very short: | **6** |
| **6**a | Leaves mostly large and heart-shaped; robust plants often with numerous flowerheads: | *Petasites, 414* |
| **6**b | Leaves not as above, generally lanceolate to elliptical; flowerheads usually solitary or few: | *Erigeron, 396* |
| **7**a | Leaves all basal; flowerheads solitary: | *Bellis, 394* |
| **7**b | Plants with leafy stems; flowerheads clustered: | **8** |
| **8**a | Flowerheads small, generally less than 15mm, borne in umbel-like heads; disk whitish or greenish-white: | *Achillea, 408* |
| **8**b | Flowerheads larger, usually more than 15mm, with a bright yellow disk: | **9** |
| **9**a | Ray-florets bent downwards; disk conical, hollow: | *Matricaria, 408* |
| **9**b | Ray-florets spreading; disk flat to somewhat domed, solid: | **10** |
| **10**a | Achenes with a prominent black spot near the top: | *Matricaria maritima, 408* |
| **10**b | Achenes without a black spot: | **11** |
| **11**a | Leaves coarsely pinnately-lobed or toothed: | *Chrysanthemum, 410* |
| **11**b | Leaves finely 2-pinnately-lobed: | **12** |
| **12**a | Annual plants, spreading to erect: | *Anthemis, 408* |
| **12**b | Perennial creeping plants: | *Chamaemelum, 408* |
| **13**a | Stem leaves opposite: | **14** |
| **13**b | Stem leaves alternate: | **16** |
| **14**a | Fruit-heads crowned by barbed spines: | *Bidens, 404* |
| **14**b | Fruit-heads not spiny: | **15** |

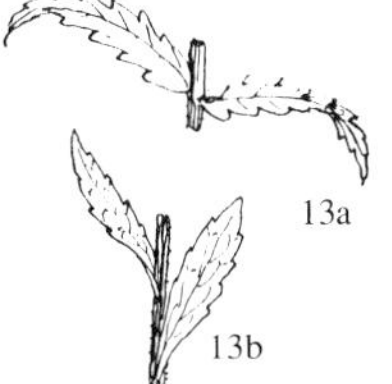


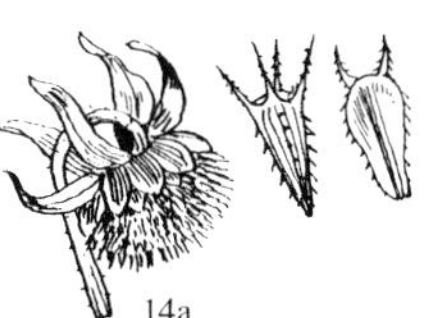

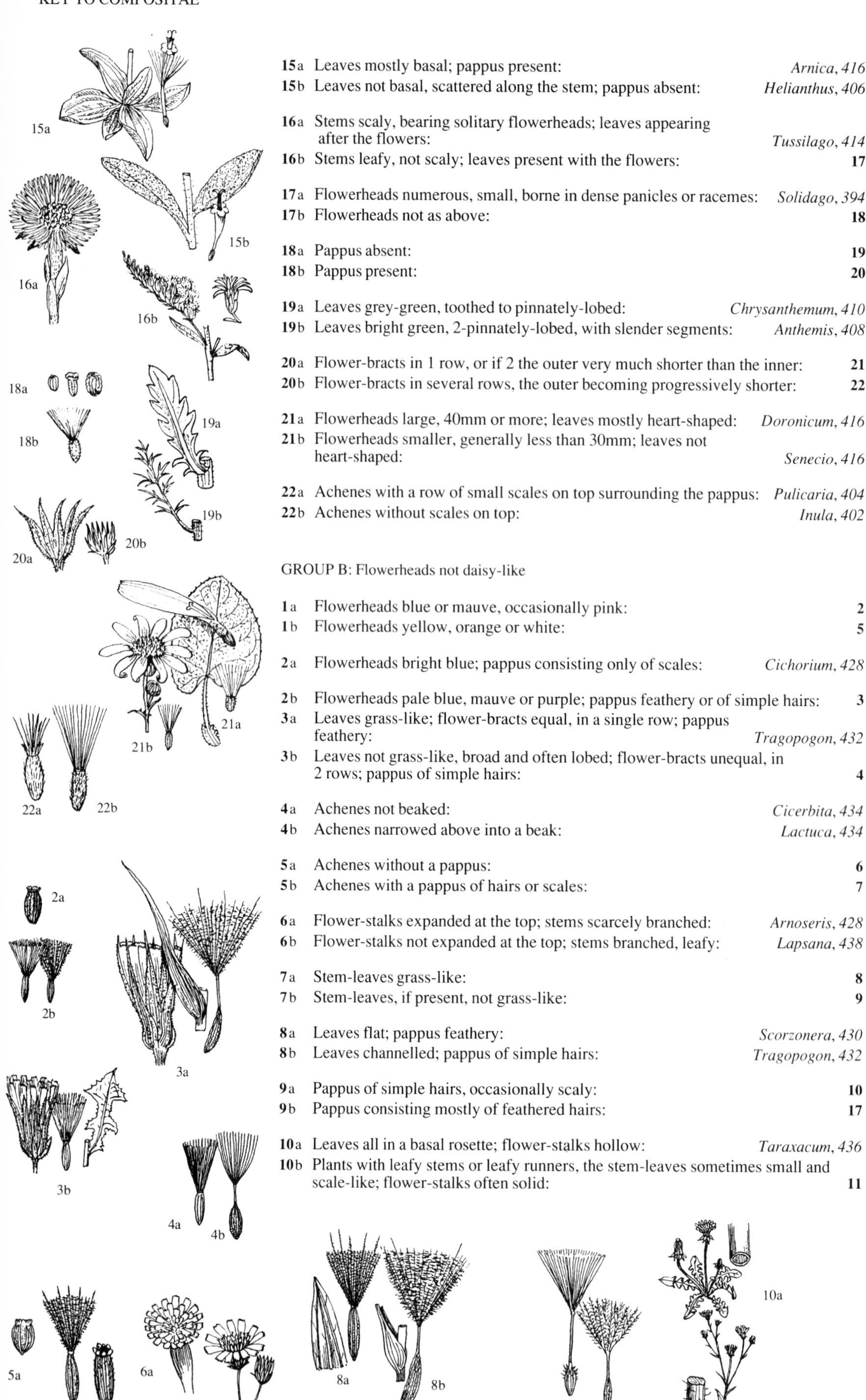


**15**a Leaves mostly basal; pappus present: *Arnica, 416*
**15**b Leaves not basal, scattered along the stem; pappus absent: *Helianthus, 406*

**16**a Stems scaly, bearing solitary flowerheads; leaves appearing after the flowers: *Tussilago, 414*
**16**b Stems leafy, not scaly; leaves present with the flowers: **17**

**17**a Flowerheads numerous, small, borne in dense panicles or racemes: *Solidago, 394*
**17**b Flowerheads not as above: **18**

**18**a Pappus absent: **19**
**18**b Pappus present: **20**

**19**a Leaves grey-green, toothed to pinnately-lobed: *Chrysanthemum, 410*
**19**b Leaves bright green, 2-pinnately-lobed, with slender segments: *Anthemis, 408*

**20**a Flower-bracts in 1 row, or if 2 the outer very much shorter than the inner: **21**
**20**b Flower-bracts in several rows, the outer becoming progressively shorter: **22**

**21**a Flowerheads large, 40mm or more; leaves mostly heart-shaped: *Doronicum, 416*
**21**b Flowerheads smaller, generally less than 30mm; leaves not heart-shaped: *Senecio, 416*

**22**a Achenes with a row of small scales on top surrounding the pappus: *Pulicaria, 404*
**22**b Achenes without scales on top: *Inula, 402*

GROUP B: Flowerheads not daisy-like

**1**a Flowerheads blue or mauve, occasionally pink: **2**
**1**b Flowerheads yellow, orange or white: **5**

**2**a Flowerheads bright blue; pappus consisting only of scales: *Cichorium, 428*

**2**b Flowerheads pale blue, mauve or purple; pappus feathery or of simple hairs: **3**
**3**a Leaves grass-like; flower-bracts equal, in a single row; pappus feathery: *Tragopogon, 432*
**3**b Leaves not grass-like, broad and often lobed; flower-bracts unequal, in 2 rows; pappus of simple hairs: **4**

**4**a Achenes not beaked: *Cicerbita, 434*
**4**b Achenes narrowed above into a beak: *Lactuca, 434*

**5**a Achenes without a pappus: **6**
**5**b Achenes with a pappus of hairs or scales: **7**

**6**a Flower-stalks expanded at the top; stems scarcely branched: *Arnoseris, 428*
**6**b Flower-stalks not expanded at the top; stems branched, leafy: *Lapsana, 438*

**7**a Stem-leaves grass-like: **8**
**7**b Stem-leaves, if present, not grass-like: **9**

**8**a Leaves flat; pappus feathery: *Scorzonera, 430*
**8**b Leaves channelled; pappus of simple hairs: *Tragopogon, 432*

**9**a Pappus of simple hairs, occasionally scaly: **10**
**9**b Pappus consisting mostly of feathered hairs: **17**

**10**a Leaves all in a basal rosette; flower-stalks hollow: *Taraxacum, 436*
**10**b Plants with leafy stems or leafy runners, the stem-leaves sometimes small and scale-like; flower-stalks often solid: **11**

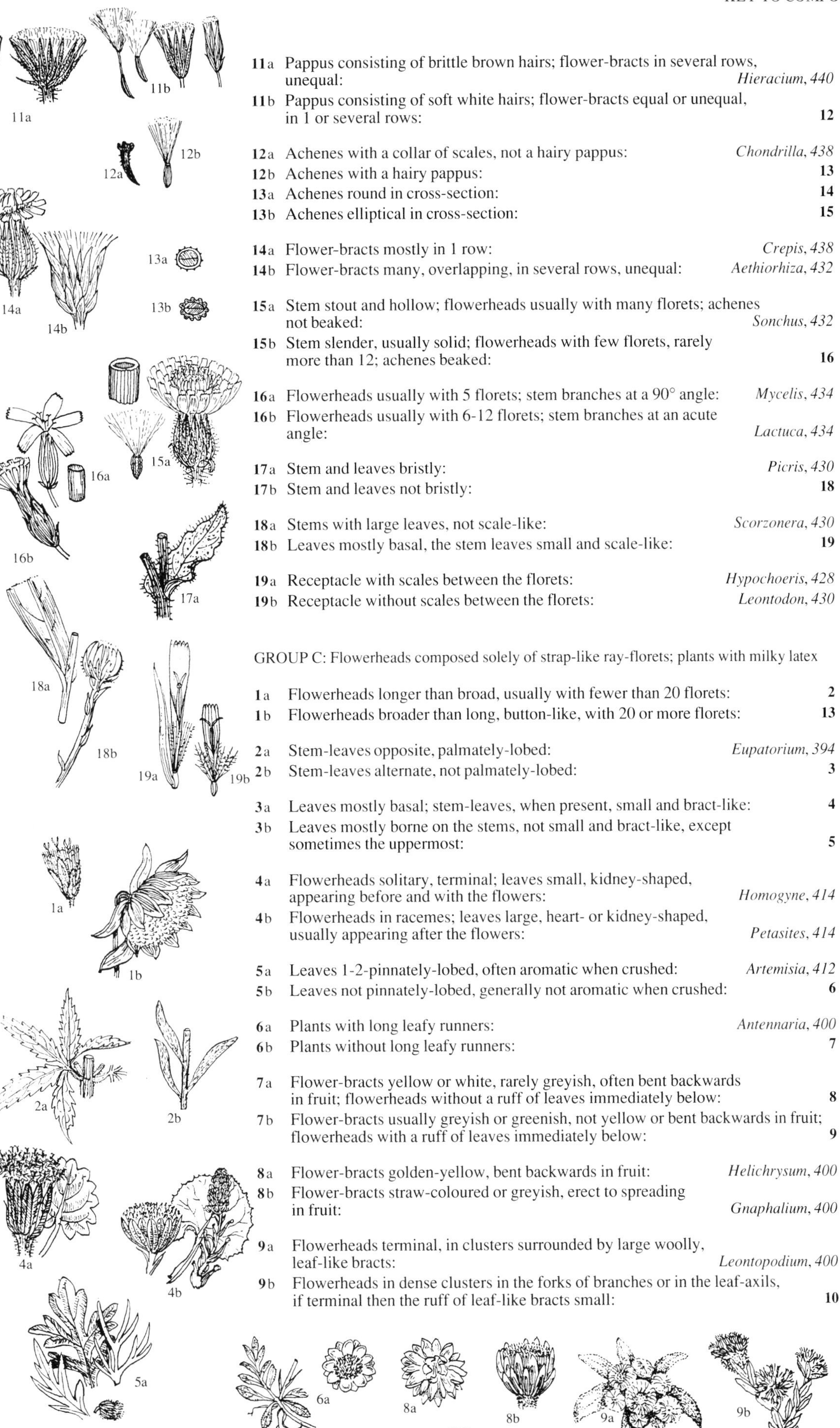


**11**a Pappus consisting of brittle brown hairs; flower-bracts in several rows, unequal: *Hieracium, 440*
**11**b Pappus consisting of soft white hairs; flower-bracts equal or unequal, in 1 or several rows: **12**

**12**a Achenes with a collar of scales, not a hairy pappus: *Chondrilla, 438*
**12**b Achenes with a hairy pappus: **13**
**13**a Achenes round in cross-section: **14**
**13**b Achenes elliptical in cross-section: **15**

**14**a Flower-bracts mostly in 1 row: *Crepis, 438*
**14**b Flower-bracts many, overlapping, in several rows, unequal: *Aethiorhiza, 432*

**15**a Stem stout and hollow; flowerheads usually with many florets; achenes not beaked: *Sonchus, 432*
**15**b Stem slender, usually solid; flowerheads with few florets, rarely more than 12; achenes beaked: **16**

**16**a Flowerheads usually with 5 florets; stem branches at a 90° angle: *Mycelis, 434*
**16**b Flowerheads usually with 6-12 florets; stem branches at an acute angle: *Lactuca, 434*

**17**a Stem and leaves bristly: *Picris, 430*
**17**b Stem and leaves not bristly: **18**

**18**a Stems with large leaves, not scale-like: *Scorzonera, 430*
**18**b Leaves mostly basal, the stem leaves small and scale-like: **19**

**19**a Receptacle with scales between the florets: *Hypochoeris, 428*
**19**b Receptacle without scales between the florets: *Leontodon, 430*

GROUP C: Flowerheads composed solely of strap-like ray-florets; plants with milky latex

**1**a Flowerheads longer than broad, usually with fewer than 20 florets: **2**
**1**b Flowerheads broader than long, button-like, with 20 or more florets: **13**

**2**a Stem-leaves opposite, palmately-lobed: *Eupatorium, 394*
**2**b Stem-leaves alternate, not palmately-lobed: **3**

**3**a Leaves mostly basal; stem-leaves, when present, small and bract-like: **4**
**3**b Leaves mostly borne on the stems, not small and bract-like, except sometimes the uppermost: **5**

**4**a Flowerheads solitary, terminal; leaves small, kidney-shaped, appearing before and with the flowers: *Homogyne, 414*
**4**b Flowerheads in racemes; leaves large, heart- or kidney-shaped, usually appearing after the flowers: *Petasites, 414*

**5**a Leaves 1-2-pinnately-lobed, often aromatic when crushed: *Artemisia, 412*
**5**b Leaves not pinnately-lobed, generally not aromatic when crushed: **6**

**6**a Plants with long leafy runners: *Antennaria, 400*
**6**b Plants without long leafy runners: **7**

**7**a Flower-bracts yellow or white, rarely greyish, often bent backwards in fruit; flowerheads without a ruff of leaves immediately below: **8**
**7**b Flower-bracts usually greyish or greenish, not yellow or bent backwards in fruit; flowerheads with a ruff of leaves immediately below: **9**

**8**a Flower-bracts golden-yellow, bent backwards in fruit: *Helichrysum, 400*
**8**b Flower-bracts straw-coloured or greyish, erect to spreading in fruit: *Gnaphalium, 400*

**9**a Flowerheads terminal, in clusters surrounded by large woolly, leaf-like bracts: *Leontopodium, 400*
**9**b Flowerheads in dense clusters in the forks of branches or in the leaf-axils, if terminal then the ruff of leaf-like bracts small: **10**

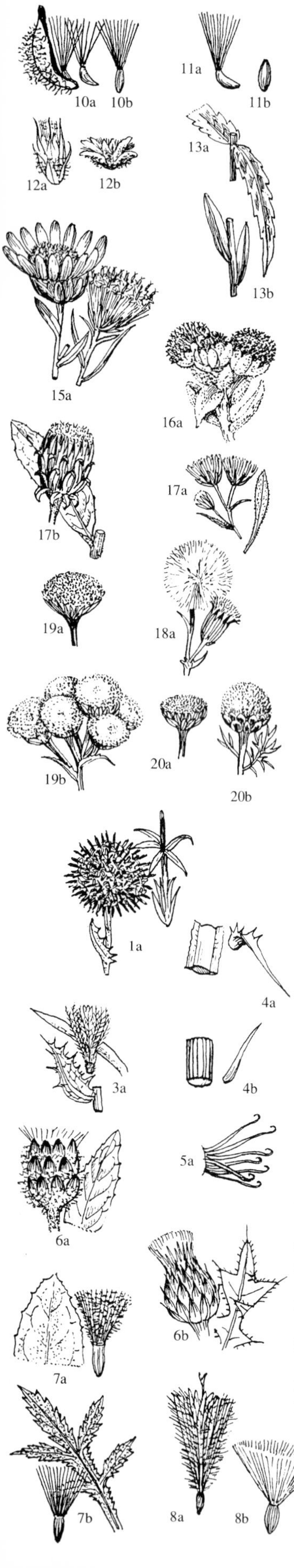

**10**a Achenes curved, falling with attached bracts: **11**
**10**b Achenes straight, not falling with the bracts: **12**

**11**a Pappus present: *Logfia, 398*
**11**b Pappus absent: *Bombycilaena, 398*

**12**a Flower-bracts erect in fruit, usually long-pointed: *Filago, 398*
**12**b Flower-bracts spreading (forming a star) in fruit, usually blunt: *Logfia, 398*

**13**a Stem-leaves opposite: *Bidens, 404*
**13**b Stem-leaves alternate: **14**

**14**a Leaves unlobed: **15**
**14**b Leaves lobed or variously dissected: **18**

**15**a Plants hairless: *Aster* (*A. linosyris* and *A. tripolium*), *396*
**15**b Plants hairy: **16**

**16**a Plants with woolly-white stems and leaves: *Otanthus, 410*
**16**b Plants not white-woolly: **17**

**17**a Flowerheads pale yellow, 3-5mm; leaves linear-lanceolate: *Conyza, 398*
**17**b Flowerheads dark yellow, 10mm or more; leaves oblong: *Inula conyza, 402*

**18**a Pappus present: *Senecio vulgaris, 418*
**18**b Pappus absent: **19**

**19**a Flowerheads solitary, domed: *Cotula, 410*
**19**b Flowerheads clustered, generally forming a flat-topped cluster: **20**

**20**a Flowerheads deep yellow, flattish; leaves aromatic: *Tanacetum, 410*
**20**b Flowerheads greenish-yellow, conical; leaves pineapple-scented: *Matricaria, 408*

GROUP D: Flowerheads composed solely of tubular disk-florets

**1**a Flowerheads in tight globose clusters, each with a single floret: *Echinops, 420*
**1**b Flowerheads not in globose clusters, each with 4 or more florets: **2**

**2**a Flowerheads yellow, cream or brownish: **3**
**2**b Flowerheads red, purple, blue or white: **5**

**3**a Leaves and flower-bracts very spiny; inner flower-bracts slender, shiny and petal-like: *Carlina, 420*
**3**b Leaves and flower-bracts hairy or with fine bristles; flower-bracts with a solitary or a few spines: **4**

**4**a Stems winged; flower-bracts with several spines: *Centaurea solstitialis, 426*
**4**b Stems unwinged; flower-bracts with one terminal spine: *Cirsium oleraceum, 424*

**5**a Flower-bracts with hooked tips: *Arctium, 420*
**5**b Flower-bracts without hooked tips: **6**

**6**a Leaves and flower-bracts not spiny: **7**
**6**b Leaves and flower-bracts spiny: **8**

**7**a Leaves unlobed; pappus feathery: *Saussurea, 420*
**7**b Leaves pinnately-lobed; pappus consisting of simple hairs: *Serratula, 424*

**8**a Pappus feathery: *Cirsium, 424*
**8**b Pappus consisting of simple hairs: **9**

**9**a Leaves with prominent white veins and/or white blotches, hairless; stems not spiny-winged: *Silybum, 424*
**9**b Leaves usually hairy, not white veined above; stems spiny-winged: **10**

**10**a Plants with white-cottony stems and leaves: *Onopordon, 424*
**10**b Plants generally with green stems and leaves: *Carduus, 422*

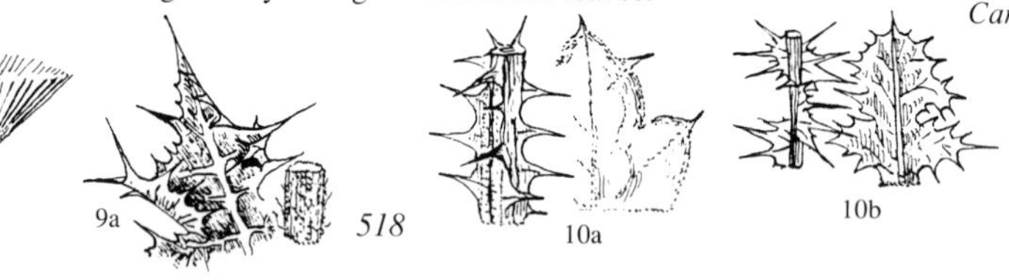

# FURTHER READING

The botanical literature covering this region is both voluminous and largely specialized; and the 'national' Floras compiled over the last century have been to a considerable extent superseded by the publication of *Flora Europaea* (Tutin, T.G. et al, 1964-1980). Here we give simply a selection of particularly important works, with some more general books which are useful and enjoyable in their different ways.


Alcenius, O. (1958). *Finlands kärlväxter.* 3rd Ed.

Christiansen, W. (1953). *Neue kritische Flora von Schleswig-Holstein.*

Clapham, A.R., Tutin, T.G. & Warburg, E.F. (1962). *Flora of the British Isles.* Ed. 2.

Clapham, A.R., Tutin, T.G. & Moore, D.M. (1987). *Flora of the British Isles.* Ed. 3.

Coste, H. (1906). *Flore descriptive et illustrée de la France, de la Corse et des Contrées limitrophes.*

Duperrex, A. (1961). *Orchids of Europe.*

Faegri, K. (1970). *Norges Planter.*

Fitter, A. (1978). *An Atlas of the Wild Flowers of Britain and Northern Europe.*

Fitter, R.S.R. (1971) *Finding Wild Flowers.*

Fitter, R.S.R., Fitter, A. & Blamey, M. (1974). *The Wild Flowers of Britain and Northern Europe.*

Fournier, P. (1961). *Les Quatre Flores de la France, Corse comprise* – first published 1934-1940.

Garcke, A. (1972). *Illustrierte Flora von Deutschland.*

Gjaervoll, O. & Jörgensen, R. *Mountain Flowers of Scandinavia.*

Grey-Wilson, C. & Blamey, M. (1979). *The Alpine Flowers of Britain and Europe.*

Grey-Wilson, C. & Blamey, M. (1974). *The Bulbous Plants of Britain and Europe and their Allies.*

Hagerup, O. & Peterson, V. (1956-60). *Botanisk Atlas.* (Denmark).

Hegi, G. (1959-71). *Flora von Mitteleuropa.*

Heimans, E. et al. (1948). *Geillustreerde Flora van Nederland.*

Heukels, H. (1962). *Flora van Nederland.* (15th Ed. by S.J. van Ooststroom).

Hultén, E. (1971). *Atlas över växternas utbredning i Norden.*

Hultén, E. (1971). *The Circumpolar Plants.*

Huxley, A. (1967). *Mountain Flowers.*

Johnson, H. (1984). *Hugh Johnson's Encyclopedia of Trees.*

Keble Martin, W. (1965). *Concise British Flora in Colour.*

Lid, J. (1962). *Norsk flora.*

Lid, J. (1963). *Norsk og svensk flora.*

Löve, Á. (1945). *Íslenzkar Jurtir.* (Iceland).

Mabey, R. & Evans, T. (1980). *The Flowering of Britain.*

McClintock, D. (1966). *Companion to Flowers.*

McClintock, D. (1975). *The Flowers of Guernsey.*

McClintock, D. & Fitter, R.S.R. (1961). *Collins Pocket Guide to Wild Flowers.*

Mitchell, A.F. (1974). *A Field Guide to the Trees of Britain and Northern Europe.*

Mitchell, A.F. & Wilkinson, J. (1982-9). *The Trees of Britain and Northern Europe.*

Mullenders, E. (1967). *Flore de la Belgique, du Nord de la France et des Régions Voisines.*

Nilsson, Ö & Nilsson, E. (1986). *Nordisk fjällflora.*

Ostenfeld, C.E.H. & Gröntved, J. (1934). *The Flora of Iceland and the Faeroes.*

Perring, F.H. & Walters, S.M. (1962). *Atlas of the British Flora.*

Phillips, R. (1977). *Wild Flowers of Britain.*

Polunin, O. (1969). *Flowers of Europe.*

Polunin, O. & Walters, S.M. (1985). *A Guide to the Vegetation of Britain and Europe.*

Raven, J.. & Walters, M. (1956). *Mountain Flowers.*

Robyns, W. (1952). *Flore générale de Belgique.*

van Rompaey, E. & Delvosalle, L. (1972). *Atlas de la Flore Belge et Luxembourgeoise.*

Rose, F. (1981). *The Wild Flower Key.*

Ross-Craig, S. (1948-1974). *Drawings of British Plants.*

Rostrup, F.G.E. (1961). *Den Danske Flora.* (19th Ed. by C.A. Jørgensen).

Rothmaler, W. (1963). *Exkursionsflora von Deutschland.*

Rouy, G.C.C. (1927). *Conspectus de la Flore de France.*

Schmeil, O. & Fitschen, J. (1965). *Flora von Deutschland.* (78th Ed. by W. Rauh).

Stefánsson, S. (1948). *Flóra Islands.* (3rd Ed. by S. Steindórsson).

Tutin, T.G. et al. (1964-1980). *Flora Europaea.*

Webb, D.A. (1967). *An Irish Flora.*

Weimarck, H. (1963). *Skånes flora.*

Wigginton, M.J. & Graham, G.G. (1981). *Guide to the Identification of some difficult Plant Groups.* Nature Conservancy Council.

# INDEX OF ENGLISH NAMES

# INDEX OF SCIENTIFIC NAMES

Entries in *italic* type are synonyms.